TABLE 3 *(continued)*

z	0.00	0.01	0.02	0.03	0.04	0.05	0.06	0.07	0.08	0.09
0.0	0.5000	0.5040	0.5080	0.5120	0.5160	0.5199	0.5239	0.5279	0.5319	0.5359
0.1	0.5398	0.5438	0.5478	0.5517	0.5557	0.5596	0.5636	0.5675	0.5714	0.5753
0.2	0.5793	0.5832	0.5871	0.5910	0.5948	0.5987	0.6026	0.6064	0.6103	0.6141
0.3	0.6179	0.6217	0.6255	0.6293	0.6331	0.6368	0.6406	0.6443	0.6480	0.6517
0.4	0.6554	0.6591	0.6628	0.6664	0.6700	0.6736	0.6772	0.6808	0.6844	0.6879
0.5	0.6915	0.6950	0.6985	0.7019	0.7054	0.7088	0.7123	0.7157	0.7190	0.7224
0.6	0.7257	0.7291	0.7324	0.7357	0.7389	0.7422	0.7454	0.7486	0.7517	0.7549
0.7	0.7580	0.7611	0.7642	0.7673	0.7704	0.7734	0.7764	0.7794	0.7823	0.7852
0.8	0.7881	0.7910	0.7939	0.7967	0.7995	0.8023	0.8051	0.8078	0.8106	0.8133
0.9	0.8159	0.8186	0.8212	0.8238	0.8264	0.8289	0.8315	0.8340	0.8365	0.8389
1.0	0.8413	0.8438	0.8461	0.8485	0.8508	0.8531	0.8554	0.8577	0.8599	0.8621
1.1	0.8643	0.8665	0.8686	0.8708	0.8729	0.8749	0.8770	0.8790	0.8810	0.8830
1.2	0.8849	0.8869	0.8888	0.8907	0.8925	0.8944	0.8962	0.8980	0.8997	0.9015
1.3	0.9032	0.9049	0.9066	0.9082	0.9099	0.9115	0.9131	0.9147	0.9162	0.9177
1.4	0.9192	0.9207	0.9222	0.9236	0.9251	0.9265	0.9279	0.9292	0.9306	0.9319
1.5	0.9332	0.9345	0.9357	0.9370	0.9382	0.9394	0.9406	0.9418	0.9429	0.9441
1.6	0.9452	0.9463	0.9474	0.9484	0.9495	0.9505	0.9515	0.9525	0.9535	0.9545
1.7	0.9554	0.9564	0.9573	0.9582	0.9591	0.9599	0.9608	0.9616	0.9625	0.9633
1.8	0.9641	0.9649	0.9656	0.9664	0.9671	0.9678	0.9686	0.9693	0.9699	0.9706
1.9	0.9713	0.9719	0.9726	0.9732	0.9738	0.9744	0.9750	0.9756	0.9761	0.9767
2.0	0.9772	0.9778	0.9783	0.9788	0.9793	0.9798	0.9803	0.9808	0.9812	0.9817
2.1	0.9821	0.9826	0.9830	0.9834	0.9838	0.9842	0.9846	0.9850	0.9854	0.9857
2.2	0.9861	0.9864	0.9868	0.9871	0.9875	0.9878	0.9881	0.9884	0.9887	0.9890
2.3	0.9893	0.9896	0.9898	0.9901	0.9904	0.9906	0.9909	0.9911	0.9913	0.9916
2.4	0.9918	0.9920	0.9922	0.9925	0.9927	0.9929	0.9931	0.9932	0.9934	0.9936
2.5	0.9938	0.9940	0.9941	0.9943	0.9945	0.9946	0.9948	0.9949	0.9951	0.9952
2.6	0.9953	0.9955	0.9956	0.9957	0.9959	0.9960	0.9961	0.9962	0.9963	0.9964
2.7	0.9965	0.9966	0.9967	0.9968	0.9969	0.9970	0.9971	0.9972	0.9973	0.9974
2.8	0.9974	0.9975	0.9976	0.9977	0.9977	0.9978	0.9979	0.9979	0.9980	0.9981
2.9	0.9981	0.9982	0.9982	0.9983	0.9984	0.9984	0.9985	0.9985	0.9986	0.9986
3.0	0.9987	0.9987	0.9987	0.9988	0.9988	0.9989	0.9989	0.9989	0.9990	0.9990
3.1	0.9990	0.9991	0.9991	0.9991	0.9992	0.9992	0.9992	0.9992	0.9993	0.9993
3.2	0.9993	0.9993	0.9994	0.9994	0.9994	0.9994	0.9994	0.9995	0.9995	0.9995
3.3	0.9995	0.9995	0.9995	0.9996	0.9996	0.9996	0.9996	0.9996	0.9996	0.9997
3.4	0.9997	0.9997	0.9997	0.9997	0.9997	0.9997	0.9997	0.9997	0.9997	0.9998

NELSON

Introduction to Probability and Statistics, Fourth Canadian Edition

by William Mendenhall, S. Ejaz Ahmed, Robert J. Beaver, and Barbara M. Beaver

Vice President, Product Solutions:
Claudine O'Donnell

Publisher, Digital and Print Content:
Jean Ford

Marketing Manager:
Kimberley Carruthers

Technical Reviewer:
Andie Burazin

Content Manager:
Courtney Thorne

Photo and Permissions Researcher:
Sandra Mark

Production Project Manager:
Jaime Smith

Production Service:
Cenveo Publisher Services

Copy Editor:
Michael Kelly

Proofreader:
Pushpa V. Giri

Indexer:
Diana Witt

Design Director:
Ken Phipps

Higher Education Design PM:
Pamela Johnston

Interior Design:
Jennifer Leung

Cover Design:
Trinh Truong

Cover Image:
Getty Images

Compositor:
Cenveo Publisher Services

Printed and bound in the United States of America
1 2 3 4 20 19 18 17

For more information contact Nelson Education Ltd., 1120 Birchmount Road, Toronto, Ontario, M1K 5G4. Or you can visit our Internet site at nelson.com

Library and Archives Canada Cataloguing in Publication

Mendenhall, William, author
Introduction to probability and statistics / William Mendenhall, University of Florida, Emeritus, Robert J. Beaver, University of California, Riverside, Barbara M. Beaver, University of California, Riverside, S. Ejaz Ahmed, Brock University. – Fourth Canadian edition.

Includes bibliographical references and index.ISBN 978-0-17-685704-2 (hardback)

1. Mathematical statistics–Textbooks.
2. Probabilities–Textbooks.
I. Ahmed, Syed Ejaz, 1957-, author
II. Beaver, RobertJ., author
III. Beaver, Barbara M., author
IV. Title. V. Title: Introduction to probability & statistics.

QA276.I58 2018 519.5
C2016-902908-5

ISBN-13: 978-0-17-685704-2
ISBN-10: 0-17-685704-4

Introduction to Probability and Statistics

FOURTH CANADIAN EDITION

William Mendenhall
University of Florida, Emeritus

S. Ejaz Ahmed
Brock University

Robert J. Beaver
University of California, Riverside

Barbara M. Beaver
University of California, Riverside

NELSON

Brief Contents

Contents

List of Applications

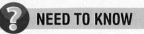

NEED TO KNOW

Preface

In our technologically advanced world, data collection and analysis are an integral part of our day-to-day activities. You likely encounter statistics every day in a newspaper, online, or on television. Consider what happens when you register with a website, like a company's Facebook page, or scan a rewards card at the check-out counter. All of that personal statistical information becomes part of a database. In order to be an educated consumer and citizen, you need to understand how statistics are used and misused in our daily lives. To that end we need to "train your brain" for statistical thinking—a theme we continue to emphasize throughout *Introduction to Probability and Statistics*, Fourth Canadian Edition.

FEATURES OF THE FOURTH CANADIAN EDITION

The fourth canadian edition retains the traditional outline for the coverage of descriptive and inferential statistics, and maintains the straightforward presentation of the previous edition. This revision continues to simplify and clarify the language and style to be more readable and "user friendly"—without sacrificing the statistical integrity of the presentation. Great effort has been taken to "train your brain" to explain not only how to apply statistical procedures, but also:

- how to meaningfully describe real sets of data
- what the results of statistical tests mean in terms of their practical applications
- how to evaluate the validity of the assumptions behind statistical tests
- what to do when statistical assumptions have been violated

Exercises

In the tradition of previous Canadian editions, the variety and number of real applications in the exercise sets is a major strength of this edition. Within the fourth canadian edition are "big picture" projects, or mini cases, added throughout the text.

These provide an opportunity for students to build on knowledge gained from previous chapters and apply it to projects. Rather than working with problems based only on the individual sections, students will be using almost all of the concepts, definitions, and techniques given in that chapter, thus bolstering students' success rate. We have also added more examples and exercises to selected chapters and a number of new and updated real data sets from applications in many interesting fields, including social media. The fourth canadian edition contains over 1300 problems, with almost 200 that are Canadian. Exercises are graduated in level of difficulty; some, involving only basic techniques, can be solved by almost all students, while others, involving practical applications and interpretation of results, will challenge students to use more sophisticated statistical reasoning and understanding.

PROJECTS

Project 9-A: Proportion of "Cured" Cancer Patients: How Does Canada Compare with Europe?[32]

Lung cancer remains the leading cause of cancer death for both Canadian men and women, responsible for the most potential years of life lost to cancer. Lung cancer alone accounts for 28% of all cancer deaths in Canada (32%. in Quebec). Most forms of lung cancer start insidiously and produce no apparent symptoms until they are too far advanced. Consequently, the chances of being cured of lung cancer are not very promising, with the five-year survival rate being less than 15%. The overall data for Europe show that the number of patients who are considered "cured" is rising steadily. For lung cancer, this proportion rose from 6% to 8%. However, there was a wide variation in the proportion of patients cured in individual European countries. For instance, the study shows that for lung cancer, less than 5% of patients were cured in Denmark, the Czech Republic, and Poland, whereas more than 10% of patients were cured in Spain.

a. Suppose a sample of 75 Quebecers was selected and it was found that 27 of them had died due to lung cancer.

Organization and Coverage

Chapters 1 to 3 present descriptive data analysis for both one and two variables, using both *MINITAB™* and *Microsoft Excel®*. We believe that Chapters 1 through 10—with the possible exception of Chapter 3—should be covered in the order presented. The remaining chapters can be covered in any order. The analysis of variance chapter precedes the regression chapter so that the instructor can present the analysis of variance as part of a regression analysis. Thus, the most effective presentation would order these three chapters as well.

Chapter 4 includes a full presentation of probability and probability distributions. Three optional sections—Counting Rules, the Total Law of Probability, and Bayes' Rule—are placed into the general flow of text, and instructors will have the option of complete or partial coverage.

The chapters on analysis of variance and linear regression include both calculational formulas and computer printouts in the basic text presentation. These chapters can be used with equal ease by instructors who wish to use the "hands-on" computational approach to linear regression and ANOVA and by those who choose to focus on the interpretation of computer-generated statistical printouts.

With the advent of computer-generated p-values, the emphasis on p-values and their use in judging statistical significance have become essential components in reporting the results of a statistical analysis. As such, the observed value of the test statistic and its p-value are presented together at the outset of our discussion of statistical hypothesis testing as equivalent tools for decision making. Statistical significance is defined in terms of preassigned values of α, and the *p-value approach* is presented as an alternative to the *critical value approach* for testing a statistical hypothesis. Examples are presented using both the *p-value* and *critical value* approaches to hypothesis testing. Discussion of the practical interpretation of statistical results, along with the difference between statistical significance and practical significance, is emphasized in the practical examples in the text.

Finally, this edition continues to address inconsistencies in the use of notation for random variables. In this edition, "X" is employed to denote a random variable and "x" or "k" to denote an observed value. The classical approach is retained for the notation of regression.

CHANGES TO THE FOURTH CANADIAN EDITION

With each revision we try to build on the strong points of previous editions, while always looking for new ways to motivate, encourage, and interest students using new technological tools.

Changes to this edition include:

- Additional exercises and diverse examples have been added throughout the book to improve and clarify topic coverage.
- Relevant coverage of social media and websites provides students with new and meaningful connections to data collection and analysis.
- New and updated data sets connect students to real and current data covering both social and scientific matters that are of interest to them.
- Streamlined content in select chapters improves readability and understanding of concepts.

Special Features of the Fourth Canadian Edition

- **NEED TO KNOW...:** A special feature of this edition are highlighted sections called "NEED TO KNOW..." and identified by this icon: NEED TO KNOW. These sections provide information consisting of definitions, procedures, or step-by-step hints on problem solving for specific questions such as "NEED TO KNOW... How to Construct a Relative Frequency Histogram."
- **Applets:** Easy access to the Internet has made it possible for students to visualize statistical concepts using an interactive webtool called an applet. Applets written by Gary McClelland, author of *Seeing Statistics*™, are found on the website for the fourth Canadian edition (http://www.nelson.com/student). Following each applet, appropriate exercises are available that provide visual reinforcement of the concepts presented in the text. Applets allow the user to perform a statistical experiment, to interact with a statistical graph, to change its form, or to access an interactive "statistical table."

TECHNOLOGY TODAY

In this edition, we continue to use the computational shortcuts and interactive visual tools that modern technology provides to give us more time to emphasize statistical reasoning as well as the understanding and interpretation of statistical results.

Students will continue to use computers for both standard statistical analyses and as a tool for reinforcing and visualizing statistical concepts. Both Microsoft Excel and *MINITAB* 16 (consistent with earlier versions of *MINITAB*) are used exclusively as the computer packages for statistical analysis. We have continued to isolate the instructions for generating computer output into individual sections called "Technology Today" at the end of most chapters. Each discussion uses numerical examples to guide the student through the *Microsoft Excel* commands and options necessary for the procedures presented in that chapter, and then presents the equivalent steps and commands needed to produce the same or similar results using *MINITAB*. We have included screen captures from both *Microsoft Excel* and *MINITAB* 16, so the student can actually work through these sections as "mini-labs."

If you do not need "hands-on" knowledge of *MINITAB* or *Microsoft Excel*, or if you are using another software package, you may choose to skip these sections and simply use the printouts as guides for the basic understanding of computer printouts.

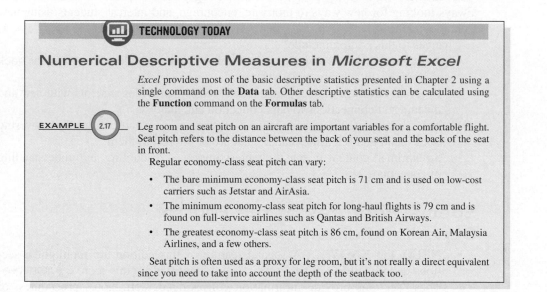

TECHNOLOGY TODAY

Numerical Descriptive Measures in *Microsoft Excel*

Excel provides most of the basic descriptive statistics presented in Chapter 2 using a single command on the **Data** tab. Other descriptive statistics can be calculated using the **Function** command on the **Formulas** tab.

EXAMPLE 2.17

Leg room and seat pitch on an aircraft are important variables for a comfortable flight. Seat pitch refers to the distance between the back of your seat and the back of the seat in front.

Regular economy-class seat pitch can vary:

- The bare minimum economy-class seat pitch is 71 cm and is used on low-cost carriers such as Jetstar and AirAsia.
- The minimum economy-class seat pitch for long-haul flights is 79 cm and is found on full-service airlines such as Qantas and British Airways.
- The greatest economy-class seat pitch is 86 cm, found on Korean Air, Malaysia Airlines, and a few others.

Seat pitch is often used as a proxy for leg room, but it's not really a direct equivalent since you need to take into account the depth of the seatback too.

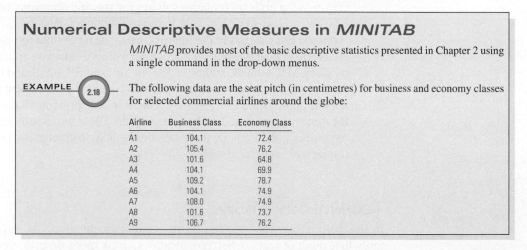

Numerical Descriptive Measures in *MINITAB*

MINITAB provides most of the basic descriptive statistics presented in Chapter 2 using a single command in the drop-down menus.

EXAMPLE 2.18

The following data are the seat pitch (in centimetres) for business and economy classes for selected commercial airlines around the globe:

Airline	Business Class	Economy Class
A1	104.1	72.4
A2	105.4	76.2
A3	101.6	64.8
A4	104.1	69.9
A5	109.2	78.7
A6	104.1	74.9
A7	108.0	74.9
A8	101.6	73.7
A9	106.7	76.2

Any student who has Internet access can use the applets found on the website for the fourth Canadian edition (http://www.nelson.com/student) to visualize a variety of statistical concepts. In addition, some of the applets can be used instead of computer software to perform simple statistical analyses. Exercises written specifically for use with these applets also appear on the text website. Students can use the applets at home or in a computer lab. They can use them as they read through the text material, once they have finished reading the entire chapter, or as a tool for exam review. Instructors can use the applets as a tool in a lab setting, or use them for visual demonstrations during lectures. We believe that these applets are a powerful tool that will increase student enthusiasm for, and understanding of, statistical concepts and procedures.

STUDY AIDS

The many and varied exercises in the text provide the best learning tool for students embarking on a first course in statistics. The answers to all odd-numbered exercises are given in the back of the text, and a detailed solution appears in the **Student Solutions Manual**, which is available as a supplement for students. Each application exercise has a title, making it easier for students and instructors to immediately identify both the context of the problem and the area of application.

3.2 EXERCISES

BASIC TECHNIQUES

3.1 Gender Differences Male and female respondents to a questionnaire about gender differences are categorized into three groups according to their answers on the first question:

	Group 1	Group 2	Group 3
Men	37	49	72
Women	7	50	31

a. Create side-by-side pie charts to describe these data.

b. Create a side-by-side bar chart to describe these data.

c. Draw a stacked bar chart to describe these data.

d. Which of the three charts best depicts the difference or similarity of the responses of men and women?

3.2 Province by Province A group of items are categorized according to a certain attribute—X, Y, Z—and according to the province in which they are produced:

APPLICATIONS

3.4 M&M'S® The colour distributions for two snack-size bags of M&M'S candies, one plain and one peanut, are displayed in the table. Choose an appropriate graphical method and compare the distributions.

	Brown	Yellow	Red	Orange	Green	Blue
Plain	15	14	12	4	5	6
Peanut	6	2	2	3	3	5

EX0305

3.5 How Much Free Time? When you were growing up, did you feel that you did not have enough free time? Parents and children have differing opinions on this subject. A research group surveyed 198 parents and 200 children and recorded their responses to the question, "How much free time does your child have?" or "How much free time do you have?" The responses are shown in the table below:[1]

Students should be encouraged to use the "**NEED TO KNOW…**" sections as they occur in the text. The placement of these sections is intended to answer questions as they would normally arise in discussions. In addition, there are numerous hints called "**NEED A TIP?**" that appear in the margins of the text. The tips are short and concise.

NEED A TIP?

x "explains" *y* or *y* "depends on" *x*.

x is the **explan**atory or independent variable.

y is the response or **depend**ent variable.

Sometimes the two variables, X and Y, are related in a particular way. the value of Y depends on the value of X; that is, the value of X in some the value of Y. For example, the cost of a home (Y) may *depend* on its a space (X); a student's grade point average (X) may *explain* her score on a test (Y). In these situations, we call Y the **dependent variable,** while **independent variable.**

If one of the two variables can be classified as the dependent variable as X, and if the data exhibit a straight-line pattern, it is possible to describ ship relating Y to X using a straight line given by the equation

Finally, sections called "**Key Concepts and Formulas**" appear in each chapter as a review in outline form of the material covered in that chapter.

CHAPTER REVIEW

Key Concepts and Formulas

I. Measures of the Centre of a Data Distribution

1. Arithmetic mean (mean) or average
   a. Population: μ
   b. Sample of n measurements: $\bar{x} = \frac{\Sigma x_i}{n}$
2. Median; **position** of the median = $0.5(n + 1)$
3. Mode
4. The median may be preferred to the mean if the data are highly skewed.

II. Measures of Variability

1. Range: R = largest − smallest
2. Variance
   a. Population of N measurements:

$$\sigma^2 = \frac{\Sigma(x_i - \mu)^2}{N}$$

   b. Sample of n measurements:

$$s^2 = \frac{\Sigma(x_i - \bar{x})^2}{n - 1} = \frac{\Sigma x_i^2 - \frac{(\Sigma x_i)^2}{n}}{n - 1}$$

3. Standard deviation
   a. Population: $\sigma = \sqrt{\sigma^2}$
   b. Sample: $s = \sqrt{s^2}$

The text website for the fourth canadian edition of *An Introduction to Probability and Statistics* (http://www.nelson.com/student) provides students with data sets for many of the text exercises saved in a variety of formats and the complete set of My Applet sections from the text. A complete set of Java Applets is also available on the text website.

INSTRUCTOR RESOURCES

The **Nelson Education Teaching Advantage (NETA)** program delivers research-based instructor resources that promote student engagement and higher-order thinking to enable the success of Canadian students and educators. Visit Nelson Education's **Inspired Instruction** website at www.nelson.com/inspired to find out more about NETA.

The following instructor resources have been created for *Introduction to Probability and Statistics*, Fourth Canadian Edition. Access these ultimate tools for customizing lectures and presentations at www.nelson.com/instructor.

NETA Test Bank

This resource was revised for the Fourth Canadian edition by Andie Burazin of University of Toronto. It includes over 700 multiple-choice questions, over 1000 true/false questions, and over 1400 problems.

cognero® Full-Circle Assessment®

The NETA Test Bank is available in a new, cloud-based platform. Nelson Testing Powered by Cognero® is a secure online testing system that allows instructors to author, edit, and manage test bank content from anywhere Internet access is available. No special installations or downloads are needed, and the desktop-inspired interface, with its drop-down menus and familiar, intuitive tools, allows instructors to create and manage tests with ease. Multiple test versions can be created in an instant, and content can be imported or exported into other systems. Tests can be delivered from a learning management system, the classroom, or wherever an instructor chooses. Nelson Testing Powered by Cognero for *Introduction to Probability and Statistics*, Fourth Canadian Edition, can be accessed through www.nelson.com/instructor.

NETA PowerPoint®

Microsoft PowerPoint® lecture slides for every chapter have been created by Cristina Anton of Grant MacEwan University. Each chapter's PowerPoint features key figures, tables, and photographs from *Introduction to Probability and Statistics*, Fourth Canadian Edition. NETA principles of clear design and engaging content have been incorporated throughout, making it simple for instructors to customize the deck for their courses.

Image Library

This resource consists of digital copies of figures, short tables, and photographs used in the book. Instructors may use these jpegs to customize the NETA PowerPoint or create their own PowerPoint presentations. An Image Library Key describes the images and lists the codes under which the jpegs are saved. Codes normally reflect the chapter number (e.g., C01 for Chapter 1), the figure or photo number (e.g., F15 for Figure 15), and the page in the textbook. For example, C01-F15-pg26 corresponds to Figure 1-15 on page 26.

TurningPoint® Slides

TurningPoint® classroom response software has been customized for *Introduction to Probability and Statistics*, Fourth Canadian Edition. Instructors can author, deliver, show, access, and grade, all in PowerPoint, with no toggling back and forth between screens. With JoinIn, instructors are no longer tied to their computers. Instead, instructors can walk about the classroom and lecture at the same time, showing slides and collecting and displaying responses with ease. Anyone who can use PowerPoint can also use JoinIn on TurningPoint.

Instructor's Solutions Manual

This manual, revised by Andie Burazin of the University of Toronto, contains step-by-step solutions to all end-of-section and end-of-chapter exercises, including case studies and projects.

WebAssign

WebAssign®, the leading homework system for math and science, has been used by more than 2.2 million students. Created by instructors for instructors, **WA** is easy to use and works with all major operating systems and browsers. **WA** adds interactive features to go far beyond simply duplicating text problems online. Students can watch solution videos, see problems solved step-by-step, and receive feedback as they complete their homework.

WebAssign allows instructors to easily assign, collect, grade, and record homework assignments via the Internet. This proven and reliable homework system uses pedagogy and content from Nelson Education's best-selling statistics textbooks, then enhances it to help students visualize the problem-solving process and reinforce statistics concepts more effectively. **WA** encourages active learning and time on task, and respects diverse ways of learning.

NEW! For applicable books only: Tutorial assistance for problems assigned most often. The authors have analyzed **WebAssign** data and identified the most frequently assigned problems, which are designated in the text by a shaded box around the problem number. Each of these key problems provides students with targeted feedback—plus either a ***Watch It*** solution video or a ***Master It*** tutorial. Problems were revised to ensure they are suitable for the online environment and easy to assign online.

STUDENT ANCILLARIES

Student Solutions Manual

The **Student Solutions Manual** (ISBN 0-17-686108-4) provides students with fully worked-out solutions to the exercises with blue exercise numbers and headings in the text.

WebAssign

WebAssign

WebAssign® is a groundbreaking homework management system that combines our exceptional statistics content with the most flexible online homework solution. **WA** will engage you with immediate feedback, rich tutorial content, and interactive features that go far beyond simply duplicating text problems online. You can watch solution videos, see problems solved step-by-step, and receive feedback as you complete your homework. Visit webassign.com/ for more information and to access **WA** today! Enter the Online Access Code from the card included with your text. If a code card is **not** provided, you can purchase instant access at **NELSONbrain.com**.

ACKNOWLEDGMENTS

I would like to express my appreciation to my colleagues, staff members, and my several graduate students for their interest and support in preparing this edition.

The Fourth Canadian Edition would not have been possible without the assistance of the incredible Nelson Education team: publisher Jean Ford, production project manager Jaime Smith, and copyeditor Michael Kelly. Thanks also to marketing manager Kim Carruthers and to the sales team at Nelson Education Ltd. for supporting this text in the market. Thank you to content manager Courtney Thorne for her continued development support and expertise and technical editor Andie Burazin for her thorough review. Many thanks to Jazib Ahmed for his work finding new and interesting data and for his assistance in revising the material. Finally, I wish to extend my gratitude to past and present reviewers whose comments and suggestions guided the development of this fourth canadian edition:

Alan Majdanac, *Douglas College*
Julie Peschke, *Athabasca University*
Jaclyn Semple, *Trent University*
Wanhua Su, *MacEwan University*
Tim Swartz, *Simon Fraser University*

Finally, a special and heartfelt thank you to my wife, Ghazala; my son, Jazib; and my daughter, Feryaal. You are a constant source of joy and a great source of encouragement and motivation. I could not have done it without you!

S. Ejaz Ahmed
Brock University

ACKNOWLEDGMENTS

Joseph Sohm/Shutterstock.com

Introduction
Train Your Brain for Statistics

What is statistics? Have you ever met a statistician? Do you know what a statistician does? Perhaps you are thinking of the person who sits in the broadcast booth at the Stanley Cup playoffs, recording the number of shots against, goals scored, or goalies' save percentages. Or perhaps the mere mention of the word *statistics* sends a shiver of fear through you. You may think you know nothing about statistics; however, it is almost inevitable that you encounter statistics in one form or another every time you pick up a daily newspaper. Here is an example:

Election Opinion Polls

Polling data in 2015 shows that the Liberal Party has taken the lead over the Conservatives. More specifically, the survey result is given below.[1]

Latest National Poll (2015)

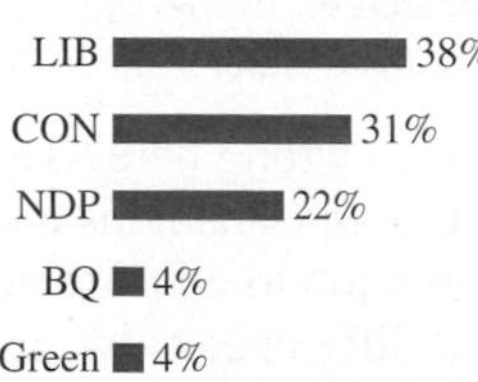

Source: © Election Almanac.

MISHELLA/Shutterstock.com.

How accurate are polling companies' reports of voter support? In the 2015 election, pollsters' predictions were between 1 and 3 percentage points off from the actual results.[2]

Articles similar to this one are commonplace in our newspapers and magazines, and in the period just prior to a national election, a new poll is reported almost every day. The language of this article is very familiar to us; however, it leaves the inquisitive reader with some unanswered questions. How were the people in the poll selected? Will these people give the same response tomorrow? Will they give the same response on election day? Will they even vote? Are these people representative of all those who will vote on election day? It is the job of a statistician to ask these questions and to find answers for them in the language of the poll.

> **Ignorance of 9/11 Is Not Fading Away Among Canadians and Americans**
> New polling data from surveys conducted by Angus Reid Global Monitor for CTV (May 05, 2009) indicate that only a small number of Canadians and Americans are aware of one specific detail related to the 9/11 attacks. According to the survey data, only 37% of Canadian respondents accurately reported that none of the terrorists who hijacked airplanes on September 11, 2001, entered the U.S. from Canada. On the other hand, just 14% of Americans reported accurately on this matter. Interestingly enough, in July 2004, the federal commission that investigated the events of 9/11 established that none of the 19 al-Qaeda operatives had entered the U.S. from either Canada or Mexico.

So what is wrong here? The survey results strongly suggest that there is a dire need to educate both Canadians and Americans on this misconception. Do you think it is a statistician's job to educate people on such issues? Further, can we say that there is a significant difference between the proportions of Canadians and Americans having this misconception? It is the job of statisticians to provide such a comparison in a convincing manner. Below are the specific responses to the survey question, "From what you

How many terrorists entered the U.S. through Canada?	Canada (%)	U.S. (%)
All of them	4	11
Some of them	38	29
None of them	37	14
Not sure	21	46

Source: © Angus Reid.

may have seen, read, or heard, how many of the terrorists who hijacked airplanes on September 11, 2001, entered the United States from Canada?"

Ontario Parents and Teachers Agree: Not Enough Classroom Time Dedicated to Learning about Agriculture

Nearly three-quarters (73%) of teachers and 7 in 10 (69%) parents feel not enough classroom time is dedicated to learning about farming or food production practices in Ontario.

These are the findings of an Ipsos Reid poll conducted on behalf of Ontario Agri-Food Education Inc. between May 2 and 12, 2006 (parents), and December 11 and 20, 2006 (teachers). The poll is based on a randomly selected sample of 704 Ontario parents of elementary or secondary school students and 211 Ontario elementary and secondary school teachers. For teachers, with a sample of this size, the results are considered accurate to within ± 6.8 percentage points, 19 times out of 20, of what they would have been had the entire population of full-time Ontario teachers (130,000 in English-speaking schools) been polled. For parents, with a sample of this size, the results are considered accurate to within ± 3.7 percentage points, 19 times out of 20, of what they would have been had the entire population of parents of elementary and secondary school students been polled. The margin of error will be larger within regions and for other subgroupings of the survey population. These data were statistically weighted to ensure the sample's regional and age/sex composition reflects that of the actual population according to 2001 Census data.

More specifically,

	Teachers (%)	Parents (%)
Not quite enough is spent on this topic	39	42
Definitely not enough is spent on this topic	34	27
Too much classroom time is spent on this topic	1	2

Further survey results showcase the following:

- Parents of elementary students (66%) are less likely than parents of secondary students (73%) to feel that not enough time is dedicated to food production practices.
- Elementary school teachers (72%) and secondary school teachers (73%) are equally likely to feel that not enough time is dedicated to food production practices.

Other agrifood topics that a large number of Ontario parents and teachers feel students do not spend enough classroom time learning about include:

Food safety practices, as they relate to farming in Ontario

- 67% of parents of elementary students and 74% of parents of secondary students feel not enough time is dedicated to food safety practices on Ontario farms.
- 74% of elementary school teachers and 71% of secondary school teachers feel not enough time is dedicated to food safety practices on Ontario farms.

Environmental practices on Ontario farms

- 70% of parents of elementary students and 77% of parents of secondary students feel not enough time is dedicated to environmental practices on Ontario farms.
- 76% of both elementary school and secondary school teachers feel not enough time is dedicated to environmental practices on Ontario farms.[3]
- The average area of liquid manure application increased, now accounting for just under one-third of the land. Farm operators are spreading manure over larger acreages, reflecting a trend towards better nutrient management planning partly due to the requirements of various provincial regulations.[4]

When you see an article like this one in a magazine, do you simply read the title and the first paragraph, or do you read further and try to understand the meaning of the numbers? How did the authors get these numbers? It is the job of the statistician to interpret the language of this study.

What Is Normal Body Temperature?

After believing for more than a century that 37°C (98.6°F) is the normal body temperature for humans, researchers now say normal is not normal anymore.

For some people at some hours of the day, 37.7°C could be fine. And readings as low as 34.8°C turn out to be highly human. Body temperature is very sensitive to hormone levels and may be higher or lower when a woman is having her menstrual period.

The 37°C standard was derived by a German doctor in 1868. Some physicians have always been suspicious of the good doctor's research. His claim: 1 million readings—in an epoch without computers.

So Mackowiak & Co. took temperature readings from 148 healthy people over a three-day period and found that the mean temperature was 36.78°C. Only 8 percent of the readings were 37.1°C.[5]

What questions come to your mind when you read this article? How did the researcher select the 148 people, and how can we be sure that the results based on these 148 people are accurate when applied to the general population? How did the researcher arrive at the normal "high" and "low" temperatures given in the article? How did the German doctor record 1 million temperatures in 1868? Again, we encounter a statistical problem with an application to everyday life.

Statistics is a branch of mathematics that has applications in almost every facet of our daily life. It is a new and unfamiliar language for most people, however, and, like any new language, statistics can seem overwhelming at first glance. We want you to "train your brain" to understand this new language *one step at a time*. Once the language of statistics is learned and understood, it provides a powerful data analytic tool in many different fields of application.

THE POPULATION AND THE SAMPLE

In the language of statistics, one of the most basic concepts is **sampling.** In most statistical problems, a specified number of measurements or data—a **sample**—is drawn from a much larger body of measurements, called the **population.**

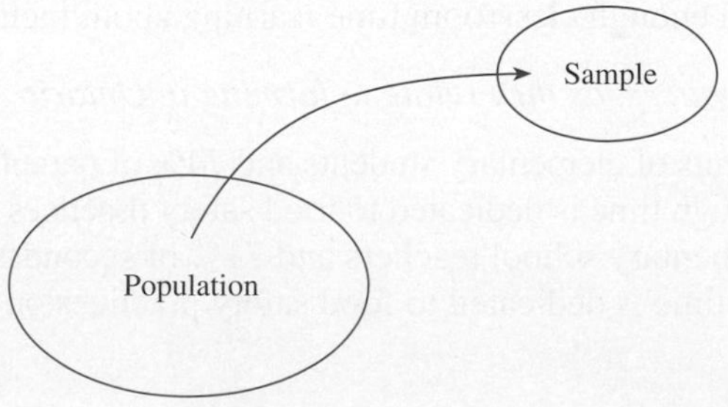

For the body-temperature experiment, the sample is the set of body-temperature measurements for the 148 healthy people chosen by the experimenter. We hope that the sample is representative of a much larger body of measurements—the population—the body temperatures of all healthy people in the world!

Which is of primary interest, the sample or the population? In most cases, we are interested primarily in the population, but the population may be difficult or impossible to enumerate. Imagine trying to record the body temperature of every healthy person on earth or the prime ministerial preference of every registered voter in Canada! Instead, **we try to describe or predict the behaviour of the population on the basis of information obtained from a representative sample from that population.**

The words *sample* and *population* have two meanings for most people. For example, you read in the newspapers that a strategic polling company poll conducted in Canada was based on a sample of 1000 people. Presumably, each person interviewed is asked a particular question, and that person's response represents a single measurement in the sample. Is the sample the set of 1000 people, or is it the 1000 responses that they give?

When we use statistical language, we distinguish between the set of objects on which the measurements are taken and the measurements themselves. To experimenters, the objects on which measurements are taken are called **experimental units.** The sample survey statistician calls them **elements of the sample.**

DESCRIPTIVE AND INFERENTIAL STATISTICS

When first presented with a set of measurements—whether a sample or a population—you need to find a way to organize and summarize it. The branch of statistics that presents techniques for describing sets of measurements is called **descriptive statistics.** You have seen descriptive statistics in many forms: bar charts, pie charts, and line charts presented by a political candidate; numerical tables in the newspaper; or the average rainfall amounts reported by the local television weather forecaster. Computer-generated graphics and numerical summaries are commonplace in our everyday communication.

DEFINITION **Descriptive statistics** consists of procedures used to summarize and describe the important characteristics of a set of measurements.

If the set of measurements is the entire population, you need only to draw conclusions based on the descriptive statistics. However, it might be too expensive or too time-consuming to enumerate the entire population. Perhaps enumerating the population would destroy it, as in the case of "time to failure" testing. For these or other reasons, you may have only a sample from the population. By looking at the sample, you want to answer questions about the population as a whole. The branch of statistics that deals with this problem is called **inferential statistics.**

DEFINITION **Inferential statistics** consists of procedures used to make inferences about population characteristics from information contained in a sample drawn from this population.

The **objective of inferential statistics** is to make inferences (that is, draw conclusions, make predictions, make decisions) about the characteristics of a population from information contained in a sample.

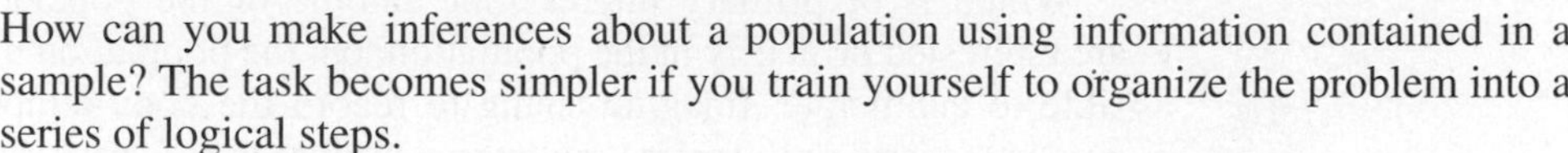

ACHIEVING THE OBJECTIVE OF INFERENTIAL STATISTICS: THE NECESSARY STEPS

How can you make inferences about a population using information contained in a sample? The task becomes simpler if you train yourself to organize the problem into a series of logical steps.

1. **Specify the questions to be answered and identify the population of interest.** In the national election poll, the objective is to determine who will get the most votes on election day. Hence, the population of interest is the set of all votes in the national election. When you select a sample, it is important that the sample be representative of *this* population, not the population of voter preferences on election day or on some other day prior to the election.
2. **Decide how to select the sample.** This is called the *design of the experiment* or the *sampling procedure*. Is the sample representative of the population of interest? For example, if a sample of registered voters is selected from the province of Quebec, will this sample be representative of all voters in Canada? Will it be the same as a sample of "likely voters"—those who are likely to actually vote in the election? Is the sample large enough to answer the questions posed in step 1 without wasting time and money on additional information? A good sampling design will answer the questions posed with minimal cost to the experimenter.
3. **Select the sample and analyze the sample information.** No matter how much information the sample contains, you must use an appropriate method of analysis to extract it. Many of these methods, which depend on the sampling procedure in step 2, are explained in the text.
4. **Use the information from step 3 to make an inference about the population.** Many different procedures can be used to make this inference, and some are better than others. For example, ten different methods might be available to estimate human response to an experimental drug, but one procedure might be more accurate than others. You should use the best inference-making procedure available (many of these are explained in the text).
5. **Determine the reliability of the inference.** Since you are using only a fraction of the population in drawing the conclusions described in step 4, you might be wrong! How can this be? If an agency conducts a statistical survey for you and estimates that your company's product will gain 34% of the market this year, how much confidence can you place in this estimate? Is this estimate accurate to within 1, 5, or 20 percentage points? Is it reliable enough to be used in setting production goals? Every statistical inference should include a measure of reliability that tells you how much confidence you have in the inference.

Now that you have learned some of the basic terms and concepts in the language of statistics, we again pose the question asked at the beginning of this discussion: Do you know what a statistician does? It is the job of the statistician to implement all of the preceding steps. This may involve questioning the experimenter to make sure that the population of interest is clearly defined, developing an appropriate sampling plan or experimental design to provide maximum information at minimum cost, correctly analyzing and drawing conclusions using the sample information, and finally measuring the reliability of the conclusions based on the experimental results.

KEYS FOR SUCCESSFUL LEARNING

As you begin to study statistics, you will find that there are many new terms and concepts to be mastered. Since statistics is an applied branch of mathematics, many of these basic concepts are mathematical—developed and based on results from calculus or higher mathematics. However, you do not have to be able to derive results in order to apply them in a logical way. In this text, we use numerical examples and commonsense arguments to explain statistical concepts, rather than more complicated mathematical arguments.

In recent years, computers have become readily available and an invaluable tool to many students. In the study of statistics, even the beginning student can use packaged programs to perform statistical analyses with a high degree of speed and accuracy. Some of the more common statistical packages available at computer facilities are *MINITAB*™, SAS (Statistical Analysis System), and SPSS (Statistical Package for the Social Sciences); personal computers will support packages such as *MINITAB*, *Microsoft Excel*®, and others. There are even online statistical programs and interactive "applets" on the Internet.

These programs, called **statistical software,** differ in the types of analyses available, the options within the programs, and the forms of printed results (called **output**). However, they are all similar. In this book, we use both *MINITAB* and *Excel* as statistical tools. Understanding the basic output of these packages will help you interpret the output from other software systems.

At the end of most chapters, you will find a section called "Technology Today." These sections present numerical examples to guide you through the *MINITAB* and *Excel* commands and options that are used for the procedures in that chapter. If you are using *MINITAB* or *Excel* in a lab or home setting, you may want to work through these sections at your own computer so that you become familiar with the hands-on methods in computer analysis. If you do not need hands-on knowledge of *MINITAB* or *Excel*, you may choose to skip this section and simply use the computer printouts for analysis as they appear in the text.

Another learning tool called statistical **applets** can be found on the textbook's website, www.probandstats3e.nelson.com. Also found on this website are explanatory sections called "Using the 4 Applets," which will help you understand how the applets can be used to visualize many of the chapter concepts. An accompanying section called "Applet APPs" provides some exercises (with solutions) that can be solved using the statistical applets. Whenever there is an applet available for a particular concept or application, you will find an icon in the left margin of the text, together with the name of the appropriate applet.

Most important, using statistics successfully requires common sense and logical thinking. For example, if we want to find the average height of all students at a particular university, would we select our entire sample from the members of the basketball team? In the body-temperature example, the logical thinker would question an 1868 average based on 1 million measurements—when computers had not yet been invented.

As you learn new statistical terms, concepts, and techniques, remember to view every problem with a critical eye and be sure that the rule of common sense applies.

Throughout the text, we will remind you of the pitfalls and dangers in the use or misuse of statistics. Benjamin Disraeli once said that there are three kinds of lies: *lies, damn lies,* and *statistics*! Our purpose is to dispel this claim—to show you how to make statistics *work* for you and not *lie* for you!

As you continue through the book, refer back to this introduction periodically. Each chapter will increase your knowledge of statistics and should, in some way, help you achieve one of the steps described here. Each of these steps is essential in attaining the overall objective of inferential statistics: to make inferences about a population using information contained in a sample drawn from that population.

Bildagentur Zoonar GmbH/Shutterstock.com

1

Describing Data with Graphs

GENERAL OBJECTIVES

Many sets of measurements are samples selected from larger populations. Other sets constitute the entire population, as in a national census. In this chapter, you will learn what a *variable* is, how to classify variables into several types, and how measurements or data are generated. You will then learn how to use graphs to describe data sets.

CHAPTER INDEX

- Data distributions and their shapes (1.1, 1.4)
- Univariate and bivariate data (1.1)
- Variables, experimental units, samples and populations data (1.1)
- Qualitative and quantitative variables—discrete and continuous (1.2)
- Pie charts, bar charts, line charts (1.3, 1.4)
- Dotplots (1.4)
- Stem and leaf plots (1.4)
- Relative frequency histograms (1.5)

NEED TO KNOW

How to Construct a Stem and Leaf Plot
How to Construct a Relative Frequency Histogram

How Is Your Blood Pressure?

Is your blood pressure normal, or is it too high or too low? The case study at the end of this chapter examines a large set of blood pressure data. You will use graphs to describe these data and to compare your blood pressure with others' of the same age and gender.

1.1 VARIABLES AND DATA

In Chapters 1 and 2, we will present some basic techniques in *descriptive statistics*—the branch of statistics concerned with describing sets of measurements, both *samples* and *populations.* In Chapter 4 we study *statistical inference*, which uses a sample to infer (i.e., to make statements) about the entire population. Once you have collected a set of measurements, how can you display this set in a clear, understandable, and readable form? First, you must be able to define what is meant by measurements or "data" and to categorize the types of data that you are likely to encounter in real life. We begin by introducing some definitions—new terms in the statistical language that you need to know.

DEFINITION A **variable** is a characteristic that changes or varies over time and/or for different individuals or objects under consideration.

For example, body temperature is a variable that changes over time within a single individual; it also varies from person to person. Religious affiliation, ethnic origin, income, height, age, and number of offspring are all variables—characteristics that vary depending on the individual chosen. A new social trend is social media where users create an online profile that extends from individual characteristics. The variables of interest of a social media profile would be number of friends, number of photos, location, and liked content.

In the Introduction, we defined an *experimental unit* or an *element of the sample* as the object on which a measurement is taken. Equivalently, we could define an experimental unit as the object on which a variable is measured. When a variable is actually measured on a set of experimental units, a set of measurements or **data** result.

DEFINITION An **experimental unit** is the individual or object on which a variable is measured. A single **measurement** or data value results when a variable is actually measured on an experimental unit.

If a measurement is generated for every experimental unit in the entire collection, the resulting data set constitutes the *population* of interest. Any smaller subset of measurements is a *sample.*

DEFINITION A **population** is the set of all measurements of interest to the investigator.

DEFINITION A **sample** is a subset of measurements selected from the population of interest.

EXAMPLE 1.1

A set of five students is selected from all undergraduates at a large Canadian university, and measurements are entered into a spreadsheet as shown in Figure 1.1. Identify the various elements involved in generating this set of measurements.

SOLUTION There are several *variables* in this example. The *experimental unit* on which the variables are measured is a particular undergraduate student on the campus, identified in column C1. Five variables are measured for each student: grade point average (GPA), gender, year in university, major, and current number of credit hours. Each of these characteristics varies from student to student. If we consider the GPAs of all students at this university to be the population of interest, the five GPAs in column C2 represent a *sample* from this population. If the GPA of each undergraduate student at the university had been measured, we would have generated the entire *population* of measurements for this variable.

FIGURE 1.1
Measurements on five undergraduate students

Worksheet 1 ***

↓	C1	C2	C3-T	C4-T	C5-T	C6
	Student	GPA	Gender	Year	Major	Number of Credit Hours
1	1	2.0	F	First	Psychology	16
2	2	2.3	F	Second	Mathematics	15
3	3	2.9	M	Second	English	17
4	4	2.7	M	First	English	15
5	5	2.6	F	Third	Business	14

The second variable measured on the students is gender, in column C3-T. This variable can take only one of two values—male (M) or female (F). It is not a numerically valued variable and hence is somewhat different from GPA. The population, if it could be enumerated, would consist of a set of Ms and Fs, one for each student at the university. Similarly, the third and fourth variables, year and major, generate nonnumerical data. Year has four categories (first, second, third, fourth), and major has one category for each undergraduate major on campus. The last variable, current number of credit hours, is numerically valued, generating a set of numbers rather than a set of qualities or characteristics.

Although we have discussed each variable individually, remember that we have measured each of these five variables on a single experimental unit: the student. Therefore, in this example, a "measurement" really consists of five observations, one for each of the five measured variables. For example, the measurement taken on student 2 produces this observation:

(2.3, F, Second, Mathematics, 15)

You can see that there is a difference between a *single* variable measured on a single experimental unit and *multiple* variables measured on a single experimental unit as in Example 1.1.

EXAMPLE 1.2

A city roads department in Ottawa would like to repair some potholes on the busiest section of a road. The best way to do the repair is when the flow of traffic is low to avoid traffic congestion and inconvenience for drivers. Practically, it would be expensive and time consuming if not impossible to monitor and record the traffic flow on every day. It was decided to record how many vehicles pass on this section of road during one particular day. Using this data information, the department decides when to repair the potholes.

Note that in this example, population constitutes all the automobiles that flow on all days on that section of the road and the sample is defined as the automobiles that passed on that section of the road on the particular day.

EXAMPLE 1.3

Analytics in the NBA has become popular for evaluating players and especially for drafting college players who are unknown commodities. A simple method would be to compare a player's profile, such as height, wingspan, accolades, stats, field goal percentage, and his related shot chart distribution, to a player in the NBA to find out how the player could potentially fare.

Note that in this example, the population includes all college players who are eligible for the NBA Draft and players who are currently or have played in the NBA. The sample may be defined as players who have played minimum of three seasons.

DEFINITION **Univariate data** result when a single variable is measured on a single experimental unit.

DEFINITION **Bivariate data** result when two variables are measured on a single experimental unit. **Multivariate data** result when more than two variables are measured.

If you measure the body temperatures of 148 people, the resulting data are *univariate.* In Example 1.1, five variables were measured on each student, resulting in *multivariate* data.

TYPES OF VARIABLES

Variables can be classified into one of two categories: **qualitative** or **quantitative.**

DEFINITION **Qualitative variables** measure a quality or characteristic on each experimental unit. **Quantitative variables** measure a numerical quantity or amount on each experimental unit.

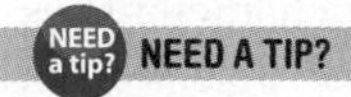

Qualitative ⇔ "quality" or characteristic
Quantitative ⇔ "quantity" or number

Qualitative variables produce data that can be categorized according to similarities or differences in kind; hence, they are often called **categorical data.** The variables gender, year, and major in Example 1.1 are qualitative variables that produce categorical data. Here are some other examples:

- Political affiliation: Liberals, Conservatives, NDP, Green, Independent
- Taste ranking: excellent, good, fair, poor
- Colour of an M&M'S® candy: brown, yellow, red, orange, green, blue
- Type of baseball pitches: fastball, breaking ball, changeup, special pitches

Quantitative variables, often represented by the letter x, produce numerical data, such as those listed here:

- $x =$ Prime interest rate
- $x =$ Number of unregistered taxicabs in a city
- $x =$ Weight of a package ready to be shipped
- $x =$ Volume of orange juice in a glass
- $x =$ Number of songs downloaded from iTunes for an album

Notice that there is a difference in the types of numerical values that these quantitative variables can assume. The number of unregistered taxicabs, for example, can take on only the values $x = 0, 1, 2, \ldots,$ whereas the weight of a package can take on any value greater than zero, or $0 < x < \infty$. To describe this difference, we define two types of quantitative variables: **discrete** and **continuous.**

DEFINITION A **discrete variable** can assume only a finite or countable number of values. A **continuous variable** can assume the infinitely many values corresponding to the points on a line interval.

The name *discrete* relates to the discrete gaps between the possible values that the variable can assume. Variables such as number of family members, number of new

NEED A TIP?

Discrete ⇔ "listable"

Continuous ⇔ "unlistable"

car sales, and number of defective tires returned for replacement are all examples of discrete variables. On the other hand, variables such as height, weight, time, distance, and volume are *continuous* because they can assume values at any point along a line interval. For any two values you pick, a third value can always be found between them!

EXAMPLE 1.4

Identify each of the following variables as qualitative or quantitative:

1. The most frequent use of your microwave oven (reheating, defrosting, warming, other)
2. The number of consumers who refuse to answer a telephone survey
3. The door chosen by a mouse in a maze experiment (A, B, or C)
4. The winning time for a horse running at the Woodbine racetrack, Toronto
5. The number of children in a Grade 5 class who are reading at or above grade level
6. The number of advertisements clicked while watching a YouTube video

NEED A TIP?

Discrete variables often involve the "number of" items in a set.

SOLUTION Variables 1 and 3 are both *qualitative* because only a quality or characteristic is measured for each individual. The categories for these two variables are shown in parentheses. The other three variables are *quantitative*. Variable 2, the number of consumers, is a *discrete* variable that can take on any of the values $x = 0, 1, 2, \ldots$, with a maximum value depending on the number of consumers called. Similarly, variable 5, the number of children reading at or above grade level, can take on any of the values $x = 0, 1, 2, \ldots$, with a maximum value depending on the number of children in the class. Variable 4, the winning time for a Woodbine horse, is the only *continuous* variable in the list. The winning time, if it could be measured with sufficient accuracy, could be 121 seconds, 121.5 seconds, 121.25 seconds, or any values between any two times we have listed.

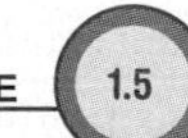

EXAMPLE 1.5

Identify each of the following quantitative variables as discrete or continuous:

a. Average daily temperature, for a small city in Quebec during a summer month
b. Number of bees on a flower
c. Driving time between Regina, Saskatchewan, and Winnipeg, Manitoba
d. Number of passengers (excluding the airline staff) on a flight from Edmonton to Vancouver
e. Amount of propane gas for a BBQ cylinder filled at Costco in Montreal
f. Guessing the number of stars in the sky

SOLUTION

a. Continuous variable: The temperature, if it could be measured with a reasonable accuracy, could be 22.5° Celsius, 26.1° Celsius, 21.9° Celsius, or any other possible values.
b. Discrete variable: It can take any value from 0 to 5 or more.
c. Continuous variable: The time could be reasonably measured, per say 6 hours 20 minutes, or 7 hours 44 minutes, or even 6 hours 11 minutes and 50 seconds, if measured in seconds too. This variable can take any possible value on a certain time interval.
d. Discrete variable: Like in part b, depending on the size of plane, it can take any value from 0 to a maximum number, depending on how many seats were sold and how many passengers showed up for this particular flight.

e. Continuous variable: A typical 13.6 kg (30 lb) cylinder holds approximately 9 kg (20 lb) of propane. This extra unfilled volume leaves some room for the liquid to expand. However, in reality, fill, if accurately measured, could be 9.12 kg, 8.93 kg, or 9.5 kg.

f. Discrete variable: It is a form of counting.

Figure 1.2 depicts the types of data we have defined. Why should you be concerned about different kinds of variables and the data that they generate? The reason is that the methods used to describe data sets depend on the type of data you have collected. For each set of data that you collect, the key will be to determine what type of data you have and how you can present them most clearly and understandably to your audience!

FIGURE 1.2
Types of data

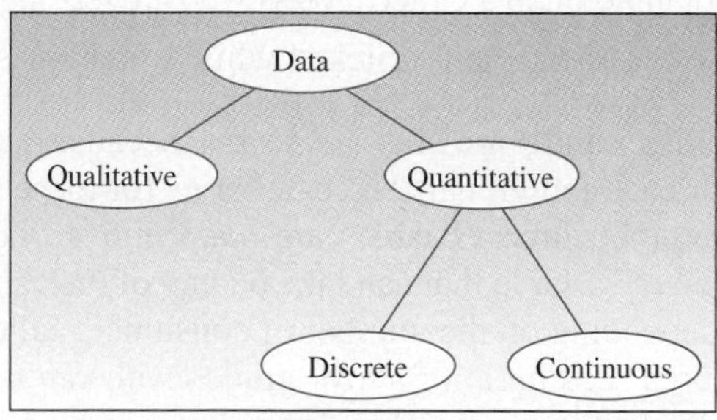

1.3 GRAPHS FOR CATEGORICAL DATA

After the data have been collected, they can be consolidated and summarized to show the following information:

- What values of the variable have been measured
- How often each value has occurred

For this purpose, you can construct a *statistical table* that can be used to display the data graphically as a data distribution. The type of graph you choose depends on the type of variable you have measured.

When the variable of interest is *qualitative*, the statistical table is a list of the categories being considered along with a measure of how often each value occurred. You can measure "how often" in three different ways:

- The **frequency,** or number of measurements in each category
- The **relative frequency,** or proportion of measurements in each category
- The **percentage** of measurements in each category

For example, if you let n be the total number of measurements in the set, you can find the relative frequency and percentage using these relationships:

$$\text{Relative frequency} = \frac{\text{Frequency}}{n}$$

$$\text{Percent} = 100 \times \text{Relative frequency}$$

You will find that the sum of the frequencies is always n, the sum of the relative frequencies is 1, and the sum of the percentages is 100%.

The categories for a qualitative variable should be chosen so that

- a measurement will belong to one and only one category
- each measurement has a category to which it can be assigned

NEED A TIP?

Three steps to a data distribution:
(1) raw data ⇒
(2) statistical table ⇒
(3) graph

For example, if you categorize meat products according to the type of meat used, you might use these categories: beef, chicken, seafood, pork, turkey, other. To categorize ranks of university faculty, you might use these categories: professor, associate professor, assistant professor, instructor, lecturer, other. The "other" category is included in both cases to allow for the possibility that a measurement cannot be assigned to one of the earlier categories.

Once the measurements have been categorized and summarized in a *statistical table*, you can use either a pie chart or a bar chart to display the distribution of the data. A **pie chart** is the familiar circular graph that shows how the measurements are distributed among the categories. A **bar chart** shows the same distribution of measurements in categories, with the height of the bar measuring how often a particular category was observed.

EXAMPLE 1.6

In a survey concerning public education, 400 school administrators were asked to rate the quality of education in Canada. Their responses are summarized in Table 1.1. Construct a pie chart and a bar chart for this set of data.

SOLUTION To construct a pie chart, assign one sector of a circle to each category. The angle of each sector should be proportional to the proportion of measurements (or *relative frequency*) in that category. Since a circle contains 360°, you can use this equation to find the angle:

$$\text{Angle} = \text{Relative frequency} \times 360^\circ$$

NEED A TIP?

Proportions add to 1.
Percents add to 100.
Sector angles add to 360°.

Table 1.2 shows the ratings along with the frequencies, relative frequencies, percentages, and sector angles necessary to construct the pie chart. Figure 1.3 shows the pie chart constructed from the values in the table. While pie charts use percentages to determine the relative sizes of the "pie slices," bar charts usually plot frequency against the categories. A bar chart for these data is shown in Figure 1.4.

The visual impact of these two graphs is somewhat different. The pie chart is used to display the relationship of the parts to the whole; the bar chart is used to emphasize the actual quantity or frequency for each category. Since the categories in this example are ordered "grades" (A, B, C, D), we would not want to rearrange the bars in the chart to change its *shape.* In a pie chart, the order of presentation is irrelevant.

TABLE 1.1 Canadian Education Rating by 400 Educators

Rating	Frequency
A	35
B	260
C	93
D	12
Total	400

TABLE 1.2 Calculations for the Pie Chart in Example 1.6

Rating	Frequency	Relative Frequency	Percent	Angle
A	35	35/400 = 0.09	9	0.09 × 360 = 32.4°
B	260	260/400 = 0.65	65	234.0°
C	93	93/400 = 0.23	23	82.8°
D	12	12/400 = 0.03	3	10.8°
Total	400	1.00	100%	360°

FIGURE 1.3
Pie chart for Example 1.6

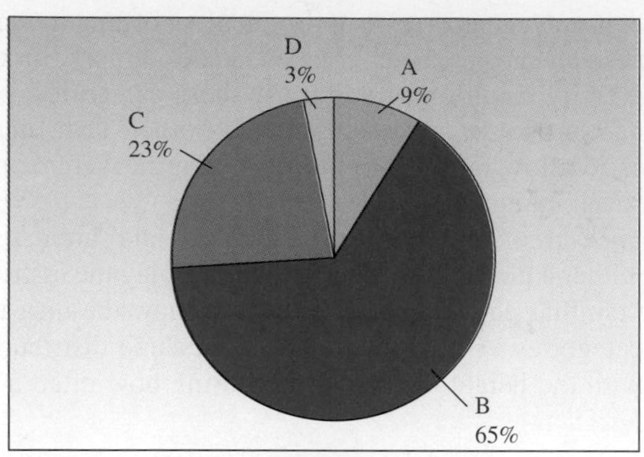

FIGURE 1.4
Bar chart for Example 1.6

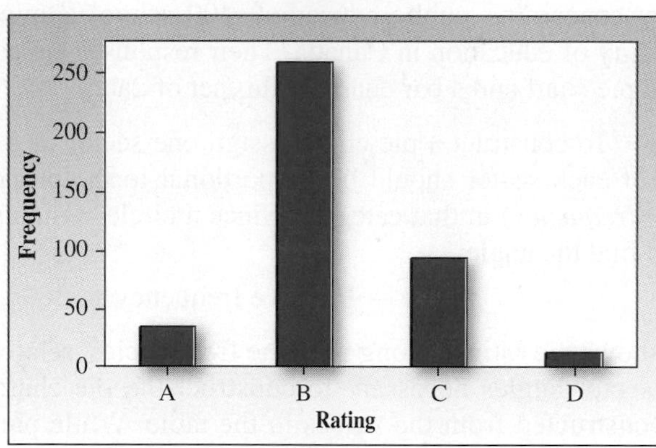

EXAMPLE 1.7

A snack size bag of peanut M&M'S® candies contains 21 candies with the colours listed in Table 1.3. The variable "colour" is *qualitative*, so Table 1.4 lists the six categories along with a tally of the number of candies of each colour. The last three columns of Table 1.4 give the three different measures of how often each category occurred. Since the categories are colours and have no particular order, you could construct bar charts with many different *shapes* just by reordering the bars. To emphasize that brown is the most frequent colour, followed by blue, green, and orange, we order the bars from largest to smallest and generate the bar chart using *MINITAB* in Figure 1.5. A bar chart in which the bars are ordered from largest to smallest is called a **Pareto chart.**

TABLE 1.3 Raw Data: Colours of 21 Candies

Brown	Green	Brown	Blue
Red	Red	Green	Brown
Yellow	Orange	Green	Blue
Brown	Blue	Blue	Brown
Orange	Blue	Brown	Orange
Yellow			

TABLE 1.4 Statistical Table: M&M'S® Data for Example 1.7

Category	Tally	Frequency	Relative Frequency	Percent
Brown	~~IIII~~ I	6	6/21	28
Green	III	3	3/21	14
Orange	III	3	3/21	14
Yellow	II	2	2/21	10
Red	II	2	2/21	10
Blue	~~IIII~~	5	5/21	24
Total		21	1	100%

FIGURE 1.5
MINITAB bar chart for Example 1.7

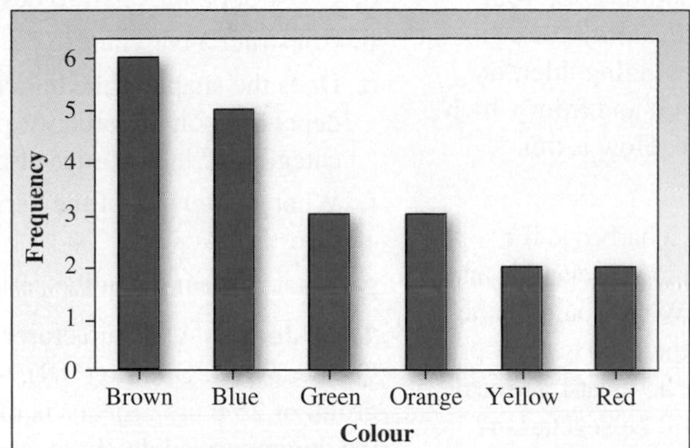

1.3 EXERCISES

UNDERSTANDING THE CONCEPTS

1.1 Experimental Units Identify the experimental units on which the following variables are measured:

a. Gender of a student

b. Number of errors on a midterm exam

c. Age of a cancer patient

d. Number of flowers on an azalea plant

e. Colour of a car entering the parking lot

1.2 Qualitative or Quantitative? Identify each variable as quantitative or qualitative:

a. Amount of time it takes to assemble a simple puzzle

b. Number of students in a Grade 1 classroom

c. Rating of a newly elected politician (excellent, good, fair, poor)

d. Province or territory in which a person lives

e. The number that lands on a dice roll

1.3 Discrete or Continuous? Identify the following quantitative variables as discrete or continuous:

a. Population in a particular area of Canada

b. Weight of newspapers recovered for recycling on a single day

c. Time to complete a sociology exam

d. Number of consumers in a poll of 1000 who consider nutritional labelling on food products to be important

e. Number of repetitions a powerlifter does on a bench press

1.4 Discrete or Continuous? Identify each quantitative variable as discrete or continuous.

a. Number of boating accidents along a 50-kilometre stretch of the St. Lawrence River

b. Time required to complete a questionnaire

c. Choice of colour for a new refrigerator

d. Number of brothers and sisters you have

e. Yield in kilograms of wheat from a 10,000-square-metre plot in a wheat field

f. The time of a rugby player's 40-yard dash

1.5 Parking on Campus Six vehicles are selected from the vehicles that are issued campus parking permits, and the following data are recorded:

Vehicle	Type	Make	Carpool?	One-way Commute Distance (kilometres)	Age of Vehicle (years)
1	Car	Honda	No	23.6	6
2	Car	Toyota	No	17.2	3
3	Truck	Toyota	No	10.1	4
4	Van	Dodge	Yes	31.7	2
5	Motor-cycle	Harley-Davidson	No	25.5	1
6	Car	Chevrolet	No	5.4	9

a. What are the experimental units?

b. What are the variables being measured? What types of variables are they?

c. Is this univariate, bivariate, or multivariate data?

1.6 Past Canadian Prime Ministers A data set consists of the ages at death for each of the 15 past prime ministers of Canada.

a. Is this set of measurements a population or a sample?

b. What is the variable being measured?

c. Is the variable in part b quantitative or qualitative?

1.7 Voter Attitudes You are a candidate for your provincial assembly, and you want to survey voter attitudes regarding your chances of winning. Identify the population that is of interest to you and from which you would like to select your sample. How is this population dependent on time?

1.8 BBQ Tonight You are hosting a barbecue for your neighbourhood block. How would you classify and identify your community members? What would be the course of action in making sure that the food is appropriate for them? For example, will there be meat-free meals for vegetarians? How many dishes will be vegetarian?

1.9 Cancer Survival Times A medical researcher wants to estimate the survival time of a patient after the onset of a particular type of cancer and after a particular regimen of radiotherapy.

a. What is the variable of interest to the medical researcher?

b. Is the variable in part a qualitative, quantitative discrete, or quantitative continuous?

c. Identify the population of interest to the medical researcher.

d. Describe how the researcher could select a sample from the population.

e. What problems might arise in sampling from this population?

1.10 New Teaching Methods An educational researcher wants to evaluate the effectiveness of a new method for teaching reading to deaf students. Achievement at the end of a period of teaching is measured by a student's score on a reading test.

a. What is the variable to be measured? What type of variable is it?

b. What is the experimental unit?

c. Identify the population of interest to the experimenter.

BASIC TECHNIQUES

1.11 Basic Statistics Fifty people are grouped into four categories—A, B, C, and D—and the number of people who fall into each category is shown in the table:

Category	Frequency
A	11
B	14
C	20
D	5

a. What is the experimental unit?

b. What is the variable being measured? Is it qualitative or quantitative?

c. Construct a pie chart to describe the data.

d. Construct a bar chart to describe the data.

e. Does the shape of the bar chart in part d change depending on the order of presentation of the four categories? Is the order of presentation important?

f. What *proportion* of the people are in category B, C, or D?

g. What *percentage* of the people are *not* in category B?

1.12 Jeans A manufacturer of jeans has plants in Quebec (QC), Ontario (ON), and Manitoba (MB). A group of 25 pairs of jeans is randomly selected from the computerized database, and the province in which each is produced is recorded:

ON	QC	QC	MB	ON
ON	ON	MB	MB	MB
QC	QC	ON	QC	MB
ON	QC	MB	MB	MB
ON	QC	QC	ON	ON

a. What is the experimental unit?

b. What is the variable being measured? Is it qualitative or quantitative?

c. Construct a pie chart to describe the data.

d. Construct a bar chart to describe the data.

e. What *proportion* of the jeans are made in Quebec?

f. What province produced the most jeans in the group?

g. If you want to find out whether the three plants produced equal numbers of jeans, or whether one produced more jeans than the others, how can you use the charts from parts c and d to help you? What conclusions can *you* draw from these data?

APPLICATIONS

1.13 Clash between Islam, West Is Political, Majority of Canadians Say The clash of civilizations is a notion popularized by Harvard University professor Samuel P. Huntington, who theorized that the principal source of conflict in the post–Cold War world will spring from cultural and religious divisions. Most Canadians reject the notion that the Islamic and Western worlds are engaged in a clash of civilizations based on culture and religion, according to an international poll published in the *Ottawa Citizen*, February 19, 2007. Pollsters interviewed 1008 Canadians in December and January as part of an international survey in 27 countries. In theory, in 19 cases out of 20, the results of the poll would not differ by more than 3 percentage points from those obtained by interviewing all Canadian adults. The GlobeScan survey found that a majority (56%) of Canadians regard the conflict between Islam and the West to be primarily about "political power and interests." Only 29% said religious and cultural differences are to blame. Thus,

Political power and interests	56%
Religious and cultural differences	29%

a. If the pollsters were planning to use these results to predict opinion for 2008, describe the population of interest to them.

b. Describe the actual population from which the sample was drawn.

c. Are all of the reasons accounted for in this table? Add another category if necessary.

d. Is the sample selected by the pollsters representative of the population described in part a? Explain.

1.14 Clash between Islam, West, continued Refer to Exercise 1.13. On the other hand, worldwide, Muslims tended to put most of the blame on politics. The data from four predominately Muslim countries is given below:

Lebanon	78%
Egypt	56%
Indonesia	56%
Turkey	55%

a. Are all of the Islamic countries accounted for in this table? Suggest a few more Islamic countries if necessary.

b. Would you use a pie chart or a bar chart to graphically describe the data? Why?

c. Draw the chart you chose in part b.

d. If you were the person conducting the opinion poll, what other types of questions might you want to investigate?

1.15 Education Attainment The data below reports educational attainment of the Aboriginal and non-Aboriginal population in Ontario, aged 15 years and over (2001 Census).

	Aboriginal Population (%)	Non-Aboriginal Population (%)
University degree	6.00	17.70
Trades, college, university certificate/diploma	28.00	27.10
Some post-secondary	12.20	11.20
High school only	12.10	14.40
Less than high school	41.70	21.60

a. Define the variable that has been measured in this table.

b. Is the variable quantitative or qualitative?

c. What do the numbers represent?

d. Construct a pie chart to describe the education levels in the Aboriginal population.

e. Construct a bar chart to describe the education levels in the non-Aboriginal population.

f. What percentage of the members of the Aboriginal population have a university degree? Is there a significant gap between the education attainment of the Aboriginal and the non-Aboriginal population?

1.16 Education Attainment, continued The educational attainment of the Aboriginal population in Canada, aged 25–64, for the years 2001 and 1996 is displayed in two bar charts (stacked), respectively.

a. Are all of the education levels accounted for in the graph? Add another category if necessary.

b. Have any improvements in Aboriginal educational attainment been made over the years?

c. Use a pie chart to describe the data. Which graph is more interesting to look at?

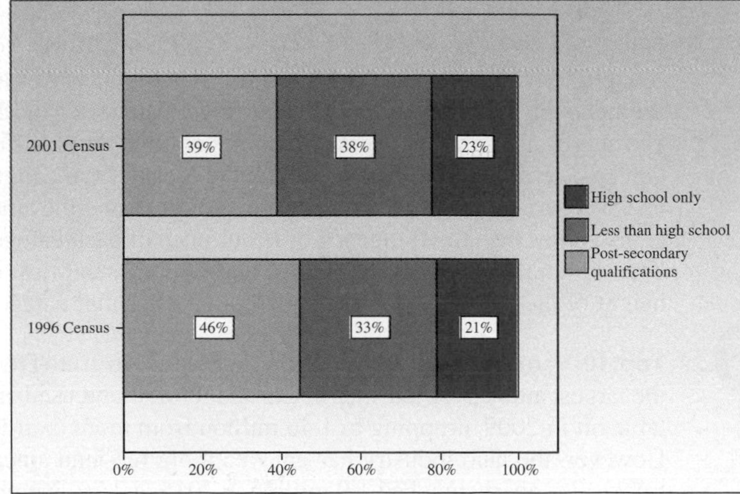

1.17 Facebook Fanatics The social networking site called Facebook has grown quickly since its inception in 2004. Social networking statistics[1] show that Facebook penetration in Canada is 50.92% compared to the country's population, and 65.55% in relation to number of Internet users. The total number of Facebook users in Canada is reaching 17,190,240.

The largest age group is currently 18–24 with a total of 4,125,658 users, followed by users in the age range of 25–34.

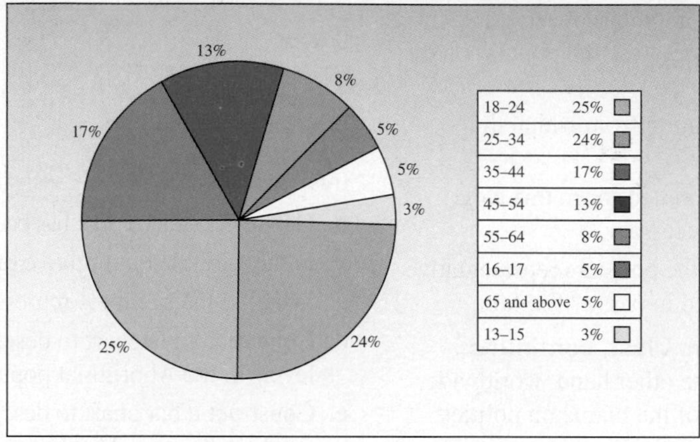

From http://www.socialbakers.com/facebook-statistics/canada © Socialbakers.

a. Define the variable that has been measured in the pie chart.

b. Is the variable quantitative or qualitative?

c. What do the percentages represent?

d. Is the pie chart drawn accurately? That is, are the sections in the correct proportion to each other?

e. Construct a bar chart to describe the age distribution.

f. Would you use a pie chart or a bar chart to graphically describe the data? Why? Which graph is more interesting to look at?

g. If you were the person collecting the data, what other types of questions might you want to investigate?

GRAPHS FOR QUANTITATIVE DATA

Quantitative variables measure an amount or quantity on each experimental unit. If the variable can take only a finite or countable number of values, it is a *discrete* variable. A variable that can assume an infinite number of values corresponding to points on a line interval is called *continuous.*

Pie Charts and Bar Charts

Sometimes information is collected for a quantitative variable measured on different segments of the population, or for different categories of classification. For example, you might measure the average incomes for people of different age groups, different genders, or living in different geographic areas of the country. In such cases, you can use pie charts or bar charts to describe the data, using the amount measured in each category rather than the frequency of occurrence of each category. The *pie chart* displays how the total quantity is distributed among the categories, and the *bar chart* uses the height of the bar to display the amount in a particular category.

EXAMPLE 1.8

Top 10 Vehicles Sold In Canada According to AutoTRADER.ca, which is one of the largest auto classified sites in Canada for new and used cars, sales figures took a notable hit in 2009, dropping to 1.46 million from more than 1.6 million the year before. However, the auto industry has enjoyed a big rebound since then: New car and light-truck sales nearly touched 1.9 million in 2015, a 2.7% increase from the previous year.

The auto industry has significantly recovered from the peak of the recession in 2008–09.[2] The Table 1.5 below shows the top 10 vehicles sold in Canada for 2015. Construct both a pie chart and a bar chart to describe the data.

TABLE 1.5 **Quantity of Vehicles Sold**

Vehicle	Units Sold
Ford F-150	118,837
Ram 1500	91,195
Honda Civic	64,950
GMC Sierra	53,727
Ford Escape	47,726
Hyundai Elantra	47,722
Toyota Corolla	47,198
Dodge Grand Caravan	46,927
Chevrolet Silverado	46,407
Toyota RAV4	42,246

Source: Chris Chase, "Canada's 25 best-selling cars in 2015," autotrader.ca, January 6, 2016. http://www.autotrader.ca/newsfeatures/20160106/canadas-25-best-selling-cars-in-2015/#TOGHLkgmDrcrf6ez.97 © autoTRADER.ca

SOLUTION The variables being measured are vehicle make (qualitative) and amount of units sold (quantitative). The bar chart in Figure 1.6 displays the categories on the horizontal axis and amount on the vertical axis.

FIGURE 1.6
Bar chart for Example 1.8

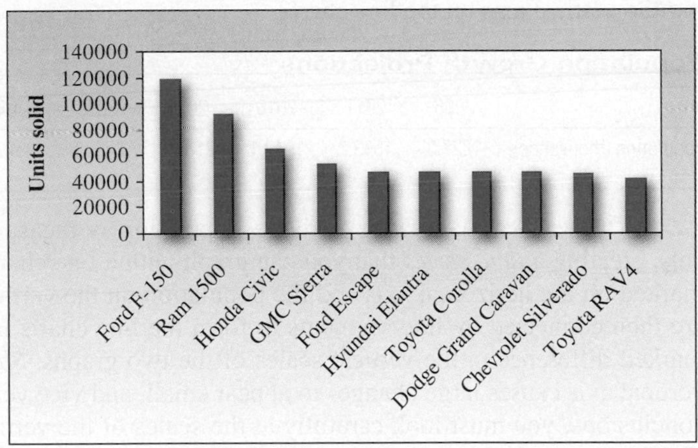

For the pie graph in Figure 1.7, each "pie slice" represents the proportion of the total top 10 vehicles sold (606,935). For example, for Toyota Corolla, the angle of the sector is

$$\frac{47,198}{606,935} \times 360° = 28.0°$$

Both graphs show that the highest number of units sold were of the Ford F-150. Since there is no inherent order to the categories, you are free to rearrange the bars or sectors of the graphs in any way you like. The *shape* of the bar chart has no bearing on its interpretation.

Line Charts

When a quantitative variable is recorded over time at equally spaced intervals (such as daily, weekly, monthly, quarterly, or yearly), the data set forms a **time series.** Time series data are most effectively presented on a **line chart** with time as the horizontal axis. The idea is to try to discern a pattern or **trend** that will likely continue into

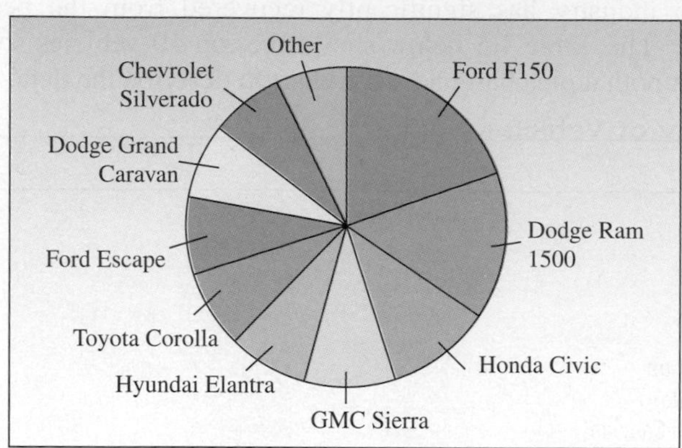

FIGURE 1.7
Pie chart for Example 1.8

the future, and then to use that pattern to make accurate predictions for the immediate future.

EXAMPLE 1.9

In the year 2030, the oldest "baby boomers" (born in 1946) will be 84 years old, and the oldest "Gen-Xers" (born in 1965) will be eligible to collect Canada Pension Plan (CPP) benefits. How will this affect consumer trends in the next 25 years? Statistics Canada gives projections for the age group 65–69 years, as shown in Table 1.6 below. Construct a line chart to illustrate the data. What is the effect of stretching and shrinking the vertical axis on the line chart?

TABLE 1.6 **Population Growth Projections**

Year	2006	2011	2016	2021	2026	2031
Population (thousands)	1227.3	1513.1	1942.1	2184.7	2466.6	2527.6

NEED A TIP?
Beware of stretching or shrinking axes when you look at a graph!

SOLUTION The quantitative variable population is measured over six time intervals, creating a *time series* that you can graph with a line chart. The time intervals are marked on the horizontal axis and the population on the vertical axis. The data points are then connected by line segments to form the line charts in Figure 1.8. Notice the marked difference in the vertical scales of the two graphs. *Shrinking* the scale on the vertical axis causes large changes to appear small, and vice versa. To avoid misleading conclusions, you must look carefully at the scales of the vertical and horizontal axes. However, from both graphs you get a clear picture of the steadily increasing numbers in the early years of the 21st century.

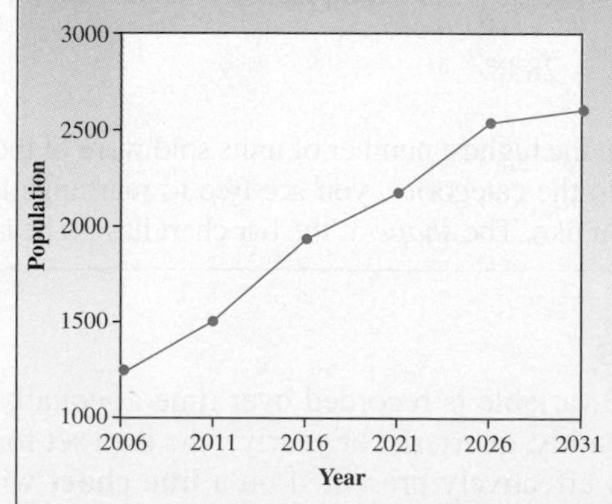

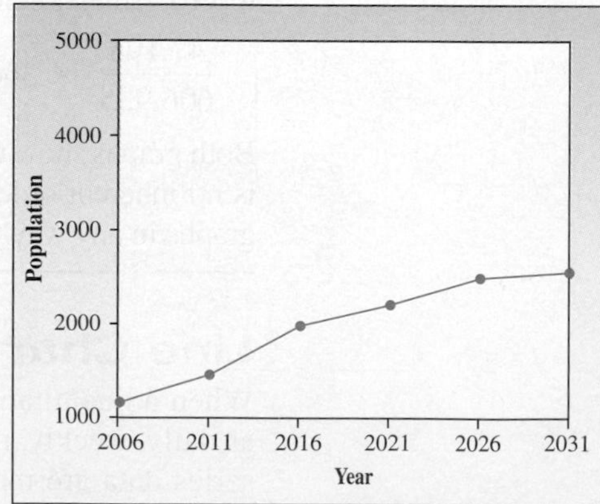

FIGURE 1.8
Line charts for Example 1.9

Dotplots

Many sets of quantitative data consist of numbers that cannot easily be separated into categories or intervals of time. You need a different way to graph this type of data!

The simplest graph for quantitative data is the **dotplot.** For a small set of measurements—for example, the set 2, 6, 9, 3, 7, 6—you can simply plot the measurements as points on a horizontal axis. This dotplot, generated by *MINITAB*, is shown in Figure 1.9(a). For a large data set, however, such as the one in Figure 1.9(b), the dotplot can be uninformative and tedious to interpret.

FIGURE 1.9
Dotplots for small and large data sets

(a)

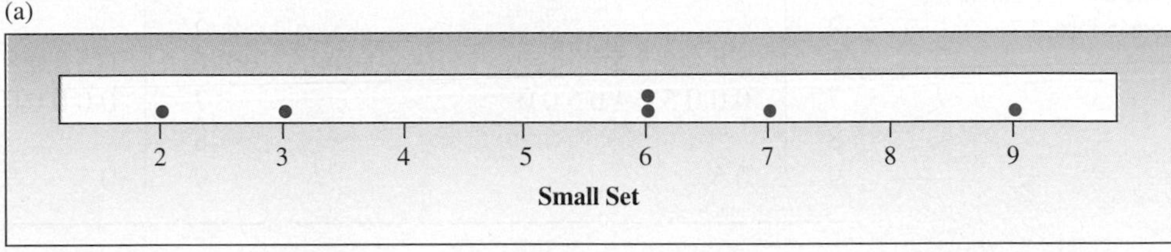

(b)

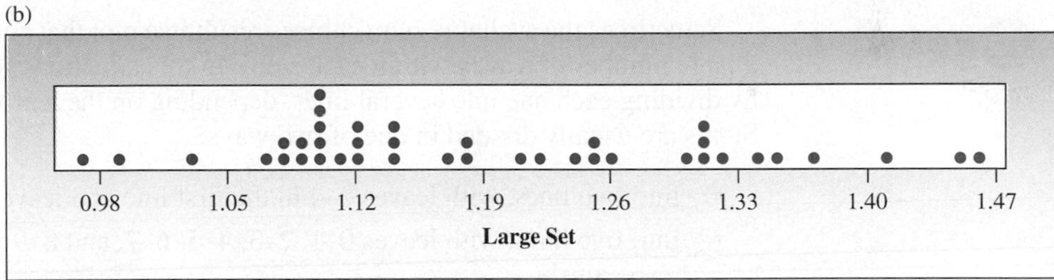

Stem and Leaf Plots

Another simple way to display the distribution of a quantitative data set is the **stem and leaf plot.** This plot presents a graphical display of the data using the actual numerical values of each data point.

NEED TO KNOW

How to Construct a Stem and Leaf Plot

1. Divide each measurement into two parts: the **stem** and the **leaf.**
2. List the stems in a column, with a vertical line to their right.
3. For each measurement, record the leaf portion in the same row as its corresponding stem.
4. Order the leaves from lowest to highest in each stem.
5. Provide a key to your stem and leaf coding so that the reader can re-create the actual measurements if necessary.

EXAMPLE  1.10

Table 1.7 lists the prices (in dollars) of 19 different brands of walking shoes. Construct a stem and leaf plot to display the distribution of the data.

TABLE 1.7 **Prices of Walking Shoes**

90	70	70	70	75	70
65	68	60	74	70	95
75	70	68	65	40	65
70					

SOLUTION To create the stem and leaf plot, you could divide each observation between the ones and the tens place. The number to the left is the stem; the number to the right is the leaf. Thus, for the shoes that cost $65, the stem is 6 and the leaf is 5. The stems, ranging from 4 to 9, are listed in Figure 1.10, along with the leaves for each of the 19 measurements. If you indicate that the leaf unit is 1, the reader will realize that the stem and leaf 6 and 8, for example, represent the number 68, recorded to the nearest dollar.

FIGURE 1.10
Stem and leaf plot for the data in Table 1.7

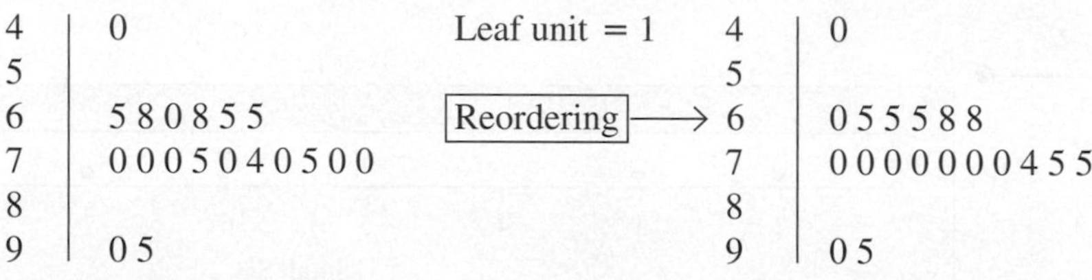

NEED A TIP?
stem | leaf

Sometimes the available stem choices result in a plot that contains too few stems and a large number of leaves within each stem. In this situation, you can stretch the stems by dividing each one into several lines, depending on the leaf values assigned to them. Stems are usually divided in one of two ways:

- Into two lines, with leaves 0–4 in the first line and leaves 5–9 in the second line
- Into five lines, with leaves 0–1, 2–3, 4–5, 6–7, and 8–9 in the five lines, respectively

EXAMPLE 1.11

The data in Table 1.8 are the GPAs of 30 first-year university students, recorded at the end of the first year. Construct a stem and leaf plot to display the distribution of the data.

TABLE 1.8 **Grade Point Averages of 30 First-Year University Students**

2.0	3.1	1.9	2.5	1.9
2.3	2.6	3.1	2.5	2.1
2.9	3.0	2.7	2.5	2.4
2.7	2.5	2.4	3.0	3.4
2.6	2.8	2.5	2.7	2.9
2.7	2.8	2.2	2.7	2.1

SOLUTION The data, though recorded to an accuracy of only one decimal place, are measurements of the continuous variable X = GPA, which can take on values in the interval 0–4.0. By examining Table 1.8, you can quickly see that the highest and lowest GPAs are 3.4 and 1.9, respectively. But how are the remaining GPAs distributed? If you use the decimal point as the dividing line between the stem and the leaf, you have only three stems, which does not produce a very good picture. Even if you divide each stem into two lines, there are only four stems, since the first line of stem 1 and the second line of stem 4 are empty! Dividing each stem into five lines produces the most descriptive plot, as shown in Figure 1.11. For these data, the leaf unit is 0.1, and the reader can infer that the stem and leaf 2 and 6, for example, represent the measurement $x = 2.6$.

FIGURE 1.11
Stem and leaf plot for the data in Table 1.8

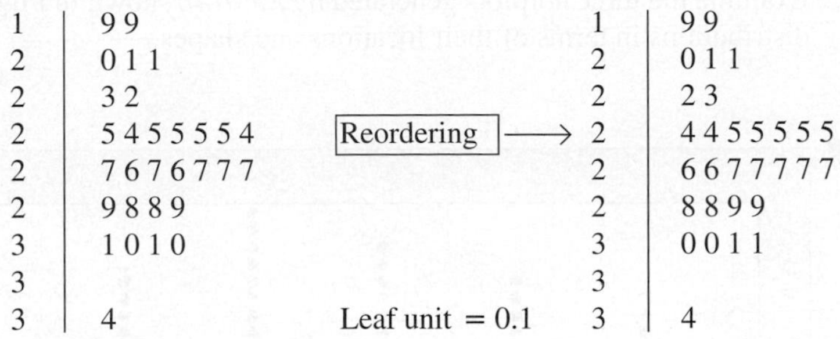

If you turn the stem and leaf plot sideways, so that the vertical line is now a horizontal axis, you can see that the data have "piled up" or been "distributed" along the axis in a pattern that can be described as "mound-shaped"—much like a pile of sand on the beach. One GPA was somewhat higher than the rest ($x = 3.4$), and the gap in the distribution shows that no GPAs were between 3.1 and 3.4.

Interpreting Graphs with a Critical Eye

Once you have created a graph or graphs for a set of data, what should you look for as you attempt to describe the data?

- First, check the horizontal and vertical **scales,** so that you are clear about what is being measured.
- Examine the **location** of the data distribution. Where on the horizontal axis is the centre of the distribution? If you are comparing two distributions, are they both centred in the same place?
- Examine the **shape** of the distribution. Does the distribution have one "peak," a point that is higher than any other? If so, this is the most frequently occurring measurement or category. Is there more than one peak? Are there an approximately equal number of measurements to the left and right of the peak?
- Look for any unusual measurements or **outliers.** That is, are any measurements much bigger or smaller than all of the others? These outliers may not be representative of the other values in the set.

Distributions are often described according to their shapes.

DEFINITION A distribution is **symmetric** if the left and right sides of the distribution, when divided at the middle value, form mirror images.

A distribution is **skewed to the right** if a greater proportion of the measurements lie to the right of the peak value. Distributions that are **skewed right** contain a few unusually large measurements.

A distribution is **skewed to the left** if a greater proportion of the measurements lie to the left of the peak value. Distributions that are **skewed left** contain a few unusually small measurements.

A distribution is **unimodal** if it has one peak; a **bimodal** distribution has two peaks. Bimodal distributions often represent a mixture of two different populations in the data set.

EXAMPLE 1.12

Examine the three dotplots generated by *MINITAB* shown in Figure 1.12. Describe these distributions in terms of their locations and shapes.

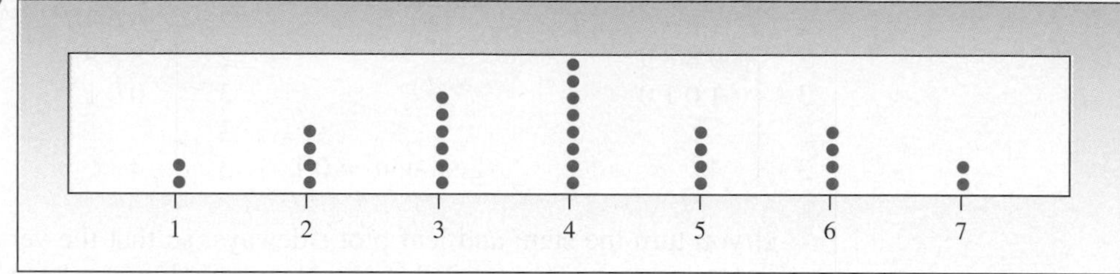

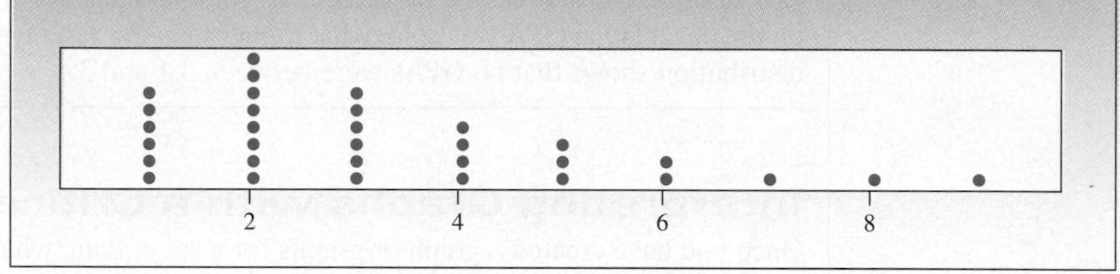

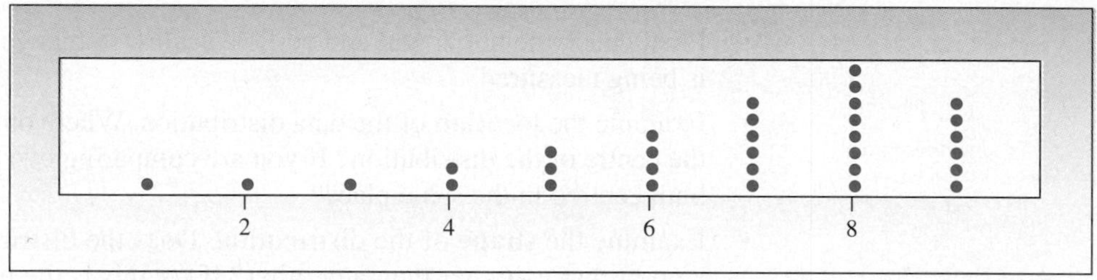

FIGURE 1.12
Shapes of data distributions for Example 1.12

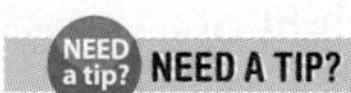

NEED A TIP?
Symmetric ⇔ mirror images
Skewed right ⇔ long right tail
Skewed left ⇔ long left tail

SOLUTION The first dotplot shows a *relatively symmetric* distribution with a single peak located at $x = 4$. If you were to fold the page at this peak, the left and right halves would *almost* be mirror images. The second dotplot, however, is far from symmetric. It has a long "right tail," meaning that there are a few unusually large observations. If you were to fold the page at the peak, a larger proportion of measurements would be on the right side than on the left. This distribution is *skewed to the right.* Similarly, the third dotplot with the long "left tail" is *skewed to the left.*

EXAMPLE 

A quality control analyst is interested in monitoring the weights of a particular style of walking sneaker. She enters the weights (in grams) of eight randomly selected shoes into the database but accidentally misplaces the decimal point in the last entry:

272 274 270 271 271 273 272 27.2

Use a dotplot to describe the data and uncover the analyst's mistake.

SOLUTION The dotplot of this small data set is shown in Figure 1.13(a). You can clearly see the *outlier* or unusual observation caused by the analyst's data entry error. Once the error has been corrected, as in Figure 1.13(b), you can see the correct distribution of the data set. Since this is a very small set, it is difficult to describe the shape of

FIGURE 1.13
Distributions of weights for Example 1.13

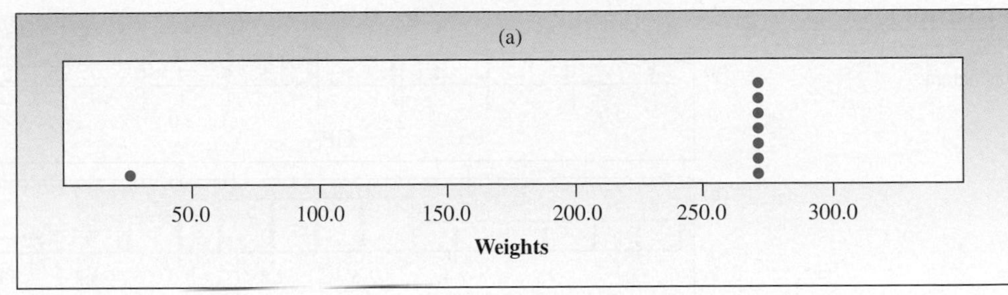

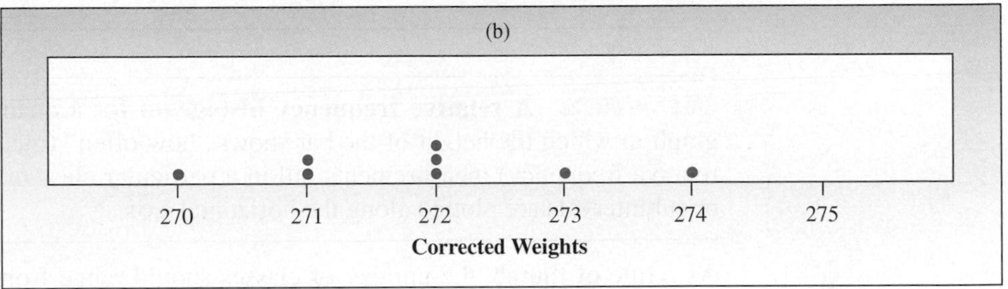

the distribution, although it seems to have a peak value around 272 and it appears to be relatively symmetric.

NEED A TIP?

Outliers lie out, away from the main body of data.

When comparing graphs created for two data sets, you should compare their *scales of measurement*, *locations*, and *shapes*, and look for unusual measurements or outliers. Remember that outliers are not always caused by errors or incorrect data entry. Sometimes they provide very valuable information that should not be ignored. You may need additional information to decide whether an outlier is a valid measurement that is simply unusually large or small, or whether there has been some sort of mistake in the data collection. If the scales differ widely, be careful about making comparisons or drawing conclusions that might be inaccurate!

1.5 RELATIVE FREQUENCY HISTOGRAMS

A relative frequency histogram resembles a bar chart, but it is used to graph quantitative rather than qualitative data. The data in Table 1.9 are the GPAs of 30 first-year university students, reproduced from Example 1.11 shown as a dotplot in Figure 1.14(a). First, divide the interval from the smallest to the largest measurements into subintervals or *classes of equal length.* If you stack up the dots in each subinterval (Figure 1.14(b)), and draw a bar over each stack, you will have created a **frequency histogram** or a **relative frequency histogram,** depending on the scale of the vertical axis.

TABLE 1.9 Grade Point Averages of 30 First-Year University Students

2.0	3.1	1.9	2.5	1.9
2.3	2.6	3.1	2.5	2.1
2.9	3.0	2.7	2.5	2.4
2.7	2.5	2.4	3.0	3.4
2.6	2.8	2.5	2.7	2.9
2.7	2.8	2.2	2.7	2.1

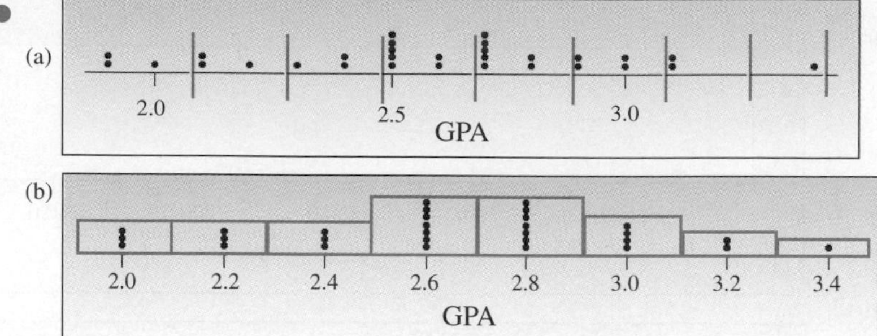

FIGURE 1.14
How to construct a histogram

DEFINITION A **relative frequency histogram** for a quantitative data set is a bar graph in which the height of the bar shows "how often" (measured as a proportion or relative frequency) measurements fall in a particular class or subinterval. The classes or subintervals are plotted along the horizontal axis.

As a rule of thumb, the number of classes should range from 5 to 12; the more data available, the more classes you need.† The classes must be chosen so that each measurement falls into one and only one class. For the GPAs in Table 1.9, we decided to use eight intervals of equal length. Since the total span of the GPAs is

$$3.4 - 1.9 = 1.5$$

the minimum class width necessary to cover the range of the data is $(1.5 \div 8) = 0.1875$. For convenience, we round this approximate width up to 0.2. Beginning the first interval at the lowest value, 1.9, we form subintervals from 1.9 up to *but not including* 2.1, 2.1 up to *but not including* 2.3, and so on. By using the **method of left inclusion,** and including the left class boundary point but not the right boundary point in the class, we eliminate any confusion about where to place a measurement that happens to fall on a class boundary point.

Table 1.10 shows the eight classes, labelled from 1 to 8 for identification. The boundaries for the eight classes, along with a tally of the number of measurements that fall in each class, are also listed in the table. As with the charts in Section 1.3, you can now measure *how often* each class occurs using *frequency* or *relative frequency.*

TABLE 1.10 **Relative Frequencies for Data of Table 1.9**

NEED A TIP?

Relative frequencies add to 1—
Frequencies add to *n*.

Class	Class Boundaries	Tally	Class Frequency	Class Relative Frequency
1	1.9 to <2.1	III	3	3/30
2	2.1 to <2.3	III	3	3/30
3	2.3 to <2.5	III	3	3/30
4	2.5 to <2.7	卌 II	7	7/30
5	2.7 to <2.9	卌 II	7	7/30
6	2.9 to <3.1	IIII	4	4/30
7	3.1 to <3.3	II	2	2/30
8	3.3 to <3.5	I	1	1/30

†You can use this table as a guide for selecting an appropriate number of classes. Remember that this is only a guide; you may use more or fewer classes than the table recommends if it makes the graph more descriptive.

Sample Size	25	50	100	200	500
Number of Classes	6	7	8	9	10

To construct the relative frequency histogram, plot the class boundaries along the horizontal axis. Draw a bar over each class interval, with height equal to the relative frequency for that class. The relative frequency histogram for the GPA data, Figure 1.15, shows at a glance how GPAs are distributed over the interval 1.9 to 3.5.

FIGURE 1.15
Relative frequency histogram

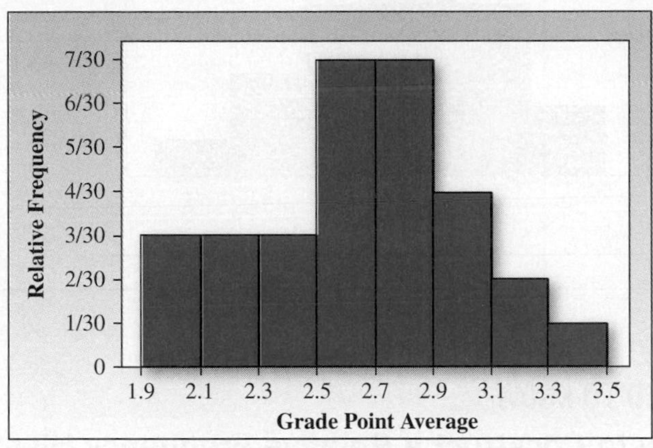

EXAMPLE 1.14

Twenty-five households are polled in a marketing survey, and Table 1.11 lists the numbers of litres of milk purchased during a particular week. Construct a relative frequency histogram to describe the data.

TABLE 1.11 Litres of Milk Purchased by 25 Households

0	3	5	4	3
2	1	3	1	2
1	1	2	0	1
4	3	2	2	2
2	2	2	3	4

SOLUTION The variable being measured is "number of litres of milk," which is a discrete variable that takes on only integer values. In this case, it is simplest to choose the classes or subintervals as the integer values over the range of observed values: 0, 1, 2, 3, 4, and 5. Table 1.12 shows the classes and their corresponding frequencies and relative frequencies. The relative frequency histogram, generated using *MINITAB*, is shown in Figure 1.16.

TABLE 1.12 Frequency Table for Example 1.14

Number of Litres	Frequency	Relative Frequency
0	2	0.08
1	5	0.20
2	9	0.36
3	5	0.20
4	3	0.12
5	1	0.04

FIGURE 1.16
MINITAB histogram for Example 1.14

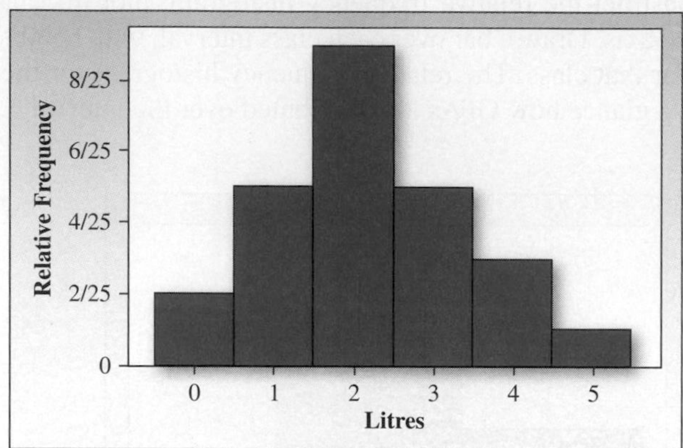

NEED TO KNOW

How to Construct a Relative Frequency Histogram

1. Choose the number of classes, usually between 5 and 12. The more data you have, the more classes you should use.
2. Calculate the approximate class width by dividing the difference between the largest and smallest values by the number of classes.
3. Round the approximate class width up to a convenient number.
4. If the data are discrete, you might assign one class for each integer value taken on by the data. For a large number of integer values, you may need to group them into classes.
5. Locate the class boundaries. The lowest class must include the smallest measurement. Then add the remaining classes using the left inclusion method.
6. Construct a statistical table containing the classes, their frequencies, and their relative frequencies.
7. Construct the histogram like a bar graph, plotting class intervals on the horizontal axis and relative frequencies as the heights of the bars.

A relative frequency histogram can be used to describe the distribution of a set of data in terms of its *location* and *shape*, and to check for *outliers* as you did with other graphs. For example, both the GPA and the "milk" data were relatively symmetric, with no unusual measurements. Since the bar constructed above each class represents the *relative frequency* or proportion of the measurements in that class, these heights can be used to give us further information:

- The proportion of the measurements that fall in a particular class or group of classes
- The probability that a measurement drawn at random from the set will fall in a particular class or group of classes

Consider the relative frequency histogram for the GPA data in Figure 1.15. What proportion of the students had GPAs of 2.7 or higher? This involves all classes beyond 2.7 in Table 1.10. Because there are 14 students in those classes, the proportion who have GPAs of 2.7 or higher is 14/30, or approximately 47%. This is also the percentage of the total area under the histogram in Figure 1.15 that lies to the right of 2.7.

Suppose you wrote each of the 30 GPAs on a piece of paper, put them in a hat, and drew one at random. What is the chance that this piece of paper contains a GPA of 2.7 or higher? Since 14 of the 30 pieces of paper fall in this category, you have 14 chances out of 30; that is, the probability is 14/30. The word *probability* is not unfamiliar to you; we will discuss it in more detail in Chapter 4.

Although we are interested in describing the set of $n = 30$ measurements, we might also be interested in the population from which the sample was drawn, which is the set of GPAs of all first-year students currently in attendance at the University of Windsor. Or, if we are interested in the academic achievement of first-year university students in general, we might consider our sample as representative of the population of GPAs for first-year students attending the University of Windsor, or other universities *similar* to the University of Windsor. A sample histogram provides valuable information about the population histogram—the graph that describes the distribution of the entire population. Remember, though, that different samples from the same population will produce *different* histograms, even if you use the same class boundaries. However, you can expect that the sample and population histograms will be *similar*. As you add more and more data to the sample, the two histograms become more and more alike. If you enlarge the sample to include the entire population, the two histograms are identical!

1.5 EXERCISES

BASIC TECHNIQUES

Data sets for all exercises marked with the data set icon are available for download from the companion website, at http://www.nelson.com/student.

EX0118

1.18 Construct a stem and leaf plot for these 50 measurements:

3.1	4.9	2.8	3.6	2.5	4.5	3.5	3.7	4.1	4.9
2.9	2.1	3.5	4.0	3.7	2.7	4.0	4.4	3.7	4.2
3.8	6.2	2.5	2.9	2.8	5.1	1.8	5.6	2.2	3.4
2.5	3.6	5.1	4.8	1.6	3.6	6.1	4.7	3.9	3.9
4.3	5.7	3.7	4.6	4.0	5.6	4.9	4.2	3.1	3.9

a. Describe the shape of the data distribution. Do you see any outliers?

b. Use the stem and leaf plot to find the smallest observation.

c. Find the eighth and ninth largest observations.

1.19 Refer to Exercise 1.18. Construct a relative frequency histogram for the data.

a. Approximately how many class intervals should you use?

b. Suppose you decide to use classes starting at 1.6 with a class width of 0.5 (i.e., 1.6 to <2.1, 2.1 to <2.6). Construct the relative frequency histogram for the data.

c. What fraction of the measurements are less than 5.1?

d. What fraction of the measurements are larger than 3.6?

e. Compare the relative frequency histogram with the stem and leaf plot in Exercise 1.18. Are the shapes similar?

Data set EX0120

1.20 Consider this set of data:

4.5	3.2	3.5	3.9	3.5	3.9
4.3	4.8	3.6	3.3	4.3	4.2
3.9	3.7	4.3	4.4	3.4	4.2
4.4	4.0	3.6	3.5	3.9	4.0

a. Construct a stem and leaf plot by using the leading digit as the stem.

b. Construct a stem and leaf plot by using each leading digit twice. Does this technique improve the presentation of the data? Explain.

1.21 A discrete variable can take on only the values 0, 1, or 2. A set of 20 measurements on this variable is shown here:

1	2	1	0	2
2	1	1	0	0
2	2	1	1	0
0	1	2	1	1

a. Construct a relative frequency histogram for the data.

b. What proportion of the measurements are greater than 1?

c. What proportion of the measurements are less than 2?

d. If a measurement is selected at random from the 20 measurements shown, what is the probability that it is a 2?

e. Describe the shape of the distribution. Do you see any outliers?

1.22 Refer to Exercise 1.21.

a. Draw a dotplot to describe the data.

b. How could you define the stem and the leaf for this data set?

c. Draw the stem and leaf plot using your decision from part b.

d. Compare the dotplot, the stem and leaf plot, and the relative frequency histogram (Exercise 1.21). Do they all convey roughly the same information?

1.23 Navigating a Maze An experimental psychologist measured the length of time it took for a rat to successfully navigate a maze on each of five days. The results are shown in the table. Create a line chart to describe the data. Do you think that any learning is taking place?

Day	1	2	3	4	5
Time (sec.)	45	43	46	32	25

Data set EX0124 **1.24 Measuring over Time** The value of a quantitative variable is measured once a year for a 10-year period. Here are the data:

Year	Measurement	Year	Measurement
1	61.5	6	58.2
2	62.3	7	57.5
3	60.7	8	57.5
4	59.8	9	56.1
5	58.0	10	56.0

a. Create a line chart to describe the variable as it changes over time.

b. Describe the measurements using the chart constructed in part a.

Data set EX0125 **1.25 Test Scores** The test scores on a 100-point test were recorded for 20 students:

61	93	91	86	55	63	86	82	76	57
94	89	67	62	72	87	68	65	75	84

a. Use an appropriate graph to describe the data.

b. Describe the shape and location of the scores.

c. Is the shape of the distribution unusual? Can you think of any reason the distribution of the scores would have such a shape?

APPLICATIONS

Data set EX0126 **1.26 A Recurring Illness** The length of time (in months) between the onset of a particular illness and its recurrence was recorded for $n = 50$ patients:

2.1	4.4	2.7	32.3	9.9	9.0	2.0	6.6	3.9	1.6
14.7	9.6	16.7	7.4	8.2	19.2	6.9	4.3	3.3	1.2
4.1	18.4	0.2	6.1	13.5	7.4	0.2	8.3	0.3	1.3
14.1	1.0	2.4	2.4	18.0	8.7	24.0	1.4	8.2	5.8
1.6	3.5	11.4	18.0	26.7	3.7	12.6	23.1	5.6	0.4

a. Construct a relative frequency histogram for the data.

b. Would you describe the shape as roughly symmetric, skewed right, or skewed left?

c. Give the fraction of recurrence times less than or equal to 10 months.

1.27 Post-secondary Education Pays Off Big—Especially for Albertans Post-secondary education is one of the best investments you can make, especially if you're living in Alberta, according to a recent interprovincial comparison of graduate salaries conducted by Alberta Learning.[3]

The average Alberta university graduate with a bachelor's degree can expect to earn more than $46,000 a year—the second highest gross salary after Ontario. When taxes and cost of living are taken into consideration, the salary of a university bachelor's graduate living in Alberta is the highest in Canada at just under $34,000, followed by Ontario at $32,000 and Saskatchewan at just under $31,000.

The average annual incomes for five different categories are shown in the table:

Educational Level	Average Annual Income ($)
High-school graduates	30,072
Certificate and diploma graduates	38,952
Bachelor degree graduates	46,565
Masters degree and earned doctorate graduates	56,974
All labour force survey categories	35,440

a. What graphical methods could you use to describe the data?

b. Select the method from part a that you think best describes the data.

c. How would you summarize the information that you see in the graph regarding educational levels and salary?

EX0128 **1.28 Preschool** The ages (in months) at which 50 children were first enrolled in a preschool are listed below:

38	40	30	35	39	40	48	36	31	36
47	35	34	43	41	36	41	43	48	40
32	34	41	30	46	35	40	30	46	37
55	39	33	32	32	45	42	41	36	50
42	50	37	39	33	45	38	46	36	31

a. Construct a stem and leaf display for the data.

b. Construct a relative frequency histogram for these data. Start the lower boundary of the first class at 30 and use a class width of 5 months.

c. Compare the graphs in parts a and b. Are there any significant differences that would cause you to choose one as the better method for displaying the data?

d. What proportion of the children were 35 months (2 years, 11 months) or older, but less than 45 months (3 years, 9 months) of age when first enrolled in preschool?

e. If one child were selected at random from this group of children, what is the probability that the child was less than 50 months old (4 years, 2 months) when first enrolled in preschool?

EX0129 **1.29 Happy in the Air?** The following table reveals complaints against Air Canada and major U.S. airlines in a given year.[4]

Airline	June 2003	Dec. 2003	Canada Total	U.S. Total	Grand Total	Passengers (millions)
Air Canada	310	176	486	36	522	20.1
Airtran Airways	0	0	0	97	97	11.7
American West Airlines	0	0	0	168	168	20.1
American Airlines	0	5	5	781	786	88.8
Continental Airlines	0	0	0	371	371	38.9
Delta	4	3	7	656	663	84.3
Northwest Airlines	1	4	5	492	497	52.0
Southwest Airlines	0	0	0	106	106	74.8
United Airlines	6	2	8	548	556	66.2
U.S. Airways	1	5	6	373	379	41.3

a. Construct a pie chart to describe the grand total number of complaints by airline.

b. Order the airlines from the smallest to the largest number of complaints. Construct a Pareto chart to describe the data. Which display is more effective?

c. Is there another variable that you could measure that might help to explain why some airlines have many more complaints than others? Explain.

EX0130 **1.30 How Long Is the Line?** To decide on the number of service counters needed for stores to be built in the future, a supermarket chain wanted to obtain information on the length of time (in minutes) required to service customers. To find the distribution of customer service times, a sample of 1000 customers' service times was recorded. Sixty of these are shown here:

3.6	1.9	2.1	0.3	0.8	0.2	1.0	1.4	1.8	1.6
1.1	1.8	0.3	1.1	0.5	1.2	0.6	1.1	0.8	1.7
1.4	0.2	1.3	3.1	0.4	2.3	1.8	4.5	0.9	0.7
0.6	2.8	2.5	1.1	0.4	1.2	0.4	1.3	0.8	1.3
1.1	1.2	0.8	1.0	0.9	0.7	3.1	1.7	1.1	2.2
1.6	1.9	5.2	0.5	1.8	0.3	1.1	0.6	0.7	0.6

a. Construct a stem and leaf plot for the data.

b. What fraction of the service times are less than or equal to 1 minute?

c. What is the smallest of the 60 measurements?

1.31 Service Times, continued Refer to Exercise 1.30. Construct a relative frequency histogram for the supermarket service times.

a. Describe the shape of the distribution. Do you see any outliers?

b. Assuming that the outliers in this data set are valid observations, how would you explain them to the management of the supermarket chain?

c. Compare the relative frequency histogram with the stem and leaf plot in Exercise 1.30. Do the two graphs convey the same information?

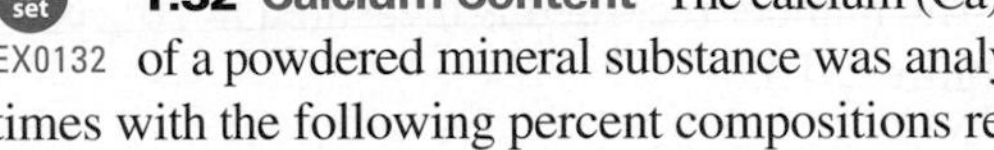

EX0132 **1.32 Calcium Content** The calcium (Ca) content of a powdered mineral substance was analyzed 10 times with the following percent compositions recorded:

0.0271	0.0282	0.0279	0.0281	0.0268
0.0271	0.0281	0.0269	0.0275	0.0276

a. Draw a dotplot to describe the data. (HINT: The scale of the horizontal axis should range from 0.0260 to 0.0290.)

b. Draw a stem and leaf plot for the data. Use the numbers in the hundredths and thousandths places as the stem.

c. Are any of the measurements inconsistent with the other measurements, indicating that the technician may have made an error in the analysis?

Data set EX0133 **1.33 Canadian Prime Ministers by Age** This is a list of prime ministers of Canada since Confederation in 1867, arranged in descending order of their age upon first taking office.[5]

Prime Minister	Age	Prime Minister	Age
C. Tupper	74	J. A. Macdonald	52
J. Abbott	70	A. Mackenzie	51
M. Bowell	69	P. Trudeau	48
L. St.-Laurent	66	J. S. Thompson	47
L. B. Pearson	66	W. L. M. King	47
P. Martin	65	S. Harper	46
J. Diefenbaker	61	K. Campbell	46
R. Bennett	60	A. Meighen	46
J. Chrétien	59	B. Mulroney	45
R. Borden	57	J. Trudeau	43
J. Turner	55	J. Clark	39
W. Laurier	54		

a. Before you graph the data, try to visualize the distribution of the ages for the prime ministers. What shape do you think it will have?

b. Construct a stem and leaf plot for the data. Describe the shape. Does it surprise you?

c. The five youngest prime ministers at the time of first taking oath appear in the lower "tail" of the distribution. Identify the five youngest prime ministers at the time of first taking oath. What common trait explains these measurements?

Data set EX0134 **1.34 RBC Counts** The red blood cell count of a healthy person was measured on each of 15 days. The number recorded is measured in 10^6 cells per microlitre (μL).

5.4	5.2	5.0	5.2	5.5
5.3	5.4	5.2	5.1	5.3
5.3	4.9	5.4	5.2	5.2

a. Use an appropriate graph to describe the data.

b. Describe the shape and location of the red blood cell counts.

c. If the person's red blood cell count is measured today as 5.7×10^6 cells/μL, would you consider this unusual? What conclusions might you draw?

Data set EX0135 **1.35 NHL Goals Against Average Leaders** Average determined by games played through 1942–43 season and by minutes played since then. A sample of goals against average leaders is listed in the table.[6]

Year	Name	Goals Against Average
1968	Gump Worsley	1.98
1967	Glenn Hall	2.38
1974	Bernie Parent	1.89
1990	Mike Liut	2.53
1961	Johnny Bower	2.50
1970	Ernie Wakely	2.11
1918	Georges Vezina	3.82
1940	Dave Kerr	1.60
1937	Norm Smith	2.13
2000	Brian Boucher	1.91
1943	John Mowers	2.47
1956	Jacques Plante	1.86
1941	Turk Broda	2.06
1952	Terry Sawchuk	1.90
1926	Alex Connell	1.17
1997	Martin Brodeur	1.88
1977	Bunny Larocque	2.09
1980	Bob Sauve	2.36
1994	Dominik Hasek	1.95
1944	Bill Durnan	2.18

Source: Goals Against Average www.infoplease.com/ipsa/A0758956.html
Permission granted by Family Education Network.

a. Construct a relative frequency histogram to describe the Goals Against Average for these 20 champions.

b. If you were to randomly choose one of the 20 names, what is the chance that you would choose a player whose average was above 2.4 for his championship year?

Data set EX0136 **1.36 Popular Movies** Described in these charts is the popularity of various genres for movies, based on various factors that make up the index where 0 to 100 rates the popularity of these genres:

Year	Genre: Action (Rating)	Genre: Comedy (Rating)	Genre: Horror (Rating)
2008	35	70	90
2009	65	50	55
2010	75	66	60
2011	80	82	70
2012	90	74	80
2013	73	75	45
2014	65	58	70
2015	73	67	50

a. Use an appropriate graph to describe the data for the horror genre.

b. Describe the shape for the comedy and action genres. Give your opinion as to why it has this particular shape.

c. Rate which genres have been performing well based on the graphs.

d. Which genre is relatively more stable than the other two?

Data set EX0137 **1.37 Organized Religion** Statistics of the world's religions are only very rough approximations, since many religions do not keep track of their membership numbers. An estimate of these numbers (in millions) is shown in the table.[7]

Religion	Members (millions)	Religion	Members (millions)
Buddhism	376	Judaism	14
Christianity	2100	Sikhism	23
Hinduism	900	Chinese traditional	394
Islam	1500	Other	61
Primal indigenous and African traditional	400		

Source: © Adherents.com.

a. Construct a pie chart to describe the total membership in the world's organized religions.

b. Construct a bar chart to describe the total membership in the world's organized religions.

c. Order the religious groups from the smallest to the largest number of members. Construct a Pareto chart to describe the data. Which of the three displays is most effective?

As you continue to work through the exercises in this chapter, you will become more experienced in recognizing different types of data and in determining the most appropriate graphical method to use. Remember that the type of graphic you use is not as important as the interpretation that accompanies the picture. Look for these important characteristics:

- Location of the centre of the data
- Shape of the distribution of data
- Unusual observations in the data set

Using these characteristics as a guide, you can interpret and compare sets of data using graphical methods, which are only the first of many statistical tools that you will soon have at your disposal.

CHAPTER REVIEW

Key Concepts

I. How Data Are Generated

1. Experimental units, variables, measurements
2. Samples and populations
3. Univariate, bivariate, and multivariate data

II. Types of Variables

1. Qualitative or categorical
2. Quantitative
   a. Discrete
   b. Continuous

III. Graphs for Univariate Data Distributions

1. Qualitative or categorical data
   a. Pie charts
   b. Bar charts
2. Quantitative data
   a. Pie and bar charts
   b. Line charts
   c. Dotplots
   d. Stem and leaf plots
   e. Relative frequency histograms
3. Describing data distributions
   a. Shapes—symmetric, skewed left, skewed right, unimodal, bimodal
   b. Proportion of measurements in certain intervals
   c. Outliers

Introduction to *Microsoft Excel*®

Microsoft Excel is a spreadsheet program in the Microsoft Office® system. It is designed for a variety of analytical applications, including statistical applications. We will assume that you are familiar with Windows®, and that you know the basic techniques necessary for executing commands from the tabs, groups, and drop-down menus at the top of the screen. If not, perhaps a lab or teaching assistant can help you to master the basics. The current version of *Excel* at the time of this printing is *Excel* 2010, used in the Windows 7 environment. When the program opens, a **spreadsheet** appears (see Figure 1.17), containing rows and columns into which you can enter data. Tabs at the bottom of the screen identify the three spreadsheets available for use; when saved as a collection, these spreadsheets are called a **workbook.**

FIGURE 1.17
Graphing with *Excel* (Microsoft screen cap)

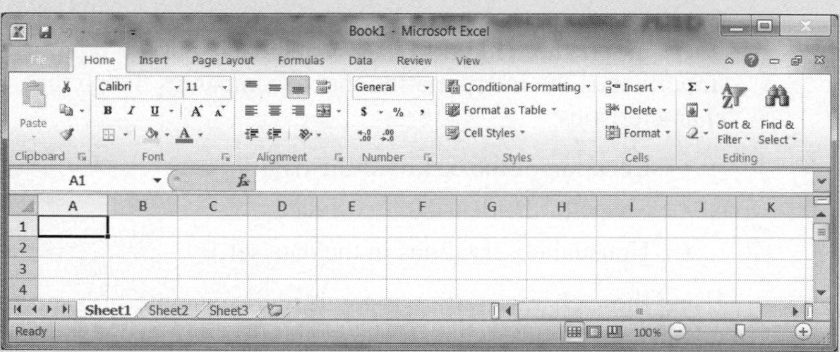

Pie charts, bar charts, and **line charts** can all be created in *Excel*. Data is entered into an *Excel* spreadsheet, including labels if needed. Highlight the data to be graphed, and then click the chart type that you want on the **Insert** tab in the **Charts** group. Once the chart has been created, it can be edited in a variety of ways to change its appearance.

Pie and Bar Charts The class statuses of 105 students in an introductory statistics class are listed in Table 1.13. The qualitative variable "class status" has been recorded for each student, and the frequencies have already been recorded.

TABLE 1.13 **Status of Students in Statistics Class**

Status	First Year	Second Year	Third Year	Fourth Year	Grad Student
Frequency	5	23	32	35	10

1. Enter the *categories* into column A of the first spreadsheet and the *frequencies* into column B. You should have two columns of data, including the labels.
2. Highlight the data, using your left mouse button to *click-and-drag* from cell A1 to cell B6 (sometimes written as **A1:B6**). Click the **Insert** tab and select **Pie** in the **Charts** group. In the drop-down list, you will see a variety of styles to choose from. Select the first option to produce the pie chart. Double-click on the title "Frequency" and change the title to "Student Status."

3. ***Editing the pie chart:*** Once the chart has been created, use your mouse to make sure that the chart is selected. You should see a green area above the tabs marked "Chart Tools." Click the **Design** tab, and look at the drop-down lists in the **Chart Layout** and **Chart Styles** groups. These lists allow you to alter the appearance of your chart. In Figure 1.18(a), the pie chart has been changed so that the percentages are shown in the appropriate sectors. By clicking on the legend, we have dragged it so that it is closer to the pie chart.

FIGURE 1.18

Two charts: (a) pie chart and (b) bar chart

(a)

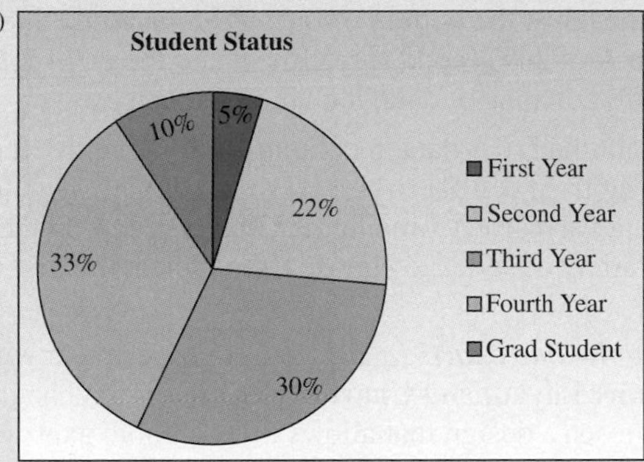

(b)

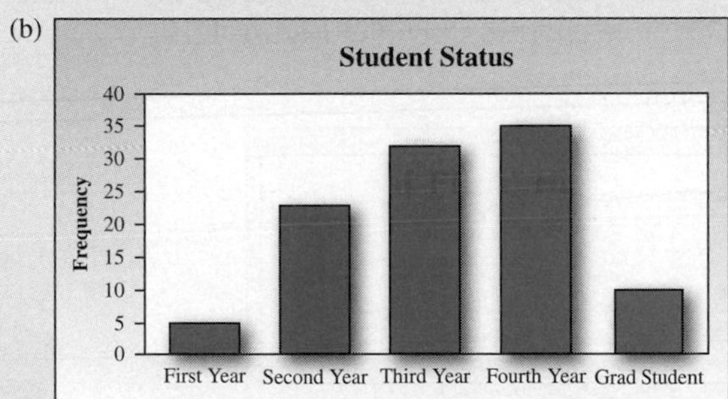

4. Click on various parts of the pie chart (legend, chart area, sector) and a box with round and/or square handles will appear. Double-click, and a dialogue box will appear. You can adjust the appearance of the selected object or region in this box and click **OK.** Click **Cancel** to exit the dialogue box without changes!
5. Still in the **Design** section, but in the **Type** group, click on **Change Chart Type** and choose the simplest **Column** type. Click **OK** to create a bar chart for the same data set, shown in Figure 1.18(b).
6. ***Editing the bar chart:*** Again, you can experiment with the various options in the **Chart Layout** and **Chart Styles** groups to change the look of the chart. You can click the entire bar chart ("Chart area") or the interior "Plot area" to stretch the chart. You can change colours by double-clicking on the appropriate region. We have chosen a design that allows axis titles and have deleted the "frequency legend entry." We have also chosen to delete the minor gridlines, by clicking the **Layout** tab in the **Chart Tools,** using the **Gridlines** drop-down list, and selecting **Primary Horizontal Gridlines ▶ Major Gridlines.** We have decreased the gaps between the bars by right-clicking on one of the bars, selecting **Format Data Series,** and changing the **Gap Width to 50%.**

EXAMPLE 1.16

Line Charts The Dow Jones Industrial Average was monitored at the close of trading for 10 days in a recent year, with the results shown in Table 1.14.

TABLE 1.14 **Dow Jones Industrial Average**

Day	1	2	3	4	5	6	7	8	9	10
DJIA	10,636	10,680	10,674	10,653	10,698	10,644	10,378	10,319	10,303	10,302

1. Click the tab at the bottom of the screen marked "Sheet 2." Enter the *Days* into column A of this second spreadsheet and the *DJIA* into column B. You should have two columns of data, including the labels.
2. Highlight the DJIA data in column B, using your left mouse button to *click-and-drag* from cell B1 to cell B11 (sometimes written as **B1:B11**). Click the **Insert** tab and select **Line** in the **Charts** group. In the drop-down list, you will see a variety of styles to choose from. Select the first option to produce the line chart.
3. ***Editing the line chart:*** Again, you can experiment with the various options in the **Chart Layout** and **Chart Styles** groups to change the look of the chart. We have chosen a design that allows titles on both axes, which we have changed to "Day" and "DJIA," and we have deleted the title and the "frequency legend entry." The line chart is shown in Figure 1.19.

FIGURE 1.19
Line graph

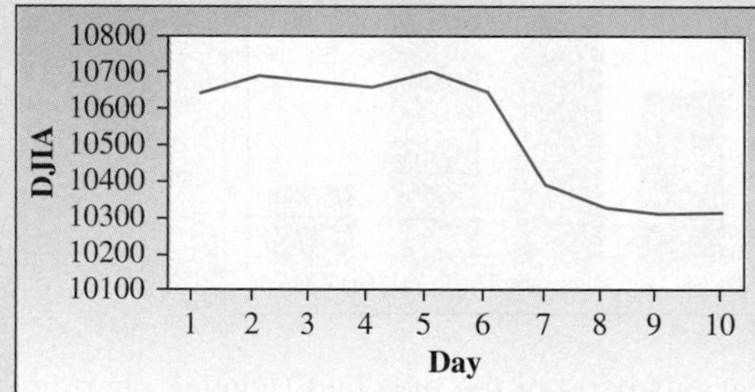

4. *Note:* If your time series involves time periods that are *not equally spaced,* it is better to use a **scatterplot** with points connected to form a line chart. This procedure is described in the "Technology Today" section in Chapter 3.

EXAMPLE 

Frequency Histograms The top 40 stocks on the over-the-counter (OTC) stocks market, listed by percentage of outstanding shares traded on a particular day, are provided in Table 1.15.

TABLE 1.15 **Percentage of OTC Stocks Traded**

11.88	6.27	5.49	4.81	4.40	3.78	3.44	3.11	2.88	2.68
7.99	6.07	5.26	4.79	4.05	3.69	3.36	3.03	2.74	2.63
7.15	5.98	5.07	4.55	3.94	3.62	3.26	2.99	2.74	2.62
7.13	5.91	4.94	4.43	3.93	3.48	3.20	2.89	2.69	2.61

1. Many of the statistical procedures that we will use in this textbook require the installation of the **Analysis ToolPak** add-in. To load this add-in, click **File ▶ Options ▶ Add-ins.** Select **Analysis ToolPak** and click **OK.**
2. Click the tab at the bottom of the screen marked "Sheet 3." Enter the data into the first column of this spreadsheet and include the label "Stocks" in the first cell.
3. *Excel* refers to the maximum value for each class interval as a **bin.** This means that *Excel* is using a **method of right inclusion,** which is slightly different from the method presented in Section 1.5. For this example, we choose to use the class intervals > 2.5–3.5, > 3.5–4.5, > 4.5–5.5, etc. Enter the *bin values* (3.5, 4.5, 5.5, … , 12.5) into the second column of the spreadsheet, labelling them as "Percent Traded" in cell **B1.**
4. Select **Data ▶ Data Analysis ▶ Histogram** and click **OK.** The Histogram dialogue box will appear, as shown in Figure 1.20.

FIGURE 1.20
How to build a histogram in *Excel*

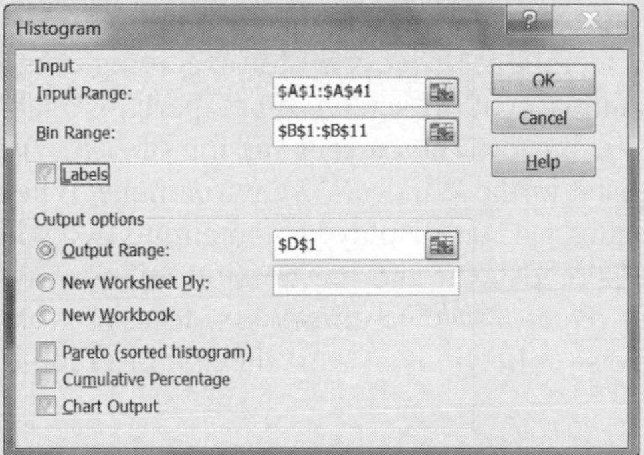

5. Highlight or type in the appropriate Input Range and Bin Range for the data. Notice that you can click the minimize button on the right of the box before you *click-and-drag* to highlight. Click the minimize button again to see the entire dialogue box. The Input Range will appear as $A$1:$A$41, with the dollar sign indicating an *absolute cell range.* Make sure to click the "Labels" and "Chart Output" check boxes. Pick a convenient cell location for the output (we picked **D1**) and click **OK.** The frequency table and histogram will appear in the spreadsheet. The histogram (Figure 1.21(a)) doesn't appear quite like we wanted.

FIGURE 1.21
(a) Histogram and (b) bar chart

(a)

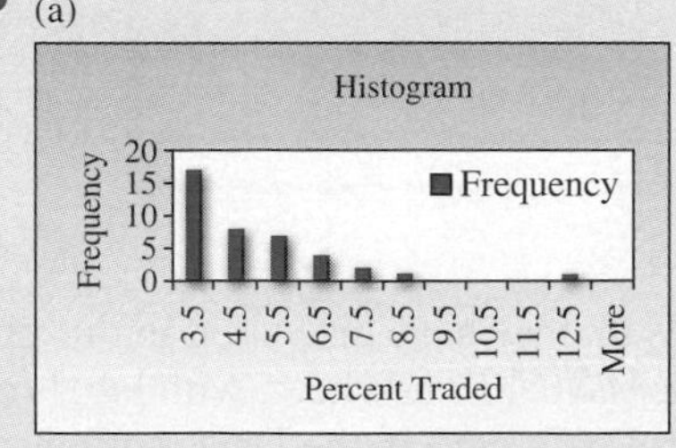

(b)

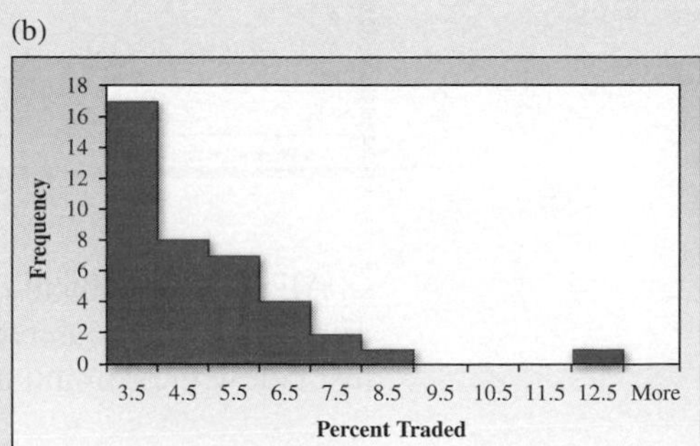

6. ***Editing the histogram:*** Click on the frequency legend entry and press the Delete key. Then select the Data Series by double-clicking on a bar. In the **Series Options** box that appears, change the **Gap Width** to 0% (no gap) and click **Close.** Stretch the graph by dragging the lower right corner, and edit the colours, title, and labels if necessary to finish your histogram, as shown in Figure 1.21(b). Remember that the numbers shown along the horizontal axis are the **bins,** the upper limit of the class interval, *not the midpoint of the interval.*
7. You can save your *Excel* workbook for use at a later time using **File ▶ Save** or **File ▶ Save As** and naming it "Chapter 1."

Introduction to *MINITAB*™

MINITAB computer software is a Windows-based program designed specifically for statistical applications. We will assume that you are familiar with Windows, and that you know the basic techniques necessary for executing commands from the tabs and drop-down menus at the top of the screen. If not, perhaps a lab or teaching assistant can help you to master the basics. The current version of *MINITAB* at the time of this printing is *MINITAB* 16, used in the Windows 7 environment. When the program opens, the main screen (see Figure 1.22) is displayed, containing two windows: the Data window, similar to an *Excel* spreadsheet, and the Session window, in which your results will appear. Just as with *Excel*, *MINITAB* allows you to save worksheets (similar to *Excel* spreadsheets), projects (collections of worksheets), or graphs.

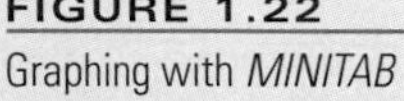

FIGURE 1.22
Graphing with *MINITAB*

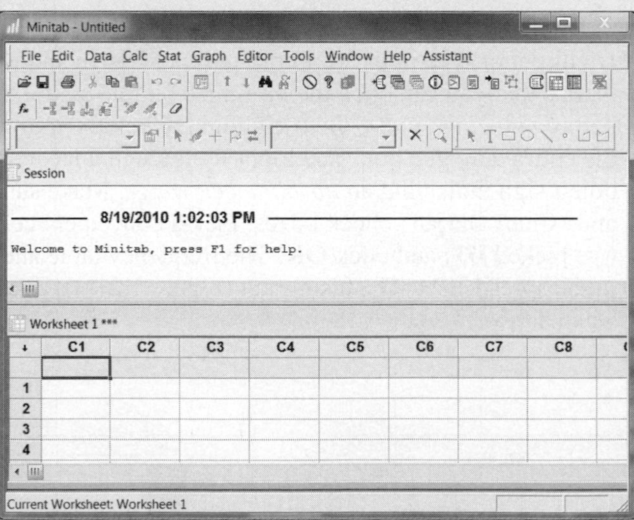

All of the graphical methods that we have discussed in this chapter can be created in *MINITAB*. Data is entered into a *MINITAB* worksheet, with labels entered in the grey cells just below the column name (C1, C2, etc.) in the Data window.

EXAMPLE 1.18

Pie and Bar Charts The class statuses of 105 students in an introductory statistics class are listed in Table 1.16. The qualitative variable "class status" has been recorded for each student, and the frequencies have already been recorded.

TABLE 1.16 Status of Students in Statistics Class

Status	First Year	Second Year	Third Year	Fourth Year	Grad Student
Frequency	5	23	32	35	10

1. Enter the *categories* into column C1, with your own descriptive name, perhaps "Status" in the grey cell. Notice that the name **C1** has changed to **C1-T** because you are entering text rather than numbers. Continue by naming column 2 (C2) "Frequency," and enter the five numerical frequencies into C2.
2. To construct a pie chart for these data, click on **Graph ▶ Pie Chart,** and a dialogue box will appear (see Figure 1.23). Click the radio button marked **Chart values from a table.** Then place your cursor in the box marked "Categorical variable." Either (1) highlight C1 in the list at the left and choose **Select,** (2) double-click on C1 in the list at the left, or (3) type C1 in the "Categorical variable" box. Similarly, place the cursor in the box marked "Summary variables" and select C2. Click **Labels** and select the tab marked **Slice Labels.** Check the boxes marked "Category names" and "Percent." When you click **OK** twice, *MINITAB* will create the pie chart in Figure 1.24(a). We have removed the legend by selecting and deleting it.

FIGURE 1.23
Screenshot of *MINITAB* (Pie Chart function)

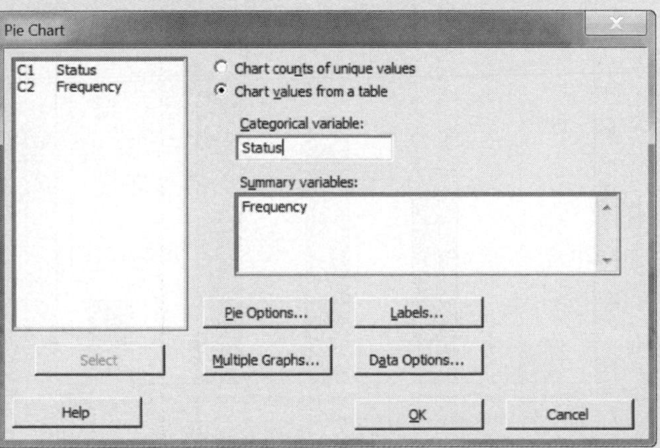

3. As you become more proficient at using the pie chart command, you may want to take advantage of some of the options available. Once the chart is created, *right-click* on the pie chart and select **Edit Pie.** You can change the colours and format of the chart, "explode" important sectors of the pie, and change the order of the categories. If you *right-click* on the pie chart and select **Update Graph Automatically,** the pie chart will automatically update when you change the data in columns C1 and C2 of the *MINITAB* worksheet.

4. If you would rather construct a bar chart, use the command **Graph ▶ Bar Chart.** In the dialogue box that appears, choose **Simple.** Choose an option in the "Bars represent" drop-down list, depending on the way that the data has been entered into the worksheet. For the data in Table 1.13, we choose "Values from a table" and click **OK.** When the dialogue box appears, place your cursor in the "Graph variables" box and select **C2.** Place your cursor in the "Categorical variable" box, and select **C1.** Click **OK** to finish the bar chart, shown in Figure 1.24(b). Once the chart is created, *right-click* on various parts of the bar chart and choose **Edit** to change the look of the chart.

FIGURE 1.24
(a) Pie Chart of Status and (b) Chart of Frequency

(a)

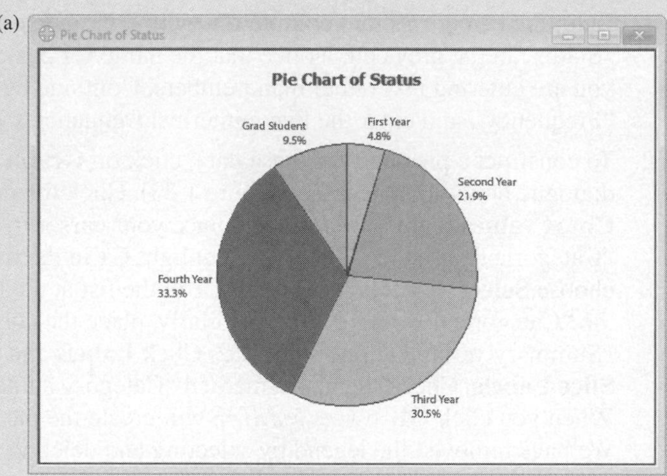

(b)

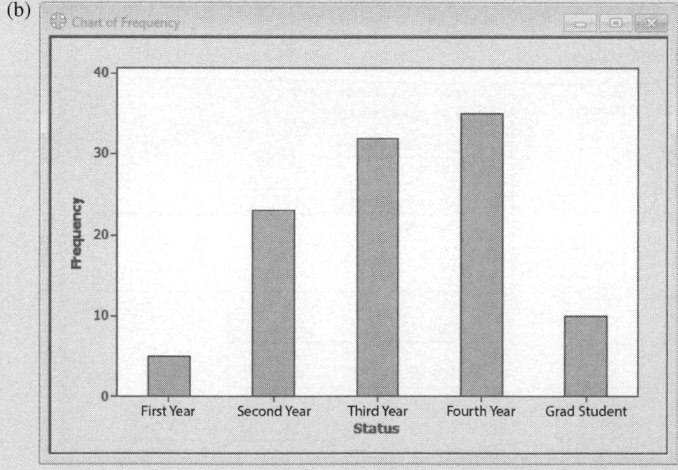

EXAMPLE 1.19

Line Charts The Dow Jones Industrial Average was monitored at the close of trading for 10 days in a recent year with the results shown in Table 1.17.

TABLE 1.17 Dow Jones Industrial Average

Day	1	2	3	4	5	6	7	8	9	10
DJIA	10,636	10,680	10,674	10,653	10,698	10,644	10,378	10,319	10,303	10,302

1. Although we could simply enter this data into third and fourth columns of the current worksheet, let's create a new worksheet using **File ▶ New ▶ Minitab Worksheet.** Enter the *Days* into column C1 of this second spreadsheet and the *DJIA* into column C2. You should have two columns of data, including the labels.
2. To create the line chart, use **Graph ▶ Time Series Plot ▶ Simple.** In the Dialog box that appears, place your cursor in the "Series" box and select "DJIA" from the list to the left. Under **Time/Scale,** choose "Stamp" and select column **C1** ("Day") in the box labelled "Stamp Columns." Click **OK** twice. You can select the numbered days shown above the line and delete them to obtain the line chart shown in Figure 1.25.

FIGURE 1.25
Time Series Plot of DJIA

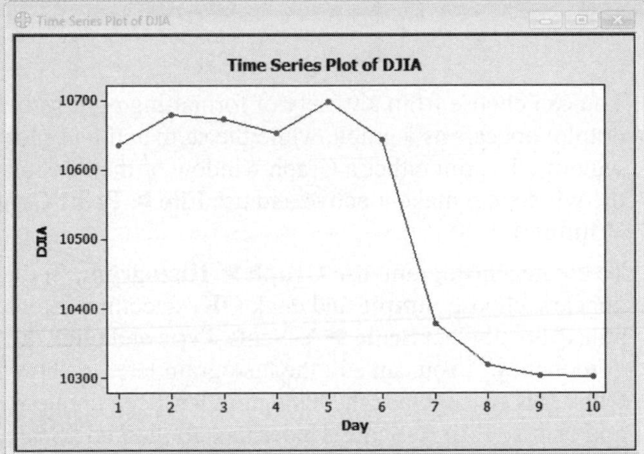

EXAMPLE  1.20

Dotplots, Stem and Leaf Plots, Histograms The top 40 stocks on the over-the-counter (OTC) market, listed by percentage of outstanding shares traded on a particular day, are provided in Table 1.18. Create a new worksheet **(File ▶ New ▶ Minitab Worksheet).** Enter the data into column C1 and name it "Stocks" in the grey cell just below the C1.

TABLE 1.18 **Percentage of OTC Stocks Traded**

11.88	6.27	5.49	4.81	4.40	3.78	3.44	3.11	2.88	2.68
7.99	6.07	5.26	4.79	4.05	3.69	3.36	3.03	2.74	2.63
7.15	5.98	5.07	4.55	3.94	3.62	3.26	2.99	2.74	2.62
7.13	5.91	4.94	4.43	3.93	3.48	3.20	2.89	2.69	2.61

1. To create a dotplot, use **Graph ▶ Dotplot.** In the dialogue box that appears, choose **One Y ▶ Simple** and click **OK.** To create a stem and leaf plot, use **Graph ▶ Stem-and-Leaf.** For either graph, place your cursor in the "Graph variables" box, and select "Stocks" from the list to the left (see Figure 1.26).

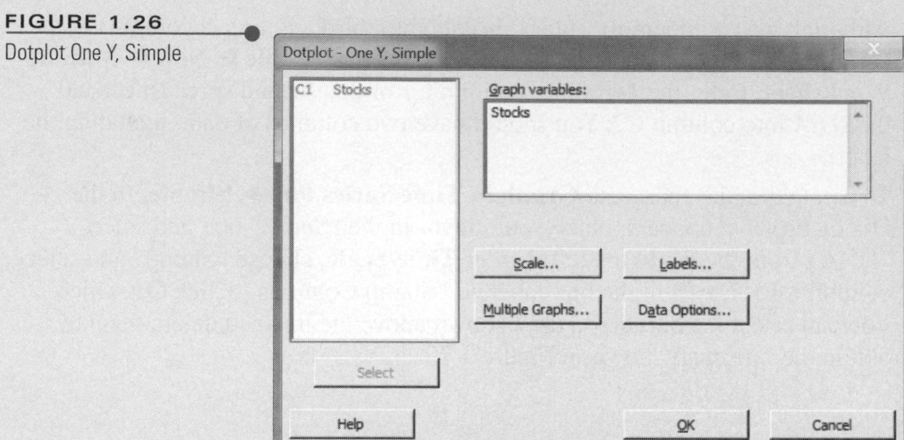

FIGURE 1.26
Dotplot One Y, Simple

2. You can choose from a variety of formatting options before clicking **OK.** The dotplot appears as a graph, while the stem and leaf plot appears in the Session window. To print either a Graph window or the Session window, click on the window to make it active and use **File ▶ Print Graph** (or **Print Session Window**).
3. To create a histogram, use **Graph ▶ Histogram.** In the Dialog box that appears, choose **Simple** and click **OK,** selecting "Stocks" for the "Graph variables" box. Select **Scale ▶ Y-Scale Type** and click the radio button marked "Frequency." (You can edit the histogram later to show relative frequencies.) Click **OK** twice. Once the histogram has been created, *right-click* on the *y*-axis and choose **Edit Y-Scale.** Under the tab marked "Scale," you can click the radio button marked "Position of ticks" and type in **0 5 10 15.** Then click the tab marked "Labels," the radio button marked "Specified," and type **0 5/40 10/40 15/40.** Click **OK.** This will reduce the number of ticks on the *y*-axis and change them to relative frequencies. Finally, double-click on the word "Frequency" along the *y*-axis. Change the box marked "Text" to read "Relative frequency" and click **OK.**
4. To adjust the type of boundaries for the histogram, *right-click* on the bars of the histogram and choose **Edit Bars.** Use the tab marked "Binning" to choose either "Cutpoints" or "Midpoints" for the histogram; you can specify the cutpoint or midpoint positions if you want. In this same **Edit** box, you can change the colours, fill type, and font style of the histogram. If you *right-click* on the bars and select **Update Graph Automatically,** the histogram will automatically update when you change the data in the "Stocks" column.

As you become more familiar with *MINITAB*, you can explore the various options available for each type of graph. It is possible to plot more than one variable at a time, to change the axes, to choose the colours, and to modify graphs in many ways. However, even with the basic default commands, it is clear that the distribution of OTC stocks in Figure 1.27 is highly skewed to the right.

FIGURE 1.27
Histogram of Stocks

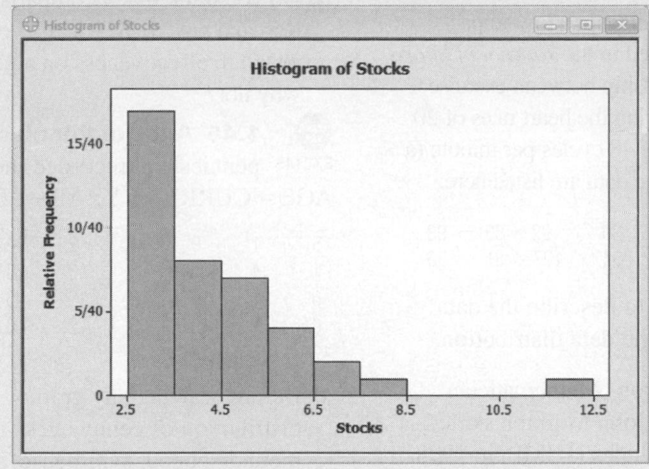

Supplementary Exercises

1.38 Quantitative or Qualitative? Identify each variable as quantitative or qualitative:

a. Ethnic origin of a candidate for public office

b. Score (0–100) on a placement examination

c. Fast-food establishment preferred by a student (McDonald's, Burger King, or Subway)

d. Mercury concentration in a sample of tuna

1.39 Symmetric or Skewed? Do you expect the distributions of the following variables to be symmetric or skewed? Explain.

a. Size in dollars of nonsecured loans

b. Size in dollars of secured loans

c. Price of a 250-gram can of peas

d. Height in centimetres of first-year women at your university

e. Number of broken taco shells in a package of 100 shells

f. Number of ticks found on each of 50 trapped cottontail rabbits

1.40 Continuous or Discrete? Identify each variable as continuous or discrete:

a. Number of homicides in Vancouver during a one-month period

b. Length of time between arrivals at an outpatient clinic

c. Number of typing errors on a page of manuscript

d. Number of defective lightbulbs in a package containing four bulbs

e. Time required to finish an examination

1.41 Continuous or Discrete, again Identify each variable as continuous or discrete:

a. Weight of two dozen shrimp

b. A person's body temperature

c. Number of people waiting for treatment at a hospital emergency room

d. Number of properties for sale by a real estate agency

e. Number of claims received by an insurance company during one day

1.42 Continuous or Discrete, again Identify each variable as continuous or discrete:

a. Number of people in line at a supermarket checkout counter

b. Depth of a snowfall

c. Length of time for a driver to respond when faced with an impending collision

d. Number of aircraft arriving at the Pierre Elliot Trudeau International airport in Montreal in a given hour

Data set EX0143 **1.43 Aqua Running** Aqua running has been suggested as a method of cardiovascular conditioning for injured athletes and others who want a low-impact aerobics program. A study reported in the *Journal of Sports Medicine* investigated the relationship between exercise cadence and heart rate by measuring the heart rates of 20 healthy volunteers at a cadence of 48 cycles per minute (a cycle consisted of two steps).[8] The data are listed here:

87	109	79	80	96	95	90	92	96	98
101	91	78	112	94	98	94	107	81	96

Construct a stem and leaf plot to describe the data. Discuss the characteristics of the data distribution.

1.44 JUMP and EQAO Scores Mathematician and award-winning playwright John Mighton started Junior Undiscovered Math Prodigies (JUMP) in 1998. Education Quality and Accountability Office (EQAO) tests are designed to assess a student's ability in reading, writing, and mathematics for Grades 3 and 6 in Ontario. The following two graphs reported in *Professionally Speaking*[9] display EQAO scores for Grades 3 and 6 before and after JUMP training at six schools in Ontario.

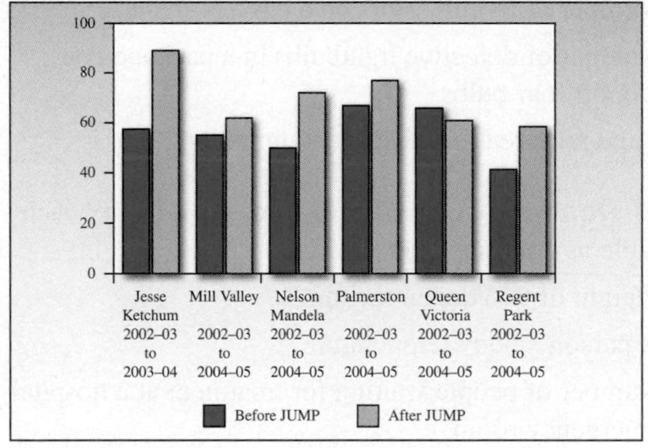

Grade 3 EQAO Score, Before and After JUMP

Source: Written by Elke Town, *Professionally Speaking* by Ontario College of Teachers. Reproduced with permission. http://professionallyspeaking.oct.ca/march_2007/math.asp

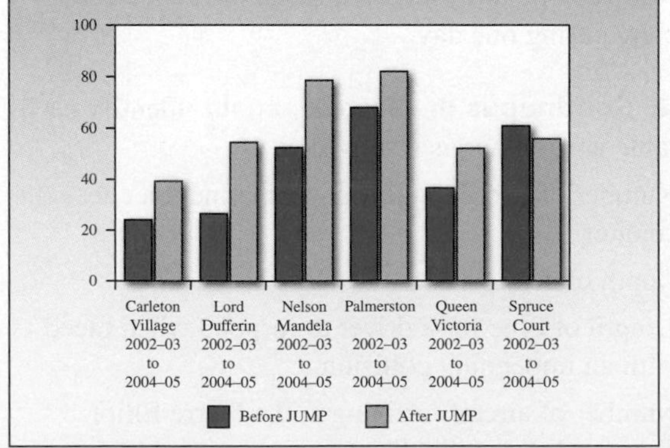

Grade 6 EQAO Score, Before and After JUMP

Source: Written by Elke Town, *Professionally Speaking* by Ontario College of Teachers. Reproduced with permission. http://professionallyspeaking.oct.ca/march_2007/math.asp

a. What graphical techniques did JUMP use?

b. Do you think that a comparison of EQAO scores pre- and post-JUMP is the best way to measure the program effectiveness on a given student? Why or why not?

Data set EX0145 **1.45 Ages of Pennies** We collected 50 pennies and recorded their ages, by calculating AGE = CURRENT YEAR − YEAR ON PENNY.

5	1	9	1	2	20	0	25	0	17
1	4	4	3	0	25	3	3	8	28
5	21	19	9	0	5	0	2	1	0
0	1	19	0	2	0	20	16	22	10
19	36	23	0	1	17	6	0	5	0

a. Before drawing any graphs, try to visualize what the distribution of penny ages will look like. Will it be mound-shaped, symmetric, skewed right, or skewed left?

b. Draw a relative frequency histogram to describe the distribution of penny ages. How would you describe the shape of the distribution?

Data set EX0146 **1.46 Ages of Pennies, continued** The data below represent the ages of a different set of 50 pennies, again calculated using AGE = CURRENT YEAR − YEAR ON PENNY.

41	9	0	4	3	0	3	8	21	3
2	10	4	0	14	0	25	12	24	19
3	1	14	7	2	4	4	5	1	20
14	9	3	5	3	0	8	17	16	0
0	7	3	5	23	7	28	17	9	2

a. Draw a relative frequency histogram to describe the distribution of penny ages. Is the shape similar to the shape of the relative frequency histogram in Exercise 1.45?

b. Draw a stem and leaf plot to describe the penny ages. Are there any unusually large or small measurements in the set?

Data set EX0147 **1.47 Canadian Federal Election** The data below shows the seats won by the Conservatives in every election in Canadian history up to 2006.[10] Use an appropriate graph to describe the number of seats won by Conservatives. Write a summary paragraph describing this set of data.

General Election (Date)	Seats	General Election (Date)	Seats
1st (1867.09.20)	101	21st (1949.06.27)	41
2nd (1872.10.12)	103	22nd (1953.08.10)	51
3rd (1874.01.22)	73	23rd (1957.06.10)	112
4th (1878.09.17)	137	24th (1958.03.31)	208
5th (1882.06.20)	139	25th (1962.06.18)	116
6th (1887.02.22)	123	26th (1963.04.08)	95

7th (1891.03.05)	123	27th (1965.11.08)	97
8th (1896.06.23)	88	28th (1968.06.25)	72
9th (1900.11.07)	81	29th (1972.10.30)	107
10th (1904.11.03)	75	30th (1974.07.08)	95
11th (1908.10.26)	85	31st (1979.05.22)	136
12th (1911.09.21)	133	32nd (1980.02.18)	103
13th (1917.12.17)	153	33rd (1984.09.04)	211
14th (1921.12.06)	50	34th (1988.11.21)	169
15th (1925.10.29)	116	35th (1993.10.25)	2
16th (1926.09.14)	91	36th (1997.06.02)	20
17th (1930.07.28)	137	37th (2000.11.27)	12
18th (1935.10.14)	39	38th (2004.06.28)	99
19th (1940.03.26)	39	39th (2006.01.23)	124
20th (1945.06.11)	67		

Source: Library of Parliament / Bibliotheque du Parlement, "Electoral Results by Party. www.lop.parl.gc.ca/Parllnfo/compilations/electionsandridings/ResultsParty.aspx Parliament of Canada. Reproduced with the permission of Library of Parliament, 2016.

Data set **1.48 Number of Wet Days** Do some cities EX0148 have more wet days? Does Vancouver deserve to be nicknamed "the wettest city"? These data list yearly averages for rainfall plus snowfall in major cities in Canada. Each city has a total of the number of days of wet weather it usually gets in a year, as well as for the normal amount of precipitation. The annual precipitation totals are averages of weather data collected from 1981 to 2010. The calculation of days includes those when precipitation totalled at least 0.2 millimetres (mm).[11]

City	Average Yearly Precipitation (mm)	Days
Abbotsford, British Columbia	1538	179
St. John's, Newfoundland	1534	212
Halifax, Nova Scotia	1468	162
Vancouver, British Columbia	1457	168
Saint John, New Brunswick	1295	158
Québec, Quebec	1184	175
Moncton, New Brunswick	1124	161
Trois-Rivières, Quebec	1123	161
Sherbrooke, Quebec	1100	187
London, Ontario	1012	168
Montréal, Quebec	1000	163
Kingston, Ontario	960	159
Windsor, Ontario	935	150
Barrie, Ontario	933	156
Saguenay, Quebec	931	198
Guelph, Ontario	931	167
Ottawa, Ontario	920	161
Kitchener-Waterloo, Ontario	916	166
Sudbury, Ontario	903	167
Hamilton, Ontario	897	149
St. Catharines-Niagara, Ontario	880	151
Oshawa, Ontario	872	146
Brantford, Ontario	867	136
Peterborough, Ontario	870	147
Toronto, Ontario	831	145
Victoria, British Columbia	705	148
Thunder Bay, Ontario	684	143
Winnipeg, Manitoba	521	125
Edmonton, Alberta	456	123
Calgary, Alberta	419	112
Regina, Saskatchewan	390	118
Saskatoon, Saskatchewan	365	87
Kelowna, British Columbia	345	120

a. Construct a relative frequency histogram for average yearly precipitation.

b. Draw a stem and leaf plot of the number of wet days. Are there any unusually large and small wet days in the data?

c. The average yearly precipitation in Abbotsford, British Columbia, is recorded as 1538. Does the geography of that city explain the observation?

d. The number wet of days in Vancouver is recorded as 168. Do you consider this unusually wet?

Data set **1.49 Queen's Plate Finish Times** History EX0149 was made at Woodbine (Toronto) as Dancethruthedawn captured the 142nd edition of the Queen's Plate Stakes. The royalty bred daughter of 1991 champion Dance Smartly emulated her dam's feat by winning both the Woodbine Oaks and the Queen's Plate. The following data set shows winning times (in seconds) for the Queen's Plate finish races from 1957 to 2006.[12]

Year	Time (seconds)	Year	Time (seconds)	Year	Time (seconds)
1957	122.6	1974	129.2	1991	123.4
1958	124.2	1975	122.6	1992	124.6
1959	124.8	1976	125.0	1993	124.2
1960	122.0	1977	126.6	1994	123.4
1961	125.0	1978	122.0	1995	123.8
1962	124.6	1979	126.6	1996	123.8
1963	124.0	1980	124.2	1997	124.2
1964	122.2	1981	124.8	1998	122.2
1965	123.8	1982	124.6	1999	123.13
1966	123.6	1983	124.2	2000	125.53
1967	123.0	1984	123.8	2001	123.78
1968	125.4	1985	124.6	2002	126.88
1969	124.2	1986	127.2	2003	122.48
1970	124.8	1987	123.6	2004	124.72
1971	123.0	1988	126.2	2005	127.37
1972	123.2	1989	123.0	2006	125.30
1973	128.0	1990	121.8		

Source: © Woodbine Entertainment Group.

a. Do you think there will be a trend in the winning times over the years? Draw a line chart to verify your answer.

b. Describe the distribution of winning times using an appropriate graph. Comment on the shape of the distribution and look for any unusual observations.

Data set EX0150 **1.50 Internet Access from Home** As Canadians become more knowledgeable about computer hardware and software, and as prices drop and installation becomes easier, home networking of PCs is rising. The table gives the number of users per 100 members of the population.[13]

Year	DSL	Cable Modem
2006	10.8	11.5
2005	9.4	9.7
2004	7.9	8.7
2003	6.3	8.0
2002	4.7	6.6
2001	3.0	5.2
2000	1.3	3.1

a. What graphical methods could you use to describe the data?

b. What trends do you expect to see in the future?

Data set EX0151 **1.51 Election Results** The 2006 election was somewhat an interesting race, in which the Conservative Party of Canada secured the largest number of House of Commons seats with respect to others. The table below lists the distribution of House of Commons seats following the 39th general election in Canada 2006.[14]

	All Seats	Conservative Party of Canada	Liberal Party of Canada	Bloc Québécois	New Democratic Party	Independent
Canada	308	124	103	51	29	1
Newfoundland & Labrador	7	3	4	0	0	0
PEI	4	0	4	0	0	0
Nova Scotia	11	3	6	0	2	0
New Brunswick	10	3	6	0	1	0
Quebec	75	10	13	51	0	1
Ontario	106	40	54	0	12	0
Manitoba	14	8	3	0	3	0
Saskatchewan	14	12	2	0	0	0
Alberta	28	28	0	0	0	0
British Columbia	36	17	9	0	10	0
Yukon	1	0	1	0	0	0
Northwest Territories	1	0	0	0	1	0
Nunavut	1	0	1	0	0	0

Source: Library of Parliament / Bibliotheque du Parlement, "Electoral Results by Party. www.lop.parl.gc.ca/ParlInfo/compilations/electionsandridings/ResultsParty.aspx Parliament of Canada. Reproduced with the permission of Library of Parliament, 2016.

a. By just looking at the table for Conservatives, what shape do you think the data distribution for seats by province/territory will have?

b. Draw a relative frequency histogram to describe the distribution of seats for Conservatives in all provinces/territories.

c. Did the histogram in part b confirm your guess in part a? Are there any outliers? How can you explain them?

Data set EX0152 **1.52 Election Results, continued** Refer to Exercise 1.51. Repeat a–c for the seat distribution in the House of Commons for the Liberal Party of Canada.

1.53 Election Results, continued Refer to Exercises 1.51 and 1.52. Construct side-by-side stem and leaf plots for the number of seats in the House of Commons for the Conservatives and the Liberals.

a. Describe the shapes of the two distributions. Are there any outliers?

b. Do the stem and leaf plots resemble the relative frequency histograms constructed in Exercises 1.51 and 1.52?

c. Explain why the distribution of the seats in the House of Commons is skewed for both parties.

1.54 Diamonds Are Forever! Much of the world's diamond industry is located in Africa, with Russia and Canada also showing large revenues from their diamond mining industries. A visual representation of the various shares of the world's diamond revenues, adapted from *Time Magazine,*[15] is shown below:

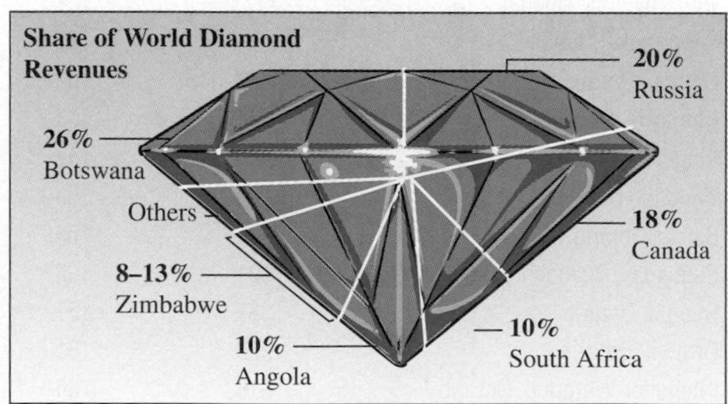

a. Draw a pie chart to describe the various shares of the world's diamond revenues.

b. Draw a bar chart to describe the various shares of the world's diamond revenues.

c. Draw a Pareto chart to describe the various shares of the world's diamond revenues.

d. Which of the charts is the most effective in describing the data?

Data set EX0155 **1.55 An Archeological Find** An article in *Archaeometry* involved an analysis of 26 samples of Romano-British pottery, found at four different kiln sites in the United Kingdom.[16] The samples were analyzed to determine their chemical composition, and the percentage of aluminum oxide in each of the 26 samples is shown in the following table.

Llanederyn		Caldicot	Island Thorns	Ashley Rails
14.4	11.6	11.8	18.3	17.7
13.8	11.1	11.6	15.8	18.3
14.6	13.4		18.0	16.7
11.5	12.4		18.0	14.8
13.8	13.1		20.8	19.1
10.9	12.7			
10.1	12.5			

Source: A. Tubb, A.J. Parker, and G. Nickless, "The Analysis of Romano-British Pottery by Atomic Absorption Spectrophotometry," *Archaeometry*, Vol. 22, Issue 2, August 1989, p. 153. *Archaeometry* by UNIVERSITY OF OXFORD. Reproduced with permission of BLACKWELL PUBLISHING LTD. in the format Book via Copyright Clearance Center.

a. Construct a relative frequency histogram to describe the aluminum oxide content in the 26 pottery samples.

b. What unusual feature do you see in this graph? Can you think of an explanation for this feature?

c. Draw a dotplot for the data, using a letter (L, C, I, or A) to locate the data point on the horizontal scale. Does this help explain the unusual feature in part b?

Data set EX0156 **1.56 Afghanistan Support Stabilizes** The following table is based on the EKOS Research Associates survey from September 2005 to October 2006, which asked the question: "Right now, the Canadian Forces are involved in a broader peace-SUPPORT operation in Afghanistan, helping to rebuild the country and maintain security with our troops fighting on the frontline if necessary. Would you say you strongly support, somewhat support, somewhat oppose or strongly oppose these contributions?"[17]

a. Draw a line chart to describe the percentages that are opposed. Use time as the horizontal axis.

b. Superimpose another line chart on the one drawn in part a to describe the percentage that is in support.

c. Use your line chart to summarize the changes during the given period. Is the support stabilizing? Why or why not?

d. The following line chart was presented on the website of EKOS. How does it differ from the graph that you drew? What characteristics of the EKOS line chart might cause distortion when the graph is interpreted, if any?

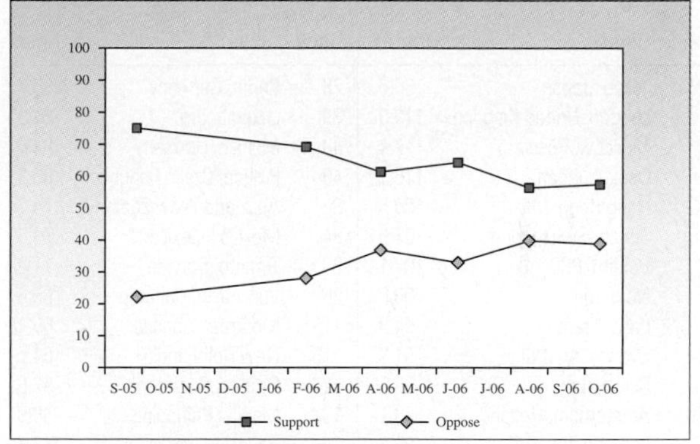

Date	Oppose (%)	Support (%)
September 2005	22	76
February 2006	28	70
April 2006	37	62
January 2006	33	65
April 2006	40	57
October 2006	39	58

Source: EKOS Research Associates: www.ekospolitics.com/article/6nov2006background.pdf.

Data set EX0157 **1.57 Pulse Rates** A group of 50 biomedical students recorded their pulse rates by counting the number of beats for 30 seconds and multiplying by 2.

80	70	88	70	84	66	84	82	66	42
52	72	90	70	96	84	96	86	62	78
60	82	88	54	66	66	80	88	56	104
84	84	60	84	88	58	72	84	68	74
84	72	62	90	72	84	72	110	100	58

a. Why are all of the measurements even numbers?

b. Draw a stem and leaf plot to describe the data, splitting each stem into two lines.

c. Construct a relative frequency histogram for the data.

d. Write a short paragraph describing the distribution of the student pulse rates.

Data set **1.58 Cost of Living in Canada— International Comparison** EX0158 On the Mercer Cost of Living Survey for 2004, which surveys 144 cities based on over 200 items in each location, Canadian cities were close to the bottom of the list. With New York being the benchmark, with an index score of 100, Toronto was 71.8, Vancouver 69.6, and Ottawa 62.6. Of the 144 cities, Toronto was 89th on the list, Vancouver was 96th, and Ottawa was 124th.[18] For further information on the Mercer Cost of Living Survey, visit www.mercerhr.com.

Rank	Cities	Index	Rank	Cities	Index
1	Tokyo, Japan	130.7	28	Berlin, Germany	85.7
2	London, United Kingdom	119.0	39	Luxembourg	84.3
3	Moscow, Russia	117.4	43	Munich, Germany	84.0
4	Osaka, Japan	116.1	49	Prague, Czech Republic	83.3
5	Hong Kong	109.5	80	Auckland, New Zealand	74.2
9	Zurich, Switzerland	101.6	84	Mexico, Mexico	73.3
11	Beijing, PRC	101.1	89	Toronto, Canada	71.8
	New York	100.0	96	Vancouver, Canada	69.6
17	Paris, France	94.8	113	Montréal, Canada	66.4
20	Sydney, Australia	91.8	116	New Delhi, India	64.7
21	Rome, Italy	90.5	124	Ottawa, Canada	62.6
26	Amsterdam, Holland	88.1	138	Manila, Philippines	48.8
27	Los Angeles, U.S.	86.6	144	Asuncion, Paraguay	42.7

a. Construct a relative frequency histogram to describe the index for the cities shown in the table. Do you see any unusual features in the histogram?

b. Construct a stem and leaf plot for the data.

c. How do you think the cities were selected for inclusion in this table?

1.59 Picture Never Lies? Want to buy a 6-pack of energy water in Alberta? Watch the price. Below you will see a visual representation of a 6-pack of energy water in Alberta and British Columbia.[19]

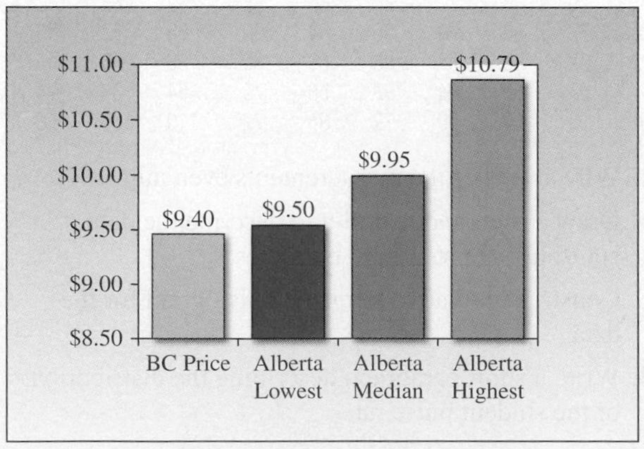

a. Comment on the accuracy of the graph shown above. Do the sizes, scale starts, and heights of the four bars accurately represent the price of a 6-pack of energy water?

b. Draw the bar graph to describe the price.

Data set **1.60 Sky Scan** EX0160 Between January and June 2001, Sky Scan, in Edmonton, conducted a number of test observations to try out meteor observing techniques. The following table shows the count of meteors detected between the hours of midnight and 6 a.m. each day during this period.[20] On two of the sample days, the computer crashed and no data was recorded.

	12:00 to 1:30	1:30 to 3:00	3:00 to 4:30	4:30 to 6:00	Total
May 27, 2001	7	5	7	8	27
May 28, 2001	6	8	7	9	30
May 29, 2001	5	5	5	7	22
May 30, 2001	8	18	8	7	41
May 31, 2001	9	11	5	12	37
June 1, 2001	8	8	6	12	34
June 2, 2001	7	7	5	8	27
June 3, 2001	10	7	2	5	24
June 4, 2001	6	7	6	8	27
June 5, 2001	7	4	1	9	21
June 6, 2001	6	5	4	10	25
June 7, 2001	3	5	9	6	23
June 8, 2001	0	0	0	0	0
June 9, 2001	10	6	6	8	30
June 10, 2001	2	10	10	8	30
June 11, 2001	16	13	9	10	48
June 12, 2001	9	6	7	12	34
June 13, 2001	0	0	0	0	0
June 14, 2001	8	9	8	13	38

Source: © Skyscan.ca. Used with permission.

The histograms below show the distribution of meteors detected per time period.

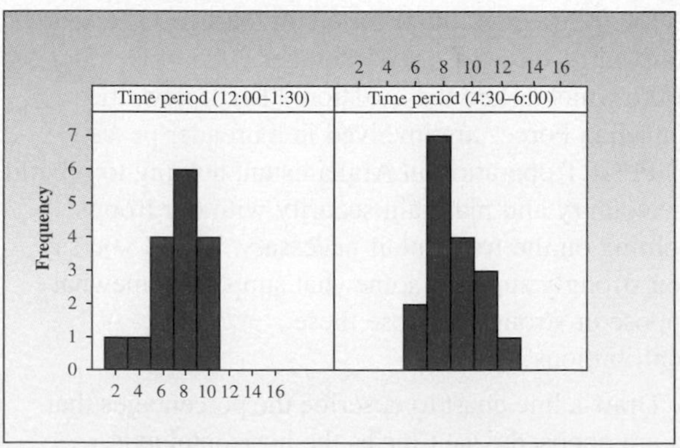

Source: © Skyscan.ca. Used with permission.

Write a summary paragraph describing and comparing the distribution of meteors detected for the time periods.

Data set EX0161 **1.61 Old Faithful** The following data are the waiting times between eruptions of the Old Faithful geyser in Yellowstone National Park.[21] Use one of the graphical methods from this chapter to describe the distribution of waiting times. If there are any unusual features in your graph, see if you can think of any practical explanation for them.

56	89	51	79	58	82	52	88	52	78
69	75	77	53	80	54	79	74	65	78
55	87	53	85	61	93	54	76	80	81
59	86	78	71	77	89	45	93	72	71
76	94	75	50	83	82	72	77	75	65
79	72	78	77	79	72	82	74	80	49
75	78	64	80	49	49	88	51	78	85
65	75	77	69	92	91	53	86	49	79
68	87	61	81	55	93	53	84	70	73
93	50	87	77	74	89	87	76	59	80

Data set EX0162 **1.62 Gasoline Tax** The table below provides a breakdown of tax components of gasoline in cents for different cities across Canada for the first six months of 2006.[22]

City	Province/Territory	Total Tax Component
St. John's	NL	39.9
Charlottetown	PEI	36.8
Halifax	NS	38.4
Fredericton	NB	37.4
Quebec	QC	38.0
Montreal	QC	39.3
Toronto	ON	30.1
Winnipeg	MB	26.9
Regina	SK	30.7
Calgary	AB	24.2
Kelowna	BC	30.4
Vancouver	BC	36.3
Victoria	BC	33.0
Whitehorse	YK	22.4
Yellowknife	NWT	26.8

Source: Table – Tax Components of Gasoline (in cents) from "Oil and Gas Prices, Taxes and Consumers" (*July 2006 Tax Bulletin*). Department of Finance Canada website http://www.fin.gc.ca/toce/2006/gas_tax-e.html. Reproduced with the permission of the Department of Finance, 2015 (www.fin.gc.ca/toc/2006/gas_tax-eng.asp). Open Gov't License: http://open.canada.ca/en/open-government-licence-canada.

a. Construct a stem and leaf display for the data.

b. How would you describe the shape of this distribution?

c. Are there cities with unusually high or low gasoline taxes? If so, which cities are they?

Data set EX0163 **1.63 Hydroelectric Plants** The following data represent the planned rated capacities in megawatts (millions of watts) for the world's 20 largest hydroelectric plants.[23]

18,200	4,500	3,000
12,600	4,200	2,940
10,000	4,200	2,715
8,370	3,840	2,700
6,400	3,444	2,541
6,300	3,300	2,512
6,000	3,100	

a. Construct a stem and leaf plot for the data.

b. How would you describe the shape of this distribution?

Data set EX0164 **1.64 Car Colours** The most popular colours for compact and sports cars in a recent year are given in the table.[24]

Colour	Percentage	Colour	Percentage
Silver	24.6	Medium Red	5.5
Black	14.3	Light Brown	4.3
Medium/Dark Blue	12.9	Gold	4.1
White	8.8	Dark Red	2.6
Bright Red	6.9	Other	9.3
Medium/Dark Grey	6.7		

Use an appropriate graphical display to describe these data.

Data set EX0165 **1.65 Tim Hortons** The number of Tim Hortons shops within 5 kilometres of city centre (downtown) in cities of southwestern Ontario is shown in the following table.[25]

City	Number	City	Number
Brampton	22	Mississauga	21
Burlington	16	Niagara Falls	10
Cambridge	12	Oakville	20
Guelph	19	St. Catharines	19
Hamilton	40	Stratford	4
Kitchener	40	Toronto	65
London	39	Waterloo	34
Markham	20	Windsor	23
Milton	6	Woodstock	7

a. Draw a dotplot to describe the data.

b. Describe the shape of the distribution.

c. Is there another variable that you could measure that might help to explain why some cities have more Tim Hortons than others? Explain.

Data set EX0166 **1.66 What's Normal?** The 37° Celsius standard for human body temperature was derived by a German doctor in 1868. In an attempt to verify his claim, Mackowiak, Wasserman, and Levine[26] took temperatures from 148 healthy people over a three-day period. A data set closely matching the one in Mackowiak's article was derived by Allen Shoemaker, and appears in the *Journal of Statistics Education*.[27]

The body temperatures for these 130 individuals are shown below:

a. Describe the shape of the distribution of temperatures.

b. Are there any unusual observations? Can you think of any explanation for these?

c. Locate the 37°C standard on the horizontal axis of the graph. Does it appear to be near the centre of the distribution?

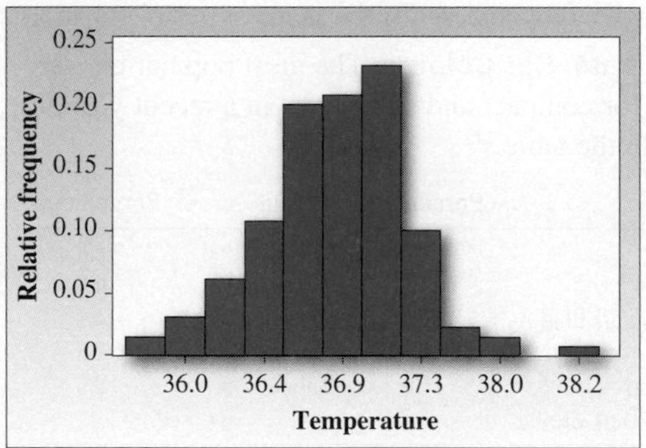

Data set EX0167 **1.67 Agriculture Demand Curves** Suppose that the four quarters of demand of these various agriculture produce products in 2015 is shown below:

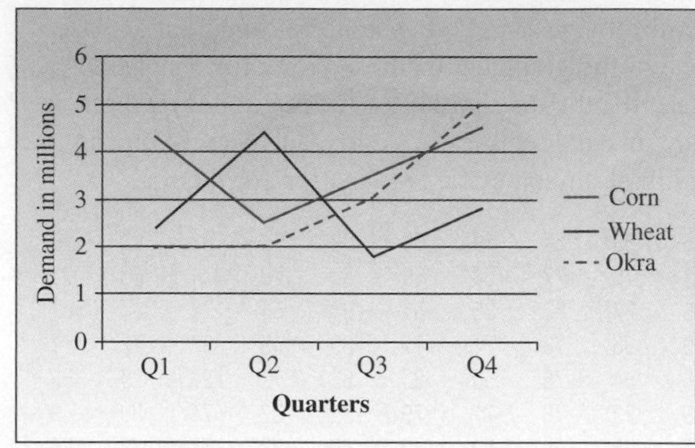

a. Describe the demand for corn, wheat, and okra during the four quarters.

b. Give your opinion as to why they have this particular shape.

c. Do you think there is any relationship between the produce? Explain.

CASE STUDY

How Is Your Blood Pressure?

Blood Pressure

Blood pressure is the pressure that the blood exerts against the walls of the arteries. When physicians or nurses measure your blood pressure, they take two readings. The systolic blood pressure is the pressure when the heart is contracting and therefore pumping. The diastolic blood pressure is the pressure in the arteries when the heart is relaxing. The diastolic blood pressure is always the lower of the two readings. Blood pressure varies from one person to another. It will also vary for a single individual from day to day and even within a given day.

If your blood pressure is too high, it can lead to a stroke or a heart attack. If it is too low, blood will not get to your extremities and you may feel dizzy. Low blood pressure is usually not serious.

So, what should *your* blood pressure be? A systolic blood pressure of 120 would be considered normal. One of 150 would be high. But since blood pressure varies with gender and increases with age, a better gauge of the relative standing of your blood pressure would be obtained by comparing it with the population of blood pressures of all persons of your gender and age in Canada. Of course, we cannot supply you with that data set, but we can show you a very large sample selected from it. The text website provides blood pressure data on 500 persons, 236 men and 264 women. This data was used as a case study in Data Analysis at the 2003 Annual Meeting of Statistical Society of Canada in Halifax. Entries for each person include that person's age and systolic blood pressure and several other variables of interest.

1. Describe the variables that have been measured in this survey. Are the variables quantitative or qualitative? Discrete or continuous? Are the data univariate, bivariate, or multivariate?
2. What types of graphical methods are available for describing this data set? What types of questions could be answered using various types of graphical techniques?

3. Using the systolic blood pressure data set, construct a relative frequency histogram for the 236 men and another for the 264 women. Use a statistical software package if you have access to one. Compare the two histograms.
4. Consider the 236 men and 264 women as the entire population of interest. Choose a sample of $n = 50$ men and $n = 50$ women, recording their systolic blood pressures and their ages. Draw two relative frequency histograms to graphically display the systolic blood pressures for your two samples. Do the shapes of the histograms resemble the population histograms from part 3?
5. How does your blood pressure compare with that of others of your same gender? Check your systolic blood pressure against the appropriate histogram in part 3 or 4 to determine whether your blood pressure is "normal" or whether it is unusually high or low.

PROJECTS

Project 1-A: Five Tips for Keeping Your Home Safe This Summer

Source: From "Keeping Your Home Safe" by Kim Fisher, *Canadian Living,* July 2008.

1. **Secure it.** Take a walk around your property. Do you see any way to get in? Purchase a good lock with a reputable brand name ($100 to $150 at a local hardware store) and lock up ladders, patio furniture, recycling bins, and anything else a thief could use to gain access to a second-storey window.
2. **Keep it living.** Your home is less likely to catch the attention of a thief if it looks like it's being lived in, so it should reflect your regular schedule and give the indication that someone is at home.
3. **Maintain it.** A home that's well cared for is a less attractive target for thieves. Trim hedges that exceed window height and cut evergreen branches up at least three feet from the ground to eliminate hiding places on your property.
4. **Shut it off.** If you're going on vacation for more than a few weeks, turn off the water to your home and drain the lines (turn on your taps for a few minutes and flush your toilets).
5. **Make a friend.** Find a friend or neighbour you trust to keep an eye on your home. Have her cut the grass, park a car in your driveway, pick up your mail, and even put out some extra garbage on garbage day.

Due to an increasing number of break-ins in a particular subdivision of North York, an insurance company specializing in property insurance is interested in knowing how many people follow the above tips. A statistical consultant was hired to provide such information. The consultant conducted a survey of 300 randomly selected households and the results of her findings are summarized in the following table.

Type of Tips	Number of Households
Secure it	74
Keep it living	36
Maintain it	31
Shut it off	17
Make a friend	42
Any combination	73
None	27

a. Identify and describe the population and sample in this problem.

b. Are the collected data based on sample or population?

c. What are the experimental units?

d. What is the variable being measured?
e. Is the variable qualitative or quantitative?
f. Is the variable of interest discrete or continuous?
g. Summarize these data in a bar chart, and explain what the chart tells us.
h. Determine the relative frequencies.
i. Construct a relative frequency bar chart.
j. Construct a pie chart to describe the data.
k. What proportion of people responded positively to tips "Secure it" or "Make a friend"?
l. Collect similar data in your neighbourhood (must get permission by appropriate authorities) and see if your data result is consistent with this data.

Project 1-B Handwashing Saves Lives: It's in Your Hands

Swine flu is a respiratory disease associated with pigs. The symptoms of swine flu in people resemble regular human flu symptoms, including fever, cough, headache, general aches, and fatigue. "The best way to limit influenza exposure in your workplace is to have sick workers stay at home and to encourage regular handwashing with hot water and soap," says Geoffrey Clark, WorkSafeBC's senior occupational hygienist. Further, it is suggested to wash hands for 20 to 30 seconds. An occupational hygienist in Moose Jaw, Saskatchewan, took a random sample of 25 students at local high school to get an idea regarding time (in seconds) for handwashing.

15	5	4	6	5	4	6	8	10	9	7	11	19
20	5	5	7	8	16	0	13	0	5	9	10	

a. What are the experimental units?
b. What is the variable being measured?
c. Is the variable qualitative or quantitative?
d. Is the variable of interest discrete or continuous?
e. Construct a dotplot and summarize the findings.
f. Based on your dotplot, what can be said about the shape of the distribution of the data? Why?
g. How will you construct a line chart to describe the data?
h. Construct a frequency histogram.
i. What proportion of students washed their hands for less than 10 seconds?
j. What proportion of students washed their hands at least 5 seconds?
k. Can you comfortably state that most students wash their hands for about 5 seconds or less? Why or why not?
l. Construct a stem and leaf plot to display the distribution of the data.
m. Would you describe the distribution of the data as symmetric, skewed to the right, or skewed to the left? Explain.
n. Do any of the observations appear to be outliers? If so, which one or ones?
o. Collect similar data at your university or college (must get permission by appropriate authorities) and see if your data result is consistent with this data.

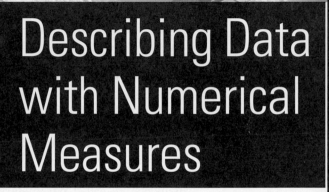

Describing Data with Numerical Measures

Shooter Bob Square Lenses/Shutterstock

The Boys of Winter

Are the superstars of the National Hockey League (NHL) of today better than those of "yesteryears"? Do the eras appear to differ in their levels of scoring? Do the goals per game appear to be more variable in some eras than others? The case study at the end of this chapter involves the goal averages of NHL superstars. Numerical descriptive measures can be used to answer these and similar questions.

GENERAL OBJECTIVES

Graphs are extremely useful for the visual description of a data set. However, they are not always the best tool when you want to make inferences about a population from the information contained in a sample. For this purpose, it is better to use numerical measures to construct a mental picture of the data.

CHAPTER INDEX

NEED TO KNOW

How to Calculate Sample Quartiles

2.1 DESCRIBING A SET OF DATA WITH NUMERICAL MEASURES

Graphs can help you describe the basic shape of a data distribution; "a picture is worth a thousand words." There are limitations, however, to the use of graphs. Suppose you need to display your data to a group of people and the bulb on the data projector blows out! Or you might need to describe your data over the telephone—no way to display the graphs! You need to find another way to convey a mental picture of the data to your audience.

A second limitation is that graphs are somewhat imprecise for use in statistical inference. For example, suppose you want to use a sample histogram to make inferences about a population histogram. How can you measure the similarities and differences between the two histograms in some concrete way? If they were identical, you could say "They are the same!" But if they are different, it is difficult to describe the "degree of difference."

One way to overcome these problems is to use **numerical measures,** which can be calculated for either a sample or a population of measurements. You can use the data to calculate a set of *numbers* that will convey a good mental picture of the frequency distribution. These measures are called **parameters** when associated with the population, and they are called **statistics** when calculated from sample measurements.

DEFINITION Numerical descriptive measures associated with a population of measurements are called **parameters;** those computed from sample measurements are called **statistics.**

2.2 MEASURES OF CENTRE

In Chapter 1, we introduced dotplots, stem and leaf plots, and histograms to describe the distribution of a set of measurements on a quantitative variable X. The horizontal axis displays the values of X, and the data are "distributed" along this horizontal line. One of the first important numerical measures is a **measure of centre**—a measure along the horizontal axis that locates the centre of the distribution.

The GPA data presented in Table 1.9 ranged from a low of 1.9 to a high of 3.4, with the centre of the histogram located in the vicinity of 2.6 (see Figure 2.1). Let's consider some rules for locating the centre of a distribution of measurements.

FIGURE 2.1
Centre of the GPA data

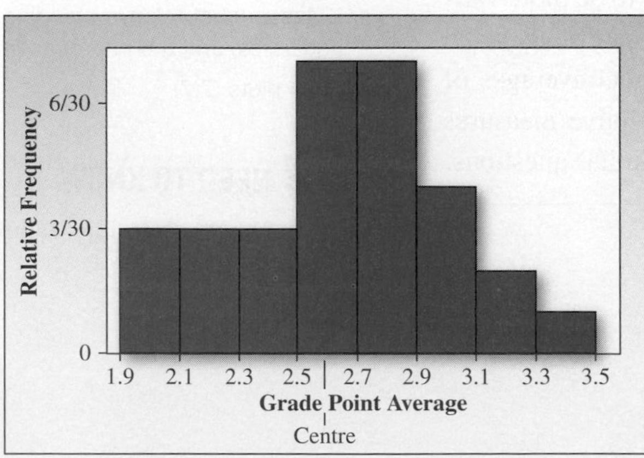

The arithmetic average of a set of measurements is a very common and useful measure of centre. This measure is often referred to as the **arithmetic mean,** or simply the **mean,** of a set of measurements. To distinguish between the mean for the sample and the mean for the population, we will use the symbol $\bar{x}$ (x-bar) for a sample mean and the symbol μ (Greek lowercase mu) for the mean of a population.

DEFINITION The **arithmetic mean** or **average** of a set of n measurements is equal to the sum of the measurements divided by n.

Since statistical formulas often involve adding or "summing" numbers, we use a shorthand symbol to indicate the process of summing. Suppose there are n measurements on the variable X—call them $x_1, x_2, \ldots, x_n$. To add the n measurements together, we use this shorthand notation:

$$\sum_{i=1}^{n} x_i \quad \text{which means } x_1 + x_2 + x_3 + \cdots + x_n$$

The Greek capital sigma (Σ) tells you to add the items that appear to its right, beginning with the number below the sigma ($i = 1$) and ending with the number above ($i = n$). However, since the typical sums in statistical calculations are almost always made on the total set of n measurements, you can use a simpler notation:

$$\sum x_i \quad \text{which means "the sum of all the } X \text{ measurements"}$$

Using this notation, we write the formula for the sample mean:

NOTATION

Sample mean: $\bar{x} = \dfrac{\sum x_i}{n}$

Population mean: μ

EXAMPLE 2.1

Draw a dotplot for the $n = 5$ measurements 2, 9, 11, 5, 6. Find the sample mean and compare its value with what you might consider the "centre" of these observations on the dotplot.

SOLUTION The dotplot in Figure 2.2 seems to be centred between 6 and 8. To find the sample mean, calculate

$$\bar{x} = \frac{\sum x_i}{n} = \frac{2 + 9 + 11 + 5 + 6}{5} = 6.6$$

FIGURE 2.2
Dotplot for Example 2.1

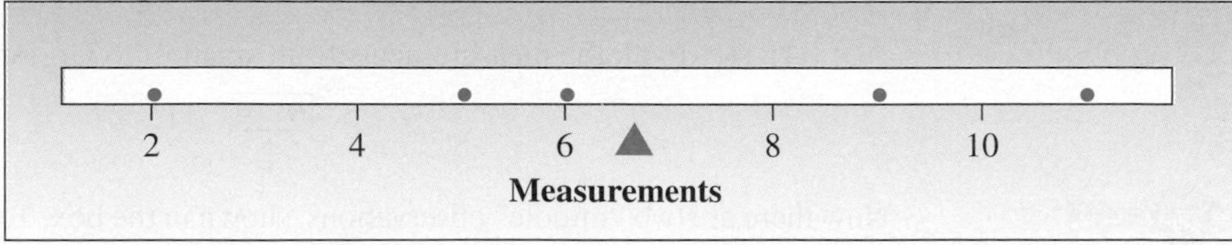

The statistic $\bar{x} = 6.6$ is the balancing point or fulcrum shown on the dotplot. It does seem to mark the centre of the data.

NEED A TIP?

mean = balancing point or fulcrum

Remember that samples are measurements drawn from a larger population that is usually unknown. An important use of the sample mean $\bar{x}$ is as an estimator of the unknown population mean μ. The GPA data in Table 1.9 are a sample from a larger population of GPAs, and the distribution is shown in Figure 2.1. The mean of the 30 GPAs is

$$\bar{x} = \frac{\sum x_i}{30} = \frac{77.5}{30} = 2.58$$

shown in Figure 2.1; it marks the balancing point of the distribution. The mean of the entire population of GPAs is unknown, but if you had to guess its value, your best estimate would be 2.58. Although the sample mean $\bar{x}$ changes from sample to sample, the population mean μ stays the same.

EXAMPLE 2.2

A survey released by Social Fresh found out from a sample size of 347 advertisers that the average cost per click (CPC) for advertising traffic to Facebook pages and apps was \$0.70.[1] Suppose that we had collected our own data from 11 advertisers where the cost per click for each was as follows:

0.80, 0.90, 0.40, 0.60, 0.70, 0.80, 0.40, 0.90, 0.60, 0.50, 0.80

Let us see how it compares to the data collected from Social Fresh:

$$\bar{x} = \frac{\sum x_i}{11} = \frac{7.4}{11} = 0.67$$

Based on the data we collected, our average CPC produced a slightly varied result, and the average value can be changed based on a given sample.

A second measure of central tendency is the **median,** which is the value in the middle position in the set of measurements ordered from smallest to largest.

DEFINITION The **median** m of a set of n measurements is the value of x that falls in the middle position when the measurements are ordered from smallest to largest.

EXAMPLE 2.3

Find the median for the set of measurements 2, 9, 11, 5, 6.

SOLUTION Rank the $n = 5$ measurements from smallest to largest:

2 5 6 9 11
↑ (under 6)

The middle observation, marked with an arrow, is in the centre of the set, or $m = 6$.

EXAMPLE 2.4

Find the median for the set of measurements 2, 9, 11, 5, 6, 27.

SOLUTION Rank the measurements from smallest to largest:

2 5 [6 9] 11 27
↑

NEED A TIP?

Roughly 50% of the measurements are smaller, 50% are larger than the median.

Now there are two "middle" observations, shown in the box. To find the median, choose a value halfway between the two middle observations:

$$m = \frac{6 + 9}{2} = 7.5$$

The value **0.5(*n* + 1)** indicates the **position of the median** in the ordered data set. If the position of the median is a number that ends in the value **0.5,** you need to average the two adjacent values.

EXAMPLE 2.5

For the $n = 5$ ordered measurements from Example 2.3, the position of the median is $0.5(n + 1) = 0.5(6) = 3$, and the median is the *3rd ordered observation*, or $m = 6$. For the $n = 6$ ordered measurements from Example 2.4, the position of the median is $0.5(n + 1) = 0.5(7) = 3.5$, and the median is the *average of the 3rd and 4th ordered observations*, or $m = (6 + 9)/2 = 7.5$.

NEED A TIP?
symmetric: mean = median
skewed right: mean > median
skewed left: mean < median

Although both the mean and the median are good measures of the centre of a distribution, the median is less sensitive to extreme values or *outliers*. For example, the value $x = 27$ in Example 2.4 is much larger than the other five measurements. The median, $m = 7.5$, is not affected by the outlier, whereas the sample average,

$$\bar{x} = \frac{\Sigma x_i}{n} = \frac{60}{6} = 10$$

is affected; its value is not representative of the remaining five observations.

When a data set has extremely small or extremely large observations, the sample mean is drawn toward the direction of the extreme measurements (see Figure 2.3).

FIGURE 2.3
Relative frequency distributions showing the effect of extreme values on the mean and median

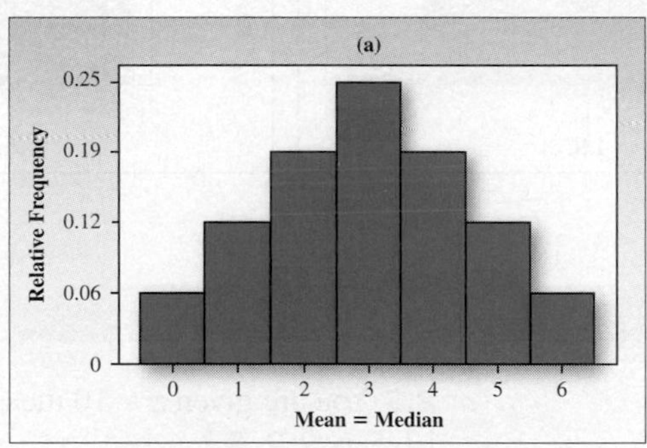

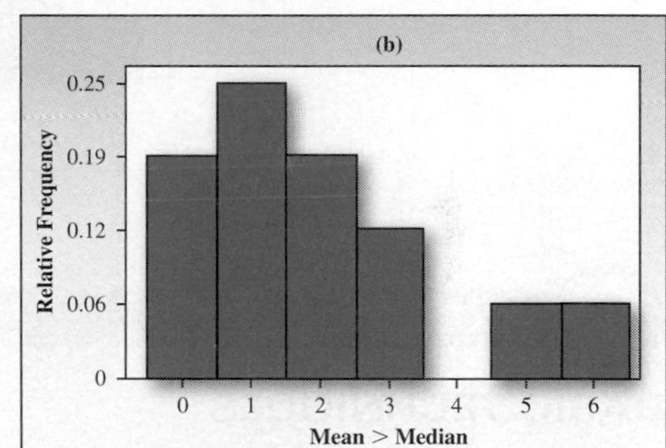

ONLINE APPLET
How Extreme Values Affect the Mean and Median

If a distribution is skewed to the right, the mean shifts to the right; if a distribution is skewed to the left, the mean shifts to the left. The median is not affected by these extreme values because the numerical values of the measurements are not used in its calculation. When a distribution is symmetric, the mean and the median are equal. If a distribution is strongly skewed by one or more extreme values, you should use the median rather than the mean as a measure of centre.

Another way to locate the centre of a distribution is to look for the value of x that occurs with the highest frequency. This measure of the centre is called the **mode.**

DEFINITION The **mode** is the category that occurs most frequently, or the most frequently occurring value of x. When measurements on a continuous variable have been grouped as a frequency or relative frequency histogram, the class with the highest peak or frequency is called the **modal class,** and the midpoint of that class is taken to be the mode.

The mode is generally used to describe large data sets, whereas the mean and median are used for both large and small data sets. From the data in Example 1.14, the mode of the distribution of the number of litres of milk purchased during one particular week is 2. The modal class and the value of x occurring with the highest frequency are the same, as shown in Figure 2.4(a).

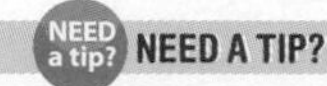

NEED A TIP?

Remember that there can be several modes or no mode (if each observation occurs only once).

For the data in Table 1.9, a GPA of 2.5 occurs five times, and therefore the mode for the distribution of GPAs is 2.5. Using the histogram to find the modal class, you find two classes that occur with equal frequency. Fortunately, these classes are side by side in the tabulation, and the choice for the value of the mode is thus 2.7, the value centred between the fourth and fifth classes. See Figure 2.4(b).

It is possible for a distribution of measurements to have more than one mode. These modes would appear as "local peaks" in the relative frequency distribution. For example, if we were to tabulate the length of fish taken from a lake during one season, we might get a *bimodal distribution*, possibly reflecting a mixture of young and old fish in the population. Sometimes bimodal distributions of sizes or weights reflect a mixture of measurements taken on males and females. In any case, a set or distribution of measurements may have more than one mode.

FIGURE 2.4

Relative frequency histograms for the milk and GPA data

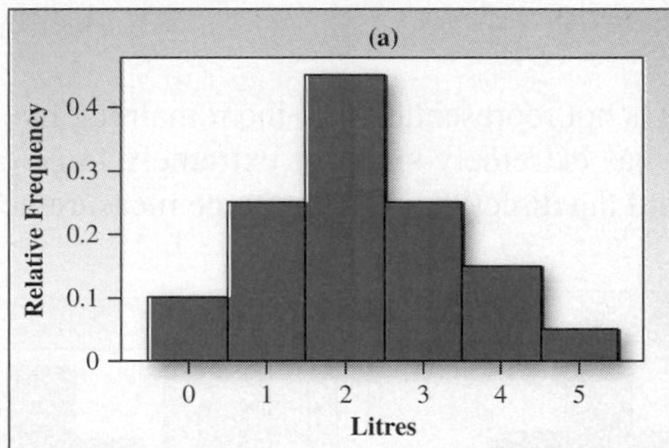

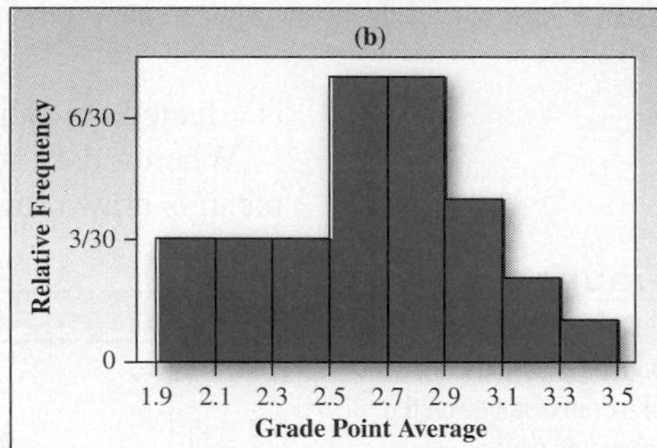

2.2 EXERCISES

BASIC TECHNIQUES

2.1 You are given $n = 5$ measurements: 0, 5, 1, 1, 3.

a. Draw a dotplot for the data. (HINT: If two measurements are the same, place one dot above the other.) Guess the approximate "centre."

b. Find the mean, median, and mode.

c. Locate the three measures of centre on the dotplot in part a. Based on the relative positions of the mean and median, are the measurements symmetric or skewed?

2.2 You are given $n = 8$ measurements: 3, 2, 5, 6, 4, 4, 3, 5.

a. Find $\bar{x}$.

b. Find m.

c. Based on the results of parts a and b, are the measurements symmetric or skewed? Draw a dotplot to confirm your answer.

2.3 You are given $n = 10$ measurements: 3, 5, 4, 6, 10, 5, 6, 9, 2, 8.

a. Calculate $\bar{x}$.

b. Find m.

c. Find the mode.

2.4 You are given $n = 11$ measurements: 5, 7, 6, 5, 5, 3, 2, 4, 6, 7, 2.

a. Calculate $\bar{x}$.

b. Find m.

c. Find the mode.

APPLICATIONS

2.5 Auto Insurance The estimated average automobile insurance premiums, 2004–2005, by province, are shown in the following table:[2]

Province	2005 ($)	2004 ($)
PEI	825	847
NS	883	833
NL	947	1014
QC	988	983
AB	1036	1126
NB	1044	1121
MB	1152	1157
SK	1197	1174
ON	1347	1396
BC	1404	1374

a. What is the average premium for the year 2005?

b. What is the average premium for the year 2004?

c. If you were a consumer, would you be interested in the average premium cost? If not, what would you be interested in?

Data set EX0206 **2.6 DVRs** The digital video recorder (DVR) is a common fixture in most Canadian households. In fact, most Canadian households have DVRs, and many have more than one. A sample of 25 households produced the following measurements on x, the number of DVRs in the household:

1	0	2	1	1
1	0	2	1	0
0	1	2	3	2
1	1	1	0	1
3	1	0	1	1

a. Is the distribution of x, the number of DVRs in a household, symmetric or skewed? Explain.

b. Guess the value of the mode, the value of x that occurs most frequently.

c. Calculate the mean, median, and mode for these measurements.

d. Draw a relative frequency histogram for the data set. Locate the mean, median, and mode along the horizontal axis. Are your answers to parts a and b correct?

2.7 More Billionaires in 2007 The world has gotten used to having less. But there's one thing the world is getting more of these days, and that's more billionaires. According to *Forbes* magazine's editors, the top 20 fortunes when combined would top the half trillion dollar mark.[3]

Name	Age	Nationality	Wealth in Billions ($)
Bill Gates	51	American	56
Warren Buffett	76	American	52
Carlos Slim Helú	67	Mexican	49
Ingvar Kamprad	80	Swedish	33
Lakshmi Mittal	56	Indian	32
Sheldon Adelson	73	American	26
Bernard Arnault	58	French	26
Amancio Ortega	71	Spanish	24
Li Ka-shing	78	Hong Kong	23
David Thomson	49	Canadian	22
Lawrence Ellison	62	American	21.5
Liliane Bettencourt	84	French	20.7
Talal Alsaud	50	Saudi	20.3
Mukesh Ambani	49	Indian	20.1
Karl Albrecht	87	German	20
Roman Abramovich	40	Russian	18.7
Stefan Persson	59	Swedish	18.4
Anil Ambani	47	Indian	18.2
Paul Allen	54	American	18
Theo Albrecht	84	German	17.5

a. Draw a stem and leaf plot for the age. Comment on the shape of the data.

b. Calculate the mean and median wealth for these 20 billionaires.

c. Which of the two measures in part b best describes the centre of the data? Explain.

2.8 Birth Order and Personality Does birth order have any effect on a person's personality? A report on a study by an MIT researcher indicates that later-born children are more likely to challenge the establishment, more open to new ideas, and more accepting of change.[4] In fact, the number of later-born children is increasing. During the Depression years of the 1930s, families averaged 2.5 children (59% later born), whereas the parents of baby boomers averaged 3 to 4 children (68% later born). What does the author mean by an average of 2.5 children?

Data set EX0209 **2.9 How to Minimize the Cost of a Wedding Cake** A wedding cake is priced by the slice. In 2015, wedding cake can cost anywhere from $1.50 to $12 a slice, depending on how complicated the cake is to make. Buttercream icing is typically cheaper than fondant icing. If you are inviting 200 guests, it would be cost effective to order cake for 125 people and serve smaller pieces, depending on the number of invited guests who attend. A quick survey of

14 bakeries gives the price per slice of cake with butter-cream icing:

3.25, 5.50, 4.75, 1.75, 3.50, 5.25, 2.50
4.25, 2.00, 4.25, 3.75, 5.75, 4.00, 6.00

a. Find the average price for the 14 different slices of wedding cake.

b. Find the median price for the 14 different slices of wedding cake.

c. Based on your findings in parts a and b, do you think that the distribution of prices is skewed? Explain.

2.10 Sports Salaries As professional sports teams become a more and more lucrative business for their owners, the salaries paid to the players have also increased. In fact, sports superstars are paid astronomical salaries for their talents. If you were asked by a sports management firm to describe the distribution of players' salaries in several different categories of professional sports, what measure of centre would you choose? Why?

2.11 Time on Task In a psychological experiment, the time on task was recorded for 10 subjects under a 5-minute time constraint. These measurements are in seconds:

175	190	250	230	240
200	185	190	225	265

a. Find the average time on task.

b. Find the median time on task.

c. If you were writing a report to describe these data, which measure of central tendency would you use? Explain.

Data set **2.12 Tim Hortons** The number of Tim
EX0212 Hortons shops within 5 kilometres (km) of city centre (downtown) in cities of southwestern Ontario is shown in the following table:[5]

22	21	39
16	10	20
12	20	6
19	19	34
40	4	23
40	65	7

a. Find the mean, the median, and the mode.

b. Compare the median and the mean. What can you say about the shape of this distribution?

c. Draw a dotplot for the data. Does this confirm your conclusion about the shape of the distribution from part b?

Data set **2.13 LED TVs** As technology improves, the
EX0213 choice of televisions becomes more complicated. Should you choose an LCD TV, an LED TV, or a plasma TV? Does the price of an LED TV depend on the size of the screen? In the table below, *Best Buy*[6] gives the prices and screen sizes for the top 10 LED TVs in the 30-inch and higher categories:

Brand	Price ($)	Size (in.)
LG UltraWide	$649.99	34.0
Dell UltraSharp	$1349.99	34.0
LG Digital Cinema	$1299.99	31.0
BenQ Widescreen 32	$999.99	32.0
BenQ Widescreen 55	$6499.99	55.0
LG 1080p	$599.95	47.0
HP Z30i	$1429.99	30.0
LG Widescreen Commercial-Grade	$1599.95	47.0
Samsung Ultra	$1749.99	31.5
HP Envy	$499.99	32.0

a. What is the average price of these 10 LED televisions?

b. What is the median price of these 10 LED televisions?

c. As a consumer, would you be interested in the average cost of an LED TV? What other variables would be important to you?

2.14 Fighting the Flu Influenza is a highly contagious respiratory disease that strikes as many as eight million Canadians in flu season, between October and April. To get an idea of people's preparation for flu season, a sample of 17 family physicians in Ancaster (a township near Hamilton, Ontario) was asked how many flu shots they had given to patients this fall. The numbers of flu shots were 8, 6, 3, 23, 2, 6, 5, 4, 1, 0, 2, 7, 11, 5, 8, 8, and 10.[7]

a. Find the sample mean.

b. Find the median number of flu shots. What is the value of the mode?

c. Based on the values of the mean and median in the previous two questions, are the measurements symmetric or skewed? Why?

d. If you were writing a report to describe these data, which measure of central tendency would you use? Explain.

Data set **2.15 NFL Draft Coverage** Rich Eisen, a
EX0215 host on the NFL Network, has been running the 40-yard dash to promote the NFL Combine coverage on the network. Below is the tabulated 40-yard dash time for the past 10 years:[8]

Year	40-Yard Dash Time
2005	6.77 seconds
2006	6.22 seconds
2007	6.43 seconds
2008	6.34 seconds
2009	6.34 seconds
2010	6.21 seconds
2011	6.18 seconds
2012	6.03 seconds
2013	6.03 seconds
2014	5.98 seconds

Source: James Brady, "Rich Eisen eyeing record 40-yard dash at NFL Combine", Feb. 23, 2015. From SB Nation www.sbnation.com/nfl/2015/2/23/8080037/nfl-combine-2015-rich-eisen-40-yard-dash. Permission courtesy of Vox Media, Inc.

a. Find the mean.

b. Is calculating the mode and the median relevant in this case? Explain your reasoning why or why not.

c. Describe the shape of the data and by looking at the shape. Explain where Rich Eisen has made the most improvement.

d. Based on the tools you have used, what would you suggest in predicting his 2015 40-yard dash?

2.3 MEASURES OF VARIABILITY

Data sets may have the same centre but look different because of the way the numbers *spread out* from the centre. Consider the two distributions shown in Figure 2.5. Both distributions are centred at $x = 4$, but there is a big difference in the way the measurements spread out, or *vary*. The measurements in Figure 2.5(a) vary from 3 to 5; in Figure 2.5(b) the measurements vary from 0 to 8.

FIGURE 2.5

Variability or dispersion of data

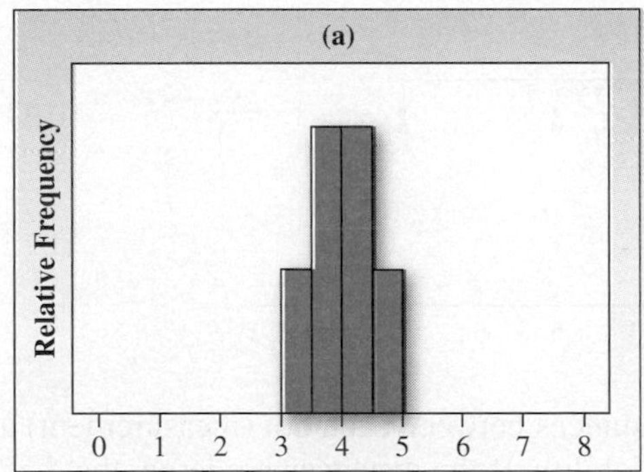

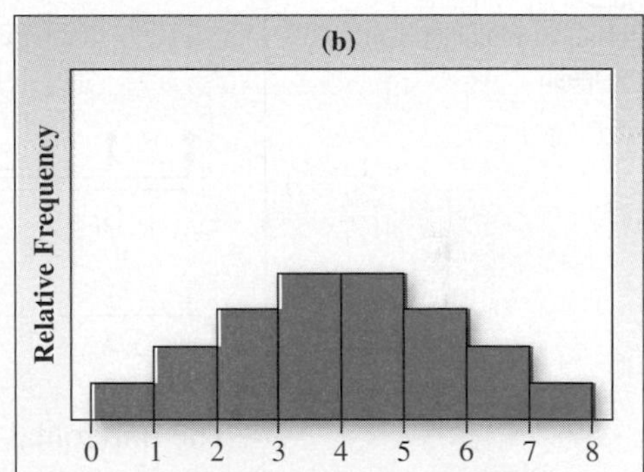

Variability or **dispersion** is a very important characteristic of data. For example, if you were manufacturing bolts, extreme variation in the bolt diameters would cause a high percentage of defective products. On the other hand, if you were trying to discriminate between good and poor accountants, you would have trouble if the examination always produced test grades with little variation, making discrimination very difficult.

Measures of variability can help you create a mental picture of the spread of the data. We will present some of the more important ones. The simplest measure of variation is the **range.**

DEFINITION The **range, *R*,** of a set of n measurements is defined as the difference between the largest and smallest measurements.

For the GPA data in Table 1.9, the measurements vary from 1.9 to 3.4. Hence, the range is $(3.4 - 1.9) = 1.5$. The range is easy to calculate, easy to interpret, and an adequate measure of variation for small sets of data. But, for large data sets, the range is not an adequate measure of variability. For example, the two relative frequency distributions in Figure 2.6 have the same range but very different shapes and variability.

FIGURE 2.6
Distributions with equal range and unequal variability

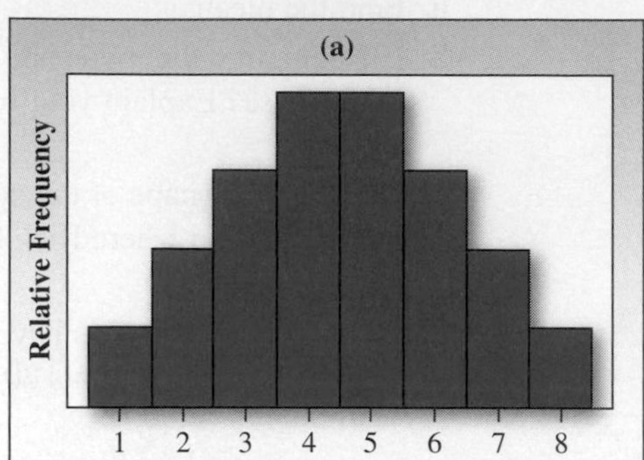

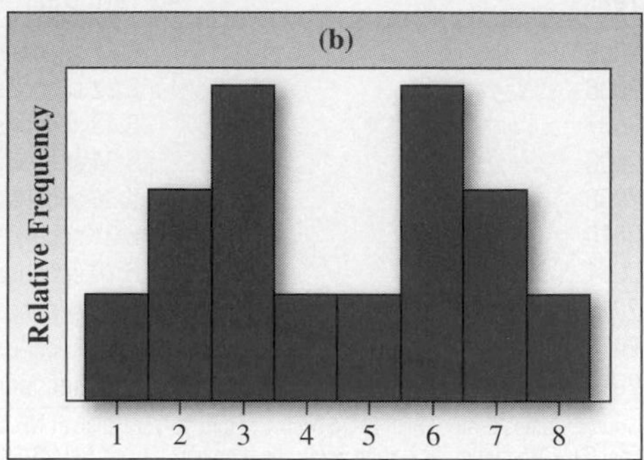

Is there a measure of variability that is more sensitive than the range? Consider, as an example, the sample measurements 5, 7, 1, 2, 4, displayed as a dotplot in Figure 2.7. The mean of these five measurements is

$$\bar{x} = \frac{\Sigma x_i}{n} = \frac{19}{5} = 3.8$$

as indicated on the dotplot.

FIGURE 2.7
Dotplot showing the deviations of points from the mean

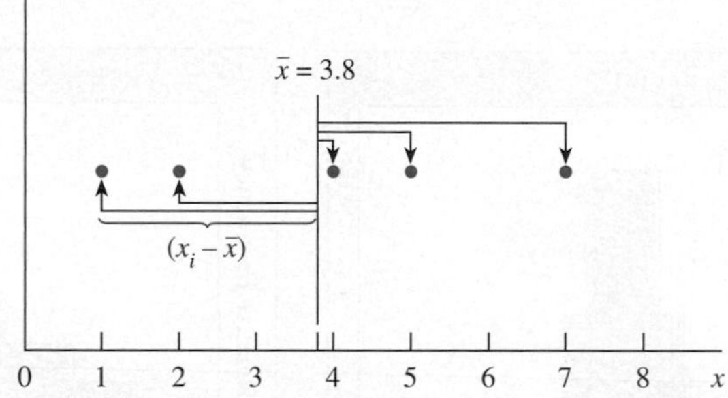

The horizontal distances between each dot (measurement) and the mean $\bar{x}$ will help you measure the variability. If the distances are large, the data are more spread out or *variable* than if the distances are small. If x_i is a particular dot (measurement), then the **deviation** of that measurement from the mean is $(x_i - \bar{x})$. Measurements to the right of the mean produce positive deviations, and those to the left produce negative deviations. The values of x and the deviations for our example are listed in the first and second columns of Table 2.1.

TABLE 2.1 **Computation of $\Sigma(x_i - \bar{x})^2$**

x	$(x_i - \bar{x})$	$(x_i - \bar{x})^2$
5	1.2	1.44
7	3.2	10.24
1	−2.8	7.84
2	−1.8	3.24
4	0.2	0.04
19	0.0	22.80

Because the deviations in the second column of the table contain information on variability, one way to combine the five deviations into one numerical measure is to average them. Unfortunately, the average will not work because some of the deviations are positive, some are negative, and the sum is always zero (unless round-off errors have been introduced into the calculations). Note that the deviations in the second column of Table 2.1 sum to zero.

Another possibility might be to disregard the signs of the deviations and calculate the average of their absolute values.† This method has been used as a measure of variability in exploratory data analysis and in the analysis of time series data. We prefer, however, to overcome the difficulty caused by the signs of the deviations by working with their sum of squares. From the sum of squared deviations, a single measure called the **variance** is calculated. To distinguish between the variance of a *sample* and the variance of a *population*, we use the symbol s^2 for a sample variance and σ^2 (Greek lowercase sigma) for a population variance. *The variance will be relatively large for highly variable data and relatively small for less variable data.*

DEFINITION The **variance of a population** of N measurements is the average of the squares of the deviations of the measurements about their mean μ. The population variance is denoted by σ^2 and is given by the formula

$$\sigma^2 = \frac{\Sigma(x_i - \mu)^2}{N}$$

Most often, you will not have all the population measurements available but will need to calculate the *variance of a sample* of n measurements.

DEFINITION The **variance of a sample** of n measurements is the sum of the squared deviations of the measurements about their mean $\bar{x}$ divided by $(n - 1)$. The sample variance is denoted by s^2 and is given by the formula

$$s^2 = \frac{\Sigma(x_i - \bar{x})^2}{n - 1}$$

NEED A TIP?

The variance and the standard deviation *cannot* be negative numbers.

For the set of $n = 5$ sample measurements presented in Table 2.1, the square of the deviation of each measurement is recorded in the third column. Adding, we obtain

$$\Sigma(x_i - \bar{x})^2 = 22.80$$

and the sample variance is

$$s^2 = \frac{\Sigma(x_i - \bar{x})^2}{n - 1} = \frac{22.80}{4} = 5.70$$

The variance is measured in terms of the square of the original units of measurement. If the original measurements are in centimetres (cm), the variance is expressed in square centimetres. Taking the square root of the variance, we obtain the **standard deviation,** which returns the measure of variability to the original units of measurement.

DEFINITION The **standard deviation** of a set of measurements is equal to the positive square root of the variance.

---

† The absolute value of a number is its magnitude, ignoring its sign. For example, the absolute value of −2, represented by the symbol |−2|, is 2. The absolute value of 2—that is, |2|—is 2.

NOTATION

n: Number of measurements in the sample

s^2: Sample variance

$s = \sqrt{s^2}$: Sample standard deviation

N: Number of measurements in the population

σ^2: Population variance

$\sigma = \sqrt{\sigma^2}$: Population standard deviation

NEED A TIP?

If you are using your calculator, make sure to choose the correct key for the sample standard deviation.

For the set of $n = 5$ sample measurements in Table 2.1, the sample variance is $s^2 = 5.70$, so the sample standard deviation is $s = \sqrt{s^2} = \sqrt{5.70} = 2.39$. The more variable the data set is, the larger the value of s.

For the small set of measurements we used, the calculation of the variance is not too difficult. However, for a larger set, the calculations can become very tedious. Most scientific calculators have built-in programs that will calculate $\bar{x}$ and s or μ and σ, so that your computational work will be minimized. The sample or population mean key is usually marked with $\bar{x}$. The sample standard deviation key is usually marked with s, s_x, or σ_{xn-1}, and the population standard deviation key with σ, σ_x, or σ_{xn}. In using any calculator with these built-in function keys, be sure you know which calculation is being carried out by each key!

If you need to calculate s^2 and s by hand, it is much easier to use the alternative computing formula given next. This computational form is sometimes called the **short-cut method for calculating s^2.**

THE COMPUTING FORMULA FOR CALCULATING s^2

$$s^2 = \frac{\sum x_i^2 - \frac{(\sum x_i)^2}{n}}{n-1}$$

The symbols $(\sum x_i)^2$ and $\sum x_i^2$ in the computing formula are shortcut ways to indicate the arithmetic operation you need to perform. You know from the formula for the sample mean that $\sum x_i$ is the sum of all the measurements. To find $\sum x_i^2$, you square each individual measurement and then add them together.

$\sum x_i^2$ = Sum of the squares of the individual measurements

$(\sum x_i)^2$ = Square of the sum of the individual measurements

The *sample standard deviation, s,* is the positive square root of s^2.

EXAMPLE 2.6

Calculate the variance and standard deviation for the five measurements in Table 2.2, which are 5, 7, 1, 2, 4. Use the computing formula for s^2 and compare your results with those obtained using the original definition of s^2.

TABLE 2.2 ● **Table for Simplified Calculation of s^2 and s**

x_i	x_i^2
5	25
7	49
1	1
2	4
4	16
19	95

Don't round off partial results as you go along!

SOLUTION The entries in Table 2.2 are the individual measurements, x_i, and their squares, x_i^2, together with their sums. Using the computing formula for s^2, you have

$$s^2 = \frac{\sum x_i^2 - \frac{(\sum x_i)^2}{n}}{n-1}$$

$$= \frac{95 - \frac{(19)^2}{5}}{4} = \frac{22.80}{4} = 5.70$$

and $s = \sqrt{s^2} = \sqrt{5.70} = 2.39$, as before.

Why Divide by $n-1$

You may wonder why you need to divide by $(n - 1)$ rather than n when computing the sample variance. Just as we used the sample mean $\bar{x}$ to estimate the population mean μ, you may want to use the sample variance s^2 to estimate the population variance σ^2. It turns out that the sample variance s^2 with $(n - 1)$ in the denominator provides better estimates of σ^2 than would an estimator calculated with n in the denominator. **For this reason, we always divide by $(n - 1)$ when computing the sample variance s^2 and the sample standard deviation s.**

Now that you have learned how to compute the variance and standard deviation of a set of measurements, remember these points:

- The value of s is always greater than or equal to zero.
- The larger the value of s^2 or s, the greater the variability of the data set.
- If s^2 or s is equal to zero, all the measurements must have the same value.
- In order to measure the variability in the same units as the original observations, we compute the standard deviation $s = \sqrt{s^2}$.

This information allows you to compare several sets of data with respect to their locations and their variability. How can you use these measures to say something more specific about a single set of data? The theorem and rule presented in the next section will help answer this question.

2.3 EXERCISES

BASIC TECHNIQUES

2.16 You are given $n = 5$ measurements: 2, 1, 1, 3, 5.

a. Calculate the sample mean, $\bar{x}$.

b. Calculate the sample variance, s^2, using the formula given by the definition.

c. Find the sample standard deviation, s.

d. Find s^2 and s using the computing formula. Compare the results with those found in parts b and c.

2.17 You are given $n = 8$ measurements: 4, 1, 3, 1, 3, 1, 2, 2.

a. Find the range.

b. Calculate $\bar{x}$.

c. Calculate s^2 and s using the computing formula.

d. Use the data entry method in your calculator to find $\bar{x}$, s, and s^2. Verify that your answers are the same as those in parts b and c.

2.18 You are given $n = 8$ measurements: 3, 1, 5, 6, 4, 4, 3, 5.

a. Calculate the range.

b. Calculate the sample mean.

c. Calculate the sample variance and standard deviation.

d. Compare the range and the standard deviation. The range is approximately how many standard deviations?

2.19 You are given $n = 6$ measurements: 6, 7, 3, 4, 2, 9.

a. Calculate the sample mean.

b. Calculate the range.

c. Calculate the sample standard deviation.

d. Compare the range with the standard deviation and comment on their relationship.

APPLICATIONS

2.20 An Archeological Find An article in *Archaeometry* involved an analysis of 26 samples of Romano-British pottery found at four different kiln sites in the United Kingdom.[9] The samples were analyzed to determine their chemical composition. The percentage of iron oxide in each of five samples collected at the Island Thorns site was:

1.28, 2.39, 1.50, 1.88, 1.51

a. Calculate the range.

b. Calculate the sample variance and the standard deviation using the computing formula.

c. Compare the range and the standard deviation. The range is approximately how many standard deviations?

Data set EX0221 **2.21 Utility Bills in Regina** The average monthly basic utility bills for an 85-square-metre apartment, which includes electricity, heating, water, and garbage collection, is \$236.56.[10] Suppose a random resident from Regina is chosen, and his or her utility bills for 12 consecutive months are recorded, beginning January 2015.

Month	Amount (\$)	Month	Amount (\$)
January	260.42	July	271.57
February	255.32	August	267.94
March	200.62	September	278.64
April	215.96	October	240.64
May	180.36	November	300.32
June	261.76	December	276.64

a. Calculate the range of the utility bills for the year 2015.

b. Calculate the average monthly utility bill for the year 2015.

c. Calculate the standard deviation for the 2015 utility bills.

2.4 ON THE PRACTICAL SIGNIFICANCE OF THE STANDARD DEVIATION

We now introduce a useful theorem developed by the Russian mathematician Tchebysheff. Proof of the theorem is not difficult, but we are more interested in its application than its proof.

Tchebysheff's Theorem Given a number k greater than or equal to 1 and a set of n measurements, at least $[1 - (1/k^2)]$ of the measurements will lie within k standard deviations of their mean.

Tchebysheff's Theorem applies to *any set of measurements* and can be used to describe either a sample or a population. We will use the notation appropriate for populations, but you should realize that we could just as easily use the mean and the standard deviation for the sample.

The idea involved in Tchebysheff's Theorem is illustrated in Figure 2.8. An interval is constructed by measuring a distance $k\sigma$ on either side of the mean μ. The number k can be any number as long as it is greater than or equal to 1. Then Tchebysheff's Theorem states that at least $1 - (1/k^2)$ of the total number n measurements lies in the constructed interval.

FIGURE 2.8
Illustrating Tchebysheff's Theorem

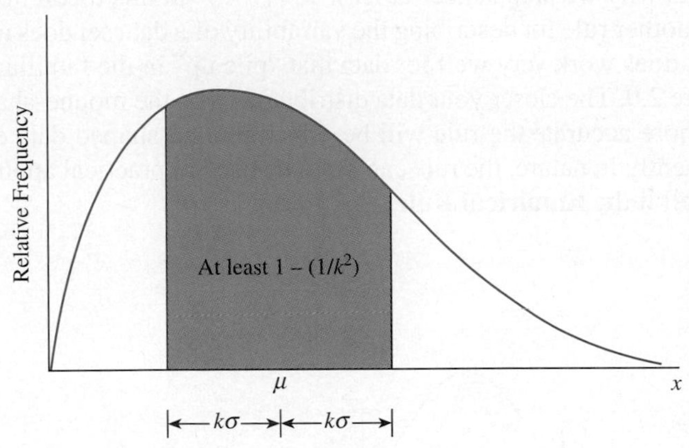

In Table 2.3, we choose a few numerical values for k and compute $[1 - (1/k^2)]$.

TABLE 2.3 **Illustrative Values of $[1 - (1/k^2)]$**

k	$1 - (1/k^2)$
1	$1 - 1 = 0$
2	$1 - 1/4 = 3/4$
3	$1 - 1/9 = 8/9$

From the calculations in Table 2.3, the theorem states:

- At least none of the measurements lie in the interval $\mu - \sigma$ to $\mu + \sigma$.
- At least 3/4 of the measurements lie in the interval $\mu - 2\sigma$ to $\mu + 2\sigma$.
- At least 8/9 of the measurements lie in the interval $\mu - 3\sigma$ to $\mu + 3\sigma$.

Although the first statement is not at all helpful, the other two values of k provide valuable information about the proportion of measurements that fall in certain intervals. The values $k = 2$ and $k = 3$ are not the only values of k you can use; for example, the proportion of measurements that fall within $k = 2.5$ standard deviations of the mean is at least $1 - [1/(2.5)^2] = 0.84$.

EXAMPLE  **2.7**

The mean and variance of a sample of $n = 25$ measurements are 75 and 100, respectively. Use Tchebysheff's Theorem to describe the distribution of measurements.

SOLUTION You are given $\bar{x} = 75$ and $s^2 = 100$. The standard deviation is $s = \sqrt{100} = 10$. The distribution of measurements is centred about $\bar{x} = 75$, and Tchebysheff's Theorem states:

- *At least* 3/4 of the 25 measurements lie in the interval $\bar{x} \pm 2s = 75 \pm 2(10)$—that is, 55 to 95.
- *At least* 8/9 of the measurements lie in the interval $\bar{x} \pm 3s = 75 \pm 3(10)$—that is, 45 to 105.

Since Tchebysheff's Theorem applies to *any* distribution, it is very conservative. This is why we emphasize "at least $1 - (1/k^2)$" in this theorem.

Another rule for describing the variability of a data set does not work for *all* data sets, but it does work very well for data that "pile up" in the familiar mound shape shown in Figure 2.9. The closer your data distribution is to the mound-shaped curve in Figure 2.9, the more accurate the rule will be. Since mound-shaped data distributions occur quite frequently in nature, the rule can often be used in practical applications. For this reason, we call it the **Empirical Rule**.

FIGURE 2.9
Mound-shaped distribution

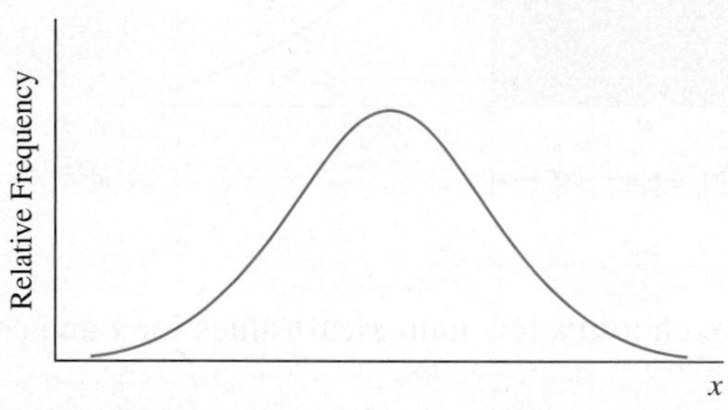

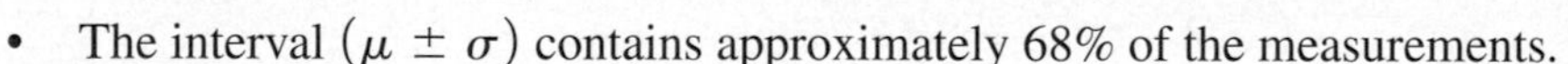

EMPIRICAL RULE Given a distribution of measurements that is approximately mound-shaped:

- The interval $(\mu \pm \sigma)$ contains approximately 68% of the measurements.
- The interval $(\mu \pm 2\sigma)$ contains approximately 95% of the measurements.
- The interval $(\mu \pm 3\sigma)$ contains approximately 99.7% of the measurements.

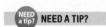

NEED A TIP?
Remember these three numbers:
68—95—99.7

The mound-shaped distribution shown in Figure 2.9 is commonly known as the **normal distribution** and will be discussed in detail in Chapter 6.

EXAMPLE 2.8

In a time study conducted at a manufacturing plant, the length of time to complete a specified operation is measured for each of $n = 40$ workers. The mean and standard deviation are found to be 12.8 and 1.7, respectively. Describe the sample data using the Empirical Rule.

SOLUTION To describe the data, calculate these intervals:

$$(\bar{x} \pm s) = 12.8 \pm 1.7 \quad \text{or} \quad 11.1 \text{ to } 14.5$$

$$(\bar{x} \pm 2s) = 12.8 \pm 2(1.7) \quad \text{or} \quad 9.4 \text{ to } 16.2$$

$$(\bar{x} \pm 3s) = 12.8 \pm 3(1.7) \quad \text{or} \quad 7.7 \text{ to } 17.9$$

According to the Empirical Rule, you expect approximately 68% of the measurements to fall into the interval from 11.1 to 14.5, approximately 95% to fall into the interval from 9.4 to 16.2, and approximately 99.7% to fall into the interval from 7.7 to 17.9.

If you doubt that the distribution of measurements is mound-shaped, or if you wish for some other reason to be conservative, you can apply Tchebysheff's Theorem and be absolutely certain of your statements. Tchebysheff's Theorem tells you that at least 3/4 of the measurements fall into the interval from 9.4 to 16.2 and at least 8/9 into the interval from 7.7 to 17.9.

EXAMPLE  2.9

Student teachers are trained to develop lesson plans, on the assumption that the written plan will help them perform successfully in the classroom. In a study to assess the relationship between written lesson plans and their implementation in the classroom, 25 lesson plans were scored on a scale of 0 to 34 according to a Lesson Plan Assessment Checklist. The 25 scores are shown in Table 2.4. Use Tchebysheff's Theorem and the Empirical Rule (if applicable) to describe the distribution of these assessment scores.

TABLE 2.4 **Lesson Plan Assessment Scores**

26.1	26.0	14.5	29.3	19.7
22.1	21.2	26.6	31.9	25.0
15.9	20.8	20.2	17.8	13.3
25.6	26.5	15.7	22.1	13.8
29.0	21.3	23.5	22.1	10.2

SOLUTION Use your calculator or the computing formulas to verify that $\bar{x} = 21.6$ and $s = 5.5$. The appropriate intervals are calculated and listed in Table 2.5. We have also referred back to the original 25 measurements and counted the actual number of measurements that fall into each of these intervals. These frequencies and relative frequencies are shown in Table 2.5.

TABLE 2.5 **Intervals $\bar{x} \pm ks$ for the Data of Table 2.4**

k	Interval $\bar{x} \pm ks$	Frequency in Interval	Relative Frequency
1	16.1–27.1	16	0.64
2	10.6–32.6	24	0.96
3	5.1–38.1	25	1.00

NEED A TIP?

Empirical Rule ⇔ mound-shaped data

Tchebysheff ⇔ any shaped data.

Is Tchebysheff's Theorem applicable? Yes, because it can be used for any set of data. According to Tchebysheff's Theorem:

- At least 3/4 of the measurements will fall between 10.6 and 32.6.
- At least 8/9 of the measurements will fall between 5.1 and 38.1.

You can see in Table 2.5 that Tchebysheff's Theorem is true for these data. In fact, the proportions of measurements that fall into the specified intervals exceed the lower bound given by this theorem.

Is the Empirical Rule applicable? You can check for yourself by drawing a graph—either a stem and leaf plot or a histogram. The *MINITAB* histogram in Figure 2.10 shows that the distribution is *relatively* mound-shaped, so the Empirical Rule should work *relatively well*:

- Approximately 68% of the measurements will fall between 16.1 and 27.1.
- Approximately 95% of the measurements will fall between 10.6 and 32.6.
- Approximately 99.7% of the measurements will fall between 5.1 and 38.1.

The relative frequencies in Table 2.5 closely approximate those specified by the Empirical Rule.

FIGURE 2.10
MINITAB histogram for Example 2.9

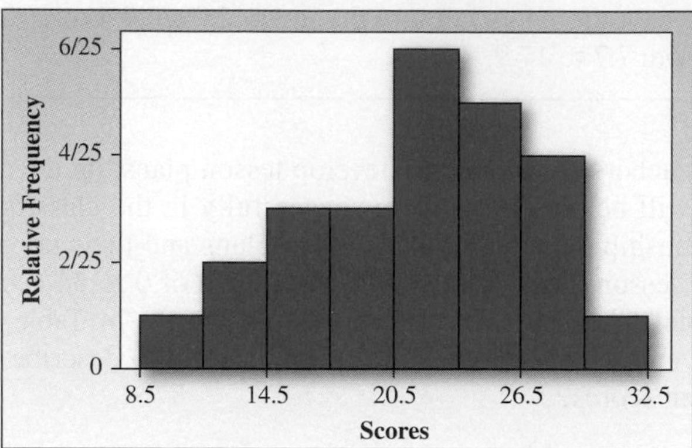

USING TCHEBYSHEFF'S THEOREM AND THE EMPIRICAL RULE

Tchebysheff's Theorem can be proven mathematically. It applies to any set of measurements—sample or population, large or small, mound-shaped or skewed.

Tchebysheff's Theorem gives a *lower bound* to the fraction of measurements to be found in an interval constructed as $\bar{x} \pm ks$. *At least* $1 - (1/k^2)$ of the measurements will fall into this interval, and probably more!

The Empirical Rule is a "rule of thumb" that can be used as a descriptive tool only when the data tend to be roughly mound-shaped (the data tend to pile up near the centre of the distribution).

When you use these two tools for describing a set of measurements, Tchebysheff's Theorem will always be satisfied, but it is a very conservative estimate of the fraction of measurements that fall into a particular interval. If it is appropriate to use the Empirical Rule (mound-shaped data), this rule will give you a more accurate estimate of the fraction of measurements that fall into the interval.

2.5 A CHECK ON THE CALCULATION OF *s*

Tchebysheff's Theorem and the Empirical Rule can be used to detect gross errors in the calculation of *s*. Roughly speaking, these two tools tell you that *most of the time*, measurements lie within *two* standard deviations of their mean. This interval is marked off in Figure 2.11, and it implies that the total range of the measurements, from smallest to largest, should be somewhere around four standard deviations. This is, of course, a very rough approximation, but it can be very useful in checking for large errors in your calculation of *s*. If the range, *R*, is about four standard deviations, or 4*s*, you can write

$$R \approx 4s \text{ or } s \approx \frac{R}{4}$$

The computed value of s using the shortcut formula should be of roughly the same order as the approximation.

FIGURE 2.11
Range approximation to s

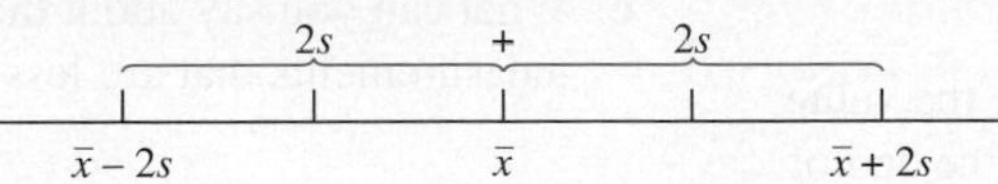

EXAMPLE 2.10

Use the range approximation to check the calculation of s for Table 2.2.

SOLUTION The range of the five measurements—5, 7, 1, 2, 4—is

$$R = 7 - 1 = 6$$

Then

$$s \approx \frac{R}{4} = \frac{6}{4} = 1.5$$

This is the same order as the calculated value $s = 2.4$.

NEED A TIP?

$s \approx R/4$ gives only an **approximate** value for s.

The range approximation is *not* intended to provide an accurate value for s. Rather, its purpose is to detect gross errors in calculating, such as the failure to divide the sum of squares of deviations by $(n - 1)$ or the failure to take the square root of s^2. If you make one of these mistakes, your answer will be many times larger than the range approximation of s.

EXAMPLE 2.11

Use the range approximation to determine an approximate value for the standard deviation for the data in Table 2.4.

SOLUTION The range $R = 31.9 - 10.2 = 21.7$. Then

$$s \approx \frac{R}{4} = \frac{21.7}{4} = 5.4$$

Since the exact value of s is 5.5 for the data in Table 2.4, the approximation is very close.

The range for a sample of n measurements will depend on the sample size, n. For larger values of n, a larger range of the x values is expected. The range for large samples (say, $n = 50$ or more observations) may be as large as $6s$, whereas the range for small samples (say, $n = 5$ or less) may be as small as or smaller than $2.5s$.

The range approximation for s can be improved if it is known that the sample is drawn from a mound-shaped distribution of data. Thus, the calculated s should not differ substantially from the range divided by the appropriate ratio given in Table 2.6.

TABLE 2.6 **Divisor for the Range Approximation of *s***

Number of Measurements	Expected Ratio of Range to s
5	2.5
10	3
25	4

2.5 EXERCISES

BASIC TECHNIQUES

2.22 A set of $n = 10$ measurements consists of the values 5, 2, 3, 6, 1, 2, 4, 5, 1, 3.

a. Use the range approximation to estimate the value of s for this set. (HINT: Use Table 2.6 at the end of Section 2.5.)

b. Use your calculator to find the actual value of s. Is the actual value close to your estimate in part a?

c. Draw a dotplot of this data set. Are the data mound-shaped?

d. Can you use Tchebysheff's Theorem to describe this data set? Why or why not?

e. Can you use the Empirical Rule to describe this data set? Why or why not?

2.23 Suppose you want to create a mental picture of the relative frequency histogram for a large data set consisting of 1000 observations, and you know that the mean and standard deviation of the data set are 36 and 3, respectively.

a. If you are fairly certain that the relative frequency distribution of the data is mound-shaped, how might you picture the relative frequency distribution? (HINT: Use the Empirical Rule.)

b. If you have no prior information concerning the shape of the relative frequency distribution, what can you say about the relative frequency histogram? (HINT: Construct intervals $\bar{x} \pm ks$ for several choices of k.)

2.24 A distribution of measurements is relatively mound-shaped with mean 50 and standard deviation 10.

a. What proportion of the measurements will fall between 40 and 60?

b. What proportion of the measurements will fall between 30 and 70?

c. What proportion of the measurements will fall between 30 and 60?

d. If a measurement is chosen at random from this distribution, what is the probability that it will be greater than 60?

2.25 A set of data has a mean of 75 and a standard deviation of 5. You know nothing else about the size of the data set or the shape of the data distribution.

a. What can you say about the proportion of measurements that fall between 60 and 90?

b. What can you say about the proportion of measurements that fall between 65 and 85?

c. What can you say about the proportion of measurements that are less than 65?

APPLICATIONS

2.26 Driving Emergencies The length of time required for an automobile driver to respond to a particular emergency situation was recorded for $n = 10$ drivers. The times (in seconds) were 0.5, 0.8, 1.1, 0.7, 0.6, 0.9, 0.7, 0.8, 0.7, 0.8.

a. Scan the data and use the procedure in Section 2.5 to find an approximate value for s. Use this value to check your calculations in part b.

b. Calculate the sample mean $\bar{x}$ and the standard deviation s. Compare with part a.

Data set EX0227 **2.27 Packaging Hamburger Meat** The data listed here are the weights (in kilograms) of 27 packages of ground beef in a supermarket meat display:

1.08	0.99	0.97	1.18	1.41	1.28	0.83
1.06	1.14	1.38	0.75	0.96	1.08	0.87
0.89	0.89	0.96	1.12	1.12	0.93	1.24
0.89	0.98	1.14	0.92	1.18	1.17	

a. Construct a stem and leaf plot or a relative frequency histogram to display the distribution of weights. Is the distribution relatively mound-shaped?

b. Find the mean and standard deviation of the data set.

c. Find the percentage of measurements in the intervals $\bar{x} \pm s$, $\bar{x} \pm 2s$, and $\bar{x} \pm 3s$.

d. How do the percentages obtained in part c compare with those given by the Empirical Rule? Explain.

e. How many of the packages weigh exactly 1 kilogram (kg)? Can you think of any explanation for this?

2.28 Breathing Rates Is your breathing rate normal? Actually, there is no standard breathing rate for humans. It can vary from as low as 4 breaths per minute to as high as 70 or 75 for a person engaged in strenuous exercise. Suppose that the resting breathing rates for university-age students have a relative frequency distribution that is mound-shaped, with a mean equal to 12 and a standard deviation of

2.3 breaths per minute. What fraction of all students would have breathing rates in the following intervals?

a. 9.7 to 14.3 breaths per minute

b. 7.4 to 16.6 breaths per minute

c. More than 18.9 or less than 5.1 breaths per minute

Data set EX0229 **2.29 Ore Samples** A geologist collected 20 different ore samples, all the same weight, and randomly divided them into two groups. She measured the titanium (Ti) content of the samples using two different methods.

Method 1					Method 2				
0.011	0.013	0.013	0.015	0.014	0.011	0.016	0.013	0.012	0.015
0.013	0.010	0.013	0.011	0.012	0.012	0.017	0.013	0.014	0.015

a. Construct stem and leaf plots for the two data sets. Visually compare their centres and their ranges.

b. Calculate the sample means and standard deviations for the two sets. Do the calculated values confirm your visual conclusions from part a?

Data set EX0230 **2.30 Social Insurance Numbers** The data below show the last digit of the Social Insurance Number for a group of 70 students:

1	6	9	1	5	9	0	2	8	4
0	7	3	4	2	3	5	8	4	2
3	2	0	0	2	1	2	7	7	4
0	0	9	9	5	3	8	4	7	4
6	6	9	0	2	6	2	9	5	8
5	1	7	7	7	8	7	5	1	8
3	4	1	9	3	8	6	6	6	6

a. The distribution of this data is relatively "flat," with each different value from 0 to 9 occurring with nearly equal frequency. Using this fact, what would be your best estimate for the mean of the data set?

b. Use the range approximation to guess the value of s for this set.

c. Use your calculator to find the actual values of $\bar{x}$ and s. Compare with your estimates in parts a and b.

2.31 Social Insurance Numbers, continued Refer to the data set in Exercise 2.30.

a. Find the percentage of measurements in the intervals $\bar{x} \pm s$, $\bar{x} \pm 2s$, and $\bar{x} \pm 3s$.

b. How do the percentages obtained in part a compare with those given by the Empirical Rule? Should they be approximately the same? Explain.

2.32 Survival Times A group of experimental animals is infected with a particular form of bacteria, and their survival time is found to average 32 days, with a standard deviation of 36 days.

a. Visualize the distribution of survival times. Do you think that the distribution is relatively mound-shaped, skewed right, or skewed left? Explain.

b. Within what limits would you expect at least 3/4 of the measurements to lie?

2.33 Survival Times, continued Refer to Exercise 2.32. You can use the Empirical Rule to see why the distribution of survival times could not be mound-shaped.

a. Find the value of x that is exactly one standard deviation below the mean.

b. If the distribution is in fact mound-shaped, approximately what percentage of the measurements should be less than the value of x found in part a?

c. Since the variable being measured is time, is it possible to find any measurements that are more than one standard deviation below the mean?

d. Use your answers to part b and c to explain why the data distribution cannot be mound-shaped.

Data set EX0234 **2.34 Timber Tracts** To estimate the amount of lumber in a tract of timber, an owner decided to count the number of trees with diameters exceeding 30 cm in randomly selected 15-by-15 m². Seventy 15-by-15 m² were chosen, and the selected trees were counted in each tract. The data are listed here:

7	8	7	10	4	8	6	8	9	10
9	6	4	9	10	9	8	8	7	9
3	9	5	9	9	8	7	5	8	8
10	2	7	4	8	5	10	7	7	7
9	6	8	8	8	7	8	9	6	8
6	11	9	11	7	7	11	7	9	13
10	8	8	5	9	9	8	5	9	8

a. Construct a relative frequency histogram to describe the data.

b. Calculate the sample mean $\bar{x}$ as an estimate of μ, the mean number of timber trees for all 15-by-15 m² in the tract.

c. Calculate s for the data. Construct the intervals $\bar{x} \pm s$, $\bar{x} \pm 2s$, and $\bar{x} \pm 3s$. Calculate the percentage of squares falling into each of the three intervals, and compare with the corresponding percentages given by the Empirical Rule and Tchebysheff's Theorem.

2.35 Wedding Cake, again Refer to Exercise 2.9 and data set EX0209. The prices (in dollars) for 14 different slices of wedding cake with buttercream icing are reproduced here:

3.25, 5.50, 4.75, 1.75, 3.50, 5.25, 2.50
4.25, 2.00, 4.25, 3.75, 5.75, 4.00, 6.00

a. Use the range approximation to find an estimate of s.

b. How does it compare to the computed value of s?

Data set EX0236 **2.36 Old Faithful** The data below are 30 waiting times between eruptions of the Old Faithful geyser in Yellowstone National Park:[11]

56 89 51 79 58 82 52 88 52 78 69 75 77 72 71
55 87 53 85 61 93 54 76 80 81 59 86 78 71 77

a. Calculate the range.

b. Use the range approximation to approximate the standard deviation of these 30 measurements.

c. Calculate the sample standard deviation s.

d. What proportion of the measurements lie within two standard deviations of the mean? Within three standard deviations of the mean? Do these proportions agree with the proportions given in Tchebysheff's Theorem?

Data set EX0237 **2.37 Twitter Usage** According to a study on Twitter, a research group examined 716,344 different users, which registered an average of 1018 followers (number of people following that person) and 227 followees (number of people a person follows).[12] The average number of tweets per user is 6.59. Below are two figures provided by the research group:

Number of tweets per user

# of tweets	# of users	Percentage
1	377,112	52.64
2	110,887	15.48
3	51,649	7.21
4	30,478	4.25
5	20,677	2.89
6	15,006	2.09
7	11,406	1.59
8	9,342	1.30
9	7,642	1.07
10	82,145	11.47

Source: Marcelo Mendoza, Barbara Poblete and Carlos Castillo, Twitter Under Crisis: Can we trust what we RT? Courtesy of the authors.

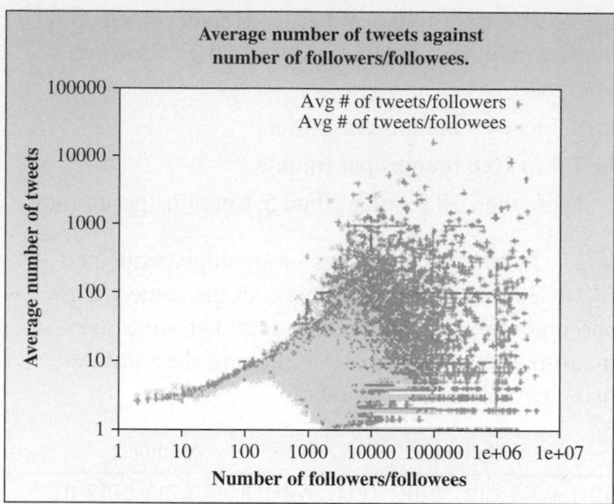

a. Using the table, how would you describe and compare the average number of tweets per user in the graph? Are there any conclusions to be made?

b. Describe the variance in the graph. Do you see any relationships of the average number of tweets against the number of followers/followees?

Data set EX0238 **2.38 Wayne Gretzky** The number of goals scored by Wayne Gretzky was recorded for seasons 1978–1999:[13]

46	51	55	92	71	87
73	52	62	40	54	40
41	31	16	38	11	23
25	23	9			

a. Draw a stem and leaf plot to describe the data.

b. Calculate the mean and standard deviation for the data.

c. What proportion of the measurements lie within two standard deviations of the mean?

2.39 Who Is Paying More at the Pump? The prices below are for regular gasoline (per litre) as of January 1, 2015:[14]

Area	Price per Litre ($)
Alberta	1.104
Ontario	1.224
New Brunswick	1.239
Manitoba	1.154
Quebec	1.329
Nova Scotia	1.270
Saskatchewan	1.175
British Columbia	1.308
Newfoundland and Labrador	1.292

Source: CFS Report - Fuel Price Update - February 2015 Reimbursement for Business Use of Personal Vehicles Study prepared for The Treasury Board of Canada Secretariat By Corporate Fleet Services.

a. Find the standard deviation of these values.

b. Find the range R. Compare the range and the standard deviation. Is range related to standard deviation, in any sense?

c. Use range to approximate the standard deviation. Is this a good approximation?

d. Subtract 0.5 from every observation and calculate the variance for the original data and the new data. What effect, if any, does subtracting 0.5 from every observation have on the sample variance?

e. What percentage of the measurements will lie within three standard deviations of the overall mean? Does the result agree with Tchebysheff's Theorem?

f. Find the percentage of measurements in the intervals $\bar{x} \pm s$ and $\bar{x} \pm 2s$. Compare these results with the Empirical Rule percentages, and comment on the shape of the distribution.

g. In how many provinces is the average gas price stabilized?

CALCULATING THE MEAN AND STANDARD DEVIATION FOR GROUPED DATA (OPTIONAL)

2.40 Suppose that some measurements occur more than once and that the data $x_1, x_2, \ldots, x_k$ are arranged in a frequency table as shown here:

Observations	Frequency f_i
x_1	f_1
x_2	f_2
.	.
.	.
.	.
x_k	f_k

The formulas for the mean and variance for grouped data are

$$\bar{x} = \frac{\Sigma x_i f_i}{n}, \quad \text{where } n = \Sigma f_i$$

and

$$s^2 = \frac{\Sigma x_i^2 f_i - \dfrac{(\Sigma x_i f_i)^2}{n}}{n - 1}$$

Notice that if each value occurs once, these formulas reduce to those given in the text. Although these formulas for grouped data are primarily of value when you have a large number of measurements, demonstrate their use for the sample 1, 0, 0, 1, 3, 1, 3, 2, 3, 0, 0, 1, 1, 3, 2.

a. Calculate $\bar{x}$ and s^2 directly, using the formulas for ungrouped data.

b. The frequency table for the $n = 15$ measurements is as follows:

x	f
0	4
1	5
2	2
3	4

Calculate $\bar{x}$ and s^2 using the formulas for grouped data. Compare with your answers to part a.

2.41 International Baccalaureate The International Baccalaureate (IB) program is an accelerated academic program offered at a growing number of high schools throughout the country. Students enrolled in this program are placed in accelerated or advanced courses and must take IB examinations in each of six subject areas at the end of Grade 11 or Grade 12. Students are scored on a scale of 1–7, with 1–2 being poor, 3 mediocre, 4 average, and 5–7 excellent. During its first year of operation at a local high school, 17 Grade 12 students attempted the IB mathematics exam, with these results:

Exam Grade	Number of Students
7	1
6	4
5	4
4	4
3	4

Calculate the mean and standard deviation for these scores.

2.42 A Skewed Distribution To illustrate the utility of the Empirical Rule, consider a distribution that is heavily skewed to the right, as shown in the accompanying figure.

a. Calculate $\bar{x}$ and s for the data shown. (NOTE: There are 10 zeros, 5 ones, and so on.)

b. Construct the intervals $\bar{x} \pm s$, $\bar{x} \pm 2s$, and $\bar{x} \pm 3s$ and locate them on the frequency distribution.

c. Calculate the proportion of the $n = 24$ measurements that fall into each of the three intervals. Compare with Tchebysheff's Theorem and the Empirical Rule. Note that, although the proportion that falls into the interval $\bar{x} \pm s$ does not agree closely with the Empirical Rule, the proportions that fall into the intervals $\bar{x} \pm 2s$ and $\bar{x} \pm 3s$ agree very well. Many times this is true, even for non-mound-shaped distributions of data.

Distribution for Exercise 2.42

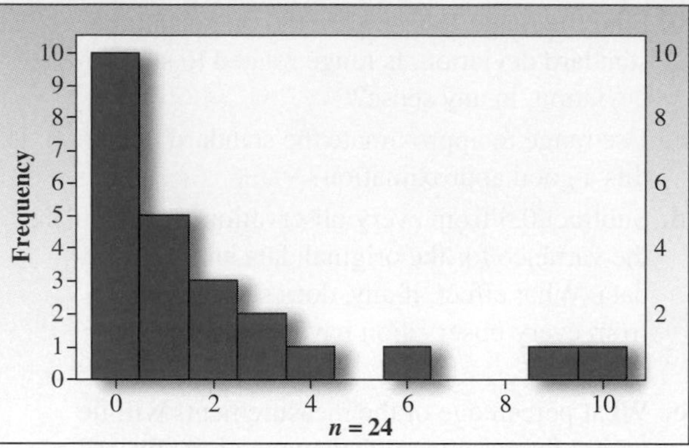

2.6 MEASURES OF RELATIVE STANDING

Sometimes you need to know the position of one observation relative to others in a set of data. For example, if you took an examination with a total of 35 points, you might want to know how your score of 30 compared to the scores of the other students in the class. The mean and standard deviation of the scores can be used to calculate a ***z*-score,** which measures the relative standing of a measurement in a data set.

DEFINITION The **sample *z*-score** is a measure of relative standing defined by

$$z\text{-score} = \frac{x - \bar{x}}{s}$$

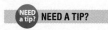

Positive *z*-score ⇔ x is above the mean.

Negative *z*-score ⇔ x is below the mean.

A *z*-score measures the distance between an observation and the mean, measured in units of standard deviation. For example, suppose that the mean and standard deviation of the test scores (based on a total of 35 points) are 25 and 4, respectively. The *z*-score for your score of 30 is calculated as follows:

$$z\text{-score} = \frac{x - \bar{x}}{s} = \frac{30 - 25}{4} = 1.25$$

Your score of 30 lies 1.25 standard deviations above the mean $(30 = \bar{x} + 1.25s)$.

The *z*-score is a valuable tool for determining whether a particular observation is likely to occur quite frequently or whether it is unlikely and might be considered an **outlier.**

According to Tchebysheff's Theorem and the Empirical Rule:

- At least 75% and more likely 95% of the observations lie within two standard deviations of their mean: their *z*-scores are between −2 and +2. *Observations with z-scores exceeding 2 in absolute value happen less than 5% of the time and are considered somewhat unlikely.*
- At least 89% and more likely 99.7% of the observations lie within three standard deviations of their mean: their *z*-scores are between −3 and +3. *Observations with z-scores exceeding 3 in absolute value happen less than 1% of the time and are considered very unlikely.*

NEED A TIP?

z-scores above 3 in absolute value are very unusual.

You should look carefully at any observation that has a z-score exceeding 3 in absolute value. Perhaps the measurement was recorded incorrectly or does not belong to the population being sampled. Perhaps it is just a highly unlikely observation, but a valid one nonetheless!

EXAMPLE 2.12

Consider this sample of $n = 10$ measurements:

1, 1, 0, 15, 2, 3, 4, 0, 1, 3

The measurement $x = 15$ appears to be unusually large. Calculate the z-score for this observation and state your conclusions.

SOLUTION Calculate $\bar{x} = 3.0$ and $s = 4.42$ for the $n = 10$ measurements. Then the z-score for the suspected outlier, $x = 15$, is calculated as

$$z\text{-score} = \frac{x - \bar{x}}{s} = \frac{15 - 3}{4.42} = 2.71$$

Hence, the measurement $x = 15$ lies 2.71 standard deviations above the sample mean, $\bar{x} = 3.0$. Although the z-score does not exceed 3, it is close enough so that you might suspect that $x = 15$ is an outlier. You should examine the sampling procedure to see whether $x = 15$ is a faulty observation.

A **percentile** is another measure of relative standing and is most often used for large data sets. (Percentiles are not very useful for small data sets.)

DEFINITION A set of n measurements on the variable x has been arranged in order of magnitude. The ***p*th percentile** is the value of x that is greater than $p\%$ of the measurements and is less than the remaining $(100 - p)\%$.

EXAMPLE 2.13

Suppose you have been notified that your score of 610 on the Verbal Graduate Record Examination placed you at the 60th percentile in the distribution of scores. Where does your score of 610 stand in relation to the scores of others who took the examination?

SOLUTION Scoring at the 60th percentile means that 60% of all the examination scores were lower than your score and 40% were higher.

In general, the 60th percentile for the variable x is a point on the *horizontal axis* of the data distribution that is greater than 60% of the measurements and less than the others. That is, 60% of the measurements are less than the 60th percentile and 40% are greater (see Figure 2.12). Since the total area under the distribution is 100%, 60% of the area is to the left and 40% of the area is to the right of the 60th percentile. Remember that the median, m, of a set of data is the middle measurement; that is, 50% of the measurements are smaller and 50% are larger than the median. Thus, the *median is the same as the 50th percentile!*

FIGURE 2.12
The 60th percentile shown on the relative frequency histogram for a data set

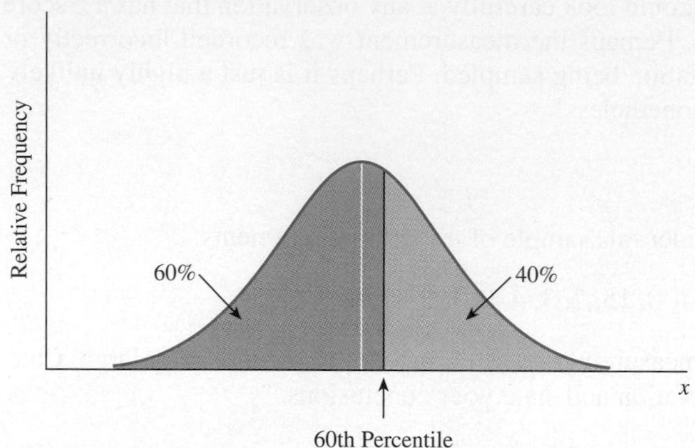

The 25th and 75th percentiles, called the **lower** and **upper quartiles,** along with the median (the 50th percentile), locate points that divide the data into four sets, each containing an equal number of measurements. Twenty-five percent of the measurements will be less than the lower (first) quartile, 50% will be less than the median (the second quartile), and 75% will be less than the upper (third) quartile. Thus, the median and the lower and upper quartiles are located at points on the *x*-axis so that the area under the relative frequency histogram for the data is partitioned into four equal areas, as shown in Figure 2.13.

FIGURE 2.13
Location of quartiles

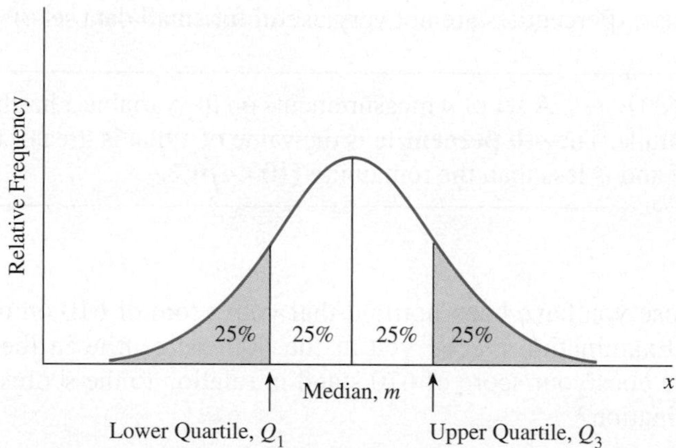

DEFINITION A set of n measurements on the variable x has been arranged in order of magnitude. The **lower quartile (first quartile),** Q_1, is the value of x that is greater than 1/4 of the measurements and is less than the remaining 3/4. The **second quartile** is the median. The **upper quartile (third quartile),** Q_3, is the value of x that is greater than 3/4 of the measurements and is less than the remaining 1/4.

For small data sets, it is often impossible to divide the set into four groups, each of which contains exactly 25% of the measurements. For example, when $n = 10$, you would need to have $2\frac{1}{2}$ measurements in each group! Even when you can perform this task (for example, if $n = 12$), there are many numbers that would satisfy the preceding definition, and could therefore be considered "quartiles." To avoid this ambiguity, we use the following rule to locate sample quartiles.

CALCULATING SAMPLE QUARTILES

- When the measurements are arranged in order of magnitude, the **lower quartile, Q_1,** is the value of x in position $0.25(n + 1)$, and the **upper quartile, Q_3,** is the value of x in position $0.75(n + 1)$.
- When $0.25(n + 1)$ and $0.75(n + 1)$ are not integers, the quartiles are found by interpolation, using the values in the two adjacent positions.[†]

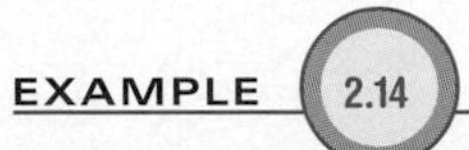

EXAMPLE 2.14

Find the lower and upper quartiles for this set of measurements:

16, 25, 4, 18, 11, 13, 20, 8, 11, 9

SOLUTION Rank the $n = 10$ measurements from smallest to largest:

4, 8, 9, 11, 11, 13, 16, 18, 20, 25

Calculate:

$$\text{Position of } Q_1 = 0.25(n + 1) = 0.25(10 + 1) = 2.75$$
$$\text{Position of } Q_3 = 0.75(n + 1) = 0.75(10 + 1) = 8.25$$

Since these positions are not integers, the lower quartile is taken to be the value 3/4 of the distance between the second and third ordered measurements, and the upper quartile is taken to be the value 1/4 of the distance between the eighth and ninth ordered measurements. Therefore,

$$Q_1 = 8 + 0.75(9 - 8) = 8 + 0.75 = 8.75$$

and

$$Q_3 = 18 + 0.25(20 - 18) = 18 + 0.5 = 18.5$$

Because the median and the quartiles divide the data distribution into four parts, each containing approximately 25% of the measurements, Q_1 and Q_3 are the upper and lower boundaries for the middle 50% of the distribution. We can measure the range of this "middle 50%" of the distribution using a numerical measure called the **interquartile range.**

DEFINITION The **interquartile range (IQR)** for a set of measurements is the difference between the upper and lower quartiles; that is, $\text{IQR} = Q_3 - Q_1$.

For the data in Example 2.14, $\text{IQR} = Q_3 - Q_1 = 18.50 - 8.75 = 9.75$. We will use the IQR along with the quartiles and the median in the next section to construct another graph for describing data sets.

[†] This definition of quartiles is consistent with the one used in the *MINITAB* package. Some textbooks use ordinary rounding when finding quartile positions, whereas other compute sample quartiles as the medians of the upper and lower halves of the data set.

EXAMPLE 2.15 Find the interquartile range for this set of measurements:

5, 2, 6, 9, 11, 2, 4, 64, 14, 47, 31

SOLUTION Rank the $n = 11$ measurements from smallest to largest:

2, 2, 4, 5, 6, 9, 11, 14, 31, 47, 64

Calculate:

$$\text{Position of } Q_1 = 0.25(11 + 1) = 0.25(12) = 3$$
$$\text{Position of } Q_3 = 0.75(11 + 1) = 0.75(12) = 9$$

Since the results are integers, we do not have to measure the distance. Therefore,

$$Q_1 = 4 \text{ and } Q_3 = 31$$

and

$$\text{IQR} = Q_3 - Q_1 = 31 - 4 = 27$$

NEED TO KNOW

How to Calculate Sample Quartiles

1. Arrange the data set in order of magnitude from smallest to largest.
2. Calculate the quartile positions:
   - Position of Q_1: $0.25(n + 1)$
   - Position of Q_3: $0.75(n + 1)$
3. If the positions are integers, then Q_1 and Q_3 are the values in the ordered data set found in those positions.
4. If the positions in step 2 are not integers, find the two measurements in positions just above and just below the calculated position. Calculate the quartile by finding a value either 1/4, 1/2, or 3/4 of the way between these two measurements.

Many of the numerical measures that you have learned are easily found using computer programs or even graphics calculators. The *MINITAB* command **Stat → Basic Statistics → Display Descriptive Statistics** (see the "Technology Today" section at the end of this chapter) produces output containing the mean, the standard deviation, the median, and the lower and upper quartiles, as well as the values of some other statistics that we have not discussed yet. The data from Example 2.14 produced the *MINITAB* output shown in Figure 2.14. Notice that the quartiles are identical to the hand-calculated values in that example.

FIGURE 2.14 *MINITAB* output for the data in Example 2.14

Descriptive Statistics: x

Variable	N	N*	Mean	SE Mean	StDev	Minimum	Q1	Median	Q3	Maximum
X	10	0	13.50	1.98	6.28	4.00	8.75	12.00	18.50	25.00

2.7 THE FIVE-NUMBER SUMMARY AND THE BOX PLOT

The median and the upper and lower quartiles shown in Figure 2.13 divide the data into four sets, each containing an equal number of measurements. If we add the largest number (Max) and the smallest number (Min) in the data set to this group, we will have a set of numbers that provide a quick and rough summary of the data distribution.

> The **five-number summary** consists of the smallest number, the lower quartile, the median, the upper quartile, and the largest number, presented in order from smallest to largest:
>
> **Min Q_1 Median Q_3 Max**
>
> By definition, 1/4 of the measurements in the data set lie between each of the four adjacent pairs of numbers.

The five-number summary can be used to create a simple graph called a **box plot** to visually describe the data distribution. From the box plot, you can quickly detect any skewness in the shape of the distribution and see whether any outliers are in the data set.

An outlier may result from transposing digits when recording a measurement, from incorrectly reading an instrument dial, from a malfunctioning piece of equipment, or from other problems. Even when there are no recording or observational errors, a data set may contain one or more valid measurements that, for one reason or another, differ markedly from the others in the set. These outliers can cause a marked distortion in commonly used numerical measures such as $\bar{x}$ and s. In fact, outliers may themselves contain important information not shared with the other measurements in the set. Therefore, isolating outliers, if they are present, is an important step in any preliminary analysis of a data set. The box plot is designed expressly for this purpose.

TO CONSTRUCT A BOX PLOT

- Calculate the median, the upper and lower quartiles, and the IQR for the data set.
- Draw a horizontal line representing the scale of measurement. Form a box just above the horizontal line with the right and left ends at Q_1 and Q_3. Draw a vertical line through the box at the location of the median.

A box plot is shown in Figure 2.15.

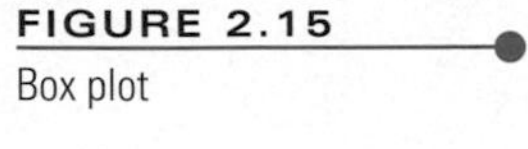

FIGURE 2.15
Box plot

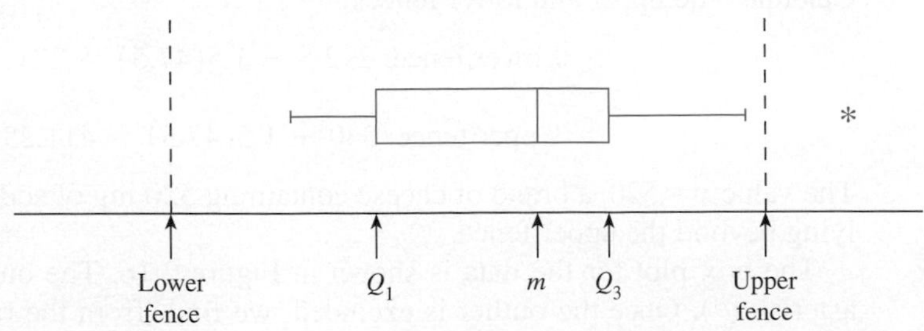

In Section 2.6, the z-score provided boundaries for finding unusually large or small measurements. You looked for z-scores greater than 2 or 3 in absolute value. The box plot uses the IQR to create imaginary "fences" to separate outliers from the rest of the data set:

DETECTING OUTLIERS—OBSERVATIONS THAT ARE BEYOND:

- Lower fence: $Q_1 - 1.5(\text{IQR})$
- Upper fence: $Q_3 + 1.5(\text{IQR})$

The upper and lower fences are shown with broken lines in Figure 2.15, but they are not usually drawn on the box plot. Any measurement beyond the upper or lower fence is an **outlier;** the rest of the measurements, inside the fences, are not unusual. Finally, the box plot marks the range of the data set using "whiskers" to connect the smallest and largest measurements (*excluding outliers*) to the box.

TO FINISH THE BOX PLOT

- Mark any **outliers** with an asterisk (*) on the graph.
- Extend horizontal lines called "whiskers" from the ends of the box to the smallest and largest observations that are *not* outliers.

EXAMPLE 2.16 As consumers become more careful about the foods they eat, food processors try to stay competitive by avoiding excessive amounts of fat, cholesterol, and sodium in the foods they sell. The following data are the amounts of sodium per slice (in milligrams) for each of eight brands of regular cheese. Construct a box plot for the data and look for outliers.

340, 300, 520, 340, 320, 290, 260, 330

SOLUTION The $n = 8$ measurements are first ranked from smallest to largest:

260, 290, 300, 320, 330, 340, 340, 520

The positions of the median, Q_1, and Q_3 are:

$$0.5(n + 1) = 0.5(9) = 4.5$$

$$0.25(n + 1) = 0.25(9) = 2.25$$

$$0.75(n + 1) = 0.75(9) = 6.75$$

so that $m = (320 + 330)/2 = 325$, $Q_1 = 290 + 0.25(10) = 292.5$, and $Q_3 = 340$. The interquartile range is calculated as

$$\text{IQR} = Q_3 - Q_1 = 340 - 292.5 = 47.5$$

Calculate the upper and lower fences:

$$\text{Lower fence: } 292.5 - 1.5(47.5) = 221.25$$

$$\text{Upper fence: } 340 + 1.5(47.5) = 411.25$$

The value $x = 520$, a brand of cheese containing 520 mg of sodium, is the only *outlier*, lying beyond the upper fence.

ONLINE APPLET

Building a Box Plot

The box plot for the data is shown in Figure 2.16. The outlier is marked with an asterisk (*). Once the outlier is excluded, we find (from the ranked data set) that the smallest and largest measurements are $x = 260$ and $x = 340$. These are the two values that form the whiskers. Since the value $x = 340$ is the same as Q_3, there is no whisker on the right side of the box.

FIGURE 2.16
Box plot for Example 2.16

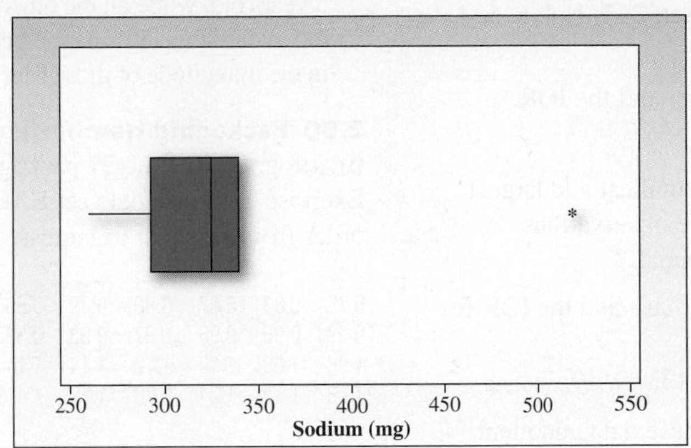

You can use the box plot to describe the shape of a data distribution by looking at the position of the median line compared to Q_1 and Q_3, the left and right ends of the box. If the median is close to the middle of the box, the distribution is fairly symmetric, providing equal-sized intervals to contain the two middle quarters of the data. If the median line is to the left of centre, the distribution is skewed to the right; if the median is to the right of centre, the distribution is skewed to the left. Also, for most skewed distributions, the whisker on the skewed side of the box tends to be longer than the whisker on the other side.

We used the *MINITAB* command **Graph → Boxplot** to draw two box plots, onc for the sodium contents of the eight brands of cheese in Example 2.16, and another for five brands of fat-free cheese with these sodium contents:

300, 300, 320, 290, 180

The two box plots are shown together in Figure 2.17. Look at the long whisker on the left side of both box plots and the position of the median lines. Both distributions are skewed to the left; that is, there are a few unusually small measurements. The regular cheese data, however, also show one brand ($x = 520$) with an unusually large amount of sodium. In general, it appears that the sodium content of the fat-free brands is lower than that of the regular brands, but the variability of the sodium content for regular cheese (excluding the outlier) is less than that of the fat-free brands.

FIGURE 2.17
MINITAB output for regular and fat-free cheese

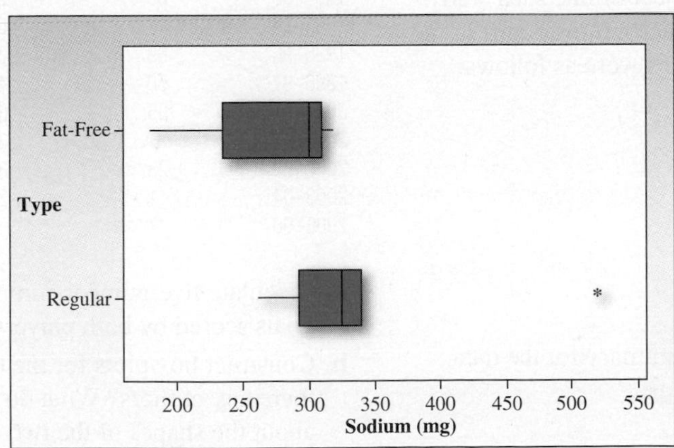

2.7 EXERCISES

BASIC TECHNIQUES

2.43 Given the following data set: 8, 7, 1, 4, 6, 6, 4, 5, 7, 6, 3, 0

a. Find the five-number summary and the IQR.

b. Calculate $\bar{x}$ and s.

c. Calculate the z-score for the smallest and largest observations. Is either of these observations unusually large or unusually small?

2.44 Find the five-number summary and the IQR for these data:

19, 12, 16, 0, 14, 9, 6, 1, 12, 13, 10, 19, 7, 5, 8

2.45 Construct a box plot for these data and identify any outliers:

25, 22, 26, 23, 27, 26, 28, 18, 25, 24, 12

2.46 Construct a box plot for these data and identify any outliers:

3, 9, 10, 2, 6, 7, 5, 8, 6, 6, 4, 9, 22

2.47 Construct a box plot for these data and identify any outliers:

21, 6, 181, 44, 32, 15, 3, 36, 25, 42

APPLICATIONS

2.48 If you scored at the 69th percentile on a placement test, how does your score compare with others?

Data set EX0249 **2.49 Mercury Concentration in Dolphins** Environmental scientists are increasingly concerned with the accumulation of toxic elements in marine mammals and the transfer of such elements to the animals' offspring. The striped dolphin (*Stenella coeruleoalba*), considered to be the top predator in the marine food chain, was the subject of one such study. The mercury concentrations (micrograms/gram) in the livers of 28 male striped dolphins were as follows:

1.70	183.00	221.00	286.00
1.72	168.00	406.00	315.00
8.80	218.00	252.00	241.00
5.90	180.00	329.00	397.00
101.00	264.00	316.00	209.00
85.40	481.00	445.00	314.00
118.00	485.00	278.00	318.00

a. Calculate the five-number summary for the data.

b. Construct a box plot for the data.

c. Are there any outliers?

d. If you knew that the first four dolphins were all less than 3 years old, while all the others were more than 8 years old, would this information help explain the difference in the magnitude of those four observations? Explain.

2.50 Packaging Hamburger Meat II The weights (in kilograms) of the 27 packages of ground beef from Exercise 2.27 (see data set EX0227) are listed here in order from smallest to largest:

0.75	0.83	0.87	0.89	0.89	0.89	0.92
0.93	0.96	0.96	0.97	0.98	0.99	1.06
1.08	1.08	1.12	1.12	1.14	1.14	1.17
1.18	1.18	1.24	1.28	1.38	1.41	

a. Confirm the values of the mean and standard deviation, calculated in Exercise 2.27 as $\bar{x} = 1.05$ and $s = 0.17$.

b. The two largest packages of meat weigh 1.38 and 1.41 kg. Are these two packages unusually heavy? Explain.

c. Construct a box plot for the package weights. What does the position of the median line and the length of the whiskers tell you about the shape of the distribution?

Data set EX0251 **2.51 Comparing NHL Superstars** How does Mario Lemieux compare to Brett Hull? The table below shows number of goals scored for each player for selected years:[15]

Season	Mario Lemieux	Brett Hull
1986–87	54	1
1987–88	70	32
1988–89	85	41
1989–90	45	72
1990–91	19	86
1991–92	44	70
1992–93	69	54
1993–94	17	57
1995–96	69	29
1996–97	50	42
2000–01	35	39
2001–02	6	30
2002–03	28	37
2003–04	1	25
2005–06	7	0

a. Calculate five-number summaries for the number of goals scored by both players.

b. Construct box plots for the two sets of data. Are there any outliers? What do the box plots tell you about the shapes of the two distributions?

c. Write a short paragraph comparing the number of goals for the two superstars.

2.52 Canadian Federal Election The data for the seats won by the Conservatives in every election in Canadian history up to 2006[16] is given in Exercise 1.47 and listed here, along with a box plot generated by *MINITAB*. Use the box plot to describe the shape of the distribution.

General Election (Date)	Seats	General Election (Date)	Seats
1st (1867.09.20)	101	21st (1949.06.27)	41
2nd (1872.10.12)	103	22nd (1953.08.10)	51
3rd (1874.01.22)	73	23rd (1957.06.10)	112
4th (1878.09.17)	137	24th (1958.03.31)	208
5th (1882.06.20)	139	25th (1962.06.18)	116
6th (1887.02.22)	123	26th (1963.04.08)	95
7th (1891.03.05)	123	27th (1965.11.08)	97
8th (1896.06.23)	88	28th (1968.06.25)	72
9th (1900.11.07)	81	29th (1972.10.30)	107
10th (1904.11.03)	75	30th (1974.07.08)	95
11th (1908.10.26)	85	31st (1979.05.22)	136
12th (1911.09.21)	133	32nd (1980.02.18)	103
13th (1917.12.17)	153	33rd (1984.09.04)	211
14th (1921.12.06)	50	34th (1988.11.21)	169
15th (1925.10.29)	116	35th (1993.10.25)	2
16th (1926.09.14)	91	36th (1997.06.02)	20
17th (1930.07.28)	137	37th (2000.11.27)	12
18th (1935.10.14)	39	38th (2004.06.28)	99
19th (1940.03.26)	39	39th (2006.01.23)	124
20th (1945.06.11)	67		

Source: Library of Parliament / Bibliotheque du Parlement, "Electoral Results by Party. www.lop.parl.gc.ca/Parllnfo/compilations/electionsandridings/ResultsParty.aspx Parliament of Canada. Reproduced with the permission of Library of Parliament, 2016.

Box plot for Exercise 2.52

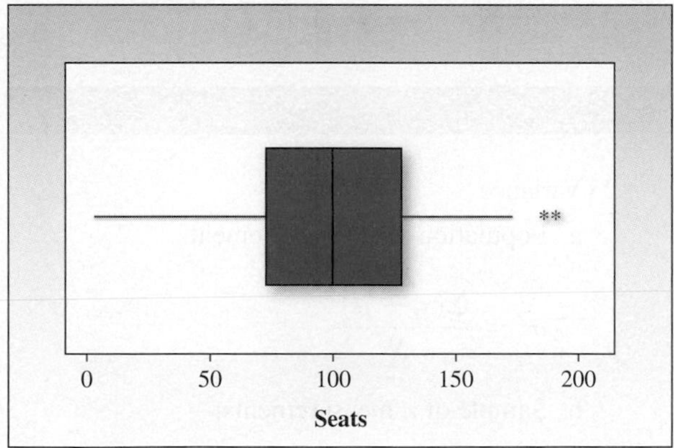

2.53 Internet Hotspots comScore, Inc. (NASDAQ: SCOR), a leader in measuring the digital world, released its monthly ranking of U.S. online activity at the top digital media properties for August 2014 based on data from the comScore Media Metrix® and Media Metrix® Multi-Platform services. Google Sites ranked as the top multi-platform property with 233.1 million unique visitors/viewers across desktop and mobile, while also ranking as the top desktop property with 190.7 million visitors.[17] The number of unique visitors at the top 26 sites are shown in the table:

Data set EX0253

Site	Unique Visitors (thousands)	Site	Unique Visitors (thousands)
Google Sites	233,143	Gannett Sites	106,760
Yahoo Sites	217,583	Wikimedia Foundation Sites	105,384
Facebook	206,881	LinkedIn	88,820
AOL, Inc.	181,216	Pandora.com	84,144
Microsoft Sites	179,454	The Weather Company	83,952
Amazon Sites	172,093	ESPN	82,420
Apple Inc.	137,758	Hearst Corporation	78,930
Turner Digital	134,957	About	78,378
Comcast NBCUniversal	132,390	Answers.com Sites	78,056
CBS Interactive	127,431	Yelp	75,354
Mode Media (formerly Glam Media)	122,971	Wal-Mart	72,159
Twitter.com	121,112	Time Inc. Network (partial)	71,699
eBay	110,635	BuzzFeed.com	68,779

Source: "comScore Ranks the Top 50 U.S. Digital Media Properties for August 2014", September 18, 2014. © 2016 comScore, Inc. Retrieved from www.comscore.com/Insights/Market-Rankings/comScore-Ranks-the-Top-50-US-Digital-Media-Properties-for-August-2014.

a. Can you tell by looking at the unique visitors whether it is roughly symmetric? Or is it skewed?

b. Calculate the mean and the median of unique visitors. Use these measures to decide whether or not the data are symmetric or skewed.

c. Draw a box plot to describe the data. Explain why the box plot confirms your conclusions in part b.

Data set EX0254

2.54 Utility Bills in Calgary The average monthly basic utility cost for an 85-square-metre apartment, which includes electricity, heating, water, and garbage collection, is \$185.75.[18] Suppose a random resident from Calgary is chosen, and his or her utility bills for 12 consecutive months are recorded, beginning January 2015:

Month	Amount (\$)	Month	Amount (\$)
January	195.65	July	184.69
February	254.32	August	154.54
March	175.67	September	289.82
April	198.65	October	231.91
May	215.12	November	267.75
June	267.34	December	255.76

a. Construct a box plot for the monthly utility costs.

b. What does the box plot tell you about the distribution of utility costs for this household?

2.55 What's Normal? Refer to Exercise 1.66 and data set EX0166. In addition to the normal body temperature in degrees Celsius for the 130 individuals, the data record the gender of the individuals. Box plots for the two groups, male and female, are shown below:[19]

Box plots for Exercise 2.55

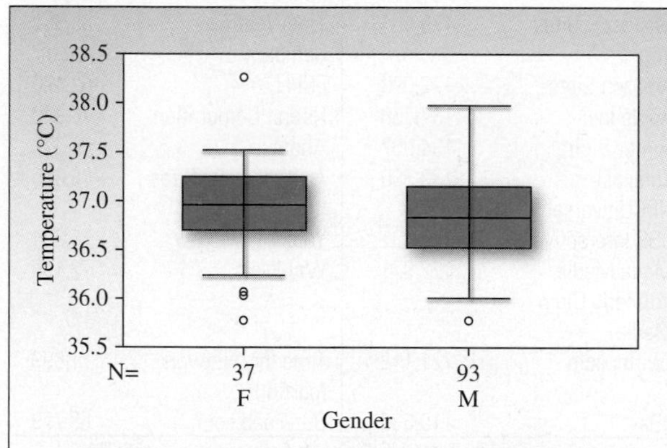

How would you describe the similarities and differences between male and female temperatures in this data set?

Data set EX0256 **2.56 Who Is Paying More at the Pump?** The prices below are the average for regular gasoline per litre as of April 2015:[20]

City	Price per Litre
Edmonton	89.227
Calgary	91.816
London	104.933
Hamilton	103.921
Toronto	105.057
Montreal	115.457
Winnipeg	90.560
Halifax	108.800
Regina	93.262
Saskatoon	96.673
Quebec City	119.458
Vancouver	121.273
Victoria	117.824

a. What is the value of the first and third quartiles?

b. What is the interquartile range?

c. Find the lower fence.

d. Find the upper fence.

e. Construct a box plot for this data.

f. Does the box plot indicate the presence of any outliers?

g. Calculate the z-score for the smallest and largest observations. Is either of these observations unusually large or unusually small?

h. Is the gas price in Hamilton relative to average gas price considered to be unusual? Why or why not? Justify your conclusion.

i. If you are an environmentally friendly person, in which city would you like to live based on the gas price? Provide a rationale.

CHAPTER REVIEW

Key Concepts and Formulas

I. Measures of the Centre of a Data Distribution

1. Arithmetic mean (mean) or average

   a. Population: μ

   b. Sample of n measurements: $\bar{x} = \dfrac{\Sigma x_i}{n}$

2. Median; **position** of the median $= 0.5(n + 1)$
3. Mode
4. The median may be preferred to the mean if the data are highly skewed.

II. Measures of Variability

1. Range: $R =$ largest $-$ smallest
2. Variance

   a. Population of N measurements:

   $$\sigma^2 = \frac{\Sigma(x_i - \mu)^2}{N}$$

   b. Sample of n measurements:

   $$s^2 = \frac{\Sigma(x_i - \bar{x})^2}{n - 1} = \frac{\Sigma x_i^2 - \dfrac{(\Sigma x_i)^2}{n}}{n - 1}$$

3. Standard deviation

   a. Population: $\sigma = \sqrt{\sigma^2}$

   b. Sample: $s = \sqrt{s^2}$

4. A rough approximation for s can be calculated as $s \approx R/4$. The divisor can be adjusted depending on the sample size.

III. Tchebysheff's Theorem and the Empirical Rule

1. Use Tchebysheff's Theorem for any data set, regardless of its shape or size.
   a. At least $1 - (1/k^2)$ of the measurements lie within k standard deviations of the mean.
   b. This is only a lower bound; there may be more measurements in the interval.
2. The Empirical Rule can be used only for relatively mound-shaped data sets. Approximately 68%, 95%, and 99.7% of the measurements are within one, two, and three standard deviations of the mean, respectively.

IV. Measures of Relative Standing

1. Sample z-score: $z = \dfrac{x - \bar{x}}{s}$
2. pth percentile; p% of the measurements are smaller, and $(100 - p)$% are larger.
3. Lower quartile, Q_1; **position** of $Q_1 = 0.25(n + 1)$
4. Upper quartile, Q_3; **position** of $Q_3 = 0.75(n + 1)$
5. Interquartile range: $\text{IQR} = Q_3 - Q_1$

V. The Five-Number Summary and Box Plots

1. The **five-number summary:**

   Min Q_1 **Median** Q_3 **Max**

   One-fourth of the measurements in the data set lie between each of the four adjacent pairs of numbers.
2. Box plots are used for detecting outliers and shapes of distributions.
3. Q_1 and Q_3 form the ends of the box. The median line is in the interior of the box.
4. Upper and lower fences are used to find outliers, observations that lie outside these fences.
   a. **Lower fence:** $Q_1 - 1.5(\text{IQR})$
   b. **Upper fence:** $Q_3 + 1.5(\text{IQR})$
5. **Outliers** are marked on the box plot with an asterisk (*).
6. **Whiskers** are connected to the box from the smallest and largest observations that are *not* outliers.
7. Skewed distributions usually have a long whisker *in the direction of the skewness*, and the median line is drawn *away from the direction of the skewness.*

TECHNOLOGY TODAY

Numerical Descriptive Measures in *Microsoft Excel*

Excel provides most of the basic descriptive statistics presented in Chapter 2 using a single command on the **Data** tab. Other descriptive statistics can be calculated using the **Function** command on the **Formulas** tab.

EXAMPLE 2.17

Leg room and seat pitch on an aircraft are important variables for a comfortable flight. Seat pitch refers to the distance between the back of your seat and the back of the seat in front.

Regular economy-class seat pitch can vary:

- The bare minimum economy-class seat pitch is 71 cm and is used on low-cost carriers such as Jetstar and AirAsia.
- The minimum economy-class seat pitch for long-haul flights is 79 cm and is found on full-service airlines such as Qantas and British Airways.
- The greatest economy-class seat pitch is 86 cm, found on Korean Air, Malaysia Airlines, and a few others.

Seat pitch is often used as a proxy for leg room, but it's not really a direct equivalent since you need to take into account the depth of the seatback too.

The following data are the seat pitch (in centimetres) for business and economy classes for selected commercial airlines around the globe:

Airline	Business Class	Economy Class
A1	104.1	72.4
A2	105.4	76.2
A3	101.6	64.8
A4	104.1	69.9
A5	109.2	78.7
A6	104.1	74.9
A7	108.0	74.9
A8	101.6	73.7
A9	106.7	76.2

1. Since the data involve two variables and a third labelling variable, enter the data into the first three columns of an *Excel* spreadsheet, using the labels in the table. Select **Data ▶ Data Analysis ▶ Descriptive Statistics**, and highlight or type the **Input range** (the data in the second and third columns) into the Descriptive Statistics dialogue box (Figure 2.18(a)). Type an Output location, make sure the check boxes for "Labels in first row" and "Summary statistics" are both checked, and click **OK**. The summary statistics (Figure 2.18(b)) will appear in the selected location in your spreadsheet.
2. You may notice that some of the cells in the spreadsheet are overlapping. To adjust this, highlight the affected columns and click the **Home** tab. In the **Cells** group, choose **Format ▶ AutoFit Column Width**. You may want to modify the appearance of the output by decreasing the decimal accuracy in certain cells. Highlight the appropriate cells and click the **Decrease Decimal** icon (**Home** tab, **Number** group) to modify the output. We have displayed the accuracy to three decimal places.

FIGURE 2.18

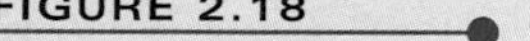

(a)

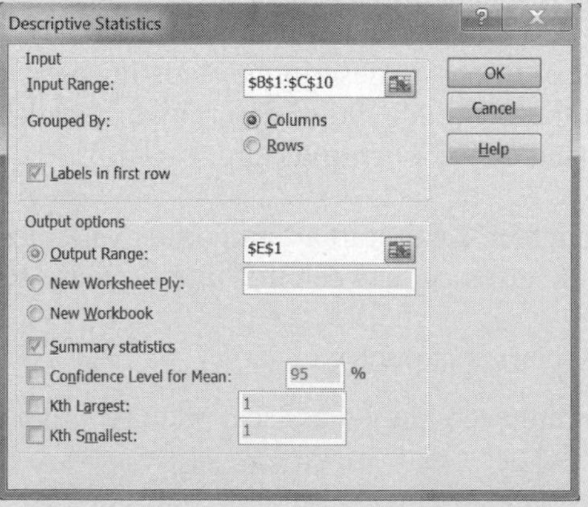

(b)

E	F	G	H
Business Class		*Economy Class*	
Mean	104.9778	Mean	73.52222
Standard Error	0.875877	Standard Error	1.371514
Median	104.1	Median	74.9
Mode	104.1	Mode	76.2
Standard Deviation	2.627631	Standard Deviation	4.114541
Sample Variance	6.904444	Sample Variance	16.92944
Kurtosis	-0.79449	Kurtosis	1.774623
Skewness	0.256476	Skewness	-1.22294
Range	7.6	Range	13.9
Minimum	101.6	Minimum	64.8
Maximum	109.2	Maximum	78.7
Sum	944.8	Sum	661.7
Count	9	Count	9

3. Notice that the sample quartiles, Q_1 and Q_3, are not given in the *Excel* output in Figure 2.18(b). You can calculate the quartiles using the function command. Place your cursor into an empty cell and select **Formulas ▶ More Functions ▶ Statistical ▶ QUARTILE.EXC**. Highlight the appropriate cells in the box marked "Array" and type an integer (0 = min,

1 = first quartile, 2 = median, 3 = third quartile, or 4 = max) in the box marked "Quart." The quartile (calculated using this textbook's method) will appear in the cell you have chosen. An alternative method for calculating the quartiles will be used if you select **Formulas ▶ More Functions ▶ Statistical ▶ QUARTILE.INC**. (NOTE: This function is called **QUARTILE** in *Excel 2007* and earlier versions.) Using the two quartiles, you can calculate the IQR and construct a box plot by hand.

Numerical Descriptive Measures in *MINITAB*

MINITAB provides most of the basic descriptive statistics presented in Chapter 2 using a single command in the drop-down menus.

EXAMPLE 2.18

The following data are the seat pitch (in centimetres) for business and economy classes for selected commercial airlines around the globe:

Airline	Business Class	Economy Class
A1	104.1	72.4
A2	105.4	76.2
A3	101.6	64.8
A4	104.1	69.9
A5	109.2	78.7
A6	104.1	74.9
A7	108.0	74.9
A8	101.6	73.7
A9	106.7	76.2

1. Since the data involve two variables and a third labelling variable, enter the data into the first three columns of a *MINITAB* worksheet, using the labels in the table. Using the drop-down menus, select **Stat ▶ Basic Statistics ▶ Display Descriptive Statistics**. The dialogue box is shown in Figure 2.19(a).

FIGURE 2.19

(a)

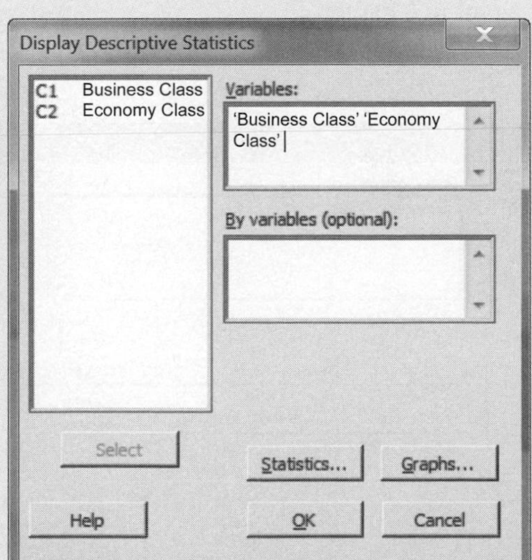

(b)

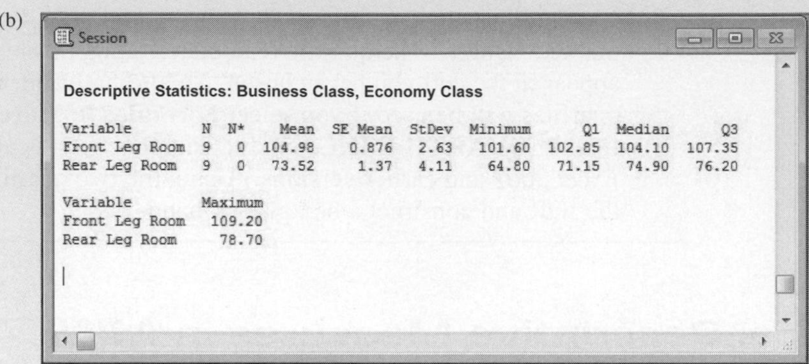
Session

Descriptive Statistics: Business Class, Economy Class

Variable	N	N*	Mean	SE Mean	StDev	Minimum	Q1	Median	Q3
Front Leg Room	9	0	104.98	0.876	2.63	101.60	102.85	104.10	107.35
Rear Leg Room	9	0	73.52	1.37	4.11	64.80	71.15	74.90	76.20

Variable	Maximum
Front Leg Room	109.20
Rear Leg Room	78.70

2. Now click on the Variables box and **select** both columns from the list on the left. (You can click on the **Graphs** option and choose one of several graphs if you like. You may also click on the **Statistics** option to select the statistics you would like to see displayed.) Click **OK**. A display of descriptive statistics for both columns will appear in the Session window (see Figure 2.19(b)). You may print this output using **File ▶ Print Session Window** if you choose.
3. To examine the distribution of the two variables and look for outliers, you can create box plots using the command **Graph ▶ Boxplot ▶ One Y ▶ Simple**. Click **OK**. Select the appropriate column of measurements in the dialogue box (see Figure 2.20(a)). You can change the appearance of the box plot in several ways. **Scale ▶ Axes and Ticks** will allow you to transpose the axes and orient the box plot horizontally, when you check the box marked "Transpose value and category scales." **Multiple Graphs** provides printing options for multiple box plots. **Labels** will let you annotate the graph with titles and footnotes. If you have entered data into the worksheet as a frequency distribution (values in one column, frequencies in another), the **Data Options** will allow the data to be read in that format. The box plot for the economy class is shown in Figure 2.20(b).
4. Save this worksheet in a file called "Seat Pitch" before exiting *MINITAB*. We will use it again in Chapter 3.

FIGURE 2.20

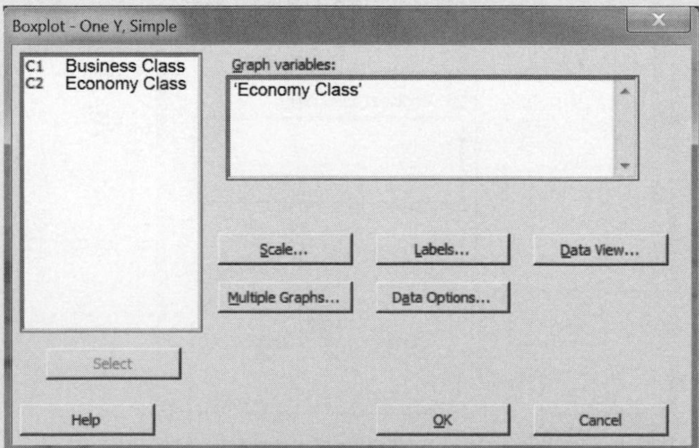

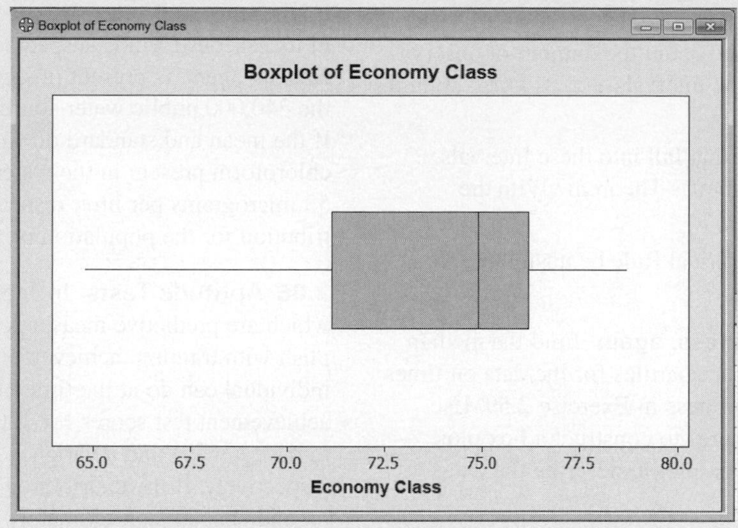

Supplementary Exercises

Data set EX0257 **2.57 Raisins** The number of raisins in each of 14 miniboxes (15 g) was counted for a generic brand and for Sunmaid brand raisins. The two data sets are shown here:

Generic Brand				Sunmaid			
25	26	25	28	25	29	24	24
26	28	28	27	28	24	28	22
26	27	24	25	25	28	30	27
26	26			28	24		

a. What are the mean and standard deviation for the generic brand?

b. What are the mean and standard deviation for the Sunmaid brand?

c. Compare the centres and variabilities of the two brands using the results of parts a and b.

Data set EX0258 **2.58 TV Viewers** The number of television viewing hours per household and the prime viewing times are two factors that affect television advertising income. A random sample of 25 households in a particular viewing area produced the following estimates of viewing hours per household:

3.0	6.0	7.5	15.0	12.0
6.5	8.0	4.0	5.5	6.0
5.0	12.0	1.0	3.5	3.0
7.5	5.0	10.0	8.0	3.5
9.0	2.0	6.5	1.0	5.0

a. Scan the data and use the range to find an approximate value for s. Use this value to check your calculations in part b.

b. Calculate the sample mean $\bar{x}$ and the sample standard deviation s. Compare s with the approximate value obtained in part a.

c. Find the percentage of the viewing hours per household that falls into the interval $\bar{x} \pm 2s$. Compare with the corresponding percentage given by the Empirical Rule.

2.59 A Recurring Illness Refer to Exercise 1.26 and data set EX0126. The lengths of time (in months) between the onset of a particular illness and its recurrence were recorded:

2.1	4.4	2.7	32.3	9.9
9.0	2.0	6.6	3.9	1.6
14.7	9.6	16.7	7.4	8.2
19.2	6.9	4.3	3.3	1.2
4.1	18.4	0.2	6.1	13.5
7.4	0.2	8.3	0.3	1.3
14.1	1.0	2.4	2.4	18.0
8.7	24.0	1.4	8.2	5.8
1.6	3.5	11.4	18.0	26.7
3.7	12.6	23.1	5.6	0.4

a. Find the range.

b. Use the range approximation to find an approximate value for s.

c. Compute s for the data and compare it with your approximation from part b.

2.60 A Recurring Illness, continued Refer to Exercise 2.59.

a. Examine the data and count the number of observations that fall into the intervals $\bar{x} \pm s$, $\bar{x} \pm 2s$, and $\bar{x} \pm 3s$.

b. Do the percentages that fall into these intervals agree with Tchebysheff's Theorem? With the Empirical Rule?

c. Why might the Empirical Rule be unsuitable for describing these data?

2.61 A Recurring Illness, again Find the median and the lower and upper quartiles for the data on times until recurrence of an illness in Exercise 2.59. Use these descriptive measures to construct a box plot for the data. Use the box plot to describe the data distribution.

2.62 Wedding Cake, again Refer to Exercise 2.9. The prices for 14 different slices of wedding cake with buttercream icing are reproduced here:

3.25, 5.50, 4.75, 1.75, 3.50, 5.25, 2.50

4.25, 2.00, 4.25, 3.75, 5.75, 4.00, 6.00

a. Calculate the five-number summary.

b. Construct a box plot for the data. Are there any outliers?

c. The value $x = 6.00$ looks large in comparison to most other prices. Use a z-score to decide whether this is an unusually expensive wedding cake.

2.63 Electrolysis An analytical chemist wanted to use electrolysis to determine the number of moles of cupric ions in a given volume of solution. The solution was partitioned into $n = 30$ portions of 0.2 millilitre each, and each of the portions was tested. The average number of moles of cupric ions for the $n = 30$ portions was found to be 0.17 mole; the standard deviation was 0.01 mole.

a. Describe the distribution of the measurements for the $n = 30$ portions of the solution using Tchebysheff's Theorem.

b. Describe the distribution of the measurements for the $n = 30$ portions of the solution using the Empirical Rule. (Do you expect the Empirical Rule to be suitable for describing these data?)

c. Suppose the chemist had used only $n = 4$ portions of the solution for the experiment and obtained the readings 0.15, 0.19, 0.17, and 0.15. Would the Empirical Rule be suitable for describing the $n = 4$ measurements? Why?

2.64 Chloroform According to the U.S. Environmental Protection Agency, chloroform, which in its gaseous form is suspected of being a cancer-causing agent, is present in small quantities in all of the 240,000 public water sources in the United States. If the mean and standard deviation of the amounts of chloroform present in the water sources are 34 and 53 micrograms per litre, respectively, describe the distribution for the population of all public water sources.

2.65 Aptitude Tests In contrast to aptitude tests, which are predictive measures of what one can accomplish with training, achievement tests tell what an individual can do at the time of the test. Mathematics achievement test scores for 400 students were found to have a mean and a variance equal to 600 and 4900, respectively. If the distribution of test scores was mound-shaped, approximately how many of the scores would fall into the interval 530 to 670? Approximately how many scores would be expected to fall into the interval 460 to 740?

2.66 Sleep and the University Student How much sleep do you get on a typical school night? A group of 10 university students were asked to report the number of hours that they slept on the previous night with the following results:

7, 6, 7.25, 7, 8.5, 5, 8, 7, 6.75, 6

a. Find the mean and the standard deviation of the number of hours of sleep for these 10 students.

b. Calculate the z-score for the largest value ($x = 8.5$). Is this an unusually sleepy university student?

c. What is the most frequently reported measurement? What is the name for this measure of centre?

d. Construct a box plot for the data. Does the box plot confirm your results in part b? (HINT: Since the z-score and the box plot are two unrelated methods for detecting outliers, and use different types of statistics, they do not necessarily have to [but usually do] produce the same results.)

Data set EX0267 **2.67 Fuel Efficiency** The litres per 100 km (L/100 km) for each of 20 medium-sized cars, selected from a production line during the month of March, are shown below:

9.7	9.9	9.5	10.1
9.8	10.3	9.4	8.9
10.2	9.4	9.5	9.6
7.9	8.1	9.8	10.0
8.9	9.9	11.3	10.9

a. Construct a relative frequency histogram for these data. How would you describe the shape of the distribution?

b. Find the mean and the standard deviation.

c. Arrange the data from smallest to largest. Find the z-scores for the largest and smallest observations. Would you consider them to be outliers? Why or why not?

d. What is the median?

e. Find the lower and upper quartiles.

2.68 Fuel Efficiency, continued Refer to Exercise 2.67. Construct a box plot for the data. Are there any outliers? Does this conclusion agree with your results in Exercise 2.67?

2.69 Polluted Seawater? Petroleum pollution in seas and oceans stimulates the growth of some types of bacteria. A count of petroleumlytic micro-organisms (bacteria per 100 millilitres) in 10 portions of seawater gave these readings:

49, 70, 54, 67, 59, 40, 61, 69, 71, 52

a. Guess the value for s using the range approximation.

b. Calculate $\bar{x}$ and s and compare with the range approximation of part a.

c. Construct a box plot for the data and use it to describe the data distribution.

2.70 Basketball Attendances at a high school's basketball games were recorded and found to have a sample mean and variance of 420 and 25, respectively. Calculate $\bar{x} \pm s$, $\bar{x} \pm 2s$, and $\bar{x} \pm 3s$ and then state the approximate fractions of measurements you would expect to fall into these intervals according to the Empirical Rule.

2.71 Standardized Test Scores A district's school board's verbal and mathematics scholastic aptitude tests are scored on a scale of 200 to 800. Although the tests were originally designed to produce mean scores of approximately 500, the mean verbal and math scores in recent years have been as low as 463 and 493, respectively, and have been trending downward. It seems reasonable to assume that a distribution of all test scores, either verbal or math, is mound-shaped. If σ is the standard deviation of one of these distributions, what is the largest value (approximately) that σ might assume? Explain.

2.72 Long-Stemmed Roses A strain of long-stemmed roses has an approximate normal distribution with a mean stem length of 38 cm and standard deviation of 5.5 cm.

a. If one accepts as "long-stemmed roses" only those roses with a stem length greater than 38 cm, what percentage of such roses would be unacceptable?

b. What percentage of these roses would have a stem length between 31 and 51 cm?

2.73 Drugs for Hypertension A pharmaceutical company wishes to know whether an experimental drug being tested in its laboratories has any effect on systolic blood pressure. Fifteen randomly selected subjects were given the drug, and their systolic blood pressures (in millimetres of mercury) are recorded:

172	148	123
140	108	152
123	129	133
130	137	128
115	161	142

a. Guess the value of s using the range approximation.

b. Calculate $\bar{x}$ and s for the 15 blood pressures.

c. Find two values, a and b, such that at least 75% of the measurements fall between a and b.

2.74 Lumber Rights A company interested in lumbering rights for a certain tract of slash pine trees is told that the mean diameter of these trees is 35 cm with a standard deviation of 7 cm. Assume the distribution of diameters is roughly mound-shaped.

a. What fraction of the trees will have diameters between 21 and 55 cm?

b. What fraction of the trees will have diameters greater than 43 cm?

Data set EX0275 **2.75 Social Ambivalence** The following data represent the social ambivalence scores for 15 people as measured by a psychological test. (The higher the score, the stronger the ambivalence.)

9	13	12
14	15	11
10	4	10
8	19	13
11	17	9

a. Guess the value of s using the range approximation.

b. Calculate $\bar{x}$ and s for the 15 social ambivalence scores.

c. What fraction of the scores actually lie in the interval $\bar{x} \pm 2s$?

2.76 TV Commercials The mean duration of television commercials on a given network is 75 seconds, with a standard deviation of 20 seconds. Assume that durations are approximately normally distributed.

a. What is the approximate probability that a commercial will last less than 35 seconds?

b. What is the approximate probability that a commercial will last longer than 55 seconds?

2.77 Parasites in Foxes A random sample of 100 foxes was examined by a team of veterinarians to determine the prevalence of a particular type of parasite. Counting the number of parasites per fox, the veterinarians found that 69 foxes had no parasites, 17 had one parasite, and so on. A frequency tabulation of the data is given here:

Number of Parasites, x	0	1	2	3	4	5	6	7	8
Number of Foxes, f	69	17	6	3	1	2	1	0	1

a. Construct a relative frequency histogram for x, the number of parasites per fox.

b. Calculate $\bar{x}$ and s for the sample.

c. What fraction of the parasite counts fall within two standard deviations of the mean? Within three standard deviations? Do these results agree with Tchebysheff's Theorem? With the Empirical Rule?

2.78 University Professors Consider a population consisting of the number of professors per university at small universities. Suppose that the number of professors per university has an average $\mu = 175$ and a standard deviation $\sigma = 15$.

a. Use Tchebysheff's Theorem to make a statement about the percentage of universities that have between 145 and 205 professors.

b. Assume that the population is normally distributed. What fraction of universities have more than 190 professors?

Data set EX0279 **2.79 Is It Accurate?** From the following data, a student calculated s to be 0.263. On what grounds might we doubt his accuracy? What is the correct value (to the nearest hundredth)?

17.2	17.1	17.0	17.1	16.9	17.0	17.1	17.0	17.3	17.2
17.1	17.0	17.1	16.9	17.0	17.1	17.3	17.2	17.4	17.1

2.80 Great Goal Scorers The number of goals scored per season by each of four NHL superstars over each player's career were recorded and shown in the box plots below:

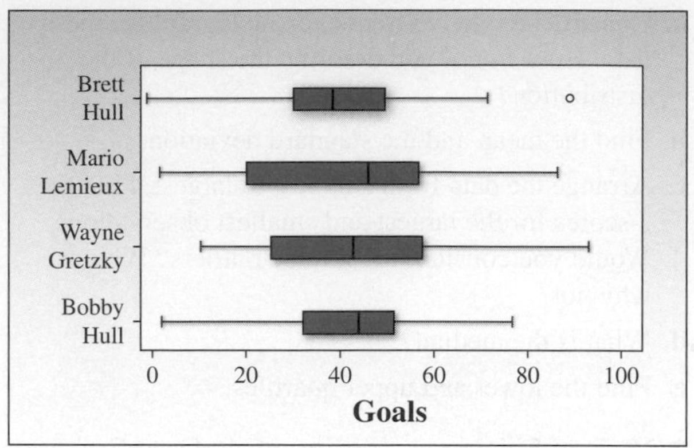

Write a short paragraph comparing the goal scoring patterns of these four players.

2.81 Bobby Hull Two box plots of Bobby Hull's goal scores are given below.[21] One is for 1957–1975, and the other includes the years 1975–1980.

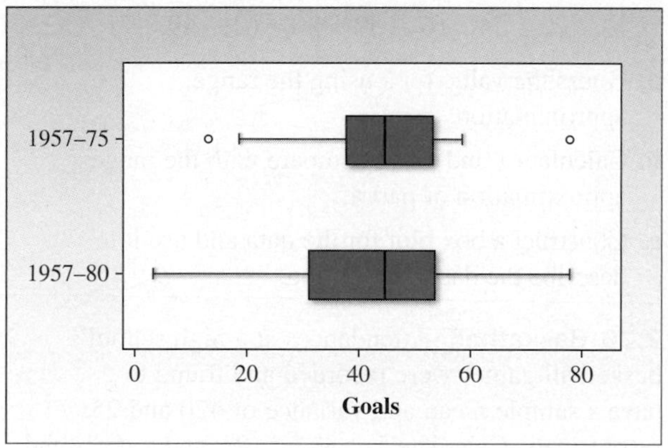

The statistics used to construct these box plots are given in the following table:

Years	Min	Q_1	Median	Q_3	IQR	Max	n
1957–80	2	31.00	44	52.00	21	77	23
1957–75	13	36.25	44	52.25	16	77	18

a. Calculate the upper fences for both of these box plots.

b. Can you explain why the record number of goals is an outlier in the 1957–1975 box plot, but not in the 1957–1980 box plot?

2.82 Ages of Pennies Here are the ages of 50 pennies from Exercise 1.45 and data set EX0145. The data have been sorted from smallest to largest.

0	0	0	0	0	0	0	0	0	0
0	0	1	1	1	1	1	1	2	2
2	3	3	3	4	4	5	5	5	5
6	8	9	9	10	16	17	17	19	19
19	20	20	21	22	23	25	25	28	36

a. What is the average age of the pennies?

b. What is the median age of the pennies?

c. Based on the results of parts a and b, how would you describe the age distribution of these 50 pennies?

d. Construct a box plot for the data set. Are there any outliers? Does the box plot confirm your description of the distribution's shape?

2.83 Environmental Factors How do Canadians rate environmental factors in terms of the threat they pose to Canada? Below are findings of the survey conducted by the Strategic Counsel:[22]

- The large majority of Canadians (61%) believe that toxic chemicals are linked to human health.
- 55% think air pollution and smog is an important factor.
- On the other hand, 52% feel that global warming and climate change is a key factor in terms of threat to Canada.

Identify the variable of interest, and any percentiles you can determine from this information.

Data set EX0284 **2.84 Breathing Patterns** Research psychologists are interested in finding out whether a person's breathing patterns are affected by a particular experimental treatment. To determine the general respiratory patterns of the $n = 30$ people in the study, the researchers collected some baseline measurements—the total ventilation in litres of air per minute adjusted for body size—for each person before the treatment. The data are shown here, along with some descriptive tools generated by *MINITAB* and *Excel*:

5.23	4.79	5.83	5.37	4.35	5.54	6.04	5.48	6.58	4.82
5.92	5.38	6.34	5.12	5.14	4.72	5.17	4.99	4.51	5.70
4.67	5.77	5.84	6.19	5.58	5.72	5.16	5.32	4.96	5.63

Descriptive Statistics: Litres

```
Variable   N    N*       Mean    SE Mean     StDev
Litres    30     0     5.3953     0.0997    0.5462

Minimum      Q1  Median      Q3  Variable  Maximum
 4.3500 4.9825  5.3750  5.7850  Litres      6.5800
```

Stem and Leaf Display: Litres

```
Stem-and-leaf of Litres  N = 30
Leaf Unit = 0.10

  1   4  3
  2   4  5
  5   4  677
  8   4  899
 12   5  1111
(4)   5  2333
 14   5  455
 11   5  6777
  7   5  889
  4   6  01
  2   6  3
  1   6  5
```

***Excel* Descriptive Statistics**

	Litres
Mean	5.3953
Standard Error	0.0997
Median	5.3750
Mode	#N/A
Standard Deviation	0.5462
Sample Variance	0.2983
Kurtosis	20.4069
Skewness	0.1301
Range	2.23
Minimum	4.35
Maximum	6.58
Sum	161.86
Count	30

a. Summarize the characteristics of the data distribution using the computer output.

b. Does the Empirical Rule provide a good description of the proportion of measurements that fall within two or three standard deviations of the mean? Explain.

c. How large or small does a ventilation measurement have to be before it is considered unusual?

Data set EX0285 **2.85 Arranging Objects** The following data are the response times in seconds for $n = 25$ Grade 1 students to arrange three objects by size:

5.2	3.8	5.7	3.9	3.7
4.2	4.1	4.3	4.7	4.3
3.1	2.5	3.0	4.4	4.8
3.6	3.9	4.8	5.3	4.2
4.7	3.3	4.2	3.8	5.4

a. Find the mean and the standard deviation for these 25 response times.

b. Order the data from smallest to largest.

c. Find the z-scores for the smallest and largest response times. Is there any reason to believe that these times are unusually large or small? Explain.

2.86 Arranging Objects, continued Refer to Exercise 2.85.

a. Find the five-number summary for this data set.

b. Construct a box plot for the data.

c. Are there any unusually large or small response times identified by the box plot?

d. Construct a stem and leaf display for the response times. How would you describe the shape of the distribution? Does the shape of the box plot confirm this result?

CASE STUDY

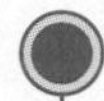

The Boys of Winter

Goals

Have goals been easier to score in the NHL in some eras than others? Many of us have heard of hockey greats such as Maurice "Rocket" Richard, Gordie Howe, Wayne Gretzky, and Mario Lemieux. But have you heard of Bill Cook, who scored 28 goals in 48 games for the New York Rangers in 1933, or Norm Ullman, who scored 42 in 70 games for the Toronto Maple Leafs in 1965? The average number of goals per game for the NHL's leading goal scorers are given on the text website. They begin in 1931 with Charlie Conacher, who averaged 0.70 goals per game when he played with the Toronto Maple Leafs.

The last entry is for the year 2006, when Jonathan Cheechoo of the San Jose Sharks averaged 0.68 goals per game. The data is divided into three different eras that hockey followers believe exhibit different goal scoring patterns:1933–1967, 1968–1993, and 1994–2006. How can we summarize the information in this data set?

1. Use *MINITAB* or *Excel* to describe the goals per game for the leading goal scorers in each of the three different eras. Generate any graphics that may help you in interpreting these data sets.
2. Do the eras appear to differ in their levels of scoring? Do the goals per game appear to be more variable in some eras than others?
3. Are there outliers in any of the eras?
4. Summarize your comparison of the three hockey eras.

PROJECTS

Project 2: Ignorance Is Not Bliss (Project 1-B continued)

An occupational hygienist believes that a two-hour training session on proper handwashing will improve time spent on handwashing. Based on a random sample of 25 high school students who had attended the session, their handwashing times (in seconds) were recorded as follows:

22	19	21	16	30	35	17	18	10	29	17	14	19
20	20	55	22	23	21	0	22	15	19	27	33	

a. Calculate the sample mean, mode, and median of this data set. Are the data mound-shaped?

b. If you were asked to select a central value to describe the data, which measure of central tendency would you use? Explain.

c. Calculate the value of the sample standard deviation (s) and the range (R), and use R to approximate s. Is this a good approximation?

d. What would happen to the mean of time spent to wash hands if all data points were raised by 4%?

e. What would happen to the standard deviation of time spent to wash hands if all data points were raised by 5%?

f. Find the percentage of measurements in the intervals $\bar{x} \pm s$ and $\bar{x} \pm 2s$. Compare these results with the Empirical Rule and Tchebysheff's Theorem percentages respectively.

g. Can you use Tchebysheff's Theorem to describe this data set? Why or why not?

h. Can you use the Empirical Rule to describe this data set? Why or why not?

i. Which, if any, of the observations appear to be outliers? Justify your answer.

j. Find the 25th, 50th, and 75th percentiles. What is the value of the interquartile range?

k. Compare range with interquartile range. What conclusion can you draw, if any?

l. Construct a box plot for this data. Does the box plot indicate the presence of any outliers?

m. Construct side-by-side box plots for this data and data given in Project 1-B (page 54). HINT: See Exercise 2.81, page 96.

n. Looking at these box plots, how would you compare the amount of time spent on washing hands?

o. Can you infer (conclude) that the training session was useful in reducing the risk of catching flu? In other words, do you support the conjecture of the occupational hygienist?

p. Now combine both data sets and make a histogram of the amount of time spent on handwashing and comment on the shape of the distribution.

3

Describing Bivariate Data

GENERAL OBJECTIVES

Sometimes the data that are collected consist of observations for two variables on the same experimental unit. Special techniques that can be used in describing these variables will help you identify possible relationships between them.

CHAPTER INDEX

NEED TO KNOW

How to Calculate the Correlation Coefficient
How to Calculate the Regression Line

Mclek/Shutterstock.com

Paying for Players

Are higher-paid hockey players actually better than lower-salaried ones? Some NHL hockey teams invest large amounts of money in signing superstars and spend significantly less on the rest of their roster, while other teams distribute their player budget more evenly but have no "superstar" talent. In the case study at the end of this chapter, we rank all 30 NHL teams by payroll in the 2005–2006 season, and then see how they perform in terms of goals scored and yielded, and more importantly, wins and points. The techniques presented in this chapter will help answer our question.

BIVARIATE DATA

Very often researchers are interested in more than just one variable that can be measured during their investigation. For example, an auto insurance company might be interested in the number of vehicles owned by a policyholder as well as the number of drivers in the household. An economist might need to measure the amount spent per week on groceries in a household and also the number of people in that household. A real estate agent might measure the selling price of a residential property and the square metres of the living area.

NEED A TIP?

"**Bi**" means "two." **Bi**variate data generate pairs of measurements.

When two variables are measured on a single experimental unit, the resulting data are called **bivariate data.** How should you display these data? Not only are both variables important when studied separately, but you also may want to explore the *relationship between the two variables.* Methods for graphing bivariate data, whether the variables are qualitative or quantitative, allow you to study the two variables together. As with *univariate data,* you use different graphs depending on the type of variables you are measuring.

3.2 GRAPHS FOR QUALITATIVE VARIABLES

When at least one of the two variables is *qualitative,* you can use either simple or more intricate pie charts, line charts, and bar charts to display and describe the data. Sometimes you will have one qualitative and one quantitative variable that have been measured in two different populations or groups. In this case, you can use two **side-by-side pie charts** or a bar chart in which the bars for the two populations are placed side by side. Another option is to use a **stacked bar chart,** in which the bars for each category are stacked on top of each other.

EXAMPLE 3.1

Are professors in the faculty of science paid more than professors in the faculty of arts? The data in Table 3.1 were collected from a sample of 400 professors whose rank, faculty, and salary were recorded. The number in each cell is the average salary (in thousands of dollars) for all professors who fell into that category. Use a graph to answer the question posed for this sample.

TABLE 3.1 **Salaries of Professors by Rank and Faculty**

	Full Professor	Associate Professor	Assistant Professor
Arts	55.8	42.2	35.2
Science	61.6	43.3	35.5

SOLUTION To display the average salaries of these 400 professors, you can use a side-by-side bar chart, as shown in Figure 3.1. The height of the bars is the average salary, with each pair of bars along the horizontal axis representing a different professorial rank. Salaries are substantially higher for full professors in the faculty of science, but there is very little difference at the lower two ranks.

FIGURE 3.1
Comparative bar charts for Example 3.1

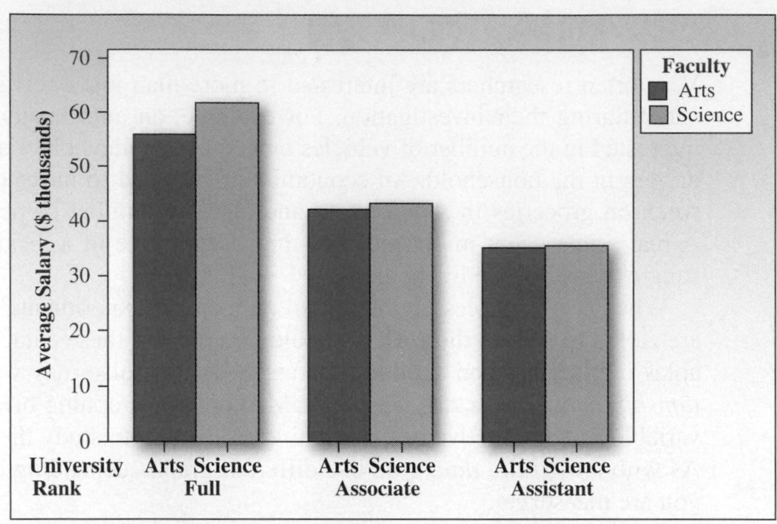

EXAMPLE 3.2

Along with the salaries for the 400 professors in Example 3.1, the researcher recorded two qualitative variables for each professor: rank and faculty. Table 3.2 shows the number of professors in each of the $2 \times 3 = 6$ categories. Use comparative charts to describe the data. Do the faculties of science employ as many high-ranking professors as the faculties of arts do?

TABLE 3.2 **Number of Professors by Rank and Faculty**

	Full Professor	Associate Professor	Assistant Professor	Total
Arts	24	57	69	150
Science	60	78	112	250

SOLUTION The numbers in the table are not quantitative measurements on a single experimental unit (the professor). They are *frequencies,* or counts, of the number of professors who fall into each category. To compare the numbers of professors in the faculty of arts and the faculty of science, you might draw two pie charts and display them side by side, as in Figure 3.2.

FIGURE 3.2
Comparative pie charts for Example 3.2

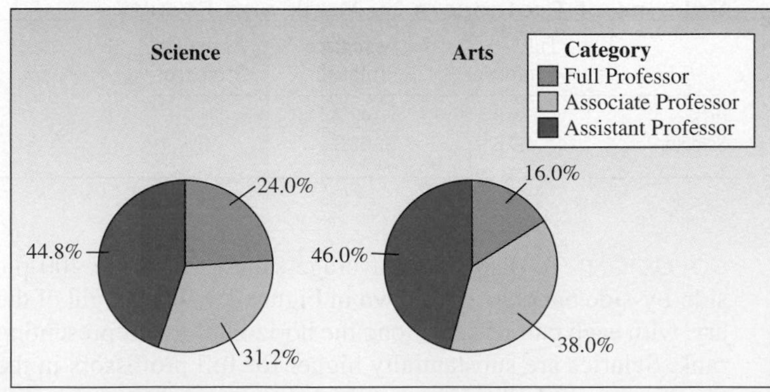

Alternatively, you could draw either a stacked or side-by-side bar chart. The stacked bar chart is shown in Figure 3.3.

FIGURE 3.3
Stacked bar chart for Example 3.2

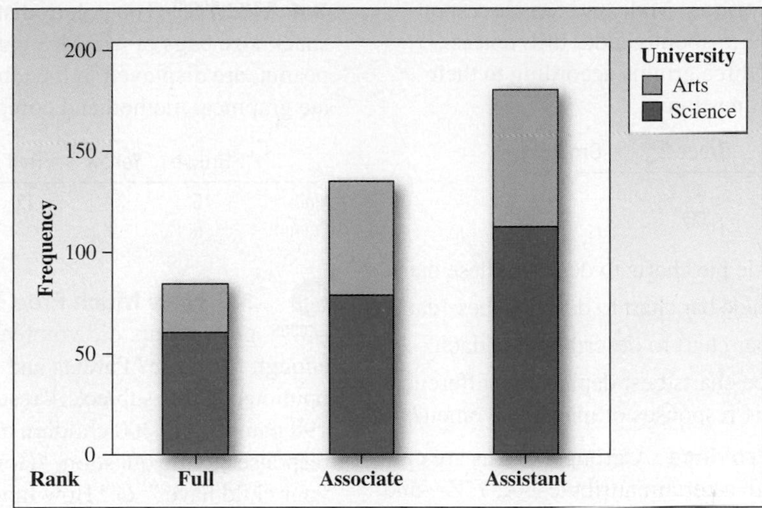

Although the graphs are not strikingly different, you can see that arts faculties have fewer full professors and relatively more (Figure 3.2) but not absolutely more (Figure 3.3) associate professors than science faculties. The reason for these differences is not clear, but you might speculate that faculties of science, with their higher salaries, are able to attract more full professors. Or perhaps faculties of arts are not as willing to promote professors to the higher-paying ranks. In any case, the graphs provide a means for comparing the two sets of data.

You can also compare the distributions for arts versus science faculties by creating *conditional data distributions.* These conditional distributions are shown in Table 3.3. One distribution shows the proportion of professors in each of the three ranks under the *condition* that the faculty is arts, and the other shows the proportions under the *condition* that the faculty is science. These *relative frequencies* are easier to compare than the *actual frequencies* and lead to the same conclusions:

- The proportion of assistant professors is roughly the same for both arts and science faculties.
- Arts faculties have a smaller proportion of full professors and a larger proportion of associate professors.

TABLE 3.3 **Proportions of Professors by Rank for Arts and Science Faculties**

	Full Professor	Associate Professor	Assistant Professor	Total
Arts	$\frac{24}{150} = 0.16$	$\frac{57}{150} = 0.38$	$\frac{69}{150} = .046$	1.00
Science	$\frac{60}{250} = 0.24$	$\frac{78}{250} = 0.31$	$\frac{112}{250} = 0.45$	1.00

3.2 EXERCISES

BASIC TECHNIQUES

3.1 Gender Differences Male and female respondents to a questionnaire about gender differences are categorized into three groups according to their answers on the first question:

	Group 1	Group 2	Group 3
Men	37	49	72
Women	7	50	31

a. Create side-by-side pie charts to describe these data.

b. Create a side-by-side bar chart to describe these data.

c. Draw a stacked bar chart to describe these data.

d. Which of the three charts best depicts the difference or similarity of the responses of men and women?

3.2 Province by Province A group of items are categorized according to a certain attribute—X, Y, Z—and according to the province in which they are produced:

	X	Y	Z
Manitoba	20	5	5
Saskatchewan	10	10	5

a. Create a comparative (side-by-side) bar chart to compare the numbers of items of each type made in Manitoba and Saskatchewan.

b. Create a stacked bar chart to compare the numbers of items of each type made in the two provinces.

c. Which of the two types of presentation in parts a and b is more easily understood? Explain.

d. What other graphical methods could you use to describe the data?

3.3 Consumer Spending The table below shows the average amounts spent per week by men and women in each of four spending categories:

	A	B	C	D
Men ($)	54	27	105	22
Women ($)	21	85	100	75

a. What possible graphical methods could you use to compare the spending patterns of women and men?

b. Choose two different methods of graphing and display the data in graphical form.

c. What can you say about the similarities or differences in the spending patterns for men and women?

d. Which of the two methods used in part b provides a better descriptive graph?

APPLICATIONS

3.4 M&M'S® The colour distributions for two snack-size bags of M&M'S candies, one plain and one peanut, are displayed in the table. Choose an appropriate graphical method and compare the distributions.

	Brown	Yellow	Red	Orange	Green	Blue
Plain	15	14	12	4	5	6
Peanut	6	2	2	3	3	5

Data set EX0305 **3.5 How Much Free Time?** When you were growing up, did you feel that you did not have enough free time? Parents and children have differing opinions on this subject. A research group surveyed 198 parents and 200 children and recorded their responses to the question, "How much free time does your child have?" or "How much free time do you have?" The responses are shown in the table below:[1]

	Just the Right Amount	Not Enough	Too Much	Don't Know
Parents	138	14	40	6
Children	130	48	16	6

a. Define the sample and the population of interest to the researchers.

b. Describe the variables that have been measured in this survey. Are the variables qualitative or quantitative? Are the data univariate or bivariate?

c. What do the entries in the cells represent?

d. Use comparative pie charts to compare the responses for parents and children.

e. What other graphical techniques could be used to describe the data? Would any of these techniques be more informative than the pie charts constructed in part d?

Data set EX0306 **3.6 Consumer Price Index** The price of living in Canada is continually increasing, as demonstrated by the consumer price indexes (CPIs) for food and shelter. These CPIs are listed in the table for the years 2005–2014.[2]

Year	2005	2006	2007	2008	2009
Food	107.0	108.9	111.8	115.7	121.4
Shelter	106.4	113.1	116.9	122.0	121.6

Year	2010	2011	2012	2013	2014
Food	123.1	127.7	130.8	132.4	135.5
Shelter	123.3	125.6	127.1	128.7	132.2

a. Create side-by-side comparative bar charts to describe the CPIs over time.

b. Draw two line charts on the same set of axes to describe the CPIs over time.

c. What conclusions can you draw using the two graphs in parts a and b? Which is the most effective?

Data set EX0307 **3.7 How Big Is the Household?** A local chamber of commerce surveyed 126 households in its city and recorded the type of residence and the number of family members in each of the households. The data are shown in the table:

	Type of Residence		
Family Members	Apartment	Duplex	Single Residence
1	8	10	2
2	15	4	14
3	9	5	24
4 or more	6	1	28

a. Use a side-by-side bar chart to compare the number of family members living in the three types of residences.

b. Use a stacked bar chart to compare the number of family members living in the three types of residences.

c. What conclusions can you draw using the graphs in parts a and b?

Data set EX0308 **3.8 Charitable Contributions** The amount of charitable donations reported by tax filers increased in 2013 over the previous year, while the actual number of donors fell 1.0%. Total donations rose 3.5% to \$8.6 billion, with gains in every province and territory except the Northwest Territories, where donations were 2.7% lower. The largest increases were in Prince Edward Island (+7.5%), Manitoba (+6.0%), and Alberta (+5.9%).

Although the total number of donors decreased, there were modest increases in donor numbers in Alberta (+0.8%), Manitoba (+0.7%), Nunavut (+0.6%), and British Columbia (+0.2%). The largest decrease in donors was in the Northwest Territories (–3.9%). The following data reports the amount (\$ thousands) for the taxation year 2012 and 2013 for eight provinces.[3]

	Amounts ($ thousands)		
Province	2012	2013	Total
Nova Scotia	173,805	177,985	351,790
New Brunswick	143,065	148,890	291,955
Quebec	858,020	883,675	1,741,695
Ontario	3,688,015	3,763,040	7,451,055
Manitoba	378,205	400,760	778,965
Saskatchewan	300,515	303,975	604,490
Alberta	1,389,605	1,472,095	2,861,700
British Columbia	1,250,720	1,319,435	2,570,155
Total	8,181,950	8,469,855	16,651,805

a. Construct a stacked bar chart to display the total donation amount provincewide given in the table.

b. Construct two comparative pie charts to display the donation amount provincewide given in the table for the years 2012 and 2013, respectively.

c. Write a short paragraph summarizing the information that can be gained by looking at these graphs. Which of the two types of comparative graphs is more effective?

Data set EX0309 **3.9 Facebook Fanatics, again** Not only is the Facebook social networking site growing rapidly in North America, but the composition of Facebook members depends on both age and gender. During a one-month period in early 2010, Facebook reported its growth by both age and gender, as shown in the table:[4]

	Growth (Number of Users)		
Age Category	Female	Male	Total
13–17	270,900	121,280	392,180
18–25	445,920	653,060	1,098,980
26–34	570,920	154,600	725,520
35–44	365,740	305,260	671,000
45–54	90,240	36,680	126,920
55–65	64,960	83,480	148,440
Total	1,808,680	1,354,360	3,163,040

Source: © Inside Network, Inc. Republished with Permission from http://www.InsideFacebook.com. 2012. All Rights Reserved.

a. Construct a stacked bar chart to display the Facebook growth given in the table.

b. Construct two comparative pie charts to display the Facebook growth given in the table.

c. Write a short paragraph summarizing the information that can be gained by looking at these graphs. Which of the two types of comparative graphs is more effective?

3.3 SCATTERPLOTS FOR TWO QUANTITATIVE VARIABLES

ONLINE APPLET
Building a Scatterplot

When both variables to be displayed on a graph are *quantitative,* one variable is plotted along the horizontal axis and the second along the vertical axis. The first variable is often called *X* and the second is called *Y,* so that the graph takes the form of a plot on the (*x, y*) axes, which is familiar to most of you. Each pair of data values is plotted as a point on this two-dimensional graph, called a **scatterplot.** It is the two-dimensional extension of the dotplot we used to graph one quantitative variable in Section 1.4.

You can describe the relationship between two variables, *X* and *Y,* using the patterns shown in the scatterplot.

- **What type of pattern do you see?** Is there a constant upward or downward trend that follows a straight-line pattern? Is there a curved pattern? Is there no pattern at all, but just a random scattering of points?
- **How strong is the pattern?** Do all of the points follow the pattern exactly, or is the relationship only weakly visible?
- **Are there any unusual observations?** An outlier is a point that is far from the cluster of the remaining points. Do the points cluster into groups? If so, is there an explanation for the observed groupings?

EXAMPLE 3.3

The number of household members, *X,* and the amount spent on groceries per week, *Y,* are measured for six households in a local area. Draw a scatterplot of these six data points:

x	2	2	3	4	1	5
y ($)	45.75	60.19	68.33	100.92	35.86	130.62

SOLUTION Label the horizontal axis *X* and the vertical axis *Y.* Plot the points using the coordinates (*x, y*) for each of the six pairs. The scatterplot in Figure 3.4 shows the six pairs marked as dots. You can see a pattern even with only six data pairs. The cost of weekly groceries increases with the number of household members in an apparent straight-line relationship.

Suppose you found that a seventh household with two members spent $115 on groceries. This observation is shown as an X in Figure 3.4. It does not fit the linear pattern of the other six observations and is classified as an outlier. Possibly these two people were having a party the week of the survey!

FIGURE 3.4
Scatterplot for Example 3.3

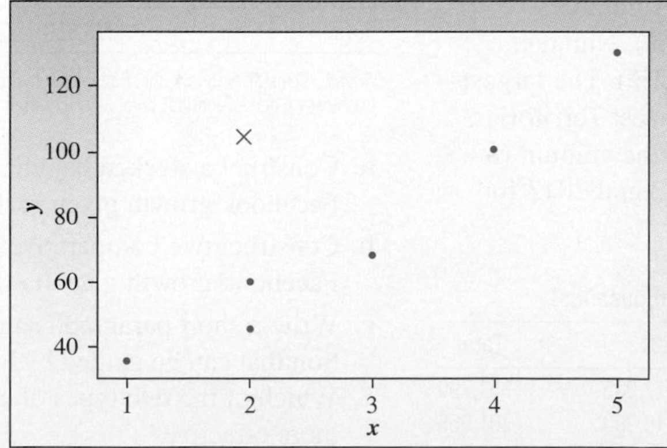

EXAMPLE 3.4

A distributor of table wines conducted a study of the relationship between price and demand using a type of wine that ordinarily sells for $10 per bottle. He sold this wine in 10 different marketing areas over a 12-month period, using five different price levels—from $10 to $14. The data are given in Table 3.4. Construct a scatterplot for the data, and use the graph to describe the relationship between price and demand.

TABLE 3.4 **Cases of Wine Sold at Five Price Levels**

Cases Sold per 10,000 Population	Price per Bottle ($)
23, 21	10
19, 18	11
15, 17	12
19, 20	13
25, 24	14

SOLUTION The 10 data points are plotted in Figure 3.5. As the price increases from $10 to $12 the demand decreases. However, as the price continues to increase, from $12 to $14, the demand begins to *increase.* The data show a curved pattern, with the relationship changing as the price changes. How do you explain this relationship? Possibly, the increased price is a signal of increased quality for the consumer, which causes the increase in demand once the cost exceeds $12. You might be able to think of other reasons, or perhaps some other variable, such as the income of people in the marketing areas, that may be causing the change.

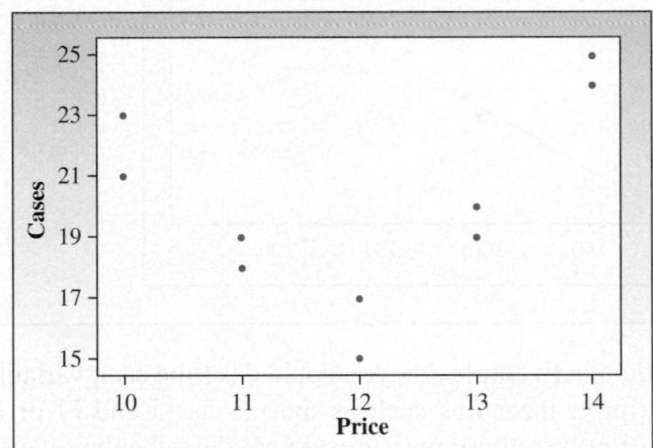

FIGURE 3.5 Scatterplot for Example 3.4

3.4 NUMERICAL MEASURES FOR QUANTITATIVE BIVARIATE DATA

A constant rate of increase or decrease is perhaps the most common pattern found in bivariate scatterplots. The scatterplot in Figure 3.4 exhibits this *linear* pattern—that is, a straight line with the data points lying both above and below the line and within a fixed distance from the line. When this is the case, we say that the two variables exhibit a *linear relationship.*

EXAMPLE 3.5

The data in Table 3.5 are the size of the living area (in m^2), *X*, and the selling price, *Y*, of 12 residential properties. The *MINITAB* scatterplot in Figure 3.6 shows a linear pattern in the data.

TABLE 3.5 **Living Area and Selling Price of 12 Properties**

Residence	x (m²)	y ($ thousands)
1	126.3	178.5
2	180.2	275.7
3	162.6	239.5
4	144.0	229.8
5	166.3	195.6
6	162.6	210.3
7	207.2	360.5
8	148.6	205.2
9	134.7	188.6
10	173.7	265.7
11	205.3	325.3
12	137.5	168.8

FIGURE 3.6

Scatterplot of *x* versus *y* for Example 3.5

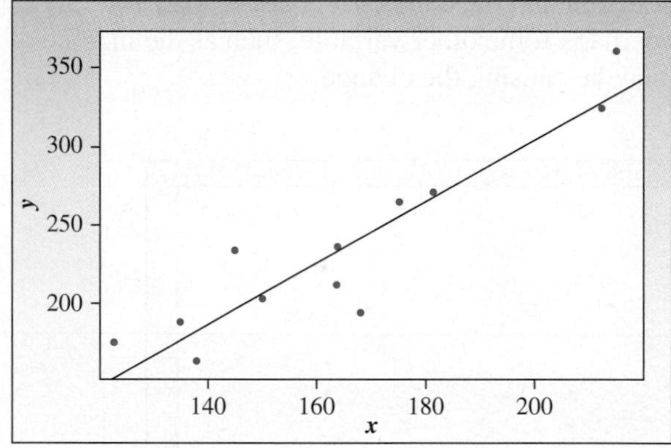

For the data in Example 3.5, you could describe each variable, *X* and *Y*, individually using descriptive measures such as the means ($\bar{x}$ and $\bar{y}$) or the standard deviations (s_x and s_y). However, these measures do not describe the relationship between *X* and *Y* for a particular residence—that is, how the size of the living space affects the selling price of the home. A simple measure that serves this purpose is called the **correlation coefficient,** denoted by *r*, and is defined as

$$r = \frac{s_{xy}}{s_x s_y}$$

The quantities s_x and s_y are the standard deviations for the variables *X* and *Y*, respectively, which can be found by using the statistics function on your calculator or the computing formula in Section 2.3. The new quantity s_{xy} is called the **covariance** between *X* and *Y* and is defined as

$$s_{xy} = \frac{\Sigma(x_i - \bar{x})(y_i - \bar{y})}{n - 1}$$

There is also a computing formula for the covariance:

$$s_{xy} = \frac{\Sigma x_i y_i - \frac{(\Sigma x_i)(\Sigma y_i)}{n}}{n - 1}$$

where $\Sigma x_i y_i$ is the sum of the products $x_i y_i$ for each of the n pairs of measurements. How does this quantity detect and measure a linear pattern in the data?

Look at the signs of the cross-products $(x_i - \bar{x})(y_i - \bar{y})$ in the numerator of r, or s_{xy}. When a data point (x, y) is in either area I or III in the scatterplot shown in Figure 3.7, the cross-product will be positive; when a data point is in area II or IV, the cross-product will be negative. We can draw these conclusions:

- If most of the points are in areas I and III (forming a positive pattern), s_{xy} and r will be positive.
- If most of the points are in areas II and IV (forming a negative pattern), s_{xy} and r will be negative.
- If the points are scattered across all four areas (forming *no* pattern), s_{xy} and r will be close to 0.

FIGURE 3.7

The signs of the cross-products $(x_i - \bar{x})(y_i - \bar{y})$ in the covariance formula

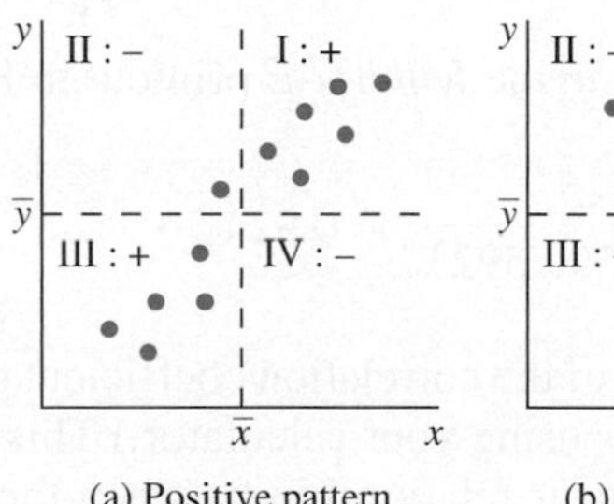

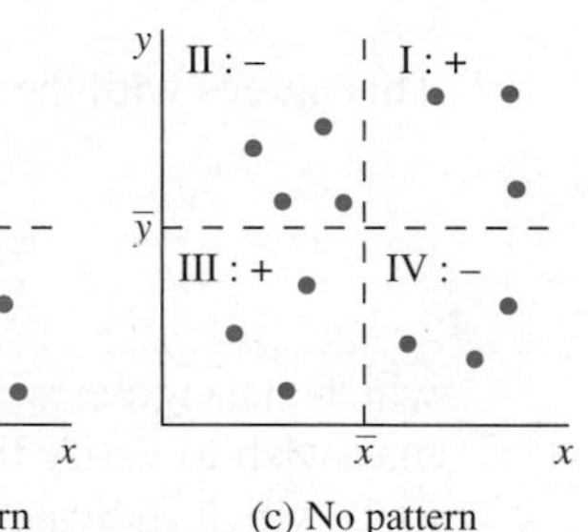

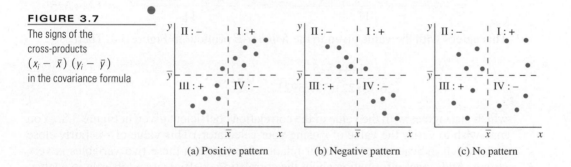

(a) Positive pattern (b) Negative pattern (c) No pattern

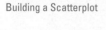

ONLINE APPLET

Building a Scatterplot

Most scientific and graphics calculators can compute the correlation coefficient, r, when the data are entered in the proper way. Check your calculator manual for the proper sequence of entry commands. Computer programs such as *MINITAB* are also programmed to perform these calculations. The *MINITAB* output in Figure 3.8 shows the covariance and correlation coefficient for x and y in Example 3.5. In the covariance table, you will find these values:

$$s_{xy} = 1444.525 \quad s_x^2 = 684.25 \quad s_y^2 = 3571.16$$

and in the correlation output, you find $r = 0.924$.

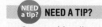

NEED A TIP?

$r > 0 \Leftrightarrow$ positive *linear* relationship.

$r < 0 \Leftrightarrow$ negative *linear* relationship.

$r \approx 0 \Leftrightarrow$ no relationship.

However you decide to calculate the correlation coefficient, it can be shown that the value of r always lies between -1 and 1. When r is positive, X increases when Y increases, and vice versa. When r is negative, X decreases when Y increases, or X increases when Y decreases. When r takes the value 1 or -1, all the points lie exactly on a straight line. If $r = 0$, then there is no apparent linear relationship between the two variables. The closer the value of r is to 1 or -1, the stronger the linear relationship between the two variables.

FIGURE 3.8
MINITAB output of covariance and *Excel* correlation output for Example 3.5

Covariances: x, y

	x	y
x	684.25	
y	1444.525	3571.16

Correlations: x, y

	x	y
x	1	
y	0.92414	1

EXAMPLE 3.6

Find the correlation coefficient for the number of square metres of living area and the selling price of a home for the data in Example 3.5.

SOLUTION Three quantities are needed to calculate the correlation coefficient. The standard deviations of the *x* and *y* variables are found using a calculator with a statistical function. You can verify that $s_x = 26.15822$ and $s_y = 59.7592$. Finally,

$$s_{xy} = \frac{\Sigma x_i y_i - \frac{(\Sigma x_i)(\Sigma y_i)}{n}}{n-1}$$

$$= \frac{477721.6 - \frac{(1949)(2843.5)}{12}}{11} = \frac{477721.6 - 461831.8}{11} = 1444.525$$

This agrees with the value given in the *MINITAB* printout in Figure 3.8. Then

$$r = \frac{s_{xy}}{s_x s_y} = \frac{1444.525}{(26.15822)(59.7592)} = 0.924$$

which also agrees with the value of the correlation coefficient given in Figure 3.8. (You may wish to verify the value of *r* using your calculator.) This value of *r* is fairly close to 1, which indicates that the linear relationship between these two variables is very strong. Additional information about the correlation coefficient and its role in analyzing linear relationships, along with alternative calculation formulas, can be found in Chapter 12.

NEED A TIP?

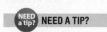

x "explains" *y* or *y* "depends on" *x*.

x is the **explan**atory or independent variable.

y is the response or **depend**ent variable.

Sometimes the two variables, *X* and *Y*, are related in a particular way. It may be that the value of *Y* depends on the value of *X*; that is, the value of *X* in some way explains the value of *Y*. For example, the cost of a home (*Y*) may *depend* on its amount of floor space (*X*); a student's grade point average (*X*) may *explain* her score on an achievement test (*Y*). In these situations, we call *Y* the **dependent variable,** while *X* is called the **independent variable.**

If one of the two variables can be classified as the dependent variable *Y* and the other as *X*, and if the data exhibit a straight-line pattern, it is possible to describe the relationship relating *Y* to *X* using a straight line given by the equation

$$Y = a + bX$$

as shown in Figure 3.9.

FIGURE 3.9
The graph of a straight line

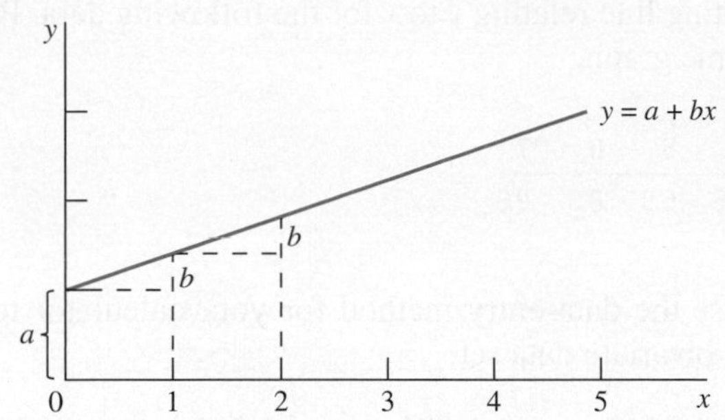

As you can see, a is where the line crosses or intersects the y-axis: a is called the *y-intercept*. You can also see that for every one-unit increase in x, y increases by an amount b. The quantity b determines whether the line is increasing ($b > 0$), decreasing ($b < 0$), or horizontal ($b = 0$) and is appropriately called the **slope** of the line.

ONLINE APPLET
How a Line Works

When plotting the (x, y) points for two variables x and y, the points generally do not fall exactly on a straight line, but they may show a trend that could be described as a linear pattern. We can describe this trend by fitting a line as best we can through the points. This best-fitting line relating Y to X, often called the **regression** or **least-squares line,** is found by minimizing the sum of the squared differences between the data points and the line itself, as shown in Figure 3.10. The formulas for computing b and a, which are derived mathematically, follow.

COMPUTING FORMULAS FOR THE LEAST-SQUARES REGRESSION LINE

$$b = r\left(\frac{s_y}{s_x}\right) \quad \text{and} \quad a = \bar{y} - b\bar{x}$$

and the least-squares regression line is: $Y = a + bX$

FIGURE 3.10
The best-fitting line

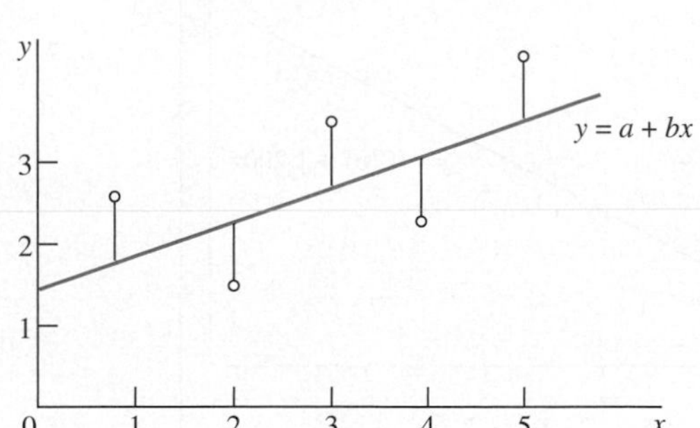

Since s_x and s_y are both positive, b and r have the same sign, so that:

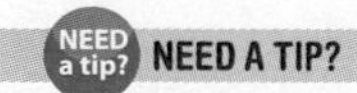
NEED A TIP?
Remember that r and b have the same sign!

- When r is positive, so is b, and the line is increasing with x.
- When r is negative, so is b, and the line is decreasing with x.
- When r is close to 0, then b is close to 0.

EXAMPLE 3.7 Find the best-fitting line relating y to x for the following data. Plot the line and the data points on the same graph.

x	2	3	4	5	6	7
y	3.0	5.0	5.5	6.0	8.0	9.5

SOLUTION Use the data entry method for your calculator to find these descriptive statistics for the bivariate data set:

NEED A TIP?

Use the regression line to predict y for a given value of x.

$$\bar{x} = 4.5 \quad \bar{y} = 6.167 \quad s_x = 1.871 \quad s_y = 2.295 \quad r = 0.978$$

Then

$$b = r\left(\frac{s_y}{s_x}\right) = 0.978\left(\frac{2.295}{1.871}\right) = 1.1996311 \cong 1.200$$

and

$$a = \bar{y} - b\bar{x} = 6.167 - 1.200(4.5) = 6.167 - 5.4 = 0.767$$

Therefore, the best-fitting line is $y = 0.767 + 1.200x$. The plot of the regression line and the actual data points are shown in Figure 3.11.

The best-fitting line can be used to estimate or predict the value of the variable Y when the value of X is known. For example, if the value $x = 3$ was observed at some time in the future, what would you predict for the value of y? From the best-fitting line in Figure 3.11, the best estimate would be

$$y = a + bx = 0.767 + 1.200(3) = 4.367$$

FIGURE 3.11

Fitted line and data points for Example 3.7

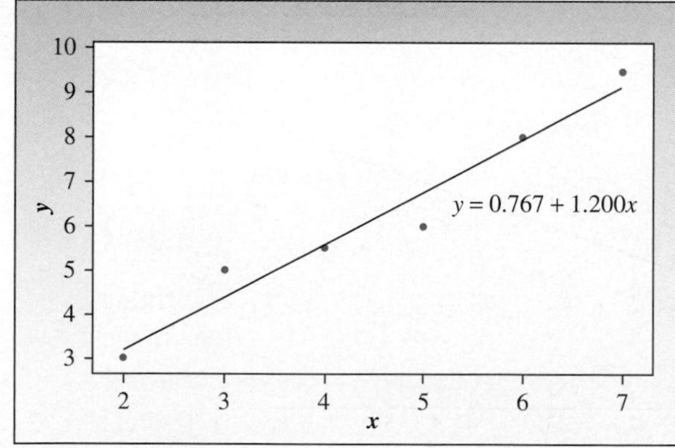

NEED TO KNOW

How to Calculate the Correlation Coefficient

1. First, create a table or use your calculator to find Σx, Σy, and Σxy.
2. Calculate the covariance, s_{xy}.
3. Use your calculator or the computing formula from Chapter 2 to calculate s_x and s_y.
4. Calculate $r = \dfrac{s_{xy}}{s_x s_y}$.

How to Calculate the Regression Line

1. First, calculate $\bar{y}$ and $\bar{x}$. Then, calculate $r = \dfrac{s_{xy}}{s_x s_y}$.
2. Find the slope, $b = r\left(\dfrac{s_y}{s_x}\right)$ and the y-intercept, $a = \bar{y} - b\bar{x}$.
3. Write the regression line by substituting the values for a and b into the equation: $Y = a + bX$.

When should you describe the linear relationship between x and y using the correlation coefficient r, and when should you use the regression line $y = a + bx$? The regression approach is used when the values of x are set in advance and then the corresponding value of y is measured. The correlation approach is used when an experimental unit is selected at random and then measurements are made on both variables x and y. This technical point will be taken up in Chapter 12 on regression analysis.

Most data analysts begin any data-based investigation by examining plots of the variables involved. If the relationship between two variables is of interest, bivariate plots are also explored in conjunction with numerical measures of location, dispersion, and correlation. Graphs and numerical descriptive measures are only the first of many statistical tools you will soon have at your disposal.

3.4 EXERCISES

BASIC TECHNIQUES

EX0310

3.10 A set of bivariate data consists of these measurements on two variables, x and y:

(3, 6) (5, 8) (2, 6) (1, 4) (4, 7) (4, 6)

a. Draw a scatterplot to describe the data.

b. Does there appear to be a relationship between x and y? If so, how do you describe it?

c. Calculate the correlation coefficient, r, using the computing formula given in this section.

d. Find the best-fitting line using the computing formulas. Graph the line on the scatterplot from part a. Does the line pass through the middle of the points?

3.11 Refer to Exercise 3.10.

a. Use the data entry method in your scientific calculator to enter the six pairs of measurements. Recall the proper memories to find the correlation coefficient, r, the y-intercept, a, and the slope, b, of the line.

b. Verify that the calculator provides the same values for r, a, and b as in Exercise 3.10.

EX0312

3.12 Consider this set of bivariate data:

x	1	2	3	4	5	6
y	5.6	4.6	4.5	3.7	3.2	2.7

a. Draw a scatterplot to describe the data.

b. Does there appear to be a relationship between x and y? If so, how do you describe it?

c. Calculate the correlation coefficient, r. Does the value of r confirm your conclusions in part b? Explain.

Data set EX0313

3.13 The value of a quantitative variable is measured once a year for a 10-year period:

Year	Measurement	Year	Measurement
1	61.5	6	58.2
2	62.3	7	57.5
3	60.7	8	57.5
4	59.8	9	56.1
5	58.0	10	56.0

a. Draw a scatterplot to describe the variable as it changes over time.

b. Describe the measurements using the graph constructed in part a.

c. Use this *MINITAB* output to calculate the correlation coefficient, r:

MINITAB output for Exercise 3.13

Covariances

```
        x         y
x    9.16667
y   -6.42222   4.84933
```

d. Find the best-fitting line using the results of part c. Verify your answer using the data entry method in your calculator.

e. Plot the best-fitting line on your scatterplot from part a. Describe the fit of the line.

APPLICATIONS

Data set EX0314

3.14 Grocery Costs These data relating the amount spent on groceries per week and the number of household members are from Example 3.3:

x	2	2	3	4	1	5
y	\$45.75	\$60.19	\$68.33	\$100.92	\$35.86	\$130.62

a. Find the best-fitting line for these data.

b. Plot the points and the best-fitting line on the same graph. Does the line summarize the information in the data points?

c. What would you estimate a household of six to spend on groceries per week? Should you use the fitted line to estimate this amount? Why or why not?

Data set EX0315

3.15 Real Estate Prices The data relating to the average monthly rate in Canada for a bachelor unit as described under the Housing Market Indicators, Canada, Provinces, and Metropolitan Areas, 1990 – 2013 by Canada Mortgage and Housing Corporation is shown below.[5] First, find the best-fitting line that describes these data, and then plot the line and the data points on the same graph. Comment on the goodness of the fitted line in describing the selling price of a residential property as a linear function of the average monthly rent over the years.

x (year)	y (\$ thousands)	x (year)	y (\$ thousands)
1990	376	2002	504
1991	388	2003	516
1992	389	2004	523
1993	398	2005	529
1994	400	2006	547
1995	409	2007	563
1996	413	2008	582
1997	420	2009	594
1998	432	2010	607
1999	448	2011	636
2000	469	2012	639
2001	490	2013	659

Data set EX0316

3.16 Students with Disabilities A social skills training program, reported in *Psychology in the Schools*, was implemented for seven students with mild disabilities in a study to determine whether the program caused improvement in pre/post measures and behaviour ratings.[6] For one such test, these are the pretest and posttest scores for the seven students:

Student	Pretest	Posttest
Earl	101	113
Ned	89	89
Jasper	112	121
Charlie	105	99
Tom	90	104
Susie	91	94
Lori	89	99

Source: Gregory K. Torrey, S.F. Vasa, J.W. Maag, and J.J. Kramer, "Social Skills Interventions Across School Settings: Case Study Reviews of Students with Mild Disabilities," *Psychology in the Schools* 29 (July 1992):248. Republished with permission of John Wiley & Sons, Inc.; permission conveyed through Copyright Clearance Center, Inc.

a. Draw a scatterplot relating the posttest score to the pretest score.

b. Describe the relationship between pretest and posttest scores using the graph in part a. Do you see any trend?

c. Calculate the correlation coefficient and interpret its value. Does it reinforce any relationship that was apparent from the scatterplot? Explain.

Data set EX0317 **3.17 Japanese Automakers in 2005** In 2005, total global production of motor vehicles was about 64 million units. The U.S. was the single largest producer with a 21% share of global output. Japan was second with 18% and Canada was the eighth largest producer with a 4% share. The Japanese overseas production by Japanese automakers steadily increased for a decade. The following table provides the production numbers at North American plants:[7]

Year	1996	1997	1998	1999	2000
Production (thousands)	2641	2665	2674	2797	2991
Year	2001	2002	2003	2004	2005
Production (thousands)	3062	3375	3487	3841	4081

a. Plot the data using a scatterplot. How would you describe the relationship between year and amount of production?

b. Find the least-squares regression line relating the production to the year being measured.

c. If you were to predict the amount of production in the year 2015, what problems might arise with your prediction?

Data set EX0318 **3.18 LED TVs, again** As technology improves, the choice of televisions becomes more complicated. Should you choose an LCD TV, an LED TV, or a plasma TV? Does the price of an LED TV depend on the size of the screen? In the table below, *Best Buy*[8] gives the prices and screen sizes for the top 10 LED TVs in the 30-inch and higher categories, as of July 2015:

Brand	Price ($)	Size
LG UltraWide	649.99	34
Dell UltraSharp	1349.99	34
LG Digital Cinema	1299.99	31
BenQ Widescreen 32	999.99	32
BenQ Widescreen 55	6499.99	55
LG 1080p	599.95	47
HP Z30i	1429.99	30
LG Widescreen Commercial-Grade	1599.95	47
Samsung Ultra	1749.99	31.5
HP Envy	499.99	32

a. Which of the two variables (price and size) is the independent variable, and which is the dependent variable?

b. Construct a scatterplot for the data. Does the relationship appear to be linear?

3.19 LED TVs, continued Refer to Exercise 3.18. Suppose we assume that the relationship between x and y is linear.

a. Find the correlation coefficient, r. What does this value tell you about the strength and direction of the relationship between size and price?

b. Refer to part a. Would it be reasonable to construct a regression line used to predict the price of an LED TV based on the size of the screen?

Data set EX0320 **3.20 Chirping Crickets** Male crickets chirp by rubbing their front wings together, and their chirping is temperature dependent. Crickets chirp faster with increasing temperature and slower with decreasing temperature. The table below shows the number of chirps per second for a cricket, recorded at 10 different temperatures:

Chirps per second	20	16	19	18	18	16	14	17	15	16
Temperature (°C)	31	23	33	30	28	24	20	27	21	29

a. Which of the two variables (temperature and number of chirps) is the independent variable, and which is the dependent variable?

b. Plot the data using a scatterplot. How would you describe the relationship between temperature and number of chirps?

c. Find the least-squares line relating the number of chirps to the temperature.

d. If a cricket is monitored at a temperature of 26°C, what would you predict his number of chirps would be?

CHAPTER REVIEW

Key Concepts

I. Bivariate Data

1. Both qualitative and quantitative variables
2. Describing each variable separately
3. Describing the relationship between the two variables

II. Describing Two Qualitative Variables

1. Side-by-side pie charts
2. Comparative line charts
3. Comparative bar charts
   a. Side-by-side
   b. Stacked
4. Relative frequencies to describe the relationship between the two variables

III. Describing Two Quantitative Variables

1. Scatterplots
   a. Linear or non-linear pattern
   b. Strength of relationship
   c. Unusual observations: clusters and outliers
2. Covariance and correlation coefficient
3. The best-fitting regression line
   a. Calculating the slope and y-intercept
   b. Graphing the line
   c. Using the line for prediction

TECHNOLOGY TODAY

Describing Bivariate Data in *Microsoft Excel*

Excel provides different graphical techniques for *qualitative* and *quantitative* bivariate data, as well as commands for obtaining bivariate descriptive measures when the data are quantitative.

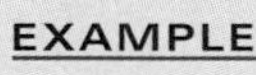

EXAMPLE 3.8

Comparative Line and Bar Charts Suppose that the 105 students whose status was tabulated in Example 1.15 were from the University of Windsor (UW), and that another 100 students from an introductory statistics class at the University of New Brunswick (UNB) were also interviewed. Table 3.6 shows the status distribution for both sets of students.

TABLE 3.6 Status of Students in a Statistics Class at UW and UNB

	First Year	Second Year	Third Year	Fourth Year	Grad Student
Frequency (UW)	5	23	32	35	10
Frequency (UNB)	10	35	24	25	6

1. Enter the data into an *Excel* spreadsheet *just as it appears in the table, including the labels.* Highlight the data in the spreadsheet, click the **Insert** tab and select **Line** in the **Charts** group. In the drop-down list, you will see a variety of styles to choose from. Select the first option to produce the line chart.

2. ***Editing the line chart***: Again, you can experiment with the various options in the **Chart Layout** and **Chart Styles** groups to change the look of the chart. We have chosen a design that allows a title on the vertical axis; we have added the title and have changed the "line style" of the UW students to a "dashed" style, by double-clicking on that line. The line chart is shown in Figure 3.12(a).

FIGURE 3.12

(a) Status of Statistics Students (line graph)
(b) Status of Statistics Students (bar chart)

(a)

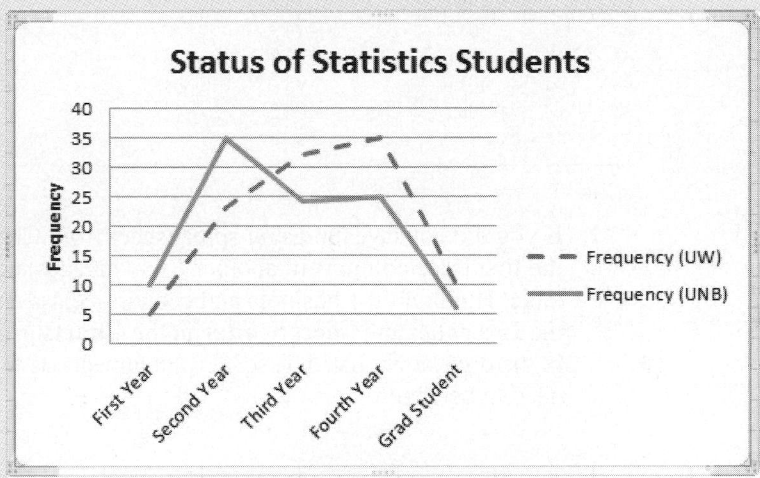

(b)

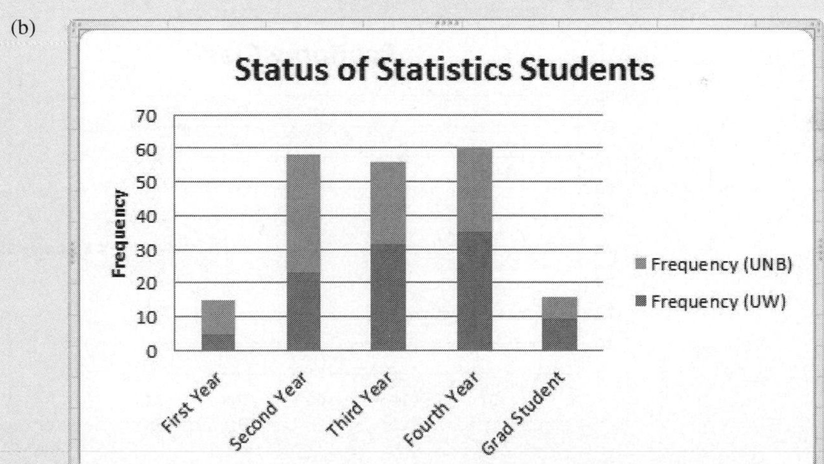

3. Once the line chart has been created, right-click on the chart area and select **Change Chart Type.** Then choose either **Stacked Column** or **Clustered Column.** The comparative bar chart (a stacked bar chart), with the same editing that you chose for the line chart, will appear as shown in Figure 3.12(b).

Scatterplots, Correlation, and the Regression Line The data from Example 2.17 give the seat pitch (in centimetres) for business and economy classes for selected commercial airlines around the globe as shown in Table 3.7:

TABLE 3.7 Leg Room and Seat Pitch on Airplanes

Airline	Business Class	Economy Class
A1	104.1	72.4
A2	105.4	76.2
A3	101.6	64.8
A4	104.1	69.9
A5	109.2	78.7
A6	104.1	74.9
A7	108.0	74.9
A8	101.6	73.7
A9	106.7	76.2

1. If you did not save the *Excel* spreadsheet from Chapter 2, enter the data into the first three columns of another *Excel* spreadsheet, using the labels in the table. Highlight the business and economy class data (columns B and C), click the **Insert** tab and select **Scatter** in the **Charts** group, and select the first option in the drop-down list. The scatterplot appears as in Figure 3.13(a), and will need to be edited!

FIGURE 3.13
(a) Economy Class
(b) Scatterplot of Business vs. Economy Class

(a)

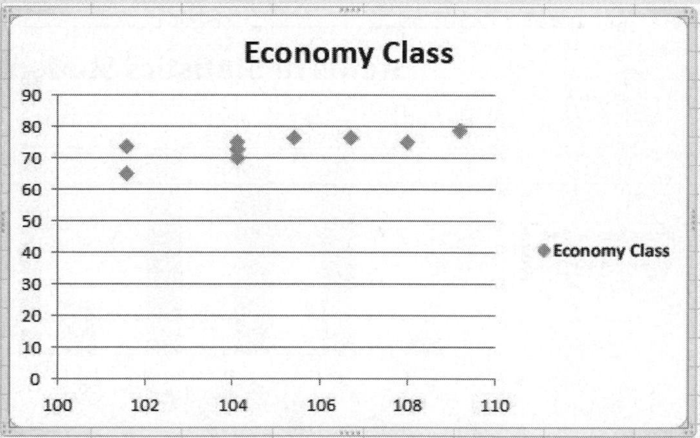

(b)

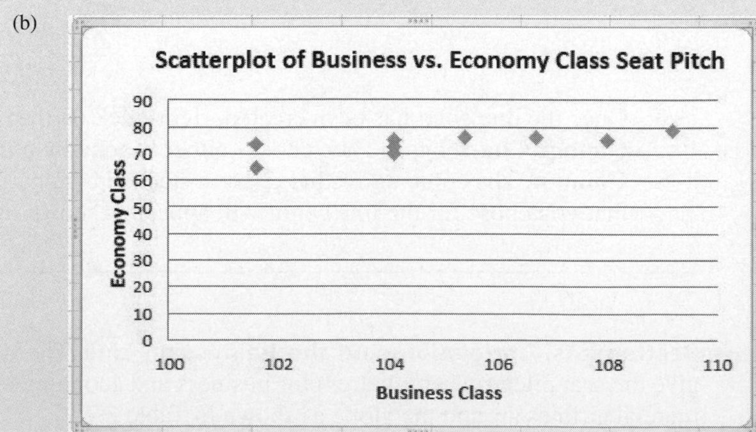

2. ***Editing the scatterplot***: With the scatterplot selected, look in the drop-down list in the **Chart Layouts** group. Find a layout that allows titles on both axes (we chose layout 1) and select it. Label the axes, remove the "legend entry," and retitle the chart as "**Scatterplot of Business vs. Economy Class Seat Pitch**." The scatterplot now appears in Figure 3.13(b). The plot is still not optimal, since *Excel* chooses to use zero as the lower limit of the vertical scale, causing the points to cluster at the top of the plot. To adjust this, double-click on the vertical axis. In the **Format Axis** dialogue box, change the **Minimum** to **Fixed**, type **63** in the box, and click **Close**. (You can make a similar adjustment to the horizontal axis if needed.)
3. To plot the best-fitting line, simply right-click on one of the data points and select **Add Trendline**. In the dialogue box that opens, make sure that the radio button marked "Linear" is selected, and check the boxes marked "Display Equation on Chart" and "Display R-squared value on Chart." The final scatterplot is shown in Figure 3.14.

FIGURE 3.14

Scatterplot of Business vs. Economy Class Seat Pitch

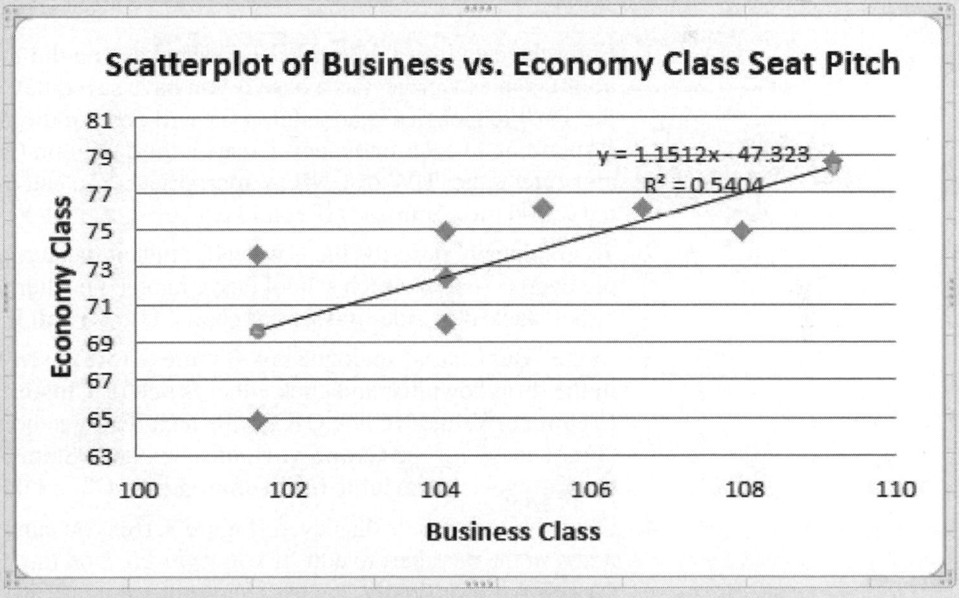

4. To find the sample correlation coefficient, *r*, you can use the command **Data ▶ Data Analysis ▶ Correlation**, selecting the two appropriate columns for the Input Range, clicking "Labels in First Row," and selecting an appropriate Output Range. When you click **OK**, the correlation matrix will appear in the spreadsheet.
5. *(ALTERNATE PROCEDURE)* You can also place your cursor in the cell in which you want the correlation coefficient to appear. Select **Formulas ▶ More Functions ▶ Statistical ▶ CORREL** or click the "Insert Function" icon at the top of the spreadsheet, choosing **CORREL** from the **Statistical** category. Highlight or type the cell ranges for the two variables in the boxes marked "Array 1" and "Array 2" and click **OK**. For our example, the value is $r = 0.735$.

Describing Bivariate Data in *MINITAB*

MINITAB provides different graphical techniques for *qualitative* and *quantitative* bivariate data, as well as commands for obtaining bivariate descriptive measures when the data are quantitative.

Comparative Line and Bar Charts Suppose that the 105 students whose status was tabulated in Example 1.18 were from the University of Windsor (UW), and that another 100 students from an introductory statistics class at the University of New Brunswick (UNB) were also interviewed. Table 3.8 shows the status distribution for both sets of students.

TABLE 3.8 Status of Students in a Statistics Class at UW and UNB

	First Year	Second Year	Third Year	Fourth Year	Grad Student
Frequency (UW)	5	23	32	35	10
Frequency (UNB)	10	35	24	25	6

1. Enter the data into a *MINITAB* worksheet as you did in Exercise 1.18, using your Chapter 1 project as a base if you have saved it. Column C1 will contain the 10 "Frequencies" and column C2 will contain the student "Status" corresponding to each frequency. Create a third column C3 called "University," and enter either **UW** or **UNB** as appropriate. You can use the familiar Windows cut-and-paste commands if you like.
2. To graphically describe the UW/UNB student data, you can use comparative pie charts—one for each school (see Chapter 1). Alternatively, you can use either stacked or side-by-side bar charts. Use **Graph ▶ Bar Chart.**
3. In the "Bar Charts" dialogue box (Figure 3.15(a)), select **Values from a Table** in the drop-down list and click either **Stack** or **Cluster** in the row marked "One Column of Values." Click **OK.** In the next dialogue box (Figure 3.15(b)), select "Frequency" for the **Graph variables** box and "Status" and "University" for the **Categorical variable for grouping** box. Click **OK.**
4. Once the bar chart is displayed (Figure 3.16), you can *right-click* on various items in the bar chart to edit. If you *right-click* on the bars and select **Update Graph Automatically,** the bar chart will automatically update when you change the data in the *MINITAB* worksheet.

(a)

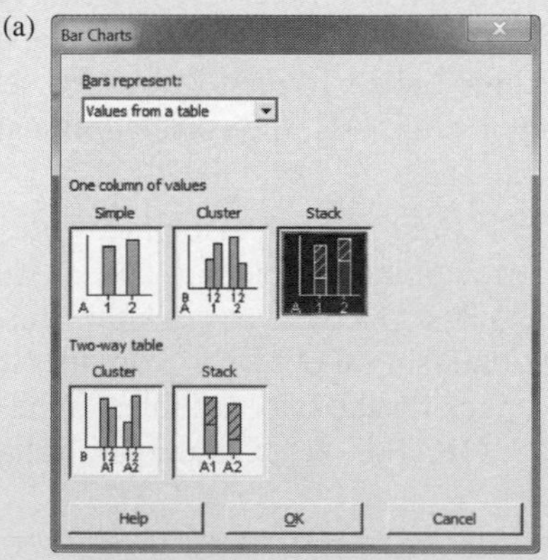

(b)

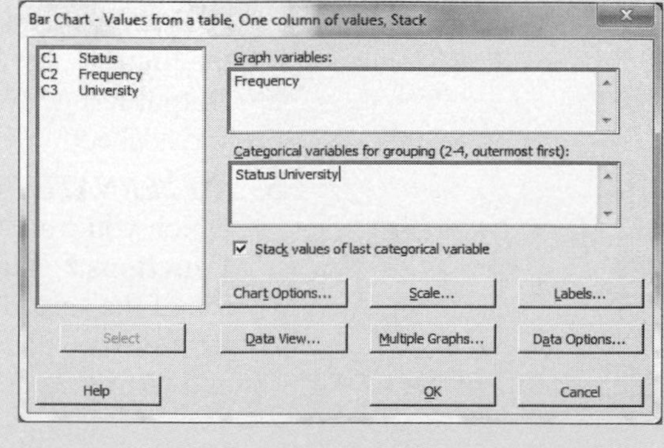

FIGURE 3.15
(a) MINITAB screen shot (Bar Charts)
(b) MINITAB screen shot (Bar Chart - Values from a table, One column of values, Stack)

FIGURE 3.16
MINITAB screen shot (Chart of Frequencey)

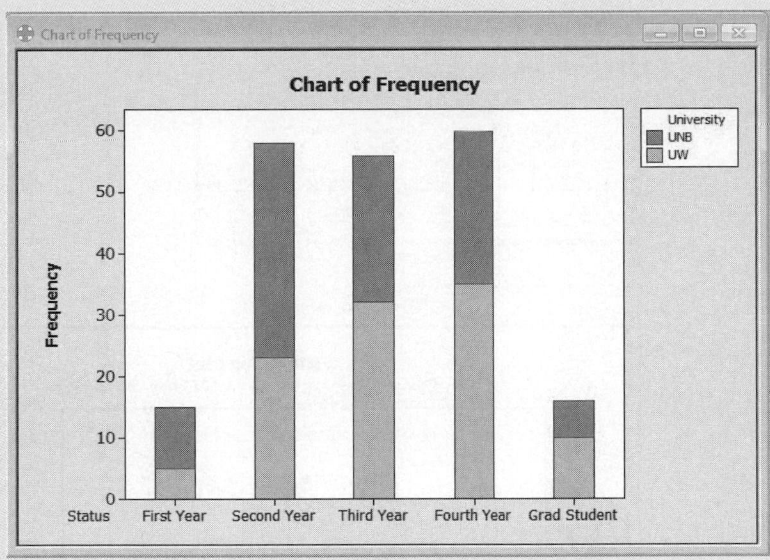

EXAMPLE 3.11

Scatterplots, Correlation, and the Regression Line The data from Example 2.18 give the seat pitch (in centimetres) for business and economy classes for selected commercial airlines around the globe:

TABLE 3.9 **Leg Room and Seat Pitch on Airplanes**

Airline	Business Class	Economy Class
A1	104.1	72.4
A2	105.4	76.2
A3	101.6	64.8
A4	104.1	69.9
A5	109.2	78.7
A6	104.1	74.9
A7	108.0	74.9
A8	101.6	73.7
A9	106.7	76.2

1. If you did not save the *MINITAB* worksheet from Chapter 2, enter the data into the first three columns of another *MINITAB* worksheet, using the labels in the table. To examine the relationship between business class and economy class, you can plot the data and numerically describe the relationship with the correlation coefficient and the best-fitting line.
2. Select **Stat ▶ Regression ▶ Fitted Line Plot**, and select "Business Class" and "Economy Class" for **Y** and **X**, respectively (see Figure 3.17(a)). Make sure that the radio button next to **Linear** is selected, and click **OK.** The plot of the nine data points and the best-fitting line will be generated as in Figure 3.17(b).

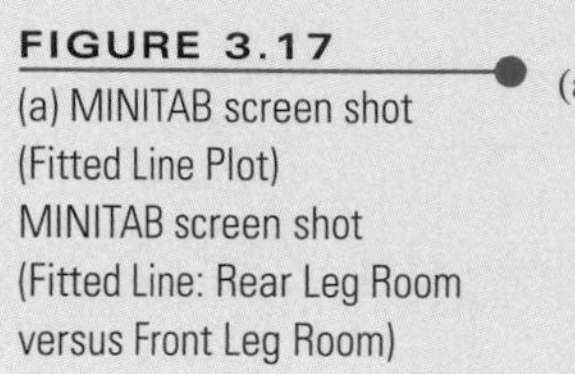

FIGURE 3.17
(a) MINITAB screen shot (Fitted Line Plot) MINITAB screen shot (Fitted Line: Rear Leg Room versus Front Leg Room)

(a)

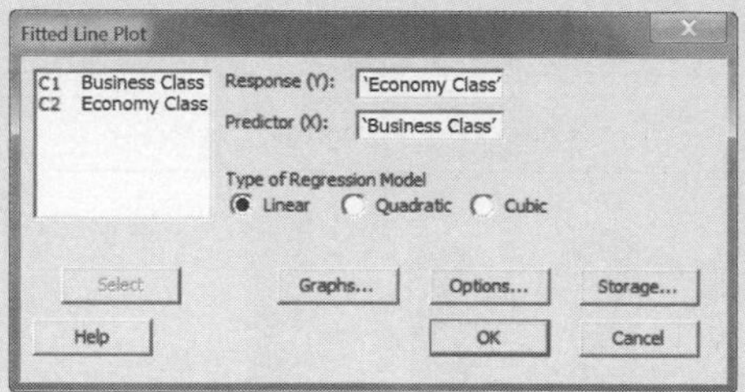

(b)

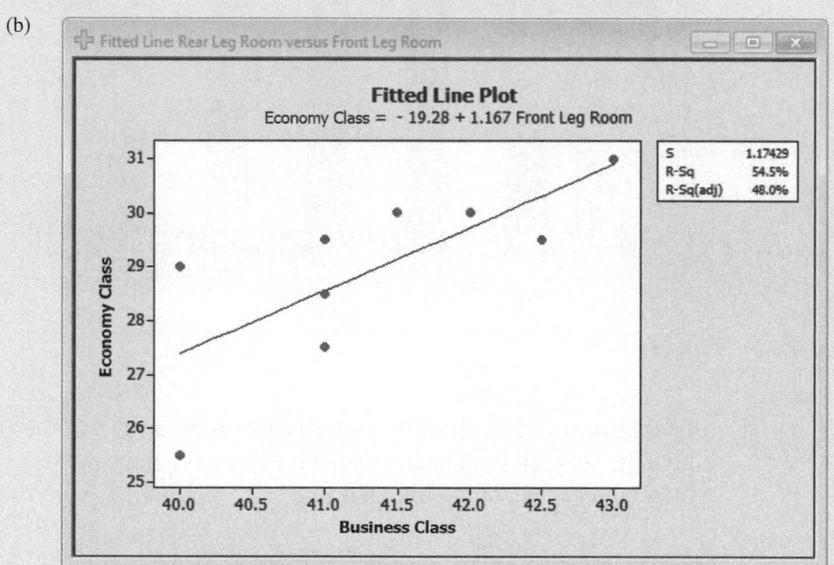

3. To calculate the correlation coefficient, use **Stat ▶ Basic Statistics ▶ Correlation**, selecting "Business Class" and "Economy Class" for the Variables box. To select both variables at once, hold the **Shift** key down as you highlight the variables and then click **Select.** Click **OK**, and the correlation coefficient will appear in the Session window (see Figure 3.18). Notice the relatively strong positive correlation and the positive slope of the regression line.

FIGURE 3.18
MINITAB sceen shot (Session)

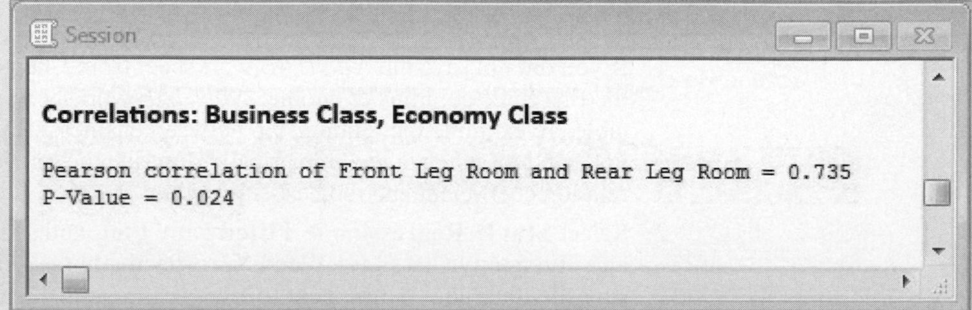

Supplementary Exercises

3.21 Professor Asimov Professor Isaac Asimov was one of the most prolific writers of all time. He wrote nearly 500 books during a 40-year career prior to his death in 1992. In fact, as his career progressed, he became even more productive in terms of the number of books written within a given period of time.[9] These data are the times (in months) required to write his books, in increments of 100:

Number of Books	100	200	300	400	490
Time (in months)	237	350	419	465	507

a. Plot the accumulated number of books as a function of time using a scatterplot.

b. Describe the productivity of Professor Asimov in light of the data set graphed in part a. Does the relationship between the two variables seem to be linear?

Data set EX0322 **3.22 Cheese, Please!** Health-conscious consumers often consult the nutritional information on food packages in an attempt to avoid foods with large amounts of fat, sodium, or cholesterol. The following information was taken from eight different brands of cheese slices:

Brand	Fat (g)	Saturated Fat (g)	Cholesterol (mg)	Sodium (mg)	Calories
Kraft Deluxe	7	4.5	20	340	80
Kraft Velveeta Slices	5	3.5	15	300	70
Private Selection	8	5.0	25	520	100
Ralphs Singles	4	2.5	15	340	60
Kraft 2% Milk Singles	3	2.0	10	320	50
Kraft Singles	5	3.5	15	290	70
Borden Singles	5	3.0	15	260	60
Lake to Lake	5	3.5	15	330	70

a. Which pairs of variables do you expect to be strongly related?

b. Draw a scatterplot for fat and saturated fat. Describe the relationship.

c. Draw a scatterplot for fat and calories. Compare the pattern to that found in part b.

d. Draw a scatterplot for fat versus sodium and another for cholesterol versus sodium. Compare the patterns. Are there any clusters or outliers?

e. For the pairs of variables that appear to be linearly related, calculate the correlation coefficients.

f. Write a paragraph to summarize the relationships you can see in these data. Use the correlations and the patterns in the four scatterplots to verify your conclusions.

3.23 Import Sale and Domestic Production in Japan The Japan Automobile Manufacturers Association of Canada, in their Annual Review 2006, provided the following two bar charts regarding import sales of automobiles and domestic automobile production in Japan, respectively:

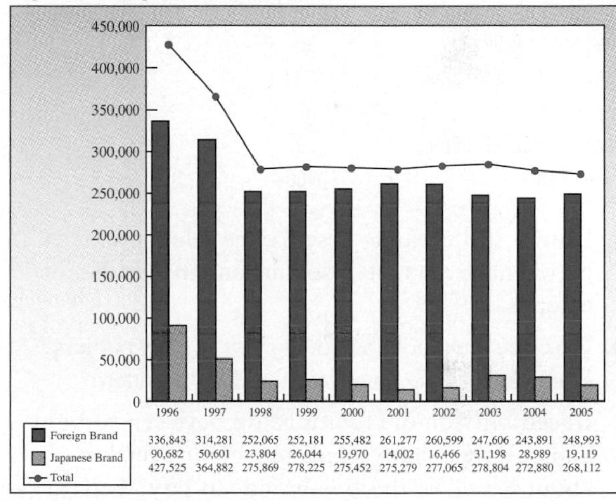

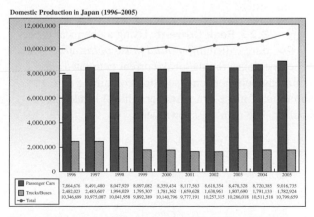

Source: © Japan Automobile Manufacturers Association of Canada.

Domestic Production in Japan (1996–2005)

	1996	1997	1998	1999	2000	2001	2002	2003	2004	2005
Passenger Cars	7,864,676	8,491,480	8,047,929	8,097,082	8,359,434	8,117,563	8,618,354	8,478,328	8,720,385	9,016,735
Trucks/Buses	2,482,023	2,483,607	1,994,029	1,795,307	1,781,362	1,659,628	1,638,961	1,807,690	1,791,133	1,782,924
Total	10,346,699	10,975,087	10,041,958	9,892,389	10,140,796	9,777,191	10,257,315	10,286,018	10,511,518	10,799,659

Source: © Japan Automobile Manufacturers Association of Canada.

a. What variables have been measured in this study? Are the variables qualitative or quantitative?

b. Describe the population of interest. Do these data represent a population or a sample drawn from the population?

c. What type of graphical presentation has been used? What other type could have been used?

d. Describe the relationship between total import sales and total domestic production over the period 1996–2005.

3.24 Cheese, again! The demand for healthy foods that are low in fats and calories has resulted in a large number of "low-fat" and "fat-free" products at the supermarket. The table shows the numbers of calories and the amounts of sodium (in milligrams) per slice for five different brands of fat-free cheese:

Brand	Sodium (mg)	Calories
Kraft Fat Free Singles	300	30
Ralphs Fat Free Singles	300	30
Borden Fat Free	320	30
Healthy Choice Fat Free	290	30
Smart Beat	180	25

a. Draw a scatterplot to describe the relationship between the amount of sodium and the number of calories.

b. Describe the plot in part a. Do you see any outliers? Do the rest of the points seem to form a pattern?

c. Based *only* on the relationship between sodium and calories, can you make a clear decision about which of the five brands to buy? Is it reasonable to base your choice on only these two variables? What other variables should you consider?

Data set EX0325 **3.25 Peak Current** Using a chemical procedure called *differential pulse polarography*, a chemist measured the peak current generated (in microamperes) when a solution containing a given amount of nickel (in parts per billion) is added to a buffer. The data are shown here:

x = Ni (ppb)	y = Peak Current (μA)
19.1	0.095
38.2	0.174
57.3	0.256
76.2	0.348
95.0	0.429
114.0	0.500
131.0	0.580
150.0	0.651
170.0	0.722

Use a graph to describe the relationship between x and y. Add any numerical descriptive measures that are appropriate. Write a paragraph summarizing your results.

Data set EX0326 **3.26 Movie Money** Does the opening weekend adequately predict the success or failure of a new movie? In 2001, 36 movies were investigated in *Entertainment Weekly*, and the following variables were recorded:[10]

- The movie's first weekend's gross earnings (in millions of dollars)
- The movie's total gross earnings (in millions of dollars)

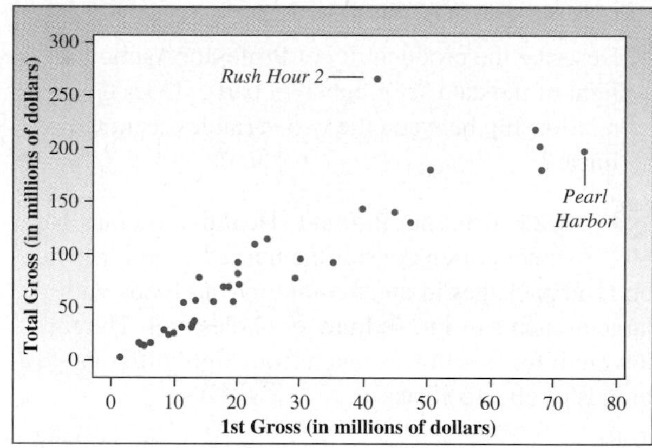

a. How would you describe the relationship between the first weekend's gross and the total gross?

b. Are there any outliers? If so, explain how they do not fit the pattern of the other movies.

c. Which dot represents the movie with the best opening weekend? Did it also have the highest total gross?

d. The film *Pearl Harbor* opened on a three-day weekend. Does that help explain its position in relation to the other data points?

3.27 Movie Money, continued The data from Exercise 3.26 were entered into a *MINITAB* worksheet, and the following output was obtained:

Covariances: 1st Gross, Total Gross

```
             1st Gross   Total Gross
1st Gross    412.528
Total Gross  1232.231    4437.109
```

a. Use the *MINITAB* output or the original data to find the correlation between first weekend and total gross.

b. Which of the two variables would you classify as the independent variable? The dependent variable?

c. If the average first weekend gross is 25.66 million dollars and the average total gross is 86.71 million dollars, find the regression line for predicting total gross as a function of the first weekend's gross.

d. If another film was released and grossed 30 million dollars on the first weekend, what would you predict that its total gross earnings will be?

Data set EX0328 **3.28 Federal Contaminated Sites (2005)** The data in the following table give the number of federal contaminated sites for provinces and territories along with the size of the province/territory, and the percentage of the total area:[11]

Province or Territory	Number of Sites	Total Area (thousands km)	Percent of Total Area
Alberta	738	405	4.1
British Columbia	2223	6	0.1
Manitoba	1062	55	0.6
New Brunswick	228	73	0.7
Newfoundland and Labrador	688	1542	15.4
Northwest Territories	281	1076	10.8
Nova Scotia	876	648	6.5
Nunavut	238	651	6.5
Ontario	2449	662	6.6
Prince Edward Island	97	945	9.5
Quebec	1851	482	4.8
Saskatchewan	432	1346	13.5
Yukon	100	2093	21.0

a. Draw a scatterplot with *MINITAB*. Is there any clear pattern in the scatterplot? Describe the relationship between number of contaminated sites and the size of the province/territory.

b. Use the *MINITAB* output to calculate the correlation coefficient. Does this confirm your answer to part a?

c. Are there any outliers or clusters in the data? If so, can you explain them?

Data set EX0329 **3.29 Aaron Rodgers** The number of passes completed and the total number of passing yards were recorded for Aaron Rodgers for each of the 15 regular season games that he played in the fall of 2010:[12]

Week	Completions	Total Yards	Week	Completions	Total Yards
1	19	188	9	27	289
2	19	255	11	22	301
3	34	316	12	26	344
4	12	181	13	21	298
5	27	293	14	7	46
6	18	313	16	25	404
7	21	295	17	19	229
8	15	170			

a. Draw a scatterplot to describe the relationship between number of completions and total passing yards for Aaron Rodgers.

b. Describe the plot in part a. Do you see any outliers? Do the rest of the points seem to form a pattern?

c. Calculate the correlation coefficient, r, between the number of completions and total passing yards.

d. What is the regression line for predicting total number of passing yards y based on the total number of completions x?

e. If Aaron Rodgers had 20 pass completions in his next game, what would you predict his total number of passing yards to be?

3.30 Pottery, continued In Exercise 1.55, we analyzed the percentage of aluminum oxide in 26 samples of Romano-British pottery found at four different kiln sites in the United Kingdom.[13] Since one of the sites provided only two measurements, that site is eliminated, and comparative box plots of aluminum oxide at the other three sites are shown.

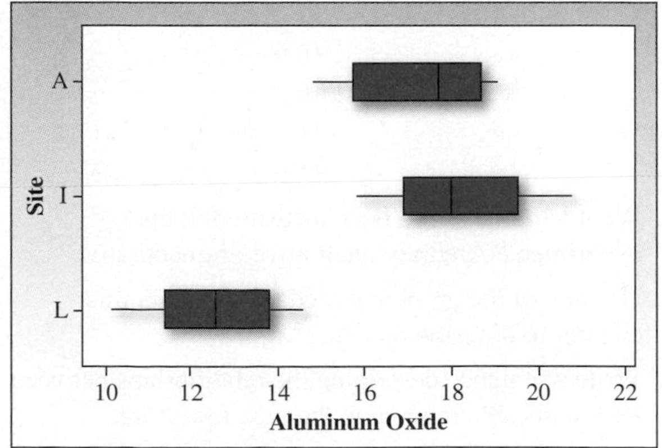

Source: A. Tubb, A. J. Parker, and G. Nickless, "The Analysis of Romano-British Pottery by Atomic Absorption Spectrophotometry," Archaeometry Vol. 22 Issue 2, August 1989, p. 153. Archaeometry by UNIVERSITY OF OXFORD. Reproduced with permission of BLACKWELL PUBLISHING LTD. in the format Book via Copyright Clearance Center.

a. What two variables have been measured in this experiment? Are they qualitative or quantitative?

b. How would you compare the amount of aluminum oxide in the samples at the three sites?

Data set EX0331 **3.31 Pottery, continued** Here is the percentage of aluminum oxide, the percentage of iron oxide, and the percentage of magnesium oxide in five samples collected at Ashley Rails in the United Kingdom:

Sample	Al	Fe	Mg
1	17.7	1.12	0.56
2	18.3	1.14	0.67
3	16.7	0.92	0.53
4	14.8	2.74	0.67
5	19.1	1.64	0.60

a. Find the correlation coefficients describing the relationships between aluminum and iron oxide content, between iron oxide and magnesium oxide, and between aluminum oxide and magnesium oxide.

b. Write a sentence describing the relationships between these three chemicals in the pottery samples.

Data set EX0332 **3.32 Internet Access from Home** The table below (see Exercise 1.50) shows DSL versus cable users per 100 members of the population.[14]

Broadband Internet Access Users per 100

Year	DSL	Cable Modem
2006	10.8	11.5
2005	9.4	9.7
2004	7.9	8.7
2003	6.3	8.0
2002	4.7	6.6
2001	3.0	5.2
2000	1.3	3.1

a. What variables have been measured in this experiment? Are they qualitative or quantitative?

b. Use one of the graphical methods given in this chapter to describe the data.

c. Write a sentence describing the relationships between DSL and cable modem in the next few years.

3.33 The Religious Cleavage in Canada Mendelsohn and Nadeau (1997), in *Canadian Journal of Political Science*, reported the strength of the religious cleavage in Canada (outside Quebec). The results are shown below:[15]

Categories	Catholics (%)	Protestants (%)
Liberal voters	36	21
Support nuclear purchase	48	55
Support unions	63	53
No restriction on abortion	36	53

a. What variables have been measured in this survey? Are they qualitative or quantitative?

b. Draw side-by-side comparative bar charts to describe the percentages of Catholics and Protestants by given categories.

c. Draw two line charts on the same set of axes to describe the same percentages of Catholic and Protestants for the given categories.

d. What conclusions can you draw using the two graphs in parts b and c? Which graph is more effective?

Data set EX0334 **3.34 Armspan and Height** Leonardo DaVinci (1452–1519) drew a sketch of a man, indicating that a person's armspan (measuring across the back with arms outstretched to make a "T") is roughly equal to the person's height. To test this claim, we measured eight people with the following results:

Person	1	2	3	4
Armspan (cm)	172.7	158.1	165.1	176.5
Height (cm)	175.3	157.5	165.1	177.8
Person	**5**	**6**	**7**	**8**
Armspan (cm)	172.7	175.3	157.5	153.0
Height (cm)	170.2	170.2	160.0	157.5

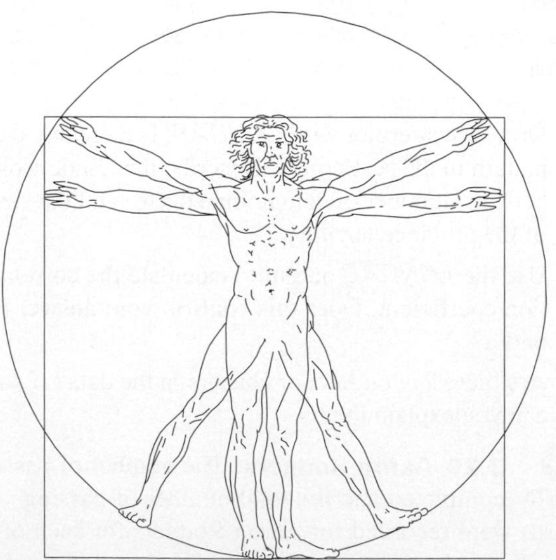

a. Draw a scatterplot for armspan and height. Use the same scale on both the horizontal and vertical axes. Describe the relationship between the two variables.

b. Calculate the correlation coefficient relating armspan and height.

c. If you were to calculate the regression line for predicting height based on a person's armspan, how would you estimate the slope of this line?

d. Find the regression line relating armspan to a person's height.

e. If a person has an armspan of 157.5 cm, what would you predict the person's height to be?

Data set EX0335 **3.35 Travellers to Canada** The number of trips, overnights x (in thousands), and the spending in foreign countries y (in millions of dollars) for the top 15 countries visited by Canadians in 2013 are given in the table.[16]

Countries	Trips	Nights (x)	Spending in Country (y)
United States	22,710	194,413	17,490
Mexico	1,598	17,146	1,639
Cuba	1,082	8,947	748
United Kingdom	908	10,955	1,056
Dominican Republic	766	6,704	674
France	729	9,068	942
Italy	375	3,897	480
Germany	345	3,501	311
Mainland China	314	6,445	521
Spain	245	2,615	284
Jamaica	243	2,189	248
Netherlands	223	1,579	158
Hong Kong	197	2,709	211
Republic of Ireland	179	2,161	205
Australia	159	3,808	329

a. Construct a scatterplot for the data (x and y).

b. Describe the form, direction, and strength of the pattern in the scatterplot.

c. Calculate the correlation coefficient between x and y and interpret its value. Does it reinforce a relation that was apparent for the scatterplot? Explain.

Data set EX0336 **3.36 Test-Interviews** Of two personnel evaluation techniques available, the first requires a two-hour test-interview while the second can be completed in less than an hour. The scores for each of the eight individuals who took both tests are given in the next table:

Applicant	Test 1 (x)	Test 2 (y)
1	75	38
2	89	56
3	60	35
4	71	45
5	92	59
6	105	70
7	55	31
8	87	52

a. Construct a scatterplot for the data (x and y).

b. Describe the form, direction, and strength of the pattern in the scatterplot.

3.37 Test-Interviews, continued Refer to Exercise 3.36.

a. Find the correlation coefficient, r, to describe the relationship between the two tests.

b. Would you be willing to use the second and quicker test rather than the longer test-interview to evaluate personnel? Explain.

Data set EX0338 **3.38 Happy in the Air? continued** The following table (data also in Exercise 1.29) reveals complaints against Air Canada and major U.S. airlines in a given year:[17]

Airline	Grand Total	Passengers (millions)
Air Canada	522	20.1
Airtran Airways	97	11.7
American West Airlines	168	20.1
American Airlines	786	88.8
Continental Airlines	371	38.9
Delta	663	84.3
Northwest Airlines	497	52.0
Southwest Airlines	106	74.8
United Airlines	556	66.2
U.S. Airways	379	41.3

a. Construct a scatterplot for the data.

b. Describe the form, direction, and strength of the pattern in the scatterplot.

c. Are there any outliers in the scatterplot? If so, which airline does this outlier represent?

d. Does the outlier from part c indicate that this airline is doing better or worse than the other airlines with respect to customer satisfaction.

3.39 Smartphones The table below shows the prices of nine smartphones with a two-year-term plan provided by Bell Mobility, along with their overall score (on a scale of 0–100) in a consumer rating survey in 2016:

EX0339

Brand and Model	Price($)	Overall Score
Samsung Galaxy S7 Edge	499.99	80
Samsung Galaxy S6 Edge+	299.99	79
iPhone 6S Plus	528.99	78
iPhone 6	98.95	78
PRIV by BlackBerry	299.99	76
Samsung Galaxy Note 5	199.99	76
BlackBerry Passport	149.95	74
Nexus 6P	49.99	72
BlackBerry Classic	49.95	68

Source: © Bell Canada; device prices effective March 24, 2016.

a. Plot the nine data points using a scatterplot. Describe the form, direction, and strength of the relationship between price and overall score.

b. Calculate r, the correlation coefficient between price and overall score.

c. Find the regression line for predicting the overall score of a smartphone based on its price.

CASE STUDY  Paying for Players

Are higher-paid hockey players actually better than lower-salaried ones? Despite the league's new salary cap, the 30 teams in the NHL in the 2005–2006 season had widely different payrolls, starting with the New Jersey Devils who paid their players a total of almost $45 million, and ranging down to the Washington Capitals, who spent a mere $19 million on salaries, less than half what they cost the Devils. The following table shows each team's payroll along with standard measures of performance, including goals scored and allowed, wins, and points earned. The information shown in the table can also be found on the text website. Use a statistical computer package to explore the relationships between various pairs of variables in the table.[18]

Team	Total Payroll ($)	Points	Wins	Losses	Overtime Losses	Goals For	Goals Against
Detroit Red Wings	39,578,300	124	58	16	8	305	209
Ottawa Senators	36,909,094	113	52	21	9	314	211
Carolina Hurricanes	35,308,700	112	52	22	8	294	260
Dallas Stars	40,651,480	112	53	23	6	265	218
Buffalo Sabres	28,515,120	110	52	24	6	281	239
Nashville Predators	31,649,440	106	49	25	8	259	227
Calgary Flames	36,589,140	103	46	25	11	218	200
New Jersey Devils	44,895,949	101	46	27	9	242	229
Philadelphia Flyers	42,566,760	101	45	26	11	267	259
New York Rangers	41,474,800	100	44	26	12	257	215
San Jose Sharks	31,005,400	99	44	27	11	266	242
Anaheim Ducks	32,060,233	98	43	27	12	254	229
Colorado Avalanche	41,044,829	95	43	30	9	283	257
Edmonton Oilers	38,469,340	95	41	28	13	256	251
Montreal Canadiens	32,994,940	93	42	31	9	243	247
Tampa Bay Lightning	39,157,379	92	43	33	6	252	260

Team	Total Payroll ($)	Points	Wins	Losses	Overtime Losses	Goals For	Goals Against
Vancouver Canucks	43,711,344	92	42	32	8	256	255
Atlanta Thrashers	37,170,200	90	41	33	8	281	275
Toronto Maple Leafs	36,796,580	90	41	33	8	257	270
Los Angeles Kings	37,856,150	89	42	35	5	249	270
Florida Panthers	26,500,510	85	37	34	11	240	257
Minnesota Wild	25,158,800	84	38	36	8	231	215
Phoenix Coyotes	30,354,345	81	38	39	5	246	271
New York Islanders	31,447,520	78	36	40	6	230	278
Boston Bruins	36,662,100	74	29	37	16	230	266
Columbus Blue Jackets	30,093,235	74	35	43	4	223	279
Washington Capitals	18,932,830	70	29	41	12	237	306
Chicago Blackhawks	30,141,200	65	26	43	13	211	285
Pittsburgh Penguins	23,122,650	58	22	46	14	244	316
St. Louis Blues	28,480,800	57	21	46	15	197	292

Note: Points determine the standings, and are calculated by the formula: Points = 2 × Wins + Overtime Losses.

1. Look at each variable individually. What can you say about symmetry? About outliers?
2. Look at the variables in pairs. Which pairs of variables are positively correlated? Which are negatively correlated? Do any pairs exhibit little or no correlation? Are some of these results counterintuitive? Can you offer an explanation for these cases?
3. Answer the following questions: Does the price of an athletic team, specifically in the NHL, convey something about its quality? Which variables did you use in arriving at your answer?

PROJECTS

Project 3-A: Child Safety Seat Survey

Canada has a Road Safety Vision of having the safest roads in the world. Yet, the leading cause of death of Canadian children remains vehicle crashes. In 2006, a national child safety seat survey was conducted by an AUTO21 research team in collaboration with Transport Canada to empirically measure Canada's progress toward achieving Road Safety Vision 2010. Child seat use was observed in parking lots and nearby intersections in 200 randomly selected sites across Canada. The following table provides a classification of a subset of children in the survey by age group and type of restraint device they were using at the time of the survey.

This research was supported in part by the AUTO21 Network of Centres of Excellence.

Reference: *T. Yi Wen, A. Snowdon, S. E. Ahmed, and A. A. Hussein (2009). Accommodating Nonrespondents in the Canadian National Child Safety Seat Survey.*

Project 3-A: Child Safety Seat Survey, continued

Table: Cross-tabulation of Age Group by Restraint Type

Category	Types of Restraint				
Age Group	Rear-Facing Infant Seat	Forward-Facing Infant Seat	Booster Seat	Seat Belt Only	**Total**
Infant (0–1 year)	181	52	1	0	**234**
Toddler (1–4 years)	49	483	117	3	**652**
School (4–9 years)	0	98	450	325	**873**
Older (>9 years)	0	0	16	627	**643**
Total	**230**	**633**	**584**	**955**	**2402**

a. What are some of the variables measured in this survey? Are they qualitative or quantitative?

b. Construct a side-by-side bar chart to describe the data in the table. Create a stacked bar chart to describe these data.

c. Construct a pie chart for each of the age groups. Which of the charts created above best depicts the difference or similarity of use of toddler restraint types?

d. Write a short paragraph summarizing the information that can be gained by looking at these graphs. Which of the three types of comparative graphs is more effective?

e. Perhaps you can suggest another graphical technique to display this data that will be the most effective one.

Based on the same survey, the following table refers to the weight (in kg), height (in cm), and gender of a random sample of 40 children chosen from the large data set.

Gender	Weight (kg)	Height (cm)	Gender	Weight (kg)	Height (cm)
F	34.02	147.50	F	52.16	147.50
M	36.29	140.00	F	34.02	132.50
F	19.73	110.23	F	8.62	70.00
F	21.09	120.08	M	14.51	95.00
F	31.75	140.75	F	16.33	92.50
F	52.39	151.25	F	56.70	145.68
F	25.17	100.00	M	33.57	139.77
M	15.88	107.50	M	20.41	120.00
M	28.12	137.50	M	41.28	145.00
F	13.61	87.50	M	29.03	116.25
F	41.28	152.50	M	36.29	152.50
F	22.45	122.50	M	23.59	117.50
M	68.95	147.50	M	31.75	133.75
F	48.08	152.50	M	28.58	130.00
M	18.60	105.00	F	34.93	132.50
M	28.80	122.50	M	38.56	138.75
F	10.66	83.75	M	27.67	135.00
F	37.42	140.75	M	29.48	120.00
F	32.66	143.70	F	26.31	130.00
M	24.49	127.50	M	41.28	147.50

a. Draw a scatterplot of weights of female children with their respective heights. Describe the relationship between the two variables.
b. Are there any outliers in the scatterplot in part a? Do the outliers indicate that the children have lower or higher height than the other children in this sample data?
c. Draw a scatterplot of male children's weights with their respective heights. Describe the form, direction, and strength of the relationship between height and weight.
d. Compare the above two scatterplots and summarize your findings including similarities and differences between sexes (i.e., difference in relationship between height and weight).
e. Create side-by-side box plots to describe the height for both female and male children. How would you compare the height in the samples for both genders?
f. Calculate the correlation coefficient relating height and weight for female data and interpret its value.
g. Does the correlation coefficient reinforce a relation that was apparent from the scatterplot in part a? Explain. If you were to calculate the regression line for predicting weight based on a female child's height, how would you estimate the slope of this line?
h. Find the regression line relating height to a female child's weight.
i. If a female child has a height of 160 cm, what would you predict her weight to be?
j. Would you use the fitted line used to predict a female child's weight for predicting a male child's weight too? Why or why not?
k. Would you collapse the gender boundaries and use the combined data to predict the child's weight regardless of their sex? Explain.

4

Probability and Probability Distributions

giulio napolitano/Shutterstock.com

GENERAL OBJECTIVES

Now that you have learned to describe a data set, how can you use sample data to draw conclusions about the sampled populations? The technique involves a statistical tool called *probability*. To use this tool correctly, you must first understand how it works. The first part of this chapter will teach you the new language of probability, presenting the basic concepts with simple examples.

The variables that we measured in Chapters 1 and 2 can now be redefined as random variables, whose values depend on the chance selection of the elements in the sample. Using probability as a tool, you can create probability distributions that serve as models for discrete random variables, and you can describe these random variables using a mean and standard deviation similar to those in Chapter 2.

CHAPTER INDEX

Probability and Decision Making in the Congo

In his exciting novel *Congo*, author Michael Crichton describes an expedition racing to find boron-coated blue diamonds in the rain forests of eastern Zaire. Can probability help the heroine Karen Ross in her search for the Lost City of Zinj? The case study at the end of this chapter involves Ross's use of probability in decision-making situations.

NEED TO KNOW

How to Calculate the Probability of an Event
The Difference between Mutually Exclusive and Independent Events

4.1 THE ROLE OF PROBABILITY IN STATISTICS

Probability and statistics are related in an important way. Probability is used as a *tool;* it allows you to evaluate the reliability of your conclusions about the population when you have only sample information. Consider these situations:

- When you toss a single coin, you will see either a head (H) or a tail (T). If you toss the coin repeatedly, you will generate an infinitely large number of Hs and Ts—the entire population. What does this population look like? If the coin is fair, then the population should contain 50% Hs and 50% Ts. Now toss the coin one more time. What is the chance of getting a head? Most people would say that the "probability" or chance is 1/2.
- Now suppose you are not sure whether the coin is fair; that is, you are not sure whether the makeup of the population is 50–50. You decide to perform a simple experiment. You toss the coin $n = 10$ times and observe 10 heads in a row. Can you conclude that the coin is fair? Probably not, because if the coin were fair, observing 10 heads in a row would be very *unlikely;* that is, the "probability" would be very small. It is more *likely* that the coin is biased.

As in the coin-tossing example, statisticians use probability in two ways. When the population is *known,* probability is used to describe the likelihood of observing a particular sample outcome. When the population is *unknown* and only a sample from that population is available, probability is used in making statements about the makeup of the population—that is, in making statistical inferences.

In Chapters 4–7, you will learn many different ways to calculate probabilities. You will assume that the population is *known* and calculate the probability of observing various sample outcomes. Once you begin to use probability for statistical inference in Chapter 8, the population will be *unknown* and you will use your knowledge of probability to make reliable inferences from sample information. We begin with some simple examples to help you grasp the basic concepts of probability.

EVENTS AND THE SAMPLE SPACE

Data are obtained by observing either uncontrolled events in nature or controlled situations in a laboratory. We use the term **random experiment** or simply **experiment** to describe either method of data collection.

DEFINITION An **experiment** is the process by which an observation (or measurement) is obtained.

The observation or measurement generated by an experiment may or may not produce a numerical value. Here are some examples of experiments:

- Recording a test grade
- Measuring daily rainfall
- Interviewing a householder to obtain his or her opinion on a greenbelt zoning ordinance
- Testing a printed circuit board to determine whether it is a defective product or an acceptable product
- Tossing a coin and observing the face that appears

When an experiment is performed, what we observe is an outcome called a **simple event,** often denoted by the capital E with a subscript.

DEFINITION A **simple event** is the outcome that is observed on a single repetition of the experiment.

EXAMPLE 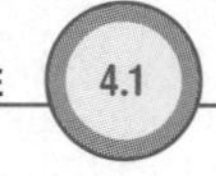

Experiment: Toss a die and observe the number that appears on the upper face. List the simple events in the experiment.

SOLUTION When the die is tossed once, there are six possible outcomes. There are the simple events, listed below.

Event E_1: Observe a 1	Event E_4: Observe a 4
Event E_2: Observe a 2	Event E_5: Observe a 5
Event E_3: Observe a 3	Event E_6: Observe a 6

We can now define an **event** as a collection of simple events, often denoted by a capital letter.

DEFINITION An **event** is a collection of simple events.

EXAMPLE continued 

We can define the events A and B for the die-tossing experiment:

A: Observe an odd number

B: Observe a number less than 4

Since event A occurs if the upper face is 1, 3, or 5, it is a collection of three simple events, and we write $A = \{E_1, E_3, E_5\}$. Similarly, the event B occurs if the upper face is 1, 2, or 3 and is defined as a collection or set of these three simple events: $B = \{E_1, E_2, E_3\}$.

Sometimes when one event occurs, it means that another event cannot.

DEFINITION Two events are **mutually exclusive** if, when one event occurs, the other cannot, and vice versa.

In the die-tossing experiment, events A and B are *not* mutually exclusive because they have two outcomes in common—if the number on the upper face of the die is a 1 or a 3. Both events A and B will occur if either E_1 or E_3 is observed when the experiment is performed. In contrast, the six simple events $E_1, E_2, \ldots, E_6$ form a set of all mutually exclusive outcomes of the experiment. When the experiment is performed once, one and only one of these simple events can occur.

DEFINITION The set of all simple events is called the sample space, S.

Sometimes it helps to visualize an experiment using a picture called a **Venn diagram,** shown in Figure 4.1. The outer box represents the *sample space,* which contains all of the *simple events,* represented by labelled points. Since an event is a collection of one or more simple events, the appropriate points are circled and labelled with the event letter. For the die-tossing experiment, the sample space is $S = \{E_1, E_2, E_3, E_4, E_5, E_6\}$ or, more simply, $S = \{1, 2, 3, 4, 5, 6\}$. The events $A = \{1, 3, 5\}$ and $B = \{1, 2, 3\}$ are circled in the Venn diagram.

FIGURE 4.1
Venn diagram for die tossing

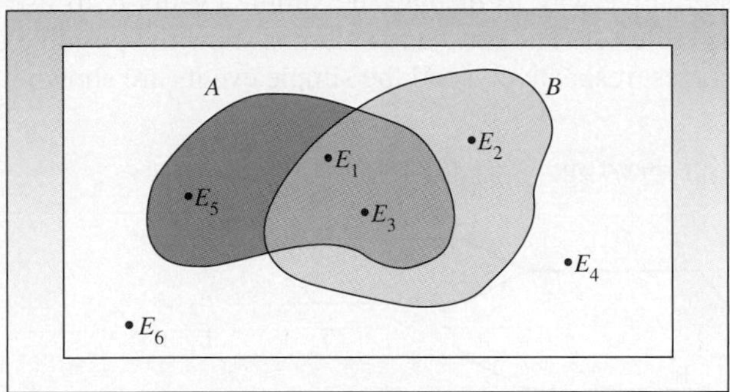

EXAMPLE 4.2

Experiment: Toss a single coin and observe the result. These are the simple events:

E_1: Observe a head (H)
E_2: Observe a tail (T)

The sample space is $S = \{E_1, E_2\}$ or more simply, $S = \{\text{H, T}\}$.

EXAMPLE 4.3

Experiment: Record a person's social media usage. The four mutually exclusive possible outcomes are these simple events:

E_1: Facebook
E_2: Twitter
E_3: LinkedIn
E_4: Instagram

The sample space is $S = \{E_1, E_2, E_3, E_4\}$, or $S =$ [Facebook, Twitter, LinkedIn, Instagram].

EXAMPLE 4.4

Experiment: Record a person's blood type. The four mutually exclusive possible outcomes are these simple events:

E_1: Blood type A
E_2: Blood type B
E_3: Blood type AB
E_4: Blood type O

The sample space is $S = \{E_1, E_2, E_3, E_4\}$, or $S = \{\text{A, B, AB, O}\}$.

Some experiments can be generated in stages, and the sample space can be displayed in a **tree diagram.** Each successive level of branching on the tree corresponds to a step required to generate the final outcome.

EXAMPLE 4.5

A medical technician records a person's blood type and Rh factor. List the simple events in the experiment.

SOLUTION For each person, a two-stage procedure is needed to record the two variables of interest. The tree diagram is shown in Figure 4.2. The eight simple events in the tree diagram form the sample space, $S = \{\text{A+, A–, B+, B–, AB+, AB–, O+, O–}\}$.

An alternative way to display the simple events is to use a **probability table,** as shown in Table 4.1. The rows and columns show the possible outcomes at the first and second stages, respectively, and the simple events are shown in the cells of the table.

FIGURE 4.2
Tree diagram for Example 4.5

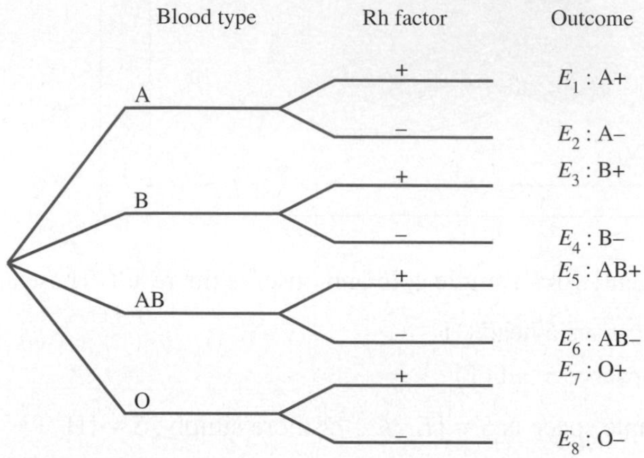

TABLE 4.1 **Probability Table for Example 4.4**

	Blood Type			
Rh Factor	A	B	AB	O
Negative	A–	B–	AB–	O–
Positive	A+	B+	AB+	O+

4.3 CALCULATING PROBABILITIES USING SIMPLE EVENTS

The probability of an event A is a measure of our belief that the event A will occur. One practical way to interpret this measure is with the concept of *relative frequency.* Recall from Chapter 1 that if an experiment is performed n times, then the relative frequency of a particular occurrence—say, A—is

$$\text{Relative frequency} = \frac{\text{Frequency}}{n}$$

where the frequency is the number of times the event A occurred. If you let n, the number of repetitions of the experiment, become larger and larger ($n \to \infty$), you will eventually generate the entire population. In this population, the relative frequency of the event A is defined as the **probability of event A;** that is,

$$P(A) = \lim_{n\to\infty} \frac{\text{Frequency}}{n}$$

Since $P(A)$ behaves like a relative frequency, $P(A)$ must be a proportion lying between 0 and 1; $P(A) = 0$ if the event A never occurs, and $P(A) = 1$ if the event A always occurs. The closer $P(A)$ is to 1, the more likely it is that A will occur.

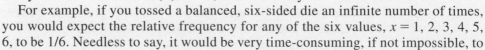

For example, if you tossed a balanced, six-sided die an infinite number of times, you would expect the relative frequency for any of the six values, $x = 1, 2, 3, 4, 5, 6$, to be 1/6. Needless to say, it would be very time-consuming, if not impossible, to repeat an experiment an infinite number of times. For this reason, there are alternative methods for calculating probabilities that make use of the relative frequency concept.

An important consequence of the relative frequency definition of probability involves the simple events. Since the simple events are mutually exclusive, their probabilities must satisfy two conditions.

REQUIREMENTS FOR SIMPLE-EVENT PROBABILITIES

- Each probability must lie between 0 and 1.
- The sum of the probabilities for all simple events in S equals 1.

When it is possible to write down the simple events associated with an experiment and to determine their respective probabilities, we can find the probability of an event A by summing the probabilities for all the simple events contained in the event A.

DEFINITION The **probability of an event** A is equal to the sum of the probabilities of the simple events contained in A.

EXAMPLE 4.6

Toss two fair coins and record the outcome. Find the probability of observing exactly one head in the two tosses.

SOLUTION To list the simple events in the sample space, you can use a tree diagram as shown in Figure 4.3. The letters H and T mean that you observed a head or a tail, respectively, on a particular toss. To assign probabilities to each of the four simple events, you need to remember that the coins are fair. Therefore, any of the four simple events is as likely as any other. Since the sum of the four simple events must be 1, each must have probability $P(E_i) = 1/4$. The simple events in the sample space are shown in Table 4.2, along with their *equally likely probabilities.* To find $P(A) = P$(observe exactly one head), you need to find all the simple events that result in event A—namely, E_2 and E_3:

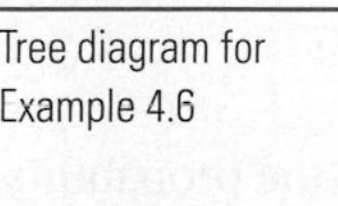

NEED A TIP?

Probabilities must lie between 0 and 1.

$$P(A) = P(E_2) + P(E_3)$$
$$= \frac{1}{4} + \frac{1}{4} = \frac{1}{2}$$

FIGURE 4.3
Tree diagram for Example 4.6

NEED A TIP?

The probabilities of all the simple events must add to 1.

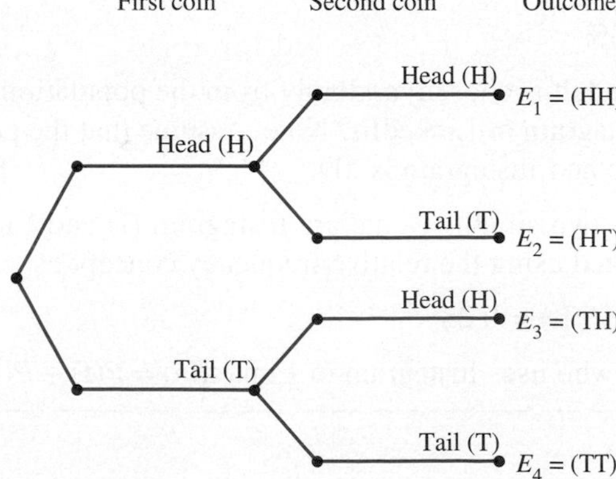

TABLE 4.2 **Simple Events and Their Probabilities**

Event	First Coin	Second Coin	$P(E_i)$
E_1	H	H	1/4
E_2	H	T	1/4
E_3	T	H	1/4
E_4	T	T	1/4

EXAMPLE 4.7

Canada is the world's second largest country by total area. Canadians adhere to a wide variety of religions. Statistics Canada (2001 census) reports that 77.1% of Canadians are identified as being Christians, about 16.5% of Canadians declare no religious affiliation, and the remaining 6.4% are affiliated with a religion other than Christianity, of which the largest was Islam. If a single Canadian is chosen randomly from the population, what is the probability he or she is Christian or has another religion?

SOLUTION The three simple events, C (Christian), N (no religious affiliation), O (other religion) do not have equally likely probabilities. Their probabilities are found using the relative frequency concept as

$$P(\text{C}) = 0.771 \qquad P(\text{N}) = 0.165 \qquad P(\text{O}) = 0.064$$

The event of interest consists of two simple events, so

$$\begin{aligned} P(\text{person is either Christian or has other religion}) &= P(\text{C}) + P(0) \\ &= 0.771 + 0.064 = 0.835 \end{aligned}$$

EXAMPLE 4.8

A survey conducted by the Pew Research Center (PRC) finds that Facebook remains by far the most popular social media site.[1] Social media has become extremely popular for its different uses, such as Twitter acting like a news feed and LinkedIn connecting professionals. According to a report done by PRC in 2014, the following list shows the percentages of adults who are online who use the named social media websites:

Facebook	71%
LinkedIn	28%
Pinterest	28%
Instagram	26%
Twitter	23%

If a single online adult is chosen randomly from the population, what is the probability he or she uses Instagram or LinkedIn? Note: Assume that the percentage of adults who use both LinkedIn and Instagram is .09.

SOLUTION The two simple events are Instagram (I) and LinkedIn (L). Their probabilities are assigned using the relative frequency concept as

$$P(\text{I}) = 0.26 \qquad P(\text{L}) = 0.28$$

$$P(\text{online adult who uses Instagram or LinkedIn}) = P(\text{I}) + P(\text{L}) = 0.26 + 0.28 = 0.54$$

A candy dish contains one yellow and two red candies. You close your eyes, choose two candies one at a time from the dish, and record their colours. What is the probability that both candies are red?

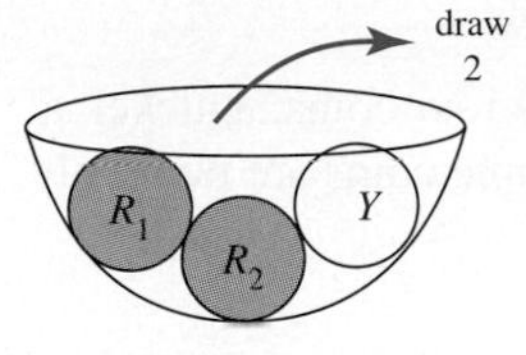

SOLUTION Since no probabilities are given, you must list the simple events in the sample space. The two-stage selection of the candies suggests a tree diagram, shown in Figure 4.4. There are two red candies in the dish, so you can use the letters R_1, R_2, and Y to indicate that you have selected the first red, the second red, or the yellow candy, respectively. Since you closed your eyes when you chose the candies, all six choices should be *equally likely* and are assigned probability 1/6. If A is the event that both candies are red, then

$$A = \{R_1R_2, R_2R_1\}$$

Thus,

$$P(A) = P(R_1R_2) + P(R_2R_1)$$
$$= \frac{1}{6} + \frac{1}{6} = \frac{1}{3}$$

FIGURE 4.4
Tree diagram for Example 4.9

First choice	Second choice	Simple event	Probability
R_1	R_2	R_1R_2	1/6
	Y	R_1Y	1/6
R_2	R_1	R_2R_1	1/6
	Y	R_2Y	1/6
Y	R_1	YR_1	1/6
	R_2	YR_2	1/6

NEED A TIP?

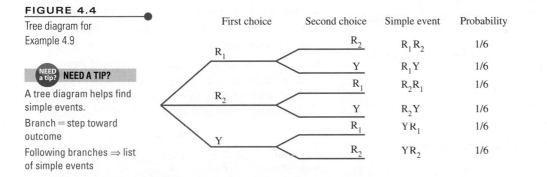

A tree diagram helps find simple events.

Branch = step toward outcome

Following branches ⇒ list of simple events

NEED TO KNOW

How to Calculate the Probability of an Event

1. List all the simple events in the sample space.
2. Assign an appropriate probability to each simple event.
3. Determine which simple events result in the event of interest.
4. Sum the probabilities of the simple events that result in the event of interest.

In your calculation, you must always be careful that you satisfy these two conditions:

- Include all simple events in the sample space.
- Assign realistic probabilities to the simple events.

One way to determine the required number of simple events is to use the counting rules presented in the next optional section. These rules can be used to solve more complex problems, which generally involve a large number of simple events. If you need to master only the basic concepts of probability, you may choose to skip the next section.

4.3 EXERCISES

BASIC TECHNIQUES

4.1 Tossing a Die An experiment involves tossing a single die. These are some events:

A: Observe a 2

B: Observe an even number

C: Observe a number greater than 2

D: Observe both *A* and *B*

E: Observe *A* or *B* or both

F: Observe both *A* and *C*

a. List the simple events in the sample space.

b. List the simple events in each of the events *A* through *F*.

c. What probabilities should you assign to the simple events?

d. Calculate the probabilities of the six events *A* through *F* by adding the appropriate simple-event probabilities.

4.2 A sample space *S* consists of five simple events with these probabilities:

$$P(E_1) = P(E_2) = 0.15 \quad P(E_3) = 0.4$$
$$P(E_4) = 2P(E_5)$$

a. Find the probabilities for simple events E_4 and E_5.

b. Find the probabilities for these two events:

$A = \{E_1, E_3, E_4\}$

$B = \{E_2, E_3\}$

c. List the simple events that are either in event *A* or event *B* or both.

d. List the simple events that are in both event *A* and event *B*.

4.3 A sample space contains 10 simple events: $E_1, E_2, \ldots, E_{10}$. If $P(E_1) = 3P(E_2) = 0.45$ and the remaining simple events are equiprobable, find the probabilities of these remaining simple events.

4.4 Free Throws A particular basketball player hits 70% of her free throws. When she tosses a pair of free throws, the four possible simple events and three of their associated probabilities are as given in the table:

Simple Event	Outcome of First Free Throw	Outcome of Second Free Throw	Probability
1	Hit	Hit	0.49
2	Hit	Miss	?
3	Miss	Hit	0.21
4	Miss	Miss	0.09

a. Find the probability that the player will hit on the first throw and miss on the second.

b. Find the probability that the player will hit on at least one of the two free throws.

4.5 Four Coins A jar contains four coins: a nickel, a dime, a quarter, and a loonie. Three coins are randomly selected from the jar.

a. List the simple events in *S*.

b. What is the probability that the selection will contain the loonie?

c. What is the probability that the total amount drawn will equal $1.10 or more?

4.6 Preschool or Not? On the first day of kindergarten, the teacher randomly selects 1 of his 25 students and records the student's gender, as well as whether or not that student had gone to preschool.

a. How would you describe the experiment?

b. Construct a tree diagram for this experiment. How many simple events are there?

c. The table below shows the distribution of the 25 students according to gender and preschool experience. Use the table to assign probabilities to the simple events in part b:

	Male	Female
Preschool	8	9
No preschool	6	2

d. What is the probability that the randomly selected student is male? What is the probability that the student is a female and did not go to preschool?

4.7 The Urn Problem An urn contains three red and two yellow balls. Two balls are randomly selected and their colours recorded. Use a tree diagram to list the 20 simple events in the experiment, keeping in mind the order in which the balls are drawn.

4.8 The Urn Problem, continued Refer to Exercise 4.7. A ball is randomly selected from the urn containing three red and two yellow balls. Its colour is noted, and the ball is returned to the urn before a second ball is selected. List the additional five simple events that must be added to the sample space in Exercise 4.7.

APPLICATIONS

4.9 Need Eyeglasses? A survey classified a large number of adults according to whether they were judged to need eyeglasses to correct their reading vision and whether they used eyeglasses when reading. The proportions falling into the four categories are shown in the table. (Note that a small proportion, 0.02, of adults used eyeglasses when in fact they were judged not to need them.)

	Used Eyeglasses for Reading	
Judged to Need Eyeglasses	Yes	No
Yes	0.44	0.14
No	0.02	0.40

If a single adult is selected from this large group, find the probability of each event:

a. The adult is judged to need eyeglasses.

b. The adult needs eyeglasses for reading but does not use them.

c. The adult uses eyeglasses for reading whether he or she needs them or not.

4.10 Roulette The game of roulette uses a wheel containing 38 pockets. Thirty-six pockets are numbered 1, 2,..., 36, and the remaining two are marked 0 and 00. The wheel is spun, and a pocket is identified as the "winner." Assume that the observance of any one pocket is just as likely as any other.

a. Identify the simple events in a single spin of the roulette wheel.

b. Assign probabilities to the simple events.

c. Let A be the event that you observe either a 0 or a 00. List the simple events in the event A and find $P(A)$.

d. Suppose you placed bets on the numbers 1 through 18. What is the probability that one of your numbers is the winner?

4.11 Jury Duty Three people are randomly selected from voter registration and driving records to report for jury duty. The gender of each person is noted by the county clerk.

a. Define the experiment.

b. List the simple events in S.

c. If each person is just as likely to be a man as a woman, what probability do you assign to each simple event?

d. What is the probability that only one of the three is a man?

e. What is the probability that all three are women?

4.12 Jury Duty, again Refer to Exercise 4.11. Suppose that there are six prospective jurors, four men and two women, who might be impanelled to sit on the jury in a criminal case. Two jurors are randomly selected from these six to fill the two remaining jury seats.

a. List the simple events in the experiment. (HINT: There are 15 simple events if you ignore the order of selection of the two jurors.)

b. What is the probability that both impanelled jurors are women?

4.13 Tea Tasters A food company plans to conduct an experiment to compare its brand of tea with that of two competitors. A single person is hired to taste and rank each of three brands of tea, which are unmarked except for identifying symbols A, B, and C.

a. Define the experiment.

b. List the simple events in S.

c. If the taster has no ability to distinguish a difference in taste among teas, what is the probability that the taster will rank tea type A as the most desirable? As the least desirable?

4.14 100-Metre Run Four equally qualified runners, John, Bill, Ed, and Dave, run a 100-metre sprint, and the order of finish is recorded.

a. How many simple events are in the sample space?

b. If the runners are equally qualified, what probability should you assign to each simple event?

c. What is the probability that Dave wins the race?

d. What is the probability that Dave wins and John places second?

e. What is the probability that Ed finishes last?

4.15 Fruit Flies In a genetics experiment, the researcher mated two *Drosophila* fruit flies and observed the traits of 300 offspring. The results are shown in the table:

	Wing Size	
Eye Colour	Normal	Miniature
Normal	140	6
Vermillion	3	151

One of these offspring is randomly selected and observed for the two genetic traits.

a. What is the probability that the fly has normal eye colour and normal wing size?

b. What is the probability that the fly has vermillion eyes?

c. What is the probability that the fly has either vermillion eyes or miniature wings, or both?

4.16 Health Care Most Canadians feel that health care is the most important issue. Research Canada: An Alliance for Health Discovery (www.rc-rc.ca) released the result of its first public opinion survey on health research in Canada. The results are shown in the table:

Opinion	Proportion
Very important	0.88
Somewhat important	0.10
Not very important	0.01
Not at all important	0.01

Suppose that one person is randomly selected and his or her opinion on this question is recorded.

a. What are the simple events in the experiment?

b. Are the simple events in part a equally likely? If not, what are the probabilities?

c. What is the probability that the person feels health care is at least somewhat important?

d. What is the probability that the person feels health care is either not very important or not at all important?

4.4 USEFUL COUNTING RULES (OPTIONAL)

Suppose that an experiment involves a large number N of simple events and you know that all the simple events are *equally likely*. Then each simple event has probability $1/N$, and the probability of an event A can be calculated as

$$P(A) = \frac{n_A}{N}$$

where n_A is the number of simple events that result in the event A. In this section, we present three simple rules that can be used to count either N, the number of simple events in the sample space; or n_A, the number of simple events in event A. Once you have obtained these counts, you can find $P(A)$ without actually listing all the simple events.

THE *mn* RULE

Consider an experiment that is performed in two stages. If the first stage can be accomplished in m ways and for each of these ways, the second stage can be accomplished in n ways, then there are mn ways to accomplish the experiment.

For example, suppose that you can order a car in one of three styles and in one of four paint colours. To find out how many options are available, you can think of first picking one of the $m = 3$ styles and then selecting one of the $n = 4$ paint colours. Using the mn Rule, as shown in Figure 4.5, you have $mn = (3)(4) = 12$ possible options.

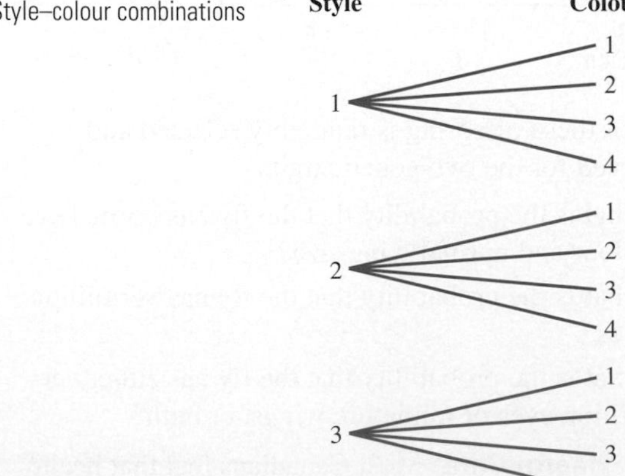

FIGURE 4.5
Style–colour combinations

EXAMPLE 4.10

Two dice are tossed. How many simple events are in the sample space S?

SOLUTION The first die can fall in one of $m = 6$ ways, and the second die can fall in one of $n = 6$ ways. Since the experiment involves two stages, forming the pairs of numbers shown on the two faces, the total number of simple events in S is

$$mn = (6)(6) = 36$$

EXAMPLE 4.11

A candy dish contains one yellow and two red candies. Two candies are selected one at a time from the dish, and their colours are recorded. How many simple events are in the sample space S?

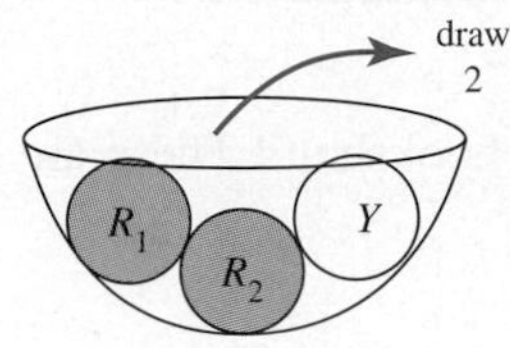

SOLUTION The first candy can be chosen in $m = 3$ ways. Since one candy is now gone, the second candy can be chosen in $n = 2$ ways. The total number of simple events is

$$mn = (3)(2) = 6$$

These six simple events were listed in Example 4.9.

We can extend the *mn* Rule for an experiment that is performed in more than two stages.

THE EXTENDED *mn* RULE

If an experiment is performed in k stages, with n_1 ways to accomplish the first stage, n_2 ways to accomplish the second stage, … , and n_k ways to accomplish the kth stage, then the number of ways to accomplish the experiment is

$$n_1 n_2 n_3 \cdots n_k$$

EXAMPLE 4.12

How many simple events are in the sample space when three coins are tossed?

SOLUTION Each coin can land in one of two ways. Hence, the number of simple events is

$$(2)(2)(2) = 8$$

EXAMPLE 4.13

A truck driver can take three routes from city A to city B, four from city B to city C, and three from city C to city D. If, when travelling from A to D, the driver must drive from A to B to C to D, how many possible A-to-D routes are available?

SOLUTION Let

n_1 = Number of routes from A to $B = 3$

n_2 = Number of routes from B to $C = 4$

n_3 = Number of routes from C to $D = 3$

Then the total number of ways to construct a complete route, taking one subroute from each of the three groups, (A to B), (B to C), and (C to D), is

$$n_1 n_2 n_3 = (3)(4)(3) = 36$$

A second useful counting rule follows from the *mn* Rule and involves **orderings** or **permutations.** For example, suppose you have three books, A, B, and C, but you have room for only two on your bookshelf. In how many ways can you select and arrange the two books? There are three choices for the two books—*A* and *B*, *A* and *C*, or *B* and *C*—but each of the pairs can be arranged in two ways on the shelf. All the permutations of the two books, chosen from three, are listed in Table 4.3. The *mn* Rule implies that there are 6 ways, because the first book can be chosen in $m = 3$ ways and the second in $n = 2$ ways, so the result is $mn = 6$.

TABLE 4.3 **Permutations of Two Books Chosen from Three**

Combinations of Two	Reordering of Combinations
AB	*BA*
AC	*CA*
BC	*CB*

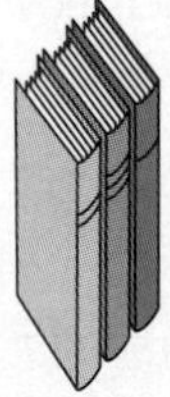

In how many ways can you arrange all three books on your bookshelf? These are the six permutations:

ABC *ACB* *BAC*
BCA *CAB* *CBA*

Since the first book can be chosen in $n_1 = 3$ ways, the second in $n_2 = 2$ ways, and the third in $n_3 = 1$ way, the total number of orderings is $n_1n_2n_3 = (3)(2)(1) = 6$.

Rather than applying the *mn* Rule each time, you can find the number of orderings using a general formula involving *factorial notation.*

A COUNTING RULE FOR PERMUTATIONS

The number of ways we can arrange *n* distinct objects, taking them *r* at a time, is

$$P_r^n = \frac{n!}{(n-r)!}$$

where $n! = n(n-1)(n-2)\cdots(3)(2)(1)$ and $0! = 1$.

Since *r* objects are chosen, this is an *r-stage* experiment. The first object can be chosen in *n* ways, the second in $(n-1)$ ways, the third in $(n-2)$ ways, and the *r*th in $(n-r+1)$ ways. We can simplify this awkward notation using the counting rule for permutations because

$$\frac{n!}{(n-r)!} = \frac{n(n-1)(n-2)\cdots(n-r+1)(n-r)\cdots(2)(1)}{(n-r)\cdots(2)(1)}$$

$$= n(n-1)\cdots(n-r+1)$$

A SPECIAL CASE: ARRANGING *n* ITEMS

The number of ways to arrange an entire set of *n* distinct items is $P_n^n = n!$

EXAMPLE 4.14

Three lottery tickets are drawn from a total of 50. If the tickets will be distributed to each of three employees in the order in which they are drawn, the order will be important. How many simple events are associated with the experiment?

SOLUTION The total number of simple events is

$$P_3^{50} = \frac{50!}{47!} = 50(49)(48) = 117{,}600$$

EXAMPLE 4.15

A piece of equipment is composed of five parts that can be assembled in any order. A test is to be conducted to determine the time necessary for each order of assembly. If each order is to be tested once, how many tests must be conducted?

SOLUTION The total number of tests equals

$$P_5^5 = \frac{5!}{0!} = 5(4)(3)(2)(1) = 120$$

When we counted the number of permutations of the two books chosen for your bookshelf, we used a systematic approach:

- First we counted the number of *combinations* or pairs of books to be chosen.
- Then we counted the number of ways to arrange the two chosen books on the shelf.

Sometimes the ordering or arrangement of the objects is not important, but only the objects that are chosen. In this case, you can use a counting rule for **combinations.** For example, you may not care in what order the books are placed on the shelf, but only which books you are able to shelve. When a five-person committee is chosen from a group of 12 students, the order of choice is unimportant because all five students will be equal members of the committee.

A COUNTING RULE FOR COMBINATIONS

The number of distinct combinations of n distinct objects that can be formed, taking them r at a time, is

$$C_r^n = \frac{n!}{r!(n-r)!}$$

The number of *combinations* and the number of *permutations* are related:

$$C_r^n = \frac{P_r^n}{r!}$$

You can see that C_r^n results when you divide the number of permutations by $r!$, the number of ways of rearranging each distinct group of r objects chosen from the total n.

EXAMPLE 4.16

A printed circuit board may be purchased from five suppliers. In how many ways can three suppliers be chosen from the five?

SOLUTION Since it is important to know only which three have been chosen, not the order of selection, the number of ways is

$$C_3^5 = \frac{5!}{3!2!} = \frac{(5)(4)}{2} = 10$$

The next example illustrates the use of counting rules to solve a probability problem.

EXAMPLE 4.17

Five manufacturers produce a certain electronic device, whose quality varies from manufacturer to manufacturer. If you were to select three manufacturers at random, what is the chance that the selection would contain exactly two of the best three?

SOLUTION The simple events in this experiment consist of all possible combinations of three manufacturers, chosen from a group of five. Of these five, three have been designated as "best" and two as "not best." You can think of a candy dish containing three red and two yellow candies, from which you will select three, as illustrated in

Figure 4.6. The total number of simple events N can be counted as the number of ways to choose three of the five manufacturers, or

$$N = C_3^5 = \frac{5!}{3!2!} = 10$$

FIGURE 4.6
Illustration for Example 4.17

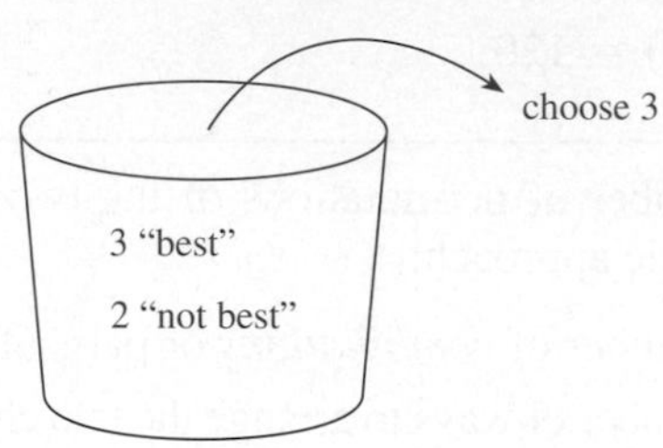

Since the manufacturers are selected at random, any of these 10 simple events will be *equally likely,* with probability 1/10. But how many of these simple events result in the event

A: Exactly two of the "best" three

You can count n_A, the number of events in A, in two steps because event A will occur when you select two of the "best" three and one of the two "not best." There are

$$C_2^3 = \frac{3!}{2!1!} = 3$$

ways to accomplish the first stage and

$$C_1^2 = \frac{2!}{1!1!} = 2$$

ways to accomplish the second stage. Applying the *mn* Rule, we find there are $n_A = (3)(2) = 6$ of the 10 simple events in event A and $P(A) = n_A/N = 6/10$.

Many other counting rules are available in addition to the three presented in this section. If you are interested in this topic, you should consult one of the many textbooks on combinatorial mathematics.

4.4 EXERCISES

BASIC TECHNIQUES

4.17 You have *two* groups of distinctly different items, 10 in the first group and 8 in the second. If you select one item from each group, how many different pairs can you form?

4.18 You have *three* groups of distinctly different items, four in the first group, seven in the second, and three in the third. If you select one item from each group, how many different triplets can you form?

4.19 Permutations Evaluate the following *permutations.* (HINT: Your scientific calculator may have a function that allows you to calculate permutations and combinations quite easily.)

a. P_3^5 **b.** P_9^{10} **c.** P_6^6 **d.** P_1^{20}

4.20 Combinations Evaluate these *combinations:*

a. C_3^5 **b.** C_9^{10} **c.** C_6^6 **d.** C_1^{20}

4.21 Choosing People In how many ways can you select five people from a group of eight if the order of selection is important?

4.22 Choosing People, again In how many ways can you select two people from a group of 20 if the order of selection is not important?

4.23 Dice Three dice are tossed. How many simple events are in the sample space?

4.24 Coins Four coins are tossed. How many simple events are in the sample space?

4.25 The Urn Problem, again Three balls are selected from an urn containing 10 balls. The order of selection is not important. How many simple events are in the sample space?

APPLICATIONS

4.26 What to Wear? You own 4 pairs of jeans, 12 clean T-shirts, and 4 wearable pairs of sneakers. How many outfits (jeans, T-shirt, and sneakers) can you create?

4.27 Itineraries A businessman in Montreal is preparing an itinerary for a visit to six major cities. The distance travelled, and hence the cost of the trip, will depend on the order in which he plans his route. How many different itineraries (and trip costs) are possible?

4.28 Vacation Plans Your family vacation involves an air flight, a rental car, and a hotel stay in Halifax. If you can choose from three air carriers, five car rental agencies, and four major hotel chains, how many options are available for your vacation accommodations?

4.29 A Card Game Three students are playing a card game. They decide to choose the first person to play by each selecting a card from the 52-card deck and looking for the highest card in value and suit. They rank the suits from lowest to highest: clubs, diamonds, hearts, and spades.

a. If the card is replaced in the deck after each student chooses, how many possible configurations of the three choices are possible?

b. How many configurations are there in which each student picks a different card?

c. What is the probability that all three students pick exactly the same card?

d. What is the probability that all three students pick different cards?

4.30 Dinner at a French Restaurant A French restaurant in Winnipeg, Manitoba, offers a special summer menu in which, for a fixed dinner cost, you can choose from one of two salads, one of two entrees, and one of two desserts. How many different dinners are available?

4.31 Playing Poker Five cards are selected from a 52-card deck for a poker hand.

a. How many simple events are in the sample space?

b. A *royal flush* is a hand that contains the A, K, Q, J, and 10, all in the same suit. How many ways are there to get a royal flush?

c. What is the probability of being dealt a royal flush?

4.32 Poker, Again Refer to Exercise 4.31. You have a poker hand containing four of a kind.

a. How many possible poker hands can be dealt?

b. In how many ways can you receive four cards of the same face value *and* one card from the other 48 available cards?

c. What is the probability of being dealt four of a kind?

4.33 A Hospital Survey A study is to be conducted in a hospital to determine the attitudes of nurses toward various administrative procedures. If a sample of 10 nurses is to be selected from a total of 90, how many different samples can be selected? (HINT: Is order important in determining the makeup of the sample to be selected for the survey?)

4.34 Traffic Problems Two city council members are to be selected from a total of five to form a subcommittee to study the city's traffic problems.

a. How many different subcommittees are possible?

b. If all possible council members have an equal chance of being selected, what is the probability that members Smith and Jones are both selected?

4.35 The NHL Professional ice hockey is very popular in Canada. National Hockey League (NHL) teams are divided into six divisions. Some information is shown in the table:

Group I Atlantic and Northwest Divisions	Group II Northeast Division
New Jersey	Boston
New York Islanders	Buffalo
New York Rangers	Montreal
Philadelphia	Ottawa
Pittsburgh	Toronto
Calgary	
Colorado	
Edmonton	
Minnesota	
Vancouver	

Two teams, one from each group, are randomly selected to play an exhibition game:

a. How many pairs of teams can be chosen?

b. What is the probability that the two teams are Calgary and Ottawa?

c. What is the probability that the Group I team is from New York?

4.36 100-Metre Run, again Refer to Exercise 4.14, in which a 100-metre sprint is run by John, Bill, Ed, and Dave. Assume that all the runners are equally qualified, so that any order of finish is equally likely. Use the *mn* Rule or permutations to answer these questions:

a. How many orders of finish are possible?

b. What is the probability that Dave wins the sprint?

c. What is the probability that Dave wins and John places second?

d. What is the probability that Ed finishes last?

4.37 Gender Bias? Consider the following case. The eight-member Human Relations Advisory Board considered the complaint of a woman who claimed discrimination, based on her gender, on the part of a local surveying company. The board, composed of five women and three men, voted 5–3 in favour of the plaintiff, the five women voting for the plaintiff and the three men against. The lawyer representing the company appealed the board's decision by claiming gender bias on the part of the board members. If the vote in favour of the plaintiff was 5–3 and the board members were not biased by gender, what is the probability that the vote would split along gender lines (five women for, three men against)?

4.38 Cramming A student prepares for an exam by studying a list of 10 problems. She can solve six of them. For the exam, the instructor selects five questions at random from the list of 10. What is the probability that the student can solve all five problems on the exam?

4.39 Monkey Business A monkey is given 12 blocks: three shaped like squares, three like rectangles, three like triangles, and three like circles. If it draws three of each kind in order—say, three triangles, then three squares, and so on—would you suspect that the monkey associates identically shaped figures? Calculate the probability of this event.

4.5 EVENT RELATIONS AND PROBABILITY RULES

Sometimes the event of interest can be formed as a combination of several other events. Let A and B be two events defined on the sample space S. Here are three important relationships between events:

DEFINITION The **union** of events A and B, denoted by $A \cup B$, is the event that either A or B or both occur.

DEFINITION The **intersection** of events A and B, denoted by $A \cap B$, is the event that both A and B occur.†

DEFINITION The **complement** of an event A, denoted by A^c, is the event that A *does not* occur.

Figures 4.7, 4.8, and 4.9 show Venn diagram representations of $A \cup B$, $A \cap B$, and A^c, respectively. Any simple event in the shaded area is a possible outcome resulting in the appropriate event. One way to find the probabilities of the union, the intersection, or the complement is to sum the probabilities of all the associated simple events.

NEED A TIP?

Intersection ⇔ "both ... and" or just "and"

Union ⇔ "either ... or ... or both" or just "or"

FIGURE 4.7

Venn diagram of $A \cup B$

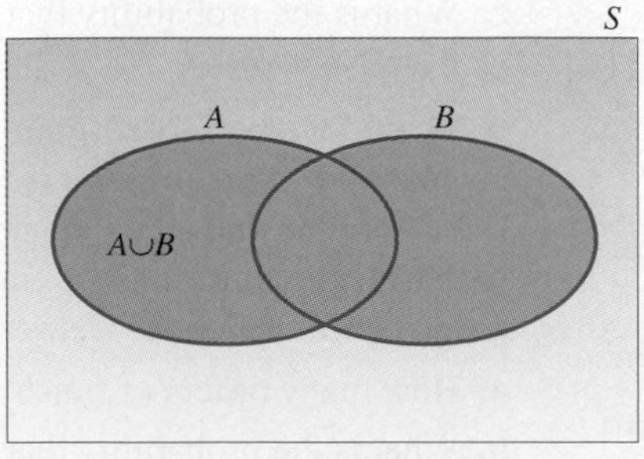

FIGURE 4.8

Venn diagram $A \cap B$

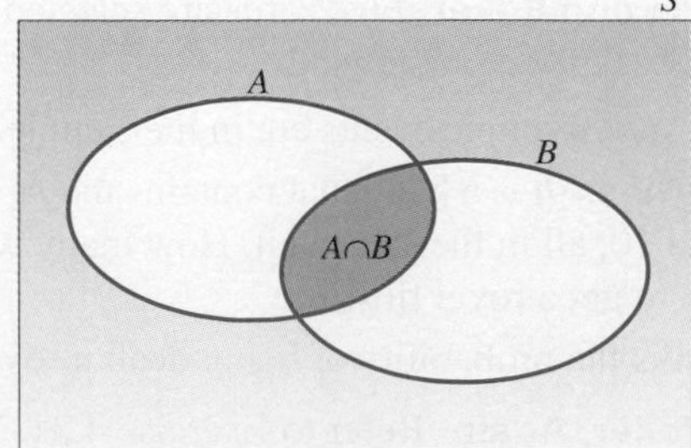

† Some authors use the notation AB.

FIGURE 4.9
The complement of an event

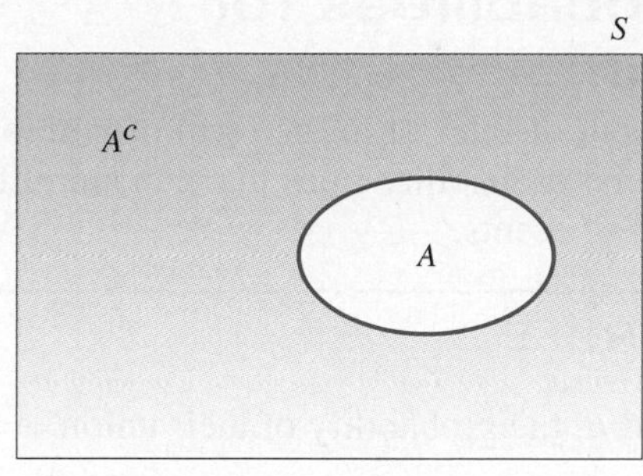

EXAMPLE 4.18

Two fair coins are tossed, and the outcome is recorded. These are the events of interest:

A: Observe at least one head
B: Observe at least one tail

Define the events A, B, $A \cap B$, $A \cup B$, and A^c as collections of simple events, and find their probabilities.

SOLUTION Recall from Example 4.6 that the simple events for this experiment are

E_1: HH (head on first coin, head on second)
E_2: HT
E_3: TH
E_4: TT

and that each simple event has probability 1/4. Event A, at least one head, occurs if E_1, E_2, or E_3 occurs, so that

$$A = \{E_1, E_2, E_3\} \qquad P(A) = \frac{3}{4}$$

and

$$A^c = \{E_4\} \qquad P(A^c) = \frac{1}{4}$$

Similarly,

$$B = \{E_2, E_3, E_4\} \qquad P(B) = \frac{3}{4}$$

$$A \cap B = \{E_2, E_3\} \qquad P(A \cap B) = \frac{1}{2}$$

$$A \cup B = \{E_1, E_2, E_3, E_4\} \qquad P(A \cup B) = \frac{4}{4} = 1$$

Note that $(A \cup B) = S$, the sample space, and is thus certain to occur.

The concept of unions and intersections can be extended to more than two events. For example, the union of three events A, B, and C, which is written as $A \cup B \cup C$, is the set of simple events that are in A or B or C or in any combination of those events. Similarly, the intersection of three events A, B, and C, which is written as $A \cap B \cap C$, is the collection of simple events common to the three events A, B, and C.

Calculating Probabilities for Unions and Complements

When we can write the event of interest in the form of a union, a complement, or an intersection, there are special probability rules that can simplify our calculations. The first rule deals with *unions* of events.

THE ADDITION RULE

Given two events, A and B, the probability of their union, $A \cup B$, is equal to

$$P(A \cup B) = P(A) + P(B) - P(A \cap B)$$

Notice in the Venn diagram in Figure 4.10 that the sum $P(A) + P(B)$ double counts the simple events common to both A and B. Subtracting $P(A \cap B)$ gives the correct result.

FIGURE 4.10
The Addition Rule

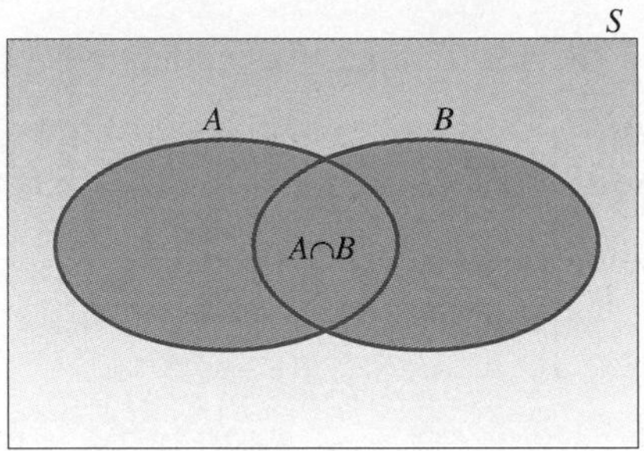

When two events A and B are **mutually exclusive** or **disjoint,** it means that when A occurs, B cannot, and vice versa. This means that the probability that they both occur, $P(A \cap B)$, must be zero. Figure 4.11 is a Venn diagram representation of two such events with no simple events in common.

FIGURE 4.11
Two disjoint events

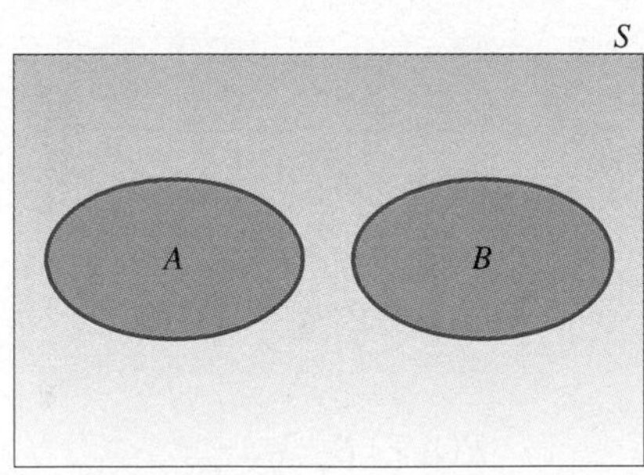

NEED A TIP?

Remember, mutually exclusive $\Leftrightarrow P(A \cap B) = 0$

When two events A and B are **mutually exclusive,** then $P(A \cap B) = 0$ and the Addition Rule simplifies to

$$P(A \cup B) = P(A) + P(B)$$

The second rule deals with *complements* of events. You can see from the Venn diagram in Figure 4.9 that A and A^c are mutually exclusive and that $A \cup A^c = S$, the entire sample space. It follows that

$$P(A) + P(A^c) = 1 \text{ and } P(A^c) = 1 - P(A)$$

RULE FOR COMPLEMENTS

$$P(A^c) = 1 - P(A)$$

EXAMPLE 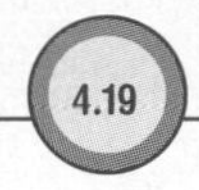 **4.19**

An oil-prospecting firm plans to drill two exploratory wells. Past evidence is used to assess the possible outcomes listed in Table 4.4.

TABLE 4.4 **Outcomes for Oil-Drilling Experiment**

Event	Description	Probability
A	Neither well produces oil or gas	0.80
B	Exactly one well produces oil or gas	0.18
C	Both wells produce oil or gas	0.02

Find $P(A \cup B)$ and $P(B \cup C)$.

SOLUTION By their definition, events *A, B,* and *C* are jointly mutually exclusive because the occurrence of one event precludes the occurrence of either of the other two. Therefore,

$$P(A \cup B) = P(A) + P(B) = 0.80 + 0.18 = 0.98$$

and

$$P(B \cup C) = P(B) + P(C) = 0.18 + 0.02 = 0.20$$

The event $A \cup B$ can be described as the event that *at most* one well produces oil or gas, and $B \cup C$ describes the event that *at least* one well produces gas or oil.

EXAMPLE  **4.20**

In a telephone survey of 1000 adults, respondents were asked about the expense of a university education and the relative necessity of some form of financial assistance. The respondents were classified according to whether they currently had a child in university and whether they thought the loan burden for most university students is too high, the right amount, or too little. The proportions responding in each category are shown in the **probability table** in Table 4.5. Suppose one respondent is chosen at random from this group.

TABLE 4.5 **Probability Table**

	Too High (*A*)	Right Amount (*B*)	Too Little (*C*)
Child in university (*D*)	0.35	0.08	0.01
No child in university (*E*)	0.25	0.20	0.11

1. What is the probability that the respondent has a child in university?
2. What is the probability that the respondent does not have a child in university?
3. What is the probability that the respondent has a child in university or thinks that the loan burden is too high?

SOLUTION Table 4.5 gives the probabilities for the six simple events in the cells of the table. For example, the entry in the top left corner of the table is the probability that a respondent has a child in university *and* thinks the loan burden is too high ($A \cap D$).

1. The event that a respondent has a child in university will occur regardless of his or her response to the question about loan burden. That is, event D consists of the simple events in the first row:

$$P(D) = 0.35 + 0.08 + 0.01 = 0.44$$

In general, the probabilities of *marginal* events such as D and A are found by summing the probabilities in the appropriate row or column.

2. The event that the respondent does not have a child in university is the complement of the event D denoted by D^c. The probability of D^c is found as

$$P(D^c) = 1 - P(D)$$

Using the result of part 1, we have

$$P(D^c) = 1 - 0.44 = 0.56$$

3. The event of interest is $P(A \cup D)$. Using the addition rule

$$\begin{aligned} P(A \cup D) &= P(A) + P(D) - P(A \cap D) \\ &= 0.60 + 0.44 - 0.35 \\ &= 0.69 \end{aligned}$$

Suppose the table below shows the probabilities of online adults who use Facebook, Twitter, and Instagram:

	Facebook (F)	Twitter (T)	Instagram (I)
Smartphone app (A)	0.50	0.12	0.21
No app on smartphone	0.05	0.06	0.06

1. What is the probability that an online adult uses the smartphone app and Facebook?
2. What is the probability that an online adult does not have any smartphone apps?
3. What is the probability that an adult uses smartphone apps or Facebook?

SOLUTION

1. The probability that an online adult uses a smartphone app, $P(A)$, and Facebook, $P(F)$, is the first cell in the table:
   $P(A \cap F) = 0.50$
2. The probability that an online adult does not use any smartphone app is
   $P(A^c) = 0.05 + 0.06 + 0.06 = 0.17$
3. The probability that an online adult uses smartphone apps or Facebook is $P(A \cup F)$:
   $P(A \cap F) = 0.50$
   $P(A) = 0.5 + 0.12 + 0.21 = 0.83$
   $P(F) = 0.50 + 0.05 = 0.55$, so
   $P(A \cup F) = P(A) + P(F) - P(A \cap F) = 0.83 + 0.55 - 0.50 = 0.88$

4.6 INDEPENDENCE, CONDITIONAL PROBABILITY, AND THE MULTIPLICATION RULE

There is a probability rule that can be used to calculate the probability of the intersection of several events. However, this rule depends on the important statistical concept of **independent** or **dependent events.**

DEFINITION Two events, *A* and *B,* are said to be **independent** if and only if the probability of event *B* is not influenced or changed by the occurrence of event *A,* or vice versa.

Colourblindness Suppose a researcher notes a person's gender and whether or not the person is colourblind to red and green. Does the probability that a person is colourblind change depending on whether the person is male or not? Define two events:

A: person is a male

B: person is colourblind

In this case, since colourblindness is a male sex-linked characteristic, the probability that a man is colourblind will be greater than the probability that a person chosen from the general population will be colourblind. The probability of event *B,* that a person is colourblind, depends on whether or not event *A,* that the person is a male, has occurred. We say that *A* and *B* are *dependent events.*

Tossing Dice On the other hand, consider tossing a single die two times, and define two events:

A: observe a 2 on the first toss

B: observe a 2 on the second toss

If the die is fair, the probability of event *A* is $P(A) = 1/6$. Consider the probability of event *B*. Regardless of whether event *A* has or has not occurred, the probability of observing a 2 on the second toss is still 1/6. We could write:

P(*B* given that *A* occurred) = 1/6

P(*B* given that *A* did not occur) = 1/6

Since the probability of event *B* is not changed by the occurrence of event *A,* we say that *A* and *B* are *independent events.*

The probability of an event *A,* given that the event *B* has occurred, is called the **conditional probability of A, given that B has occurred,** denoted by $P(A|B)$. The vertical bar is read "given" and the events appearing to the right of the bar are those that you know have occurred. We will use these probabilities to calculate the probability that *both A and B* occur when the experiment is performed.

THE GENERAL MULTIPLICATION RULE

The probability that *both A and B* occur when the experiment is performed is

$$P(A \cap B) = P(A)P(B|A)$$

or

$$P(A \cap B) = P(B)P(A|B)$$

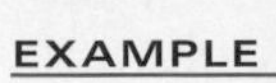

EXAMPLE 4.22

Suppose a group of 40 first-year science students at Brock University were classified according to their gender and drinking habits, as shown in Table 4.6. One student is selected at random from that group of 40 people.

TABLE 4.6 Gender and Drinking Habits of First-Year Students

	Drinking Habits		
Gender	Drinker (D)	Non-drinker (N)	Total
Male (M)	2	24	26
Female (F)	6	8	14
Total	8	32	40

1. What is the probability the student drinks?
2. What is the probability the student does not drink?
3. What is the probability the student is female and does not drink?
4. What is the probability the student is male?
5. What is the probability the student is female?
6. What is the probability the student is male and drinks?
7. What is the probability that a student is a drinker or a male student?

SOLUTION Using the information in Table 4.6, we can calculate all these probabilities.

1. $P(\text{D}) = 8/40 = 0.20$, since the total number of students who drinks are 8 among 40 students.
2. $P(\text{N}) = 32/40 = 0.80$; alternatively $P(\text{N}) = P(\text{D}^c) = 1 - P(\text{N}) = 1 - 0.20 = 0.80$.
3. $P(\text{female and does not drink}) = P(\text{F} \cap \text{N}) = 8/40 = 0.20$.
4. $P(\text{M}) = 26/40 = 0.65$.
5. $P(\text{F}) = 14/40 = 0.35$; alternatively $P(\text{F}) = P(\text{M}^c) = 1 - P(\text{M}) = 1 - 0.35 = 0.65$.
6. $P(\text{male and drinks}) = P(\text{M} \cap \text{D}) = 2/40 = 0.05$.
7. $P(\text{drinks or male}) = P(\text{D} \cup \text{M}) = P(\text{D}) + P(\text{M}) - P(\text{D} \cap \text{M})$

$$= P(D) + P(M) - P(M \cap D)$$
$$= 0.20 + 0.65 - 0.05 = 0.80$$

EXAMPLE 4.23

In a colour preference experiment, eight toys are placed in a container. The toys are identical except for colour—two are red, and six are green. A child is asked to choose two toys *at random.* What is the probability that the child chooses the two red toys?

SOLUTION You can visualize the experiment using a tree diagram as shown in Figure 4.12. Define the following events:

R: red toy is chosen

G: green toy is chosen

The event A (both toys are red) can be constructed as the intersection of two events:

$$A = (\text{R on first choice}) \cap (\text{R on second choice})$$

Since there are only two red toys in the container, the probability of choosing red on the first choice is 2/8. However, once this red toy has been chosen, the probability of red on the second choice is *dependent* on the outcome of the first choice (see Figure 4.12).

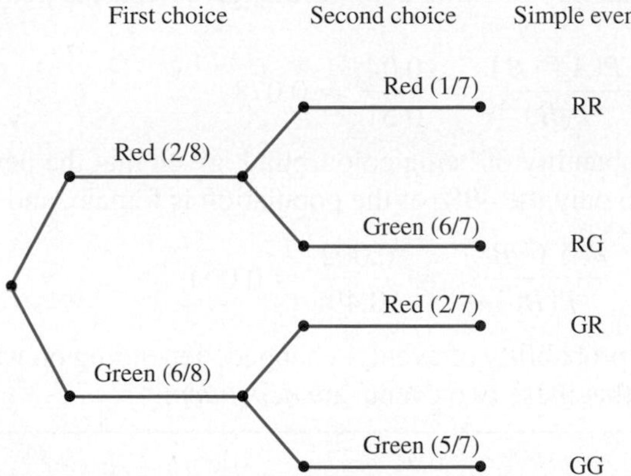

FIGURE 4.12
Tree diagram for Example 4.23

If the first choice was red, the probability of choosing a second red toy is only 1/7 because there is only one red toy among the seven remaining. If the first choice was green, the probability of choosing red on the second choice is 2/7 because there are two red toys among the seven remaining. Using this information and the Multiplication Rule, you can find the probability of event *A*.

$$P(A) = P(\text{R on first choice} \cap \text{R on second choice})$$
$$= P(\text{R on first choice})\ P(\text{R on second choice})|\text{R on first})$$
$$= \left(\frac{2}{8}\right)\left(\frac{1}{7}\right) = \frac{2}{56} = \frac{1}{28}$$

Sometimes you may need to use the Multiplication Rule in a slightly different form, so that you can calculate the **conditional probability, $P(A|B)$.** Just rearrange the terms in the Multiplication Rule.

CONDITIONAL PROBABILITIES

The conditional probability of event *A*, given that event *B* has occurred is

$$P(A|B) = \frac{P(A \cap B)}{P(B)} \quad \text{if} \quad P(B) \neq 0$$

The conditional probability of event *B*, given that event *A* has occurred is

$$P(B|A) = \frac{P(A \cap B)}{P(A)} \quad \text{if} \quad P(A) \neq 0$$

Colourblindness, continued Suppose that in the general population, there are 51% men and 49% women, and that the proportions of colourblind men and women are shown in the probability table below:

	Men (B)	Women (B^C)	Total
Colourblind (A)	0.04	0.002	0.042
Not colourblind (A^C)	0.47	0.488	0.958
Total	0.51	0.49	1.00

If a person is drawn at random from this population and is found to be a man (event *B*), what is the probability that the man is colourblind (event *A*)? If we know that the event *B* has occurred, we must restrict our focus to only the 51% of the population that is male. The probability of being colourblind, given that the person is male, is 4% of the 51% or

$$P(A \mid B) = \frac{P(A \cap B)}{P(B)} = \frac{0.04}{0.51} = 0.078$$

What is the probability of being colourblind, given that the person is female? Now we are restricted to only the 49% of the population is female, and

$$P(A \mid B^C) = \frac{P(A \cap B^C)}{P(B^C)} = \frac{0.002}{0.49} = 0.004$$

Notice that the probability of event A changed, depending on whether event B occurred. This indicates that these two events are *dependent*.

When two events are **independent**—that is, if the probability of event B is the same, whether or not event A has occurred, then event A does not affect event B and

$$P(B \mid A) = P(B)$$

The Multiplication Rule can now be simplified.

THE MULTIPLICATION RULE FOR INDEPENDENT EVENTS

If two events A and B are independent, the probability that *both A and B* occur is

$$P(A \cap B) = P(A)P(B)$$

Similarly, if *A, B,* and *C* are mutually independent events (all pairs of events are independent), then the probability that *A, B, and C* all occur is

$$P(A \cap B \cap C) = P(A)P(B)P(C)$$

Coin Tosses at Football Games A football team is involved in two overtime periods during a given game, so that there are three coin tosses. If the coin is fair, what is the probability that they lose all three tosses?

SOLUTION If the coin is fair, the event can be described in three steps:

A: Lose the first toss

B: Lose the second toss

C: Lose the third toss

Since the tosses are independent, and since $P(\text{win}) = P(\text{lose}) = 0.5$ for any of the three tosses,

$$P(A \cap B \cap C) = P(A)P(B)P(C) = (.5)(.5)(.5) = 0.125$$

How can you check to see if two events are independent or dependent? The easiest solution is to redefine the concept of **independence** in a more formal way.

CHECKING FOR INDEPENDENCE

Two events A and B are said to be **independent** if and only if either

$$P(A \cap B) = P(A)P(B)$$

or

$$P(B \mid A) = P(B)$$

Otherwise, the events are said to be **dependent.**

EXAMPLE 4.24

Toss two coins and observe the outcome. Define these events:

A: Head on the first coin

B: Tail on the second coin

Are events A and B independent?

SOLUTION From previous examples, you know that $S = \{HH, HT, TH, TT\}$. Use these four simple events to find

$$P(A) = \frac{1}{2}, P(B) = \frac{1}{2}, \text{ and } P(A \cap B) = \frac{1}{4}$$

NEED A TIP?

Remember, independence $\Leftrightarrow$ $P(A \cap B) = P(A)P(B)$

Since $P(A)P(B) = \left(\frac{1}{2}\right)\left(\frac{1}{2}\right) = \frac{1}{4}$ and $P(A \cap B) = \frac{1}{4}$, we have $P(A)P(B) = P(A \cap B)$ and the two events must be independent.

EXAMPLE 4.25

Refer to the probability table in Example 4.20, which is reproduced below.

	Too High (A)	Right Amount (B)	Too Little (C)
Child in university (D)	0.35	0.08	0.01
No child in university (E)	0.25	0.20	0.11

Are events D and A independent? Explain.

SOLUTION

1. Use the probability table to find $P(A \cap D) = 0.35$, $P(A) = 0.60$, and $P(D) = 0.44$. Then

$$P(A)P(D) = (.60)(.44) = 0.264 \text{ and } P(A \cap D) = 0.35$$

Since these two probabilities are not the same, events A and D are *dependent.*

2. Alternately, calculate

$$P(A|D) = \frac{P(A \cap D)}{P(D)} = \frac{0.35}{0.44} = 0.80$$

Since $P(A|D) = 0.80$ and $P(A) = 0.60$, we are again led to the conclusion that events A and D are *dependent.*

NEED TO KNOW

The Difference between Mutually Exclusive and Independent Events

Many students find it hard to tell the difference between *mutually exclusive* and *independent events.*

- When two events are *mutually exclusive* or *disjoint,* they cannot both happen when the experiment is performed. Once the event B has occurred, event A cannot occur, so that $P(A | B) = 0$, or vice versa. The occurrence of event B certainly affects the probability that event A can occur.
- Therefore, mutually exclusive events must be *dependent.*
- When two events are *mutually exclusive* or *disjoint,* $P(A \cap B) = 0$ and $P(A \cup B) = P(A) + P(B)$
- When two events are *independent,* $P(A \cap B) = P(A)P(B)$, and $P(A \cup B) = P(A) + P(B) - P(A)P(B)$

Using probability rules to calculate the probability of an event requires some experience and ingenuity. You need to express the event of interest as a union or intersection (or the combination of both) of two or more events whose probabilities are known or easily calculated. Often you can do this in different ways; the key is to find the right combination.

EXAMPLE 4.26

Two cards are drawn from a deck of 52 cards. Calculate the probability that the draw includes an ace and a 10.

SOLUTION Consider the event of interest:

A: draw an ace and a 10

Then $A = B \cup C$, where

B: draw the ace on the first draw and the 10 on the second

C: draw the 10 on the first draw and the ace on the second

Refer to Figure 4.13 below.

FIGURE 4.13

Tree diagram for Example 4.26

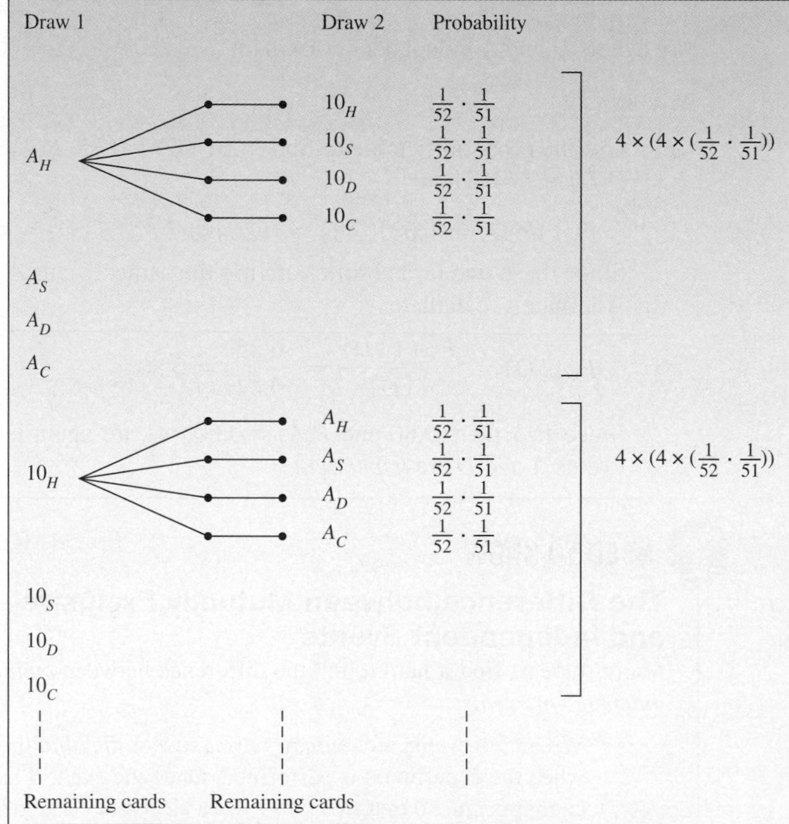

Events B and C were chosen to be mutually exclusive and also to be intersections of events with known probabilities; that is,

$$B = B_1 \cap B_2 \text{ and } C = C_1 \cap C_2$$

where

B_1: Draw an ace on the first draw
B_2: Draw a 10 on the second draw
C_1: Draw a 10 on the first draw
C_2: Draw an ace on the second draw

Applying the Multiplication Rule, you get

$$P(B_1 \cap B_2) = P(B_1)P(B_2|B_1) = \left(\frac{4}{52}\right)\left(\frac{4}{51}\right)$$

and

$$P(C_1 \cap C_2) = \left(\frac{4}{52}\right)\left(\frac{4}{51}\right)$$

Then, applying the Addition Rule,

$$P(A) = P(B) + P(C) = \left(\frac{4}{52}\right)\left(\frac{4}{51}\right) + \left(\frac{4}{52}\right)\left(\frac{4}{51}\right) = \frac{8}{663}$$

Check each composition carefully to be certain that it is actually equal to the event of interest.

4.6 EXERCISES

BASIC TECHNIQUES

4.40 An experiment can result in one of five equally likely simple events, $E_1, E_2, \ldots, E_5$. Events *A, B,* and *C* are defined as follows:

$A: E_1, E_3$	$P(A) = 0.4$
$B: E_1, E_2, E_4, E_5$	$P(B) = 0.8$
$C: E_3, E_4$	$P(C) = 0.4$

Find the probabilities associated with these compound events by listing the simple events in each.

a. A^c **b.** $A \cap B$ **c.** $B \cap C$
d. $A \cup B$ **e.** $B|C$ **f.** $A|B$
g. $A \cup B \cup C$ **h.** $(A \cap B)^c$

4.41 Refer to Exercise 4.40. Use the definition of a complementary event to find these probabilities:

a. $P(A^c)$ **b.** $P((A \cap B)^c)$

Do the results agree with those obtained in Exercise 4.40?

4.42 Refer to Exercise 4.40. Use the definition of conditional probability to find these probabilities:

a. $P(A|B)$ **b.** $P(B|C)$

Do the results agree with those obtained in Exercise 4.40?

4.43 Refer to Exercise 4.40. Use the Addition and Multiplication Rules to find these probabilities:

a. $P(A \cup B)$ **b.** $P(A \cap B)$ **c.** $P(B \cap C)$

Do the results agree with those obtained in Exercise 4.40?

4.44 Refer to Exercise 4.40.

a. Are events *A* and *B* independent?
b. Are events *A* and *B* mutually exclusive?

4.45 Dice An experiment consists of tossing a single die and observing the number of dots that show on the upper face. Events *A, B,* and *C* are defined as follows:

A: Observe a number less than 4

B: Observe a number less than or equal to 2

C: Observe a number greater than 3

Find the probabilities associated with the events below using either the simple event approach or the rules and definitions from this section.

a. S **b.** $A|B$ **c.** B
d. $A \cap B \cap C$ **e.** $A \cap B$ **f.** $A \cap C$
g. $B \cap C$ **h.** $A \cup C$ **i.** $B \cup C$

4.46 Refer to Exercise 4.45.

a. Are events *A* and *B* independent? Mutually exclusive?

b. Are events *A* and *C* independent? Mutually exclusive?

4.47 Suppose that $P(A) = 0.4$ and $P(B) = 0.2$. If events *A* and *B* are independent, find these probabilities:

a. $P(A \cap B)$ **b.** $P(A \cup B)$

4.48 Suppose that $P(A) = 0.3$ and $P(B) = 0.5$. If events *A* and *B* are mutually exclusive, find these probabilities:

a. $P(A \cap B)$ **b.** $P(A \cup B)$

4.49 Suppose that $P(A) = 0.4$ and $P(A \cap B) = 0.12$.

a. Find $P(B|A)$.

b. Are events *A* and *B* mutually exclusive?

c. If $P(B) = 0.3$, are events *A* and *B* independent?

4.50 An experiment can result in one or both of events *A* and *B* with the probabilities shown in this probability table:

	A	A^c
B	0.34	0.46
B^c	0.15	0.05

Find the following probabilities:

a. $P(A)$ **b.** $P(B)$ **c.** $P(A \cap B)$
d. $P(A \cup B)$ **e.** $P(A|B)$ **f.** $P(B|A)$

4.51 Refer to Exercise 4.50.

a. Are events *A* and *B* mutually exclusive? Explain.

b. Are events *A* and *B* independent? Explain.

APPLICATIONS

4.52 Drug Testing Some companies are testing prospective employees for drug use, with the intent of improving efficiency and reducing absenteeism, accidents, and theft. Opponents claim that this procedure is creating a class of unhirables and that some persons may be placed in this class because the tests themselves are not 100% reliable. Suppose a company uses a test that is 98% accurate—that is, it correctly identifies a person as a drug user or non-user with probability 0.98—and to reduce the chance of error, each job applicant is required to take two tests. If the outcomes of the two tests on the same person are independent events, what are the probabilities of these events?

a. A non-user fails both tests.

b. A drug user is detected (i.e., he or she fails at least one test).

c. A drug user passes both tests.

4.53 Grant Funding Whether a grant proposal is funded quite often depends on the reviewers. Suppose a group of research proposals was evaluated by a group of experts as to whether the proposals were worthy of funding. When these same proposals were submitted to a second independent group of experts, the decision to fund was reversed in 30% of the cases. If the probability that a proposal is judged worthy of funding by the first peer review group is 0.2, what are the probabilities of these events?

a. A worthy proposal is approved by both groups.

b. A worthy proposal is disapproved by both groups.

c. A worthy proposal is approved by one group.

4.54 Drug Offenders A study of the behaviour of a large number of drug offenders after treatment for drug abuse suggests that the likelihood of conviction within a two-year period after treatment may depend on the offender's education. The proportions of the total number of cases that fall into four education/conviction categories are shown in the table below:

	Status within Two Years after Treatment		
Education	Convicted	Not Convicted	Total
10 years or more	0.10	0.30	0.40
9 years or less	0.27	0.33	0.60
Total	0.37	0.63	1.00

Suppose a single offender is selected from the treatment program. Here are the events of interest:

A: The offender has 10 or more years of education.

B: The offender is convicted within two years after completion of treatment.

Find the appropriate probabilities for these events:

a. A **b.** B **c.** $A \cap B$
d. $A \cup B$ **e.** A^c **f.** $(A \cup B)^c$
g. $(A \cap B)^c$ **h.** A given that B has occurred
i. B given that A has occurred

4.55 Drug Offenders, continued Use the probabilities of Exercise 4.54 to show that these equalities are true:

a. $P(A \cap B) = P(A)P(B \mid A)$

b. $P(A \cap B) = P(B)P(A \mid B)$

c. $P(A \cup B) = P(A) + P(B) - P(A \cap B)$

4.56 The Birthday Problem Two people enter a room and their birthdays (ignoring years) are recorded.

a. Identify the nature of the simple events in *S*.

b. What is the probability that the two people have a specific pair of birthdates?

c. Identify the simple events in event *A*: Both people have the same birthday.

d. Find $P(A)$. **e.** Find $P(A^c)$.

4.57 The Birthday Problem, continued If *n* people enter a room, find these probabilities:

A: None of the people have the same birthday.

B: At least two of the people have the same birthday.

Solve for

a. $n = 3$ **b.** $n = 4$

(NOTE: Surprisingly, $P(B)$ increases rapidly as *n* increases. For example, for $n = 20$, $P(B) = 0.411$; for $n = 40$, $P(B) = 0.891$.)

4.58 Starbucks or Tim Hortons? A university student frequents one of two coffee houses on campus, choosing Tim Hortons 70% of the time and Starbucks 30% of the time. Regardless of where she goes, she buys a decaffeinated coffee on 60% of her visits.

a. The next time she goes into a coffee house on campus, what is the probability that she goes to Tim Hortons and orders a decaffeinated coffee?

b. Are the two events in part a independent? Explain.

c. If she goes into a coffee house and orders a decaffeinated coffee, what is the probability that she is at Starbucks?

d. What is the probability that she goes to Tim Hortons or orders a decaffeinated coffee or both?

4.59 Alcohol Consumption and Smoking among Canadian Medical Students A team of researchers at the Centre for Addiction and Mental Health, Toronto, conducted a survey to quantify the extent, and to assess student perception, of alcohol and tobacco use among medical students at the University of Calgary. For this purpose, a questionnaire was distributed to first-, second-, and third-year medical students attending the University of Calgary medical school. Of the 327 students enrolled, 175 of students responded to the questionnaire. Six percent of the students currently smoke, while 86% currently drink, with a majority drinking fewer than 11 drinks per week.[2] Suppose a group of 100 medical students at another Canadian university were classified according to their gender, smoking habits, and alcohol consumption habits, as shown in the table below:

	Smoking Habits		
Alcohol Consumption Habits	Smoker (S)	Non-smoker (N)	Total
No	3	14	17
Yes	6	77	83
Total	9	91	100

Source: S. Thakore, Z. Ismail, S. Jarvis, E. Payne, S. Keetbaas, R. Payne, L. Rothenburg, "The Perceptions and Habits of Alcohol Consumption and Smoking Among Canadian Medical Students," *Academic Psychiatry*, May 2009, Volume 33, Issue 3, pp. 193–197.

Convert the frequency table shown above into a probability table. If one student is selected at random from that group of 100 students, find the following probabilities:

a. What is the probability the student consumes alcohol? Is your answer consistent with the survey result? Why or why not?

b. What is the probability the student smokes? Is your answer consistent with the survey result? Why or why not?

c. What is the probability the selected student is a smoker and does not consume alcohol?

d. What is the probability the student does not drink?

e. What is the probability the selected student consumes alcohol and smokes?

f. If the student is a smoker, what is the probability that student consumes alcohol?

g. If the selected student consumes alcohol, what is the probability the student does not smoke?

4.60 Inspection Lines A certain manufactured item is visually inspected by two different inspectors. When a defective item comes through the line, the probability that it gets by the first inspector is 0.1. Of those that get past the first inspector, the second inspector will "miss" 5 out of 10. What fraction of the defective items get by both inspectors?

4.61 Smoking and Cancer A survey of people in a given region showed that 20% were smokers. The probability of death due to lung cancer, given that a person smoked, was roughly 10 times the probability of death due to lung cancer, given that a person did not smoke. If the probability of death due to lung cancer in the region is 0.006, what is the probability of death due to lung cancer given that a person is a smoker?

4.62 Canadians Are Huge Online Users According to comScore, a company that provides digital marketing intelligence, Canadians spend more time online when compared to a group of countries that includes the United States, France, and the United Kingdom. In Canada, 68% of the population is online, as compared to 62% in

France and the United Kingdom, 60% in Germany, 59% in the United States, 57% in Japan, and 36% in Italy. In fact, users in Canada spent an average of almost 42 hours a month online (or more than 2500 minutes).[3] Five hundred online users were selected from St. Catharines, Ontario, and were categorized by age and gender as shown in the probability table below:

	Age Group			
Gender	Under 18 Years (A)	18–35 Years (B)	Over 35 Years (C)	Total
Male (M)	0.21	0.16	0.10	0.47
Female (F)	0.24	0.18	0.11	0.53
Total	0.45	0.34	0.21	1.00

If one online user is selected at random from that group of 500 online users, find the following probabilities:

a. What is the probability the randomly selected online user is under 18 years of age?

b. What is the probability the randomly selected online user is female?

c. What is the probability the randomly selected online user is female and under 18 years of age?

d. If the randomly selected online user is male, what is the probability he is over 35 years of age?

e. If the randomly selected online user is under 18 years of age, what is the probability the online user is female?

f. What is the probability the randomly selected online user is male or over 35 years of age?

g. If the randomly selected online user is female, what is the probability that she is 18 to 35 years old?

h. Are the gender and age of the online user independent events? Why or why not?

i. Are the gender and age of the online user mutually exclusive events? Explain.

4.63 Smoke Detectors A smoke-detector system uses two devices, *A* and *B*. If smoke is present, the probability that it will be detected by device *A* is 0.95; by device *B*, 0.98; and by both devices, 0.94.

a. If smoke is present, find the probability that the smoke will be detected by device *A* or device *B* or both devices.

b. Find the probability that the smoke will not be detected.

4.64 Social Media and Canadians A Vancouver-based Internet marketing firm called 6S Marketing conducted a survey of its database to track the use of social media in Canada. The firm polled 10,000 Canadians and uncovered some interesting data:

- Seventy percent of Canadians say they use social media.
- Facebook is the most popular social networking site, with 70% of people surveyed currently having an account.
- Forty-seven percent of Canadians use Twitter, and the majority of users are 19–25 years of age.

However, a social media blogger challenges the claim that 47% of Canadians use Twitter, arguing that this number is inflated. Perhaps 6S Marketing's database consists of people who have a high likelihood of using Twitter, and hence, the survey may not be representative of the Canadian population.

The results of another survey of 500 Canadians using Twitter are shown in the following probability table. They are classified by education level and age in the probability table.[4]

	Age					
Education	19–25 (*A*)	26–32 (*B*)	33–39 (*C*)	40–46 (*D*)	Other (*E*)	Total
High School (*H*)	0.17	0.11	0.09	0.04	0.01	0.42
Post-secondary (*G*)	0.22	0.13	0.11	0.05	0.01	0.52
Other (*O*)	0.03	0.02	0.01	0	0	0.06
Total	0.42	0.26	0.21	0.09	0.02	1.00

If one individual is selected at random from that group of 500 individuals, compute the following probabilities:

a. $P(A)$ **b.** $P(A \cap G)$ **c.** $P(A|G)$

d. $P(G)$ **e.** $P(A \cup G)$ **f.** $P(D|O)$

g. Are events *A* and *G* independent? Justify your answer.

h. Are events *D* and *O* mutually exclusive? Why or why not?

i. Are events *D* and *O* independent? Explain.

j. Are events *A* and *G* mutually exclusive? Why or why not?

4.65 Plant Genetics Gregor Mendel was a monk who suggested in 1865 a theory of inheritance based on the science of genetics. He identified heterozygous individuals for flower colour that had two alleles (one r = recessive white colour allele and one R = dominant red colour allele). When these individuals were mated, 3/4 of the offspring were observed to have red flowers and 1/4 had white flowers. The table summarizes this mating; each parent gives one of its alleles to form the gene of the offspring.

	Parent 2	
Parent 1	r	R
r	rr	rR
R	Rr	RR

We assume that each parent is equally likely to give either of the two alleles and that, if either one or two of the alleles in a pair is dominant (R), the offspring will have red flowers.

a. What is the probability that an offspring in this mating has at least one dominant allele?

b. What is the probability that an offspring has at least one recessive allele?

c. What is the probability that an offspring has one recessive allele, given that the offspring has red flowers?

4.66 Online Dating According to a BBC poll, nearly one-third of all Web users surf the Internet for a boyfriend or girlfriend. The BBC poll surveyed close to 11,000 Internet users in 19 countries. In Canada, according to another poll, only one-quarter of Internet users have tried online dating; 69% said that they would most likely not. Further, 64% of online daters say common interests are the most important factor in finding a potential partner online, while 49% say physical characteristics (from photos and videos) are most important.[5] Let A be the event an online dater will find a potential partner based on common interest. On the other hand, let B be the event an online dater will find a potential partner based on physical characteristics. An online dater is selected at random, and the following probabilities are given:

$$P(A \cap B) = 0.23, P(A \cap B^c) = 0.13,$$
$$P(A^c \cap B) = 0.11, \text{ and } P(A^c \cap B^c) = 0.53.$$

a. What is the probability the randomly selected online dater will find a partner based on common interests?

b. What is the probability the randomly selected online dater will find a partner on at least one of the two factors?

c. If the randomly selected online dater finds a partner based on common interests, what is the probability he or she will also find a partner with regards to physical characteristics?

d. Are the events A and B mutually exclusive? Explain.

e. Are the events A and B independent? Justify your answer.

4.67 Student Well-Being Survey This exercise uses a collection of responses to a survey question by students in the Toronto District School Board.[6] The question asked the students about their social and emotional well-being. The table summarizes the results:

	Grades 7–8 (G_1)	Grades 9–12 (G_2)
Good about yourself (A)	0.80	0.70
Reasonably happy (B)	0.78	0.67
Able to enjoy daily activities (C)	0.76	0.64
Hopeful about the future (D)	0.72	0.62
You like the way you look (E)	0.67	0.58

Assuming that the percentage of participation from both groups is equal, if one individual is drawn at random, find the following probabilities:

a. $P(A)$ **b.** $P(C)$ **c.** $P(A \cap G_2)$
d. $P(G_2|A)$ **e.** $P(G_2|B)$ **f.** $P(G_2|C)$
g. $P(C|G_1)$ **h.** $P(B^c)$

4.68 Choosing a Mate Men and women often disagree on how they think about selecting a mate. Suppose that a poll of 1000 individuals in their twenties gave the following responses to the question of whether it is more important for their future mate to be able to communicate their feelings (F) than it is for that person to make a good living (G).

	Feelings (F)	Good Living (G)	Total
Men (M)	0.35	0.20	0.55
Women (W)	0.36	0.09	0.45
Total	0.71	0.29	1.00

If an individual is selected at random from this group of 1000 individuals, calculate the following probabilities:

a. $P(F)$ **b.** $P(G)$ **c.** $P(F|M)$
d. $P(F|W)$ **e.** $P(M|F)$ **f.** $P(W|G)$

4.69 Wage Losses and the "Motherhood Gap" According to a national survey by TD Economics, 81% of working mothers in Canada re-enter the workforce after leaving their jobs to have kids.[7] Three moms were randomly selected:

a. What is the probability that the first selected mom has not re-entered the workforce?

b. What is the probability that all three moms re-entered the workforce?

c. What is the probability that none of them has re-entered the workforce?

d. What is the probability that at least two have re-entered the workforce?

4.70 Alex and Steven The top two 2014–2015 NHL shooting percentage leaders are Alex Tanguay and Steven Stamkos. Hockeyreference.com reports that the shooting percentage for Alex is 18.66, while for Steven it is 17.66[8]. Assume that the shots are independent, and that each player takes two shots during a particular game.

a. What is the probability that Alex makes both of his shots?

b. What is the probability that Steven makes exactly one of his two shots?

c. What is the probability that Steven makes both of his shots, and Alex makes neither of his?

4.71 Golfing Player A has entered a golf tournament but it is not certain whether player B will enter. Player A has a probability of 1/6 of winning the tournament if player B enters, and a probability of 3/4 of winning if player B does not enter the tournament. If the probability that player B enters is 1/3, find the probability that player A wins the tournament.

4.7 BAYES' RULE (OPTIONAL)

Colourblindness Let us reconsider the experiment involving colourblindness from Section 4.6. Notice that the two events

B: person selected is a man

B^c: person selected is a woman

taken together make up the sample space S, consisting of both men and women. Since colourblind people can be either male or female, the event A, which is that a person is colourblind, consists of both those simple events that are in A **and** B and those simple events that are in A **and** B^c. Since these two *intersections* are *mutually exclusive,* you can write the event A as

$$A = (A \cap B) \cup (A \cap B^c)$$

and

$$P(A) = P(A \cap B) + P(A \cap B^c)$$
$$= 0.04 + 0.002 = 0.042$$

Suppose now that the sample space can be partitioned into k subpopulations, $S_1, S_2, S_3, \ldots, S_k$, that, as in the colourblindness example, are **mutually exclusive and exhaustive;** that is, taken together they make up the entire sample space. In a similar way, you can express an event A as

$$A = (A \cap S_1) \cup (A \cap S_2) \cup (A \cap S_3) \cup \cdots \cup (A \cap S_k)$$

Then

$$P(A) = P(A \cap S_1) + P(A \cap S_2) + P(A \cap S_3) + \cdots + P(A \cap S_k)$$

This is illustrated for $k = 3$ in Figure 4.14.

FIGURE 4.14
Decomposition of event A

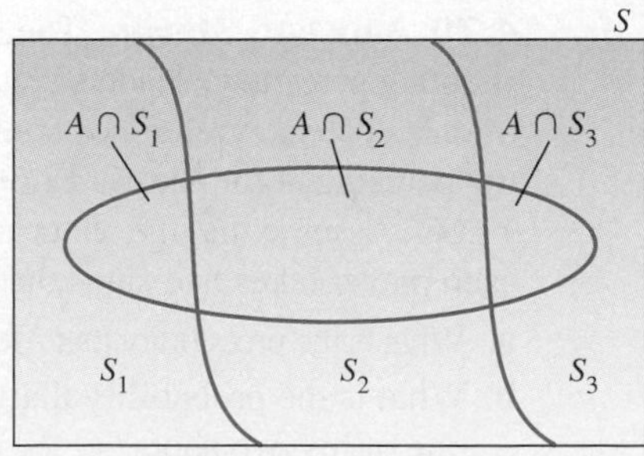

You can go one step further and use the Multiplication Rule to write $P(A \cap S_i)$ as $P(S_i)P(A|S_i)$, for $i = 1, 2, \ldots, k$. The result is known as the **Law of Total Probability.**

LAW OF TOTAL PROBABILITY

Given a set of events $S_1, S_2, S_3, \ldots, S_k$ that are mutually exclusive and exhaustive and an event A, the probability of the event A can be expressed as

$$P(A) = P(S_1)P(A|S_1) + P(S_2)P(A|S_2) + P(S_3)P(A|S_3) + \cdots + P(S_k)P(A|S_k)$$

EXAMPLE 4.27

From the 2006 population estimates by Statistics Canada, Table 4.7 gives the fraction of Canadians 15 years of age and older who are employed full-time in each of three age groups by gender.[9] The table also shows the fraction of Canadians 15 years of age and older in each age group.[10] Use the Law of Total Probability to determine the unconditional probability of a Canadian 15 years and older having a full-time job.

TABLE 4.7 **Probability Table**

	Male			Female		
	G_1 15–24	G_2 25–44	G_3 >45	G_4 15–24	G_5 25–44	G_6 >45
Fraction of Canadians employed full-time	0.36	0.80	0.50	0.28	0.61	0.33
Fraction of Canadians 15 years and older	0.08	0.18	0.23	0.08	0.18	0.25

SOLUTION Let A be the event that an individual chosen at random from the Canadian population 15 years and older is employed full-time. Let $G_1, G_2, \ldots, G_6$ represent the event that the individual selected belongs to each of the six age groups, respectively. Since the six age groups are exhaustive, you can write the event A as

$$A = (A \cap G_1) \cup (A \cap G_2) \cup (A \cap G_3) \cup (A \cap G_4) \cup (A \cap G_5) \cup (A \cap G_6)$$

Using the Law of Total Probability, you can find the probability of A as:

$$\begin{aligned} P(A) &= P(A \cap G_1) + P(A \cap G_2) + P(A \cap G_3) + P(A \cap G_4) + P(A \cap G_5) + P(A \cap G_6) \\ &= P(G_1)P(A|G_1) + P(G_2)P(A|G_2) + P(G_3)P(A|G_3) + P(G_4)P(A|G_4) \\ &\quad + P(G_5)P(A|G_5) + P(G_6)P(A|G_6) \end{aligned}$$

From the probabilities in Table 4.7,

$$\begin{aligned} P(A) &= (0.08)(0.36) + (0.18)(0.80) + (0.23)(0.50) + (0.08)(0.28) + (0.18)(0.61) \\ &\quad + (0.25)(0.33) \\ &= 0.5025 \end{aligned}$$

The *unconditional probability* that an individual selected at random from the population of Canadians 15 years of age and older is employed full-time is about 0.50. Notice that the Law of Total Probability is a weighted average of the probabilities within each group, with weights 0.08, 0.18, 0.23, 0.08, 0.18, and 0.25, which reflect the relative sizes of the groups.

Often you need to find the conditional probability of an event B, given that an event A has occurred. One such situation occurs in screening tests, which used to be associated primarily with medical diagnostic tests but are now finding applications in a variety of fields. Automatic test equipment is routinely used to inspect parts in high-volume production processes. Steroid testing of athletes, home pregnancy tests, and AIDS testing are some other applications. Screening tests are evaluated on the probability of a false negative or a false positive, and both of these are *conditional probabilities.*

A **false positive** is the event that the test is positive for a given condition, given that the person does not have the condition. A **false negative** is the event that the test is negative for a given condition, given that the person has the condition. You can evaluate these conditional probabilities using a formula derived by the probabilist Thomas Bayes.

The experiment involves selecting a sample from one of k subpopulations that are mutually exclusive and exhaustive. Each of these subpopulations, denoted by $S_1, S_2, \ldots, S_k$, has a selection probability $P(S_1), P(S_2), P(S_3), \ldots, P(S_k)$, called *prior probabilities.* An event A is observed in the selection. What is the probability that the sample came from subpopulation S_i, given that A has occurred?

You know from Section 4.6 that $P(S_i|A) = [P(A \cap S_i)]/P(A)$, which can be rewritten as $P(S_i|A) = [P(S_i)P(A|S_i)]/P(A)$. Using the Law of Total Probability to rewrite $P(A)$, you have

$$P(S_i|A) = \frac{P(S_i)P(A|S_i)}{P(S_1)P(A|S_1) + P(S_2)P(A|S_2) + P(S_3)P(A|S_3) + \cdots + P(S_k)P(A|S_k)}$$

These new probabilities are often referred to as *posterior probabilities*—that is, probabilities of the subpopulations (also called *states of nature*) that have been updated after observing the sample information contained in the event A. Bayes suggested that if the prior probabilities are unknown, they can be taken to be $1/k$, which implies that each of the events S_1 through S_k is equally likely.

BAYES' RULE

Let $S_1, S_2, \ldots, S_k$ represent k mutually exclusive and exhaustive subpopulations with prior probabilities $P(S_1), P(S_2), \ldots, P(S_k)$. If an event A occurs, the posterior probability of S_i given A is the conditional probability

$$P(S_i|A) = \frac{P(S_i)P(A|S_i)}{\sum_{j=1}^{k} P(S_j)P(A|S_j)}$$

for $i = 1, 2, \ldots, k$

EXAMPLE 4.28

The XL-2100 lamp unit is designed to be used with the Sony LCD projection TV "Grand WEGA." An electronics company purchases the lamp unit from three manufacturers located in Japan, China, and Korea respectively. The company located in Japan delivers the shipment on time 90% of the time, the company in Korea makes the shipment on time 85% of the time, whereas the company in China is 95% on time. The past record shows that the electronics company placed the orders with the company in China 70% of the time and to the company in Japan about 20% of the time. If a lamp unit was selected at random and identified as being shipped in time, find the probability that the selected unit was manufactured by the company located in China.

SOLUTION Let A be the event that a lamp unit was shipped in time. Let S_1, S_2, and S_3 represent the event that the lamp unit was manufactured in Japan, China, and Korea, respectively. Here we are interested in finding posterior probability, that is, $P(S_2|A)$. Note that the three locations are exhaustive and form a partition of the sample space for the experiment of selecting one manufacturing location. Then,

$$P(S_2|A) = P(S_2 \text{ and } A)/P(A) = P(A|S_2)P(S_2)/P(A)$$

$$P(A) = P(A \text{ and } S_1) + P(A \text{ and } S_2) + P(A \text{ and } S_3).$$

Note that S_1, S_2, and S_3 are mutually exclusive and exhaustive.

$$P(A) = P(A|S_1)P(S_1) + P(A|S_2)P(S_2) + P(A|S_3)P(S_3)$$
$$= (0.90)(0.20) + (0.95)(0.70) + (0.85)(0.10)$$
Note: $P(S_1) = 1 - 0.20 - 0.70 = 0.10$
$$= 0.180 + 0.665 + 0.085 = 0.93.$$
$$P(S_2|A) = P(A|S_2)P(S_2)/P(A) = (0.95)(0.70)/0.93 = 0.715.$$

Note that the posterior probability 0.715 is slightly greater than the prior probability of 0.70, since the plant in China has the highest probability of shipping the item on time. The updated (posterior) probability incorporates this fact!

EXAMPLE 4.29

Refer to Project 3-A. Canada has a Road Safety Vision of having the safest roads in the world. Yet, the leading cause of death of Canadian children remains vehicle crashes. In 2006, a national child seat safety survey was conducted by an AUTO21 research team in collaboration with Transport Canada to empirically measure Canada's progress toward achieving Road Safety Vision 2010. Child seat use was observed in parking lots and nearby intersections in 200 randomly selected sites across Canada. The restraints were classified in four categories: rear-facing infant seat, forward-facing, booster seat, and seat belt only. Further, the ages were divided in four groups: infant, toddler, school years, and over 9 years. Their findings show that 21.3% of toddlers were in a rear-facing infant seat, 76.3% of toddlers were in a forward-facing infant seat, 20% of toddlers were in a booster seat, whereas only 0.3% of toddlers were wearing a seat belt only. Further, the survey results show that among all the children 9.5% were in a rear-facing infant seat, 26.3% were in a forward-facing seat, 24.5% were in a booster seat, and 39.7% were wearing a seat belt only. If a child is selected at random and observed to be a toddler, find the probability that the child was in a rear-facing infant seat.

SOLUTION Let A be the event that a child selected at random is a toddler. Further, let S_1, S_2, S_3, and S_4 represent the event that the selected child used each of the four types of restraint.

Here, we will be calculating a posterior probability using Bayes' rule. In other words, we are supposed to calculate a conditional probability. Thus, we are interested in calculating $P(S_1|A)$, which is given by

$$P(S_1|A) = P(S_1 \text{ and } A)/P(A) = P(A|S_1)P(S_1)/P(A)$$
$$P(A) = P(A \text{ and } S_1) + P(A \text{ and } S_2) + P(A \text{ and } S_3) + P(A \text{ and } S_4)$$

Note that S_1, S_2, S_3, and S_4 are mutually exclusive and exhaustive.

$$P(A) = P(A|S_1)P(S_1) + P(A|S_2)P(S_2) + P(A|S_3)P(S_3) + P(A|S_4)P(S_4)$$
$$= (0.213)(0.095) + (0.763)(0.263) + (0.200)(0.245) + (0.003)(0.397)$$
$$= 0.020 + 0.201 + 0.049 + 0.001 = 0.271.$$
$$P(S_1|A) = P(S_1 \text{ and } A)/P(A) = P(A|S_1)P(S_1)/P(A) = (0.213)(0.095)/0.271$$
$$= 0.020/0.271 = 0.075$$

EXAMPLE 4.30

What are the deal breakers for you on the dating scene? Several questions were asked of respondents in an online survey conducted by the *Globe and Mail* on Friday, June 25, 2009. More than 7500 Canadians coast to coast participated in this poll. One of the questions was: Which of the following is the biggest dating deal breaker for you? The percentages of affirmative responses given by females and males to each category of deal breaker are below:

Has kids: Female 4%; Male 9%
Uneducated: Female 21%; Male 27%
Lack of financial stability: Female 13%; Male 7%

Suppose that in this survey about 69% of respondents were female. In a follow-up study, a person is selected at random.

1. What is the probability that the chosen person will respond that "has kids" is the biggest dating deal breaker given that the individual is a female?
2. What is the probability that the respondent is a female and thinks "uneducated" is the biggest dating deal breaker?
3. What is the probability that a selected person is a female given that she considers that "has kids" is the biggest dating deal breaker factor?

SOLUTION 1. You need to find the conditional probability. First, let A be the event that the selected individual responds with "has kids" being the greatest deal breaker and F be the event that the selected person is a female. In other words, we are interested in finding $P(A|F)$.

The poll results provide $P(A \text{ and } F) = 0.04$ and

$$P(A|F) = P(A \text{ and } F)/P(F) = (0.04)/(0.69) = 0.0580.$$

2. Let B be the event "uneducated." You are supposed to find $P(B \text{ and } F)$.

$$P(B \text{ and } F) = 0.21.$$

3. In this question you are asked to calculate the conditional probability (posterior probability), that is, $P(F|A)$. Thus,

$$P(F|A) = P(A \text{ and } F)/P(A),$$
$$P(A) = P(A \text{ and } F) + P(A \text{ and } M) = 0.04 + 0.09 = 0.13$$
$$P(F|A) = P(A \text{ and } F)/P(A) = 0.04/0.13 = 0.3077.$$

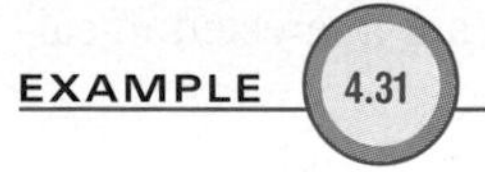
EXAMPLE 4.31

Refer to Example 4.27. Find the probability that the person selected is 45 years of age or older, and female, given that the individual is employed full-time.

SOLUTION You need to find the conditional probability given by

$$P(G_6|A) = \frac{P(A \cap G_6)}{P(A)}$$

You have already calculated $P(A) = 0.5025$ using the Law of Total Probability, therefore,

$$P(G_6|A) = \frac{P(G_6)P(A|G_6)}{\sum_{j=1}^{6} P(G_j)P(A|G_j)}$$

$$= \frac{(0.25)(0.33)}{(0.08)(0.36) + (0.18)(.80) + (0.23)(0.50) + (0.08)(0.28) + (0.18)(0.61) + (0.25)(0.33)}$$

$$= \frac{0.0825}{0.5025} = 0.1642$$

In this case, the posterior probability of 0.16 is somewhat less than the prior probability of 0.25 (from Table 4.7). This group *a priori* was the highest, and only a small proportion of this segment was employed full-time.

What is the posterior probability that the individual selected is a male in the 25–44 age range, given that the individual is employed full-time?

$$P(G_2|A) = \frac{(0.18)(0.80)}{(0.08)(0.36) + (0.18)(0.80) + (0.23)(0.50) + (0.08)(0.28) + (0.18)(0.61) + (.25)(.33)}$$

$$= \frac{0.144}{0.5025} = 0.2866$$

The posterior probability of 0.29 is substantially greater than the prior probability of 0.18. In effect this age group was *a priori* the second smallest segment in the population sampled, but at the same time, the proportion of individuals in this group who were employed full-time had the highest probability of any age groups. These two facts taken together cause an upward adjustment of around a third in the prior probability of 0.18.

Bayesian and Frequentist Perspectives in Statistics

There are two schools of thought in statistical science literature: the classical (Frequentist) and Bayesian methodologies. The main difference between Frequentist and Bayesian statistics boils down to different interpretations of probability. A Frequentist defines probability in the repeated sampling context, or using the notion of "what happens in long run." For example, the probability of having baby girls (or tossing a coin that turns heads) is 1/2. A Frequentist will argue that if the probability of having baby girls is 1/2, which suggests if we repeat this experiment of having babies, it is highly probable that approximately the same number of boys and girls will be born. In other words, the chances of having human boys and girls in the world are 50/50. However, Bayesian point of view is much different, in that assigning a probability to an outcome of the experiment is rather personal and is based on the experimenter's judgment. In some instances this judgment may be some sort of conjecture about the outcome of the experiment. For this reason Bayesian perspective is commonly known as a subjectivist point of view. The Bayesian will argue that the probability of having baby girls being 1/2 means that one can safely guess that the probability for having either girls or boys is 1/2; however, the gender of a baby is not important, and this fact can be used for further study. In some sense, the Bayesian elucidation of probability can be viewed as an extension of logic that provides reasoning with uncertainty. In a Bayesian framework, some prior probability is specified, which is then updated using current available data. The Bayesian probability calculation may incorporate (but not necessarily) both sample and non-sample information (NSI) in terms of prior information (prior probability).

The Bayesian probability calculation is already defined and discussed in the previous section. Here we summarize some terminology and definitions in the calculation of Bayesian probability. Let S be a conjecture or hypothesis, and let A be the data, then the Bayesian inference uses Bayes' formula for conditional probability:

$$P(S|A) = P(A|S)P(S)/P(A)$$

$P(S)$ is the *prior probability* of S. In other words, the probability that S is correct before the data A was observed.

$P(A|S)$ is the *conditional probability* of observed data A given that the conjecture S is true.

$P(A|S)$ is also termed as the *likelihood*.

$P(A)$ is called the *marginal probability* of data A.

Thus, given exhaustive set of mutually exclusive hypotheses S_i, $i = 1,2, \ldots k$.

$$P(A) = P(A|S_1)P(S_1) + P(A|S_2)P(S_2) + P(A|S_3)P(S_3) + \ldots + P(A|S_k)P(S_k)$$

The outcome $P(S|A)$ is called the *posterior probability*, the probability that the hypothesis is true, given the data and the previous belief about the hypothesis.

4.7 EXERCISES

BASIC TECHNIQUES

4.72 Bayes' Rule A sample is selected from one of two populations, S_1 and S_2, with probabilities $P(S_1) = 0.7$ and $P(S_2) = 0.3$. If the sample has been selected from S_1, the probability of observing an event A is $P(A|S_1) = 0.2$. Similarly, if the sample has been selected from S_2, the probability of observing A is $P(A|S_2) = 0.3$.

a. If a sample is randomly selected from one of the two populations, what is the probability that event A occurs?

b. If the sample is randomly selected and event A is observed, what is the probability that the sample was selected from population S_1? From population S_2?

4.73 Bayes' Rule II If an experiment is conducted, one and only one of three mutually exclusive events S_1, S_2, and S_3 can occur, with these probabilities:

$$P(S_1) = 0.2 \qquad P(S_2) = 0.5 \qquad P(S_3) = 0.3$$

The probabilities of a fourth event A occurring, given that event S_1, S_2, or S_3 occurs, are

$$P(A|S_1) = 0.2 \quad P(A|S_2) = 0.1 \quad P(A|S_3) = 0.3$$

If event A is observed, find $P(S_1|A)$, $P(S_2|A)$, and $P(S_3|A)$.

4.74 Law of Total Probability A population can be divided into two subgroups that occur with probabilities 60% and 40%, respectively. An event A occurs 30% of the time in the first subgroup and 50% of the time in the second subgroup. What is the unconditional probability of the event A, regardless of which subgroup it comes from?

APPLICATIONS

4.75 Violent Crime City crime records show that 20% of all crimes are violent and 80% are non-violent, involving theft, forgery, and so on. Ninety percent of violent crimes are reported versus 70% of non-violent crimes.

a. What is the overall reporting rate for crimes in the city?

b. If a crime in progress is reported to the police, what is the probability that the crime is violent? What is the probability that it is non-violent?

c. Refer to part b. If a crime in progress is reported to the police, why is it more likely that it is a non-violent crime? Wouldn't violent crimes be more likely to be reported? Can you explain these results?

4.76 Worker Error A worker-operated machine produces a defective item with probability 0.01 if the worker follows the machine's operating instructions exactly, and with probability 0.03 if the worker does not. If the worker follows the instructions 90% of the time, what proportion of all items produced by the machine will be defective?

4.77 Airport Security Suppose that, in a particular city, airport A handles 50% of all airline traffic, and airports B and C handle 30% and 20%, respectively. The detection rates for weapons at the three airports are 0.9, 0.5, and 0.4, respectively. If a passenger at one of the airports is found to be carrying a weapon through the boarding gate, what is the probability that the passenger is using airport A? Airport C?

4.78 Football Strategies A particular football team is known to run 30% of its plays to the left and 70% to the right. A linebacker on an opposing team notes that the right guard shifts his stance most of the time (80%) when plays go to the right and that he uses a balanced stance the remainder of the time. When plays go to the left, the guard takes a balanced stance 90% of the time and the shift stance the remaining 10%. On a particular play, the linebacker notes that the guard takes a balanced stance.

a. What is the probability that the play will go to the left?

b. What is the probability that the play will go to the right?

c. If you were the linebacker, which direction would you prepare to defend if you saw the balanced stance?

4.79 No Pass, No Play Many public schools are implementing a "no pass, no play" rule for athletes. Under this system, a student who fails a course is disqualified from participating in extracurricular activities during the next grading period. Suppose the probability that an athlete who has not previously been disqualified will be disqualified is 0.15 and the probability that an athlete who has been disqualified will be disqualified again in the next time period is 0.5. If 30% of the athletes have been disqualified before, what is the unconditional probability that an athlete will be disqualified during the next grading period?

4.80 Medical Diagnostics Medical case histories indicate that different illnesses may produce identical symptoms. Suppose a particular set of symptoms, which we will denote as event *H*, occurs only when any one of three illnesses—*A*, *B*, or *C*—occurs. (For the sake of simplicity, we will assume that illnesses *A*, *B*, and *C* are mutually exclusive.) Studies show these probabilities of getting the three illnesses:

$$P(A) = 0.01$$
$$P(B) = 0.005$$
$$P(C) = 0.02$$

The probabilities of developing the symptoms *H*, given a specific illness, are

$$P(H|A) = 0.90$$
$$P(H|B) = 0.95$$
$$P(H|C) = 0.75$$

Assuming that an ill person shows the symptoms *H*, what is the probability that the person has illness *A*?

4.81 Cheating on Your Taxes? Suppose 5% of all people filing the income tax form seek deductions that they know are illegal, and an additional 2% incorrectly list deductions because they are unfamiliar with income tax regulations. Of the 5% who are guilty of cheating, 80% will deny knowledge of the error if confronted by an investigator. If the filer of the long form is confronted with an unwarranted deduction and he or she denies the knowledge of the error, what is the probability that he or she is guilty?

4.82 Screening Tests Suppose that a certain disease is present in 10% of the population, and that there is a screening test designed to detect this disease if present. The test does not always work perfectly. Sometimes the test is negative when the disease is present, and sometimes it is positive when the disease is absent. The table below shows the proportion of times that the test produces various results:

	Test Is Positive (*P*)	Test Is Negative (*N*)
Disease present (*D*)	0.08	0.02
Disease absent (D^c)	0.05	0.85

a. Find the following probabilities from the table: $P(D)$, $P(D^c)$, $P(N|D^c)$, $P(N|D)$.

b. Use Bayes' Rule and the results of part a to find $P(D|N)$.

c. Use the definition of conditional probability to find $P(D|N)$. (Your answer should be the same as the answer to part b.)

d. Find the probability of a false positive, that the test is positive, given that the person is disease-free.

e. Find the probability of a false negative, that the test is negative, given that the person has the disease.

f. Are either of the probabilities in parts d or e large enough that you would be concerned about the reliability of this screening method? Explain.

4.8 DISCRETE RANDOM VARIABLES AND THEIR PROBABILITY DISTRIBUTIONS

In Chapter 1, *variables* were defined as characteristics that change or vary over time and/or for different individuals or objects under consideration. *Quantitative variables* generate numerical data, whereas *qualitative variables* generate categorical data.

However, even qualitative variables can generate numerical data if the categories are numerically coded to form a scale. For example, if you toss a single coin, the qualitative outcome could be recorded as "0" if a head and "1" if a tail.

Random Variables

A numerically valued variable X will vary or change depending on the particular outcome of the experiment being measured. For example, suppose you toss a die and measure X, the number observed on the upper face. The variable X can take on any of six values—1, 2, 3, 4, 5, 6—depending on the *random* outcome of the experiment. For this reason, we refer to the variable X as a **random variable.**

DEFINITION A variable x is a **random variable** if the value that it assumes, corresponding to the outcome of an experiment, is a chance or random event.

You can think of many examples of random variables:

- X = Number of defects on a *randomly selected* piece of furniture
- X = GPA score for a *randomly selected* university student
- X = Number of telephone calls received by a crisis intervention hotline during a *randomly selected* time period

As in Chapter 1, quantitative random variables are classified as either *discrete* or *continuous,* according to the values that X can assume. It is important to distinguish between discrete and continuous random variables because different techniques are used to describe their distributions. We focus on discrete random variables in the remainder of this chapter; continuous random variables are the subject of Chapter 6.

Probability Distributions

In Chapters 1 and 2, you learned how to construct the *relative frequency distribution* for a set of numerical measurements on a variable X. The distribution gave this information about X:

- What values of X occurred
- How often each value of X occurred

You also learned how to use the mean and standard deviation to measure the centre and variability of this data set.

In this chapter, we defined *probability* as the limiting value of the relative frequency as the experiment is repeated over and over again. Now we define the **probability distribution** for a random variable X as the *relative frequency distribution* constructed for the entire population of measurements.

DEFINITION The **probability distribution** for a discrete random variable is a formula, table, or graph that gives the possible values of X, and the probability $p(x)$ associated with each value of x.

The values of X represent mutually exclusive numerical events. Summing $p(x)$ over all values of x is equivalent to adding the probabilities of all simple events and therefore equals 1.

REQUIREMENTS FOR A DISCRETE PROBABILITY DISTRIBUTION

- $0 \leq p(x) \leq 1$
- $\Sigma\, p(x) = 1$

EXAMPLE 4.32 Toss two fair coins and let X equal the number of heads observed. Find the probability distribution for X.

SOLUTION The simple events for this experiment with their respective probabilities are listed in Table 4.8.

TABLE 4.8 **Simple Events and Probabilities in Tossing Two Coins**

Simple Event	Coin 1	Coin 2	$P(E_i)$	x
E_1	H	H	1/4	2
E_2	H	T	1/4	1
E_3	T	H	1/4	1
E_4	T	T	1/4	0

Since E_1 = HH results in two heads, this simple event results in the value $x = 2$. Similarly, the value $x = 1$ is assigned to E_2, and so on. For each value of x, you can calculate $p(x)$ by adding the probabilities of the simple events in that event. For example, when $x = 0$,

$$p(0) = P(E_4) = \frac{1}{4}$$

and when $x = 1$,

$$p(1) = P(E_2) + P(E_3) = \frac{1}{2}$$

The values of x and their respective probabilities, $p(x)$, are listed in Table 4.9. Notice that the probabilities add to 1.

TABLE 4.9 **Probability Distribution for X (X = Number of Heads)**

x	Simple Events in x	$p(x)$
0	E_4	1/4
1	E_2, E_3	1/2
2	E_1	1/4
		$\Sigma\, p(x) = 1$

ONLINE APPLET
Flipping Coins

The probability distribution in Table 4.9 can be graphed using the methods of Section 1.5 to form the **probability histogram** in Figure 4.15.† The three values of the random variable X are located on the horizontal axis, and the probabilities $p(x)$ are located on the vertical axis (replacing the relative frequencies used in Chapter 1). Since the width of each bar is 1, the area under the bar is the probability of observing the particular value of x and the total area equals 1.

† The probability distribution in Table 4.9 can also be presented using a formula, which is given in Section 5.2.

FIGURE 4.15
Probability histogram for Example 4.32

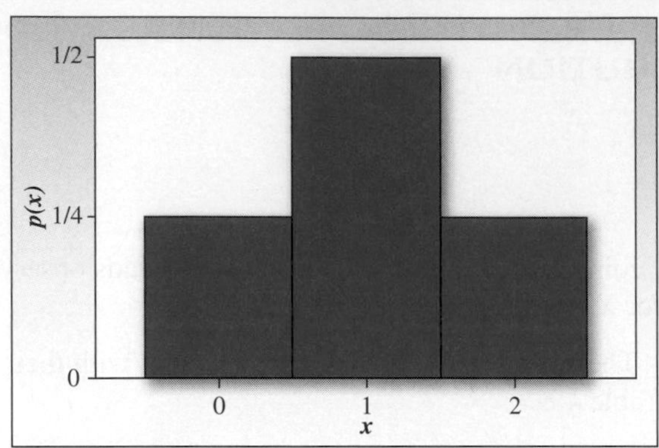

The Mean and Standard Deviation for a Discrete Random Variable

The probability distribution for a discrete random variable looks very similar to the relative frequency distribution discussed in Chapter 1. The difference is that the relative frequency distribution describes a *sample* of n measurements, whereas the probability distribution is constructed as a model for the *entire population* of measurements. Just as the mean $\bar{x}$ and the standard deviation s measured the centre and spread of the sample data, you can calculate similar measures to describe the centre and spread of the population.

The population mean, which measures the average value of x in the population, is also called the **expected value** of the random variable X. It is the value that you would *expect* to observe on *average* if the experiment is repeated over and over again. The formula for calculating the population mean is easier to understand by example. Toss those two fair coins again, and let x be the number of heads observed. We constructed this probability distribution for x:

x	0	1	2
$p(x)$	1/4	1/2	1/4

Suppose the experiment (tossing coins) is repeated a large number of times—say, $n = 4{,}000{,}000$ times. Intuitively, you would expect to observe approximately 1 million zeros, 2 million ones, and 1 million twos. Then the average value of x would equal

$$\frac{\text{Sum of measurements}}{n} = \frac{1{,}000{,}000(0) + 2{,}000{,}000(1) + 1{,}000{,}000(2)}{4{,}000{,}000}$$

$$= \left(\frac{1}{4}\right)(0) + \left(\frac{1}{2}\right)(1) + \left(\frac{1}{4}\right)(2)$$

Note that the first term in this sum is $(0)p(0)$, the second is equal to $(1)p(1)$, and the third is $(2)p(2)$. The average value of X, then, is

$$\Sigma xp(x) = 0 + \frac{1}{2} + \frac{2}{4} = 1$$

This result provides some intuitive justification for the definition of the expected value of a discrete random variable X.

DEFINITION Let X be a discrete random variable with probability distribution $p(x)$. The mean or **expected value of** X is given as

$$\mu = E(X) = \Sigma xp(x)$$

where the elements are summed over all values of the random variable x.

We could use a similar argument to justify the formulas for the **population variance** σ^2 and the **population standard deviation** σ. These numerical measures describe the spread or variability of the random variable using the "average" or "expected value" of the squared deviations of the x-values from their mean μ.

DEFINITION Let X be a discrete random variable with probability distribution $p(x)$ and mean μ. The **variance of** x is

$$\sigma^2 = E[(X - \mu)^2] = \Sigma(x - \mu)^2 p(x)$$

where the summation is over all values of the random variable x.†

DEFINITION The **standard deviation σ of a random variable** X is equal to the positive square root of its variance.

EXAMPLE  4.33

An electronics store sells a particular model of computer laptop. There are only four laptops in stock, and the manager wonders what today's demand for this particular model will be. She learns from the marketing department that the probability distribution for X, the daily demand for the laptop, is as shown in the table. Find the mean, variance, and standard deviation of X. Is it likely that five or more customers will want to buy a laptop today?

x	0	1	2	3	4	5
$p(x)$	0.10	0.40	0.20	0.15	0.10	0.05

SOLUTION Table 4.10 shows the values of x and $p(x)$, along with the individual terms used in the formulas for μ and σ^2. The sum of the values in the third column is

$$\mu = \Sigma xp(x) = (0)(.10) + (1)(.40) + \cdots + (5)(.05) = 1.90$$

while the sum of the values in the fifth column is

$$\sigma^2 = \Sigma(x - \mu)^2 p(x)$$
$$= (0 - 1.9)^2(.10) + (1 - 1.9)^2(.40) + \cdots + (5 - 1.9)^2(.05) = 1.79$$

and

$$\sigma = \sqrt{\sigma^2} = \sqrt{1.79} = 1.34$$

† It can be shown (proof omitted) that

$$\sigma^2 = \Sigma(x - \mu)^2 p(x) = \Sigma x^2 p(x) - \mu^2$$

This result is analogous to the computing formula for the sum of squares of deviations given in Chapter 2.

TABLE 4.10 **Calculations for Example 4.33**

x	$p(x)$	$xp(x)$	$(x-\mu)^2$	$(x-\mu)^2\, p(x)$
0	0.10	0.00	3.61	0.361
1	0.40	0.40	0.81	0.324
2	0.20	0.40	0.01	0.002
3	0.15	0.45	1.21	0.1815
4	0.10	0.40	4.41	0.441
5	0.05	0.25	9.61	0.4805
Total	1.00	$\mu = 1.90$		$\sigma^2 = 1.79$

The graph of the probability distribution is shown in Figure 4.16. Since the distribution is approximately mound-shaped, approximately 95% of all measurements should lie within *two* standard deviations of the mean—that is,

$$\mu \pm 2\sigma \Rightarrow 1.90 \pm 2(1.34) \quad \text{or} \; -0.78 \text{ to } 4.58$$

Since $x = 5$ lies outside this interval, you can say it is unlikely that five or more customers will want to buy a laptop today. In fact, $P(X \geq 5)$ is exactly 0.05, or 1 time in 20.

FIGURE 4.16
Probability distribution for Example 4.33

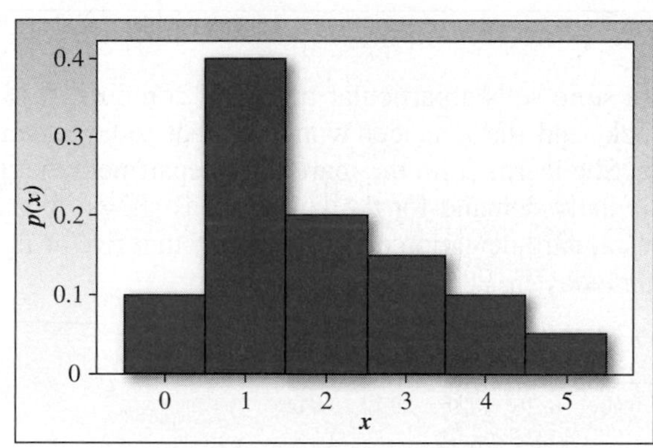

EXAMPLE 4.34

In a lottery conducted to benefit the local fire station, 8000 tickets are to be sold at $5 each. The prize is $12,000. If you purchase two tickets, what is your expected gain?

SOLUTION Your gain x may take one of two values. You will either lose $10 (i.e., your "gain" will be −$10) or win $11,990, with probabilities 7998/8000 and 2/8000, respectively. The probability distribution for the gain x is shown in the table:

x	$p(x)$
−$10	7998/8000
$11,990	2/8000

The expected gain will be

$$\mu = E(X) = \Sigma xp(x)$$

$$= (-\$10)\left(\frac{7998}{8000}\right) + (\$11{,}900)\left(\frac{2}{8000}\right) = -\$7$$

Recall that the expected value of X is the average of the theoretical population that would result if the lottery were repeated an infinitely large number of times. If this were done, your average or expected gain per lottery ticket would be a loss of \$7.

EXAMPLE 4.35 Determine the yearly premium for a \$1000 insurance policy covering an event that, over a long period of time, has occurred at the rate of 2 times in 100. Let X equal the yearly financial gain to the insurance company resulting from the sale of the policy, and let C equal the unknown yearly premium. Calculate the value of C such that the expected gain $E(X)$ will equal zero. Then C is the premium required to break even. To this, the company would add administrative costs and profit.

SOLUTION The first step in the solution is to determine the values that the gain x may take and then to determine $p(x)$. If the event does not occur during the year, the insurance company will gain the premium of $X = C$ dollars. If the event does occur, the gain will be negative; that is, the company will lose \$1000 less the premium of C dollars already collected. Then $x = -(1000 - C)$ dollars. The probabilities associated with these two values of x are 98/100 and 2/100, respectively. The probability distribution for the gain is shown in the table:

x = Gain	$p(x)$
C	98/100
$-(1000 - C)$	2/100

Since the company wants the insurance premium C such that, in the long run (for many similar policies), the mean gain will equal zero, you can set the expected value of x equal to zero and solve for C. Then

$$\mu = E(X) = \Sigma xp(x)$$

$$= C\left(\frac{98}{100}\right) + [-(1000 - C)]\left(\frac{2}{100}\right) = 0$$

or

$$\frac{98}{100}C + \frac{2}{100}C - 20 = 0$$

Solving this equation for C, you obtain $C = \$20$. Therefore, if the insurance company charged a yearly premium of \$20, the average gain calculated for a large number of similar policies would equal zero. The actual premium would equal \$20 plus administrative costs and profit.

The method for calculating the expected value of X for a continuous random variable is similar to what you have done, but in practice it involves the use of calculus. Nevertheless, the basic results concerning expectations are the same for continuous and discrete random variables. For example, regardless of whether X is continuous or discrete, $\mu = E(X)$ and $\sigma^2 = E[(X - \mu)^2]$.

4.8 EXERCISES

BASIC TECHNIQUES

4.83 Discrete or Continuous? Identify the following as discrete or continuous random variables:

a. Total number of points scored in a football game
b. Shelf life of a particular drug
c. Height of the ocean's tide at a given location
d. Length of a two-year-old black bass
e. Number of aircraft near-collisions in a year

4.84 Discrete or Continuous II Identify the following as discrete or continuous random variables:

a. Increase in length of life attained by a cancer patient as a result of surgery
b. Tensile breaking strength (in kilograms per square centimetre) of steel cable with a diameter of 2.5 cm
c. Number of deer killed per year in a state wildlife preserve
d. Number of overdue accounts in a department store at a particular time
e. Your blood pressure

4.85 Probability Distribution I A random variable *X* has this probability distribution:

x	0	1	2	3	4	5
$p(x)$	0.1	0.3	0.4	0.1	?	0.05

a. Find $p(4)$.
b. Construct a probability histogram to describe $p(x)$.
c. Find μ, σ^2, and σ.
d. Locate the interval $\mu \pm 2\sigma$ on the *x*-axis of the histogram. What is the probability that *X* will fall into this interval?
e. If you were to select a very large number of values of *X* from the population, would most fall into the interval $\mu \pm 2\sigma$? Explain.

4.86 Probability Distribution II A random variable *X* can assume five values: 0, 1, 2, 3, 4. A portion of the probability distribution is shown here:

x	0	1	2	3	4
$p(x)$	0.1	0.3	0.3	?	0.1

a. Find $p(3)$.
b. Construct a probability histogram for $p(x)$.
c. Calculate the population mean, variance, and standard deviation.
d. What is the probability that *X* is greater than 2?
e. What is the probability that *X* is 3 or less?

4.87 Dice Let *X* equal the number observed on the throw of a single balanced die.

a. Find and graph the probability distribution for *X*.
b. What is the average or expected value of *X*?
c. What is the standard deviation of *X*?
d. Locate the interval $\mu \pm 2\sigma$ on the *x*-axis of the graph in part a. What proportion of all the measurements would fall into this range?

4.88 Grocery Visits Let *X* represent the number of times a customer visits a grocery store in a one-week period. Assume this is the probability distribution of *X*:

x	0	1	2	3
$p(x)$	0.1	0.4	0.4	0.1

Find the expected value of *X*, the average number of times a customer visits the store.

APPLICATIONS

4.89 Canadians Welcome Refugees Following the terrorist attacks in Paris, France, in 2015, a poll conducted by IPSOS Public Affairs stated that 54% of Canadians supported the federal government's plan to accept 25,000 Syrian refugees by February 2016.[11] Suppose that you randomly select three Canadians and ask them if they support the federal government's plan

a. Find the probability distribution for *X*, the number of people in the population of three who support the federal government's plan.
b. Construct the probability histogram for $p(x)$.
c. What is the probability that exactly one of the three would support the federal government's plan?
d. What are the population mean and standard deviation for the random variable *X*?

4.90 Which Key Fits? A key ring contains four office keys that are identical in appearance, but only one will open your office door. Suppose you randomly select one key and try it. If it does not fit, you randomly select one of the three remaining keys. If it does not fit, you randomly select one of the last two. Each different sequence that could occur in selecting the keys represents one of a set of equiprobable simple events.

a. List the simple events in *S* and assign probabilities to the simple events.

b. Let X equal the number of keys that you try before you find the one that opens the door ($x = 1, 2, 3, 4$). Then assign the appropriate value of X to each simple event.

c. Calculate the values of $p(x)$ and display them in a table.

d. Construct a probability histogram for $p(x)$.

4.91 Roulette Exercise 4.10 described the game of roulette. Suppose you bet \$5 on a single number—say, the number 18. The payoff on this type of bet is usually 35 to 1. What is your expected gain?

4.92 Gender Bias? A company has five applicants for two positions: two women and three men. Suppose that the five applicants are equally qualified and that no preference is given for choosing either gender. Let x equal the number of women chosen to fill the two positions.

a. Find $p(x)$.

b. Construct a probability histogram for x.

4.93 Defective Equipment A piece of electronic equipment contains six computer chips, two of which are defective. Three chips are selected at random, removed from the piece of equipment, and inspected. Let X equal the number of defectives observed, where $x = 0$, 1, or 2. Find the probability distribution for X. Express the results graphically as a probability histogram.

4.94 Drilling Oil Wells Past experience has shown that, on the average, only 1 in 10 wells drilled hits oil. Let X be the number of drillings until the first success (oil is struck). Assume that the drillings represent independent events.

a. Find $p(1)$, $p(2)$, and $p(3)$.

b. Give a formula for $p(x)$.

c. Graph $p(x)$.

4.95 Tennis, Anyone? Two tennis professionals, A and B, are scheduled to play a match; the winner is the first player to win three sets in a total that cannot exceed five sets. The event that A wins any one set is independent of the event that A wins any other, and the probability that A wins any one set is equal to 0.6. Let X equal the total number of sets in the match; that is, $x = 3$, 4, or 5. Find $p(x)$.

4.96 Tennis, again The probability that a tennis player A can win a set from tennis player B is one measure of the comparative abilities of the two players. In Exercise 4.95 you found the probability distribution for X, the number of sets required to play a best-of-five-sets match, given that the probability that A wins any one set—call this $P(A)$—is 0.6.

a. Find the expected number of sets required to complete the match for $P(A) = 0.6$.

b. Find the expected number of sets required to complete the match when the players are of equal ability—that is, $P(A) = 0.5$.

c. Find the expected number of sets required to complete the match when the players differ greatly in ability—that is, say, $P(A) = 0.9$.

4.97 The PGA One professional golfer plays best on short-distance holes. Experience has shown that the numbers x of shots required for 3-, 4-, and 5-par holes have the probability distributions shown in the table:

Par-3 Holes		Par-4 Holes		Par-5 Holes	
x	$p(x)$	x	$p(x)$	x	$p(x)$
2	0.12	3	0.14	4	0.04
3	0.80	4	0.80	5	0.80
4	0.06	5	0.04	6	0.12
5	0.02	6	0.02	7	0.04

What is the golfer's expected score on these holes?

a. A par-3 hole

b. A par-4 hole

c. A par-5 hole

4.98 Insuring Your Diamonds You can insure a \$50,000 diamond for its total value by paying a premium of D dollars. If the probability of theft in a given year is estimated to be 0.01, what premium should the insurance company charge if it wants the expected gain to equal \$1000?

4.99 Health Canada Testing The maximum patent life for a new drug is, in most cases, 20 years. Subtracting the length of time required by Health Canada for testing and approval of the drug provides the actual patent life of the drug—that is, the length of time that a company has to recover research and development costs and make a profit. Suppose the distribution of the lengths of patent life for new drugs is as shown here:

Years, x	3	4	5	6	7	8
$p(x)$	0.03	0.05	0.07	0.10	0.14	0.20

Years, x	9	10	11	12	13
$p(x)$	0.18	0.12	0.07	0.03	0.01

a. Find the expected number of years of patent life for a new drug.

b. Find the standard deviation of X.

c. Find the probability that X falls into the interval $\mu \pm 2\sigma$.

4.100 No Time for Vegetables In a survey conducted by Ipsos Canada, 41% of Canadians say that because of their busy lifestyles, they find it hard to prepare and eat enough vegetables.[12] Suppose you had conducted your own telephone survey at the same time. You randomly called people and asked them whether they find time to eat vegetables. Assume that the percentage given in the Ipsos survey can be taken to approximate the percentage of all adult Canadians who find time to eat vegetables.

a. Find the probability for X, the number of calls until you find the first person who *does not* find time to eat vegetables.

b. What problems might arise as you randomly call people and ask them to take part in your survey? How would this affect the reliability of the probabilities calculated in part a?

4.101 Shipping Charges From experience, a shipping company knows that the cost of delivering a small package within 24 hours is \$14.80. The company charges \$15.50 for shipment but guarantees to refund the charge if delivery is not made within 24 hours. If the company fails to deliver only 2% of its packages within the 24-hour period, what is the expected gain per package?

4.102 Actuaries A manufacturing representative is considering taking out an insurance policy to cover possible losses incurred by marketing a new product. If the product is a complete failure, the representative feels that a loss of \$80,000 would be incurred; if it is only moderately successful, a loss of \$25,000 would be incurred. Insurance actuaries have determined from market surveys and other available information that the probabilities that the product will be a failure or only moderately successful are 0.01 and 0.05, respectively. Assuming that the manufacturing representative is willing to ignore all other possible losses, what premium should the insurance company charge for a policy in order to break even?

CHAPTER REVIEW

Key Concepts and Formulas

I. Experiments and the Sample Space

1. Experiments, events, mutually exclusive events, simple events
2. The sample space
3. Venn diagrams, tree diagrams, probability tables

II. Probabilities

1. Relative frequency definition of probability
2. Properties of probabilities
   a. Each probability lies between 0 and 1
   b. Sum of all simple-event probabilities equals 1
3. $P(A)$, the sum of the probabilities for all simple events in A

III. Counting Rules

1. mn Rule; extended mn Rule
2. Permutations: $P_r^n = \dfrac{n!}{(n-r)!}$
3. Combinations: $C_r^n = \dfrac{n!}{r!(n-r)!}$

IV. Event Relations

1. Unions and intersections
2. Events
   a. Disjoint or mutually exclusive: $P(A \cap B) = 0$
   b. Complementary: $P(A) = 1 - P(A^c)$
3. Conditional probability: $P(A|B) = \dfrac{P(A \cap B)}{P(B)}$
4. Independent and dependent events
5. Addition Rule: $P(A \cup B) = P(A) + P(B) - P(A \cap B)$
6. Multiplication Rule: $P(A \cap B) = P(A)P(B|A)$
7. Law of Total Probability
8. Bayes' Rule

V. Discrete Random Variables and Probability Distributions

1. Random variables, discrete and continuous
2. Properties of probability distributions
   a. $0 \le p(x) \le 1$
   b. $\Sigma p(x) = 1$
3. Mean or expected value of a discrete random variable: $\mu = \Sigma x p(x)$
4. Variance and standard deviation of a discrete random variable: $\sigma^2 = \Sigma(x-\mu)^2 p(x)$ and $\sigma = \sqrt{\sigma^2}$

TECHNOLOGY TODAY

Discrete Probability Distributions in *Microsoft Excel*

Although *Excel* cannot help you solve the types of general probability problems presented in this chapter, it is useful for calculating the mean, variance, and standard deviation of the random variable x. In Chapters 5 and 6, we will use *Excel* to calculate exact probabilities for three special cases: the binomial, the Poisson, and the normal random variables.

EXAMPLE 4.36 Suppose you have this general discrete probability distribution:

x	0	1	3	5
$p(x)$	0.25	0.35	0.25	0.15

1. Enter the values of x and $p(x)$ into columns A and B of a new *Excel* spreadsheet. Then create two columns—column C (named "$x^*p(x)$") and column D (named "$xsq*p(x)$). You can now use the **Function** command to fill in columns C and D. In *Excel*, an "equals" sign indicates that you are going to type an equation (or insert a function). Hence, in cell C2, we type: =A2*B2.
   Then, to copy this formula to the remaining three cells in column C, simply click on cell C2, grab the square in the lower right corner of the cell with your mouse, and drag to cell C5 to copy.
2. To fill in column D, type the following equation into cell D2: = A2*A2*B2 and then copy this formula to the remaining three cells in column D as explained above.
3. Finally, use the first three cells in column F to type the names "Mean," "Variance," and "Std Dev." Again, use the equation (or insert function) commands. In cell G1 (Mean), type: =SUM(C2:C5); in cell G2 (Variance), type: =SUM(D2:D5)–(G1*G1); and in cell G3 (Standard Deviation), type: =SQRT(G2). The resulting spreadsheet is shown in Figure 4.17.

FIGURE 4.17
MINITAB screen shot

	A	B	C	D	E	F	G
1	x	p(x)	x*p(x)	xsq*p(x)		Mean	1.85
2	0	0.25	0	0		Variance	2.9275
3	1	0.35	0.35	0.35		Std Dev	1.710994
4	3	0.25	0.75	2.25			
5	5	0.15	0.75	3.75			

Discrete Probability Distributions in *MINITAB*

Although *MINITAB* cannot help you solve the types of general probability problems presented in this chapter, it is useful for graphing the probability distribution $p(x)$ for a general discrete random variable X when the probabilities are known, and for calculating the mean, variance, and standard deviation of the random variable X. In Chapters 5 and 6, we will use *MINITAB* to calculate exact probabilities for three special cases: the binomial, the Poisson, and the normal random variables.

EXAMPLE 4.37 Suppose you have this general probability distribution:

x	0	1	3	5
$p(x)$	0.25	0.35	0.25	0.15

1. Enter the values of x and $p(x)$ into columns C1 and C2 of a new *MINITAB* worksheet. In the grey boxes just below C3, C4, and C5, respectively, type the names "Mean," "Variance," and "Std Dev." You can now use the **Calc ▶ Calculator** command to calculate μ, σ^2, and σ and to store the results in columns C3–C5 of the worksheet.
2. Use the same approach for all three parameters. In the Calculator dialogue box, select "Mean" as the column in which to store μ. In the Expression box, use the Functions list, the calculator keys, and the variables list on the left to highlight, select, and create the expression for the mean (see Figure 4.18(a)):

   SUM('x'*'p(x)')

   MINITAB will multiply each row element in C1 times the corresponding row element in C2, sum the resulting products, and store the result in C3! You can check the result by hand if you like.
3. The formulas for the variance and standard deviation are selected in a similar way:

   Variance: SUM(('x' – 'Mean')**2*'p(x)')
   Std Dev: SQRT('Variance')
4. To see the tabular form of the probability distribution and the three parameters, use **Data ▶ Display Data** and select all five columns. Click **OK** and the results will be displayed in the Session window, as shown in Figure 4.18(b).

FIGURE 4.18
MINITAB screen shot
(a) Calculator
MINITAB screen shot
(b) Session

(a)

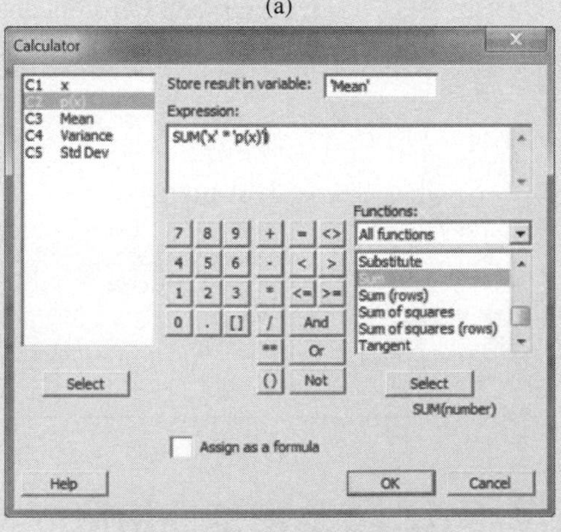

(b)

Session

Data Display

Row	x	p(x)	Mean	Variance	Std Dev
1	0	0.25	1.85	2.9275	1.71099
2	1	0.35			
3	3	0.25			
4	5	0.15			

The probability histogram can be plotted using the *MINITAB* command **Graph ▶ Scatterplot ▶ Simple ▶ OK.** In the Scatterplot dialogue box, select 'p(x)' for **Y variables** and 'x' for **X variables.** To display the discrete probability bars, click on **Data View,** uncheck the box marked "Symbols," and check the box marked "Project Lines." Click **OK** twice to see the plot. You will see a single straight line projected at each of the four values of x. If you want the plot to look more like the discrete probability histograms in Section 4.8, position your cursor on one of the lines, right-click the mouse, and choose "Edit Project Lines." Under the "Attributes" tab, select **Custom** and change the line size to **75.** Click **OK.** If the bar width is not satisfactory, you can readjust the line size. Finally, right-click on the *x*-axis, choose "Edit X Scale," and select **−.5** and **5.5** for the minimum and maximum **Scale Ranges.** Click **OK.** The probability histogram is shown in Figure 4.19.

Locate the mean on the graph. Is it at the centre of the distribution? If you mark off two standard deviations on either side of the mean, do most of the possible values of *x* fall into this interval?

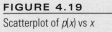

FIGURE 4.19
Scatterplot of $p(x)$ vs x

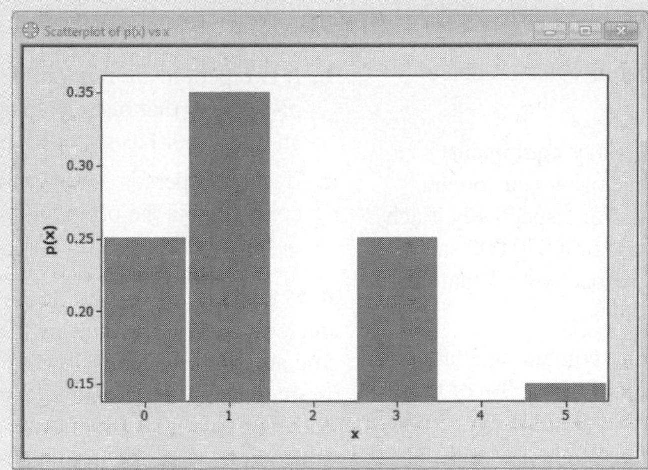

Supplementary Exercises

4.103 Playing the Slots A slot machine has three slots; each will show a cherry, a lemon, a star, or a bar when spun. The player wins if all three slots show the same three items. If each of the four items is equally likely to appear on a given spin, what is your probability of winning?

4.104 Whistle Blowers "Whistle blowers" is the name given to employees who report corporate fraud, theft, and other unethical and perhaps criminal activities by fellow employees or by their employer. Although there is legal protection for whistle blowers, it has been reported that approximately 23% of those who reported fraud suffered reprisals such as demotion or poor performance ratings. Suppose the probability that an employee will fail to report a case of fraud is 0.69. Find the probability that a worker who observes a case of fraud will report it and will subsequently suffer some form of reprisal.

4.105 Aspirin Two cold tablets are accidentally placed in a box containing two aspirin tablets. The four tablets are identical in appearance. One tablet is selected at random from the box and is swallowed by the first patient. A tablet is then selected at random from the three remaining tablets and is swallowed by the second patient. Define the following events as specific collections of simple events:

a. The sample space *S*

b. The event *A* that the first patient obtained a cold tablet

c. The event *B* that exactly one of the two patients obtained a cold tablet

d. The event *C* that neither patient obtained a cold tablet

4.106 Refer to Exercise 4.105. By summing the probabilities of simple events, find $P(A)$, $P(B)$, $P(A \cap B)$, $P(A \cup B)$, $P(C)$, $P(A \cap C)$, and $P(A \cup C)$.

4.107 DVRs A retailer sells two styles of high-priced digital video recorders (DVR) that experience indicates are in equal demand. (Fifty percent of all potential customers prefer style 1, and 50% favour style 2.) If the retailer stocks four of each, what is the probability that the first four customers seeking a DVR all purchase the same style?

4.108 Boxcars A boxcar contains seven complex electronic systems. Unknown to the purchaser, three are defective. Two of the seven are selected for thorough testing and are then classified as defective or non-defective. What is the probability that no defectives are found?

4.109 Heavy Equipment A heavy-equipment salesperson can contact either one or two customers per day with probability 1/3 and 2/3, respectively. Each contact will result in either no sale or a $50,000 sale with probability 9/10 and 1/10, respectively. What is the expected value of the daily sales?

4.110 Fire Insurance A county containing a large number of rural homes is thought to have 60% of those homes insured against fire. Four rural homeowners are chosen at random from the entire population, and x are found to be insured against fire. Find the probability distribution for x. What is the probability that at least three of the four will be insured?

4.111 Fire Alarms A fire-detection device uses three temperature-sensitive cells acting independently of one another in such a manner that any one or more can activate the alarm. Each cell has a probability $p = 0.8$ of activating the alarm when the temperature reaches 38°C or higher. Let x equal the number of cells activating the alarm when the temperature reaches 38°C.

a. Find the probability distribution of x.

b. Find the probability that the alarm will function when the temperature reaches 38°C.

c. Find the expected value and the variance for the random variable x.

4.112 Catching a Cold Is your chance of getting a cold influenced by the number of social contacts you have? A study by Sheldon Cohen, a psychology professor at Carnegie Mellon University, seems to show that the more social relationships you have, the *less susceptible* you are to colds. A group of 276 healthy men and women were grouped according to their number of relationships (such as parent, friend, church member, neighbour). They were then exposed to a virus that causes colds. An adaptation of the results is shown in the table:[13]

	Number of Relationships		
	Three or Fewer	Four or Five	Six or More
Cold	49	43	34
No cold	31	57	62
Total	80	100	96

a. If one person is selected at random from the 276 people in the study, what is the probability that the person got a cold?

b. If two people are randomly selected, what is the probability that one has four or five relationships and the other has six or more relationships?

c. If a single person is randomly selected and has a cold, what is the probability that he or she has three or fewer relationships?

4.113 Plant Genetics Refer to the experiment conducted by Gregor Mendel in Exercise 4.65. Suppose you are interested in following two independent traits in snap peas—seed texture (S = smooth, s = wrinkled) and seed colour (Y = yellow, y = green)—in a second-generation cross of heterozygous parents. Remember that the capital letter represents the dominant trait. Complete the table with the gene pairs for both traits. All possible pairings are equally likely.

	Seed Colour			
Seed Texture	yy	yY	Yy	YY
ss	(ss yy)	(ss yY)		
sS				
Ss				
SS				

a. What proportion of the offspring from this cross will have smooth yellow peas?

b. What proportion of the offspring will have smooth green peas?

c. What proportion of the offspring will have wrinkled yellow peas?

d. What proportion of the offspring will have wrinkled green peas?

e. Given that an offspring has smooth yellow peas, what is the probability that this offspring carries one s allele? One s allele *and* one y allele?

4.114 Profitable Stocks An investor has the option of investing in three of five recommended stocks. Only two will show a substantial profit within the next five years, though it is unknown which ones. If the investor selects the three stocks at random (giving every combination of three stocks an equal chance of selection), what is the probability that the two profitable stocks are selected? What is the probability that only one of the two profitable stocks is selected?

4.115 Racial Bias? Four union members, two from a minority group, are assigned to four distinctly different one-person jobs, which can be ranked in order of desirability.

a. Define the experiment.

b. List the simple events in *S*.

c. If the assignment to the jobs is unbiased—that is, if any one ordering of assignments is as probable as any other—what is the probability that the two people from the minority group are assigned to the least desirable jobs?

4.116 A Reticent Salesman A salesperson figures that the probability of consummating a sale during the first contact with a client is 0.4 but improves to 0.55 on the second contact if the client did not buy during the first contact. Suppose this salesperson makes one and only one callback to any client. If she contacts a client, calculate the probabilities for these events:

a. The client will buy.

b. The client will not buy.

4.117 Bus or Subway A man takes either a bus or the subway to work with probabilities 0.3 and 0.7, respectively. When he takes the bus, he is late 30% of the days. When he takes the subway, he is late 20% of the days. If the man is late for work on a particular day, what is the probability that he took the bus?

4.118 Guided Missiles The failure rate for a guided missile control system is 1 in 1000. Suppose that a duplicate, but completely independent, control system is installed in each missile so that, if the first fails, the second can take over. The reliability of a missile is the probability that it does not fail. What is the reliability of the modified missile?

4.119 Rental Trucks A rental truck agency services its vehicles on a regular basis, routinely checking for mechanical problems. Suppose that the agency has six moving vans, two of which need to have new brakes. During a routine check, the vans are tested one at a time.

a. What is the probability that the last van with brake problems is the fourth van tested?

b. What is the probability that no more than four vans need to be tested before both brake problems are detected?

c. Given that one van with bad brakes is detected in the first two tests, what is the probability that the remaining van is found on the third or fourth test?

4.120 Winning the Lottery On Wednesday, October 26, 2005, a single winning Lotto 6/49 lottery ticket was sold in Camrose, Alberta. The ticket belonged to 17 oil industry workers. According to the Alberta Lottery Fund, the winning numbers were 5, 11, 20, 30, 37, and 43. The prize was $54,000,000.

a. How many ways can six numbers be picked from 49, without replacement?

b. If exactly one ticket is purchased, what is the probability of being a winner?

c. How many different orders of the numbers 5, 11, 20, 30, 37, and 43 are possible?

4.121 Winning the Lottery II On Thursday, September 28, 2006, an 85-year-old lady in Etobicoke, Ontario, woke from a dream about winning a lottery, and wrote down the six numbers on a piece of paper. She bought one Lotto 6/49 lottery ticket that day using the numbers 1, 10, 18, 24, 31, and 46. On Friday, she bought another ticket with the same numbers! On Saturday, the real numbers were chosen and her two tickets were both winners. There was one other winning ticket sold in another province. The total jackpot was $24,000,000. Because the lady had two of the three winning tickets, she received 2/3 of the jackpot.

a. What is the probability of getting the six correct numbers with one ticket?

b. If two tickets with the same numbers on both tickets are purchased, what is the probability that the purchaser would get the correct six numbers?

c. If two tickets with different numbers are purchased, what is the probability of one of the two tickets having all six winning numbers?

d. Is it a good strategy to select the same numbers twice (if one is only guessing)?

e. If two 6/49 tickets are selected randomly, what is the probability that none of the 12 numbers are repeated?

4.122 The Match Game Two people each toss a coin. They obtain a "match" if either both coins are heads or both are tails. Suppose the tossing is repeated three times.

a. What is the probability of three matches?

b. What is the probability that all six tosses (three for each participant) result in tails?

c. Coin tossing provides a model for many practical experiments. Suppose that the coin tosses represent the answers given by two students for three specific true–false questions on an examination. If the two students gave three matches for answers, would the low probability found in part a suggest collusion?

4.123 Contract Negotiations Experience has shown that, 50% of the time, a particular union–management contract negotiation led to a contract settlement within a two-week period, 60% of the time the union strike fund was adequate to support a strike, and 30% of the time both conditions were satisfied. What is the probability of a contract settlement given that the union strike fund is adequate to support a strike? Is settlement of a contract within a two-week period dependent on whether the union strike fund is adequate to support a strike?

4.124 Work Tenure Suppose the probability of remaining with a particular company 10 years or longer is 1/6. A man and a woman start work at the company on the same day.

a. What is the probability that the man will work there less than 10 years?

b. What is the probability that both the man and the woman will work there less than 10 years? (Assume they are unrelated and their lengths of service are independent of each other.)

c. What is the probability that one or the other or both will work 10 years or longer?

4.125 Accident Insurance Accident records collected by an automobile insurance company give the following information: The probability that an insured driver has an automobile accident is 0.15; if an accident has occurred, the damage to the vehicle amounts to 20% of its market value with probability 0.80, 60% of its market value with probability 0.12, and a total loss with probability 0.08. What premium should the company charge on a $22,000 car so that the expected gain by the company is zero?

4.126 Waiting Times Suppose that at a particular supermarket the probability of waiting five minutes or longer for checkout at the cashier's counter is 0.2. On a given day, a husband and wife decide to shop individually at the market, each checking out at different cashier counters. They both reach cashier counters at the same time.

a. What is the probability that the man will wait less than five minutes for checkout?

b. What is probability that both of them will be checked out in less than five minutes? (Assume that the checkout times for the two are independent events.)

c. What is the probability that one or the other or both will wait five minutes or longer?

4.127 Quality Control A quality-control plan calls for accepting a large lot of crankshaft bearings if a sample of seven is drawn and none are defective. What is the probability of accepting the lot if none in the lot are defective? If 1/10 are defective? If 1/2 are defective?

4.128 Mass Transit Only 40% of all people in a community favour the development of a mass transit system. If four citizens are selected at random from the community, what is the probability that all four favour the mass transit system? That none favours the mass transit system?

4.129 Blood Pressure Meds A research physician compared the effectiveness of two blood pressure drugs *A* and *B* by administering the two drugs to each of four pairs of identical twins. Drug *A* was given to one member of a pair; drug *B* to the other. If, in fact, there is no difference in the effects of the drugs, what is the probability that the drop in the blood pressure reading for drug *A* exceeds the corresponding drop in the reading for drug *B* for all four pairs of twins? Suppose drug *B* created a greater drop in blood pressure than drug *A* for each of the four pairs of twins. Do you think this provides sufficient evidence to indicate that drug *B* is more effective in lowering blood pressure than drug *A*?

4.130 Blood Tests To reduce the cost of detecting a disease, blood tests are conducted on a pooled sample of blood collected from a group of *n* people. If no indication of the disease is present in the pooled blood sample (as is usually the case), none have the disease. If analysis of the pooled blood sample indicates that the disease is present, each individual must submit to a blood test. The individual tests are conducted in sequence. If, among a group of five people, one person has the disease, what is the probability that six blood tests (including the pooled test) are required to detect the single diseased person? If two people have the disease, what is the probability that six tests are required to locate both diseased people?

4.131 Tossing a Coin How many times should a coin be tossed to obtain a probability equal to or greater than 0.9 of observing at least one head?

4.132 Flextime The number of companies offering flexible work schedules has increased as companies try to help employees cope with the demands of home and work. One flextime schedule is to work four 10-hour shifts. A survey provided the following information for 220 firms located in two cities in Quebec:

	Flextime Schedule		
City	Available	Not Available	Total
A	39	75	114
B	25	81	106
Total	64	156	220

A company is selected at random from this pool of 220 companies.

a. What is the probability that the company is located in city *A*?

b. What is the probability that the company is located in city *B* and offers flextime work schedules?

c. What is the probability that the company does not have flextime schedules?

d. What is the probability that the company is located in city *B*, given that the company has flextime schedules available?

4.133 A Colour Recognition Experiment An experiment is run as follows—the colours red, yellow, and blue are each flashed on a screen for a short period of time. A subject views the colours and is asked to choose which one was flashed for the longest time. The experiment is repeated three times with the same subject.

a. If all the colours were flashed for the same length of time, find the probability distribution for x, the number of times that the subject chose the colour red. Assume that the three choices are independent.

b. Construct the probability histogram for the random variable x.

4.134 Pepsi or Coke? A taste-testing experiment is conducted at a local supermarket, where passing shoppers are asked to taste two soft-drink samples—one Pepsi and one Coke—and state their preference. Suppose that four shoppers are chosen at random and asked to participate in the experiment, and that there is actually no difference in the taste of the two brands.

a. What is the probability that all four shoppers choose Pepsi?

b. What is the probability that exactly one of the four shoppers chooses Pepsi?

4.135 MRIs An article in *The American Journal of Sports Medicine* compared the results of magnetic resonance imaging (MRI) evaluation with arthroscopic surgical evaluation of cartilage tears at two sites in the knees of 35 patients. The $2 \times 35 = 70$ examinations produced the classifications shown in the table.[14] Actual tears were confirmed by arthroscopic surgical examination.

	Tears	No Tears	Total
MRI Positive	27	0	27
MRI Negative	4	39	43
Total	31	39	70

a. What is the probability that a site selected at random has a tear and has been identified by MRI as having a tear?

b. What is the probability that a site selected at random has no tear but has been identified by MRI as having a tear?

c. What is the probability that a site selected at random has a tear that has not been identified by MRI?

d. What is the probability of a positive MRI, given that there is a tear?

e. What is the probability of a false negative—that is, a negative MRI, given that there is a tear?

4.136 Viruses A certain virus afflicted the families in three adjacent houses in a row of 12 houses. If three houses were randomly chosen from a row of 12 houses, what is the probability that the three houses would be adjacent? Is there reason to believe that this virus is contagious?

4.137 Orchestra Politics The board of directors of a major symphony orchestra has voted to create a players' committee for the purpose of handling employee complaints. The council will consist of the president and vice-president of the symphony board and two orchestra representatives. The two orchestra representatives will be randomly selected from a list of six volunteers, consisting of four men and two women.

a. Find the probability distribution for X, the number of women chosen to be orchestra representatives.

b. Find the mean and variance for the random variable X.

c. What is the probability that both orchestra representatives will be women?

4.138 Independence and Mutually Exclusive Suppose that $P(A) = 0.3$ and $P(B) = 0.4$.

a. If $P(A \cap B) = 0.12$, are A and B independent? Justify your answer.

b. If $P(A \cup B) = 0.7$, what is $P(A \cap B)$? Justify your answer.

c. If A and B are independent, what is $P(A|B)$?

d. If A and B are mutually exclusive, what is $P(A|B)$?

4.139 Bringing Home the Bacon The following information reflects the results of a survey reported by Mya Frazier in an *Ad Age Insights* white paper.[15] Working spouses were asked "Who is the household breadwinner?" Suppose that one person is selected at random from these 200 individuals.

	You	Spouse or Significant Other	About Equal	Total
Men	64	16	20	100
Women	32	45	23	100
Total	96	61	43	200

a. What is the probability that this person will identify himself or herself as the household breadwinner?

b. What is the probability that the person selected will be a man who indicates that he and his spouse/significant other are equal breadwinners?

c. If the person selected indicates that the spouse or significant other is the breadwinner, what is the probability that the person is a man?

CASE STUDY  Probability and Decision Making in the Congo

In his exciting novel *Congo,* Michael Crichton describes a search by Earth Resources Technology Service (ERTS), a geological survey company, for deposits of boron-coated blue diamonds, diamonds that ERTS believes to be the key to a new generation of optical computers.[16] In the novel, ERTS is racing against an international consortium to find the Lost City of Zinj, a city that thrived on diamond mining and existed several thousand years ago (according to African fable), deep in the rain forests of eastern Zaire.

After the mysterious destruction of its first expedition, ERTS launches a second expedition under the leadership of Karen Ross, a 24-year-old computer genius who is accompanied by Professor Peter Elliot, an anthropologist; Amy, a talking gorilla; and the famed mercenary and expedition leader, "Captain" Charles Munro. Ross's efforts to find the city are blocked by the consortium's offensive actions, by the deadly rain forest, and by hordes of "talking" killer gorillas whose perceived mission is to defend the diamond mines. Ross overcomes these obstacles by using space-age computers to evaluate the probabilities of success for all possible circumstances and all possible actions that the expedition might take. At each stage of the expedition, she is able to quickly evaluate the chances of success.

At one stage in the expedition, Ross is informed by her Houston headquarters that their computers estimate that she is 18 hours and 20 minutes behind the competing Euro-Japanese team, instead of 40 hours ahead. She changes plans and decides to have the 12 members of her team—Ross, Elliot, Munro, Amy, and eight native porters—parachute into a volcanic region near the estimated location of Zinj. As Crichton relates, "Ross had double-checked outcome probabilities from the Houston computer, and the results were unequivocal. The probability of a successful jump was 0.7980, meaning that there was approximately one chance in five that someone would be badly hurt. However, given a successful jump, the probability of expedition success was 0.9943, making it virtually certain that they would beat the consortium to the site."

Keeping in mind that this is an excerpt from a novel, let us examine the probability, 0.7980, of a successful jump. If you were one of the 12-member team, what is the probability that you would successfully complete your jump? In other words, if the probability of a successful jump by all 12 team members is 0.7980, what is the probability that a single member could successfully complete the jump?

PROJECTS

Project 4-A: Child Safety Seat Survey, Part 2 (Continued from Project 3-A)

Canada has a Road Safety Vision of having the safest roads in the world. Yet, the leading cause of death of Canadian children remains vehicle crashes. In 2006, a national child seat safety survey was conducted by an AUTO21 research team in collaboration with Transport Canada to empirically measure Canada's progress towards achieving Road Safety Vision 2010. Child seat use was observed in parking lots and nearby intersections in 200 randomly selected sites across Canada. The following table provides a classification of a subset of children in the survey by age groups and type of restraint device they were using at the time of the survey.

Table: Cross-Tabulation of Age Group by Restraint Type

Category	Types of Restraints				
Age Group	Rear-Facing Infant Seat	Forward-Facing Infant Seat	Booster Seat	Seat Belt Only	Total
Infant (0–1 year)	181	52	1	0	234
Toddler (1–4 years)	49	483	117	3	652
School (4–9 years)	0	98	450	325	873
Older (>9 years)	0	0	16	627	643
Total	230	633	584	955	2402

a. Convert the frequency table shown above into a probability table.

b. What is the probability that the randomly selected child is in a rear-facing infant seat?

c. What is the probability that the randomly selected child is an infant?

d. If the randomly selected child is a toddler, what is the probability that the child is in a booster seat?

e. If the randomly selected child is in a rear-facing infant seat, what is the probability that the child is an infant?

f. What is the probability that the randomly selected child is a toddler or in a forward-facing infant seat?

g. If the randomly selected child is a toddler, what is the probability that the child is in a forward-facing infant seat?

h. Are the types of restraint the child uses and age mutually exclusive events? Explain.

i. Are the types of restraint the child uses and age independent events? Explain.

j. If a randomly selected child was using a forward-facing infant seat, find the probability that the child is a toddler.

Project 4-B: False Results in Medical Testing

A laboratory test for a rare disease affecting 2% of the population is either positive, indicating the rare disease is present, or negative, indicating the disease is not present. However, when people who have the rare disease are tested in the laboratory, 90% of the tests results were positive and 10% were falsely negative (a "false negative" result). On the other hand, for those people who don't have the rare disease are tested, 11% of the tests are positive (a "false positive" result).

a. What is the probability that a randomly selected person's test results would come back positive?

b. Are you surprised with the result in part a? Did you expect this number to be higher? Can you suggest an alternative way to calculate the probability in part a?

c. What is the probability that the selected person has the rare disease given that the test result is positive?

d. What is the probability that the selected person has the rare disease given that the test result is negative?

e. Interpret the events "false positive" and "false negative," respectively.

Project 4-C: Selecting Condiments

A box of condiments has ten small packages, in which three are ketchup, three are mustard, and four are relish. A sample of three packages is randomly selected (without replacement) from the box.

a. Find the probability distribution for X, the number of mustard packages in the sample.

b. What are mean and variance of X?

c. What is the probability that at most one mustard package is selected?

d. What is the probability that at least one mustard package is selected?

e. What is the probability that X is within one standard deviation from its mean?

f. Now, suppose that you win \$25 for each package of ketchup chosen and lose \$15 otherwise. Let y denote the total winnings.

(i) Find the probability distribution for y.

(ii) Verify that this distribution satisfies the axioms of a probability distribution.

(iii) Calculate the expected loss.

5

Several Useful Discrete Distributions

Nikolay Litov/Shutterstock.com

How Safe Is Plastic Surgery? Myth versus Fact!

A popular belief is that plastic surgery increases the frequency of cancer-related deaths. A Canadian study suggests that women with breast implants and those who have had other forms of plastic surgery actually have lower rates of cancer than the general population, but have higher rates of suicide.

Although the researchers offered no definitive explanation for the increased suicide rates, they suggest greater attention be paid to the mental state of potential cosmetic surgery patients. Previous international studies have reported similar results but small sample sizes have limited their impact. The study, funded by Health Canada and carried out jointly by the Public Health Agency of Canada, the University of Toronto, Cancer Care Ontario, and the University of Laval, is the largest of its kind to date, according to the researchers.[1] The case study at the end of this chapter examines how this question can be answered using one of the discrete probability distributions presented here.

GENERAL OBJECTIVES

Discrete random variables are used in many practical applications. Three important discrete random variables—the binomial, the Poisson, and the hypergeometric—are presented in this chapter. These random variables are often used to describe the number of occurrences of a specified event in a fixed number of trials or a fixed unit of time or space.

CHAPTER INDEX

NEED TO KNOW

How to Use Table 1 in Appendix I to Calculate Binomial Probabilities
How to Use Table 2 in Appendix I to Calculate Poisson Probabilities

5.1

INTRODUCTION

Examples of *discrete random variables* can be found in a variety of everyday situations and across most academic disciplines. However, there are three discrete probability distributions that serve as *models* for a large number of these applications. In this chapter we study the binomial, the Poisson, and the hypergeometric probability distributions and discuss their usefulness in different physical situations. However, first we briefly present the uniform and Bernoulli probability distributions.

The Uniform Probability Distribution

Recall the experiment of rolling a die. The probability of each value of x, where X represents the number that appeared and which has possible values $x = 1, 2, 3, 4, 5, 6$ is equally likely with $p(x) = 1/6$ for all values of x. Therefore, the graph of the probability distribution will have a flat shape, called the *discrete uniform probability distribution*.

FIGURE 5.1
Discrete uniform probability distribution

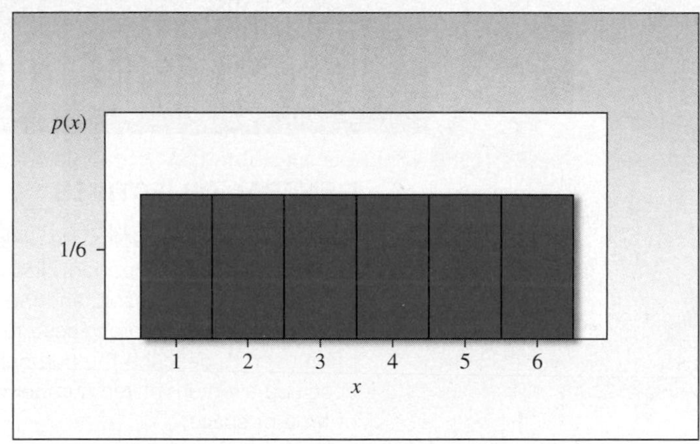

EXAMPLE 5.1

A poker hand consists of five cards from a deck of 52 ordinary cards. A player received five cards and did not look at them. Suppose that among these five cards, one card is the ace of spades, but this is not known to the player. Let X be the number of cards in the player's hand that are turned over until we observe the ace of spades. Assuming that turning over each card was done randomly without replacement, what is the probability function of X?

SOLUTION One way to calculate the probabilities is by using the Multiplication Rule, that is

$$P(X = 1) = p(1) = P \quad \text{(Observing the ace of spades on the first trial)} = \frac{1}{5}$$

$$P(X = 2) = p(2) = P \quad \text{(Observing any other card on the first trial and observing the ace of spades on the second trial)}$$

$$= \left(\frac{4}{5}\right)\left(\frac{1}{4}\right) = \frac{1}{5}$$

By similar arguments,

$$p(3) = \left(\frac{4}{5}\right)\left(\frac{3}{4}\right)\left(\frac{1}{3}\right) = \frac{1}{5}$$

$$p(4) = \left(\frac{4}{5}\right)\left(\frac{3}{4}\right)\left(\frac{2}{3}\right)\left(\frac{1}{2}\right) = \frac{1}{5}$$

$$p(5) = \left(\frac{4}{5}\right)\left(\frac{3}{4}\right)\left(\frac{2}{3}\right)\left(\frac{1}{2}\right)\left(\frac{1}{1}\right) = \frac{1}{5}$$

This is another example of the uniform distribution. That is, the probability remains constant regardless of what values are taken by the random variable *X*. Thus,

$$p(x) = \frac{1}{5}, \quad x = 1, 2, 3, 4, 5$$

In general, the probability distribution or probability mass function of a uniform random variable *X* is given by

$$p(x) = \frac{1}{k}, \quad x = 1, 2, \ldots, k$$

Clearly, this is a probability distribution.

The Bernoulli Probability Model

A single coin-tossing experiment is a simple example of an important discrete random variable called the *Bernoulli random variable*. Many practical experiments result in data similar to the head or tail outcomes of the coin toss. In many experiments in daily life, there are only two outcomes. For instance:

- Flipping a coin
- Rolling a die to determine whether it is a 3 or not
- Writing the MCAT exam and waiting for the result—pass or fail
- Taking a painkiller medicine once and waiting to see whether it is effective or not
- Results of a pregnancy—a girl or a boy
- Hitting a target; the outcome is hit or miss
- Recording your own opinion about Canadian involvement in the Afghanistan war—the result is support or no support
- In an NHL hockey game, the outcome of a penalty shot—goal or no goal

Regardless of the situation, assume there are only two outcomes in a given experiment. We call such an experiment a Bernoulli trial, and conveniently or conventionally/historically we refer to the two outcomes as a *success, 1,* or *failure, 0.* In other words, let *X* be the random variable that denotes the outcome. Then *X* will either take on the value 0 or 1. The next step is to assign probabilities to these two outcomes, which is simple in this case. For example, suppose the probability of the success of an outcome is p, where $0 < p < 1$ and the probability of failure (not a success) will be $1 - p$. Thus, the Bernoulli trial is an experiment with only two possible outcomes with positive probabilities p and $1 - p$. Thus, a Bernoulli random variable can be expressed as an indicator variable:

$X = 1, 0$ if a success occurred
otherwise, in other words, a failure occurred

Further, the probabilities are:

$$p(0) = P(X = 0) = 1 - p$$
$$p(1) = P(X = 1) = p$$

Hence,

$$p(x) = p^x(1 - p)^{1-x}, \quad x = 0, 1$$

Here, $0 <= p(x) < = 1$ for $x = 0,1$, and $\Sigma p(x) = 1$.

The quantity p is called the *parameter* of the distribution. Here, the value p can be specified by the experimenter for the problem at hand.

EXAMPLE 5.2

The Islamic calendar is lunar. The beginning and ending of the calendar are determined by the sighting of the crescent moon (new moon). Muslims are supposed to sight the crescent everywhere they live. It is a purely lunar calendar, having 12 lunar months in a year of about 354 (12 × 29.53 = 354.36) days, so the months rotate backward through the seasons and are not fixed to the Gregorian calendar. Muslims around the globe fast during the month of Ramadan (the ninth month) and after completing the fast and sighting the new moon, the next day they celebrate an event called Eid-ul-fitre. The moon may appear either on the 29th or 30th day of the month. However, 60% of Muslims believe that the new moon will be sighted on the 29th day of Ramadan, and so the fasting will be completed in 29 days. What is the probability that the moon will not be sighted on the 29th day of the month?

SOLUTION Clearly, this problem involves the Bernoulli random variable with $p = 0.6$. Therefore, desired probability is given by

$$p(0) = 0.6^0(1 - 0.6)^{1-0} = 0.4$$

EXAMPLE  5.3

As of July 2015, according to the Vegas Insider, where many sports odds are determined, the Toronto Maple Leafs had 1/100 chance to win the Stanley Cup and 1/45 of winning their conference for the 2015/2016 year.[2] What is the probability that the Maple Leafs will win the Stanley Cup?

SOLUTION Since there is one trial and two outcomes, this follows the Bernoulli random variable with $p = 0.01$. Thus,

$$P(1) = 0.01^1(1 - 0.01)^{1-1} = 0.01$$

5.2 THE BINOMIAL PROBABILITY DISTRIBUTION

The Bernoulli trial can be generalized to n independent trials. A coin-tossing experiment is a simple example of an important discrete random variable called the **binomial random variable.** As mentioned above, many everyday events feature an outcome similar to a head or tail result. For example, consider the political polls used to predict voter preferences in elections. Each sampled voter can be compared to a coin because the voter may be in favour of our candidate—a "head"—or not—a "tail." In most cases, the proportion of voters who favour our candidate does not equal 1/2; that is, the coin is not fair. In fact, the proportion of voters who favour our candidate is exactly what the poll is designed to measure!

Here are some other situations that are similar to the coin-tossing experiment:

- A sociologist is interested in the proportion of elementary school teachers who are men.
- A soft-drink marketer is interested in the proportion of cola drinkers who prefer her brand.
- A geneticist is interested in the proportion of the population who possess a gene linked to Alzheimer's disease.

Each sampled person is analogous to tossing a coin, but the probability of a "head" is not necessarily equal to 1/2. Although these situations have different practical objectives, they all exhibit the common characteristics of the **binomial experiment.**

DEFINITION A **binomial experiment** is one that has these five characteristics:

1. The experiment consists of n identical trials.
2. Each trial results in one of two outcomes. For lack of a better name, the one outcome is called a success, *S,* and the other a failure, *F.*
3. The probability of success on a single trial is equal to p and remains the same from trial to trial. The probability of failure is equal to $(1 - p) = q$.
4. The trials are independent.
5. We are interested in x, the number of successes observed during the n trials, for $x = 0, 1, 2, \ldots, n$.

EXAMPLE 5.4

Suppose there are approximately 1,000,000 adults in a prairie province (i.e., Alberta, Manitoba, Saskatchewan) and an unknown proportion, p, favour term limits for politicians. A sample of 1000 adults will be chosen in such a way that every one of the 1,000,000 adults has an equal chance of being selected, and each adult is asked whether he or she favours term limits. (The ultimate objective of this survey is to estimate the unknown proportion p, a problem that we will discuss in Chapter 8.) Is this a binomial experiment?

SOLUTION Does the experiment have the five binomial characteristics?

1. A "trial" is the choice of a single adult from the 1,000,000 adults in a prairie province. This sample consists of $n = 1000$ identical trials.
2. Since each adult will either favour or not favour term limits, there are two outcomes that represent the "successes" and "failures" in the binomial experiment.†
3. The probability of success, p, is the probability that an adult favours term limits. Does this probability remain the same for each adult in the sample? For all practical purposes, the answer is *yes.* For example, if 500,000 adults in the population favour term limits, then the probability of a "success" when the first adult is chosen is $500{,}000/1{,}000{,}000 = 1/2$. When the second adult is chosen, the probability p changes slightly, depending on the first choice. That is, there will be either 499,999 or 500,000 successes left among the 999,999 adults. In either case, p is still approximately equal to 1/2.

†Although it is traditional to call the two possible outcomes of a trial "success" and "failure," they could have been called "head" and "tail," "red" and "white," or any other pair of words. Consequently, the outcome called a "success" does not need to be viewed as a success in the ordinary use of the word.

4. The independence of the trials is guaranteed because of the large group of adults from which the sample is chosen. The probability of an adult favouring term limits does not change depending on the responses of previously chosen people.
5. The random variable X is the number of adults in the sample who favour term limits.

Because the survey satisfies the five characteristics reasonably well, for all practical purposes it can be viewed as a binomial experiment.

EXAMPLE 5.5

Suppose there are 7000 adults who use Instagram in Regina, Saskatchewan, and an unknown proportion p, prefer Instagram for sharing photos. A sample of 5000 adults will be chosen in such a way that every one of the 5000 adults has an equal chance of being selected, and each adult is asked whether he or she favours Instagram. (The ultimate objective of this survey is to estimate the unknown proportion p, a problem that we will discuss in Chapter 8.) Is this a binomial experiment?

SOLUTION Does the experiment have the five binomial characteristics?

1. A "trial" is the choice of a single adult from the sample of 5000 adults in Regina, Saskatchewan.
2. Since each adult will either favour Instagram or not, there are two outcomes that represent "success" and "failure" in the binomial experiment.
3. The probability of success, p, is the probability that an adult favours Instagram. Does this probability remain the same for each adult in the sample? For all practical purposes, the answer is yes. For example, if 3000 adults in the population favour Instagram, then the probability of a "success" when the first adult is chosen is 3000/5000 = 3/5. When the second adult is chosen, the probability p changes slightly, depending on the first choice. That is, there will be either 2999 or 3000 successes left among the 4999 adults. In either case, p is still approximately equal to 3/5.

EXAMPLE 5.6

A patient fills a prescription for a 10-day regimen of two pills daily. Unknown to the pharmacist and the patient, the 20 tablets consist of 18 pills of the prescribed medication and two pills that are the generic equivalent of the prescribed medication. The patient selects two pills at random for the first day's dosage. If we check the selection and record the number of pills that are generic, is this a binomial experiment?

SOLUTION Again, check the sampling procedure for the characteristics of a binomial experiment.

1. A "trial" is the selection of a pill from the 20 in the prescription. This experiment consists of $n = 2$ trials.
2. Each trial results in one of two outcomes. Either the pill is generic (call this a "success") or not (a "failure").
3. Since the pills in a prescription bottle can be considered randomly "mixed," the unconditional probability of drawing a generic pill on a given trial would be 2/20.
4. The condition of independence between trials is *not* satisfied, because the probability of drawing a generic pill on the second trial is dependent on the first trial. For example, if the first pill drawn is generic, then there is only one generic pill in the remaining 19. Therefore,

$$P(\text{generic on trial 2}|\text{generic on trial 1}) = \frac{1}{19}$$

If the first selection *does not* result in a generic pill, then there are still two generic pills in the remaining 19, and the probability of a "success" (a generic pill) changes to

$$P(\text{generic on trial } 2|\text{no generic on trial } 1) = \frac{2}{19}$$

Therefore, the trials are dependent and the sampling does not represent a binomial experiment.

Think about the difference between these examples. When the sample (the n identical trials) came from a large population, the probability of success p stayed about the same from trial to trial. When the population size N was small, the probability of success p changed quite dramatically from trial to trial, and the experiment *was not* binomial.

RULE OF THUMB

If the sample size is large relative to the population size—in particular, if $n/N \geq 0.05$—then the resulting experiment is not binomial.

In Chapter 4, we tossed two fair coins and constructed the probability distribution for X, the number of heads—a binomial experiment with $n = 2$ and $p = 0.5$. The general binomial probability distribution is constructed in the same way, but the procedure gets complicated as n gets larger. Fortunately, the probabilities $p(x)$ follow a general pattern. This allows us to use a single formula to find $p(x)$ for any given value of x.

THE BINOMIAL PROBABILITY DISTRIBUTION

A binomial experiment consists of n identical trials with probability of success p on each trial. The probability of k successes in n trials is

$$P(X = k) = C_k^n p^k q^{n-k} = \frac{n!}{k!(n-k)!} p^k q^{n-k}$$

for values of $k = 0, 1, 2, \ldots, n$ and where $q = 1 - p$. The symbol

$$C_k^n = \frac{n!}{k!(n-k)!} \quad \text{where } n! = n(n-1)(n-2)\ldots(2)(1) \text{ and } 0! \equiv 1$$

The general formulas for μ, σ^2, and σ given in Chapter 4 can be used to derive the following simpler formulas for the binomial mean and standard deviation.

MEAN AND STANDARD DEVIATION FOR THE BINOMIAL RANDOM VARIABLE

The random variable X, the number of successes in n trials, has a probability distribution with this centre and spread:

Mean: $\mu = np$

Variance: $\sigma^2 = npq$

Standard deviation: $\sigma = \sqrt{npq}$

EXAMPLE 5.7

Find $P(X = 2)$ for a binomial random variable with $n = 10$ and $p = 0.1$.

SOLUTION $P(X = 2)$ is the probability of observing two successes and eight failures in a sequence of 10 trials. You might observe the two successes first, followed by eight consecutive failures:

S, S, F, F, F, F, F, F, F, F

NEED A TIP?

$n! = n(n-1)(n-2) \ldots (2)(1)$
For example,
$5! = 5(4)(3)(2)(1) = 120$
and $0! \equiv 1$.

Since p is the probability of success and q is the probability of failure, this particular sequence has probability

$ppqqqqqqqq = p^2q^8$

However, many *other* sequences also result in $X = 2$ successes. The binomial formula uses C_2^{10} to count the number of sequences and gives the exact probability when you use the binomial formula with $k = 2$:

$$P(X = 2) = C_2^{10}(0.1)^2(0.9)^{10-2}$$

$$= \frac{10!}{2!(10-2)!}(0.1)^2(0.9)^8 = \frac{10(9)}{2(1)}(0.01)(0.430467) = 0.1937$$

You could repeat the procedure in Example 5.7 for each value of x—0, 1, 2, . . . , 10—and find all the values of $p(x)$ necessary to construct a probability histogram for X. This would be a long and tedious job, but the resulting graph would look like Figure 5.2(a). You can check the height of the bar for $X = 2$ and find $p(2) = P(X = 2) = 0.1937$. The graph is skewed right; that is, most of the time you will observe small values of X. The mean or "balancing point" is around $X = 1$; in fact, you can use the formula to find the exact mean:

$$\mu = np = 10(0.1) = 1$$

Figures 5.2(b) and 5.2(c) show two other binomial distributions with $n = 10$ but with different values of p. Look at the shapes of these distributions. When $p = 0.5$, the distribution is exactly symmetric about the mean, $\mu = np = 10(0.5) = 5$. When $p = 0.9$, the distribution is the "mirror image" of the distribution for $p = 0.1$ and is skewed to the left.

FIGURE 5.2
Binomial probability distributions

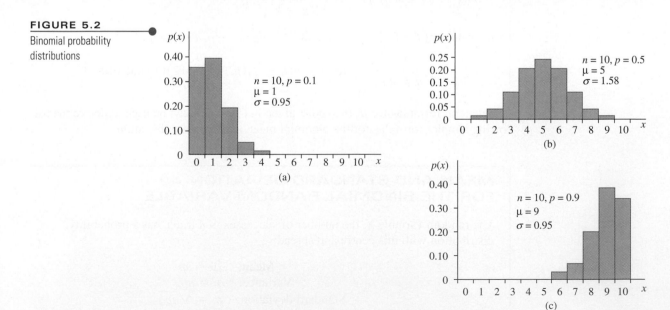

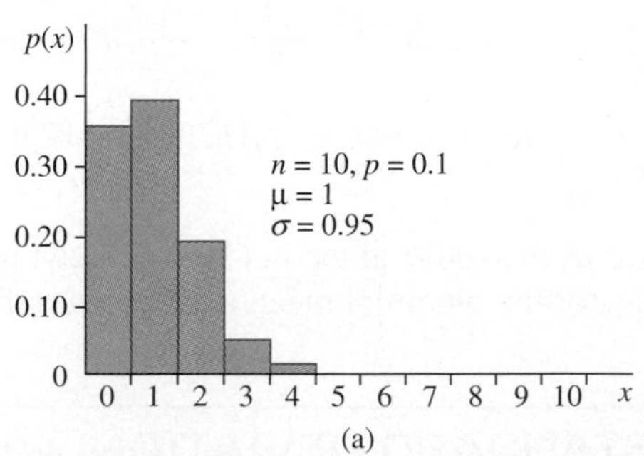

p(x)
0.25
0.20
0.15
0.10
0.05
0
0 1 2 3 4 5 6 7 8 9 10 x
n = 10, p = 0.5
μ = 5
σ = 1.58
(b)

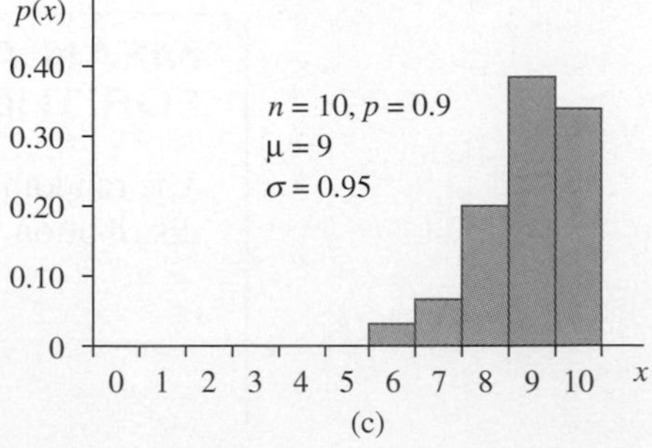

EXAMPLE 5.8

Jonas Valanciunas of the NBA's Toronto Raptors is a 77.8% free throw shooter.[3] Suppose the Raptors are up by four points in the fourth quarter against their opponent. The opponent decides to hack Valanciunas and send him to the foul line. They do this strategy twice in a row and the Raptors do not score in this situation. What is the probability that Valanciunas makes four free throws in a row?

SOLUTION $P(X = 4)$ is the probability of observing four successes and zero failures in a sequence of four trials.

$$P(X = 4) = C_4^4(0.778)^4(0.222)^{4-4}$$
$$= 0.3664$$

In this case, Jonas Valanciunas has a 36.64% chance of making four free throws in a row. This provides context to the idea that if Valanciunas has a relatively high free-throw percentage, it does not translate to 77.8% for each consecutive free throw made.

EXAMPLE 

Over a long period of time it has been observed that a given marksman can hit a target on a single trial with probability equal to 0.8. Suppose he fires four shots at the target.

1. What is the probability that he will hit the target exactly two times?
2. What is the probability that he will hit the target at least once?

SOLUTION A "trial" is a single shot at the target, and you can define a "success" as a hit and a "failurc" as a miss, so that $n = 4$ and $p = 0.8$. If you assume that the marksman's chance of hitting the target does not change from shot to shot, then the number X of times he hits the target is a *binomial random variable*.

1. $P(X = 2) = p(2) = C_2^4(0.8)^2(0.2)^{4-2}$

$$= \frac{4!}{2!2!}(0.64)(0.04) = \frac{4(3)(2)(1)}{2(1)(2)(1)}(0.64)(0.04) = 0.1536$$

The probability is 0.1536 that he will hit the target exactly two times.

2. $P(\text{at least once}) = P(X \geq 1) = p(1) + p(2) + p(3) + p(4)$

$$= 1 - p(0)$$
$$= 1 - C_0^4(0.8)^0(0.2)^4$$
$$= 1 - 0.0016 = 0.9984$$

Although you could calculate $P(X = 1)$, $P(X = 2)$, $P(X = 3)$, and $P(X = 4)$ to find this probability, using the complement of the event makes your job easier; that is,

$$P(X \geq 1) = 1 - P(X < 1) = 1 - P(X = 0)$$

Can you think of any reason your assumption of independent trials might be wrong? If the marksman learns from his previous shots (that is, he notices the location of his previous shot and adjusts his aim), then his probability p of hitting the target may increase from shot to shot. The trials would *not* be independent, and the experiment would *not* be binomial.

Calculating binomial probabilities can become tedious even for relatively small values of n. As n gets larger, it becomes almost impossible without the help of a calculator or computer. Fortunately, both of these tools are available to us. Computer-generated tables of **cumulative binomial probabilities** are given in Table 1 of Appendix I for values of n ranging from 2 to 25 and for selected values of p. These probabilities can also be generated using *MINITAB* or the Java applets on the text website.

NEED A TIP?

Use Table 1 rather than the binomial formula whenever possible. *This is an easier way!*

Cumulative binomial probabilities differ from the *individual* binomial probabilities that you calculated with the binomial formula. Once you find the column of probabilities for the correct values of n and p in Table 1, the row marked k gives the sum of all the binomial probabilities from $x = 0$ to $x = k$. Table 5.1 shows part of Table 1 for $n = 5$ and $p = 0.6$. If you look in the row marked $k = 3$, you will find

$$P(X \le 3) = p(0) + p(1) + p(2) + p(3) = 0.663$$

TABLE 5.1 Portion of Table 1 in Appendix I for $n = 5$

	p													
k	0.01	0.05	0.10	0.20	0.30	0.40	0.50	0.60	0.70	0.80	0.90	0.95	0.99	k
0	—	—	—	—	—	—	—	0.010	—	—	—	—	—	0
1	—	—	—	—	—	—	—	0.087	—	—	—	—	—	1
2	—	—	—	—	—	—	—	0.317	—	—	—	—	—	2
3	—	—	—	—	—	—	—	0.663	—	—	—	—	—	3
4	—	—	—	—	—	—	—	0.922	—	—	—	—	—	4
5	—	—	—	—	—	—	—	1.000	—	—	—	—	—	5

If the probability you need to calculate is not in this form, you will need to think of a way to rewrite your probability to make use of the tables!

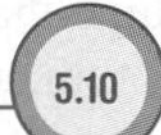

EXAMPLE 5.10

Use the cumulative binomial table for $n = 5$ and $p = 0.6$ to find the probabilities of these events:

1. Exactly three successes
2. Three or more successes

SOLUTION

1. If you find $k = 3$ in Table 5.1, the tabled value is

$$P(X \le 3) = p(0) + p(1) + p(2) + p(3)$$

Since you want only $P(X = 3) = p(3)$, you must subtract out the unwanted probability:

$$P(X \le 2) = p(0) + p(1) + p(2)$$

which is found in Table 5.1 with $k = 2$. Then

$$P(X = 3) = P(X \le 3) - P(X \le 2)$$
$$= 0.663 - 0.317 = 0.346$$

2. To find $P(\text{three or more successes}) = P(X \ge 3)$ using Table 5.1, you must use the complement of the event of interest. Write

$$P(X \ge 3) = 1 - P(X < 3) = 1 - P(X \le 2)$$

You can find $P(X \le 2)$ in Table 5.1 with $k = 2$. Then

$$P(X \ge 3) = 1 - P(X \le 2)$$
$$= 1 - 0.317 = 0.683$$

EXAMPLE 5.11

Refer to Example 5.10 and the binomial random variable X with $n = 5$ and $p = 0.6$. Use the cumulative binomial table in Table 5.1 to find the remaining binomial probabilities: $p(0), p(1), p(2), p(4)$, and $p(5)$. Construct the probability histogram for the random variable X and describe its shape and location.

SOLUTION You can find $P(X = 0)$ directly from Table 5.1 with $k = 0$. That is, $p(0) = 0.010$. The other probabilities can be found by subtracting successive entries in Table 5.1:

$$P(X = 1) = P(X \leq 1) - P(X = 0) = 0.087 - 0.010 = 0.077$$

$$P(X = 2) = P(X \leq 2) - P(X \leq 1) = 0.317 - 0.087 = 0.230$$

$$P(X = 4) = P(X \leq 4) - P(X \leq 3) = 0.922 - 0.663 = 0.259$$

$$P(X = 5) = P(X \leq 5) - P(X \leq 4) = 1.000 - 0.922 = 0.078$$

The probability histogram is shown in Figure 5.3. The distribution is relatively mound-shaped, with a centre around 3.

FIGURE 5.3
Binomial probability distribution for Example 5.11.

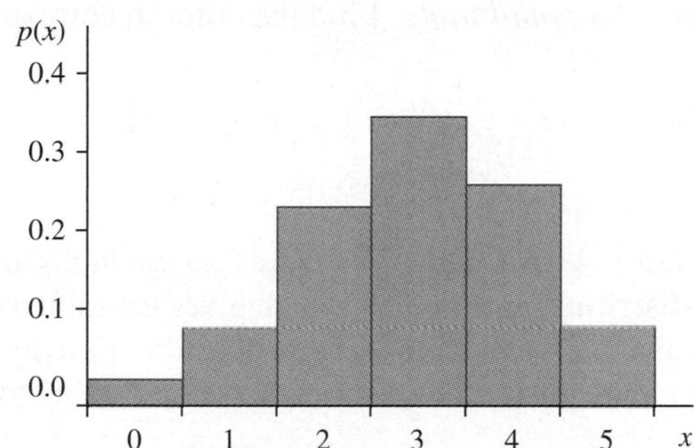

NEED TO KNOW

How to Use Table 1 to Calculate Binomial Probabilities

1. Find the necessary values of n and p. Isolate the appropriate column in Table 1.
2. Table 1 gives $P(X \leq k)$ in the row marked k. Rewrite the probability you need so that it is in this form.
   - List the values of x in your event.
   - From the list, write the event as either the difference of two probabilities:

     $P(X \leq a) - P(X \leq b)$ for $a > b$

     or the complement of the event:

     $1 - P(X \leq a)$

     or just the event itself:

     $P(X \leq a)$ or $P(X < a) = P(X \leq a - 1)$

EXAMPLE 5.12 

A regimen consisting of a daily dose of vitamin C was tested to determine its effectiveness in preventing the common cold. Ten people who were following the prescribed regimen were observed for a period of one year. Eight survived the winter without a cold.

Suppose the probability of surviving the winter without a cold is 0.5 when the vitamin C regimen is not followed. What is the probability of observing eight or more survivors, given that the regimen is ineffective in increasing resistance to colds?

SOLUTION If you assume that the vitamin C regimen is ineffective, then the probability p of surviving the winter without a cold is 0.5. The probability distribution for X, the number of survivors, is

$$p(x) = C_x^{10}(0.5)^x(0.5)^{10-x}$$

You have learned three ways to find P(8 or more survivors) $= P(X \geq 8)$. You will get the same results with any of the three; choose the most convenient method for your particular problem.

1. *The binomial formula:*

$$\begin{aligned} P(8 \text{ or more}) &= p(8) + p(9) + p(10) \\ &= C_8^{10}(0.5)^{10} + C_9^{10}(0.5)^{10} + C_{10}^{10}(0.5)^{10} \\ &= 0.055 \end{aligned}$$

2. *The cumulative binomial table:* Find the column corresponding to $p = 0.5$ in the table for $n = 10$:

$$\begin{aligned} P(8 \text{ or more}) &= P(X \geq 8) = 1 - P(X \leq 7) \\ &= 1 - 0.945 = 0.055 \end{aligned}$$

3. *Output from MINITAB or Excel:* The outputs shown in Figure 5.4(a) and (b) give the **cumulative distribution function,** which gives the same probabilities you found in the cumulative binomial tables. The **probability density function** gives the individual binomial probabilities, which you found using the binomial formula.

FIGURE 5.4 (a) *MINITAB* output for Example 5.12

Cumulative Distribution Function

Binomial with n = 10 and p = 0.5

x	P(X <= x)
0	0.00098
1	0.01074
2	0.05469
3	0.17187
4	0.37695
5	0.62305
6	0.82813
7	0.94531
8	0.98926
9	0.99902
10	1.00000

Probability Density Function

Binomial with n = 10 and p = 0.5

x	P(X = x)
0	0.000977
1	0.009766
2	0.043945
3	0.117188
4	0.205078
5	0.246094
6	0.205078
7	0.117188
8	0.043945
9	0.009766
10	0.000977

FIGURE 5.4(b) *Excel* output for Example 5.12

x	P(X <= x)	P(X = x)
0	0.000977	0.000977
1	0.010742	0.009766
2	0.054688	0.043945
3	0.171875	0.117188
4	0.376953	0.205078
5	0.623047	0.246094
6	0.828125	0.205078
7	0.945313	0.117188
8	0.989258	0.043945
9	0.999023	0.009766
10	1.000000	0.000977

Using the cumulative distribution function, calculate

$$P(X \geq 8) = 1 - P(X \leq 7)$$
$$= 1 - 0.94531 = 0.05469$$

Or, using the probability density function, calculate

$$P(X \geq 8) = p(8) + p(9) + p(10)$$
$$= 0.043945 + 0.009766 + 0.000977 = 0.05469$$

EXAMPLE  5.13

Would you rather take a multiple-choice or a full recall test? If you have absolutely no knowledge of the material, you will score zero on a full recall test. However, if you are given five choices for each question, you have at least one chance in five of guessing correctly! If a multiple-choice exam contains 100 questions, each with five possible answers, what is the expected score for a student who is guessing on each question? Within what limits will the "no-knowledge" scores fall?

SOLUTION If X is the number of correct answers on the 100-question exam, the probability of a correct answer, p, is one in five, so that $p = 0.2$. Since the student is randomly selecting answers, the $n = 100$ answers are independent, and the expected score for this binomial random variable is

$$\mu = np = 100(0.2) = 20 \qquad \text{correct answers}$$

To evaluate the spread or variability of the scores, you can calculate

$$\sigma = \sqrt{npq} = \sqrt{100(0.2)(0.8)} = 4$$

Then, using your knowledge of variation from Tchebysheff's Theorem and the Empirical Rule, you can make these statements:

- A large proportion of the scores will lie within two standard deviations of the mean, or from $20 - 8 = 12$ to $20 + 8 = 28$.
- Almost all the scores will lie within three standard deviations of the mean, or from $20 - 12 = 8$ to $20 + 12 = 32$.

The "guessing" option gives the student a better score than the zero score on the full recall test, but the student still will not pass the exam. What other options does the student have?

5.2 EXERCISES

BASIC TECHNIQUES

5.1 Consider a binomial random variable with $n = 8$ and $p = 0.7$. Let X be the number of successes in the sample.

a. Find the probability that X is 3 or less.

b. Find the probability that X is 3 or more.

c. Find $P(X < 3)$.

d. Find $P(X = 3)$.

e. Find $P(3 \leq X \leq 5)$.

5.2 Consider a binomial random variable with $n = 9$ and $p = 0.3$. Let X be the number of successes in the sample.

a. Find the probability that X is exactly 2.

b. Find the probability that X is less than 2.

c. Find $P(X > 2)$.

d. Find $P(2 \leq X \leq 4)$.

5.3 The Urn Problem A jar contains five balls: three red and two white. Two balls are randomly selected without replacement from the jar, and the number X of red balls is recorded. Explain why X is or is not a binomial random variable. (HINT: Compare the characteristics of this experiment with the characteristics of a binomial experiment given in this section.) If the experiment is binomial, give the values of n and p.

5.4 The Urn Problem, continued Refer to Exercise 5.3. Assume that the sampling was conducted with replacement. That is, assume that the first ball was selected from the jar, observed, and then replaced, and that the balls were then mixed before the second ball was selected. Explain why X, the number of red balls observed, is or is not a binomial random variable. If the experiment is binomial, give the values of n and p.

5.5 Evaluate these binomial probabilities:

a. $C_2^8(0.3)^2(0.7)^6$ **b.** $C_0^4(0.05)^0(0.95)^4$
c. $C_3^{10}(0.5)^3(0.5)^7$ **d.** $C_1^7(0.2)^1(0.8)^6$

5.6 Evaluate these binomial probabilities:

a. $C_0^8(0.2)^0(0.8)^8$ **b.** $C_1^8(0.2)^1(0.8)^7$
c. $C_2^8(0.2)^2(0.8)^6$ **d.** $P(X \leq 1)$ when $n = 8, p = 0.2$
e. P(two or fewer successes)

5.7 Let X be a binomial random variable with $n = 7$, $p = 0.3$. Find these values:

a. $P(X = 4)$ **b.** $P(X \leq 1)$ **c.** $P(X > 1)$
d. $\mu = np$ **e.** $\sigma = \sqrt{npq}$

5.8 Use the formula for the binomial probability distribution to calculate the values of $p(x)$, and construct the probability histogram for X when $n = 6$ and $p = 0.2$. [HINT: Calculate $P(X = k)$ for seven different values of k.]

5.9 Refer to Exercise 5.8. Construct the probability histogram for a binomial random variable X with $n = 6$ and $p = 0.8$. Use the results of Exercise 5.8; do not recalculate all the probabilities.

5.10 If X has a binomial distribution with $p = 0.5$, will the shape of the probability distribution be symmetric, skewed to the left, or skewed to the right?

5.11 Let X be a binomial random variable with $n = 10$ and $p = 0.4$. Find these values:

a. $P(X = 4)$ **b.** $P(X \geq 4)$ **c.** $P(X > 4)$
d. $P(X \leq 4)$ **e.** $\mu = np$ **f.** $\sigma = \sqrt{npq}$

5.12 Use Table 1 in Appendix I to find the sum of the binomial probabilities from $X = 0$ to $X = k$ for these cases:

a. $n = 10, p = 0.1, k = 3$
b. $n = 15, p = 0.6, k = 7$
c. $n = 25, p = 0.5, k = 14$

5.13 Use Table 1 in Appendix I to evaluate the following probabilities for $n = 6$ and $p = 0.8$:

a. $P(X \geq 4)$ **b.** $P(X = 2)$
c. $P(X < 2)$ **d.** $P(X > 1)$

Verify these answers using the values of $p(x)$ calculated in Exercise 5.9.

5.14 Find $P(X \leq k)$ in each case:

a. $n = 20, p = 0.05, k = 2$
b. $n = 15, p = 0.7, k = 8$
c. $n = 10, p = 0.9, k = 9$

5.15 Use Table 1 in Appendix I to find the following:

a. $P(X < 12)$ for $n = 20, p = 0.5$
b. $P(X \leq 6)$ for $n = 15, p = 0.4$
c. $P(X > 4)$ for $n = 10, p = 0.4$
d. $P(X \geq 6)$ for $n = 15, p = 0.6$
e. $P(3 < X < 7)$ for $n = 10, p = 0.5$

5.16 Find the mean and standard deviation for a binomial distribution with these values:

a. $n = 1000, p = 0.3$ **b.** $n = 400, p = 0.01$
c. $n = 500, p = 0.5$ **d.** $n = 1600, p = 0.8$

5.17 Find the mean and standard deviation for a binomial distribution with $n = 100$ and these values of p:

a. $p = 0.01$ **b.** $p = 0.9$ **c.** $p = 0.3$
d. $p = 0.7$ **e.** $p = 0.5$

5.18 In Exercise 5.17, the mean and standard deviation for a binomial random variable were calculated for a fixed sample size, $n = 100$, and for different values of p. Graph the values of the standard deviation for the five values of p given in Exercise 5.17. For what value of p does the standard deviation seem to be a maximum?

5.19 Let X be a binomial random variable with $n = 20$ and $p = 0.1$.

a. Calculate $P(X \leq 4)$ using the binomial formula.
b. Calculate $P(X \leq 4)$ using Table 1 in Appendix I.
c. Use the *Excel* output below to calculate $P(X \leq 4)$. Compare the results of parts a, b, and c.
d. Calculate the mean and standard deviation of the random variable X.
e. Use the results of part d to calculate the intervals $\mu \pm \sigma$, $\mu \pm 2\sigma$, and $\mu \pm 3\sigma$. Find the probability that an observation will fall into each of these intervals.
f. Are the results of part e consistent with Tchebysheff's Theorem? With the Empirical Rule? Why or why not?

Excel Output for Exercise 5.19: Binomial with $n = 20$ and $p = 0.1$

x	p(x)	x	p(x)
0	0.1216	11	0.0000
1	0.2702	12	0.0000
2	0.2852	13	0.0000
3	0.1901	14	0.0000
4	0.0898	15	0.0000
5	0.0319	16	0.0000
6	0.0089	17	0.0000
7	0.0020	18	0.0000
8	0.0004	19	0.0000
9	0.0001	20	0.0000
10	0.0000		

APPLICATIONS

5.20 Calgary Weather A meteorologist in Calgary recorded the number of days of rain during a 30-day period. If the random variable X is defined as the number of days of rain, does X have a binomial distribution? If not, why not? If so, are both values of n and p known?

5.21 Telemarketers A market research firm hires operators to conduct telephone surveys. The computer randomly dials a telephone number, and the operator asks the respondent whether or not he or she has time to answer some questions. Let X be the number of telephone calls made until the first respondent is willing to answer the operator's questions. Is this a binomial experiment? Explain.

5.22 MCAT Scores In 2015 the average combined LSAT score for students in Canada was 161.[4] Suppose that approximately 45% of all students took this test, and that 100 students are randomly selected from throughout Canada. Which of the following random variables has an approximate binomial distribution? If possible, give the values for n and p.

a. The number of students who took the LSAT

b. The scores of the 100 students on the LSAT

c. The number of students who scored above average on the LSAT

5.23 Security Systems A home security system is designed to have a 99% reliability rate. Suppose that nine homes equipped with this system experience an attempted burglary. Find the probabilities of these events:

a. At least one of the alarms is triggered.

b. More than seven of the alarms are triggered.

c. Eight or fewer alarms are triggered.

5.24 Blood Types In a certain population, 85% of the people have Rh-positive blood. Suppose that two people from this population get married. What is the probability that they are both Rh-negative, thus making it inevitable that their children will be Rh-negative?

5.25 Car Colours Car colour preferences change over the years and according to the particular model that the customer selects. In a recent year, 10% of all luxury cars sold were black. If 25 cars of that year and type are randomly selected, find the following probabilities:

a. At least five cars are black.

b. At most six cars are black.

c. More than four cars are black.

d. Exactly four cars are black.

e. Between three and five cars (inclusive) are black.

f. More than 20 cars are not black.

5.26 Harry Potter Of all the Harry Potter books purchased in a recent year, about 60% were purchased for readers 14 or older.[5] If 12 Harry Potter fans who bought books that year are surveyed, find the following probabilities:

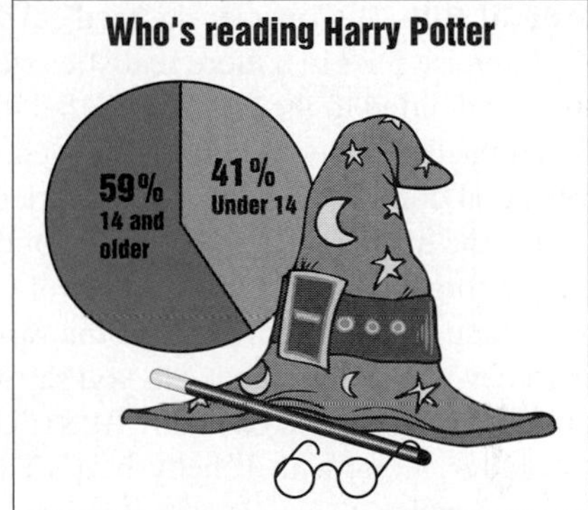

a. At least five of them are 14 or older.

b. Exactly nine of them are 14 or older.

c. Less than three of them are 14 or older.

5.27 O Canada! The National Hockey League (NHL) has 80% of its players born outside the United States, and of those born outside the United States, 50% are born in Canada.[6] Suppose that $n = 12$ NHL players were selected at random. Let X be the number of players in the sample who were born outside of the United States so that $p = 0.8$. Find the following probabilities:

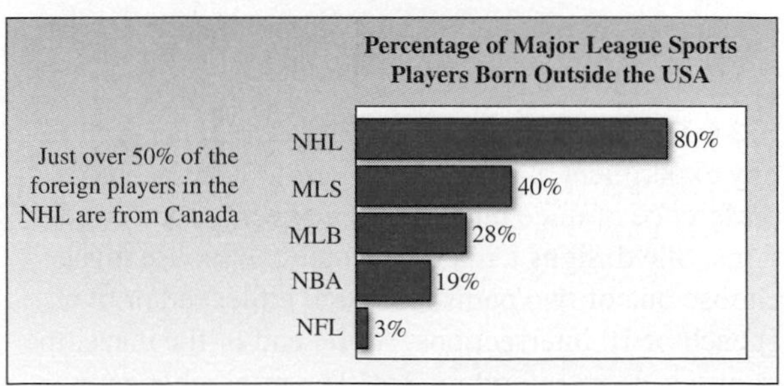

a. At least five of the sampled players were born outside the United States.

b. Exactly seven of the players were born outside the United States.

c. Fewer than six were born outside the United States.

5.28 Medical Bills Records show that 30% of all patients admitted to an alternative medicine clinic fail to pay their bills and that eventually the bills are forgiven. Suppose $n = 4$ new patients represent a random selection from the large set of prospective patients served by the clinic. Find these probabilities:

a. All the patients' bills will eventually have to be forgiven.

b. One will have to be forgiven.

c. None will have to be forgiven.

5.29 Medical Bills II Consider the medical payment problem in Exercise 5.28 in a more realistic setting. Of all patients admitted to the clinic, 30% fail to pay their bills and the debts are eventually forgiven. If the clinic treats 2000 different patients over a period of 1 year, what is the mean (expected) number of debts that have to be forgiven? If X is the number of forgiven debts in the group of 2000 patients, find the variance and standard deviation of X. What can you say about the probability that X will exceed 700? (HINT: Use the values of μ and σ, along with Tchebysheff's Theorem, to answer this question.)

5.30 Whitefly Infestation Suppose that 10% of the fields in a given agricultural area are infested with the sweet potato whitefly. One hundred fields in this area are randomly selected and checked for whitefly.

a. What is the average number of fields sampled that are infested with whitefly?

b. Within what limits would you expect to find the number of infested fields, with probability approximately 95%?

c. What might you conclude if you found that $x = 25$ fields were infested? Is it possible that one of the characteristics of a binomial experiment is not satisfied in this experiment? Explain.

5.31 Colour Preferences in Mice In a psychology experiment, a researcher plans to test the colour preference of mice under certain experimental conditions. She designs a maze in which the mouse must choose one of two paths, coloured either red or blue, at each of 10 intersections. At the end of the maze, the mouse is given a food reward. The researcher counts the number of times the mouse chooses the red path. If you were the researcher, how would you use this count to decide whether the mouse has any preference for colour?

5.32 Pet Peeves Across the board, 22% of car leisure travellers rank "traffic and other drivers" as their pet peeve while travelling. Of car leisure travellers in the densely populated U.S. Northeast, 33% list this as their pet peeve.[7] A random sample of $n = 8$ such travellers in the Northeast were asked to state their pet peeve while travelling. The *MINITAB* printout shows the *cumulative* and *individual* probabilities.

MINITAB Output for Exercise 5.32

Cumulative Distribution Function

Binomial with n = 8 and p = 0.33

x	P(X <= x)
0	0.04061
1	0.20061
2	0.47644
3	0.74814
4	0.91543
5	0.98134
6	0.99758
7	0.99986
8	1.00000

Probability Density Function

Binomial with n = 8 and p = 0.033

x	P(X = x)
0	0.040607
1	0.160003
2	0.275826
3	0.271709
4	0.167283
5	0.065915
6	0.016233
7	0.002284
8	0.000141

a. Use the binomial formula to find the probability that all eight give "traffic and other drivers" as their pet peeve.

b. Confirm the results of part a using the *MINITAB* printout.

c. What is the probability that at most seven give "traffic and other drivers" as their pet peeve?

5.33 Fast Food and Gas Stations Forty percent of all Canadians who travel by car look for gas stations and food outlets that are close to or visible from the highway. Suppose a random sample of $n = 25$ Canadians who travel by car are asked how they determine where to stop for food and gas. Let x be the number in the sample who respond that they look for gas stations and food outlets that are close to or visible from the highway.

a. What are the mean and variance of x?

b. Calculate the interval $\mu \pm 2\sigma$. What values of the binomial random variable x fall into this interval?

c. Find $P(6 \leq x \leq 14)$. How does this compare with the fraction in the interval $\mu \pm 2\sigma$ for any distribution? For mound-shaped distributions?

5.34 Taste Test for PTC The taste test for PTC (phenylthiocarbamide) is a favourite exercise for every human genetics class. It has been established that a single gene determines the characteristic, and that 70% of Canadians are "tasters," while 30% are "non-tasters." Suppose that 20 Canadians are randomly chosen and are tested for PTC.

a. What is the probability that 17 or more are "tasters"?

b. What is the probability that 15 or fewer are "tasters"?

5.35 Less Vegetable Servings Many Canadians report consuming fewer servings of vegetables in the winter than in the summer, according to an Ipsos Reid/Campbell Company of Canada survey.[8] Specifically, almost half of Canadians indicate they consume fewer vegetable servings on a typical winter day than on a typical summer day. Suppose that the 50% figure is correct and that 15 Canadians are randomly selected for the survey.

a. What is the probability that exactly eight of the respondents have fewer vegetable servings?

b. What is the probability that at most four of the respondents have fewer vegetable servings?

c. What is the probability that more than 10 respondents have fewer vegetable servings?

5.3 THE POISSON PROBABILITY DISTRIBUTION

Another discrete random variable that has numerous practical applications is the **Poisson random variable.** Its probability distribution provides a good model for data that represent the number of occurrences of a specified event in a given unit of time or space. Here are some examples of experiments for which the random variable X can be modelled by the Poisson random variable:

- The number of calls received by a switchboard during a given period of time
- The number of bacteria per small volume of fluid
- The number of customer arrivals at a checkout counter during a given minute
- The number of machine breakdowns during a given day
- The number of traffic accidents at a given intersection during a given time period

In each example, **X represents the number of events that occur in a period of time or space during which an average of μ such events can be expected to occur.** The only assumptions needed when one uses the Poisson distribution to model experiments such as these are that the counts or events occur **randomly and independently** of one another. The formula for the Poisson probability distribution, as well as its mean and variance, are given next.

THE POISSON PROBABILITY DISTRIBUTION

Let μ be the average number of times that an event occurs in a certain period of time or space. The probability of k occurrences of this event is

$$P(X = k) = \frac{\mu^k e^{-\mu}}{k!}$$

for values of $k = 0, 1, 2, 3, \ldots$. The mean and standard deviation of the Poisson random variable X are

Mean: μ

Standard deviation: $\sigma = \sqrt{\mu}$

The symbol $e = 2.71828 \ldots$ is evaluated using your scientific calculator, which should have a function such as e^x. For each value of k, you can obtain the individual probabilities for the Poisson random variable, just as you did for the binomial random variable.

EXAMPLE 5.14

The average number of traffic accidents on a certain section of the Trans-Canada Highway is two per week. Assume that the number of accidents follows a Poisson distribution with $\mu = 2$.

1. Find the probability of no accidents on this section of highway during a one-week period.
2. Find the probability of at most three accidents on this section of highway during a two-week period.

SOLUTION

1. The average number of accidents per week is $\mu = 2$. Therefore, the probability of no accidents on this section of highway during a given week is

$$P(X = 0) = p(0) = \frac{2^0 e^{-2}}{0!} = e^{-2} = 0.135335$$

2. During a two-week period, the average number of accidents on this section of highway is $2(2) = 4$. The probability of at most three accidents during a two-week period is

$$P(X \leq 3) = p(0) + p(1) + p(2) + p(3)$$

where

$$p(0) = \frac{4^0 e^{-4}}{0!} = 0.018316 \qquad p(2) = \frac{4^2 e^{-4}}{2!} = 0.146525$$

$$p(1) = \frac{4^1 e^{-4}}{1!} = 0.073263 \qquad p(3) = \frac{4^3 e^{-4}}{3!} = 0.195367$$

Therefore,

$$P(X \leq 3) = 0.018316 + 0.073263 + 0.146525 + 0.195367 = 0.433471$$

NEED A TIP?

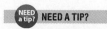

Use either the Poisson formula or Table 2 to calculate Poisson probabilities.

Once the values for $p(x)$ have been calculated, you can use them to construct a probability histogram for the random variable X. Graphs of the Poisson probability distribution for $\mu = 0.5$, 2, and 4 are shown in Figure 5.5.

Alternatively, you can use **cumulative Poisson tables** (Table 2 in Appendix I) or the cumulative or individual probabilities generated by *MINITAB* or *Excel*. All of these options are usually more convenient than hand calculation. The procedures are similar to those used for the binomial random variable.

FIGURE 5.5

Poisson probability distributions for $\mu = 0.5$, 2, and 4

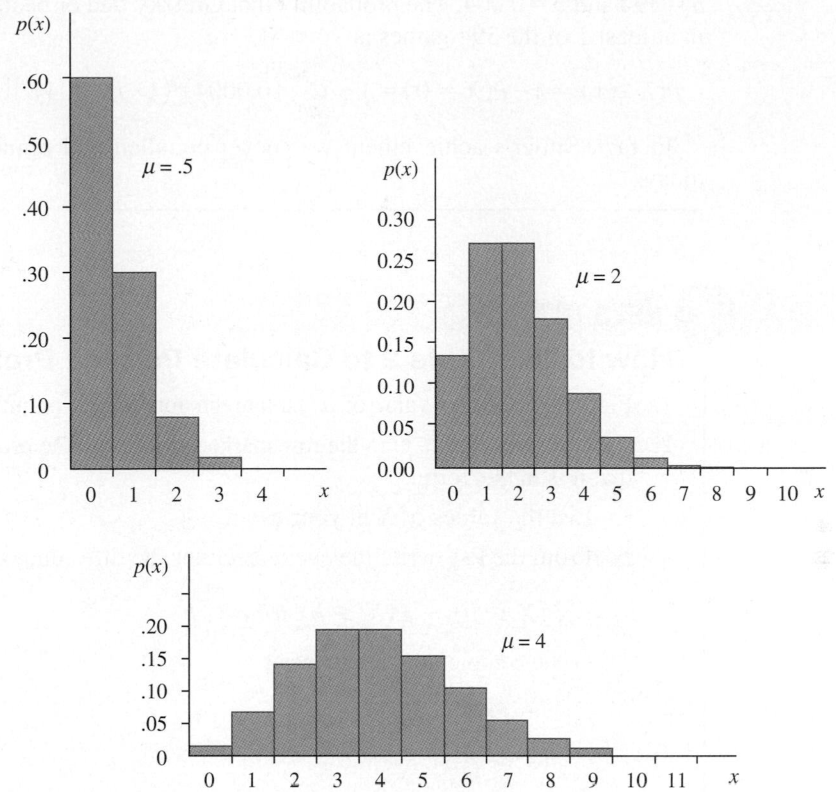

EXAMPLE  5.15

On February 7, 1976, Darryl Sittler scored 10 points in a hockey game while playing for the Toronto Maple Leafs. Wayne Gretzky has the most points of any hockey player in history, averaging 2.629 points per game in the 394 regular season games played in the five seasons 1981–82 to 1985–86. Goals in hockey games follow a Poisson distribution.[9] From just this information, compute the probability that Gretzky had at least 10 points in at least one of the 394 games.

SOLUTION Let Y be the random variable representing the number of goals that Gretzky scores in a single game. Then $\mu = E(Y) = 2.629$. So

$$P(Y \geq 10) = \sum_{y=10}^{\infty} \frac{\mu^y e^{-\mu}}{y!} = 1 - \sum_{y=0}^{9} \frac{\mu^y e^{-\mu}}{y!} = 1 - 0.9996 = 0.0004$$

Next let X be the number of games (among the 394) in which Gretzky scored at least 10 goals. Since the mean 2.629 was computed from the 394 games, our use of this information means that the points in the different games are not really independent but the numbers of goals in different hockey games are generally independent and we will assume independence here as a good approximation to reality. Thus X is binomial with

$n = 394$ and $p = 0.004$. The probability that Gretzky tied or beat Sittler's 10-point game in at least 1 of the 394 games is

$$P(X \geq 1) = 1 - P(X = 0) = 1 - C_0^{394}(0.0004)^0(1 - 0.0004)^{394} = 1 - 0.8542 = 0.1458$$

In fact, Sittler's achievement was never equalled and remains as an NHL record today.

NEED TO KNOW

How to Use Table 2 to Calculate Poisson Probabilities

1. Find the necessary value of μ. Isolate the appropriate column in Table 2.
2. Table 2 gives $P(X \leq k)$ in the row marked k. Rewrite the probability you need so that it is in this form.
   - List the values of X in your event.
   - From the list, write the event as either the difference of two probabilities:

     $P(X \leq a) - P(X \leq b)$ *for* $a > b$

     or the complement of the event:

     $1 - P(X \leq a)$

     or just the event itself:

     $P(X \leq a)$ *or* $P(X < a - 1)$

EXAMPLE 5.16

Refer to Example 5.14, where we calculated probabilities for a Poisson distribution with $\mu = 2$ and $\mu = 4$. Use the cumulative Poisson table to find the probabilities of these events:

1. No accidents during a one-week period
2. At most three accidents during a two-week period

SOLUTION A portion of Table 2 in Appendix I is shown in Figure 5.6.

FIGURE 5.6
Portion of Table 2 in Appendix I

	μ				
k	2.0	2.5	3.0	3.5	4.0
0	0.135	0.082	0.055	0.033	0.018
1	0.406	0.287	0.199	0.136	0.092
2	0.677	0.544	0.423	0.321	0.238
3	0.857	0.758	0.647	0.537	0.433
4	0.947	0.891	0.815	0.725	0.629
5	0.983	0.958	0.916	0.858	0.785
6	0.995	0.986	0.966	0.935	0.889
7	0.999	0.996	0.988	0.973	0.949
8	1.000	0.999	0.996	0.990	0.979
9		1.000	0.999	0.997	0.992
10			1.000	0.999	0.997
11				1.000	0.999
12					1.000

1. From Example 5.14, the average number of accidents in a one-week period is $\mu = 2.0$. Therefore, the probability of no accidents in a one-week period can be read directly from Table 2 in the column marked "2.0" as $P(X = 0) = p(0) = 0.135$.
2. The average number of accidents in a two-week period is $2(2) = 4$. Therefore, the probability of at most three accidents in a two-week period is found in Table 2, indexing $\mu = 4.0$ and $k = 3$ as $P(X \leq 3) = 0.433$.

Both of these probabilities match the calculations done in Example 5.14, correct to three decimal places.

In Section 5.2, we used the cumulative binomial tables to simplify the calculation of binomial probabilities. Unfortunately, in practical situations, n is often large and no tables are available.

NEED A TIP?

You can estimate binomial probabilities with the Poisson when n is large and p is small.

THE POISSON APPROXIMATION TO THE BINOMIAL DISTRIBUTION

The Poisson probability distribution provides a simple, easy-to-compute, and accurate approximation to binomial probabilities when n is large and $\mu = np$ is small, preferably with $np < 7$. An approximation suitable for larger values of $\mu = np$ will be given in Chapter 6.

EXAMPLE 5.17

Suppose a life insurance company insures the lives of 5000 men aged 42. If actuarial studies show the probability that any 42-year-old man will die in a given year to be 0.001, find the exact probability that the company will have to pay $X = 4$ claims during a given year.

SOLUTION The exact probability is given by the binomial distribution as

$$P(X = 4) = p(4) = \frac{5000!}{4!4996!}(0.001)^4(0.999)^{4996}$$

for which binomial tables are not available. To compute $P(X = 4)$ without the aid of a computer would be very time-consuming, but the Poisson distribution can be used to provide a good approximation to $P(X = 4)$. Computing $\mu = np = (5000)(0.001) = 5$ and substituting into the formula for the Poisson probability distribution, we have

$$p(4) \approx \frac{\mu^4 e^{-\mu}}{4!} = \frac{5^4 e^{-5}}{4!} = \frac{(625)(0.006738)}{24} = 0.175$$

The value of $p(4)$ could also be obtained using Table 2 in Appendix I with $\mu = 5$ as

$$p(4) = P(X \leq 4) - P(X \leq 3) = 0.440 - 0.265 = 0.175$$

EXAMPLE 5.18

In 2001, an 82-year-old man in Ontario claimed that a convenience store clerk defrauded him of $250,000 by telling him that his ticket was not a winner and keeping it. The story received national attention in 2006 with a news story aired on the CBC. Analysis of Ontario Lottery Commission winning tickets indicated that 214 major lottery prizes had been won by clerks. A statistician computed that the expected number should have been 57. The following numbers are fictitious but indicate how such an analysis could be done. Suppose that there were 10,000 major lottery prizes of $25,000 or more during a 20-year period. Suppose that clerks who sell lottery tickets bought 0.57% of all tickets sold (i.e., 5.7 of every 1000 tickets sold). Then each of the 10,000 prizes can be viewed as a trial in a binomial experiment with $n = 10{,}000$ and p is the probability of a winning ticket being won by a clerk, so $p = 0.0057$. What is the probability of having at least 214 clerks as prize winners?

SOLUTION Let X be the number of prizes won by salesclerks among the 10,000 prizes. Then X is binomial with $n = 10{,}000$ and $p = 0.0057$. So

$$P(X \geq 214) = C_x^{10{,}000}(0.0057)^x\ (1 - 0.0057)^{10000-x}$$

This probability is hard to compute so we use a normal approximation (Chapter 6, p. 251) to a binomial. Here $\mu = np = 10{,}000(0.0057) = 57$ and $\sigma^2 = npq = 10{,}000(0.0057)(0.9943) = 56.6751$. Using the results of Section 6.4, we have

$$P(X \geq 214) = P(X \geq 213.5) = P\left(\frac{X - \mu}{\sigma} \geq \frac{213.5 - 57}{\sqrt{56.6751}}\right) \approx P(Z > 20.78826)$$

This probability is so small that it is beyond the limits of the normal table and it is much less than 0.00001.

EXAMPLE 5.19

A manufacturer of power lawn mowers buys one-horsepower, two-cycle engines in lots of 1000 from a supplier. She then equips each of the mowers produced by her plant with one of the engines. History shows that the probability of any one engine from that supplier proving unsatisfactory is 0.001. In a shipment of 1000 engines, what is the probability that none is defective? Three are? Four are?

SOLUTION This is a binomial experiment with $n = 1000$ and $p = 0.001$. The expected number of defectives in a shipment of $n = 1000$ engines is $\mu = np = (1000)(0.001) = 1$. Since this is a binomial experiment with $np < 7$, the probability of X defective engines in the shipment may be approximated by

$$P(X = k) = p(k) = \frac{\mu^k e^{-\mu}}{k!} = \frac{1^k e^{-1}}{k!} = \frac{e^{-1}}{k!}$$

Therefore,

$$p(0) \approx \frac{e^{-1}}{0!} = \frac{0.368}{1} = 0.368$$

$$p(3) \approx \frac{e^{-1}}{3!} = \frac{0.368}{6} = 0.061$$

$$p(4) \approx \frac{e^{-1}}{4!} = \frac{0.368}{24} = 0.015$$

The individual Poisson probabilities for $\mu = 1$ along with the individual binomial probabilities for $n = 1000$ and $p = 0.001$ were generated by *Excel* and are shown in Figure 5.7. The individual probabilities, even though they are computed with totally different formulas, are almost the same. The exact binomial probabilities are in the left section of Figure 5.7, and the Poisson approximations are on the right.

FIGURE 5.7
Excel output of binomial and Poisson probabilities

x	Binomial p(x)	x	Poisson p(x)
0	0.3677	0	0.3679
1	0.3681	1	0.3679
2	0.1840	2	0.1839
3	0.0613	3	0.0613
4	0.0153	4	0.0153
5	0.0030	5	0.0031
6	0.0005	6	0.0005
7	0.0001	7	0.0001
8	0.0000	8	0.0000
9	0.0000	9	0.0000
10	0.0000	10	0.0000

5.3 EXERCISES

BASIC TECHNIQUES

5.36 Consider a Poisson random variable X with $\mu = 2.5$. Use the Poisson formula to calculate the following probabilities:

a. $P(X = 0)$ **b.** $P(X = 1)$
c. $P(X = 2)$ **d.** $P(X \leq 2)$

5.37 Consider a Poisson random variable X with $\mu = 3$. Use the Poisson formula to calculate the following probabilities:

a. $P(X = 0)$ **b.** $P(X = 1)$ **c.** $P(X > 1)$

5.38 Consider a Poisson random variable X with $\mu = 3$. Use Table 2 in Appendix I to find the following probabilities:

a. $P(X \leq 3)$ **b.** $P(X > 3)$
c. $P(X = 3)$ **d.** $P(3 \leq X \leq 5)$

5.39 Let X be a Poisson random variable with mean $\mu = 2$. Calculate these probabilities:

a. $P(X = 0)$ **b.** $P(X = 1)$
c. $P(X > 1)$ **d.** $P(X = 5)$

5.40 Let X be a Poisson random variable with mean $\mu = 2.5$. Use Table 2 in Appendix I to calculate these probabilities:

a. $P(X \geq 5)$ **b.** $P(X > 6)$
c. $P(X = 2)$ **d.** $P(1 \leq X \leq 4)$

5.41 Poisson vs. Binomial Let X be a binomial random variable with $n = 20$ and $p = 0.1$.

a. Calculate $P(X \leq 2)$ using Table 1 in Appendix I to obtain the exact binomial probability.
b. Use the Poisson approximation to calculate $P(X \leq 2)$.
c. Compare the results of parts a and b. Is the approximation accurate?

5.42 Poisson vs. Binomial II To illustrate how well the Poisson probability distribution approximates the binomial probability distribution, calculate the Poisson approximate values for $p(0)$ and $p(1)$ for a binomial probability distribution with $n = 25$ and $p = 0.05$. Compare the answers with the exact values obtained from Table 1 in Appendix I.

APPLICATIONS

5.43 Airport Safety The increased number of small commuter planes in major airports has heightened concern over air safety. An eastern airport has recorded a monthly average of five near-misses on landings and takeoffs in the past five years.

a. Find the probability that during a given month there are no near-misses on landings and takeoffs at the airport.
b. Find the probability that during a given month there are five near-misses.
c. Find the probability that there are at least five near-misses during a particular month.

5.44 Intensive Care The number X of people entering the intensive care unit at a particular hospital on any one day has a Poisson probability distribution with mean equal to five people per day.

a. What is the probability that the number of people entering the intensive care unit on a particular day is two? Less than or equal to two?

b. Is it likely that X will exceed 10? Explain.

5.45 Cross-Border Drinking Alcohol-related crashes increased in Windsor after bar hours extended, the *Windsor Star* reported on November 7, 2005.[10] The number of injuries and fatalities from alcohol-related car accidents rose by 45% in Windsor since the Ontario government extended drinking hours to 2 a.m., according to a study by the University of Western Ontario. There were an average of 1.7 alcohol-related traffic casualties a month in Windsor between the hours of 11 p.m. and 3 a.m. before the drinking hours were extended. After the change, the number increased to 2.5. Assume that the average number of fatalities for the current year is still 2.5. What are the probabilities of the following events?

a. There will be two fatalities during the year.

b. Two or more fatalities

c. At most one fatality

d. At least one fatality

5.46 Cross-Border Drinking, continued Refer to Exercise 5.45.

a. Calculate the mean and standard deviation for X, the number of fatalities per year.

b. Within what limits would you expect the number of fatalities per year to fall?

5.47 Bacteria in Water Samples If a drop of water is placed on a slide and examined under a microscope, the number X of a particular type of bacteria present has been found to have a Poisson probability distribution. Suppose the maximum permissible count per water specimen for this type of bacteria is five. If the mean count for your water supply is two and you test a single specimen, is it likely that the count will exceed the maximum permissible count? Explain.

5.48 *E. coli* Outbreak Increased research and discussion have focused on the number of illnesses involving the organism *Escherichia coli* (01257: H7), which causes a breakdown of red blood cells and intestinal hemorrhages in its victims.[11] Suppose that sporadic outbreaks of *E. coli* have occurred in Alberta at a rate of 2.5 per 100,000 for a period of two years. Let us suppose that this rate has not changed.

a. What is the probability that at most five cases of *E. coli* per 100,000 are reported in Alberta in a given year?

b. What is the probability that more than five cases of *E. coli* per 100,000 are reported in a given year?

c. Approximately 95% of occurrences of *E. coli* involve at most how many cases?

5.4 THE HYPERGEOMETRIC PROBABILITY DISTRIBUTION

Suppose you are selecting a sample of elements from a population and you record whether or not each element possesses a certain characteristic. You are recording the typical "success" or "failure" data found in the binomial experiment. The sample survey of Example 5.4 is a practical illustration of these sampling situations.

If the number of elements in the population is large relative to the number in the sample (as in Example 5.4), the probability of selecting a success on a single trial is equal to the proportion p of successes in the population. Because the population is large in relation to the sample size, this probability will remain constant (for all practical purposes) from trial to trial, and the number X of successes in the sample will follow a binomial probability distribution. However, if the number of elements in the population is small in relation to the sample size ($n/N \geq 0.05$), the probability of a success for a given trial is dependent on the outcomes of preceding trials. Then the number X of successes follows what is known as a **hypergeometric probability distribution.**

It is easy to visualize the **hypergeometric random variable** X by thinking of a bowl containing M red balls and $N - M$ white balls, for a *total of* N balls in the bowl. You select n balls from the bowl and record X, the number of red balls that you see. If you now define a "success" to be a red ball, you have an example of the hypergeometric random variable X.

The formula for calculating thc probability of exactly k successes in n trials is given next.

THE HYPERGEOMETRIC PROBABILITY DISTRIBUTION

A population contains M successes and $N - M$ failures. The probability of exactly k successes in a random sample of size n is

$$P(X = k) = \frac{C_k^M C_{n-k}^{N-M}}{C_n^N}$$

for values of k that depend on N, M, and n with

$$C_n^N = \frac{N!}{n!(N - n)!}$$

The mean and variance of a hypergeometric random variable are very similar to those of a binomial random variable with a correction for the finite population size:

$$\mu = n\left(\frac{M}{N}\right)$$

$$\sigma^2 = n\left(\frac{M}{N}\right)\left(\frac{N - M}{N}\right)\left(\frac{N - n}{N - 1}\right)$$

EXAMPLE  5.20

A case of wine has 12 bottles, three of which contain spoiled wine. A sample of four bottles is randomly selected from the case.

1. Find the probability distribution for X, the number of bottles of spoiled wine in the sample.
2. What are the mean and variance of X?

SOLUTION For this example, $N = 12$, $n = 4$, $M = 3$, and $(N - M) = 9$. Then

$$p(x) = \frac{C_x^3 C_{4-x}^9}{C_4^{12}}$$

1. The possible values for x are 0, 1, 2, and 3, with probabilities

NEED A TIP?

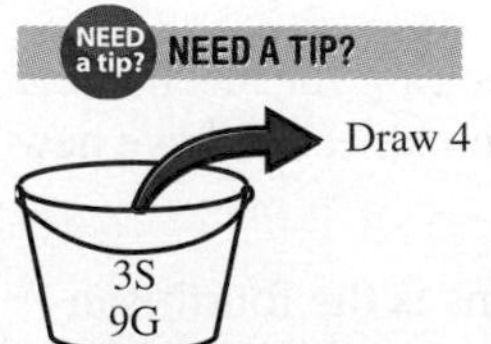

$$p(0) = \frac{C_0^3 C_4^9}{C_4^{12}} = \frac{1(126)}{495} = 0.25$$

$$p(1) = \frac{C_1^3 C_3^9}{C_4^{12}} = \frac{3(84)}{495} = 0.51$$

$$p(2) = \frac{C_2^3 C_2^9}{C_4^{12}} = \frac{3(36)}{495} = 0.22$$

$$p(3) = \frac{C_3^3 C_1^9}{C_4^{12}} = \frac{1(9)}{495} = 0.02$$

2. The mean is given by

$$\mu = 4\left(\frac{3}{12}\right) = 1$$

and the variance is

$$\sigma^2 = 4\left(\frac{3}{12}\right)\left(\frac{9}{12}\right)\left(\frac{12-4}{11}\right) = 0.55$$

EXAMPLE 5.21

A particular industrial product is shipped in lots of 20. Testing to determine whether each item is defective is costly; hence, the manufacturer samples production rather than using a 100% inspection plan. A sampling plan constructed to minimize the number of defectives shipped to customers calls for sampling five items from each lot and rejecting the lot if more than one defective is observed. (If the lot is rejected, each item in the lot is then tested.) If a lot contains four defectives, what is the probability that it will be accepted?

SOLUTION Let X be the number of defectives in the sample. Then $N = 20$, $M = 4$, $(N - M) = 16$, and $n = 5$. The lot will be rejected if $X = 2$, 3, or 4. Then

$$P(\text{accept the lot}) = P(X \le 1) = p(0) + p(1) = \frac{C_0^4 C_5^{16}}{C_5^{20}} + \frac{C_1^4 C_4^{16}}{C_5^{20}}$$

$$= \frac{\left(\frac{4!}{0!4!}\right)\left(\frac{16!}{5!11!}\right)}{\frac{20!}{5!15!}} + \frac{\left(\frac{4!}{1!3!}\right)\left(\frac{16!}{4!12!}\right)}{\frac{20!}{5!15!}}$$

$$= \frac{91}{323} + \frac{455}{969} = 0.2817 + 0.4696 = 0.7513$$

EXAMPLE  **5.22**

A rental truck agency services its vehicles on a regular basis, checking for mechanical problems. Suppose that the agency has six moving vans, two of which need to have new brakes. During a routine check, the vans are tested one at a time.

1. What is the probability that the last van with brake problems is the fourth van tested?

2. What is the probability that no more than four vans need to be tested before both brake problems are detected?
3. Given that one van with bad brakes is detected in the first two tests, what is the probability that the remaining van is found faulty on the third or fourth test?

SOLUTION

1. If the fourth van tested is the last van with brake problems, then in the first three tests, we must find one van with brake problems and two without. That is, in choosing three out of the six vans, we must find one that is faulty and two that are not faulty. Think of this problem as choosing three balls—one white and two red from a total of six. Since the experiment is modelled using the hypergeometric distribution, the probability can be calculated as

$$P_{(\text{one faulty and two not})} = \frac{\binom{2}{1}\binom{4}{2}}{\binom{6}{3}} = \frac{3}{5}$$

Once this is accomplished, the second van with brake problems must be chosen on the fourth test. Using the Multiplication Rule, the probability that the fourth van tested is the last with faulty brakes is $\frac{3}{5}\left(\frac{1}{3}\right) = \frac{1}{5}$.

2. If it is known that the first faulty van is found in the first two tests, there are four vans left from which to select those tested third and fourth. Of those four, only one is faulty. Hence,

$$\begin{aligned} &P\text{ (one faulty found on third or fourth test} \mid \text{one faulty found on first two tests)} \\ &\quad = P\text{ (one faulty} \mid \text{one faulty in first two tests)} \\ &\quad + P\text{ (one not faulty, then one faulty} \mid \text{one faulty in first two tests)} \end{aligned}$$

$$= \frac{\binom{1}{1}\binom{4-1}{1-1}}{\binom{4}{1}} + \frac{\binom{1}{0}\binom{4-1}{1-0}\binom{1}{1}\binom{3-1}{1-1}}{\binom{4}{1}\binom{3}{1}} = \frac{1}{4} + \frac{3}{4} \times \frac{1}{3} = \frac{1}{2}$$

This can be solved in another way by stating the problem in an alternative way:

$$P_{(1\text{ faulty among 2 tests on 4 items})} = \frac{\binom{1}{1}\binom{3}{1}}{\binom{4}{2}} = \frac{1}{2}$$

EXAMPLE 5.23 An eight-cylinder automobile engine has two misfiring spark plugs. The mechanic removes all four plugs from one side of the engine. What is the probability the two misfiring spark plugs are among those removed?

SOLUTION The random variable X, the number of misfiring spark plugs, has a hypergeometric distribution. Therefore,

$$P(X = 2) = \frac{\binom{2}{2}\binom{6}{2}}{\binom{8}{4}} = \frac{3}{14}$$

5.4 EXERCISES

BASIC TECHNIQUES

5.49 Evaluate these probabilities:

a. $\frac{C_1^3 C_1^2}{C_2^5}$ **b.** $\frac{C_2^4 C_1^3}{C_3^7}$ **c.** $\frac{C_4^5 C_0^3}{C_4^8}$

5.50 Let X be the number of successes observed in a sample of $n = 4$ items selected from a population of $N = 8$. Suppose that of the $N = 8$ items, 5 are considered "successes."

a. Find the probability of observing all successes.

b. Find the probability of observing one success.

c. Find the probability of observing at most two successes.

5.51 Let X be the number of successes observed in a sample of $n = 5$ items selected from $N = 10$. Suppose that, of the $N = 10$ items, 6 are considered "successes."

a. Find the probability of observing no successes.

b. Find the probability of observing at least two successes.

c. Find the probability of observing exactly two successes.

5.52 Let X be a hypergeometric random variable with $N = 15$, $n = 3$, and $M = 4$.

a. Calculate $p(0)$, $p(1)$, $p(2)$, and $p(3)$.

b. Construct the probability histogram for x.

c. Use the formulas given in Section 5.4 to calculate μ and σ^2.

d. What proportion of the population of measurements fall into the interval $(\mu \pm 2\sigma)$? Into the interval $(\mu \pm 3\sigma)$? Do these results agree with those given by Tchebysheff's Theorem?

5.53 Candy Choices A candy dish contains five blue and three red candies. A child reaches up and selects three candies without looking.

a. What is the probability that there are two blue and one red candies in the selection?

b. What is the probability that the candies are all red?

c. What is the probability that the candies are all blue?

APPLICATIONS

5.54 Defective Computer Chips A piece of electronic equipment contains six computer chips, two of which are defective. Three computer chips are randomly chosen for inspection, and the number of defective chips is recorded. Find the probability distribution for X, the number of defective computer chips. Compare your results with the answers obtained in Exercise 4.93.

5.55 Gender Bias? A company has five applicants for two positions: two women and three men. Suppose that the five applicants are equally qualified and that no preference is given for choosing either gender. Let x equal the number of women chosen to fill the two positions.

a. Write the formula for $p(x)$, the probability distribution of X.

b. What are the mean and variance of this distribution?

5.56 Teaching Credentials In Southern Ontario, a growing number of persons pursuing a teaching credential are choosing paid internships over traditional student teaching programs. A group of eight candidates for three local teaching positions consisted of five candidates who had enrolled in paid internships and three candidates who had enrolled in traditional student teaching programs. Let us assume that all eight candidates are equally qualified for the positions. Let X represent the number of internship-trained candidates who are hired for these three positions.

a. Does X have a binomial distribution or a hypergeometric distribution? Support your answer.

b. Find the probability that three internship-trained candidates are hired for these positions.

c. What is the probability that none of the three hired was internship-trained?

d. Find $P(X \leq 1)$.

5.57 Seed Treatments Seeds are often treated with a fungicide for protection in poor-draining, wet environments. In a small-scale trial prior to a large-scale experiment to determine what dilution of the fungicide to apply, five treated seeds and five untreated seeds were planted in clay soil and the number of plants emerging from the treated and untreated seeds were recorded. Suppose the dilution was not effective and only four plants emerged. Let X represent the number of plants that emerged from treated seeds.

a. Find the probability that $X = 4$.

b. Find $P(X \leq 3)$.

c. Find $P(2 \leq X \leq 3)$.

CHAPTER REVIEW

Key Concepts and Formulas

I. The Binomial Random Variable

1. **Five characteristics:** n identical independent trials, each resulting in either *success S* or *failure F*; probability of success is p and remains constant from trial to trial; and X is the number of successes in n trials
2. **Calculating binomial probabilities**
   a. Formula: $P(X = k) = C_k^n p^k q^{n-k}$
   b. Cumulative binomial tables
   c. Individual and cumulative probabilities using *MINITAB* and *Excel*
3. Mean of the binomial random variable: $\mu = np$
4. Variance and standard deviation: $\sigma^2 = npq$ and $\sigma = \sqrt{npq}$

II. The Poisson Random Variable

1. The number of events that occur in a period of time or space, during which an average of μ such events are expected to occur
2. **Calculating Poisson probabilities**
   a. Formula: $P(X = k) = \dfrac{\mu^k e^{-\mu}}{k!}$
   b. Cumulative Poisson tables
   c. Individual and cumulative probabilities using *MINITAB* and *Excel*
3. Mean of the Poisson random variable: μ
4. Variance and standard deviation: $\sigma^2 = \mu$ and $\sigma = \sqrt{\mu}$
5. Binomial probabilities can be approximated with Poisson probabilities when $np < 7$, using $\mu = np$.

III. The Hypergeometric Random Variable

1. The number of successes in a sample of size n from a finite population containing M successes and $N - M$ failures
2. Formula for the probability of k successes in n trials:
$$P(X = k) = \frac{C_k^M C_{n-k}^{N-M}}{C_n^N}$$
3. Mean of the hypergeometric random variable:
$$\mu = n\left(\frac{M}{N}\right)$$
4. Variance and standard deviation:
$$\sigma^2 = n\left(\frac{M}{N}\right)\left(\frac{N-M}{N}\right)\left(\frac{N-n}{N-1}\right)$$
and $\quad \sigma = \sqrt{\sigma^2}$

Binomial and Poisson Probabilities in *Microsoft Excel*

For a random variable that has either a binomial or a Poisson probability distribution, *Excel* has been programmed to calculate either exact probabilities—$P(X = k)$—for a given value of k or cumulative probabilities—$P(X \leq k)$—for a given value of k. You must specify which distribution you are using and the necessary parameters: n and p for the binomial distribution and μ for the Poisson distribution.

Binomial Probabilities

1. Consider a binomial distribution with $n = 10$ and $p = 0.25$. The *value* of p does not appear in Table 1 of Appendix I, but you can use *Excel* to generate the entire probability distribution as well as the cumulative probabilities by entering the numbers 0–10 in column A.

   One way to quickly enter a set of consecutive integers in a column is to do the following:

   - Name columns A, B, and C as "*X*," "P(*X* = k)," and "P(*X* <= k)," respectively.
   - Enter the first two values of *X*—0 and 1—to create a pattern in column A.
   - Use your mouse to highlight the first two integers. Then grab the square handle in the lower right corner of the highlighted area. Drag the handle down to continue the pattern.
   - As you drag, you will see an integer appear in a small rectangle. Release the mouse when you have the desired number of integers—in this case, 10.

2. Once the necessary values of x have been entered, place your cursor in the cell corresponding to $p(0)$, cell B2 in the spreadsheet. Select the **Insert Function** icon *fx*. In the drop-down list, select the **Statistical** category, select the **BINOM.DIST** function, and click **OK.** (NOTE: This function is called **BINOMDIST** in ***Excel 2007*** and earlier versions.) The dialogue box shown in Figure 5.8 will appear.

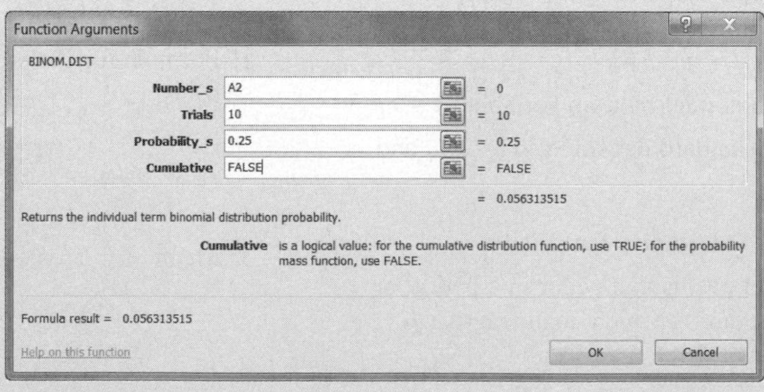

FIGURE 5.8
MINITAB screenshot - BINOMDIST

3. You must type in or select numbers or cell locations for each of the four boxes. When you place your cursor in the box, you will see a description of the necessary input for that box. Enter the address of the cell corresponding to $X = 0$ (cell A2) in the first box, the value of n in the second box, the value of p in the third box, and the word FALSE in the fourth box to calculate $P(X = k)$.

4. The resulting probability is marked as "Formula result = 0.056313515" at the bottom of the box, and when you click **OK**, the probability $P(X = 0)$ will appear in cell B2. To obtain the other probabilities, simply place your cursor in cell B2, grab the square handle in the lower right corner of the cell and drag the handle down to copy the formula into the other nine cells. *Excel* will automatically adjust the cell location in the formula as you copy.

5. If you want to generate the cumulative probabilities, $P(X \leq k)$, place your cursor in the cell corresponding to $P(X \leq 0)$, cell C2 in the spreadsheet. Then select **Insert Function ▶ Statistical ▶ BINOM.DIST,** and click **OK.** Continue as in steps 3 and 4, but type TRUE in the fourth line of the dialogue box to calculate $P(X \leq k)$. The resulting output is shown in Figure 5.9.

FIGURE 5.9

MINITAB screenshot - BINOMDIST con't

	A	B	C
1	x	P(x = k)	P(x <= k)
2	0	0.0563	0.0563
3	1	0.1877	0.2440
4	2	0.2816	0.5256
5	3	0.2503	0.7759
6	4	0.1460	0.9219
7	5	0.0584	0.9803
8	6	0.0162	0.9965
9	7	0.0031	0.9996
10	8	0.0004	1.0000
11	9	0.0000	1.0000
12	10	0.0000	1.0000

6. What value k is such that only 5% of the values of X exceed this value (and 95% are less than or equal to k)? Place your cursor in an empty cell, select **Insert Function ▶ Statistical ▶ BINOM.INV**, and click **OK.** (NOTE: This function is new to ***Excel 2010***.) The resulting dialogue box will calculate what is sometimes called the **inverse cumulative probability.** Type **10** in the first box, **0.25** in the second box, and **0.95** in the third box. When you click **OK,** the number 5 will appear in the empty cell. This is the smallest value of k for which $P(X \leq k)$ is greater than or equal to 0.95. Refer to Figure 5.9 and notice that $P(X \leq 5) = 0.9803$ so that $P(X > 5) = 1 - 0.9803 = 0.0197$. Hence, if you observed a value of $x = 5$, this would be an unusual observation.

Poisson Probabilities

1. The procedures for calculating individual or cumulative probabilities and probability distributions for the Poisson random variable are similar to those used for the binomial distribution.
2. To find Poisson probabilities $P(X = k)$ or $P(X \leq k)$ select **Insert Function ▶ Statistical ▶ POISSON.DIST,** and click **OK.** (NOTE: This function is called **POISSON** in *Excel 2007* and earlier versions.) Enter the values for k, μ, and **FALSE/TRUE** before clicking **OK.**
3. There is no **inverse cumulative probability** as there was for the binomial distribution.

Binomial and Poisson Probabilities in *MINITAB*

For a random variable that has either a binomial or a Poisson probability distribution, *MINITAB* has been programmed to calculate either exact probabilities—$P(X = x)$—for a given value of x or the cumulative probabilities—$P(X \le x)$—for a given value of x. (NOTE: *MINITAB* uses the notation "X" for the random variable and "x" for a particular value of the random variable.) You must specify which distribution you are using and the necessary parameters: n and p for the binomial distribution and μ for the Poisson distribution.

Binomial Probabilities

1. Consider a binomial distribution with $n = 10$ and $p = 0.25$. The *value* of p does not appear in Table 1 of Appendix I, but you can use *MINITAB* to generate the entire probability distribution as well as the cumulative probabilities by entering the numbers 0 to 10 in column A.
2. One way to quickly enter a set of consecutive integers in a column is to do the following:
   - Name columns C1, C2, and C3 as "x," "P(X = x)," and "P(X<=x)," respectively.
   - Enter the first two values of x—0 and 1—to create a pattern in column C1.
   - Use your mouse to highlight the first two integers. Then grab the square handle in the lower right corner of the highlighted area. Drag the handle down to continue the pattern.
   - As you drag, you will see an integer appear in a small rectangle. Release the mouse when you have the desired number of integers—in this case, 10.
3. Once the necessary values of x have been entered, use **Calc ▶ Probability Distributions ▶ Binomial** to generate the dialogue box shown in Figure 5.10.

FIGURE 5.10
MINITAB screenshot - Bionomial Probability Distribution

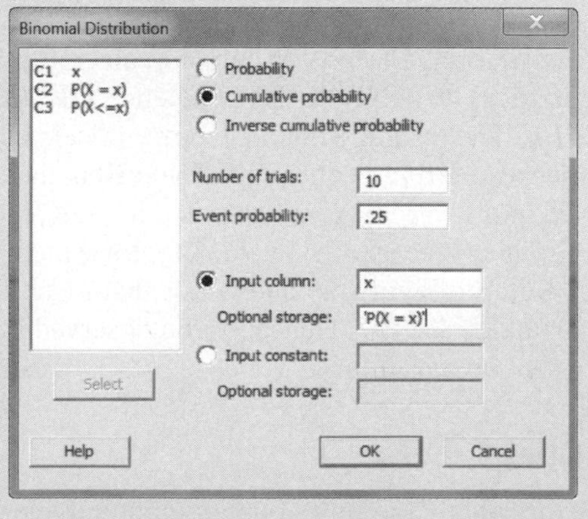

4. Type the number of trials and the value of p (Event probability) in the appropriate boxes, select C1 ('x') for the input column, and select C2 ('P(X = x)') for the Optional storage. Make sure that the radio button marked "Probability" is selected. The probability distribution for x will appear in column C2 when you click **OK.** (NOTE: If you do not select a column for the Optional storage, the results will be displayed in the Session window.)
5. If you want to generate the cumulative probabilities, $P(X \le k)$, again use **Calc ▶Probability Distributions ▶ Binomial** to generate the dialogue box. This time,

select the radio button marked "Cumulative probability" and select C3 (P(X<=x)) for the Optional storage in the dialogue box (Figure 5.10). The cumulative probability distribution will appear in column C3 when you click **OK.** You can display both distributions in the Session window using **Data ▶ Display Data,** selecting columns C1–C3 and clicking **OK.** The results are shown in Figure 5.11(a).

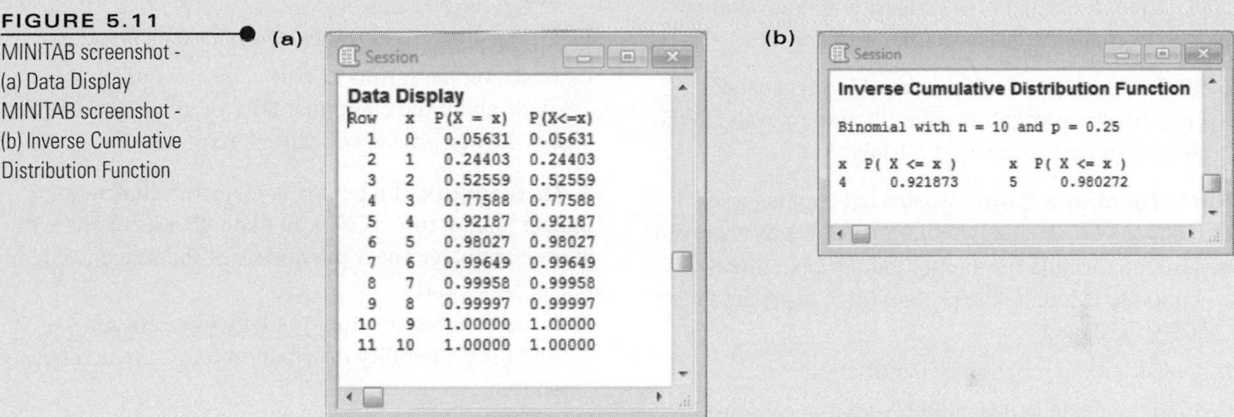

(a) Session

Data Display

Row	x	P(X = x)	P(X<=x)
1	0	0.05631	0.05631
2	1	0.24403	0.24403
3	2	0.52559	0.52559
4	3	0.77588	0.77588
5	4	0.92187	0.92187
6	5	0.98027	0.98027
7	6	0.99649	0.99649
8	7	0.99958	0.99958
9	8	0.99997	0.99997
10	9	1.00000	1.00000
11	10	1.00000	1.00000

(b) Session

Inverse Cumulative Distribution Function

Binomial with n = 10 and p = 0.25

x	P(X <= x)	x	P(X <= x)
4	0.921873	5	0.980272

FIGURE 5.11
MINITAB screenshot - (a) Data Display
MINITAB screenshot - (b) Inverse Cumulative Distribution Function

6. What value x is such that only 5% of the values of the random variable X exceed this value (and 95% are less than or equal to x)? Again, use **Calc ▶ Probability Distributions ▶ Binomial** to generate the dialogue box. This time, select the radio button marked "Inverse cumulative probability" and enter the probability **0.95** into the "Input constant" box (Figure 5.10). When you click **OK,** the values of x on either side of the "0.95 mark" will appear in the Session window as shown in Figure 5.11(b). Hence, if you observed a value of $x = 5$, this would be an unusual observation, because $P(X > 5) = 1 - 0.980272 = 0.019728$.

Poisson Probabilities

1. The procedures for calculating individual or cumulative probabilities and probability distributions for the Poisson random variable are similar to those used for the binomial distribution.
2. To find Poisson probabilities $P(X = x)$ or $P(X \leq x)$, use **Calc ▶ Probability Distributions ▶ Poisson** to generate the dialogue box. Enter the value for the mean μ, choose the appropriate radio button, and input either a column or a constant to indicate the value(s) of X for which you want to calculate a probability before clicking **OK.**
3. The **inverse cumulative probability** calculates the values of x such that $P(X \leq x) = C$, where C is a constant probability, between 0 and 1. Follow the steps described for the binomial random variable in step 6 above.

Supplementary Exercises

5.58 List the five identifying characteristics of the binomial experiment.

5.59 Under what conditions can the Poisson random variable be used to approximate the probabilities associated with the binomial random variable? What application does the Poisson distribution have other than to estimate certain binomial probabilities?

5.60 Under what conditions would you use the hypergeometric probability distribution to evaluate the probability of x successes in n trials?

5.61 Tossing a Coin A balanced coin is tossed three times. Let X equal the number of heads observed.

a. Use the formula for the binomial probability distribution to calculate the probabilities associated with $x = 0$, 1, 2, and 3.

b. Construct the probability distribution.

c. Find the mean and standard deviation of X, using these formulas:

$$\mu = np$$

$$\sigma = \sqrt{npq}$$

d. Using the probability distribution in part b, find the fraction of the population measurements lying within one standard deviation of the mean. Repeat for two standard deviations. How do your results agree with Tchebysheff's Theorem and the Empirical Rule?

5.62 Coins, continued Refer to Exercise 5.61. Suppose the coin is definitely unbalanced and the probability of a head is equal to $p = 0.1$. Follow the instructions in parts a, b, c, and d. Note that the probability distribution loses its symmetry and becomes skewed when p is not equal to 1/2.

5.63 Cancer Survivor Rates The 10-year survival rate for bladder cancer is approximately 50%. If 20 people who have bladder cancer are properly treated for the disease, what is the probability that:

a. At least 1 will survive for 10 years?

b. At least 10 will survive for 10 years?

c. At least 15 will survive for 10 years?

5.64 Garbage Collection A city administrator claims that 80% of all people in the city favour garbage collection by contract to a private concern (in contrast to collection by city employees). To check the theory that the proportion of people in the city favouring private collection is 0.8, you randomly sample 25 people and find that X, the number of people who support the commissioner's claim, is 22.

a. What is the probability of observing at least 22 people who support the administrator's claim if, in fact, $p = 0.8$?

b. What is the probability that x is exactly equal to 22?

c. Based on the results of part a, what would you conclude about the claim that 80% of all people in the city favour private collection? Explain.

5.65 Integers If a person is given the choice of an integer from 0 to 9, is it more likely that he or she will choose an integer near the middle of the sequence than one at either end?

a. If the integers are equally likely to be chosen, find the probability distribution for X, the number chosen.

b. What is the probability that the person will choose a 4, 5, or 6?

c. What is the probability that the person will not choose a 4, 5, or 6?

5.66 Integers II Refer to Exercise 5.65. Twenty people are asked to select a number from 0 to 9. Eight of them choose a 4, 5, or 6.

a. If the choice of any one number is as likely as any other, what is the probability of observing eight or more choices of the numbers 4, 5, or 6?

b. What conclusions would you draw from the results of part a?

5.67 Income Splitting According to an Ipsos Reid survey (February 27, 2007), conducted on behalf of CanWest/Global News, most Canadians (77%) are in favour of "income splitting" for couples.[12] Suppose that we randomly select $n = 15$ Canadians and let x be the number who are in favour of income splitting for couples. You may round up 77% to 80% for the calculation.

a. What is the probability distribution for x?

b. What is $P(X \le 8)$?

c. What is the probability that X exceeds 8?

d. What is the largest value of c for which $P(X \le c) \le 0.10$?

5.68 Vacation Homes Approximately 60% of Canadians rank "owning a vacation home nestled on a beach or near a mountain resort" as their number one choice for a status symbol. A sample of $n = 400$ Canadians is randomly selected.

a. What is the average number in the sample who would rank owning a vacation home number one?

b. What is the standard deviation of the number in the sample who would rank owning a vacation home number one?

c. Within what range would you expect to find the number in the sample who would rank having a vacation home as the number one status symbol?

d. If only 200 in a sample of 400 people ranked owning a vacation home as the top status symbol, would you consider this unusual? Explain. What conclusions might you draw from this sample information?

5.69 Reality TV Reality TV (*Survivor*, *The Amazing Race*, etc.) is a relatively new phenomenon in television programming, with contestants escaping to remote locations, taking dares, breaking world records, or racing across the country. Of those who watch reality TV, 50% say that their favourite reality show involves escaping to remote locations.[13] If 20 reality-TV fans are randomly selected, find the following probabilities:

a. Exactly 16 say that their favourite reality show involves escaping to remote locations.

b. From 15 to 18 say that their favourite reality show involves escaping to remote locations.

c. Five or fewer say that their favourite reality show involves escaping to remote locations. Would this be an unlikely occurrence?

5.70 Psychosomatic Problems A psychiatrist believes that 80% of all people who visit doctors have problems of a psychosomatic nature. She decides to select 25 patients at random to test her theory.

a. Assuming that the psychiatrist's theory is true, what is the expected value of X, the number of the 25 patients who have psychosomatic problems?

b. What is the variance of X, assuming that the theory is true?

c. Find $P(X \leq 14)$. (Use tables and assume that the theory is true.)

d. Based on the probability in part c, if only 14 of the 25 sampled had psychosomatic problems, what conclusions would you make about the psychiatrist's theory? Explain.

5.71 Student Fees A student union states that 80% of all students favour an increase in student fees to subsidize a new recreational area. A random sample of $n = 25$ students produced 15 in favour of increased fees. What is the probability that 15 or fewer in the sample would favour the issue if the student union is correct? Do the data support the student union's assertion, or does it appear that the percentage favouring an increase in fees is less than 80%?

5.72 Grey Hair on Campus University campuses are greying! According to a recent article, one in four college students is aged 30 or older. Many of these students are women updating their job skills. Assume that the 25% figure is accurate, that your university is representative of universities at large, and that you sample $n = 200$ students, recording X, the number of students age 30 or older.

a. What are the mean and standard deviation of X?

b. If there are 35 students in your sample who are age 30 or older, would you be willing to assume that the 25% figure is representative of your campus? Explain.

5.73 Probability of Rain Most weather forecasters protect themselves very well by attaching probabilities to their forecasts, such as "The probability of rain today is 40%." Then, if a particular forecast is incorrect, you are expected to attribute the error to the random behaviour of the weather rather than to the inaccuracy of the forecaster. To check the accuracy of a particular forecaster, records were checked only for those days when the forecaster predicted rain "with 30% probability." A check of 25 of those days indicated that it rained on 10 of the 25.

a. If the forecaster is accurate, what is the appropriate value of p, the probability of rain on one of the 25 days?

b. What are the mean and standard deviation of X, the number of days on which it rained, assuming that the forecaster is accurate?

c. Calculate the z-score for the observed value, $X = 10$. [HINT: Recall from Section 2.6 that z-score $= (X - \mu)/\sigma$.]

d. Do these data disagree with the forecast of a "30% probability of rain"? Explain.

5.74 What's for Breakfast? A packaging experiment is conducted by placing two different package designs for a breakfast food side by side on a supermarket shelf. The objective of the experiment is to see whether buyers indicate a preference for one of the two package designs. On a given day, 25 customers purchased a package from the supermarket. Let x equal the number of buyers who choose the second package design.

a. If there is no preference for either of the two designs, what is the value of p, the probability that a buyer chooses the second package design?

b. If there is no preference, use the results of part a to calculate the mean and standard deviation of x.

c. If 5 of the 25 customers choose the first package design and 20 choose the second design, what do you conclude about the customers' preference for the second package design?

5.75 Plant Density One model for plant competition assumes that there is a zone of resource depletion around each plant seedling. Depending on the size of the zones and the density of the plants, the zones of resource depletion may overlap with those of other seedlings in the vicinity. When the seeds are randomly dispersed over a wide area, the number of neighbours that a seedling may have usually follows a Poisson distribution with a mean equal to the density of seedlings per unit area. Suppose that the density of seedlings is four per square metre (m^2).

a. What is the probability that a given seedling has no neighbours within 1 m^2?

b. What is the probability that a seedling has at most three neighbours per m^2?

c. What is the probability that a seedling has five or more neighbours per m^2?

d. Use the fact that the mean and variance of a Poisson random variable are equal to find the proportion of neighbours that would fall into the interval $\mu \pm 2\sigma$. Comment on this result.

5.76 Plant Genetics A peony plant with red petals was crossed with another plant having streaky petals. The probability that an offspring from this cross has red flowers is 0.75. Let X be the number of plants with red petals resulting from 10 seeds from this cross that were collected and germinated.

a. Does the random variable X have a binomial distribution? If not, why not? If so, what are the values of n and p?

b. Find $P(X \geq 9)$.

c. Find $P(X \leq 1)$.

d. Would it be unusual to observe one plant with red petals and the remaining nine plants with streaky petals? If these experimental results actually occurred, what conclusions could you draw?

5.77 Dominant Traits The alleles for black (B) and white (b) feather colour in chickens show incomplete dominance; individuals with the gene pair Bb have "blue" feathers. When one individual that is homozygous dominant (BB) for this trait is mated with an individual that is homozygous recessive (bb) for this trait, 1/4 of the offspring will carry the gene pair BB, 1/2 will carry the gene pair Bb, and 1/4 will carry the gene pair bb. Let X be the number of chicks with "blue" feathers in a sample of $n = 20$ chicks resulting from crosses involving homozygous dominant chickens (BB) with homozygous recessive chickens (bb).

a. Does the random variable X have a binomial distribution? If not, why not? If so, what are the values of n and p?

b. What is the mean number of chicks with "blue" feathers in the sample?

c. What is the probability of observing fewer than five chicks with "blue" feathers?

d. What is the probability that the number of chicks with "blue" feathers is greater than or equal to 10 but less than or equal to 12?

5.78 Football Coin Tosses During the 1992 football season, the Los Angeles Rams had a bizarre streak of coin-toss losses. In fact, they lost the call 11 weeks in a row.[14]

a. The Rams' computer system manager said that the odds against losing 11 straight tosses are 2047 to 1. Is he correct?

b. After these results were published, the Rams lost the call for the next two games, for a total of 13 straight losses. What is the probability of this happening if, in fact, the coin was fair?

5.79 Diabetes in Children Insulin-dependent diabetes (IDD) is a common chronic disorder of children. This disease occurs most frequently in persons of northern European descent, but the incidence ranges from a low of 1–2 cases per 100,000 per year to a high of more than 40 per 100,000 in parts of Finland.[15] Let us assume that an area in Europe has an incidence of 5 cases per 100,000 per year.

a. Can the distribution of the number of cases of IDD in this area be approximated by a Poisson distribution? If so, what is the mean?

b. What is the probability that the number of cases of IDD in this area is less than or equal to 3 per 100,000?

c. What is the probability that the number of cases is greater than or equal to 3 but less than or equal to 7 per 100,000?

d. Would you expect to observe 10 or more cases of IDD per 100,000 in this area in a given year? Why or why not?

5.80 Problems with Your New Smartphone? A new study by Square Trade indicates that smartphones are 50% more likely to malfunction than simple phones over a three-year period.[16] Of smartphone failures, 30% are related to internal components not working, and overall, there is a 31% chance of having your smartphone fail over three years. Suppose that smartphones are shipped in cartons of $N = 50$ phones. Before shipment, $n = 10$ phones are selected from each carton, and the carton is shipped if none of the selected phones are defective. If one or more are found to be defective, the whole carton is tested.

a. What is the probability distribution of X, the number of defective phones related to internal components not working in the sample of $n = 10$ phones?

b. What is the probability that the carton will be shipped if two of the $N = 50$ smartphones in the carton have defective internal components?

c. What is the probability that the carton will be shipped if it contains four defectives? Six defectives?

5.81 Dark Chocolate Despite reports that dark chocolate is beneficial to the heart, 47% of adults still prefer milk chocolate to dark chocolate.[17] Suppose a random sample of $n = 5$ adults is selected and asked whether they prefer milk chocolate to dark chocolate.

a. What is the probability that all five adults say that they prefer milk chocolate to dark chocolate?

b. What is the probability that exactly three of the five adults say they prefer milk chocolate to dark chocolate?

c. What is the probability that at least one adult prefers milk chocolate to dark chocolate?

5.82 Defective Flash Drives A manufacturer of flash drives ships them in lots of 1200 drives per lot. Before shipment, 20 drives are randomly selected from each lot and tested. If none is defective, the lot is shipped. If one or more are defective, every drive in the lot is tested.

a. What is the probability distribution for x, the number of defective drives in the sample of 20?

b. What distribution can be used to approximate probabilities for the random variable X in part a?

c. What is the probability that a lot will be shipped if it contains 10 defectives? 20 defectives? 30 defectives?

5.83 Tobacco Control An Ipsos Reid poll conducted on behalf of the Canadian Cancer Society finds that the majority of British Columbians support implementing tougher tobacco control measures in the province.[18] Overall, 71% of British Columbians support a ban on all smoking in workplaces and public establishments, including the elimination of designated smoking areas. The majority also support establishing regulations that would prohibit the sale of cigarettes and tobacco products in "pharmacies" (66%) and "on university and college campuses" (60%). Suppose that in random sample $n = 5$ British Columbians is taken, X is the number who support the prohibition of the sale of cigarettes and tobacco products in pharmacies.

a. What is the probability that all five persons will support prohibiting the sale of cigarettes and tobacco products in pharmacies?

b. What is the probability that three of the five British Columbians will support prohibiting the sale of cigarettes and tobacco products in pharmacies?

c. What is the probability that one British Columbian will support prohibiting the sale of cigarettes and tobacco products in pharmacies?

5.84 Tay–Sachs Disease Tay–Sachs disease is a genetic disorder that is usually fatal in young children. If both parents are carriers of the disease, the probability that their offspring will develop the disease is approximately 0.25. Suppose a husband and wife are both carriers of the disease and the wife is pregnant on three different occasions. If the occurrence of Tay–Sachs in any one offspring is independent of the occurrence in any other, what are the probabilities of these events?

a. All three children will develop Tay–Sachs disease.

b. Only one child will develop Tay–Sachs disease.

c. The third child will develop Tay–Sachs disease, given that the first two did not.

5.85 Wildlife Refuge There is significant opposition among Canadians toward U.S. plans to approve drilling in the Arctic Wildlife Refuge, according to a survey commissioned by World Wildlife Fund Canada, carried out by Ekos Research Associates.[19] The survey results are based on a telephone survey using a random sample of 1020 Canadians, 18 years and older. Surveying took place between July 6 and 12, 2005. When asked how strongly they support or oppose U.S. plans to approve drilling in the Alaska Wildlife Refuge, 90% of Canadians voice opposition (either oppose or strongly oppose). Suppose a group of 12 Canadians, 18 years and older, is randomly selected and asked whether they opposed (or strongly opposed) U.S. drilling in the Arctic.

a. What is the probability that more than six of the respondents will oppose the drilling?

b. What is the probability that fewer than five of the respondents will oppose the drilling?

c. What is the probability that exactly ten of the respondents will oppose the drilling?

5.86 The Triangle Test A procedure often used to control the quality of name-brand food products utilizes a panel of five "tasters." Each member of the panel tastes three samples, two of which are from batches of the product known to have the desired taste and the other from the latest batch. Each taster selects the sample that is different from the other two. Assume that the latest batch does have the desired taste, and that there is no communication between the tasters.

a. If the latest batch tastes the same as the other two batches, what is the probability that the taster picks it as the one that is different?

b. What is the probability that exactly one of the tasters picks the latest batch as different?

c. What is the probability that at least one of the tasters picks the latest batch as different?

5.87 Do You Return Your Questionnaires? The president of a company specializing in public opinion surveys claims that approximately 70% of all people to whom the agency sends questionnaires respond by filling out and returning the questionnaire. Twenty such questionnaires are sent out. Assume that the president's claim is correct.

a. What is the probability that exactly 10 of the questionnaires are filled out and returned?

b. What is the probability that at least 12 of the questionnaires are filled out and returned?

c. What is the probability that at most 10 of the questionnaires are filled out and returned?

5.88 Questionnaires, continued Refer to Exercise 5.87. $n = 20$ questionnaires are sent out.

a. What is the average number of questionnaires that will be returned?

b. What is the standard deviation of the number of questionnaires that will be returned?

c. If $x = 10$ of the 20 questionnaires are returned to the company, would you consider this to be an unusual response? Explain.

5.89 Poultry Problems A preliminary investigation reported that approximately 30% of locally grown poultry were infected with an intestinal parasite that, though not harmful to those consuming the poultry, decreases the usual weight growth rates in the birds. A diet supplement believed to be effective against this parasite was added to the bird's food. Twenty-five birds were examined after having the supplement for at least two weeks, and three birds were still found to be infested with the parasite.

a. If the diet supplement is ineffective, what is the probability of observing three or fewer birds infected with the intestinal parasite?

b. If in fact the diet supplement was effective and reduced the infection rate to 10%, what is the probability of observing three or fewer infected birds?

5.90 Machine Breakdowns In a food processing and packaging plant, there are, on average, two packaging machine breakdowns per week. Assume the weekly machine breakdowns follow a Poisson distribution.

a. What is the probability that there are no machine breakdowns in a given week?

b. Calculate the probability that there are no more than two machine breakdowns in a given week.

5.91 Safe Drivers? Evidence shows that the probability that a driver will be involved in a serious auotmobile accident during a given year is 0.01. A particular corporation employs 100 full-time travelling sales reps. Based on this evidence, use the Poisson approximation to the binomial distribution to find the probability that exactly two of the sales reps will be involved in a serious automobile accident during the coming year.

5.92 Stressed Out A subject is taught to do a task in two different ways. Studies have shown that when subjected to mental strain and asked to perform the task, the subject most often reverts to the method first learned, regardless of whether it was easier or more difficult. If the probability that a subject returns to the first method learned is 0.8 and six subjects are tested, what is the probability that at least five of the subjects revert to their first learned method when asked to perform their task under stress?

5.93 Enrolling in University A west coast university has found that about 90% of its accepted applicants for enrollment in first year will actually enroll. In 2012, 1360 applicants were accepted to the university. Within what limits would you expect to find the size of the first-year class at this university in the fall of 2012?

5.94 Earthquakes! Suppose that one out of every 10 homeowners in the city of Kobe in western Japan has invested in earthquake insurance. If 15 homeowners are randomly chosen to be interviewed,

a. What is the probability that at least one has earthquake insurance?

b. What is the probability that four or more have earthquake insurance?

c. Within what limits would you expect the number of homeowners insured against earthquakes to fall?

5.95 Bad Wiring Improperly wired control panels were mistakenly installed on two of eight large automated machine tools. It is uncertain which of the machine tools have the defective panels, and a sample of four tools is randomly chosen for inspection. What is the probability that the sample will include no defective panels? Both defective panels?

5.96 High Gas Prices How do you cope with the high gas prices? Do you drive less or do you have other money-wise or even environmentally friendly innovative ideas? As a result of soaring fuel prices, most Canadians are either cutting down on the amount they drive, making other significant changes to their daily lives, or doing both. An Ekos poll revealed that 58% of Canadians say they have reacted in some substantial way; 46% of Canadians say they are driving less; and 23% say they have made other adjustments.[20] Alternatively, some people are both driving less and making other adjustments. A random sample of 100 people was selected.

a. What is the average number of people who say they are driving less?

b. What is the standard deviation for the number of people who say they are driving less?

c. If 59 of the respondents in the sample said they are driving less, would this be an unusual occurrence? Explain.

5.97 Successful Surgeries A new surgical procedure is said to be successful 80% of the time. Suppose the operation is performed five times and the results are assumed to be independent of one another. What are the probabilities of these events?

a. All five operations are successful.

b. Exactly four are successful.

c. Less than two are successful.

5.98 Surgery, continued Refer to Exercise 5.97. If less than two operations were successful, how would you feel about the performance of the surgical team?

5.99 Engine Failure Suppose the four engines of a commercial aircraft are arranged to operate independently and that the probability of in-flight failure of a single engine is 0.01. What is the probability of these events on a given flight?

a. No failures are observed.

b. No more than one failure is observed.

5.100 McDonald's or Burger King? Suppose that 50% of all young adults prefer McDonald's to Burger King when asked to state a preference. A group of 100 young adults were randomly selected and their preferences recorded.

a. What is the probability that more than 60 preferred McDonald's?

b. What is the probability that between 40 and 60 (inclusive) preferred McDonald's?

c. What is the probability that between 40 and 60 (inclusive) preferred Burger King?

5.101 After Graduation Most of today's university graduates want to start earning money as soon as they graduate from university. However, for some decades now, the enrollment in graduate programs at Canadian universities has been increasing. About 20% of senior undergrad students are likely to pursue a graduate program after graduating. Suppose that 50 senior undergrad students were randomly selected.

a. What is the average value of x, the number of senior undergrad students in the group who say they will pursue a graduate program after graduation? What is the standard deviation of x?

b. Would it be unlikely to find 15 or more in the group who say they will pursue a graduate program after graduation?

c. How many standard deviations from the mean is the value $x = 15$? Does this confirm your answer in part b?

CASE STUDY 

How Safe Is Plastic Surgery? Myth versus Fact![21]

How safe is plastic surgery? Do breast implants reduce cancer risk? Do they increase the risk of suicide?

First we consider the question: *Do breast implants increase the suicide rate?* Researchers combed vital statistics and death certificates of 24,558 women in Ontario and Quebec who underwent breast implant surgery from 1974 to 1989. They identified 58 suicides among the 480 breast-implant recipients who died. In a comparable general female population without implants, they would have expected 33 suicide deaths. How unusual was this number of suicide deaths? That is, statistically speaking, is 58 a highly improbable number of suicide deaths in this group of women with breast implants? If the answer is yes, then either some external factor caused this unusually large number, or we have observed a very rare event!

The Poisson probability distribution provides a good approximation to the distribution of variables, such as the number of deaths in a region due to a rare disease, the number of accidents in a manufacturing plant per month, or the number of airline crashes per month. Therefore it is reasonable to assume that Poisson distribution provides an appropriate model for the number of suicides in this instance.

1. Here, the 58 reported cases represented a rate of 75% higher than is typically expected (can you verify from the above information that 75% is a correct percentage?). What is a reasonable estimate of μ, the average number of such suicide cases for Ontario and Quebec?
2. Based on your estimate μ, what is the estimated standard deviation of the number of suicide cases in both provinces?
3. What is the z-score for the $x = 58$ observed cases of suicide deaths? How do you interpret this z-score in light of following comments?

"That's not a very large number," said Dr. Jacques Brisson of the department of social and preventive medicine at Laval University in Quebec. "It's an increase in the number of deaths related to suicide but the number itself is not (large). So the risk of suicide among women with breast implants is not great. It's actually small, relatively speaking, but it still is larger than the general population."

Now, let us consider a second question: *Do breast implants reduce cancer risk?* The research study discovered that "overall, women who underwent breast implantation had lower rates of mortality than the general population, at comparable ages over the same period. They also had lower rates of cancer, contrary to popular belief, with 229 cancer deaths among the nearly 25,000 women. About 303 would be expected in the general female population."

1. Based on the above study, the 229 cancer deaths reported among women who had breast implants represents 24.5% less than the expected number among the general female population. Verify that 24.5% is the correct percentage.
2. What is a reasonable estimate of μ, the average number of such death cases for Ontario and Quebec?
3. Based on your estimate μ, what is the estimated standard deviation of the number of cancer death cases in both provinces?
4. What is the z-score for the $x = 229$ observed cases of cancer deaths? How do you interpret this z-score?

PROJECTS 

Project 5: Relations among Useful Discrete Probability Distributions

A Bernoulli experiment consists of only one trial with two outcomes (success/failure) with probability of success p, then the Bernoulli distribution is

$$P(X = k) = p^k q^{n-k},\ k = 0, 1$$

The sum of the n independent Bernoulli trials forms a binomial experiment. The binomial probability distribution provides a simple, easy-to-compute approximation with a reasonable accuracy to hypergeometric probabilities when n/N is less than or equal to 0.10. Further, the Poisson distribution gives an accurate approximation to binomial probabilities when n is large and μ is small.

Child-abuse victims and developing cancer: Truth or myth Is physical childhood abuse somehow related to the development of cancer later in life? A recent survey revealed that people who have been physically abused as children were 49% more likely to develop cancer as adults.[22]

a. Suppose in a region in Saskatchewan, among a group of 20 adults with cancer, seven were physically abused during their childhood. A random sample of five adult persons is taken from this group. Assume that sampling occurs without replacement, and the random variable x represents the number of adults abused during their childhood period in the sample.

   (i) Write the formula for $p(x)$, the probability distribution of X.

   (ii) What are the mean and variance of X?

   (iii) What proportion of the population of measurements falls into the interval $(\mu \pm 2\sigma)$? Into the interval $(\mu \pm 3\sigma)$? Do these results agree with those given by Tchebysheff's Theorem?

   (iv) What is the probability that at least one person was abused during childhood?

b. Suppose another survey in British Columbia reveals that among 180 adults with cancer, only 80 adults were abused in their childhood. If five of the adults are selected for further study, find the probability that only two of the five adults with childhood abuse will be selected in the sample by using

(i) the formula for the hypergeometric distribution;

(ii) the formula for the binomial distribution with as $p = \dfrac{8}{180}$ an approximation. Is this approximation close enough? Why or why not?

(iii) Calculate the mean and variance using both binomial and hypergeometric distributions, respectively. Provide a comparison and summarize your findings.

c. According to a report from the Public Health Agency of Canada, approximately 850 Canadian children aged 0–14 develop cancer each year. Childhood cancers account for less than 1% of all cancers diagnosed in Canada.[23]

Assuming that in some part of Quebec the probability is 0.00007 that a child will develop cancer, use the Poisson distribution to approximate the binomial probabilities that among 28,572 children

(i) none will develop cancer.

(ii) at most two will develop cancer.

(iii) Calculate the probability that at least seven in a sample of ten children will not develop cancer.

FUNFUNPHOTO/Shutterstock.com

6

The Normal Probability Distribution

GENERAL OBJECTIVES

In Chapters 4 and 5, you learned about discrete random variables and their probability distributions. In this chapter, you will learn about continuous random variables and their probability distributions and about one very important continuous random variable—the normal. You will learn how to calculate normal probabilities and, under certain conditions, how to use the normal probability distribution to approximate the binomial probability distribution. Then, in Chapter 7 and in the chapters that follow, you will see how the normal probability distribution plays a central role in statistical inference.

CHAPTER INDEX

- Probability distributions for continuous random variables (6.1)
- The normal probability distribution (6.2)
- Calculation of areas associated with the normal probability distribution (6.3)
- The normal approximation to the binomial probability distribution (6.4)

NEED TO KNOW

How to Use Table 3 in Appendix I to Calculate Probabilities under the Standard Normal Curve

How to Calculate Binomial Probabilities Using the Normal Approximation

"Are You Going to Curve the Grades?"

"Curving the grades" doesn't necessarily mean that you will receive a higher grade on a test, although many students would like to think so! Curving actually refers to a method of assigning the letter grades A, B, C, D, or F using fixed proportions of the grades corresponding to each of the letter grades. One such curving technique assumes that the distribution of the grades is approximately normal and uses these proportions:

Letter grade	A	B	C	D	F
Proportion of grades	10%	20%	40%	20%	10%

In the case study at the end of this chapter, we will examine this and other assigned proportions for curving grades.

6.1 PROBABILITY DISTRIBUTIONS FOR CONTINUOUS RANDOM VARIABLES

When a random variable x is discrete, you can assign a positive probability to each value that x can take and get the probability distribution for X. The sum of all the probabilities associated with the different values of X is 1. However, not all experiments result in random variables that are discrete. **Continuous random variables,** such as heights and weights, length of life of a particular product, or experimental laboratory error, can assume the infinitely many values corresponding to points on a line interval. If you try to assign a positive probability to each of these uncountable values, the probabilities will no longer sum to 1, as with discrete random variables. Therefore, you must use a different approach to generate the probability distribution for a continuous random variable.

Suppose you have a set of measurements on a continuous random variable, and you create a relative frequency histogram to describe their distribution. For a small number of measurements, you could use a small number of classes; then as more and more measurements are collected, you can use more classes and reduce the class width. The outline of the histogram will change slightly, for the most part becoming less and less irregular, as shown in Figure 6.1. As the number of measurements becomes very large and the class widths become very narrow, the relative frequency histogram appears more and more like the smooth curve shown in Figure 6.1(d). This smooth curve describes the **probability distribution of the continuous random variable.**

FIGURE 6.1
Relative frequency histograms for increasingly large sample sizes

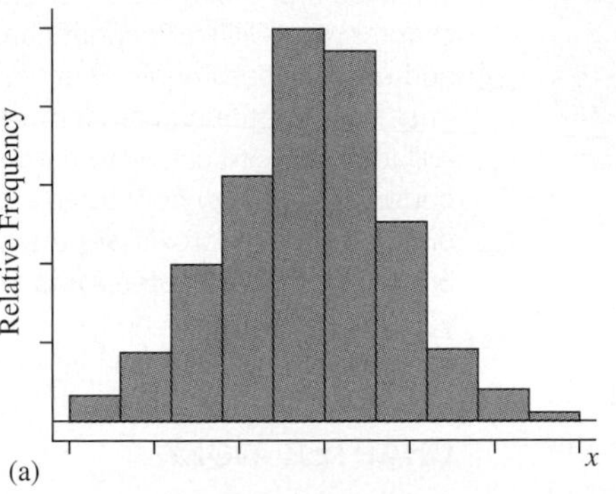

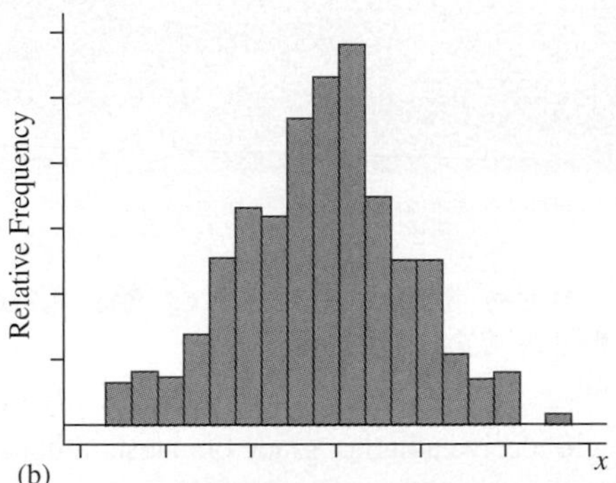

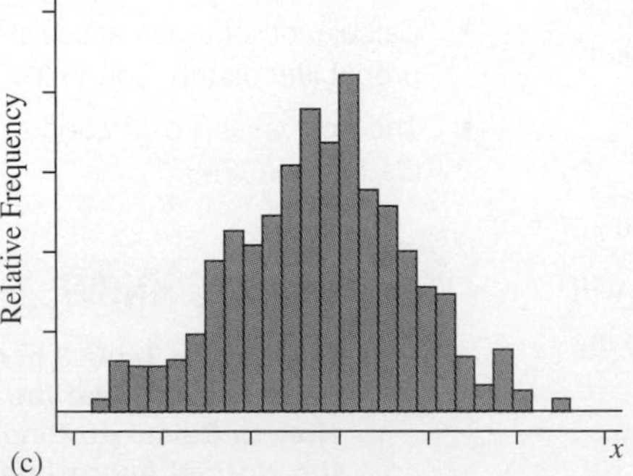

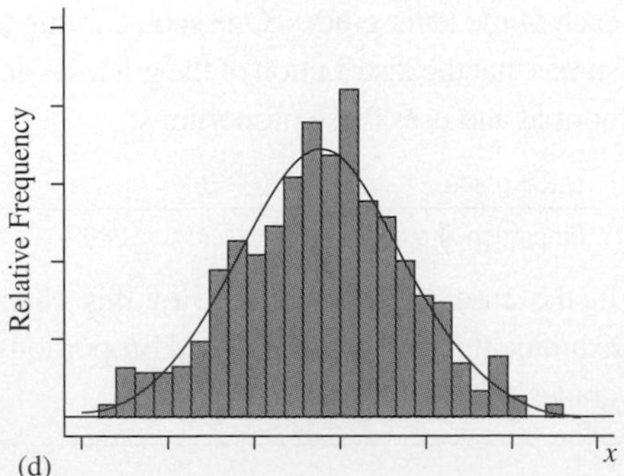

How can you create a model for this probability distribution? A continuous random variable can take on any of an infinite number of values on the real line, much like the infinite number of grains of sand on a beach. The probability distribution is created by distributing one unit of probability along the line, much as you might distribute a handful of sand. The probability—grains of sand or measurements—will pile up in certain places, and the result is the probability distribution shown in Figure 6.2. The depth or **density** of the probability, which varies with x, may be described by a mathematical formula $f(x)$, called the **probability distribution** or **probability density function** for the random variable X.

FIGURE 6.2
The probability distribution $f(x)$; $P(a < x < b)$ is equal to the shaded area under the curve

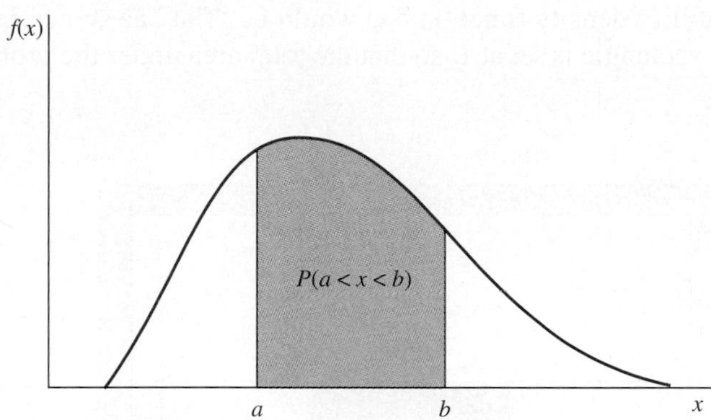

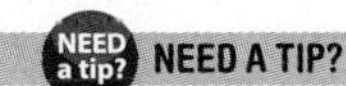

For continuous random variables, **area = probability.**

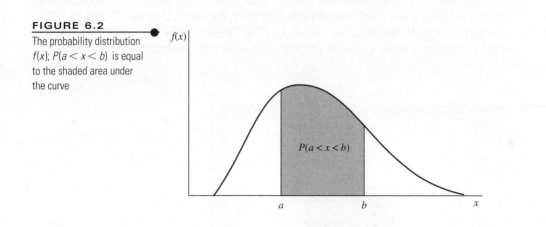

Several important properties of continuous probability distributions parallel their discrete counterparts. Just as the sum of discrete probabilities (or the sum of the relative frequencies) is equal to 1, and the probability that X falls into a certain interval can be found by summing the probabilities in that interval, continuous probability distributions have the characteristics listed next.

- The area under a continuous probability distribution is equal to 1.
- The probability that X will fall into a particular interval—say, from a to b—is equal to the area under the curve between the two points a and b. This is the shaded area in Figure 6.2.

NEED A TIP?

Area under the curve equals 1.

There is also one important difference between discrete and continuous random variables. Consider the probability that X equals some particular value—say, a. Since there is no area above a single point—say, $X = a$—in the probability distribution for a continuous random variable, our definition implies that the probability is 0.

- $P(X = a) = 0$ for continuous random variables.
- This implies that $P(X \geq a) = P(X > a)$ and $P(X \leq a) = P(X < a)$.
- This is *not* true in general for discrete random variables.

How do you choose the model—that is, the probability distribution $f(x)$—appropriate for a given experiment? Many types of continuous curves are available for modelling.

Some are mound-shaped, like the one in Figure 6.1(d), but others are not. In general, try to pick a model that meets these criteria:

- It fits the accumulated body of data.
- It allows you to make the best possible inferences using the data.

Uniform Distribution

The *uniform random variable* is used to model the behaviour of a continuous random variable whose values are uniformly or evenly distributed over a given interval. For example, the error X introduced by rounding an observation to the nearest centimetre would probably have a uniform distribution over the interval from −0.5 to 0.5. The probability density function $f(x)$ would be "flat" as shown in Figure 6.3. The height of the rectangle is set at 1, so that the total area under the probability distribution is 1.

FIGURE 6.3

A uniform probability distribution

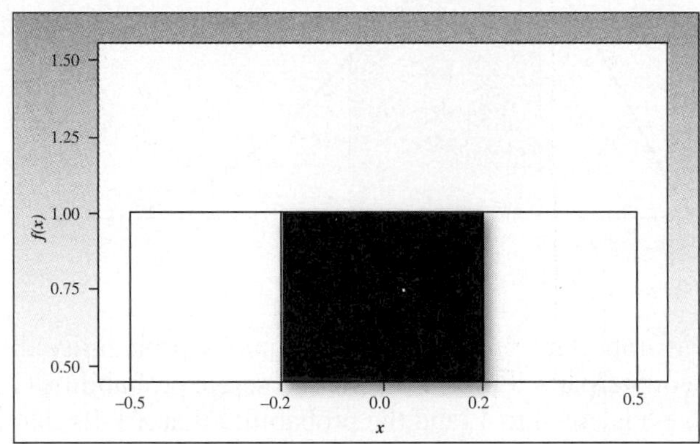

What is the probability that the rounding error is less than 0.2 in magnitude?

SOLUTION This probability corresponds to the area under the distribution between $x = -0.2$ and $x = 0.2$. Since the height of the rectangle is 1,

$$P(-0.2 < X < 0.2) = [0.2 - (-0.2)] \times 1 = 0.4$$

Exponential Distribution

The exponential random variable is used to model continuous random variables such as waiting times, or more importantly, lifetimes (life length) associated with electronic components. In general, the exponential probability density function is given by

$$f(x) = \frac{1}{\mu} e^{-x/\mu}, \quad 0 \leq x < \infty, \quad \mu > 0$$

It is the only parameter of the exponential distribution. The parameter μ is actually the mean of the distribution. An exponential probability distribution with a mean $\mu = 1$ is plotted in Figure 6.4. Generally speaking, the exponential distributions have the form shown in Figure 6.4.

FIGURE 6.4
Exponential probability distribution

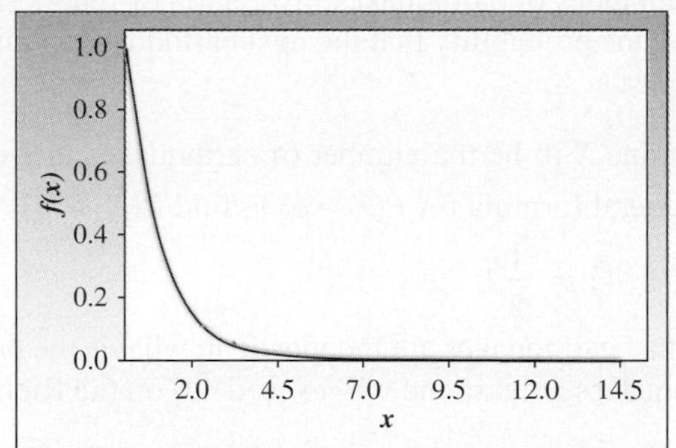

The exponential probability distribution is arguably easy to work with by direct integration. For a given value of μ, the probability of a given interval can be easily computed so there is no need for tables. Also, it can be shown that

$$P(X > a) = e^{-a/\mu}, \quad \text{for } a > 0.$$

EXAMPLE 6.1

The waiting time at a Canadian supermarket checkout has an exponential distribution with an average time of five minutes. Thus, the probability density function is

$$f(x) = \frac{1}{5}e^{-\frac{1}{5}x},\ x > 0.$$

The function is shown in Figure 6.5.

FIGURE 6.5
An exponential probability distribution for Example 6.1

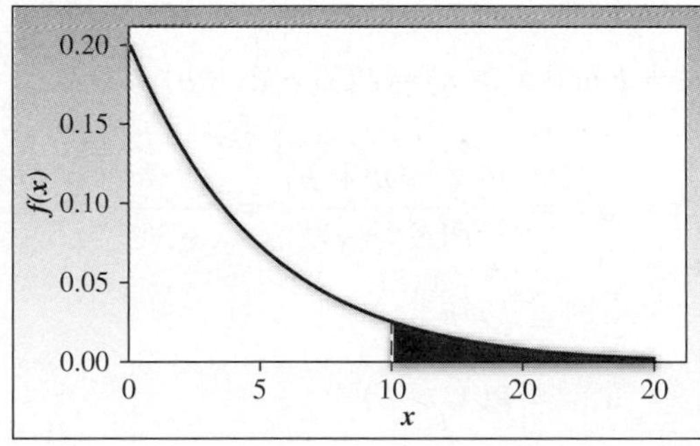

To find the area under a curve, you can use the fact that $P(X > a) = e^{-\frac{1}{5}a}$ for $a > 0$. What is the probability that you have to wait 10 minutes or more at the checkout?

SOLUTION The probability to be calculated is the area shaded in Figure 6.5. Use the general formula for $P(X > a)$ to find that

$$P(X > 10) = e^{-\frac{1}{5}(10)} = 0.135$$

EXAMPLE 6.2

The magnitude of most earthquakes is measured on the Richter scale. It was invented by Charles F. Richter in 1934. For example, using this scale, a magnitude 5–5.9 is termed as a moderate earthquake (slight damage to buildings and other structures).

Suppose the magnitudes of earthquakes in a region of Indonesia can be modelled with a mean of 4. Find the probability that the next earthquake to hit this region will exceed 4 on the Richter scale.

SOLUTION Define X to be the number of earthquakes that exceed 4 on the Richter scale. Use the general formula for $P(X > a)$ to find $P(X > 4) = e^{-\frac{4}{4}} = e^{-1} = 0.37$.

Noting that $P(X > 4) \neq \frac{1}{2}$!

Now, assuming that earthquakes are independent, what is the probability that out of the next seven earthquakes at least one will exceed 4.0 on the Richter scale?

SOLUTION This is a binomial problem now, with $n = 7$, $p = P(X > 4) = 0.37$.

Hence
P (at least one of the seven earthquakes will exceed 4) $= P(X \geq 1) = 1 - P(X = 0) = 1 - C^7_0\,(0.37)^0\,(1 - 0.37)^{7-0} = 1 - (0.63)^7 = 1 - 0.0394 = 0.96$.

EXAMPLE 6.3

Suppose X has an exponential probability density function with mean μ. Show that

$$P(X > a + b \mid X > a) = P(X > b),\ a > 0,\ b > 0$$

SOLUTION $P(X > a + b \mid X > a)$ is a conditional probability, so by the definition of conditional probability

$$P(X > a + b|X > a) = \frac{P(X > a + b \text{ and } X > a)}{P(X > a)}$$

Note that $P(X > a + b \text{ and } X > a) = P(X > a + b)$

$$\begin{aligned} P(X > a + b|X > a) &= \frac{P(X > a + b)}{P(X > a)} = \frac{e^{-(a+b)/\mu}}{e^{-a/\mu}} \\ &= e^{-b/\mu} \\ &= P(X > b) \end{aligned}$$

This remarkable property of an exponential distribution is often called the memoryless property of the distribution.

Remark 1: Let $a = 20$ and $b = 15$, then by the memoryless property of an exponential distribution

$$P(X > 20 + 15 = 35|X > 20) = P(X > 15)$$

So $P(X > 35 \mid X > 20) \neq P(X > 20)$.

Thus, the events $X > 35$ and $X > 20$ are not independent, a common misunderstanding and misinterpretation of the memoryless property.

Remark 2: The failure rate (hazard rate) function is another function that is important in analyzing lifetime data. The exponential distribution has a constant failure rate.

Remark 3: The commonly known exponential distribution and geometric distribution (discrete) are the only memoryless probability distributions.

Your model may not always fit the experimental situation perfectly, but you should try to choose a model that *best fits* the population relative frequency histogram. The better the model approximates reality, the better your inferences will be. Fortunately, many continuous random variables have mound-shaped frequency distributions, such as the data in Figure 6.1(d). The **normal probability distribution** provides a good model for describing this type of data.

6.2 THE NORMAL PROBABILITY DISTRIBUTION

Continuous probability distributions can assume a variety of shapes. However, a large number of random variables observed in nature possess a frequency distribution that is approximately mound-shaped or, as the statistician would say, is approximately a normal probability distribution. The formula that generates this distribution is shown next.

NORMAL PROBABILITY DISTRIBUTION

$$f(x) = \frac{1}{\sigma\sqrt{2\pi}} e^{-(x-\mu)^2/(2\sigma^2)} \quad -\infty < x < \infty$$

The symbols e and π are mathematical constants given approximately by 2.7183 and 3.1416, respectively; μ and σ ($\sigma > 0$) are parameters that represent the population mean and standard deviation, respectively.

The graph of a normal probability distribution with mean μ and standard deviation σ is shown in Figure 6.6. The mean μ locates the *centre* of the distribution, and the distribution is *symmetric* about its mean μ. Since the total area under the normal probability distribution is equal to 1, the symmetry implies that the area to the right of μ is 0.5 and the area to the left of μ is also 0.5. The *shape* of the distribution is determined by σ, the population standard deviation. As you can see in Figure 6.7, large values of σ reduce the height of the curve and increase the spread; small values of σ increase the height of the curve and reduce the spread. Figure 6.7 shows three normal probability distributions with different means and standard deviations. Notice the differences in shape and location.

FIGURE 6.6
Normal probability distribution

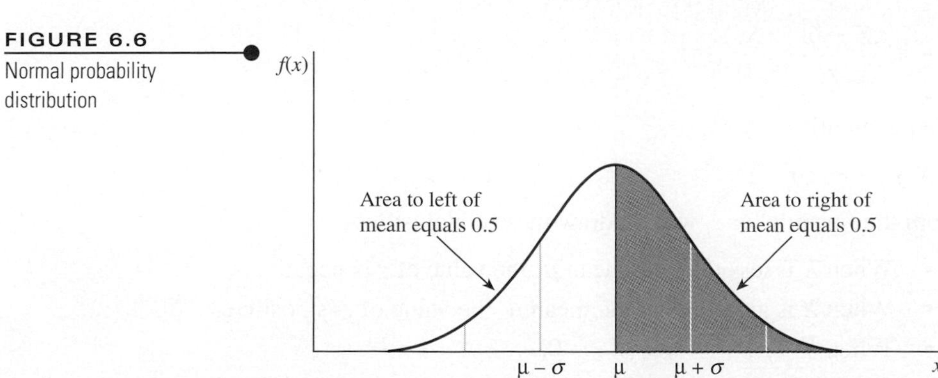

FIGURE 6.7
Normal probability distributions with differing values of μ and σ

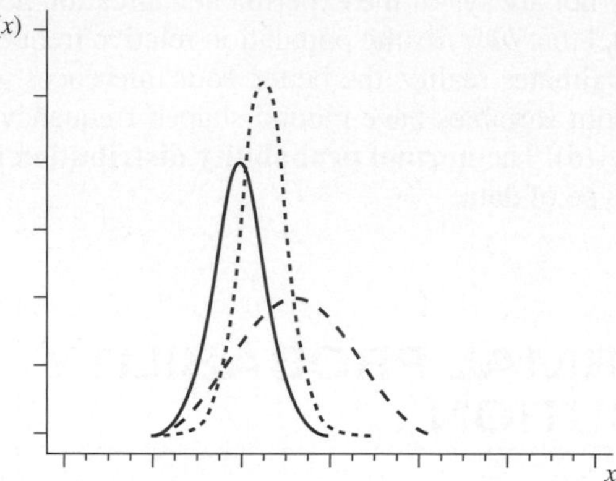

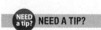

ONLINE APPLET
Visualizing Normal Curves

You rarely find a variable with values that are infinitely small ($-\infty$) or infinitely large ($+\infty$). Even so, many *positive* random variables (such as heights, weights, and times) have distributions that are well approximated by a normal distribution. According to the Empirical Rule, almost all values of a normal random variable lie in the interval $\mu \pm 3\sigma$. As long as the values within three standard deviations of the mean are *positive*, the normal distribution provides a good model to describe the data.

6.3 TABULATED AREAS OF THE NORMAL PROBABILITY DISTRIBUTION

To find the probability that a normal random variable X lies in the interval from a to b, we need to find the area under the normal curve between the points a and b (see Figure 6.2). However (see Figure 6.7), there are an infinitely large number of normal distributions—one for each different mean and standard deviation. A separate table of areas for each of these curves is obviously impractical. Instead, we use a standardization procedure that allows us to use the same table for all normal distributions.

The Standard Normal Random Variable

A normal random variable X is **standardized** by expressing its value as the number of standard deviations (σ) it lies to the left or right of its mean μ. This is really just a change in the units of measure that we use, as if we were measuring in centimetres rather than in metres! The standardized normal random variable, z, is defined as

NEED A TIP?
Area under the z-curve equals 1.

$$z = \frac{X - \mu}{\sigma}$$

or equivalently,

$$X = \mu + z\sigma$$

From the formula for z, we can draw these conclusions:

- When X is less than the mean μ, the value of z is negative.
- When X is greater than the mean μ, the value of z is positive.
- When $X = \mu$, the value of $z = 0$.

The probability distribution for z, shown in Figure 6.8, is called the **standardized normal distribution** because its mean is 0 and its standard deviation is 1. Values of z on the left side of the curve are negative, while values on the right side are positive. The area under the standard normal curve to the left of a specified value of z—say, z_0—is the probability $P(z \leq z_0)$. This **cumulative area** is recorded in Table 3 of Appendix I and is shown as the shaded area in Figure 6.8. An abbreviated version of Table 3 is given in Table 6.1. Notice that the table contains both positive and negative values of z. The left-hand column of the table gives the value of z correct to the tenth place; the second decimal place for z, corresponding to hundredths, is given across the top row.

FIGURE 6.8
Standardized normal distribution

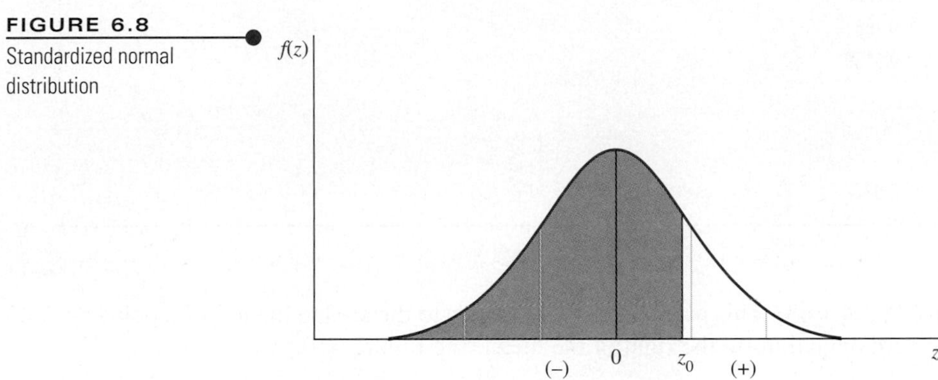

TABLE 6.1

Abbreviated Version of Table 3 in Appendix I

Table 3. Areas Under the Normal Curve

z	0.00	0.01	0.02	0.03	...	0.09
−3.4	0.0003	0.0003	0.0003	0.0003		
−3.3	0.0005	0.0005	0.0005	0.0004		
−3.2	0.0007	0.0007	0.0006	0.0006		
−3.1	0.0010	0.0009	0.0009	0.0009		
−3.0	0.0013	0.0013	0.0013	0.0012	...	0.0010
−2.9	0.0019	.	.	.		
−2.8	0.0026	.	.	.		
−2.7	0.0035	.	.	.		
−2.6	0.0047					
−2.5	0.0062					
.	.					
.	.					
.	.					
−2.0	0.0228					
.	.					
.	.					
.	.					

Table 3. Areas Under the Normal Curve (*continued*)

z	0.00	0.01	0.02	0.03	0.04	...	0.09
0.0	0.5000	0.5040	0.5080	0.5120	0.5160		
0.1	0.5398	0.5438	0.5478	0.5517	0.5557		
0.2	0.5793	0.5832	0.5871	0.5910	0.5948		
0.3	0.6179	0.6217	0.6255	0.6293	0.6331		
0.4	0.6554	0.6591	0.6628	0.6664	0.6700	...	0.6879
0.5	0.6915	.	.	.			
0.6	0.7257	.	.	.			
0.7	0.7580	.	.	.			
0.8	0.7881						
0.9	0.8159						
.	.						
.	.						
.	.						
2.0	0.9772						

EXAMPLE 6.4

Find $P(z \leq 1.63)$. This probability corresponds to the area to the left of a point $z = 1.63$ standard deviations to the right of the mean (see Figure 6.9).

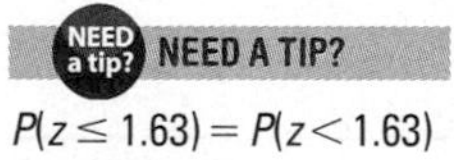

NEED A TIP?

$P(z \leq 1.63) = P(z < 1.63)$

SOLUTION The area is shaded in Figure 6.9. Since Table 3 in Appendix I gives areas under the normal curve to the left of a specified value of z, you simply need to find the tabled value for $z = 1.63$. Proceed down the left-hand column of the table to $z = 1.6$ and across the top of the table to the column marked 0.03. The intersection of this row and column combination gives the area 0.9484, which is $P(z \leq 1.63)$.

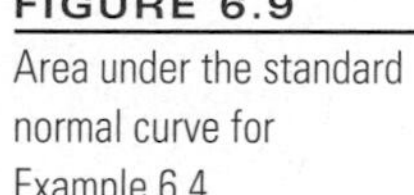

FIGURE 6.9

Area under the standard normal curve for Example 6.4

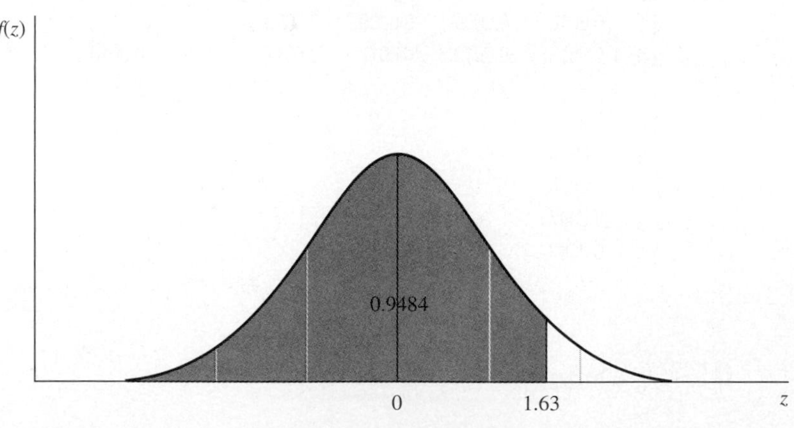

Areas to the left of $z = 0$ are found using negative values of z.

EXAMPLE 6.5

Find $P(z \geq -0.5)$. This probability corresponds to the area to the *right* of a point $z = -0.5$ standard deviation to the left of the mean (see Figure 6.10).

FIGURE 6.10
Area under the standard normal curve for Example 6.5

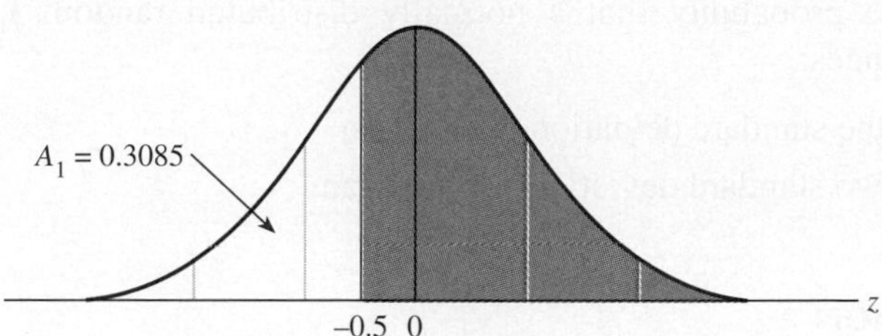

SOLUTION The area given in Table 3 is the area to the left of a specified value of z. Indexing $z = -0.5$ in Table 3, we can find the area A_1 to the *left* of -0.5 to be 0.3085. Since the area under the curve is 1, we find $P(z \geq -0.5) = 1 - A_1 = 1 - 0.3085 = 0.6915$.

EXAMPLE 6.6

Find $P(-0.5 \leq z \leq 1.0)$. This probability is the area between $z = -0.5$ and $z = 1.0$, as shown in Figure 6.11.

FIGURE 6.11
Area under the standard normal curve for Example 6.6

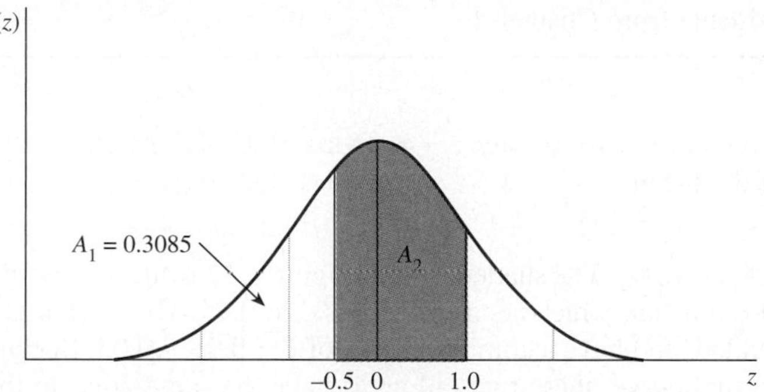

SOLUTION The area required is the shaded area A_2 in Figure 6.11. From Table 3 in Appendix I, you can find the area to the left of $z = -0.5$ ($A_1 = 0.3085$) and the area to the left of $z = 1.0$ ($A_1 + A_2 = 0.8413$). To find the area marked A_2, we subtract the two entries:

$$A_2 = (A_1 + A_2) - A_1 = 0.8413 - 0.3085 = 0.5328$$

That is, $P(-0.5 \leq z \leq 1.0) = 0.5328$.

NEED TO KNOW

How to Use Table 3 to Calculate Probabilities under the Standard Normal Curve

- To calculate the area to the left of a z-value, find the area directly from Table 3.
- To calculate the area to the right of a z-value, find the area in Table 3, and subtract from 1.
- To calculate the area between two values of z, find the two areas in Table 3, and subtract one area from the other.

EXAMPLE 6.7

Find the probability that a normally distributed random variable will fall within these ranges:

1. One standard deviation of its mean
2. Two standard deviations of its mean

SOLUTION

1. Since the standard normal random variable z measures the distance from the mean in units of standard deviations, you need to find

$$P(-1 \le z \le 1) = 0.8413 - 0.1587 = 0.6826$$

Remember that you calculate the area between two z-values by subtracting the tabled entries for the two values.

2 As in part 1, $P(-2 \le z \le 2) = 0.9772 - 0.0228 = 0.9544$.

These probabilities agree with the approximate values of 68% and 95% in the Empirical Rule from Chapter 2.

EXAMPLE 6.8

Find the value of z—say, z_0—such that 0.95 of the area is within $\pm z_0$ standard deviations of the mean.

We know the area. Work from the inside of the table out.

SOLUTION The shaded area in Figure 6.12 is the area within $\pm z_0$ standard deviations of the mean, which needs to be equal to 0.95. The "tail areas" under the curve are not shaded, and have a combined area of $1 - 0.95 = 0.05$. Because of the symmetry of the normal curve, these two tail areas have the same area, so that $A_1 = 0.05/2 = 0.025$ in Figure 6.12. Thus, the entire *cumulative area* to the left of z_0 to equal $A_1 + A_2 = 0.95 + 0.025 = 0.9750$. This area is found in the interior of Table 3 in Appendix I in the row corresponding to $z = 1.9$ and the 0.06 column. Hence, $z_0 = 1.96$. Note that this result is very close to the approximate value, $z = 2$, used in the Empirical Rule.

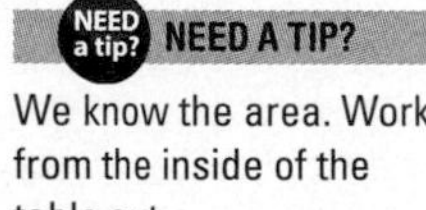

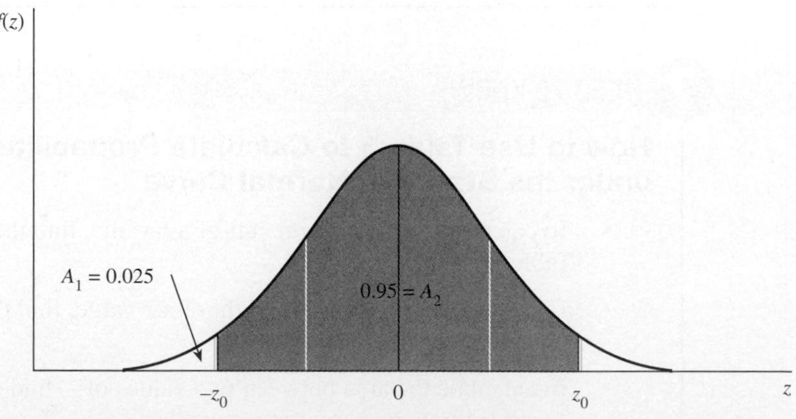

FIGURE 6.12
Area under the standard normal curve for Example 6.8

Calculating Probabilities for a General Normal Random Variable

Most of the time, the probabilities you are interested in will involve X, a normal random variable with mean μ and standard deviation σ. You must then *standardize* the interval of interest, writing it as the equivalent interval in terms of z, the standard normal random variable. Once this is done, the probability of interest is the area that you find using the *standard normal probability distribution.*

EXAMPLE 6.9

Let X be a normally distributed random variable with a mean of 10 and a standard deviation of 2. Find the probability that x lies between 11 and 13.6.

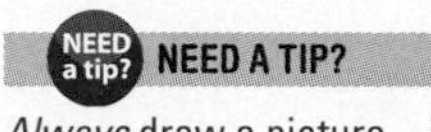

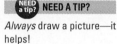

NEED A TIP?

Always draw a picture—it helps!

SOLUTION The interval from $X = 11$ to $X = 13.6$ must be standardized using the formula for z. When $X = 11$,

$$z = \frac{X - \mu}{\sigma} = \frac{11 - 10}{2} = 0.5$$

and when $X = 13.6$,

$$z = \frac{X - \mu}{\sigma} = \frac{13.6 - 10}{2} = 1.8$$

The desired probability is therefore $P(0.5 \le z \le 1.8)$, the area lying between $z = 0.5$ and $z = 1.8$, as shown in Figure 6.13. From Table 3 in Appendix I, you find that the area to the left of $z = 0.5$ is 0.6915, and the area to the left of $z = 1.8$ is 0.9641. The desired probability is the difference between these two probabilities, or

$$P(0.5 \le z \le 1.8) = 0.9641 - 0.6915 = 0.2726$$

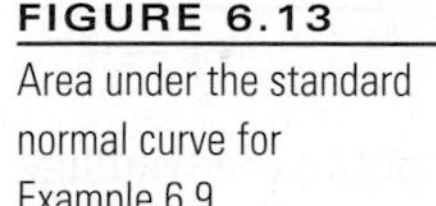

FIGURE 6.13

Area under the standard normal curve for Example 6.9

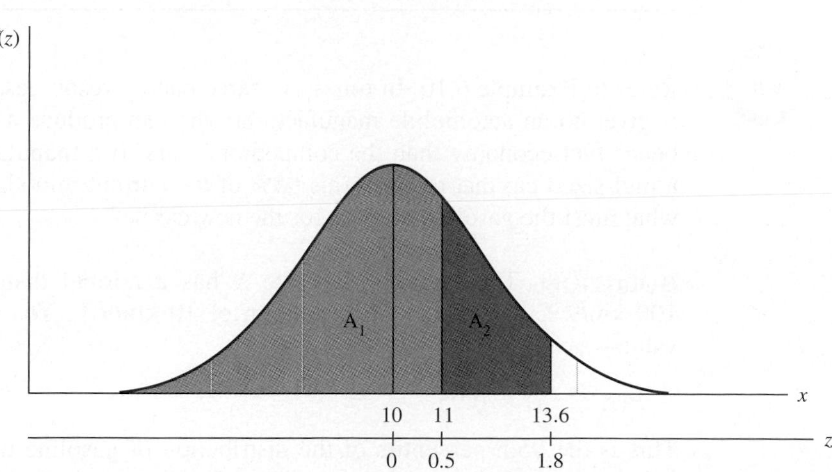

EXAMPLE 6.10

Studies show that gasoline use for mid-sized cars sold in Canada is normally distributed, with a mean of 100 km per 9 L (km/L) and a standard deviation of 10 km/9 L. What percentage of mid-sized cars get 110 km/9 L or more?

Normal Probability Distributions and Normal Probabilities and z-scores

SOLUTION The proportion of mid-sized cars that get 110 km/9 L or more is given by the shaded area in Figure 6.14. To solve this problem, you must first find the z-value corresponding to $X = 110$. Substituting into the formula for z, you get

$$z = \frac{X - \mu}{\sigma} = \frac{110 - 100}{10} = 1.0$$

The area A_1 to the left of $z = 1.0$, is 0.8413 (from Table 3 in Appendix I). Then the proportion of mid-sized cars that get 110 km/9 L or more is equal to:

$$P(X \geq 110) = 1 - P(z < 1) = 1 - 0.8413 = 0.1587$$

The percentage exceeding 110 km/9 L is

$$100(0.1587) = 15.87\%.$$

FIGURE 6.14

Area under the standard normal curve for Example 6.10

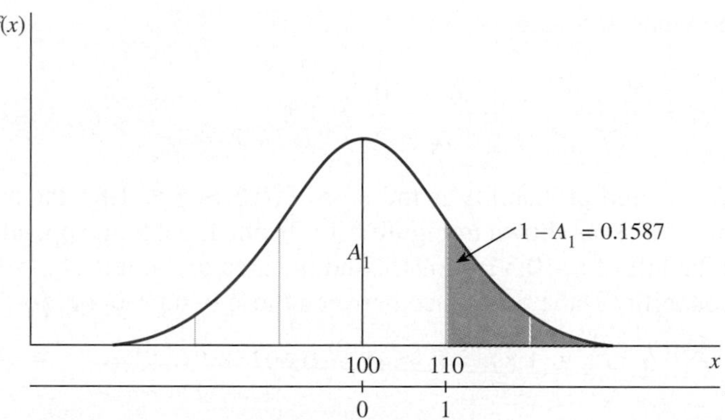

EXAMPLE 6.11

Refer to Example 6.10. In times of scarce energy resources, a competitive advantage is given to an automobile manufacturer who can produce a car that has substantially better fuel economy than the competitors' cars. If a manufacturer wishes to develop a mid-sized car that outperforms 95% of the current mid-sized cars in fuel economy, what must the gasoline use rate for the new car be?

SOLUTION The gasoline use rate X has a normal distribution with a mean of 100 km/9 L and a standard deviation of 10 km/9 L. You need to find a particular value—say, x_0—such that

$$P(X \leq x_0) = 0.95$$

This is the 95th percentile of the distribution of gasoline use rate X. Since the only information you have about normal probabilities is in terms of the standard normal random variable z, start by standardizing the value of x_0:

$$z_0 = \frac{x_0 - 100}{10}$$

Since the value of z_0 corresponds to x_0, it must *also* have area 0.95 to its left, as shown in Figure 6.15. If you look in the interior of Table 3 in Appendix I, you will find that the area 0.9500 is exactly halfway between the areas for $z = 1.64$ and $z = 1.65$. Thus, z_0 must be exactly halfway between 1.64 and 1.65, or

$$z_0 = \frac{x_0 - 100}{10} = 1.645$$

Solving for x_0, you obtain

$$x_0 = \mu + z_0\sigma = 100 + (1.645)(10) = 116.45$$

FIGURE 6.15
Area under the standard normal curve for Example 6.11

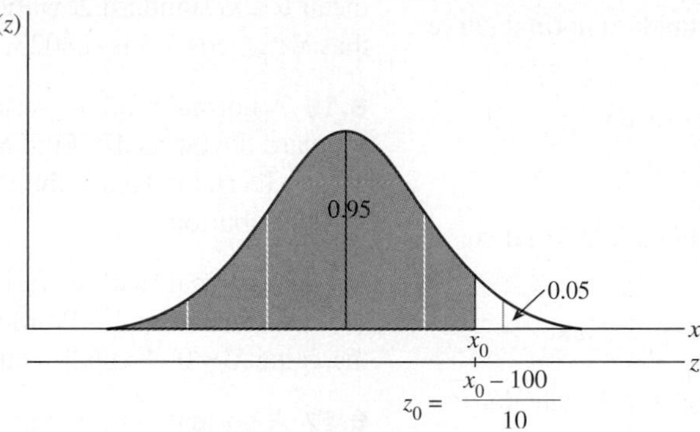

The manufacturer's new mid-sized car must therefore get 116.45 km/9 L to outperform 95% of the mid-sized cars available on the Canadian market.

EXAMPLE 6.12

According to research published at the International Conference on Human Computer Interaction (HCI) in 2015, researchers determined the age when both male and female participants began to use social media from the ages 9 to 17.[1] The average age that females started to use social media is 12.263, with a standard deviation of 1.627. Would a 9-year-old female be too young to use social media? How common is it to be using social media at that age?

SOLUTION $z_9 = 9–12.263/1.627 = –2.0$

From the z-score table, the area under the normal curve, –2.0 is 0.0228, or 2.28%. It is uncommon for a 9-year-old female to start using social media as it represents only 2.28% of the data, with an average of 12.263 and a standard deviation of 1.627.

6.3 EXERCISES

BASIC TECHNIQUES

6.1 Consider a standard normal random variable with $\mu = 0$ and standard deviation $\sigma = 1$. Use Table 3 to find the following probabilities:

a. $P(z < 2)$ **b.** $P(z > 1.16)$
c. $P(-2.33 < z < 2.33)$ **d.** $P(z < 1.88)$

6.2 Find these probabilities associated with the standard normal random variable z:

a. $P(z > 5)$ **b.** $P(-3 < z < 3)$
c. $P(z < 2.81)$ **d.** $P(z > 2.81)$

6.3 Calculate the area under the standard normal curve to the left of these values:

a. $z = 1.6$ **b.** $z = 1.83$
c. $z = 0.90$ **d.** $z = 4.18$

6.4 Calculate the area under the standard normal curve between these values:

a. $z = -1.4$ and $z = 1.4$ **b.** $z = -3.0$ and $z = 3.0$

6.5 Find the following probabilities for the standard normal random variable z:

a. $P(-1.43 < z < 0.68)$ **b.** $P(0.58 < z < 1.74)$
c. $P(-1.55 < z < -0.44)$ **d.** $P(z > 1.34)$
e. $P(z < -4.32)$

6.6 Find these probabilities for the standard normal random variable z:

a. $P(z < 2.33)$ **b.** $P(z < 1.645)$
c. $P(z > 1.96)$ **d.** $P(-2.58 < z < 2.58)$

6.7 a. Find a z_0 such that $P(z > z_0) = 0.025$.

b. Find a z_0 such that $P(z < z_0) = 0.9251$.

6.8 Find a z_0 such that $P(-z_0 < z < z_0) = 0.8262$.

6.9 a. Find a z_0 that has area 0.9505 to its left.

b. Find a z_0 that has area 0.05 to its left.

6.10 a. Find a z_0 such that $P(-z_0 < z < z_0) = 0.90$.

b. Find a z_0 such that $P(-z_0 < z < z_0) = 0.99$.

6.11 Find the following *percentiles* for the standard normal random variable z:

a. 90th percentile **b.** 95th percentile
c. 98th percentile **d.** 99th percentile

6.12 A normal random variable x has mean $\mu = 10$ and standard deviation $\sigma = 2$. Find the probabilities of these X-values:

a. $X > 13.5$ **b.** $X < 8.2$ **c.** $9.4 < X < 10.6$

6.13 A normal random variable x has mean $\mu = 1.20$ and standard deviation $\sigma = 0.15$. Find the probabilities of these X-values:

a. $1.00 < X < 1.10$ **b.** $X > 1.38$
c. $1.35 < X < 1.50$

6.14 A normal random variable x has an unknown mean μ and standard deviation $\sigma = 2$. If the probability that X exceeds 7.5 is 0.8023, find μ.

6.15 A normal random variable X has mean 35 and standard deviation 10. Find a value of X that has area 0.01 to its right. This is the *99th percentile* of this normal distribution.

6.16 A normal random variable X has mean 50 and standard deviation 15. Would it be unusual to observe the value $X = 0$? Explain your answer.

6.17 A normal random variable X has an unknown mean and standard deviation. The probability that X exceeds 4 is 0.9772, and the probability that X exceeds 5 is 0.9332. Find μ and σ.

APPLICATIONS

6.18 Ground Beef The meat department at a local supermarket specifically prepares its "1-kilogram" packages of ground beef so that there will be a variety of weights, some slightly more and some slightly less than 1 kilogram (kg). Suppose that the weights of these "1-kilogram" packages are normally distributed with a mean of 1 kg and a standard deviation of 0.15 kg.

a. What proportion of the packages will weigh more than 1 kg?

b. What proportion of the packages will weigh between 0.95 and 1.05 kg?

c. What is the probability that a randomly selected package of ground beef will weigh less than 0.80 kg?

d. Would it be unusual to find a package of ground beef that weighs 1.45 kg? How would you explain such a large package?

6.19 Human Heights Human heights are one of the many biological random variables that can be modelled by the normal distribution. The average height of Canadian women aged 18 and older is 163 centimetres (cm), while the average height for men is 177 cm. Assume the standard deviation for the Canadian men is 8 cm.

a. What proportion of all men will be taller than 185 cm?

b. What is the probability that a randomly selected man will be between 170 and 185 cm tall?

c. The height of the current prime minister is 188 cm tall. Is this an unusual height?

6.20 Christmas Trees The diameters of Douglas firs grown at a Christmas tree farm are normally distributed with a mean of 10 cm and a standard deviation of 3 cm.

a. What proportion of the trees will have diameters between 8 and 12 cm?

b. What proportion of the trees will have diameters less than 7 cm?

c. Your Christmas tree stand will expand to a diameter of 14 cm. What proportion of the trees will not fit in your Christmas tree stand?

6.21 Cerebral Blood Flow Cerebral blood flow (CBF) in the brains of healthy people is normally distributed with a mean of 74 and a standard deviation of 16.

a. What proportion of healthy people will have CBF readings between 60 and 80?

b. What proportion of healthy people will have CBF readings above 100?

c. If a person has a CBF reading below 40, that person is classified as at risk for a stroke. What proportion of healthy people will mistakenly be diagnosed as "at risk"?

6.22 Braking Distances For a car travelling 50 kilometres per hour (km/h), the distance required to brake to a stop is normally distributed with a mean of 15 metres (m) and a standard deviation of 2.5 m. Suppose you are travelling 50 km/h in a residential area and a car moves abruptly into your path at a distance of 18 m.

a. If you apply your brakes, what is the probability that you will brake to a stop within 12 m or less? Within 15 m or less?

b. If the only way to avoid a collision is to brake to a stop, what is the probability that you will avoid the collision?

6.23 Elevator Capacities Suppose you must establish regulations concerning the maximum number of people who can occupy an elevator. Suppose a study of elevator occupancies indicates that, if eight people occupy the elevator, the probability distribution of the total weight of the eight people has a mean equal to 550 kg and a variance equal to 445 kg. What is the probability that the total weight of eight people exceeds 590 kg? 680 kg? (Assume that the probability distribution is approximately normal.)

6.24 A Phosphate Mine The discharge of suspended solids from a phosphate mine is normally distributed, with a mean daily discharge of 27 milligrams per litre (mg/L) and a standard deviation of 14 mg/L. What proportion of days will the daily discharge exceed 50 mg/L?

6.25 Sunflowers An experimenter publishing in the *Annals of Botany* investigated whether the stem diameters of the dicot sunflower would change depending on whether the plant was left to sway freely in the wind or was artificially supported.[2] Suppose that the unsupported stem diameters at the base of a particular species of sunflower plant have a normal distribution with an average diameter of 35 millimetres (mm) and a standard deviation of 3 mm.

a. What is the probability that a sunflower plant will have a stem diameter of more than 40 mm?

b. If two sunflower plants are randomly selected, what is the probability that both plants will have a stem diameter of more than 40 mm?

c. Within what limits would you expect the stem diameters to lie, with probability 0.95?

d. What diameter represents the 90th percentile of the distribution of diameters?

6.26 Breathing Rates The number of times, X, an adult human breathes per minute when at rest depends on the age of the human and varies greatly from person to person. Suppose the probability distribution for X is approximately normal, with the mean equal to 16 and the standard deviation equal to 4. If a person is selected at random and the number X of breaths per minute while at rest is recorded, what is the probability that X will exceed 22?

6.27 Economic Forecasts One method of arriving at economic forecasts is to use a consensus approach. A forecast is obtained from each of a large number of analysts, and the average of these individual forecasts is the consensus forecast. Suppose the individual

2013 January prime interest rate forecasts of all economic analysts are approximately normally distributed, with the mean equal to 4.5% and the standard deviation equal to 0.1%. If a single analyst is randomly selected from among this group, what is the probability that the analyst's forecast of the prime interest rate will take on these values?

a. Exceed 4.75%

b. Be less than 4.375%

6.28 Tax Audit How does the Canada Revenue Agency decide on the percentage of income tax returns to audit for each province? Suppose they do it by randomly selecting 50 values from a normal distribution with a mean equal to 1.55% and a standard deviation equal to 0.45%. (Computer programs are available for this type of sampling.)

a. What is the probability that a particular province will have more than 2.5% of its income tax returns audited?

b. What is the probability that a province will have less than 1% of its income tax returns audited?

6.29 Bacteria in Drinking Water Suppose the numbers of a particular type of bacteria in samples of 1 millilitre (mL) of drinking water tend to be approximately normally distributed, with a mean of 85 and a standard deviation of 9. What is the probability that a given 1-mL sample will contain more than 100 bacteria?

6.30 Loading Grain A grain loader can be set to discharge grain in amounts that are normally distributed, with mean μ kg and standard deviation equal to 700 kg. If a company wishes to use the loader to fill containers that hold 54,440 kg of grain and wants to overfill only one container in 100, at what value of μ should the company set the loader?

6.31 How Many Words? A publisher has discovered that the numbers of words contained in a new manuscript are normally distributed, with a mean equal to 20,000 words in excess of that specified in the author's contract and a standard deviation of 10,000 words. If the publisher wants to be almost certain (say, with a probability of 0.95) that the manuscript will have less than 100,000 words, what number of words should the publisher specify in the contract?

6.32 Tennis Anyone? A stringer of tennis racquets has found that the actual string tension achieved for any individual racquet stringing will vary as much as 3 kilograms per square centimetre (km/cm^2) from the desired tension set on the stringing machine. If the stringer wishes to string at a tension lower than that specified by a customer only 5% of the time, how much above or below the customer's specified tension should the stringer set the stringing machine? (NOTE: Assume that the distribution of string tensions produced by the stringing machine is normally distributed, with a mean equal to the tension set on the machine and a standard deviation equal to 1 km/cm^2.)

6.33 Sunday Shopping When Nova Scotia lifted its Sunday shopping ban, retail stores of all kinds could finally open on Sundays and other holidays, the only exception being Remembrance Day. Nova Scotia was one of the last provinces to place Sunday restrictions on retailers. Suppose that the amount of money spent at shopping centres on Sundays has a normal distribution with a mean of $85 and with a standard deviation of $10. A shopper is randomly selected on a Sunday and asked about his spending patterns.

a. What is the probability that he has spent more than $90 at the mall?

b. What is the probability that he has spent between $90 and $100 at the mall?

c. If two shoppers are randomly selected, what is the probability that both shoppers have spent more than $100 at the mall?

6.34 Pulse Rates Your pulse rate is a measure of the number of heartbeats per minute. It can be measured in several places on your body, where an artery passes close to the skin. Once you find the pulse, count the number of beats per minute, or, count for 30 seconds and multiply by two. What's a *normal* pulse rate? That depends on a variety of factors. Pulse rates between 60 and 100 beats per minute are considered normal for children over 10 and adults.[3] Suppose that these pulse rates are approximately normally distributed with a mean of 78 and a standard deviation of 12.

a. What proportion of adults will have pulse rates between 60 and 100?

b. What is the 95th percentile for the pulse rates of adults?

c. Would a pulse rate of 110 be considered unusual? Explain.

6.4 THE NORMAL APPROXIMATION TO THE BINOMIAL PROBABILITY DISTRIBUTION (OPTIONAL)

In Chapter 5, you learned three ways to calculate probabilities for the binomial random variable x:

- Using the binomial formula, $P(X = k) = C^n_k p^k q^{n-k}$
- Using the cumulative binomial tables
- Using *MINITAB* and *Excel*

The binomial formula produces lengthy calculations, and the tables are available for only certain values of n and p. There is another option available when $np < 7$; the Poisson probabilities can be used to approximate $P(X = k)$. When this approximation *does not work* and n is large, the normal probability distribution provides another approximation for binomial probabilities.

THE NORMAL APPROXIMATION TO THE BINOMIAL PROBABILITY DISTRIBUTION

Let X be a binomial random variable with n trials and probability p of success. The probability distribution of X is approximated using a normal curve with

$$\mu = np \quad \text{and} \quad \sigma = \sqrt{npq}$$

This approximation is adequate as long as n is large and p is not too close to 0 or 1.

Since the normal distribution is continuous, the area under the curve at any single point is equal to 0. Keep in mind that this result applies only to continuous random variables. Because the binomial random variable X is a discrete random variable, the probability that X takes some specific value—say, $X = 11$—will not necessarily equal 0.

Figures 6.16 and 6.17 show the binomial probability histograms for $n = 25$ with $p = 0.5$ and $p = 0.1$, respectively. The distribution in Figure 6.16 is exactly symmetric. If you superimpose a normal curve with the same mean, $\mu = np$, and the same standard deviation, $\sigma = \sqrt{npq}$, over the top of the bars, it "fits" quite well; that is, the areas under the curve are almost the same as the areas under the bars. However, when the probability of success, p, gets small and the distribution is skewed, as in Figure 6.17, the symmetric normal curve no longer fits very well. If you try to use the normal curve areas to approximate the area under the bars, your approximation will not be very good.

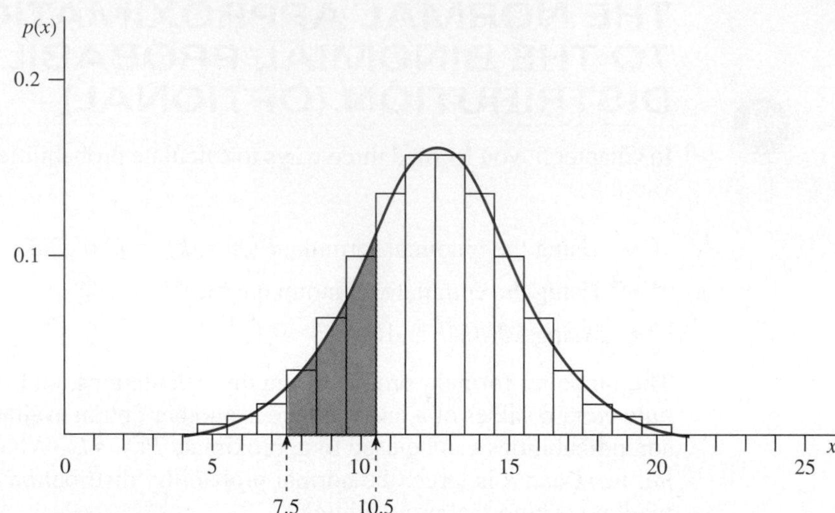

FIGURE 6.16

The binomial probability distribution for $n = 25$ and $p = 0.5$ and the approximating normal distribution with $\mu = 12.5$ and $\sigma = 2.5$

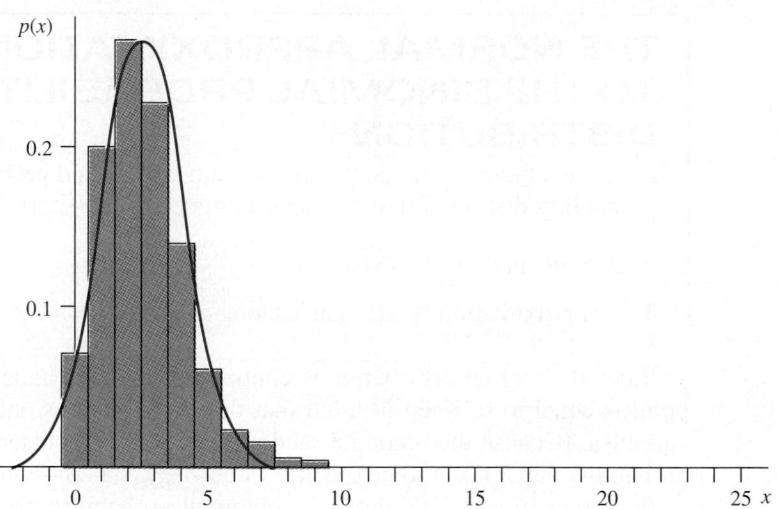

FIGURE 6.17

The binomial probability distribution and the approximating normal distribution for $n = 25$ and $p = 0.1$

EXAMPLE 6.13

Use the normal curve to approximate the probability that $X = 8$, 9, or 10 for a binomial random variable with $n = 25$ and $p = 0.5$. Compare this approximation to the exact binomial probability.

SOLUTION You can find the exact binomial probability for this example because there are cumulative binomial tables for $n = 25$. From Table 1 in Appendix I,

$$P(X = 8, 9, \text{or } 10) = P(X \le 10) - P(X \le 7) = 0.212 - 0.022 = 0.190$$

To use the normal approximation, first find the appropriate mean and standard deviation for the normal curve:

$$\mu = np = 25(0.5) = 12.5$$

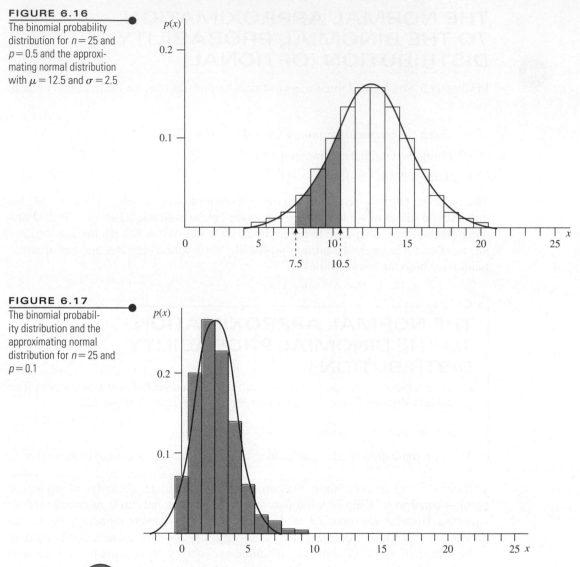

$$\sigma = \sqrt{npq} = \sqrt{25(0.5)(0.5)} = 2.5$$

NEED A TIP?

Use the continuity correction *only* if x has a **binomial** distribution!

ONLINE APPLET

Normal Approximation to Binomial Probabilities

The probability that you need corresponds to the area of the three rectangles lying over $X = 8$, 9, and 10. The equivalent area under the normal curve lies between $X = 7.5$ (the lower edge of the rectangle for $X = 8$) and $X = 10.5$ (the upper edge of the rectangle for $X = 10$). This area is shaded in Figure 6.16.

To find the normal probability, follow the procedures of Section 6.3. First you standardize each interval endpoint:

$$z = \frac{X - \mu}{\sigma} = \frac{7.5 - 12.5}{2.5} = -2.0$$

$$z = \frac{X - \mu}{\sigma} = \frac{10.5 - 12.5}{2.5} = -0.8$$

Then the approximate probability (shaded in Figure 6.18) is found from Table 3 in Appendix I:

$$P(-2.0 < z < -0.8) = 0.2119 - 0.0228 = 0.1891$$

You can compare the approximation, 0.1891, to the actual probability, 0.190. They are quite close!

FIGURE 6.18

Area under the normal curve for Example 6.13

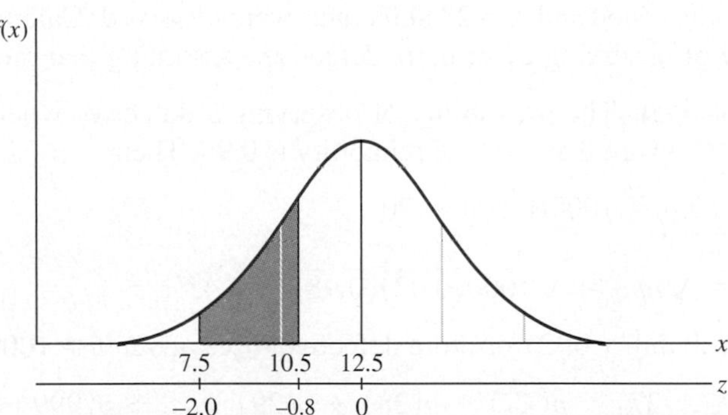

You must be careful not to exclude half of the two extreme probability rectangles when you use the normal approximation to the binomial probability distribution. This adjustment, called the **continuity correction,** helps account for the fact that you are approximating a *discrete random variable* with a *continuous* one. If you forget the correction, your approximation will not be very good! Use this correction only for *binomial probabilities*; do not try to use it when the random variable is already continuous, such as a height or weight.

How can you tell when it is appropriate to use the normal approximation to binomial probabilities? The normal approximation works well when the binomial histogram is roughly symmetric. This happens when the binomial distribution is not "bunched up" near 0 or n—that is, when it can spread out at least two standard deviations from its mean without exceeding its limits, 0 and n. Using this criterion, you can derive this simple rule of thumb:

RULE OF THUMB

The normal approximation to the binomial probabilities will be adequate if both $np > 5$ and $nq > 5$.

NEED TO KNOW

How to Calculate Binomial Probabilities Using the Normal Approximation

- Find the necessary values of n and p. Calculate $\mu = np$ and $\sigma = \sqrt{npq}$.
- Write the probability you need in terms of X and locate the appropriate area on the curve.
- Correct the value of x by ± 0.5 to include the entire block of probability for that value. This is the *continuity correction.*
- Convert the necessary X-values to z-values using

$$z = \frac{x \pm 0.5 - np}{\sqrt{npq}}$$

- Use Table 3 in Appendix I to calculate the approximate probability.

EXAMPLE 6.14

The reliability of an electrical fuse is the probability that a fuse, chosen at random from production, will function under its designed conditions. A random sample of 1000 fuses was tested and $X = 27$ defectives were observed. Calculate the approximate probability of observing 27 or more defectives, assuming that the fuse reliability is 0.98.

NEED A TIP?

If np and nq are both greater than 5, you can use the normal approximation.

SOLUTION The probability of observing a defective when a single fuse is tested is $p = 0.02$, given that the fuse reliability is 0.98. Then

$$\mu = np = 1000(0.02) = 20$$

$$\sigma = \sqrt{npq} = \sqrt{1000(0.02)(0.98)} = 4.43$$

The probability of 27 or more defective fuses, given $n = 1000$, is

$$P(X \geq 27) = p(27) + p(28) + p(29) + \ldots + p(999) + p(1000)$$

It is appropriate to use the normal approximation to the binomial probability because

$$np = 1000(0.02) = 20 \text{ and } nq = 1000(0.98) = 980$$

are both greater than 5. The normal area used to approximate $P(X \geq 27)$ is the area under the normal curve to the right of 26.5, so that the entire rectangle for $X = 27$ is included. Then, the z-value corresponding to $X = 26.5$ is

$$z = \frac{X - \mu}{\sigma} = \frac{26.5 - 20}{4.43} = \frac{6.5}{4.43} = 1.47$$

and the area to the left of $z = 1.47$ is equal to 0.9292, as shown in Figure 6.19. Since the total area under the curve is 1, you have

$$P(X \geq 27) \approx P(z \geq 1.47) = 1 - 0.9292 = 0.0708$$

FIGURE 6.19
Normal approximation to the binomial for Example 6.14

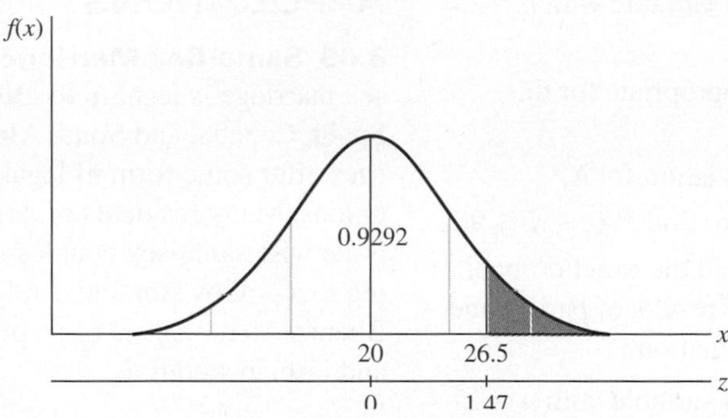

EXAMPLE 6.15

A producer of soft drinks was fairly certain that her brand had a 10% share of the soft drink market. In a market survey involving 2500 consumers of soft drinks, $X = 211$ expressed a preference for her brand. If the 10% figure is correct, find the probability of observing 211 or fewer consumers who prefer her brand of soft drink.

SOLUTION If the producer is correct, then the probability that a consumer prefers her brand of soft drink is $p = 0.10$. Then

$$\mu = np = 2500(0.10) = 250$$

$$\sigma = \sqrt{npq} = \sqrt{2500(0.10)(0.90)} = 15$$

The probability of observing 211 or fewer who prefer her brand is

$$P(X \leq 211) = p(0) + p(1) + \ldots + p(210) + p(211)$$

The normal approximation to this probability is the area to the left of 211.5 under a normal curve with a mean of 250 and a standard deviation of 15. First calculate

$$z = \frac{X - \mu}{\sigma} = \frac{211.5 - 250}{15} = -2.57$$

Then

$$P(X \leq 211) \approx P(z < -2.57) = 0.0051$$

The probability of observing a sample value of 211 or less when $p = 0.10$ is so small that you can conclude that one of two things has occurred: Either you have observed an unusual sample even though really $p = 0.10$, *or* the sample reflects that the actual value of p is less than 0.10 and perhaps closer to the observed sample proportion, $211/2500 = 0.08$.

6.4 EXERCISES

6.35 Consider a binomial random variable X with $n = 25$ and $p = 0.6$.

a. Can the normal approximation be used to approximate probabilities in this case? Why or why not?

b. What are the mean and standard deviation of X?

c. Using the correction for continuity, approximate $P(X > 9)$.

6.36 Consider a binomial random variable X with $n = 45$ and $p = 0.05$.

a. Are np and nq both larger than 5?

b. Based on your answer to part a, can we use the normal approximation to approximate the binomial probabilities associated with X? If not, is there another possible approximation we could use?

6.37 Let X be a binomial random variable with $n = 25$ and $p = 0.3$.

a. Is the normal approximation appropriate for this binomial random variable?

b. Find the mean and standard deviation for X.

c. Use the normal approximation to find $P(6 \le X \le 9)$.

d. Use Table 1 in Appendix I to find the exact probability $P(6 \le X \le 9)$. Compare the results of parts c and d. How close was your approximation?

6.38 Let X be a binomial random variable with $n = 15$ and $p = 0.5$.

a. Is the normal approximation appropriate?

b. Find $P(X \ge 6)$ using the normal approximation.

c. Find $P(X > 6)$ using the normal approximation.

d. Find the exact probabilities for parts b and c, and compare these with your approximations.

6.39 Let X be a binomial random variable with $n = 100$ and $p = 0.2$. Find approximations to these probabilities:

a. $P(X > 22)$ **b.** $P(X \ge 22)$

c. $P(20 < X < 25)$ **d.** $P(X \le 25)$

6.40 Let X be a binomial random variable for $n = 25$, $p = 0.2$.

a. Use Table 1 in Appendix I to calculate $P(4 \le X \le 6)$.

b. Find μ and σ for the binomial probability distribution, and use the normal distribution to approximate the probability $P(4 \le X \le 6)$. Note that this value is a good approximation to the exact value of $P(4 \le X \le 6)$ even though $np = 5$.

6.41 Suppose the random variable X has a binomial distribution corresponding to $n = 20$ and $p = 0.30$. Use Table 1 of Appendix I to calculate these probabilities:

a. $P(X = 5)$ **b.** $P(X \ge 7)$

6.42 Refer to Exercise 6.41. Use the normal approximation to calculate $P(X = 5)$ and $P(X \ge 7)$. Compare with the exact values obtained from Table 1 in Appendix I.

6.43 Consider a binomial experiment with $n = 20$ and $p = 0.4$. Calculate $P(X \ge 10)$ using each of these methods:

a. Table 1 in Appendix I

b. The normal approximation to the binomial probability distribution

6.44 Find the normal approximation to $P(355 \le X \le 360)$ for a binomial probability distribution with $n = 400$ and $p = 0.9$.

APPLICATIONS

6.45 Same-Sex Marriage in Argentines Same-sex marriage is legal in the Netherlands, Belgium, Spain, Canada, and South Africa, and at least 18 countries offer some form of legal recognition to same-sex unions. Many residents of Argentina's capital have no issue with same-sex couples getting married, according to a poll by Analogias released by Página 12.[4] In Buenos Aires, 73.1% of respondents agreed with gay and lesbian weddings.

Does this hold true for you and your statistics classmates? Assume that it does and that your class contains 50 students. What are the approximate probabilities for these events?

a. More than 30 students support same-sex marriage.

b. Fewer than 40 students support same-sex marriage.

c. Fewer than 15 students *do not* support same-sex marriage.

d. Are you willing to assume that you and your classmates are a representative sample of all Canadians when it comes to this question? How does your answer affect the probabilities in parts a to c?

6.46 Genetic Defects Data collected over a long period of time show that a particular genetic defect occurs in 1 of every 1000 children. The records of a medical clinic show $x = 60$ children with the defect in a total of 50,000 examined. If the 50,000 children were a random sample from the population of children represented by past records, what is the probability of observing a value of x equal to 60 or more? Would you say that the observation of $x = 60$ children with genetic defects represents a rare event?

6.47 No Shows Airlines and hotels often grant reservations in excess of capacity to minimize losses due to no-shows. Suppose the records of a hotel show that, on the average, 10% of its prospective guests will not claim their reservations. If the hotel accepts 215 reservations and there are only 200 rooms in the hotel, what is the probability that all guests who arrive to claim a room will receive one?

6.48 Lung Cancer Compilation of large masses of data on lung cancer shows that approximately 1 of every 40 adults acquires the disease. Workers in a certain occupation are known to work in an air-polluted environment that may cause an increased rate of lung cancer. A random sample of $n = 400$ workers shows 19 with identifiable cases of lung cancer. Do the data provide sufficient evidence to indicate a higher rate of lung cancer for these workers than for the national average?

6.49 Death Penalty in Lima An overwhelming majority of adults in Peru's capital believe capital punishment would be suitable in specific circumstances, according to a poll by Apoyo published in *El Comercio*.[5] In Lima, 81% of respondents would allow the death penalty for people convicted of raping young children. The above finding is based on interviews with 503 Peruvian adults in Lima conducted on January 11 and 12, 2007.

a. Find the approximate probability that exactly 420 randomly selected Peruvian adults in Lima would allow the death penalty for people convicted of raping young children.

b. Find the approximate probability that fewer than 420 randomly selected Peruvian adults in Lima would allow the death penalty for people convicted of raping young children.

c. Find the approximate probability that more than 420 randomly selected Peruvian adults in Lima would allow the death penalty for people convicted of raping young children.

d. Can you assume that Peruvian adults in Lima are a representative sample of all adults in Peru? Explain.

6.50 The Rh Factor In a certain population, 15% of the people have Rh-negative blood. A blood bank serving this population receives 92 blood donors on a particular day.

a. What is the probability that 10 or fewer are Rh-negative?

b. What is the probability that 15 to 20 (inclusive) of the donors are Rh-negative?

c. What is the probability that more than 80 of the donors are Rh-positive?

6.51 Pepsi's Market Share Two of the biggest soft drink rivals, Pepsi and Coke, are very concerned about their market shares. The pie chart that follows claims that PepsiCo's share of the beverage market is 24.2%.[6] Assume that this proportion will be close to the probability that a person selected at random indicates a preference for a Pepsi product when choosing a soft drink.

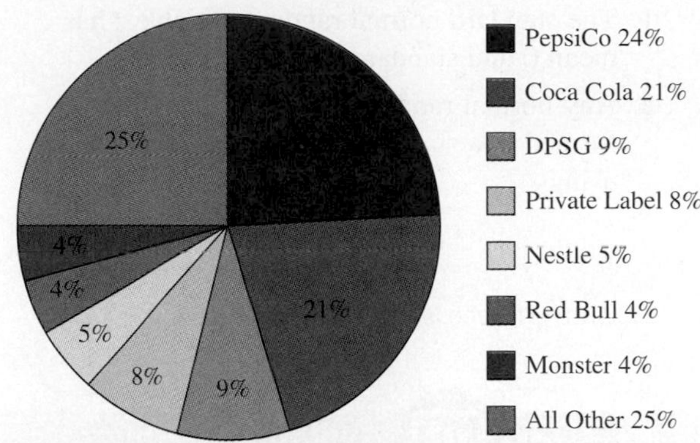

© PepsiCo, Inc., www.pepsico.com/docs/album/default-document-library/pepsico-2014-annual-report_final.pdf?sfvrsn=0.

A group of $n = 500$ consumers is selected and the number preferring a Pepsi product is recorded. Use the normal curve to approximate the following binomial probabilities.

a. Exactly 150 consumers prefer a Pepsi product.

b. Between 120 and 150 consumers (inclusive) prefer a Pepsi product.

c. Fewer than 150 consumers prefer a Pepsi product.

d. Would it be unusual to find that 232 of the 500 consumers preferred a Pepsi product? If this were to occur, what conclusions would you draw?

6.52 Trying to Be More Frugal? Assume that 37% of Canadians are concerned they won't be able to afford their current lifestyle in 2017, and that a random sample 100 Canadians is selected.

a. What is the average number of Canadians who felt they will have to cut back on spending?

b. What is the standard deviation of Canadians who felt they will have to cut back on spending?

c. Suppose that in our sample of 100 Canadians, there are 50 who felt they will have to cut back on spending. Would you consider this to be an unusual occurrence? Explain.

CHAPTER REVIEW

Key Concepts and Formulas

I. Continuous Probability Distributions

1. Continuous random variables
2. Probability distributions or probability density functions
   a. Curves are smooth.
   b. Area under the curve equals 1.
   c. The area under the curve between a and b represents the probability that x falls between a and b.
   d. $P(X = a) = 0$ for continuous random variables.

II. The Normal Probability Distribution

1. Symmetric about its mean μ
2. Shape determined by its standard deviation σ

III. The Standard Normal Distribution

1. The standard normal random variable z has mean 0 and standard deviation 1.
2. Any normal random variable X can be transformed to a standard normal random variable using

$$z = \frac{X - \mu}{\sigma}$$

3. Convert necessary values of x to z.
4. Use Table 3 in Appendix I to compute standard normal probabilities.
5. Several important z-values have right-tail areas as follows:

Right-Tail Area	0.005	0.01	0.025	0.05	0.10
z-Value	2.58	2.33	1.96	1.645	1.28

TECHNOLOGY TODAY

Normal Probabilities in *Microsoft Excel*

When the random variable of interest has a normal probability distribution, you can generate the following probabilities using the following functions:

1. **NORM.DIST** and **NORM.S.DIST:** Generate cumulative probabilities—$P(X \le x_0)$ for a general normal random variable or $P(z \le z_0)$ for a standard normal random variable. (NOTE: These functions are called **NORMDIST** and **NORMSDIST** in *Excel 2007* and earlier versions.)
2. **NORM.INV** and **NORM.S.INV:** Generate inverse cumulative probabilities—the value x_0 such that the area to its left under the general normal probability distribution is equal to a, or the value z_0 such that the area to its left under the standard normal probability distribution is equal to a. (NOTE: These functions are called **NORMINV** and **NORMSINV** in *Excel 2007* and earlier versions.)

You must specify which normal distribution you are using and, if it is a general normal random variable, the values for the mean μ and the standard deviation σ. As in Chapter 5, you must also specify the values for x_0, z_0, or a, depending on the function you are using.

EXAMPLE 6.16

For a standard normal random variable z, find $P(1.2 < z < 1.96)$. Find the value z_0 with area 0.025 to its right.

1. Name columns A and B of a spreadsheet as "z0," and "P(z <= z0)," respectively. Then enter the two values for z_0 (1.2 and 1.96) in cells A2 and A3. To generate cumulative probabilities for these two values, first place your cursor in cell B2. Select **Insert Function** ▶ **Statistical** ▶ **NORM.S.DIST** and click **OK.** The dialogue box shown in Figure 6.20 will appear.
2. Enter the location of first value of z_0 (cell A2) into the first box and the word TRUE into the second box. The resulting probability is marked as "Formula result = 0.8849" at the bottom of the box, and when you click **OK,** the probability $P(z \le 1.2)$ will appear in cell B2. To obtain the other probability, simply place your cursor in cell B2, grab the square handle in the lower right corner of the cell and drag the handle down to copy the formula into the other cell and obtain $P(z \le 1.96) = 0.9750$. *Excel* has automatically adjusted the cell location in the formula as you copied.

FIGURE 6.20

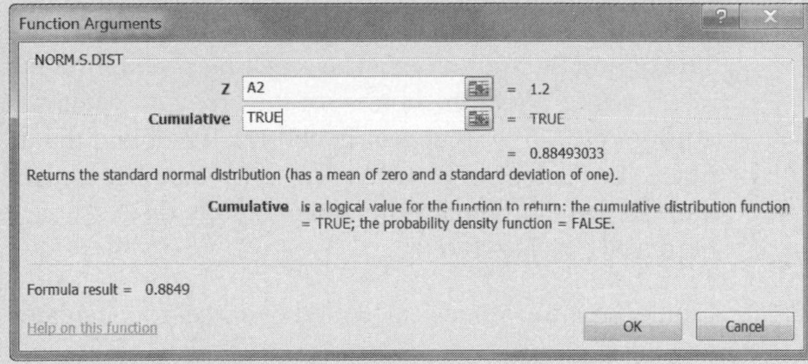
Function Arguments
NORM.S.DIST
Z A2 = 1.2
Cumulative TRUE = TRUE
= 0.88493033
Returns the standard normal distribution (has a mean of zero and a standard deviation of one).
Cumulative is a logical value for the function to return: the cumulative distribution function = TRUE; the probability density function = FALSE.
Formula result = 0.8849
Help on this function
OK
Cancel

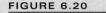

3. To find $P(1.2 < z < 1.96)$, remember that the cumulative probability is the area to the left of the given value of z. Hence,

$$P(1.2 < z < 1.96) = P(z < 1.96) - P(z < 1.2) = 0.9750 - 0.8849 = 0.0901$$

You can check this calculation using Table 3 in Appendix I if you wish!

4. To calculate inverse cumulative probabilities, place your cursor in an empty cell, select **Insert Function ▶ Statistical ▶ NORM.S.INV** and click **OK.** We need a value z_0 with area 0.025 to its right, or area 0.975 to its left. Enter **.975** in the box marked "Probability" and notice the "Formula Result = 1.959963985," which will appear in the empty cell when you click **OK.** This value, when rounded to two decimal places, is the familiar $z_0 = 1.96$ used in Example 6.8.

EXAMPLE
6.17

Suppose that the average birth weights of babies born at hospitals owned by a major health maintenance organization (HMO) are approximately normal with mean 6.75 pounds and standard deviation 0.54 pounds. What proportion of babies born at these hospitals weigh between 6 and 7 pounds? Find the 95th percentile of these birth weights.

1. Name columns A and B of an *Excel* spreadsheet as "x0," and "P(x<=x0)," respectively. Then enter the two values for x_0 (6 and 7) in cells A2 and A3. Proceed as in Example 6.15, this time selecting **Insert Function ▶ Statistical ▶ NORM. DIST** and clicking **OK.** The dialogue box shown in Figure 6.21 will appear.

FIGURE 6.21

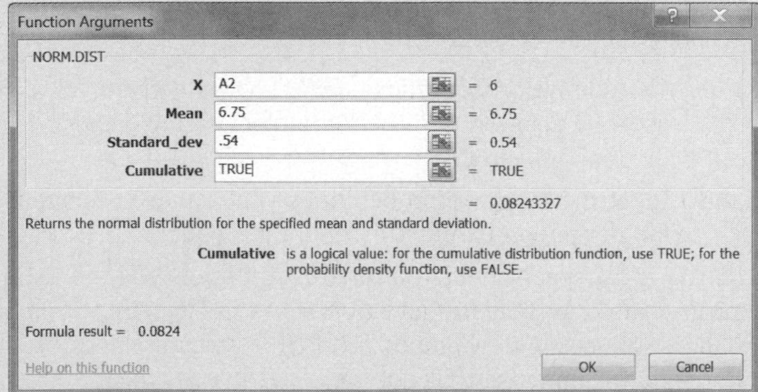
Function Arguments
NORM.DIST
X A2 = 6
Mean 6.75 = 6.75
Standard_dev .54 = 0.54
Cumulative TRUE = TRUE
= 0.08243327
Returns the normal distribution for the specified mean and standard deviation.
Cumulative is a logical value: for the cumulative distribution function, use TRUE; for the probability density function, use FALSE.
Formula result = 0.0824
Help on this function
OK
Cancel

2. Enter the location of first value of x_0 (cell A2) into the first box, the appropriate mean and standard deviation in the second and third boxes, and the word TRUE into the fourth box. The resulting probability is marked as "Formula result = 0.0824" at the bottom of the box, and when you click **OK**, the probability $P(x \leq 6)$ will appear in cell B2. To obtain the other probability, simply place your cursor in cell B2, grab the square handle in the lower right corner of the cell and drag the handle down to copy the formula into the other cell and obtain $P(x \leq 7) = 0.6783$.

3. Finally, use the values calculated by *Excel* to calculate

$$P(6 < X < 7) = P(X < 7) - P(X < 6) = 0.6783 - 0.0824 = 0.5959$$

4. To calculate the 95th percentile, place your cursor in an empty cell, select **Insert Function ▶ Statistical ▶ NORM.INV** and click **OK**. We need a value x_0 with area 0.95 to its left. Enter **.95** in the box marked "Probability," the appropriate mean and standard deviation (see Figure 6.22), and notice the "Formula Result = 7.638220959," which will appear in the empty cell when you click **OK.**

FIGURE 6.22

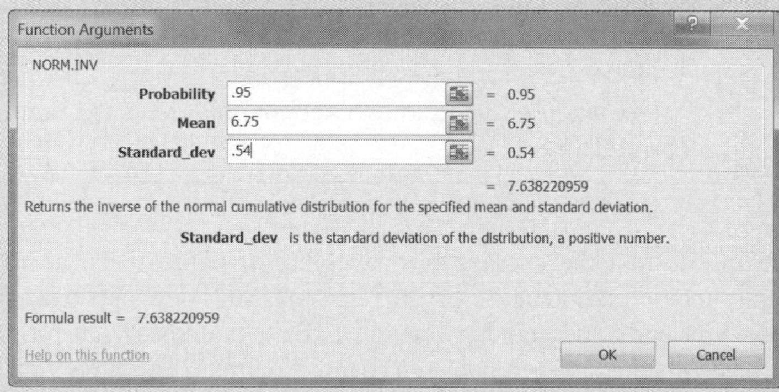
Function Arguments
NORM.INV
Probability .95 = 0.95
Mean 6.75 = 6.75
Standard_dev .54 = 0.54
= 7.638220959
Returns the inverse of the normal cumulative distribution for the specified mean and standard deviation.
Standard_dev is the standard deviation of the distribution, a positive number.
Formula result = 7.638220959
Help on this function
OK
Cancel

That is, 95% of all babies born at these hospitals weigh 7.638 pounds or less. Would you consider a baby who weighs 9 pounds to be unusually large?

Normal Probabilities in *MINITAB*

When the random variable of interest has a normal probability distribution, you can generate the following probabilities:

1. Cumulative probabilities—$P(X \leq x)$ for a given value of x. (NOTE: *MINITAB* uses the notation "X" for the random variable and "x" for a particular value of the random variable.)
2. Inverse cumulative probabilities—the value X such that the area to its left under the normal probability distribution is equal to a.

You must specify which normal distribution you are using and the values for the mean μ and the standard deviation σ. As in Chapter 5, you have the option of specifying only one single value of X (or a) or several values of X (or a), which should be stored in a column (say, C1) of the *MINITAB* worksheet.

For a standard normal random variable z, find $P(1.2 < z < 1.96)$. Find the value z_0 with area 0.025 to its right.

1. Name columns C1 and C2 of a *MINITAB* worksheet as "x" and "P(X<=x)," respectively. Then enter the two values for x (1.2 and 1.96) in the first two cells of column C1. To generate cumulative probabilities for these two values, select **Calc ▶ Probability Distributions ▶ Normal** and the dialogue box shown in Figure 6.23 will appear.

FIGURE 6.23

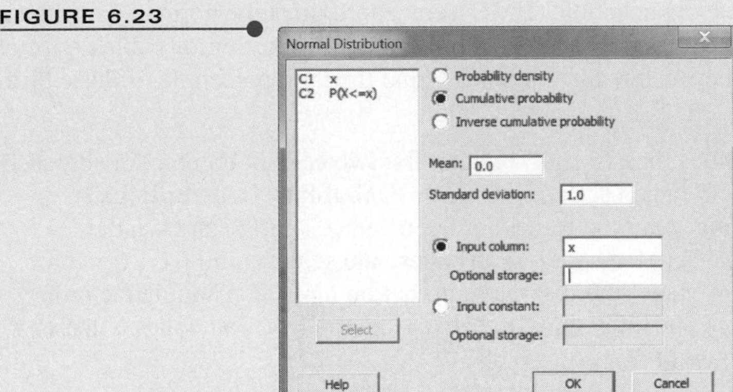

2. By default, *MINITAB* chooses $\mu = 0$ and $\sigma = 1$ as the mean and standard deviation of the standard normal z distribution, so you need only to enter the Input column (C1) and make sure that the radio button marked "Cumulative probability" is selected. If you do not specify a column for "Optional storage," *MINITAB* will display the results in the Session window, shown in Figure 6.24(a).

FIGURE 6.24

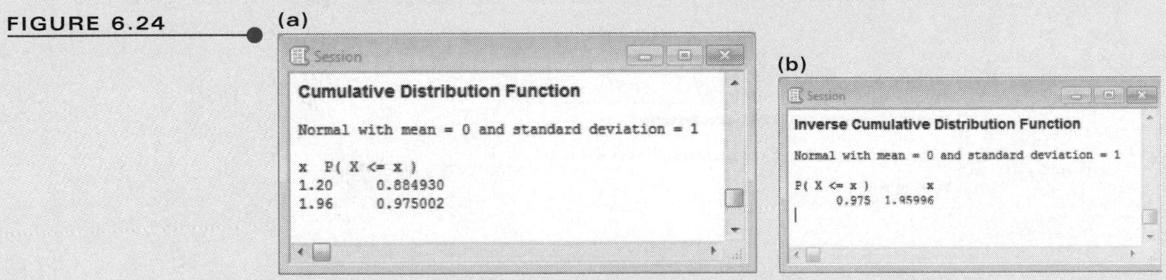

3. To find $P(1.2 < z < 1.96)$, remember that the cumulative probability is the area to the left of the given value of z. Hence,

$$P(1.2 < z < 1.96) = P(z < 1.96) - P(z < 1.2)$$
$$= 0.975002 - 0.884930 = 0.090072$$

You can check this calculation using Table 3 in Appendix I if you wish!

4. To calculate inverse cumulative probabilities, select **Calc ▶ Probability Distributions ▶ Normal**, and click the radio button marked "Inverse cumulative probability," shown in Figure 6.23. We need a value z_0 with area 0.025 to its right, or area 0.975 to its left. Enter **.975** in the box marked "Input constant" and click **OK.** The value of z_0 will appear in the Session window, shown in Figure 6.24(b). This value, when rounded to two decimal places, is the familiar $z_0 = 1.96$ used in Example 6.8.

EXAMPLE 6.19

Suppose that the average birth weights of babies born at hospitals owned by a major health maintenance organization (HMO) are approximately normal with mean 6.75 pounds and standard deviation 0.54 pounds. What proportion of babies born at these hospitals weigh between 6 and 7 pounds? Find the 95th percentile of these birth weights.

1. Enter the two values for x (6 and 7) in the first two cells of column C1. Proceed as in Example 6.17, again selecting **Calc ▶ Probability Distributions ▶ Normal.** This time, enter the values for the mean ($\mu = 6.75$) and standard deviation ($\sigma = 0.54$) in the appropriate boxes, and select column C1 ("x") for the Input column. Make sure that the radio button marked "Cumulative probability" is selected and click **OK.** In the Session window, you will see that $P(X \leq 7) = 0.678305$ and $P(X \leq 6) = 0.082433$.

2. Finally, use the values calculated by *MINITAB* to calculate

$$P(6 < X < 7) = P(X < 7) - P(X < 6) = 0.678305 - 0.082433 = 0.595872$$

3. To calculate the 95th percentile, selecting **Calc ▶ Probability Distributions ▶ Normal,** enter the values for the mean ($\mu = 6.75$) and standard deviation ($\sigma = 0.54$) in the appropriate boxes, and make sure that the radio button marked "Inverse cumulative probability" is selected. We need a value x_0 with area 0.95 to its left. Enter **.95** in the box marked "Input constant" and click **OK.** In the Session window, you will see the 95th percentile, as shown in Figure 6.25.

FIGURE 6.25

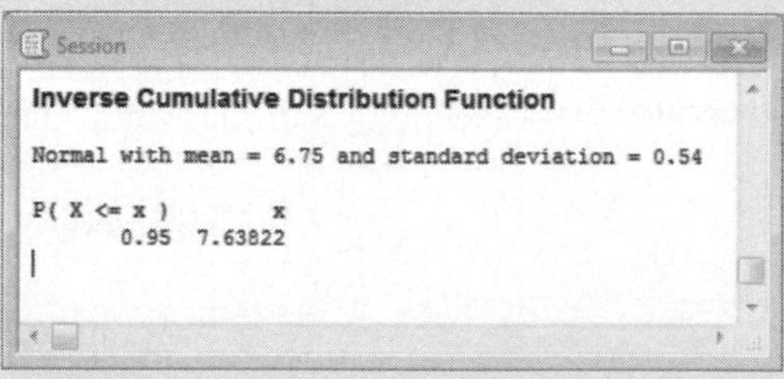

That is, 95% of all babies born at these hospitals weigh 7.63822 pounds or less. Would you consider a baby who weighs 9 pounds to be unusually large?

Supplementary Exercises

6.53 Calculate the area under the standard normal curve to the left of these values:
a. $z=-0.90$ **b.** $z=2.34$ **c.** $z=5.4$

6.54 Calculate the area under the standard normal curve between these values:
a. $z=-2.0$ and $z=2.00$ **b.** $z=-2.3$ and $z=-1.5$

6.55 Find the following probabilities for the standard normal random variable z:
a. $P(-1.96 \leq z \leq 1.96)$ **b.** $P(z > 1.96)$
c. $P(z < -1.96)$

6.56 Find a z_0 such that
a. $P(z > z_0) = 0.9750$ **b.** $P(z > z_0) = 0.3594$

6.57 Find a z_0 such that
a. $P(-z_0 \leq z \leq z_0) = 0.95$ **b.** $P(-z_0 \leq z \leq z_0) = 0.98$
c. $P(-z_0 \leq z \leq z_0) = 0.90$ **d.** $P(-z_0 \leq z \leq z_0) = 0.99$

6.58 A normal random variable x has mean $\mu = 5$ and standard deviation $\sigma = 2$. Find the probabilities associated with the following intervals:
a. $1.2 < X < 10$ **b.** $X > 7.5$ **c.** $X \leq 0$

6.59 Let X be a binomial random variable with $n = 36$ and $p = 0.54$. Use the normal approximation to find:
a. $P(X \leq 25)$ **b.** $P(15 \leq X \leq 20)$ **c.** $P(X > 30)$

6.60 Using Table 3 in Appendix I, calculate the area under the standard normal curve to the left of the following:
a. $z = 1.2$ **b.** $z = -0.99$
c. $z = 1.46$ **d.** $z = -0.42$

6.61 Find the following probabilities for the standard normal random variable:
a. $P(0.3 < z < 1.56)$ **b.** $P(-0.2 < z < 0.2)$

6.62
a. Find the probability that z is greater than -0.75.
b. Find the probability that z is less than 1.35.

6.63 Find z_0 such that $P(z > z_0) = 0.5$.

6.64 Find the probability that z lies between $z = -1.48$ and $z = 1.48$.

6.65 Find z_0 such that $P(-z_0 < z < z_0) = 0.5$. What percentiles do $-z_0$ and z_0 represent?

6.66 Drill Bits It is estimated that the mean life span of oil-drilling bits is 75 hours. Suppose an oil exploration company purchases drill bits that have a life span that is approximately normally distributed with a mean equal to 75 hours and a standard deviation equal to 12 hours.
a. What proportion of the company's drill bits will fail before 60 hours of use?
b. What proportion will last at least 60 hours?
c. What proportion will have to be replaced after more than 90 hours of use?

6.67 Faculty Ages The influx of new ideas into a university, introduced primarily by new young faculty, is becoming a matter of concern because of the increasing ages of faculty members; that is, the distribution of faculty ages is shifting upward due most likely to a shortage of vacant positions and an oversupply of PhDs. Thus, faculty members are reluctant to move and give up a secure position. If the retirement age at most universities is 65, would you expect the distribution of faculty ages to be normal? Explain.

6.68 Bearing Diameters A machine operation produces bearings whose diameters are normally distributed, with mean and standard deviation equal to 1.265 and 0.005, respectively. If specifications require that the bearing diameter equal 1.270 cm $\pm$ 0.101 cm, what fraction of the production will be unacceptable?

6.69 Used Cars A used-car dealership has found that the length of time before a major repair is required on the cars it sells is normally distributed, with a mean equal to 10 months and a standard deviation of 3 months. If the dealer wants only 5% of the cars to fail before the end of the guarantee period, for how many months should the cars be guaranteed?

6.70 Restaurant Sales The daily sales total (excepting Saturday) at a small restaurant has a probability distribution that is approximately normal, with a mean μ equal to \$1230 per day and a standard deviation σ equal to \$120.
a. What is the probability that the sales will exceed \$1400 for a given day?
b. The restaurant must have at least \$1000 in sales per day to break even. What is the probability that on a given day the restaurant will not break even?

6.71 Washers The life span of a type of automatic washer is approximately normally distributed, with mean and standard deviation equal to 3.1 and 1.2 years, respectively. If this type of washer is guaranteed for 1 year, what fraction of original sales will require replacement?

6.72 Garage Door Openers Most users of automatic garage door openers activate their openers at distances that are normally distributed, with a mean of 10 m and a standard deviation of 2.7 m. To minimize interference with other remote-controlled devices, the manufacturer is required to limit the operating distance to 15 m. What percentage of the time will users attempt to operate the opener outside its operating limit?

6.73 How Long Is the Test? The average length of time required to complete a college achievement test was found to equal 70 minutes, with a standard deviation of 12 minutes. When should the test be terminated if you wish to allow sufficient time for 90% of the students to complete the test? (Assume that the time required to complete the test is normally distributed.)

6.74 Servicing Automobiles The length of time required for the periodic maintenance of an automobile will usually have a probability distribution that is mound-shaped and, because some long service times will occur occasionally, is skewed to the right. The length of time required to run a 8000 km check and to service an automobile has a mean equal to 1.4 hours and a standard deviation of 0.7 hour. Suppose that the service department plans to service 50 automobiles per 8-hour day and that, in order to do so, it must spend no more than an average of 1.6 hours per automobile. What proportion of all days will the service department have to work overtime?

6.75 TV Viewers An advertising agency has stated that 20% of all television viewers watch a particular program. In a random sample of 1000 viewers, $x = 184$ viewers were watching the program. Do these data present sufficient evidence to contradict the advertiser's claim?

6.76 Forecasting Earnings A researcher notes that senior corporation executives are not very accurate forecasters of their own annual earnings. He states that his studies of a large number of company executive forecasts "showed that the average estimate missed the mark by 15%."

a. Suppose the distribution of these forecast errors has a mean of 15% and a standard deviation of 10%. Is it likely that the distribution of forecast errors is approximately normal?

b. Suppose the probability is 0.5 that a corporate executive's forecast error exceeds 15%. If you were to sample the forecasts of 100 corporate executives, what is the probability that more than 60 would be in error by more than 15%?

6.77 Filling Pop Cups A pop machine can be regulated to discharge an average of μ millilitres per cup. If the millilitres of fill are normally distributed, with standard deviation equal to 10 mL, give the setting for μ so that 250 mL cups will overflow only 1% of the time.

6.78 Light Bulbs A manufacturing plant uses 3000 electric light bulbs whose life spans are normally distributed, with mean and standard deviation equal to 500 and 50 hours, respectively. In order to minimize the number of bulbs that burn out during operating hours, all the bulbs are replaced after a given period of operation. How often should the bulbs be replaced if we wish no more than 1% of the bulbs to burn out between replacement periods?

6.79 The First-Year Class The admissions office of a small university is asked to accept deposits from a number of qualified prospective first-year students so that, with probability about 0.95, the size of the first-year class will be less than or equal to 120. Suppose the applicants constitute a random sample from a population of applicants, 80% of whom would actually enter the first-year class if accepted.

a. How many deposits should the admissions counsellor accept?

b. If applicants in the number determined in part a are accepted, what is the probability that the first-year class size will be less than 105?

6.80 No Shows, again An airline finds that 5% of the persons making reservations on a certain flight will not show up for the flight. If the airline sells 160 tickets for a flight that has only 155 seats, what is the probability that a seat will be available for every person holding a reservation and planning to fly?

6.81 Long Distance It is known that 30% of all calls coming into a telephone exchange are long-distance calls. If 200 calls come into the exchange, what is the probability that at least 50 will be long-distance calls?

6.82 Plant Genetics In Exercise 5.76, a cross between two peony plants—one with red petals and one with streaky petals—produced offspring plants with red petals 75% of the time. Suppose that 100 seeds from this cross were collected and germinated, and x, the number of plants with red petals, was recorded.

a. What is the exact probability distribution for x?

b. Is it appropriate to approximate the distribution in part a using the normal distribution? Explain.

c. Use an appropriate method to find the approximate probability that between 70 and 80 (inclusive) offspring plants have red flowers.

d. What is the probability that 53 or fewer offspring plants had red flowers? Is this an unusual occurrence?

e. If you actually observed 53 of 100 offspring plants with red flowers, and if you were certain that the genetic ratio 3:1 was correct, what other explanation could you give for this unusual occurrence?

6.83 Suppliers A or B? A purchaser of electric relays buys from two suppliers, A and B. Supplier A supplies two of every three relays used by the company. If 75 relays are selected at random from those in use by the company, find the probability that at most 48 of these relays come from supplier A. Assume that the company uses a large number of relays.

6.84 Snacking and TV Is television dangerous to your diet? Psychologists believe that excessive eating may be associated with emotional states (being upset or bored) and environmental cues (watching television, reading, and so on). To test this theory, suppose you randomly selected 60 overweight persons and matched them by weight and gender in pairs. For a period of two weeks, one of each pair is required to spend evenings reading novels of interest to him or her. The other member of each pair spends each evening watching television. The calorie count for all snack and drink intake for the evenings is recorded for each person, and you record $x = 19$, the number of pairs for which the television watchers' calorie intake exceeded the intake of the readers. If there is no difference in the effects of television and reading on calorie intake, the probability p that the calorie intake of one member of a pair exceeds that of the other member is 0.5. Do these data provide sufficient evidence to indicate a difference between the effects of television watching and reading on calorie intake? (HINT: Calculate the z-score for the observed value, $x = 19$.)

6.85 Gestation Times *The Biology Data Book* reports that the gestation time for human babies averages 278 days with a standard deviation of 12 days.[7] Suppose that these gestation times are normally distributed.

a. Find the upper and lower quartiles for the gestation times.

b. Would it be unusual to deliver a baby after only six months of gestation? Explain.

6.86 Tax Audits In Exercise 6.28 we suggested that the Canada Revenue Agency assign auditing rates per province by randomly selecting 50 auditing percentages from a normal distribution with a mean equal to 1.55% and a standard deviation of 0.45%.

a. What is the probability that a particular province would have more than 2% of its tax returns audited?

b. What is the expected value of x, the number of provinces that will have more than 2% of their income tax returns audited?

c. Is it likely that as many as 4 of the 13 provinces will have more than 2% of their income tax returns audited?

6.87 Your Favourite Sport There is a difference in sports preferences between men and women, according to a recent survey. Among the 10 most popular sports, men include competition-type sports—pool and billiards, basketball, and softball—whereas women include aerobics, running, hiking, and calisthenics. However, the top recreational activity for men was still the relaxing sport of fishing, with 41% of those surveyed indicating that they had fished during the year. Suppose 180 randomly selected men are asked whether they had fished in the past year.

a. What is the probability that fewer than 50 had fished?

b. What is the probability that between 50 and 75 had fished?

c. If the 180 men selected for the interview were selected by the marketing department of a sporting-goods company based on information obtained from their mailing lists, what would you conclude about the reliability of their survey results?

6.88 Introvert or Extrovert? A psychological introvert–extrovert test produced scores that had a normal distribution with a mean and standard deviation of 75 and 12, respectively. If we wish to designate the *highest* 15% as extroverts, what would be the proper score to choose as the cutoff point?

6.89 Normal Temperatures In Exercise 1.66, Allen Shoemaker derived a distribution of human body temperatures, which has a distinct mound-shape.[8] Suppose we assume that the temperatures of healthy humans is approximately normal with a mean of 37.1 °C and a standard deviation of 0.2 degrees.

a. If a healthy person is selected at random, what is the probability that the person has a temperature above 37.22 °C?

b. What is the 95th percentile for the body temperatures of healthy humans?

6.90 Stamps Philatelists (stamp collectors) often buy stamps at or near retail prices, but, when they sell, the price is considerably lower. For example, it may be reasonable to assume that (depending on the mix of a collection, condition, demand, economic conditions, etc.) a collection will sell at $x\%$ of the retail price, where X is normally distributed with a mean equal to 45% and a standard deviation of 4.5%. If a philatelist has a collection to sell that has a retail value of \$30,000, what is the probability that the philatelist receives these amounts for the collection?

a. More than \$15,000 **b.** Less than \$15,000

c. Less than \$12,000

6.91 Test Scores The scores on a national achievement test were approximately normally distributed, with a mean of 540 and a standard deviation of 110.

a. If you achieved a score of 680, how far, in standard deviations, did your score depart from the mean?

b. What percentage of those who took the examination scored higher than you?

6.92 Transplanting Cells Briggs and King developed the technique of nuclear transplantation, in which the nucleus of a cell from one of the later stages of the development of an embryo is transplanted into a zygote (a single-cell fertilized egg) to see whether the nucleus can support normal development. If the probability that a single transplant from the early gastrula stage will be successful is 0.65, what is the probability that more than 70 transplants out of 100 will be successful?

CASE STUDY  "Are You Going to Curve the Grades?"

Very often, at the end of an exam that seemed particularly difficult, students will ask the professor, "Are you going to curve the grades?" Unfortunately, "curving the grades" doesn't necessarily mean that you will receive a higher grade on a test, although you might like to think so! Curving grades is actually a technique whereby a fixed proportion of the highest grades receive As (even if the highest grade is a failing grade on a percentage basis), and a fixed proportion of the lowest grades receive Fs (even if the lowest score is a passing grade on a percentage basis). The B, C, and D grades are also assigned according to fixed proportion. One such allocation uses the following proportions:

Letter grade	A	B	C	D	F
Proportion of grade	10%	20%	40%	20%	10%

1. If the average C grade is centred on the average grade for all students, and if we assume that the grades are normally distributed, how many standard deviations on each side of the mean will designate the C grades?
2. How many standard deviations on either side of the mean will be the cutoff points for the B and D grades?

A histogram of the grades for an introductory statistics class together with summary statistics follows.

Descriptive Statistics: Grades

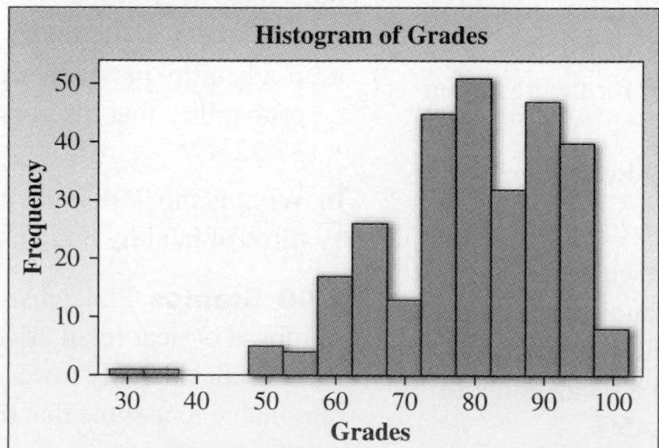

Variable	N	N*	Mean	StDev	Minimum	Q1	Median	Q3	Maximum
Grades	290	0	79.972	12.271	31.000	73.000	82.000	88.000	100.000

For ease of calculation, round the number of standard deviations for C grades to ±0.5 standard deviations and for B and D grades to ±1.5 standard deviations.

3. Find the cutoff points for A, B, C, D, and F grades corresponding to these rounded values.
4. If you had a score of 92 on the exam and you had the choice of curving the grades or using the absolute standard of 90–100 for an A, 80–89 for a B, 70–79 for a C, and so on, what would be your choice? Explain your reasoning. Is the skewness of the distribution of grades a problem?

PROJECTS

Project 6-A: The Spectrum of Prematurity[9]

Moderate prematurity refers to babies who are born between 28 and 32 completed weeks gestational age with a birth weight range between 1000 and 1500 grams.

The length of time a baby has spent in the womb, or more specifically the number of completed weeks of pregnancy, is called the gestational age. Based on their gestational age and birth weight, premature babies are placed into categories of mild, moderate, and extreme prematurity.

- *Mild prematurity* refers to babies who are born between 33 and 36 completed weeks of gestation and/or have a birth weight between 1500 and 2500 grams.
- *Moderate prematurity* refers to babies who are born between 28 and 32 completed weeks of gestation with a birth weight range between 1000 and 1500 grams.

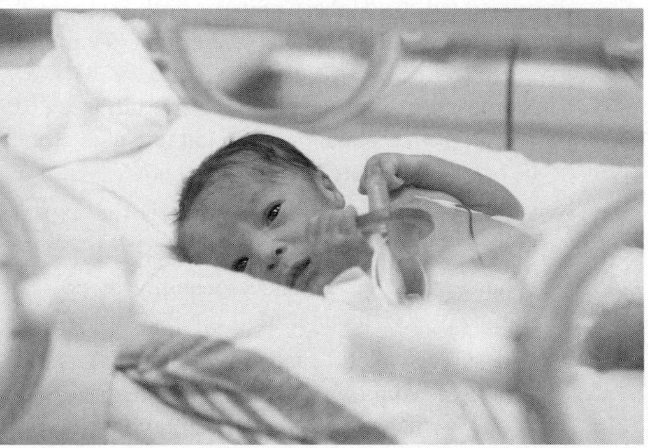

Fuse/Thinkstock

- *Extreme prematurity* refers to babies who are born before 28 completed weeks of gestation or have a birth weight of less than 1000 grams.

a. Generally speaking, the gestational time for human babies is approximately normally distributed with an average of 40 weeks and a standard deviation of 2 weeks.
   (i) Calculate the probability of having a birth with mild prematurity.
   (ii) What is the probability of having a birth with extreme prematurity?
   (iii) Find the upper and lower quartiles for the gestation times.

(iv) Would it be unusual to deliver a baby after only 24 weeks of gestation? Explain.

(v) A randomly selected baby would be an age of less than X weeks to be one of the bottom 20% in gestational age. What is the value of x?

(vi) Before what gestational time does 83.4% of the gestational time occur?

b. The birth weight of a baby is approximately normally distributed with an average of 3.4 kg and a standard deviation of 800 grams.

(i) Calculate the probability of having a birth with moderate prematurity.

(ii) What is the probability of having a birth with extreme prematurity?

(iii) What is the probability of having a baby weighing at least 6 kg? Do you think it is highly unlikely to have a baby with this weight? Explain.

(iv) A randomly selected baby will weigh more than X kg to be one of the top 5% in weight. What is the value of x?

(v) Above what weight do 87.70% of the weights occur?

(vi) Suppose that on another planet the baby (may not be human) birth weight X follows the normal distribution. The probability that X exceeds 4 kg is 0.975, and the probability that X exceeds 5 kg is 0.95. Find μ and σ. Are these values comparable with the expected weight of a human baby? Explain.

Project 6-B: Premature Babies in Canada[10]

In 2006–2007, the Canadian in-hospital preterm birth (prematurity, that is, birth before 37 weeks of gestation) and SGA (small for their gestational age) rates were approximately 8.1% and 8.3%, respectively, accounting for more than 54,000 live births combined. Among the provinces, Alberta and Newfoundland and Labrador demonstrated the highest preterm birth rates, at 8.7% and 8.6%, respectively. The highest provincial SGA rates occurred in Ontario (8.9%) and Alberta (8.7%).

a. Suppose a random sample of 25 live births is taken in Newfoundland and Labrador. Would the normal approximation for some probability calculation for preterm birth be appropriate here? Why or why not?

b. Based on this information, what is the expected number of SGA births if we randomly sample 200 births in Alberta?

c. What is the approximate probability of 60 or more SGA births from a random sample of 200 in Ontario?

d. If a survey of 500 randomly selected Canadian in-hospital births is conducted, what is the approximate probability that at most 50 of these births will be declared SGA? Further, what is the probability that more than 10% of the sampled births will be termed SGA?

e. Suppose a survey of 25 randomly selected Canadian in-hospital births is conducted. Calculate the exact and approximate probability that at least one of these births will be termed SGA. Compare the results. How close or far was your approximation? Explain.

Nata789/Shutterstock.com

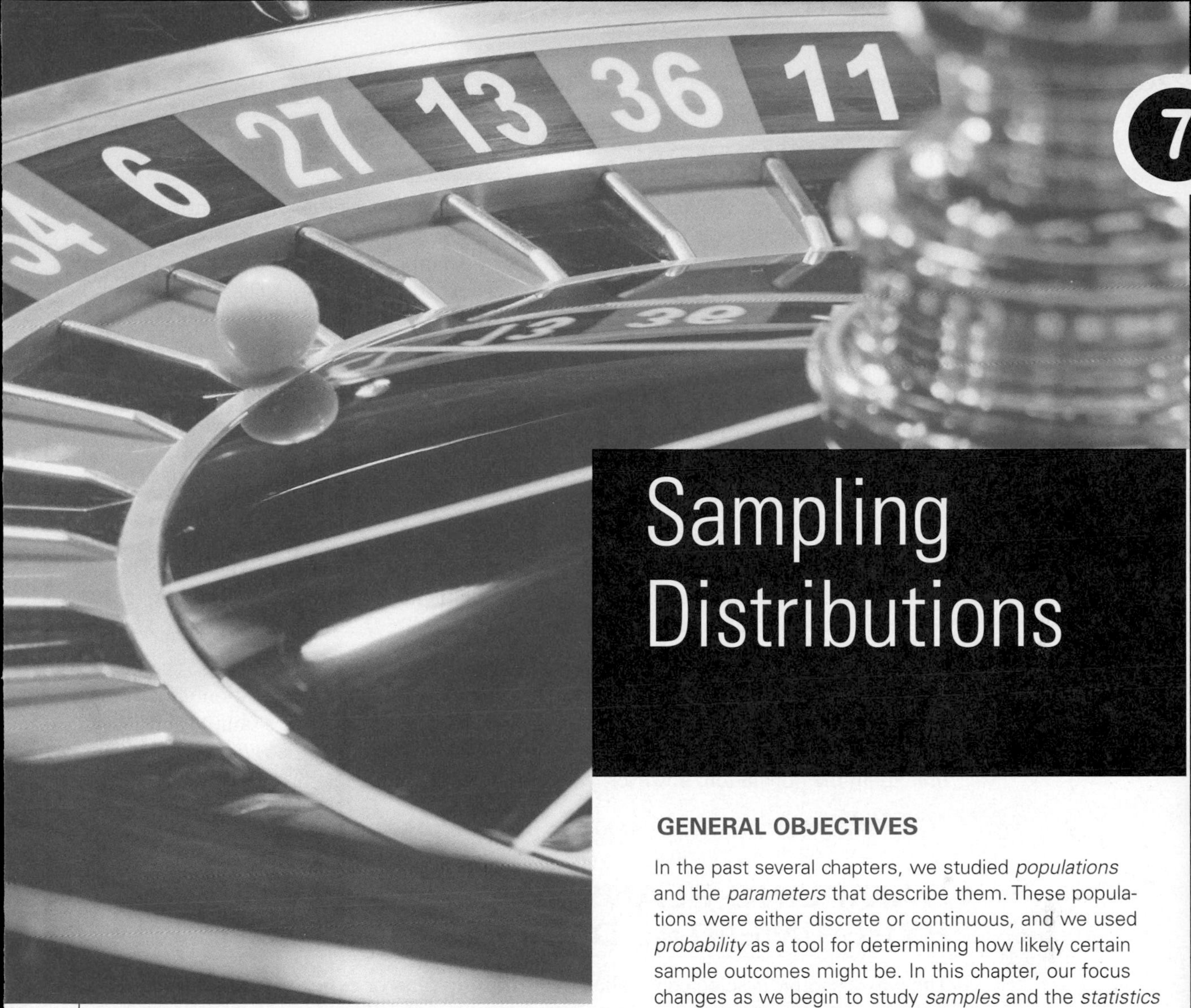

7

Sampling Distributions

GENERAL OBJECTIVES

In the past several chapters, we studied *populations* and the *parameters* that describe them. These populations were either discrete or continuous, and we used *probability* as a tool for determining how likely certain sample outcomes might be. In this chapter, our focus changes as we begin to study *samples* and the *statistics* that describe them. These sample statistics are used to make inferences about the corresponding population parameters. This chapter involves sampling and sampling distributions, which describe the behaviour of sample statistics in repeated sampling.

CHAPTER INDEX

- Random samples (7.2)
- Sampling plans and experimental designs (7.2)
- Statistics and sampling distributions (7.3)
- The Central Limit Theorem (7.4)
- The sampling distribution of the sample mean, $\bar{x}$ (7.5)
- The sampling distribution of the sample proportion, $\hat{p}$ (7.6)
- Statistical process control: $\bar{x}$ and p charts (7.7)

Sampling the Roulette at Monte Carlo

How would you like to try your hand at gambling without the risk of losing? You could do it by simulating the gambling process, making imaginary bets, and observing the results. This technique, called a Monte Carlo procedure, is the topic of the case study at the end of this chapter.

NEED TO KNOW

When the Sample Size Is Large Enough to Use the Central Limit Theorem

How to Calculate Probabilities for the Sample Mean $\bar{X}$

How to Calculate Probabilities for the Sample Proportion $\hat{p}$

INTRODUCTION

NEED A TIP?

Parameter ⇔ Population
Statistic ⇔ Sample

In the previous three chapters, you have learned a lot about probability distributions, such as the binomial and normal distributions. The shape of the normal distribution is determined by its mean μ and its standard deviation σ, whereas the shape of the binomial distribution is determined by the success probability p. These numerical descriptive measures—called **parameters**—are needed to calculate the probability of observing sample results.

In practical situations, you may be able to decide which *type* of probability distribution to use as a model, but the values of the *parameters* that specify its *exact form* are unknown. Here are two examples:

- A pollster is sure that the responses to his "agree/disagree" questions will follow a binomial distribution, but p, the proportion of those who "agree" in the population, is unknown.
- An agronomist believes that the yield per hectare of a variety of wheat is approximately normally distributed, but the mean μ and standard deviation σ of the yields are unknown.

In these cases, you must rely on the *sample* to learn about these parameters. The proportion of those who "agree" in the pollster's sample provides information about the actual value of p. The mean and standard deviation of the agronomist's sample approximate the actual values of μ and σ. If you want the sample to provide *reliable information* about the population, however, you must select your sample in a certain way!

7.2

SAMPLING PLANS AND EXPERIMENTAL DESIGNS

The way a sample is selected is called the **sampling plan** or **experimental design,** and it determines the quantity of information in the sample. Knowing the sampling plan used in a particular situation will often allow you to measure the reliability or goodness of your inference.

Simple random sampling is a commonly used sampling plan in which every sample of size n has the same chance of being selected. For example, suppose you want to select a sample of size $n = 2$ from a population containing $N = 4$ objects. If the four objects are identified by the symbols x_1, x_2, x_3, and x_4, there are six distinct pairs that could be selected, as listed in Table 7.1. If the sample of $n = 2$ observations is selected so that each of these six samples has the same chance of selection, given by 1/6, then the resulting sample is called a **simple random sample,** or just a **random sample.**

TABLE 7.1 Ways of Selecting a Sample of Size 2 from 4 Objects

Sample	Observations in Sample
1	x_1, x_2
2	x_1, x_3
3	x_1, x_4
4	x_2, x_3
5	x_2, x_4
6	x_3, x_4

DEFINITION If a sample of n elements is selected from a population of N elements using a sampling plan in which each of the possible samples has the same chance of selection, then the sampling is said to be **random** and the resulting sample is a **simple random sample.**

Perfect random sampling is difficult to achieve in practice. If the size of the population N is small, you might write each of N numbers on a poker chip, mix the chips, and select a sample of n chips. The numbers that you select correspond to the n measurements that appear in the sample. Since this method is not always very practical, a simpler and more reliable method uses **random numbers**—digits generated so that the values 0 to 9 occur randomly and with equal frequency. These numbers can be generated by computer or may even be available on your scientific calculator. Alternatively, Table 10 in Appendix I is a table of random numbers that you can use to select a *random sample.*

EXAMPLE 7.1

A computer database at a downtown law firm contains files for $N = 1000$ clients. The firm wants to select $n = 5$ files for review. Select a simple random sample of five files from this database.

SOLUTION You must first label each file with a number from 1 to 1000. Perhaps the files are stored alphabetically, and the computer has already assigned a number to each. Then generate a sequence of ten three-digit random numbers. If you are using Table 10 of Appendix I, select a random starting point and use a portion of the table similar to the one shown in Table 7.2. The random starting point ensures that you will not use the same sequence over and over again. The first three digits of Table 7.2 indicate the number of the first file to be reviewed. The random number 001 corresponds to file #1, and the last file, #1000, corresponds to the random number 000 (The file 1000 is labelled as "000."). Using Table 7.2, you would choose the five files numbered 155, 450, 32, 882, and 350 for review. Alternately, you might choose to read across the lines, and choose files 155, 350, 989, 450, and 369 for review.

TABLE 7.2 **Portion of a Table of Random Numbers**

15574	35026	98924
45045	36933	28630
03225	78812	50856
88292	26053	21121

The situation described in Example 7.1 is called an **observational study** because the data already existed before you decided to *observe* or describe their characteristics. Most sample surveys, in which information is gathered with a questionnaire, fall into this category. Computer databases make it possible to assign identification numbers to each element even when the population is large and to select a simple random sample. You must be careful when conducting a *sample survey,* however, to watch for these frequently occurring problems:

- **Non-response:** You have carefully selected your random sample and sent out your questionnaires, but only 50% of those surveyed return their questionnaires. Are the responses you received still representative of the entire population, or are they **biased** because only those people who were particularly opinionated about the subject chose to respond?

- **Undercoverage:** You have selected your random sample using telephone records as a database. Does the database you used systematically exclude certain segments of the population—perhaps those who do not have telephones?
- **Wording bias:** Your questionnaire may have questions that are too complicated or tend to confuse the reader. Possibly the questions are sensitive in nature—for example, "Have you ever used drugs?" or "Have you ever cheated on your income tax?"—and the respondents will not answer truthfully.

Methods have been devised to solve some of these problems, but only if you know that they exist. If your survey is *biased* by any of these problems, then your conclusions will not be very reliable, even though you did select a random sample!

Some research involves **experimentation,** in which an experimental condition or *treatment* is imposed on the *experimental units.* Selecting a simple random sample is more difficult in this situation.

EXAMPLE 7.2

A research chemist is testing a new method for measuring the amount of titanium (Ti) in ore samples. She chooses 10 ore samples of the same weight for her experiment. Five of the samples will be measured using a standard method, and the other five using the new method. Use random numbers to assign the 10 ore samples to the new and standard groups. Do these data represent a simple random sample from the population?

SOLUTION There are really two populations in this experiment. They consist of titanium measurements, using either the new or standard method, for *all possible* ore samples of this weight. These populations do not exist in fact; they are **hypothetical populations,** envisioned in the mind of the researcher. Thus, it is impossible to select a simple random sample using the methods of Example 7.1. Instead, the researcher selects what she believes are 10 *representative* ore samples and hopes that these samples will *behave as if* they had been randomly selected from the two populations.

The researcher can, however, randomly select the five samples to be measured with each method. Number the samples from 1 to 10. The five samples selected for the new method may correspond to five one-digit random numbers. Use this sequence of random digits generated on a scientific calculator:

948247817184610

Since you cannot select the same ore sample twice, you must skip any digit that has already been chosen. Ore samples 9, 4, 8, 2, and 7 will be measured using the new method. The other samples—1, 3, 5, 6, and 10—will be measured using the standard method.

In addition to *simple random sampling,* there are other sampling plans that involve randomization and therefore provide a probabilistic basis for inference making. Three such plans are based on *stratified, cluster,* and *systematic sampling.*

When the population consists of two or more subpopulations, called **strata,** a sampling plan that ensures that each subpopulation is represented in the sample is called a **stratified random sample.**

DEFINITION **Stratified random sampling** involves selecting a simple random sample from each of a given number of subpopulations, or **strata.**

Citizens' opinions about the construction of a performing arts centre could be collected using a stratified random sample with city voting wards as strata. National polls usually involve some form of stratified random sampling with provinces as strata.

A stratified random sample is often selected by taking individual or separate random samples from every strata in such a way that the size of the random samples may vary with the importance of different strata. A stratified sample can be relatively more efficient than a simple random sample. A stratified random sample is useful in the following situations:

- The sampled population contains two or more mutually exclusive and clearly distinguishable subpopulations or strata that are intentionally highly homogenous.
- The sampled population contains two or more mutually exclusive strata or strata that differ greatly from one another with respect to some characteristic of interest.

Keep in mind that a stratified random sample can be chosen only when a complete list of the population to be sampled is readily available to researchers.

Another form of random sampling is used when the available sampling units are groups of elements, called **clusters.** For example, a household is a *cluster* of individuals living together. A city block or a neighbourhood might be a convenient sampling unit and might be considered a *cluster* for a given sampling plan.

DEFINITION A **cluster sample** is a simple random sample of clusters from the available clusters in the population.

When a particular cluster is included in the sample, a census of every element in the cluster is taken. Recall that in a stratified sample, a simple random sample is taken from each of the strata. Cluster sampling is usually less precise than a simple random sample or stratified random sample. Cluster sampling is, however, more cost effective in many cases. For that reason, cluster sampling is preferred when it can be economically justified. The cluster sample is used when the population consists of natural clusters, such as city blocks, subdivisions, and school districts.

Sometimes the population to be sampled is ordered, such as an alphabetized list of people with driver's licences, a list of utility users arranged by service addresses, or a list of customers by account numbers. In these and other situations, one element is chosen at random from the first k elements, and then every kth element thereafter is included in the sample.

DEFINITION A **1-in-k systematic random sample** involves the random selection of one of the first k elements in an ordered population, and then the systematic selection of every kth element thereafter.

NEED A TIP?

All sampling plans used for making inferences must involve *randomization*!

Not all sampling plans, however, involve random selection. You have probably heard of the non-random telephone polls in which those people who wish to express support for a question call one "900 number" and those opposed call a second "900 number." Each person must pay for his or her call. It is obvious that those people who call do not represent the population at large. This type of sampling plan is one form of a **convenience sample**—a sample that can be easily and simply obtained without random selection. Advertising for subjects who will be paid a fee for participating in an experiment produces a convenience sample. **Judgment sampling** allows the sampler to decide who will or will not be included in the sample. **Quota sampling,** in which the makeup of the sample must reflect the makeup of the population on some preselected characteristic, often has a non-random component in the selection process. **Remember that non-random samples can be described but cannot be used for making inferences!**

EXAMPLE 7.3

Identify the sampling design for each of the following:

1. The faculty of mathematics and science at Brock University consists of six academic departments and two centres. Dr. S. Ejaz Ahmed, the academic dean of the faculty, decides which research proposals submitted to him by each department and centre will be submitted for Canada Foundation for Innovation funding consideration.
2. Every student in an introductory biology class at the University of Regina is assigned a number. The professor then randomly selects five numbers and interviews the sampled students.
3. Class rosters list student names in alphabetical order. A professor wants to select some students to participate in a research project. The professor randomly selects a student from the list and then selects every fifth student thereafter.
4. An apartment complex executive in Toronto randomly selects five buildings from the complex of 25 buildings, and then interviews all tenants from every apartment in the five selected buildings.
5. The alumni association of an Ontarian university is comprised of 85% in-province residents and 15% out-of-province residents. The executive of the association wishes to contact 100 alumni from the university. However, they were instructed to contact 70 in-province alumni and 30 out-of-province alumni.
6. The list of students in a Niagara Region high school was divided by grade. A simple random sample of students was selected from each grade. Each student selected for the sample was asked how much time she or he spends for math homework per week, on average.
7. The list of students in a Niagara Region high school was divided by grade. A systematic random sample of students was selected from each grade. Each student selected for the sample was asked how much time he or she spends for math homework per week, on average.

SOLUTION

1. Judgment sampling
2. Simple random sampling
3. 1-in-5 systematic sampling
4. Cluster sampling
5. Quota sampling
6. Stratified sampling
7. Stratified sampling

7.2 EXERCISES

BASIC TECHNIQUES

7.1 A population consists of $N = 500$ experimental units. Use a random number table to select a random sample of $n = 20$ experimental units. (HINT: Since you need to use three-digit numbers, you can assign two three-digit numbers to each of the sampling units in the manner shown in the table.) What is the probability that each experimental unit is selected for inclusion in the sample?

Experimental Units	Random Numbers
1	001, 501
2	002, 502
3	003, 503
4	004, 504
.	.
.	.
.	.
499	499, 999
500	500, 000

7.2 A political analyst wishes to select a sample of $n = 20$ people from a population of 2000. Use the random number table to identify the people to be included in the sample.

7.3 A population contains 50,000 voters. Use the random number table to identify the voters to be included in a random sample of $n = 15$.

7.4 A small city contains 20,000 voters. Use the random number table to identify the voters to be included in a random sample of $n = 15$.

7.5 Every 10th Person A random sample of public opinion in a small town was obtained by selecting every 10th person who passed by the busiest corner in the downtown area. Will this sample have the characteristics of a random sample selected from the town's citizens? Explain.

7.6 Parks and Recreation A questionnaire was mailed to 1000 registered municipal voters selected at random. Only 500 questionnaires were returned, and of the 500 returned, 360 respondents were strongly opposed to a surcharge proposed to support the city Parks and Recreation Department. Are you willing to accept the 72% figure as a valid estimate of the percentage in the city who are opposed to the surcharge? Why or why not?

7.7 MPAC Lists and Jury Selection Juries are selected in Ontario through a process described in the *Juries Act*. Enumeration lists are obtained from the Municipal Property Assessment Corporation (MPAC). These lists are updated only once every four years, and persons at the households contacted by MPAC provide their date of birth and citizenship information. As a result, MPAC lists do not reflect changes in personal circumstances within that time period. The random selection process for juries proceeds as follows: a request for jury selection is made and MPAC selects persons for each district who have indicated by the returns (enumeration form) that they are over 18 years of age and are Canadian citizens. Questionnaires are subsequently mailed out to the randomly selected individuals to determine whether they may be considered as potential jurors.[1] In what ways might this list not cover certain sectors of the population adequately?

7.8 Sex and Violence One question on a survey questionnaire is phrased as follows: "Don't you agree that there is too much sex and violence during prime TV viewing hours?" Comment on possible problems with the responses to this question. Suggest a better way to pose the question.

APPLICATIONS

7.9 Omega-3 Fats Contrary to current thought about omega-3 fatty acids, new research shows that the beneficial fats may not help reduce second heart attacks in heart attack survivors. The study included 4837 men and women being treated for heart disease. The experimental group received an additional 400 mg of the fats daily.[2] Suppose that this experiment was repeated with 50 individuals in the control group and 50 individuals in the experimental group. Determine a randomization scheme to assign the 100 individuals to the two groups.

7.10 Cancer in Rats The *Press Enterprise* identified a byproduct of chlorination called MX that has been linked to cancer in rats.[3] A scientist wants to conduct a validation study using 25 rats in the experimental group, each to receive a fixed dose of MX, and 25 rats in a control group that will receive no MX. Determine a randomization scheme to assign the 50 individual rats to the two groups.

7.11 Racial Bias? Does the race of an interviewer matter? This question was investigated by Chris Gilberg and colleagues and reported in an issue of *Chance* magazine.[4] The interviewer asked, "Do you feel that affirmative action should be used as an occupation selection criteria?" with possible answers of yes or no.

a. What problems might you expect with responses to this question when asked by interviewers of different ethnic origins?

b. When people were interviewed by an African American, the response was about 70% in favour of affirmative action, approximately 35% when interviewed by an Asian, and approximately 25% when interviewed by a Caucasian. Do these results support your answer in part a?

7.12 MRIs In a study described in the *American Journal of Sports Medicine,* Peter D. Franklin and colleagues reported on the accuracy of using magnetic resonance imaging (MRI) to evaluate ligament sprains and tears on 35 patients.[5] Consecutive patients with acute or chronic knee pain were selected from the clinical practice of one of the authors and agreed to participate in the study.

a. Describe the sampling plan used to select study participants.

b. What chance mechanism was used to select this sample of 35 individuals with knee pain?

c. Can valid inferences be made using the results of this study? Why or why not?

d. Devise an alternative sampling plan. What would you change?

7.13 Tai Chi and Fibromyalgia A small new study shows that tai chi, an ancient Chinese practice of exercise and meditation, may relieve symptoms of chronic painful fibromyalgia. The study assigned 66 fibromyalgia patients to take either a 12-week tai chi class or attend a wellness education class.[6]

a. Provide a randomization scheme to assign 66 subjects to the two groups.

b. Will your randomization scheme result in equal-sized groups? Explain.

7.14 Blood Thinner A study of an experimental blood thinner was conducted to determine whether it works better than the simple aspirin tablet in warding off heart attacks and strokes.[7] The study reported in the *Press Enterprise* involved 19,185 people who had suffered heart attacks, strokes, or pain from clogged arteries. Each person was randomly assigned to take either aspirin or the experimental drug for one to three years. Assume that each person was equally likely to be assigned one of the two medications.

a. Devise a randomization plan to assign the medications to the patients.

b. Will there be an equal number of patients in each treatment group? Explain.

7.15 Health Care: Canada Speaks Two different polls were conducted by two different organizations, both of which involved people's feelings about national priorities/important issues. Here is a question from each, along with the response of the sampled Canadians:[8]

National Priorities

Environics Research Group for Research Canada (September 19, 2006, Nationwide, n = 1000, margin of error ± 3.1)

"Thinking about Canada's national priorities, which of the following would you say is important: health care, education, jobs, the environment, national security, or other?"

	Response
Health Care	98%
Education	98%
Jobs	96%

The Strategic Council finding of the Globe and Mail/CTV August Polling Program (August 7, 2005, Nationwide, n = 1000, margin of error ± 3.1)

"In your view what is the most important issue facing Canada today—that is, one about which you are most concerned?"

	Response
Health Care	16%
Other Social Issue (except health care)	12%
Infrastructure/Environmental Issues	9%

a. Read the two poll questions. Which of the two readings is more unbiased? Explain.

b. Look at the responses for the two different polls. How would you explain the large difference in the percentages for the health care category?

7.16 Ask Canada A nationwide policy survey was sent by the Conservative Party Election Committee to voters asking for opinions on a variety of political issues. Here are some questions from the survey:

- In recent years has the federal government grown more or less intrusive in your personal and business affairs?
- Is Prime Minister Harper right in trying to rein in the size and scope of the federal government against the wishes of the big government Liberals?
- Do you agree that the obstructionist NDP should not be allowed to gain control of the Parliament in forthcoming elections?

Comment on the effect of wording bias on the responses gathered using this survey.

7.3 STATISTICS AND SAMPLING DISTRIBUTIONS

When you select a random sample from a population, the numerical descriptive measures you calculate from the sample are called **statistics.** These statistics vary or change for each different random sample you select; that is, they are *random variables.* The probability distributions for statistics are called **sampling distributions** because, in repeated sampling, they provide this information:

- What values of the statistic can occur
- How often each value occurs

DEFINITION The **sampling distribution of a statistic** is the probability distribution for the possible values of the statistic that results when random samples of size n are repeatedly drawn from the population.

There are three ways to find the sampling distribution of a statistic:

1. Derive the distribution *mathematically* using the laws of probability.
2. Use a *simulation* to approximate the distribution. That is, draw a large number of samples of size n, calculating the value of the statistic for each sample, and tabulate the results in a relative frequency histogram. When the number of samples is large, the histogram will be very close to the theoretical sampling distribution.
3. Use *statistical theorems* to derive exact or approximate sampling distributions.

The next example demonstrates how to derive the sampling distributions of two statistics for a very small population.

EXAMPLE 7.4

A population consists of $N = 5$ numbers: 3, 6, 9, 12, 15. If a random sample of size $n = 3$ is selected without replacement, find the sampling distributions for the sample mean $\bar{x}$ and the sample median m.

SOLUTION You are sampling from the population shown in Figure 7.1. It contains five distinct numbers and each is equally likely, with probability $p(x) = 1/5$. You can easily find the population mean and median as

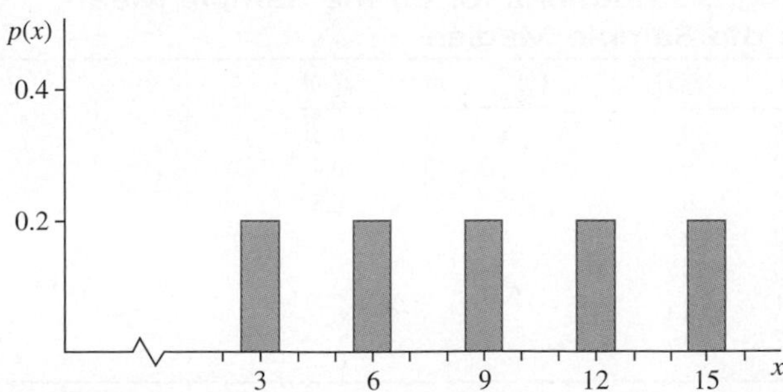

FIGURE 7.1 Probability histogram for the $N = 5$ population values in Example 7.4

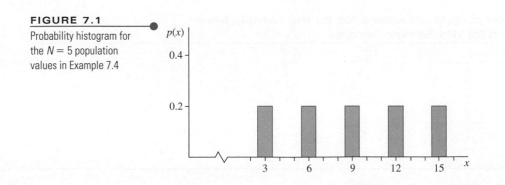

NEED A TIP?

Sampling distributions can be either discrete or continuous.

$$\mu = \frac{3 + 6 + 9 + 12 + 15}{5} = 9 \quad \text{and} \quad M = 9$$

There are 10 possible random samples of size $n = 3$ and each is equally likely, with probability 1/10. These samples, along with the calculated values of $\bar{x}$ and m for each, are listed in Table 7.3. You will notice that some values of $\bar{x}$ are more likely than others because they occur in more than one sample. For example,

$$P(\bar{x} = 8) = \frac{2}{10} = 0.2 \quad \text{and} \quad P(m = 6) = \frac{3}{10} = 0.3$$

The values in Table 7.3 are tabulated, and the sampling distributions for $\bar{x}$ and m are shown in Table 7.4 and Figure 7.2.

Since the population of $N = 5$ values is symmetric about the value $x = 9$, both the *population mean* and the *median* equal 9. It would seem reasonable, therefore, to consider using either $\bar{x}$ or m as a possible estimator of $M = \mu = 9$. Which estimator would you choose? From Table 7.3, you see that, in using m as an estimator, you would be in error by $9 - 6 = 3$ with probability 0.3 or by $9 - 12 = -3$ with probability 0.3. That is, the error in estimation using m would be 3 with probability 0.6. In using $\bar{x}$, however, an error of 3 would occur with probability only 0.2. On these grounds alone, you may wish to use $\bar{x}$ as an estimator in preference to m.

TABLE 7.3 **Values of $\bar{x}$ and m for Simple Random Sampling When $n = 3$ and $N = 5$**

Sample	Sample Values	$\bar{x}$	m
1	3, 6, 9	6	6
2	3, 6, 12	7	6
3	3, 6, 15	8	6
4	3, 9, 12	8	9
5	3, 9, 15	9	9
6	3, 12, 15	10	12
7	6, 9, 12	9	9
8	6, 9, 15	10	9
9	6, 12, 15	11	12
10	9, 12, 15	12	12

TABLE 7.4 **Sampling Distributions for (a) the Sample Mean and (b) the Sample Median**

(a)	$\bar{x}$	$p(\bar{x})$	(b)	m	$p(m)$
	6	0.1		6	0.3
	7	0.1		9	0.4
	8	0.2		12	0.3
	9	0.2			
	10	0.2			
	11	0.1			
	12	0.1			

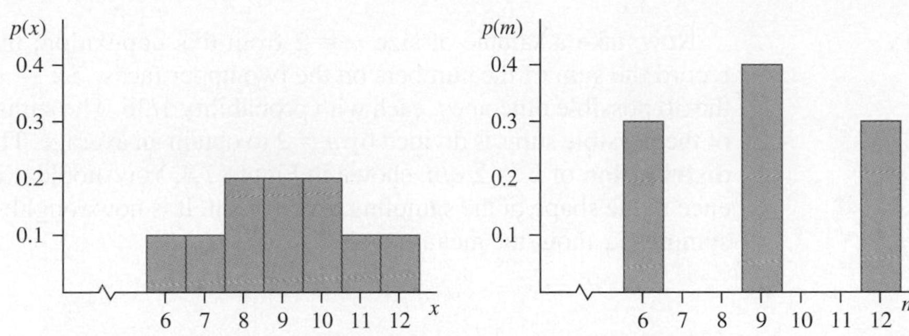

FIGURE 7.2
Probability histograms for the sampling distributions of the sample mean, $\bar{x}$, and the sample median, *m*, in Example 7.4

NEED A TIP?

Almost every statistic has a mean and a standard deviation (or *standard error*) describing its centre and spread.

It was not too difficult to derive these sampling distributions in Example 7.4 because the number of elements in the population was very small. When this is not the case, you may need to use one of these methods:

- Use a simulation to approximate the sampling distribution empirically.
- Rely on statistical theorems and theoretical results.

One important statistical theorem that describes the sampling distribution of statistics that are sums or averages is presented in the next section. This theorem plays a central role in statistical theory and applications since it provides the sampling distributions of average or sum when samples are taken from an infinite population.

7.4 THE CENTRAL LIMIT THEOREM

The **Central Limit Theorem** states that, under rather general conditions, sums and means of random samples of measurements drawn from a population tend to have an approximately normal distribution. Suppose you toss a balanced die $n = 1$ time. The random variable X is the number observed on the upper face. This familiar random variable can take six values, each with probability 1/6, and its probability distribution is shown in Figure 7.3. The shape of the distribution is *flat* or *uniform* and symmetric about the mean $\mu = 3.5$.

FIGURE 7.3
Probability distribution for *x*, the number appearing on a single toss of a die

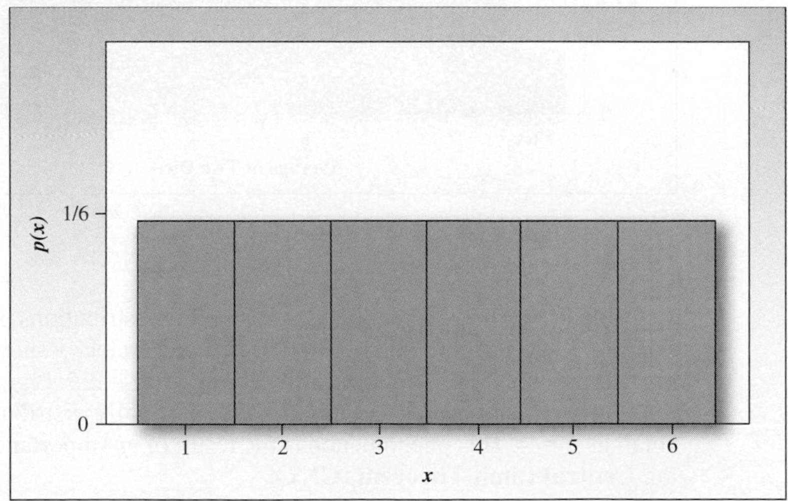

Now, take a sample of size $n = 2$ from this population; that is, toss two dice and record the sum of the numbers on the two upper faces, $\Sigma x_i = x_1 + x_2$. Table 7.5 shows the 36 possible outcomes, each with probability 1/36. The sums are tabulated, and each of the possible sums is divided by $n = 2$ to obtain an average. The result is the **sampling distribution** of $\bar{x} = \Sigma x_i/n$, shown in Figure 7.4. You should notice the dramatic difference in the shape of the sampling distribution. It is now roughly mound-shaped but still symmetric about the mean $\mu = 3.5$.

TABLE 7.5 **Sums of the Upper Faces of Two Dice**

	First Die					
Second Die	**1**	**2**	**3**	**4**	**5**	**6**
1	2	3	4	5	6	7
2	3	4	5	6	7	8
3	4	5	6	7	8	9
4	5	6	7	8	9	10
5	6	7	8	9	10	11
6	7	8	9	10	11	12

FIGURE 7.4
Sampling distribution of $\bar{x}$ for $n = 2$ dice

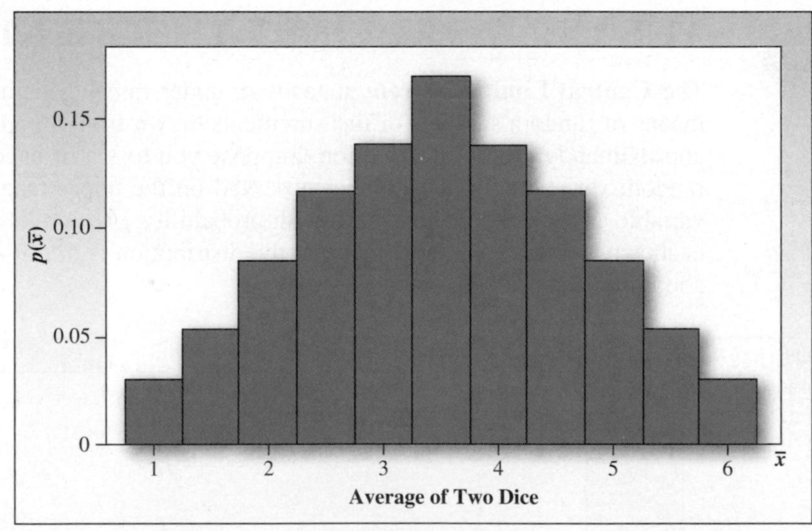

Using *MINITAB*, we generated the sampling distributions of $\bar{x}$ when $n = 3$ and $n = 4$. For $n = 3$, the sampling distribution in Figure 7.5 clearly shows the mound shape of the normal probability distribution, still centred at $\mu = 3.5$. Figure 7.6 dramatically shows that the distribution of $\bar{x}$ is approximately normally distributed based on a sample as small as $n = 4$. This phenomenon is the result of an important statistical theorem called the **Central Limit Theorem (CLT).**

FIGURE 7.5
MINITAB sampling distribution of $\bar{x}$ for $n = 3$ dice

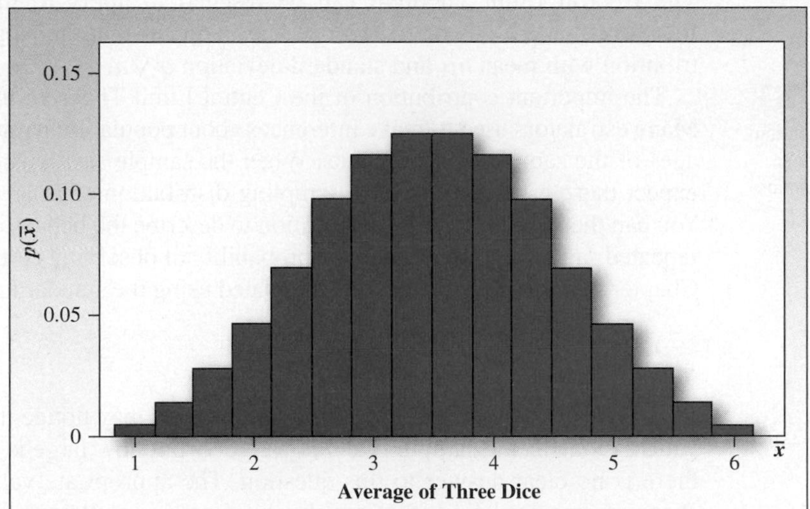

FIGURE 7.6
MINITAB sampling distribution of $\bar{x}$ for $n = 4$ dice

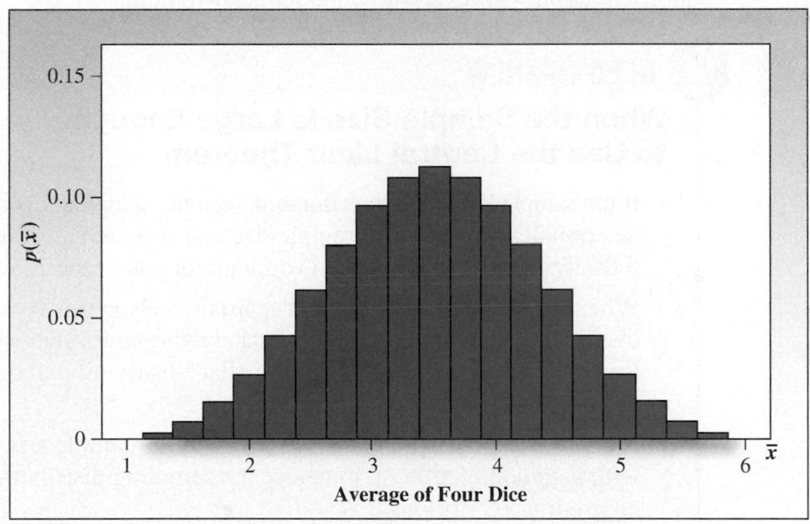

Central Limit Theorem

If random samples of n observations are drawn from a non-normal population with finite mean μ and standard deviation σ, then, when n is large, the sampling distribution of the sample mean $\bar{x}$ is approximately normally distributed, with mean μ and standard deviation

$$\frac{\sigma}{\sqrt{n}}$$

Central Limit Theorem

The approximation becomes more accurate as n becomes large. The approximation is exact for normal populations.

Regardless of shape of the population and the sample size n, the sampling distribution of $\bar{x}$ always has a mean identical to the mean of the sampled population and a standard deviation equal to the population standard deviation σ divided by $\sqrt{n}$. Consequently, *the spread of the distribution of sample means is considerably less than the spread of the sampled population. The variability depends on the size of the sample n; the larger the value of n, the smaller the variability.*

In some applications, practitioners are interested in obtaining information about the sum of the sample measurements instead of information about the sample means.

The Central Limit Theorem can be restated to apply to the **sum of the sample measurements** Σx_i, which, as n becomes large, also has an approximately normal distribution with mean $n\mu$ and standard deviation $\sigma\sqrt{n}$.

NEED A TIP?

The sampling distribution of $\bar{x}$ always has a mean σ and standard deviation $\sigma\sqrt{n}$. The CLT helps describe its **shape**.

$\sigma = 18$

n	$\sigma\sqrt{n}$
1	18
4	9
9	6
36	3

The important contribution of the Central Limit Theorem is in statistical inference. Many estimators used to make inferences about population parameters are sums or averages of the sample measurements. When the sample size is sufficiently large, you can expect these estimators to have sampling distributions that are approximately normal. You can then use the normal distribution to describe the behaviour of these estimators in repeated sampling and evaluate the probability of observing certain sample results. As in Chapter 6, these probabilities are calculated using the standard normal random variable

$$z = \frac{\text{Estimator} - \text{Mean}}{\text{Standard deviation}}$$

As you reread the Central Limit Theorem, you may notice that the approximation is valid as long as the sample size n is "large"—but how large is "large"? Unfortunately, there is no clear answer to this question. The appropriate value of n depends on the shape of the population from which you sample as well as on how you want to use the approximation. However, these guidelines will help:

NEED TO KNOW

When the Sample Size Is Large Enough to Use the Central Limit Theorem

- If the sampled population is **normal,** then the sampling distribution of $\bar{x}$ will also be normal, no matter what sample size you choose. This result can be proven theoretically, but it should not be too difficult for you to accept without proof.
- When the sampled population is approximately **symmetric,** the sampling distribution of $\bar{x}$ becomes approximately normal for relatively small values of n. Remember how rapidly ($n = 3$) the "flat" distribution in the dice example became mound-shaped.
- When the sampled population is **skewed,** the sample size n must be larger, with n equalling at least 30 before the sampling distribution of $\bar{x}$ becomes approximately normal.

These guidelines suggest that, for many populations, the sampling distribution of $\bar{x}$ will be approximately normal for moderate sample sizes; an exception to this rule occurs in sampling a binomial population when either p or $q = (1 - p)$ is very small. As specific applications of the Central Limit Theorem arise, we will give you the appropriate sample size n.

7.5 THE SAMPLING DISTRIBUTION OF THE SAMPLE MEAN

If the population mean μ is unknown, you might choose one of several *statistics* as an estimator; the sample mean $\bar{x}$ and the sample median m are two that readily come to mind. Which should you use? Consider these criteria in choosing the estimator for μ:

- Is it easy or hard to calculate?
- Does it produce estimates that are consistently too high or too low?
- Is it more or less variable than other possible estimators?

The sampling distributions for $\bar{x}$ and m with $n = 3$ for the small population in Example 7.4 showed that, in terms of these criteria, the sample mean performed better than the sample median as an estimator of μ. The sample mean can be easily computed and is a consistent estimator of the population mean. Furthermore, it is an unbiased estimator of μ; that is, its expected value is equal to population mean, μ, or $E(\bar{x}) = \mu$. In many situations, the sample mean $\bar{x}$ has desirable properties as an estimator that are not shared by other competing estimators; therefore, it is more widely used.

The Sampling Distribution of the Sample Mean, $\bar{x}$

- If a random sample of n measurements is selected from a population with mean μ and standard deviation σ, the sampling distribution of the sample mean $\bar{x}$ will have mean μ and standard deviation †

$$\frac{\sigma}{\sqrt{n}}$$

- If the population has a *normal* distribution, the sampling distribution of $\bar{x}$ will be *exactly* normally distributed, *regardless of the sample size, n.*
- If the population distribution is *non-normal,* the sampling distribution of $\bar{x}$ will be *approximately* normally distributed for large samples (by the Central Limit Theorem).

Standard Error

DEFINITION The standard deviation of a statistic used as an estimator of a population parameter is also called the **standard error of the estimator** (abbreviated **SE**) because it refers to the precision of the estimator. Therefore, the standard deviation of $\bar{x}$—given by $\sigma/\sqrt{n}$—is referred to as the **standard error of the mean** (abbreviated as $SE(\bar{x})$ or just SE).

† When repeated samples of size n are randomly selected from a *finite* population with N elements whose mean is μ and whose variance is σ^2, the standard deviation of $\bar{x}$ is

$$\frac{\sigma}{\sqrt{n}}\sqrt{\frac{N-n}{N-1}}$$

where σ^2 is the population variance. When N is large relative to the sample size n, $\sqrt{(N-n)/(N-1)}$ is approximately equal to 1, and the standard deviation of $\bar{x}$ is

$$\frac{\sigma}{\sqrt{n}}$$

NEED TO KNOW

How to Calculate Probabilities for the Sample Mean $\bar{x}$

If you know that the sampling distribution of $\bar{x}$ is *normal* or *approximately normal,* you can describe the behaviour of the sample mean $\bar{x}$ by calculating the probability of observing certain values of $\bar{x}$ in repeated sampling.

1. Find μ and calculate SE $(\bar{x}) = \sigma/\sqrt{n}$.
2. Write down the event of interest in terms of $\bar{x}$, and locate the appropriate area on the normal curve.
3. Convert the necessary values of $\bar{x}$ to z-values using
$$z = \frac{\bar{x} - \mu}{\sigma/\sqrt{n}}$$
4. Use Table 3 in Appendix I to calculate the probability.

EXAMPLE  **7.5**

The duration of Alzheimer's disease from the onset of symptoms until death ranges from 3 to 20 years; the average is 8 years with a standard deviation of 4 years. The administrator of a large medical centre randomly selects the medical records of 30 deceased Alzheimer's patients from the medical centre's database and records the average duration. Find the approximate probabilities for these events:

1. The average duration is less than seven years.
2. The average duration exceeds seven years.
3. The average duration lies within one year of the population mean $\mu = 8$.

SOLUTION Since the administrator has selected a random sample from the database at this medical centre, he can draw conclusions about only past, present, or future patients with Alzheimer's disease at this medical centre. If, on the other hand, this medical centre can be considered representative of other medical centres in the country, it may be possible to draw more far-reaching conclusions.

NEED A TIP?

If x is normal, $\bar{x}$ is normal for any n.

If x is not normal, $\bar{x}$ is approximately normal for large n.

What can you say about the shape of the sampled population? It is not symmetric because the mean $\mu = 8$ does not lie halfway between the maximum and minimum values. Since the mean is closer to the minimum value, the distribution is skewed to the right, with a few patients living a long time after the onset of the disease. Regardless of the shape of the population distribution, however, the sampling distribution of $\bar{x}$ has a mean $\mu = 8$ and standard deviation $\sigma/\sqrt{n} = 4/\sqrt{30} = 0.73$. In addition, because the sample size is $n = 30$, the Central Limit Theorem ensures the approximate normality of the sampling distribution of $\bar{x}$.

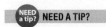

NEED A TIP?

Remember that for continuous random variables, there is no probability assigned to a single point. Therefore, $p(\bar{x} \leq 7) = p(\bar{x} < 7)$.

1. The probability that $\bar{x}$ is less than 7 is given by the shaded area in Figure 7.7. To find this area, you need to calculate the value of z corresponding to $\bar{x} = 7$:
$$z = \frac{\bar{x} - \mu}{\sigma/\sqrt{n}} = \frac{7 - 8}{0.73} = -1.37$$
From Table 3 in Appendix I, you can find the cumulative area corresponding to $z = -1.37$ and
$$P(\bar{x} < 7) = P(z < -1.37) = 0.0853$$

FIGURE 7.7
The probability that $\bar{x}$ is less than 7 for Example 7.5

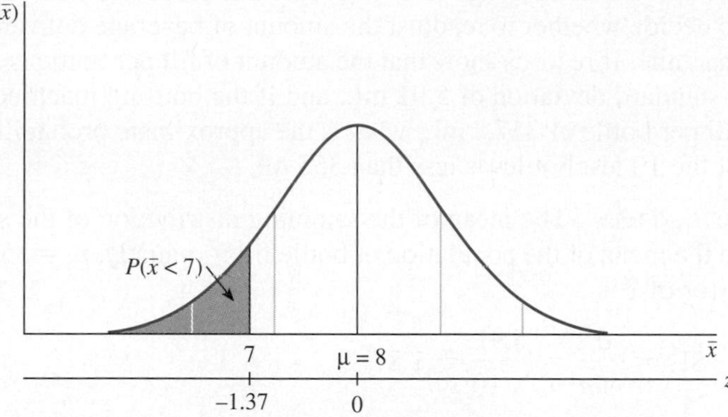

NOTE: You must use $\sigma/\sqrt{n}$ (not σ) in the formula for z because you are finding an area under the sampling distribution for $\bar{x}$, not under the probability distribution for x.

2. The event that $\bar{x}$ exceeds 7 is the complement of the event that $\bar{x}$ is less than 7. Thus, the probability that $\bar{x}$ exceeds 7 is

$$\begin{aligned} P(\bar{x} > 7) &= 1 - P(\bar{x} \leq 7) \\ &= 1 - 0.0853 = 0.9147 \end{aligned}$$

3. The probability that $\bar{x}$ lies within one year of $\mu = 8$ is the shaded area in Figure 7.8. The z-value corresponding to $\bar{x} = 7$ is $z = -1.37$, from step 1, and the z-value for $\bar{x} = 9$ is

$$z = \frac{\bar{x} - \mu}{\sigma/\sqrt{n}} = \frac{9 - 8}{0.73} = 1.37$$

The probability of interest is

$$\begin{aligned} P(7 < \bar{x} < 9) &= P(-1.37 < z < 1.37) \\ &= 0.9147 - 0.0853 = 0.8294 \end{aligned}$$

FIGURE 7.8
The probability that $\bar{x}$ lies within one year of $\mu = 8$ for Example 7.5

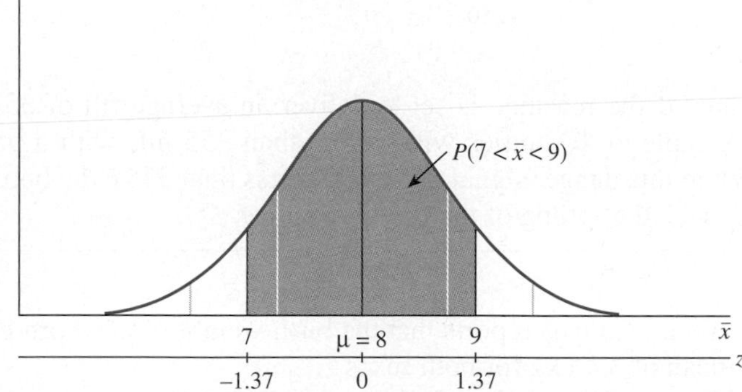

EXAMPLE 7.6

To avoid difficulties with the federal or provincial and local consumer protection agencies, a beverage bottler must make reasonably certain that 355 millilitre (mL) bottles actually contain 355 mL of beverage. To determine whether a bottling machine is working satisfactorily, one bottler randomly samples 10 bottles per hour and measures the amount of beverage in each bottle. The mean $\bar{x}$ of the 10 fill measurements is used to decide whether to readjust the amount of beverage delivered per bottle by the filling machine. If records show that the amount of fill per bottle is normally distributed, with a standard deviation of 5.91 mL, and if the bottling machine is set to produce a mean fill per bottle of 357.8 mL, what is the approximate probability that the sample mean $\bar{x}$ of the 10 test bottles is less than 355 mL?

SOLUTION The mean of the sampling distribution of the sample mean $\bar{x}$ is identical to the mean of the population of bottle fills—namely, $\mu = 357.8$ mL—and the standard error of $\bar{x}$ is

$$\text{SE} = \frac{\sigma}{\sqrt{n}} = \frac{5.91}{\sqrt{10}} = 1.87$$

(NOTE: σ is the standard deviation of the population of bottle fills, and n is the number of bottles in the sample.) Since the amount of fill is normally distributed, $\bar{x}$ is also normally distributed, as shown in Figure 7.9.

To find the probability that $\bar{x}$ is less than 355 mL, express the value $\bar{x} = 355$ in units of standard deviations:

$$z = \frac{\bar{x} - \mu}{\sigma/\sqrt{n}} = \frac{355 - 357.8}{1.87} = -1.50$$

Then

$$P(\bar{x} < 355) = P(z < -1.50) = 0.0668$$

FIGURE 7.9
Probability distribution of $\bar{x}$, the mean of the $n = 10$ bottle fills, for Example 7.6

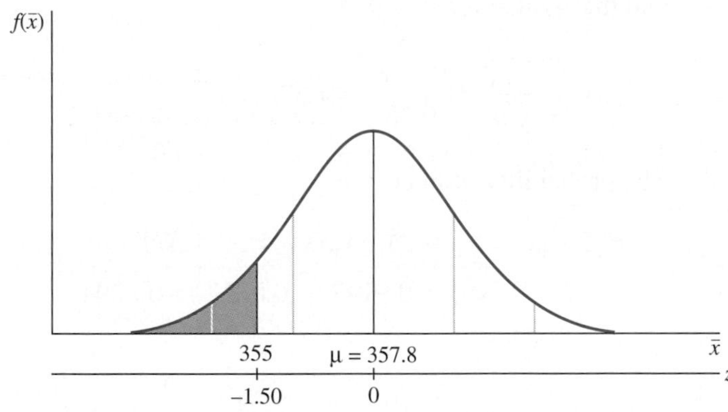

Thus, if the machine is set to deliver an average fill of 357.8 mL, the mean fill $\bar{x}$ of a sample of 10 bottles will be less than 355 mL with a probability equal to 0.0668. When this danger signal occurs ($\bar{x}$ is less than 355), the bottler takes a larger sample to recheck the setting of the filling machine.

EXAMPLE 7.7

Statistics Canada reports that the birth weight of newborn babies in Saskatchewan has a mean of 3.45 kg for both sexes.[9]

Suppose the standard deviation is 0.70 kg. Further, we randomly sample 49 birth certificates in Saskatchewan and record the birth weights of sampled babies. Find the mean and standard deviation of the sampling distribution of $\bar{x}$. What is the probability that the sample mean birth weight as recorded on the birth certificates will be less than 3.25 kg?

SOLUTION The mean: $\mu_{\bar{x}} = \mu = 3.45$ kg and the standard deviation: $\sigma/\sqrt{n} = 0.10$ kg

The probability that the sample mean birth weight as recorded on the birth certificates will be less than 3.25 kg is

$$P(\bar{x} < 3.25) = P(z < -2) = 0.0228$$

EXAMPLE 7.8 In September 2015, Statistics Canada reported that the average hourly wage for Ontario workers aged 15 to 24 was \$13.64.[10] A random sample of 64 individuals aged 15 to 24 in Ontario was taken, and the individuals' wages were recorded. Suppose the standard deviation is known to be \$6.00. What is the probability that the sample mean hourly wage is greater than \$12.00?

SOLUTION The mean is $\mu_{\bar{x}} = \mu = 13.64\text{kg}$, and the standard deviation is $\sigma/\sqrt{n} = \$0.75$. The probability that the sample hourly wage is greater than \$12.00 is

$$P(\bar{x} < 12) = P(z < -2.19) = 0.9857$$

7.5 EXERCISES

BASIC TECHNIQUES

7.17 Random samples of size n were selected from populations with the means and variances given here. Find the mean and standard deviation of the sampling distribution of the sample mean in each case:

a. $n = 36, \mu = 10, \sigma^2 = 9$
b. $n = 100, \mu = 5, \sigma^2 = 4$
c. $n = 8, \mu = 120, \sigma^2 = 1$

7.18 Refer to Exercise 7.17.

a. If the sampled populations are normal, what is the sampling distribution of $\bar{x}$ for parts a, b, and c?

b. According to the Central Limit Theorem, if the sampled populations are *not* normal, what can be said about the sampling distribution of $\bar{x}$ for parts a, b, and c?

7.19 Refer to Exercise 7.17, part b.

a. Sketch the sampling distribution for the sample mean and locate the mean and the interval $\mu \pm 2\sigma/\sqrt{n}$ along the x-axis.

b. Shade the area under the curve that corresponds to the probability that $\bar{x}$ lies within 0.15 unit of the population mean μ.

c. Find the probability described in part b.

7.20 A population consists of $N = 5$ numbers: 1, 3, 5, 6, and 7. It can be shown that the mean and standard deviation for this population are $\mu = 4.5$ and $\sigma = 2.15$, respectively.

a. Construct a probability histogram for this population.

b. Use the random number table, Table 10 in Appendix I, to select a random sample of size $n = 10$ with replacement from the population. Calculate the sample mean, $\bar{x}$. Repeat this procedure, calculating the sample mean $\bar{x}$ for your second sample. (HINT: Assign the random digits 0 and 1 to the measurement $x = 1$, assign digits 2 and 3 to the measurement $x = 3$, and so on.)

c. To simulate the sampling distribution of $\bar{x}$, we have selected 50 more samples of size $n = 10$ with replacement, and have calculated the corresponding sample means. Construct a relative frequency histogram for these 50 values of $\bar{x}$. What is the shape of this distribution?

4.8	4.2	4.2	4.5	4.3	4.3	5.0	4.0	3.3	4.7
3.0	5.9	5.7	4.2	4.4	4.8	5.0	5.1	4.8	4.2
4.6	4.1	3.4	4.9	4.1	4.0	3.7	4.3	4.3	4.5
5.0	4.6	4.1	5.1	3.4	5.9	5.0	4.3	4.5	3.9
4.4	4.2	4.2	5.2	5.4	4.8	3.6	5.0	4.5	4.9

7.21 Refer to Exercise 7.20.

a. Use the data entry method in your calculator to find the mean and standard deviation of the 50 values of $\bar{x}$ given in Exercise 7.20, part c.

b. Compare the values calculated in part a to the theoretical mean μ and the theoretical standard deviation $\sigma/\sqrt{n}$ for the sampling distribution of $\bar{x}$. How close do the values calculated from the 50 measurements come to the theoretical values?

7.22 A random sample of n observations is selected from a population with standard deviation $\sigma = 1$. Calculate the standard error of the mean (SE) for these values of n:

a. $n = 1$ **b.** $n = 2$ **c.** $n = 4$
d. $n = 9$ **e.** $n = 16$ **f.** $n = 25$
g. $n = 100$

7.23 Refer to Exercise 7.22. Plot the standard error of the mean (SE) versus the sample size n and connect the points with a smooth curve. What is the effect of increasing the sample size on the standard error?

7.24 A random sample of n observations is selected from a population with standard deviation $\sigma = 5$. Calculate the standard error of the mean (SE) for these values of n:

a. $n = 1$ **b.** $n = 2$ **c.** $n = 4$
d. $n = 9$ **e.** $n = 16$ **f.** $n = 25$
g. $n = 100$

7.25 Refer to Exercise 7.24. Plot the standard error of the mean (SE) versus the sample size n and connect the points with a smooth curve. What is the effect of increasing the sample size on the standard error?

7.26 A random sample of size $n = 49$ is selected from a population with mean $\mu = 53$ and standard deviation $\sigma = 21$.

a. What will be the approximate shape of the sampling distribution of $\bar{x}$?

b. What will be the mean and standard deviation of the sampling distribution of $\bar{x}$?

7.27 Refer to Exercise 7.26. Find the probability that the sample mean is greater than 55.

7.28 A random sample of size $n = 40$ is selected from a population with mean $\mu = 100$ and standard deviation $\sigma = 20$.

a. What will be the approximate shape of the sampling distribution of $\bar{x}$?

b. What will be the mean and standard deviation of the sampling distribution of $\bar{x}$?

7.29 Refer to Exercise 7.28. Find the probability that the sample mean is between 105 and 110.

7.30 Suppose a random sample of $n = 25$ observations is selected from a population that is normally distributed, with mean equal to 106 and standard deviation equal to 12.

a. Give the mean and the standard deviation of the sampling distribution of the sample mean $\bar{x}$.

b. Find the probability that $\bar{x}$ exceeds 110.

c. Find the probability that the sample mean deviates from the population mean $\mu = 106$ by no more than 4.

APPLICATIONS

7.31 Measurement Error When research chemists perform experiments, they may obtain slightly different results on different replications, even when the experiment is performed identically each time. These differences are due to a phenomenon called "measurement error."

a. List some variables in a chemical experiment that might cause some small changes in the final response measurement.

b. If you want to make sure that your measurement error is small, you can replicate the experiment and take the sample average of all the measurements. To decrease the amount of variability in your average measurement, should you use a large or a small number of replications? Explain.

7.32 Tomatoes Explain why the weight of a package of one dozen tomatoes should be approximately normally distributed if the dozen tomatoes represent a random sample.

7.33 Bacteria in Water Use the Central Limit Theorem to explain why a Poisson random variable—say, the number of a particular type of bacteria in a cubic metre of water—has a distribution that can be approximated by a normal distribution when the mean μ is large.

7.34 Faculty Salaries Suppose that university faculty with the rank of assistant professor earn an average of \$74,000 per year with a standard deviation of \$6000. In an attempt to verify this salary level, a random sample of 60 assistant professors was selected from a personnel database for all universities in Canada.

a. Describe the sampling distribution of the sample mean $\bar{x}$.

b. Within what limits would you expect the sample average to lie, with probability 0.95?

c. Calculate the probability that the sample mean $\bar{x}$ is greater than \$78,000.

d. If your random sample actually produced a sample mean of \$78,000, would you consider this unusual? What conclusion might you draw?

7.35 Tax Savings An important expectation of a federal income tax reduction is that consumers will reap a substantial portion of the tax savings. Suppose estimates of the portion of total tax saved, based on a random sampling of 35 economists, have a mean of 26% and a standard deviation of 12%.

a. What is the approximate probability that a sample mean, based on a random sample of $n = 35$ economists, will lie within 1% of the mean of the population of the estimates of all economists?

b. Is it necessarily true that the mean of the population of estimates of all economists is equal to the percentage of tax savings that will actually be achieved? Why?

7.36 Paper Strength A manufacturer of paper used for packaging requires a minimum strength of 1400 g/cm^2. To check on the quality of the paper, a random sample of 10 pieces of paper is selected each hour from the previous hour's production and a strength measurement is recorded for each. The standard deviation σ of the strength measurements, computed by pooling the sum of squares of deviations of many samples, is known to equal 140 g/cm^2, and the strength measurements are normally distributed.

a. What is the approximate sampling distribution of the sample mean of $n = 10$ test pieces of paper?

b. If the mean of the population of strength measurements is 1450 g/cm^2, what is the approximate probability that, for a random sample of $n = 10$ test pieces of paper, $\bar{x} < 1400$?

c. What value would you select for the mean paper strength μ in order that $P(\bar{x} < 1400)$ be equal to 0.001?

7.37 Potassium Levels The normal daily human potassium requirement is in the range of 2000 to 6000 milligrams (mg), with larger amounts required during hot summer weather. The amount of potassium in food varies, depending on the food. For example, there are approximately 7 mg in a cola drink, 46 mg in a beer, 630 mg in a banana, 300 mg in a carrot, and 440 mg in a glass of orange juice. Suppose the distribution of potassium in a banana is normally distributed, with mean equal to 630 mg and standard deviation equal to 40 mg per banana. You eat $n = 3$ bananas per day, and T is the total number of milligrams of potassium you receive from them.

a. Find the mean and standard deviation of T.

b. Find the probability that your total daily intake of potassium from the three bananas will exceed 2000 mg. (HINT: Note that T is the sum of three random variables, x_1, x_2, and x_3, where x_1 is the amount of potassium in banana number 1, etc.)

7.38 Deli Sales The total daily sales, x, in the deli section of a local market is the sum of the sales generated by a fixed number of customers who make purchases on a given day.

a. What kind of probability distribution do you expect the total daily sales to have? Explain.

b. For this particular market, the average sale per customer in the deli section is \$8.50 with $\sigma = \$2.50$. If 30 customers make deli purchases on a given day, give the mean and standard deviation of the probability distribution of the total daily sales, x.

7.39 Normal Temperatures In Exercise 1.66, Allen Shoemaker derived a distribution of human body temperatures with a distinct mound shape.[11] Suppose we assume that the temperatures of healthy humans is approximately normal with a mean of 37° Celsius and a standard deviation of 0.2 degrees.

a. If 130 healthy people are selected at random, what is the probability that the average temperature for these people is 36.81°C or lower?

b. Would you consider an average temperature of 36.81°C to be an unlikely occurrence, given that the true average temperature of healthy people is 37°C? Explain.

7.40 Sports and Achilles Tendon Injuries Some sports that involve a significant amount of running, jumping, or hopping put participants at risk for Achilles tendinopathy (AT), an inflammation and thickening of the Achilles tendon. A study in the *American Journal of Sports Medicine* looked at the diameter (in mm) of the affected and non-affected tendons for patients who participated in these types of sports activities.[12] Suppose that the Achilles tendon diameters in the general population have a mean of 5.97 millimetres (mm) with a standard deviation of 1.95 mm.

a. What is the probability that a randomly selected sample of 31 patients would produce an average diameter of 6.5 mm or less for the non-affected tendon?

b. When the diameters of the affected tendon were measured for a sample of 31 patients, the average diameter was 9.80 mm. If the average tendon diameter in the population of patients with AT is no different than the average diameter of the non-affected tendons (5.97 mm), what is the probability of observing an average diameter of 9.80 mm or higher?

c. What conclusions might you draw from the results of part b?

7.6 THE SAMPLING DISTRIBUTION OF THE SAMPLE PROPORTION

There are many practical examples of the binomial random variable x. One common application involves consumer preference or opinion polls, in which we use a random sample of n people to estimate the proportion p of people in the population who have a specified characteristic. If x of the sampled people have this characteristic, then the sample proportion

$$\hat{p} = \frac{x}{n}$$

can be used to estimate the population proportion p (Figure 7.10).[†]

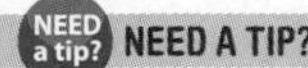

NEED A TIP?

Q: How do you know if it's binomial or not?
A: Look to see if the measurement taken on a single experimental unit in the sample is a "success/failure" type. If so, it's probably binomial.

The binomial random variable x has a probability distribution $p(x)$, described in Chapter 5, with mean np and standard deviation $\sqrt{npq}$. Since $\hat{p}$ is simply the value of x, expressed as a proportion $\left(\hat{p} = \frac{x}{n}\right)$, the sampling distribution of $\hat{p}$ is identical in shape to the probability distribution of x, except that it has a new scale along the horizontal axis.

FIGURE 7.10
Sampling distribution of the binomial random variable x and the sample proportion $\hat{p}$

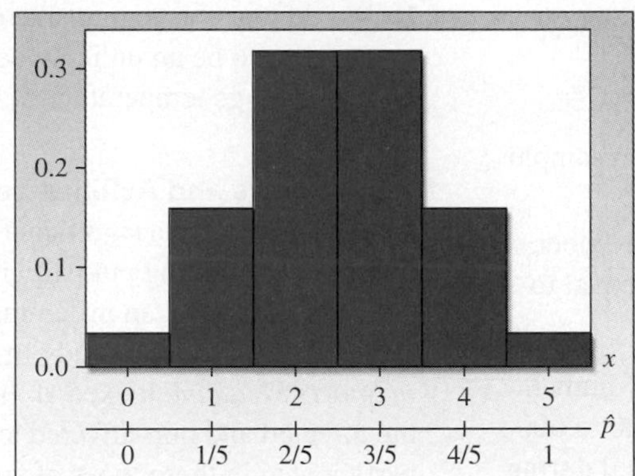

Because of this change of scale, the mean and standard deviation of $\hat{p}$ are also rescaled, so that the mean of the sampling distribution of $\hat{p}$ is p, and its standard error is

$$SE(\hat{p}) = \sqrt{\frac{pq}{n}} \quad \text{where } q = 1 - p$$

Finally, just as we can approximate the probability distribution of a sample mean with a normal distribution when the sample size n is large, we can do the same with the sampling distribution of $\hat{p}$. Recall that the sample proportion is actually a sample mean with $x = 1$ or 0.

[†]A "hat" placed over the symbol of a population parameter denotes a statistic used to estimate the population parameter. For example, the symbol $\hat{p}$ denotes the sample proportion.

Properties of the Sampling Distribution of the Sample Proportion, $\hat{p}$

- If a random sample of n observations is selected from a binomial population with parameter p, then the sampling distribution of the sample proportion

$$\hat{p} = \frac{x}{n}$$

will have a mean

$$p$$

and a standard deviation

$$SE(\hat{p}) = \sqrt{\frac{pq}{n}} \quad \text{where } q = 1 - p$$

- When the sample size n is large, the sampling distribution of $\hat{p}$ can be approximated by a normal distribution. The approximation will be adequate if $np > 5$ and $nq > 5$.

EXAMPLE 7.9

A sports facility located in Prince Edward Island recently conducted a survey on the importance of sports for children. In the survey, 500 mothers and fathers were asked about the importance of sports for boys and girls. Of the parents interviewed, 60% agreed that the genders are equal and should have equal opportunities to participate in sports. Describe the sampling distribution of the sample proportion $\hat{p}$ of parents who agree that the genders are equal and should have equal opportunities.

SOLUTION You can assume that the 500 parents represent a random sample of the parents of all boys and girls in Canada and that the true proportion in the population is equal to some unknown value that you can call p. The sampling distribution of $\hat{p}$ can be approximated by a normal distribution,† with mean equal to p (see Figure 7.11) and standard error

$$SE(\hat{p}) = \sqrt{\frac{pq}{n}}$$

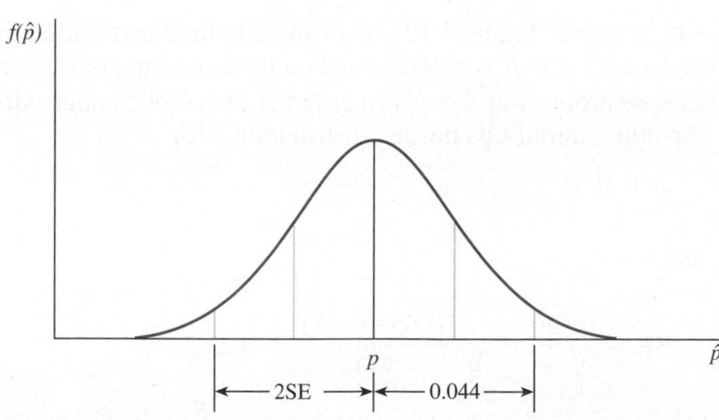

FIGURE 7.11
The sampling distribution for $\hat{p}$ based on a sample of $n = 500$ parents for Example 7.9

† Checking the conditions that allow the normal approximation to the distribution of $\hat{p}$, you can see that $n = 500$ is adequate for values of p near 0.60 because $n\hat{p} = 300$ and $n\hat{q} = 200$ are both greater than 5.

You can see from Figure 7.11 that the sampling distribution of $\hat{p}$ is centred over its mean p. Even though you do not know the exact value of p (the sample proportion $\hat{p} = 0.60$ may be larger or smaller than p), an approximate value for the standard deviation of the sampling distribution can be found using the sample proportion $\hat{p} = 0.60$ to approximate the unknown value of p. Thus,

$$\text{SE} = \sqrt{\frac{pq}{n}} \approx \sqrt{\frac{\hat{p}\hat{q}}{n}}$$

$$= \sqrt{\frac{(0.60)(0.40)}{500}} = 0.022$$

Therefore, approximately 95% of the time, $\hat{p}$ will fall within $2\text{SE} \approx 0.044$ of the (unknown) value of p.

NEED TO KNOW

How to Calculate Probabilities for the Sample Proportion $\hat{p}$

1. Find the necessary values of n and p.
2. Check whether the normal approximation to the binomial distribution is appropriate ($np > 5$ and $nq > 5$).
3. Write down the event of interest in terms of $\hat{p}$, and locate the appropriate area on the normal curve.
4. Convert the necessary values of $\hat{p}$ to z-values using

$$z = \frac{\hat{p} - p}{\sqrt{\frac{pq}{n}}}$$

5. Use Table 3 in Appendix I to calculate the probability.

EXAMPLE  **7.10**

Refer to Example 7.9. Suppose the proportion p of parents in the population is actually equal to 0.55. What is the probability of observing a sample proportion as large as or larger than the observed value $\hat{p} = 0.60$?

SOLUTION Figure 7.12 shows the sampling distribution of $\hat{p}$ when $p = 0.55$, with the observed value $\hat{p} = 0.60$ located on the horizontal axis. The probability of observing a sample proportion $\hat{p}$ equal to or larger than 0.60 is approximated by the shaded area in the upper tail of this normal distribution with

$$p = 0.55$$

and

$$\text{SE} = \sqrt{\frac{pq}{n}} = \sqrt{\frac{(0.55)(0.45)}{500}} = 0.0222$$

FIGURE 7.12
The sampling distribution of $\hat{p}$ for $n = 500$ and $p = 0.55$ for Example 7.10

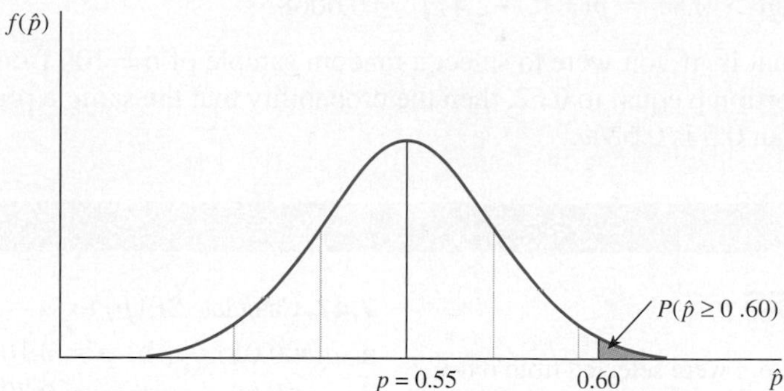

To find this shaded area, first calculate the z-value corresponding to $\hat{p} = 0.60$:

$$z = \frac{\hat{p} - p}{\sqrt{pq/n}} = \frac{0.60 - 0.55}{0.0222} = 2.25$$

Using Table 3 in Appendix I, you find

$$P(\hat{p} > 0.60) \approx P(z > 2.25) = 1 - 0.9878 = 0.0122$$

That is, if you were to select a random sample of $n = 500$ observations from a population with proportion p equal to 0.55, the probability that the sample proportion $\hat{p}$ would be as large as or larger than 0.60 is only 0.0122.

When the normal distribution was used in Chapter 6 to approximate the binomial probabilities associated with x, a correction of ± 0.5 was applied to improve the approximation. The equivalent correction here is $\pm(0.5/n)$. For example, for $\hat{p} = 0.60$ the value of z with the correction is

$$z_1 = \frac{(0.60 - 0.001) - 0.55}{\sqrt{\dfrac{(0.55)(0.45)}{500}}} = 2.20$$

with $P(\hat{p} > 0.60) \approx 0.0139$. To two-decimal-place accuracy, this value agrees with the earlier result. When n is large, the effect of using the correction is generally negligible. You should solve problems in this and the remaining chapters *without* the correction factor unless you are specifically instructed to use it.

EXAMPLE 7.11

A study done by Statistics Canada showed that 62% of the population in Nunavut are smokers.[13] A smoker is an individual aged 12 or older who regularly smokes. Suppose that a random sample of 100 individuals from Nunavut was chosen, and 50 of them were smokers. What is the probability of observing a sample proportion less than 0.5?

SOLUTION

$$\hat{p} = \frac{50}{100} = 0.5$$

$$z = \frac{\hat{p} - p}{\sqrt{\dfrac{pq}{n}}}$$

$$z = \frac{0.5 - 0.62}{\sqrt{\dfrac{(0.62)(0.38)}{100}}} = -2.47$$

$$P(\hat{p} > 0.5) = p(z < -2.47) = 0.0068$$

That is, if you were to select a random sample of n = 100 from a population with proportion p equal to 0.62, then the probability that the sample proportion $\hat{p}$ would be less than 0.5 is 0.68%.

7.6 EXERCISES

BASIC TECHNIQUES

7.41 Random samples of size n were selected from binomial populations with population parameters p given here. Find the mean and the standard deviation of the sampling distribution of the sample proportion $\hat{p}$ in each case:

a. $n = 100, p = 0.3$

b. $n = 400, p = 0.1$

c. $n = 250, p = 0.6$

7.42 Sketch each of the sampling distributions in Exercise 7.41. For each, locate the mean p and the interval $p \pm 2$ SE along the $\hat{p}$-axis of the graph.

7.43 Refer to the sampling distribution in Exercise 7.41, part a.

a. Sketch the sampling distribution for the sample proportion and shade the area under the curve that corresponds to the probability that $\hat{p}$ lies within 0.08 of the population proportion p.

b. Find the probability described in part a.

7.44 Is it appropriate to use the normal distribution to approximate the sampling distribution of $\hat{p}$ in the following circumstances?

a. $n = 50, p = 0.05$

b. $n = 75, p = 0.1$

c. $n = 250, p = 0.99$

7.45 Random samples of size $n = 75$ were selected from a binomial population with $p = 0.4$. Use the normal distribution to approximate the following probabilities:

a. $P(\hat{p} \leq 0.43)$

b. $P(0.35 \leq \hat{p} \leq 0.43)$

7.46 Random samples of size $n = 500$ were selected from a binomial population with $p = 0.1$.

a. Is it appropriate to use the normal distribution to approximate the sampling distribution of $\hat{p}$? Check to make sure the necessary conditions are met.

Using the results of part a, find these probabilities:

b. $\hat{p} > 0.12$

c. $\hat{p} < 0.10$

d. $\hat{p}$ lies within 0.02 of p

7.47 Calculate SE($\hat{p}$) for $n = 100$ and these values of p:

a. $p = 0.01$ **b.** $p = 0.10$ **c.** $p = 0.30$

d. $p = 0.50$ **e.** $p = 0.70$ **f.** $p = 0.90$

g. $p = 0.99$

h. Plot SE($\hat{p}$) versus p on graph paper and sketch a smooth curve through the points. For what value of p is the standard deviation of the sampling distribution of $\hat{p}$ a maximum? What happens to the standard error when p is near 0 or near 1.0?

7.48 A random sample of size $n = 50$ is selected from a binomial distribution with population proportion $p = 0.7$.

a. What will be the approximate shape of the sampling distribution of $\hat{p}$?

b. What will be the mean and standard deviation (or standard error) of the sampling distribution of $\hat{p}$?

c. Find the probability that the sample proportion $\hat{p}$ is less than 0.8.

7.49 A random sample of size $n = 80$ is selected from a binomial distribution with population proportion $p = 0.25$.

a. What will be the approximate shape of the sampling distribution of $\hat{p}$?

b. What will be the mean and standard deviation (or standard error) of the sampling distribution of $\hat{p}$?

c. Find the probability that the sample proportion $\hat{p}$ is between 0.18 and 0.44.

7.50 **a.** Is the normal approximation to the sampling distribution of $\hat{p}$ appropriate when $n = 400$ and $p = 0.8$?

b. Use the results of part a to find the probability that $\hat{p}$ is greater than 0.83.

c. Use the results of part a to find the probability that $\hat{p}$ lies between 0.76 and 0.84.

APPLICATIONS

7.51 Road Trip! Parents with children list a GPS system (28%) and a DVD player (28%) as "must have" accessories for a road trip.[14] Suppose a sample of $n = 1000$ parents is randomly selected, and parents are asked what devices they would like to have for a family road trip. Let $\hat{p}$ be the proportion of parents in the sample who choose either a GPS system or a DVD player.

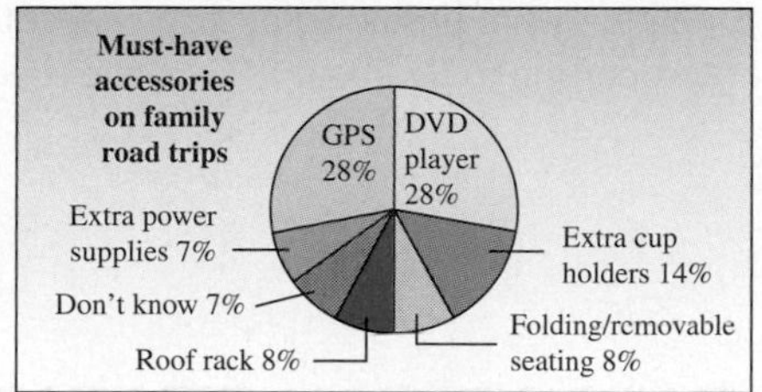

To survive road trips with the family, parents consider a GPS navigation device and a DVD player essential.

a. If $p = 0.28 + 0.28 = 0.56$, how can you approximate the distribution of $\hat{p}$?

b. What is the probability that $\hat{p}$ exceeds 0.6?

c. What is the probability that $\hat{p}$ lies between 0.5 and 0.6?

d. Would a sample percentage of $\hat{p} = 0.7$ contradict the reported value of 0.56?

7.52 Fats and Sweets According to a study in the *American Journal of Public Health,* diets high in fat, sugar, and grains were associated with lower diet costs after adjustment for energy intakes, gender, and age.[15] For most levels of energy intake, each additional 100 g of fat and sweets was associated with a 0.05−0.40 per day reduction in diet costs. In contrast, each additional 100 g of fruit and vegetables was associated with 0.18−0.29 per day increase in diet costs. The study was based on freely chosen diets of 837 French adults that were assessed by a dietary history method. Mean national food prices for 57 foods were used to estimate diet costs. Suppose a random sample of 100 French adults was selected and 46% of the adults eat fat and sweets due to cost factors.

a. Does the distribution of $\hat{p}$, the sample proportion of French adults who eat fats and sweets, have an approximately normal distribution? If so, what are its mean and standard deviation?

b. What is the probability that the sample proportion, $\hat{p}$, exceeds 0.5?

c. What is the probability that $\hat{p}$ lies within the interval 0.35 to 0.55?

d. What might you conclude if the sample proportion were as small as 30%?

7.53 Surfing the Net Do you use the Internet to gather information for a project? A survey reports that the percentage of students who used the Internet as their major resource for a school project in a recent year was 66%.[16] Suppose that you take a sample of $n = 1000$ students, and record the number of students who used the Internet as their major resource for their school project during the past year. Let $\hat{p}$ be the proportion of students surveyed who used the Internet as a major resource in the past year.

a. How can you approximate the distribution of $\hat{p}$?

b. What is the probability that the sample proportion $\hat{p}$ exceeds 68%?

c. What is the probability that the sample proportion lies between 64% and 68%?

d. Would a sample proportion of 70% contradict the reported value of 66%?

7.54 M&M'S® According to the M&M'S® website, the average percentage of brown M&M'S® candies in a package of milk chocolate M&M's® is 13%.[17] (This percentage varies, however, among the different types of packaged M&M'S®.) Suppose you randomly select a package of milk chocolate M&M'S® that contains 55 candies and determine the proportion of brown candies in the package.

a. What is the approximate distribution of the sample proportion of brown candies in a package that contains 55 candies?

b. What is the probability that the sample proportion of brown candies is less than 20%?

c. What is the probability that the sample proportion exceeds 35%?

d. Within what range would you expect the sample proportion to lie about 95% of the time?

7.55 Measured Obesity Over the past several years, the prevalence of overweight and obese children and adolescents has risen, with the most substantial increases observed in economically developed countries.[18] According to the results of a study by Statistics Canada, a substantial share of Canadian youth are part of this trend. In 2014, 23.1% of Canadian children and adolescents aged 12 to 17 were overweight or obese.[19] Assume that the percentage is decreased to 21.5% now and a random sample of 100 children is selected.

a. What is the probability that the sample proportion of overweight children exceeds 25%?

b. What is the probability that the sample proportion of overweight children is less than 12%?

c. Would it be unusual to find that 30% of the sampled children were overweight? Explain.

7.56 Oh, Nuts! Are you a chocolate "purist," or do you like other ingredients in your chocolate? *American Demographics* reports that almost 75% of consumers like traditional ingredients such as nuts or caramel in their chocolate. They are less enthusiastic about the taste of mint or coffee, which provide more distinctive flavours.[20] A random sample of 200 consumers is selected and the number who like nuts or caramel in their chocolate is recorded.

a. What is the approximate sampling distribution for the sample proportion $\hat{p}$? What are the mean and standard deviation for this distribution?

b. What is the probability that the sample proportion is greater than 80%?

c. Within what limits would you expect the sample proportion to lie about 95% of the time?

7.7 A SAMPLING APPLICATION: STATISTICAL PROCESS CONTROL (OPTIONAL)

Statistical process control (SPC) methodology was developed to monitor, control, and improve products and services. Steel bearings must conform to size and hardness specifications, industrial chemicals must have a low prespecified level of impurities, and accounting firms must minimize and ultimately eliminate incorrect bookkeeping entries. It is often said that statistical process control consists of 10% statistics and 90% engineering and common sense. We can statistically monitor a process mean and tell when the mean falls outside preassigned limits, but we cannot tell *why* it is out of control. Answering this last question requires knowledge of the process and problem-solving ability—the other 90%!

Product quality is usually monitored using statistical control charts. Measurements on a process variable to be monitored change over time. The cause of a change in the variable is said to be *assignable* if it can be found and corrected. Other variation—small haphazard changes due to alteration in the production environment—that is not controllable is regarded as *random variation.* If the variation in a process variable is solely random, the process is said to be *in control.* The first objective in statistical process control is to eliminate assignable causes of variation in the process variable and then get the process in control. The next step is to reduce variation and get the measurements on the process variable within *specification limits,* the limits within which the measurements on usable items or services must fall.

Once a process is in control and is producing a satisfactory product, the process variables are monitored with **control charts.** Samples of n items are drawn from the process at specified intervals of time, and a sample statistic is computed. These statistics are plotted on the control chart, so that the process can be checked for shifts in the process variable that might indicate control problems.

A Control Chart for the Process Mean: The $\bar{x}$ Chart

Assume that n items are randomly selected from the production process at equal intervals and that measurements are recorded on the process variable. If the process is in control, the sample means should vary about the population mean μ in a random manner. Moreover, according to the Central Limit Theorem, the sampling distribution of $\bar{x}$ should be approximately normal, so that almost all of the values of fall into the interval $(\mu \pm 3\text{SE}) = \mu \pm 3(\sigma/\sqrt{n})$. Although the exact values of μ and σ are unknown, you can obtain accurate estimates by using the sample measurements.

Every control chart has a *centreline* and *control limits.* The centreline for the $\bar{x}$ **chart** is the estimate of μ, the grand average of all the sample statistics calculated from the measurements on the process variable. The upper and lower *control limits* are placed three standard deviations above and below the centreline. If you monitor the process mean based on k samples of size n taken at regular intervals, the centreline is $\bar{\bar{x}}$, the average of the sample means, and the control limits are at $\bar{\bar{x}} \pm 3(\sigma/\sqrt{n})$, with σ estimated by s, the standard deviation of the nk measurements.

EXAMPLE 7.12

A statistical process control monitoring system samples the inside diameters of $n = 4$ bearings each hour. Table 7.6 provides the data for $k = 25$ hourly samples. Construct an $\bar{x}$ chart for monitoring the process mean.

SOLUTION The sample mean was calculated for each of the $k = 25$ samples. For example, the mean for sample 1 is

$$\bar{x} = \frac{0.992 + 1.007 + 1.016 + 0.991}{4} = 1.0015$$

TABLE 7.6 **25 Hourly Samples of Bearing Diameters, $n = 4$ Bearings per Sample**

Sample	Sample Measurements				Sample Mean, $\bar{x}$
1	0.992	1.007	1.016	0.991	1.00150
2	1.015	0.984	0.976	1.000	0.99375
3	0.988	0.993	1.011	0.981	0.99325
4	0.996	1.020	1.004	0.999	1.00475
5	1.015	1.006	1.002	1.001	1.00600
6	1.000	0.982	1.005	0.989	0.99400
7	0.989	1.009	1.019	0.994	1.00275
8	0.994	1.010	1.009	0.990	1.00075
9	1.018	1.016	0.990	1.011	1.00875
10	0.997	1.005	0.989	1.001	0.99800
11	1.020	0.986	1.002	0.989	0.99925
12	1.007	0.986	0.981	0.995	0.99225
13	1.016	1.002	1.010	0.999	1.00675
14	0.982	0.995	1.011	0.987	0.99375
15	1.001	1.000	0.983	1.002	0.99650
16	0.992	1.008	1.001	0.996	0.99925
17	1.020	0.988	1.015	0.986	1.00225
18	0.993	0.987	1.006	1.001	0.99675
19	0.978	1.006	1.002	0.982	0.99200
20	0.984	1.009	0.983	0.986	0.99050
21	0.990	1.012	1.010	1.007	1.00475
22	1.015	0.983	1.003	0.989	0.99750
23	0.983	0.990	0.997	1.002	0.99300
24	1.011	1.012	0.991	1.008	1.00550
25	0.987	0.987	1.007	0.995	0.99400

The sample means are shown in the last column of Table 7.6. The centreline is located at the average of the sample means, or

$$\bar{\bar{x}} = \frac{24.9675}{25} = 0.9987$$

The calculated value of s, the sample standard deviation of all $nk = 4(25) = 100$ observations, is $s = 0.011458$, and the estimated standard error of the mean of $n = 4$ observations is

$$\frac{s}{\sqrt{n}} = \frac{0.011458}{\sqrt{4}} = 0.005729$$

The upper and lower control limits are found as

$$\text{UCL} = \bar{\bar{x}} + 3\frac{s}{\sqrt{n}} = 0.9987 + 3(0.005729) = 1.015887$$

and

$$\text{LCL} = \bar{\bar{x}} - 3\frac{s}{\sqrt{n}} = 0.9987 - 3(0.005729) = 0.981513$$

Figure 7.13 shows a *MINITAB* printout of the $\bar{x}$ chart constructed from the data. If you assume that the samples used to construct the $\bar{x}$ chart were collected when the process was in control, the chart can now be used to detect changes in the process mean. Sample means are plotted periodically, and if a sample mean falls outside the control limits, a warning should be conveyed. The process should be checked to locate the cause of the unusually large or small mean.

FIGURE 7.13

MINITAB $\bar{x}$ chart for Example 7.12

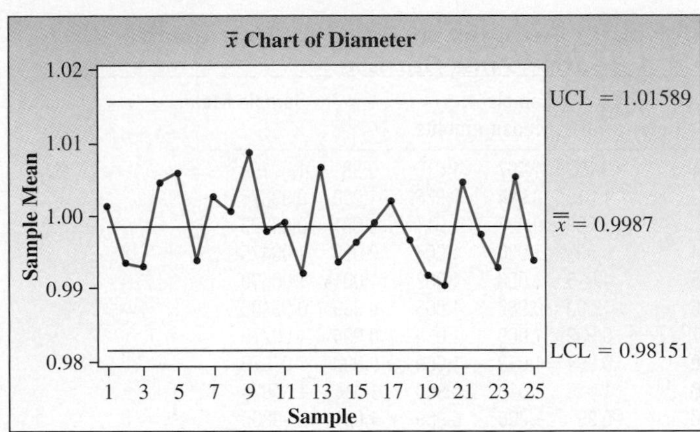

A Control Chart for the Proportion Defective: The *p* Chart

Sometimes the observation made on an item is simply whether or not it meets specifications; thus, it is judged to be defective or non-defective. If the fraction defective produced by the process is p, then x, the number of defectives in a sample of n items, has a binomial distribution.

To monitor a process for defective items, samples of size n are selected at periodic intervals and the sample proportion $\hat{p}$ is calculated. When the process is in control, $\hat{p}$ should fall into the interval $p \pm 3\text{SE}$, where p is the proportion of defectives in the population (or the process fraction defective) with standard error

$$\text{SE} = \sqrt{\frac{pq}{n}} = \sqrt{\frac{p(1-p)}{n}}$$

The process fraction defective is unknown but can be estimated by the average of the k sample proportions

$$\bar{p} = \frac{\Sigma \hat{p}_i}{k}$$

and the standard error is estimated by

$$\text{SE} = \sqrt{\frac{\bar{p}(1-\bar{p})}{n}}$$

The centreline for the ***p* chart** is located at $\bar{p}$, and the upper and lower control limits are

$$\text{UCL} = \bar{p} + 3\sqrt{\frac{\bar{p}(1-\bar{p})}{n}}$$

and

$$\text{LCL} = \bar{p} - 3\sqrt{\frac{\bar{p}(1-\bar{p})}{n}}$$

A manufacturer of ballpoint pens randomly samples 400 pens per day and tests each to see whether the ink flow is acceptable. The proportions of pens judged defective each day over a 40-day period are listed in Table 7.7. Construct a control chart for the proportion $\hat{p}$ defective in samples of $n = 400$ pens selected from the process.

TABLE 7.7 **Proportions of Defectives in Samples of $n = 400$ Pens**

Day	Proportion	Day	Proportion	Day	Proportion	Day	Proportion
1	0.0200	11	0.0100	21	0.0300	31	0.0225
2	0.0125	12	0.0175	22	0.0200	32	0.0175
3	0.0225	13	0.0250	23	0.0125	33	0.0225
4	0.0100	14	0.0175	24	0.0175	34	0.0100
5	0.0150	15	0.0275	25	0.0225	35	0.0125
6	0.0200	16	0.0200	26	0.0150	36	0.0300
7	0.0275	17	0.0225	27	0.0200	37	0.0200
8	0.0175	18	0.0100	28	0.0250	38	0.0150
9	0.0200	19	0.0175	29	0.0150	39	0.0150
10	0.0250	20	0.0200	30	0.0175	40	0.0225

SOLUTION The estimate of the process proportion defective is the average of the $k = 40$ sample proportions in Table 7.7. Therefore, the centreline of the control chart is located at

$$\bar{p} = \frac{\Sigma\hat{p}_i}{k} = \frac{0.0200 + 0.0125 + \cdots + 0.0225}{40} = \frac{0.7600}{40} = 0.019$$

An estimate of SE, the standard error of the sample proportions, is

$$\sqrt{\frac{\bar{p}(1-\bar{p})}{n}} = \sqrt{\frac{(0.019)(0.981)}{400}} = 0.00683$$

and $3\text{SE} = (3)(0.00683) = 0.0205$. Therefore, the upper and lower control limits for the p chart are located at

$$\text{UCL} = \bar{p} + 3\text{SE} = 0.0190 + 0.0205 = 0.0395$$

and

$$\text{LCL} = \bar{p} - 3\text{SE} = 0.0190 - 0.0205 = -0.0015$$

Or, since p cannot be negative, $\text{LCL} = 0$.

The p control chart is shown in Figure 7.14. Note that all 40 sample proportions fall within the control limits. If a sample proportion collected at some time in the future falls outside the control limits, the manufacturer should be concerned about an increase in the defective rate. He should take steps to look for the possible causes of this increase.

FIGURE 7.14
MINITAB p chart for Example 7.13

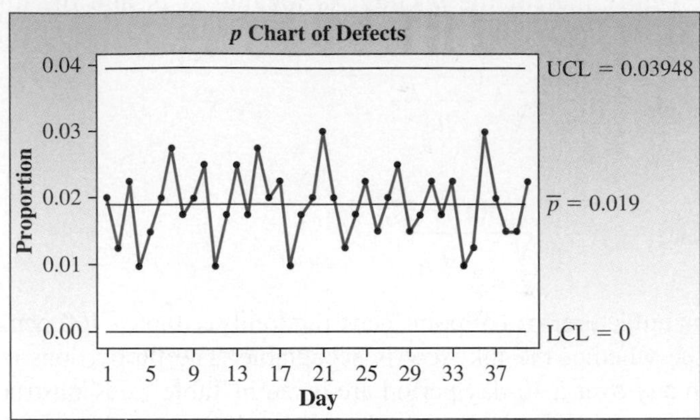

Other commonly used control charts are the *R chart,* which is used to monitor variation in the process variable by using the sample range, and the *c chart,* which is used to monitor the number of defects per item.

7.7 EXERCISES

BASIC TECHNIQUES

7.57 The sample means were calculated for 30 samples of size $n = 10$ for a process that was judged to be in control. The means of the 30 $\bar{x}$-values and the standard deviation of the combined 300 measurements were $\bar{\bar{x}} = 20.74$ and $s = 0.87$, respectively.

a. Use the data to determine the upper and lower control limits for an $\bar{x}$ chart.

b. What is the purpose of an $\bar{x}$ chart?

c. Construct an $\bar{x}$ chart for the process and explain how it can be used.

7.58 The sample means were calculated for 40 samples of size $n = 5$ for a process that was judged to be in control. The means of the 40 values and the standard deviation of the combined 200 measurements were $\bar{\bar{x}} = 155.9$ and $s = 4.3$, respectively.

a. Use the data to determine the upper and lower control limits for an $\bar{x}$ chart.

b. Construct an $\bar{x}$ chart for the process and explain how it can be used.

7.59 Explain the difference between an $\bar{x}$ chart and a *p* chart.

7.60 Samples of $n = 100$ items were selected hourly over a 100-hour period, and the sample proportion of defectives was calculated each hour. The mean of the 100 sample proportions was 0.035.

a. Use the data to find the upper and lower control limits for a *p* chart.

b. Construct a *p* chart for the process and explain how it can be used.

7.61 Samples of $n = 200$ items were selected hourly over a 100-hour period, and the sample proportion of defectives was calculated each hour. The mean of the 100 sample proportions was 0.041.

a. Use the data to find the upper and lower control limits for a *p* chart.

b. Construct a *p* chart for the process and explain how it can be used.

APPLICATIONS

7.62 Blackjack A gambling casino records and plots the mean daily gain or loss from five blackjack tables on an $\bar{x}$ chart. The overall mean of the sample means and the standard deviation of the combined data over 40 weeks were $\bar{\bar{x}} = \$10,752$ and $s = \$1605$, respectively.

a. Construct an $\bar{x}$ chart for the mean daily gain per blackjack table.

b. How can this $\bar{x}$ chart be of value to the manager of the casino?

7.63 Brass Rivets A producer of brass rivets randomly samples 400 rivets each hour and calculates the proportion of defectives in the sample. The mean sample proportion calculated from 200 samples was equal to 0.021. Construct a control chart for the proportion of defectives in samples of 400 rivets. Explain how the control chart can be of value to a manager.

Data set EX0764 **7.64 Lumber Specs** The manager of a building-supplies company randomly samples incoming lumber to see whether it meets quality specifications. From each shipment, 100 pieces of 2 × 4 lumber are inspected and judged according to whether they are first (acceptable) or second (defective) grade. The proportions of second-grade 2 × 4s recorded for 30 shipments were as follows:

0.14	0.21	0.19	0.18	0.23	0.20	0.25	0.19	0.22	0.17
0.21	0.15	0.23	0.12	0.19	0.22	0.15	0.26	0.22	0.21
0.14	0.20	0.18	0.22	0.21	0.13	0.20	0.23	0.19	0.26

Construct a control chart for the proportion of second-grade 2 × 4s in samples of 100 pieces of lumber. Explain how the control chart can be of use to the manager of the building-supplies company.

7.65 Coal-Burning Power Plant A coal-burning power plant tests and measures three specimens of coal each day to monitor the percentage of ash in the coal. The overall mean of 30 daily sample means and the combined standard deviation of all the data were $\bar{\bar{x}} = 7.24$ and $s = 0.07$, respectively. Construct an $\bar{x}$ chart for the process and explain how it can be of value to the manager of the power plant.

Data set EX0766 **7.66 Nuclear Power Plant** The data in the table are measures of the radiation in air particulates at a nuclear power plant. Four measurements were recorded at weekly intervals over a 26-week period. Use the data to construct an $\bar{x}$ chart and plot the 26 values of $\bar{x}$. Explain how the chart can be used.

Week	Radiation			
1	0.031	0.032	0.030	0.031
2	0.025	0.026	0.025	0.025
3	0.029	0.029	0.031	0.030
4	0.035	0.037	0.034	0.035
5	0.022	0.024	0.022	0.023
6	0.030	0.029	0.030	0.030
7	0.019	0.019	0.018	0.019
8	0.027	0.028	0.028	0.028
9	0.034	0.032	0.033	0.033
10	0.017	0.016	0.018	0.018
11	0.022	0.020	0.020	0.021
12	0.016	0.018	0.017	0.017
13	0.015	0.017	0.018	0.017
14	0.029	0.028	0.029	0.029
15	0.031	0.029	0.030	0.031
16	0.014	0.016	0.016	0.017
17	0.019	0.019	0.021	0.020
18	0.024	0.024	0.024	0.025
19	0.029	0.027	0.028	0.028
20	0.032	0.030	0.031	0.030
21	0.041	0.042	0.038	0.039
22	0.034	0.036	0.036	0.035
23	0.021	0.022	0.024	0.022
24	0.029	0.029	0.030	0.029
25	0.016	0.017	0.017	0.016
26	0.020	0.021	0.020	0.022

7.67 Baseball Bats A hardwoods manufacturing plant has several different production lines to make baseball bats of different weights. One such production line is designed to produce bats weighing 910 grams. During a period of time when the production process was known to be in statistical control, the average bat weight was found to be 900 grams. The observed data were gathered from 50 samples, each consisting of five measurements. The standard deviation of all samples was found to be $s = 6$ grams. Construct an $\bar{x}$-chart to monitor the 910-gram bat production process.

7.68 Baseball Bats, continued Refer to Exercise 7.67 and suppose that during a day when the state of the 910-gram bat production process was unknown, the following measurements were obtained at hourly intervals:

Hour	$\bar{x}$	Hour	$\bar{x}$
1	985	4	938
2	921	5	985
3	946	6	901

Each measurement represents a statistic computed from a sample of five bat weights selected from the production process during a certain hour. Use the control chart constructed in Exercise 7.67 to monitor the process.

CHAPTER REVIEW

Key Concepts and Formulas

I. Sampling Plans and Experimental Designs

1. Simple random sampling
    a. Each possible sample of size n is equally likely to occur.
    b. Use a computer or a table of random numbers.
    c. Problems are non-response, undercoverage, and wording bias.
2. Other sampling plans involving randomization
    a. Stratified random sampling
    b. Cluster sampling
    c. Systematic 1-in-k sampling
3. Non-random sampling
    a. Convenience sampling
    b. Judgment sampling
    c. Quota sampling

II. Statistics and Sampling Distributions

1. Sampling distributions describe the possible values of a statistic and how often they occur in repeated sampling.
2. Sampling distributions can be derived mathematically, approximated empirically, or found using statistical theorems.
3. The Central Limit Theorem states that sums and averages of measurements from a non-normal population with finite mean μ and standard deviation σ have approximately normal distributions for large samples of size n.

III. Sampling Distribution of the Sample Mean

1. When samples of size n are randomly drawn from a normal population with mean μ and variance σ^2, the sample mean $\bar{x}$ has a normal distribution with mean μ and standard deviation $\sigma/\sqrt{n}$.
2. When samples of size n are randomly drawn from a non-normal population with mean μ and variance σ^2, the Central Limit Theorem ensures that the sample mean $\bar{x}$ will have an approximately normal distribution with mean μ and standard deviation $\sigma/\sqrt{n}$ when n is large ($n \geq 30$).
3. Probabilities involving the sample mean can be calculated by standardizing the value of $\bar{x}$ using z:

$$z = \frac{\bar{x} - \mu}{\sigma/\sqrt{n}}$$

IV. Sampling Distribution of the Sample Proportion

1. When samples of size n are drawn from a binomial population with parameter p, the sample proportion $\hat{p}$ will have an approximately normal distribution with mean p and standard deviation $\sqrt{pq/n}$ as long as $np > 5$ and $nq > 5$.
2. Probabilities involving the sample proportion can be calculated by standardizing the value $\hat{p}$ using z:

$$z = \frac{\hat{p} - p}{\sqrt{\dfrac{pq}{n}}}$$

V. Statistical Process Control

1. To monitor a quantitative process, use an $\bar{x}$ chart. Select k samples of size n and calculate the overall mean $\bar{\bar{x}}$ and the standard deviation s of all nk measurements. Create upper and lower control limits as

$$\bar{\bar{x}} \pm 3\frac{s}{\sqrt{n}}$$

If a sample mean exceeds these limits, the process is out of control.

2. To monitor a *binomial* process, use a p chart. Select k samples of size n and calculate the average of the sample proportions as

$$\bar{p} = \frac{\Sigma \hat{p}_i}{k}$$

Create upper and lower control limits as

$$\bar{p} \pm 3\sqrt{\frac{\bar{p}(1-\bar{p})}{n}}$$

If a sample proportion exceeds these limits, the process is out of control.

The Central Limit Theorem at Work—*Microsoft Excel*

Excel can be used to explore the way the Central Limit Theorem works in practice. Remember that, according to the Central Limit Theorem, if random samples of size n are drawn from a non-normal population with mean μ and standard deviation σ, then when n is large, the sampling distribution of the sample mean $\bar{x}$ will be approximately normal with the same mean μ and with standard error $\sigma/\sqrt{n}$. Let's try sampling from a non-normal population using *Excel*.

In a new spreadsheet, generate 100 samples of size $n = 30$ from a continuous uniform distribution (Chapter 6, p. 236) over the interval (0, 10). Label column A as "Sample" and enter the numbers 1 to 100 in that column. Then select **Data ▶ Data Analysis ▶ Random Number Generation** to obtain the dialogue box in Figure 7.15(a). Type **30** for the number of variables and 100 for the number of random numbers. In the drop-down "Distribution" list, choose **Uniform,** with parameters between **0** and **10.** We will leave the first row of our spreadsheet empty, starting the "Output Range" at cell B2. Press **OK** to see the 100 random samples of size $n = 30$. You can look at the distribution of the entire set of data using **Data ▶ Data Analysis ▶ Histogram,** choosing bins 1, 2, … , 9, 10 and using the procedures described in the "Technology Today" section in Chapter 2. For our data, the distribution, shown in Figure 7.15(b) is not mound-shaped, but is fairly flat, as expected for the uniform distribution.

For the uniform distribution that we have used, the mean and standard deviation are $\mu = 5$ and $\sigma = 2.89$, respectively. Check the descriptive statistics for the $30 \times 100 = 3000$ measurements (use the functions = **AVERAGE(B2:AE101)** and = **STDEV(B2:AE101)**), and you will find that the 100 observations have a sample mean and standard deviation *close to* but not exactly equal to $\mu = 5$ and $\sigma = 2.89$, respectively.

FIGURE 7.15
MINITAB screen shots (a) Random Number Generation (b) Histogram of Uniform Data

(a)

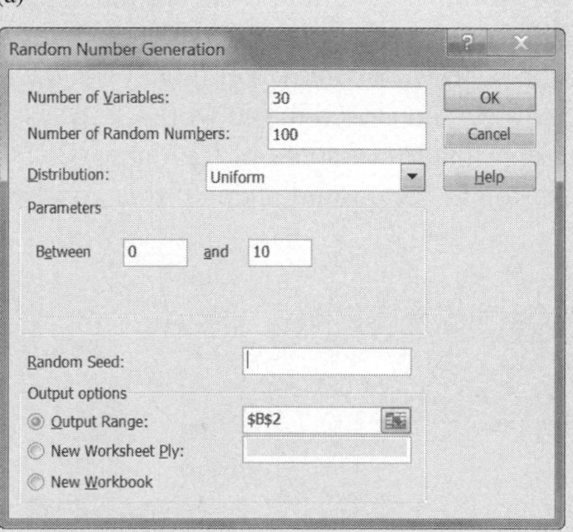

(b)

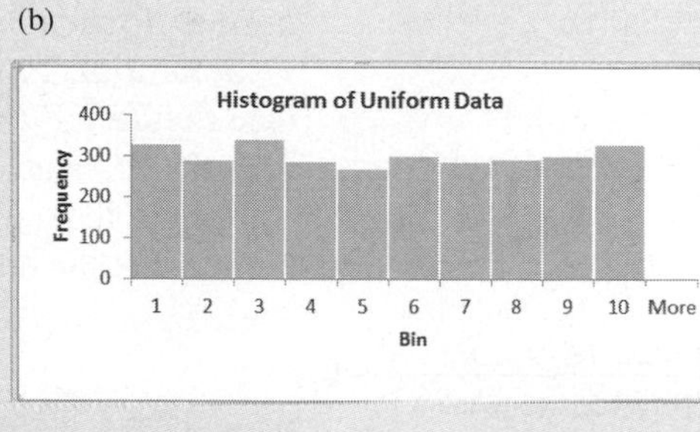

Now, generate 100 values of $\bar{x}$ based on samples of size $n = 30$ by creating a column of means for the 100 rows. First, label column AF as "x-bar" and place your cursor in cell AF2. Use **Insert Function ▶ Statistical ▶ Average** (or type = **AVERAGE(B2:AE2)**) to obtain the first average. Then copy the formula into the other 99 cells in column AF. You can now look at the distribution of these 100 sample means using **Data ▶ Data Analysis ▶ Histogram** and choosing bins 1, 1.5, 2, 2.5, … , 9, 9.5, 10. The distribution for our 100 sample means is shown in Figure 7.16.

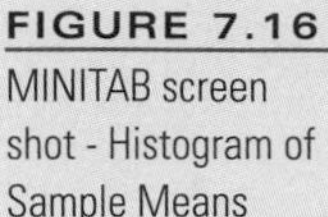

FIGURE 7.16
MINITAB screen shot - Histogram of Sample Means

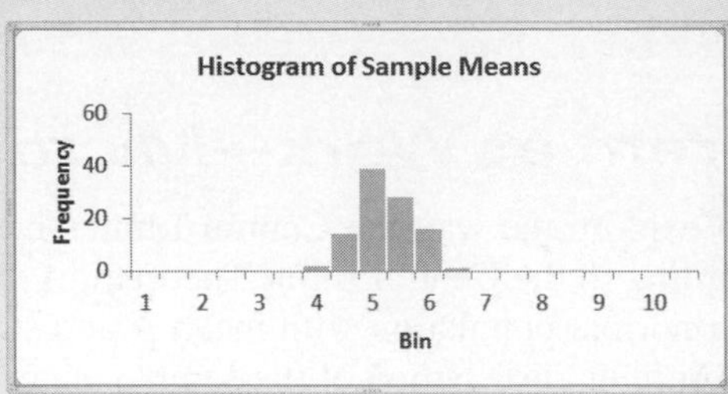

Notice the distinct mound shape of the distribution in Figure 7.16 compared to the original distribution in Figure 7.15(b). Also, if you check the mean and standard deviation for the 100 sample means in column AE, you will find that they are not too different from the theoretical values, $\mu = 5$ and $\sigma/\sqrt{n} = 2.89/\sqrt{30} = 0.53$. (For our data, the sample mean is 4.98 and the standard deviation is 0.49.) Since we had only 100 samples, our results are not *exactly* equal to the theoretical values. If we had generated an *infinite* number of samples, we would have had an exact match. This is the Central Limit Theorem at work!

The Central Limit Theorem at Work—*MINITAB*

MINITAB provides a perfect tool for exploring the way the Central Limit Theorem works in practice. Remember that, according to the Central Limit Theorem, if random samples of size n are drawn from a non-normal population with mean μ and standard deviation σ, then when n is large, the sampling distribution of the sample mean $\bar{x}$ will be approximately normal with the same mean μ and with standard error $\sigma/\sqrt{n}$. Let's try sampling from a non-normal population with the help of *MINITAB*.

In a new *MINITAB* worksheet, generate 100 samples of size $n = 30$ from a non-normal distribution called the exponential distribution. Use **Calc ▶ Random Data ▶ Exponential.** Type **100** for the number of rows of data, and store the results in C1–C30 (see Figure 7.17(a)). Leave the mean at the default of 1.0, the threshold at 0.0, and click **OK.** The data are generated and stored in the worksheet. Use **Graph ▶ Histogram ▶ Simple** to look at the distribution of some of the data—say, C1 (as in Figure 7.17(b)). Notice that the distribution is not mound-shaped; it is highly skewed to the right.

FIGURE 7.17
MINITAB screen shots
(a) Exponential Distribution
(b) Histogram of C1

(a)

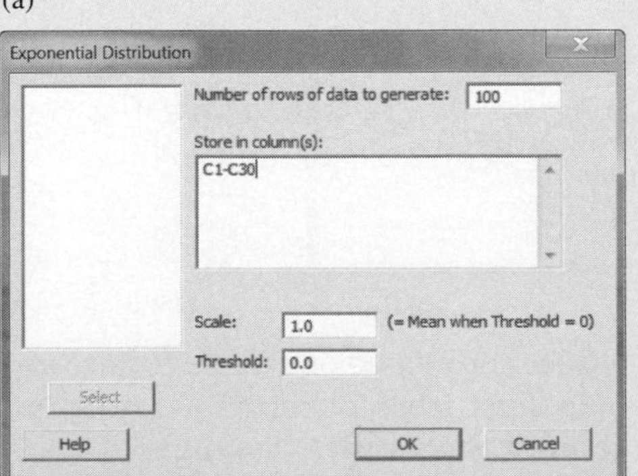

(b)

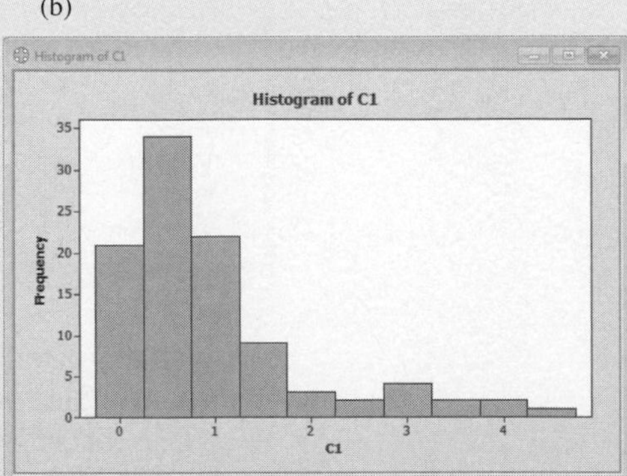

For the exponential distribution that we have used, the mean and standard deviation are $\mu = 1$ and $\sigma = 1$, respectively. Check the descriptive statistics for one of the columns (use **Stat ▶ Basic Statistics ▶ Display Descriptive Statistics**), and you will find that the 100 observations have a sample mean and standard deviation that are both *close to* but not exactly equal to 1. Now, generate 100 values of $\bar{x}$ based on samples of size $n = 30$ by creating a column of means for the 100 rows. Use **Calc ▶ Row Statistics,** and select **Mean.** To average the entries in all 30 columns, select or type **C1–C30** in the Input variables box, and store the results in **C31** (see Figure 7.18(a)). You can now look at the distribution of the sample means using **Graph ▶ Histogram ▶ Simple,** selecting **C31,** and clicking **OK.** The distribution of the 100 sample means generated for our example is shown in Figure 7.18(b).

FIGURE 7.18
MINITAB screen shots
(a) Row Statistics
(b) Histogram of C31

(a)

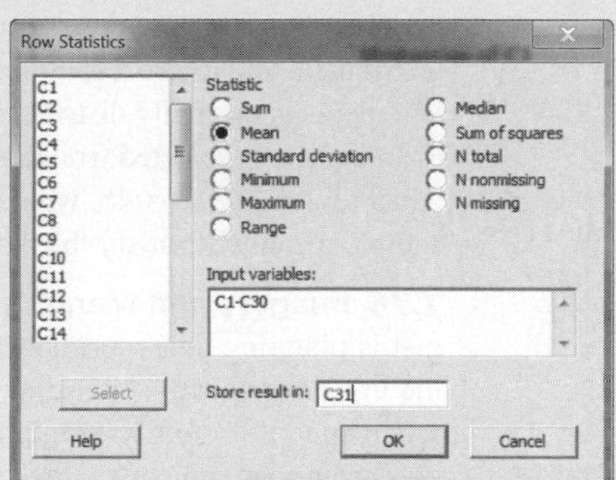

(b)

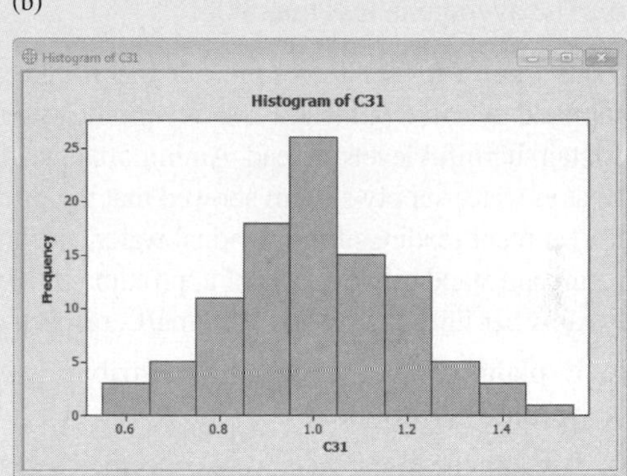

Notice the distinct mound shape of the distribution in Figure 7.18(b) compared to the original distribution in Figure 7.17(b). Also, if you check the descriptive statistics for C31, you will find that the mean and standard deviation of our 100 sample means are not too different from the theoretical values, $\mu = 1$ and $\sigma/\sqrt{n} = 1/\sqrt{30} = 0.18$. (For our data, the sample mean is 1.0024 and the standard deviation is 0.1813.) Since we had only 100 samples, our results are not *exactly* equal to the theoretical values. If we had generated an *infinite* number of samples, we would have had an exact match. This is the Central Limit Theorem at work!

Supplementary Exercises

7.69 A finite population consists of four elements: 6, 1, 3, 2.

a. How many different samples of size $n = 2$ can be selected from this population if you sample without replacement? (Sampling is said to be *without replacement* if an element cannot be selected twice for the same sample.)

b. List the possible samples of size $n = 2$.

c. Compute the sample mean for each of the samples given in part b.

d. Find the sampling distribution of $\bar{x}$. Use a probability histogram to graph the sampling distribution of $\bar{x}$.

e. If all four population values are equally likely, calculate the value of the population mean μ. Do any of the samples listed in part b produce a value of $\bar{x}$ exactly equal to μ?

7.70 Refer to Exercise 7.69. Find the sampling distribution for $\bar{x}$ if random samples of size $n = 3$ are selected *without replacement.* Graph the sampling distribution of $\bar{x}$.

7.71 Suppose a random sample of $n = 5$ observations is selected from a population that is normally distributed, with mean equal to 1 and standard deviation equal to 0.36.

a. Give the mean and standard deviation of the sampling distribution of $\bar{x}$.

b. Find the probability $\bar{x}$ that exceeds 1.3.

c. Find the probability that the sample mean $\bar{x}$ is less than 0.5.

d. Find the probability that the sample mean deviates from the population mean $\mu = 1$ by more than 0.4.

7.72 Batteries A certain type of automobile battery is known to last an average of 1110 days with a standard deviation of 80 days. If 400 of these batteries are selected, find the following probabilities for the average length of life of the selected batteries:

a. The average is between 1100 and 1110.

b. The average is greater than 1120.

c. The average is less than 900.

7.73 Lead Pipes Studies indicate that drinking water supplied by some old lead-lined city piping systems may contain harmful levels of lead. An important study of the Boston water supply system showed that the distribution of lead content readings for individual water specimens had a mean and standard deviation of approximately 0.033 milligrams per litre (mg/L) and 0.10 mg/L, respectively.[21]

a. Explain why you believe this distribution is or is not normally distributed.

b. Because the researchers were concerned about the shape of the distribution in part a, they calculated the average daily lead levels at 40 different locations on each of 23 randomly selected days. What can you say about the shape of the distribution of the average daily lead levels from which the sample of 23 days was taken?

c. What are the mean and standard deviation of the distribution of average lead levels in part b?

7.74 Biomass The total amount of vegetation held by the earth's forests is important to both ecologists and politicians because green plants absorb carbon dioxide. An underestimate of the earth's vegetative mass, or biomass, means that much of the carbon dioxide emitted by human activities (primarily fossil-burning fuels) will not be absorbed, and a climate-altering buildup of carbon dioxide will occur. Studies[22] indicate that the biomass for tropical woodlands, thought to be about 35 kilograms per square metre (kg/m^2), may in fact be too high and that tropical biomass values vary regionally—from about 5 to 55 kg/m^2. Suppose you measure the tropical biomass in 400 randomly selected square-metre plots.

a. Approximate σ, the standard deviation of the biomass measurements.

b. What is the probability that your sample average is within two units of the true average tropical biomass?

c. If your sample average is $\bar{x} = 31.75$, what would you conclude about the overestimation that concerns the scientists?

7.75 Hard Hats The safety requirements for hard hats worn by construction workers and others, established by the American National Standards Institute (ANSI), specify that each of three hats pass the following test. A hat is mounted on an aluminum head form. A 4-kilogram (kg) steel ball is dropped on the hat from a height of 1.5 metres, and the resulting force is measured at the bottom of the head form. The force exerted on the head form by each of the three hats must be less than 455 kg, and the average of the three must be less than 386 kg. (The relationship between this test and actual human head damage is unknown.) Suppose the exerted force is normally distributed, and hence a sample mean of three force measurements is normally distributed. If a random sample of three hats is selected from a shipment with a mean equal to 410 and $\sigma = 45$, what is the probability that the sample mean will satisfy the ANSI standard?

7.76 Imagery and Memory A research psychologist is planning an experiment to determine whether the use of imagery—picturing a word in your mind—affects people's ability to memorize. He wants to use two groups of subjects: a group that memorizes a set of 20 words using the imagery technique, and a control group that does not use imagery.

a. Use a randomization technique to divide a group of 20 subjects into two groups of equal size.

b. How can the researcher randomly select the group of 20 subjects?

c. Suppose the researcher offers to pay subjects \$50 each to participate in the experiment and uses the first 20 students who apply. Would this group behave as if it were a simple random sample of size $n = 20$?

7.77 Child Abuse A study of nearly 2000 women included questions dealing with child abuse and its effect on the women's adult life.[23] The study reported on the likelihood that a woman who was abused as a child would suffer either physical abuse or physical problems arising from depression, anxiety, low self-esteem, and drug abuse as an adult.

a. Is this an observational study or a designed experiment?

b. What problems might arise because of the sensitive nature of this study? What kinds of biases might occur?

7.78 Sprouting Radishes A biology experiment was designed to determine whether sprouting radish seeds inhibit the germination of lettuce seeds.[24] Three 10-centimetre petri dishes were used. The first contained 26 lettuce seeds, the second contained 26 radish seeds, and the third contained 13 lettuce seeds and 13 radish seeds.

a. Assume that the experimenter had a package of 50 radish seeds and another of 50 lettuce seeds. Devise a plan for randomly assigning the radish and lettuce seeds to the three treatment groups.

b. What assumptions must the experimenter make about the packages of 50 seeds in order to assure randomness in the experiment?

7.79 Canadian Identity Do Canadians see themselves as Canadians, as a resident of a province, or equally as a Canadian and a resident of a province? According to Strategic Counsel, who presented the findings of the *Globe and Mail*/CTV 2005 polling program, 41% said they see themselves as Canadians.[25] The result is based on interviews conducted by telephone among a national sample of 1000 adult Canadians 18 years of age or older.

a. Is this an observational study or a designed experiment?

b. What problems might or could have occurred because of the sensitive nature of the subject? What kinds of biases might have occurred?

7.80 Telephone Service Suppose a telephone company executive wishes to select a random sample of $n = 20$ (a small number is used to simplify the exercise) out of 7000 customers for a survey of customer attitudes concerning service. If the customers are numbered for identification purposes, indicate the customers whom you will include in your sample. Use the random number table and explain how you selected your sample.

7.81 Rh-Positive The proportion of individuals with an Rh-positive blood type is 85%. You have a random sample of $n = 500$ individuals.

a. What are the mean and standard deviation of $\hat{p}$, the sample proportion with Rh-positive blood type?

b. Is the distribution of $\hat{p}$ approximately normal? Justify your answer.

c. What is the probability that the sample proportion $\hat{p}$ exceeds 82%?

d. What is the probability that the sample proportion lies between 83% and 88%?

e. The sample proportion would lie between what two limits 99% of the time?

7.82 What survey design is used in each of these situations?

a. A random sample of $n = 50$ city blocks is selected, and a census is done for each single-family dwelling on each block.

b. The highway patrol stops every 10th vehicle on a given city artery between 9:00 A.M. and 3:00 P.M. to perform a routine traffic safety check.

c. One hundred households in each of four city wards are surveyed concerning a pending city tax relief referendum.

d. Every 10th tree in a managed slash pine plantation is checked for pine needle borer infestation.

e. A random sample of $n = 1000$ taxpayers from the city of Halifax is selected by the Canada Revenue Agency and their tax returns are audited.

7.83 Elevator Loads The maximum load (with a generous safety factor) for the elevator in an office building is 900 kg. The relative frequency distribution of the weights of all men and women using the elevator is mound-shaped (slightly skewed to the heavy weights), with mean μ equal to 65 kg and standard deviation σ equal to 16 kg. What is the largest number of people you can allow on the elevator if you want their total weight to exceed the maximum weight with a small probability (say, near 0.01)? (HINT: If $x_1, x_2, \ldots, x_n$ are independent observations made on a random variable x, and if x has mean μ and variance σ^2, then the mean and variance of Σx_i are $n\mu$ and $n\sigma^2$, respectively. This result was given in Section 7.4.)

7.84 Wiring Packages The number of wiring packages that can be assembled by a company's employees has a normal distribution, with a mean equal to 16.4 per hour and a standard deviation of 1.3 per hour.

a. What are the mean and standard deviation of the number x of packages produced per worker in an eight-hour day?

b. Do you expect the probability distribution for x to be mound-shaped and approximately normal? Explain.

c. What is the probability that a worker will produce at least 135 packages per eight-hour day?

7.85 Wiring Packages, continued Refer to Exercise 7.84. Suppose the company employs 10 assemblers of wiring packages.

a. Find the mean and standard deviation of the company's daily (eight-hour day) production of wiring packages.

b. What is the probability that the company's daily production is less than 1280 wiring packages per day?

Data set EX0786 **7.86 Defective Light Bulbs** The table lists the number of defective 60-watt light bulbs found in samples of 100 bulbs selected over 25 days from a manufacturing process. Assume that during these 25 days, the manufacturing process was not producing an excessively large fraction of defectives.

Day	1	2	3	4	5	6	7	8	9	10
Defectives	4	2	5	8	3	4	4	5	6	1

Day	11	12	13	14	15	16	17	18	19	20
Defectives	2	4	3	4	0	2	3	1	4	0

Day	21	22	23	24	25
Defectives	2	2	3	5	3

a. Construct a p chart to monitor the manufacturing process, and plot the data.

b. How large must the fraction of defective items be in a sample selected from the manufacturing process before the process is assumed to be out of control?

c. During a given day, suppose a sample of 100 items is selected from the manufacturing process and 15 defective bulbs are found. If a decision is made to shut down the manufacturing process in an attempt to locate the source of the implied controllable variation, explain how this decision might lead to erroneous conclusions.

7.87 Light Bulbs, continued A hardware store chain purchases large shipments of light bulbs from the manufacturer described in Exercise 7.86 and specifies that each shipment must contain no more than 4% defectives. When the manufacturing process is in control, what is the probability that the hardware store's specifications are met?

7.88 Light Bulbs, again Refer to Exercise 7.86. During a given week, the number of defective bulbs in each of five samples of 100 were found to be 2, 4, 9, 7, and 11. Is there reason to believe that the production process has been producing an excessive proportion of defectives at any time during the week?

Data set EX0789 **7.89 Canned Tomatoes** During long production runs of canned tomatoes, the average weights (in mL) of samples of five cans of standard-grade tomatoes in puree form were taken at 30 control points during an 11-day period. These results are shown in the table.[26] When the machine is performing normally, the average weight per can is 621 mL with a standard deviation of 35.5 mL.

a. Compute the upper and lower control limits and the centreline for the $\bar{x}$ chart.

b. Plot the sample data on the $\bar{x}$ chart and determine whether the performance of the machine is in control.

Sample Number	Average Weight	Sample Number	Average Weight
1	654.9	16	606.7
2	603.8	17	578.3
3	623.7	18	646.4
4	606.7	19	598.2
5	618.0	20	586.8
6	584.0	21	612.3
7	569.8	22	635.0
8	606.7	23	603.8
9	609.5	24	598.2
10	572.7	25	569.8
11	575.5	26	601.0
12	569.8	27	564.2
13	615.2	28	598.2
14	595.3	29	612.3
15	612.3	30	603.8

Source: Peter Hackl and Johannes Ledolter, "A Control Chart Based on Ranks," *Journal of Quality Technology*, Vol. 23, Iss. 2, April 1991. Reprinted and adapted with permission from the *Journal of Quality Technology* © 1991 ASQ, www.asq.org. No further distribution allowed without permission.

7.90 Pepsi or Coke? The battle for consumer preference continues between Pepsi and Coke. How can you make your preferences known? There is a webpage where you can vote for one of these colas if you click on the link that says PAY CASH for your opinion. Explain why the respondents do not represent a random sample of the opinions of purchasers or drinkers of these drinks. Explain the types of distortions that could creep into an Internet opinion poll.

7.91 Strawberries An experimenter wants to find an appropriate temperature at which to store fresh strawberries to minimize the loss of ascorbic acid. There are 20 storage containers, each with controllable temperature, in which strawberries can be stored. If two storage temperatures are to be used, how would the experimenter assign the 20 containers to one of the two storage temperatures?

7.92 Filling Pop Cans A bottler of soft drinks packages cans in six-packs. Suppose that the fill per can has an approximate normal distribution with a mean of 355 mL and a standard deviation of 5.91 mL.

a. What is the distribution of the total fill for a case of 24 cans?

b. What is the probability that the total fill for a case is less than 8457 mL?

c. If a six-pack of pop can be considered a random sample of size $n = 6$ from the population, what is the probability that the average fill per can for a six-pack of pop is less than 348.93 mL?

7.93 Total Packing Weight Packages of food whose average weight is 454 grams with a standard deviation of 17 grams are shipped in boxes of 24 packages. If the package weights are approximately normally distributed, what is the probability that a box of 24 packages will weigh more than 11,113 grams (11.113 kg)?

7.94 Electronic Components A manufacturing process is designed to produce an electronic component for use in small portable television sets. The components are all of standard size and need not conform to any measurable characteristic, but are sometimes inoperable when emerging from the manufacturing process. Fifteen samples were selected from the process at times when the process was known to be in statistical control. Fifty components were observed within each sample, and the number of inoperable components was recorded.

6, 7, 3, 5, 6, 8, 4, 5, 7, 3, 1, 6, 5, 4, 5

Construct a p chart to monitor the manufacturing process.

CASE STUDY  Sampling the Roulette at Monte Carlo

Goals

The technique of simulating a process that contains random elements and repeating the process over and over to see how it behaves is called a **Monte Carlo procedure.** It is widely used in business and other fields to investigate the properties of an operation that is subject to random effects, such as weather, human behaviour, and so on. For example, you could model the behaviour of a manufacturing company's inventory by creating, on paper, daily arrivals and departures of manufactured products from the company's warehouse. Each day a random number of items produced by the company would be received into inventory. Similarly, each day a random number of orders of varying random sizes would be shipped. Based on the input and output of items, you could calculate the inventory—that is, the number of items on hand at the end of each day. The values of the random variables, the number of items produced, the number of orders, and the number of items per order needed for each day's simulation would be obtained from theoretical distributions of observations that closely model the corresponding distributions of the variables that have been observed over time in the manufacturing operation. By repeating the simulation of the supply, the shipping, and the calculation of daily inventory for a large number of days (a sampling of what might really happen), you can observe the behaviour of the plant's daily inventory. The Monte Carlo procedure is particularly valuable because it enables the manufacturer to see how the daily inventory would behave when certain changes are made in the supply pattern or in some other aspect of the operation that could be controlled.

In an article entitled "The Road to Monte Carlo," Daniel Seligman comments on the Monte Carlo method, noting that, although the technique is widely used in business schools to study capital budgeting, inventory planning, and cash flow management, no one seems to have used the procedure to study how well we might do if we were to gamble at Monte Carlo.[27]

To follow up on this thought, Seligman programmed his personal computer to simulate the game of roulette. Roulette involves a wheel with its rim divided into 38 pockets. Thirty-six of the pockets are numbered 1 to 36 and are alternately coloured red and black. The two remaining pockets are coloured green and are marked 0 and 00. To play the game, you bet a certain amount of money on one or more pockets. The wheel is spun and turns until it stops. A ball falls into a slot on the wheel to indicate the winning number. If you have money on that number, you win a specified amount. For example, if you were to play the number 20, the payoff is 35 to 1. If the wheel does not stop at that number, you lose your bet. Seligman decided to see how his nightly gains (or losses) would fare if he were to bet \$5 on each turn of the wheel and repeat the process 200 times each night. He did this 365 times, thereby simulating the outcomes of 365 nights at the casino. Not surprisingly, the mean "gain" per \$1000 evening for the 365 nights was a *loss* of \$55, the average of the winnings retained by the gambling house. The surprise, according to Seligman, was the extreme variability of the nightly "winnings." Seven times out of the 365 evenings, the fictitious gambler lost the \$1000 stake, and only once did he win a maximum of \$1160. On 141 nights, the loss exceeded \$250.

1. To evaluate the results of Seligman's Monte Carlo experiment, first find the probability distribution of the gain x on a single \$5 bet.
2. Find the expected value and variance of the gain x from part 1.
3. Find the expected value and variance for the evening's gain, the sum of the gains or losses for the 200 bets of \$5 each.
4. Use the results of part 2 to evaluate the probability of 7 out of 365 evenings resulting in a loss of the total \$1000 stake.
5. Use the results of part 3 to evaluate the probability that the largest evening's winnings were as great as \$1160.

PROJECTS

Project 7-A: Canada's Average IQ Just Jumped a Bunch—Stephen Hawking Came to Canada!

[*Source:* The Canadian Press, "Physicist Stephen Hawking accepts post at Waterloo Institute," *Toronto Star*, 27 November 2008, www.thestar.com/sciencetech/article/544641. Reproduced by permission of The Canadian Press.]

Renowned physicist Stephen Hawking recently accepted a research position at the Perimeter Institute for Theoretical Physics in Waterloo, Ontario. In doing so, the famed cosmologist has heightened Canada's profile in the international physics community. The ratio of mental age (MA) divided by chronological age (CA) and multiplied 100 is called "Intelligence Quotient" (IQ). It is suggested that the average IQ of top civil servants, professors, and research scientists is 140. Suppose the standard deviation is 5.

a. Suppose a full professor from a Canadian university is selected at random. What is the probability that the IQ of the selected Canadian professor is below 130? State any necessary assumptions you have made to compute this probability.

b. Suppose the assumption(s) made in part a was not justifiable. A researcher decided to take a random sample of 81 full professors from the Canadian university system.

AP Photo/Zero Gravity Corp./FILE

(i) What is the sampling distribution of the sample mean $\bar{x}$? Explain.

(ii) Find the mean and standard deviation of the sampling distribution of $\bar{x}$.

(iii) What is the probability that the sample average IQ of this sample is less than 130? Compare your answer with answer given in a. Summarize your findings.

(iv) Suppose that Stephen Hawking's score was 187. If you include this score in the sample of 81 Canadian professors, do you think Canadian professors' average IQ will improve significantly?

(v) Will the probability in (iii) change if we change the wording from "less than" to "less than or equal to"? Why or why not?

(vi) Can you calculate the probability that the sample average IQ is exactly 135? Justify your answer.

(vii) If the sample mean $\bar{x}$ is actually 130, what can be said about the claim that $\mu = 140$? What conclusion might you draw?

(viii) What is the probability that the sample mean differs from the population mean by more than 2?

(ix) Within what limits would you expect the sample average to be, with probability 0.95?

c. The total IQ score Σx_i is the sum of the individual 81 selected professors.

(i) What kind of sampling probability distribution do you expect the total scores to have? Explain.

(ii) Provide the mean and standard deviation of the probability distribution of the total score Σx_i.

(iii) Find the probability that the total score will be between 11,000 and 11,400, inclusively.

Project 7-B: Test the Nation on CBC

Where do Canada's most clever people live? Who wins the IQ battle of the sexes? We found out on March 18, 2007.

"Test the Nation" was the biggest survey ever conducted to see just how smart Canadians are. In this live two-hour special, 1.5 million Canadians participated in a real-time interactive IQ test. Viewers took the test in the comfort of their own homes, on the Internet or with pen and paper, while seven teams—tattoo artists, millionaires, fitness instructors, surgeons, mayors, talk jocks, and celebrities—exercised their grey matter in CBC's Toronto studio. Suppose that 20% of Canadians scored above average (that is, scoring 110 or more). You took a random sample of $n = 64$ and recorded their IQ scores.

a. What are the mean and standard deviation of sample proportion $\hat{p}$, the sample proportion of individuals who scored above average?

b. Is the distribution of $\hat{p}$ approximately normal? Justify your answer.

c. What is the probability that the sample proportion $\hat{p}$ exceeds 18%?

d. What is the probability that the sample proportion $\hat{p}$ lies between 19% and 23%?

e. Ninety-nine percent of the time, the sample proportion $\hat{p}$ would lie between what two limits?

f. What might you conclude if the sample proportion were as small as 10%?

g. Would a value of $\hat{p} = 0.35$ be considered unusual? Justify your answer.

8

Large-Sample Estimation

pio3/Shutterstock.com

GENERAL OBJECTIVES

In previous chapters, you learned about the probability distributions of random variables and the sampling distributions of several statistics that, for large sample sizes, can be approximated by a normal distribution according to the Central Limit Theorem. This chapter presents a method for estimating population parameters and illustrates the concept with practical examples. The Central Limit Theorem and the sampling distributions presented in Chapter 7 play a key role in evaluating the reliability of the estimates.

CHAPTER INDEX

NEED TO KNOW

How to Estimate a Population Mean or Proportion
How to Choose the Sample Size

How Reliable Is That Poll?

Do the national polls conducted by various polling organizations, the news media, and others provide accurate estimates of the percentages of people in Canada who favour various propositions? The case study at the end of this chapter examines the reliability of a poll on "The New Canada," conducted by Ipsos Reid using the theory of large-sample estimation.

8.1 WHERE WE'VE BEEN AND WHERE WE'RE GOING

The first seven chapters of this book have given you the building blocks you will need to understand statistical inference and how it can be applied in practical situations. The first three chapters were concerned with using descriptive statistics, both graphical and numerical, to describe and interpret sets of measurements. In the next three chapters, you learned about probability and probability distributions—the basic tools used to describe *populations* of measurements. The binomial and the normal distributions were emphasized as important for practical applications. The seventh chapter provided the link between probability and statistical inference. Many statistics are either sums or averages calculated from sample measurements. The Central Limit Theorem states that, even if the sampled populations are not normal, the sampling distributions of these *statistics* will be approximately normal when the sample size n is large. These statistics are the tools you use for *inferential statistics*—making inferences about a population using information contained in a sample.

Statistical Inference

Inference—specifically, decision making and prediction—is centuries old and plays a very important role in most peoples' lives. Here are some applications:

- The government needs to predict short- and long-term interest rates.
- A broker wants to forecast the behaviour of the stock market.
- A metallurgist wants to decide whether a new type of steel is more resistant to high temperatures than the old type.
- A consumer wants to estimate the selling price of her house before putting it on the market.

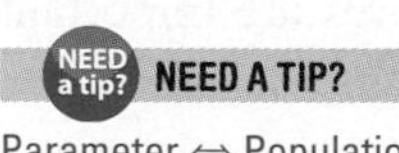

Parameter ⇔ Population
Statistic ⇔ Sample

There are many ways to make these decisions or predictions, some subjective and some more objective in nature. How good will your predictions or decisions be? Although you may feel that your own built-in decision-making ability is quite good, experience suggests that this may not be the case. It is the job of the mathematical statistician to provide methods of statistical inference making that are better and more reliable than just subjective guesses.

Statistical inference is concerned with making decisions or predictions about **parameters**—the numerical descriptive measures that characterize a population. Three parameters you encountered in earlier chapters are the population mean μ, the population standard deviation σ, and the binomial proportion p. In statistical inference, a practical problem is restated in the framework of a population with a specific parameter of interest. For example, the metallurgist could measure the *average* coefficients of expansion for both types of steel and then compare their values.

Methods for making inferences about population parameters fall into one of two categories:

- **Estimation:** Estimating or predicting the value of the parameter
- **Hypothesis testing:** Making a decision about the value of a parameter based on some preconceived idea about what its value might be

EXAMPLE 8.1

The circuits in computers and other electronics equipment consist of one or more printed circuit boards (PCB), and computers are often repaired by simply replacing one or more defective PCBs. In an attempt to find the proper setting of a plating process applied to one side of a PCB, a production supervisor might *estimate* the average thickness of copper plating on PCBs using samples from several days of operation. Since he has no knowledge of the average thickness μ before observing the production process, his is an *estimation* problem.

EXAMPLE 8.2

The supervisor in Example 8.1 is told by the plant owner that the thickness of the copper plating must not be less than 0.003 centimetre (cm) in order for the process to be in control. To decide whether or not the process is in control, the supervisor might formulate a test. He could *hypothesize* that the process is in control—that is, assume that the average thickness of the copper plating is 0.003 cm or greater—and use samples from several days of operation to decide whether or not his hypothesis is correct. The supervisor's decision-making approach is called a *test of hypothesis.*

Which method of inference should be used? That is, should the parameter be estimated, or should you test a hypothesis concerning its value? The answer is dictated by the practical question posed and is often determined by personal preference. Since both estimation and tests of hypotheses are used frequently in scientific literature, we include both methods in this and the next chapter.

A statistical problem, which involves planning, analysis, and inference making, is incomplete without a measure of the **goodness of the inference.** That is, how accurate or reliable is the method you have used? If a stockbroker predicts that the price of a stock will be \$80 next Monday, will you be willing to take action to buy or sell your stock without knowing how reliable her prediction is? Will the prediction be within \$1, \$2, or \$10 of the actual price next Monday? Statistical procedures are important because they provide two types of information:

- Methods for making the inference
- A numerical measure of the goodness or reliability of the inference

8.2 TYPES OF ESTIMATORS

To estimate the value of a population parameter, you can use information from the sample in the form of an **estimator.** Estimators are calculated using information from the sample observations, and hence, by definition they are also *statistics.*

DEFINITION An **estimator** is a rule, usually expressed as a formula, that tells us how to calculate an estimate based on information in the sample.

Estimators are used in two different ways:

- **Point estimation:** Based on sample data, a single number is calculated to estimate the population parameter. The rule or formula that describes this calculation is called the **point estimator,** and the resulting number is called a **point estimate.**

- **Interval estimation:** Based on sample data, two numbers are calculated to form an interval within which the parameter is expected to lie. The rule or formula that describes this calculation is called the **interval estimator,** and the resulting pair of numbers is called an **interval estimate** or **confidence interval.**

EXAMPLE 8.3

A veterinarian wants to estimate the average weight gain per month of four-month-old golden retriever pups that have been placed on a lamb and rice diet. The *population* consists of the weight gains per month of all four-month-old golden retriever pups that are given this particular diet. The veterinarian wants to estimate the unknown parameter μ, the average monthly weight gain for this *hypothetical* population. One possible *estimator* based on sample data is the sample mean, $\bar{x} = \Sigma x_i/n$. It could be used in the form of a single number or *point estimate*—for instance, 1.72 kg—or you could use an *interval estimate* and estimate that the average weight gain will be between 1.22 and 2.22 kg.

Both point and interval estimation procedures use information provided by the sampling distribution of the specific estimator you have chosen to use. We will begin by discussing *point estimation* and its use in estimating population means and proportions.

8.3 POINT ESTIMATION

In a practical situation, there may be several statistics that could be used as point estimators for a population parameter. To decide which of several choices is best, you need to know how the estimator behaves in repeated sampling, described by its *sampling distribution.*

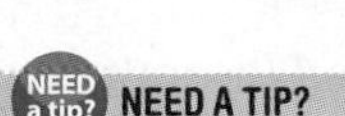

NEED A TIP?

Parameter = Target's bull's-eye
Estimator = Bullet or arrow

By way of analogy, think of firing a revolver at a target. The parameter of interest is the bull's-eye at which you are aiming. Each bullet represents a single sample estimate, fired at the board, which represents the estimator. Suppose your friend fires a single bullet and hits the bull's-eye. Can you conclude that he has excellent aim? Would you stand next to the target while he fires a second bullet? Probably not, because you have no measure of how well he performs in repeated trials. Does he always hit the bull's-eye, or is he consistently too high or too low? Do his shots cluster closely around the target, or do they consistently miss the target by a wide margin? Figure 8.1 shows several target configurations. Which target would you pick as belonging to the best shot?

FIGURE 8.1
Which shooter is best?

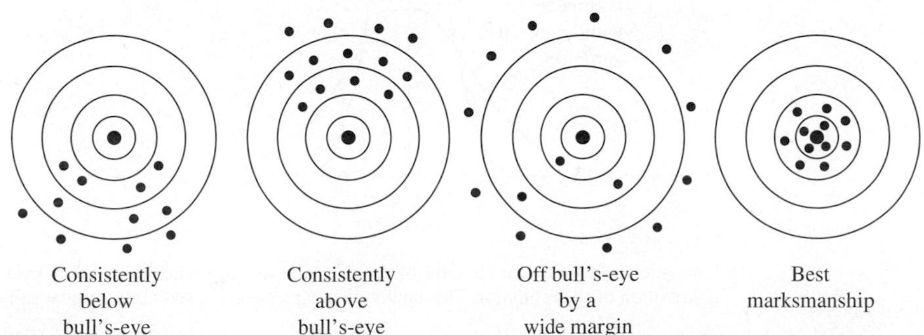

Sampling distributions provide information that can be used to select the **best estimator.** What characteristics would be valuable? First, the **sampling distribution of the point estimator should be centred over the true value of the parameter to be estimated.** That is, the estimator should not consistently underestimate or overestimate the parameter of interest. Such an estimator is said to be **unbiased.**

DEFINITION An estimator of a parameter is said to be **unbiased** if the mean of its distribution is equal to the true value of the parameter. Otherwise, the estimator is said to be **biased.**

The sampling distributions for an unbiased estimator and a biased estimator are shown in Figure 8.2. The sampling distribution for the biased estimator is shifted to the right of the true value of the parameter. This biased estimator is more likely than an unbiased one to overestimate the value of the parameter.

FIGURE 8.2
Distributions for biased and unbiased estimators

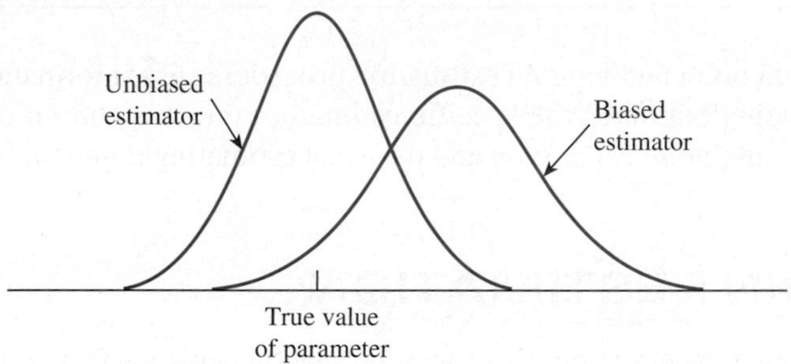

The second desirable characteristic of an estimator is that **the spread (as measured by the variance) of the sampling distribution should be as small as possible.** This ensures that, with a high probability, an individual estimate will fall close to the true value of the parameter. The sampling distributions for two unbiased estimators, one with a small variance[†] and the other with a larger variance, are shown in Figure 8.3.

FIGURE 8.3
Comparison of estimator variability

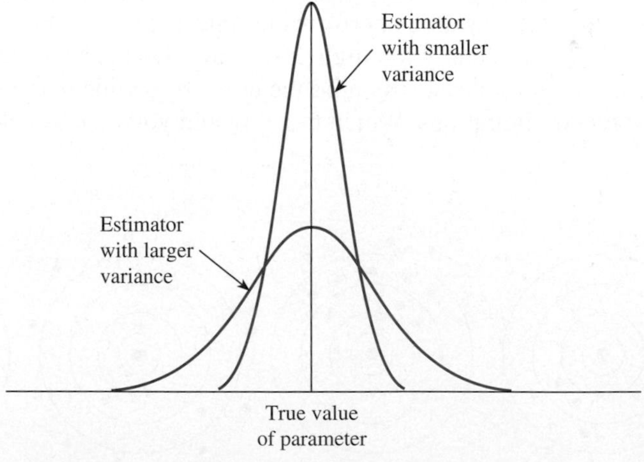

[†] Statisticians usually use the term *variance of an estimator* when in fact they mean the variance of the sampling distribution of the estimator. This contractive expression is used almost universally.

Naturally, you would prefer the estimator with the smaller variance because the estimates tend to lie closer to the true value of the parameter than in the distribution with the larger variance.

In real-life sampling situations, you may know that the sampling distribution of an estimator centres about the parameter that you are attempting to estimate, but all you have is the estimate computed from the n measurements contained in the sample. How far from the true value of the parameter will your estimate lie? How close is the shooter's bullet to the bull's-eye?

DEFINITION The distance between an estimate and the estimated parameter is called the **error of estimation.**

In this chapter, you may assume that the sample sizes are always large and, therefore, that the *unbiased* estimators you will study have sampling distributions that can be approximated by a normal distribution (because of the Central Limit Theorem). Remember that, for any point estimator with a normal distribution, the Empirical Rule states that approximately 95% of all the point estimates will lie within two (or more exactly, 1.96) standard deviations of the mean of that distribution. For *unbiased* estimators, this implies that the difference between the point estimator and the true value of the parameter will be less than 1.96 standard deviations or 1.96 standard errors (SE). This quantity, called the **95% margin of error** (or simply the **margin of error**), provides a practical upper bound for the error of estimation (see Figure 8.4). It is possible that the error of estimation will exceed this margin of error, but that is very unlikely.

FIGURE 8.4
Sampling distribution of an unbiased estimator

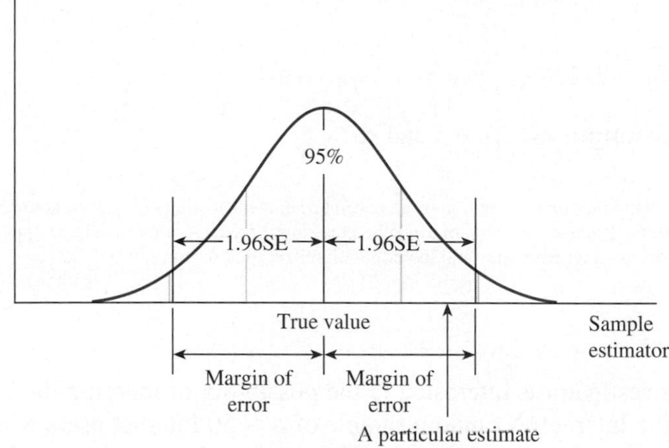

POINT ESTIMATION OF A POPULATION PARAMETER

- Point estimator: a statistic calculated using sample measurements
- 95% margin of error: 1.96 × Standard error of the estimator

NEED A TIP?
95% margin of error = 1.96 × standard error

The sampling distributions for two unbiased point estimators (sample mean $\bar{x}$ and sample proportion) were discussed in Chapter 7. It can be shown that both of these point estimators have the *minimum variability* of all unbiased estimators and are thus the *best estimators* you can find in each situation. A detailed discussion may be found in advanced statistics books.

The variability of the estimator is measured using its standard error. However, you might have noticed that the standard error usually depends on unknown parameters such as σ or p. These parameters must be estimated using sample statistics such as s and $\hat{p}$. Although not exactly correct, experimenters generally refer to the estimated standard error as *the standard error.*

NEED TO KNOW

How to Estimate a Population Mean or Proportion

1. To estimate the population mean μ for a quantitative population, the point estimator $\bar{x}$ is *unbiased* with standard error estimated as

$$\text{SE} = \frac{s}{\sqrt{n}}^{\dagger}$$

The 95% margin of error when $n \geq 30$ is estimated as

$$\pm 1.96\left(\frac{s}{\sqrt{n}}\right)$$

2. To estimate the population proportion p for a binomial population, the point estimator $\hat{p} = x/n$ is *unbiased,* with standard error estimated as

$$\text{SE} = \sqrt{\frac{\hat{p}\hat{q}}{n}}$$

The 95% margin of error is estimated as

$$\pm 1.96\sqrt{\frac{\hat{p}\hat{q}}{n}}$$

Assumptions: $n\hat{p} > 5$ and $n\hat{q} > 5$.

† When you sample from a normal distribution, the statistic $(\bar{x} - \mu)/(s/\sqrt{n})$ has a t distribution, which will be discussed in Chapter 10. When the sample is *large,* this statistic is approximately normally distributed whether the sampled population is normal or non-normal.

EXAMPLE 8.4

An investigator is interested in the possibility of merging the capabilities of television and the Internet. A random sample of $n = 50$ Internet users who were polled about the time they spend watching television produced an average of 11.5 hours per week, with a standard deviation of 3.5 hours. Use this information to estimate the population mean time Internet users spend watching television.

SOLUTION The random variable measured is the time spent watching television per week. This is a quantitative random variable best described by its mean μ. The point estimate of μ, the average time Internet users spend watching television, is $\bar{x} = 11.5$ hours. The margin of error is estimated as

$$1.96\text{SE} = 1.96\left(\frac{s}{\sqrt{n}}\right) = 1.96\left(\frac{3.5}{\sqrt{50}}\right) = 0.97 \approx 1$$

You can feel fairly confident that the sample estimate of 11.5 hours of television watching for Internet users is within ± 1 hour of the population mean.

In reporting research results, investigators often attach either the sample standard deviation s (sometimes called SD) or the standard error $s/\sqrt{n}$ (usually called SE or SEM) to the estimates of population means. You should always look for an explanation somewhere in the text of the report that tells you whether the investigator is reporting $\bar{x} \pm$ SD or $\bar{x} \pm$ SE. In addition, the sample means and standard deviations or standard errors are often presented as "error bars" using the graphical format shown in Figure 8.5.

FIGURE 8.5
Plot of treatment means and their standard errors

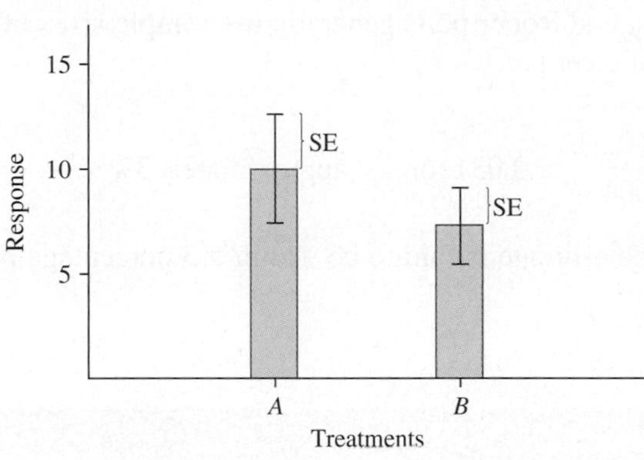

EXAMPLE 8.5

In addition to the average time Internet users spend watching television, the researcher from Example 8.4 is interested in estimating the proportion of individuals in the population at large who want to purchase a television that also acts as a computer. In a random sample of $n = 100$ adults, 45% in the sample indicated that they might buy one. Estimate the true population proportion of adults who are interested in buying a television that also acts as a computer, and find the margin of error for the estimate.

SOLUTION The parameter of interest is now p, the proportion of individuals in the population who want to purchase a television that also acts as a computer. The best estimator of p is the sample proportion, $\hat{p}$, which for this sample is $\hat{p} = 0.45$. In order to find the margin of error, you can approximate the value of p with its estimate $\hat{p} = 0.45$:

$$1.96\,\text{SE} = 1.96\sqrt{\frac{\hat{p}\hat{q}}{n}} = 1.96\sqrt{\frac{0.45(0.55)}{100}} = 0.10$$

With this margin of error, you can be fairly confident that the estimate of 0.45 is within $\pm$ 0.10 of the true value of p. Hence, you can conclude that the true value of p could be as small as 0.35 or as large as 0.55. This margin of error is quite large when compared to the estimate itself and reflects the fact that large samples are required to achieve a small margin of error when estimating p.

Table 8.1 shows how the numerator of the standard error of $\hat{p}$ changes for various values of p. Notice that, for most values of p—especially when p is between 0.3 and 0.7—there is very little change in $\sqrt{pq}$, the numerator of SE, reaching its maximum value when $p = 0.5$. This means that the margin of error using the estimator $\hat{p}$ will also be a maximum when $p = 0.5$. Some pollsters routinely use the maximum margin of error—often called the **sampling error**—when estimating p, in which case they calculate

$$1.96\,\text{SE} = 1.96\sqrt{\frac{0.5(0.5)}{n}} \quad \text{or sometimes} \quad 2\,\text{SE} = 2\sqrt{\frac{0.5(0.5)}{n}}$$

TABLE 8.1 Some Calculated Values of $\sqrt{pq}$

p	pq	$\sqrt{pq}$	p	pq	$\sqrt{pq}$
0.1	0.09	0.30	0.6	0.24	0.49
0.2	0.16	0.40	0.7	0.21	0.46
0.3	0.21	0.46	0.8	0.16	0.40
0.4	0.24	0.49	0.9	0.09	0.30
0.5	0.25	0.50			

Gallup, Harris, and Roper polls generally use sample sizes of approximately 1000, so their margin of error is

$$1.96\sqrt{\frac{0.5(0.5)}{1000}} = 0.031 \text{ or } \quad \text{approximately } 3\%$$

In this case, the estimate is said to be within ± 3 percentage points of the true population proportion.

8.3 EXERCISES

BASIC TECHNIQUES

8.1 Explain what is meant by "margin of error" in point estimation.

8.2 What are two characteristics of the best point estimator for a population parameter?

8.3 Calculate the margin of error in estimating a population mean μ for these values:

a. $n = 30, \sigma^2 = 0.2$

b. $n = 30, \sigma^2 = 0.9$

c. $n = 30, \sigma^2 = 1.5$

8.4 Refer to Exercise 8.3. What effect does a larger population variance have on the margin of error?

8.5 Calculate the margin of error in estimating a population mean μ for these values:

a. $n = 50, s^2 = 4$

b. $n = 500, s^2 = 4$

c. $n = 5000, s^2 = 4$

8.6 Refer to Exercise 8.5. What effect does an increased sample size have on the margin of error?

8.7 Calculate the margin of error in estimating a binomial proportion for each of the following values of n. Use $p = 0.5$ to calculate the standard error of the estimator.

a. $n = 30$ **b.** $n = 100$

c. $n = 400$ **d.** $n = 1000$

8.8 Refer to Exercise 8.7. What effect does increasing the sample size have on the margin of error?

8.9 Calculate the margin of error in estimating a binomial proportion p using samples of size $n = 100$ and the following values for p:

a. $p = 0.1$ **b.** $p = 0.3$ **c.** $p = 0.5$

d. $p = 0.7$ **e.** $p = 0.9$

f. Which of the values of p produces the largest margin of error?

8.10 Suppose you are writing a questionnaire for a sample survey involving $n = 100$ individuals. The questionnaire will generate estimates for several different binomial proportions. If you want to report a single margin of error for the survey, which margin of error from Exercise 8.9 is the correct one to use?

8.11 A random sample of $n = 900$ observations from a binomial population produced $x = 655$ successes. Estimate the binomial proportion p and calculate the margin of error.

8.12 A random sample of $n = 50$ observations from a quantitative population produced $\bar{x} = 56.4$ and $s^2 = 2.6$. Give the best point estimate for the population mean μ, and calculate the margin of error.

8.13 A random sample of $n = 500$ observations from a binomial population produced $x = 450$ successes. Estimate the binomial proportion p and calculate the margin of error.

8.14 A random sample of $n = 75$ observations from a quantitative population produced $\bar{x} = 29.7$ and $s^2 = 10.8$. Give the best point estimate for the population mean μ and calculate the margin of error.

8.15 In a binomial experiment, $n = 120$ trials yielded 95 successes.

a. What is the best point estimator for the population proportion p?

b. Calculate the standard error for this estimator. What will be the margin of error?

APPLICATIONS

8.16 The San Andreas Fault Geologists are interested in shifts and movements of the earth's surface indicated by fractures (cracks) in the earth's crust. One of the most famous large fractures is the San Andreas fault in California. A geologist attempting to study the movement of the relative shifts in the earth's crust at a particular location found many fractures in the local rock structure. In an attempt to determine the mean angle of the breaks, she sampled $n = 50$ fractures and found the sample mean and standard deviation to be 39.8° and 17.2°, respectively. Estimate the mean angular direction of the fractures and find the margin of error for your estimate.

8.17 Biomass Estimates of the earth's biomass, the total amount of vegetation held by the earth's forests, are important in determining the amount of unabsorbed carbon dioxide that is expected to remain in the earth's atmosphere.[1] Suppose a sample of 75 one-square-metre plots, randomly chosen in North America's boreal (northern) forests, produced a mean biomass of 4.2 kilograms per square metre (kg/m^2), with a standard deviation of 1.5 kg/m^2. Estimate the average biomass for the boreal forests of North America and find the margin of error for your estimate.

8.18 Consumer Confidence An increase in the rate of consumer savings is frequently tied to a lack of confidence in the economy and is said to be an indicator of a recessional tendency in the economy. A random sampling of $n = 200$ savings accounts in a local community showed a mean increase in savings account values of 7.2% over the past 12 months, with a standard deviation of 5.6%. Estimate the mean percent increase in savings account values over the past 12 months for depositors in the community. Find the margin of error for your estimate.

8.19 Multimedia Kids Do our children spend as much time enjoying the outdoors and playing with family and friends as previous generations did? Or are our children spending more and more time glued to the television, computer, and other multimedia equipment? A random sample of 250 children between the ages of 8 and 18 showed that 170 children had a TV in their bedroom and that 120 of them had a video game console in their bedroom.

a. Estimate the proportion of all 8- to 18-year-olds who have a TV in their bedroom, and calculate the margin of error for your estimate.

b. Estimate the proportion of all 8- to 18-year-olds who have a video game console in their bedroom, and calculate the margin of error for your estimate.

8.20 Hotel Costs Even within a particular chain of hotels, lodging during the summer months can vary substantially depending on the type of room and the amenities offered.[2] Suppose that we randomly select 50 billing statements from each of the computer databases of the Marriott, Westin, and Doubletree hotel chains, and record the nightly room rates.

	Marriott	Westin	Doubletree
Sample Average ($)	150	165	125
Sample Standard Deviation ($)	17.2	22.5	12.8

a. Describe the sampled population(s).

b. Find a point estimate for the average room rate for the Marriott hotel chain. Calculate the margin of error.

c. Find a point estimate for the average room rate for the Westin hotel chain. Calculate the margin of error.

d. Find a point estimate for the average room rate for the Doubletree hotel chain. Calculate the margin of error.

e. Display the results of parts b, c, and d graphically, using the form shown in Figure 8.5. Use this display to compare the average room rates for the three hotel chains.

8.21 Male Teachers Although most school districts do not specifically recruit men to be elementary school teachers, those men who do choose a career in elementary education are highly valued and find the career very rewarding.[3] If there were 40 men in a random sample of 250 elementary school teachers, estimate the proportion of male elementary school teachers in the entire population. Give the margin of error for your estimate.

8.22 Fear Factor In a survey of 1275 Canadians aged 18 and over conducted by EKOS Research, 69% of Canadians said they thought the world is more dangerous today than it was 25 years ago.[4]

a. Find a point estimate for the proportion of Canadian adults who fear that the world is more dangerous today than it was 25 years ago. Calculate the margin of error.

b. The poll reports a margin of error of plus or minus 2.7%. Does this agree with your results in part a? If not, what value of p produces the margin of error given in the poll?

8.23 Mosaic or Melting Pot? The concept of the mosaic—where cultural differences within society are deemed valuable and regarded as something that should be preserved—is often used to distinguish Canada from the United States.

Based on the online survey of a representative national sample of 1006 Canadian adults, 54% believe Canada should be a melting pot, while 33% of Canadians endorse the concept of the mosaic. The melting pot is particularly attractive for Quebecers (64%), Albertans (60%), and respondents over the age of 55. The mosaic gets its best marks among British Columbians (42%) and respondents aged 18 to 34 (47%).[5]

a. Based on the survey result, give a point estimate for p for those Canadians adults who believe Canada should be a melting pot.

b. Find the margin of error of your point estimate.

8.24 "900" Numbers Radio and television stations often air controversial issues during broadcast time and ask viewers to indicate their agreement or disagreement with a given stand on the issue. A poll is conducted by asking those viewers who *agree* to call a certain 900 telephone number and those who *disagree* to call a second 900 telephone number. All respondents pay a fee for their calls.

a. Does this polling technique result in a random sample?

b. What can be said about the validity of the results of such a survey? Do you need to worry about a margin of error in this case?

8.25 Hungry Rats In an experiment to assess the strength of the hunger drive in rats, 30 previously trained animals were deprived of food for 24 hours. At the end of the 24-hour period, each animal was put into a cage where food was dispensed if the animal pressed a lever. The length of time the animal continued pressing the bar (although receiving no food) was recorded for each animal. If the data yielded a sample mean of 19.3 minutes with a standard deviation of 5.2 minutes, estimate the true mean time and calculate the margin of error.

8.4 INTERVAL ESTIMATION

An *interval estimator* is a rule for calculating two numbers—say, a and b—to create an interval that you are fairly certain contains the parameter of interest. The concept of "fairly certain" means "with high probability." We measure this probability using the **confidence coefficient,** designated by $1 - \alpha$.

DEFINITION The probability that a confidence interval will contain the estimated parameter is called the **confidence coefficient.**

Like Lariat Roping:
Parameter = Fence post
Interval estimate = Lariat

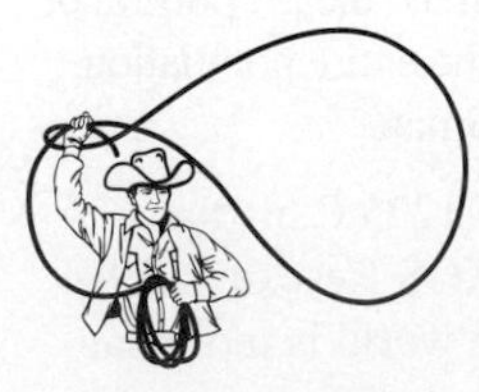

For example, experimenters often construct 95% confidence intervals. This means that the confidence coefficient, or the probability that the interval will contain the estimated parameter, is 0.95. You can increase or decrease your amount of certainty by changing the confidence coefficient. Some values typically used by experimenters are 0.90, 0.95, 0.98, and 0.99.

Consider an analogy—this time, throwing a lariat at a fence post. The fence post represents the parameter that you wish to estimate, and the loop formed by the lariat represents the confidence interval. Each time you throw your lariat, you hope to rope the fence post; however, sometimes your lariat misses. In the same way, each time you draw a sample and construct a confidence interval for a parameter, you hope to include the parameter in your interval, but, just like the lariat, sometimes you miss. Your "success rate"—the proportion of intervals that "rope the post" in repeated sampling—is the confidence coefficient.

Constructing a Confidence Interval

When the sampling distribution of a point estimator is approximately normal, an interval estimator or **confidence interval** can be constructed using the following reasoning. For simplicity, assume that the confidence coefficient is 0.95 and refer to Figure 8.6.

FIGURE 8.6
Parameter ±1.96 SE

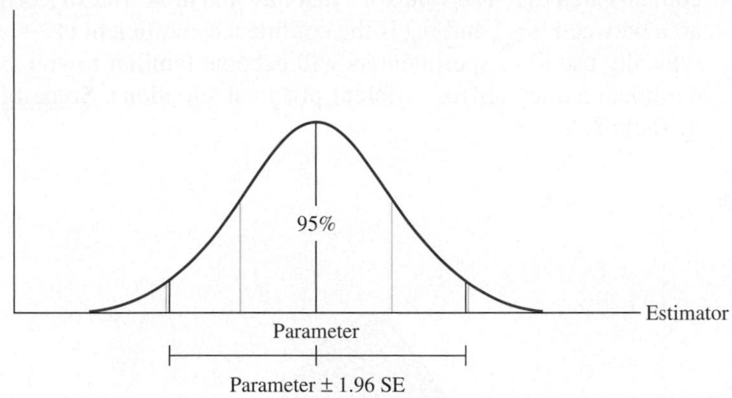

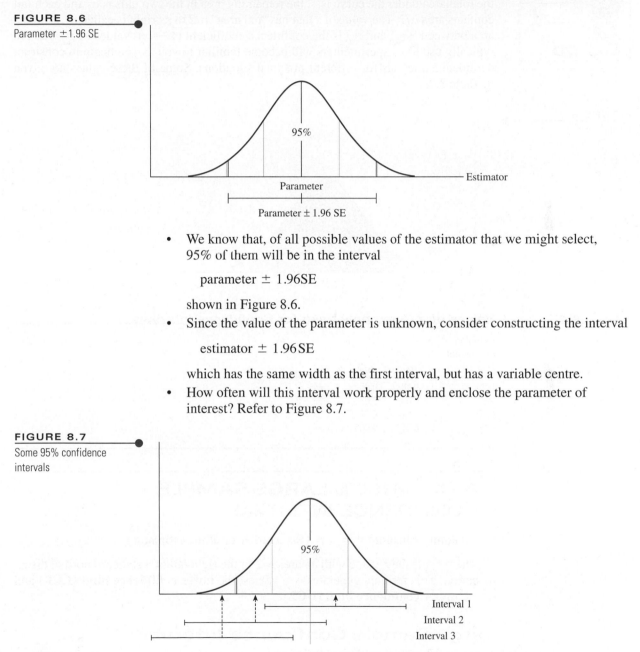

- We know that, of all possible values of the estimator that we might select, 95% of them will be in the interval

  parameter ± 1.96SE

  shown in Figure 8.6.
- Since the value of the parameter is unknown, consider constructing the interval

  estimator ± 1.96SE

  which has the same width as the first interval, but has a variable centre.
- How often will this interval work properly and enclose the parameter of interest? Refer to Figure 8.7.

FIGURE 8.7
Some 95% confidence intervals

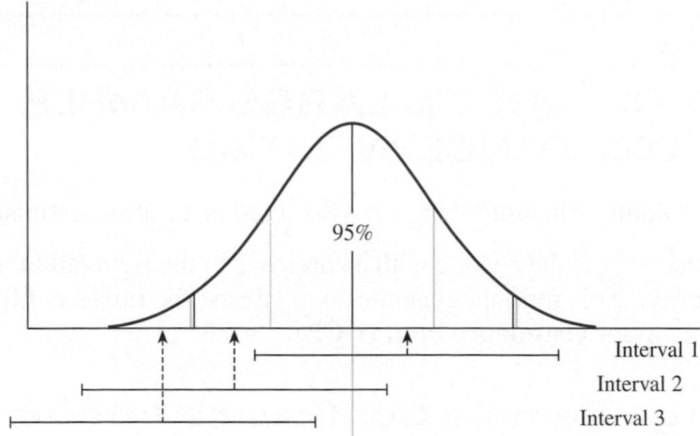

NEED A TIP?

Like a game of ring toss:
Parameter = Peg
Interval estimate = Ring

The first two intervals work properly—the parameter (marked with the lower light grey line) is contained within both intervals. The third interval does not work, since it fails to enclose the parameter. This happened because the value of the estimator at the centre of the interval was too far away from the parameter. Fortunately, values of the estimator only fall this far away 5% of the time—our procedure will work properly 95% of the time!

You may want to change the *confidence coefficient* from $(1 - \alpha) = 0.95$ to another confidence level $(1 - \alpha)$. To accomplish this, you need to change the value $z = 1.96$, which locates an area 0.95 in the centre of the standard normal curve, to a value of z that locates the area $(1 - \alpha)$ in the centre of the curve, as shown in Figure 8.8. Since the total area under the curve is 1, the remaining area in the two tails is α, and each tail contains area $\alpha/2$. The value of z that has "tail area" $\alpha/2$ to its right is called $z_{\alpha/2}$, and the area between $-z_{\alpha/2}$ and $z_{\alpha/2}$ is the confidence coefficient $(1 - \alpha)$. Values of $z_{\alpha/2}$ that are typically used by experimenters will become familiar to you as you begin to construct confidence intervals for different practical situations. Some of these values are given in Table 8.2.

FIGURE 8.8
Location of $z_{\alpha/2}$

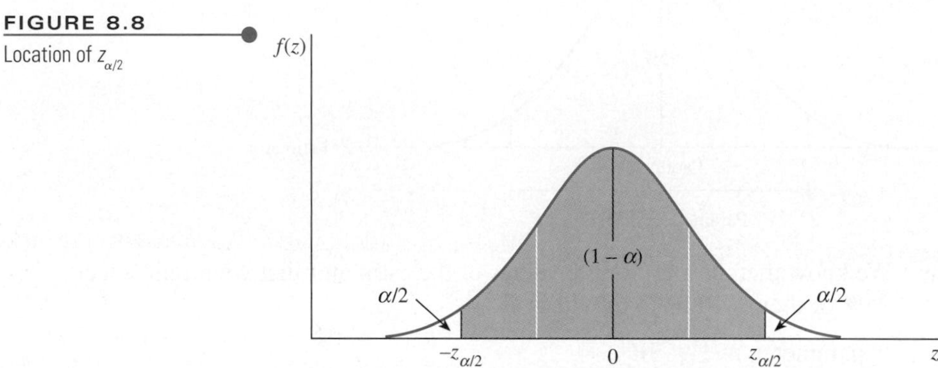

TABLE 8.2 **Values of z Commonly Used for Confidence Intervals**

Confidence coefficient, $(1 - \alpha)$	α	$\alpha/2$	$z_{\alpha/2}$
0.90	0.10	0.05	1.645
0.95	0.05	0.025	1.96
0.98	0.02	0.01	2.33
0.99	0.01	0.005	2.58

A $(1 - \alpha)100\%$ LARGE-SAMPLE CONFIDENCE INTERVAL

$$(\text{Point estimator}) \pm z_{\alpha/2} \times (\text{Standard error of the estimator})$$

where $z_{\alpha/2}$ is the z-value with an area $\alpha/2$ in the right tail of a standard normal distribution. This formula generates two values; the **lower confidence limit (LCL)** and the **upper confidence limit (UCL).**

Large-Sample Confidence Interval for a Population Mean μ

Practical problems very often lead to the estimation of μ, the mean of a population of quantitative measurements. Here are some examples:

- The average achievement of students at a particular university
- The average strength of a new type of steel

- The average number of deaths per age category
- The average demand for a new cosmetics product

When the sample size n is large, the sample mean $\bar{x}$ is the best point estimator for the population mean μ. Since its sampling distribution is approximately normal, it can be used to construct a confidence interval according to the general approach given earlier.

A $(1 - \alpha)100\%$ LARGE-SAMPLE CONFIDENCE INTERVAL FOR A POPULATION MEAN μ

$$\bar{x} \pm z_{\alpha/2}\frac{\sigma}{\sqrt{n}}$$

where $z_{\alpha/2}$ is the z-value corresponding to an area $\alpha/2$ in the upper tail of a standard normal z distribution, and

n = Sample size
σ = Standard deviation of the sampled population

If σ is unknown, it can be approximated by the sample standard deviation s when the sample size is large ($n \geq 30$) and the approximate confidence interval is

$$\bar{x} \pm z_{\alpha/2}\frac{s}{\sqrt{n}}$$

Another way to find the large-sample confidence interval for a population mean μ is to begin with the statistic

$$z = \frac{\bar{x} - \mu}{\sigma/\sqrt{n}}$$

which has a standard normal distribution. If you write $z_{\alpha/2}$ as the value of z with area $\alpha/2$ to its right, then you can write

$$P\left(-z_{\alpha/2} < \frac{\bar{x} - \mu}{\sigma/\sqrt{n}} < z_{\alpha/2}\right) = 1 - \alpha$$

You can rewrite this inequality as

$$-z_{\alpha/2}\frac{\sigma}{\sqrt{n}} < \bar{x} - \mu < z_{\alpha/2}\frac{\sigma}{\sqrt{n}}$$

$$-\bar{x} - z_{\alpha/2}\frac{\sigma}{\sqrt{n}} < -\mu < -\bar{x} + z_{\alpha/2}\frac{\sigma}{\sqrt{n}}$$

so that

$$P\left(\bar{x} - z_{\alpha/2}\frac{\sigma}{\sqrt{n}} < \mu < \bar{x} + z_{\alpha/2}\frac{\sigma}{\sqrt{n}}\right) = 1 - \alpha$$

Both $\bar{x} - z_{\alpha/2}(\sigma/\sqrt{n})$ and $\bar{x} + z_{\alpha/2}(\sigma/\sqrt{n})$, the lower and upper confidence limits, are actually random quantities that depend on the sample mean $\bar{x}$. Therefore, in repeated sampling, the random interval, $\bar{x} \pm z_{\alpha/2}(\sigma/\sqrt{n})$, will contain the population mean μ with probability $(1 - \alpha)$.

EXAMPLE 8.6

A scientist interested in monitoring chemical contaminants in food, and thereby the accumulation of contaminants in human diets, selected a random sample of $n = 50$ male adults. It was found that the average daily intake of dairy products was $\bar{x} = 756$ grams per day with a standard deviation of $s = 35$ grams per day. Use this sample information to construct a 95% confidence interval for the mean daily intake of dairy products for men.

SOLUTION Since the sample size of $n = 50$ is large, the distribution of the sample mean $\bar{x}$ is approximately normally distributed with mean μ and standard error estimated by $s/\sqrt{n}$. The approximate 95% confidence interval is

$$\bar{x} \pm 1.96\left(\frac{s}{\sqrt{n}}\right)$$

$$756 \pm 1.96\left(\frac{35}{\sqrt{50}}\right)$$

$$756 \pm 9.70$$

Hence, the 95% confidence interval for μ is from 746.30 to 765.70 grams per day.

NEED A TIP?

A 95% confidence interval tells you that, if you were to construct many of these intervals (all of which would have slightly different endpoints), 95% of them would enclose the population mean.

Interpreting the Confidence Interval

What does it mean to say you are "95% confident" that the true value of the population mean μ is within a given interval? If you were to construct 20 such intervals, each using different sample information, your intervals might look like those shown in Figure 8.9(a). Of the 20 intervals, you might expect that 95% of them, or 19 out of 20, will perform as planned and contain μ within their upper and lower bounds. If you constructed 100 such intervals (Figure 8.9(b)), you would expect about 95 of them to perform as planned. Remember that you cannot be absolutely sure that any one particular interval contains the mean μ. You will never know whether your particular interval is one of the 19 that "worked," or whether it is the one interval that "missed." Your confidence in the estimated interval follows from the fact that when repeated intervals are calculated, 95% of these intervals will contain μ.

ONLINE APPLET

Interpreting Confidence Intervals

FIGURE 8.9
Interpreting confidence intervals

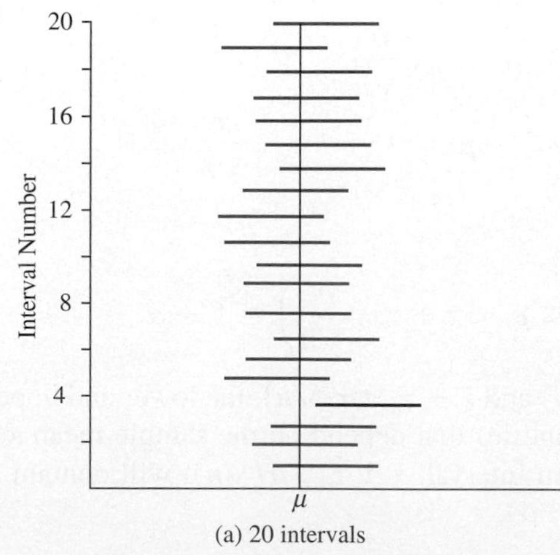

(a) 20 intervals

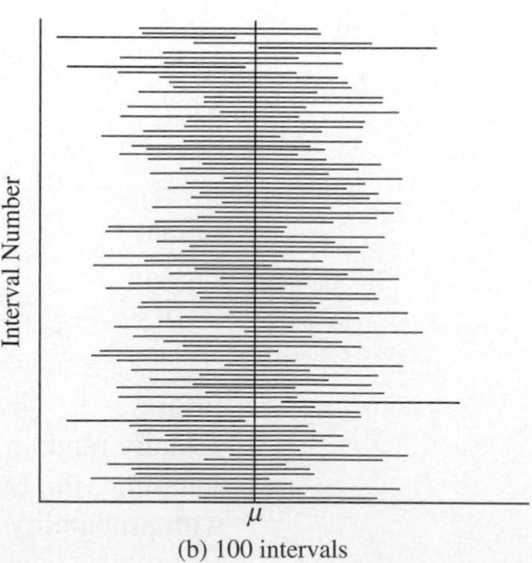

(b) 100 intervals

A good confidence interval has two desirable characteristics:

- It is as narrow as possible. The narrower the interval, the more exactly you have located the estimated parameter.
- It has a large confidence coefficient, near 1. The larger the confidence coefficient, the more likely it is that the interval will contain the estimated parameter.

EXAMPLE 8.7

Construct a 99% confidence interval for the mean daily intake of dairy products for adult men in Example 8.6.

SOLUTION To change the confidence level to 0.99, you must find the appropriate value of the standard normal z that puts area $(1 - \alpha) = 0.99$ in the centre of the curve. This value, with tail area $\alpha/2 = 0.005$ to its right, is found from Table 8.2 to be $z = 2.58$ (see Figure 8.10). The 99% confidence interval is then

$$\bar{x} \pm 2.58\left(\frac{s}{\sqrt{n}}\right)$$

$$756 \pm 2.58(4.95)$$

$$756 \pm 12.77$$

FIGURE 8.10

Standard normal values for a 99% confidence interval

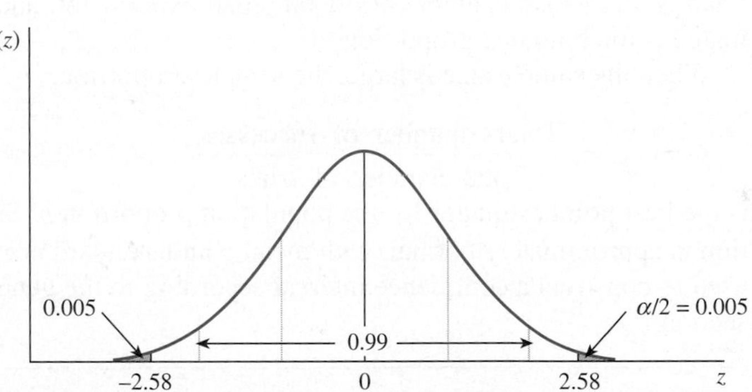

or 743.23 to 768.77 grams per day. This confidence interval is *wider* than the 95% confidence interval in Example 8.6. The increased width is necessary to increase the confidence, just as you might want a wider loop on your lariat to ensure roping the fence post! The only way to *increase the confidence* without increasing the width of the interval is to *increase the sample size, n.*

The standard error of $\bar{x}$,

$$SE = \frac{\sigma}{\sqrt{n}}$$

measures the variability or spread of the values of $\bar{x}$. The more variable the population data, measured by σ, the more variable will be $\bar{x}$, and the standard error will be larger. On the other hand, if you increase the sample size n, more information is available for estimating μ. The estimates should fall closer to μ and the standard error will be smaller.

The confidence intervals of Examples 8.6 and 8.7 are approximate because you substituted s as an approximation for σ. That is, instead of the confidence coefficient being 0.95, the value specified in the example, the true value of the coefficient may be

0.92, 0.94, or 0.97. But this discrepancy is of little concern from a practical point of view; as far as your "confidence" is concerned, there is little difference among these confidence coefficients. Most interval estimators used in statistics yield approximate confidence intervals because the assumptions upon which they are based are not satisfied exactly. Having made this point, we will not continue to refer to confidence intervals as "approximate." It is of little practical concern as long as the actual confidence coefficient is near the value specified.

Large-Sample Confidence Interval for a Population Proportion *p*

Many research experiments or sample surveys have as their objective the estimation of the proportion of people or objects in a large group that possess a certain characteristic. Here are some examples:

- The proportion of sales that can be expected in a large number of customer contacts
- The proportion of seeds that germinate
- The proportion of "likely" voters who plan to vote for a particular political candidate

Each is a practical example of the binomial experiment, and the parameter to be estimated is the binomial proportion p.

When the sample size is large, the sample proportion,

$$\hat{p} = \frac{x}{n} = \frac{\text{Total number of successes}}{\text{Total number of trials}}$$

is the best point estimator for the population proportion p. Since its sampling distribution is approximately normal, with mean p and standard error $\text{SE} = \sqrt{pq/n}$, $\hat{p}$ can be used to construct a confidence interval according to the general approach given in this section.

Right Tail Area	z-Value
0.05	1.645
0.025	1.96
0.01	2.33
0.005	2.58

A $(1 - \alpha)$100% LARGE-SAMPLE CONFIDENCE INTERVAL FOR A POPULATION PROPORTION *p*

$$\hat{p} \pm z_{\alpha/2}\sqrt{\frac{pq}{n}}$$

where $z_{\alpha/2}$ is the z-value corresponding to an area $\alpha/2$ in the right tail of a standard normal z distribution. Since p and q are unknown, they are estimated using the best point estimators: $\hat{p}$ and $\hat{q}$. The sample size is considered large when the normal approximation to the binomial distribution is adequate—namely, when $n\hat{p} > 5$ and $n\hat{q} > 5$.

A random sample of 985 "likely" voters—those who are likely to vote in the upcoming election—were polled during a phone-athon conducted by the Liberal Party. Of those surveyed, 592 indicated that they intended to vote for the Liberal candidate in the upcoming election. Construct a 90% confidence interval for p, the proportion of likely voters in the population who intend to vote for the Liberal candidate. Based on this information, can you conclude that the candidate will win the election?

SOLUTION The point estimate for p is

$$\hat{p} = \frac{x}{n} = \frac{592}{985} = 0.601$$

and the standard error is

$$\sqrt{\frac{\hat{p}\hat{q}}{n}} = \sqrt{\frac{(0.601)(0.399)}{985}} = 0.016$$

The z-value for a 90% confidence interval is the value that has area $\alpha/2 = 0.05$ in the upper tail of the z distribution, or $z_{0.05} = 1.645$ from Table 8.2. The 90% confidence interval for p is thus

$$\hat{p} \pm 1.645\sqrt{\frac{\hat{p}\hat{q}}{n}}$$

$$0.601 \pm 0.026$$

or $0.575 < p < 0.627$. You estimate that the percentage of likely voters who intend to vote for the Liberal candidate is between 57.5% and 62.7%. Will the candidate win the election? Assuming that she needs more than 50% of the vote to win, and since both the upper and lower confidence limits exceed this minimum value, you can say with 90% confidence that the candidate will win.

There are some problems, however, with this type of sample survey. What if the voters who consider themselves "likely to vote" do not actually go to the polls? What if a voter changes his or her mind between now and election day? What if a surveyed voter does not respond truthfully when questioned by the campaign worker? The 90% confidence interval you have constructed gives you 90% confidence only if you have selected a *random sample from the population of interest.* You can no longer be assured of "90% confidence" if your sample is biased, or if the population of voter responses changes before the day of the election!

You may have noticed that the point estimator with its 95% margin of error looks very similar to a 95% confidence interval for the same parameter. This close relationship exists for most of the parameters estimated in this book, but it is not true in general. Sometimes the best point estimator for a parameter *does not* fall in the middle of the best confidence interval; the best confidence interval may not even be a function of the best point estimator. Although this is a theoretical distinction, you should remember that there is a difference between point and interval estimation, and that the choice between the two depends on the preference of the experimenter.

8.4 EXERCISES

BASIC TECHNIQUES

8.26 Find and interpret a 95% confidence interval for a population mean μ for these values:

a. $n = 36, \bar{x} = 13.1, s^2 = 3.42$

b. $n = 64, \bar{x} = 2.73, s^2 = 0.1047$

8.27 Find a 90% confidence interval for a population mean μ for these values:

a. $n = 125, \bar{x} = 0.84, s^2 = 0.086$

b. $n = 50, \bar{x} = 21.9, s^2 = 3.44$

c. Interpret the intervals found in parts a and b.

8.28 Find a $(1 - \alpha)100\%$ confidence interval for a population mean μ for these values:

a. $\alpha = 0.01, n = 38, \bar{x} = 34, s^2 = 12$

b. $\alpha = 0.10, n = 65, \bar{x} = 1049, s^2 = 51$

c. $\alpha = 0.05, n = 89, \bar{x} = 66.3, s^2 = 2.48$

8.29 A random sample of $n = 300$ observations from a binomial population produced $x = 263$ successes. Find a 90% confidence interval for p and interpret the interval.

8.30 Suppose the number of successes observed in $n = 500$ trials of a binomial experiment is 27. Find a 95% confidence interval for p. Why is the confidence interval narrower than the confidence interval in Exercise 8.29?

8.31 A random sample of n measurements is selected from a population with unknown mean μ and known standard deviation $\sigma = 10$. Calculate the width of a 95% confidence interval for μ for these values of n:

a. $n = 100$ **b.** $n = 200$ **c.** $n = 400$

8.32 Compare the confidence intervals in Exercise 8.31. What effect does each of these actions have on the width of a confidence interval?

a. Double the sample size

b. Quadruple the sample size

8.33 Refer to Exercise 8.32.

a. Calculate the width of a 90% confidence interval for μ when $n = 100$.

b. Calculate the width of a 99% confidence interval for μ when $n = 100$.

c. Compare the widths of 90%, 95%, and 99% confidence intervals for μ. What effect does increasing the confidence coefficient have on the width of the confidence interval?

8.34 A binomial experiment yielded 150 successes from a trial of 350.

a. Calculate a 90% confidence interval for the population proportion p.

b. Calculate a 95% confidence interval for the population proportion p.

c. Why is the confidence interval in part a narrower than in part b?

APPLICATIONS

8.35 A Chemistry Experiment Due to a variation in laboratory techniques, impurities in materials, and other unknown factors, the results of an experiment in a chemistry laboratory will not always yield the same numerical answer. In an electrolysis experiment, a class measured the amount of copper precipitated from a saturated solution of copper sulphate over a 30-minute period. The $n = 30$ students calculated a sample mean and standard deviation equal to 0.145 and 0.0051 mole, respectively. Find a 90% confidence interval for the mean amount of copper precipitated from the solution over a 30-minute period.

8.36 Acid Rain Acid rain, caused by the reaction of certain air pollutants with rainwater, appears to be a growing problem in Eastern Canada. (Acid rain affects the soil and causes corrosion on exposed metal surfaces.) Pure rain falling through clean air registers a pH value of 5.7 (pH is a measure of acidity: 0 is acid; 14 is alkaline). Suppose water samples from 40 rainfalls are analyzed for pH, and $\bar{x}$ and s are equal to 3.7 and 0.5, respectively. Find a 99% confidence interval for the mean pH in rainfall and interpret the interval. What assumption must be made for the confidence interval to be valid?

8.37 'Tweens When it comes to advertising, "'tweens" (kids aged 10 to 13) are not ready for the hardline messages that advertisers often use to reach teenagers. One study found that 78% of 'tweens understand and enjoy ads that are silly in nature. Unlike teenagers, 'tweens would much rather see dancing (69%) and "boyfriends and girlfriends" (63%) than "sexy looking people" or "kissing."[6] Suppose that these results are based on a sample of size $n = 1030$ 'tweens.

a. Construct a 95% confidence interval estimate of the proportion of 'tweens who understand and enjoy ads that are silly in nature.

b. Construct a 95% confidence interval for the proportion of 'tweens who would rather see dancing.

8.38 Ground Beef The meat department of a local supermarket chain packages ground beef using meat trays of two sizes: one designed to hold approximately 0.454 kilograms (kg) of meat, and one that holds approximately 1.36 kg. A random sample of 35 packages in the smaller meat trays produced weight measurements with an average of 0.458 kg and a standard deviation of 0.082 kg.

a. Construct a 99% confidence interval for the average weight of all packages sold in the smaller meat trays by this supermarket chain.

b. What does the phrase "99% confident" mean?

c. Suppose that the quality control department of this supermarket chain intends that the amount of ground beef in the smaller trays should be 0.454 kg on average. Should the confidence interval in part a concern the quality control department? Explain.

8.39 Kids' Take on Media To provide current and factual information on the role media plays in children's lives, the Canadian Teachers' Federation (CTF) commissioned Erin Research to conduct definitive research on Canadian children's experience with communications media.[7] The research involved 5756 students in Grades 3 to 10 in every province and territory. This is the first survey of its size to question young people on all the media they use, what satisfaction they get from media, and whether violence in their daily media consumption influences their behaviour and values.

According to survey results, in Grades 3 to 6, roughly 30% of kids claim they never have adult input concerning what TV shows they can watch. By Grade 6, 50% report no adult input as to how long they can watch. In Grade 8, the figures for those who experience no parental supervision of their TV viewing rises to approximately 60%. On the other hand, close to 90% of children report that they watch TV with their family either "most of the time" or "sometimes," and this pattern remains fairly constant for Grades 3 to 10. Construct a 90% confidence interval for the overall proportion of children who say that they watch TV with their family either "most of the time" or "sometimes."

8.40 SUVs A sample survey is designed to estimate the proportion of sports utility vehicles being driven in PEI. A random sample of 500 registrations are selected from a PEI database, and 68 are classified as sports utility vehicles.

a. Use a 95% confidence interval to estimate the proportion of sports utility vehicles in PEI.

b. How can you estimate the proportion of sports utility vehicles in PEI with a higher degree of accuracy? (HINT: There are two answers.)

8.41 e-Shopping In a report of why e-shoppers abandon their online sales transactions, Alison Stein Wellner[8] found that "pages took too long to load" and "site was so confusing that I couldn't find the product" were the two complaints heard most often. Based on customers' responses, the average time to complete an online order form will take 4.5 minutes. Suppose that $n = 50$ customers responded and that the standard deviation of the time to complete an online order is 2.7 minutes.

a. Do you think that *X*, the time to complete the online order form, has a mound-shaped distribution? If not, what shape would you expect?

b. If the distribution of the completion times is not normal, you can still use the standard normal distribution to construct a confidence interval for μ, the mean completion time for online shoppers. Why?

c. Construct a 95% confidence interval for μ, the mean completion time for online orders.

8.42 What's Normal? What *is* normal, when it comes to people's body temperatures? A random sample of 130 human body temperatures, provided by Allen Shoemaker[9] in the *Journal of Statistical Education,* had a mean of 36.81° Celsius and a standard deviation of 0.73°.

a. Construct a 99% confidence interval for the average body temperature of healthy people.

b. Does the confidence interval constructed in part a contain the value 37° C, the usual average temperature cited by physicians and others? If not, what conclusions can you draw?

8.43 Multiculturalism? Based on the online survey of a representative national sample of 1006 Canadian adults, conducted by Angus Reid, 55% think multiculturalism has been "very good" or "good" for Canada. Further, the survey found 54% want Canada to be a melting pot where immigrants assimilate and blend into Canadian society. The poll reported a margin of error—which measures sampling variability—is +/− 3.1%, 19 times out of 20.[10]

a. Construct a 90% confidence interval for the proportion of Canadian adults who think the multiculturalism has been "very good" or "good" for Canada.

b. Construct a 90% confidence interval for the proportion of Canadian adults who want Canada to be a melting pot where immigrants assimilate and blend into Canadian society.

c. How did the researchers calculate the margin of error for this survey? Confirm that their margin of error is correct.

8.44 A Tolerant Society? Refer to Exercise 8.43. Respondents across the country were asked whether Canada is tolerant or intolerant toward nine different minority groups. One-third of respondents (33%) think Canadian society is intolerant toward Muslims, 30% say it is intolerant toward Aboriginal Canadians, and 24% believe it is intolerant toward immigrants from South Asia.

a. Construct a 99% confidence interval for the proportion of respondents who think Canadian society is intolerant toward Muslims.

b. Construct a 99% confidence interval for the proportion of respondents who think Canadian society is intolerant toward immigrants from South Asia.

c. Construct a 99% confidence interval for the proportion of respondents who think Canadian society is intolerant toward Aboriginal Canadians.

8.45 Voter Turnout in Canada How likely are you to vote in the next federal general election? A random sample of 300 adults was taken, and 192 of them said that they always vote in federal general elections.

a. Construct a 95% confidence interval for the proportion of adult Canadians who say they always vote in federal general elections.

b. In the 36 federal general elections and three referendums held in Canada from 1867 to 1997, an average of 71% of registered electors have turned out to vote.[11] Voter turnout ranged from a low of 44% in the prohibition plebiscite of 1898 to a high of 79% at the general election of 1958. Based on the interval constructed in part a, would you disagree with their reported percentage? Explain.

c. Can we use the interval estimate from part a to estimate the actual proportion of adult Canadians who vote in the next federal general election? Why or why not?

8.5 ESTIMATING THE DIFFERENCE BETWEEN TWO POPULATION MEANS

A problem equally as important as the estimation of a single population mean μ for a quantitative population is the comparison of two population means. You may want to make comparisons such as these:

- The average scores on the Medical College Admission Test (MCAT) for students whose major was biochemistry and those whose major was biology
- The average yields in a chemical plant using raw materials furnished by two different suppliers
- The average stem diameters of plants grown on two different types of nutrients

For each of these examples, there are two populations: the first with mean and variance μ_1 and σ_1^2 and the second with mean and variance μ_2 and σ_2^2. A random sample of n_1 measurements is drawn from population 1 and a second random sample of size n_2 is independently drawn from population 2. Finally, the estimates of the population parameters are calculated from the sample data using the estimators $\bar{x}_1$, s_1^2, $\bar{x}_2$, and s_2^2 as shown in Table 8.3.

TABLE 8.3 **Samples from Two Quantitative Populations**

	Population 1	Population 2
Mean	μ_1	μ_2
Variance	σ_1^2	σ_2^2

	Sample 1	Sample 2
Mean	$\bar{x}_1$	$\bar{x}_2$
Variance	s_1^2	s_2^2
Sample size	n_1	n_2

Intuitively, the difference between two sample means would provide the maximum information about the actual difference between two population means, and this is in fact the case. The best point estimator of the difference $(\mu_1 - \mu_2)$ between the population means is $(\bar{x}_1 - \bar{x}_2)$. The sampling distribution of this estimator is not difficult to derive, but we state it here without proof.

PROPERTIES OF THE SAMPLING DISTRIBUTION OF $(\bar{x}_1 - \bar{x}_2)$, THE DIFFERENCE BETWEEN TWO SAMPLE MEANS

When independent random samples of n_1 and n_2 observations have been selected from populations with means μ_1 and μ_2 and variances σ_1^2 and σ_2^2, respectively, the sampling distribution of the difference $(\bar{x}_1 - \bar{x}_2)$ has the following properties:

1. The mean of $(\bar{x}_1 - \bar{x}_2)$ is

   $$\mu_1 - \mu_2$$

   and the standard error is

   $$SE = \sqrt{\frac{\sigma_1^2}{n_1} + \frac{\sigma_2^2}{n_2}}$$

   which can be estimated as

   $$SE = \sqrt{\frac{s_1^2}{n_1} + \frac{s_2^2}{n_2}} \text{ when the sample sizes are large.}$$

2. **If the sampled populations are normally distributed,** then the sampling distribution of $(\bar{x}_1 - \bar{x}_2)$ is **exactly** normally distributed, regardless of the sample size.
3. **If the sampled populations are not normally distributed,** then the sampling distribution of $(\bar{x}_1 - \bar{x}_2)$ is **approximately** normally distributed when n_1 and n_2 are both 30 or more, due to the Central Limit Theorem.

Since $(\mu_1 - \mu_2)$ is the mean of the sampling distribution, it follows that $(\bar{x}_1 - \bar{x}_2)$ is an unbiased estimator of $(\mu_1 - \mu_2)$ with an approximately normal distribution when n_1 and n_2 are large. That is, the statistic

$$z = \frac{(\bar{x}_1 - \bar{x}_2) - (\mu_1 - \mu_2)}{\sqrt{\frac{s_1^2}{n_1} + \frac{s_2^2}{n_2}}}$$

has an approximately standard normal z distribution, and the general procedures of Section 8.4 can be used to construct point and interval estimates. Although the choice between point and interval estimation depends on your personal preference, most experimenters choose to construct confidence intervals for two-sample problems. The appropriate formulas for both methods are given next.

NEED a tip? NEED A TIP?

Right Tail Area	z-Value
0.05	1.645
0.025	1.96
0.01	2.33
0.005	2.58

LARGE-SAMPLE POINT ESTIMATION OF $(\mu_1 - \mu_2)$

Point estimator: $(\bar{x}_1 - \bar{x}_2)$

95% margin of error: $\pm 1.96SE = \pm 1.96\sqrt{\frac{s_1^2}{n_1} + \frac{s_2^2}{n_2}}$

A $(1 - \alpha)100\%$ LARGE-SAMPLE CONFIDENCE INTERVAL FOR $(\mu_1 - \mu_2)$

$$(\bar{x}_1 - \bar{x}_2) \pm z_{\alpha/2}\sqrt{\frac{s_1^2}{n_1} + \frac{s_2^2}{n_2}}$$

EXAMPLE  8.9

The wearing qualities of two types of automobile tires were compared by road-testing samples of $n_1 = n_2 = 100$ tires for each type. The number of kilometres until wearout was defined as a specific amount of tire wear. The test results are given in Table 8.4. Estimate $(\mu_1 - \mu_2)$, the difference in mean kilometres to wearout, using a 99% confidence interval. Is there a difference in the average wearing quality for the two types of tires?

TABLE 8.4 **Sample Data Summary for Two Types of Tires**

Tire 1	Tire 2
$\bar{x}_1 = 42{,}500$ *km*	$\bar{x}_2 = 40{,}400$ *km*
$s_1^2 = 2{,}317{,}455$	$s_2^2 = 3{,}154{,}314$

SOLUTION The point estimate of $(\mu_1 - \mu_2)$ is

$$(\bar{x}_1 - \bar{x}_2) = 42{,}500 - 40{,}400 = 2100\,\text{km}$$

and the standard error of $(\bar{x}_1 - \bar{x}_2)$ is estimated as

$$SE = \sqrt{\frac{s_1^2}{n_1} + \frac{s_2^2}{n_2}} = \sqrt{\frac{2{,}317{,}455}{100} + \frac{3{,}154{,}314}{100}} = 233.92\,\text{km}$$

The 99% confidence interval is calculated as

$$(\bar{x}_1 - \bar{x}_2) \pm 2.58\sqrt{\frac{s_1^2}{n_1} + \frac{s_2^2}{n_2}}$$

$$2100 \pm 2.58(233.92)$$

$$2100 \pm 603.51$$

or $1496.49 < (\mu_1 - \mu_2) < 2703.52$. The difference in the average kilometres to wearout for the two types of tires is estimated to lie between LCL = 1496.49 and UCL = 2703.52 kilometres (km) of wear.

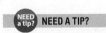

NEED A TIP?

If 0 is not in the interval, you *can* conclude that there is a difference in the population means.

Based on this confidence interval, can you conclude that there is a difference in the average kilometres to wearout for the two types of tires? If there were no difference in the two population means, then μ_1 and μ_2 would be equal and $(\mu_1 - \mu_2) = 0$. If you look at the confidence interval you constructed, you will see that 0 is not one of the possible values for $(\mu_1 - \mu_2)$. Therefore, it is not likely that the means are the same; you can conclude that there is a difference in the average kilometres to wearout for the two types of tires. The confidence interval has allowed you to *make a decision* about the equality of the two population means.

EXAMPLE 8.10

The scientist in Example 8.6 wondered whether there was a difference in the average daily intakes of dairy products between men and women. He took a sample of $n = 50$ adult women and recorded their daily intakes of dairy products in grams per day. He did the same for adult men. A summary of his sample results is listed in Table 8.5. Construct a 95% confidence interval for the difference in the average daily intakes of dairy products for men and women. Can you conclude that there is a difference in the average daily intakes for men and women?

TABLE 8.5 **Sample Values for Daily Intakes of Dairy Products**

	Men	Women
Sample size	50	50
Sample mean	756	762
Sample standard deviation	35	30

SOLUTION The confidence interval is constructed using a value of z with tail area $\alpha/2 = 0.025$ to its right; that is, $z_{0.025} = 1.96$. Using the sample standard deviations to approximate the unknown population standard deviations, the 95% confidence interval is

$$(\bar{x}_1 - \bar{x}_2) \pm 1.96\sqrt{\frac{s_1^2}{n_1} + \frac{s_2^2}{n_2}}$$

$$(756 - 762) \pm 1.96\sqrt{\frac{35^2}{50} + \frac{30^2}{50}}$$

$$-6 \pm 12.78$$

or $-18.78 < (\mu_1 - \mu_2) < 6.78$. Look at the possible values for $(\mu_1 - \mu_2)$ in the confidence interval. It is possible that the difference $(\mu_1 - \mu_2)$ could be negative (indicating that the average for women exceeds the average for men), it could be positive (indicating that men have the higher average), or it could be 0 (indicating no difference between the averages). Based on this information, you *should not be willing to conclude* that there is a difference in the average daily intakes of dairy products for men and women.

Examples 8.9 and 8.10 deserve further comment with regard to using sample estimates in place of unknown parameters. The sampling distribution of

$$\frac{(\bar{x}_1 - \bar{x}_2) - (\mu_1 - \mu_2)}{\sqrt{\frac{\sigma_1^2}{n_1} + \frac{\sigma_2^2}{n_2}}}$$

has a standard normal distribution for all sample sizes when both sampled populations are normal. On the other hand, it will have an *approximate* standard normal distribution when the sampled populations are not normal but the sample sizes are large (≥ 30). When σ_1^2 and σ_2^2 are not known and are estimated by the sample estimates s_1^2 and s_2^2, the resulting statistic will still have an approximate standard normal distribution when the sample sizes are large. The behaviour of this statistic when the population variances are unknown and the sample sizes are small will be discussed in Chapter 10.

8.5 EXERCISES

BASIC TECHNIQUES

8.46 Independent random samples were selected from populations 1 and 2. The sample sizes, means, and variances are as follows:

	Population	
	1	2
Sample size	35	49
Sample mean	12.7	7.4
Sample variance	1.38	4.14

a. Find a 95% confidence interval for estimating the difference in the population means ($\mu_1 - \mu_2$).

b. Based on the confidence interval in part a, can you conclude that there is a difference in the means for the two populations? Explain.

8.47 Independent random samples were selected from populations 1 and 2. The sample sizes, means, and variances are as follows:

	Population	
	1	2
Sample size	64	64
Sample mean	2.9	5.1
Sample variance	0.83	1.67

a. Find a 90% confidence interval for the difference in the population means. What does the phrase "90% confident" mean?

b. Find a 99% confidence interval for the difference in the population means. Can you conclude that there is a difference in the two population means? Explain.

8.48 Independent random samples of size $n_1 = n_2 = 100$ were selected from each of two populations. The mean and standard deviations for the two samples were $\bar{x}_1 = 125.2$, $\bar{x}_2 = 123.7$, $s_1 = 5.6$, and $s_2 = 6.8$.

a. Construct a 99% confidence interval for estimating the difference in the two population means.

b. Does the confidence interval in part a provide sufficient evidence to conclude that there is a difference in the two population means? Explain.

8.49 Independent random samples of size $n_1 = n_2 = 500$ were selected from each of two populations. The mean and standard deviations for the two samples were $\bar{x}_1 = 125.2$, $\bar{x}_2 = 123.7$, $s_1 = 5.6$, and $s_2 = 6.8$.

a. Find a point estimate for the difference in the two population means. Calculate the margin of error.

b. Based on the results in part a, can you conclude that there is a difference in the two population means? Explain.

APPLICATIONS

8.50 Selenium A small amount of the trace element selenium, 50–200 micrograms (μg) per day, is considered essential to good health. Suppose that random samples of $n_1 = n_2 = 30$ adults were selected from two regions of Canada and that a day's intake of selenium, from both liquids and solids, was recorded for each person. The mean and standard deviation of the selenium daily intakes for the 30 adults from region 1 were $\bar{x}_1 = 167.1$ and $s_1 = 24.3$ μg, respectively. The corresponding statistics for the 30 adults from region 2 were $\bar{x}_2 = 140.9$ and $s_2 = 17.6$. Find a 95% confidence interval for the difference in the mean selenium intakes for the two regions. Interpret this interval.

8.51 9-1-1 A study was conducted to compare the mean numbers of police emergency calls per eight-hour shift in two districts of a large city. Samples of 100 eight-hour shifts were randomly selected from the police records for each of the two regions, and the number of emergency calls was recorded for each shift. The sample statistics are listed here:

	Region	
	1	2
Sample size	100	100
Sample mean	2.4	3.1
Sample variance	1.44	2.64

Find a 90% confidence interval for the difference in the mean numbers of police emergency calls per shift between the two districts of the city. Interpret the interval.

8.52 Teaching Biology In developing a standard for assessing the teaching of precollege sciences in the United States, an experiment was conducted to evaluate a teacher-developed curriculum, "Biology: A Community Context" (BACC), that was standards-based, activity-oriented, and inquiry-centred. This approach was compared to the historical presentation through lecture, vocabulary, and memorized facts. Students were tested on biology concepts that featured biological knowledge and process skills in the traditional sense. The perhaps not-so-startling results from a test on biology concepts, published in the *American Biology Teacher,* are shown in the following table:[12]

	Mean	Sample Size	Standard Deviation
Pretest: All BACC classes	13.38	372	5.59
Pretest: All traditional	14.06	368	5.45
Posttest: All BACC classes	18.5	365	8.03
Posttest: All traditional	16.5	298	6.96

Source: William H. Leonard, Barbara J. Speziale, and John E. Pernick, "Performance Assessment of a Standards-based High School Biology Curriculum," *The American Biology Teacher* 63, no. 5 (2001): 310–316. Reprinted by permission of University of California Press.

a. Find a 95% confidence interval for the mean score for the posttest for all BACC classes.

b. Find a 95% confidence interval for the mean score for the posttest for all traditional classes.

c. Find a 95% confidence interval for the difference in mean scores for the posttest BACC classes and the posttest traditional classes.

d. Does the confidence interval in part c provide evidence that there is a real difference in the posttest BACC and traditional class scores? Explain.

8.53 Are You Dieting? An experiment was conducted to compare two diets—A and B—designed for weight reduction. Two groups of 30 overweight dieters each were randomly selected. One group was placed on diet A and the other on diet B, and their weight losses were recorded over a 90-day period. The means and standard deviations of the weight-loss measurements for the two groups are shown in the table. Find a 95% confidence interval for the difference in mean weight loss for the two diets. Interpret your confidence interval.

Diet A	Diet B
$\bar{x}_A = 21.3$	$\bar{x}_B = 13.4$
$S_A = 2.6$	$S_B = 1.9$

8.54 Starting Salaries In an attempt to compare the starting salaries of university graduates majoring in education and social sciences, random samples of 50 recent university graduates in each major were selected and the following information was obtained:

Major	Mean	SD
Education ($)	40,554	2225
Social sciences ($)	38,348	2375

a. Find a point estimate for the difference in the average starting salaries of university students majoring in education and the social sciences. What is the margin of error for your estimate?

b. Based on the results of part a, do you think that there is a significant difference in the means for the two groups in the general population? Explain.

8.55 Biology Skills Refer to Exercise 8.52. In addition to tests involving biology concepts, students were also tested on process skills. The results of pretest and posttest scores, published in the *American Biology Teacher,* are given below:[13]

	Mean	Sample Size	Standard Deviation
Pretest: All BACC classes	10.52	395	4.79
Pretest: All traditional	11.97	379	5.39
Posttest: All BACC classes	14.06	376	5.65
Posstest: All traditional	12.96	308	5.93

Source: William H. Leonard, Barbara J. Speziale, and John E. Pernick, "Performance Assessment of a Standards-based High School Biology Curriculum," *The American Biology Teacher 63,* no. 5 (2001): 310-316. Reprinted by permission of University of California Press.

a. Find a 95% confidence interval for the mean score on process skills for the posttest for all BACC classes.

b. Find a 95% confidence interval for the mean score on process skills for the posttest for all traditional classes.

c. Find a 95% confidence interval for the difference in mean scores on process skills for the posttest BACC classes and the posttest traditional classes.

d. Does the confidence interval in part c provide evidence that there is a real difference in the mean process skills scores between posttest BACC and traditional class scores? Explain.

8.56 Hotel Costs Refer to Exercise 8.20. The means and standard deviations for 50 billing statements from each of the computer databases of each of the three hotel chains are given in the table:[14]

	Marriott	Westin	Doubletree
Sample average ($)	150	165	125
Sample standard deviation	17.2	22.5	12.8

a. Find a 95% confidence interval for the difference in the average room rates for the Marriott and the Doubletree hotel chains.

b. Find a 99% confidence interval for the difference in the average room rates for the Westin and the Doubletree hotel chains.

c. Do the intervals in parts a and b contain the value $(\mu_1 - \mu_2) = 0$? Why is this of interest to the researcher?

d. Do the data indicate a difference in the average room rates between the Marriott and the Doubletree chains? Between the Westin and the Doubletree chains?

8.57 Noise and Stress To compare the effect of stress in the form of noise on the ability to perform a simple task, 70 subjects were divided into two groups. The first group of 30 subjects acted as a control, while the second group of 40 were the experimental group. Although each subject performed the task in the same control room, each of the experimental group subjects had to perform the task while loud rock music was played. The time to finish the task was recorded for each subject and the following summary was obtained:

	Control	Experimental
n	30	40
$\bar{x}$	15 minutes	23 minutes
s	4 minutes	10 minutes

a. Find a 99% confidence interval for the difference in mean completion times for these two groups.

b. Based on the confidence interval in part a, is there sufficient evidence to indicate a difference in the average time to completion for the two groups? Explain.

8.58 What's Normal II Of the 130 people in Exercise 8.42, 65 were female and 65 were male.[15] The means and standard deviation of their temperatures are shown below:

	Men	Women
Sample mean	36.7°C	36.98°C
Standard deviation	0.70°C	0.75°C

Find a 95% confidence interval for the difference in the average body temperatures for males versus females. Based on this interval, can you conclude that there is a difference in the average temperatures for males versus females? Explain.

ESTIMATING THE DIFFERENCE BETWEEN TWO BINOMIAL PROPORTIONS

A simple extension of the estimation of a binomial proportion p is the estimation of the difference between two binomial proportions. You may wish to make comparisons such as these:

- The proportion of defective items manufactured in two production lines
- The proportion of female voters and the proportion of male voters who favour an equal rights amendment
- The germination rates of untreated seeds and seeds treated with a fungicide

These comparisons can be made using the difference $(p_1 - p_2)$ between two binomial proportions, p_1 and p_2. Independent random samples consisting of n_1 and n_2 trials are drawn from populations 1 and 2, respectively, and the sample estimates $\hat{p}_1$ and $\hat{p}_2$ are calculated. The unbiased estimator of the difference $(p_1 - p_2)$ is the sample difference $(\hat{p}_1 - \hat{p}_2)$.

PROPERTIES OF THE SAMPLING DISTRIBUTION OF THE DIFFERENCE $(\hat{p}_1 - \hat{p}_2)$ BETWEEN TWO SAMPLE PROPORTIONS

Assume that independent random samples of n_1 and n_2 observations have been selected from binomial populations with parameters p_1 and p_2, respectively. The sampling distribution of the difference between sample proportions

$$(\hat{p}_1 - \hat{p}_2) = \left(\frac{x_1}{n_1} - \frac{x_2}{n_2}\right)$$

has these properties:

1. The mean of $(\hat{p}_1 - \hat{p}_2)$ is

$$p_1 - p_2$$

and the standard error is

$$\text{SE} = \sqrt{\frac{p_1 q_1}{n_1} + \frac{p_2 q_2}{n_2}}$$

which is estimated as

$$\text{SE} = \sqrt{\frac{\hat{p}_1 \hat{q}_1}{n_1} + \frac{\hat{p}_2 \hat{q}_2}{n_2}}$$

2. The sampling distribution of $(\hat{p}_1 - \hat{p}_2)$ can be approximated by a normal distribution when n_1 and n_2 are large, due to the Central Limit Theorem.

Although the range of a single proportion is from 0 to 1, the difference between two proportions ranges from -1 to 1. To use a normal distribution to approximate the distribution of $(\hat{p}_1 - \hat{p}_2)$, both $\hat{p}_1$ and $\hat{p}_2$ should be approximately normal; that is, $n_1\hat{p}_1 > 5$, $n_1\hat{q}_1 > 5$ and $n_2\hat{p}_2 > 5$, $n_2\hat{q}_2 > 5$.

The appropriate formulas for point and interval estimation are given next.

LARGE-SAMPLE POINT ESTIMATION OF $(p_1 - p_2)$

Point estimator: $(\hat{p}_1 - \hat{p}_2)$

95% margin of error: $\pm 1.96\,\text{SE} = \pm 1.96\sqrt{\frac{\hat{p}_1 \hat{q}_1}{n_1} + \frac{\hat{p}_2 \hat{q}_2}{n_2}}$

A $(1 - \alpha)100\%$ LARGE-SAMPLE CONFIDENCE INTERVAL FOR $(p_1 - p_2)$

$$(\hat{p}_1 - \hat{p}_2) \pm z_{\alpha/2}\sqrt{\frac{\hat{p}_1 \hat{q}_1}{n_1} + \frac{\hat{p}_2 \hat{q}_2}{n_2}}$$

Assumption: n_1 and n_2 must be sufficiently large so that the sampling distribution of $(\hat{p}_1 - \hat{p}_2)$ can be approximated by a normal distribution—namely, if $n_1\hat{p}_1$, $n_1\hat{q}_1$, $n_2\hat{p}_2$, and $n_2\hat{q}_2$ are all greater than 5.

EXAMPLE  **8.11**

A bond proposal for school construction will be submitted to the voters at the next municipal election. A major portion of the money derived from this bond issue will be used to build schools in a rapidly developing section of the city, and the remainder will be used to renovate and update school buildings in the rest of the city. To assess the viability of the bond proposal, a random sample of $n_1 = 50$ residents in the developing section and $n_2 = 100$ residents from the other parts of the city were asked whether they plan to vote for the proposal. The results are tabulated in Table 8.6.

TABLE 8.6 **Sample Values for Opinion on Bond Proposal**

	Developing Section	Rest of the City
Sample size	50	100
Number favouring proposal	38	65
Proportion favouring proposal	0.76	0.65

1. Estimate the difference in the true proportions favouring the bond proposal with a 99% confidence interval.
2. If both samples were pooled into one sample of size $n = 150$, with 103 in favour of the proposal, provide a point estimate of the proportion of city residents who will vote for the bond proposal. What is the margin of error?

SOLUTION

1. The best point estimate of the difference $(p_1 - p_2)$ is given by

   $$(\hat{p}_1 - \hat{p}_2) = 0.76 - 0.65 = 0.11$$

   and the standard error of $(\hat{p}_1 - \hat{p}_2)$ is estimated as

   $$\sqrt{\frac{\hat{p}_1\hat{q}_1}{n_1} + \frac{\hat{p}_2\hat{q}_2}{n_2}} = \sqrt{\frac{(0.76)(0.24)}{50} + \frac{(0.65)(0.35)}{100}} = 0.077$$

   For a 99% confidence interval, $z_{.005} = 2.58$, and the approximate 99% confidence interval is found as

   $$(\hat{p}_1 - \hat{p}_2) \pm z_{0.005}\sqrt{\frac{\hat{p}_1\hat{q}_1}{n_1} + \frac{\hat{p}_2\hat{q}_2}{n_2}}$$
   $$0.11 \pm (2.58)(0.077)$$
   $$0.11 \pm 0.199$$

   or $(-0.089, 0.309)$. Since this interval contains the value $(p_1 - p_2) = 0$, it is possible that $p_1 = p_2$, which implies that there may be no difference in the proportions favouring the bond issue in the two sections of the city.

2. If there is no difference in the two proportions, then the two samples are not really different and might well be combined to obtain an overall estimate of the proportion of the city residents who will vote for the bond issue. If both samples are pooled, then $n = 150$ and

   $$\hat{p} = \frac{103}{150} = 0.69$$

   Therefore, the point estimate of the overall value of p is 0.69, with a margin of error given by

   $$\pm 2.58\sqrt{\frac{(0.69)(0.31)}{150}} = \pm 2.58(.0378) = \pm 0.097$$

   Notice that 0.69 ± 0.097 produces the interval 0.59 to 0.79, which includes only proportions greater than 0.5. Therefore, if voter attitudes do not change adversely prior to the election, the bond proposal should pass by a reasonable majority.

8.6 EXERCISES

BASIC TECHNIQUES

8.59 Independent random samples of $n_1 = 500$ and $n_2 = 500$ observations were selected from binomial populations 1 and 2, and $x_1 = 120$ and $x_2 = 147$ successes were observed.

a. What is the best point estimator for the difference $(p_1 - p_2)$ in the two binomial proportions?

b. Calculate the approximate standard error for the statistic used in part a.

c. What is the margin of error for this point estimate?

8.60 Independent random samples of $n_1 = 800$ and $n_2 = 640$ observations were selected from binomial populations 1 and 2, and $x_1 = 337$ and $x_2 = 374$ successes were observed.

a. Find a 90% confidence interval for the difference $(p_1 - p_2)$ in the two population proportions. Interpret the interval.

b. What assumptions must you make for the confidence interval to be valid? Are these assumptions met?

8.61 Independent random samples of $n_1 = 1265$ and $n_2 = 1688$ observations were selected from binomial populations 1 and 2, and $x_1 = 849$ and $x_2 = 910$ successes were observed.

a. Find a 99% confidence interval for the difference $(p_1 - p_2)$ in the two population proportions. What does "99% confidence" mean?

b. Based on the confidence interval in part a, can you conclude that there is a difference in the two binomial proportions? Explain.

8.62 Suppose that two independent random samples were taken from a binomial population and the results are tabulated below:

Population	Proportion	Sample Size	Successes
A	p_A	140	79
B	p_B	88	63

a. What is the best point estimator for the difference $(p_B - p_A)$ in the two binomial proportions?

b. Calculate the approximate standard error for the statistic used in part a.

c. What is the margin of error for this point estimate?

d. Find a 90% confidence interval for the difference $(p_B - p_A)$ in the two population proportions. Interpret the interval.

e. Find a 99% confidence interval for the difference $(p_B - p_A)$ in the two population proportions.

f. Compare the width of 90% and 99% confidence intervals for $(p_B - p_A)$. What effect does increasing the confidence coefficient have on the width of the confidence interval?

g. What assumptions must you make for the confidence interval to be valid? Are these assumptions met?

APPLICATIONS

8.63 M&M'S® Does the maker of M&M'S® (Mars, Inc.) use the same proportion of red candies in its plain and peanut varieties? A random sample of 56 plain M&M'S® contained 12 red candies, and another random sample of 32 peanut M&M'S® contained 8 red candies.

a. Construct a 95% confidence interval for the difference in the proportions of red candies for the plain and peanut varieties.

b. Based on the confidence interval in part a, can you conclude that there is a difference in the proportions of red candies for the plain and peanut varieties? Explain.

8.64 Consumer Outlook According to the 14th Annual RBC Homeownership Survey conducted by Ipsos Reid in 2007, most Canadians continued to think purchasing a home was a good investment.[16] Additionally, there was less concern about interest and/or mortgage rate hikes than at the same time the year before: 51% were concerned about interest rate increases in 2007 versus 56% in 2006; 43% thought mortgage rates would go up in 2007 versus 70% in 2006. Suppose that these results are based on 1000 randomly selected adult Canadians.

a. Construct a 99% confidence interval for the difference in the proportion of Canadians regarding the interest increase in 2006 versus 2007.

b. Does the data indicate that the true proportion of 2007 is higher than in 2006? Explain.

8.65 Consumer Outlook, continued Refer to Exercise 8.64. In a 2007 opinion poll, suppose that 430 adults identified those who thought mortgage rates would go up. The following question was posed: *Do you think Bank of Canada may start hiking the base rate in coming years*? Suppose that 390 who thought mortgage rates would go up and 100 of those who thought mortgage rates would not go up answered yes to this question.

a. Construct an 80% confidence interval for the difference in the true proportion of Canadians regarding their opinion on Bank of Canada rate increases in the future.

b. Does the data indicate that there is a difference in true proportion in part a? Why or why not?

8.66 Catching a Cold Do well-rounded people get fewer colds? A study in the *Chronicle of Higher Education* was conducted by scientists at Carnegie Mellon University, the University of Pittsburgh, and the University of Virginia. They found that people who have only a few social outlets get more colds than those who are involved in a variety of social activities.[17] Suppose that of the 276 healthy men and women tested, $n_1 = 96$ had only a few social outlets and $n_2 = 105$ were busy with six or more activities. When these people were exposed to a cold virus, the following results were observed:

	Few Social Outlets	Many Social Outlets
Sample size	96	105
Percent with colds	62%	35%

a. Construct a 99% confidence interval for the difference in the two population proportions.

b. Does there appear to be a difference in the population proportions for the two groups?

c. You might think that coming into contact with more people would lead to more colds, but the data show the opposite effect. How can you explain this unexpected finding?

8.67 Union, Yes! A sampling of political candidates—200 randomly chosen from the West and 200 from the East—was classified according to whether the candidate received backing by a national labour union and whether the candidate won. In the West, 120 winners had union backing, and in the East, 142 winners were backed by a national union. Find a 95% confidence interval for the difference between the proportions of union-backed winners in the West versus the East. Interpret this interval.

8.68 Birth Order and University Success In a study of the relationship between birth order and university success, an investigator found that 126 in a sample of 180 university graduates were firstborn or only children. In a sample of 100 non-graduates of comparable age and socio-economic background, the number of firstborn or only children was 54. Estimate the difference between the proportions of firstborn or only children in the two populations from which these samples were drawn. Use a 90% confidence interval and interpret your results.

8.69 Ads in Outer Space? Do you think that we should let The Source film a commercial in outer space? The commercialism of our space program is a topic of great interest since Dennis Tito paid $20 million to ride along with the Russians on the space shuttle.[18] In a survey of 500 men and 500 women, 20% of the men and 26% of the women responded that space should remain commercial-free.

a. Construct a 98% confidence interval for the difference in the proportions of men and women who think that space should remain commercial-free.

b. What does it mean to say that you are "98% confident"?

c. Based on the confidence interval in part a, can you conclude that there is a difference in the proportions of men and women who think space should remain commercial-free?

8.70 Generation Next Born between 1980 and 1990, Generation Next is engaged with technology, and the vast majority is dependent upon it.[19] Suppose that in a survey of 500 female and 500 male students in Generation Next, 345 of the females and 365 of the males reported that they decided to attend university in order to make more money.

a. Construct a 98% confidence interval for the difference in the proportions of female and male students who decided to attend university in order to make more money.

b. What does it mean to say that you are "98% confident"?

c. Based on the confidence interval in part a, can you conclude that there is a difference in the proportions of female and male students who decided to attend university in order to make more money?

8.71 Excedrin or Tylenol? In a study to compare the effects of two pain relievers, it was found that of $n_1 = 200$ randomly selected individuals instructed to use the first pain reliever, 93% indicated that it relieved their pain. Of $n_2 = 450$ randomly selected individuals instructed to use the second pain reliever, 96% indicated that it relieved their pain.

a. Find a 99% confidence interval for the difference in the proportions experiencing relief from pain for these two pain relievers.

b. Based on the confidence interval in part a, is there sufficient evidence to indicate a difference in the proportions experiencing relief for the two pain relievers? Explain.

8.72 Auto Accidents Last year's records of auto accidents occurring on a given section of highway were classified according to whether the resulting damage was \$1000 or more and to whether a physical injury resulted from the accident. The data follows:

	Under \$1000	\$1000 or more
Number of accidents	32	41
Number involving injuries	10	23

a. Estimate the true proportion of accidents involving injuries when the damage was \$1000 or more for similar sections of highway and find the margin of error.

b. Estimate the true difference in the proportion of accidents involving injuries for accidents with damage under \$1000 and those with damage of \$1000 or more. Use a 95% confidence interval.

8.7 ONE-SIDED CONFIDENCE BOUNDS

The confidence intervals discussed in Sections 8.4 to 8.6 are sometimes called **two-sided confidence intervals** because they produce both an upper (UCL) and a lower (LCL) bound for the parameter of interest. Sometimes, however, an experimenter is interested in only one of these limits; that is, the experimenter needs only an upper bound (or possibly a lower bound) for the parameter of interest. In this case, you can construct a **one-sided confidence bound** for the parameter of interest, such as μ, p, $\mu_1 - \mu_2$ or $p_1 - p_2$.

When the sampling distribution of a point estimator is approximately normal, an argument similar to the one in Section 8.4 can be used to show that one-sided confidence bounds, constructed using the following equations *when the sample size is large,* will contain the true value of the parameter of interest $(1 - \alpha)100\%$ of the time in repeated sampling.

A $(1 - \alpha)100\%$ LOWER CONFIDENCE BOUND (LCB)

(Point estimator) $- z_\alpha \times$ (Standard error of the estimator)

A $(1 - \alpha)100\%$ UPPER CONFIDENCE BOUND (UCB)

(Point estimator) $+ z_\alpha \times$ (Standard error of the estimator)

The z-value used for a $(1 - \alpha)100\%$ one-sided confidence bound, z_α, locates an area α in a single tail of the normal distribution as shown in Figure 8.11.

FIGURE 8.11

z-value for a one-sided confidence bound

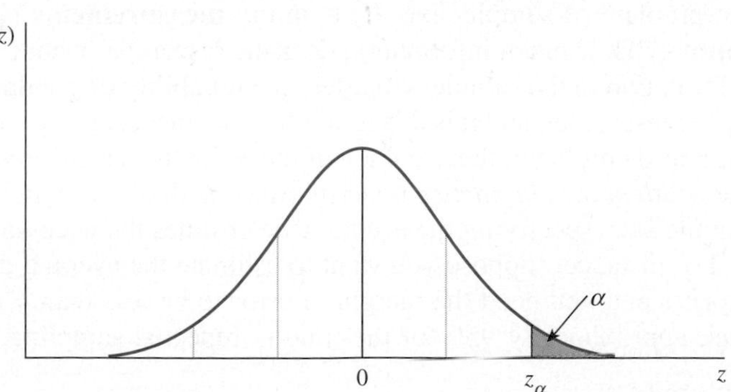

EXAMPLE 8.12

A corporation plans to issue some short-term notes and is hoping that the interest it will have to pay will not exceed 11.5%. To obtain some information about this problem, the corporation marketed 40 notes, one through each of 40 brokerage firms. The mean and standard deviation for the 40 interest rates were 10.3% and 0.31%, respectively. Since the corporation is interested in only an upper limit on the interest rates, find a 95% upper confidence bound for the mean interest rate that the corporation will have to pay for the notes.

SOLUTION Since the parameter of interest is μ, the point estimator is $\bar{x}$ with standard error SE $\approx \frac{s}{\sqrt{n}}$. The confidence coefficient is 0.95, so that $\alpha = 0.05$ and $z_{.05} = 1.645$. Therefore, the 95% upper confidence bound is

$$\text{UCB} = \bar{x} + 1.645\left(\frac{s}{\sqrt{n}}\right) = 10.3 + 1.645\left(\frac{0.31}{\sqrt{40}}\right) = 10.3 + 0.0806 = 10.4$$

Thus, you can estimate that the mean interest rate that the corporation will have to pay on its notes will be less than 10.4%. The corporation should not be concerned about its interest rates exceeding 11.5%. How confident are you of this conclusion? Fairly confident, because intervals constructed in this manner contain μ 95% of the time.

8.8 CHOOSING THE SAMPLE SIZE

Designing an experiment is essentially a plan for buying a certain amount of information. Just as the price you pay for a video game varies depending on where and when you buy it, the price of statistical information varies depending on how and where the information is collected. As when you buy any product, you should buy as much statistical information as you can for the minimum possible cost.

The total amount of relevant information in a sample is controlled by two factors:

- The **sampling plan** or **experimental design:** the procedure for collecting the information
- The **sample size** $\boldsymbol{n}$**:** the amount of information you collect

You can increase the amount of information you collect by *increasing* the sample size, or perhaps by *changing* the type of sampling plan or experimental design you are using. We will discuss the simplest sampling plan—random sampling from a relatively large population—and focus on ways to choose the sample size n needed to purchase a given amount of information.

A researcher makes little progress in planning an experiment before encountering the problem of sample size. **How many measurements should be included in the sample?** How much information does the researcher want to buy? The total amount of information in the sample will affect the reliability or goodness of the inferences made by the researcher, and it is this reliability that the researcher must specify. In a statistical estimation problem, the accuracy of the estimate is measured by the *margin of error* or the *width of the confidence interval.* Since both of these measures are a function of the sample size, specifying the accuracy determines the necessary sample size.

For instance, suppose you want to estimate the average daily yield μ of a chemical process and you need the margin of error to be less than 4 metric tonnes. This means that, approximately 95% of the time in repeated sampling, the distance between the sample mean $\bar{x}$ and the population mean μ should be less than 1.96 SE. You want this quantity to be less than 4. That is,

$$1.96\,\text{SE} < 4 \quad \text{or} \quad 1.96\left(\frac{\sigma}{\sqrt{n}}\right) < 4$$

Solving for *n,* you obtain

$$n > \left(\frac{1.96}{4}\right)^2 \sigma^2 \quad \text{or} \quad n > 0.24\sigma^2$$

If you know σ, the population standard deviation, you can substitute its value into the formula and solve for *n*. If σ is unknown—which is usually the case—you can use the best approximation available:

- An estimate *s* obtained from a previous sample
- A range estimate based on knowledge of the largest and smallest possible measurements: $\sigma \approx \text{Range}/4$

For this example, suppose that a prior study of the chemical process produced a sample standard deviation of $s = 21$ metric tonnes. Then

$$n > 0.24\sigma^2 = 0.24(21)^2 = 105.8$$

Using a sample of size $n = 106$ or larger, you could be reasonably certain (with probability approximately equal to 0.95) that your estimate of the average yield will be within $\pm$ 4 metric tonnes of the actual average yield.

The solution $n = 106$ is only approximate because you had to use an approximate value for σ to calculate the standard error of the mean. Although this may bother you, it is the best method available for selecting the sample size, and it is certainly better than guessing!

Sometimes researchers request a different confidence level than the 95% confidence specified by the margin of error. In this case, the half-width of the confidence interval provides the accuracy measure for your estimate; that is, the bound *B* on the error of your estimate is

$$z_{\alpha/2}\left(\frac{\sigma}{\sqrt{n}}\right) < B$$

This method for choosing the sample size can be used for all four estimation procedures presented in this chapter. The general procedure is described next.

NEED TO KNOW

How to Choose the Sample Size

Determine the parameter to be estimated and the standard error of its point estimator. Then proceed as follows:

1. Choose *B,* the bound on the error of your estimate, and a confidence coefficient $(1 - \alpha)$.
2. For a one-sample problem, solve this equation for the sample size *n*:

   $$z_{\alpha/2} \times (\text{Standard error of the estimator}) \le B$$

   where $z_{\alpha/2}$ is the value of *z* having area $\alpha/2$ to its right.
3. For a two-sample problem, set $n_1 = n_2 = n$ and solve the equation in step 2.

NOTE: For most estimators (all presented in this textbook), the standard error is a function of the sample size *n.*

EXAMPLE 8.13

Producers of polyvinyl plastic pipe want to have a supply of pipes sufficient to meet marketing needs. They wish to survey wholesalers who buy polyvinyl pipe in order to estimate the proportion who plan to increase their purchases next year. What sample size is required if they want their estimate to be within 0.04 of the actual proportion with probability equal to 0.90?

SOLUTION For this particular example, the bound B on the error of the estimate is 0.04. Since the confidence coefficient is $(1 - \alpha) = 0.90$, α must equal 0.10 and $\alpha/2$ is 0.05. The z-value corresponding to an area equal to 0.05 in the upper tail of the z distribution is $z_{0.05} = 1.645$. You then require

$$1.645\,\text{SE} = 1.645\sqrt{\frac{pq}{n}} = 0.04$$

In order to solve this equation for n, you must substitute an approximate value of p into the equation. If you want to be certain that the sample is large enough, you should use $p = 0.5$ (substituting $p = 0.5$ will yield the largest possible solution for n because the maximum value of pq occurs when $p = q = 0.5$). Then

$$1.645\sqrt{\frac{(0.5)(0.5)}{n}} \leq 0.04$$

or

$$\sqrt{n} \geq \frac{(1.645)(0.5)}{0.04} = 20.56$$

$$n \geq (20.56)^2 = 422.8$$

Therefore, the producers must include at least 423 wholesalers in its survey if it wants to estimate the proportion p correct to within 0.04.

EXAMPLE 8.14

A personnel director wishes to compare the effectiveness of two methods of training industrial employees to perform a certain assembly operation. A number of employees are to be divided into two equal groups: the first receiving training method 1 and the second training method 2. Each will perform the assembly operation, and the length of assembly time will be recorded. It is expected that the measurements for both groups will have a range of approximately eight minutes. For the estimate of the difference in mean times to assemble to be correct to within one minute with a probability equal to 0.95, how many workers must be included in each training group?

SOLUTION Letting $B = 1$ minute, you get

$$1.96\sqrt{\frac{\sigma_1^2}{n_1} + \frac{\sigma_2^2}{n_2}} \leq 1$$

Since you wish n_1 to equal n_2, you can let $n_1 = n_2 = n$ and obtain the equation

$$1.96\sqrt{\frac{\sigma_1^2}{n} + \frac{\sigma_2^2}{n}} \leq 1$$

As noted above, the variability (range) of each method of assembly is approximately the same, and hence $\sigma_1^2 = \sigma_2^2 = \sigma^2$. Since the range, equal to eight minutes, is approximately equal to 4σ, you have

$$4\sigma \approx 8 \text{ or } \sigma \approx 2$$

Substituting this value for σ_1 and σ_2 in the earlier equation, you get

$$1.96\sqrt{\frac{(2)^2}{n} + \frac{(2)^2}{n}} \leq 1$$

$$1.96\sqrt{\frac{8}{n}} \leq 1$$

$$\sqrt{n} \geq 1.96\sqrt{8}$$

Solving, you have $n \geq 31$. Thus, each group should contain at least $n = 31$ workers.

Table 8.7 provides a summary of the formulas used to determine the sample sizes.

TABLE 8.7 **Sample Size Formulas**

Parameter	Estimator	Sample Size	Assumptions
μ	$\bar{x}$	$n \geq \frac{z_{\alpha/2}^2\sigma^2}{B^2}$	
$\mu_1 - \mu_2$	$\bar{x}_1 - \bar{x}_2$	$n \geq \frac{z_{\alpha/2}^2(\sigma_1^2 + \sigma_2^2)}{B^2}$	$n_1 = n_2 = n$
p	$\hat{p}$	$n \geq \frac{z_{\alpha/2}^2 pq}{B^2}$ or $n \geq \frac{(0.25)z_{\alpha/2}^2}{B^2}$	$p = 0.5$
$p_1 - p_2$	$\hat{p}_1 - \hat{p}_2$	$n \geq \frac{z_{\alpha/2}^2(p_1q_1 + p_2q_2)}{B^2}$ or $n \geq \frac{2(0.25)z_{\alpha/2}^2}{B^2}$	$n_1 = n_2 = n$; $n_1 = n_2 = n$ and $p_1 = p_2 = 0.5$

8.8 EXERCISES

BASIC TECHNIQUES

8.73 Find a 90% one-sided upper confidence bound for the population mean μ for these values:

a. $n = 40, s^2 = 65, \bar{x} = 75$

b. $n = 100, s = 2.3, \bar{x} = 1.6$

8.74 Find a 99% lower confidence bound for the binomial proportion p when a random sample of $n = 400$ trials produced $x = 196$ successes.

8.75 Independent random samples of size 50 are drawn from two quantitative populations, producing the sample information in the table. Find a 95% upper confidence bound for the difference in the two population means.

	Sample 1	Sample 2
Sample size	50	50
Sample mean	12	10
Sample standard deviation	5	7

8.76 Suppose you wish to estimate a population mean based on a random sample of n observations, and prior experience suggests that $\sigma = 12.7$. If you wish to estimate μ correct to within 1.6, with probability equal to 0.95, how many observations should be included in your sample?

8.77 Suppose you wish to estimate a binomial parameter p correct to within 0.04, with probability equal to 0.95. If you suspect that p is equal to some value between 0.1 and 0.3 and you want to be certain that your sample is large enough, how large should n be? (HINT: When calculating the standard error, use the value of p in the interval $0.1 < p < 0.3$ that will give the largest sample size.)

8.78 Independent random samples of $n_1 = n_2 = n$ observations are to be selected from each of two populations 1 and 2. If you wish to estimate the difference between the two population means correct to within 0.17, with probability equal to 0.90, how large should n_1 and n_2 be? Assume that you know $\sigma_1^2 \approx \sigma_2^2 \approx 27.8$.

8.79 Independent random samples of $n_1 = n_2 = n$ observations are to be selected from each of two binomial populations 1 and 2. If you wish to estimate the difference in the two population proportions correct to within 0.05, with probability equal to 0.98, how large should n be? Assume that you have no prior information on the values of p_1 and p_2, but you want to make certain that you have an adequate number of observations in the samples.

8.80 Suppose that two independent random samples of equal sizes are to be taken from two binomial populations such that

Population	Prior Proportion
I	$p_I = 0.5$
II	$p_{II} = 0.6$

If you wish to estimate the difference in the two population proportions correct within 0.10, with probability equal to 0.90, how large should n be?

APPLICATIONS

8.81 Operating Expenses A random sampling of a company's monthly operating expenses for $n = 36$ months produced a sample mean of \$5474 and a standard deviation of \$764. Find a 90% upper confidence bound for the company's mean monthly expenses.

8.82 Fear Factor II Exercise 8.22 discussed a research poll done by EKOS Research to determine whether Canadians think the world is more dangerous today than 25 years ago. Suppose you were designing a poll of this type.

a. Explain how you would select your sample. What problems might you encounter in this process?

b. If you wanted to estimate the percentage of the population who agree with a particular statement in your survey questionnaire correct to within 1%, with probability 0.95, approximately how many people would have to be polled?

8.83 Political Corruption A questionnaire is designed to investigate attitudes about political corruption in government. The experimenter would like to survey two different groups—Liberals and Conservatives—and compare the responses to various "yes–no" questions for the two groups. The experimenter requires that the sampling error for the difference in the proportion of yes responses for the two groups is no more than ±3 percentage points. If the two samples are the same size, how large should the samples be?

8.84 Less Red Meat! Canadians are becoming more conscious of the importance of good nutrition, and some researchers believe that we may be altering our diets to include less red meat and more fruits and vegetables. To test this theory, a researcher decides to select hospital nutritional records for subjects surveyed 10 years ago and to compare the average amount of beef consumed per year to the amounts consumed by an equal number of subjects she will interview this year. She knows that the amount of beef consumed annually by Canadians ranges from 0 to approximately 47 kg. How many subjects should the researcher select for each group if she wishes to estimate the difference in the average annual per-capita beef consumption correct to within 2.3 kg with 99% confidence?

8.85 Red Meat, continued Refer to Exercise 8.84. The researcher selects two groups of 400 subjects each and collects the following sample information on the annual beef consumption now and 10 years ago:

	Ten Years Ago	This Year
Sample mean (kg)	33.1	28.6
Sample standard deviation (kg)	11.3	12.7

a. The researcher would like to show that per-capita beef consumption has decreased in the last 10 years, so she needs to show that the difference in the averages is greater than 0. Find a 99% lower confidence bound for the difference in the average per-capita beef consumptions for the two groups.

b. What conclusions can the researcher draw using the confidence bound from part a?

8.86 Hunting Season If a wildlife service wishes to estimate the mean number of days of hunting per hunter for all hunters licensed in the province during a given season, with a bound on the error of estimation equal to two hunting days, how many hunters must be included in the survey? Assume that data collected in earlier surveys have shown σ to be approximately equal to 10.

8.87 Polluted Rain Suppose you wish to estimate the mean pH of rainfalls in an area that suffers heavy pollution due to the discharge of smoke from a power plant. You know that σ is in the neighbourhood of 0.5 pH, and you wish your estimate to lie within 0.1 of μ, with a probability near 0.95. Approximately how many rainfalls must be included in your sample (one pH reading per rainfall)? Would it be valid to select all of your water specimens from a single rainfall? Explain.

8.88 pH in Rainfall Refer to Exercise 8.87. Suppose you wish to estimate the difference between the mean acidity for rainfalls at two different locations, one in a relatively unpolluted area along the ocean and the other in an area subject to heavy air pollution. If you wish your estimate to be correct to the nearest 0.1 pH, with probability near 0.90, approximately how many rainfalls (pH values) would have to be included in each sample? (Assume that the variance of the pH measurements is approximately 0.25 at both locations and that the samples will be of equal size.)

8.89 Selenium, again Refer to the comparison of the daily adult intake of selenium in two different regions of Canada in Exercise 8.50. Suppose you wish to estimate the difference in the mean daily intakes between the two regions correct to within 5 micrograms, with probability equal to 0.90. If you plan to select an equal number of adults from the two regions (i.e., $n_1 = n_2$), how large should n_1 and n_2 be?

CHAPTER REVIEW

Key Concepts and Formulas

I. Types of Estimators

1. Point estimator: a single number is calculated to estimate the population parameter.
2. Interval estimator: two numbers are calculated to form an interval that, with a certain amount of confidence, contains the parameter.

II. Properties of Good Estimators

1. Unbiased: the average value of the estimator equals the parameter to be estimated.
2. Minimum variance: of all the unbiased estimators, the best estimator has a sampling distribution with the smallest standard error.
3. The margin of error measures the maximum distance between the estimator and the true value of the parameter.

III. Large-Sample Point Estimators

To estimate one of four population parameters when the sample sizes are large, use the following point estimators with the appropriate margins of error.

Parameter	Point Estimator	95% Margin of Error
μ	$\bar{x}$	$\pm 1.96\left(\frac{s}{\sqrt{n}}\right)$
p	$\hat{p} = \frac{x}{n}$	$\pm 1.96\sqrt{\frac{\hat{p}\hat{q}}{n}}$
$\mu_1 - \mu_2$	$\bar{x}_1 - \bar{x}_2$	$\pm 1.96\sqrt{\frac{s_1^2}{n_1} + \frac{s_2^2}{n_2}}$
$p_1 - p_2$	$(\hat{p}_1 - \hat{p}_2) = \left(\frac{x_1}{n_1} - \frac{x_2}{n_2}\right)$	$\pm 1.96\sqrt{\frac{\hat{p}_1\hat{q}_1}{n_1} + \frac{\hat{p}_2\hat{q}_2}{n_2}}$

IV. Large-Sample Interval Estimators

To estimate one of four population parameters when the sample sizes are large, use the following interval estimators.

Parameter	$(1 - \alpha)100\%$ Confidence Interval
μ	$\bar{x} \pm z_{\alpha/2}\left(\frac{s}{\sqrt{n}}\right)$
p	$\hat{p} \pm z_{\alpha/2}\sqrt{\frac{\hat{p}\hat{q}}{n}}$
$\mu_1 - \mu_2$	$(\bar{x}_1 - \bar{x}_2) \pm z_{\alpha/2}\sqrt{\frac{s_1^2}{n_1} + \frac{s_2^2}{n_2}}$
$p_1 - p_2$	$(\hat{p}_1 - \hat{p}_2) \pm z_{\alpha/2}\sqrt{\frac{\hat{p}_1\hat{q}_1}{n_1} + \frac{\hat{p}_2\hat{q}_2}{n_2}}$

1. All values in the interval are possible values for the unknown population parameter.
2. Any values outside the interval are unlikely to be the value of the unknown parameter.
3. To compare two population means or proportions, look for the value 0 in the confidence interval. If 0 is in the interval, it is possible that the two population means are equal, and you should not declare a difference. If 0 is not in the interval, it is unlikely that the two means are equal, and you can confidently declare a difference.

V. One-Sided Confidence Bounds

Use either the upper (+) or lower (−) two-sided bound, with the critical value of z changed from $z_{\alpha/2}$ to z_α.

VI. Choosing the Sample Size

1. Determine the size of the margin of error, B, that you are willing to tolerate.
2. Choose the sample size by solving for n or $n = n_1 = n_2$ in the inequality: $z_{\alpha/2}$ SE $\leq B$, where SE is a function of the sample size n.
3. For quantitative populations, estimate the population standard deviation using a previously calculated value of s or the range approximation $\sigma \approx$ Range/4.
4. For binomial populations, use the conservative approach and approximate p using the value $p = 0.5$.

Supplementary Exercises

8.90 State the Central Limit Theorem. Of what value is the Central Limit Theorem in large-sample statistical estimation?

8.91 A random sample of $n = 64$ observations has a mean $\bar{x} = 29.1$ and a standard deviation $s = 3.9$.

a. Give the point estimate of the population mean μ and find the margin of error for your estimate.

b. Find a 90% confidence interval for μ. What does "90% confident" mean?

c. Find a 90% lower confidence bound for the population mean μ. Why is this bound different from the lower confidence limit in part b?

d. How many observations do you need to estimate μ to within 0.5, with probability equal to 0.95?

8.92 Independent random samples of $n_1 = 50$ and $n_2 = 60$ observations were selected from populations 1 and 2, respectively. The sample sizes and computed sample statistics are given in the table:

	Population 1	Population 2
Sample size	50	60
Sample mean	100.4	96.2
Sample standard deviation	0.8	1.3

Find a 90% confidence interval for the difference in population means and interpret the interval.

8.93 Refer to Exercise 8.92. Suppose you wish to estimate $(\mu_1 - \mu_2)$ correct to within 0.2, with probability equal to 0.95. If you plan to use equal sample sizes, how large should n_1 and n_2 be?

8.94 A random sample of $n = 500$ observations from a binomial population produced $x = 240$ successes.

a. Find a point estimate for *p*, and find the margin of error for your estimator.

b. Find a 90% confidence interval for *p*. Interpret this interval.

8.95 Refer to Exercise 8.94. How large a sample is required if you wish to estimate *p* correct to within 0.025, with probability equal to 0.90?

8.96 Independent random samples of $n_1 = 40$ and $n_2 = 80$ observations were selected from binomial populations 1 and 2, respectively. The number of successes in the two samples were $x_1 = 17$ and $x_2 = 23$. Find a 99% confidence interval for the difference between the two binomial population proportions. Interpret this interval.

8.97 Refer to Exercise 8.96. Suppose you wish to estimate $(p_1 - p_2)$ correct to within 0.06, with probability equal to 0.99, and you plan to use equal sample sizes—that is, $n_1 = n_2$. How large should n_1 and n_2 be?

8.98 Ethnic Cuisine Ethnic groups in North America buy differing amounts of various food products because of their ethnic cuisine. Asians buy fewer canned vegetables than do other groups, and Hispanics purchase more cooking oil. A researcher interested in market segmentation for these two groups would like to estimate the proportion of households that select certain brands for various products. If the researcher wishes these estimates to be within 0.03 with probability 0.95, how many households should she include in the samples?

8.99 Women on Wall Street Women on Wall Street can earn large salaries, but may need to make sacrifices in their personal lives. In fact, many women in the securities industry have to make significant personal sacrifices. A survey of 482 women and 356 men found that only half of the women have children, compared to three-quarters of the men surveyed.[20]

a. What are the values of $\hat{p}_1$ and $\hat{p}_2$ for the women and men in this survey?

b. Find a 95% confidence interval for the difference in the proportion of women and men on Wall Street who have children.

c. What conclusions can you draw regarding the groups compared in part b?

8.100 Smoking and Blood Pressure An experiment was conducted to estimate the effect of smoking on the blood pressure of a group of 35 cigarette smokers. The difference for each participant was obtained by taking the difference in the blood pressure readings at the beginning of the experiment and again five years later. The sample mean increase, measured in millimetres of mercury, was $\bar{x} = 9.7$. The sample standard deviation was $s = 5.8$. Estimate the mean increase in blood pressure that one would expect for cigarette smokers over the time span indicated by the experiment. Find the margin of error. Describe the population associated with the mean that you have estimated.

8.101 Blood Pressure, continued Using a confidence coefficient equal to 0.90, place a confidence interval on the mean increase in blood pressure for Exercise 8.100.

8.102 Iodine Concentration Based on repeated measurements of the iodine concentration in a solution, a chemist reports the concentration as 4.614 moles/litre, with an "error margin of 0.006."

a. How would you interpret the chemist's "error margin"?

b. If the reported concentration is based on a random sample of $n = 30$ measurements, with a sample standard deviation $s = 0.017$, would you agree that the chemist's "error margin" is 0.006?

8.103 Heights If it is assumed that the heights of men are normally distributed, with a standard deviation of 6 cm, how large a sample should be taken to be fairly sure (probability 0.95) that the sample mean does not differ from the true mean (population mean) by more than 0.50 in absolute value?

8.104 Chicken Feed An experimenter fed different rations, A and B, to two groups of 100 chicks each. Assume that all factors other than rations are the same for both groups. Of the chicks fed ration A, 13 died, and of the chicks fed ration B, 6 died.

a. Construct a 98% confidence interval for the true difference in mortality rates for the two rations.

b. Can you conclude that there is a difference in the mortality rates for the two rations?

8.105 Antibiotics You want to estimate the mean hourly yield for a process that manufactures an antibiotic. You observe the process for 100 hourly periods chosen at random, with the results $\bar{x} = 34$ ounces per hour and $s = 3$. Estimate the mean hourly yield for the process using a 95% confidence interval.

8.106 Cheese and Beer The average European has become accustomed to eating away from home, especially at fast-food restaurants. Partly as a result of this fast-food habit, the per-capita consumption of cheese (the main ingredient in pizza) and beer has risen dramatically from a decade ago. A study in *Ecological*

Economics reports that the average European consumes 14.2 kg of cheese and drinks 83.9 litres of beer per year.[21] To test the accuracy of these reported averages, a random sample of 40 consumers is selected, and these summary statistics are recorded:

	Cheese (kg/yr)	Beer (L/yr)
Sample mean	16.1	81
Sample standard deviation	2.1	10.2

Use your knowledge of statistical estimation to estimate the average per-capita annual consumption for these two products. Does this sample cause you to support or to question the accuracy of the reported averages? Explain.

8.107 Sunflowers In an article in the *Annals of Botany,* a researcher reported the basal stem diameters of two groups of dicot sunflowers: those that were left to sway freely in the wind and those that were artificially supported.[22] A similar experiment was conducted for monocot maize plants. Although the authors measured other variables in a more complicated experimental design, assume that each group consisted of 64 plants (a total of 128 sunflower and 128 maize plants). The values shown in the table are the sample means plus or minus the standard error.

	Sunflower	Maize
Free-standing	35.3 ± 0.72	16.2 ± 0.41
Supported	32.1 ± 0.72	14.6 ± 0.40

Use your knowledge of statistical estimation to compare the free-standing and supported basal stem diameters for the two plants. Write a paragraph describing your conclusions, making sure to include a measure of the accuracy of your inference.

8.108 Working in Retirement Research released by Investors Group shows that non-retired Canadians overwhelmingly say their physical health is better than their financial health (67%)—and many may be relying on continued good health to enable them to keep working during their retirement years.[23] The majority of working Canadians (56%) agree that they think they would not have enough money to live on if they stopped working entirely, but maintaining social connections and gaining new experiences also appear to be on the minds of Canadians as they envision their retirement lifestyle. From survey respondents, 30% said the opportunity to maintain connections with other people was a benefit of working in retirement. The research given in the table showed interesting gender differences. Results are based on a sample of 2170 Canadians.

	Men	Women	All
Physical health is better than financial health	64%	70%	67%
Social connection	27%	37%	36%
Money is important	28%	34%	–

a. Construct a 95% confidence interval for the true proportion who agreed that physical health is better than financial health.

b. If a women's group consisted of $n = 160$ individuals, find a 95% confidence interval for the proportion of Canadian women who agreed that money is more important.

c. Find a 98% confidence interval for the true difference in proportions of women and men who said social connection is important.

8.109 University Costs A university administrator wishes to estimate the average cost of the first year at a particular university correct to within \$500, with a probability of 0.95. If a random sample of first-year students is to be selected and each asked to keep financial data, how many must be included in the sample? Assume that the dean knows only that the range of expenditures will vary from approximately \$4800 to \$13,000.

8.110 Quality Control A quality-control engineer wants to estimate the fraction of defectives in a large lot of film cartridges. From previous experience, he feels that the actual fraction of defectives should be somewhere around 0.05. How large a sample should he take if he wants to estimate the true fraction to within 0.01, using a 95% confidence interval?

8.111 Circuit Boards Samples of 400 printed circuit boards were selected from each of two production lines A and B. Line A produced 40 defectives, and line B produced 80 defectives. Estimate the difference in the actual fractions of defectives for the two lines with a confidence coefficient of 0.90.

8.112 Circuit Boards II Refer to Exercise 8.111. Suppose 10 samples of $n = 400$ printed circuit boards were tested and a confidence interval was constructed for p for each of the 10 samples. What is the probability that exactly one of the intervals will not contain the true value of p? That at least one interval will not contain the true value of p?

8.113 Ice Hockey The ability to accelerate rapidly is an important attribute for an ice hockey player.

G. Wayne Marino investigated some of the variables related to the acceleration and speed of a hockey player from a stopped position.[24] Sixty-nine hockey players, varsity and intramural, were included in the experiment. Each player was required to move as rapidly as possible from a stopped position to cover a distance of 6 metres (m). The means and standard deviations of some of the variables recorded for each of the 69 skaters are shown in the table:

	Mean	SD
Weight (kilograms)	75.270	9.470
Stride length (metres)	1.110	0.205
Stride rate (strides/second)	3.310	0.390
Average acceleration (metres/second^2)	2.962	0.529
Instantaneous velocity (metres/second)	5.753	0.892
Time to skate (seconds)	1.953	0.131

a. Give the formula that you would use to construct a 95% confidence interval for one of the population means (e.g., mean time to skate the 6 m distance).

b. Construct a 95% confidence interval for the mean time to skate. Interpret this interval.

8.114 Ice Hockey II Exercise 8.113 presented statistics from a study of fast starts by ice hockey skaters. The mean and standard deviation of the 69 individual average acceleration measurements over the 6 m distance were 2.962 and 0.529 metres per second (m/sec), respectively.

a. Find a 95% confidence interval for this population mean. Interpret the interval.

b. Suppose you were dissatisfied with the width of this confidence interval and wanted to cut the interval in half by increasing the sample size. How many skaters (total) would have to be included in the study?

8.115 Ice Hockey III The mean and standard deviation of the speeds of the sample of 69 skaters at the end of the 6 m distance in Exercise 8.113 were 5.753 and 0.892 (m/sec) respectively.

a. Find a 95% confidence interval for the mean velocity at the 6 m mark. Interpret the interval.

b. Suppose you wanted to repeat the experiment and you wanted to estimate this mean velocity correct to within 0.1 second (sec), with probability 0.99. How many skaters would have to be included in your sample?

8.116 Student Loan Debt In 2014, average Canadian student debt estimates hovered in the mid- to high-$20,000 range. The Canadian Federation of Students pegged it at $27,000, which is close to the nearly $26,300 that many students said they expected to owe after graduation in a recent BMO survey.[25] A group of $n = 49$ university students are randomly selected, and the average student loan debt is found to be $25,500 with a standard deviation of 700. Construct a 99% confidence interval for the true average student loan debt for university students in Canada. Does this contradict the reported average of $26,300? Explain.

8.117 Recidivism An experimental rehabilitation technique was used on released convicts. It was shown that 79 of 121 men subjected to the technique pursued useful and crime-free lives for a three-year period following prison release. Find a 95% confidence interval for p, the probability that a convict subjected to the rehabilitation technique will follow a crime-free existence for at least three years after prison release.

8.118 Specific Gravity If 36 measurements of the specific gravity of aluminum had a mean of 2.705 and a standard deviation of 0.028, construct a 98% confidence interval for the actual specific gravity of aluminum.

8.119 Audiology Research In a study to establish the absolute threshold of hearing, 70 male first-year university students were asked to participate. Each subject was seated in a soundproof room and a 150 Hz tone was presented at a large number of stimulus levels in a randomized order. The subject was instructed to press a button if he detected the tone; the experimenter recorded the lowest stimulus level at which the tone was detected. The mean for the group was 21.6 db with $s = 2.1$. Estimate the mean absolute threshold for all first-year university students and calculate the margin of error.

8.120 Right- or Left-Handed A researcher classified his subjects as innately right-handed or left-handed by comparing thumbnail widths. He took a sample of 400 men and found that 80 men could be classified as left-handed according to his criterion. Estimate the proportion of all males in the population who would test to be left-handed using a 95% confidence interval.

8.121 The Citrus Red Mite An entomologist wishes to estimate the average development time of the citrus red mite correct to within 0.5 day. From previous experiments it is known that σ is in the neighbourhood of four days. How large a sample should the entomologist take to be 95% confident of her estimate?

8.122 The Citrus Red Mite, continued A grower believes that one in five of his citrus trees are infected with the citrus red mite, mentioned in Exercise 8.121. How large a sample should be taken if the grower wishes to estimate the proportion of his trees that are infected with citrus red mite to within 0.08?

CASE STUDY

How Reliable Is That Poll?

What elements, if any, go into a Canadian "national identity" has long been a question of interest to scholars and ordinary citizens alike. The Centre for Research and Information on Canada (CRIC) and the *Globe and Mail* newspaper attempted to answer this question in their public opinion poll on "The New Canada," conducted by Ipsos Reid between April 21 and May 4, 2003, involving $n = 2000$ adults 18 years or older.[26] The survey included a set of questions asking Canadians how proud (or not) 18 highly visible things or events made them to be Canadian. Do national polls conducted by the Ipsos Reid and Decima organizations, the news media, and so on, really provide accurate estimates of the percentages of people in Canada who have various opinions? Let's look at some of the results of the poll survey methods that were used.

Respondents to the poll were asked the following question about each of the 18 things or events: *I will read you a list of things and events that some people say make them proud to be Canadian. I would like you to tell me whether each of these makes you feel proud to be a Canadian. Please use a scale of 0–10, where 0 means it does not make you feel proud at all, and 10 means it makes you feel very proud. You can use any number between 0 and 10. How about…?*

The following table shows how the public responded to this question for each thing or event, in descending order of the percentage indicating it made them proud to be Canadian, represented by a rating of 6 or higher on the 10-point scale.

The CRIC-*Globe and Mail* Survey on "The New Canada"

Things or Event	Proud (Rated 6–10) %	Neutral (Rated 5) %	Not Proud (Rated 0–4) %
The vastness and beauty of the land	95	3	2
When the United Nations ranks Canada as the best country in the world in which to live	92	4	4
Canada's politeness and civility	87	7	6
The fact that people from different cultural groups in Canada get along and live in peace	87	8	5
Canadian scientific inventions, like the Canadarm	87	7	6
When Canadian airports took in American planes that were diverted on September 11th, 2001	87	7	6
Canada's participation in peacekeeping activities around the world	86	7	7
The Charter of Rights and Freedoms	83	8	9
Canada's participation in key battles of World War I or World War II	80	10	9
Multiculturalism	79	11	11
Canadian Olympic hockey team victories	79	10	11
The success of Canadian musicians or actors or artists	78	12	10
Canada's health care system	71	12	17
The CBC	67	16	17
Pierre Trudeau	62	14	24
Having two official languages, English and French	60	16	25
When Canada decided to not participate in the war on Iraq	59	11	30
The Queen	43	15	42

Source: Andrew Parkin and Matthew Mendelsohn, "Table 5: Proud to Be Canadian," *The CRIC Papers: A New Canada: An Identity Shaped by Diversity*, October 2003, p. 11, http://130.15.161.246:82/docs/Government%20surveys/cric/cricgmnc03-eng/cricpapers_oct3003.pdf. Permission courtesy of Carleton University.

For several questions in the present poll, the sample of 2000 Canadians 18 years and older was split into two groups of 1000 adults each. For the questions on pride in the previous table, the full sample was asked about six of the things or events, while six each were asked about only in the first or second split sample groups. For example, the full sample of 2000 was asked to rate multiculturalism, while only the first split sample was asked to rate Pierre Trudeau, and only the second split sample was asked to rate "the fact that people from different cultural groups in Canada get along and live in peace."

The following description of the "Methodology" that appears on the website of the Canadian Opinion Research Archive, the poll's distributor, explains how the results of the poll should be viewed.

> The CRIC-*Globe and Mail* Survey on "The New Canada" was designed by the Centre for Research and Information on Canada, the *Globe and Mail,* and the Canadian Opinion Research Archive. The survey was carried out between April 21 and May 4, 2003, by Ipsos Reid. A representative sample of 2000 randomly selected Canadians were interviewed by telephone. A survey of this size has a margin of error of plus or minus 2.2%, 19 times out of 20.

1. Verify the margin of error of ±2.2 percentage points given by the survey designers for the full sample of 2000 adults. Find the margin of error for the two split samples of 1000 adults.
2. Do the numbers reported in the table represent the number of people who fell into those categories? If not, what do those numbers represent?
3. When the set of questions on pride in Canada were asked, the pollster rotated the order of things and events given to the respondent. Why do you suppose this technique was used?
4. Construct 95% confidence intervals for the proportion of Canadians who:
   a. report that they were "proud" when Canada decided to not participate in the war on Iraq.
   b. report that they were "not proud" about this decision.
5. Compare the percentage of people in the full sample of 2000 who say that multiculturalism makes them proud to be Canadian with the percentage of those in the second split sample who say that "the fact that people from different cultural groups in Canada get along and live in peace" makes them proud. Is this a surprising difference?
6. If these questions were asked today, would you expect the responses to be similar to those reported here or would you expect them to differ substantially?

PROJECTS

Project 8-A: Saving Time and Making Patients Safer

North York General Hospital, a multi-site community teaching hospital, gets approximately 75,000 visits to its emergency department each year. With that many patients, the emergency department was often very overcrowded and patients weren't flowing through care as they should. North York General Hospital used "Lean Tools," a quality improvement method that emphasizes reducing waste in the work environment in its many forms to create a productive work environment and a journey through the system for the patient that is as smooth and efficient as possible. Between November 2006 and September 2008, North York General Hospital achieved a reduction in the average time to see a physician. More specifically, the average time to see a physician in sub-acute care decreased from 3.3 hours to 2.8 hours.[27] Suppose a sample of 100 waiting times before 2006, and between 2006 and September 2008 were randomly selected from the hospital records for each of the two time periods. The sample statistics are listed below:

Sample Statistic	Waiting Time Before Quality Improvement	Waiting Time After Quality Improvement
Sample size	100	100
Sample mean	3.5	3.1
Sample variance	2.82	1.68

a. Give a point estimate for μ_2 the true mean waiting time after the quality improvement and the margin of error.

b. Construct a 90% confidence interval for μ_2 the true mean waiting time after quality improvement. Interpret your confidence interval.

c. Based on the interval in part b, can the hospital's claim be rejected? Justify your answer.

d. Suppose that the quality control department of this hospital intends to have the waiting time of 2.7 hours on average to see a physician in sub-acute care. Should the confidence interval in part b concern the quality control department? Explain.

e. If the distribution of the waiting time is not normal, you can still use the standard normal distribution to construct a confidence interval for μ_2. Why?

f. Now, suppose for a future survey the hospital administrator wishes to estimate μ_2. How many records should be sampled in order to estimate μ_2? The administrator would like the margin of error to be less than 0.5 hours with a confidence of 0.95, using standard deviation of 1.7.

g. Let μ_1 be the true mean waiting time before the quality improvement. What is the best point estimator for $(\mu_1 - \mu_2)$, the difference in average waiting time before and after the quality improvement?

h. Describe completely the sampling distribution of the point estimator for $(\mu_1 - \mu_2)$ and calculate the margin of error.

i. Find a 98% confidence interval for $(\mu_1 - \mu_2)$.

j. Based on the interval in part i, can one conclude there is a difference in the true average waiting time before quality improvement and after? Justify your answer.

k. What does the phrase "98% confident" mean?

Project 8-B: Attitudes of Canadian Women Toward Birthing Centres and Midwife Care for Childbirth

[*Source:* From "Attitudes of Canadian women toward birthing centres and midwife care for childbirth," Shi Wu Wen, Leslie S. Mery, Michael S. Kramer, Vanie Jimenez, Konia Trouton, Pearl Herbert, and Beverly Chalmers, *CMAJ*, Volume 161, Issue 6.

A study reveals that interest among Canadian women in home and birthing centre delivery and in midwife care for women at low risk has been growing recently. In the past few years in Canada, several provincial governments have legislated midwife care, and other provinces are considering doing so. A study was carried out under the auspices of the Canadian Prenatal Surveillance System and the result of the study was published in *CMAJ,* 21 Sept. 1999; 161 (6). These results demonstrate that a substantial proportion of Canadian women of reproductive age would be willing to deliver at a birthing centre and to receive childbirth and postpartum care from a nurse or midwife. Just under one-third of respondents (31%, 95% confidence interval [CI] 28% to 33%) answered yes to the question, "Would you go to a birthing centre, rather than a hospital, to have a baby?" To investigate the recent trend a province-wide survey was conducted and the sample statistics are listed below.

Province or Region of Residence	Atlantic Provinces	Ontario	Quebec	Western Provinces
Sample Size	49	360	169	225
Sample Proportion	0.28	0.31	0.35	0.30

Source: From "Attitudes of Canadian women toward birthing centres and midwife care for childbirth," Shi Wu Wen, Leslie S. Mery, Michael S. Kramer, Vanie Jimenez, Konia Trouton, Pearl Herbert, and Beverly Chalmers, *CMAJ,* Volume 161, Issue 6.

a. Give a point estimate for p_2, the proportion of Ontario women who prefer delivery in a birthing centre. Find the estimated margin of error.

b. Give a point estimate for p_3, the proportion of Quebec women who prefer delivery in a birthing centre. Find the estimated margin of error for your point estimate.

c. Estimate the true proportions, p_3 with a 99% confidence interval. Interpret this interval.

d. A health care researcher believes that 35% of the women in the Atlantic provinces prefer delivery in a birthing centre. How many women must you include in a simple random sample if you want to be 90% confident that the true population proportion lies within 0.01 of your sample proportion?

e. A health care researcher in the western provinces wishes to estimate the true proportion of women who prefer delivery in a birthing centre with a margin of error of no more than 0.1 and with probability 0.99. How many observations does the researcher need to include in the sample to achieve his goal?

f. Estimate the difference between the true proportions of women who prefer delivery in a birthing centre residing in Ontario and Quebec. Use a 98% confidence interval and interpret your results.

g. Based on the interval in part f, can one conclude whether there is a difference in the proportion of women in their attitudes toward having a delivery in a birthing centre? Justify your answer.

h. Find the point estimate of the difference between the true proportions of women who prefer delivery in a birthing centre residing in Atlantic and western provinces. Calculate the margin of error of your point estimator.

i. Is it possible to compare the true proportions based on all four provinces/regions of residence simultaneously? Explain.

9

Large-Sample Tests of Hypotheses

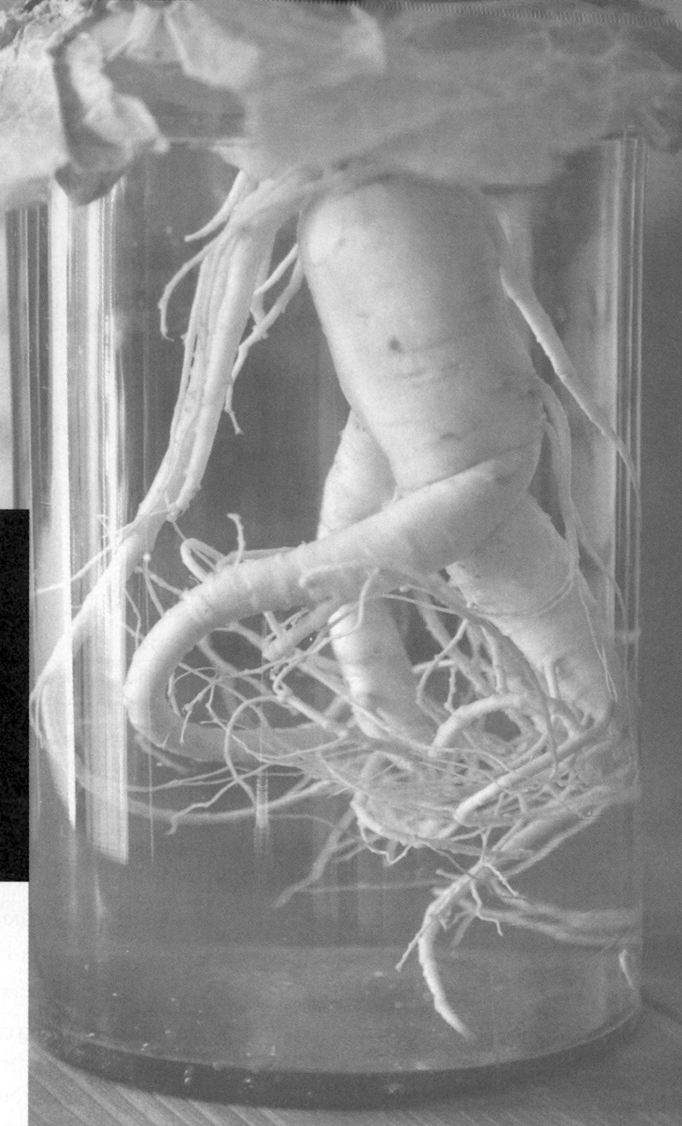

Anna Jurkovska/Shutterstock.com

GENERAL OBJECTIVES

In this chapter, the concept of a statistical test of hypothesis is formally introduced. The sampling distributions of statistics presented in Chapters 7 and 8 are used to construct large-sample tests concerning the values of population parameters of interest to the experimenter.

CHAPTER INDEX

NEED TO KNOW

Rejection Regions, *p*-Values, and Conclusions
How to Calculate β

Cure for the Cold—Pooling Data: Making Sense or Folly?

Can ginseng prevent colds? Edmonton company CV Technologies Inc. has conducted clinical trials, with results published in the *Journal of the American Geriatrics Society*, that it claims prove that its proprietary ginseng extract can do exactly that. Later, an article was published in the *Vancouver Sun* in which two professors from the University of British Columbia criticized the claims. They suggested that trials do not provide definite evidence that the product had any effect. What's going on here? The case study at the end of this chapter looks at how the trials were conducted, and you will analyze the data using large sample strategies.

TESTING HYPOTHESES ABOUT POPULATION PARAMETERS

In practical situations, statistical inference can involve either estimating a population parameter or making decisions about the value of the parameter. For example, if a pharmaceutical company is fermenting a vat of antibiotic, samples from the vat can be used to *estimate* the mean potency μ for all of the antibiotic in the vat. In contrast, suppose that the company is not concerned about the exact mean potency of the antibiotic, but is concerned only that it meet the minimum government potency standards. Then the company can use samples from the vat to decide between these two possibilities:

- The mean potency μ does not exceed the minimum allowable potency.
- The mean potency μ exceeds the minimum allowable potency.

The pharmaceutical company's problem illustrates a **statistical test of hypothesis.**

The reasoning used in a statistical test of hypothesis is similar to the process in a court trial. In trying a person for theft, the court must decide between innocence and guilt. As the trial begins, the accused person is assumed to be *innocent.* The prosecution collects and presents all available evidence in an attempt to contradict the innocent hypothesis and hence obtain a conviction. If there is enough evidence against innocence, the court will reject the innocence hypothesis and declare the defendant *guilty*. If the prosecution does not present enough evidence to prove the defendant guilty, the court will find the defendant *not guilty.* Notice that this does not prove that the defendant is innocent, but merely that there was not enough evidence to conclude that the defendant was guilty.

We use this same type of reasoning to explain the basic concepts of hypothesis testing. These concepts are used to test the four population parameters discussed in Chapter 8: a single population mean or proportion (μ or p) and the difference between two population means or proportions ($\mu_1 - \mu_2$ or $p_1 - p_2$). When the sample sizes are large, the point estimators for each of these four parameters have normal sampling distributions, so that all four large-sample statistical tests follow the same general pattern.

A STATISTICAL TEST OF HYPOTHESIS

A statistical test of hypothesis consists of five parts:

1. The null hypothesis, denoted by H_0
2. The alternative hypothesis, denoted by H_a
3. The test statistic and its p-value
4. The rejection region
5. The conclusion

When you specify these five elements, you define a particular test; changing one or more of the parts creates a new test. Let's look at each part of the statistical test of hypothesis in more detail. Note that these numbered labels will be used in most examples throughout this chapter for reference.

DEFINITION The two competing hypotheses are the **alternative hypothesis H_a,** generally the hypothesis that the researcher wishes to support, and the **null hypothesis H_0,** a contradiction of the alternative hypothesis.

1–2

As you will soon see, it is easier to show support for the alternative hypothesis by proving that the null hypothesis is false. Hence, the statistical researcher always begins by assuming that the null hypothesis H_0 is true. The researcher then uses the sample data to decide whether the evidence favours H_a rather than H_0 and draws one of these two **conclusions:**

- Reject H_0 and conclude that H_a is true.
- Accept (do not reject) H_0 as true.

EXAMPLE 9.1

You wish to show that the average hourly wage of clerical workers in the province of New Brunswick is different from \$14, which is the national average. This is the alternative hypothesis, written as

2 $H_a: \mu \neq 14$

The null hypothesis is

1 $H_0: \mu = 14$

You would like to reject the null hypothesis, thus concluding that the New Brunswick mean is not equal to \$14.

EXAMPLE 9.2

A milling process currently produces an average of 3% defectives. You are interested in showing that a simple adjustment on a machine will decrease p, the proportion of defectives produced in the milling process. Thus, the alternative hypothesis is

2 $H_a: p < 0.03$

and the null hypothesis is

1 $H_0: p = 0.03$

If you can reject H_0, you can conclude that the adjusted process produces fewer than 3% defectives.

EXAMPLE 9.3

According to Statistics Canada, the mean birth weight of newborn babies in Canada is 3.372 kg for both sexes.[1] However, a pediatrician in Alberta believes that mean birth of newborn babies for both sexes in Alberta is much higher than the national average. In this case, the alternative hypothesis is

2 $H_a: \mu > 3.372$

The null hypothesis is

1 $H_0: \mu = 3.372$

If you fail to reject the null hypothesis, then you can conclude that the mean birth weight of newborn babies in the province of Alberta is not higher than 3.372 kg. In other words, it is same as the national average.

EXAMPLE 9.4

Based on the online survey of a representative national sample of 1006 Canadian adults, 54% respondents believe Canada should be a melting pot.[2] A researcher thinks that the melting pot is particularly attractive for Quebecers and the proportion of Quebecers who support the idea of melting pot is higher than 54%.

Thus,

2 $H_a\!:p > 0.54$

1 $H_0\!:p = 0.54$

If the null hypothesis is rejected, then you may draw the conclusion that the proportion of residents of Quebecers who support the idea of melting pot is higher than the national proportion of 54%.

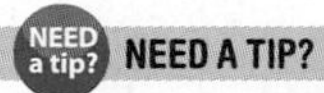

NEED A TIP?

Two-tailed ⇔ Look for a $\neq$ sign in H_a.

One-tailed ⇔ Look for a $>$ or $<$ sign in H_a.

There is a difference in the forms of the alternative hypotheses given in Examples 9.1 and 9.2. In Example 9.1, no directional difference is suggested for the value of μ; that is, μ might be either larger or smaller than $14 if H_a is true. This type of test is called a **two-tailed test of hypothesis.** In Example 9.2, however, you are specifically interested in detecting a directional difference in the value of p; that is, if H_a is true, the value of p is less than 0.03. This type of test is called a **one-tailed test of hypothesis.**

The decision to reject or accept the null hypothesis is based on information contained in a sample drawn from the population of interest. This information takes these forms:

3
- **Test statistic:** a single number calculated from the sample data
- ***p*-value:** a probability calculated using the test statistic

Either or both of these measures act as decision makers for the researcher in deciding whether to reject or accept H_0.

EXAMPLE 9.5

For the test of hypothesis in Example 9.1, the average hourly wage $\bar{x}$ for a random sample of 100 New Brunswick clerical workers might provide a good *test statistic* for testing

$$H_0\!:\mu = 14 \text{ versus } H_a\!:\mu \neq 14$$

If the null hypothesis H_0 is true, then the sample mean should not be too far from the population mean $\mu = 14$. Suppose that this sample produces a sample mean $\bar{x} = 15$ with standard deviation $s = 2$. Is this sample evidence likely or unlikely to occur, if in fact H_0 is true? You can use two measures to find out. Since the sample size is large, the sampling distribution of $\bar{x}$ is approximately normal with mean $\mu = 14$ and standard error $\sigma/\sqrt{n}$, estimated as

$$\text{SE} = \frac{s}{\sqrt{n}} = \frac{2}{\sqrt{100}} = 0.2$$

- The **test statistic** $\bar{x} = 15$ lies

3 $$z = \frac{\bar{x} - \mu}{\sigma/\sqrt{n}} = \frac{15 - 14}{0.2} = 5$$

standard deviations from the population mean μ.

- The ***p*-value** is the probability of observing a test statistic as extreme as or more extreme than the observed value, if in fact H_0 is true. For this example, we define "extreme" as far below or far above what we would have expected. That is,

$$p\text{-value} = P(z > 5) + P(z < -5) \approx 0$$

The *large value of the test statistic* and the *small p-value* mean that you have observed a very unlikely event, if indeed H_0 is true and $\mu = 14$.

How do you decide whether to reject or accept H_0? The entire set of values that the test statistic may assume is divided into two sets, or regions. One set, consisting of values that support the alternative hypothesis and lead to rejecting H_0, is called the

4 **rejection region.** The other, consisting of values that support the null hypothesis, is called the **acceptance region.**

For example, in Example 9.1, you would be inclined to believe that New Brunswick's average hourly wage was different from \$14 if the sample mean is either much less than \$14 or much greater than \$14. The two-tailed rejection region consists of very small and very large values of $\bar{x}$, as shown in Figure 9.1. In Example 9.2, since you want to prove that the percentage of defectives has *decreased,* you would be inclined to reject H_0 for values of $\hat{p}$ that are much smaller than 0.03. Only *small* values of $\hat{p}$ belong in the left-tailed rejection region shown in Figure 9.2. When the rejection region is in the left tail of the distribution, the test is called a **left-tailed test.** A test with its rejection region in the right tail is called a **right-tailed test.**

FIGURE 9.1
Rejection and acceptance regions for Example 9.1

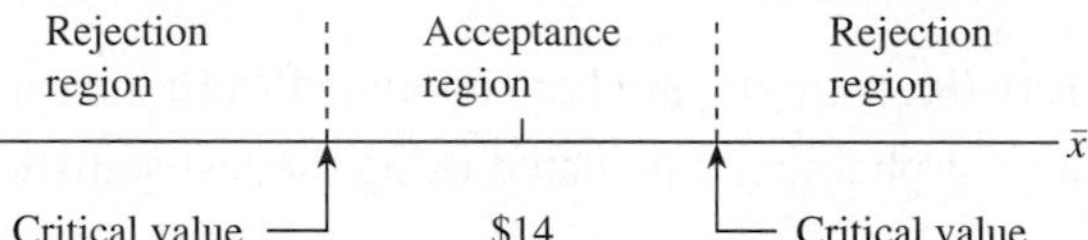

FIGURE 9.2
Rejection and acceptance regions for Example 9.2

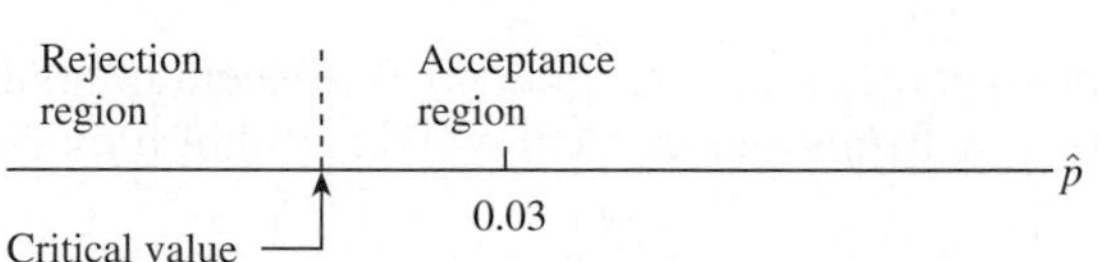

5 If the test statistic falls into the rejection region, then the null hypothesis is rejected. If the test statistic falls into the acceptance region, then either the null hypothesis is accepted or the test is judged to be inconclusive. We will clarify the different types of conclusions that are appropriate as we consider several practical examples of hypothesis tests.

Finally, how do you decide on the **critical values** that separate the acceptance and rejection regions? That is, how do you decide how much statistical evidence you need before you can reject H_0? This depends on the amount of confidence that you, the researcher, want to attach to the test conclusions and the **significance level α,** the risk you are willing to take of making an incorrect decision.

DEFINITION A **Type I error** for a statistical test is the error of rejecting the null hypothesis when it is true. The **level of significance (significance level)** for a statistical test of hypothesis is

$$\alpha = P(\text{Type I error}) = P(\text{falsely rejecting } H_0) = P(\text{rejecting } H_0 \text{ when it is true})$$

This value α represents the *maximum tolerable risk* of incorrectly rejecting H_0. Once this significance level is fixed, the rejection region can be set to allow the researcher to reject H_0 with a fixed degree of confidence in the decision. It is important that a significance value should be chosen prior to hypothesis testing, which cannot be changed afterward.

In the next section, we will show you how to use a test of hypothesis to test the value of a population mean μ. As we continue, we will clarify some of the computational details and add some additional concepts to complete your understanding of hypothesis testing.

9.3 A LARGE-SAMPLE TEST ABOUT A POPULATION MEAN

Consider a random sample of n measurements drawn from a population that has mean μ and standard deviation σ. You want to test a hypothesis of the form†

1 $H_0: \mu = \mu_0$

where μ_0 is some hypothesized value for μ, versus a one-tailed alternative hypothesis:

2 $H_a: \mu > \mu_0$

The subscript zero indicates the value of the parameter specified by H_0. Notice that H_0 provides an exact value for the parameter to be tested, whereas H_a gives a range of possible values for μ.

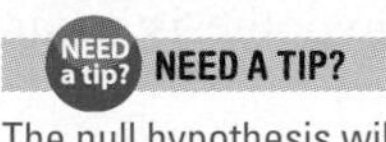

NEED A TIP?

The null hypothesis will *always* have an "equals" sign attached.

The Essentials of the Test

The sample mean $\bar{x}$ is the best estimate of the actual value of μ, which is presently in question. What values of $\bar{x}$ would lead you to believe that H_0 is false and μ is, in fact, greater than the hypothesized value? The values of $\bar{x}$ that are extremely *large* would imply that μ is larger than hypothesized. Hence, you should reject H_0 if $\bar{x}$ is too large.

The next problem is to define what is meant by "too large." Values of $\bar{x}$ that lie too many standard deviations to the right of the mean are not very likely to occur. Those values have very little area to their right. Hence, you can define "too large" as being too many standard deviations away from μ_0. But what is "too many"? This question can be answered using the *significance level* α, the probability of rejecting H_0 when H_0 is *true*.

Remember that the standard error of $\bar{x}$ is estimated as

$$SE = \frac{s}{\sqrt{n}}$$

Since the sampling distribution of the sample mean $\bar{x}$ is approximately normal when n **is large,** the number of standard deviations that $\bar{x}$ lies from μ_0 can be measured using the **standardized test statistic:**

3 $$z = \frac{\bar{x} - \mu_0}{s/\sqrt{n}}$$

which has an approximate standard normal distribution when H_0 is true and $\mu = \mu_0$.

4 The significance level α is equal to the area under the normal curve lying above the rejection region. Thus, if you want $\alpha = 0.01$, you will reject H_0 when $\bar{x}$ is more than 2.33 standard deviations to the right of μ_0. Equivalently, you will reject H_0 if the standardized test statistic z is greater than 2.33 (see Figure 9.3).

†Note that if the test rejects the null hypothesis $\mu = \mu_0$ in favour of the alternative hypothesis $\mu > \mu_0$, then it will certainly reject a null hypothesis that includes $\mu < \mu_0$, since this is even more contradictory to the alternative hypothesis. For this reason, in this text we state the null hypothesis for a one-tailed test as $\mu = \mu_0$ rather than $\mu \leq \mu_0$.

FIGURE 9.3

The rejection region for a right-tailed test with $\alpha = 0.01$

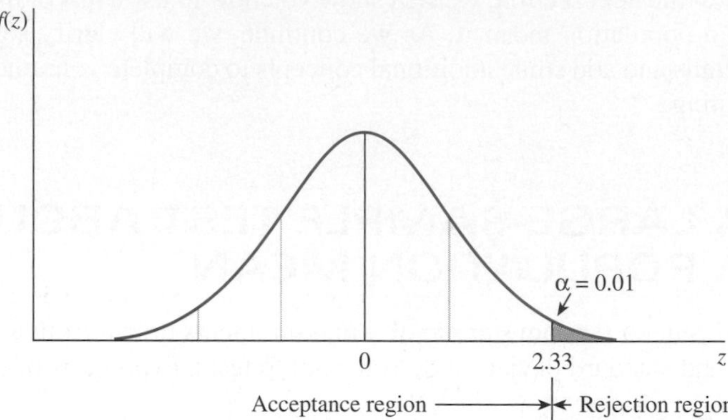

EXAMPLE 9.6

The average weekly earnings for women in managerial and professional positions is \$670. Do men in the same positions have average weekly earnings that are higher than those for women? A random sample of $n = 40$ men in managerial and professional positions showed $\bar{x} = \$725$ and $s = \$102$. Test the appropriate hypothesis using $\alpha = 0.01$.

NEED A TIP?

For one-tailed tests, look for directional words such as "greater," "less than," "higher," "lower," etc.

SOLUTION You would like to show that the average weekly earnings for men are higher than \$670, the women's average. Hence, if μ is the average weekly earnings in managerial and professional positions for men, you can set out the formal test of hypothesis in steps:

1–2 **Null and alternative hypotheses:**

$$H_0: \mu = 670 \text{ versus } H_a: \mu > 670$$

3 **Test statistic:** Using the sample information, with s as an estimate of the population standard deviation, calculate

$$z = \frac{\bar{x} - 670}{s/\sqrt{n}} = \frac{725 - 670}{102/\sqrt{40}} = 3.41$$

4 **Rejection region:** For this one-tailed test, values of $\bar{x}$ much larger than 670 would lead you to reject H_0; or, equivalently, values of the *standardized test statistic* z in the right tail of the standard normal distribution. To control the risk of making an incorrect decision as $\alpha = 0.01$, you must set the **critical value** separating the rejection and acceptance regions so that the area in the right tail is exactly $\alpha = 0.01$. This value is found in Table 3 of Appendix I to be $z = 2.33$, as shown in Figure 9.3. The null hypothesis will be rejected if the observed value of the test statistic, z, is greater than 2.33.

5 **Conclusion:** Compare the observed value of the test statistic, $z = 3.41$, with the critical value necessary for rejection, $z = 2.33$. Since the observed value of the test statistic falls in the rejection region, you can reject H_0 and conclude that the average weekly earnings for men in managerial and professional positions are higher than the average for women. The probability that you have made an incorrect decision is $\alpha = 0.01$.

NEED A TIP?

If the test is two-tailed, you will not see any directional words. The experimenter is only looking for a "difference" from the hypothesized value.

If you wish to detect departures either greater or less than μ_0, then the alternative hypothesis is *two-tailed*, written as

$$H_a: \mu \neq \mu_0$$

which implies either $\mu > \mu_0$ or $\mu < \mu_0$. Values of $\bar{x}$ that are either "too large" or "too small" in terms of their distance from μ_0 are placed in the rejection region. If you choose $\alpha = 0.01$, the area in the rejection region is equally divided between the two tails of the normal distribution, as shown in Figure 9.4. Using the standardized test statistic z, you can reject H_0 if $z > 2.58$ or $z < -2.58$. For different values of α, the critical values of z that separate the rejection and acceptance regions will change accordingly.

FIGURE 9.4
The rejection region for a two-tailed test with $\alpha = 0.01$

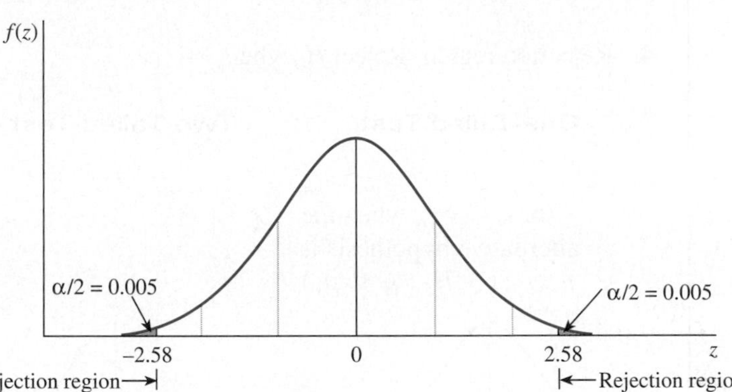

EXAMPLE 9.7

The daily yield for a local chemical plant has averaged 880 metric tonnes for the last several years. The quality control manager would like to know whether this average has changed in recent months. She randomly selects 50 days from the computer database and computes the average and standard deviation of the $n = 50$ yields as $\bar{x} = 871$ metric tonnes and $s = 21$ metric tonnes, respectively. Test the appropriate hypothesis using $\alpha = 0.05$.

SOLUTION

1–2 **Null and alternative hypotheses:**

$$H_0: \mu = 880 \text{ versus } H_a: \mu \neq 880$$

3 **Test statistic:** The point estimate for μ is $\bar{x}$. Therefore, the test statistic is

$$z = \frac{\bar{x} - \mu_0}{s/\sqrt{n}} = \frac{871 - 880}{21/\sqrt{50}} = -3.03$$

4 **Rejection region:** For this two-tailed test, you use values of z in both the right and left tails of the standard normal distribution. Using $\alpha = 0.05$, the **critical values** separating the rejection and acceptance regions cut off areas of $\alpha/2 = 0.025$ in the right and left tails. These values are $z = \pm 1.96$ and the null hypothesis will be rejected if $z > 1.96$ or $z < -1.96$.

5 **Conclusion:** Since $z = -3.03$ and the calculated value of z falls in the rejection region, the manager can reject the null hypothesis that $\mu = 880$ metric tonnes and conclude that it has changed. The probability of rejecting H_0 when H_0 is true and $\alpha = 0.05$, a fairly small probability. Hence, she is reasonably confident that her finding is correct.

Large-Sample Statistical Test for μ

1. Null hypothesis: $H_0: \mu = \mu_0$
2. Alternative hypothesis:

One-Tailed Test	**Two-Tailed Test**
$H_a : \mu > \mu_0$	$H_a : \mu \neq \mu_0$
(or, $H_a : \mu < \mu_0$)	

3. Test statistic: $z = \dfrac{\bar{x} - \mu_0}{\sigma/\sqrt{n}}$ estimated as $z = \dfrac{\bar{x} - \mu_0}{s/\sqrt{n}}$

4. Rejection region: Reject H_0 when

One-Tailed Test	**Two-Tailed Test**

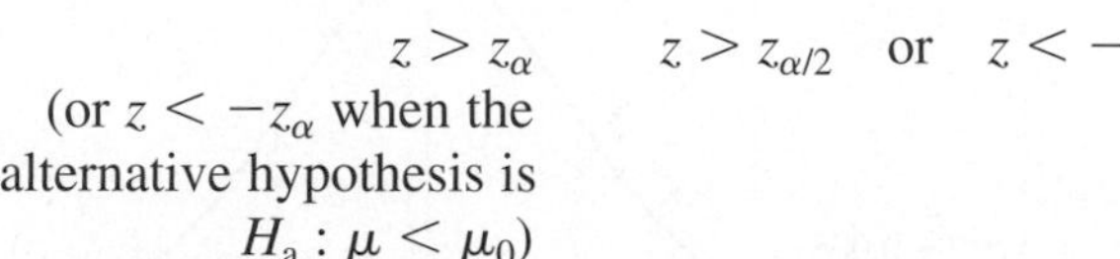

$z > z_\alpha$ (or $z < -z_\alpha$ when the alternative hypothesis is $H_a : \mu < \mu_0$)	$z > z_{\alpha/2}$ or $z < -z_{\alpha/2}$

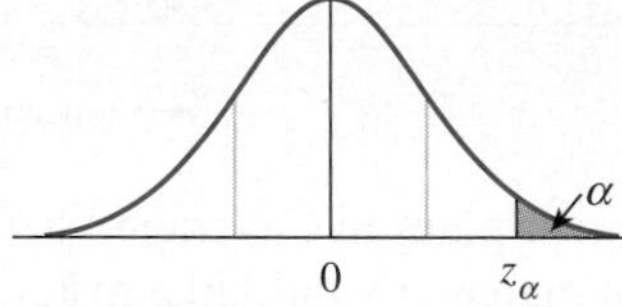

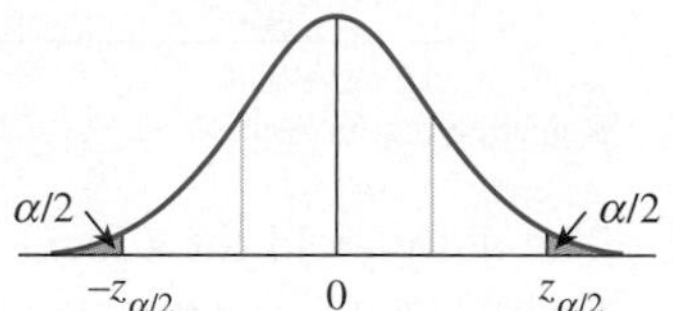

Assumptions: The n observations in the sample are randomly selected from the population and n is large—say, $n \geq 30$.

Calculating the *p*-Value

In the previous examples, the decision to reject or accept H_0 was made by comparing the calculated value of the test statistic with a critical value of z based on the significance level α of the test. However, different significance levels may lead to different conclusions. For example, if in a right-tailed test, the test statistic is $z = 2.03$, you can reject H_0 at the 5% level of significance because the test statistic exceeds $z = 1.645$. However, you cannot reject H_0 at the 1% level of significance because the test statistic is less than $z = 2.33$ (see Figure 9.5). To avoid any ambiguity in their conclusions, some experimenters prefer to use a variable level of significance called the ***p*-value** for the test.

DEFINITION The ***p*-value** or observed significance level of a statistical test is the smallest value of α for which H_0 can be rejected. It is the *actual risk* of committing a Type I error, if H_0 is rejected based on the observed value of the test statistic. The p-value measures the strength of the evidence against H_0.

In the right-tailed test with observed test statistic $z = 2.03$, the smallest critical value you can use and still reject, H_0 is $z = 2.03$. For this critical value, the risk of an incorrect decision is

$$P(z \geq 2.03) = 1 - 0.9788 = 0.0212$$

This probability is the *p-value* for the test. *Notice that it is actually the area to the right of the calculated value of the test statistic.*

FIGURE 9.5
Variable rejection regions

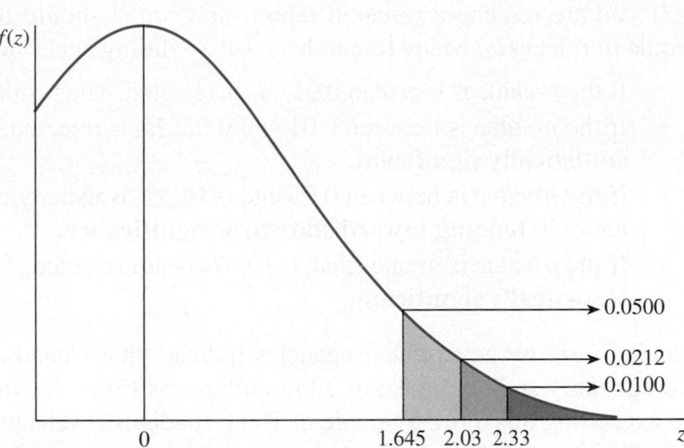

NEED A TIP?

p-value = tail area (one or two tails) "beyond" the observed value of the test statistic.

A *small p-value* indicates that the observed value of the test statistic lies far away from the hypothesized value of μ. This presents strong evidence that H_0 is false and should be rejected. *Large p-values* indicate that the observed test statistic is not far from the hypothesized mean and does not support rejection of H_0. How small does the *p*-value need to be before H_0 can be rejected?

DEFINITION If the *p*-value is less than or equal to a preassigned significance level α, then the null hypothesis can be rejected, and you can report that the results are **statistically significant** at level α.

In the previous instance, if you choose $\alpha = 0.05$ as your significance level, H_0 can be rejected because the *p*-value is less than 0.05. However, if you choose $\alpha = 0.01$ as your significance level, the *p*-value (0.0212) is not small enough to allow rejection of H_0. The results are significant at the 5% level, but not at the 1% level. You might see these results reported in professional journals as *significant* ($p < 0.05$).†

EXAMPLE 9.8

Refer to Example 9.7. The quality control manager wants to know whether the daily yield at a local chemical plant—which has averaged 880 metric tonnes for the last several years—has changed in recent months. A random sample of 50 days gives an average yield of 871 metric tonnes with a standard deviation of 21 metric tonnes. Calculate the *p*-value for this two-tailed test of hypothesis. Use the *p*-value to draw conclusions regarding the statistical test.

SOLUTION The rejection region for this two-tailed test of hypothesis is found in both tails of the normal probability distribution. Since the observed value of the test statistic is $z = -3.03$, the smallest rejection region that you can use and still reject H_0 is $|z| > 3.03$. For this rejection region, the value of α is the *p*-value:

$$p\text{-value} = P(z > 3.03) + P(z < -3.03) = (1 - 0.9988) + 0.0012 = 0.0024$$

Notice that the two-tailed p-value is actually twice the tail area corresponding to the calculated value of the test statistic. If this *p*-value = 0.0024 is less than or equal to the preassigned level of significance α, H_0 can be rejected. For this test, you can reject H_0 at either the 1% or the 5% level of significance.

---

†In reporting statistical significance, many researchers write ($p < 0.05$) or ($P < 0.05$) to mean that the *p*-value of the test was smaller than 0.05, making the results significant at the 5% level. The symbol *p* or *P* in the expression has no connection with our notation for probability or with the binomial parameter *p*.

If you are reading a research report, how small should the p-value be before you decide to reject H_0? Many researchers use a "sliding scale" to classify their results:

- If the p-value is less than 0.01, H_0 is rejected. The results are **highly significant.**
- If the p-value is between 0.01 and 0.05, H_0 is rejected. The results are **statistically significant.**
- If the p-value is between 0.05 and 0.10, H_0 is usually not rejected. The results are only **tending toward statistical significance.**
- If the p-value is greater than 0.10, H_0 is not rejected. The results are **not statistically significant.**

EXAMPLE 9.9

Standards set by government agencies indicate that Canadians should not exceed an average daily sodium intake of 3300 milligrams (mg). To find out whether Canadians are exceeding this limit, a sample of 100 Canadians is selected, and the mean and standard deviation of daily sodium intake are found to be 3400 mg and 1100 mg, respectively. Use $\alpha = 0.05$ to conduct a test of hypothesis.

SOLUTION The hypotheses to be tested are

$$H_0: \mu = 3300 \quad \text{versus} \quad H_a: \mu > 3300$$

and the test statistic is

$$z = \frac{\bar{x} - \mu_0}{s/\sqrt{n}} = \frac{3400 - 3300}{1100/\sqrt{100}} = 0.9091$$

NEED A TIP?

Small p-value ⇔ large z-value.

Small p-value ⇒ reject H_0.

How small? p-value $\leq \alpha$.

The two approaches developed in this section yield the same conclusions:

- **The critical value approach:** Since the significance level is $\alpha = 0.05$ and the test is one-tailed, the rejection region is determined by a critical value with tail area equal to $\alpha = 0.05$; that is, H_0 can be rejected if $z > 1.645$. Since $z = 0.9091$ is not greater than the critical value, H_0 is not rejected (see Figure 9.6).
- **The p-value approach:** Calculate the p-value, the probability that z is greater than or equal to $z = 0.9091$:

$$p\text{-value} = P(z > 0.9091) = 1 - 0.8186 = 0.1814$$

The null hypothesis can be rejected only if the *p-value is less than or equal to the specified 5% significance level.* Therefore, H_0 is not rejected and the results are *not statistically significant* (see Figure 9.6). There is not enough evidence to indicate that the average daily sodium intake exceeds 3300 mg.

FIGURE 9.6

Rejection region and p-value for Example 9.9

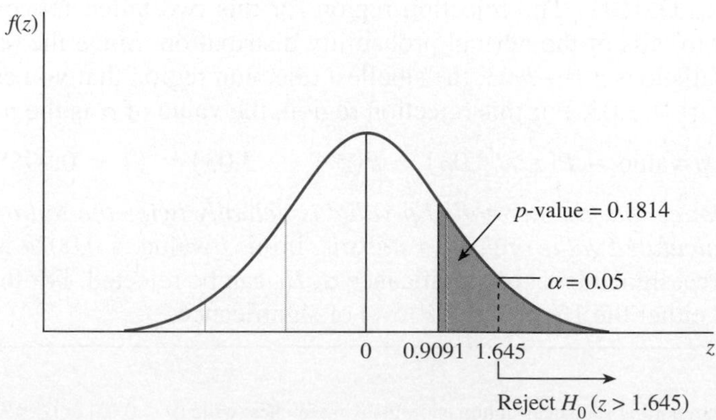

ONLINE APPLET

Large-Sample Test of a Population Mean

Notice that these two approaches are actually the same, as shown in Figure 9.6. As soon as the calculated value of the test statistic z becomes *larger than* the critical value, z_α, the p-value becomes *smaller than* the significance level α. You can use the most convenient of the two methods; the conclusions you reach will always be the same! The p-value approach does have two advantages, however:

- Statistical output from packages such as *MINITAB* usually reports the p-value of the test.
- Based on the p-value, your test results can be evaluated using any significance level you wish to use. Many researchers report the smallest possible significance level for which their results are *statistically significant.*

Sometimes it is easy to confuse the significance level α with the p-value (or observed significance level). They are both probabilities calculated as areas in the tails of the sampling distribution of the test statistic. However, the significance level α is preset by the experimenter before collecting the data. The p-value is linked directly to the data and actually describes how likely or unlikely the sample results are, assuming that H_0 is true. *The smaller the p-value, the more unlikely it is that H_0 is true!*

NEED TO KNOW

Rejection Regions, *p*-Values, and Conclusions

The significance level, α, lets you set the risk that you are willing to take of making an incorrect decision in a test of hypothesis.

- To set a rejection region, choose a **critical value** of z so that the area in the tail(s) of the z-distribution is (are) either α for a one-tailed test or $\alpha/2$ for a two-tailed test. Use the right tail for an upper-tailed test and the left tail for a lower-tailed test. Reject H_0 when the test statistic exceeds the critical value and falls in the rejection region.
- To find a ***p*-value,** find the area in the tail "beyond" the test statistic. If the test is one-tailed, this is the p-value. If the test is two-tailed, this is only half the p-value and must be doubled. Reject H_0 when the p-value is less than α.

Two Types of Errors

You might wonder why, when H_0 was not rejected in the previous example, we did not say that H_0 was definitely true and $\mu = 3300$. This is because, if we choose to *accept* H_0, we must have a measure of the probability of error associated with this decision.

Since there are two choices in a statistical test, there are also two types of errors that can be made. In the courtroom trial, a defendant could be judged not guilty when he's really guilty, or vice versa—the same is true in a statistical test. In fact, the null hypothesis may be either true or false, regardless of the decision the experimenter makes. These two possibilities, along with the two decisions that can be made by the researcher, are shown in Table 9.1.

TABLE 9.1 **Decision Table**

Decision	Null Hypothesis: True	False
Reject H_0	Type I error	Correct decision
Accept H_0	Correct decision	Type II error

In addition to the Type I error with probability α defined earlier in this section, it is possible to commit a second error, called a **Type II error,** which has probability β.

DEFINITION A **Type I error** for a statistical test is the error of rejecting the null hypothesis when it is true. The probability of making a Type I error is denoted by the symbol α.

A **Type II error** for a statistical test is the error of accepting the null hypothesis when it is false and some alternative hypothesis is true. The probability of making a Type II error is denoted by the symbol β.

Notice that the probability of a Type I error is exactly the same as the **level of significance** α and is therefore controlled by the researcher. When H_0 is rejected, you have an accurate measure of the reliability of your inference; the probability of an incorrect decision is α. However, the probability β of a Type II error is not always controlled by the experimenter. In fact, when H_0 is false and H_a is true, you may not be able to specify an exact value for μ but only a range of values. This makes it difficult, if not impossible, to calculate β. Without a measure of reliability, it is not wise to conclude that H_0 is true. Rather than risk an incorrect decision, you should withhold judgment, concluding that you *do not have enough evidence to reject* H_0. Instead of *accepting* H_0, you should *not reject or fail to reject* H_0.

NEED A TIP?

$\alpha = P(\text{reject } H_0 \text{ when } H_0 \text{ true})$.

$\beta = P(\text{accept } H_0 \text{ when } H_0 \text{ false})$.

Keep in mind that *"accepting" a particular hypothesis means deciding in its favour.* Regardless of the outcome of a test, you are never *certain* that the hypothesis you "accept" is true. *There is always a risk of being wrong (measured by α or β).* Consequently, you never "accept" H_0 if β is unknown or its value is unacceptable to you. When this situation occurs, you should withhold judgment and collect more data.

The Power of a Statistical Test

The goodness of a statistical test is measured by the size of the two error rates: α, the probability of rejecting H_0 when it is true, and β, the probability of accepting H_0 when H_0 is false and H_a is true. A "good" test is one for which both of these error rates are small. The experimenter begins by selecting α, the probability of a Type I error. If he or she also decides to control the value of β, the probability of accepting H_0 when H_a is true, then an appropriate sample size is chosen.

Another way of evaluating a test is to look at the complement of a Type II error—that is, rejecting H_0 when H_a is true—which has probability

$$1 - \beta = P(\text{reject } H_0 \text{ when } H_a \text{ is true})$$

The quantity $(1 - \beta)$ is called the **power** of the test because it measures the probability of taking the action that we wish to have occur—that is, rejecting the null hypothesis when it is false and H_a is true.

DEFINITION The **power of a statistical test,** given as

$$1 - \beta = P(\text{reject } H_0 \text{ when } H_a \text{ is true})$$

measures the ability of the test to perform as required.

A graph of $(1 - \beta)$, the probability of rejecting H_0 when in fact H_0 is false, as a function of the true value of the parameter of interest is called the **power curve** for the statistical test. Ideally, you would like α to be small and the *power* $(1 - \beta)$ to be large.

For a fixed α, when testing the usual null hypothesis, clearly the power of the test should be as high as possible. Rejecting a test with a very low power may not be fruitful. In performing statistical tests, make sure that the power of the test will be sufficiently high to detect the departure from the null hypothesis under consideration. Fail to do so, and the conclusions drawn from a low powerful test will be meaningless. The statistical tests used in this book are usually more powerful than others.

EXAMPLE 9.10

Refer to Example 9.7. Calculate β and the power of the test $(1 - \beta)$ when μ is actually equal to 870 metric tonnes.

SOLUTION The acceptance region for the test of Example 9.7 is located in the interval $[\mu_0 \pm 1.96(s/\sqrt{n})]$. Substituting numerical values, you get

$$880 \pm 1.96\left(\frac{21}{\sqrt{50}}\right) \quad \text{or} \quad 874.18 \text{ to } 885.82$$

The probability of accepting H_0, given $\mu = 870$, is equal to the area under the sampling distribution for the test statistic $\bar{x}$ in the interval from 874.18 to 885.82. Since $\bar{x}$ is normally distributed with a mean of 870 and $SE = 21/\sqrt{50} = 2.97$, β is equal to the area under the normal curve with $\mu = 870$ located between 874.18 and 885.82 (see Figure 9.7). Calculating the z-values corresponding to 874.18 and 885.82, you get

FIGURE 9.7
Calculating β in Example 9.10

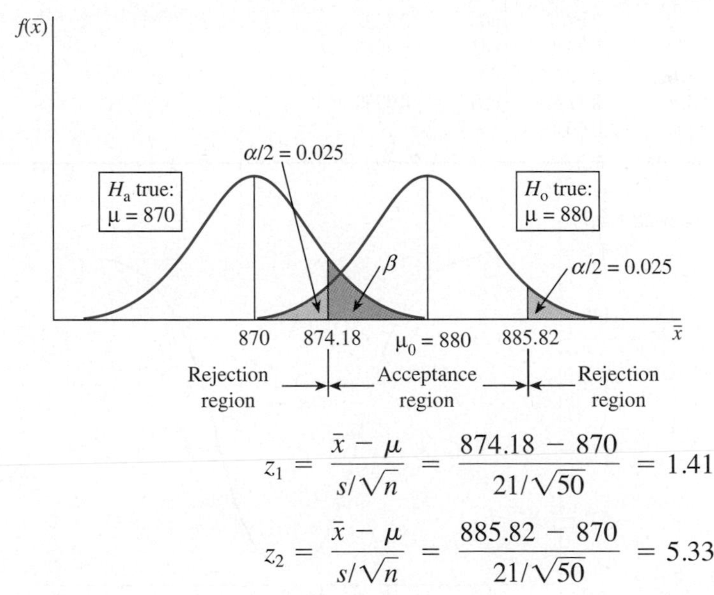

$$z_1 = \frac{\bar{x} - \mu}{s/\sqrt{n}} = \frac{874.18 - 870}{21/\sqrt{50}} = 1.41$$

$$z_2 = \frac{\bar{x} - \mu}{s/\sqrt{n}} = \frac{885.82 - 870}{21/\sqrt{50}} = 5.33$$

Then

$$\beta = P(\text{accept}\, H_0 \,\text{when}\;\; \mu = 870) = P(874.18 < \bar{x} < 885.82 \text{ when } \mu = 870)$$
$$= P(1.41 < z < 5.33)$$

You can see from Figure 9.7 that the area under the normal curve with $\mu = 870$ above $\bar{x} = 885.82$ (or $z = 5.33$) is negligible. Therefore,

$$\beta = P(z > 1.41)$$

From Table 3 in Appendix I you can find

$$\beta = 1 - 0.9207 = 0.0793$$

Hence, the power of the test is

$$1 - \beta = 1 - 0.0793 = 0.9207$$

The probability of correctly rejecting H_0, given that μ is really equal to 870, is 0.9207, or approximately 92 chances in 100.

Power of a *z*-Test

Values of $(1 - \beta)$ can be calculated for various values of μ_a different from $\mu_0 = 880$ to measure the power of the test. For example, if $\mu_a = 885$,

$$\begin{aligned}\beta &= P(874.18 < \bar{x} < 885.82 \text{ when } \mu = 885)\\ &= P(-3.64 < z < 0.28)\\ &= 0.6103 - 0 = 0.6103\end{aligned}$$

and the power is $(1 - \beta) = 0.3897$. Table 9.2 shows the power of the test for various values of μ_a, and a power curve is graphed in Figure 9.8. Note that the power of the test increases as the distance between μ_a and μ_0 increases. The result is a U-shaped curve for this two-tailed test.

TABLE 9.2 **Value of $(1 - \beta)$ for Various Values of μ_a for Example 9.10**

μ_a	$(1 - \beta)$	μ_a	$(1 - \beta)$
865	0.9990	883	0.1726
870	0.9207	885	0.3897
872	0.7673	888	0.7673
875	0.3897	890	0.9207
877	0.1726	895	0.9990
880	0.0500		

FIGURE 9.8 Power curve for Example 9.10

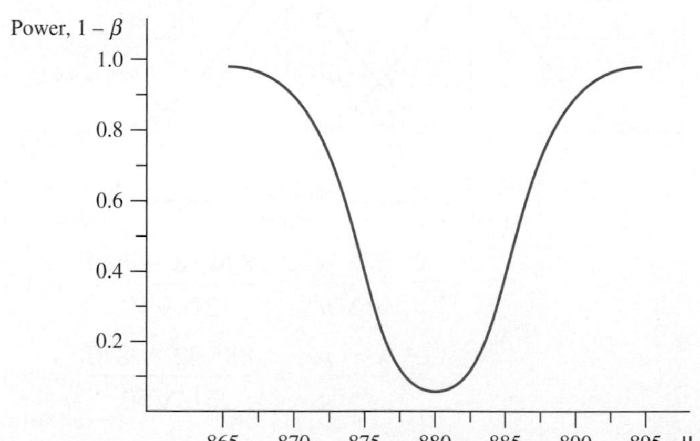

There are many important links among the two error rates, α and β, the power, $(1 - \beta)$, and the sample size, n. Look at the two curves shown in Figure 9.7.

- If α (the sum of the two tail areas in the curve on the right) is increased, the shaded area corresponding to β decreases, and vice versa.
- The only way to decrease β for a fixed α is to "buy" more information—that is, increase the sample size n.

What would happen to the area β as the curve on the left is moved closer to the curve on the right ($\mu = 880$)? With the rejection region in the right curve fixed, the value of β will *increase.* What effect does this have on the power of the test? Look at Figure 9.8. You may also want to note the following:

- As the distance between the true (μ_a) and hypothesized (μ_0) values of the mean increases, the power $(1 - \beta)$ increases. The test is better at detecting *differences* when the distance is *large.*
- The closer the true value (μ_a) gets to the hypothesized value (μ_0), the less power $(1 - \beta)$ the test has to detect the difference.
- The only way to increase the power $(1 - \beta)$ for a fixed α is to "buy" more information—that is, increase the sample size, n.

The experimenter must decide on the values of α and β—measuring the risks of the possible errors he or she can tolerate. He or she also must decide how much power is needed to detect differences that are practically important in the experiment. Once these decisions are made, the sample size can be chosen by consulting the power curves corresponding to various sample sizes for the chosen test.

NEED TO KNOW

How to Calculate β

1. Find the critical value or values of $\bar{x}$ used to separate the acceptance and rejection regions.
2. Using one or more values for μ consistent with the alternative hypothesis H_a, calculate the probability that the sample mean $\bar{x}$ falls in the *acceptance region.* This produces the value $\beta = P(\text{accept } H_a \text{ when } \mu = \mu_a)$.
3. Remember that the **power** of the test is $(1 - \beta)$.

9.3 EXERCISES

BASIC TECHNIQUES

9.1 Find the appropriate rejection regions for the large-sample test statistic z in these cases:

a. A right-tailed test with $\alpha = 0.01$

b. A two-tailed test at the 5% significance level

9.2 Refer to Exercise 9.1. Suppose that the observed value of the test statistic was $z = 2.16$. For the rejection regions constructed in parts a and b of Exercise 9.1, draw the appropriate conclusion for the tests. If appropriate, give a measure of the reliability of your conclusion.

9.3 Find the appropriate rejection regions for the large-sample test statistic z in these cases:

a. A left-tailed test at the 1% significance level

b. A two-tailed test with $\alpha = 0.01$

c. Suppose that the observed value of the test statistic was $z = -2.41$. For the rejection regions constructed in parts a and b, draw the appropriate conclusion for the tests. If appropriate, give a measure of the reliability of your conclusion.

9.4 Find the p-value for the following large-sample z tests:

a. A right-tailed test with observed $z = 1.15$

b. A two-tailed test with observed $z = -2.78$

c. A left-tailed test with observed $z = -1.81$

9.5 For the three tests given in Exercise 9.4, use the p-value to determine the significance of the results. Explain what "statistically significant" means in terms of rejecting or accepting H_0 and H_a.

9.6 A random sample of $n = 35$ observations from a quantitative population produced a mean $\bar{x} = 2.4$ and a standard deviation $s = 0.29$. Suppose your research objective is to show that the population mean μ exceeds 2.3.

a. Give the null and alternative hypotheses for the test.

b. Locate the rejection region for the test using a 5% significance level.

c. Find the standard error of the mean.

d. Before you conduct the test, use your intuition to decide whether the sample mean $\bar{x} = 2.4$ is likely or unlikely, assuming that $\mu = 2.3$. Now conduct the test. Do the data provide sufficient evidence to indicate that $\mu > 2.3$?

9.7 Refer to Exercise 9.6.

a. Calculate the *p*-value for the test statistic in part d.

b. Use the *p*-value to draw a conclusion at the 5% significance level.

c. Compare the conclusion in part b with the conclusion reached in part d of Exercise 9.6. Are they the same?

9.8 Refer to Exercise 9.6. You want to test $H_0 : \mu = 2.3$ against $H_a : \mu > 2.3$.

a. Find the critical value of $\bar{x}$ used for rejecting H_0.

b. Calculate $\beta = P(\text{accept } H_0 \text{ when } \mu = 2.4)$.

c. Repeat the calculation of β for $\mu = 2.3$, 2.5, and 2.6.

d. Use the values of β from parts b and c to graph the power curve for the test.

9.9 A random sample of 100 observations from a quantitative population produced a sample mean of 26.8 and a sample standard deviation of 6.5. Use the *p*-value approach to determine whether the population mean is different from 28. Explain your conclusions.

9.10 Find the *p*-value for the following large-sample *z* tests:

a. A right-tailed test with observed $z = 2.73$

b. A two-tailed test with observed $z = 1.32$

c. A left-tailed test with observed $z = -0.89$

9.11 For the three tests given in Exercise 9.10, use the *p*-value to determine the significance of the results. Explain what "statistically significant" means in terms of rejecting or accepting H_0 and H_a.

APPLICATIONS

9.12 Airline Occupancy Rates High airlineoc-cupancy rates on scheduled flights are essential to corporate profitability. Suppose a scheduled flight must average at least 60% occupancy in order to be profitable, and an examination of the occupancy rate for 120 10:00 A.M. flights from Toronto to Calgary showed a mean occupancy per flight of 58% and a standard deviation of 11%.

a. If μ is the mean occupancy per flight and if the company wishes to determine whether or not this scheduled flight is unprofitable, give the alternative and the null hypotheses for the test.

b. Does the alternative hypothesis in part a imply a one- or two-tailed test? Explain.

c. Do the occupancy data for the 120 flights suggest that this scheduled flight is unprofitable? Test using $\alpha = 0.05$.

9.13 Ground Beef The meat department of a local supermarket chain packages ground beef in trays of two sizes. The smaller tray is intended to hold 1 kilogram (kg) of meat. A random sample of 35 packages in the smaller meat tray produced weight measurements with an average of 1.01 kg and a standard deviation of 20 grams.

a. If you were the quality control manager and wanted to make sure that the average amount of ground beef was indeed 1 kg, what hypotheses would you test?

b. Find the *p*-value for the test and use it to perform the test in part a.

c. How would you, as the quality control manager, report the results of your study to a consumer interest group?

9.14 Invasive Species In a study of the pernicious giant hogweed, Jan Pergl[3] and associates compared the density of these plants in two different sites within the Caucasus region of Russia. In its native area, the average density was found to be 5 plants/m^2. In an invaded area in the Czech Republic, a sample of $n = 50$ plants produced an average density of 11.17 plants/m^2 with a standard deviation of 3.9 plants/m^2.

a. Does the invaded area in the Czech Republic have an average density of giant hogweed that is different from $\mu = 5$ at the $\alpha = 0.05$ level of significance?

b. What is the *p*-value associated with the test in part a? Can you reject H_0 at the 5% level of significance using the *p*-value?

9.15 Advertising at the Movies "Welcome to the new movie pre-show!" Before you can see the newly released movie you have just paid to see, you must sit through a variety of trivia slides, snack bar ads, paid product advertising, and movie trailers. Although the total barrage of advertising may last up to 20 minutes or more, a particular theatre chain claims that the average length of any one advertisement is no more than 3 minutes.[4] To test this claim, 50 theatre advertisements were randomly selected and found to have an average duration of 3 minutes 15 seconds with a standard deviation of 30 seconds. Do the data provide sufficient evidence to indicate that the average duration of theatre ads is more than that claimed by the theatre? Test at the 1% level of significance. (HINT: Change "seconds" to fractions of a "minute.")

9.16 Potency of an Antibiotic A drug manufacturer claimed that the mean potency of one of its antibiotics was 80%. A random sample of $n = 100$ capsules were tested and produced a sample mean of $\bar{x} = 79.7$, with a standard deviation of $s = 0.8\%$. Do the data present sufficient evidence to refute the manufacturer's claim? Let $\alpha = 0.05$.

a. State the null hypothesis to be tested.

b. State the alternative hypothesis.

c. Conduct a statistical test of the null hypothesis and state your conclusion.

9.17 Flextime Many companies are becoming involved in *flextime,* in which a worker schedules his or her own work hours or compresses work weeks. A company that was contemplating the installation of a flextime schedule estimated that it needed a minimum mean of seven hours per day per assembly worker in order to operate effectively. Each of a random sample of 80 of the company's assemblers was asked to submit a tentative flextime schedule. If the mean number of hours per day for Monday was 6.7 hours and the standard deviation was 2.7 hours, do the data provide sufficient evidence to indicate that the mean number of hours worked per day on Mondays, for all of the company's assemblers, will be less than seven hours? Test using $\alpha = 0.05$.

9.18 Raise Your MCAT Test Scores! There are many books and crash courses available to prepare for the MCAT examination. An organization that offers such courses claims the average score improvement for its crash-course participants is between 29 to 37 points. Are the claims made by this organization exaggerated? That is, is the average score improvement less than 29, the minimum claimed in the advertising flyer? A random sample of 36 students who took the crash course achieved an average score of 27 with a standard deviation of 5.2 points.

a. Use the p-value approach to test the claim. At which significance levels can you reject H_0?

b. If you were a competitor of this organization, how would you state your conclusions to put your company in the best possible light?

c. If you worked for this organization, how would you state your conclusions to protect your company's reputation?

9.19 What's Normal? What *is* normal, when it comes to people's body temperatures? A random sample of 130 human body temperatures, provided by Allen Shoemaker[5] in the *Journal of Statistical Education,* had a mean of 36.81° Celsius and a standard deviation of 0.73°. Does the data indicate that the average body temperature for healthy humans is different from 37°, the usual average temperature cited by physicians and others? Test using both methods given in this section.

a. Use the p-value approach with $\alpha = 0.05$.

b. Use the critical value approach with $\alpha = 0.05$.

c. Compare the conclusions from parts a and b. Are they the same?

d. The 37 °C standard was derived by a German doctor in 1868, who claimed to have recorded 1 million temperatures in the course of his research.[6] What conclusions can you draw about his research in light of your conclusions in parts a and b?

9.20 Sports and Achilles Tendon Injuries Some sports that involve a significant amount of running, jumping, or hopping put participants at risk for Achilles tendinopathy (AT), an inflammation and thickening of the Achilles tendon. A study in The *American Journal of Sports Medicine* looked at the diameter (in mm) of the affected tendons for patients who participated in these types of sports activities.[7] Suppose that the Achilles tendon diameters in the general population have a mean of 5.97 millimetres (mm). When the diameters of the affected tendon were measured for a random sample of 31 patients, the average diameter was 9.80 with a standard deviation of 1.95 mm. Is there sufficient evidence to indicate that the average diameter of the tendon for patients with AT is greater than 5.97 mm? Test at the 5% level of significance.

9.21 Time Spent on Social Media Social media has become extremely popular. According to the data provided by comScore via a new mobile measurement report, the average time per month spent by users on Twitter is 170 minutes.[8] Suppose that a random sample is taken of 120 users whose average time spent per month on Twitter is 173 minutes and standard deviation is 15 minutes. Does the data represent sufficient evidence to reject comScore's claim? Test using $\alpha = 0.05$.

a. State the null hypothesis to be tested.

b. State the alternate hypothesis.

c. Conduct a statistical test of the null hypothesis and state your conclusions.

d. Find the p-value for the test and use it to perform the test in part a.

e. If you are a Twitter user, do you find the results surprising?

9.22 Life Expectancy of Canadians As our quality of life improves, so does our lifespan. Statistics from the World Bank show that, in 2015, Canada's average life expectancy is 81 years of age.[9] Suppose that a random sample is taken of 240 Canadians who have recently passed away, and their average lifespan was calculated to be 79 years old with a standard deviation of 16 years. Does the data represent sufficient evidence to support the World Bank's claim? Test using $\alpha = 0.05$.

a. Clearly state the respective null and alternate hypotheses.

b. Conduct a statistical test of the null hypothesis.

c. What are your conclusions?

d. Calculate the p-value for the test and use it to perform the test in part a.

e. As a Canadian, what are your thoughts on living to the age of 81?

9.4 A LARGE-SAMPLE TEST OF HYPOTHESIS FOR THE DIFFERENCE BETWEEN TWO POPULATION MEANS

In many situations, the statistical question to be answered involves a comparison of two population means. For example, a federal agency is interested in reducing its massive 31,646.5 million litres/year gasoline bill by replacing gasoline-powered trucks with electric-powered trucks. To determine whether significant savings in operating costs are achieved by changing to electric-powered trucks, a pilot study should be undertaken using, say, 100 conventional gasoline-powered trucks and 100 electric-powered trucks operated under similar conditions.

The statistic that summarizes the sample information regarding the difference in population means $(\mu_1 - \mu_2)$ is the difference in sample means $(\bar{x}_1 - \bar{x}_2)$. Therefore, in testing whether the difference in sample means indicates that the true difference in population means differs from a specified value, $(\mu_1 - \mu_2) = D_0$, you can use the standard error of $(\bar{x}_1 - \bar{x}_2)$:

$$\sqrt{\frac{\sigma_1^2}{n_1} + \frac{\sigma_2^2}{n_2}} \quad \text{estimated by} \quad SE = \sqrt{\frac{s_1^2}{n_1} + \frac{s_2^2}{n_2}}$$

in the form of a z-statistic to measure how many standard deviations the difference $(\bar{x}_1 - \bar{x}_2)$ lies from the hypothesized difference D_0. The formal testing procedure is described next.

Large-Sample Statistical Test for $(\mu_1 - \mu_2)$

1. Null hypothesis: $H_0 : (\mu_1 - \mu_2) = D_0$, where D_0 is some specified difference that you wish to test. For many tests, you will hypothesize that there is no difference between μ_1 and μ_2; that is, $D_0 = 0$.
2. Alternative hypothesis:

One-Tailed Test	Two-Tailed Test
$H_a : (\mu_1 - \mu_2) > D_0$ [or $H_a : (\mu_1 - \mu_2) < D_0$]	$H_a : (\mu_1 - \mu_2) \neq D_0$

3. Test statistic: $z \approx \dfrac{(\bar{x}_1 - \bar{x}_2) - D_0}{\text{SE}} = \dfrac{(\bar{x}_1 - \bar{x}_2) - D_0}{\sqrt{\dfrac{s_1^2}{n_1} + \dfrac{s_2^2}{n_2}}}$

4. Rejection region: Reject H_0 when

One-Tailed Test	**Two-Tailed Test**
$z > z_\alpha$ [or $z < -z_\alpha$ when the alternative hypothesis is $H_a: (\mu_1 - \mu_2) < D_0$] or when p-value $< \alpha$	$z > z_{\alpha/2}$ or $z < -z_{\alpha/2}$

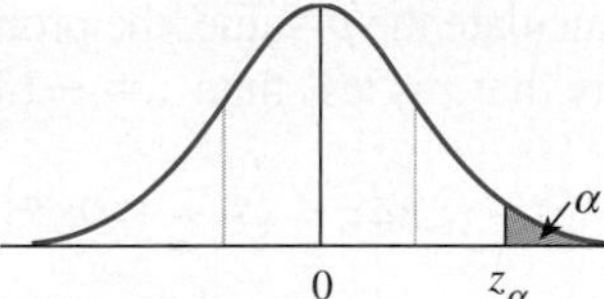

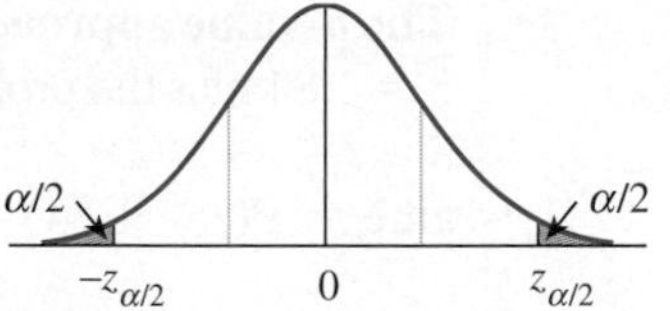

Assumptions: The samples are randomly and independently selected from the two populations and $n_1 \geq 30$ and $n_2 \geq 30$.

EXAMPLE 9.11

To determine whether car ownership affects a student's academic achievement, two random samples of 100 male students were each drawn from the student body. The grade point average for the $n_1 = 100$ non-owners of cars had an average and variance equal to $\bar{x}_1 = 2.70$ and $s_1^2 = 0.36$, while $\bar{x}_2 = 2.54$ and $s_2^2 = 0.40$ for the $n_2 = 100$ car owners. Do the data present sufficient evidence to indicate a difference in the mean achievements between car owners and non-owners of cars? Test using $\alpha = 0.05$.

SOLUTION To detect a difference, if it exists, between the mean academic achievements for non-owners of cars μ_1 and car owners μ_2, you will test the null hypothesis that there is no difference between the means against the alternative hypothesis that $(\mu_1 - \mu_2) \neq 0$; that is,

$$H_0: (\mu_1 - \mu_2) = D_0 = 0 \quad \text{versus} \quad H_a: (\mu_1 - \mu_2) \neq 0$$

Substituting into the formula for the test statistic, you get

$$z = \frac{(\bar{x}_1 - \bar{x}_2) - D_0}{\sqrt{\dfrac{s_1^2}{n_1} + \dfrac{s_2^2}{n_2}}} = \frac{2.70 - 2.54}{\sqrt{\dfrac{0.36}{100} + \dfrac{0.40}{100}}} = 1.84$$

- **The critical value approach:** Using a two-tailed test with significance level $\alpha = 0.05$, you place $\alpha/2 = 0.025$ in each tail of the z distribution and reject H_0 if $z > 1.96$ or $z < -1.96$. Since $z = 1.84$ does not exceed 1.96 and is not less than -1.96, H_0 cannot be rejected (see Figure 9.9). That is, there is insufficient evidence to declare a difference in the average academic achievements for the two groups. Remember that you should not be willing to *accept* H_0—declare the two means to be the same—until β is evaluated for some meaningful values of $(\mu_1 - \mu_2)$.

NEED A TIP?

|test statistic| > |critical value| ⇔ reject H_0.

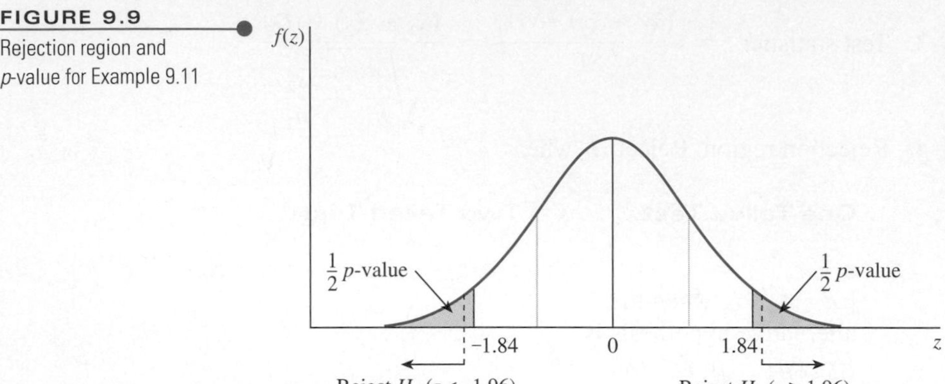

FIGURE 9.9
Rejection region and *p*-value for Example 9.11

- **The *p*-value approach:** Calculate the *p*-value, the probability that z is greater than $z = 1.84$ plus the probability that z is less than $z = -1.84$, as shown in Figure 9.9:

$$p\text{-value} = P(z > 1.84) + P(z < -1.84) = (1 - 0.9671) + 0.0329 = 0.0658$$

The *p*-value lies between 0.10 and 0.05, so you can reject H_0 at the 0.10 level but not at the 0.05 level of significance. Since the *p*-value of 0.0658 exceeds the specified significance level $\alpha = 0.05$, H_0 cannot be rejected. Again, you should not be willing to *accept* H_0 until β is evaluated for some meaningful values of $(\mu_1 - \mu_2)$.

Hypothesis Testing and Confidence Intervals

Whether you use the critical value or the *p*-value approach for testing hypotheses about $(\mu_1 - \mu_2)$, you will always reach the same conclusion because the calculated value of the test statistic and the critical value are related *exactly* in the same way that the *p*-value and the significance level α are related. You might remember that the confidence intervals constructed in Chapter 8 could also be used to answer questions about the difference between two population means. In fact, for a two-tailed test, the $(1 - \alpha)100\%$ confidence interval for the parameter of interest can be used to test its value, just as you did informally in Chapter 8. The value of α indicated by the confidence coefficient in the confidence interval is equivalent to the significance level α in the statistical test. For a one-tailed test, the equivalent confidence interval approach would use the one-sided confidence bounds in Section 8.8 with confidence coefficient α. In addition, by using the confidence interval approach, you gain a range of possible values for the parameter of interest, regardless of the outcome of the test of hypothesis.

- If the confidence interval you construct *contains* the value of the parameter specified by H_0, then that value is one of the likely or possible values of the parameter and H_0 should not be rejected.
- If the hypothesized value *lies outside* of the confidence limits, the null hypothesis is rejected at the α level of significance.

Construct a 95% confidence interval for the difference in average academic achievements between car owners and non-owners. Using the confidence interval, can you conclude that there is a difference in the population means for the two groups of students?

SOLUTION For the large-sample statistics discussed in Chapter 8, the 95% confidence interval is given as

$$\text{Point estimator} \pm 1.96 \times (\text{Standard error of the estimator})$$

For the difference in two population means, the confidence interval is approximated as

$$(\bar{x}_1 - \bar{x}_2) \pm 1.96\sqrt{\frac{s_1^2}{n_1} + \frac{s_2^2}{n_2}}$$

$$(2.70 - 2.54) \pm 1.96\sqrt{\frac{0.36}{100} + \frac{0.40}{100}}$$

$$0.16 \pm 0.17$$

or $-0.01 < (\mu_1 - \mu_2) < 0.33$. This interval gives you a range of possible values for the difference in the population means. Since the hypothesized difference, $(\mu_1 - \mu_2) = 0$, is contained in the confidence interval, you should not reject H_0. Look at the signs of the possible values in the confidence interval. You cannot tell from the interval whether the difference in the means is negative (−), positive (+), or zero (0)—the latter of the three would indicate that the two means are the same. Hence, you can really reach no conclusion in terms of the question posed. There is not enough evidence to indicate that there is a difference in the average achievements for car owners versus non-owners. The conclusion is the same onc reached in Example 9.11.

EXAMPLE 9.13

Football games and rugby games are similar in the sense that a ball is used to score points. Suppose that a random sample of 30 CFL footballs was chosen, and their average weight was 450 grams with a sample standard deviation of 8, while a random sample of 40 rugby balls was chosen, and their average weight was 445 grams with a sample variance of 81. Does the data indicate that rugby balls weigh less than footballs? Test using $\alpha = 0.05$.

$$H0: \mu_1 - \mu_2 = 0 \quad \text{versus} \quad H_a: \mu_1 > \mu_2$$

SOLUTION

$$z = \frac{449 - 447}{\sqrt{\frac{64}{30} + \frac{81}{40}}} = 0.98$$

The critical value approach: $z_{.05} = 1.645$, which is greater than 0.98. Thus, the null hypothesis is not rejected. This means that there is no statistical evidence to conclude that the weight of a CFL football is heavier than the weight of a rugby ball.

9.4 EXERCISES

BASIC TECHNIQUES

9.23 Independent random samples of 80 measurements were drawn from two quantitative populations, 1 and 2. Here is a summary of the sample data:

	Sample 1	Sample 2
Sample size	80	80
Sample mean	11.6	9.7
Sample variance	27.9	38.4

a. If your research objective is to show that μ_1 is larger than μ_2, state the alternative and the null hypotheses that you would choose for a statistical test.

b. Is the test in part a one- or two-tailed?

c. Calculate the test statistic that you would use for the test in part a. Based on your knowledge of the standard normal distribution, is this a likely or unlikely observation, assuming that H_0 is true and the two population means are the same?

d. *p-value approach:* Find the p-value for the test. Test for a significant difference in the population means at the 1% significance level.

e. *Critical value approach:* Find the rejection region when $\alpha = 0.01$. Do the data provide sufficient evidence to indicate a difference in the population means?

9.24 Independent random samples of 36 and 45 observations are drawn from two quantitative populations, 1 and 2, respectively. The sample data summary is shown here:

	Sample 1	Sample 2
Sample size	36	45
Sample mean	1.24	1.31
Sample variance	0.0560	0.0540

Do the data present sufficient evidence to indicate that the mean for population 1 is smaller than the mean for population 2? Use one of the two methods of testing presented in this section, and explain your conclusions.

9.25 Suppose you wish to detect a difference between μ_1 and μ_2 (either $\mu_1 > \mu_2$ or $\mu_1 < \mu_2$) and, instead of running a two-tailed test using $\alpha = 0.05$, you use the following test procedure. You wait until you have collected the sample data and have calculated $\bar{x}_1$ and $\bar{x}_2$. If $\bar{x}_1$ is larger than $\bar{x}_2$, you choose the alternative hypothesis $H_a : \mu_1 > \mu_2$ and run a one-tailed test placing $\alpha_1 = 0.05$ in the upper tail of the z distribution. If, on the other hand, $\bar{x}_2$ is larger than $\bar{x}_1$, you reverse the procedure and run a one-tailed test, placing $\alpha_2 = 0.05$ in the lower tail of the z distribution. If you use this procedure and if μ_1 actually equals μ_2, what is the probability α that you will conclude that μ_1 is not equal to μ_2 (i.e., what is the probability α that you will incorrectly reject H_0 when H_0 is true)? This exercise demonstrates why statistical tests should be formulated *prior* to observing the data.

9.26 For the three tests given in Exercise 9.25, use the p-value to determine the significance of the results. Explain what "statistically significant" means in terms of rejecting or accepting H_0 and H_a.

APPLICATIONS

9.27 Cure for the Common Cold? An experiment was planned to compare the mean time (in days) required to recover from a common cold for persons given a daily dose of 4 mg of vitamin C versus those who were not given a vitamin supplement. Suppose that 35 adults were randomly selected for each treatment category and that the mean recovery times and standard deviations for the two groups were as follows:

	No Vitamin Supplement	4 mg Vitamin C
Sample size	35	35
Sample mean	6.9	5.8
Sample standard deviation	2.9	1.2

a. Suppose your research objective is to show that the use of vitamin C reduces the mean time required to recover from a common cold and its complications. Give the null and alternative hypotheses for the test. Is this a one- or a two-tailed test?

b. Conduct the statistical test of the null hypothesis in part a and state your conclusion. Test using $\alpha = 0.05$.

9.28 Healthy Eating Canadians are becoming more conscious about the importance of good nutrition, and some researchers believe we may be altering our diets to include less red meat and more fruits and vegetables. To test the theory that the consumption of red meat has decreased over the last 10 years, a researcher decides to select hospital nutrition records for 400 subjects surveyed 10 years ago and to compare their average amount of beef consumed per year to amounts consumed by an equal number of subjects interviewed this year. The data (in kg) are given in the table:

	Ten Years Ago	This Year
Sample mean	33.1	28.6
Sample standard deviation	11.3	12.7

a. Do the data present sufficient evidence to indicate that per-capita beef consumption has decreased in the last 10 years? Test at the 1% level of significance.

b. Find a 99% lower confidence bound for the difference in the average per-capita beef consumptions for the two groups. (This calculation was done as part of Exercise 8.85.) Does your confidence bound confirm your conclusions in part a? Explain. What additional information does the confidence bound give you?

9.29 Lead Levels in Drinking Water Analyses of drinking water samples for 100 homes in each of two different sections of a city gave the following means and standard deviations of lead levels (in parts per million):

	Section 1	Section 2
Sample size	100	100
Mean	34.1	36.0
Standard deviation	5.9	6.0

a. Calculate the test statistic and its p-value (observed significance level) to test for a difference in the two population means. Use the p-value to evaluate the statistical significance of the results at the 5% level.

b. Use a 95% confidence interval to estimate the difference in the mean lead levels for the two sections of the city.

c. Suppose that the city environmental engineers will be concerned only if they detect a difference of more than 5 parts per million in the two sections of the city. Based on your confidence interval in part b, is the statistical significance in part a of *practical significance* to the city engineers? Explain.

9.30 Starting Salaries, again In an attempt to compare the starting salaries for university graduates who majored in education and the social sciences (see Exercise 8.54), random samples of 50 recent university graduates in each major were selected and the following information was obtained:

Major	Mean	SD
Education ($)	40,554	2225
Social sciences ($)	38,348	2375

a. Do the data provide sufficient evidence to indicate a difference in average starting salaries for university graduates who majored in education and the social sciences? Test using $\alpha = 0.05$.

b. Compare your conclusions in part a with the results of part b in Exercise 8.54. Are they the same? Explain.

9.31 Hotel Costs In Exercise 8.20, we explored the average cost of lodging at three different hotel chains.[10] We randomly select 50 billing statements from the computer databases of the Marriott, Westin, and Doubletree hotel chains, and record the nightly room rates. A portion of the sample data is shown in the table:

	Marriott	Westin
Sample average ($)	150	165
Sample standard deviation ($)	17.2	22.5

a. Before looking at the data, would you have any preconceived idea about the direction of the difference between the average room rates for these two hotels? If not, what null and alternative hypotheses should you test?

b. Use the *critical value* approach to determine if there is a significant difference in the average room rates for the Marriott and the Westin hotel chains. Use $\alpha = 0.01$.

c. Find the p-value for this test. Does this p-value confirm the results of part b?

9.32 Hotel Costs II Refer to Exercise 9.31. The table below shows the sample data collected to compare the average room rates at the Westin and Doubletree hotel chains:[11]

	Westin	Doubletree
Sample average ($)	165	125
Sample standard deviation ($)	22.5	12.8

a. Do the data provide sufficient evidence to indicate a difference in the average room rates for the Westin and the Doubletree hotel chains? Use $\alpha = .05$.

b. Construct a 95% confidence interval for the difference in the average room rates for the two chains. Does this interval confirm your conclusions in part a?

9.33 MMT in Gasoline The addition of MMT, a compound containing manganese (Mn), to gasoline as an octane enhancer has caused concern about human exposure to Mn because high intakes have been linked to serious health effects. In a study of ambient air concentrations of fine Mn, Wallace and Slonecker (*Journal of the Air and Waste Management Association*) presented the accompanying summary information about the amounts of fine Mn (in nanograms per cubic metre) in mostly rural national park sites and in mostly urban California sites:[12]

	National Parks	California
Mean	0.94	2.8
Standard deviation	1.2	2.8
Number of sites	36	26

a. Is there sufficient evidence to indicate that the mean concentrations differ for these two types of sites at the $\alpha = 0.05$ level of significance? Use the large-sample z-test. What is the p-value of this test?

b. Construct a 95% confidence interval for $(\mu_1 - \mu_2)$. Does this interval confirm your conclusions in part a?

9.34 Noise and Stress In Exercise 8.57, you compared the effect of stress in the form of noise on the ability to perform a simple task. Seventy subjects were divided into two groups; the first group of 30 subjects acted as a control, while the second group of 40 was the experimental group. Although each subject performed the task in the same control room, each of the experimental group subjects had to perform the task while loud rock music was played. The time to finish the task was recorded for each subject and the following summary was obtained:

	Control	Experimental
n	30	40
$\bar{x}$	15 minutes	23 minutes
s	4 minutes	10 minutes

a. Is there sufficient evidence to indicate that the average time to complete the task was longer for the experimental "rock music" group? Test at the 1% level of significance.

b. Construct a 99% one-sided upper bound for the difference (control – experimental) in average times for the two groups. Does this interval confirm your conclusions in part a?

9.35 What's Normal II Of the 130 people in Exercise 9.19, 65 were female and 65 were male.[13] The means and standard deviations of their temperatures are shown below.

	Men	Women
Sample mean (°C)	36.72	36.88
Standard deviation	0.70	0.74

a. Use the *p*-value approach to test for a significant difference in the average temperatures for males versus females.

b. Are the results significant at the 5% level? At the 1% level?

9.36 Gender Differences in Lifespan According to a sample collected of 150 males, their average lifespan was 79 years old with a standard deviation of 12.5 years. A sample collected of 100 females showed that their average lifespan was 83 years old with a sample standard deviation of 9.3 years.

a. Is there sufficient evidence to conclude that women on average live longer than men? Let alpha = 0.05.

b. Construct a 95% confidence interval for the difference of average lifespan between both genders.

9.5 A LARGE-SAMPLE TEST OF HYPOTHESIS FOR A BINOMIAL PROPORTION

When a random sample of n identical trials is drawn from a binomial population, the sample proportion $\hat{p}$ has an approximately normal distribution when n is large, with mean p and standard error

$$\text{SE} = \sqrt{\frac{pq}{n}}$$

When you test a hypothesis about p, the proportion in the population possessing a certain attribute, the test follows the same general form as the large-sample tests in Sections 9.3 and 9.4. To test a hypothesis of the form

$$H_0: p = p_0$$

versus a one- or two-tailed alternative

$$H_a: p > p_0 \quad \text{or} \quad H_a: p < p_0 \quad \text{or} \quad H_a: p \neq p_0$$

the test statistic is constructed using $\hat{p}$, the best estimator of the true population proportion p. The sample proportion $\hat{p}$ is standardized, using the hypothesized mean and standard error, to form a test statistic z, which has a standard normal distribution if H_0 is true. This large-sample test is summarized next.

Large-Sample Statistical Test for *p*

1. Null hypothesis: $H_0: p = p_0$
2. Alternative hypothesis:

One-Tailed Test	Two-Tailed Test
$H_a: p > p_0$ (or, $H_a: p < p_0$)	$H_a: p \neq p_0$

3. Test statistic: $z = \dfrac{\hat{p} - p_0}{\text{SE}} = \dfrac{\hat{p} - p_0}{\sqrt{\dfrac{p_0 q_0}{n}}}$ with $\hat{p} = \dfrac{x}{n}$

where x is the number of successes in n binomial trials.†

4. Rejection region: Reject H_0 when

One-Tailed Test	**Two-Tailed Test**
$z > z_\alpha$ (or $z < -z_\alpha$ when the alternative hypothesis is $H_a: p < p_0$)	$z > z_{\alpha/2}$ or $z < -z_{\alpha/2}$

or when p-value $< \alpha$

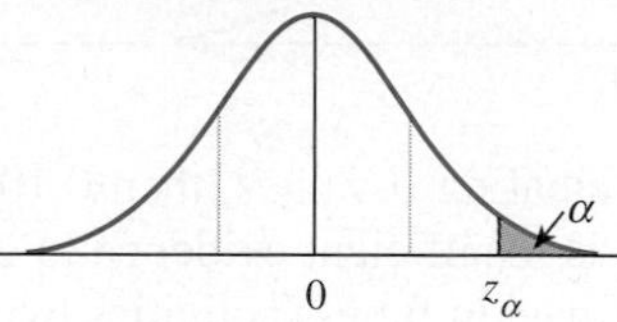

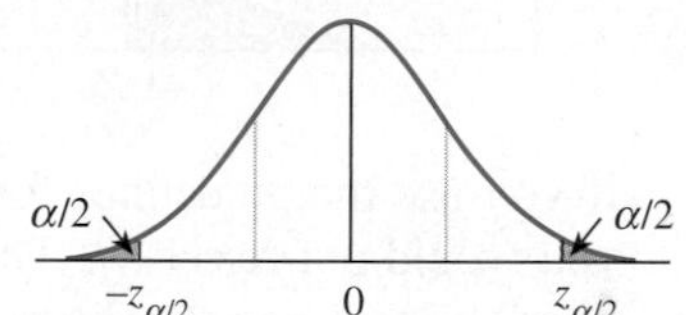

Assumption: The sampling satisfies the assumptions of a binomial experiment (see Section 5.2), and n is large enough so that the sampling distribution of $\hat{p}$ can be approximated by a normal distribution ($np_0 > 5$ and $nq_0 > 5$).

EXAMPLE 9.14

Regardless of age, about 20% of Canadian adults participate in fitness activities at least twice a week. However, these fitness activities change as the people get older, and occasionally participants become non-participants as they age. In a local survey of $n = 100$ adults over 40 years old, a total of 15 people indicated that they participated in a fitness activity at least twice a week. Do these data indicate that the participation rate for adults over 40 years of age is significantly less than the 20% figure? Calculate the p-value and use it to draw the appropriate conclusions.

SOLUTION Assuming that the sampling procedure satisfies the requirements of a binomial experiment, you can answer the question posed using a one-tailed test of hypothesis:

$$H_0: p = 0.2 \quad \text{versus} \quad H_a: p < 0.2$$

Begin by assuming that H_0 is true—that is, the true value of p is $p_0 = 0.2$. Then $\hat{p} = x/n$ will have an approximate normal distribution with mean p_0 and standard error $\sqrt{p_0 q_0/n}$. (This is different from the estimation procedure in which the unknown standard error is estimated by $\sqrt{\hat{p}\hat{q}/n}$.) The observed value of $\hat{p}$ is $15/100 = 0.15$ and the test statistic is

p-value $\leq \alpha \Leftrightarrow$ reject H_0

p-value $> \alpha \Leftrightarrow$ do not reject H_0.

$$z = \frac{\hat{p} - p_0}{\sqrt{\dfrac{p_0 q_0}{n}}} = \frac{0.15 - 0.20}{\sqrt{\dfrac{(0.20)(0.80)}{100}}} = -1.25$$

†An equivalent test statistic can be found by multiplying the numerator and denominator of z by n to obtain

$$z = \frac{x - np_0}{\sqrt{np_0 q_0}}$$

The p-value associated with this test is found as the area under the standard normal curve to the left of $z = -1.25$ as shown in Figure 9.10. Therefore,

$$p\text{-value} = P(z < -1.25) = 0.1056$$

FIGURE 9.10
p-value for Example 9.14

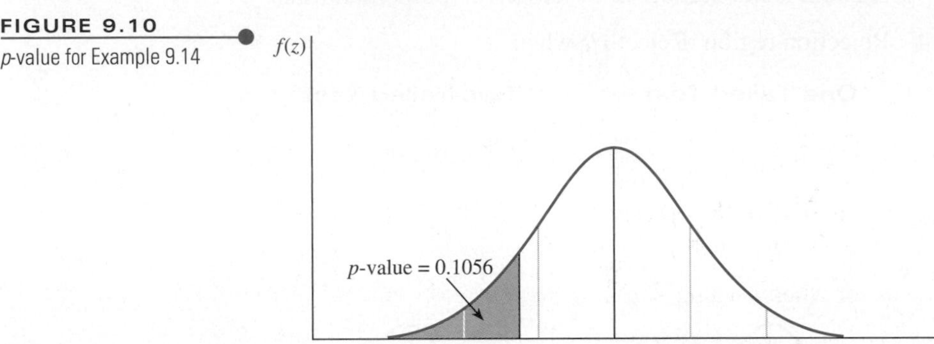

If you use the guidelines for evaluating p-values, then 0.1056 is greater than 0.10, and you would not reject H_0. There is insufficient evidence to conclude that the percentage of adults over age 40 who participate in fitness activities twice a week is less than 20%.

Statistical Significance and Practical Importance

It is important to understand the difference between results that are "significant" and results that are practically "important." In statistical language, the word *significant* does not necessarily mean "important," but only that the results could not have occurred by chance. For example, suppose that in Example 9.14, the researcher had used $n = 400$ adults in her experiment and had observed the same sample proportion. The test statistic is now

$$z = \frac{\hat{p} - p_0}{\sqrt{\dfrac{p_0 q_0}{n}}} = \frac{0.15 - 0.20}{\sqrt{\dfrac{(0.20)(0.80)}{n = 400}}} = -2.50$$

with

$$p\text{-value} = P(z < -2.50) = 0.0062$$

Now the results are *highly significant:* H_0 is rejected, and there is sufficient evidence to indicate that the percentage of adults over age 40 who participate in physical fitness activities is less than 20%. However, is this drop in activity really *important*? Suppose that physicians would be concerned only about a drop in physical activity of more than 10%. If there had been a drop of more than 10% in physical activity, this would imply that the true value of p was less than 0.10. What is the largest possible value of p? Using a 95% upper one-sided confidence bound, you have

$$\hat{p} + 1.645\sqrt{\frac{\hat{p}\hat{q}}{n}}$$

$$0.15 + 1.645\sqrt{\frac{(0.15)(0.85)}{400}}$$

$$0.15 + 0.029$$

or $p < 0.179$. The physical activity for adults aged 40 and older has dropped from 20%, but you cannot say that it has dropped below 10%. So, the results, although *statistically significant*, are not *practically important*.

In this book, you will learn how to determine whether results are statistically significant. When you use these procedures in a practical situation, however, you must also make sure the results are practically important.

9.5 EXERCISES

BASIC TECHNIQUES

9.37 A random sample of $n = 1000$ observations from a binomial population produced $x = 279$.

a. If your research hypothesis is that p is less than 0.3, what should you choose for your alternative hypothesis? Your null hypothesis?

b. What is the critical value that determines the rejection region for your test with $\alpha = 0.05$?

c. Do the data provide sufficient evidence to indicate that p is less than 0.3? Use a 5% significance level.

9.38 A random sample of $n = 1400$ observations from a binomial population produced $x = 529$.

a. If your research hypothesis is that p differs from 0.4, what hypotheses should you test?

b. Calculate the test statistic and its p-value. Use the p-value to evaluate the statistical significance of the results at the 1% level.

c. Do the data provide sufficient evidence to indicate that p is different from 0.4?

9.39 A random sample of 120 observations was selected from a binomial population, and 72 successes were observed. Do the data provide sufficient evidence to indicate that p is greater than 0.5? Use one of the two methods of testing presented in this section, and explain your conclusions.

APPLICATIONS

9.40 R and M Entertainment On behalf of the Canadian Teachers Federation (CTF), Erin Research conducted a study on Canadian children and their experience with communication media. The research indicates that 75% of students say R-movies should be open to kids age 12 and up, but less than 50% say the same about M-rated games. The research involved 5756 students in Grades 3 to 10. Suppose that $n = 100$ students in Grades 3 to 10 are sampled, and of those sampled, 68 support R-movies open to kids ages 12 and up, and of those sampled 47 say the same about M-rated games.

a. Two claims can be tested using the sample information. What are the two sets of hypotheses to be tested?

b. Do the data represent sufficient evidence to contradict the claim that more than half of students say that M-rated games should be open to kids age 12 and up?

c. Do the data present sufficient evidence to show that the 75% figure claimed in the research article is incorrect?

9.41 Plant Genetics A peony plant with red petals was crossed with another plant having streaky petals. A geneticist states that 75% of the offspring resulting from this cross will have red flowers. To test this claim, 100 seeds from this cross were collected and germinated and 58 plants had red petals.

a. What hypothesis should you use to test the geneticist's claim?

b. Calculate the test statistic and its p-value. Use the p-value to evaluate the statistical significance of the results at the 1% level.

9.42 Early Detection of Breast Cancer Of those women who are diagnosed to have early-stage breast cancer, one-third eventually die of the disease. Suppose a community public health department instituted a screening program to provide for the early detection of breast cancer and to increase the survival rate p of those diagnosed to have the disease. A random sample of 200 women was selected from among those who were periodically screened by the program and who were diagnosed to have the disease. Let x represent the number of those in the sample who survive the disease.

a. If you wish to detect whether the community screening program has been effective, state the null hypothesis that should be tested.

b. State the alternative hypothesis.

c. If 164 women in the sample of 200 survive the disease, can you conclude that the community screening program was effective? Test using $\alpha = 0.05$ and explain the practical conclusions from your test.

d. Find the p-value for the test and interpret it.

9.43 Sweet Potato Whitefly Suppose that 10% of the fields in a given agricultural area are infested with the sweet potato whitefly. One hundred fields in this area are randomly selected, and 25 are found to be infested with whitefly.

a. Assuming that the experiment satisfies the conditions of the binomial experiment, do the data indicate that the proportion of infested fields is greater than expected? Use the p-value approach, and test using a 5% significance level.

b. If the proportion of infested fields is found to be significantly greater than 0.10, why is this of practical significance to the agronomist? What practical conclusions might she draw from the results?

9.44 Taste Testing In a head-to-head taste test of store-brand foods versus national brands, *Consumer Reports* found that it was hard to find a taste difference in the two.[14] If the national brand is indeed better tasting than the store brand, it should be judged as better more than 50% of the time.

a. State the null and alternative hypothesis to be tested. Is this a one- or a two-tailed test?

b. Suppose that, of the 35 food categories used for the taste test, the national brand was found to be better than the store brand in eight of the taste comparisons. Use this information to test the hypothesis in part a. Use $\alpha = 0.01$. What practical conclusions can you draw from the results?

9.45 Avian Flu Avian flu is clearly on the radar of Canadians. Given the amount of media coverage on the avian flu issue, it is not surprising to find that about 60% are concerned about it. However, the concern is fairly modest, only 19% being "very concerned," but this suggests that concern has the potential to grow as events unfold. The above result is based on a report to the *Globe and Mail* and CTV from The Strategic Counsel, 2005.[15] Suppose a random sample of 90 people is selected in 2007. Of these people, 20 are very concerned and 50 are concerned about the issue.

a. Do the sample data provide sufficient evidence to indicate that the proportion of very concerned is different from 19%? Use $\alpha = 0.05$.

b. Do the sample data provide sufficient evidence to indicate that the proportion of concerned is different from 60%? Use $\alpha = 0.05$.

c. Is there any reason to conduct a one-tailed test for either part a or b? Explain.

9.46 A Cure for Insomnia An experimenter has prepared a drug-dose level that he claims will induce sleep for at least 80% of people suffering from insomnia. After examining the dosage, we feel that his claims regarding the effectiveness of his dosage are inflated. In an attempt to disprove his claim, we administer his prescribed dosage to 50 insomniacs and observe that 37 of them have had sleep induced by the drug dose. Is there enough evidence to refute his claim at the 5% level of significance?

9.47 Avian Flu, continued Refer to Exercise 9.45. The research report further indicates evidence that the public is somewhat concerned about whether or not Canadian health authorities are prepared to deal with this emerging problem. The survey showed that 45% believe that the Canadian health authorities are not prepared. To test this survey, a sample of 300 adults was taken, and 140 of them said they are concerned about the preparedness. Is there sufficient evidence to dispute the claim regarding the preparedness? Use $\alpha = 0.01$.

9.48 Aboriginal People in Canada According to the most recent census data, Aboriginal people account for approximately 3.3% of the Canadian population.[16] However, the female-to-male ratio is slightly higher among Aboriginal peoples (51.2% female and 48.8% male) than in the total population (50.9% female and 49.1% male). Most of the Aboriginal people live in the territories. For example, in Nunavut the proportion of Aboriginal people is about 85%. In a random sample of 300 people in Nunavut, 261 people identified themselves as Aboriginal. Does this data provide sufficient evidence to indicate that the proportion of Aboriginal people in Nunavut is different from that reported than in census data? Test using $\alpha = 0.01$.

9.49 Love for Pets in Canada In 2014, a nationwide survey conducted by Ipsos of over 4200 pet-owning households showed that Canadians own a total of 13.4 million dogs and cats. This total breaks down to approximately 6.4 million dogs and 7 million cats in the country.[17]

In a random sample of 300 households, 114 households said that they owned at least one pet. Does this data provide sufficient evidence to indicate that the proportion of households with at least one pet is different from that reported by the Statistics Canada? Test using $\alpha = 0.05$.

9.6 A LARGE-SAMPLE TEST OF HYPOTHESIS FOR THE DIFFERENCE BETWEEN TWO BINOMIAL PROPORTIONS

When random and independent samples are selected from two *binomial* populations, the focus of the experiment may be the difference $(p_1 - p_2)$ in the proportions of individuals or items possessing a specified characteristic in the two populations. In this situation, you can use the difference in the sample proportions $(\hat{p}_1 - \hat{p}_2)$ along with its standard error,

$$SE = \sqrt{\frac{p_1 q_1}{n_1} + \frac{p_2 q_2}{n_2}}$$

in the form of a z-statistic to test for a significant difference in the two population proportions. The null hypothesis to be tested is usually of the form

$$H_0: p_1 = p_2 \text{ or } H_0: (p_1 - p_2) = 0$$

versus either a one- or two-tailed alternative hypothesis. The formal test of hypothesis is summarized in the next display. In estimating the standard error for the z-statistic, you should use the fact that when H_0 is true, the two population proportions are equal to some common value—say, p. To obtain the best estimate of this common value, the sample data are "pooled" and the estimate of p is

NEED A TIP?

Remember: Each trial results in one of two outcomes (S or F).

$$\hat{p} = \frac{\text{Total number of successes}}{\text{Total number of trials}} = \frac{x_1 + x_2}{n_1 + n_2}$$

Remember that, in order for the difference in the sample proportions to have an approximately normal distribution, the sample sizes must be large and the proportions should not be too close to 0 or 1.

Large-Sample Statistical Test for $(p_1 - p_2)$

1. Null hypothesis: $H_0: (p_1 - p_2) = 0$ or equivalently $H_0: p_1 = p_2$
2. Alternative hypothesis:

One-Tailed Test	Two-Tailed Test
$H_a: (p_1 - p_2) > 0$ [or $H_a: (p_1 - p_2) < 0$]	$H_a: (p_1 - p_2) \neq 0$

3. Test statistic: $z = \dfrac{(\hat{p}_1 - \hat{p}_2) - 0}{SE} = \dfrac{\hat{p}_1 - \hat{p}_2}{\sqrt{\dfrac{p_1 q_1}{n_1} + \dfrac{p_2 q_2}{n_2}}} = \dfrac{\hat{p}_1 - \hat{p}_2}{\sqrt{\dfrac{pq}{n_1} + \dfrac{pq}{n_2}}}$

   where $\hat{p}_1 = x_1/n_1$ and $\hat{p}_2 = x_2/n_2$. Since the common value of $p_1 = p_2 = p$ (used in the standard error) is unknown, it is estimated by

$$\hat{p} = \frac{x_1 + x_2}{n_1 + n_2}$$

and the test statistic is

$$z = \frac{(\hat{p}_1 - \hat{p}_2) - 0}{\sqrt{\dfrac{\hat{p}\hat{q}}{n_1} + \dfrac{\hat{p}\hat{q}}{n_2}}} \quad \text{or} \quad z = \frac{\hat{p}_1 - \hat{p}_2}{\sqrt{\hat{p}\hat{q}\left(\dfrac{1}{n_1} + \dfrac{1}{n_2}\right)}}$$

4. Rejection region: Reject H_0 when

One-Tailed Test	**Two-Tailed Test**
$z > z_\alpha$ [or $z < -z_\alpha$ when the alternative hypothesis is $H_a : (p_1 - p_2) < 0$]	$z > z_{\alpha/2}$ or $z < -z_{\alpha/2}$

or when p-value $< \alpha$

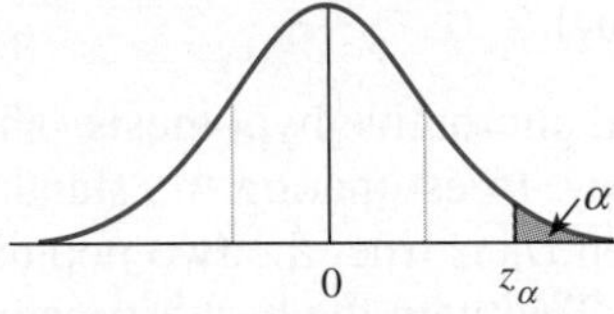

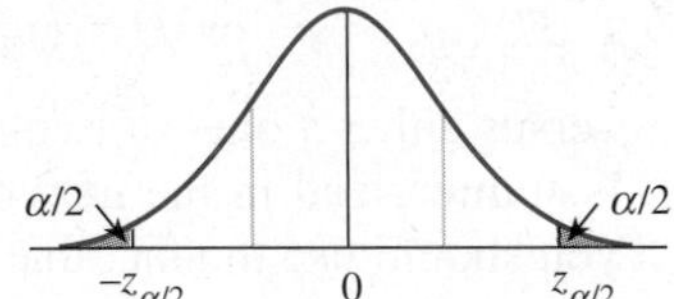

Assumptions: Samples are selected in a random and independent manner from two binomial populations, and n_1 and n_2 are large enough so that the sampling distribution of $(\hat{p}_1 - \hat{p}_2)$ can be approximated by a normal distribution. That is, $n_1\hat{p}_1$, $n_1\hat{q}_1$, $n_2\hat{p}_2$, and $n_2\hat{q}_2$ should all be greater than 5.

EXAMPLE 9.15

The records of a hospital show that 52 men in a sample of 1000 men versus 23 women in a sample of 1000 women were admitted because of heart disease. Do these data present sufficient evidence to indicate a higher rate of heart disease among men admitted to the hospital? Use $\alpha = 0.05$.

SOLUTION Assume that the number of patients admitted for heart disease has an approximate binomial probability distribution for both men and women with parameters p_1 and p_2, respectively. Then, since you wish to determine whether $p_1 > p_2$, you will test the null hypothesis $p_1 = p_2$—that is, $H_0 : (p_1 - p_2) = 0$—against the alternative hypothesis $H_a : p_1 > p_2$ or, equivalently, $H_a : (p_1 - p_2) > 0$. To conduct this test, use the z-test statistic and approximate the standard error using the pooled estimate of p. Since H_a implies a one-tailed test, you can reject H_0 only for large values of z. Thus, for $\alpha = 0.05$, you can reject H_0 if $z > 1.645$ (see Figure 9.11).

The pooled estimate of p required for the standard error is

$$\hat{p} = \frac{x_1 + x_2}{n_1 + n_2} = \frac{52 + 23}{1000 + 1000} = 0.0375$$

and the test statistic is

$$z = \frac{\hat{p}_1 - \hat{p}_2}{\sqrt{\hat{p}\hat{q}\left(\dfrac{1}{n_1} + \dfrac{1}{n_2}\right)}} = \frac{0.052 - 0.023}{\sqrt{(0.0375)(0.9625)\left(\dfrac{1}{1000} + \dfrac{1}{1000}\right)}} = 3.41$$

FIGURE 9.11
Location of the rejection region in Example 9.15

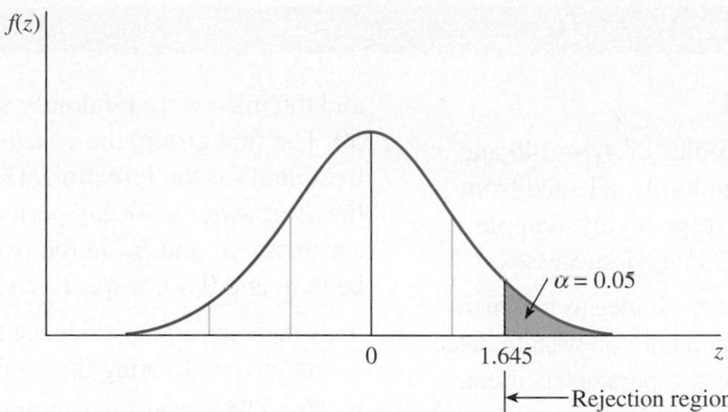

Since the computed value of z falls in the rejection region, you can reject the hypothesis that $p_1 = p_2$. The data present sufficient evidence to indicate that the percentage of men entering the hospital because of heart disease is higher than that of women. (NOTE: This does not imply that the *incidence* of heart disease is higher in men. Perhaps fewer women enter the hospital when afflicted with the disease!)

How *much higher* is the proportion of men than women entering the hospital with heart disease? A 95% lower one-sided confidence bound will help you find the lowest likely value for the difference.

$$(\hat{p}_1 - \hat{p}_2) - 1.645\sqrt{\frac{\hat{p}_1\hat{q}_1}{n_1} + \frac{\hat{p}_2\hat{q}_2}{n_2}}$$

$$(0.052 - 0.023) - 1.645\sqrt{\frac{0.052(0.948)}{1000} + \frac{0.023(0.977)}{1000}}$$

$$0.029 - 0.014$$

or $(p_1 - p_2) > 0.015$. The proportion of men is roughly 1.5% higher than women. Is this of *practical importance*? This is a question for the researcher to answer.

In some situations, you may need to test for a difference D_0 (other than 0) between two binomial proportions. If this is the case, the test statistic is modified for testing $H_0 : (p_1 - p_2) = D_0$, and a pooled estimate for a common p is no longer used in the standard error. The modified test statistic is

$$z = \frac{(\hat{p}_1 - \hat{p}_2) - D_0}{\sqrt{\frac{\hat{p}_1\hat{q}_1}{n_1} + \frac{\hat{p}_2\hat{q}_2}{n_2}}}$$

Although this test statistic is not used often, the procedure is no different from other large-sample tests you have already mastered!

9.6 EXERCISES

BASIC TECHNIQUES

9.50 Independent random samples of $n_1 = 140$ and $n_2 = 140$ observations were randomly selected from binomial populations 1 and 2, respectively. Sample 1 had 74 successes, and sample 2 had 81 successes.

a. Suppose you have no preconceived idea as to which parameter, p_1 or p_2, is the larger, but you want to detect only a difference between the two parameters if one exists. What should you choose as the alternative hypothesis for a statistical test? The null hypothesis?

b. Calculate the standard error of the difference in the two sample proportions, $(\hat{p}_1 - \hat{p}_2)$. Make sure to use the pooled estimate for the common value of p.

c. Calculate the test statistic that you would use for the test in part a. Based on your knowledge of the standard normal distribution, is this a likely or unlikely observation, assuming that H_0 is true and the two population proportions are the same?

d. *p-value approach:* Find the p-value for the test. Test for a significant difference in the population proportions at the 1% significance level.

e. *Critical value approach:* Find the rejection region when $\alpha = 0.01$. Do the data provide sufficient evidence to indicate a difference in the population proportions?

9.51 Refer to Exercise 9.50. Suppose, for practical reasons, you know that p_1 cannot be larger than p_2.

a. Given this knowledge, what should you choose as the alternative hypothesis for your statistical test? The null hypothesis?

b. Does your alternative hypothesis in part a imply a one- or two-tailed test? Explain.

c. Conduct the test and state your conclusions. Test using $\alpha = 0.05$.

9.52 Independent random samples of 280 and 350 observations were selected from binomial populations 1 and 2, respectively. Sample 1 had 132 successes, and sample 2 had 178 successes. Do the data present sufficient evidence to indicate that the proportion of successes in population 1 is smaller than the proportion in population 2? Use one of the two methods of testing presented in this section, and explain your conclusions.

APPLICATIONS

9.53 Treatment versus Control An experiment was conducted to test the effect of a new drug on a viral infection. The infection was induced in 100 mice, and the mice were randomly split into two groups of 50. The first group, the *control group*, received no treatment for the infection. The second group received the drug. After a 30-day period, the proportions of survivors, $\hat{p}_1$ and $\hat{p}_2$, in the two groups were found to be 0.36 and 0.60, respectively.

a. Is there sufficient evidence to indicate that the drug is effective in treating the viral infection? Use $\alpha = 0.05$.

b. Use a 95% confidence interval to estimate the actual difference in the cure rates for the treated versus the control groups.

9.54 Tai Chi and Fibromyalgia A new study (Exercise 7.13) indicates that tai chi, an ancient Chinese practice of exercise and meditation, may relieve symptoms of chronic painful fibromyalgia. The study assigned 66 fibromyalgia patients to take either a 12-week tai chi class ($n_1 = 33$) or a wellness education class ($n_2 = 33$).[18] The results of the study are shown in the following table:

	Tai Chi	Wellness Education
Number who felt better at end of course	26	13

a. Is there a significant difference in the proportion of all fibromyalgia patients who would admit to feeling better after taking the tai chi class compared to the wellness education class? Use $\alpha = 0.01$.

b. Find the p-value for the test in part a. How would you describe the significance or non-significance of the test?

9.55 Movie Marketing Marketing to targeted age groups has become a standard method of advertising, even in movie theatre advertising. Advertisers use computer software to track the demographics of moviegoers and then decide on the type of products to advertise before a particular movie.[19] One statistic that might be of interest is how frequently adults with children under 18 attend movies as compared to those without children. Suppose that a theatre database is used to randomly select 1000 adult ticket purchasers. These adults are then surveyed and asked whether they were frequent moviegoers—that is, do they attend movies 12 or more times a year? The results are shown in the table:

	With Children under 18	Without Children
Sample size	440	560
Number who attend 12+ times per year	123	145

a. Is there a significant difference in the population proportions of frequent moviegoers in these two demographic groups? Use $\alpha = 0.01$.

b. Why would a statistically significant difference in these population proportions be of *practical importance* to the advertiser?

9.56 M&M'S® In Exercise 8.63, you investigated whether Mars, Inc., uses the same proportion of red M&M'S® in its plain and peanut varieties. Random samples of plain and peanut M&M'S® provide the following sample data for the experiment:

	Plain	Peanut
Sample size	56	32
Number of red M&M'S®	12	8

Use a test of hypothesis to determine whether there is a significant difference in the proportions of red candies for the two types of M&M'S® Let $\alpha = 0.05$ and compare your results with those of Exercise 8.63.

9.57 Hormone Therapy and Alzheimer's Disease In the last few years, many research studies have shown that the purported benefits of hormone replacement therapy (HRT) do not exist, and in fact, that hormone replacement therapy actually increases the risk of several serious diseases. A four-year experiment involving 4532 women, reported in *The Press Enterprise*, was conducted at 39 medical centres. Half of the women took placebos and half took *Prempro*, a widely prescribed type of hormone replacement therapy. There were 40 cases of dementia in the hormone group and 21 in the placebo group.[20] Is there sufficient evidence to indicate that the risk of dementia is higher for patients using *Prempro*? Test at the 1% level of significance.

9.58 HRT, continued Refer to Exercise 9.57. Calculate a 99% lower one-sided confidence bound for the difference in the risk of dementia for women using hormone replacement therapy versus those who do not. Would this difference be of *practical importance* to a woman considering HRT? Explain.

9.59 Clopidogrel and Aspirin A large study was conducted to test the effectiveness of clopidogrel in combination with aspirin in warding off heart attacks and strokes.[21] The trial involved more than 15,500 people 45 years of age or older from 32 countries. These people had been diagnosed with cardiovascular disease or had multiple risk factors. The subjects were randomly assigned to one of two groups. After two years, there was no difference in the risk of heart attack, stroke, or dying from heart disease between those who took clopidogrel and low-dose aspirin daily and those who took low-dose aspirin plus a dummy pill. The two-drug combination actually increased the risk of dying (5.4% versus 3.8%) or dying specifically from cardiovascular disease (3.9% versus 2.2%).

a. The subjects were randomly assigned to one of the two groups. Explain how you could use the random number table to make these assignments.

b. No sample sizes were given in the article: however, let us assume that the sample sizes for each group were $n_1 = 7720$ and $n_2 = 7780$. Determine whether the risk of dying was significantly different for the two groups.

c. What do the results of the study mean in terms of *practical significance*?

9.7 SOME COMMENTS ON TESTING HYPOTHESES

A statistical test of hypothesis is a fairly clear-cut procedure that enables an experimenter to either reject or accept the null hypothesis H_0, with measured risks α and β. The experimenter can control the risk of falsely rejecting H_0 by selecting an appropriate value of α. On the other hand, the value of β depends on the sample size and the values of the parameter under test that are of practical importance to the experimenter. When this information is not available, an experimenter may decide to select an affordable sample size, in the hope that the sample will contain sufficient information to reject the null hypothesis. The chance that this decision is in error is given by α, whose value has been set in advance. If the sample does not provide sufficient evidence to reject H_0, the experimenter may wish to state the results of the test as "The data do not support the rejection of H_0" rather than accepting H_0 without knowing the chance of error β.

Some experimenters prefer to use the observed p-value of the test to evaluate the strength of the sample information in deciding to reject H_0. These values can usually be generated by computer and are often used in reports of statistical results:

- If the p-value is greater than 0.05, the results are reported as NS—not significant at the 5% level.
- If the p-value lies between 0.05 and 0.01, the results are reported as $P < 0.05$—significant at the 5% level.
- If the p-value lies between 0.01 and 0.001, the results are reported as $P < 0.01$—"highly significant" or significant at the 1% level.
- If the p-value is less than 0.001, the results are reported as $P < 0.001$—"very highly significant" or significant at the 0.1% level.

Still other researchers prefer to construct a confidence interval for a parameter and perform a test informally. If the value of the parameter specified by H_0 is included within the upper and lower limits of the confidence interval, then "H_0 is not rejected." If the value of the parameter specified by H_0 is not contained within the interval, then "H_0 is rejected." These results will agree with a two-tailed test; one-sided confidence bounds are used for one-tailed alternatives.

Finally, consider the choice between a one- and two-tailed test. In general, experimenters wish to know whether a treatment causes what could be a beneficial increase in a parameter or what might be a harmful decrease in a parameter. Therefore, most tests are two-tailed unless a one-tailed test is strongly dictated by practical considerations. For example, assume you will sustain a large financial loss if the mean μ is greater than μ_0 but not if it is less. You will then want to detect values larger than μ_0 with a high probability and thereby use a right-tailed test. In the same vein, if pollution levels higher than μ_0 cause critical health risks, then you will certainly wish to detect levels higher than μ_0 with a right-tailed test of hypothesis. In any case, the choice of a one- or two-tailed test should be dictated by the practical consequences that result from a decision to reject or not reject H_0 in favour of the alternative.

CHAPTER REVIEW

Key Concepts and Formulas

I. Parts of a Statistical Test

1. **Null hypothesis:** a contradiction of the alternative hypothesis
2. **Alternative hypothesis:** the hypothesis the researcher wants to support
3. **Test statistic** and its ***p*-value:** sample evidence calculated from the sample data
4. **Rejection region—critical values** and **significance levels:** values that separate rejection and non-rejection of the null hypothesis
5. **Conclusion:** reject or do not reject the null hypothesis, stating the practical significance of your conclusion

II. Errors and Statistical Significance

1. The **significance level** α is the probability of rejecting H_0 when it is in fact true.
2. The ***p*-value** is the probability of observing a test statistic as extreme as or more extreme than the one observed; also, the smallest value of α for which H_0 can be rejected.
3. When the ***p*-value** is less than the **significance level** α, the null hypothesis is rejected. This happens when the **test statistic** exceeds the **critical value.**
4. In a **Type II error,** β is the probability of accepting H_0 when it is in fact false. The **power of the test** is $(1 - \beta)$, the probability of rejecting H_0 when it is false.

III. Large-Sample Test Statistics Using the z Distribution

To test one of the four population parameters when the sample sizes are large, use the following test statistics:

Parameter	Test Statistic
μ	$z = \dfrac{\bar{x} - \mu_0}{s/\sqrt{n}}$
p	$z = \dfrac{\hat{p} - p_0}{\sqrt{\dfrac{p_0 q_0}{n}}}$
$\mu_1 - \mu_2$	$z = \dfrac{(\bar{x}_1 - \bar{x}_2) - D_0}{\sqrt{\dfrac{s_1^2}{n_1} + \dfrac{s_2^2}{n_2}}}$
$p_1 - p_2$	$z = \dfrac{\hat{p}_1 - \hat{p}_2}{\sqrt{\hat{p}\hat{q}\left(\dfrac{1}{n_1} + \dfrac{1}{n_2}\right)}}$ or $z = \dfrac{(\hat{p}_1 - \hat{p}_2) - D_0}{\sqrt{\dfrac{\hat{p}_1\hat{q}_1}{n_1} + \dfrac{\hat{p}_2\hat{q}_2}{n_2}}}$

Supplementary Exercises

Starred (*) exercises are optional.

9.60 a. Define α and β for a statistical test of hypothesis.

b. For a fixed sample size n, if the value of α is decreased, what is the effect on β?

c. In order to decrease both α and β for a particular alternative value of μ, how must the sample size change?

9.61 What is the p-value for a test of hypothesis? How is it calculated for a large-sample test?

9.62 What conditions must be met so that the z test can be used to test a hypothesis concerning a population mean μ?

9.63 Define the power of a statistical test. As the alternative value of μ gets farther from μ_0, how is the power affected?

9.64 Acidity in Rainfall Refer to Exercise 8.36 and the collection of water samples to estimate the mean acidity (in pH) of rainfalls in Eastern Canada. As noted, the pH for pure rain falling through clean air is approximately 5.7. The sample of $n = 40$ rainfalls produced pH readings with $\bar{x} = 3.7$ and $s = 0.5$. Do the data provide sufficient evidence to indicate that the mean pH for rainfalls is more acidic ($H_a : \mu < 5.7$ pH) than pure rainwater? Test using $\alpha = 0.05$. Note that this inference is appropriate only for the area in which the rainwater specimens were collected.

9.65 Washing Machine Colours A manufacturer of automatic washers provides a particular model in one of three colours. Of the first 1000 washers sold, it is noted that 400 were of the first colour. Can you conclude that more than one-third of all customers have a preference for the first colour?

a. Find the p-value for the test.

b. If you plan to conduct your test using $\alpha = 0.05$, what will be your test conclusions?

9.66 Commercials in Space The commercialism of our space program[22] was the topic of Exercise 8.69. In a survey of 500 men and 500 women, 20% of the men and 26% of the women responded that space should remain commercial-free.

a. Is there a significant differencc in the population proportions of men and women who think that space should remain commercial-free? Use $\alpha = 0.01$.

b. Can you think of any reason why a statistically significant difference in these population proportions might be of *practical importance* to the administrators of the space program? To the advertisers? To the politicians?

9.67 Bass Fishing The pH factor is a measure of the acidity or alkalinity of water. A reading of 7.0 is neutral; values in excess of 7.0 indicate alkalinity; those below 7.0 imply acidity. Loren Hill states that the best chance of catching bass occurs when the pH of the water is in the range 7.5 to 7.9.[23] Suppose you suspect that acid rain is lowering the pH of your favourite fishing spot and you wish to determine whether the pH is less than 7.5.

a. State the alternative and null hypotheses that you would choose for a statistical test.

b. Does the alternative hypothesis in part a imply a one- or a two-tailed test? Explain.

c. Suppose that a random sample of 30 water specimens gave pH readings with $\bar{x} = 7.3$ and $s = 0.2$. Just glancing at the data, do you think that the difference $\bar{x} - 7.5 = -0.2$ is large enough to indicate that the mean pH of the water samples is less than 7.5? (Do *not* conduct the test.)

d. Now conduct a statistical test of the hypotheses in part a and state your conclusions. Test using $\alpha = 0.05$. Compare your statistically based decision with your intuitive decision in part c.

9.68 Traffic Tickets An Ontario website dealing with traffic tickets claims "We fully dismiss at least 94% of all cases."

a. Suppose that we examine 25 cases and exactly 15 were fully dismissed. Would you reject the claim of 94% or more made by the website?

b. Suppose that the records for the past year indicate that of 442 cases, 260 cases were fully dismissed. If this year is typical of all years, find a 95% confidence interval for p, the true proportion of fully dismissed cases.

c. Using the result of part b, are you willing to reject a figure of 94% or greater for the true proportion of fully dismissed cases? Explain.

9.69 White-Tailed Deer In an article entitled "A Strategy for Big Bucks," Charles Dickey discusses studies of the habits of white-tailed deer that indicate that they live and feed within very limited ranges—approximately 607,028 to 829,605 square metres (m^2).[24] To determine whether there was a difference between the ranges of deer located in two different geographic areas, 40 deer were caught, tagged, and fitted with small radio transmitters. Several months later, the deer were tracked and identified, and the distance x from the release point was recorded. The mean and standard deviation of the distances from the release point were as follows:

	Location 1	Location 2
Sample size	40	40
Sample mean	908 m	976 m
Sample standard deviation	347 m	293 m

a. If you have no preconceived reason for believing one population mean is larger than another, what would you choose for your alternative hypothesis? Your null hypothesis?

b. Does your alternative hypothesis in part a imply a one- or a two-tailed test? Explain.

c. Do the data provide sufficient evidence to indicate that the mean distances differ for the two geographic locations? Test using $\alpha = 0.05$.

9.70 Female Models In a study to assess various effects of using a female model in automobile advertising, 100 men were shown photographs of two automobiles matched for price, colour, and size, but of different makes. One of the automobiles was shown with a female model to 50 of the men (group A), and both automobiles were shown without the model to the other 50 men (group B). In group A, the automobile shown with the model was judged as more expensive by 37 men; in group B, the same automobile was judged as the more expensive by 23 men. Do these results indicate that using a female model influences the perceived cost of an automobile? Use a one-tailed test with $\alpha = 0.05$.

9.71 Bolts Random samples of 200 bolts manufactured by a type A machine and 200 bolts manufactured by a type B machine showed 16 and 8 defective bolts, respectively. Do these data present sufficient evidence to suggest a difference in the performance of the machine types? Use $\alpha = 0.05$.

9.72 Biomass Exercise 7.74 reported that the biomass for tropical woodlands, thought to be about 35 kilograms per square metre (kg/m^2), may in fact be too high and that tropical biomass values vary regionally—from about 5 to 55 kg/m^2.[25] Suppose you measure the tropical biomass in 400 randomly selected square-metre plots and obtain $\bar{x} = 31.75$ and $s = 10.5$. Do the data present sufficient evidence to indicate that scientists are overestimating the mean biomass for tropical woodlands and that the mean is in fact lower than estimated?

a. State the null and alternative hypotheses to be tested.

b. Locate the rejection region for the test with $\alpha = 0.01$.

c. Conduct the test and state your conclusions.

9.73 Anti-Terrorism Bill C-36 The Anti-Terrorism Bill was passed by the House of Commons on November 28, 2001. Many people argued that the law increased the burden for charities. A researcher believed that the fraction p_1 of Conservatives in favour of the Anti-Terrorism Bill was greater than fraction p_2 of New Democrats. The researcher acquired an independent random sample of 200 Conservatives and 200 New Democrats and found 136 Conservatives and 124 New Democrats favouring the Anti-Terrorism Bill. Do these data support the researcher's belief?

a. Find the p-value for the test.

b. If you plan to conduct your test using $\alpha = 0.05$, what will be your test conclusions?

9.74 Anti-Terrorism Bill C-36, continued Refer to Exercise 9.73. Some thought should have been given to designing a test for which β is tolerably low when p_1 exceeds p_2 by an important amount. For example, find a common sample size n for a test with $\alpha = 0.05$ and $\beta \le 0.20$, when in fact p_1 exceeds p_2 by 0.1. (HINT: The maximum value of $p(1 - p)$ is 0.25.)

9.75 Losing Weight In a comparison of the mean one-month weight losses for women aged 20 to 30 years, these sample data were obtained for each of two diets:

	Diet I	Diet II
Sample size n	40	40
Sample mean $\bar{x}$	4.54 kg	3.63 kg
Sample variance s^2	1.95 kg	2.59 kg

Do the data provide sufficient evidence to indicate that diet I produces a greater mean weight loss than diet II? Use $\alpha = 0.05$.

9.76 Increased Yield An agronomist has shown experimentally that a new irrigation/fertilization regimen produces an increase of 17,635 cm^3 per quadrant (significant at the 1% level) when compared with the regimen currently in use. The cost of implementing and using the new regimen will not be a factor if the increase in yield exceeds 26,453 cm^3 per quadrant. Is statistical significance the same as practical importance in this situation? Explain.

9.77 Breaking Strengths of Cables A test of the breaking strengths of two different types of cables was conducted using samples of $n_1 = n_2 = 100$ pieces of each type of cable.

Cable I	Cable II
$\bar{x}_1 = 1925$	$\bar{x}_2 = 1905$
$s_1 = 40$	$s_2 = 30$

Do the data provide sufficient evidence to indicate a difference between the mean breaking strengths of the two cables? Use $\alpha = 0.05$.

9.78 Put On the Brakes The braking ability was compared for two 2012 automobile models. Random samples of 64 automobiles were tested for each type. The recorded measurement was the distance (in metres) required to stop when the brakes were applied at 65 kilometres per hour (km/hr). These are the computed sample means and variances:

Model I	Model II
$\bar{x}_1 = 36$	$\bar{x}_2 = 33$
$s_1^2 = 31$	$s_2^2 = 27$

Do the data provide sufficient evidence to indicate a difference between the mean stopping distances for the two models?

9.79 Spraying Fruit Trees A fruit grower wants to test a new spray that a manufacturer claims will *reduce* the loss due to insect damage. To test the claim, the grower sprays 200 trees with the new spray and 200 other trees with the standard spray. The following data were recorded:

	New Spray	Standard Spray
Mean yield per tree $\bar{x}$ (kg)	109	103
Variance s^2	445	372

a. Do the data provide sufficient evidence to conclude that the mean yield per tree treated with the new spray exceeds that for trees treated with the standard spray? Use $\alpha = 0.05$.

b. Construct a 95% confidence interval for the difference between the mean yields for the two sprays.

9.80 Actinomycin D A biologist hypothesizes that high concentrations of actinomycin D inhibit RNA synthesis in cells and hence the production of proteins as well. An experiment conducted to test this theory compared the RNA synthesis in cells treated with two concentrations of actinomycin D: 0.6 and 0.7 microgram per millilitre. Cells treated with the lower concentration (0.6) of actinomycin D showed that 55 out of 70 developed normally, whereas only 23 out of 70 appeared to develop normally for the higher concentration (0.7). Do these data provide sufficient evidence to indicate a difference between the rates of normal RNA synthesis for cells exposed to the two different concentrations of actinomycin D?

a. Find the p-value for the test.

b. If you plan to conduct your test using $\alpha = 0.05$, what will be your test conclusions?

9.81 A Maze Experiment In a maze-running study, a rat is run in a T maze and the result of each run recorded. A reward in the form of food is always placed at the right exit. If learning is taking place, the rat will choose the right exit more often than the left. If no learning is taking place, the rat should randomly choose either exit. Suppose that the rat is given $n = 100$ runs in the maze and that he chooses the right exit $x = 64$ times. Would you conclude that learning is taking place? Use the p-value approach, and make a decision based on this p-value.

9.82 PCBs Polychlorinated biphenyls (PCBs) have been found to be dangerously high in some game birds found along the marshlands of the southeastern coast of North America. A concentration of PCBs higher than 5 parts per million (ppm) in these game birds is considered to be dangerous for human consumption. A sample of 38 game birds produced an average of 7.2 ppm with a standard deviation of 6.2 ppm. Is there sufficient evidence to indicate that the mean ppm of PCBs in the population of game birds exceeds the U.S. FDA's recommended limit of 5 ppm? Use $\alpha = 0.01$.

9.83* PCBs, continued Refer to Exercise 9.82.

a. Calculate β and $1 - \beta$ if the true mean ppm of PCBs is 6 ppm.

b. Calculate β and $1 - \beta$ if the true mean ppm of PCBs is 7 ppm.

c. Find the power, $1 - \beta$, when $\mu = 8, 9, 10$, and 12. Use these values to construct a power curve for the test in Exercise 9.82.

d. For what values of μ does this test have power greater than or equal to 0.90?

9.84 Stricter Emission Standards An Ipsos Reid survey conducted on behalf of Can West/Global Television shows that 68% of Canadians support the view that as Canada's largest producer of oil and gas, Alberta should be subjected to stricter emission standards even if it means a significant increase to the cost of producing oil and gas.[26] To test this claim, a random sample of 100 Canadians is selected by a researcher, and 65 of them support stricter emission standards in Alberta. Is there significant evidence to indicate that the percentage reported by the survey is correct? Test at the 5% level. Interestingly, another Ipsos Reid survey conducted on behalf of Can West/Global Television reveals that 23% of Canadians would consider relocating to Alberta for a 25% pay increase.[27](Source: Copyright © 2016 Ipsos in Canada)

9.85 Heights and Gender It is a well-accepted fact that males are taller on average than females. But how much taller? The heights and genders of 105 biomedical students were recorded and the data are summarized below:

	Males	Females
Sample size	48	77
Sample mean (cm)	176.7	163.7
Sample standard deviation	6.7	6.6

a. Perform a test of hypothesis to either confirm or refute our initial claim that males are taller on average than females. Use $\alpha = 0.01$.

b. If the results of part a show that our claim was correct, construct a 99% confidence one-sided lower confidence bound for the average difference in heights between male and female students. How much taller are males than females?

9.86 Right Number of Immigrants The Strategic Counsel conducted a survey in 2005 for the *Globe and Mail*/CTV August polling program. The following question was posed to randomly selected Canadians. "*Does Canada accept the right number of immigrants per year?*" The results are shown in the table:[28]

	Total % $n = 1000$	Rest of Canada $n_1 = 753$	Ontario $n_2 = 380$	Quebec $n_3 = 247$
Too many	32	33	38	30
About the right number	46	45	42	48
Too few	10	9	7	14
Other	12	13	13	8

Does this data provide sufficient statistical evidence to indicate that the percentage of Quebecers who said "too many" differ from the opinion of the rest of Canada on the same question? Test using $\alpha = 0.01$.

9.87 Breaststroke Swimmers How much training time does it take to become a world-class breaststroke swimmer? A survey published in *The American Journal of Sports Medicine* reported the number of metres per week swum by two groups of swimmers—those who competed only in breaststroke and those who competed in the individual medley (which includes breaststroke). The number of metres per week practising the breaststroke swim was recorded and the summary statistics are shown below.[29]

	Breaststroke	Individual Medley
Sample size	130	80
Sample mean (m)	9017	5853
Sample standard deviation	7162	1961

Is there sufficient evidence to indicate a difference in the average number of metres swum by these two groups of swimmers? Test using $\alpha = 0.01$.

9.88 Breaststroke, continued Refer to Exercise 9.87.

a. Construct a 99% confidence interval for the difference in the average number of metres swum by breaststroke versus individual medley swimmers.

b. How much longer do pure breaststroke swimmers practise that stroke than individual medley swimmers? What is the practical reason for this difference?

CASE STUDY 

Cure for the Cold—Pooling Data: Making Sense or Folly?

Edmonton company CV Technologies Inc. has conducted clinical trials, with results published in the *Journal of the American Geriatrics Society,* showing that its proprietary ginseng extract can prevent colds. It obtained results that indicated a reduction in laboratory-confirmed respiratory illness (colds and flu) of 89%, and the results were statistically significant.[30] However, on February 25, 2006, the *Vancouver Sun* published an article in which two professors from the University of British Columbia criticized the claims, accusing the article's authors of "data-mining," and saying that the trials were not definitive evidence that the product had any effect.[31]

The study consisted of two randomized clinical trials, one conducted in 2000, the second in 2001, with nursing-home patients as subjects. In each trial, the subjects were randomly assigned to take either 200 mg of the ginseng extract or a placebo twice daily. The trials were conducted as double-blind studies, in which neither the participants nor the investigators responsible for following the participants knew to which group a participant belonged. The results are given in the following table:

	Trial 1		Trial 2	
	Ginseng	Placebo	Ginseng	Placebo
Sample size	40	49	57	52
Symptoms of respiratory illness	15	18	18	18
Laboratory-confirmed influenza or RSV	0	3	1	6

The original purpose of the studies was to see whether the ginseng extract would reduce the incidence of respiratory illnesses as defined by symptoms such as cough, sore throat, and runny nose. A secondary purpose of the studies was to measure the difference in the incidence of laboratory-confirmed respiratory illness (influenza or respiratory syncytial virus) between the two groups. The researchers, reporting the results in the *Journal of the American Geriatrics Society*, found "no significant difference between the placebo and the [ginseng extract] groups for the number of [acute respiratory illnesses] defined by symptoms." They also found "no significant difference in the severity or duration of symptoms related to [acute respiratory illnesses] between the two groups in either study."

However, when they looked at laboratory-confirmed illness, there did appear to be a difference in the placebo groups—6 and 12% of the subjects in the two studies contracted laboratory-confirmed illnesses. In the ginseng groups, the percentages were 0 and 2%. These results were not statistically significant. However, when the researchers pooled the data from the two studies, they did get statistically significant results.

The professors had two main criticisms of the studies. First, they argued that symptomatic respiratory illness was a more "relevant endpoint," saying of laboratory-confirmed illness that "we don't look for that in clinical practice… That's why they [the researchers] [originally] picked symptoms-based outcomes as their endpoint."

Their second criticism was that combining the two studies erodes the credibility of the results: "Taking two studies that don't show a benefit and then adding them together to get a positive result is a form of data-mining. It's torturing the data until it confesses." The professors argued that if the original intent had been to combine the results of the two studies, then it would be a legitimate technique, but if not, it might seem that "the researchers did a second study because they didn't like the initial results."

1. Test whether each study, analyzed separately, supports the hypothesis that the ginseng extract reduces the likelihood of laboratory-confirmed influenza.
2. Repeat your analysis using the pooled data from the two studies.
3. Should it matter to the analysis whether the original endpoint was symptomatic or laboratory-confirmed respiratory illness? How about whether or not the researchers originally intended to conduct two studies and pool the results? Does the data care what the researchers were thinking?

PROJECTS

Project 9-A: Proportion of "Cured" Cancer Patients: How Does Canada Compare with Europe?[32]

Lung cancer remains the leading cause of cancer death for both Canadian men and women, responsible for the most potential years of life lost to cancer. Lung cancer alone accounts for 28% of all cancer deaths in Canada (32%. in Quebec). Most forms of lung cancer start insidiously and produce no apparent symptoms until they are too far advanced. Consequently, the chances of being cured of lung cancer are not very promising, with the five-year survival rate being less than 15%. The overall data for Europe show that the number of patients who are considered "cured" is rising steadily. For lung cancer, this proportion rose from 6% to 8%. However, there was a wide variation in the proportion of patients cured in individual European countries. For instance, the study shows that for lung cancer, less than 5% of patients were cured in Denmark, the Czech Republic, and Poland, whereas more than 10% of patients were cured in Spain.

a. Suppose a sample of 75 Quebecers was selected and it was found that 27 of them had died due to lung cancer.

   (i) Perform the appropriate test of hypothesis to determine whether the proportion of deaths in Quebec due to the leading cause (lung cancer) has changed. Test using $\alpha = 0.05$.

   (ii) Find the p-value for this test and interpret it.

   (iii) Construct a 95% confidence interval estimate of the population proportion of the premature cancer death in Quebec.

   (iv) Explain how to use this confidence interval to test the hypotheses.

b. Suppose two independent samples were taken. The following data were recorded:

Quebec:	$n_1 = 150$	Number of deaths due to cancer $= x_1 = 47$
Rest of Canada:	$n_2 = 1000$	Number of deaths due to cancer $= x_2 = 291$

   (i) Suppose the scientists have no preconceived theory concerning which proportion parameter is the larger and they wish to detect only a difference between the two parameters, if it exists. What should they choose as the null and alternative hypotheses for a statistical test?

   (ii) What type of error could occur in testing the null hypothesis in (i), if H_0 is false?

   (iii) Calculate the standard error of difference between the two sample proportions. Make sure to use the pooled estimate for the common value of true proportion.

(iv) Calculate the test statistic that you would use for the test in (i). Based on your knowledge of the standard normal distribution, is this a likely or unlikely observation, assuming that H_0 is true and the two population proportions are the same?

(v) Find the p-value for the test. Test for a significant difference between the population proportions at the 1% significance level.

(vi) Find the rejection region when $\alpha = 0.01$. Using critical value approach, determine whether the data provide sufficient evidence to indicate a difference between the population proportions.

(vii) Use a 95% confidence interval to estimate the actual difference between the cancer death proportions for the people in Quebec versus rest of the Canada. Summarize your findings.

c. To test the claim regarding proportion of cured cancer patients in Spain, a random sample of 500 patients yielded 39 who were cured.

(i) State the appropriate null and alternative hypotheses.

(ii) Calculate the value of the test statistic.

(iii) Calculate the p-value and write your conclusion.

Project 9-B: Walking and Talking: My Favourite Sport[33]

Walking is perhaps the simplest exercise you can do anytime it fits in your busy schedule. Further, it can be viewed as one of the most economical methods of transportation. More importantly, walking reduces the risk of coronary disease. Walking speed varies quite a bit, depending on stride length, terrain or walking surface, and even one's own physical condition, age, and sex. On average, women walk at 4.8 kilometres per hour (km/h) on reasonably flat terrain. At this pace, the average woman should be able to easily carry on a conversation. For the average man, the speed is a little quicker, at 5.6 km/h, mainly due to his longer legs and thus longer stride length. Also, men typically have more muscle mass, which increases speed because it is muscles which do the work to move the body. If you walk regularly (three times a week for 30 minutes or more), you will gradually increase your personal average speed because you will increase your cardiovascular endurance and you will build strength (or muscle) particularly in your legs. Both of these factors will lead to walking faster. A digital pedometer will tell you your walking speed so you don't have to do the math! But you need to do statistics!

a. A random sample of 40 average Canadian women is selected and their walking distance for a one-hour period is recorded. The mean and standard deviation are found to be 4.5 and 0.4 km/h, respectively.

(i) Perform the appropriate test of hypothesis to determine whether the average walking speed is less than the stated average for women. Use a 0.01 level of significance.

(ii) What effect, if any, would there be on the conclusion of the test of hypothesis in the first question if you changed α to 0.20?

(iii) Find the p-value for this test in part (i).

(iv) Describe the Type I and Type II errors for this testing problem.

(v) Compute the power of the test if the actual mean walking distance is 4.7 km/h. Interpret the results.

(vi) Is the assumption of normality required to perform the test in part (i) or to calculate the power in part (iv)? Explain.

b. Two independent samples of average Canadian men and women are taken and their walking speeds (km/h) are observed. The results of random sampling are recorded below:

Men	$n_1 = 81$	$\bar{x}_1 = 4.9$	$s_1 = 0.5$
Women	$n_2 = 64$	$\bar{x}_2 = 5.4$	$s_2 = 0.2$

(i) Perform the appropriate test of hypothesis to determine whether there is a significant difference in average walking speed (km/h) between Canadian men and women. Use the *p*-value approach and the critical value approach and explain your conclusion.

(ii) Find a 99% lower confidence bound for the difference in the average walking speed (km/h) for the two groups. Does your confidence bound confirm your conclusions in the previous question? Explain. What additional information does the confidence bound give you?

Inference from Small Samples

Mike Norkum/Shutterstock.com

Does Bait Type Affect the Visit of the American Marten in Ontario?

The American marten has become a focal species for the conservation of forested landscapes throughout North America primarily due to its association with older forests and its sensitivity to human disturbance. Are the martens more attracted to chicken? Or lured by peanut butter? The answers to these questions are posed in the case study at the end of this chapter.

GENERAL OBJECTIVES

The basic concepts of large-sample statistical estimation and hypothesis testing for practical situations involving population means and proportions were introduced in Chapters 8 and 9. Because all of these techniques rely on the Central Limit Theorem to justify the normality of the estimators and test statistics, they apply only when the samples are large. This chapter supplements the large-sample techniques by presenting small-sample tests and confidence intervals for population means and variances. Unlike their large-sample counterparts, these small-sample techniques require the sampled populations to be normal, or approximately so.

CHAPTER INDEX

NEED TO KNOW

How to Decide Which Test to Use

INTRODUCTION

Suppose you need to run an experiment to estimate a population mean or the difference between two means. The process of collecting the data may be very expensive or very time-consuming. If you cannot collect a *large sample,* the estimation and test procedures of Chapters 8 and 9 are of no use to you.

This chapter introduces some equivalent statistical procedures that can be used when the *sample size is small.* The estimation and testing procedures involve these familiar parameters:

- A single population mean, μ
- The difference between two population means, $(\mu_1 - \mu_2)$
- A single population variance, σ^2
- The comparison of two population variances, σ_1^2 and σ_2^2

Small-sample tests and confidence intervals for binomial proportions will be omitted from our discussion.

STUDENT'S *t* DISTRIBUTION

In conducting an experiment to evaluate a new but very costly process for producing synthetic diamonds, you are able to study only six diamonds generated by the process. How can you use these six measurements to make inferences about the average weight μ of diamonds from this process?

In discussing the sampling distribution of $\bar{x}$ in Chapter 7, we made these points:

- When the original sampled population is normal, $\bar{x}$ and $z = (\bar{x} - \mu)/(\sigma/\sqrt{n})$ both have normal distributions, *for any sample size.*
- When the original sampled population is *not* normal, $\bar{x}$, $z = (\bar{x} - \mu)/(\sigma/\sqrt{n})$, and $z \approx (\bar{x} - \mu)/(\sigma/\sqrt{n})$ all have approximately normal distributions, if the sample size is *large.*

NEED A TIP?

When $n < 30$, the Central Limit Theorem will not guarantee that

$$\frac{\bar{x} - \mu}{s/\sqrt{n}}$$

is approximately normal.

Unfortunately, when the sample size n is small, the statistic $(\bar{x} - \mu)/(s/\sqrt{n})$ *does not* have a normal distribution. Therefore, all the critical values of z that you used in Chapters 8 and 9 are no longer correct. For example, you *cannot say* that $\bar{x}$ will lie within 1.96 standard errors of μ 95% of the time.

This problem is not new; it was studied by statisticians and experimenters in the early 1900s. To find the sampling distribution of this statistic, there are two ways to proceed:

- Use an empirical approach. Draw repeated samples and compute $(\bar{x} - \mu)/(s/\sqrt{n})$ for each sample. The relative frequency distribution that you construct using these values will approximate the shape and location of the sampling distribution.
- Use a mathematical approach to derive the actual density function or curve that describes the sampling distribution.

This second approach was used by an Englishman named W.S. Gosset in 1908. He derived a complicated formula for the density function of

$$t = \frac{\bar{x} - \mu}{s/\sqrt{n}}$$

for random samples of size n from a normal population, and he published his results under the pen name "Student." Ever since, the statistic has been known as **Student's *t*.** It has the following characteristics:

- It is mound-shaped and symmetric about $t = 0$, just like z.
- It is more variable than z, with "heavier tails"; that is, the t curve does not approach the horizontal axis as quickly as z does. This is because the t statistic involves two random quantities, $\bar{x}$ and s, whereas the z statistic involves only the sample mean, $\bar{x}$. You can see this phenomenon in Figure 10.1.
- The shape of the t distribution depends on the sample size n. As n increases, the variability of t decreases because the estimate s of σ is based on more and more information. Eventually, when n is infinitely large, the t and z distributions are identical!

FIGURE 10.1
Standard normal z and the t distribution with 5 degrees of freedom

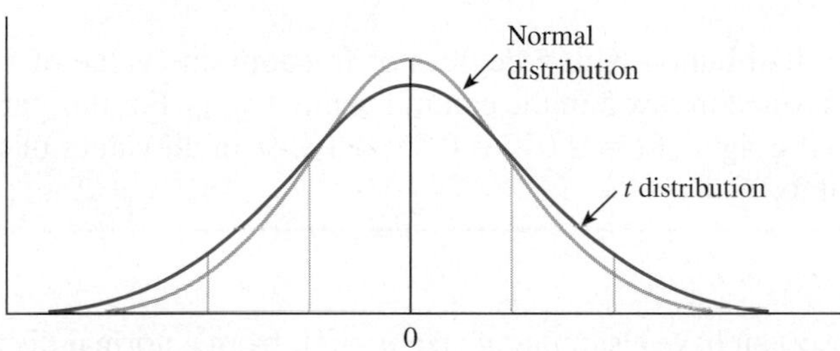

NEED A TIP?

For a one-sample t, $df = n - 1$.

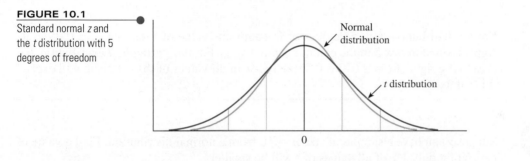

The divisor $(n - 1)$ in the formula for the sample variance s^2 is called the **number of degrees of freedom (*df*) associated with s^2.** It determines the *shape* of the t distribution. The origin of the term *degrees of freedom* is theoretical and refers to the number of independent squared deviations in s^2 that are available for estimating σ^2. These degrees of freedom may change for different applications, and since they specify the correct t distribution to use, you need to remember to calculate the correct degrees of freedom for each application.

The table of probabilities for the standard normal z distribution is no longer useful in calculating critical values or p-values for the t statistic. Instead, you will use Table 4 in Appendix I, which is partially reproduced in Table 10.1. When you index a particular number of degrees of freedom, the table records t_α, a value of t that has tail area α to its right, as shown in Figure 10.2.

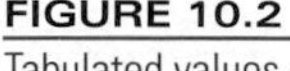

FIGURE 10.2
Tabulated values of Student's t

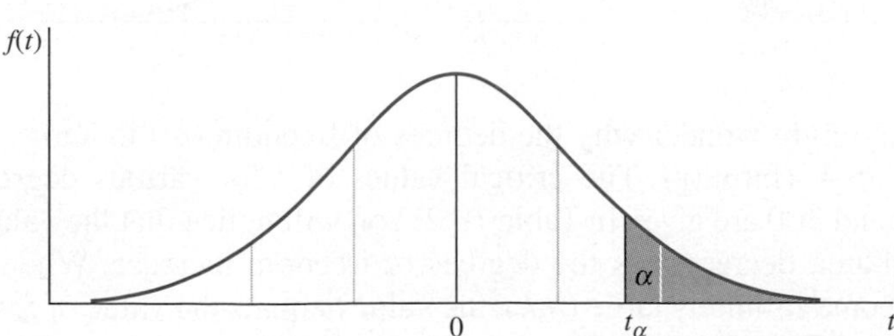

TABLE 10.1 **Format of the Student's *t* Table from Table 4 in Appendix I**

ONLINE APPLET
Student's *t* Probabilities
Comparing *t* and *z*

df	$t_{0.100}$	$t_{0.050}$	$t_{0.025}$	$t_{0.010}$	$t_{0.005}$	*df*
1	3.078	6.314	12.706	31.821	63.657	1
2	1.886	2.920	4.303	6.965	9.925	2
3	1.638	2.353	3.182	4.541	5.841	3
4	1.533	2.132	2.776	3.747	4.604	4
5	1.476	2.015	2.571	3.365	4.032	5
6	1.440	1.943	2.447	3.143	3.707	6
7	1.415	1.895	2.365	2.998	3.499	7
8	1.397	1.860	2.306	2.896	3.355	8
9	1.383	1.833	2.262	2.821	3.250	9
.	.	.	.	.	.	.
.	.	.	.	.	.	.
.	.	.	.	.	.	.
26	1.315	1.706	2.056	2.479	2.779	26
27	1.314	1.703	2.052	2.473	2.771	27
28	1.313	1.701	2.048	2.467	2.763	28
29	1.311	1.699	2.045	2.462	2.756	29
inf.	1.282	1.645	1.960	2.326	2.576	inf.

EXAMPLE 
10.1

For a t distribution with 5 degrees of freedom, the value of t that has area 0.05 to its right is found in row 5 in the column marked $t_{0.050}$. For this particular t distribution, the area to the right of $t = 2.015$ is 0.05; only 5% of all values of the t statistic will exceed this value.

EXAMPLE 
10.2

Suppose you have a sample of size $n = 10$ from a normal distribution. Find a value of t such that only 1% of all values of t will be smaller.

SOLUTION The degrees of freedom that specify the correct t distribution are $df = n - 1 = 9$, and the necessary t-value must be in the lower portion of the distribution, with area 0.01 to its left, as shown in Figure 10.3. Since the t distribution is symmetric about 0, this value is simply the negative of the value on the right-hand side with area 0.01 to its right, or $-t_{0.01} = -2.821$.

FIGURE 10.3
t Distribution for Example 10.2

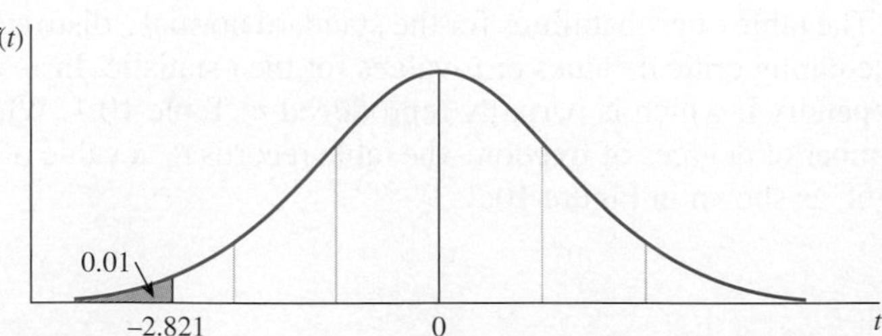

You might wonder why the degrees of freedom (df) in Table 10.1 jump from $df = 29$ to $df =$ (infinity). The critical values of t for various degrees of freedom between 29 and 300 are given in Table 10.2. You will notice that the value of t for the same right-tail area decreases as the degrees of freedom increase. When the degrees of freedom become infinitely large (*inf.*), the value t equals the value of z, which is given in the last row of Table 10.2.

TABLE 10.2 ● **Critical Values of Student's *t* for Degrees of Freedom between *df* = 29 and *df* = *infinity***

	Right-Tail Area		
df	0.05	0.025	0.01
29	1.699	2.045	2.462
49	1.677	2.010	2.405
69	1.667	1.995	2.382
100	1.660	1.984	2.364
200	1.653	1.972	2.345
300	1.650	1.968	2.339
inf.	1.645	1.96	2.326

At the same time, as the degrees of freedom increase, the shape of the t distribution becomes less variable until it ultimately looks like (and is) the standard normal distribution. Notice that when the degrees of freedom with t are $df = 300$, there is almost no difference. When $df = 29$ and $n = 30$, the critical values of t are quite close to their normal counterparts; this may explain why we use the arbitrary dividing line between large and small sample as $n = 30$. Rather than produce a t-table with many more critical values, the critical values of z are sufficient when n reaches 30.

Assumptions behind Student's *t* Distribution

The critical values of t allow you to make reliable inferences *only if* you follow all the rules; that is, your sample must meet these requirements specified by the t distribution:

Assumptions for one-sample *t*:
- Random sample
- Normal distribution

- The sample must be randomly selected.
- The population from which you are sampling must be normally distributed.

These requirements may seem quite restrictive. How can you possibly know the shape of the probability distribution for the entire population if you have only a sample? If this were a serious problem, however, the t statistic could be used in only very limited situations. Fortunately, the shape of the t distribution is not affected very much as long as the sampled population has an *approximately mound-shaped* distribution. Statisticians say that the t statistic is **robust,** meaning that the distribution of the statistic does not change significantly when the normality assumption is violated.

How can you tell whether your sample is from a normal population? Although there are statistical procedures designed for this purpose, the easiest and quickest way to check for normality is to use the graphical techniques of Chapter 2: Draw a dotplot or construct a stem and leaf plot. As long as your plot tends to "mound up" in the centre, you can be fairly safe in using the t statistic for making inferences.

The random sampling requirement, on the other hand, is quite critical if you want to produce reliable inferences. If the sample is not random, or if it does not at *least behave as* a random sample, then your sample results may be affected by some unknown factor and your conclusions may be incorrect. When you design an experiment or read about experiments conducted by others, look critically at the way the data have been collected!

10.2 EXERCISES

BASIC TECHNIQUES

10.1 What are the assumptions to use a t-distribution?

10.2 Let $n = 9$, which is taken from a normal population. Find the value of t such that only 5% of all values of t will be the smallest.

10.3 Comment on the slope of the t-distribution and its implication if assumption is violated.

10.4 For $n = 5$, use Table 4 in Appendix I to find $t_{0.100}$, $t_{0.025}$, and $t_{0.005}$. Comment on the magnitude of the values. Why are the values increasing? What assumptions are necessary to use the table?

10.5 Comment on the shape of the t-distribution when the sample size increases beyond 50.

10.6 Use Table 4 of Appendix I to find the t-value that has 0.1 probability above it.

a. 10 degrees of freedom

b. 20 degrees of freedom

c. 30 degrees of freedom

10.3 SMALL-SAMPLE INFERENCES CONCERNING A POPULATION MEAN

As with large-sample inference, small-sample inference can involve either **estimation** or **hypothesis testing,** depending on the preference of the experimenter. We explained the basics of these two types of inference in the earlier chapters, and we use them again now, with a different sample statistic, $t = (\bar{x} - \mu)/(s/\sqrt{n})$, and a different sampling distribution, the Student's t, with $(n - 1)$ degrees of freedom.

SMALL-SAMPLE HYPOTHESIS TEST FOR μ

1. Null hypothesis: $H_0 : \mu = \mu_0$
2. Alternative hypothesis:

One-Tailed Test	Two-Tailed Test
$H_a : \mu > \mu_0$ (or, $H_a : \mu < \mu_0$)	$H_a : \mu \neq \mu_0$

3. Test statistic: $t = \left(\dfrac{\bar{x} - \mu_0}{s/\sqrt{n}}\right)$
4. Rejection region: Reject H_0 when

One-Tailed Test	Two-Tailed Test
$t > t_\alpha$ (or $t < -t_\alpha$ when the alternative hypothesis is $H_a : \mu < \mu_0$) or when p-value $< \alpha$	$t > t_{\alpha/2}$ or $t < -t_{\alpha/2}$

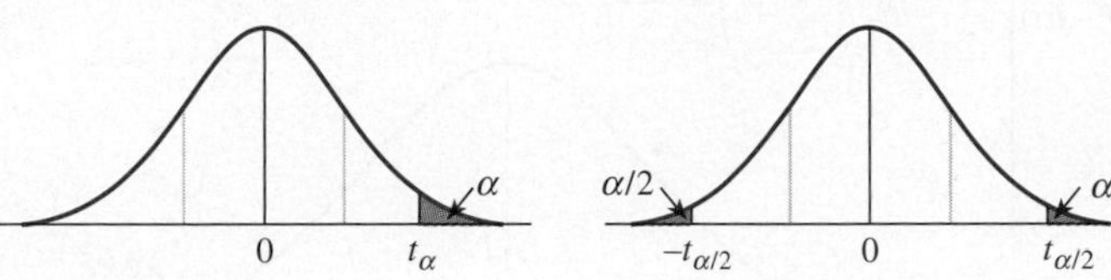

The critical values of t, t_α, and $t_{\alpha/2}$ are based on $(n - 1)$ degrees of freedom. These tabulated values can be found using Table 4 of Appendix I.

Assumption: The sample is randomly selected from a normally distributed population.

SMALL-SAMPLE $(1 - \alpha)100\%$ CONFIDENCE INTERVAL FOR μ

$$\bar{x} \pm t_{\alpha/2}\frac{s}{\sqrt{n}}$$

where $s/\sqrt{n}$ is the estimated standard error of $\bar{x}$, often referred to as the **standard error of the mean.**

EXAMPLE 10.3

A new process for producing synthetic diamonds can be operated at a profitable level only if the average weight of the diamonds is greater than 0.5 karat. To evaluate the profitability of the process, six diamonds are generated, with recorded weights 0.46, 0.61, 0.52, 0.48, 0.57, and 0.54 karat. Do the six measurements present sufficient evidence to indicate that the average weight of the diamonds produced by the process is in excess of 0.5 karat?

SOLUTION The population of diamond weights produced by this new process has mean μ, and you can set out the formal test of hypothesis in steps, as you did in Chapter 9:

1–2 **Null and alternative hypotheses:**

$$H_0: \mu = 0.5 \qquad \text{versus} \qquad H_a: \mu > 0.5$$

3 **Test statistic:** You can use your calculator to verify that the mean and standard deviation for the six diamond weights are 0.53 and 0.0559, respectively. The test statistic is a t statistic, calculated as

$$t = \frac{\bar{x} - \mu_0}{s/\sqrt{n}} = \frac{0.53 - 0.5}{0.0559/\sqrt{6}} = 1.32$$

As with the large-sample tests, the test statistic provides evidence for either rejecting or accepting H_0 depending on how far from the centre of the t distribution it lies.

4 **Rejection region:** If you choose a 5% level of significance ($\alpha = 0.05$), the right-tailed rejection region is found using the critical values of t from Table 4 of Appendix I. With $df = n - 1 = 5$, you can reject H_0 if $t > t_{0.05} = 2.015$, as shown in Figure 10.4.

5 **Conclusion:** Since the calculated value of the test statistic, 1.32, does not fall in the rejection region, you cannot reject H_0. The data do not present sufficient evidence to indicate that the mean diamond weight exceeds 0.5 karat.

FIGURE 10.4
Rejection region for Example 10.3

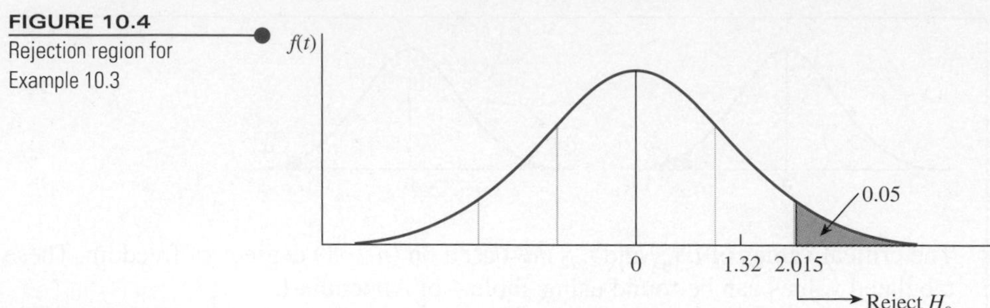

NEED A TIP?

A 95% confidence interval tells you that, if you were to construct many of these intervals (all of which would have slightly different endpoints), 95% of them would enclose the population mean.

As in Chapter 9, the conclusion to *accept* H_0 would require the difficult calculation of β, the probability of a Type II error. To avoid this problem, we choose to *not reject* H_0. We can then calculate the lower bound for μ using a small-sample lower one-sided confidence bound. This bound is similar to the large-sample one-sided confidence bound, except that the critical z_α is replaced by a critical t_α from Table 4. For this example, a 95% lower one-sided confidence bound for μ is:

$$\bar{x} - t_\alpha \frac{s}{\sqrt{n}}$$

$$0.53 - 2.015\frac{0.0559}{\sqrt{6}}$$

$$0.53 - 0.046$$

The 95% lower bound for μ is $\mu > 0.484$. The range of possible values includes mean diamond weights both smaller and greater than 0.5; this confirms the failure of our test to show that μ exceeds 0.5.

Remember from Chapter 9 that there are two ways to conduct a test of hypothesis:

- **The critical value approach:** Set up a rejection region based on the critical values of the statistic's sampling distribution. If the test statistic falls in the rejection region, you can reject H_0.
- **The *p*-value approach:** Calculate the *p*-value based on the observed value of the test statistic. If the *p*-value is smaller than the significance level, α, you can reject H_0. If there is no *preset* significance level, use the guidelines in Section 9.3 to judge the statistical significance of your sample results.

We used the first approach in the solution to Example 10.3. We use the second approach to solve Example 10.4.

EXAMPLE 10.4

Labels on 3.79 litre (L) cans of paint usually indicate the drying time and the area that can be covered in one coat. Most brands of paint indicate that, in one coat, 3.79 L will cover between 23.2 and 46.4 square metres (m^2), depending on the texture of the surface to be painted. One manufacturer, however, claims that 3.79 L of its paint will cover 37.2 m^2 of surface area. To test this claim, a random sample of ten 3.79 L cans of white paint were used to paint ten identical areas using the same kind of equipment. The actual areas (in m^2) covered by these 3.79 L of paint are given here:

28.8	28.9	38.3	34.2	41.5
34.9	28.2	38.1	33.9	32.5

Do the data present sufficient evidence to indicate that the average coverage differs from 37.2 m^2? Find the p-value for the test, and use it to evaluate the statistical significance of the results.

NEED A TIP?

Remember from Chapter 2 how to calculate $\bar{x}$ and s using the data entry method on your calculator.

SOLUTION To test the claim, the hypotheses to be tested are

$$H_0 : \mu = 37.2 \quad \text{versus} \quad H_a : \mu \neq 37.2$$

The sample mean and standard deviation for the recorded data are

$$\bar{x} = 33.93 \quad s = 4.49$$

and the test statistic is

$$t = \frac{\bar{x} - \mu_0}{s/\sqrt{n}} = \frac{33.93 - 37.2}{4.49\sqrt{10}} = -2.30$$

The p-value for this test is the probability of observing a value of the t statistic as contradictory to the null hypothesis as the one observed for this set of data—namely, $t = -2.30$. Since this is a two-tailed test, the p-value is the probability that either $t \leq -2.30$ or $t \geq 2.30$.

Unlike the z-table, the table for t gives the values of t corresponding to upper-tail areas equal to 0.100, 0.050, 0.025, 0.010, and 0.005. Consequently, you can only approximate the upper-tail area that corresponds to the probability that $t > 2.30$. Since the t statistic for this test is based on 9 df, we refer to the row corresponding to $df = 9$ in Table 4. The five critical values for various tail areas are shown in Figure 10.5, an enlargement of the tail of the t distribution with 9 degrees of freedom. The value $t = 2.30$ falls between $t_{0.025} = 2.262$ and $t_{0.010} = 2.821$. Therefore, the right-tail area corresponding to the probability that $t > 2.30$ lies between 0.01 and 0.025. Since this area represents only half of the p-value, you can write

$$0.01 < \frac{1}{2}(p\text{-value}) < 0.025\text{—or—}0.02 < p\text{-value} < 0.05$$

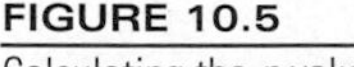

FIGURE 10.5
Calculating the p-value for Example 10.4 (shaded area $= \frac{1}{2}$ p-value)

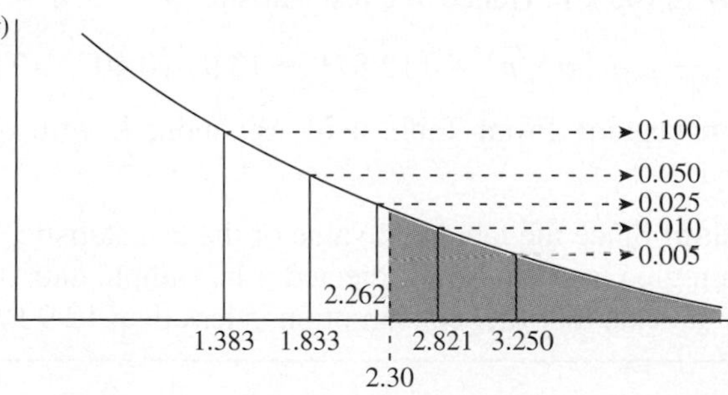

What does this tell you about the significance of the statistical results? For you to reject H_0, the p-value must be less than the specified significance level, α. Hence, you could reject H_0 at the 5% level, but not at the 2% or 1% level. Therefore, the p-value for this test would typically be reported by the experimenter as

$$p\text{-value} < 0.05 \quad (\text{or sometimes } P < 0.05)$$

For this test of hypothesis, H_0 is rejected at the 5% significance level. There is sufficient evidence to indicate that the average coverage differs from 37.2 m^2.

Within what limits does this average coverage *really* fall? A 95% confidence interval gives the upper and lower limits for μ as:

$$\bar{x} \pm t_{\alpha/2}\left(\frac{s}{\sqrt{n}}\right)$$

$$33.93 \pm 2.262\left(\frac{4.49}{\sqrt{10}}\right)$$

$$33.93 \pm 3.21$$

$$(30.7, 37.1)$$

Thus, you can estimate that the average area covered by 3.79 L of this brand of paint lies in the interval 30.72 to 37.14. A more precise interval estimate (a shorter interval) can generally be obtained by increasing the sample size. Notice that the upper limit of this interval is very close to the value of 37.2 m^2, the coverage claimed on the label. This coincides with the fact that the observed value of $t = -2.30$ is just slightly less than the left-tail critical value of $t_{0.025} = -2.262$, making the p-value just slightly less than 0.05.

EXAMPLE 10.5

Suppose the average fuel consumption of an SUV manufactured at an automobile plant in Windsor is 12.9 L/100 km in city. The average fuel consumption (in L/100 km) for seven randomly selected SUVs is 13.5, 13.0, 12.6, 12.2, 12.8, 12.9, and 13.1.

Assume that the distribution of fuel consumption distribution is normal. Do the data provide sufficient evidence to indicate the average fuel consumption is less than 12.9 L/100 km? Test using $\alpha = 0.10$.

SOLUTION You can set out the formal test of hypothesis in the following steps:

1–2 **Null and alternative hypotheses:**

$$H_o : \mu = 12.9 \text{ versus } H_a: \mu < 12.9$$

3 **Test statistic:** The sample mean is $\bar{x} = 12.871$, and the sample standard deviation is $s = 0.407$ L/100 km. Hence, the test statistic is

$$t = (\bar{x} - \mu_0)/(s/\sqrt{n}) = (12.871 - 12.9)/(0.407/\sqrt{7}) = -0.189$$

4 **Rejection region:** From Table 4 of Appendix I, with $df = 7 - 1 = 6$, we have $t_{0.10} = -1.440$

5 **Conclusion:** Since the tabulated value of the test statistic, -0.189, does not fall in the rejection region, and H_0 is not rejected. The sample data do not present sufficient evidence to indicate that fuel consumption is less than 12.9 L/100 km.

EXAMPLE 10.6

A manufacturer claims that the average lifetime of an electrical component produced at its plant is 75 weeks. A consumer agency reports that the lifetime (in weeks) of nine randomly selected components were 74.5, 75.0, 72.3, 76.0, 75.2, 75.1, 75.3, 74.9, and 74.8.

Assume the component lifetime distribution is normal. Do the sample data suggest the average lifetime is smaller than 75 weeks? Find the approximate p-value for the test. Does the sample data support the alternative hypothesis at the $\alpha = 0.05$ level?

SOLUTION To test the claim, the hypotheses to be tested are

$$H_0 : \mu = 75 \text{ versus } H_a : \mu < 75$$

The sample mean and standard deviation are

$\bar{x} = 74.789 \quad s = 1.020$

The test statistic is

$t = (\bar{x} - \mu_0)/(s/\sqrt{n}) = (74.789 - 75.0)/(1.02/\sqrt{9}) = -0.62$

The p-value for this test is the probability of observing a value of the t statistic as contradictory to the null hypothesis as the one observed for this set of data. Hence,

$p \text{ value} = P(t < -0.62) = P(t > 0.62) > 0.10$

No, the sample data does not support the alternative hypothesis at the 0.05 level, since p-value $> \alpha$ and H_0 is not rejected.

EXAMPLE  10.7

"Tar" is the term used to describe the toxic chemicals found in cigarettes. The concentration of tar in a cigarette determines its rating: High-tar cigarettes contain at least 22 milligrams (mg) of tar; medium-tar cigarettes from 15 mg to 21 mg; and low-tar cigarettes 7 mg or less of tar. In Canada, tobacco companies are currently required by law to print the levels of tar, nicotine, carbon monoxide, hydrogen cyanide, formaldehyde, and benzene in smoke from their cigarettes.[1]

From a particular Canadian brand, 10 cigarettes were randomly selected and the content of tar (in mg) was measured. The results were 13.5, 14.0, 13.9, 14.2, 15.1, 14.6, 13.8, 14.0, 14.1, and 14.7 mg per cigarette. Assume these measurements were taken from a population with a normal distribution. Construct a 90% confidence interval for the mean tar content of any cigarette of this brand.

SOLUTION The sample mean and standard deviation are

$\bar{x} = 14.19 \quad s = 0.477$

The 90% confidence interval is

$$\bar{x} \pm t_{\alpha/2} s/\sqrt{n}$$

$$14.19 \pm (1.833)(0.477)/\sqrt{10}$$

$$14.19 \pm 0.276,$$

$$\text{or } 13.9 < \mu < 14.5$$

EXAMPLE  10.8

A red blood cell (RBC) count is a useful blood test that can provide information about how the number of red blood cells in a person's blood. This test might be done as one component of a complete blood cell (CBC) count. A RBC count is the number of red blood cells per volume of blood, and is reported in either millions in a microlitre or millions in a litre of blood.[2]

Here are the red blood cell counts (in 10^6 cells per microlitre) of a healthy Canadian male measured on each of 15 days: 5.6, 5.4, 5.2, 5.4, 5.7, 5.5, 5.6, 5.4, 5.3, 5.5, 5.5, 5.1, 5.6, 5.4, and 5.4. Find a 95% confidence interval estimate of the true mean red blood cell count for this Canadian male during the period of testing.

SOLUTION

$$\bar{x} = \Sigma x_i/n = 5.44$$

$$s^2 = \left[\Sigma x_i^2 - \frac{(\Sigma x_i)^2}{n}\right]/(n-1) = 0.02543$$

$$s = 0.15946.$$

A 95% confidence interval for the population mean μ is

$$\bar{x} \pm t_{\alpha/2}s/\sqrt{n}$$

$$5.44 \pm (2.145)0.15946/\sqrt{15}$$

$$5.44 \pm 0.0883$$

or $5.3517 < \mu < 5.5283$

You can estimate with 95% confidence that the true mean red blood cell count is roughly between 5.35 and 5.53. Intervals constructed using this procedure will enclose the true mean 95% of the time in repeated sampling.

Most statistical computing packages contain programs that will implement the Student's t test or construct a confidence interval for μ when the data are properly entered into the computer's database. Most of these programs will calculate and report the *exact p-value* of the test, allowing you to quickly and accurately draw conclusions about the statistical significance of the results. The results of the *MINITAB* one-sample t test and confidence interval procedures are given in Figure 10.6. Besides the observed value of $t = -2.30$ and the confidence interval (30.72, 37.14), the output gives the sample mean, the sample standard deviation, the standard error of the mean (SE Mean $= s/\sqrt{n}$), and the exact p-value of the test ($P = 0.047$). This is consistent with the range for the p-value that we found using Table 4 in Appendix I:

$$0.02 < p\text{-value} < 0.05$$

FIGURE 10.6
MINITAB output for Example 10.4

One Sample T: Area

```
Test of mu = 37.2 vs not = 37.2

Variable          N       Mean    StDev    SE Mean
Area             10      33.93     4.49       1.42

Variable        95% CI               T         P
Area       (30.72, 37.14)        -2.30     0.047
```

You can see the value of using the computer output to evaluate statistical results:

- The exact p-value eliminates the need for tables and critical values.
- All of the numerical calculations are done for you.

The most important job—which is left for the experimenter—is to *interpret* the results in terms of their practical significance!

10.3 EXERCISES

BASIC TECHNIQUES

10.7 Find the following t-values in Table 4 of Appendix I:

a. $t_{0.05}$ for 5 *df*
b. $t_{0.025}$ for 8 *df*
c. $t_{0.10}$ for 18 *df*
d. $t_{0.025}$ for 30 *df*

10.8 Find the critical value(s) of t that specify the rejection region in these situations:

a. A two-tailed test with $\alpha = 0.01$ and 12 *df*
b. A right-tailed test with $\alpha = 0.05$ and 16 *df*
c. A two-tailed test with $\alpha = 0.05$ and 25 *df*
d. A left-tailed test with $\alpha = 0.01$ and 7 *df*

10.9 Use Table 4 in Appendix I to approximate the p-value for the t statistic in each situation:

a. A two-tailed test with $t = 2.43$ and 12 df

b. A right-tailed test with $t = 3.21$ and 16 df

c. A two-tailed test with $t = -1.19$ and 25 df

d. A left-tailed test with $t = -8.77$ and 7 df

Data set EX1010 **10.10 Test Scores** The test scores on a 100-point test were recorded for 20 students:

71	93	91	86	75
73	86	82	76	57
84	89	67	62	72
77	68	65	75	84

a. Can you reasonably assume that these test scores have been selected from a normal population? Use a stem and leaf plot to justify your answer.

b. Calculate the mean and standard deviation of the scores.

c. If these students can be considered a random sample from the population of all students, find a 95% confidence interval for the average test score in the population.

10.11 The following $n = 10$ observations are a sample from a normal population:

7.4 7.1 6.5 7.5 7.6 6.3 6.9 7.7 6.5 7.0

a. Find the mean and standard deviation of these data.

b. Find a 99% upper one-sided confidence bound for the population mean μ.

c. Test $H_0 : \mu = 7.5$ versus $H_a : \mu < 7.5$. Use $\alpha = 0.01$.

d. Do the results of part b support your conclusion in part c?

APPLICATIONS

Data set EX1012 **10.12 Healing Properties of Herbs** Parsley, cilantro, mint, and basil are herbs used as flavouring in cooking. But did you know that these herbs have long been valued around the globe for their healing properties? Mint has been used to help with indigestion, basil has been used to help with kidney problems, and cilantro has antiseptic properties.

Based on a recent market survey, the prices of these herbs with approximately the same weight were recorded:

Parsley		Cilantro	Mint	Basil	
0.99	0.53	1.27	1.49	2.56	0.62
1.92	1.41	1.22	1.29	1.92	0.66
1.23	1.12	1.19	1.27	1.30	0.62
0.85	0.63	1.22	1.35	1.79	0.65
0.65	0.67		1.29	1.23	0.60
0.69	0.60		1.00		0.67
0.60	0.66		1.27		
			1.28		

Assume that the prices included in this survey represent a random sample of the prices of these herbs across North America.

a. Find a 95% confidence interval for the average price for parsley. Interpret this interval. That is, what does the "95%" refer to?

b. Find a 95% confidence interval for the average price for cilantro. How does the width of this interval compare to the width of the interval in part a? Can you explain why?

c. Find 95% confidence intervals for the other two samples (mint and basil). Plot the four treatment means and their standard errors in a two-dimensional plot similar to Figure 8.5. What kind of broad comparisons can you make about the four treatments? (We will discuss the procedure for comparing more than two population means in Chapter 11.)

10.13 Dissolved O_2 Content Industrial wastes and sewage dumped into our rivers and streams absorb oxygen and thereby reduce the amount of dissolved oxygen available for fish and other forms of aquatic life. One provincial agency requires a minimum of 5 parts per million (ppm) of dissolved oxygen in order for the oxygen content to be sufficient to support aquatic life. Six water specimens taken from a river at a specific location during the low-water season (July) gave readings of 4.9, 5.1, 4.9, 5.0, 5.0, and 4.7 ppm of dissolved oxygen. Do the data provide sufficient evidence to indicate that the dissolved oxygen content is less than 5 ppm? Test using $\alpha = 0.05$.

10.14 Lobsters In a study of the infestation of the *Thenus orientalis* lobster by two types of barnacles, *Octolasmis tridens* and *O. lowei*, the carapace lengths (in millimetres mm) of 10 randomly selected lobsters caught in the seas near Singapore are measured:[3]

78 66 65 63 60 60 58 56 52 50

Find a 95% confidence interval for the mean carapace length of the *T. orientalis* lobsters.

Data set EX1015 **10.15 Smoking and Lung Capacity** It is recognized that cigarette smoking has a deleterious effect on lung function. In a study of the effect of cigarette smoking on the carbon monoxide diffusing capacity (DL) of the lung, researchers found that current smokers had DL readings significantly lower than those of either ex-smokers or non-smokers. The carbon monoxide diffusing capacities for a random sample of $n = 20$ current smokers are listed here:

103.768	88.602	73.003	123.086	91.052
92.295	61.675	90.677	84.023	76.014
100.615	88.017	71.210	82.115	89.222
102.754	108.579	73.154	106.755	90.479

a. Do these data indicate that the mean DL reading for current smokers is significantly lower than 100 DL, the average for non-smokers? Use $\alpha = 0.01$.

b. Find a 99% upper one-sided confidence bound for the mean DL reading for current smokers. Does this bound confirm your conclusions in part a?

Data set EX1016 **10.16 Wayne Gretzky** In Exercise 2.38 (EX0238), the number of goals scored by Wayne Gretzky were recorded for seasons 1978–1999:[4]

46	51	55	92	71	87
73	52	62	40	54	40
41	31	16	38	11	23
25	23	9			

a. Construct a stem and leaf plot of the $n = 21$ observations.

Based on this plot, is it reasonable to assume that the underlying population is approximately normal, as required for the one-sample t test? Explain.

b. Calculate the mean and standard deviation for Wayne Gretzky's goals scored.

c. Construct a 95% confidence interval to estimate the goals scored for Wayne Gretzky.

10.17 Purifying Organic Compounds Organic chemists often purify organic compounds by a method known as fractional crystallization. An experimenter wanted to prepare and purify 4.85 g of aniline. Ten 4.85 g quantities of aniline were individually prepared and purified to acetanilide. The following dry yields were recorded:

3.85	3.80	3.88	3.85	3.90
3.36	3.62	4.01	3.72	3.82

Estimate the mean grams of acetanilide that can be recovered from an initial amount of 4.85 g of aniline. Use a 95% confidence interval.

10.18 Organic Compounds, continued Refer to Exercise 10.17. Approximately how many 4.85 g specimens of aniline are required if you wish to estimate the mean number of grams of acetanilide correct to within 0.06 g with probability equal to 0.95?

10.19 Bulimia Although there are many treatments for bulimia nervosa, some subjects fail to benefit from treatment. In a study to determine which factors predict who will benefit from treatment, an article in the *British Journal of Clinical Psychology* indicates that self-esteem was one of these important predictors.[5] The table gives the mean and standard deviation of self-esteem scores prior to treatment, at posttreatment, and during a follow-up:

	Pretreatment	Posttreatment	Follow-up
Sample mean $\bar{x}$	20.3	26.6	27.7
Standard deviation s	5.0	7.4	8.2
Sample size n	21	21	20

a. Use a test of hypothesis to determine whether there is sufficient evidence to conclude that the true pretreatment mean is less than 25.

b. Construct a 95% confidence interval for the true posttreatment mean.

c. In Section 10.4, we will introduce small-sample techniques for making inferences about the difference between two population means. Without the formality of a statistical test, what are you willing to conclude about the differences among the three sampled population means represented by the results in the table?

Data set EX1020 **10.20 RBC Counts** Here are the red blood cell counts (in 10^6 cells per microlitre) of a healthy person measured on each of 15 days:

5.4	5.2	5.0	5.2	5.5
5.3	5.4	5.2	5.1	5.3
5.3	4.9	5.4	5.2	5.2

Find a 95% confidence interval estimate of μ, the true mean red blood cell count for this person during the period of testing.

Data set EX1021 **10.21 Ground Beef** These data are the weights (in kilograms) of 27 packages of ground beef in a supermarket meat display:

1.08	0.99	0.97	1.18	1.41	1.28	0.83
1.06	1.14	1.38	0.75	0.96	1.08	0.87
0.89	0.89	0.96	1.12	1.12	0.93	1.24
0.89	0.98	1.14	0.92	1.18	1.17	

a. Interpret the accompanying *MINITAB* printouts for the one-sample test and estimation procedures.

MINITAB output for Exercise 10.21

One-Sample T: Weights

```
Test of mu = 1 vs not = 1
Variable N              Mean      StDev    SE Mean
Weights 27           1.05222    0.16565    0.03188

Variable              95% CI          T          P
Weights (0.98669, 1.11775)         1.64      0.113
```

b. Verify the calculated values of t and the upper and lower confidence limits.

10.22 Cholesterol The serum cholesterol levels of 50 subjects randomly selected from the L.A. Heart Data, data from an epidemiological heart disease study on employees, follow:[6]

EX1022

148	304	300	240	368	139	203	249	265	229
303	315	174	209	253	169	170	254	212	255
262	284	275	229	261	239	254	222	273	299
278	227	220	260	221	247	178	204	250	256
305	225	306	184	242	282	311	271	276	248

a. Construct a histogram for the data. Are the data approximately mound-shaped?

b. Use a t distribution to construct a 95% confidence interval for the average serum cholesterol levels for employees.

10.23 Cholesterol, continued Refer to Exercise 10.22. Since $n > 30$, use the methods of Chapter 8 to create a large-sample 95% confidence interval for the average serum cholesterol level for employees. Compare the two intervals. (HINT: The two intervals should be quite similar. This is the reason we choose to approximate the sample distribution of $\frac{\bar{x} - \mu}{s/\sqrt{n}}$ with a z-distribution when $n > 30$.)

10.4 SMALL-SAMPLE INFERENCES FOR THE DIFFERENCE BETWEEN TWO POPULATION MEANS: INDEPENDENT RANDOM SAMPLES

The physical setting for the problem considered in this section is the same as the one in Section 8.5, except that the sample sizes are no longer large. Independent random samples of n_1 and n_2 measurements are drawn from two populations, with means and variances μ_1, σ_1^2, μ_2, and σ_2^2, and your objective is to make inferences about $(\mu_1 - \mu_2)$, the difference between the two population means.

When the sample sizes are small, you can no longer rely on the Central Limit Theorem to ensure that the sample means will be normal. If the original populations *are normal,* however, then the sampling distribution of the difference in the sample means, $(\bar{x}_1 - \bar{x}_2)$, will be normal (even for small samples) with mean $(\mu_1 - \mu_2)$ and standard error

$$\sqrt{\frac{\sigma_1^2}{n_1} + \frac{\sigma_2^2}{n_2}}$$

In Chapters 7 and 8, you used the sample variances, s_1^2 and s_2^2, to calculate an *estimate* of the standard error, which was then used to form a large-sample confidence interval or a test of hypothesis based on the large-sample z statistic:

$$z \approx \frac{(\bar{x}_1 - \bar{x}_2) - (\mu_1 - \mu_2)}{\sqrt{\frac{s_1^2}{n_1} + \frac{s_2^2}{n_2}}}$$

NEED A TIP?

Assumptions for the two-sample (independent) t-test:

- Random independent samples
- Normal distributions
- $\sigma_1 = \sigma_2$

Unfortunately, when the sample sizes are small, this statistic does not have an approximately normal distribution—nor does it have a Student's t distribution. In order to form a statistic with a sampling distribution that can be derived theoretically, you must make one more assumption.

Suppose that the variability of the measurements in the two normal populations is the same and can be measured by a common variance σ^2. That is, *both populations have exactly the same shape,* and $\sigma_1^2 = \sigma_2^2 = \sigma^2$. Then the standard error of the difference in the two sample means is

$$\sqrt{\frac{\sigma_1^2}{n_1} + \frac{\sigma_2^2}{n_2}} = \sqrt{\sigma^2\left(\frac{1}{n_1} + \frac{1}{n_2}\right)}$$

It can be proven mathematically that, if you use the appropriate sample estimate s^2 for the population variance σ^2, then the resulting test statistic,

$$t = \frac{(\bar{x}_1 - \bar{x}_2) - (\mu_1 - \mu_2)}{\sqrt{s^2\left(\frac{1}{n_1} + \frac{1}{n_2}\right)}}$$

has a *Student's t distribution.* The only remaining problem is to find the sample estimate s^2 and the appropriate number of *degrees of freedom* for the t statistic.

Remember that the population variance σ^2 describes the shape of the normal distributions from which your samples come, so that either s_1^2 or s_2^2 would give you an estimate of σ^2. But why use just one when information is provided by both? A better procedure is to combine the information in both sample variances using a *weighted average,* in which the weights are determined by the relative amount of information (the number of measurements) in each sample. For example, if the first sample contained twice as many measurements as the second, you might consider giving the first sample variance twice as much weight. To achieve this result, use this formula:

$$s^2 = \frac{(n_1 - 1)(n_2 - 1)s^2}{df_1 + df_2}$$

Remember from Section 10.3 that the degrees of freedom for the one-sample t statistic are $(n - 1)$, the denominator of the sample estimate s^2. Since s_1^2 has $(n_1 - 1)$ df and s_2^2 has $(n_2 - 1)$ df, the total number of degrees of freedom is the sum:

$$(n_1 - 1) + (n_2 - 1) = n_1 + n_2 - 2$$

Thus,

$$s^2 = \frac{(n_1 - 1)(n_2 - 1)s^2}{n_1 + n_2 - 2}$$

CALCULATION OF S^2

- If you have a scientific calculator, calculate each of the two sample standard deviations s_1 and s_2 separately, using the data entry procedure for your particular calculator. These values are squared and used in this formula:

$$s^2 = \frac{(n_1 - 1)s_1^2 + (n_2 - 1)s_2^2}{n_1 + n_2 - 2}$$

NEED A TIP?

For the two-sample (independent) t-test, $df = n_1 + n_2 - 2$

It can be shown that s^2 is an unbiased estimator of the common population variance σ^2. If s^2 is used to estimate σ^2 and if the samples have been randomly and independently drawn from normal populations with a common variance, then the statistic

$$t = \frac{(\bar{x}_1 - \bar{x}_2) - (\mu_1 - \mu_2)}{\sqrt{s^2\left(\frac{1}{n_1} + \frac{1}{n_2}\right)}}$$

has a Student's t distribution with $(n_1 + n_2 - 2)$ degrees of freedom. The small-sample estimation and test procedures for the difference between two means are given next.

TEST OF HYPOTHESIS CONCERNING THE DIFFERENCE BETWEEN TWO MEANS: INDEPENDENT RANDOM SAMPLES

1. Null hypothesis: $H_0 : (\mu_1 - \mu_2) = D_0$, where D_0 is some specified difference that you wish to test. For many tests, you will hypothesize that there is no difference between μ_1 and μ_2; that is, $D_0 = 0$.
2. Alternative hypothesis:

One-Tailed Test	**Two-Tailed Test**
$H_a : (\mu_1 - \mu_2) > D_0$ [or $H_a : (\mu_1 - \mu_2) < D_0$]	$H_a : (\mu_1 - \mu_2) \neq D_0$

3. Test statistic: $$t = \frac{(\bar{x}_1 - \bar{x}_2) - D_0}{\sqrt{s^2\left(\frac{1}{n_1} + \frac{1}{n_2}\right)}}$$ where $$s^2 = \frac{(n_1 - 1)s_1^2 + (n_2 - 1)s_2^2}{n_1 + n_2 - 2}$$
4. Rejection region: Reject H_0 when

One-Tailed Test	**Two-Tailed Test**
$t > t_\alpha$ [or $t < -t_\alpha$ when the alternative hypothesis is $H_a : (\mu_1 - \mu_2) < D_0$]	$t > t_{\alpha/2}$ or $t < -t_{\alpha/2}$

   or when p-value $< \alpha$

The critical values of t, t_α, and $t_\alpha/2$ are based on $(n_1 + n_2 - 2)$ *df.* The tabulated values can be found using Table 4 of Appendix I.

Assumptions: The samples are randomly and independently selected from normally distributed populations. The variances of the populations σ_1^2 and σ_2^2 are equal.

SMALL-SAMPLE $(1 - \alpha)100\%$ CONFIDENCE INTERVAL FOR $(\mu_1 - \mu_2)$ BASED ON INDEPENDENT RANDOM SAMPLES

$$(\bar{x}_1 - \bar{x}_2) \pm t_{\alpha/2}\sqrt{s^2\left(\frac{1}{n_1} + \frac{1}{n_2}\right)}$$

where s^2 is the pooled estimate of σ^2.

EXAMPLE 10.9

A course can be taken for credit either by attending lecture sessions at fixed times and days, or by doing online sessions that can be done at the student's own pace and schedule. The course coordinator wants to determine if these two ways of taking the course resulted in a significant difference in achievement as measured by the final exam for the course. Table 10.3 gives the scores on an examination with 45 possible points for one group of $n_1 = 9$ students who took the course online, and a second group of $n_2 = 9$ students who took the course with conventional lectures. Do these data present sufficient evidence to indicate that the average grade for students who take the course online is significantly higher than for those who attend a conventional class?

TABLE 10.3 Test Scores for Online and Classroom Presentations

Online	Classroom
32	35
37	31
35	29
28	25
41	34
44	40
35	27
31	32
34	31

SOLUTION Let μ_1 and μ_2 be the mean scores for the online group and the classroom group, respectively. Then, since you seek evidence to support the theory that $\mu_1 > \mu_2$, you can test the null hypothesis

$$H_0: \mu_1 = \mu_2 \text{ [or } H_0: (\mu_1 - \mu_2) = 0]$$

versus the alternative hypothesis

$$H_a: \mu_1 > \mu_2 \text{ [or } H_a: (\mu_1 - \mu_2) > 0]$$

To conduct the t-test for these two independent samples, you must assume that the sampled populations are both normal and have the same variance σ^2. Is this reasonable? Stem and leaf plots of the data in Figure 10.7 show at least a "mounding" pattern, so that the assumption of normality is not unreasonable.

FIGURE 10.7
Stem and leaf plots for Example 10.9

Online	Classroom
2 | 8	2 | 579
3 | 124	3 | 1124
3 | 557	3 | 5
4 | 14	4 | 0

NEED A TIP?

Stem and leaf plots can help you decide if the normality assumption is reasonable.

Furthermore, the standard deviations of the two samples, calculated as

$$s_1 = 4.9441 \text{ and } s_2 = 4.4752$$

are not different enough for us to doubt that the two distributions may have the same shape. If you make these two assumptions and calculate (using full accuracy) the pooled estimate of the common variance as

$$s^2 = \frac{(n_1 - 1)s_1^2 + (n_2 - 1)s_2^2}{n_1 + n_2 - 2} = \frac{8(4.9441)^2 + 8(4.4752)^2}{9 + 9 - 2} = 22.2361$$

you can then calculate the test statistic

$$t = \frac{\bar{x}_1 - \bar{x}_2}{\sqrt{s^2\left(\frac{1}{n_1} + \frac{1}{n_2}\right)}} = \frac{35.22 - 31.56}{\sqrt{22.2361\left(\frac{1}{9} + \frac{1}{9}\right)}} = 1.65$$

NEED A TIP?

If you are using a calculator, don't round off until the final step!

The alternative hypothesis $H_a : \mu_1 > \mu_2$ or, equivalently $H_a : (\mu_1 - \mu_2) > 0$, implies that you should use a one-tailed test in the upper tail of the t distribution with $(n_1 + n_2 - 2) = 16$ degrees of freedom. You can find the appropriate critical value for a rejection region with $\alpha = 0.05$ in Table 4 of Appendix I, and H_0 will be rejected if $t > 1.746$. Comparing the observed value of the test statistic $t = 1.65$ with the critical value $t_{.05} = 1.746$, you cannot reject the null hypothesis (see Figure 10.8). There is insufficient evidence to indicate that the average online course grade is higher than the average conventional course grade at the 5% level of significance.

FIGURE 10.8

Rejection region for Example 10.9

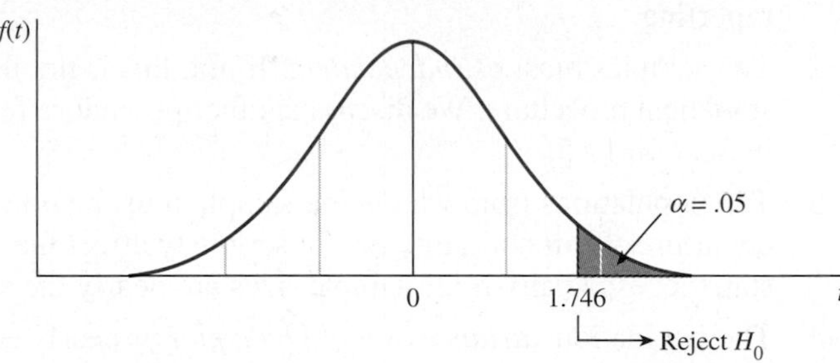

EXAMPLE 10.10

Find the p-value that would be reported for the statistical test in Example 10.9.

SOLUTION The observed value of t for this one-tailed test is $t = 1.65$. Therefore,

$$p - \text{value} = P(t > 1.65)$$

for a t statistic with 16 degrees of freedom. Remember that you cannot obtain this probability directly from Table 4 in Appendix I; you can only *bound* the p-value using the critical values in the table. Since the observed value, $t = 1.65$, lies between $t_{0.100} = 1.337$ and $t_{0.050} = 1.746$, the tail area to the right of 1.65 is between 0.05 and 0.10. The p-value for this test would be reported as

$$0.05 < p - \text{value} < 0.10$$

ONLINE APPLET

Two-Sample t Test: Independent Samples

Because the p-value is greater than 0.05, most researchers would report the results as *not significant.*

EXAMPLE 10.11

Use a lower 95% confidence bound to estimate the difference $(\mu_1 - \mu_2)$ in Example 10.9. Does the lower confidence bound indicate that the online test score average is significantly higher than the classroom test score average?

SOLUTION The lower confidence bound takes a familiar form—the point estimator $(\bar{x}_1 - \bar{x}_2)$ minus an amount equal to t_α times the standard error of the estimator. Substituting into the formula, you can calculate the 95% lower confidence bound:

$$(\bar{x}_1 - \bar{x}_2) - t_\alpha\sqrt{s^2\left(\frac{1}{n_1} + \frac{1}{n_2}\right)}$$

$$(35.22 - 31.56) - 1.746\sqrt{22.2361\left(\frac{1}{9} + \frac{1}{9}\right)}$$

$$3.66 - 3.88$$

or $(\mu_1 - \mu_2) > -0.22$. Since the value $(\mu_1 - \mu_2) = 0$ is included in the confidence interval, it is possible that the two means are equal. There is insufficient evidence to indicate that the online average is higher than the classroom average.

NEED A TIP?

larger s^2/smaller $s^2 < 3$

$\Leftrightarrow$

variance assumption is reasonable

The two-sample procedure that uses a pooled estimate of the common variance σ^2 relies on four important assumptions:

1. The samples must be *randomly selected.* Samples not randomly selected may introduce bias into the experiment and thus alter the significance levels you are reporting.
2. The samples must be *independent.* If not, this is not the appropriate statistical procedure. We discuss another procedure for dependent samples in Section 10.5.
3. The populations from which you sample must be *normal.* However, moderate departures from normality do not seriously affect the distribution of the test statistic, especially if the sample sizes are nearly the same.
4. The population *variances should be equal* or nearly equal to ensure that the procedures are valid.

If the population variances are far from equal, there is an alternative procedure for estimation and testing that has an *approximate t* distribution in repeated sampling. As a rule of thumb, you should use this procedure if the ratio of the two sample variances,

$$\frac{\text{Larger } s^2}{\text{Smaller } s^2} > 3$$

Since the population variances are not equal, the pooled estimator s^2 is no longer appropriate, and each population variance must be estimated by its corresponding sample variance. The resulting test statistic is

$$\frac{(\bar{x}_1 - \bar{x}_2) - D_0}{\sqrt{\dfrac{s_1^2}{n_1} + \dfrac{s_2^2}{n_2}}}$$

When the sample sizes are *small,* critical values for this statistic are found using degrees of freedom approximated by the formula

$$df \approx \frac{\left(\dfrac{s_1^2}{n_1} + \dfrac{s_2^2}{n_2}\right)^2}{\dfrac{(s_1^2/n_1)^2}{(n_1 - 1)} + \dfrac{(s_2^2/n_2)^2}{(n_2 - 1)}}$$

The degrees of freedom are taken to be the integer part of this result.

Computer packages such as *MINITAB* and *Excel* can be used to implement this procedure, sometimes called *Satterthwaite's approximation,* as well as the *pooled method* described earlier. In fact, some experimenters choose to analyze their data using *both* methods. As long as both analyses lead to the same conclusions, you need not concern yourself with the equality or inequality of variances.

The *MINITAB* and *Excel* outputs resulting from the pooled method of analysis for the data of Example 10.9 are shown in Figure 10.9(a) and (b). Notice that the ratio of the two sample variances, $(4.94/4.48)^2 = 1.22$, is less than 3, which makes the pooled method appropriate. The calculated value of $t = 1.65$ and the exact p-value $= 0.059$ with 16 degrees of freedom are shown in both of the outputs. The exact p-value makes it quite easy for you to determine the significance or non-significance of the sample results. You will find instructions for generating this output in the section "Technology Today" at the end of this chapter.

FIGURE 10.9(a)
MINITAB output for Example 10.9

Two-Sample T-Test and CI: Online, Classroom

Two-sample T for Online vs Classroom

	N	Mean	StDev	SE Mean
Online	9	35.22	4.94	1.6
Classroom	9	31.56	4.48	1.5

Difference = mu (Online) − mu (Classroom)
Estimate for difference: 3.67
95% lower bound for difference: −0.21
T-Test of difference = 0 (vs >): T-Value = 1.65 P-Value = 0.059 *DF* = 16
Both use Pooled StDev = 4.7155

FIGURE 10.9(b)
Excel output for Example 10.9

D	E	F
t-Test: Two-Sample Assuming Equal Variances		
	Online	*Classroom*
Mean	35.222	31.556
Variance	24.444	20.028
Observations	9	9
Pooled Variance	22.236	
Hypothesized Mean Difference	0	
df	16	
t Stat	1.649	
P(T<=t) one-tail	0.059	
t Critical one-tail	1.746	
P(T<=t) two-tail	0.119	
t Critical two-tail	2.120	

If there is reason to believe that the normality assumptions have been violated, you can test for a shift in location of two population distributions using the nonparametric Wilcoxon rank sum test of Chapter 15. This test procedure, which requires fewer assumptions concerning the nature of the population probability distributions, is almost as sensitive in detecting a difference in population means when the conditions necessary for the t test are satisfied. It may be more sensitive when the normality assumption is not satisfied.

10.4 EXERCISES

BASIC TECHNIQUES

10.24 Give the number of degrees of freedom for s^2, the pooled estimator of σ^2, in these cases:

a. $n_1 = 16, n_2 = 8$

b. $n_1 = 10, n_2 = 12$

c. $n_1 = 15, n_2 = 3$

10.25 Calculate s^2, the pooled estimator for σ^2, in these cases:

a. $n_1 = 10, n_2 = 4, s_1^2 = 3.4, s_2^2 = 4.9$

b. $n_1 = 12, n_2 = 21, s_1^2 = 18, s_2^2 = 23$

10.26 Two independent random samples of sizes $n_1 = 4$ and $n_2 = 5$ are selected from each of two normal populations:

Population 1	12	3	8	5	
Population 2	14	7	7	9	6

a. Calculate s^2, the pooled estimator of σ^2.

b. Find a 90% confidence interval for $(\mu_1 - \mu_2)$, the difference between the two population means.

c. Test $H_0: (\mu_1 - \mu_2) = 0$ against $H_a: (\mu_1 - \mu_2) < 0$ for $\alpha = 0.05$. State your conclusions.

10.27 Independent random samples of $n_1 = 16$ and $n_2 = 13$ observations were selected from two normal populations with equal variances:

	Population	
	1	2
Sample size	16	13
Sample mean	34.6	32.2
Sample variance	4.8	5.9

a. Suppose you wish to detect a difference between the population means. State the null and alternative hypotheses for the test.

b. Find the rejection region for the test in part a for $\alpha = 0.01$.

c. Find the value of the test statistic.

d. Find the approximate p-value for the test.

e. Conduct the test and state your conclusions.

10.28 Refer to Exercise 10.27. Find a 99% confidence interval for $(\mu_1 - \mu_2)$.

10.29 The *MINITAB* printout shows a test for the difference in two population means.

MINITAB output for Exercise 10.29

Two-Sample T-Test and CI: Sample 1, Sample 2

```
Two-sample T for Sample 1 vs Sample 2
           N     Mean     StDev     SE Mean
Sample 1   6     29.00    4.00          1.6
Sample 2   7     28.86    4.67          1.8
Difference = mu (Sample 1) - (Sample 2)
Estimate for difference: 0.14
95% CI for difference: (-5.2, 5.5)
T-Test of difference = 0 (vs not =) :
T-Value = 0.06 P-Value = 0.95 DF = 11
Both use Pooled StDev = 4.38
```

a. Do the two sample standard deviations indicate that the assumption of a common population variance is reasonable?

b. What is the observed value of the test statistic? What is the p-value associated with this test?

c. What is the pooled estimate s^2 of the population variance?

d. Use the answers to part b to draw conclusions about the difference in the two population means.

e. Find the 95% confidence interval for the difference in the population means. Does this interval confirm your conclusions in part d?

10.30 The *Excel* printout shows a test for the difference in two population means.

Excel output for Exercise 10.30

D	E	F
t-Test: Two-Sample Assuming Equal Variances		
	Sample 1	*Sample 2*
Mean	28.667	28.286
Variance	5.067	2.238
Observations	6	7
Pooled Variance	3.524	
Hypothesized Mean Difference	0	
df	11	
t Stat	0.365	
P(T<=t) one-tail	0.361	
t Critical one-tail	1.796	
P(T<=t) two-tail	0.722	
t Critical two-tail	2.201	

a. Do the two sample variances indicate that the assumption of a common population variance is reasonable?

b. What is the observed value of the test statistic? If this is a two-tailed test, what is the p-value associated with the test?

c. What is the pooled estimate s^2 of the population variance?

d. Use the answers to part b to draw conclusions about the difference in the two population means.

e. Use the information in the printout to construct a 95% confidence interval for the difference in the population means. Does this interval confirm your conclusions in part d?

APPLICATIONS

10.31 Healthy Teeth Jan Lindhe conducted a study on the effect of an oral antiplaque rinse on plaque buildup on teeth.[7] Fourteen people whose teeth were thoroughly cleaned and polished were randomly assigned to two groups of seven subjects each. Both groups were assigned to use oral rinses (no brushing) for a two-week period. Group 1 used a rinse that contained an antiplaque agent. Group 2, the control group, received a similar rinse except that, unknown to the subjects, the rinse contained no antiplaque agent. A plaque index x, a measure of plaque buildup, was recorded at 4, 7, and 14 days. The mean and standard deviation for the 14-day plaque measurements are shown in the table for the two groups.

	Control Group	Antiplaque Group
Sample size	7	7
Mean	1.26	0.78
Standard deviation	0.32	0.32

a. State the null and alternative hypotheses that should be used to test the effectiveness of the antiplaque oral rinse.

b. Do the data provide sufficient evidence to indicate that the oral antiplaque rinse is effective? Test using $\alpha = 0.05$.

c. Find the approximate p-value for the test.

Data set EX1032 **10.32 Herbs, again** In Exercise 10.12, we presented data on the prices of parsley, cilantro, mint, and basil of approximately the same weight. A portion of the data is reproduced in the table below. Use the *MINITAB* printout to answer the questions.

Parsley		Basil	
0.99	0.53	2.56	0.62
1.92	1.41	1.92	0.66
1.23	1.12	1.30	0.62
0.85	0.63	1.79	0.65
0.65	0.67	1.23	0.60
0.69	0.60		0.67
0.60	0.66		

MINITAB output for Exercise 10.32

Two-Sample T-Test and CI: Parsley, Basil

```
Two sample T for Parsley vs Basil

         N      Mean     StDev    SE Mean
Water   14     0.896     0.400       0.11
Oil     11     1.147     0.679       0.20

Difference = mu (Parsley) - mu (Basil)
Estimate for difference: -0.250844
95% CI for difference: (-0.700004, 0.198316)
T-Test of difference = 0 (vs not =):
T-Value = -1.16 P-Value = 0.260 DF = 23
Both use Pooled StDev = 0.5389
```

a. Do the data in the table present sufficient evidence to indicate a difference in the average prices of parsley versus basil? Test using $\alpha = 0.05$.

b. What is the p-value for the test?

c. The *MINITAB* analysis uses the pooled estimate of σ^2. Is the assumption of equal variances reasonable? Why or why not?

10.33 Runners and Cyclists Chronic anterior compartment syndrome is a condition characterized by exercise-induced pain in the lower leg. Swelling and impaired nerve and muscle function also accompany this pain, which is relieved by rest. Susan Beckham and colleagues conducted an experiment involving 10 healthy runners and 10 healthy cyclists to determine whether there are significant differences in pressure measurements within the anterior muscle compartment for runners and cyclists.[8] The data summary—compartment pressure in millimetres of mercury (Hg)—is as follows:

	Runners		Cyclists	
Condition	Mean	Standard Deviation	Mean	Standard Deviation
Rest	14.5	3.92	11.1	3.98
80% maximal O_2 consumption	12.2	3.49	11.5	4.95
Maximal O_2 consumption	19.1	16.9	12.2	4.47

a. Test for a significant difference in compartment pressure between runners and cyclists under the resting condition. Use $\alpha = 0.05$.

b. Construct a 95% confidence interval estimate of the difference in means for runners and cyclists under the condition of exercising at 80% of maximal oxygen consumption.

c. To test for a significant difference in compartment pressure at maximal oxygen consumption, should you use the pooled or unpooled t test? Explain.

10.34 Disinfectants An experiment published in *The American Biology Teacher* studied the efficacy of using 95% ethanol or 20% bleach as a disinfectant in removing bacterial and fungal contamination when culturing plant tissues. The experiment was repeated 15 times with each disinfectant, using eggplant as the plant tissue being cultured.[9] Five cuttings per plant were placed on a petri dish for each disinfectant and stored at 25 °C for four weeks. The observation reported was the number of uncontaminated eggplant cuttings after the four-week storage.

Disinfectant	95% Ethanol	20% Bleach
Mean	3.73	4.80
Variance	2.78095	0.17143
n	15	15

Pooled variance 1.47619

a. Are you willing to assume that the underlying variances are equal?

b. Using the information from part a, are you willing to conclude that there is a significant difference in the mean numbers of uncontaminated eggplants for the two disinfectants tested?

Data set EX1035 **10.35 Titanium** A geologist collected 20 different ore samples, all of the same weight, and randomly divided them into two groups. The titanium contents of the samples, found using two different methods, are listed in the table:

Method 1					Method 2				
0.011	0.013	0.013	0.015	0.014	0.011	0.016	0.013	0.012	0.015
0.013	0.010	0.013	0.011	0.012	0.012	0.017	0.013	0.014	0.015

a. Use an appropriate method to test for a significant difference in the average titanium contents using the two different methods.

b. Determine a 95% confidence interval estimate for $(\mu_1 - \mu_2)$. Does your interval estimate substantiate your conclusion in part a? Explain.

Data set EX1036 **10.36 Raisins** The numbers of raisins in each of 14 miniboxes (15 g) were counted for a generic brand and for Sunmaid brand raisins:

Generic Brand				Sunmaid			
25	26	25	28	25	29	24	24
26	28	28	27	28	24	28	22
26	27	24	25	25	28	30	27
26	26			28	24		

a. Although counts cannot have a normal distribution, do these data have approximately normal distributions? (HINT: Use a histogram or stem and leaf plot.)

b. Are you willing to assume that the underlying population variances are equal? Why?

c. Use the p-value approach to determine whether there is a significant difference in the mean numbers of raisins per minibox. What are the implications of your conclusion?

10.37 Dissolved O_2 Content, continued Refer to Exercise 10.13, in which we measured the dissolved oxygen content in river water to determine whether a stream had sufficient oxygen to support aquatic life. A pollution control inspector suspected that a river community was releasing amounts of semitreated sewage into a river. To check his theory, he drew five randomly selected specimens of river water at a location above the town, and another five below. The dissolved oxygen readings (in parts per million) are as follows:

Above town	4.8	5.2	5.0	4.9	5.1
Below town	5.0	4.7	4.9	4.8	4.9

a. Do the data provide sufficient evidence to indicate that the mean oxygen content below the town is less than the mean oxygen content above? Test using $\alpha = 0.05$.

b. Suppose you prefer estimation as a method of inference. Estimate the difference in the mean dissolved oxygen contents for locations above and below the town. Use a 95% confidence interval.

Data set EX1038 **10.38 Freestyle Swimmers** In an effort to compare the average swimming times for two swimmers, each swimmer was asked to swim freestyle for a distance of 100 metres (m) at randomly selected times. The swimmers were thoroughly rested between laps and did not race against each other, so that each sample of times was an independent random sample. The times for each of 10 trials are shown for the two swimmers:

Swimmer 1		Swimmer 2	
59.62	59.74	59.81	59.41
59.48	59.43	59.32	59.63
59.65	59.72	59.76	59.50
59.50	59.63	59.64	59.83
60.01	59.68	59.86	59.51

Suppose that swimmer 2 was last year's winner when the two swimmers raced. Does it appear that the average time for swimmer 2 is still faster than the average time for swimmer 1 in the 100 m freestyle? Find the approximate p-value for the test and interpret the results.

10.39 Freestyle Swimmers, continued Refer to Exercise 10.38. Construct a lower 95% one-sided confidence bound for the difference in the average times for the two swimmers. Does this interval confirm your conclusions in Exercise 10.38?

Data set EX1040 **10.40 Comparing NHL Superstars** How does Mario Lemieux compare to Brett Hull? The table below shows number of goals scored for each player for selected years:[10]

Season	Mario Lemieux	Brett Hull
1986–87	54	1
1987–88	70	32
1988–89	85	41
1989–90	45	72

Season	Mario Lemieux	Brett Hull
1990–91	19	86
1991–92	44	70
1992–93	69	54
1993–94	17	57
1995–96	69	29
1996–97	50	42
2000–01	35	39
2001–02	6	30
2002–03	28	37
2003–04	1	25
2005–06	7	0

a. Does the data indicate that there is a difference in the average number of goals scored for the two players? Test using $\alpha = 0.05$.

b. Construct a 95% confidence interval for the difference in the average number of goals scored for the two players. Does the confidence interval confirm your conclusion in part a? Explain.

10.41 Comparing NFL Quarterbacks How does Aaron Rodgers, quarterback for the 2011 Super Bowl Champion Green Bay Packers, compare to Drew Brees, quarterback for the 2010 Super Bowl Champion New Orleans Saints? The table below shows the number of completed passes for each athlete during the 2010 NFL football season.[11] Use the *Excel* printout to answer the questions that follow.

Aaron Rodgers			Drew Brees		
19	21	7	27	37	25
19	15	25	28	34	29
34	27	19	30	27	35
12	22		33	29	22
27	26		24	23	
18	21		21	24	

Excel printout for Exercise 10.41

E	F	G
t-Test: Two-Sample Assuming Equal Variances		
	Rodgers	*Brees*
Mean	20.800	28.000
Variance	44.029	23.333
Observations	15.000	16.000
Pooled Variance	33.324	
Hypothesized Mean Difference	0.000	
df	29.000	
t Stat	-3.470	
P(T<=t) one-tail	0.001	
t Critical one-tail	1.699	
P(T<=t) two-tail	0.002	
t Critical two-tail	2.045	

a. The *Excel* analysis uses the pooled estimate of σ^2. Is the assumption of equal variances reasonable? Why or why not?

b. Do the data indicate that there is a difference in the average number of completed passes for the two quarterbacks? Test using $\alpha = 0.05$.

c. What is the p-value for the test?

d. Use the information given in the printout to construct a 95% confidence interval for the difference in the average number of completed passes for the two quarterbacks. Does the confidence interval confirm your conclusion in part b? Explain.

Data set EX1042 **10.42 An Archeological Find** An article in *Archaeometry* involved an analysis of 26 samples of Romano-British pottery, found at four different kiln sites in the United Kingdom.[12] The samples were analyzed to determine their chemical composition and the percentage of aluminum oxide in each of 10 samples at two sites is shown below:

Island Thorns	Ashley Rails
18.3	17.7
15.8	18.3
18.0	16.7
18.0	14.8
20.8	19.1

Does the data provide sufficient information to indicate that there is a difference in the average percentage of aluminum oxide at the two sites? Test at the 5% level of significance.

10.5 SMALL-SAMPLE INFERENCES FOR THE DIFFERENCE BETWEEN TWO MEANS: A PAIRED-DIFFERENCE TEST

To compare the wearing qualities of two types of automobile tires, A and B, a tire of type A and one of type B are randomly assigned and mounted on the rear wheels of each of five automobiles. The automobiles are then operated for a specified number of kilometres (km), and the amount of wear is recorded for each tire. These measurements appear in Table 10.4. Do the data present sufficient evidence to indicate a difference in the average wear for the two tire types?

TABLE 10.4 **Average Wear for Two Types of Tires**

Automobile	Tire A	Tire B
1	10.6	10.2
2	9.8	9.4
3	12.3	11.8
4	9.7	9.1
5	8.8	8.3
	$\bar{x}_1 = 10.24$	$\bar{x}_2 = 9.76$
	$s_1 = 1.316$	$s_2 = 1.328$

Table 10.4 shows a difference of $(\bar{x}_1 - \bar{x}_2) = (10.24 - 9.76) = 0.48$ between the two sample means, while the standard deviations of both samples are approximately 1.3. Given the variability of the data and the small number of measurements, this is a rather small difference, and you would probably not suspect a difference in the average wear for the two types of tires. Let's check your suspicions using the methods of Section 10.4.

Look at the *MINITAB* analysis in Figure 10.10. The two-sample *pooled t* test is used for testing the difference in the means based on two independent random samples. The calculated value of t used to test the null hypothesis $H_0 : \mu_1 = \mu_2$ is $t = 0.57$ with p-value $=$ 0.582, a value that is not nearly small enough to indicate a significant difference in the two population means. The corresponding 95% confidence interval, given as

$$-1.448 < (\mu_1 - \mu_2) < 2.408$$

is quite wide and also does not indicate a significant difference in the population means.

FIGURE 10.10
MINITAB output using *t* test for independent samples for the tire data

Two-Sample T-Test and CI: Tire A, Tire B

```
Two-sample T for Tire A vs Tire B
         N        Mean       StDev    SE Mean
Tire A   5       10.24        1.32       0.59
Tire B   5        9.76        1.33       0.59

Difference = mu (Tire A) - mu (Tire B)
Estimate for difference: 0.480000
95% CI for difference: (-1.448239, 2.408239)
T-Test of difference = 0 (vs not =): T-Value = 0.57  P-Value = 0.582 DF = 8
Both use Pooled StDev = 1.3221
```

Take a second look at the data and you will notice that the wear measurement for type A is greater than the corresponding value for type B for *each* of the five automobiles. Wouldn't this be unlikely, if there's really no difference between the two tire types?

Consider a simple intuitive test, based on the binomial distribution of Chapter 5. If there is no difference in the mean tire wear for the two types of tires, then it is just as likely as not that tire A shows more wear than tire B. The five automobiles then correspond to five binomial trials with $p = P$(tire A shows more wear than tire B) $= 0.5$. Is the observed value of $x = 5$ positive differences unusual? The probability of observing $x = 5$ or the equally unlikely value $x = 0$ can be found in Table 1 in Appendix I to be $2(0.031) = 0.062$, which is quite small compared to the likelihood of the more powerful t test, which had a p-value of 0.58. Isn't it peculiar that the t test, which uses more information (the actual sample measurements) than the binomial test, fails to supply sufficient information for rejecting the null hypothesis?

TABLE 10.5 **Differences in Tire Wear, Using the Data of Table 10.4**

Automobile	A	B	$d = A - B$
1	10.6	10.2	0.4
2	9.8	9.4	0.4
3	12.3	11.8	0.5
4	9.7	9.1	0.6
5	8.8	8.3	0.5
			$\bar{d} = 0.48$

There is an explanation for this inconsistency. The t test described in Section 10.4 is *not* the proper statistical test to be used for our example. The statistical test procedure of Section 10.4 requires that the two samples be *independent and random.* Certainly, the independence requirement is violated by the manner in which the experiment was conducted. The (pair of) measurements, an A and a B tire, for a particular automobile are definitely related. A glance at the data shows that the readings have approximately the same magnitude for a particular automobile but vary markedly from one automobile to another. This, of course, is exactly what you might expect. Tire wear is largely determined by driver habits, the balance of the wheels, and the road surface. Since each automobile has a different driver, you would expect a large amount of variability in the data from one automobile to another.

In designing the tire wear experiment, the experimenter realized that the measurements would vary greatly from automobile to automobile. If the tires (five of type A and five of type B) were randomly assigned to the ten wheels, resulting in *independent* random samples, this variability would result in a large standard error and make it difficult to detect a difference in the means. Instead, he chose to "pair" the measurements, comparing the wear for type A and type B tires on each of the five automobiles. This experimental design, sometimes called a **paired-difference** or **matched pairs** design, allows us to eliminate the car-to-car variability by looking at only the five difference measurements shown in Table 10.5. These five differences form a single random sample of size $n = 5$.

Notice that in Table 10.5 the sample mean of the differences, $d = \text{A} - \text{B}$, is calculated as

$$\bar{d} = \frac{\Sigma d_i}{n} = 0.48$$

and is exactly the same as the difference of the sample means: $(\bar{x}_1 - \bar{x}_2) = (10.24 - 9.76) = 0.48$. It should not surprise you that this can be proven to be true in general, and also that the same relationship holds for the population means. That is, the average of the population differences is

$$\mu_d = (\mu_1 - \mu_2)$$

Because of this fact, you can use the sample differences to test for a significant difference in the two population means, $(\mu_1 - \mu_2) = \mu_d$. The test is a single-sample t test of the difference measurements to test the null hypothesis

$$H_0: \mu_d = 0 \;\; [\text{or } H_0: (\mu_1 - \mu_2) = 0]$$

versus the alternative hypothesis

$$H_\text{a}: \mu_d \neq 0 \;\; [\text{or } H_\text{a}: (\mu_1 - \mu_2) \neq 0]$$

The test procedures take the same form as the procedures used in Section 10.3 and are described next.

PAIRED-DIFFERENCE TEST OF HYPOTHESIS FOR $(\mu_1 - \mu_2) = \mu_d$: DEPENDENT SAMPLES

1. Null hypothesis: $H_0 : \mu_d = 0$
2. Alternative hypothesis:

One-Tailed Test	Two-Tailed Test
$H_a : \mu_d > 0$ (or $H_a : \mu_d < 0$)	$H_a : \mu_d \neq 0$

3. Test statistic: $t = \dfrac{\bar{d} - 0}{s_d/\sqrt{n}} = \dfrac{\bar{d}}{s_d/\sqrt{n}}$

   where n = Number of paired differences
   $\bar{d}$ = Mean of the sample differences
   s_d = Standard deviation of the sample differences

4. Rejection region: Reject H_0 when

One-Tailed Test	Two-Tailed Test
$t > t_\alpha$ (or $t < -t_\alpha$ when the alternative hypothesis is $H_a : \mu_d < 0$)	$t > t_{\alpha/2}$ or $t < -t_{\alpha/2}$

   or when p-value $< \alpha$

The critical values of t, t_α, and $t_{\alpha/2}$ are based on $(n - 1)$ *df*. These tabulated values can be found using Table 4 in Appendix I.

$(1 - \alpha)100\%$ SMALL-SAMPLE CONFIDENCE INTERVAL FOR $(\mu_1 - \mu_2) = \mu_d$, BASED ON A PAIRED-DIFFERENCE EXPERIMENT

$$\bar{d} \pm t_{\alpha/2}\left(\frac{s_d}{\sqrt{n}}\right)$$

Assumptions: The experiment is designed as a paired-difference test so that the n differences represent a random sample from a normal population.

Do the data in Table 10.4 provide sufficient evidence to indicate a difference in the mean wear for tire types A and B? Test using $\alpha = 0.05$.

SOLUTION You can verify using your calculator that the average and standard deviation of the five difference measurements are

$$\bar{d} = 0.48 \text{ and } s_d = 0.0837$$

Then

$$H_0 : \mu_d = 0 \text{ and } H_a : \mu_d \neq 0$$

and

$$t = \frac{\bar{d} - 0}{s_d/\sqrt{n}} = \frac{0.48}{0.0837/\sqrt{5}} = 12.8$$

The critical value of t for a two-tailed statistical test, $\alpha = 0.05$ and 4 *df*, is 2.776. Certainly, the observed value of $t = 12.8$ is extremely large and highly significant. Hence, you can conclude that there is a difference in the mean wear for tire types A and B.

EXAMPLE  10.13

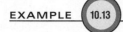

Find a 95% confidence interval for $(\mu_1 - \mu_2) = \mu_d$ using the data in Table 10.4.

SOLUTION A 95% confidence interval for the difference between the mean wears is

$$\bar{d} \pm t_{\alpha/2}\left(\frac{s_d}{\sqrt{n}}\right)$$

$$0.48 \pm 2.776\left(\frac{0.0837}{\sqrt{5}}\right)$$

$$0.48 \pm 0.10$$

or $0.38 < (\mu_1 - \mu_2) < 0.58$. How does the width of this interval compare with the width of an interval you might have constructed *if* you had designed the experiment in an unpaired manner? It probably would have been of the same magnitude as the interval calculated in Figure 10.10, where the observed data were *incorrectly* analyzed using the unpaired analysis. This interval, $-1.45 < (\mu_1 - \mu_2) < 2.41$, is much wider than the paired interval, which indicates that the paired difference design increased the accuracy of our estimate, and we have gained valuable information by using this design.

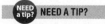

NEED a tip? **NEED A TIP?**

Confidence intervals are always interpreted in the same way! In repeated sampling, intervals constructed in this way enclose the true value of the parameter $(1 - \alpha)$ 100% of the time.

The *paired-difference test* or *matched pairs design* used in the tire wear experiment is a simple example of an experimental design called a **randomized block design.** When there is a great deal of variability among the experimental units, even before any experimental procedures are implemented, the effect of this variability can be minimized by **blocking**—that is, comparing the different procedures within groups of relatively similar experimental units called **blocks.** In this way, the "noise" caused by the large variability does not mask the true differences between the procedures. We will discuss randomized block designs in more detail in Chapter 11.

Paired difference test: $df = n - 1$

It is important for you to remember that the *pairing* or *blocking* occurs when the experiment is planned, and not after the data are collected. An experimenter may choose to use pairs of identical twins to compare two learning methods. A physician may record a patient's blood pressure before and after a particular medication is given. Once you have used a paired design for an experiment, you no longer have the option of using the unpaired analysis of Section 10.4. The independence assumption has been purposely violated, and your only choice is to use the paired analysis described here!

Although pairing was very beneficial in the tire wear experiment, this may not always be the case. In the paired analysis, the degrees of freedom for the t test are cut in half—from $(n + n - 2) = 2(n - 1)$ to $(n - 1)$. This reduction *increases* the critical value of t for rejecting H_0 and also increases the width of the confidence interval for the difference in the two means. If pairing is not effective, this increase is not offset by a *decrease* in the variability, and you may in fact lose rather than gain information by pairing. This, of course, did not happen in the tire experiment—the large reduction in the standard error more than compensated for the loss in degrees of freedom.

Except for notation, the paired-difference analysis is the same as the single-sample analysis presented in Section 10.3. However, *MINITAB* provides a single procedure called **Paired** *t* to analyze the differences, as shown in Figure 10.11. The *p*-value for the paired analysis, 0.000, indicates a *highly significant* difference in the means. You will find instructions for generating this *MINITAB* output in the "Technology Today" section at the end of this chapter.

FIGURE 10.11
MINITAB output for paired-difference analysis of tire wear data

Paired T-Test and CI: Tire A, Tire B

```
Paired T for Tire A - Tire B
              N       Mean       StDev   SE Mean
Tire A        5    10.2400      1.3164    0.5887
Tire B        5     9.7600      1.3278    0.5938
Difference    5   0.480000    0.083666  0.037417

95% CI for mean difference: (0.376115, 0.583885)
T-Test of mean difference = 0 (vs not = 0): T-Value = 12.83  P-Value = 0.000
```

10.5 EXERCISES

BASIC TECHNIQUES

10.43 A paired-difference experiment was conducted using $n = 10$ pairs of observations.

a. Test the null hypothesis $H_0 : (\mu_1 - \mu_2) = 0$ against $H_a : (\mu_1 - \mu_2) \neq 0$ for $\alpha = 0.05$, $\bar{d} = 0.3$, and $s_d^2 = 0.16$. Give the approximate *p*-value for the test.

b. Find a 95% confidence interval for $(\mu_1 - \mu_2)$.

c. How many pairs of observations do you need if you want to estimate $(\mu_1 - \mu_2)$ correct to within 0.1 with probability equal to 0.95?

10.44 A paired-difference experiment consists of $n = 18$ pairs, $\bar{d} = 5.7$, and $s_d^2 = 256$. Suppose you wish to detect $\mu_d > 0$.

a. Give the null and alternative hypotheses for the test.

b. Conduct the test and state your conclusions.

10.45 A paired-difference experiment was conducted to compare the means of two populations:

	Pairs				
Population	1	2	3	4	5
1	1.3	1.6	1.1	1.4	1.7
2	1.2	1.5	1.1	1.2	1.8

a. Do the data provide sufficient evidence to indicate that μ_1 differs from μ_2? Test using $\alpha = 0.05$.

b. Find the approximate *p*-value for the test and interpret its value.

c. Find a 95% confidence interval for $(\mu_1 - \mu_2)$. Compare your interpretation of the confidence interval with your test results in part a.

d. What assumptions must you make for your inferences to be valid?

APPLICATIONS

EX1046 **10.46 Auto Insurance** The cost of automobile insurance has become a sore subject because the rates are dependent on so many variables, such as the province in which you live, the number of cars you insure, and the company with which you are insured. The table is adapted from a publication of Fraser Institute Digital Publication, February 2007, and lists estimated average auto insurance premiums in 10 provinces, 2004–2005.[13]

Province	Avg. Earned Premium in 2005	Avg. Earned Premium in 2004
BC	$1404	$1374
ON	$1374	$1396
SK	$1197	$1174
MB	$1152	$1157
NB	$1044	$1121
AB	$1036	$1126
QC	$988	$983
NL	$947	$1014
NS	$871	$883
PE	$825	$847

Source: The Fraser Institute, "Table 1: Estimated Average Automobile Insurance Premiums, 2004–05, by Province, Straight Dollars," from *The False Promise of Government Auto Insurance*, p. 5, by Brett J. Skinner, February 2007. http://www.fraserinstitute.ca. Available at http://www.fraserinstitute.org/studies/false-promise-government-auto-insurance.

a. Why would you expect these pairs of observations to be dependent?

b. Do the data provide sufficient evidence to indicate that there is a difference in the average premiums between 2004 and 2005? Test using $\alpha = 0.01$.

c. Find the approximate p-value for the test and interpret its value.

d. Find a 99% confidence interval for the difference in the average premiums for the years 2004 and 2005.

e. Can we use the information in the table to make valid comparisons for Saskatchewan (SK) for the given years? Why or why not?

10.47 Runners and Cyclists II Refer to Exercise 10.33. In addition to the compartment pressures, the level of creatine phosphokinase (CPK) in blood samples, a measure of muscle damage, was determined for each of 10 runners and 10 cyclists before and after exercise.[14] The data summary—CPK values in units/litre—is as follows:

	Runners		Cyclists	
Condition	Mean	Standard Deviation	Mean	Standard Deviation
Before exercise	255.63	115.48	173.8	60.69
After exercise	284.75	132.64	177.1	64.53
Difference	29.13	21.01	3.3	6.85

a. Test for a significant difference in mean CPK values for runners and cyclists before exercise under the assumption that $\sigma_1^2 \neq \sigma_2^2$; use $\alpha = 0.05$. Find a 95% confidence interval estimate for the corresponding difference in means.

b. Test for a significant difference in mean CPK values for runners and cyclists after exercise under the assumption that $\sigma_1^2 \neq \sigma_2^2$; use $\alpha = 0.05$. Find a 95% confidence interval estimate for the corresponding difference in means.

c. Test for a significant difference in mean CPK values for runners before and after exercise.

d. Find a 95% confidence interval estimate for the difference in mean CPK values for cyclists before and after exercise. Does your estimate indicate that there is no significant difference in mean CPK levels for cyclists before and after exercise?

Data set EX1048 **10.48 Canada's Food Basics** An advertisement for Food Basics claims that Food Basics has had consistently lower prices than four other full-service supermarkets. As part of a survey conducted by an "independent market basket price-checking company," the average weekly total, based on the prices (in \$) of approximately 95 items, is given for two different supermarket chains recorded during four consecutive weeks in a particular month:

Week	Food Basics	Zehrs
1	254.26	256.03
2	240.62	255.65
3	231.90	255.12
4	234.13	261.18

a. Is there a significant difference in the average prices for these two different supermarket chains?

b. What is the approximate p-value for the test conducted in part a?

c. Construct a 99% confidence interval for the difference in the average prices for the two supermarket chains. Interpret this interval.

10.49 No Left Turn An experiment was conducted to compare the mean reaction times to two types of traffic signs: prohibitive (No Left Turn) and permissive (Left Turn Only). Ten drivers were included in the experiment. Each driver was presented with 40 traffic signs, 20 prohibitive and 20 permissive, in random order. The mean time to reaction (in milliseconds) was recorded for each driver and is shown here:

Driver	Prohibitive	Permissive
1	824	702
2	866	725
3	841	744
4	770	663
5	829	792
6	764	708
7	857	747
8	831	685
9	846	742
10	759	610

Excel printout for Exercise 10.49

D	E	F
t-Test: Paired Two Sample for Means		
	Prohibitive	*Permissive*
Mean	818.7	711.8
Variance	1573.344444	2596.4
Observations	10	10
Pearson Correlation	0.693852753	
Hypothesized Mean Difference	0	
df	9	
t Stat	9.14983257	
P(T<=t) one-tail	3.72891E-06	
t Critical one-tail	1.833112933	
P(T<=t) two-tail	7.45782E-06	
t Critical two-tail	2.262157163	

a. Explain why this is a paired-difference experiment, and give reasons why the pairing should be useful in increasing information on the difference between the mean reaction times to prohibitive and permissive traffic signs.

b. Use the *Excel* printout to determine whether there is a significant difference in mean reaction times to prohibitive and permissive traffic signs. Use the *p*-value approach.

10.50 Healthy Teeth II Exercise 10.31 describes a dental experiment conducted to investigate the effectiveness of an oral rinse used to inhibit the growth of plaque on teeth. Subjects were divided into two groups: One group used a rinse with an antiplaque ingredient, and the control group used a rinse containing inactive ingredients. Suppose that the plaque growth on each person's teeth was measured after using the rinse after four hours and then again after eight hours. If you wish to estimate the difference in plaque growth from four to eight hours, should you use a confidence interval based on a paired or an unpaired analysis? Explain.

10.51 Ground or Air? The earth's temperature (which affects seed germination, crop survival in bad weather, and many other aspects of agricultural production) can be measured using either ground-based sensors or infrared-sensing devices mounted in aircraft or space satellites. Ground-based sensoring is tedious, requiring many replications to obtain an accurate estimate of ground temperature. On the other hand, airplane or satellite sensoring of infrared waves appears to introduce a bias in the temperature readings. To determine the bias, readings were obtained at five different locations using both ground- and air-based temperature sensors. The readings (in degrees Celsius) are listed here:

Location	Ground	Air
1	46.9	47.3
2	45.4	48.1
3	36.3	37.9
4	31.0	32.7
5	24.7	26.2

a. Do the data present sufficient evidence to indicate a bias in the air-based temperature readings? Explain.

b. Estimate the difference in mean temperatures between ground- and air-based sensors using a 95% confidence interval.

c. How many paired observations are required to estimate the difference between mean temperatures for ground- versus air-based sensors correct to within 0.2 °C, with probability approximately equal to 0.95?

Data set EX1052 **10.52 Red Dye** To test the comparative brightness of two red dyes, nine samples of cloth were taken from a production line and each sample was divided into two pieces. One of the two pieces in each sample was randomly chosen and red dye 1 applied; red dye 2 was applied to the remaining piece. The following data represent a "brightness score" for each piece. Is there sufficient evidence to indicate a difference in mean brightness scores for the two dyes? Use $\alpha = 0.05$.

Sample	1	2	3	4	5	6	7	8	9
Dye 1	10	12	9	8	15	12	9	10	15
Dye 2	8	11	10	6	12	13	9	8	13

Data set EX1053 **10.53 Tax Assessors** In response to a complaint that a particular tax assessor (A) was biased, an experiment was conducted to compare the assessor named in the complaint with another tax assessor (B) from the same office. Eight properties were selected, and each was assessed by both assessors. The assessments (in thousands of dollars) are shown in the table:

Property	Assessor A	Assessor B
1	76.3	75.1
2	88.4	86.8
3	80.2	77.3
4	94.7	90.6
5	68.7	69.1
6	82.8	81.0
7	76.1	75.3
8	79.0	79.1

Use the *MINITAB* printout to answer the questions.

MINITAB output for Exercise 10.53

Paired T-Test and CI: Assessor A, Assessor B

```
Paired T for Assessor A - Assessor B

              N      Mean     StDev   SE Mean
Assessor A    8   80.7750    7.9939    2.8263
Assessor B    8   79.2875    6.8510    2.4222
Difference    8   1.48750   1.49134   0.52727

95% lower bound for mean difference: 0.48855
T-Test of mean difference = 0 (vs > 0):
T-Value = 2.82 P-value = 0.013
```

a. Do the data provide sufficient evidence to indicate that assessor A tends to give higher assessments than assessor B?

b. Estimate the difference in mean assessments for the two assessors.

c. What assumptions must you make in order for the inferences in parts a and b to be valid?

d. Suppose that assessor A had been compared with a more stable standard—say, the average $\bar{x}$ of the assessments given by four assessors selected from the tax office. Thus, each property would be assessed by A and also by each of the four other assessors and $(x_A - \bar{x})$ would be calculated. If the test in part a is valid, can you use the paired-difference t test to test the hypothesis that the bias, the mean difference between A's assessments and the mean of the assessments of the four assessors, is equal to 0? Explain.

Data set EX1054 **10.54 Memory Experiments** A psychology class performed an experiment to compare whether a recall score in which instructions to form images of 25 words were given is better than an initial recall score for which no imagery instructions were given. Twenty students participated in the experiment with the following results:

Student	With Imagery	Without Imagery	Student	With Imagery	Without Imagery
1	20	5	11	17	8
2	24	9	12	20	16
3	20	5	13	20	10
4	18	9	14	16	12
5	22	6	15	24	7
6	19	11	16	22	9
7	20	8	17	25	21
8	19	11	18	21	14
9	17	7	19	19	12
10	21	9	20	23	13

Does it appear that the average recall score is higher when imagery is used?

Data set EX1055 **10.55 Music in the Workplace** Before contracting to have stereo music piped into each of his suites of offices, an executive had his office manager randomly select seven offices in which to have the system installed. The average time (in minutes) spent outside these offices per excursion among the employees involved was recorded before and after the music system was installed with the following results:

Office number	1	2	3	4	5	6	7
No music	8	9	5	6	5	10	7
Music	5	6	7	5	6	7	8

Would you suggest that the executive proceed with the installation? Conduct an appropriate test of hypothesis. Find the approximate p-value and interpret your results.

10.6 INFERENCES CONCERNING A POPULATION VARIANCE

You have seen in the preceding sections that an estimate of the population variance σ^2 is usually needed before you can make inferences about population means. Sometimes, however, the population variance σ^2 is the primary objective in an experimental investigation. It may be *more* important to the experimenter than the population mean! Consider these examples:

- Scientific measuring instruments must provide unbiased readings with a very small error of measurement. An aircraft altimeter that measures the correct altitude on the *average* is fairly useless if the measurements are in error by as much as 300 metres above or below the correct altitude.
- Machined parts in a manufacturing process must be produced with minimum variability in order to reduce out-of-size and hence defective parts.
- Aptitude tests must be designed so that scores *will* exhibit a reasonable amount of variability. For example, an 800-point test is not very discriminatory if all students score between 601 and 605.

In previous chapters, you have used

$$s^2 = \frac{\Sigma(x_i - \bar{x})^2}{n - 1}$$

as an unbiased estimator of the population variance σ^2. This means that, in repeated sampling, the average of all your sample estimates will equal the target parameter, σ^2. But how close or far from the target is your estimator s^2 likely to be? To answer this question, we use the sampling distribution of s^2, which describes its behaviour in repeated sampling.

Consider the distribution of s^2 based on repeated *random* sampling from a *normal* distribution with a specified mean and variance. We can show theoretically that the distribution begins at $s^2 = 0$ (since the variance cannot be negative) with a mean equal to σ^2. Its shape is *non-symmetric* and changes with each different sample size and each different value of σ^2. Finding critical values for the sampling distribution of s^2 would be quite difficult and would require separate tables for each population variance. Fortunately, we can simplify the problem by *standardizing,* as we did with the z distribution.

DEFINITION The standardized statistic

$$\chi^2 = \frac{(n - 1)s^2}{\sigma^2}$$

is called a **chi-square variable** and has a sampling distribution called the **chi-square probability distribution,** with $n - 1$ degrees of freedom.

The equation of the density function for this statistic is quite complicated to look at, but it traces the curve shown in Figure 10.12.

FIGURE 10.12
A chi-square distribution

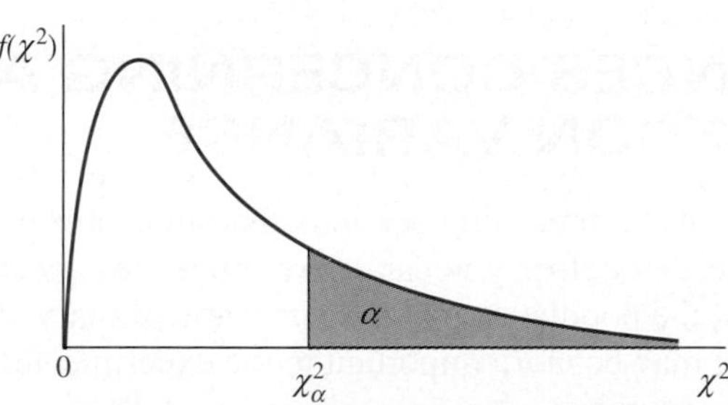

Certain critical values of the chi-square statistic, which are used for making inferences about the population variance, have been tabulated by statisticians and appear in Table 5 of Appendix I. Since the shape of the distribution varies with the sample size n or, more precisely, the degrees of freedom, $n - 1$, associated with s^2, Table 5, partially reproduced in Table 10.6, is constructed in exactly the same way as the t table, with the degrees of freedom in the first and last columns. The symbol χ^2_α indicates that the tabulated χ^2-value has an area α to its right (see Figure 10.12).

TABLE 10.6 **Format of the Chi-Square Table from Table 5 in Appendix I**

df	$\chi^2_{0.995}$	...	$\chi^2_{0.950}$	$\chi^2_{0.900}$	$\chi^2_{0.100}$	$\chi^2_{0.050}$	...	$\chi^2_{0.005}$	df
1	0.0000393		0.0039321	0.0157908	2.70554	3.84146		7.87944	1
2	0.0100251		0.102587	0.210720	4.60517	5.99147		10.5966	2
3	0.0717212		0.351846	0.584375	6.25139	7.81473		12.8381	3
4	0.206990		0.710721	1.063623	7.77944	9.48773		14.8602	4
5	0.411740		1.145476	1.610310	9.23635	11.0705		16.7496	5
6	0.0675727		1.63539	2.204130	10.6446	12.5916		18.5476	6
.	.		.	.	.	.		.	.
.	.		.	.	.	.		.	.
.	.		.	.	.	.		.	.
15	4.60094		7.26094	8.54675	22.3072	24.9958		32.8013	15
16	5.14224		7.96164	9.31223	23.5418	26.2962		34.2672	16
17	5.69724		8.67176	10.0852	24.7690	27.5871		35.7185	17
18	6.26481		9.39046	10.8649	25.9894	28.8693		37.1564	18
19	6.84398		10.1170	11.6509	27.2036	30.1435		38.5822	19
.	.		.	.	.	.		.	.
.	.		.	.	.	.		.	.
.	.		.	.	.	.		.	.

NEED A TIP?

Testing one variance: $df = n = 1$

ONLINE APPLET

Chi-Square Probabilities

You can see in Table 10.6 that, because the distribution is non-symmetric and starts at 0, both upper and lower tail areas must be tabulated for the chi-square statistic. For example, the value $\chi^2_{0.95}$ is the value that has 95% of the area under the curve to its right and 5% of the area to its left. This value cuts off an area equal to 0.05 in the lower tail of the chi-square distribution.

EXAMPLE 10.14

Check your ability to use Table 5 in Appendix I by verifying the following statements:

1. The probability that χ^2, based on $n = 16$ measurements ($df = 15$), exceeds 24.9958 is 0.05.
2. For a sample of $n = 6$ measurements, 95% of the area under the χ^2 distribution lies to the right of 1.145476.

These values are shaded in Table 10.6.

The statistical test of a null hypothesis concerning a population variance

$$H_0 : \sigma^2 = \sigma_0^2$$

uses the test statistic

$$\chi^2 = \frac{(n-1)s^2}{\sigma_0^2}$$

Notice that when H_0 is true, s^2/σ_0^2 should be near 1, so χ^2 should be close to $(n - 1)$, the degrees of freedom. If σ^2 is really greater than the hypothesized value σ_0^2, the test statistic will tend to be larger than $(n - 1)$ and will probably fall toward the upper tail of the distribution. If $\sigma^2 < \sigma_0^2$, the test statistic will tend to be smaller than $(n - 1)$ and will probably fall toward the lower tail of the chi-square distribution. As in other testing situations, you may use either a one- or a two-tailed statistical test, depending on the alternative hypothesis. This test of hypothesis and the $(1 - \alpha)100\%$ confidence interval for σ^2 are both based on the chi-square distribution and are described next.

TEST OF HYPOTHESIS CONCERNING A POPULATION VARIANCE

1. Null hypothesis: $H_0 : \sigma^2 = \sigma_0^2$
2. Alternative hypothesis:

One-Tailed Test

$H_a : \sigma^2 > \sigma_0^2$
(or $H_a : \sigma^2 < \sigma_0^2$)

Two-Tailed Test

$H_a : \sigma^2 \neq \sigma_0^2$

3. Test statistic: $\chi^2 = \dfrac{(n-1)s^2}{\sigma_0^2}$
4. Rejection region: Reject H_0 when

One-Tailed Test

$\chi^2 > \chi^2_\alpha$
(or $\chi^2 < \chi^2_{(1-\alpha)}$ when the alternative hypothesis is $H_a{:}\sigma^2 < \sigma_0^2$), where χ^2_α and $\chi^2_{(1-\alpha)}$ are, respectively, the upper- and lower-tail values of χ^2 that place α in the tail areas

Two-Tailed Test

$\chi^2 > \chi^2_{\alpha/2}$ or $\chi^2 < \chi^2_{(1-\alpha/2)}$, where $\chi^2_{\alpha/2}$ and $\chi^2_{(1-\alpha/2)}$ are, respectively, the upper- and lower-tail values of χ^2 that place $\alpha/2$ in the tail areas

or when p-value $< \alpha$

The critical values of χ^2 are based on $(n - 1)$ *df*. These tabulated values can be found using Table 5 of Appendix I.

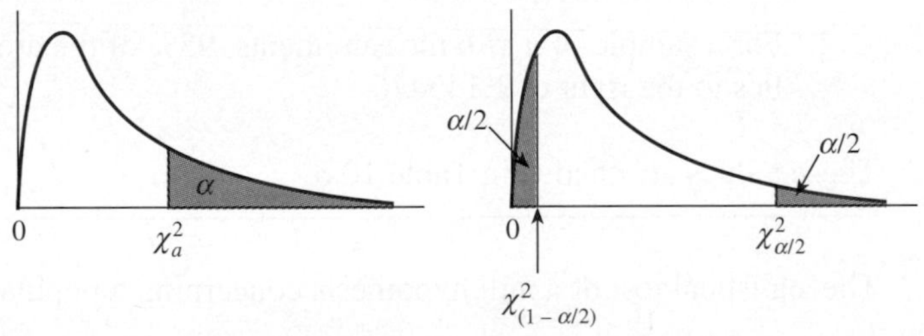

(1 − α)100% CONFIDENCE INTERVAL FOR σ²

$$\frac{(n-1)s^2}{\chi^2_{\alpha/2}} < \sigma^2 < \frac{(n-1)s^2}{\chi^2_{(1-\alpha/2)}}$$

where $\chi^2_{\alpha/2}$ and $\chi^2_{(1-\alpha/2)}$ are the upper and lower χ^2-values, which locate one-half of α in each tail of the chi-square distribution.

Assumption: The sample is randomly selected from a normal population.

EXAMPLE 10.15

A cement manufacturer claims that concrete prepared from its product has a relatively stable compressive strength and that the strength measured in kilograms per square centimetre (kg/cm^2) lies within a range of 40 kg/cm^2. A sample of $n = 10$ measurements produced a mean and variance equal to, respectively,

$$\bar{x} = 312 \text{ and } s^2 = 195$$

Do these data present sufficient evidence to reject the manufacturer's claim?

SOLUTION In Section 2.5, you learned that the range of a set of measurements should be approximately four standard deviations. The manufacturer's claim that the range of the strength measurements is within 40 kg/cm^2 must mean that the standard deviation of the measurements is roughly 10 kg/cm^2 or less. To test the claim, the appropriate hypotheses are

$$H_0: \sigma^2 = 10^2 = 100 \text{ versus } H_a: \sigma^2 > 100$$

If the sample variance is much larger than the hypothesized value of 100, then the test statistic

$$\chi^2 = \frac{(n-1)s^2}{\sigma_0^2} = \frac{1755}{100} = 17.55$$

will be unusually large, favouring rejection of H_0 and acceptance of H_a. There are two ways to use the test statistic to make a decision for this test:

- **The critical value approach:** The appropriate test requires a one-tailed rejection region in the right tail of the χ^2 distribution. The critical value for $\alpha = 0.05$ and $(n - 1) = 9$ *df* is $\chi^2_{0.05} = 16.9190$ from Table 5 in Appendix I. Figure 10.13 shows the rejection region; you can reject H_0 if the test statistic exceeds 16.9190. Since the observed value of the test statistic is $\chi^2 = 17.55$, you can conclude that the null hypothesis is false and that the range of concrete strength measurements exceeds the manufacturer's claim.

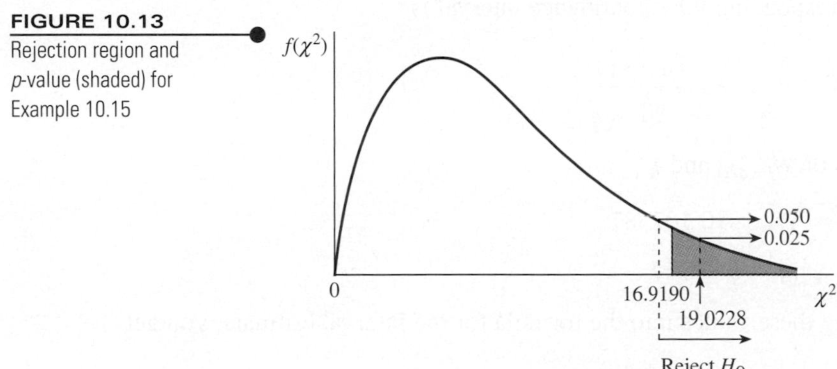

FIGURE 10.13 Rejection region and *p*-value (shaded) for Example 10.15

- **The *p*-value approach:** The *p*-value for a statistical test is the smallest value of α for which H_0 can be rejected. It is calculated, as in other one-tailed tests, as the area in the tail of the χ^2 distribution to the right of the observed value, $\chi^2 = 17.55$. Although computer packages allow you to calculate this area exactly, Table 5 in Appendix I allows you only to bound the *p*-value. Since the value 17.55 lies between $\chi^2_{0.050} = 16.9190$ and $\chi^2_{0.025} = 19.0228$, the *p*-value lies between 0.025 and 0.05. Most researchers would reject H_0 and report these results as significant at the 5% level, or $P < 0.05$. Again, you can reject H_0 and conclude that the range of measurements exceeds the manufacturer's claim.

EXAMPLE 10.16

An experimenter is convinced that her measuring instrument had a variability measured by standard deviation $\sigma = 2$. During an experiment, she recorded the measurements 4.1, 5.2, and 10.2. Do these data confirm or disprove her assertion? Test the appropriate hypothesis, and construct a 90% confidence interval to estimate the true value of the population variance.

SOLUTION Since there is no preset level of significance, you should choose to use the p-value approach in testing these hypotheses:

$$H_0 : \sigma^2 = 4 \text{ versus } H_a : \sigma^2 \neq 4$$

Use your scientific calculator to verify that the sample variance is $s^2 = 10.57$ and the test statistic is

$$\chi^2 = \frac{(n-1)s^2}{\sigma_0^2} = \frac{2(10.57)}{4} = 5.29$$

Since this is a two-tailed test, the rejection region is divided into two parts, half in each tail of the χ^2 distribution. If you approximate the area to the right of the observed test statistic, $\chi^2 = 5.29$, you will have only *half* of the p-value for the test. Since an equally unlikely value of χ^2 might occur in the lower tail of the distribution, with equal probability, you must *double* the upper area to obtain the p-value. With 2 *df,* the observed value, 5.29, falls between $\chi^2_{0.10}$ and $\chi^2_{0.05}$ so that

$$0.05 < \frac{1}{2}(p - \text{value}) < 0.10 \text{ or } 0.10 < p - \text{value} < 0.20$$

Since the p-value is greater than 0.10, the results are not statistically significant. There is insufficient evidence to reject the null hypothesis $H_0 : \sigma^2 = 4$.

The corresponding 90% confidence interval is

$$\frac{(n-1)s^2}{\chi^2_{\alpha/2}} < \sigma^2 < \frac{(n-1)s^2}{\chi^2_{(1-\alpha/2)}}$$

The values of $\chi^2_{(1-\alpha/2)}$ and $\chi^2_{\alpha/2}$ are

$$\chi^2_{(1-\alpha/2)} = \chi^2_{0.95} = 0.102587$$

$$\chi^2_{(\alpha/2)} = \chi^2_{0.05} = 5.99147$$

Substituting these values into the formula for the interval estimate, you get

$$\frac{2(10.57)}{5.99147} < \sigma^2 < \frac{2(10.57)}{0.102587} \quad \text{or} \quad 3.53 < \sigma^2 < 206.07$$

Thus, you can estimate the population variance to fall into the interval 3.53 to 206.07. This very wide confidence interval indicates how little information on the population variance is obtained from a sample of only three measurements. Consequently, it is not surprising that there is insufficient evidence to reject the null hypothesis $\sigma^2 = 4$. To obtain more information on σ^2, the experimenter needs to increase the sample size.

Although *Excel* does not have a single command to implement these procedures, you can use the function tool in *Excel* to find the test statistic, the p-value, and/or the upper and lower confidence limits yourself. If you use *MINITAB,* the command **Stat ▶ Basic Statistics ▶ 1 Variance** allows you to enter either raw data or summary statistics to perform the chi-square test for a single variance, and calculate a confidence interval. The pertinent part of the *MINITAB 16* printout for Example 10.16 is shown in Figure 10.14.

FIGURE 10.14
MINITAB output for Example 10.16

Test and CI for One Variance: Measurements

```
Null hypothesis               Sigma-squared = 4
Alternative hypothesis        Sigma-squared not = 4

Statistics
Variable       N    StDev     Variance
Measurements   3     3.25     10.6
90% Confidence Intervals
CI for         CI for         StDev              Variance
Variable       Method
Measurements   Chi-Square     (1.88, 14.36)      (3.5, 206.1)

Test
Variable       Method         Statistic          DF  P-Value
Measurements   Chi-Square     5.28               2   0.142
```

10.6 EXERCISES

BASIC TECHNIQUES

10.56 A random sample of $n = 25$ observations from a normal population produced a sample variance equal to 21.4. Do these data provide sufficient evidence to indicate that $\sigma^2 > 15$? Test using $\alpha = 0.05$.

10.57 A random sample of $n = 15$ observations was selected from a normal population. The sample mean and variance were $\bar{x} = 3.91$ and $s^2 = 0.3214$. Find a 90% confidence interval for the population variance σ^2.

10.58 A random sample of size $n = 7$ from a normal population produced these measurements: 1.4, 3.6, 1.7, 2.0, 3.3, 2.8, 2.9.

a. Calculate the sample variance, s^2.

b. Construct a 95% confidence interval for the population variance, σ^2.

c. Test $H_0: \sigma^2 = 0.8$ versus $H_a: \sigma^2 \neq 0.8$ using $\alpha = 0.05$. State your conclusions.

d. What is the approximate p-value for the test in part c?

APPLICATIONS

10.59 Instrument Precision A precision instrument is guaranteed to read accurately to within 2 units. A sample of four instrument readings on the same object yielded the measurements 353, 351, 351, and 355. Test the null hypothesis that $\sigma = 0.7$ against the alternative $\sigma > 0.7$. Use $\alpha = 0.05$.

10.60 Instrument Precision, continued Find a 90% confidence interval for the population variance in Exercise 10.59.

10.61 Drug Potency To properly treat patients, drugs prescribed by physicians must have a potency that is accurately defined. Consequently, not only must the distribution of potency values for shipments of a drug have a mean value as specified on the drug's container, but also the variation in potency must be small. Otherwise, pharmacists would be distributing drug prescriptions that could be harmfully potent or have a low potency and be ineffective. A drug manufacturer claims that its drug is marketed with a potency of 5 ± 0.1 milligram per cubic centimetre (mg/cc). A random sample of four containers gave potency readings equal to 4.94, 5.09, 5.03, and 4.90 mg/cc.

a. Do the data present sufficient evidence to indicate that the mean potency differs from 5 mg/cc?

b. Do the data present sufficient evidence to indicate that the variation in potency differs from the error limits specified by the manufacturer? (HINT: It is sometimes difficult to determine exactly what is meant by limits on potency as specified by a manufacturer. Since it implies that the potency values will fall into the interval 5.0 ± 0.1 mg/cc with very high probability—the implication is *always*—let us assume that the range 0.2; or (4.9 to 5.1), represents 6σ, as suggested by the Empirical Rule. Note that letting the range equal 6σ rather than 4σ places a stringent interpretation on the manufacturer's claim. We want the potency to fall into the interval 5.0 ± 0.1 with very high probability.)

10.62 Drug Potency, continued Refer to Exercise 10.61. Testing of 60 additional randomly selected containers of the drug gave a sample mean and variance equal to 5.04 and 0.0063 (for the total of $n = 64$ containers). Using a 95% confidence interval, estimate the variance of the manufacturer's potency measurements.

10.63 Hard Hats A manufacturer of hard safety hats for construction workers is concerned about the mean and the variation of the forces helmets transmit to wearers when subjected to a standard external force. The manufacturer desires the mean force transmitted by helmets to be 363 kg (or less), well under the legal 454 kg limit, and σ to be less than 18.1. A random sample of $n = 40$ helmets was tested, and the sample mean and variance were found to be equal to 374 kg and 350 kg^2, respectively.

a. If $\mu = 363$ and $\sigma = 18.1$, is it likely that any helmet, subjected to the standard external force, will transmit a force to a wearer in excess of 454 kg? Explain.

b. Do the data provide sufficient evidence to indicate that when the helmets are subjected to the standard external force, the mean force transmitted by the helmets exceeds 454 kg?

10.64 Hard Hats, continued Refer to Exercise 10.63. Do the data provide sufficient evidence to indicate that σ exceeds 18.1?

EX1065 **10.65 Light Bulbs** A manufacturer of industrial light bulbs likes its bulbs to have a mean life that is acceptable to its customers and a variation in life that is relatively small. If some bulbs fail too early in their life, customers become annoyed and shift to competitive products. Large variations above the mean reduce replacement sales, and variation in general disrupts customers' replacement schedules. A sample of 20 bulbs tested produced the following lengths of life (in hours):

2100	2302	1951	2067	2415	1883	2101	2146	2278	2019
1924	2183	2077	2392	2286	2501	1946	2161	2253	1827

The manufacturer wishes to control the variability in length of life so that σ is less than 150 hours. Do the data provide sufficient evidence to indicate that the manufacturer is achieving this goal? Test using $\alpha = 0.01$.

10.7 COMPARING TWO POPULATION VARIANCES

Just as a single population variance is sometimes important to an experimenter, you might also need to compare two population variances. You might need to compare the precision of one measuring device with that of another, the stability of one manufacturing process with that of another, or even the variability in the grading procedure of one university professor with that of another.

One way to compare two population variances, σ_1^2 and σ_2^2, is to use the ratio of the sample variances, s_1^2/s_2^2. If s_1^2/s_2^2 is nearly equal to 1, you will find little evidence to indicate that σ_1^2 and σ_2^2 are unequal. On the other hand, a very large or very small value for s_1^2/s_2^2 provides evidence of a difference in the population variances.

How large or small must s_1^2/s_2^2 be for sufficient evidence to exist to reject the following null hypothesis?

$$H_0 : \sigma_1^2 = \sigma_2^2$$

The answer to this question may be found by studying the distribution of s_1^2/s_2^2 in repeated sampling.

When independent random samples are drawn from two *normal* populations with *equal variances*—that is, $\sigma_1^2 = \sigma_2^2$—then s_1^2/s_2^2 has a probability distribution in repeated sampling that is known to statisticians as an ***F* distribution,** shown in Figure 10.15.

FIGURE 10.15
An *F* distribution with $df_1 = 10$ and $df_2 = 10$

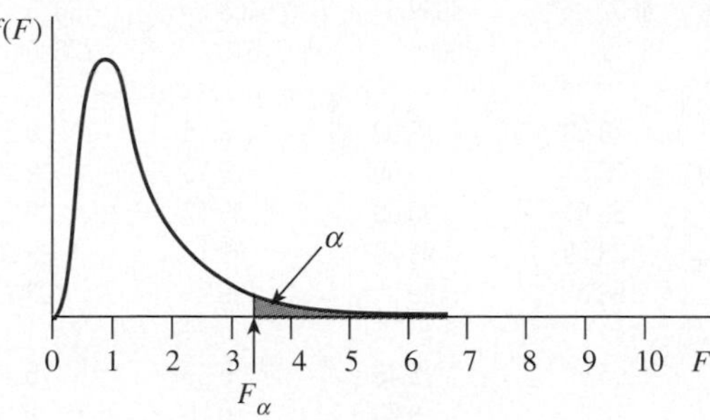

ASSUMPTIONS FOR s_1^2/s_2^2 to HAVE AN *F* DISTRIBUTION

- Random and independent samples are drawn from each of two normal populations.
- The variability of the measurements in the two populations is the same and can be measured by a common variance, σ^2; that is, $\sigma_1^2 = \sigma_2^2 = \sigma^2$.

NEED A TIP?
Testing two variances: $df_1 = n_1 - 1$ and $df_2 = n_2 - 1$

It is not important for you to know the complex equation of the density function for *F*. For your purposes, you need only to use the well-tabulated critical values of *F* given in Table 6 in Appendix I.

Like the χ^2 distribution, the shape of the *F* distribution is non-symmetric and depends on the number of degrees of freedom associated with s_1^2 and s_2^2, represented as $df_1 = (n_1 - 1)$ and $df_2 = (n_2 - 1)$, respectively. This complicates the tabulation of critical values of the *F* distribution because a table is needed for each different combination of df_1, df_2, and α.

In Table 6 in Appendix I, critical values of *F* for right-tailed areas corresponding to $\alpha = 0.100, 0.050, 0.025, 0.010$, and 0.005 are tabulated for various combinations of df_1 numerator degrees of freedom and df_2 denominator degrees of freedom. A portion of Table 6 is reproduced in Table 10.7. The numerator degrees of freedom df_1 are listed across the top margin, and the denominator degrees of freedom df_2 are listed along the side margin. The values of α are listed in the second column. For a fixed combination of df_1 and df_2, the appropriate critical values of *F* are found in the line indexed by the value of α required.

EXAMPLE 10.17

Check your ability to use Table 6 in Appendix I by verifying the following statements:

1. The value of *F* with area 0.05 to its right for $df_1 = 6$ and $df_2 = 9$ is 3.37.
2. The value of *F* with area 0.05 to its right for $df_1 = 5$ and $df_2 = 10$ is 3.33.
3. The value of *F* with area 0.01 to its right for $df_1 = 6$ and $df_2 = 9$ is 5.80.

These values are shaded in Table 10.7.

TABLE 10.7 ● **Format of the *F* Table from Table 6 in Appendix I**

		df_1					
df_2	α	1	2	3	4	5	6
1	0.100	39.86	49.50	53.59	55.83	57.24	58.20
	0.050	161.4	199.5	215.7	224.6	230.2	234.0
	0.025	647.8	799.5	864.2	899.6	921.8	937.1
	0.010	4052	4999.5	5403	5625	5764	5859
	0.005	16211	20000	21615	22500	23056	23437
2	0.100	8.53	9.00	9.16	9.24	9.29	9.33
	0.050	18.51	19.00	19.16	19.25	19.30	19.33
	0.025	38.51	39.00	39.17	39.25	39.30	39.33
	0.010	98.50	99.00	99.17	99.25	99.30	99.33
	0.005	198.5	199.0	199.2	199.2	199.3	199.3
3	0.100	5.54	5.46	5.39	5.34	5.31	5.28
	0.050	10.13	9.55	9.28	9.12	9.01	8.94
	0.025	17.44	16.04	15.44	15.10	14.88	14.73
	0.010	34.12	30.82	29.46	28.71	28.24	27.91
	0.005	55.55	49.80	47.47	46.19	45.39	44.84
.	.	.		.			.
.	.	.		.			.
.	.	.		.			.
9	0.100	3.36	3.01	2.81	2.69	2.61	2.55
	0.050	5.12	4.26	3.86	3.63	3.48	3.37
	0.025	7.21	5.71	5.08	4.72	4.48	4.32
	0.010	10.56	8.02	6.99	6.42	6.06	5.80
	0.005	13.61	10.11	8.72	7.96	7.47	7.13
10	0.100	3.29	2.92	2.73	2.61	2.52	2.46
	0.050	4.96	4.10	3.71	3.48	3.33	3.22
	0.025	6.94	5.46	4.83	4.47	4.24	4.07
	0.010	10.04	7.56	6.55	5.99	5.64	5.39
	0.005	12.83	9.43	8.08	7.34	6.87	6.54

The statistical test of the null hypothesis

$$H_0 : \sigma_1^2 = \sigma_2^2$$

uses the test statistic

$$F = \frac{s_1^2}{s_2^2}$$

When the alternative hypothesis implies a one-tailed test—that is,

$$H_a : \sigma_1^2 > \sigma_2^2$$

you can find the right-tailed critical value for rejecting H_0 directly from Table 6 in Appendix I. However, when the alternative hypothesis requires a two-tailed test—that is,

$$H_0 : \sigma_1^2 \neq \sigma_2^2$$

the rejection region is divided between the upper and lower tails of the F distribution. These left-tailed critical values are *not given* in Table 6 for the following reason: You are free to decide which of the two populations you want to call "Population 1." If you always choose to call the population with the *larger* sample variance "Population 1," then the observed value of your test statistic will always be in the right tail of the F distribution. Even though half of the rejection region, the area $\alpha/2$ to its left, will be in the lower tail of the distribution, you will never need to use it! Remember these points, though, for a two-tailed test:

- The area in the right tail of the rejection region is only $\alpha/2$.
- The area to the right of the observed test statistic is only 1/2(p-value).

The formal procedures for a test of hypothesis and a $(1 - \alpha)100\%$ confidence interval for two population variances are shown next.

TEST OF HYPOTHESIS CONCERNING THE EQUALITY OF TWO POPULATION VARIANCES

1. Null hypothesis: $H_0 : \sigma_1^2 = \sigma_2^2$
2. Alternative hypothesis:

One-Tailed Test	**Two-Tailed Test**
$H_a : \sigma_1^2 > \sigma_2^2$ (or $H_a : \sigma_1^2 < \sigma_2^2$)	$H_a : \sigma_1^2 \neq \sigma_2^2$

3. Test statistic:

One-Tailed Test	**Two-Tailed Test**
$F = \dfrac{s_1^2}{s_2^2}$	$F = \dfrac{s_1^2}{s_2^2}$

where s_1^2 is the larger sample variance

4. Rejection region: Reject H_0 when

One-Tailed Test	**Two-Tailed Test**
$F > F_\alpha$	$F > F_{\alpha/2}$

or when p-value $< \alpha$

The critical values of F_α and $F_{\alpha/2}$ are based on $df_1 = (n_1 - 1)$ and $df_2 = (n_2 - 1)$. These tabulated values, for $\alpha = 0.100, 0.050, 0.025, 0.010$, and 0.005, can be found using Table 6 in Appendix I.

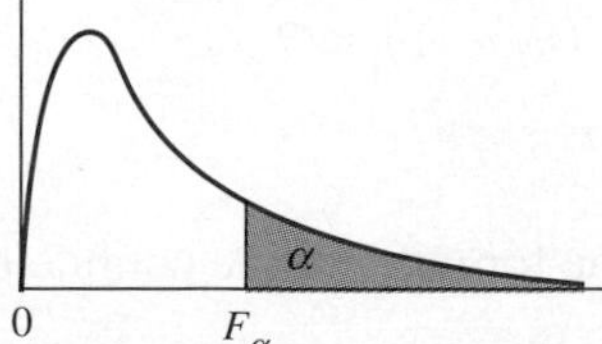

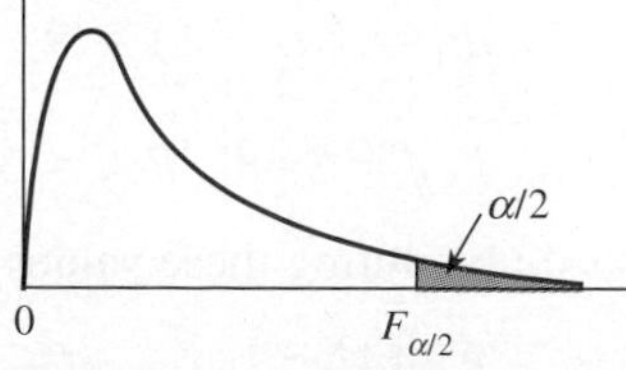

Assumptions: The samples are randomly and independently selected from normally distributed populations.

CONFIDENCE INTERVAL FOR $\sigma_1^2 \sigma_2^2$

$$\left(\frac{s_1^2}{s_2^2}\right)\frac{1}{F_{df_1,df_2}} < \frac{\sigma_1^2}{\sigma_2^2} < \left(\frac{s_1^2}{s_2^2}\right)F_{df_2,df_1}$$

where $df_1 = (n_1 - 1)$ and $df_2 = (n_2 - 1)$. F_{df_1, df_2} is the tabulated critical value of F corresponding to df_1 and df_2 degrees of freedom in the numerator and denominator of F, respectively, with area $\alpha/2$ to its right.

Assumptions: The samples are randomly and independently selected from normally distributed populations.

EXAMPLE 10.18

An experimenter is concerned that the variability of responses using two different experimental procedures may not be the same. Before conducting his research, he conducts a prestudy with random samples of 10 and 8 responses and gets $s_1^2 = 7.14$ and $s_2^2 = 3.21$, respectively. Do the sample variances present sufficient evidence to indicate that the population variances are unequal?

SOLUTION Assume that the populations have probability distributions that are reasonably mound-shaped and hence satisfy, for all practical purposes, the assumption that the populations are normal. You wish to test these hypotheses:

$$H_0 : \sigma_1^2 = \sigma_2^2 \quad \text{versus} \quad H_a : \sigma_1^2 \neq \sigma_2^2$$

Using Table 6 in Appendix I for $\alpha/2 = 0.025$, you can reject H_0 when $F > 4.82$ with $\alpha = 0.05$. The calculated value of the test statistic is

$$F = \frac{s_1^2}{s_2^2} = \frac{7.14}{3.21} = 2.22$$

Because the test statistic does not fall into the rejection region, you cannot reject $H_0: \sigma_1^2 = \sigma_2^2$. Thus, there is insufficient evidence to indicate a difference in the population variances.

EXAMPLE 10.19

Refer to Example 10.18 and find a 90% confidence interval for σ_1^2/σ_2^2.

SOLUTION The 90% confidence interval for σ_1^2/σ_2^2 is

$$\left(\frac{s_1^2}{s_2^2}\right)\frac{1}{F_{df_1,df_2}} < \frac{\sigma_1^2}{\sigma_2^2} < \left(\frac{s_1^2}{s_2^2}\right)F_{df_2,df_1}$$

where

$$s_1^2 = 7.14 \quad s_2^2 = 3.21$$

$$df_1 = (n_1 - 1) = 9 \quad df_2 = (n_2 - 1) = 7$$

$$F_{9,7} = 3.68 \quad F_{7,9} = 3.29$$

Substituting these values into the formula for the confidence interval, you get

$$\left(\frac{7.14}{3.21}\right)\frac{1}{3.68} < \frac{\sigma_1^2}{\sigma_2^2} < \left(\frac{7.14}{3.21}\right)3.29 \quad \text{or} \quad .604 < \frac{\sigma_1^2}{\sigma_2^2} < 7.32$$

The calculated interval estimate 0.604 to 7.32 includes 1.0, the value hypothesized in H_0. This indicates that it is quite possible that $\sigma_1^2 = \sigma_2^2$ and therefore agrees with the test conclusions. Do not reject $H_0: \sigma_1^2 = \sigma_2^2$.

The *Excel* function called **F.TEST (FTEST** in *Excel 2007* and earlier versions) can be used to perform the F test for the equality of variances when you have entered the raw data into the spreadsheet. The *MINITAB* command **Stat ▶ Basic Statistics ▶ 2 Variances** is a little more flexible, since it allows you to enter either raw data or summary statistics to perform the F test. In addition, *MINITAB 16* calculates confidence intervals for the ratio of two variances or two standard deviations (which we have not discussed). The *MINITAB 16* printout for Example 10.18, containing the F statistic and its p-value, is shown in Figure 10.16.

FIGURE 10.16
MINITAB output for Example 10.18

Test and CI for Two Variances

```
Null hypothesis           Sigma (1) / Sigma (2) = 1
Alternative hypothesis    Sigma (1) / Sigma (2) not = 1
Significance level        Alpha = 0.05

Statistics
Sample     N    StDev    Variance
1         10    2.672       7.140
2          8    1.792       3.210
Ratio of standard deviations = 1.491
Ratio of variances = 2.224

95% Confidence Intervals
CI for
Distribution      CI for StDev        Variance
of Data
Normal           (0.679, 3.055)     (0.461, 9.335)
Test
Method               DF1     DF2     Statistic   P-Value
F Test (normal)        9       7          2.22     0.304
```

EXAMPLE  10.20

The variability in the amount of impurities present in a batch of chemical used for a particular process depends on the length of time the process is in operation. A manufacturer using two production lines 1 and 2 has made a slight adjustment to line 2, hoping to reduce the variability as well as the average amount of impurities in the chemical. Samples of $n_1 = 25$ and $n_2 = 25$ measurements from the two batches yield these means and variances:

$$\bar{x}_1 = 3.2 \quad s_1^2 = 1.04$$

$$\bar{x}_2 = 3.0 \quad s_2^2 = 0.51$$

Do the data present sufficient evidence to indicate that the process variability is less for line 2?

SOLUTION The experimenter believes that the average levels of impurities are the same for the two production lines but that her adjustment may have decreased the variability of the levels for line 2, as illustrated in Figure 10.17. This adjustment would be good for the company because it would decrease the probability of producing shipments of the chemical with unacceptably high levels of impurities.

FIGURE 10.17
Distributions of impurity measurements for two production lines

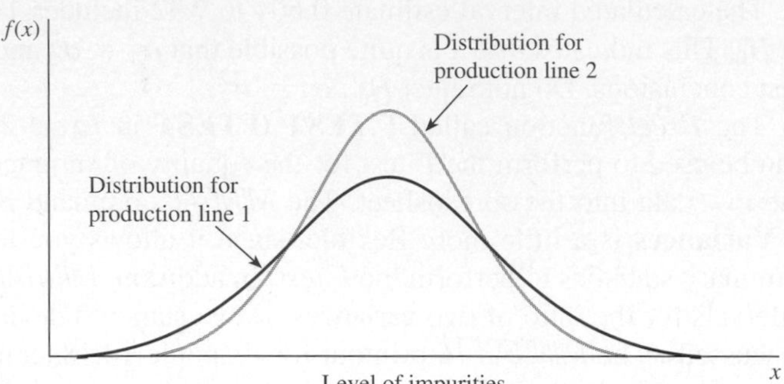

To test for a decrease in variability, the test of hypothesis is

$$H_0 : \sigma_1^2 = \sigma_2^2 \quad \text{versus} \quad H_a : \sigma_1^2 > \sigma_2^2$$

and the observed value of the test statistic is

$$F = \frac{s_1^2}{s_2^2} = \frac{1.04}{0.51} = 2.04$$

Using the p-value approach, you can bound the one-tailed p-value using Table 6 in Appendix I with $df_1 = df_2 = (25 - 1) = 24$. The observed value of F falls between $F_{0.050} = 1.98$ and $F_{0.025} = 2.27$, so that $0.025 < p\text{-value} < 0.05$. The results are judged significant at the 5% level, and H_0 is rejected. You can conclude that the variability of line 2 is less than that of line 1.

The F test for the difference in two population variances completes the battery of tests you have learned in this chapter for making inferences about population parameters under these conditions:

- The sample sizes are small.
- The sample or samples are drawn from normal populations.

You will find that the F and χ^2 distributions, as well as the Student's t distribution, are very important in other applications in the chapters that follow. They will be used for different estimators designed to answer different types of inferential questions, but the basic techniques for making inferences remain the same.

In the next section, we review the assumptions required for all of these inference tools, and discuss options that are available when the assumptions do not seem to be reasonably correct.

10.7 EXERCISES

BASIC TECHNIQUES

10.66 Independent random samples from two normal populations produced the variances listed here:

Sample Size	Sample Variance
16	55.7
20	31.4

a. Do the data provide sufficient evidence to indicate that σ_1^2 differs from σ_2^2? Test using $\alpha = 0.05$.

b. Find the approximate p-value for the test and interpret its value.

10.67 Refer to Exercise 10.66 and find a 95% confidence interval for σ_1^2/σ_2^2.

10.68 Independent random samples from two normal populations produced the given variances:

Sample Size	Sample Variance
13	18.3
13	7.9

a. Do the data provide sufficient evidence to indicate that $\sigma_1^2 > \sigma_2^2$? Test using $\alpha = 0.05$.

b. Find the approximate p-value for the test and interpret its value.

APPLICATIONS

10.69 MCAT Scores The MCAT consists of four sections: Physical Science (PS), Verbal Reasoning (VR), Biological Sciences (BS), and Writing (WS). The student receives a separate score for each section. The PS, VR, and BS sections are scored on a scale of 1 to 15. On the other hand, the WS is scored on a scale J (low) to T (high). Suppose for 2012 medical school matriculants, the results for PS and BS were given in the table:

MCAT PS	MCAT VR	MCAT BS
$\bar{x} = 9.94$	$\bar{x} = 9.8$	$\bar{x} = 10.51$
$s = 0.71$	$s = 0.95$	$s = 0.69$
$n = 15$	$n = 15$	$n = 15$

To use the two-sample t test with a pooled estimate of σ^2, you must assume that the two population variances are equal. Test this assumption for PS and BS using the F test for equality of variances. What is the approximate p-value for the test?

10.70 SAT Scores The SAT subject tests in chemistry and physics[15] for two groups of 15 students each electing to take these tests are given below:

Chemistry	Physics
$\bar{x} = 644$	$\bar{x} = 658$
$s = 114$	$s = 103$
$n = 15$	$n = 15$

To use the two-sample t test with a pooled estimate of σ^2, you must assume that the two population variances are equal. Test this assumption using the F test for equality of variances. What is the approximate p-value for the test?

10.71 Construct a 90% confidence interval for the variance ratio in Exercise 10.70.

10.72 Herbs III In Exercise 10.32 and dataset EX1032, you conducted a test to detect a difference in the average prices of bunches of parsley versus bunches of basil.

a. What assumption had to be made concerning the population variances so that the test would be valid?

b. Do the data present sufficient evidence to indicate that the variances violate the assumption in part a? Test using $\alpha = 0.05$.

Data set EX1073 **10.73 Roethlisberger and Rodgers** Quarterbacks not only need to have a good passing percentage, but they need to be consistent. That is, the variability in the number of passes completed per game should be small. The table below gives the number of passes completed for Ben Roethlisberger and Aaron Rodgers, quarterbacks for the Pittsburgh Steelers and Green Bay Packers, respectively, during the 2010 NFL season.[16]

Aaron Rodgers			Ben Roethlisberger	
19	21	7	16	20
19	15	25	19	22
34	27	19	17	21
12	22		17	23
27	26		30	22
18	21		18	15

a. Does the data indicate that there is a difference in the variability in the number of passes completed for the two quarterbacks? Use $\alpha = 0.01$.

b. If you were going to test for a difference in the two population means, would it be appropriate to use the two-sample t test that assumes equal variances? Explain.

10.74 Runners and Cyclists III Refer to Exercise 10.33. Susan Beckham and colleagues conducted an experiment involving 10 healthy runners and 10 healthy cyclists to determine if there are significant differences in pressure measurements within the anterior muscle compartment for runners and cyclists.[17] The data—compartment pressure, in millimetres of mercury (Hg)—are reproduced here:

	Runners		Cyclists	
Condition	Mean	Standard Deviation	Mean	Standard Deviation
Rest	14.5	3.92	11.1	3.98
80% maximal O_2 consumption	12.2	3.49	11.5	4.95
Maximal O_2 consumption	19.1	16.9	12.2	4.47

For each of the three variables measured in this experiment, test to see whether there is a significant difference in the variances for runners versus cyclists. Find the approximate p-values for each of these tests. Will a two-sample t test with a pooled estimate of σ^2 be appropriate for all three of these variables? Explain.

10.75 Impurities A pharmaceutical manufacturer purchases a particular material from two different suppliers. The mean level of impurities in the raw material is approximately the same for both suppliers, but the manufacturer is concerned about the variability of the impurities from shipment to shipment. If the level of impurities tends to vary excessively for one source of supply, it could affect the quality of the pharmaceutical product. To compare the variation in percentage impurities for the two suppliers, the manufacturer selects 10 shipments from each of the two suppliers and measures the percentage of impurities in the raw material for each shipment. The sample means and variances are shown in the table:

Supplier A	Supplier B
$\bar{x}_1 = 1.89$	$\bar{x}_2 = 1.85$
$s_1^2 = 0.273$	$s_2^2 = 0.094$
$n_1 = 10$	$n_2 = 10$

a. Do the data provide sufficient evidence to indicate a difference in the variability of the shipment impurity levels for the two suppliers? Test using $\alpha = 0.01$. Based on the results of your test, what recommendation would you make to the pharmaceutical manufacturer?

b. Find a 99% confidence interval for σ_2^2 and interpret your results.

NEED TO KNOW

How to Decide Which Test to Use

Are you interested in testing means? If the design involves:

a. One random sample, use the one-sample t statistic.

b. Two independent random samples, are the population variances equal?

   i. If equal, use the two-sample t statistic with pooled s^2.

   ii. If unequal, use the unpooled t with estimated *df*.

c. Two paired samples with random pairs, use a one-sample t for analyzing differences.

Are you interested in testing variances? If the design involves:

a. One random sample, use the χ^2 test for a single variance.

b. Two independent random samples, use the F test to compare two variances.

10.8

REVISITING THE SMALL-SAMPLE ASSUMPTIONS

All of the tests and estimation procedures discussed in this chapter require that the data satisfy certain conditions in order that the error probabilities (for the tests) and the confidence coefficients (for the confidence intervals) be equal to the values you have specified. For example, if you construct what you believe to be a 95% confidence interval, you want to be certain that, in repeated sampling, 95% (and not 85% or 75% or less) of all such intervals will contain the parameter of interest. These conditions are summarized in these assumptions:

ASSUMPTIONS

1. For all tests and confidence intervals described in this chapter, it is assumed that **samples are randomly selected from normally distributed populations.**
2. When two samples are selected, it is assumed that they are **selected in an independent manner** except in the case of the paired-difference experiment.
3. For tests or confidence intervals concerning the difference between two population means μ_1 and μ_2 based on independent random samples, it is assumed that $\sigma_1^2 = \sigma_2^2$.

In reality, you will never know everything about the sampled population. If you did, there would be no need for sampling or statistics. It is also highly unlikely that a population will *exactly* satisfy the assumptions given in the box. Fortunately, the procedures presented in this chapter give good inferences even when the data exhibit moderate departures from the necessary conditions.

A statistical procedure that is not sensitive to departures from the conditions on which it is based is said to be **robust.** The Student's t tests are quite robust for moderate departures from normality. Also, as long as the sample sizes are nearly equal, there is not much difference between the pooled and unpooled t statistics for the difference in two population means. However, if the sample sizes are not nearly equal, and if the population variances are unequal, the pooled t statistic provides inaccurate conclusions.

If you are concerned that your data do not satisfy the assumptions, other options are available:

- If you can select relatively large samples, you can use one of the large-sample procedures of Chapters 8 and 9, which do not rely on the normality or equal variance assumptions.
- You may be able to use a *nonparametric test* to answer your inferential questions. These tests have been developed specifically so that few or no distributional assumptions are required for their use. Tests that can be used to compare the locations or variability of two populations are presented in Chapter 15.

CHAPTER REVIEW

Key Concepts and Formulas

I. Experimental Designs for Small Samples

1. **Single random sample:** The sampled population must be normal.
2. **Two independent random samples:** Both sampled populations must be normal.
   a. Populations have a common variance σ^2.
   b. Populations have different variances: σ_1^2 and σ_2^2.
3. **Paired-difference** or **matched pairs** design: The samples are not independent.

II. Statistical Tests of Significance

1. Based on the t, F, and χ^2 distributions
2. Use the same procedure as in Chapter 9
3. **Rejection region—critical values** and **significance levels:** based on the t, F, or χ^2 distributions with the appropriate degrees of freedom
4. **Tests of population parameters:** a single mean, the difference between two means, a single variance, and the ratio of two variances

III. Small-Sample Test Statistics

To test one of the population parameters when the sample sizes are small, use the following test statistics:

Parameter	Test Statistic	Degrees of Freedom
μ	$t = \dfrac{\bar{x} - \mu_0}{s/\sqrt{n}}$	$n - 1$
$\mu_1 - \mu_2$ (equal variances)	$t = \dfrac{(\bar{x}_1 - \bar{x}_2) - (\mu_1 - \mu_2)}{\sqrt{s^2\left(\dfrac{1}{n_1} + \dfrac{1}{n_2}\right)}}$	$n_1 + n_2 - 2$
$\mu_1 - \mu_2$ (unequal variances)	$t \approx \dfrac{(\bar{x}_1 - \bar{x}_2) - (\mu_1 - \mu_2)}{\sqrt{\dfrac{s_1^2}{n_1} + \dfrac{s_2^2}{n_2}}}$	Satterthwaite's approximation
$\mu_1 - \mu_2$ (paired samples)	$t = \dfrac{\bar{d} - \mu_d}{s_d/\sqrt{n}}$	$n - 1$
σ^2	$\chi^2 = \dfrac{(n-1)s^2}{\sigma_0^2}$	$n - 1$
σ_1^2/σ_2^2	$F = s_1^2/s_2^2$	$n_1 - 1$ and $n_2 - 1$

TECHNOLOGY TODAY

Small-Sample Testing—*Microsoft Excel*

The tests of hypotheses for two population means based on the Student's t distribution and the F test for the ratio of two variances can be found using the *Excel* command **Data ▶ Data Analysis.** Remember that you need to have loaded the *Excel* add-ins called **Analysis ToolPak** (see the instructions in the "Technology Today" section of Chapter 1). You will find three choices for the two-sample t tests and one F test in the list of "Analysis Tools." To choose the proper t tests, you must first decide whether the samples are independent or paired; for the independent samples test, you must decide whether or not the population variances can be assumed equal.

EXAMPLE 10.21

Two-Sample *t* Test Assuming Equal Variances The test scores on the same algebra test were recorded for nine students randomly selected from a classroom taught by teacher A and eight students randomly selected from a classroom taught by teacher B. Is there a difference in the average scores for students taught by these two teachers?

Teacher A	65	88	93	95	80	76	79	77	
Teacher B	91	85	70	82	92	68	86	87	75

Enter the data into columns A and B of an *Excel* spreadsheet.

1. Use **Data ▶ Data ▶ Analysis ▶ Descriptive Statistics** or ▶ **Statistical ▶ STDEV.S** (**STDEV** in *Excel 2007* and earlier versions) to find the standard deviations for the two samples, $s_1 = 9.913$ and $s_2 = 8.800$. Since the ratio of the two variances is s_1^2/s_2^2 (less than 3), you are safe in assuming that the population variances are the same.
2. Select **Data ▶ Data Analysis ▶ t-Test: Two-Sample Assuming Equal Variances** to generate the dialogue box in Figure 10.18(a). Highlight or type the **Variable 1 Range** and **Variable 2 Range** (the data in the first and second columns) into the first two boxes. In the box marked "Hypothesized Mean Difference" type **0** (since we are testing $H_0: \mu_1 - \mu_2\ 0$) and check "Labels" if necessary.
3. The default significance level is $\alpha = 0.05$ in *Excel.* Change this significance level if necessary. Enter a cell location for the **Output Range** and click **OK.** The output will appear in the selected cell location, and should be adjusted using **Format ▶ AutoFit Column Width** on the **Home** tab in the **Cells** group while it is still highlighted. You can decrease the decimal accuracy if you like, using on the **Home** tab in the **Number** group.
4. The observed value of the test statistic $t = -0.0337$ is found in Figure 10.18(b) in the row labelled "*t Stat*" followed by the one-tailed p-value "*P(T < = t) one-tail*" and the critical value marking the rejection region for a one-tailed test with $a = 0.05$. The last two rows of output give the p-value and critical t-value for a two-tailed test.
5. For this example, the p-value =0 .9736 indicates that there is no significant difference in the average scores for students taught by the two teachers.

FIGURE 10.18
a) *MINITAB* screen shot
b) *Excel* screen shot

(a)

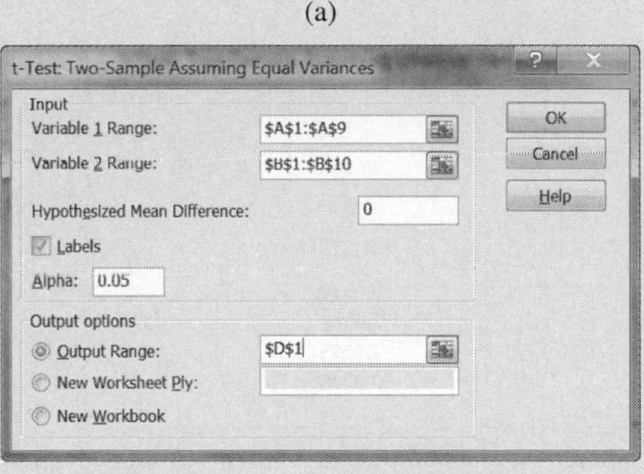

(b)

D	E	F
t-Test: Two-Sample Assuming Equal Variances		
	A	*B*
Mean	81.625	81.778
Variance	98.268	77.444
Observations	8	9
Pooled Variance	87.162	
Hypothesized Mean Difference	0	
df	15	
t Stat	-0.0337	
P(T<=t) one-tail	0.4868	
t Critical one-tail	1.7531	
P(T<=t) two-tail	0.9736	
t Critical two-tail	2.1314	

6. In Section 10.7, we presented a formal test of hypothesis for the equality of two variances using the F test. To implement this test using *Excel,* select **Data ▶ Data Analysis ▶ F-Test: Two-Sample for Variances**. Follow the directions for the Equal Variances t test, but replace the "Alpha" valuc with 0.025, and you will generate the output in Figure 10.19.

FIGURE 10.19
Excel screen shot

F-Test Two-Sample for Variances		
	A	*B*
Mean	81.625	81.78
Variance	98.268	77.44
Observations	8	9
df	7	8
F	1.2689	
P(F<=f) one-tail	0.3701	
F Critical one-tail	4.5286	

Notice that only the one-tailed p-value and critical value are given in the output, which is why we specified the single tail to be 0.025. Hence, for our two-tailed test, $\alpha = 0.05$ and p-value = 0.7402. There is no significant difference in the two variances.

EXAMPLE 

Two-Sample *t* Test Assuming Unequal Variances

1. Refer to Example 10.21. If the ratio of the two sample variances had been so large that you could not assume equal variances (we use "greater than 3" as a rule of thumb), you should select **Data ▶ Data Analysis ▶ t-Test: Two-Sample Assuming Unequal Variances.**
2. Follow the directions for the equal variances t test in Example 10.21, and you will generate similar output. If we use this test on the data from Example 10.21, the following output results (Figure 10.20).

FIGURE 10.20
Excel screen shot

H	I	J
t-Test: Two-Sample Assuming Unequal Variances		
	A	*B*
Mean	81.625	81.778
Variance	98.268	77.444
Observations	8	9
Hypothesized Mean Difference	0	
df	14	
t Stat	-0.0334	
P(T<=t) one-tail	0.4869	
t Critical one-tail	1.7613	
P(T<=t) two-tail	0.9738	
t Critical two-tail	2.1448	

3. You will see slight differences in the observed value of the test statistic, the degrees of freedom, and the p-values for the test, but the conclusions did not change.

NOTE: When calculating the degrees of freedom for *Satterthwaite's Approximation,* the **Data Analysis Tool** in *Excel* rounds to the nearest integer. An alternative *Excel* function for calculating the p-value for this test (*fx* **Statistical ▶ T.TEST**) uses the exact value of *df* given by Satterthwaite's formula. Because of these different approaches to determining the degrees of freedom, the results of **T.TEST** and the t test tool will differ slightly in the unequal variances case, and will *also* differ slightly from the *MINITAB* output.

EXAMPLE  **10.23**

Paired *t* Test Refer to the tire wear data from Table 10.4 (page 426).

1. To perform a paired-difference test for these dependent samples, enter the data into the first two columns of an *Excel* spreadsheet and select **Data ▶ Data Analysis ▶ t-Test: Paired Two Sample for Means**.
2. Follow the directions for the equal variances t test in Example 10.21, and you will generate similar output. For the data in Table 10.4, you obtain the output in Figure 10.21. Again, you can decrease the decimal accuracy if you like, using .00 on the **Home** tab in the **Number** group.
3. Using the observed value of the test statistic ($t = 12.83$) with two-tailed p-value = 0.0002, there is strong evidence to indicate a difference in the two population means.

FIGURE 10.21
Excel screen shot

D	E	F
t-Test: Paired Two Sample for Means		
	Tire A	*Tire B*
Mean	10.24	9.76
Variance	1.733	1.763
Observations	5	5
Pearson Correlation	0.998	
Hypothesized Mean Difference	0	
df	4	
t Stat	12.8285	
P(T<=t) one-tail	0.0001	
t Critical one-tail	2.1318	
P(T<=t) two-tail	0.0002	
t Critical two-tail	2.7764	

Small-Sample Testing and Estimation—*MINITAB*

The tests of hypotheses for two population means based on the Student's t distribution and the F test for the ratio of two variances can be found using the *MINITAB* command **Stat ▶ Basic Statistics.** You will find choices for **1-Sample t, 2-Sample t, Paired t,** and **2 Variances,** which will perform the tests and estimation procedures of Sections 10.3, 10.4, 10.5, and 10.7. To choose the proper two sample t tests, you must first decide whether the samples are independent or paired; for the independent samples test, you must decide whether or not the population variances can be assumed equal.

EXAMPLE 10.24

One-Sample *t* Test Refer to Example 10.3, in which the average weight of diamonds using a new process was compared to an average weight of 0.5 karat.

1. Enter the six recorded weights—0.46, 0.61, 0.52, 0.48, 0.57, 0.54—in column C1 and name them "Weights." Use **Stat ▶ Basic Statistics ▶ 1-Sample t** to generate the dialogue boxes in Figure 10.22.

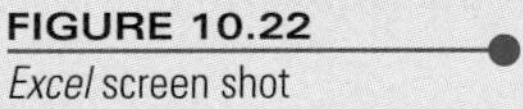

FIGURE 10.22
Excel screen shot

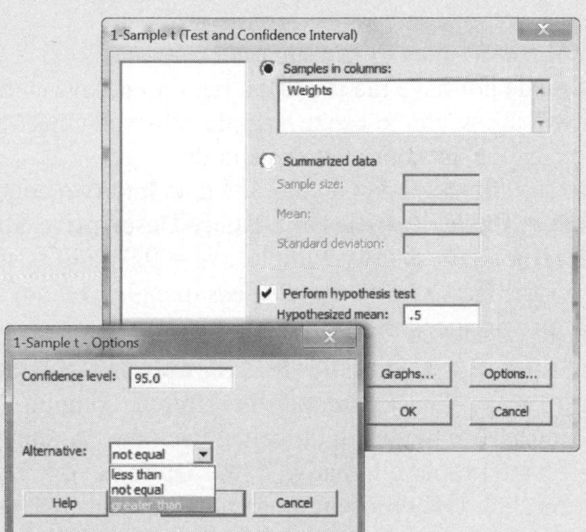

2. To test $H_0: \mu = 0.5$ versus $H_a: \mu > 0.5$, use the list on the left to select "Weights" for the box marked "Samples in Columns." Check the box marked "Perform hypothesis test." Then, place your cursor in the box marked "Hypothesized mean:" and enter **.5** as the test value. Finally, use **Options** and the drop-down menu marked "Alternative" to select "greater than." You can change the default confidence coefficient of **.95** if you wish. Click **OK** twice to obtain the output in Figure 10.23.

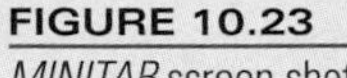

FIGURE 10.23
MINITAB screen shot

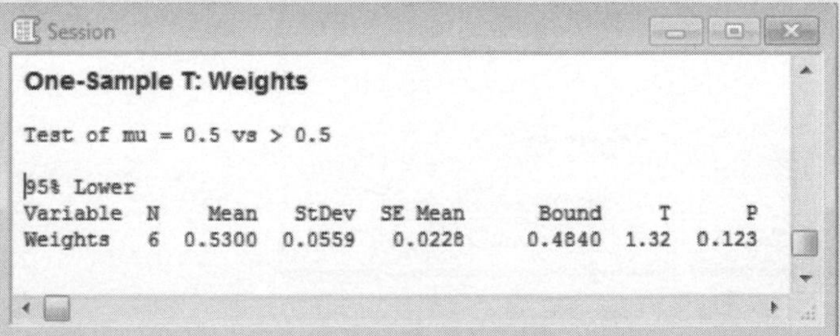

Session
One-Sample T: Weights
Test of mu = 0.5 vs > 0.5
95% Lower
Variable N Mean StDev SE Mean Bound T P
Weights 6 0.5300 0.0559 0.0228 0.4840 1.32 0.123

3. Notice that *MINITAB* produces a one- or a two-sided confidence interval for the single population mean, consistent with the alternative hypothesis you have chosen.

Two-Sample *t* Test The test scores on the same algebra test were recorded for nine students randomly selected from a classroom taught by teacher A and eight students randomly selected from a classroom taught by teacher B. Is there a difference in the average scores for students taught by these two teachers?

Teacher A	65	88	93	95	80	76	79	77	
Teacher B	91	85	70	82	92	68	86	87	75

1. The data can be entered into the worksheet in one of three ways:
   - Enter measurements from both samples into a single column and enter letters (A or B) in a second column to identify the sample from which the measurement comes.
   - Enter the samples in two separate columns.
   - If you do not have the raw data, but rather have summary statistics, *MINITAB 16* will allow you to use these values by selecting "Summarized data" and entering the appropriate values in the boxes.
2. Use the second method and enter the data into two columns of the worksheet. Use **Stat ▶ Basic Statistics ▶ Display Descriptive Statistics** to find the standard deviations for the two samples, $s_1 = 9.91$ and $s_2 = 8.80$. Since the ratio of the two variances is $s_1^2/s_2^2 = 1.27$ (less than 3), you are safe in assuming that the population variances are the same.
3. Select **Stat ▶ Basic Statistics ▶ 2-Sample t** to generate the dialogue box in Figure 10.24(a). Check "samples in different columns," selecting the appropriate columns from the list at the left. Check the "Assume equal variances" box and select the proper alternative in the **Options** box. The two-sample output when you click **OK** twice automatically contains a 95% one- or two-sided confidence interval as well as the test statistic and *p*-value (you can change the confidence coefficient if you wish). The output is shown in Figure 10.24(b).

FIGURE 10.24
a) *MINITAB* screen shot
b) *Excel* screen shot

(a)

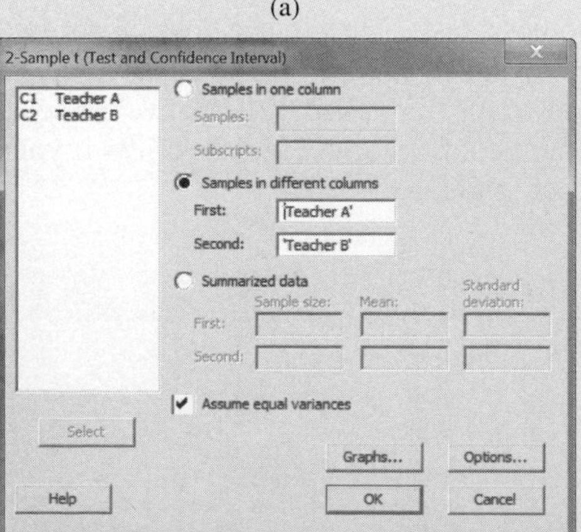

(b)

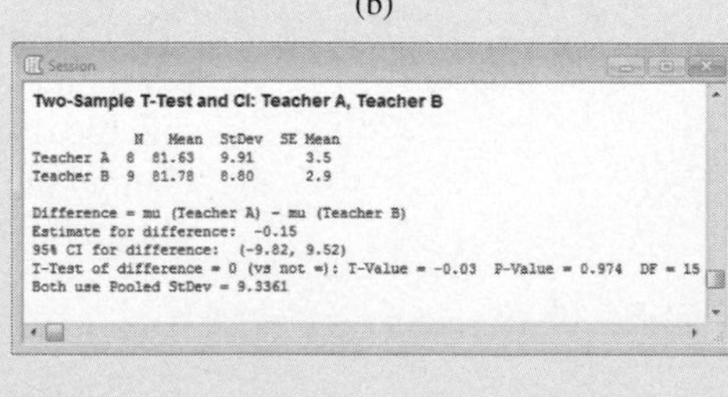

```
Session
Two-Sample T-Test and CI: Teacher A, Teacher B

           N   Mean  StDev  SE Mean
Teacher A  8  81.63   9.91      3.5
Teacher B  9  81.78   8.80      2.9

Difference = mu (Teacher A) - mu (Teacher B)
Estimate for difference:  -0.15
95% CI for difference:  (-9.82, 9.52)
T-Test of difference = 0 (vs not =): T-Value = -0.03  P-Value = 0.974  DF = 15
Both use Pooled StDev = 9.3361
```

4. The observed value of the test statistic $t = -0.03$ is labelled "*T-Value*" followed by the two-tailed "*P-Value.*" For this example, the *p*-value = 0.974 indicates that there is no significant difference in the average scores for students taught by the two teachers.
5. In Section 10.7, we presented a formal test of hypothesis for the equality of two variances using the *F* test. To implement this test using *MINITAB*, select **Stat ▶ Basic Statistics ▶ 2 Variances.** In the drop-down list, select "Samples in two columns" and enter the appropriate columns from the list on the left. The pertinent portion of the output is shown in Figure 10.25. For our two-tailed test with a 0.05, the test statistic is $F = 1.27$ and the *p*-value = 0.740. There is no significant difference in the two variances.

FIGURE 10.25
MINITAB screen shot

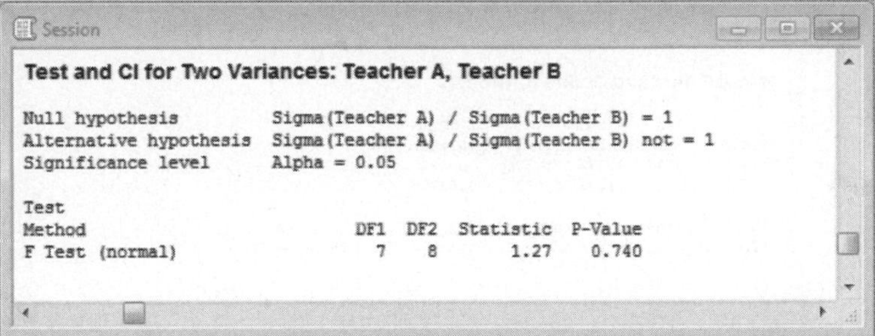

EXAMPLE 10.26 Two-Sample *t* Test Assuming Unequal Variances

1. Refer to Example 10.25. If the ratio of the two sample variances had been so large that you could not assume equal variances (we use "greater than 3" as a rule of thumb), you should select **Stat ▶ Basic Statistics ▶ 2-Sample t,** but DO NOT check the box marked "Assume Equal Variances." If we use this test on the data from Example 10.25, the following output results (Figure 10.26).

FIGURE 10.26
MINITAB screen shot

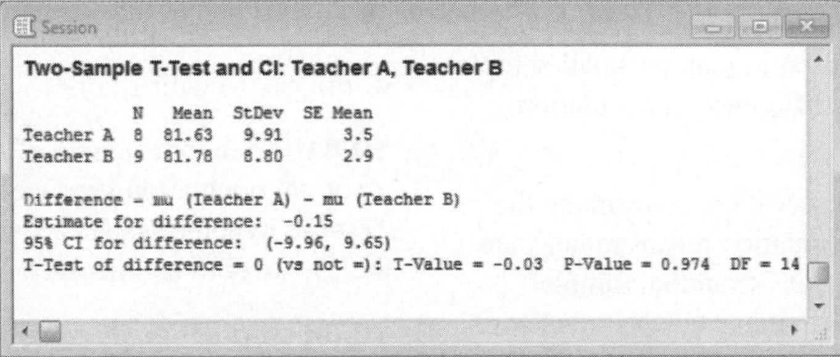

2. You will see slight differences in the degrees of freedom; there is no "Pooled StDev" listed, but the conclusions did not change.

NOTE: When calculating the degrees of freedom for *Satterthwaite's Approximation, MINITAB* uses the integer part of the calculated value, which is different from the procedures used in *Excel.* Because of these different approaches to determining the degrees of freedom, the results of the outputs from *Excel* and *MINITAB* will differ slightly.

EXAMPLE 10.27 **Paired-Difference Test** Refer to the tire wear data from Table 10.4 on page 426.

1. To perform a paired-difference test for these dependent samples, enter the data into the first two columns of a *MINITAB* worksheet and select **Stat ▶ Basic Statistics ▶ Paired t.**
2. Follow the directions for the independent samples t test, and you will generate similar output. For the data in Table 10.4, you obtain the output in Figure 10.27.
3. Using the observed value of the test statistic ($t = 12.83$) with two-tailed p-value $= 0.000$, there is strong evidence to indicate a difference in the two population means.

FIGURE 10.27
MINITAB screen shot

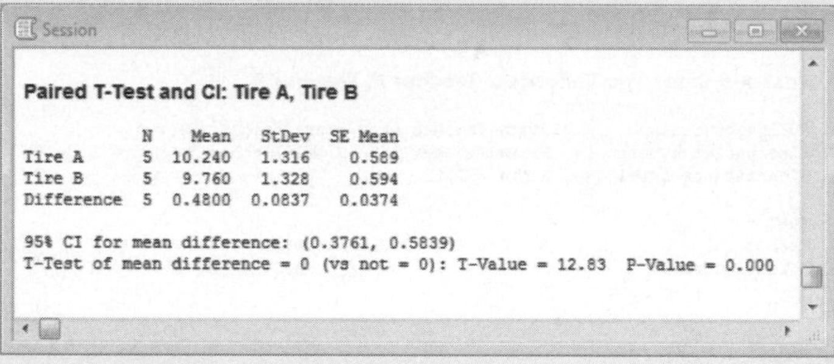

Session
Paired T-Test and CI: Tire A, Tire B
N Mean StDev SE Mean
Tire A 5 10.240 1.316 0.589
Tire B 5 9.760 1.328 0.594
Difference 5 0.4800 0.0837 0.0374
95% CI for mean difference: (0.3761, 0.5839)
T-Test of mean difference = 0 (vs not = 0): T-Value = 12.83 P-Value = 0.000

Supplementary Exercises

10.76 What assumptions are made when Student's t test is used to test a hypothesis concerning a population mean?

10.77 What assumptions are made about the populations from which random samples are obtained when the t distribution is used in making small-sample inferences concerning the difference in population means?

10.78 Why use paired observations to estimate the difference between two population means rather than estimation based on independent random samples selected from the two populations? Is a paired experiment always preferable? Explain.

10.79 Use Table 4 in Appendix I to find the following critical values:

a. An upper one-tailed rejection region with $\alpha = 0.05$ and 11 *df*

b. A two-tailed rejection region with $\alpha = 0.05$ and 7 *df*

c. A lower one-tailed rejection region with $\alpha = 0.01$ and 15 *df*

10.80 Use Table 4 in Appendix I to bound the following p-values:

a. $P(t > 1.2)$ with 5 *df*

b. $P(t > 2) + P(t < -2)$ with 10 *df*

c. $P(t < -3.3)$ with 8 *df*

d. $P(t > 0.6)$ with 12 *df*

10.81 A random sample of $n = 12$ observations from a normal population produced $\bar{x} = 47.1$ and $s^2 = 4.7$. Test the hypothesis $H_0: \mu = 48$ against $H_0: \mu \neq 48$ at the 5% level of significance.

10.82 Impurities II A manufacturer can tolerate a small amount (0.05 milligrams per litre [mg/L]) of impurities in a raw material needed for manufacturing its product. Because the laboratory test for the impurities is subject to experimental error, the manufacturer tests each batch 10 times. Assume that the mean value of the experimental error is 0 and hence that the mean value of the 10 test readings is an unbiased estimate of the true amount of the impurities in the batch. For a particular batch of the raw material, the mean of the 10 test readings is 0.058 mg/L, with a standard deviation of 0.012 mg/L. Do the data provide sufficient evidence to indicate that the amount of impurities in the batch exceeds 0.05 mg/L? Find the p-value for the test and interpret its value.

10.83 Red Pine The main stem growth measured for a sample of 17 four-year-old red pine trees produced a mean and standard deviation equal to 25 and 7 centimetres (cm), respectively. Find a 90% confidence interval for the mean growth of a population of six-year-old red pine trees subjected to similar environmental conditions.

10.84 Sodium Hydroxide The object of a general chemistry experiment is to determine the amount (in millilitres [mL]) of sodium hydroxide (NaOH) solution needed to neutralize 1 g of a specified acid. This will be an exact amount, but when the experiment is run in the laboratory, variation will occur as the result of experimental error. Three titrations are made using phenolphthalein as an indicator of the neutrality of the solution (pH equals 7 for a neutral solution). The three volumes of NaOH required to attain a pH of 7 in each of the three titrations are as follows: 82.10, 75.75, and 75.44 mL. Use a 99% confidence interval to estimate the mean number of millilitres required to neutralize 1 g of the acid.

10.85 Sodium Chloride Measurements of water intake, obtained from a sample of 17 rats that had been injected with a sodium chloride solution, produced a mean and standard deviation of 31.0 and 6.2 cubic centimetres (cm^3), respectively. Given that the average water intake for non-injected rats observed over a comparable period of time is 22.0 cm^3, do the data indicate that injected rats drink more water than non-injected rats? Test at the 5% level of significance. Find a 90% confidence interval for the mean water intake for injected rats.

10.86 Sea Urchins An experimenter was interested in determining the mean thickness of the cortex of the sea urchin egg. The thickness was measured for $n = 10$ sea urchin eggs. These measurements were obtained:

4.5	6.1	3.2	3.9	4.7
5.2	2.6	3.7	4.6	4.1

Estimate the mean thickness of the cortex using a 95% confidence interval.

10.87 Fabricating Systems A production plant has two extremely complex fabricating systems; one system is twice as old as the other. Both systems are checked, lubricated, and maintained once every two weeks. The number of finished products fabricated daily by each of the systems is recorded for 30 working days. The results are given in the table. Do these data present sufficient evidence to conclude that the variability in daily production warrants increased maintenance of the older fabricating system? Use the p-value approach.

New System	Old System
$\bar{x}_2 = 246$	$\bar{x}_2 = 240$
$s_1 = 15.6$	$s_2 = 28.2$

Data set EX1088

10.88 Fossils The data in the table are the diameters and heights of 10 fossil specimens of a species of small shellfish, *Rotularia (Annelida) fallax,* that were unearthed in a mapping expedition near the Antarctic Peninsula.[18] The table gives an identification symbol for the fossil specimen, the fossil's diameter and height in millimetres, and the ratio of diameter to height (D/H).

Specimen	Diameter	Height	D/H
OSU 36651	185	78	2.37
OSU 36652	194	65	2.98
OSU 36653	173	77	2.25
OSU 36654	200	76	2.63
OSU 36655	179	72	2.49
OSU 36656	213	76	2.80
OSU 36657	134	75	1.79
OSU 36658	191	77	2.48
OSU 36659	177	69	2.57
OSU 36660	199	65	3.06
$\bar{x}$:	184.5	73	2.54
s:	21.5	5	0.37

a. Find a 95% confidence interval for the mean diameter of the species.

b. Find a 95% confidence interval for the mean height of the species.

c. Find a 95% confidence interval for the mean ratio of diameter to height.

d. Compare the three intervals constructed in parts a, b, and c. Is the average of the ratios the same as the ratio of the average diameter to average height?

10.89 Fossils, continued Refer to Exercise 10.88 and data set EX1088. Suppose you want to estimate the mean diameter of the fossil specimens correct to within 5 mm with probability equal to 0.95. How many fossils do you have to include in your sample?

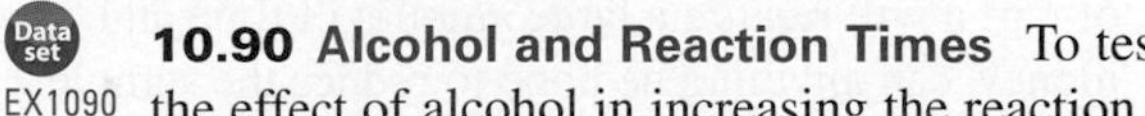

Data set EX1090

10.90 Alcohol and Reaction Times To test the effect of alcohol in increasing the reaction time to respond to a given stimulus, the reaction times of seven people were measured. After consuming 89 mL of 40% alcohol, the reaction time for each of the seven people was measured again. Do the following data indicate that the mean reaction time after consuming alcohol was greater than the mean reaction time before consuming alcohol? Use $\alpha = 0.05$.

Person	1	2	3	4	5	6	7
Before	4	5	5	4	3	6	2
After	7	8	3	5	4	5	5

Data set EX1091 **10.91 Cheese, Please** Here are the prices per 28 g of $n = 13$ different brands of individually wrapped cheese slices:

29.0	24.1	23.7	19.6	27.5
28.7	28.0	23.8	18.9	23.9
21.6	25.9	27.4		

Construct a 95% confidence interval estimate of the underlying average price per gram of individually wrapped cheese slices.

10.92 Drug Absorption An experiment was conducted to compare the mean lengths of time required for the bodily absorption of two drugs A and B. Ten people were randomly selected and assigned to receive one of the drugs. The length of time (in minutes) for the drug to reach a specified level in the blood was recorded, and the data summary is given in the table:

Drug A	Drug B
$\bar{x}_1 = 27.2$	$\bar{x}_2 = 33.5$
$s_1^2 = 16.36$	$s_2^2 = 18.92$

a. Do the data provide sufficient evidence to indicate a difference in mean times to absorption for the two drugs? Test using $\alpha = 0.05$.

b. Find the approximate p-value for the test. Does this value confirm your conclusions?

c. Find a 95% confidence interval for the difference in mean times to absorption. Does the interval confirm your conclusions in part a?

10.93 Drug Absorption, continued Refer to Exercise 10.92. Suppose you wish to estimate the difference in mean times to absorption correct to within one minute with probability approximately equal to 0.95.

a. Approximately how large a sample is required for each drug (assume that the sample sizes are equal)?

b. If conducting the experiment using the sample sizes of part a will require a large amount of time and money, can anything be done to reduce the sample sizes and still achieve the one-minute margin of error for estimation?

Data set EX1094 **10.94 Ring-Necked Pheasants** The weights in grams of 10 males and 10 female juvenile ring-necked pheasants are given below:

Males		Females	
1384	1672	1073	1058
1286	1370	1053	1123
1503	1659	1038	1089
1627	1725	1018	1034
1450	1394	1146	1281

a. Use a statistical test to determine if the population variance of the weights of the male birds differs from that of the females.

b. Test whether the average weight of juvenile male ring-necked pheasants exceeds that of the females by more than 300 grams. (HINT: The procedure that you use should take into account the results of the analysis in part a.)

Data set EX1095 **10.95 Bees** Insects hovering in flight expend enormous amounts of energy for their size and weight. The data shown here were taken from a much larger body of data collected by T. M. Casey and colleagues.[19] They show the wing stroke frequencies (in hertz) for two different species of bees, $n_1 = 4$ *Euglossa mandibularis* Friese and $n_2 = 6$ *Euglossa imperialis* Cockerell.

E. mandibularis Friese	*E. imperialis* Cockerell
235	180
225	169
190	180
188	185
	178
	182

a. Based on the observed ranges, do you think that a difference exists between the two population variances?

b. Use an appropriate test to determine whether a difference exists.

c. Explain why a Student's t test with a pooled estimator s^2 is unsuitable for comparing the mean wing stroke frequencies for the two species of bees.

Data set EX1096 **10.96 Calcium** The calcium (Ca) content of a powdered mineral substance was analyzed 10 times with the following percent compositions recorded:

0.0271	0.0282	0.0279	0.0281	0.0268
0.0271	0.0281	0.0269	0.0275	0.0276

a. Find a 99% confidence interval for the true calcium content of this substance.

b. What does the phrase "99% confident" mean?

c. What assumptions must you make about the sampling procedure so that this confidence interval will be valid? What does this mean to the chemist who is performing the analysis?

10.97 Sun or Shade? Karl Niklas and T.G. Owens examined the differences in a particular plant, *Plantago Major L.*, when grown in full sunlight versus shade conditions.[20] In this study, shaded plants received direct sunlight for less than two hours each day, whereas full-sun plants were never shaded. A partial summary of the data based on $n_1 = 16$ full-sun plants and $n_2 = 15$ shade plants is shown here:

	Full Sun		Shade	
	$\bar{x}$	s	$\bar{x}$	s
Leaf area (cm^2)	128.00	43.00	78.70	41.70
Overlap area (cm^2)	46.80	2.21	8.10	1.26
Leaf number	9.75	2.27	6.93	1.49
Thickness (mm)	0.90	0.03	0.50	0.02
Length (cm)	8.70	1.64	8.91	1.23
Width (cm)	5.24	0.98	3.41	0.61

a. What assumptions are required in order to use the small-sample procedures given in this chapter to compare full-sun versus shade plants? From the summary presented, do you think that any of these assumptions have been violated?

b. Do the data present sufficient evidence to indicate a difference in mean leaf area for full-sun versus shade plants?

c. Do the data present sufficient evidence to indicate a difference in mean overlap area for full-sun versus shade plants?

10.98 Orange Juice A comparison of the precisions of two machines developed for extracting juice from oranges is to be made using the following data:

Machine A	Machine B
$s^2 = 49.9\ g^2$	$s^2 = 33.5\ g^2$
$n = 25$	$n = 25$

a. Is there sufficient evidence to indicate that there is a difference in the precision of the two machines at the 5% level of significance?

b. Find a 95% confidence interval for the ratio of the two population variances. Does this interval confirm your conclusion from part a? Explain.

10.99 At Home or at School? Four sets of identical twins (pairs A, B, C, and D) were selected at random from a computer database of identical twins. One child was selected at random from each pair to form an "experimental group." These four children were sent to school. The other four children were kept at home as a control group. At the end of the school year, the following IQ scores were obtained:

Pair	Experimental Group	Control Group
A	110	111
B	125	120
C	139	128
D	142	135

Does this evidence justify the conclusion that lack of school experience has a depressing effect on IQ scores? Use the p-value approach.

Data set EX10100 **10.100 Dieting** Eight obese persons were placed on a diet for one month, and their weights (in kilograms), at the beginning and at the end of the month, were recorded:

	Weights	
Subjects	Initial	Final
1	141	119
2	134	114
3	130	113
4	138	117
5	122	106
6	147	121
7	126	110
8	136	120

Estimate the mean weight loss for obese persons when placed on the diet for a one-month period. Use a 95% confidence interval and interpret your results. What assumptions must you make so that your inference is valid?

Data set EX10101 **10.101 Repair Costs** Car manufacturers try to design the bumpers of their automobiles to prevent costly damage in parking-lot type accidents. To compare two models of automobiles, the cars were purposely subject to a series of four front and rear impacts at 8 kilometres per hour (km/h), and the repair costs were recorded.[21]

Impact Type	Honda Civic	Hyundai Elantra
Front into barrier	$403	$247
Rear into barrier	447	0
Front into angle barrier	404	407
Rear into pole	227	185

Do the data provide sufficient evidence to indicate a difference in the average cost of repair for the Honda Civic and the Hyundai Elantra? Test using $\alpha = 0.05$.

Data set EX10102 **10.102 Breathing Patterns** Research psychologists measured the baseline breathing patterns—the total ventilation (in litres of air per minute) adjusted for body size—for each of $n = 30$ patients, so that they could estimate the average total ventilation for patients before any experimentation was done. The data, along with some *MINITAB* output, are presented here:

5.23	5.72	5.77	4.99	5.12	4.82
5.54	4.79	5.16	5.84	4.51	5.14
5.92	6.04	5.83	5.32	6.19	5.70
4.72	5.38	5.48	5.37	4.96	5.58
4.67	5.17	6.34	6.58	4.35	5.63

MINITAB output for Exercise 10.102

Stem-and-Leaf Display: Ltrs/min

```
Stem-and-leaf of Ltrs/min N = 30
Leaf Unit = 0.10

  1   4   3
  2   4   5
  5   4   677
  8   4   899
 12   5   1111
 (4)  5   2333
 14   5   455
 11   5   6777
  7   5   889
  4   6   01
  2   6   3
  1   6   5
```

What information does the stem and leaf plot give you about the data? Wht is this important?

10.103 Reaction Times A comparison of reaction times (in seconds) for two different stimuli in a psychological word-association experiment produced the following results when applied to a random sample of 16 people:

Stimulus 1	1	3	2	1	2	1	3	2
Stimulus 2	4	2	3	3	1	2	3	3

Do the data present sufficient evidence to indicate a difference in mean reaction times for the two stimuli? Test using $\alpha = 0.05$.

10.104 Reaction Times II Refer to Exercise 10.103. Suppose that the word-association experiment is conducted using eight people as blocks and making a comparison of reaction times within each person; that is, each person is subjected to both stimuli in a random order. The reaction times (in seconds) for the experiment are as follows:

Person	Stimulus 1	Stimulus 2
1	3	4
2	1	2
3	1	3
4	2	1
5	1	2
6	2	3
7	3	3
8	2	3

Do the data present sufficient evidence to indicate a difference in mean reaction times for the two stimuli? Test using $\alpha = 0.05$.

10.105 Refer to Exercises 10.103 and 10.104. Calculate a 95% confidence interval for the difference in the two population means for each of these experimental designs. Does it appear that blocking increased the amount of information available in the experiment?

Data set EX10106 **10.106 Impact Strength** The following data are readings (in metre-kilograms) of the impact strengths of two kinds of packaging material:

A	B
1.25	0.89
1.16	1.01
1.33	0.97
1.15	0.95
1.23	0.94
1.20	1.02
1.32	0.98
1.28	1.06
1.21	0.98

Excel output for Exercise 10.106

D	E	F
t-Test: Two-Sample Assuming Equal Variances		
	A	B
Mean	1.2367	0.9778
Variance	0.0042	0.0024
Observations	9	9
Pooled Variance	0.0033	
Hypothesized Mean Difference	0	
df	16	
t Stat	9.5641	
P(T<=t) one-tail	0.0000	
t Critical one-tail	1.7459	
P(T<=t) two-tail	0.0000	
t Critical two-tail	2.1199	

a. Use the *Excel* printout to determine whether there is evidence of a difference in the mean strengths for the two kinds of material.

b. Are there practical implications to your results?

10.107 Cake Mixes An experiment was conducted to compare the densities (in grams per cubic centimetre) of cakes prepared from two different cake mixes. Six cake pans were filled with batter A, and six were filled with batter B. Expecting a variation in oven temperature, the experimenter placed a pan filled with batter A and another with batter B *side by side* at six different locations in the oven. The six paired observations of densities are as follows:

Batter A	0.135	0.102	0.098	0.141	0.131	0.144
Batter B	0.129	0.120	0.112	0.152	0.135	0.163

a. Do the data present sufficient evidence to indicate a difference between the average densities of cakes prepared using the two types of batter?

b. Construct a 95% confidence interval for the difference between the average densities for the two mixes.

10.108 Under what assumptions can the F distribution be used in making inferences about the ratio of population variances?

10.109 Got Milk? A dairy is in the market for a new container-filling machine and is considering two models, manufactured by company A and company B. Ruggedness, cost, and convenience are comparable in the two models, so the deciding factor is the variability of fills. The model that produces fills with the smaller variance is preferred. If you obtain samples of fills for each of the two models, an F test can be used to test for the equality of population variances. Which type of rejection region would be most favoured by each of these individuals?

a. The manager of the dairy—Why?

b. A sales representative for company A—Why?

c. A sales representative for company B—Why?

10.110 Got Milk II Refer to Exercise 10.109. Wishing to demonstrate that the variability of fills is less for her model than for her competitor's, a sales representative for company A acquired a sample of 30 fills from her company's model and a sample of 10 fills from her competitor's model. The sample variances were $s_A^2 = 0.027$ and $s_B^2 = 0.065$, respectively. Does this result provide statistical support at the 0.05 level of significance for the sales representative's claim?

10.111 Chemical Purity A chemical manufacturer claims that the purity of its product never varies by more than 2%. Five batches were tested and given purity readings of 98.2, 97.1, 98.9, 97.7, and 97.9%.

a. Do the data provide sufficient evidence to contradict the manufacturer's claim? (HINT: To be generous, let a range of 2% equal 4σ.)

b. Find a 90% confidence interval for σ^2.

10.112 454-Gram Cans? A cannery prints "weight 454 grams" on its label. The quality control supervisor selects nine cans at random and weighs them. He finds $\bar{x} = 445$ and $s = 14.2$. Do the data present sufficient evidence to indicate that the mean weight is less than that claimed on the label?

10.113 Reaction Time III A psychologist wishes to verify that a certain drug increases the reaction time to a given stimulus. The following reaction times (in tenths of a second) were recorded before and after injection of the drug for each of four subjects:

	Reaction Time	
Subject	Before	After
1	7	13
2	2	3
3	12	18
4	12	13

Test at the 5% level of significance to determine whether the drug significantly increases reaction time.

Data set EX10114 **10.114 Food Production** At a time when energy conservation is so important, some scientists think closer scrutiny should be given to the cost (in energy) of producing various forms of food. Suppose you wish to compare the mean amount of oil required to produce 4047 square metres (m^2) of corn versus 4047 m^2 of cauliflower. The readings (in barrels of oil per 4047 m^2), based on 80,937 m^2 plots, seven for each crop, are shown in the table. Use these data to find a 90% confidence interval for the difference between the mean amounts of oil required to produce these two crops:

Corn	Cauliflower
5.6	15.9
7.1	13.4
4.5	17.6
6.0	16.8
7.9	15.8
4.8	16.3
5.7	17.1

10.115 Alcohol and Altitude The effect of alcohol consumption on the body appears to be much greater at high altitudes than at sea level. To test this theory, a scientist randomly selects 12 subjects and randomly divides them into two groups of six each. One group is put into a chamber that simulates conditions at an altitude of 3658 m, and each subject ingests a drink containing 100 cubic centimetres (cc) of alcohol. The second group receives the same drink in a chamber that simulates conditions at sea level. After two hours, the amount of alcohol in the blood (grams per 100 cc) for each subject is measured. The data are shown in the table. Do the data provide sufficient evidence to support the theory that retention of alcohol in the blood is greater at high altitudes?

Sea Level	3658 m
0.07	0.13
0.10	0.17
0.09	0.15
0.12	0.14
0.09	0.10
0.13	0.14

10.116 Stock Risks The closing prices of two common stocks were recorded for a period of 15 days. The means and variances are as follows:

$\bar{x}_1 = 40.33$ $\qquad$ $\bar{x}_2 = 42.54$

$s_1^2 = 1.54$ $\qquad$ $s_2^2 = 2.96$

a. Do these data present sufficient evidence to indicate a difference between the variabilities of the closing prices of the two stocks for the populations associated with the two samples? Give the p-value for the test and interpret its value.

b. Construct a 99% confidence interval for the ratio of the two population variances.

10.117 Auto Design An experiment is conducted to compare two new automobile designs. Twenty people are randomly selected, and each person is asked to rate each design on a scale of 1 (poor) to 10 (excellent). The resulting ratings will be used to test the null hypothesis that the mean level of approval is the same for both designs against the alternative hypothesis that one of the automobile designs is preferred. Do these data satisfy the assumptions required for the Student's t test of Section 10.4? Explain.

10.118 Safety Programs The data shown here were collected on lost-time accidents (the figures given are mean work-hours lost per month over a period of one year) before and after an industrial safety program was put into effect. Data were recorded for six industrial plants. Do the data provide sufficient evidence to indicate whether the safety program was effective in reducing lost-time accidents? Test using $\alpha = 0.01$.

	Plant Number					
	1	2	3	4	5	6
Before program	38	64	42	70	58	30
After program	31	58	43	65	52	29

Data set EX10119 **10.119 Two Different Entrees** To compare the demand for two different entrees, the manager of a cafeteria recorded the number of purchases of each entree on seven consecutive days. The data are shown in the table. Do the data provide sufficient evidence to indicate a greater mean demand for one of the entrees? Refer to the *Excel* printout.

Day	A	B
Monday	420	391
Tuesday	374	343
Wednesday	434	469
Thursday	395	412
Friday	637	538
Saturday	594	521
Sunday	679	625

Excel output for Exercise 10.119

E	F	G
t-Test: Paired Two Sample for Means		
	A	B
Mean	504.714	471.286
Variance	16191.238	9495.571
Observations	7	7
Pearson Correlation	0.945	
Hypothesized Mean Difference	0	
df	6	
t Stat	1.862	
P(T<=t) one-tail	0.056	
t Critical one-tail	1.943	
P(T<=t) two-tail	0.112	
t Critical two-tail	2.447	

10.120 Pollution Control The U.S. Environmental Protection Agency limit on the allowable discharge of suspended solids into rivers and streams is 60 milligrams per litre (mg/L) per day. A study of water samples selected from the discharge at a phosphate mine shows that over a long period, the mean daily discharge of suspended solids is 48 mg/L, but day-to-day discharge readings are variable. Inspectors measured the discharge rates of suspended solids for $n = 20$ days and found $s^2 = 39$ (mg/L)^2. Find a 90% confidence interval for σ^2. Interpret your results.

10.121 Enzymes Two methods were used to measure the specific activity (in units of enzyme activity per milligram of protein) of an enzyme. One unit of enzyme activity is the amount that catalyzes the formation of one micromole of product per minute under specified conditions. Use an appropriate test or estimation procedure to compare the two methods of measurement. Comment on the validity of any assumptions you need to make.

Method 1	125	137	130	151	142
Method 2	137	143	151	156	149

10.122 Connector Rods A producer of machine parts claimed that the diameters of the connector rods produced by its plant had a variance of at most 0.173 cm^2. A random sample of 15 connector rods from the plant produced a sample mean and variance of 1.4 cm and 0.58 cm^2, respectively.

a. Is there sufficient evidence to reject the producer's claim at the $\alpha = 0.05$ level of significance?

b. Find a 95% confidence interval for the variance of the rod diameters.

10.123 Sleep and the University Student How much sleep do you get on a typical study night? Ten university students were asked to report the number of hours that they slept on the previous night with the following results:

7, 6, 7.25, 7, 8.5, 5, 8, 7, 6.75, 6

a. Find a 99% confidence interval for the average number of hours that university students sleep.

b. What assumptions are required in order for this confidence interval to be valid?

Data set EX10124 **10.124 Arranging Objects** The following data are the response times in seconds for $n = 25$ first graders to arrange three objects by size:

5.2	3.8	5.7	3.9	3.7
4.2	4.1	4.3	4.7	4.3
3.1	2.5	3.0	4.4	4.8
3.6	3.9	4.8	5.3	4.2
4.7	3.3	4.2	3.8	5.4

Find a 95% confidence interval for the average response time for first graders to arrange three objects by size. Interpret this interval.

Data set EX10125 **10.125 The NBA Finals** Want to attend a pro-basketball finals game? The average prices for the NBA rematch of the Boston Celtics and the L.A. Lakers in 2010 compared to the average ticket prices in 2008 are given in the following table:[22]

Game	2008 ($)	2010 ($)
1	593	532
2	684	855
3	727	541
4	907	458
5	769	621
6	753	681
7	533	890

a. If we were to assume that the prices given in the table have been randomly selected, test for a significant difference between the 2008 and 2010 prices. Use $\alpha = 0.01$.

b. Find a 98% confidence interval for the mean difference, $\mu_d = \mu_{08} - \mu_{10}$. Does this estimate confirm the results of part a?

10.126 Finger-Lickin' Good! Maybe too good, according to tests performed by the consumer testing division of *Good Housekeeping*. Nutritional information provided by KFC claims that each small bag of potato wedges contains 136 g of food, for a total of 280 calories. A sample of 10 orders from KFC restaurants averaged 358 calories.[23] If the standard deviation of this sample was $s = 54$, is there sufficient evidence to indicate that the average number of calories in small bags of KFC potato wedges is greater than advertised? Test at the 1% level of significance.

10.127 Mall Rats One study investigated consumer habits at the mall. We tend to spend the most money shopping on the weekends, and, in particular, on Sundays from 4 to 6 P.M. Wednesday morning shoppers spend the least![24] Suppose that a random sample of 20 weekend shoppers and a random sample of 20 weekday shoppers were selected, and the amount spent per trip to the mall was recorded:

	Weekends	Weekdays
Sample size	20	20
Sample mean	$78	$67
Sample standard deviation	$22	$20

a. Is it reasonable to assume that the two population variances are equal? Use the F test to test this hypothesis with $\alpha = 0.05$.

b. Based on the results of part a, use the appropriate test to determine whether there is a difference in the average amount spent per trip on weekends versus weekdays. Use $\alpha = 0.05$.

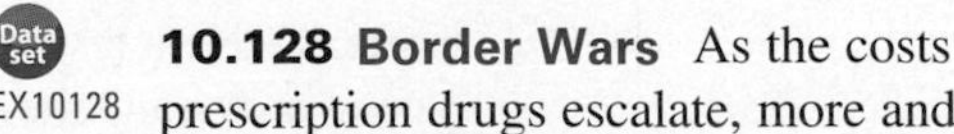

Data set EX10128 **10.128 Border Wars** As the costs of prescription drugs escalate, more and more

American senior citizens are ordering prescriptions from Canada, or actually crossing the border to buy prescription drugs. The price of a typical prescription for nine best-selling drugs was recorded at randomly selected stores in both the United States and in Canada:[25]

Drug	U.S.	Canada
Lipitor	$290	$179
Zocor	412	211
Prilosec	117	72
Norvasc	139	125
Zyprexa	571	396

Drug	U.S.	Canada
Paxil	276	171
Prevacid	484	196
Celebrex	161	67
Zoloft	235	156

a. Is there sufficient evidence to indicate that the average cost of prescription drugs in the United States is different from the average cost in Canada? Use $\alpha = 0.01$.

b. What is the approximate the p-value for this test? Does this confirm your conclusions in part a?

CASE STUDY

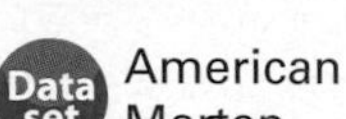

American Marten

How Does Bait Type Affect the Visit of the American Marten in Ontario?

The American marten has become a focal species for the conservation of forested landscapes throughout North America primarily due to its association with older forests and its sensitivity to human disturbance. Information on their diet is vital to understanding the habitat requirements of this species. Are the marten more attracted to chicken? Or lured by peanut butter? A research group in Ancient Forest Exploration & Research used sooted track plate boxes in order to compare marten bait types. During September 2003, eight arrays of track boxes were placed in forested areas of Rabbit Lake watershed in the Temagami region of Central Ontario. Arrays were separated by at least 4 km, located at least 1 km from a paved road and marked with Gusto, a commercial marten lure. Each array consisted of four track boxes with different bait types, one with chicken, one with a jam-lard-fish oil mixture, one with peanut butter, and a control with no bait. Track boxes were checked for a maximum of seven days, and the number of days until a track box was visited was recorded. The data for five of the arrays is available and listed in the following table:[26]

	Number of Days until Bait Taken			
Array	Chicken	Jam-Lard-Fish Oil	Peanut Butter	Control
1	2	1	7	7
2	4	4	4	5
3	6	4	4	6
4	4	4	4	4
5	6	6	6	6

Source: Adapted from Peter Quinby, James Hodson, and Mike Henry, "Forest Landscape Baselines: Evaluating Track Plate Independence and Bait Type for Detecting Marten in Temagami, Ontario," No. 25, Table 2, p. 3, 2005. http://www.ancientforest.org/wp-content/uploads/2014/12/flb25.pdf

1. American martens feed on a variety of small mammals, birds, insects, fish, carrion, and vegetation.[27] Marten food habits differ substantially in different geographic areas. Do the data indicate that the jam-lard-fish oil mixture attracts the martens in the Temagami region more efficiently than chicken does? Use the p-value approach to testing to reach your conclusion.

2. Peanut butter is one of the favourite foods of North Americans. Do American martens have a taste for peanut butter as well? Answer by comparing the effect of peanut butter and the control.

3. Construct a 95% confidence interval to estimate the average difference in days taken for the marten to visit a chicken track box and a peanut butter track box. What do you conclude about the difference between the average number of days taken for the marten to visit each of these track boxes?

4. Based on your answers to parts 1, 2, and 3, what can you conclude about the order of the effectiveness of the four different baits?

PROJECTS 

Project 10: Watch Your Sugar Level!

The most common signs and symptoms of diabetes include excessive thirst, frequent urination, increased appetite, unexplained weight loss, fatigue, blurred vision, tingling and numbness in the hands or feet, slow healing wounds, and frequent infections.[28] If your fasting blood glucose level is between 5.5 mmol/L (100 mg/dL) and 6.9 mmol/L (126 mg/dL), you are considered to have prediabetes (impaired fasting glucose), and you have an increased chance of developing diabetes.[29] The Canadian Diabetes Association (CDA) criteria for diagnosing diabetes are met when any of the following results have been repeated on at least two different days:

- A fasting blood glucose level is 7.0 mmol/L (126 mg/dL) or higher.
- A 2-hour oral glucose tolerance test result is 11.1 mmol/L (200 mg/dL) or higher.
- Symptoms of diabetes are present and a random blood glucose test is 11.1 mmol/L (200 mg/dL) or higher.

The fasting blood glucose level of a randomly selected person was measured for seven days and the following readings were made: 7.0, 7.7, 6.8, 7.1, 7.1, 6.1, 7.3. It is reasonable to assume the glucose level distribution is normal.

a. It is of interest to know if the sample data suggest the average blood glucose level is higher than 7.0 mmol/L.
   (i) State the appropriate hypotheses.
   (ii) Compute the test statistic for the hypotheses.
   (iii) Compute the approximate p-value associated with the test statistic in part (ii). Do the sample data support the alternative hypothesis at the $\alpha = 0.05$ level?
   (iv) Construct a 95% confidence interval for the mean fasting blood glucose. Based on this interval, can the null hypothesis in (i) be rejected? Justify your answer.

b. A diabetes researcher would like to study the variability in the fasting blood glucose level readings for the selected person in part a. The researcher believes that fasting blood glucose never varies more than 0.8 mmol/L. It is of interest to determine if the sample data present sufficient evidence to reject the researcher's claim.
   (i) State the appropriate null and alternative hypotheses.
   (ii) What is the appropriate conclusion at the 0.05 level of significance?
   (iii) What is the approximate p-value for the test statistic?
   (iv) Estimate the population variance using a 90% confidence interval.
   (v) Can we infer at the 90% confidence that the population variance is significantly less than 1?

c. At a diabetic clinic, a group of people considered to have prediabetes (impaired fasting glucose) were randomly selected. Then they were randomly assigned in two groups, namely treatment group (subject to a strict diet plan) and placebo (normal diet), for a specified number of days.

After the completion of the clinical trial, their blood glucose level readings provided the following statistics:

Placebo	$n_1 = 10$	$\bar{x}_1 = 6.5$	$s_1^2 = 0.21$
Treatment	$n_2 = 10$	$\bar{x}_2 = 5.4$	$s_1^2 = 0.17$

Assume the assumptions of normality and equal variances are satisfied.

(i) Is it reasonable to assume equality of variances in this problem? Justify your answer.

(ii) Do the data provide sufficient evidence at the 5% significance level that the strict diet plan lowers the blood glucose level? What is the appropriate conclusion?

(iii) Approximate the p-value for the test.

(iv) Using the p-value approach and $\alpha = 0.10$, what conclusion can be drawn about the difference in average blood glucose level readings between treatment and placebo groups?

(v) Estimate with 95% confidence the difference in mean blood glucose levels between the two groups. Interpret the interval estimate.

d. The researcher was not too sure about the assumption of equality of variances and was under the impression that the population variance of the placebo group is greater than the variance of the treatment group.

(i) Do the sample variances present sufficient evidence to indicate that the population variance of the placebo group is greater than the variance of the treatment group? Test the appropriate hypotheses using a significance level of 0.10. Interpret your results.

(ii) Estimate with 90% confidence the ratio of the two population variances.

(iii) Describe what the interval estimate tells you and briefly explain how to use the interval estimate to test the hypotheses.

e. Ten adults with symptoms of diabetes were randomly selected and a random blood glucose test was administered before and after a medication. Their blood glucose level readings were gathered before and after the medication; the results are shown below.

Adult	1	2	3	4	5	6	7	8	9	10
After medication	10.5	11.1	9.8	8.4	7.9	8.1	6.9	11.0	8.3	8.6
Before medication	11.7	12.9	10.2	11.4	10.1	11.3	10.9	13.3	11.2	11.0

(i) Are the samples independent? Justify your answer.

(ii) The consulting statistician suggested using a paired-difference t test for the analysis. Do you agree with the statistician's opinion? Justify your answer.

(iii) State the appropriate null and alternative hypotheses to test whether the medication has been effective in decreasing blood glucose level.

(iv) Calculate the value of the test statistic.

(v) Set up the appropriate rejection region for the hypotheses assuming $\alpha = 0.05$.

(vi) What is the appropriate conclusion?

(vii) Find the approximate p-value.

(viii) Find a 90% confidence interval for the difference in average blood glucose levels between the two groups. Based on this interval, can one conclude that there is a significant difference in average blood glucose level between the two groups? Justify your answer. Is your finding consistent with that in (vi)?

11

The Analysis of Variance

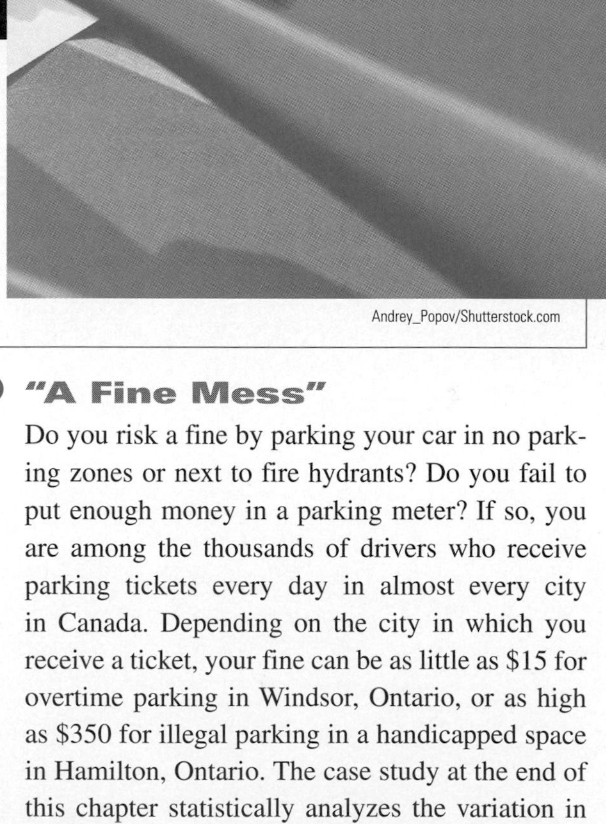

Andrey_Popov/Shutterstock.com

GENERAL OBJECTIVES

The quantity of information contained in a sample is affected by various factors that the experimenter may or may not be able to control. This chapter introduces three different *experimental designs*, two of which are direct extensions of the unpaired and paired designs of Chapter 10. A new technique called the *analysis of variance* is used to determine how the different experimental factors affect the average response.

CHAPTER INDEX

- The analysis of variance (11.2)
- The completely randomized design (11.3)
- Tukey's method for paired comparisons (11.4)
- The randomized block design (11.5, 11.6)
- Factorial experiments (11.7, 11.8)

NEED TO KNOW

How to Determine Whether Calculations Are Accurate

"A Fine Mess"

Do you risk a fine by parking your car in no parking zones or next to fire hydrants? Do you fail to put enough money in a parking meter? If so, you are among the thousands of drivers who receive parking tickets every day in almost every city in Canada. Depending on the city in which you receive a ticket, your fine can be as little as $15 for overtime parking in Windsor, Ontario, or as high as $350 for illegal parking in a handicapped space in Hamilton, Ontario. The case study at the end of this chapter statistically analyzes the variation in parking fines in cities in the province of Ontario.

11.1 THE DESIGN OF AN EXPERIMENT

The way that a sample is selected is called the *sampling plan* or *experimental design* and determines the amount of information in the sample. Some research involves an **observational study,** in which the researcher does not actually produce the data but only *observes* the characteristics of data that already exist. Most sample surveys, in which information is gathered with a questionnaire, fall into this category. The researcher forms a plan for collecting the data—called the *sampling plan*—and then uses the appropriate statistical procedures to draw conclusions about the population or populations from which the sample comes.

Other research involves **experimentation.** The researcher may deliberately impose one or more experimental conditions on the experimental units in order to determine their effect on the response. Here are some new terms we will use to discuss the design of a statistical experiment.

DEFINITION An **experimental unit** is the object on which a measurement (or measurements) is taken.

A **factor** is an independent variable whose values are controlled and varied by the experimenter.

A **level** is the intensity setting of a factor.

A **treatment** is a specific combination of factor levels.

The **response** is the variable being measured by the experimenter.

EXAMPLE 11.1

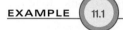

A group of people is randomly divided into an experimental group and a control group. The control group is given an aptitude test after having eaten a full breakfast. The experimental group is given the same test without having eaten any breakfast. What are the factors, levels, and treatments in this experiment?

SOLUTION The *experimental units* are the people on which the *response* (test score) is measured. The *factor* of interest could be described as "meal" and has two *levels:* "breakfast" and "no breakfast." Since this is the only factor controlled by the experimenter, the two levels—"breakfast" and "no breakfast"—also represent the *treatments* of interest in the experiment.

EXAMPLE 11.2

Suppose that the experimenter in Example 11.1 began by randomly selecting 20 men and 20 women for the experiment. These two groups were then randomly divided into 10 each for the experimental and control groups. What are the factors, levels, and treatments in this experiment?

SOLUTION Now there are two *factors* of interest to the experimenter, and each factor has two *levels:*

- "Gender" at two levels: men and women
- "Meal" at two levels: breakfast and no breakfast

In this more complex experiment, there are four *treatments,* one for each specific combination of factor levels: men without breakfast, men with breakfast, women without breakfast, and women with breakfast.

In this chapter, we will concentrate on experiments that have been designed in three different ways, and we will use a technique called the *analysis of variance* to judge the effects of various factors on the experimental response. Two of these *experimental designs* are extensions of the unpaired and paired designs from Chapter 10.

11.2 WHAT IS AN ANALYSIS OF VARIANCE?

The responses that are generated in an experimental situation always exhibit a certain amount of *variability*. In an **analysis of variance,** you divide the total variation in the response measurements into portions that may be attributed to various *factors* of interest to the experimenter. If the experiment has been properly designed, these portions can then be used to answer questions about the effects of the various factors on the response of interest.

You can better understand the logic underlying an analysis of variance by looking at a simple experiment. Consider two sets of samples randomly selected from populations 1 (◆) and 2 (○), each with identical pairs of means, $\bar{x}_1$ and $\bar{x}_2$ respectively. The two sets are shown in Figure 11.1. Is it easier to detect the difference in the two means when you look at set A or set B? You will probably agree that set A shows the difference more clearly. In set A, the variability of the measurements *within* the groups (◆s and ○s) is much smaller than the variability *between* the two groups. In set B, there is more variability *within* the groups (◆s and ○s), causing the two groups to "mix" together and making it more difficult to see the *identical* difference in the means.

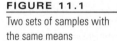

FIGURE 11.1
Two sets of samples with the same means

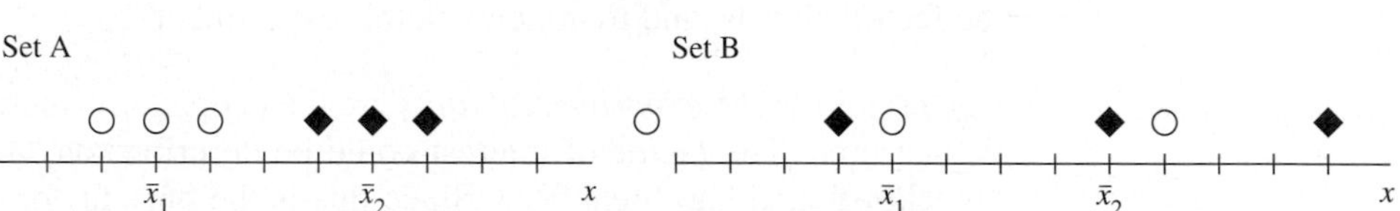

The comparison you have just done intuitively is formalized by the analysis of variance. Moreover, the analysis of variance can be used not only to compare two means but also to make comparisons of *more than two* population means and to determine the effects of various factors in more complex experimental designs. The analysis of variance relies on statistics with sampling distributions that are modelled by the F distribution of Section 10.7.

The Assumptions for an Analysis of Variance

The assumptions required for an analysis of variance are similar to those required for the Student's t and F statistics of Chapter 10. Regardless of the experimental design used to generate the data, you must assume that the observations within each treatment group are **normally distributed** with a **common variance** σ^2. As in Chapter 10, the analysis of variance procedures are fairly **robust** when the sample sizes are equal and when the data are fairly mound-shaped. Violating the assumption of a common variance is more serious, especially when the sample sizes are not nearly equal.

Assumptions for Analysis of Variance Test and Estimation Procedures

- The observations within each population are normally distributed with a common variance σ^2.
- Assumptions regarding the sampling procedure are specified for each design in the sections that follow.

This chapter describes the analysis of variance for three different experimental designs. The first design is based on independent random sampling from several populations and is an extension of the *unpaired t-test* of Chapter 10. The second is an extension of the *paired-difference* or *matched pairs* design and involves a random assignment of treatments within matched sets of observations. The third is a design that allows you to judge the effect of two experimental factors on the response. The sampling procedures necessary for each design will be restated in their respective sections.

11.3 THE COMPLETELY RANDOMIZED DESIGN: A ONE-WAY CLASSIFICATION

One of the simplest experimental designs is the **completely randomized design,** in which random samples are selected independently from each of k populations. This design involves only one *factor,* the population from which the measurement comes—hence its designation as a **one-way classification.** There are k different *levels* corresponding to the k populations, which are also the *treatments* in this one-way classification. Are the k population means all the same, or is at least one mean different from the others?

Why do you need a new procedure, the *analysis of variance,* to compare the population means when you already have the Student's t-test available? In comparing $k = 3$ means, you could test each of three pairs of hypotheses:

$$H_0: \mu_1 = \mu_2 \quad H_0: \mu_1 = \mu_3 \quad H_0: \mu_2 = \mu_3$$

to find out where the differences lie. However, you must remember that each test you perform is subject to the possibility of error. To compare $k = 4$ means, you would need six tests, and you would need 10 tests to compare $k = 5$ means. The more tests you perform on a set of measurements, the more likely at least one of your conclusions will be incorrect. The analysis of variance procedure provides one overall test to judge the equality of the k population means. Once you have determined whether there is *actually* a difference in the means, you can use another procedure to find out where the differences lie.

How can you select these k random samples? Sometimes the populations actually exist in fact, and you can use a computerized random number generator or a random number table to randomly select the samples. For example, in a study to compare the average sizes of health insurance claims in four different states, you could use a computer database provided by the health insurance companies to select random samples from the four states. In other situations, the populations may be *hypothetical,* and responses can be generated only after the experimental treatments have been applied.

EXAMPLE 11.3

A researcher is interested in the effects of five types of insecticides in controlling boll weevils in cotton fields. Explain how to implement a completely randomized design to investigate the effects of the five insecticides on crop yield.

SOLUTION The only way to generate the equivalent of five random samples from the hypothetical populations corresponding to the five insecticides is to use a method called a **randomized assignment.** A fixed number of cotton plants are chosen for treatment, and each is assigned a random number. Suppose that each sample is to have an equal number of measurements. Using a randomization device, you can assign the first n plants chosen to receive insecticide 1, the second n plants to receive insecticide 2, and so on, until all five treatments have been assigned.

Whether you use *random selection* or *random assignment,* both of these methods result in a completely randomized design, or one-way classification, for which the analysis of variance is used.

The Analysis of Variance for a Completely Randomized Design

Suppose you want to compare k population means, $\mu_1, \mu_2, \ldots, \mu_k$, based on independent random samples of size $n_1, n_2, \ldots, n_k$ from normal populations with a common variance σ^2. That is, each of the normal populations has the same shape, but their locations might be different, as shown in Figure 11.2.

FIGURE 11.2
Normal populations with a common variance but different means

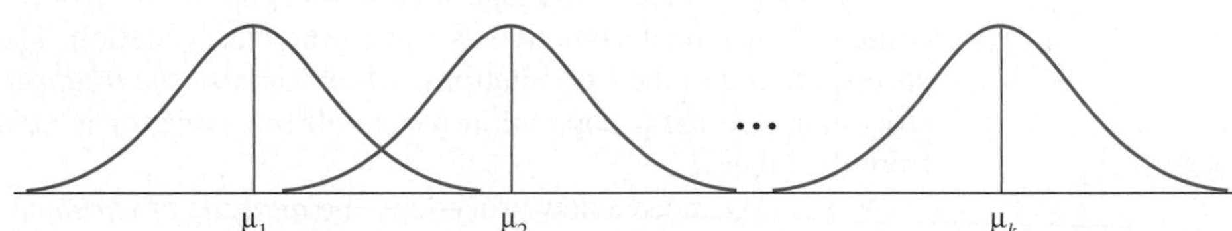

Partitioning the Total Variation in an Experiment

Let x_{ij} be the jth measurement ($j = 1, 2, \ldots, n_i$) in the ith sample $(i = 1, 2, \ldots, k)$. The analysis of variance procedure begins by considering the total variation in the experiment, which is measured by a quantity called the **total sum of squares (Total SS):**

$$\text{Total SS} = \Sigma(x_{ij} - \bar{x})^2 = \Sigma x_{ij}^2 - \frac{(\Sigma x_{ij})^2}{n}$$

This is the familiar numerator in the formula for the sample variance for the entire set of $n = n_1 + n_2 + \ldots + n_k$ measurements. The second part of the calculational formula is sometimes called the **correction for the mean (CM).** If we let G represent the *grand total* of all n observations, then

$$\text{CM} = \frac{(\Sigma x_{ij})^2}{n} = \frac{G^2}{n}$$

This Total SS is partitioned into two components. The first component, called the **sum of squares for treatments (SST),** measures the variation among the k sample means:

$$\text{SST} = \Sigma n_i(\bar{x}_i - \bar{x})^2 = \Sigma\frac{T_i^2}{n_i} - \text{CM}$$

where T_i is the total of the observations for treatment *i*. The second component, called the **sum of squares for error (SSE),** is used to measure the pooled variation within the *k* samples:

$$\text{SSE} = (n_1 - 1)s_1^2 + (n_2 - 1)s_2^2 + \cdots + (n_k - 1)s_k^2$$

This formula is a direct extension of the numerator in the formula for the pooled estimate of σ^2 from Chapter 10. We can show algebraically that, in the analysis of variance,

$$\text{Total SS} = \text{SST} + \text{SSE}$$

Therefore, you need to calculate only two of the three sums of squares—Total SS, SST, and SSE—and the third can be found by subtraction.

Each of the sources of variation, when divided by its appropriate **degrees of freedom,** provides an estimate of the variation in the experiment. Since Total SS involves *n* squared observations, its degrees of freedom are $df = (n - 1)$. Similarly, the sum of squares for treatments involves *k* squared observations, and its degrees of freedom are $df = (k - 1)$. Finally, the sum of squares for error, a direct extension of the pooled estimate in Chapter 10, has

$$df = (n_1 - 1) + (n_2 - 1) + \cdots + (n_k - 1) = n - k$$

Notice that the degrees of freedom for treatments and error are additive—that is,

$$df(\text{total}) = df(\text{treatments}) + df(\text{error})$$

These two sum of squares and their respective degrees of freedom are combined to form the **mean squares** as MS = SS/*df*. The total variation in the experiment is then displayed in an **analysis of variance** (or **ANOVA**) **table.**

The column labelled "SS" satisfies: Total SS = SST + SSE

ANOVA Table for *k* Independent Random Samples: Completely Randomized Design

Source	*df*	SS	MS	*F*
Treatments	$k - 1$	SST	MST = SST/$(k - 1)$	MST/MSE
Error	$n - k$	SSE	MSE = SSE/$(n - k)$	
Total	$n - 1$	Total SS		

where

$$\text{Total SS} = \Sigma x_{ij}^2 - \text{CM}$$
$$= (\text{Sum of squares of all } x\text{-values}) - \text{CM}$$

with

$$\text{CM} = \frac{(\Sigma x_{ij})^2}{n} = \frac{G^2}{n}$$

The column labelled "*df*" always adds up to $n - 1$.

$$\text{SST} = \Sigma\frac{T_i^2}{n_i} - \text{CM} \qquad \text{MST} = \frac{\text{SST}}{k - 1}$$

$$\text{SSE} = \text{Total SS} - \text{SST} \qquad \text{MSE} = \frac{\text{SSE}}{n - k}$$

and

G = Grand total of all n observations
T_i = Total of all observations in sample i
n_i = Number of observations in sample i
$n = n_1 + n_2 + \cdots + n_k$

EXAMPLE 

In an experiment to determine the effect of nutrition on the attention spans of elementary school students, a group of 15 students were randomly assigned to each of three meal plans: no breakfast, light breakfast, and full breakfast. Their attention spans (in minutes) were recorded during a morning reading period and are shown in Table 11.1. Construct the analysis of variance table for this experiment.

TABLE 11.1 **Attention Spans of Students After Three Meal Plans**

No Breakfast	Light Breakfast	Full Breakfast
8	14	10
7	16	12
9	12	16
13	17	15
10	11	12
$T_1 = 47$	$T_2 = 70$	$T_3 = 65$

SOLUTION To use the calculational formulas, you need the $k = 3$ treatment totals together with $n_1 = n_2 = n_3 = 5$, $n = 15$, and $\Sigma x_{ij} = 182$. Then

$$\text{CM} = \frac{(182)^2}{15} = 2208.2667$$

$$\text{Total SS} = (8^2 + 7^2 + \cdots + 12^2) - \text{CM} = 2338 - 2208.2667 = 129.7333$$

with $(n - 1) = (15 - 1) = 14$ degrees of freedom,

$$\text{SST} = \frac{47^2 + 70^2 + 65^2}{5} - \text{CM} = 2266.8 - 2208.2667 = 58.5333$$

with $(k - 1) = (3 - 1) = 2$ degrees of freedom, and by subtraction,

$$\text{SSE} = \text{Total SS} - \text{SST} = 129.7333 - 58.5333 = 71.2$$

with $(n - k) = (15 - 3) = 12$ degrees of freedom. These three sources of variation, their degrees of freedom, sums of squares, and mean squares are shown in the shaded area of the ANOVA tables generated by *MINITAB* and *Excel*, given in Figure 11.3. You will find instructions for generating this output in the "Technology Today" section at the end of this chapter.

FIGURE 11.3(a) *MINITAB* output for Example 11.4

One-way ANOVA: Span versus Meal

```
Source  DF      SS     MS     F      P
Meal     2   58.53  29.27  4.93  0.027
Error   12   71.20   5.93
Total   14  129.73

S = 2.436   R-Sq = 45.12%   R-Sq(adj) = 35.97%
                            Individual 95% CIs For Mean
                            Based on Pooled StDev

Level  N    Mean  StDev  --+---------+---------+---------+-------
1      5   9.400  2.302  (---------*--------)
2      5  14.000  2.550                      (--------*--------)
3      5  13.000  2.449                  (--------*--------)
                         --+---------+---------+---------+-------
                         7.5      10.0      12.5      15.0

Pooled StDev = 2.436
```

FIGURE 11.3(b)
Excel output for Example 11.4

Anova: Single Factor

SUMMARY

Groups	*Count*	*Sum*	*Average*	*Variance*
None	5	47	9.4	5.3
Light	5	70	14	6.5
Full	5	65	13	6

ANOVA

Source of Variation	*SS*	*df*	*MS*	*F*	*P-value*	*F crit*
Between Groups	58.533	2	29.267	4.933	0.027	3.885
Within Groups	71.2	12	5.933			
Total	129.733	14				

The computer outputs give some additional information about the variation in the experiment. The lower section in *MINITAB* and the upper section in *Excel* show the means and standard deviations (or variances) for the three meal plans. More importantly, you can see in the upper section in *MINITAB* and the lower section in *Excel* two columns marked "F" and "P" ("*F*" and "*P*-value" in *Excel*). We can use these values to test a hypothesis concerning the equality of the three treatment means.

Testing the Equality of the Treatment Means

The *mean squares* in the analysis of variance table can be used to test the null hypothesis

NEED A TIP?
MS = SS/*df*

$$H_0 : \mu_1 = \mu_2 = \cdots = \mu_k$$

versus the alternative hypothesis

H_a : At least one of the means is different from the others

using the following theoretical argument:

- Remember that σ^2 is the common variance for all k populations. The quantity

$$\text{MSE} = \frac{\text{SSE}}{n - k}$$

is a pooled estimate of σ^2, a weighted average of all k sample variances, whether or not H_0 is true.

- If H_0 is true, then the variation in the sample means, measured by MST = [SST/(k − 1)], also provides an unbiased estimate of σ^2. However, if H_0 is false and the population means are different, then MST—which measures the variation in the sample means—will be unusually *large,* as shown in Figure 11.4.

FIGURE 11.4
Sample means drawn from identical versus different populations

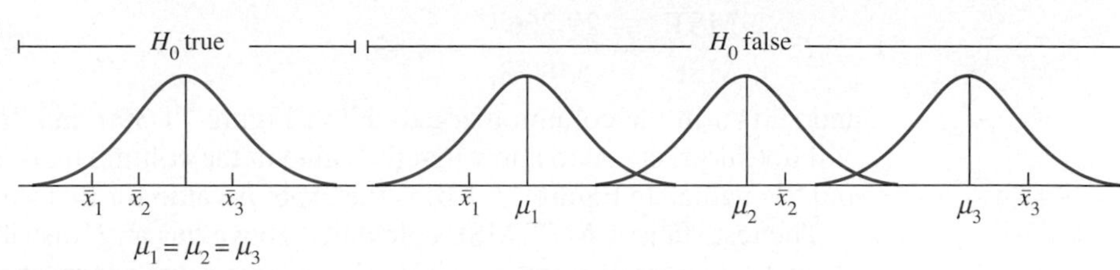

- The test statistic

$$F = \frac{\text{MST}}{\text{MSE}}$$

F tests for ANOVA tables are **always** upper (right) tailed.

tends to be larger than usual if H_0 is false. Hence, you can reject H_0 for large values of F, using a *right-tailed* statistical test. When H_0 is true, this test statistic has an F distribution with $df_1 = (k - 1)$ and $df_2 = (n - k)$ degrees of freedom, and *right-tailed* critical values of the F distribution (from Table 6 in Appendix I) or computer-generated *p*-values can be used to draw statistical conclusions about the equality of the population means.

F Test for Comparing *k* Population Means

1. Null hypothesis: $H_0 : \mu_1 = \mu_2 = \ldots = \mu_k$
2. Alternative hypothesis: H_a : One or more pairs of population means differ
3. Test statistic: $F = \text{MST/MSE}$, where F is based on $df_1 = (k - 1)$ and $df_2 = (n - k)$
4. Rejection region: Reject H_0 if $F > F_\alpha$, where F_α lies in the upper tail of the F distribution (with $df_1 = k - 1$ and $df_2 = n - k$) or if the *p*-value $< \alpha$.

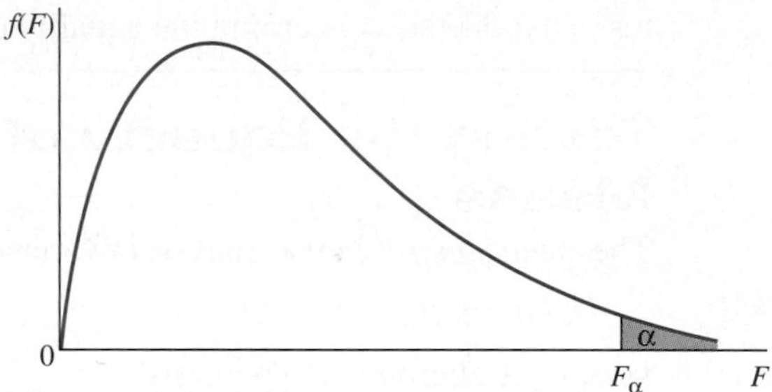

Assumptions

- The samples are randomly and independently selected from their respective populations.
- The populations are normally distributed with means $\mu_1, \mu_2, \ldots, \mu_k$ and equal variances, $\sigma_1^2 = \sigma_2^2 = \cdots = \sigma_k^2 = \sigma^2$.

EXAMPLE 11.5

Do the data in Example 11.4 provide sufficient evidence to indicate a difference in the average attention spans depending on the type of breakfast eaten by the student?

SOLUTION To test $H_0 : \mu_1 = \mu_2 = \mu_3$ versus the alternative hypothesis that the average attention span is different for at least one of the three treatments, you use the analysis of variance F statistic, calculated as

$$F = \frac{\text{MST}}{\text{MSE}} = \frac{29.2667}{5.9333} = 4.93$$

and shown in the column marked "F" in Figure 11.3(a) and "*F*" in Figure 11.3(b). It will not surprise you to know that the value in the column marked "P" in Figure 11.3(a) and "*P*-Value" in Figure 11.3(b) is the exact *p*-value for this statistical test.

The test statistic MST/MSE calculated above has an F distribution with $df_1 = 2$ and $df_2 = 12$ degrees of freedom. Using the critical value approach with $\alpha = 0.05$, you can reject H_0 if $F > F_{0.05} = 3.89$ from Table 6 in Appendix I (see Figure 11.5). Since the observed value, $F = 4.93$, exceeds the critical value, you reject H_0. There is sufficient evidence to indicate that at least one of the three average attention spans is different from at least one of the others.

FIGURE 11.5
Rejection region for Example 11.5

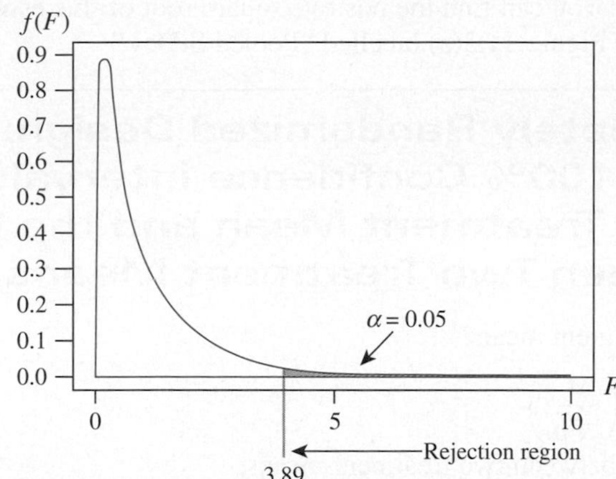

NEED A TIP?

Computer printouts give the exact p-value—use the p-value to make your decision.

You could have reached this same conclusion using the exact p-value, $P = 0.027$, given in Figure 11.3. Since the p-value is less than $\alpha = 0.05$, the results are statistically significant at the 5% level. You still conclude that at least one of the three average attention spans is different from at least one of the others.

Estimating Differences in the Treatment Means

The next obvious question you might ask involves the nature of the differences in the population means. Which means are different from the others? How can you estimate the difference, or possibly the individual means for each of the three treatments? In Section 11.4, we will present a procedure that you can use to compare all possible pairs of treatment means simultaneously. However, if you have a special interest in a particular mean or pair of means, you can construct confidence intervals using the small-sample procedures of Chapter 10, based on the Student's t distribution. For a single population mean, μ_i, the confidence interval is

$$\bar{x}_i \pm t_{\alpha/2}\left(\frac{s}{\sqrt{n_i}}\right)$$

where $\bar{x}_i$ is the sample mean for the ith treatment. Similarly, for a comparison of two population means—say, μ_i and μ_j—the confidence interval is

$$(\bar{x}_i - \bar{x}_j) \pm t_{\alpha/2}\sqrt{s^2\left(\frac{1}{n_i} + \frac{1}{n_j}\right)}$$

Before you can use these confidence intervals, however, two questions remain:

- How do you calculate s or s^2, the best estimate of the common variance σ^2?
- How many degrees of freedom are used for the critical value of t?

To answer these questions, remember that in an analysis of variance, the mean square for error, MSE, always provides an unbiased estimator of σ^2 and uses information from the entire set of measurements. Hence, it is the best available estimator of σ^2, regardless of what test or estimation procedure you are using. You should *always* use

$$s^2 = \text{MSE} \quad \text{with } df = (n - k)$$

to estimate σ^2! You can find the positive square root of this estimator, $s = \sqrt{MSE}$, on the last line of Figure 11.3(a) labelled "Pooled StDev."

NEED A TIP?

Degrees of freedom for confidence intervals are the *df* for **error**.

Completely Randomized Design: (1 − α)100% Confidence Intervals for a Single Treatment Mean and the Difference between Two Treatment Means

Single treatment mean:

$$\bar{x}_i \pm t_{\alpha/2}\left(\frac{s}{\sqrt{n_i}}\right)$$

Difference between two treatment means:

$$(\bar{x}_i - \bar{x}_j) \pm t_{\alpha/2}\sqrt{s^2\left(\frac{1}{n_i} + \frac{1}{n_j}\right)}$$

with

$$s = \sqrt{s^2} = \sqrt{MSE} = \sqrt{\frac{SSE}{n-k}}$$

where $n = n_1 + n_2 + \ldots + n_k$ and $t_{\alpha/2}$ is based on $(n - k)$ *df*.

EXAMPLE 11.6

The researcher in Example 11.4 believes that students who eat no breakfast will have significantly shorter attention spans, but that there may be no difference between those who eat a light and a full breakfast. Find a 95% confidence interval for the average attention span for students who eat no breakfast, as well as a 95% confidence interval for the difference in the average attention spans for light versus full breakfast eaters.

SOLUTION For $s^2 = MSE = 5.9333$ so that $s = \sqrt{5.9333} = 2.436$ with $df = (n - k) = 12$, you can calculate the two confidence intervals:

- For no breakfast:

$$\bar{x}_1 \pm t_{\alpha/2}\left(\frac{s}{\sqrt{n_1}}\right)$$

$$9.4 \pm 2.179\left(\frac{2.436}{\sqrt{5}}\right)$$

$$9.4 \pm 2.37$$

or between 7.03 and 11.77 minutes (min).

- For light versus full breakfast:

$$(\bar{x}_2 - \bar{x}_3) \pm t_{\alpha/2}\sqrt{s^2\left(\frac{1}{n_2} + \frac{1}{n_3}\right)}$$

$$(14 - 13) \pm 2.179\sqrt{5.9333\left(\frac{1}{5} + \frac{1}{5}\right)}$$

$$1 \pm 3.36$$

a difference of between −2.36 and 4.36 min.

You can see that the second confidence interval does not indicate a difference in the average attention spans for students who ate light versus full breakfasts, as the researcher suspected. If the researcher, because of prior beliefs, wishes to test the other two possible pairs of means—none versus light breakfast, and none versus full breakfast—the methods given in Section 11.4 should be used for testing all three pairs.

EXAMPLE 11.7

A mechanical engineer at a manufacturing plant in Hamilton, Ontario, keeps a close watch on the performance and condition of the machines, inspecting for wear due to friction. The following data are the weight losses (in milligrams) of certain machine parts due to friction when used with three different lubricants:

	Lubricant A	Lubricant B	Lubricant C
	12	10	12
	9	8	8
	8	9	14
	11	13	11
	10	8	12
	8	7	13
	7	5	7
	6	11	8
Total (T_i)	71	71	85
Means ($\bar{x}_i$)	8.875	8.875	10.625
Standard deviation (S_i)	2.031	2.475	2.615

$\Sigma x_{ij}^2 = 2283.0$, and $\Sigma x_{ij} = 227$

1. State the null and alternative hypotheses to test whether there is a significant difference in mean weight losses among the three lubricants.
2. How many degrees of freedom are associated with the F test statistic?
3. Calculate the value of the test statistic F.
4. Set up the ANOVA table.
5. What is the appropriate conclusion?

SOLUTION

1. $H_0: \mu_A = \mu_B = \mu_C$ versus H_a: At least one of the population means is different from the others.
2. The test statistic F has an F distribution with degrees of freedom given by: $df_1 - k-1 = 2$, and $df_2 = n - k = 24 - 3 = 21$
3. $\text{CM} = \left(\Sigma x_{ij}\right)^2/n = (227)^2/24 = 2147.042$

   $\text{Total SS} = \Sigma x_{ij}^2 - \text{CM} = 2283 - 2147.042 = 135.958$

   $\text{SST} = [\text{C}]\Sigma \text{T}_i^2/\text{n}_i - \text{CM} = 2163.375 - 2147.042 = 16.333$

   $\text{MST} = \text{SST}/(k - 1) = 16.333/2 = 8.1665$

   $\text{SSE} = \text{Total SS} - \text{SST} = 119.625$

   $\text{MSE} = \text{SSE}/(n - k) = 119.625/21 = 5.6964$

   The value of the test statistic is

   $F = \text{MST}/\text{MSE} = 8.1665/5.6964 = 1.4336$

4.

Source of Variation	*df*	SS	MS	*F*
Treatments	2	16.333	8.1665	1.4336
Error	21	119.625	5.6964	
Total	23	135.958		

5. With df_1, = 2, $df_2 = 21$, and $\alpha = 0.01$, the rejection region is

$$\text{Reject } H_0 \text{ if } F > F_{0.01} = 5.78$$

Since $F < 5.78$, do not reject H_0; therefore, we *cannot* conclude that there is a significant difference in average weight loss among the three lubricants.

NEED TO KNOW

How to Determine Whether Calculations Are Accurate

The following suggestions apply to all the analyses of variance in this chapter:

1. When calculating sums of squares, be certain to carry at least six significant figures before performing subtractions.
2. Remember, sums of squares can never be negative. If you obtain a negative sum of squares, you have made a mistake in arithmetic.
3. Always check your analysis of variance table to make certain that the degrees of freedom sum to the total degrees of freedom $(n - 1)$ and that the sums of squares sum to Total SS.

11.3 EXERCISES

BASIC TECHNIQUES

11.1 Suppose you wish to compare the means of six populations based on independent random samples, each of which contains 10 observations. Insert, in an ANOVA table, the sources of variation and their respective degrees of freedom.

11.2 The values of Total SS and SSE for the experiment in Exercise 11.1 are Total SS = 21.4 and SSE = 16.2.

a. Complete the ANOVA table for Exercise 11.1.

b. How many degrees of freedom are associated with the F statistic for testing $H_0: \mu_1 = \mu_2 = \cdots = \mu_6$?

c. Give the rejection region for the test in part b for $\alpha = 0.05$.

d. Do the data provide sufficient evidence to indicate differences among the population means?

e. Approximate the p-value for the test. Does this value confirm your conclusions in part d?

11.3 The sample means corresponding to populations 1 and 2 in Exercise 11.1 are $\bar{x}_1 = 3.07$ and $\bar{x}_2 = 2.52$.

a. Find a 95% confidence interval for μ_1.

b. Find a 95% confidence interval for the difference $(\mu_1 - \mu_2)$.

11.4 Suppose you wish to compare the means of four populations based on independent random samples, each of which contains six observations. Insert, in an ANOVA table, the sources of variation and their respective degrees of freedom.

11.5 The values of Total SS and SST for the experiment in Exercise 11.4 are Total SS = 473.2 and SST = 339.8.

a. Complete the ANOVA table for Exercise 11.4.

b. How many degrees of freedom are associated with the F statistic for testing $H_0: \mu_1 = \mu_2 = \mu_3 = \mu_4$?

c. Give the rejection region for the test in part b for $\alpha = 0.05$.

d. Do the data provide sufficient evidence to indicate differences among the population means?

e. Approximate the p-value for the test. Does this confirm your conclusions in part d?

11.6 The sample means corresponding to populations 1 and 2 in Exercise 11.4 are $\bar{x}_1 = 88.0$ and $\bar{x}_2 = 83.9$.

a. Find a 90% confidence interval for μ_1.

b. Find a 90% confidence interval for the difference $(\mu_1 - \mu_2)$.

Data set EX1107 **11.7** These data are observations collected using a completely randomized design:

Sample 1	Sample 2	Sample 3
3	4	2
2	3	0
4	5	2
3	2	1
2	5	

a. Calculate CM and Total SS.

b. Calculate SST and MST.

c. Calculate SSE and MSE.

d. Construct an ANOVA table for the data.

e. State the null and alternative hypotheses for an analysis of variance F test.

f. Use the p-value approach to determine whether there is a difference in the three population means.

11.8 Refer to Exercise 11.7 and data set EX1107. Do the data provide sufficient evidence to indicate a difference between μ_2 and μ_3? Test using the t test of Section 10.4 with $\alpha = 0.05$.

11.9 Refer to Exercise 11.7 and data set EX1107.

a. Find a 90% confidence interval for μ_1.

b. Find a 90% confidence interval for the difference $(\mu_1 - \mu_3)$.

APPLICATIONS

11.10 Reducing Hostility A clinical psychologist wished to compare three methods for reducing hostility levels in university students using a certain psychological test (HLT). High scores on this test were taken to indicate great hostility. Eleven students who got high and nearly equal scores were used in the experiment. Five were selected at random from among the 11 problem cases and treated by method A, three were taken at random from the remaining six students and treated by method B, and the other three students were treated by method C. All treatments continued throughout a semester, when the HLT test was given again. The results are shown in the table:

Method	Scores on the HLT Test				
A	73	83	76	68	80
B	54	74	71		
C	79	95	87		

a. Perform an analysis of variance for this experiment.

b. Do the data provide sufficient evidence to indicate a difference in mean student response to the three methods after treatment?

11.11 Hostility, continued Refer to Exercise 11.10. Let μ_A and μ_B, respectively, denote the mean scores at the end of the semester for the populations of extremely hostile students who were treated throughout that semester by method A and method B.

a. Find a 95% confidence interval for μ_A.

b. Find a 95% confidence interval for μ_B.

c. Find a 95% confidence interval for $(\mu_A - \mu_B)$.

d. Is it correct to claim that the confidence intervals found in parts a, b, and c are jointly valid?

Data set EX1112 **11.12 Assembling Electronic Equipment** An experiment was conducted to compare the effectiveness of three training programs, A, B, and C, in training assemblers of a piece of electronic equipment. Fifteen employees were randomly assigned, five each, to the three programs. After completion of the courses, each person was required to assemble four pieces of the equipment, and the average length of time required to complete the assembly was recorded. Several of the employees resigned during the course of the program; the remainder were evaluated, producing the data shown in the accompanying table. Use the *MINITAB* printout to answer the questions:

Training Program	Average Assembly Time (min)				
A	59	64	57	62	
B	52	58	54		
C	58	65	71	63	64

a. Do the data provide sufficient evidence to indicate a difference in mean assembly times for people trained by the three programs? Give the p-value for the test and interpret its value.

b. Find a 99% confidence interval for the difference in mean assembly times between persons trained by programs A and B.

c. Find a 99% confidence interval for the mean assembly times for persons trained by program A.

d. Do you think the data will satisfy (approximately) the assumption that they have been selected from normal populations? Why?

MINITAB output for Exercise 11.12

One-way ANOVA: Time versus Program

```
Source     DF       SS      MS      F      P
Program     2    170.5    85.2   5.70  0.025
Error       9    134.5    14.9
Total      11    304.9
S = 3.865   R-Sq = 55.90%   R-Sq(adj) = 46.10%
                                  Individual 95% CIs For Mean
                                  Based on Pooled StDev
Level       N     Mean   StDev  -+---------+---------+---------+-----
1           4   60.500   3.109              (--------*--------)
2           3   54.667   3.055   (---------*---------)
3           5   64.200   4.658                        (------*-------)
                                 -+---------+---------+---------+-----
Pooled StDev =   3.865          50.0      55.0      60.0      65.0
```

11.13 Swampy Sites Data set EX1113 An ecological study was conducted to compare the rates of growth of vegetation at four swampy undeveloped sites and to determine the cause of any differences that might be observed. Part of the study involved measuring the leaf lengths of a particular plant species on a preselected date in May. Six plants were randomly selected at each of the four sites to be used in the comparison. The data in the table are the mean leaf length per plant (in centimetres) for a random sample of 10 leaves per plant. The *MINITAB* analysis of variance computer printout for these data is also provided:

Location	Mean Leaf Length (cm)					
1	5.7	6.3	6.1	6.0	5.8	6.2
2	6.2	5.3	5.7	6.0	5.2	5.5
3	5.4	5.0	6.0	5.6	4.9	5.2
4	3.7	3.2	3.9	4.0	3.5	3.6

MINITAB output for Exercise 11.13

One-way ANOVA: Length versus Location

```
Source     DF      SS      MS      F      P
Location    3  19.740   6.580  57.38  0.000
Error      20   2.293   0.115
Total      23  22.033
S = 0.3386   R-Sq = 89.59%   R-Sq(adj) = 88.03%
                                 Individual 95% CIs For Mean
                                 Based on Pooled StDev
Level       N     Mean   StDev  --------+---------+---------+---------+-
1           6   6.0167  0.2317                                (--*---)
2           6   5.6500  0.3937                           (---*--)
3           6   5.3500  0.4087                       (---*--)
4           6   3.6500  0.2881  (---*--)
                                --------+---------+---------+---------+-
Pooled StDev =  0.3386               4.00      4.80      5.60      6.40
```

a. You will recall that the test and estimation procedures for an analysis of variance require that the observations be selected from normally distributed (at least, roughly so) populations. Why might you feel reasonably confident that your data satisfy this assumption?

b. Do the data provide sufficient evidence to indicate a difference in mean leaf length among the four locations? What is the p-value for the test?

c. Suppose, prior to seeing the data, you decided to compare the mean leaf lengths of locations 1 and 4. Test the null hypothesis $\mu_1 = \mu_4$ against the alternative $\mu_1 \neq \mu_4$.

d. Refer to part c. Construct a 99% confidence interval for $(\mu_1 - \mu_4)$.

e. Rather than use an analysis of variance F test, it would seem simpler to examine one's data, select the two locations that have the smallest and largest sample mean lengths, and then compare these two means using a Student's t test. If there is evidence to indicate a difference in these means, there is clearly evidence of a difference among the four. (If you were to use this logic, there would be no need for the analysis of variance F test.) Explain why this procedure is invalid.

11.14 Dissolved O_2 Content Data set EX1114 Water samples were taken at four different locations in a river to determine whether the quantity of dissolved oxygen, a measure of water pollution, varied from one location to another. Locations 1 and 2 were selected above an industrial plant, one near the shore and the other in midstream; location 3 was adjacent to the industrial water discharge for the plant; and location 4 was slightly downriver in midstream. Five water specimens were randomly selected at each location, but one specimen, corresponding to location 4, was lost in the laboratory. The data and an *MINITAB* analysis of variance computer printout are provided here (the greater the pollution, the lower the dissolved oxygen readings):

Location	Mean Dissolved Oxygen Content				
1	5.9	6.1	6.3	6.1	6.0
2	6.3	6.6	6.4	6.4	6.5
3	4.8	4.3	5.0	4.7	5.1
4	6.0	6.2	6.1	5.8	

MINITAB output for Exercise 11.14

One-way ANOVA: Oxygen versus Location

```
Source     DF      SS      MS      F      P
Location    3  7.8361  2.6120  63.66  0.000
Error      15  0.6155  0.0410
Total      18  8.4516
S = 0.2026   R-Sq = 92.72%   R-Sq(adj) = 91.26%
                                 Individual 95% CIs For Mean
                                 Based on Pooled StDev
Level       N     Mean   StDev  ----+---------+---------+---------+--
1           5   6.0800  0.1483                       (--*---)
2           5   6.4400  0.1140                             (--*---)
3           5   4.7800  0.3114  (---*--)
4           4   6.0250  0.1708                      (--*---)
                                ----+---------+---------+---------+--
Pooled StDev =  0.2026            4.80      5.40      6.00      6.60
```

a. Do the data provide sufficient evidence to indicate a difference in the mean dissolved oxygen contents for the four locations?

b. Compare the mean dissolved oxygen content in midstream above the plant with the mean content adjacent to the plant (location 2 versus location 3). Use a 95% confidence interval.

Data set EX1115 **11.15 Calcium** The calcium content of a powdered mineral substance was analyzed five times by each of three methods, with similar standard deviations:

Method	Percent Calcium				
1	0.0279	0.0276	0.0270	0.0275	0.0281
2	0.0268	0.0274	0.0267	0.0263	0.0267
3	0.0280	0.0279	0.0282	0.0278	0.0283

Use an appropriate test to compare the three methods of measurement. Comment on the validity of any assumptions you need to make.

Data set EX1116 **11.16 Healing Herbs, Again** In Exercise 10.12, we reported the estimated average prices for different bunches of herbs with approximately the same weight:

Parsley		Cilantro	Mint	Basil	
0.99	0.53	1.27	1.49	2.56	0.62
1.92	1.41	1.22	1.29	1.92	0.66
1.23	1.12	1.19	1.27	1.30	0.62
0.85	0.63	1.22	1.35	1.79	0.65
0.65	0.67		1.29	1.23	0.60
0.69	0.60		1.00		0.67
0.60	0.66		1.27		
			1.28		

a. Use an analysis of variance for a completely randomized design to determine if there are significant differences in the prices of these four herbs. Can you reject the hypothesis of no difference in average price for these packages at the $\alpha = 0.05$ level of significance? At the $\alpha = 0.01$ level of significance?

b. Find a 95% confidence interval estimate of the difference in price between parsley and basil. Does there appear to be a significant difference in the price of these two kinds of herbs?

c. Find a 95% confidence interval estimate of the difference in price between mint and cilantro. Does there appear to be a significant difference in the price of these two kinds of herbs?

d. What other confidence intervals might be of interest to the researcher who conducted this experiment?

Data set EX1117 **11.17 The Cost of Lumber** A national home builder wants to compare the prices per 2.4 cubic metres of standard or better grade Douglas fir framing lumber. He randomly selects five suppliers in each of the four provinces where the builder is planning to begin construction. The prices are given in the table:

Province			
1	2	3	4
\$241	\$216	\$230	\$245
235	220	225	250
238	205	235	238
247	213	228	255
250	220	240	255

a. What type of experimental design has been used?

b. Construct the analysis of variance table for this data.

c. Do the data provide sufficient evidence to indicate that the average price per 2.4 cubic metres of Douglas fir differs among the four provinces? Test using $\alpha = 0.05$.

Data set EX1118 **11.18 Good at Math?** Twenty Grade 3 students were randomly separated into four equal groups, and each group was taught a mathematical concept using a different teaching method. At the end of the teaching period, progress was measured by a unit test. The scores are shown below (one child in Group 3 was absent on the day that the test was administered):

Group			
1	2	3	4
112	111	140	101
92	129	121	116
124	102	130	105
89	136	106	126
97	99		119

a. What type of design has been used in this experiment?

b. Construct an ANOVA table for the experiment.

c. Do the data present sufficient evidence to indicate a difference in the average scores for the four teaching methods? Test using $\alpha = 0.05$.

11.4 RANKING POPULATION MEANS

Many experiments are exploratory in nature. You have no preconceived notions about the results and have not decided (before conducting the experiment) to make specific treatment comparisons. Rather, you want to rank the treatment means, determine which means differ, and identify sets of means for which no evidence of difference exists.

One option might be to order the sample means from the smallest to the largest and then to conduct *t* tests for adjacent means in the ordering. If two means differ by more than

$$t_{\alpha/2}\sqrt{s^2\left(\frac{1}{n_i}+\frac{1}{n_j}\right)}$$

you conclude that n_i *and* n_j differ for *i* not equal to *j*. The problem with this procedure is that the probability of making a Type I error—that is, concluding that two means differ when, in fact, they are equal—is α for each test. If you compare a large number of pairs of means, the probability of detecting at least one difference in means, when in fact none exists, is quite large.

A simple way to avoid the high risk of declaring differences when they do not exist is to use the **studentized range,** the difference between the smallest and the largest in a set of *k* sample means, as the yardstick for determining whether there is a difference in a pair of population means. This method, often called **Tukey's method for paired comparisons,** makes the probability of declaring that a difference exists between at least one pair in a set of *k* treatment means, when no difference exists, equal to α.

Tukey's method for making paired comparisons is based on the usual analysis of variance assumptions. **In addition, it assumes that the sample means are independent and based on samples of equal size.** The yardstick that determines whether a difference exists between a pair of treatment means is the quantity ω (Greek letter omega), which is presented next.

Yardstick for Making Paired Comparisons

$$\omega = q_\alpha(k,df)\left(\frac{s}{\sqrt{n_t}}\right)$$

where

k = Number of treatments

s^2 = MSE = Estimator of the common variance σ^2 and $s=\sqrt{s^2}$

df = Number of degrees of freedom for s^2

n_t = Common sample size—that is, the number of observations in each of the *k* treatment means

$q_\alpha(k, df)$ = Tabulated value from Tables 11(a) and 11(b) in Appendix I, for $\alpha = 0.05$ and 0.01, respectively, and for various combinations of *k* and *df*

Rule: Two population means are judged to differ if the corresponding sample means differ by ω or more.

Table 11(a) and 11(b) in Appendix I list the values of $q_\alpha(k, df)$ for $\alpha = 0.05$ and 0.01, respectively. To illustrate the use of the tables, refer to the portion of Table 11(a) reproduced in Table 11.2. Suppose you want to make pairwise comparisons of $k = 5$ means with $\alpha = 0.05$ for an analysis of variance, where s^2 possesses 9 *df*. The tabulated value for $k = 5$, $df = 9$, and $\alpha = 0.05$, shaded in Table 11.2, is $q_{0.05}(5, 9) = 4.76$.

TABLE 11.2 A Partial Reproduction of Table 11(a) in Appendix I; Upper 5% Points

df	2	3	4	5	6	7	8	9	10	11	12
1	17.97	26.98	32.82	37.08	40.41	43.12	45.40	47.36	49.07	50.59	51.96
2	6.08	8.33	9.80	10.88	11.74	12.44	13.03	13.54	13.99	14.39	14.75
3	4.50	5.91	6.82	7.50	8.04	8.48	8.85	9.18	9.46	9.72	9.95
4	3.93	5.04	5.76	6.29	6.71	7.05	7.35	7.60	7.83	8.03	8.21
5	3.64	4.60	5.22	5.67	6.03	6.33	6.58	6.80	6.99	7.17	7.32
6	3.46	4.34	4.90	5.30	5.63	5.90	6.12	6.32	6.49	6.65	6.79
7	3.34	4.16	4.68	5.06	5.36	5.61	5.82	6.00	6.16	6.30	6.43
8	3.26	4.04	4.53	4.89	5.17	5.40	5.60	5.77	5.92	6.05	6.18
9	3.20	3.95	4.41	4.76	5.02	5.24	5.43	5.59	5.74	5.87	5.98
10	3.15	3.88	4.33	4.65	4.91	5.12	5.30	5.46	5.60	5.72	5.83
11	3.11	3.82	4.26	4.57	4.82	5.03	5.20	5.35	5.49	5.61	5.71
12	3.08	3.77	4.20	4.51	4.75	4.95	5.12	5.27	5.39	5.51	5.61

EXAMPLE 11.8

Refer to Example 11.4, in which you compared the average attention spans for students given three different "meal" treatments in the morning: no breakfast, a light breakfast, or a full breakfast. The ANOVA F test in Example 11.5 indicated a significant difference in the population means. Use Tukey's method for paired comparisons to determine which of the three population means differ from the others.

SOLUTION For this example, there are $k = 3$ treatment means, with $s = \sqrt{\text{MSE}} = 2.436$. Tukey's method can be used, with each of the three samples containing $n_t = 5$ measurements and $(n - k) = 12$ degrees of freedom. Consult Table 11 in Appendix I to find $q_{0.05}(k, df) = q_{0.05}(3, 12) = 3.77$ and calculate the "yardstick" as

$$\omega = q_{0.05}(3,12)\left(\frac{s}{\sqrt{n_t}}\right) = 3.77\left(\frac{2.436}{\sqrt{5}}\right) = 4.11$$

The three treatment means are arranged in order from the smallest, 9.4, to the largest, 14.0, in Figure 11.6. The next step is to check the difference between every pair of means. The only difference that exceeds $\omega = 4.11$ is the difference between no breakfast and a light breakfast. These two treatments are thus declared significantly different. You cannot declare a difference between the other two pairs of treatments. To indicate this fact visually, Figure 11.6 shows a line under those pairs of means that are not significantly different.

FIGURE 11.6 Ranked means for Example 11.8

None	Full	Light
9.4	13.0	14.0

The results here may seem confusing. However, it usually helps to think of ranking the means and interpreting non-significant differences as our inability to distinctly rank those means underlined by the same line. For this example, the light breakfast definitely ranked higher than no breakfast, but the full breakfast could not be ranked higher than no breakfast, or lower than the light breakfast. The probability that we make at least one error among the three comparisons is at most $\alpha = 0.05$.

NEED A TIP?

If zero is not in the interval, there is evidence of a difference between the two methods.

Most computer programs provide an option to perform **paired comparisons,** including Tukey's method. The *MINITAB* output in Figure 11.7 shows its form of Tukey's test, which differs slightly from the method we have presented. The three intervals that you see in the printout marked "Lower" and "Upper" represent the difference in the two sample means plus or minus the yardstick ω. If the interval contains the value 0, the two means are judged to be not significantly different. You can see that only means 1 and 2 (none versus light) show a significant difference.

FIGURE 11.7

MINITAB output for Example 11.8

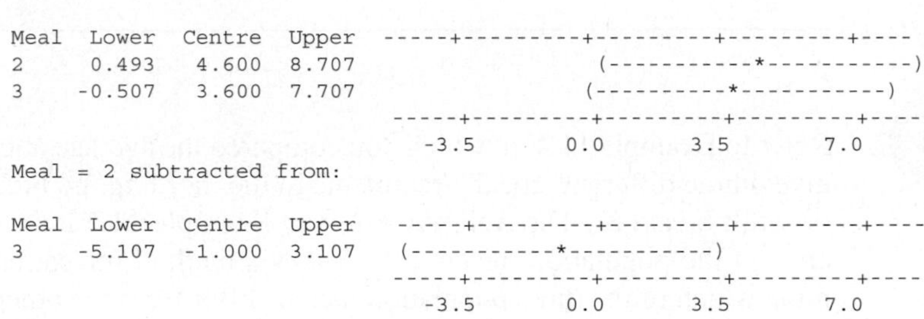

```
Tukey's 95% Simultaneous Confidence Intervals
All Pairwise Comparisons among Levels of Meal
Individual confidence level = 97.94%

Meal = 1 subtracted from:

Meal  Lower  Centre  Upper  -----+---------+---------+---------+----
2     0.493   4.600  8.707                 (-----------*-----------)
3    -0.507   3.600  7.707               (-----------*-----------)
                            -----+---------+---------+---------+----
                              -3.5       0.0       3.5       7.0
Meal = 2 subtracted from:

Meal  Lower   Centre  Upper  -----+---------+---------+---------+----
3    -5.107  -1.000  3.107  (-----------*-----------)
                             -----+---------+---------+---------+----
                               -3.5       0.0       3.5       7.0
```

As you study two more experimental designs in the next sections of this chapter, remember that once you have found a factor to be significant, you should use Tukey's method or another method of paired comparisons to find out exactly where the differences lie!

11.4 EXERCISES

BASIC TECHNIQUES

11.19 Suppose you wish to use Tukey's method for paired comparisons to rank a set of population means. In addition to the analysis of variance assumptions, what other property must the treatment means satisfy?

11.20 Consult Tables 11(a) and 11(b) in Appendix I and find the values of $q_\alpha(k, df)$ for these cases:

a. $\alpha = 0.05, k = 5, df = 7$
b. $\alpha = 0.05, k = 3, df = 10$
c. $\alpha = 0.01, k = 4, df = 8$
d. $\alpha = 0.01, k = 7, df = 5$

11.21 If the sample size for each treatment is n_t and if s^2 is based on 12 df, find ω in these cases:

a. $\alpha = 0.05, k = 4, n_t = 5$
b. $\alpha = 0.01, k = 6, n_t = 8$

11.22 An independent random sampling design was used to compare the means of six treatments based on samples of four observations per treatment. The pooled estimator of σ^2 is 9.12, and the sample means follow:

$\bar{x}_1 = 101.6 \quad \bar{x}_2 = 98.4 \quad \bar{x}_3 = 112.3$

$\bar{x}_4 = 92.9 \quad \bar{x}_5 = 104.2 \quad \bar{x}_6 = 113.8$

a. Give the value of ω that you would use to make pairwise comparisons of the treatment means for $\alpha = 0.05$.
b. Rank the treatment means using pairwise comparisons.

APPLICATIONS

11.23 Swampy Sites, again Refer to Exercise 11.13 and data set EX1113. Rank the mean leaf growth for the four locations. Use $\alpha = 0.01$.

11.24 Calcium Refer to Exercise 11.15 and data set EX1115. The paired comparisons option in *MINITAB* generated the output provided here. What do these results tell you about the differences in the population means? Does this confirm your conclusions in Exercise 11.15?

MINITAB output for Exercise 11.24

```
Tukey's 95% Simultaneous Confidence Intervals
All Pairwise Comparisons among Levels of Method
Individual confidence level = 97.94%

Method = 1 subtracted from:

Method       Lower       Centre       Upper
2        -0.0014377  -0.0008400  -0.0002423
3        -0.0001777   0.0004200   0.0010177
Method   --------+---------+---------+---------+
2           (-----*-----)
3                  (-----*-----)
         --------+---------+---------+---------+
           -0.0010    0.0000    0.0010    0.0020

Method = 2 subtracted from:

Method       Lower       Centre       Upper
3         0.0006623   0.0012600   0.0018577
Method   --------+---------+---------+---------+
3                                (-----*-----)
         --------+---------+---------+---------+
           -0.0010    0.0000    0.0010    0.0020
```

Data set EX1125 **11.25 Glucose Tolerance** Physicians depend on laboratory test results when managing medical problems such as diabetes or epilepsy. In a uniformity test for glucose tolerance, three different laboratories were each sent $n_t = 5$ identical blood samples from a person who had drunk 50 milligrams (mg) of glucose dissolved in water. The laboratory results (in mg/dL) are listed here:

Lab 1	Lab 2	Lab 3
120.1	98.3	103.0
110.7	112.1	108.5
108.9	107.7	101.1
104.2	107.9	110.0
100.4	99.2	105.4

a. Do the data indicate a difference in the average readings for the three laboratories?

b. Use Tukey's method for paired comparisons to rank the three treatment means. Use $\alpha = 0.05$.

11.26 The Cost of Lumber, continued The analysis of variance F test in Exercise 11.17 (and data set EX1117) determined that there was indeed a difference in the average cost of lumber for the four provinces. The following information from Exercise 11.17 is given in the table:

Sample Means			
	$\bar{x}_1 = 242.2$	MSE	41.25
	$\bar{x}_2 = 214.8$	Error *df*:	16
	$\bar{x}_3 = 231.6$	n_i:	5
	$\bar{x}_4 = 248.6$	k:	4

Use Tukey's method for paired comparisons to determine which means differ significantly from the others at the $\alpha = 0.01$ level.

Data set EX1127 **11.27 GRE Scores** The Graduate Record Examination (GRE) scores were recorded for students admitted to three different graduate programs at a local university:

Graduate Program		
1	2	3
532	670	502
548	590	607
619	640	549
509	710	524
627	690	542

a. Do these data provide sufficient evidence to indicate a difference in the mean GRE scores for applicants admitted to the three programs?

b. Find a 95% confidence interval for the difference in mean GRE scores for programs 1 and 2.

c. If you find a significant difference in the average GRE scores for the three programs, use Tukey's method for paired comparisons to determine which means differ significantly from the others. Use $\alpha = 0.05$.

11.5 THE RANDOMIZED BLOCK DESIGN: A TWO-WAY CLASSIFICATION

The *completely randomized design* introduced in Section 11.3 is a generalization of the *two independent samples* design presented in Section 10.4. It is meant to be used when the experimental units are quite similar or *homogeneous* in their makeup and when there is only one factor—the *treatment*—that might influence the response. Any other variation in the response is due to random variation or *experimental error*. Sometimes it is clear to the researcher that the experimental units are *not homogeneous*. Experimental subjects or animals, agricultural fields, days of the week, and other experimental units often add their own variability to the response. Although the researcher is not really interested in this source of variation, but rather in some chosen *treatment*, the researcher may be able to increase the information by isolating this source of variation using the **randomized block design**—a direct extension of the *matched pairs* or *paired-difference design* in Section 10.5.

NEED A TIP?

b = blocks
k = treatments
$n = bk$

In a randomized block design, the experimenter is interested in comparing k treatment means. The design uses *blocks* of k experimental units that are relatively similar, or *homogeneous*, with one unit within each block *randomly* assigned to each treatment. If the randomized block design involves k treatments within each of b blocks, then the total number of observations in the experiment is $n = bk$.

For example, a production supervisor wants to compare the mean times for assembly-line operators to assemble an item using one of three methods: A, B, or C. Expecting variation in assembly times from operator to operator, the supervisor uses a randomized block design to compare the three methods. Five assembly-line operators are selected to serve as blocks, and each is assigned to assemble the item three times, once for each of the three methods. Since the sequence in which the operator uses the three methods may be important (fatigue or increasing dexterity may be factors affecting the response), each operator should be assigned a random sequencing of the three methods. For example, operator 1 might be assigned to perform method C first, followed by A and B. Operator 2 might perform method A first, then C and B.

To compare four different teaching methods, a group of students might be divided into blocks of size 4, so that the groups are most nearly *matched* according to academic achievement. To compare the average costs for three different cellular phone companies, costs might be compared at each of three usage levels: low, medium, and high. To compare the average yields for three species of fruit trees when a variation in yield is expected because of the field in which the trees are planted, a researcher uses five fields. Each field is divided into three *plots* on which the three species of fruit trees are planted.

Matching or *blocking* can take place in many different ways. Comparisons of treatments are often made within blocks of time, within blocks of people, or within similar external environments. The purpose of blocking is to remove or isolate the *block-to-block* variability that might otherwise hide the effect of the treatments. You will find more examples of the use of the randomized block design in the exercises at the end of the next section.

11.6 THE ANALYSIS OF VARIANCE FOR A RANDOMIZED BLOCK DESIGN

The randomized block design identifies two factors: **treatments** and **blocks**—both of which affect the response.

Partitioning the Total Variation in the Experiment

Let x_{ij} be the response when the ith treatment ($i = 1, 2, \ldots, k$) is applied in the jth block ($j = 1, 2, \ldots, b$). The total variation in the $n = bk$ observations is

$$\text{Total SS} = \Sigma(x_{ij} - \bar{x})^2 = \Sigma x_{ij}^2 - \frac{(\Sigma x_{ij})^2}{n}$$

This is partitioned into *three* (rather than two) parts in such a way that

Total SS = SSB + SST + SSE

where

- SSB (sum of squares for blocks) measures the variation among the block means.
- SST (sum of squares for treatments) measures the variation among the treatment means.
- SSE (sum of squares for error) measures the variation of the differences among the treatment observations *within* blocks, which measures the experimental error.

The calculational formulas for the four sums of squares are similar in form to those you used for the completely randomized design in Section 11.3. Although you can simplify your work by using a computer program to calculate these sums of squares, the formulas are given next.

Calculating the Sums of Squares for a Randomized Block Design, *k* Treatments in *b* Blocks

$$\text{CM} = \frac{G^2}{n}$$

where

$$G = \Sigma x_{ij} = \text{Total of all } n = bk \text{ observations}$$

$$\text{Total SS} = \Sigma x_{ij}^2 - \text{CM}$$

$$= (\text{Sum of squares of all } x\text{-values}) - \text{CM}$$

Total SS = SST + SSB + SSE

$$\text{SST} = \Sigma \frac{T_i^2}{b} - \text{CM}$$

$$\text{SSB} = \Sigma \frac{B_j^2}{k} - \text{CM}$$

$$\text{SSE} = \text{Total SS} - \text{SST} - \text{SSB}$$

with

T_i = Total of all observations receiving treatment i, $i = 1, 2, \ldots, k$
B_j = Total of all observations in block j, $j = 1, 2, \ldots, b$

Each of the three **sources of variation,** when divided by the appropriate **degrees of freedom,** provides an estimate of the variation in the experiment. Since Total SS involves $n = bk$ squared observations, its degrees of freedom are $df = (n - 1)$. Similarly, SST involves k squared totals, and its degrees of freedom are $df = (k - 1)$, while SSB involves b squared totals and has $(b - 1)$ degrees of freedom. Finally, since the degrees of freedom are additive, the remaining degrees of freedom associated with SSE can be shown algebraically to be $df = (b - 1)(k - 1)$.

These three sources of variation and their respective degrees of freedom are combined to form the **mean squares** as MS = SS/df, and the total variation in the experiment is then displayed in an **analysis of variance** (or **ANOVA**) **table** as shown here:

NEED A TIP?

Degrees of freedom are additive.

ANOVA Table for a Randomized Block Design, k Treatments and b Blocks

Source	*df*	SS	MS	*F*
Treatments	$k - 1$	SST	MST = SST/$(k - 1)$	MST/MSE
Blocks	$b - 1$	SSB	MSB = SSB/$(b - 1)$	MSB/MSE
Error	$(b - 1)(k - 1)$	SSE	MSE = SSE/$(b - 1)(k - 1)$	
Total	$n - 1 = bk - 1$			

EXAMPLE 11.9

The cellular phone industry is involved in a fierce battle for customers, with each company devising its own complex pricing plan to lure customers. Since the cost of a cellphone minute varies drastically depending on the number of minutes per month used by the customer, a consumer watchdog group decided to compare the average costs for four cellular phone companies using three different usage levels as blocks. The monthly costs (in dollars) computed by the cellphone companies for peak-time callers at low (20 min per month), middle (150 min per month), and high (1000 min per month) usage levels are given in Table 11.3. Construct the analysis of variance table for this experiment.

TABLE 11.3 **Monthly Phone Costs of Four Companies at Three Usage Levels**

	Company				
Usage Level	A	B	C	D	Totals
Low	27	24	31	23	$B_1 = 105$
Middle	68	76	65	67	$B_2 = 276$
High	308	326	312	300	$B_3 = 1246$
Totals	$T_1 = 403$	$T_2 = 426$	$T_3 = 408$	$T_4 = 390$	$G = 1627$

NEED A TIP?

Blocks contain experimental units that are *relatively the same.*

SOLUTION The experiment is designed as a *randomized block design* with $b = 3$ usage levels (blocks) and $k = 4$ companies (treatments), so there are $n = bk = 12$ observations and $G = 1627$. Then

$$\text{CM} = \frac{G^2}{n} = \frac{1627^2}{12} = 220{,}594.0833$$

$$\text{Total SS} = (27^2 + 24^2 + \cdots + 300^2) - \text{CM} = 189{,}798.9167$$

$$\text{SST} = \frac{403^2 + \cdots + 390^2}{3} - \text{CM} = 222.25$$

$$\text{SSB} = \frac{105^2 + 276^2 + 1246^2}{4} - \text{CM} = 189{,}335.1667$$

and by subtraction,

$$\text{SSE} = \text{Total SS} - \text{SST} - \text{SSB} = 241.5$$

These four sources of variation, their degrees of freedom, sums of squares, and mean squares are shown in the shaded areas of the analysis of variance tables, generated by *MINITAB* and *Excel* and given in Figures 11.8(a) and 11.8(b). You will find instructions for generating this output in the "Technology Today" section at the end of this chapter.

FIGURE 11.8(a)
MINITAB output for Example 11.9

Two-way ANOVA: Dollars versus Usage, Company

Source	DF	SS	MS	F	P
Usage	2	189335	94667.6	2351.99	0.000
Company	3	222	74.1	1.84	0.240
Error	6	242	40.3		
Total	11	189799			

S = 6.344   R-Sq = 99.87%   R-Sq(adj) = 99.77%

FIGURE 11.8(b)
Excel output for Example 11.9

Anova: Two-Factor Without Replication

SUMMARY	*Count*	*Sum*	*Average*	*Variance*
Low	4	105	26.25	12.917
Middle	4	276	69	23.333
High	4	1246	311.5	118.333
A	3	403	134.333	23040.333
B	3	426	142	26068
C	3	408	136	23521
D	3	390	130	22159

ANOVA

Source of Variation	*SS*	*df*	*MS*	*F*	*P-value*	*F crit*
Usage	189335.167	2	94667.583	2351.990	0.000	5.143
Company	222.25	3	74.083	1.841	0.240	4.757
Error	241.5	6	40.25			
Total	189798.917	11				

Notice that both the ANOVA tables show two different F statistics and p-values. It will not surprise you to know that these statistics are used to test hypotheses concerning the equality of both the *treatment* and *block* means.

Testing the Equality of the Treatment and Block Means

The *mean squares* in the analysis of variance table can be used to test the null hypotheses

H_0 : No difference among the k treatment means

or

H_0 : No difference among the b block means

versus the alternative hypothesis

H_a : At least one of the means is different from at least one other

using a theoretical argument similar to the one we used for the completely randomized design.

- Remember that σ^2 is the common variance for the observations in all bk block-treatment combinations. The quantity

$$\text{MSE} = \frac{\text{SSE}}{(b-1)(k-1)}$$

is an unbiased estimate of σ^2, whether or not H_0 is true.

- The two mean squares, MST and MSB, estimate σ^2 only if H_0 is true and tend to be unusually *large* if H_0 is false and either the treatment or block means are different.
- The test statistics

$$F = \frac{\text{MST}}{\text{MSE}} \quad \text{and} \quad F = \frac{\text{MSB}}{\text{MSE}}$$

are used to test the equality of treatment and block means, respectively. Both statistics tend to be larger than usual if H_0 is false. Hence, you can reject H_0 for large values of F, using *right-tailed* critical values of the F distribution with the appropriate degrees of freedom (see Table 6 in Appendix I) or computer-generated p-values to draw statistical conclusions about the equality of the population means.

Tests for a Randomized Block Design

For comparing treatment means:

1. Null hypothesis: H_0 : The treatment means are equal
2. Alternative hypothesis: H_a : At least two of the treatment means differ
3. Test statistic: $F = \text{MST/MSE}$, where F is based on $df_1 = (k - 1)$ and $df_2 = (b - 1)(k - 1)$
4. Rejection region: Reject if $F > F_\alpha$, where F_α lies in the upper tail of the F distribution (see the figure), or when the p-value $< \alpha$

For comparing block means:

1. Null hypothesis: H_0 : The block means are equal
2. Alternative hypothesis: H_a : At least two of the block means differ
3. Test statistic: $F = \text{MSB/MSE}$, where F is based on $df_1 = (b - 1)$ and $df_2 = (b - 1)(k - 1)$
4. Rejection region: Reject if $F > F_\alpha$, where F_α lies in the upper tail of the F distribution (see the figure), or when the p-value $< \alpha$

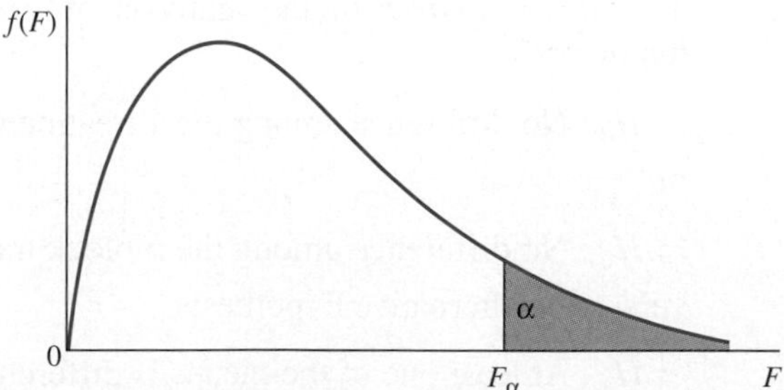

EXAMPLE 11.10 Do the data in Example 11.9 provide sufficient evidence to indicate a difference in the average monthly cellphone cost depending on the company the customer uses?

SOLUTION The cellphone companies represent the *treatments* in this randomized block design, and the differences in their average monthly costs are of primary interest to the researcher. To test

H_0 : No difference in the average cost among companies

versus the alternative that the average cost is different for at least one of the four companies, you use the analysis of variance F statistic, calculated as

$$F = \frac{\text{MST}}{\text{MSE}} = \frac{74.1}{40.3} = 1.84$$

and shown in the column marked "F" and the row marked "Company" in Figures 11.8(a) and 11.8(b). The exact p-value for this statistical test is also given in Figures 11.8(a) and 11.8(b) as 0.240, which is too large to allow rejection of H_0. The results do not show a significant difference in the treatment means. That is, there is insufficient evidence to indicate a difference in the average monthly costs for the four companies.

The researcher in Example 11.10 was fairly certain in using a *randomized block design* that there would be a significant difference in the block means—that is, a significant difference in the average monthly costs depending on the usage level. This suspicion is justified by looking at the test of equality of block means. Notice that the observed test statistic is $F = 2351.99$ with $P = 0.000$, showing a highly significant difference, as expected, in the block means.

Identifying Differences in the Treatment and Block Means

Once the overall F test for equality of the treatment or block means has been performed, what more can you do to identify the nature of any differences you have found? As in Section 11.3, you can use Tukey's method of paired comparisons to determine which pairs of treatment or block means are significantly different from one another. However, if the F test does not indicate a significant difference in the means, there is no reason to use Tukey's procedure. If you have a special interest in a particular *pair* of treatment or block means, you can estimate the difference using a $(1 - \alpha)100\%$ confidence interval.† The formulas for these procedures, shown next, follow a pattern similar to the formulas for the completely randomized design. Remember that MSE always provides an unbiased estimator of σ^2 and uses information from the entire set of measurements. Hence, it is the best available estimator of σ^2, regardless of what test or estimation procedure you are using. You will again use

$$s^2 = \text{MSE} \text{ with } df = (b - 1)(k - 1)$$

to estimate σ^2 in comparing the treatment and block means.

NEED A TIP?

Degrees of freedom for Tukey's test and for confidence intervals are **error *df*.**

Comparing Treatment and Block Means

Tukey's yardstick for comparing block means:

$$\omega = q_\alpha(b, df)\left(\frac{s}{\sqrt{k}}\right)$$

Tukey's yardstick for comparing treatment means:

$$\omega = q_\alpha(k, df)\left(\frac{s}{\sqrt{b}}\right)$$

†You cannot construct a confidence interval for a single mean unless the blocks have been randomly selected from among the population of all blocks. The procedure for constructing intervals for single means is beyond the scope of this book.

$(1 - \alpha)100\%$ confidence interval for the difference in two block means:

$$(\overline{B}_i - \overline{B}_j) \pm t_{\alpha/2}\sqrt{s^2\left(\frac{1}{k} + \frac{1}{k}\right)}$$

where $\overline{B}_i$ is the average of all observations in block i.

$(1 - \alpha)$ 100% confidence interval for the difference in two treatment means:

$$(\overline{T}_i - \overline{T}_j) \pm t_{\alpha/2}\sqrt{s^2\left(\frac{1}{b} + \frac{1}{b}\right)}$$

where $\overline{T}_i$ is the average of all observations in treatment i.

Note: The values $q_\alpha(*, df)$ from Table 11 in Appendix I, $t_{\alpha/2}$ from Table 4 in Appendix I, and $s^2 = \text{MSE}$ all depend on $df = (b - 1)(k - 1)$ degrees of freedom.

EXAMPLE 11.11

Identify the nature of any differences you found in the average monthly cellphone costs from Example 11.9.

SOLUTION Since the F test did not show any significant differences in the average costs for the four companies, there is no reason to use Tukey's method of paired comparisons. Suppose, however, that you are an executive for company B and your major competitor is company C. Can you claim a significant difference in the two average costs? Using a 95% confidence interval, you can calculate

$$(\overline{T}_2 - \overline{T}_3) \pm t_{0.025}\sqrt{\text{MSE}\left(\frac{2}{b}\right)}$$

$$\left(\frac{426}{3} - \frac{408}{3}\right) \pm 2.447\sqrt{40.3\left(\frac{2}{3}\right)}$$

$$6 \pm 12.68$$

so the difference between the two average costs is estimated as between $-\$6.68$ and \$18.68. Since 0 is contained in the interval, you do not have evidence to indicate a significant difference in your average costs. Sorry!

NEED A TIP?

You **cannot** form a confidence interval or test an hypothesis about a single treatment mean in a randomized block design!

Some Cautionary Comments on Blocking

Here are some important points to remember:

- A randomized block design should not be used when treatments and blocks both correspond to **experimental** factors of interest to the researcher. In designating one factor as a *block,* you may assume that the effect of the treatment will be the same, regardless of which block you are using. If this is *not* the case, the two factors—blocks and treatments—are said to **interact,** and your analysis could lead to incorrect conclusions regarding the relationship between the treatments and the response. When an *interaction* is suspected between two factors, you should analyze the data as a **factorial experiment,** which is introduced in the next section.
- Remember that blocking may not always be beneficial. When SSB is removed from SSE, the number of degrees of freedom associated with SSE gets smaller.

For blocking to be beneficial, the information gained by isolating the block variation must outweigh the loss of degrees of freedom for error. Usually, though, if you suspect that the experimental units are not homogeneous and you can group the units into blocks, it pays to use the *randomized block design!*

- Finally, remember that you cannot construct confidence intervals for individual treatment means unless it is reasonable to assume that the b blocks have been randomly selected from a population of blocks. If you construct such an interval, the sample treatment mean will be biased by the positive and negative effects that the blocks have on the response.

11.6 EXERCISES

BASIC TECHNIQUES

11.28 A randomized block design was used to compare the means of three treatments within six blocks. Construct an ANOVA table showing the sources of variation and their respective degrees of freedom.

11.29 Suppose that the analysis of variance calculations for Exercise 11.28 are SST = 11.4, SSB = 17.1, and Total SS = 42.7. Complete the ANOVA table, showing all sums of squares, mean squares, and pertinent F-values.

11.30 Do the data of Exercise 11.28 provide sufficient evidence to indicate differences among the treatment means? Test using $\alpha = 0.05$.

11.31 Refer to Exercise 11.28. Find a 95% confidence interval for the difference between a pair of treatment means A and B if $\bar{x}_A = 21.9$ and $\bar{x}_B = 24.2$.

11.32 Do the data of Exercise 11.28 provide sufficient evidence to indicate that blocking increased the amount of information in the experiment about the treatment means? Justify your answer.

Data set EX1133 **11.33** The data that follow are observations collected from an experiment that compared four treatments, A, B, C, and D, within each of three blocks, using a randomized block design:

	Treatment				
Block	A	B	C	D	Total
1	6	10	8	9	33
2	4	9	5	7	25
3	12	15	14	14	55
Total	22	34	27	30	113

a. Do the data present sufficient evidence to indicate differences among the treatment means? Test using $\alpha = 0.05$.

b. Do the data present sufficient evidence to indicate differences among the block means? Test using $\alpha = 0.05$.

c. Rank the four treatment means using Tukey's method of paired comparisons with $\alpha = 0.01$.

d. Find a 95% confidence interval for the difference in means for treatments A and B.

e. Does it appear that the use of a randomized block design for this experiment was justified? Explain.

Data set EX1134 **11.34** The data shown here are observations collected from an experiment that compared three treatments, A, B, and C, within each of five blocks, using a randomized block design:

	Block					
Treatment	1	2	3	4	5	Total
A	2.1	2.6	1.9	3.2	2.7	12.5
B	3.4	3.8	3.6	4.1	3.9	18.8
C	3.0	3.6	3.2	3.9	3.9	17.6
Total	8.5	10.0	8.7	11.2	10.5	48.9

Excel output for Exercise 11.34

Anova: Two-Factor without Replication

SUMMARY	*Count*	*Sum*	*Average*	*Variance*
A	5	12.5	2.5	0.265
B	5	18.8	3.76	0.073
C	5	17.6	3.52	0.167
1	3	8.5	2.833	0.443
2	3	10	3.333	0.413
3	3	8.7	2.9	0.79
4	3	11.2	3.733	0.223
5	3	10.5	3.5	0.48

ANOVA

Source of Variation	*SS*	*df*	*MS*	*F*	*P-value*	*F crit*
Rows	4.476	2	2.238	79.929	0.000	4.459
Columns	1.796	4	0.449	16.036	0.001	3.838
Error	0.224	8	0.028			
Total	6.496	14				

Use the *Excel* ouput to analyze the experiment. Investigate possible differences in the block and/or treatment means and, if any differences exist, use an appropriate method to specifically identify where the differences lie. Has blocking been effective in this experiment? Present your results in the form of a report.

11.35 The partially completed ANOVA table for a randomized block design is presented here:

Source	*df*	SS	MS	*F*
Treatments	4	14.2		
Blocks		18.9		
Error	24			
Total	34	41.9		

a. How many blocks are involved in the design?

b. How many observations are in each treatment total?

c. How many observations are in each block total?

d. Fill in the blanks in the ANOVA table.

e. Do the data present sufficient evidence to indicate differences among the treatment means? Test using $\alpha = 0.05$.

f. Do the data present sufficient evidence to indicate differences among the block means? Test using $\alpha = 0.05$.

APPLICATIONS

Data set EX1136 **11.36 Fuel Efficiency** A study was conducted to compare automobile fuel efficiency for three brands of gasoline, A, B, and C. Four automobiles, all of the same make and model, were used in the experiment, and each gasoline brand was tested in each automobile. Using each brand in the same automobile has the effect of eliminating (blocking out) automobile-to-automobile variability. The data (in litres per 100 kilometres (L/100 km)) are as follows:

	Automobile			
Gasoline Brand	1	2	3	4
A	7	9	9	8
B	9	10	9	9
C	8	9	8	9

a. Do the data provide sufficient evidence to indicate a difference in mean fuel efficiency (L/100 km) for the three brands of gasoline?

b. Is there evidence of a difference in mean fuel efficiency (L/100 km) for the four automobiles?

c. Suppose that *prior to looking at the data*, you had decided to compare the mean fuel efficiency (L/100 km) for gasoline brands A and B. Find a 90% confidence interval for this difference.

d. Use an appropriate method to identify the pairwise differences, if any, in the average fuel efficiency (L/100 km) for the three brands of gasoline.

Data set EX1137 **11.37 Water Resistance in Textiles** An experiment was conducted to compare the effects of four different chemicals, A, B, C, and D, in producing water resistance in textiles. A strip of material, randomly selected from a bolt, was cut into four pieces, and the four pieces were randomly assigned to receive one of the four chemicals, A, B, C, or D. This process was replicated three times, thus producing a randomized block design. The design, with moisture-resistance measurements, is shown in the figure (low readings indicate low moisture penetration). Analyze the experiment using a method appropriate for this randomized block design. Identify the blocks and treatments, and investigate any possible differences in treatment means. If any differences exist, use an appropriate method to specifically identify where the differences lie. What are the practical implications for the chemical producers? Has blocking been effective in this experiment? Present your results in the form of a report.

Illustration for Exercise 11.37

Blocks (bolt samples)

1	2	3
C 9.9	D 13.4	B 12.7
A 10.1	B 12.9	D 12.9
B 11.4	A 12.2	C 11.4
D 12.1	C 12.3	A 11.9

11.38 Glare in Rearview Mirrors An experiment was conducted to compare the glare characteristics of four types of automobile rearview mirrors. Forty drivers were randomly selected to participate in the experiment. Each driver was exposed to the glare produced by a headlight located nine metres behind the rear window of the experimental automobile. The driver then rated the glare produced by the rearview mirror on a scale of 1 (low) to 10 (high). Each of the four mirrors was tested by each driver; the mirrors were assigned to a driver in random order. An analysis of variance of the data produced this ANOVA table:

Source	*df*	SS	MS	*F*
Mirrors		46.98		
Drivers			8.42	
Error				
Total		638.61		

a. Fill in the blanks in the ANOVA table.

b. Do the data present sufficient evidence to indicate differences in the mean glare ratings of the four rearview mirrors? Calculate the approximate *p*-value and use it to make your decision.

c. Do the data present sufficient evidence to indicate that the level of glare perceived by the drivers varied from driver to driver? Use the *p*-value approach.

d. Based on the results of part b, what are the practical implications of this experiment for the manufacturers of the rearview mirrors?

Data set EX1139 **11.39 Slash Pine Seedings** An experiment was conducted to determine the effects of three methods of soil preparation on the first-year growth of slash pine seedlings. Four locations (provincial forest lands) were selected, and each location was divided into three plots. Since it was felt that soil fertility within a location was more homogeneous than between locations, a randomized block design was employed using locations as blocks. The methods of soil preparation were A (no preparation), B (light fertilization), and C (burning). Each soil preparation was randomly applied to a plot within each location. On each plot, the same number of seedlings were planted and the average first-year growth of the seedlings was recorded on each plot. Use the *MINITAB* printout to answer the questions.

	Location			
Soil Preparation	1	2	3	4
A	11	13	16	10
B	15	17	20	12
C	10	15	13	10

a. Conduct an analysis of variance. Do the data provide evidence to indicate a difference in the mean growths for the three soil preparations?

b. Is there evidence to indicate a difference in mean growths for the four locations?

c. Use Tukey's method for paired comparisons to rank the mean growths for the three soil preparations. Use $\alpha = 0.01$.

d. Use a 95% confidence interval to estimate the difference in mean growths for methods A and B.

MINITAB output for Exercise 11.39

Two-way ANOVA: Growth versus Soil Prep, Location

```
Source      DF       SS        MS      F      P
Soil Prep    2   38.000   19.0000  10.06  0.012
Location     3   61.667   20.5556  10.88  0.008
Error        6   11.333    1.8889
Total       11  111.000

S = 1.374   R-Sq = 89.79%   R-Sq(adj) = 81.28%

                   Individual 95% CIs For Mean Based on
                   Pooled StDev
Soil Prep   Mean   ---------+---------+---------+---------+--
1           12.5      (-------*-------)
2           16.0                        (-------*-------)
3           12.0    (-------*-------)
                   ---------+---------+---------+---------+--
                          12.0      14.0      16.0      18.0

                   Individual 95% CIs For Mean Based on
                   Pooled StDev
Location    Mean   -----+---------+---------+---------+-----
1        12.0000          (-------*-------)
2        15.0000                    (-------*-------)
3        16.3333                          (-------*-------)
4        10.6667   (-------*------)
                   -----+---------+---------+---------+-----
                      10.0      12.5      15.0      17.5
```

Data set EX1140 **11.40 Digitalis and Calcium Uptake** A study was conducted to compare the effects of three levels of digitalis on the levels of calcium in the heart muscles of dogs. Because general level of calcium uptake varies from one animal to another, the tissue for a heart muscle was regarded as a block, and comparisons of the three digitalis levels (treatments) were made within a given animal. The calcium uptakes for the three levels of digitalis, A, B, and C, were compared based on the heart muscles of four dogs and the results are given in the table. Use the *Excel* printout to answer the questions.

Dogs			
1	2	3	4
A	C	B	A
1342	1698	1296	1150
B	B	A	C
1608	1387	1029	1579
C	A	C	B
1881	1140	1549	1319

a. How many degrees of freedom are associated with SSE?

b. Do the data present sufficient evidence to indicate a difference in the mean uptakes of calcium for the three levels of digitalis?

c. Use Tukey's method of paired comparisons with $\alpha = 0.01$ to rank the mean calcium uptakes for the three levels of digitalis.

d. Do the data indicate a difference in the mean uptakes of calcium for the four heart muscles?

e. Use Tukey's method for paired comparisons with $\alpha = 0.01$ to rank the mean calcium uptakes for the heart muscles of the four dogs used in the experiment. Are these results of any practical value to the researcher?

f. Give the standard error of the difference between the mean calcium uptakes for two levels of digitalis.

g. Find a 95% confidence interval for the difference in mean responses between treatments A and B.

Excel output for Exercise 11.40

ANOVA: Two-Factor without Replication

SUMMARY	Count	Sum	Average	Variance
A	4	4661	1165.25	16891.583
B	4	5610	1402.5	20261.667
C	4	6707	1676.75	72681.583
1	3	4831	1610.333	72634.333
2	3	4225	1408.333	78182.333
3	3	3874	1291.333	67616.333
4	3	4048	1349.333	46700.333

ANOVA

Source of Variation	SS	df	MS	F	P-value	F crit
Digitalis	524177.167	2	262088.583	258.237	0.000	5.143
Dogs	173415	3	57805	56.955	0.000	4.757
Error	6089.5	6	1014.917			
Total	703681.667	11				

Data set EX1141 **11.41 Bidding on Construction Jobs** A building contractor employs three construction engineers, A, B, and C, to estimate and bid on jobs. To determine whether one tends to be a more conservative (or liberal) estimator than the others, the contractor selects four projected construction jobs and has each estimator independently estimate the cost (in dollars per square foot) of each job. The data are shown in the table:

	Construction Job				
Estimator	1	2	3	4	Total
A	35.10	34.50	29.25	31.60	130.45
B	37.45	34.60	33.10	34.40	139.55
C	36.30	35.10	32.45	32.90	136.75
Total	108.85	104.20	94.80	98.90	406.75

Analyze the experiment using the appropriate methods. Identify the blocks and treatments, and investigate any possible differences in treatment means. If any differences exist, use an appropriate method to specifically identify where the differences lie. Has blocking been effective in this experiment? What are the practical implications of the experiment? Present your results in the form of a report.

11.7 THE $a \times b$ FACTORIAL EXPERIMENT: A TWO-WAY CLASSIFICATION

Suppose the manager of a manufacturing plant suspects that the output (in number of units produced per shift) of a production line depends on two factors:

- Which of two supervisors is in charge of the line
- Which of three shifts—day, swing, or night—is being measured

That is, the manager is interested in two *factors:* "supervisor" at two levels and "shift" at three levels. Can you use a randomized block design, designating one of the two factors as a block factor? In order to do this, you would need to assume that the effect of the two supervisors is the same, regardless of which shift you are considering. This may not be the case; maybe the first supervisor is most effective in the morning, and the second is more effective at night. You cannot generalize and say that one supervisor is better than the other or that the output of one particular shift is best. You need to investigate not only the average output for the two supervisors and the average output for the three shifts, but also the **interaction** or relationship between the two factors.

Consider two different examples that show the effect of *interaction* on the responses in this situation.

EXAMPLE 11.12

Suppose that the two supervisors are each observed on three randomly selected days for each of the three different shifts. The average outputs for the three shifts are shown in Table 11.4 for each of the supervisors. Look at the relationship between the two factors in the line chart for these means, shown in Figure 11.9. Notice that supervisor 2 always produces a higher output, regardless of the shift. The two factors behave *independently;* that is, the output is always about 100 units higher for supervisor 2, no matter which shift you look at.

TABLE 11.4 Average Outputs for Two Supervisors on Three Shifts

	Shift		
Supervisor	Day	Swing	Night
1	487	498	550
2	602	602	637

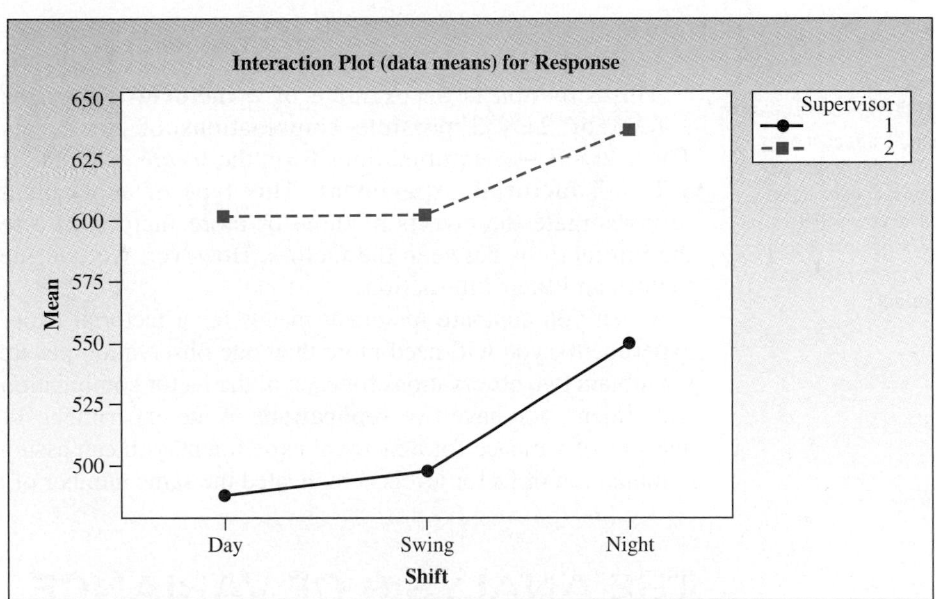

FIGURE 11.9 Interaction plot for means in Table 11.4

Now consider another set of data for the same situation, shown in Table 11.5. There is a definite difference in the results, depending on which shift you look at, and the *interaction* can be seen in the crossed lines of the chart in Figure 11.10.

TABLE 11.5 Average Outputs for Two Supervisors on Three Shifts

	Shift		
Supervisor	Day	Swing	Night
1	602	498	450
2	487	602	657

FIGURE 11.10
Interaction plot for means in Table 11.5

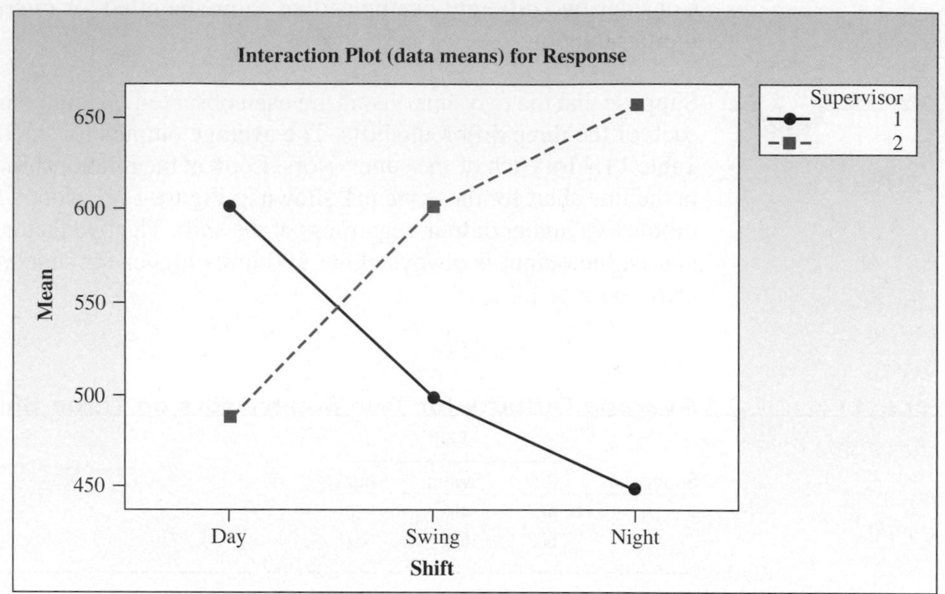

NEED A TIP?
When the effect of one factor on the response changes, depending on the level at which the other factor is measured, the two factors are said to **interact**.

This situation is an example of a **factorial experiment** in which there are a total of 2×3 possible combinations of the levels for the two factors. These $2 \times 3 = 6$ combinations form the *treatments*, and the experiment is called a **2 × 3 factorial experiment.** This type of experiment can actually be used to investigate the effects of three or more factors on a response and to explore the interactions between the factors. However, we confine our discussion to two factors and their interaction.

When you compare treatment means for a factorial experiment (or for any other experiment), you will need more than one observation per treatment. For example, if you obtain two observations for each of the factor combinations of a complete factorial experiment, you have two **replications** of the experiment. In the next section on the analysis of variance for a factorial experiment, you can assume that each treatment or combination of factor levels is replicated the same number of times, r.

11.8 THE ANALYSIS OF VARIANCE FOR AN $a \times b$ FACTORIAL EXPERIMENT

An analysis of variance for a two-factor factorial experiment replicated r times follows the same pattern as the previous designs. If the letters A and B are used to identify the two factors, the total variation in the experiment

$$\text{Total SS} = \Sigma(x - \bar{x})^2 = \Sigma x^2 - \text{CM}$$

is partitioned into *four* parts in such a way that

$$\text{Total SS} = \text{SSA} + \text{SSB} + \text{SS (AB)} + \text{SSE}$$

where

- SSA (sum of squares for factor A) measures the variation among the factor A means.
- SSB (sum of squares for factor B) measures the variation among the factor B means.
- SS(AB) (sum of squares for interaction) measures the variation *among* the different combinations of factor levels.
- SSE (sum of squares for error) measures the variation of the differences among the observations *within* each combination of factor levels—the experimental error.

Sums of squares SSA and SSB are often called the **main effect** sums of squares, to distinguish them from the **interaction** sum of squares. Although you can simplify your work by using a computer program to calculate these sums of squares, the calculational formulas are given next. You can assume that there are:

- a levels of factor A
- b levels of factor B
- r replications of each of the ab factor combinations
- A total of $n = abr$ observations

Calculating the Sums of Squares for a Two-Factor Factorial Experiment

$$\text{CM} = \frac{G^2}{n} \qquad \text{Total SS} = \Sigma x^2 - \text{CM}$$

$$\text{SSA} = \Sigma\frac{A_i^2}{br} - \text{CM} \quad \text{SSB} = \Sigma\frac{B_j^2}{ar} - \text{CM}$$

$$\text{SS(AB)} = \Sigma\frac{(AB)_{ij}^2}{r} - \text{CM} - \text{SSA} - \text{SSB}$$

where

G = Sum of all $n = abr$ observations
A_i = Total of all observations at the ith level of factor A, $i = 1, 2, \ldots, a$
B_j = Total of all observations at the jth level of factor B, $j = 1, 2, \ldots, b$
$(AB)_{ij}$ = Total of the r observations at the ith level of factor A and the jth level of factor B

Each of the five **sum of squares,** when divided by the appropriate **degrees of freedom,** provides an estimate of the variation in the experiment. These estimates are called **mean squares**—MS = SS/df—and are displayed along with their respective sums of squares and df in the **analysis of variance** (or **ANOVA**) **table.**

ANOVA Table for *r* Replications of a Two-Factor Factorial Experiment: Factor A at *a* Levels and Factor B at *b* Levels

Source	*df*	SS	MS	*F*
A	$a-1$	SSA	$MSA = \frac{SSA}{a-1}$	$\frac{MSA}{MSE}$
B	$b-1$	SSB	$MSB = \frac{SSB}{b-1}$	$\frac{MSB}{MSE}$
AB	$(a-1)(b-1)$	SS(AB)	$MS(AB) = \frac{SS(AB)}{(a-1)(b-1)}$	$\frac{MS(AB)}{MSE}$
Error	$ab(r-1)$	SSE	$MSE = \frac{SSE}{ab(r-1)}$	
Total	$abr-1$	Total SS		

Finally, the equality of means for various levels of the factor combinations (the interaction effect) and for the levels of both main effects, A and B, can be tested using the ANOVA *F* tests, as shown next.

Tests for a Factorial Experiment

- **For interaction:**

1. Null hypothesis: H_0 : Factors A and B do not interact
2. Alternative hypothesis: H_a : Factors A and B interact
3. Test statistic: $F =$ MS(AB)/MSE, where F is based on $df_1 = (a-1)(b-1)$ and $df_2 = ab(r-1)$
4. Rejection region: Reject H_0 when $F > F_\alpha$, where F_α lies in the upper tail of the F distribution (see the figure), or when the p-value $< \alpha$

- **For main effects, factor A:**

1. Null hypothesis: H_0 : There are no differences among the factor A means
2. Alternative hypothesis: H_a : At least two of the factor A means differ
3. Test statistic: $F =$ MSA/MSE, where F is based on $df_1 = (a-1)$ and $df_2 = ab(r-1)$
4. Rejection region: Reject H_0 when $F > F_\alpha$ (see the figure) or when the p-value $< \alpha$

- **For main effects, factor B:**

1. Null hypothesis: H_0 : There are no differences among the factor B means
2. Alternative hypothesis: H_a : At least two of the factor B means differ
3. Test statistic: $F =$ MSB/MSE, where F is based on $df_1 = (b-1)$ and $df_2 = ab(r-1)$
4. Rejection region: Reject H_0 when $F > F_\alpha$ (see the figure) or when the p-value $< \alpha$

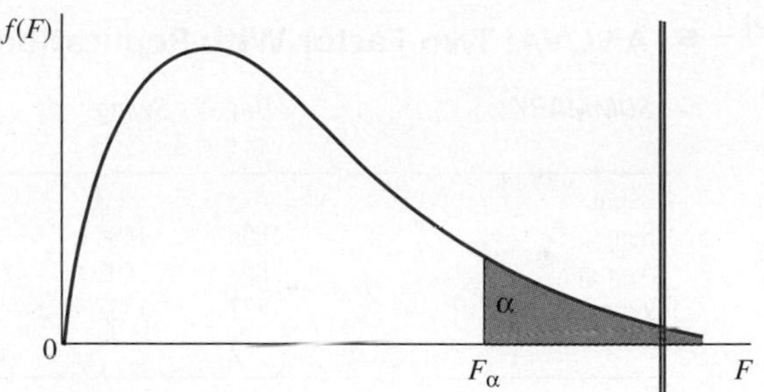

EXAMPLE 11.13 Table 11.6 shows the original data used to generate Table 11.5 in Example 11.12. That is, the two supervisors were each observed on three randomly selected days for each of the three different shifts, and the production outputs were recorded. Analyze these data using the appropriate analysis of variance procedure.

TABLE 11.6 **Outputs for Two Supervisors on Three Shifts**

	Shift		
Supervisor	Day	Swing	Night
1	571	480	470
	610	474	430
	625	540	450
2	480	625	630
	516	600	680
	465	581	661

SOLUTION The computer output in Figure 11.11(a) was generated using the two-way analysis of variance procedure in the *MINITAB* software package. Figure 11.11(b) provides the *Excel* output for Example 11.13. You can verify the quantities in the ANOVA table using the calculational formulas presented earlier, or you may choose just to use the results and interpret their meaning.

FIGURE 11.11(a)
MINITAB output for Example 11.13

Two-way ANOVA: Output versus Supervisor, Shift

```
Source       DF       SS        MS      F      P
Supervisor    1    19208   19208.0  26.68  0.000
Shift         2      247     123.5   0.17  0.844
Interaction   2    81127   40563.5  56.34  0.000
Error        12     8640     720.0
Total        17   109222

S = 26.83   R-Sq = 92.09%   R-Sq(adj) = 88.79%

                      Individual 95% CIs For Mean Based on
                      Pooled StDev
Supervisor     Mean   ----+---------+---------+---------+-----
1           516.667   (-------*------)
2           582.000                          (-------*-------)
                      ----+---------+---------+---------+-----
                        510       540       570       600

                      Individual 95% CIs For Mean Based on
                      Pooled StDev
Shift          Mean   ---+---------+---------+---------+-------
Day           544.5   (---------------*---------------)
Swing         550.0       (---------------*---------------)
Night         553.5         (---------------*---------------)
                      ---+---------+---------+---------+-------
                        525       540       555       570
```

FIGURE 11.11(b)
Excel output for Example 11.13

ANOVA: Two-Factor With Replication

SUMMARY	*Day*	*Swing*	*Night*	*Total*
	1			
Count	3	3	3	9
Sum	1806	1494	1350	4650
Average	602	498	450	516.667
Variance	777	1332	400	5155.25
	2			
Count	3	3	3	9
Sum	1461	1806	1971	5238
Average	487	602	657	582
Variance	687	487	637	6096.5
	Total			
Count	6	6	6	
Sum	3267	3300	3321	
Average	544.5	550	553.5	
Variance	4553.1	3972.4	13269.5	

ANOVA

Source of Variation	*SS*	*df*	*MS*	*F*	*P-value*	*F crit*
Supervisor	19208	1	19208	26.678	0.000	4.747
Shift	247	2	123.5	0.172	0.844	3.885
Interaction	81127	2	40563.5	56.338	0.000	3.885
Within	8640	12	720			
Total	109222	17				

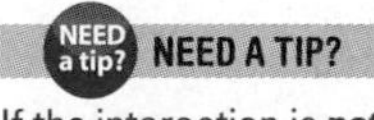

NEED A TIP?

If the interaction is **not significant,** test each of the factors individually.

At this point, you have undoubtedly discovered the familiar pattern in testing the significance of the various experimental factors with the F statistic and its p-value. The small p-value ($P = 0.000$) in the row marked "Supervisor" means that there is sufficient evidence to declare a difference in the mean levels for factor A—that is, a difference in mean outputs per supervisor. This fact is visually apparent in the non-overlapping confidence intervals for the supervisor means shown in the printout. But this is overshadowed by the fact that there is strong evidence ($P = 0.000$) of an *interaction* between factors A and B. This means that the average output for a given shift depends on the supervisor on duty. You saw this effect clearly in Figure 11.11. The three largest mean outputs occur when supervisor 1 is on the day shift and when supervisor 2 is on either the swing or night shift. As a practical result, the manager should schedule supervisor 1 for the day shift and supervisor 2 for the night shift.

If the interaction effect *is* significant, the differences in the treatment means can be further studied, *not* by comparing the means for factor A or B individually but rather by looking at comparisons for the 2×3 (AB) factor-level combinations. If the interaction effect is *not significant*, then the significance of the main effect means should be investigated, first with the overall F test and next with Tukey's method for paired comparisons and/or specific confidence intervals. Remember that these analysis of variance procedures always use $s^2 = \text{MSE}$ as the best estimator of σ^2 with degrees of freedom equal to $df = ab(r - 1)$.

For example, using Tukey's yardstick to compare the average outputs for the two supervisors on each of the three shifts, you could calculate

$$\omega = q_{0.05}(6,12)\left(\frac{s}{\sqrt{r}}\right) = 4.75\left(\frac{\sqrt{720}}{\sqrt{3}}\right) = 73.59$$

Since all three pairs of means—602 and 487 on the day shift, 498 and 602 on the swing shift, and 450 and 657 on the night shift—differ by more than ω, our practical conclusions have been confirmed statistically.

11.8 EXERCISES

BASIC TECHNIQUES

11.42 Suppose you were to conduct a two-factor factorial experiment, factor A at four levels and factor B at five levels, with three replications per treatment.

a. How many treatments are involved in the experiment?

b. How many observations are involved?

c. List the sources of variation and their respective degrees of freedom.

11.43 The analysis of variance table for a 3 × 4 factorial experiment, with factor A at three levels and factor B at four levels, and with two observations per treatment, is shown here:

Source	*df*	SS	MS	*F*
	2	5.3		
	3	9.1		
	6			
	12	24.5		
Total	23	43.7		

a. Fill in the missing items in the table.

b. Do the data provide sufficient evidence to indicate that factors A and B interact? Test using $\alpha = 0.05$. What are the practical implications of your answer?

c. Do the data provide sufficient evidence to indicate that factors A and B affect the response variable x? Explain.

11.44 Refer to Exercise 11.43. The means of two of the factor-level combinations—say, A_1B_1 and A_2B_1—are $\bar{x}_1 = 8.3$ and $\bar{x}_2 = 6.3$, respectively. Find a 95% confidence interval for the difference between the two corresponding population means.

11.45 The table gives data for a 3 × 3 factorial experiment, with two replications per treatment:

EX1145

	Levels of Factor A		
Levels of Factor B	1	2	3
1	5, 7	9, 7	4, 6
2	8, 7	12, 13	7, 10
3	14, 11	8, 9	12, 15

a. Perform an analysis of variance for the data, and present the results in an analysis of variance table.

b. What do we mean when we say that factors A and B interact?

c. Do the data provide sufficient evidence to indicate interaction between factors A and B? Test using $\alpha = 0.05$.

d. Find the approximate p-value for the test in part c.

e. What are the practical implications of your results in part c? Explain your results using a line graph similar to the one in Figure 11.9.

Data set **11.46 2 × 2 Factorial** The table gives data for a 2 × 2 factorial experiment, with four replications per treatment:

EX1146

	Levels of Factor A	
Levels of Factor B	1	2
1	2.1, 2.7, 2.4, 2.5	3.7, 3.2, 3.0, 3.5
2	3.1, 3.6, 3.4, 3.9	2.9, 2.7, 2.2, 2.5

a. The accompanying graph was generated by *MINITAB*. Verify that the four points that connect the two lines are the means of the four observations within each factor-level combination. What does the graph tell you about the interaction between factors A and B?

MINITAB interaction plot for Exercise 11.46

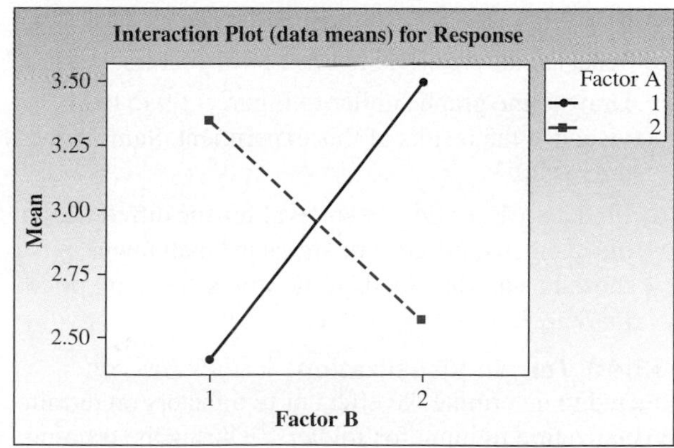

b. Use the *MINITAB* output to test for a significant interaction between A and B. Does this confirm your conclusions in part a?

MINITAB output for Exercise 11.46

Two-way ANOVA: Response versus Factor A, Factor B

```
Source       DF      SS        MS       F      P
Factor A      1  0.0000   0.00000    0.00  1.000
Factor B      1  0.0900   0.09000    1.00  0.338
Interaction   1  3.4225   3.42250   37.85  0.000
Error        12  1.0850   0.09042
Total        15  4.5975

S = 0.3007   R-Sq = 76.40%   R-Sq(adj) = 70.50%
```

c. Considering your results in part b, how can you explain the fact that neither of the main effects is significant?

d. If a significant interaction is found, is it necessary to test for significant main effect differences? Explain.

e. Write a short paragraph summarizing the results of this experiment.

APPLICATIONS

Data set EX1147 **11.47 Demand for Diamonds** A chain of jewellery stores conducted an experiment to investigate the effect of price and location on the demand for its diamonds. Six small-town stores were selected for the study, as well as six stores located in large suburban malls. Two stores in each of these locations were assigned to each of three item percentage markups. The percentage gain (or loss) in sales for each store was recorded at the end of one month. The data are shown in the accompanying table:

	Markup		
Location	1	2	3
Small towns	10	−3	−10
	4	7	−24
Suburban malls	14	8	−4
	18	3	3

a. Do the data provide sufficient evidence to indicate an interaction between markup and location? Test using $\alpha = 0.05$.

b. What are the practical implications of your test in part a?

c. Draw a line graph similar to Figure 11.9 to help visualize the results of this experiment. Summarize the results.

d. Find a 95% confidence interval for the difference in mean change in sales for stores in small towns versus those in suburban malls if the stores are using price markup 3.

11.48 Terrain Visualization A study was conducted to determine the effect of two factors on terrain visualization training for soldiers.[1] During the training programs, participants viewed contour maps of various terrains and then were permitted to view a computer reconstruction of the terrain as it would appear from a specified angle. The two factors investigated in the experiment were the participants' spatial abilities (abilities to visualize in three dimensions) and the viewing procedures (active or passive). Active participation permitted participants to view the computer-generated reconstructions of the terrain from any and all angles. Passive participation gave the participants a set of preselected reconstructions of the terrain. Participants were tested according to spatial ability, and from the test scores 20 were categorized as possessing high spatial ability, 20 medium, and 20 low. Then 10 participants within each of these groups were assigned to each of the two training modes, active or passive. The accompanying tables are the ANOVA table computed by the researchers and the table of the treatment means:

Source	*df*	MS	Error *df*	*F*	*p*
Main effects:					
Training condition	1	103.7009	54	3.66	0.0610
Ability	2	760.5889	54	26.87	0.0005
Interaction:					
Training condition × Ability	2	124.9905	54	4.42	0.0167
Within cells	54	28.3015			

	Training Condition	
Spatial Ability	Active	Passive
High	17.895	9.508
Medium	5.031	5.648
Low	1.728	1.610

Note: Maximum score = 36.

a. Explain how the authors arrived at the degrees of freedom shown in the ANOVA table.

b. Are the *F*-values correct?

c. Interpret the test results. What are their practical implications?

d. Use Table 6 in Appendix I to approximate the *p*-values for the F statistics shown in the ANOVA table.

Source: Barsam, H.F., Simutis, Z.M. (1984). Computer-Based Graphics for Terrain Visualization Training. *Human Factors* 1984, Vol. 26(6), pp. 659–665.

Data set EX1149 **11.49 Canadian Flight Attendants** The national average salary for a flight attendant is \$42,643. The median annual salary for flight attendants in Ontario is \$72,751, as of February 22, 2016, with a range usually between $58,386 and $90,208.[2] However, the salary for a flight attendant may vary depending on a number of factors including industry, company size, location, years of experience, and level of education. The table reports the average national salary (in $ thousands) for four Canadian airlines at four different levels of experience:

Airline	Entry Level	Mid-Career	Experienced	Late Career
A	43	55	53	62
B	40	50	49	58
C	37	47	55	67
D	49	57	64	71

Data set EX1150 **11.50 Animation Helps?** To explore ways to increase the educational experience using animation versus static images in a learning environment, Cyril Rebetez and colleagues[3] ran a factorial experiment that measured the retention of information under four factorial conditions: with animation or without animation; and reinforcement through snapshots or without snapshots of the major frames in the animation. It was expected that the animation would lead to better retention of information and that having the snapshots available would also help with retention of information. The following data are based on the results of their experiment:

	Learning Setting			
	Static		Animated	
Snapshots	Without	With	Without	With
	58.9	42.0	57.7	64.3
	48.9	53.9	55.9	66.4
	51.8	54.4	57.2	63.1
	53.0	47.6	65.1	55.8
	51.3	50.5	59.3	57.9
	49.8	50.2	65.7	61.5
	61.5	47.0	60.8	61.2
	47.8	52.4	59.3	61.9

Use the *MINITAB* output to analyze the experiment with the appropriate method. Identify the two factors, and investigate any possible effects due to their interaction or the main effects. What are the practical implications of these results? Why do the interaction plots seem counterintuitive to the analysis? If the interaction effect is real, what might you do as an experimenter to show that the interaction is, in fact, significant? Explain your conclusions in the form of a report.

MINITAB output for Exercise 11.50

ANOVA: Retention versus Setting, Snapshots

```
Factor      Type   Levels   Values
Setting     fixed       2   Animated, Static
Snapshots   fixed       2   With, Without
Analysis of Variance for Retention

Source      DF      SS       MS        F       P
Setting      1  722.95   722.95    44.56   0.000
Snapshots    1    6.04     6.04     0.37   0.547
Setting*
  Snapshots  1   40.73    40.73     2.51   0.124
Error       28  454.28    16.22
Total       31 1223.99
S = 4.02793   R-Sq = 62.89%   R-Sq(adj) = 58.91%
```

MINITAB plots for Exercise 11.50

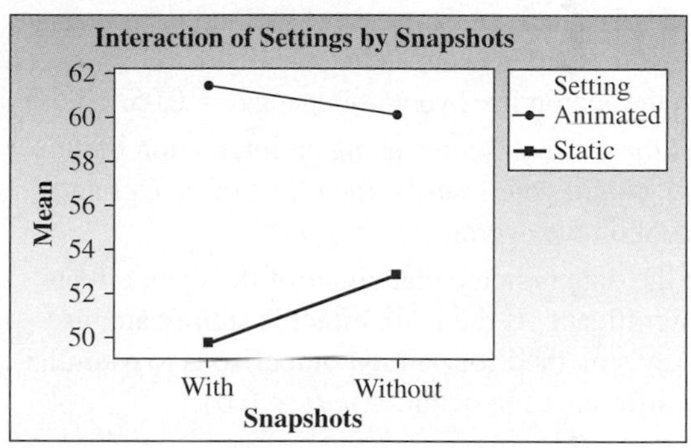

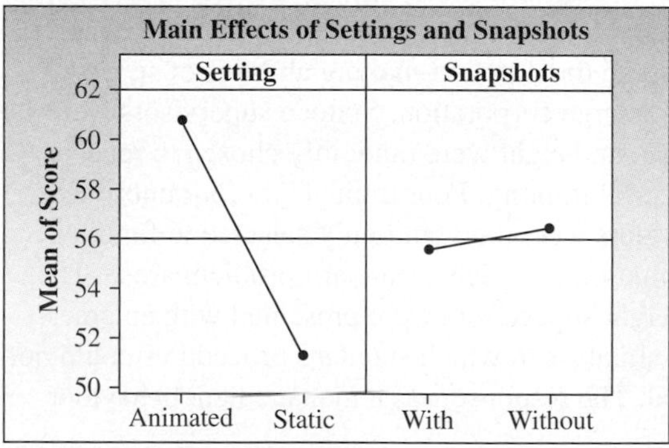

Data set EX1151 **11.51 Standardized Test Scores** A local school board was interested in comparing test scores on a standarized reading test for Grade 3 students in its district. It selected a random sample of five male and five female Grade 3 students at each of four different elementary schools in the district and recorded the test scores. The results are shown in the table below:

Gender	School 1	School 2	School 3	School 4
Male	631	642	651	350
	566	710	611	565
	620	649	755	543
	542	596	693	509
	560	660	620	494
Female	669	722	709	505
	644	769	545	498
	600	723	657	474
	610	649	722	470
	559	766	711	463

a. What type of experimental design is this? What are the experimental units? What are the factors and levels of interest to the school board?

b. Perform the appropriate analysis of variance for this experiment.

c. Do the data indicate that effect of gender on the average test score is different depending on the student's school? Test the appropriate hypothesis using $\alpha = 0.05$.

d. Plot the average scores using an interaction plot. How would you describe the effect of gender and school on the average test scores?

e. Do the data indicate that either of the main effects is significant? If the main effect is significant, use Tukey's method for paired comparisons to examine the differences in detail. Use $\alpha = 0.01$.

11.52 Management Training An experiment was conducted to investigate the effect of management training on the decision-making abilities of supervisors in a large corporation. Sixteen supervisors were selected, and eight were randomly chosen to receive managerial training. Four trained and four untrained supervisors were then randomly selected to function in a situation in which a standard problem arose. The other eight supervisors were presented with an emergency situation in which standard procedures could not be used. The response was a management behaviour rating for each supervisor as assessed by a rating scheme devised by the experimenter.

a. What are the experimental units in this experiment?

b. What are the two factors considered in the experiment?

c. What are the levels of each factor?

d. How many treatments are there in the experiment?

e. What type of experimental design has been used?

Data set **11.53 Management Training, continued**

EX1153 Refer to Exercise 11.52. The data for this experiment are shown in the table:

	Training (A)		
Situation (B)	Trained	Not Trained	Totals
Standard	85	53	519
	91	49	
	80	38	
	78	45	
Emergency	76	40	473
	67	52	
	82	46	
	71	39	
Totals	630	362	992

a. Construct the ANOVA table for this experiment.

b. Is there a significant interaction between the presence or absence of training and the type of decision-making situation? Test at the 5% level of significance.

c. Do the data indicate a significant difference in behaviour ratings for the two types of situations at the 5% level of significance?

d. Do behaviour ratings differ significantly for the two types of training categories at the 5% level of significance.

e. Plot the average scores using an interaction plot. How would you describe the effect of training and emergency situation on the decision-making abilities of the supervisors?

11.9 REVISITING THE ANALYSIS OF VARIANCE ASSUMPTIONS

In Section 11.2, you learned that the assumptions and test procedures for the analysis of variance are similar to those required for the t and F tests in Chapter 10—namely, that observations within a treatment group must be normally distributed with common variance σ^2. You also learned that the analysis of variance procedures are fairly robust when the sample sizes are equal and the data are fairly mound-shaped. If this is the case, one way to protect yourself from inaccurate conclusions is to try, when possible, to select samples of equal sizes!

There are some quick and simple ways to check the data for violation of assumptions. Look first at the type of response variable you are measuring. You might immediately see a problem with either the normality or common variance assumption. It may be that the data you have collected cannot be measured *quantitatively*. For example, many responses, such as product preferences, can be ranked only as "A is better than B" or "C is the least preferable." Data that are *qualitative* cannot have a normal distribution. If the response variable is *discrete* and can assume only three values—say, 0, 1, or 2—then it is again unreasonable to assume that the response variable is normally distributed.

Suppose that the response variable is binomial—say, the proportion p of people who favour a particular type of investment. Although binomial data can be approximately mound-shaped under certain conditions, they violate the equal variance assumption. The variance of a sample proportion is

$$\sigma^2 = \frac{pq}{n} = \frac{p(1-p)}{n}$$

so that the variance changes depending on the value of p. As the treatment means change, the value of p changes and so does the variance σ^2. A similar situation occurs when the response variable is a Poisson random variable—say, the number of industrial accidents per month in a manufacturing plant. Since the variance of a Poisson random variable is $\sigma^2 = \mu$, the variance changes exactly as the treatment mean changes.

If you cannot see any flagrant violations in the type of data being measured, look at the range of the data within each treatment group. If these ranges are nearly the same, then the common variance assumption is probably reasonable. To check for normality, you might make a quick dotplot or stem and leaf plot for a particular treatment group. However, quite often you do not have enough measurements to obtain a reasonable plot.

If you are using a computer program to analyze your experiment, there are some valuable **diagnostic tools** you can use. These procedures are too complicated to be performed using hand calculations, but they are easy to use when the computer does all the work!

Residual Plots

In the analysis of variance, the total variation in the data is partitioned into several parts, depending on the factors identified as important to the researcher. Once the effects of these sources of variation have been removed, the "leftover" variability in each observation is called the **residual** for that data point. These residuals represent **experimental error,** the basic variability in the experiment, and should have an approximately *normal distribution* with a mean of 0 and the *same variation* for each treatment group. Most computer packages will provide options for plotting these residuals:

- The **normal probability plot of residuals** is a graph that plots the residuals for each observation against the expected value of that residual *had it come from a normal distribution*. If the residuals are approximately normal, the plot will closely resemble a *straight line*, sloping upward to the right.
- The **plot of residuals versus fit** or **residuals versus variables** is a graph that plots the residuals against the expected value of that observation *using the experimental design we have used*. If no assumptions have been violated and there are no "leftover" sources of variation other than experimental error, this plot should show a *random scatter* of points around the horizontal "zero error line" for each treatment group, with approximately the same vertical spread.

EXAMPLE 11.14

The data from Example 11.4 involving breakfast and the attention spans of three groups of elementary students were analyzed using *MINITAB*. The graphs in Figure 11.12, generated by *MINITAB*, are the normal probability plot and the residuals versus fit plot for this experiment. Look at the straight-line pattern in the normal probability plot, which indicates a normal distribution in the residuals. In the other plot, the residuals are plotted against the estimated expected values, which are the sample averages for each of the three treatments in the completely randomized design. The random scatter around the horizontal "zero error line" and the constant spread indicate *no violations* in the constant variance assumption.

FIGURE 11.12
MINITAB diagnostic plots for Example 11.14

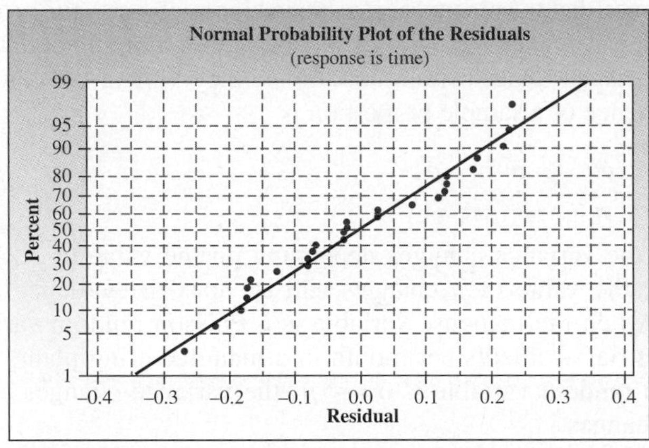

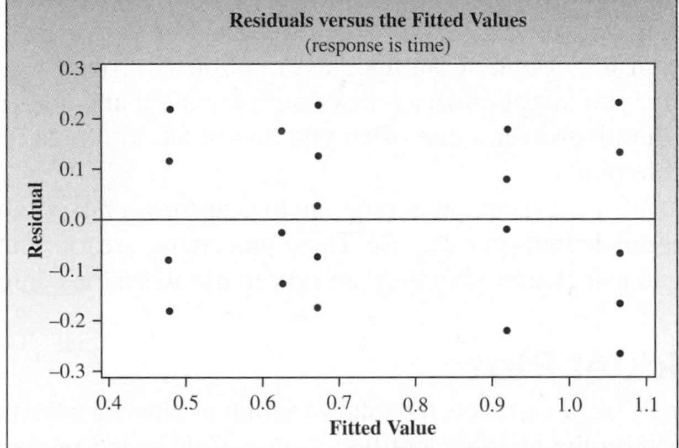

EXAMPLE 11.15

A company plans to promote a new product by using one of three advertising campaigns. To investigate the extent of product recognition from these three campaigns, 15 market areas were selected and 5 were randomly assigned to each advertising plan. At the end of the ad campaigns, random samples of 400 adults were selected in each area and the proportions who were familiar with the new product were recorded, as in Table 11.7. Have any of the analysis of variance assumptions been violated in this experiment?

TABLE 11.7 **Proportions of Product Recognition for Three Advertising Campaigns**

Campaign 1	Campaign 2	Campaign 3
0.33	0.28	0.21
0.29	0.41	0.30
0.21	0.34	0.26
0.32	0.39	0.33
0.25	0.27	0.31

SOLUTION The experiment is designed as a *completely randomized design*, but the response variable is a binomial sample proportion. This indicates that both the normality and the common variance assumptions might be invalid. Look at the normal probability plot of the residuals and the plot of residuals versus fit generated as an option in the *MINITAB* analysis of variance procedure and shown in Figure 11.13. The curved pattern in the normal probability plot indicates that the residuals *do not have a normal distribution*. In the residual versus fit plot, you can see three vertical lines of residuals, one for each of the three ad campaigns. Notice that two of the lines (campaigns 1 and 3) are close together and have similar spread. However, the third line (campaign 2) is farther to the right, which indicates a larger sample proportion and consequently a *larger variance* in this group. Both analysis of variance assumptions are suspect in this experiment.

FIGURE 11.13
MINITAB diagnostic plots for Example 11.15

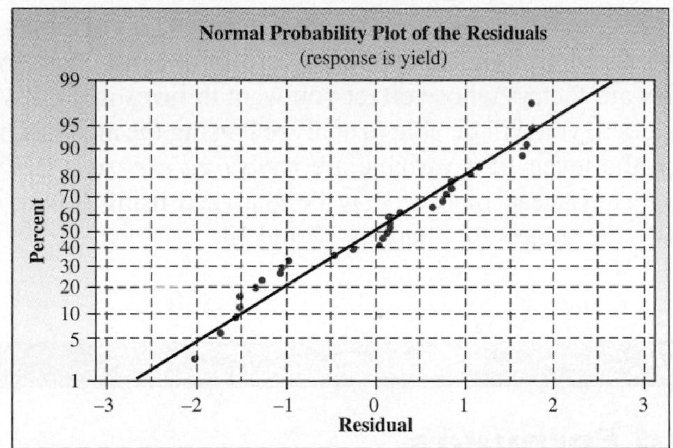

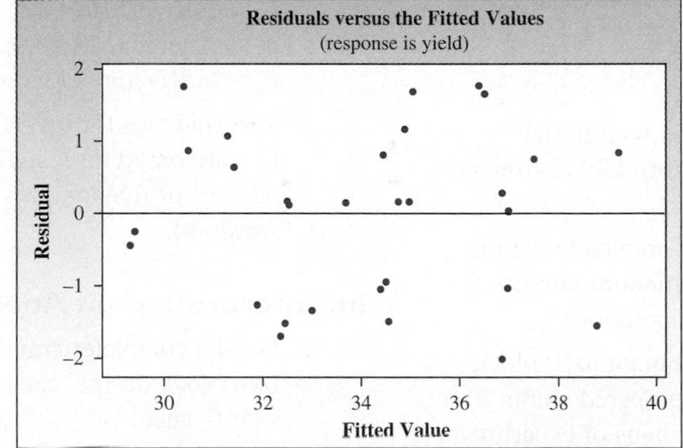

What can you do when the ANOVA assumptions are not satisfied? The *constant variance* assumption can often be remedied by **transforming** the response measurements. That is, instead of using the original measurements, you might use their square roots, logarithms, or some other function of the response. Transformations that tend to stabilize the variance of the response also tend to make their distributions more nearly normal.

When nothing can be done to even *approximately* satisfy the ANOVA assumptions or if the data are rankings, you should use **nonparametric** testing and estimation procedures, presented in Chapter 15. We have mentioned these procedures before; they are almost as powerful in detecting treatment differences as the tests presented in this chapter when the data are normally distributed. When the parametric ANOVA assumptions are violated, the nonparametric tests are generally more powerful.

A BRIEF SUMMARY

We presented three different experimental designs in this chapter, each of which can be analyzed using the analysis of variance procedure. The objective of the analysis of variance is to detect differences in the mean responses for experimental units that have received different treatments—that is, different combinations of the experimental factor levels. Once an overall test of the differences is performed, the nature of these differences (if any exist) can be explored using methods of paired comparisons and/or interval estimation procedures.

The three designs presented in this chapter represent only a brief introduction to the subject of analyzing designed experiments. Designs are available for experiments that involve several design variables, as well as more than two treatment factors and other more complex designs. Remember that **design variables** are factors whose effect you want to control and hence remove from experimental error, whereas **treatment variables** are factors whose effect you want to investigate. If your experiment is properly designed, you will be able to analyze it using the analysis of variance. Experiments in which the levels of a variable are *measured experimentally* rather than *controlled* or *preselected* ahead of time may be analyzed using **linear** or **multiple regression analysis**—the subject of Chapters 12 and 13.

CHAPTER REVIEW

Key Concepts and Formulas

I. Experimental Designs

1. Experimental units, factors, levels, treatments, response variables.
2. Assumptions: Observations within each treatment group must be normally distributed with a common variance σ^2.
3. One-way classification—completely randomized design: Independent random samples are selected from each of k populations.
4. Two-way classification—randomized block design: k treatments are compared within b relatively homogeneous groups of experimental units called *blocks*.
5. Two-way classification—$a \times b$ factorial experiment: Two factors, A and B, are compared at several levels. Each factor-level combination is replicated r times to allow for the investigation of an interaction between the two factors.

II. Analysis of Variance

1. The total variation in the experiment is divided into variation (sums of squares) explained by the various experimental factors and variation due to experimental error (unexplained).
2. If there is an effect due to a particular factor, its mean square ($MS = SS/df$) is usually large and $F = MS(\text{factor})/MSE$ is large.
3. Test statistics for the various experimental factors are based on F statistics, with appropriate degrees of freedom (df_2 = Error degrees of freedom).

III. Interpreting an Analysis of Variance

1. For the completely randomized and randomized block design, each factor is tested for significance.
2. For the factorial experiment, first test for a significant interaction. If the interaction is significant, main effects need not be tested. The nature of the differences in the factor-level combinations should be further examined.
3. If a significant difference in the population means is found, Tukey's method for pairwise comparisons or a similar method can be used to further identify the nature of the differences.
4. If you have a special interest in one population mean or the difference between two population means, you can use a confidence interval estimate. (For a randomized block design,

confidence intervals do not provide unbiased estimates for single population means.)

IV. Checking the Analysis of Variance Assumptions

1. To check for normality, use the normal probability plot for the residuals. The residuals should exhibit a straight-line pattern, increasing upwards toward the right.
2. To check for equality of variance, use the residuals versus fit plot. The plot should exhibit a random scatter, with the same vertical spread around the horizontal "zero error line."

TECHNOLOGY TODAY

Analysis of Variance Procedures—*Microsoft Excel*

The statistical procedures to perform the analysis of variance for the three experimental designs in this chapter can be found using the *Excel* command **Data ▶ Data Analysis.** You will see choices for **Single Factor, Two-Factor Without Replication,** and **Two-Factor With Replication** that will generate dialogue boxes used for the completely randomized, randomized block, and factorial designs, respectively.

Completely Randomized Design Refer to the breakfast study in Example 11.4, in which the effect of nutrition on attention span (in minutes) was studied.

No Breakfast	Light Breakfast	Full Breakfast
8	14	10
7	16	12
9	12	16
13	17	15
10	11	12

1. Enter the data into columns A, B, and C of an *Excel* spreadsheet with one sample per column.
2. Use **Data ▶ Data Analysis ▶ Anova: Single Factor** to generate the dialogue box in Figure 11.14(a). Highlight or type the **Input Range** (the data in the first three columns) into the first box. In the section marked "Grouped by" choose the radio button for **Columns** and check "Labels" if necessary.
3. The default significance level is $\alpha = 0.05$ in *Excel.* Change this significance level if necessary. Enter a cell location for the **Output Rang**e and click **OK.** The output will appear in the selected cell location, and should be adjusted using **Format ▶ AutoFit Column Width** on the **Home** tab in the **Cells** group while it is still highlighted. You can decrease the decimal accuracy if you like, using on the **Home** tab in the **Number** group (see Figure 11.14(b)).
4. The observed value of the test statistic $F = 4.93$ is found in the row labelled "*Between Groups*" followed by the "*P-value*" and the critical value marking the rejection region for a one-tailed test with $\alpha = 0.05$. For this example, the p-value $= 0.0273$ indicates that there is a significant difference in the average attention spans depending on the type of breakfast.

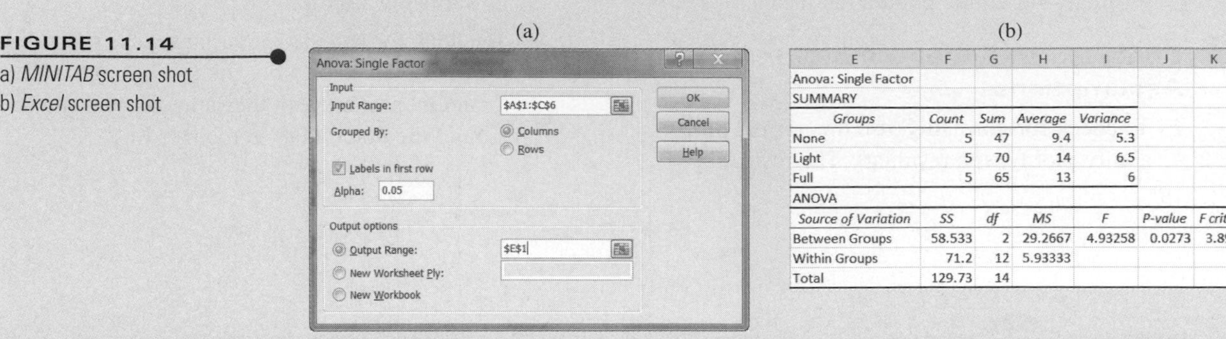

E	F	G	H	I	J	K
Anova: Single Factor						
SUMMARY						
Groups	*Count*	*Sum*	*Average*	*Variance*		
None	5	47	9.4	5.3		
Light	5	70	14	6.5		
Full	5	65	13	6		
ANOVA						
Source of Variation	*SS*	*df*	*MS*	*F*	*P-value*	*F crit*
Between Groups	58.533	2	29.2667	4.93258	0.0273	3.89
Within Groups	71.2	12	5.93333			
Total	129.73	14				

FIGURE 11.14
a) *MINITAB* screen shot
b) *Excel* screen shot

EXAMPLE 11.17

Randomized Block Design Refer to the cellphone study in Example 11.9, in which the effect of usage level on cost (in dollars) was studied for four different companies.

	Company			
Usage Level	A	B	C	D
Low	27	24	31	23
Middle	68	76	65	67
High	308	326	312	300

1. Enter the data into columns A–E of an *Excel* spreadsheet, using column A for usage labels and row 1 for company labels, just as shown in the table above.
2. Use **Data ▶ Data Analysis ▶ Anova: Two-Factor Without Replication** to generate the dialogue box in Figure 11.15(a). Highlight or type the **Input Range** (the data in the first five columns) into the first box and check "Labels" if necessary. Change the significance level if needed, and click **OK.** You can adjust the output, possibly changing the labels "Rows" and "Columns" to "Usage" and "Company," as shown in Figure 11.15(b).
3. The observed value of the test statistic for treatments (companies) is $F = 1.84$ with p-value $= 0.240$ indicating that there is no significant difference among the four companies. The test for blocks (usage) is highly significant, with p-value $= 0.000$.

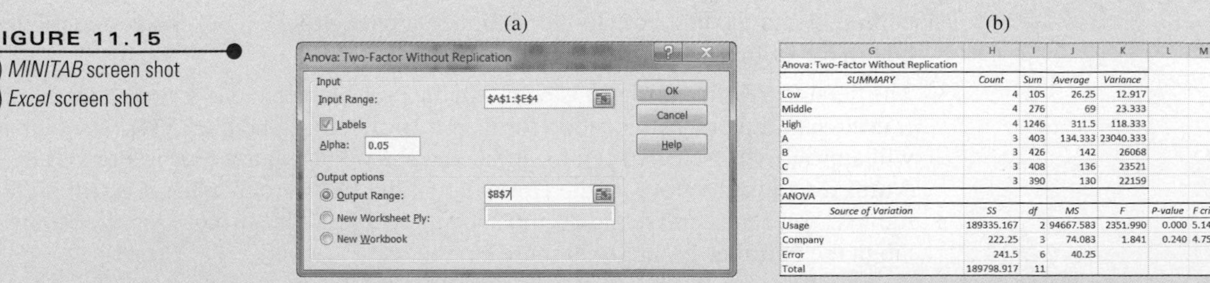

G	H	I	J	K	L	M
Anova: Two-Factor Without Replication						
SUMMARY	*Count*	*Sum*	*Average*	*Variance*		
Low	4	105	26.25	12.917		
Middle	4	276	69	23.333		
High	4	1246	311.5	118.333		
A	3	403	134.333	23040.333		
B	3	426	142	26068		
C	3	408	136	23521		
D	3	390	130	22159		
ANOVA						
Source of Variation	*SS*	*df*	*MS*	*F*	*P-value*	*F crit*
Usage	189335.167	2	94667.583	2351.990	0.000	5.143
Company	222.25	3	74.083	1.841	0.240	4.757
Error	241.5	6	40.25			
Total	189798.917	11				

FIGURE 11.15
a) *MINITAB* screen shot
b) *Excel* screen shot

EXAMPLE 11.18

Factorial Experiment Refer to the production output study in Example 11.12, in which the effect of supervisor and shift on production output was studied.

Supervisor	Shift: Day	Swing	Night
1	571	480	470
	610	474	430
	625	540	450
2	480	625	630
	516	600	680
	465	581	661

1. Enter the data into columns A–D of an *Excel* spreadsheet, using column A for supervisor labels and row 1 for shift labels, just as shown in the table above.
2. Use **Data ▶ Data Analysis ▶ Anova: Two-Factor With Replication** to generate the dialogue box in Figure 11.16(a). Highlight or type the **Input Range** (the data in the first four columns) into the first box. Enter the number of replications (3) into "Rows per Sample," change the significance level if needed, and click **OK.** You can adjust the output, possibly changing the labels "Sample" and "Columns" to "Supervisor" and "Shift," as shown in Figure 11.16(b).
3. Refer to the ANOVA table at the bottom of the printout. There is a significant interaction between shift and supervisor (p-value = 0.000). The differences in the treatment means can now be studied by looking at comparisons for the $3 \times 2 = 6$ factor level combinations.

FIGURE 11.16

a) *MINITAB* screen shot
b) *Excel* screen shot

(a)

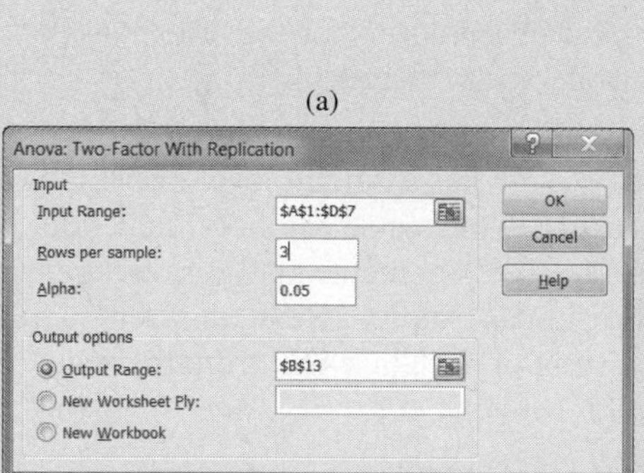

(b)

G	H	I	J	K	L	M
Anova: Two-Factor With Replication						
SUMMARY	Day	Swing	Night	Total		
1						
Count	3	3	3	9		
Sum	1806	1494	1350	4650		
Average	602	498	450	516.667		
Variance	777	1332	400	5155.25		
2						
Count	3	3	3	9		
Sum	1461	1806	1971	5238		
Average	487	602	657	582		
Variance	687	487	637	6096.5		
Total						
Count	6	6	6			
Sum	3267	3300	3321			
Average	544.5	550	553.5			
Variance	4553.1	3972.4	13269.5			
ANOVA						
Source of Variation	*SS*	*df*	*MS*	*F*	*P-value*	*F crit*
Supervisor	19208	1	19208	26.678	0.000	4.747
Shift	247	2	123.5	0.172	0.844	3.885
Interaction	81127	2	40563.5	56.338	0.000	3.885
Within	8640	12	720			
Total	109222	17				

NOTE: *Excel* does not provide options for performing Tukey's test or for generating diagnostic plots.

Analysis of Variance Procedures—*MINITAB*

The statistical procedures to perform the analysis of variance for the three experimental designs in this chapter can be found using the *MINITAB* command **Stat ▶ ANOVA.** You will see choices for **One-Way, One-Way (Unstacked),** and **Two-Way** that will generate dialogue boxes used for the completely randomized, randomized block, and factorial designs, respectively.

Completely Randomized Design Refer to the breakfast study in Example 11.4, in which the effect of nutrition on attention span was studied.

No Breakfast	Light Breakfast	Full Breakfast
8	14	10
7	16	12
9	12	16
13	17	15
10	11	12

1. Enter the 15 recorded attention spans in column C1 of a *MINITAB* worksheet and name them "Span." Next, enter the integers 1, 2, and 3 into a second column C2 to identify the meal assignment (*treatment*) for each observation. You can let *MINITAB* set this pattern for you using **Calc ▶ Make Patterned Data ▶ Simple Set of Numbers** and entering the appropriate numbers, as shown in Figure 11.17.

FIGURE 11.17
MINITAB screen shot

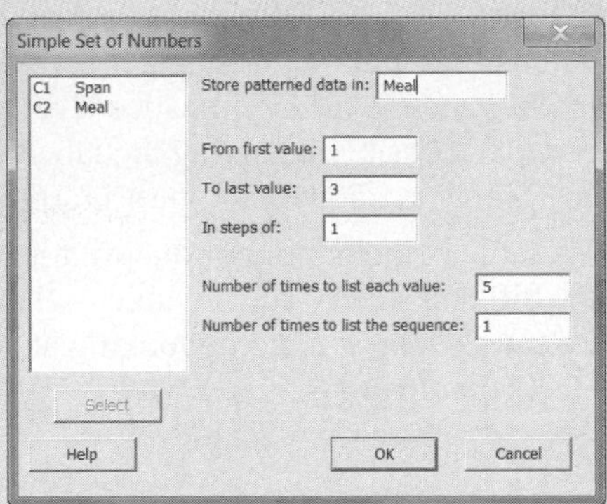

2. Use **Stat ▶ ANOVA ▶ One-Way** to generate the dialogue box in Figure 11.18(a).† Select the column of observations for the "Response" box and the column of treatment indicators for the "Factor" box.
3. Now you have several options. Under **Comparisons,** you can select "Tukey's family error rate" (which has a default level of 5%) to obtain paired comparisons output. Under **Graphs,** you can select individual value plots and/or box plots to compare the three meal assignments, and you can generate residual plots (use "Normal plot of residuals" and/or "Residuals versus fits") to verify the validity of the ANOVA assumptions. Click **OK** from the main dialogue box to obtain the output in Figure 11.18(b).

FIGURE 11.18
a) *MINITAB* screen shot
b) *MINITAB* screen shot

(a)

(b)

```
Session
One-way ANOVA: Span versus Meal
Source  DF      SS     MS     F      P
Meal     2   58.53  29.27  4.93  0.027
Error   12   71.20   5.93
Total   14  129.73
S = 2.436   R-Sq = 45.12%   R-Sq(adj) = 35.97%

Individual 95% CIs For Mean Based on
                         Individual 95% CIs For Mean Based on
                         Pooled StDev
Level  N    Mean  StDev  --+---------+---------+---------+-------
1      5   9.400  2.302  (---------*--------)
2      5  14.000  2.550                   (--------*--------)
3      5  13.000  2.449               (--------*--------)
                         --+---------+---------+---------+-------
                         7.5      10.0      12.5      15.0

Pooled StDev = 2.436
```

†If you had entered each of the three samples into separate columns, the proper command would have been **Stat ▶ ANOVA ▶ One-Way (Unstacked).**

4. The observed value of the test statistic $F = 4.93$ is found in the row labelled "*Meal*" followed by the p-value $= .0273$. With $\alpha = .05$, there is a significant difference in the average attention spans depending on the type of breakfast.

The **Stat ▶ ANOVA ▶ Two-Way** command can be used for both the randomized block and the factorial designs. You must first enter all of the observations into a single column and then integers or descriptive names to indicate either of these cases:

- The *block* and *treatment* for each of the measurements in a randomized block design
- The levels of *factors A and B* for the factorial experiment

MINITAB will recognize a number of replications within each factor level combination in the factorial experiment and will break out the sum of squares for interaction as long as you do not check the box "Fit additive model." Since these two designs involve the same sequence of commands, we will use the data from Example 11.12 to generate the ANOVA.

Two-Way Classification Refer to the production output study in Example 11.12, in which the effect of supervisor and shift on production output was studied.

Supervisor	Shift: Day	Swing	Night
1	571	480	470
	610	474	430
	625	540	450
2	480	625	630
	516	600	680
	465	581	661

1. Enter the data into the worksheet as shown in Figure 11.19(a). See if you can use the **Calc ▶ Make Patterned Data ▶ Simple Set of Numbers** to enter the data in columns C2–C3.
2. Use **Stat ▶ ANOVA ▶ Two-Way** to generate the dialogue box in Figure 11.19(b). Choose "Output" for the "Response" box, and "Supervisor" and "Shift" for the "Row Factor" and "Column Factor," respectively. You may choose to display the main effect means along with 95% confidence intervals by checking "Display means," and you may select residual plots if you wish. Click **OK** to obtain the ANOVA printout in Figure 11.11(a) on page 503.

FIGURE 11.19
a) *MINITAB* screen shot
b) *MINITAB* screen shot

(a)

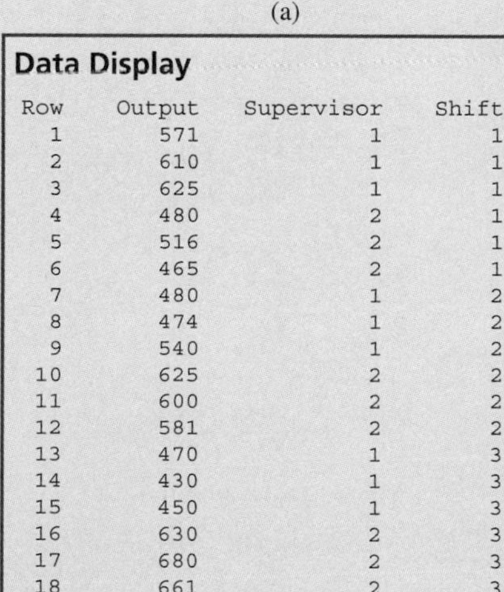

Data Display

Row	Output	Supervisor	Shift
1	571	1	1
2	610	1	1
3	625	1	1
4	480	2	1
5	516	2	1
6	465	2	1
7	480	1	2
8	474	1	2
9	540	1	2
10	625	2	2
11	600	2	2
12	581	2	2
13	470	1	3
14	430	1	3
15	450	1	3
16	630	2	3
17	680	2	3
18	661	2	3

(b)

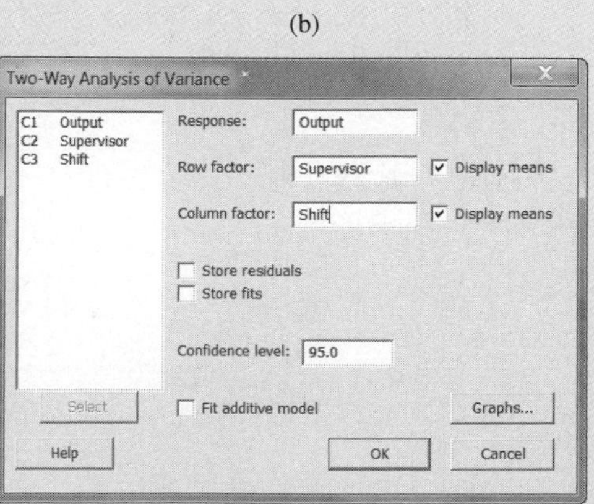

3. Since the interaction between supervisors and shifts is highly significant, you may want to explore the nature of this interaction by plotting the average output for each supervisor at each of the three shifts. Use **Stat ▶ ANOVA ▶ Interaction Plot** and choose the appropriate response and factor variables. The plot is shown in Figure 11.20. You can see the strong difference in the behaviours of the mean outputs for the two supervisors, indicating a strong interaction between the two factors.

FIGURE 11.20
MINITAB screen shot

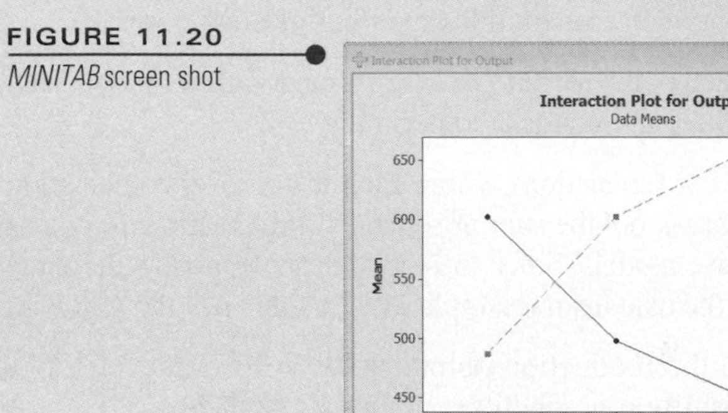

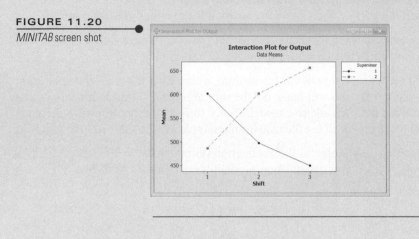

Supplementary Exercises

11.54 Reaction Times vs. Stimuli EX1154 Twenty-seven people participated in an experiment to compare the effects of five different stimuli on reaction time. The experiment was run using a completely randomized design, and, regardless of the results of the analysis of variance, the experimenters wanted to compare stimuli A and D. The results of the experiment are given here. Use the *MINITAB* printout to complete the exercise:

Stimulus	Reaction Time (sec)							Total	Mean
A	0.8	0.6	0.6	0.5				2.5	0.625
B	0.7	0.8	0.5	0.5	0.6	0.9	0.7	4.7	0.671
C	1.2	1.0	0.9	1.2	1.3	0.8		6.4	1.067
D	1.0	0.9	0.9	1.1	0.7			4.6	0.920
E	0.6	0.4	0.4	0.7	0.3			2.4	0.480

MINITAB output for Exercise 11.54

One-way ANOVA: Time versus Stimulus

```
Source    DF      SS      MS      F      P
Stimulus   4  1.2118  0.3030  11.67  0.000
Error     22  0.5711  0.0260
Total     26  1.7830

S = 0.1611   R-Sq = 67.97%   R-Sq(adj) = 62.14%

                                Individual 95% CIs For Mean Based on
                                Pooled StDev
Level    N    Mean   StDev  -------+---------+---------+---------+--
A        4  0.6250  0.1258       (------*------)
B        7  0.6714  0.1496           (----*----)
C        6  1.0667  0.1966                          (----*----)
D        5  0.9200  0.1483                     (-----*-----)
E        5  0.4800  0.1643  (-----*-----)
                            -------+---------+---------+---------+--
Pooled StDev =  0.1611           0.50      0.75      1.00      1.25
```

a. Conduct an analysis of variance and test for a difference in the mean reaction times due to the five stimuli.

b. Compare stimuli A and D to see if there is a difference in mean reaction times.

11.55 Refer to Exercise 11.54. Use this *MINITAB* output to identify the differences in the treatment means:

MINITAB output for Exercise 11.55

```
Tukey 95% Simultaneous Confidence Intervals
All Pairwise Comparisons among Levels of Stimulus

Individual confidence level = 99.29%

Stimulus = A subtracted from:

Stimulus   Lower   Centre    Upper  --------+---------+---------+---------+-
B        -0.2535   0.0464   0.3463                (-----*-----)
C         0.1328   0.4417   0.7505                      (-----*-----)
D        -0.0260   0.2950   0.6160                    (-----*-----)
E        -0.4660  -0.1450   0.1760             (-----*-----)
                                    --------+---------+---------+---------+-
                                        -0.50      0.00      0.50      1.00

Stimulus = B subtracted from:

Stimulus   Lower   Centre    Upper  --------+---------+---------+---------+-
C         0.1290   0.3952   0.6615                      (-----*-----)
D        -0.0316   0.2486   0.5288                    (-----*-----)
E        -0.4716  -0.1914   0.0888             (-----*-----)
                                    --------+---------+---------+---------+-
                                        -0.50      0.00      0.50      1.00

Stimulus = C subtracted from:

Stimulus   Lower   Centre    Upper  --------+---------+---------+---------+-
D        -0.4364  -0.1467   0.1431             (-----*-----)
E        -0.8764  -0.5867  -0.2969  (-----*-----)
                                    --------+---------+---------+---------+-
                                        -0.50      0.00      0.50      1.00

Stimulus = D subtracted from:

Stimulus   Lower   Centre    Upper  --------+---------+---------+---------+-
E        -0.7426  -0.4400  -0.1374    (-----*-----)
                                    --------+---------+---------+---------+-
                                        -0.50      0.00      0.50      1.00
```

11.56 Refer to Exercise 11.54. What do the normal probability plot and the residuals versus fit plot tell you about the validity of your analysis of variance results?

MINITAB diagnostic plots for Exercise 11.56

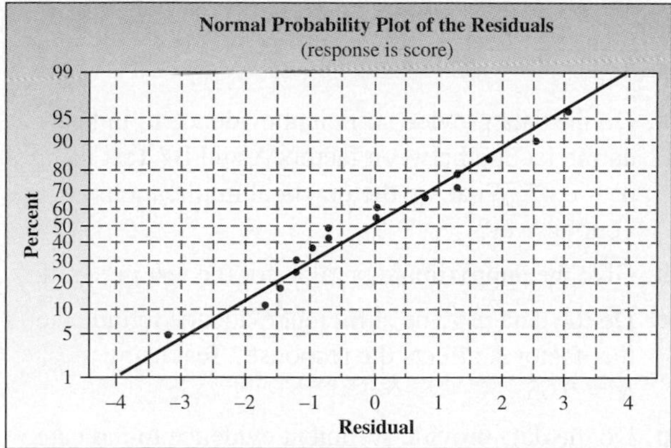

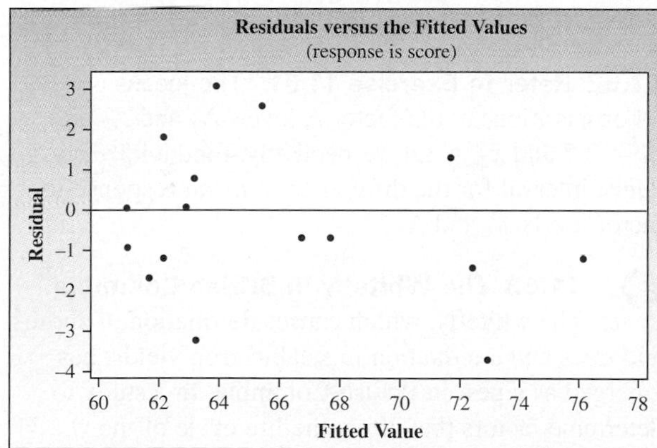

Data set EX1157 **11.57 Reaction Times II** The experiment in Exercise 11.54 might have been conducted more effectively using a randomized block design with people as blocks, since you would expect mean reaction time to vary from one person to another. Hence, four people were used in a new experiment, and each person was subjected to each of the five stimuli in a random order. The reaction times (in seconds) are listed here:

	Stimulus				
Subject	A	B	C	D	E
1	0.7	0.8	1.0	1.0	0.5
2	0.6	0.6	1.1	1.0	0.6
3	0.9	1.0	1.2	1.1	0.6
4	0.6	0.8	0.9	1.0	0.4

MINITAB output for Exercise 11.57

Two-way ANOVA: Time versus Subject, Stimulus

```
Source     DF     SS        MS        F      P
Subject     3  0.140  0.046667     6.59  0.007
Stimulus    4  0.787  0.196750    27.78  0.000
Error      12  0.085  0.007083
Total      19  1.012

S = 0.08416   R-Sq = 91.60%    R-Sq(adj) = 86.70%

                        Individual 95% CIs For Mean Based on
                        Pooled StDev
Stimulus     Mean     --------+---------+---------+---------+-
A           0.700          (----*----)
B           0.800                (----*----)
C           1.050                            (---*----)
D           1.025                          (---*----)
E           0.525     (---*----)
                      --------+---------+---------+---------+-
                           0.60      0.80      1.00      1.20
```

a. Use the *MINITAB* printout to analyze the data and test for differences in treatment means.

b. Use Tukey's method for paired comparisons to identify the significant pairwise differences in the stimuli.

c. Does it appear that blocking was effective in this experiment?

Data set EX1158 **11.58 Heart Rate and Exercise** An experiment was conducted to examine the effect of age on heart rate when a person is subjected to a specific amount of exercise. Ten male subjects were randomly selected from four age groups: 10–19, 20–39, 40–59, and 60–69. Each subject walked on a treadmill at a fixed grade for a period of 12 minutes, and the increase in heart rate, the difference before and after exercise, was recorded (in beats per minute):

	10–19	20–39	40–59	60–69
	29	24	37	28
	33	27	25	29
	26	33	22	34
	27	31	33	36
	39	21	28	21
	35	28	26	20
	33	24	30	25
	29	34	34	24
	36	21	27	33
	22	32	33	32
Total	309	275	295	282

Use an appropriate computer program to answer these questions:

a. Do the data provide sufficient evidence to indicate a difference in mean increase in heart rate among the four age groups? Test by using $\alpha = 0.05$.

b. Find a 90% confidence interval for the difference in mean increase in heart rate between age groups 10–19 and 60–69.

c. Find a 90% confidence interval for the mean increase in heart rate for the age group 20–39.

d. Approximately how many people would you need in each group if you wanted to be able to estimate a group mean correct to within two beats per minute with probability equal to 0.95?

Data set EX1159 **11.59 Learning to Sell** A company wished to study the effects of four training programs on the sales abilities of their sales personnel. Thirty-two people were randomly divided into four groups of equal size, and each group was then subjected to one of the different sales training programs. Because there were some dropouts during the training programs due to illness, vacations, and so on, the number of trainees completing the programs varied from group to group. At the end of the training programs, each salesperson was randomly assigned a sales area from a group of sales areas that were judged to have equivalent sales potentials. The sales made by each of the four groups of salespeople during the first week after completing the training program are listed in the table:

	Training Program			
	1	2	3	4
	78	99	74	81
	84	86	87	63
	86	90	80	71
	92	93	83	65
	69	94	78	86
	73	85		79
		97		73
		91		70
Total	482	735	402	588

Analyze the experiment using the appropriate method. Identify the treatments or factors of interest to the researcher and investigate any significant effects. What are the practical implications of this experiment? Write a paragraph explaining the results of your analysis.

11.60 4 × 2 Factorial Suppose you were to conduct a two-factor factorial experiment, factor A at four levels and factor B at two levels, with r replications per treatment.

a. How many treatments are involved in the experiment?

b. How many observations are involved?

c. List the sources of variation and their respective degrees of freedom.

11.61 2 × 3 Factorial The analysis of variance table for a 2 × 3 factorial experiment, factor A at two levels and factor B at three levels, with five observations per treatment, is shown in the table:

Source	*df*	SS	MS	*F*
A		1.14		
B		2.58		
AB		0.49		
Error				
Total		8.41		

a. Do the data provide sufficient evidence to indicate an interaction between factors A and B? Test using $\alpha = 0.05$. What are the practical implications of your answer?

b. Give the approximate p-value for the test in part a.

c. Do the data provide sufficient evidence to indicate that factor A affects the response? Test using $\alpha = 0.05$.

d. Do the data provide sufficient evidence to indicate that factor B affects the response? Test using $\alpha = 0.05$.

11.62 Refer to Exercise 11.61. The means of all observations at the factor A levels A_1 and A_2 are $\bar{x}_1 = 3.7$ and $\bar{x}_2 = 1.4$, respectively. Find a 95% confidence interval for the difference in mean response for factor levels A_1 and A_2.

Data set EX1163 **11.63 The Whitefly in British Columbia** The whitefly, which causes defoliation of shrubs and trees and a reduction in salable crop yields, has emerged as a pest in British Columbia. In a study to determine factors that affect the life cycle of the whitefly, an experiment was conducted in which whiteflies were placed on two different types of plants at three different temperatures. The observation of interest was the total number of eggs laid by caged females under one of the six possible treatment combinations. Each treatment combination was run using five cages:

	Temperature		
Plant	21.1°C	25°C	27.8°C
Cotton	37	34	46
	21	54	32
	36	40	41
	43	42	36
	31	16	38
Cucumber	50	59	43
	53	53	62
	25	31	71
	37	69	49
	48	51	59

MINITAB output for Exercise 11.63

Two-way ANOVA: Eggs versus Temperature, Plant

```
Source        DF       SS       MS      F      P
Temperature    2   487.47   243.73   1.98  0.160
Plant          1  1512.30  1512.30  12.29  0.002
Interaction    2   111.20    55.60   0.45  0.642
Error         24  2952.40   123.02
Total         29  5063.37

S = 11.09   R-Sq = 41.69%   R-Sq(adj) = 29.54%
```

a. What type of experimental design has been used?

b. Do the data provide sufficient evidence to indicate that the effect of temperature on the number of eggs laid is different depending on the type of plant? Use the *MINITAB* printout to test the appropriate hypothesis.

c. Plot the treatment means for cotton as a function of temperature. Plot the treatment means for cucumber as a function of temperature. Comment on the similarity or difference in these two plots.

d. Find the mean number of eggs laid on cotton and cucumber based on 15 observations each. Calculate a 95% confidence interval for the difference in the underlying population means.

Data set EX1164 **11.64 Pollution from Chemical Plants** Four chemical plants, producing the same product and owned by the same company, discharge effluents into streams in the vicinity of their locations. To check on the extent of the pollution created by the effluents and to determine whether this varies from plant to plant, the company collected random samples of liquid waste, five specimens for each of the four plants. The data are shown in the table:

Plant	Polluting Effluents (400 g/4 L of waste)				
A	1.65	1.72	1.50	1.37	1.60
B	1.70	1.85	1.46	2.05	1.80
C	1.40	1.75	1.38	1.65	1.55
D	2.10	1.95	1.65	1.88	2.00

a. Do the data provide sufficient evidence to indicate a difference in the mean amounts of effluents discharged by the four plants?

b. If the maximum mean discharge of effluents is 600 g/4 L, do the data provide sufficient evidence to indicate that the limit is exceeded at plant A?

c. Estimate the difference in the mean discharge of effluents between plants A and D, using a 95% confidence interval.

Data set EX1165 **11.65 Canada's Food Basics** Exercise 10.48 examined an advertisement for Food Basics, a supermarket chain in Canada. The advertiser claims that Food Basics has consistently had lower prices than four other full-service supermarkets. The average weekly total based on the prices of approximately 95 items is given for five different supermarket chains recorded during four consecutive weeks:

Week	Food Basics	Zehrs	Loblaws	IGA	Price Chopper
1	$254.26	$256.03	$267.92	$260.71	$258.84
2	240.62	255.65	251.55	251.80	242.14
3	231.90	255.12	245.89	246.77	246.80
4	234.13	261.18	254.12	249.45	248.99

a. What type of design has been used in this experiment?

b. Conduct an analysis of variance for the data.

c. Is there sufficient evidence to indicate a difference in the average weekly totals for the five supermarkets? Use $\alpha = 0.05$.

d. Use Tukey's method for paired comparisons to determine which of the means are significantly different from each other. Use $\alpha = 0.05$.

Data set EX1166 **11.66 Yield of Wheat** The yields of wheat (in kilogram per hectare) were compared for five different varieties, A, B, C, D, and E, at six different locations. Each variety was randomly assigned to a plot at each location. The results of the experiment are shown in the accompanying table, along with a *MINITAB* printout of the analysis of variance. Analyze the experiment using the appropriate method. Identify the treatments or factors of interest to the researcher and investigate any effects that exist. Use the diagnostic plots to comment on the validity of the analysis of variance assumptions. What are the practical implications of this experiment? Write a paragraph explaining the results of your analysis.

	Location					
Variety	1	2	3	4	5	6
A	35.3	31.0	32.7	36.8	37.2	33.1
B	30.7	32.2	31.4	31.7	35.0	32.7
C	38.2	33.4	33.6	37.1	37.3	38.2
D	34.9	36.1	35.2	38.3	40.2	36.0
E	32.4	28.9	29.2	30.7	33.9	32.1

MINITAB output for Exercise 11.66

Two-way ANOVA: Yield versus Varieties, Location

```
Source     DF       SS       MS      F      P
Varieties   4  142.670  35.6675  18.61  0.000
Locations   5   68.142  13.6283   7.11  0.001
Error      20   38.303   1.9165
Total      29  249.142

S = 1.384   R-Sq = 84.62%   R-Sq(adj) = 77.69%
```

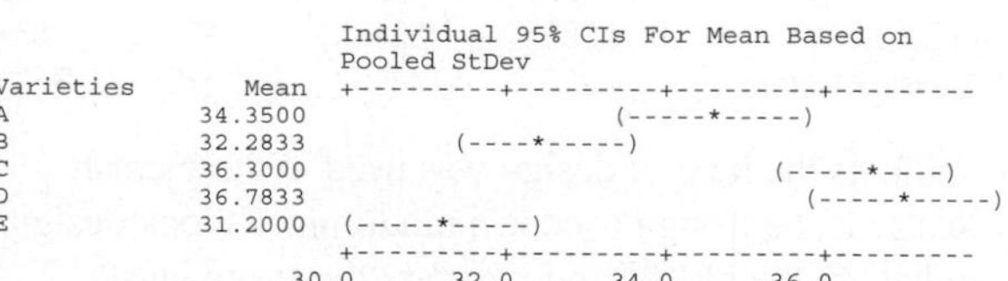

```
                   Individual 95% CIs For Mean Based on
                   Pooled StDev
Varieties    Mean  ---------+---------+---------+---------+---------
A         34.3500                        (-----*-----)
B         32.2833              (----*-----)
C         36.3000                                  (-----*----)
D         36.7833                                    (-----*-----)
E         31.2000  (-----*-----)
                   ---------+---------+---------+---------+---------
                         30.0      32.0      34.0      36.0
```

MINITAB diagnostic plot for Exercise 11.66

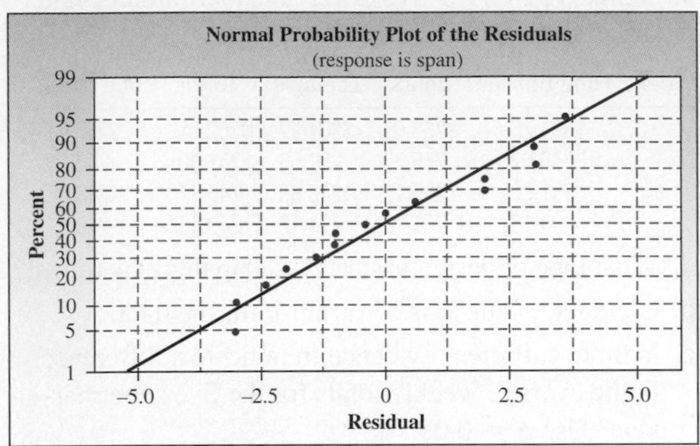

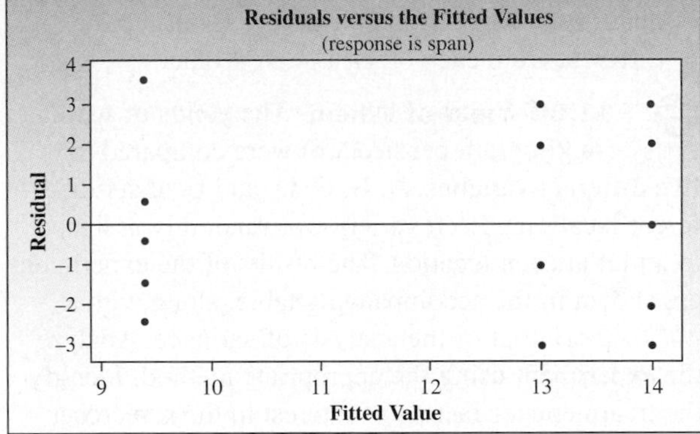

Data set EX1167 **11.67 Crash Tests** Information on crash tests conducted by the Insurance Institute for Highway Safety regarding bumper repair costs for damage sustained in front and rear crashes of vehicles into barriers/poles at approximately 8 km/h are given in the following table. These types of crash tests evaluate how well the bumpers prevent costly damage to vehicles in parking lot–type impacts.[4]

	Types of Crash			
Autos	Front into Barrier	Rear into Barrier	Front into Angle Barrier	Rear into Pole
Hyundai Elantra	$ 247	$ 0	$ 407	$ 185
Ford Focus	31	1,137	507	939
Honda Civic	403	447	404	227
Dodge Stratus/ Chrysler Cirrus	278	174	626	1,473
Lexus LS 30	75	395	1,526	765
Dodge Grand Caravan	329	822	703	2,268
Isuzu Rodeo	1,769	924	1,932	552
Mitsubishi Montero	1,210	2,495	2,525	2,831

a. What is the type of design was used in these crash tests? If the design used is a randomized block design, what are the blocks and what are the treatments?

b. Are there significant differences in the cost of crashes for the vehicles considered here?

c. Are there significant differences among the four types of crashes?

d. Use Tukey's pairwise procedure to investigate the differences in average repair costs for the eight vehicles. Comment on the results found using this procedure. Use $\alpha = 0.05$.

Data set EX1168 **11.68 Physical Fitness** Researchers Russell R. Pate and colleagues analyzed the results of the National Health and Nutrition Examination Survey to assess cardiorespiratory fitness levels in youth aged 12 to 19 years.[5] Estimated maximum oxygen uptake (VO_{2max}) was used to measure a person's cardiorespiratory level. The focus of the study investigated the relationship between levels of physical activity (more than others, same as others, or less than others) and gender on VO_{2max}. The data that follows are based on this study:

	Physical Activity		
	More	Same	Less
Males	50.1	45.7	40.9
	47.2	44.2	41.3
	49.7	46.8	39.2
	50.4	44.9	40.9
	More	Same	Less
Females	41.2	37.2	36.5
	39.8	39.4	35.0
	41.5	38.6	37.2
	38.2	37.8	35.4

a. Is this a factorial experiment or a randomized block design? Explain.

b. Is there a significant interaction between levels of physical activity and gender? Are there significant differences between males and females? Levels of physical activity?

c. If the interaction is significant, use Tukey's pairwise procedure to investigate differences among the six cell means. Comment on the results found using this procedure. Use $\alpha = 0.05$.

Data set EX1169 **11.69 Professor's Salaries** In a study of starting salaries of assistant professors,[6] five male assistant professors and five female assistant professors at each of three types of faculties granting doctoral degrees were polled and their initial starting salaries were recorded. The results of the survey in $1000 are given in the following table:

Gender	Faculty of Arts	Faculty of Sciences	Nursing
	49.3	81.8	66.9
	49.9	71.2	57.3
Males	48.5	62.9	57.7
	68.5	69.0	46.2
	54.0	69.0	52.2
	45.4	55.1	50.4
	54.7	62.1	52.1
Females	67.0	59.5	49.8
	61.2	54.8	61.9
	63.3	69.7	51.6

a. What type of design was used in collecting these data?

b. Use an analysis of variance to test if there are significant differences in gender, in type of institution, and to test for a significant interaction of gender × type of institution.

c. Find a 95% confidence interval estimate for the difference in starting salaries for male assistant professors and female assistant professors. Interpret this interval in terms of a gender difference in starting salaries.

d. Use Tukey's procedure to investigate differences in assistant professor salaries for the three types of institutions. Use $\alpha = 0.01$.

e. Summarize the results of your analysis.

Data set EX1170 **11.70 Pottery in the United Kingdom** An article in *Archaeometry* involved an analysis of 26 samples of Romano-British pottery, found at four different kiln sites in the United Kingdom.[7] Since one site only yielded two samples, consider the samples found at the other three sites. The samples were analyzed to determine their chemical composition and the percentage of iron oxide is shown below:

Llanederyn		Island Thorns	Ashley Rails
7.00	5.78	1.28	1.12
7.08	5.49	2.39	1.14
7.09	6.92	1.50	0.92
6.37	6.13	1.88	2.74
7.06	6.64	1.51	1.64
6.26	6.69		
4.26	6.44		

Source: A. Tubb, A. J. Parker, and G. Nickless, "The Analysis of Romano-British Pottery by Atomic Absorption Spectrophotometry," *Archaeometry*, Vol. 22, Issue 2, August 1989, p. 153. *Archaeometry* by UNIVERSITY OF OXFORD. Reproduced with permission of BLACKWELL PUBLISHING LTD. in the format Book via Copyright Clearance Center.

a. What type of experimental design is this?

b. Use an analysis of variance to determine if there is a difference in the average percentage of iron oxide at the three sites. Use $\alpha = 0.01$.

c. If you have access to a computer program, generate the diagnostic plots for this experiment. Does it appear that any of the analysis of variance assumptions have been violated? Explain.

Data set EX1171 **11.71 Cellphones** How satisfied are you with your current cellphone service provider? Surveys indicate that there is a high level of dissatisfaction among consumers, resulting in high customer turnover rates. The table shows the overall satisfaction scores, based on a maximum score of 100, for four service providers in four different cities:

	Vancouver	Halifax	Toronto	Regina
Bell	63	66	61	64
Rogers	67	67	64	60
Telus	60	68	60	61
Virgin Mobile	71	75	73	73

a. What type of experimental design was used in this article? If the design used is a randomized block design, what are the blocks and what are the treatments?

b. Conduct an analysis of variance for the data.

c. Are there significant differences in the average satisfaction scores for the four wireless providers considered here?

d. Are there significant differences in the average satisfaction scores for the four cities?

11.72 Cellphones, continued Refer to Exercise 11.71. The diagnostic plots for this experiment are shown below. Does it appear that any of the analysis of variance assumptions have been violated? Explain.

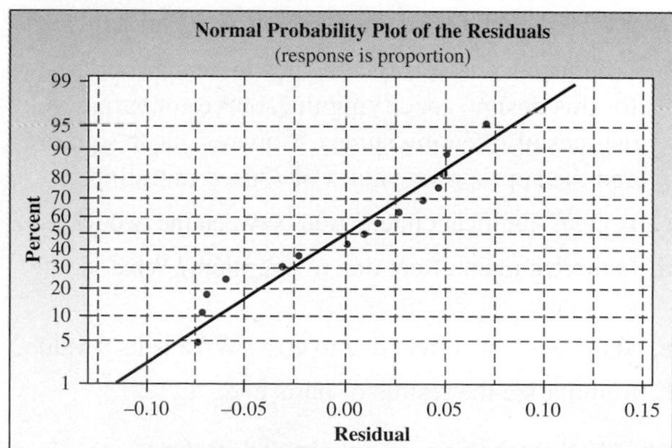

Source: William Mendenhall, R.J. Beaver, and B.M. Beaver, *Introduction to Probability and Statistics*, 13USe, p. 500. Copyright © Cengage Learning, 2009.

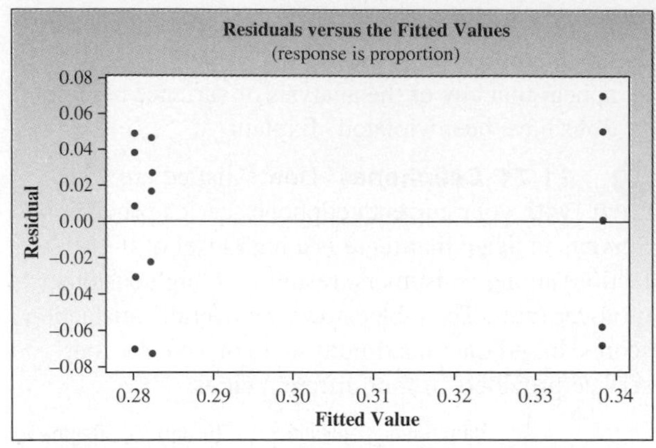

Source: William Mendenhall, R.J. Beaver, and B.M. Beaver, *Introduction to Probability and Statistics*, 13USe, p. 500. Copyright © Cengage Learning, 2009.

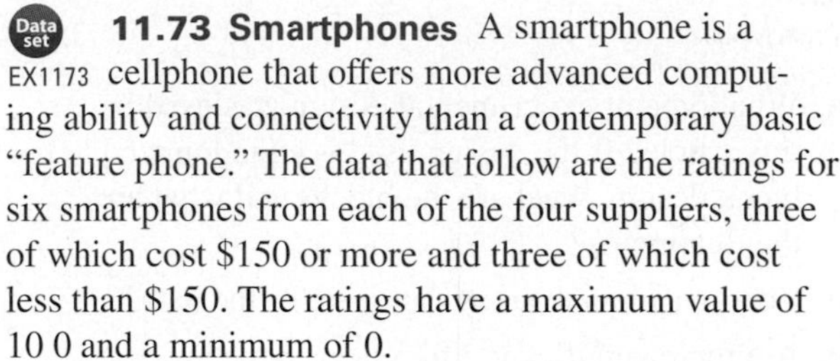

Data set EX1173 **11.73 Smartphones** A smartphone is a cellphone that offers more advanced computing ability and connectivity than a contemporary basic "feature phone." The data that follow are the ratings for six smartphones from each of the four suppliers, three of which cost \$150 or more and three of which cost less than \$150. The ratings have a maximum value of 10 0 and a minimum of 0.

	Supplier			
	Bell	Rogers	Telus	Virgin Mobile
Cost ≥ \$150	76	74	72	75
	74	69	71	73
	69	68	71	73
Cost < \$150	69	69	71	72
	67	64	71	71
	64	60	70	70

a. What type of experiment was used to evaluate these smartphones? What are the factors? How many levels of each factor are used in the experiment?

b. Produce an analysis of variance table appropriate for this design, specifying the sources of variation, degrees of freedom, sums of squares, mean squares, and the appropriate values of F used in testing.

c. Is there significant interaction between the two factors?

d. Is there a main effect due to suppliers? What is its p-value?

e. Is there a main effect due to cost? What is its p-value?

f. Summarize the results of parts c–e.

11.74 Smartphones, continued Refer to Exercise 11.73. The diagnostic plots for this experiment are shown below. Does it appear that any of the analysis of variance assumptions have been violated? Explain.

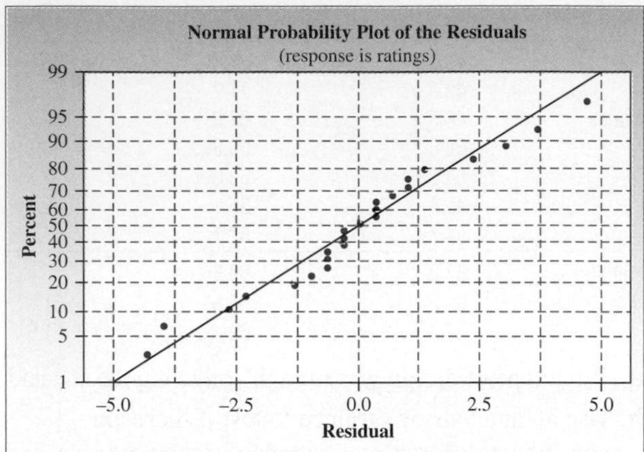

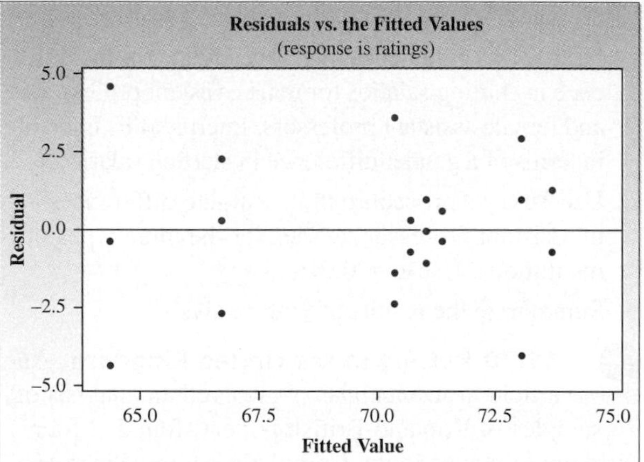

Data set EX1175 **11.75 Professor's Salaries II** Each year, Statistics Canada reports on salaries of academic professors at universities and colleges in Canada. The following data (in dollars), adapted from this report, are based on samples of $n = 10$ in each of three professorial ranks, for both male and female professors:[8]

Full Professor		Associate Professor		Assistant Professor	
Male	Female	Male	Female	Male	Female
155690.6	151430.2	93966.7	89348.0	76616.3	74441.1
139550.3	110820.4	93075.8	89281.3	78979.3	70006.6
132050.2	146290.4	94580.4	89649.6	73854.5	78930.3
116450.4	145680.6	95055.5	96927.2	77283.0	72367.4
154240.4	59900.6	95920.4	95696.8	81067.0	74048.9
144320.5	115600.6	93846.1	92245.1	80577.5	74858.1
91050.7	90560.1	91900.4	93694.7	76894.1	69644.6
82730.1	78130.0	94707.1	93408.8	77405.5	71568.6
141250.9	163270.3	97582.2	93612.1	78028.6	75303.6
139200.9	104420.4	96316.7	90471.8	80937.8	71846.3

a. Identify the design used in this survey.

b. Use the appropriate analysis of variance for these data.

c. Do the data indicate that the salary at the different ranks vary by gender?

d. If there is no interaction, determine whether there are differences in salaries by rank, and whether there are differences by gender. Discuss your results.

e. Plot the average salaries using an interaction plot. If the main effect of ranks is significant, use Tukey's method for pairwise comparisons to determine if there are significant differences among the ranks.

CASE STUDY  "A Fine Mess"

Do you risk a parking ticket by parking where you shouldn't or forgetting how much time you have left on the parking meter? Do the fines associated with various parking infractions vary depending on the city in which you receive a parking ticket? To look at this issue, the fines imposed for overtime parking, parking in a no parking zone, and parking next to a fire route were recorded for 11 cities in the province of Ontario.[9]

City	Expired Meter	Fire Route	No Parking Zone
Guelph	$15	$60	$20
Cambridge	15	60	15
Kitchener	15	45	20
Waterloo	18	40	20
London	15	75	30
Mississauga	25	100	25
Newmarket	25	75	30
Whitby	15	45	15
Peterborough	15	15	15
Windsor	15	85	30
Hamilton	18	30	30

1. Identify the design used for the data collection in this case study.
2. Analyze the data using the appropriate analysis. What can you say about the variation among the cities in this study? Among fines for the three types of violations? Can Tukey's procedure be of use in further delineating any significant differences you may find? Would confidence interval estimates be useful in your analysis?
3. Summarize the results of your analysis of these data.

PROJECTS  Project 11: *Globe and Mail* Series Exposes the Pervasive Health Risks Associated with Canada's Excessive Salt Consumption[10]

(*Source*: © 2015, Heart and Stroke Foundation of Canada. Reproduced with the permission of the Heart and Stroke Foundation of Canada. www.heartandstroke.ca)

Each day, the average Canadian consumes excessive amounts of sodium—an average of 3,100 milligrams—more than double the adequate intake. There is no recommended daily intake (RDI) for sodium… but there are adequate daily intake and tolerable upper intake levels. 3100 is more than double the 1200 to 1500 AI.

The **adequate daily intake** for a healthy adult is 1200 mg to 1500 mg of sodium according to Health Canada and the U.S. National Academy of Sciences (Institute of

Medicine) and a **tolerable upper intake level** is 2300 mg. Statistics Canada estimates the average Canadian consumes more than 3100 mg of sodium daily, making the goal of between 1200 mg and 2300 mg a significant reduction.

The health implications are extremely serious and include high blood pressure, heart attacks, and a range of other major health problems. Perhaps most disturbing is the fact most Canadians aren't even aware of this hidden health threat. About 80% of the salt we consume comes from processed foods, including fast foods, prepared meals, processed meats such as hot dogs and lunchmeats, canned soups, bottled dressings, packaged sauces, condiments such as ketchup and pickles, and salty snacks like potato chips.

a. A group of researchers were interested in comparing salt intake among four Canadian provinces. Five people were randomly selected from each of the four provinces and their sodium intake was measured in grams in a given day. The sample data is given in the following table:

Quebec	Ontario	Alberta	British Columbia
2.5	2.9	2.8	2.6
3.1	2.7	3.2	2.7
2.6	3.2	3.2	2.9
2.8	3.0	2.9	2.9
2.4	2.2	3.3	3.1

(i) Identify the experimental design.

(ii) Create the appropriate ANOVA table.

(iii) Test the null hypothesis of no difference in the true mean amount of sodium for the four provinces. Use $\alpha = 0.05$. What conclusion can be drawn? Find and interpret the approximate p-value.

(iv) Develop and interpret a 90% confidence interval for $\mu_2 - \mu_3$ (Quebec versus Alberta).

(v) Do the data provide sufficient evidence to indicate a difference between μ_2 and μ_3? Test using the two-sample independent t test with $\alpha = 0.05$.

(vi) Find a 90% confidence interval for μ_1 (Ontario).

(vii) Rank the province means using pair-wise comparisons.

(viii) Use Tukey's method of comparison to determine which of the four provinces' means differ from the others. Use $\alpha = 0.05$.

(ix) Use Tukey's multiple comparison method to determine which means differ.

The following strategies help reduce added, unnecessary salt:

Strategy 1: Cut down on prepared and processed foods.

Strategy 2: Eat more fresh vegetables and fruits.

Strategy 3: Reduce the amount of salt you add while cooking, baking, or at the table.

Strategy 4: Experiment with other seasonings, such as garlic, lemon juice, and fresh or dried herbs.

Strategy 5: Look for products with claims such as low sodium, sodium reduced, or no salt added.

Strategy 6: When eating out, ask for nutrient information for the menu items and select meals lower in sodium.

b. A number of people from the selected provinces were randomized to the above strategies for lowering the salt intake. The data summarized in the following table were collected.

Strategies	Province			
	Quebec	Ontario	Alberta	British Columbia
Strategy 1	2.5	2.7	2.9	2.4
Strategy 2	2.6	2.7	2.8	2.2
Strategy 3	2.3	2.5	2.7	2.3
Strategy 4	2.6	2.4	2.9	2.5
Strategy 5	2.9	2.6	2.6	2.6
Strategy 6	2.2	2.3	2.4	1.9

(i) What is the appropriate experimental design?

(ii) What are the blocks?

(iii) What are the treatments?

(iv) Is blocking necessary in this problem? Justify your answer. Let $\alpha = 0.05$.

(v) Perform an analysis of variance treating the provinces as blocks. Use $\alpha = 0.05$.

(vi) Why is it necessary to treat the provinces as blocks?

(vii) Use the *p*-value approach to determine whether there is a significant difference among the five strategies in reducing added unnecessary salt. What does this tell you about the appropriateness of the experimental design?

(viii) Use the *p*-value approach to determine whether there is a significant difference among the provinces' amount of salt intake at the 0.05 level of significance.

(ix) Find a 95% confidence interval for the difference in means for Strategies 1 and 6.

(x) Does it appear that the use of a randomized block design for this experiment was justified? Explain.

c. A researcher suspects that the effectiveness of the different strategies for lowering the amount of salt intake is affected by the education level of the subjects. Therefore, she used a 6×4 factorial design with three replicates in each cell to examine her conjecture. The data is summarized in the following table:

Strategies	Education level			
	Less Than High School	High School	University/ College	Postgraduate
Strategy 1	2.5	2.7	2.9	2.4
	2.7	2.8	2.1	2.3
	3.2	2.9	2.5	2.5
Strategy 2	2.6	2.7	2.8	2.2
	2.3	2.5	2.7	2.3
	2.3	2.4	2.3	2.4
Strategy 3	2.9	2.7	2.4	2.5
	3.1	2.9	2.6	2.2
	3.0	2.4	2.3	2.1
Strategy 4	2.6	2.4	2.9	2.5
	2.9	2.5	2.3	2.1
	2.8	2.7	2.4	2.0
Strategy 5	3.0	2.5	2.2	2.4
	2.9	2.4	2.3	1.9
	2.6	2.7	2.6	2.5
Strategy 6	2.9	2.5	2.2	1.9
	3.2	2.3	2.1	1.7
	3.0	2.0	2.0	1.4

(i) Identify the experimental design.

(ii) Test at the 5% significance level whether there are differences among the means of each of the six strategies.

(iii) Do the data provide sufficient evidence to indicate an interaction between strategy and education level? Test using $\alpha = 0.05$.

(iv) What are the practical implications of the test results on the previous question?

(v) Test at 10% significance level to determine if differences exist among the means of the education levels of the subjects.

(vi) Find a 95% confidence interval for the difference in mean between high school and postgraduate education levels if the subjects were using Strategy 6.

Jeff Whyte/Shutterstock.com

12

Linear Regression and Correlation

GENERAL OBJECTIVES

In this chapter, we consider the situation in which the mean value of a random variable *y* is related to another variable *x*. By measuring both *y* and *x* for each experimental unit, thereby generating bivariate data, you can use the information provided by *x* to estimate the average value of *y* and to predict values of *y* for preassigned values of *x*.

CHAPTER INDEX

NEED TO KNOW

How to Ensure That Calculations Are Correct

Are Foreign Companies "Buying Up the Canadian Economy"?

Canada's for sale! We're losing our sovereignty! Dire statements such as these frequently appear in the media when news breaks about a well-known Canadian company being purchased by a foreign competitor, with the implication that there will soon be no Canadian-owned companies left. In the case study at the end of this chapter, we explore the relationship between the percentage of foreign ownership and year, using a simple linear regression analysis.

INTRODUCTION

High-school students, their parents, and university administrations are concerned about the academic achievements of students after they have enrolled in a university. Can you estimate or predict a student's first-year university grade point average (GPA) at the end of Grade 12 before the student enrolls in university? At first glance this might seem like a difficult problem. However, you would expect highly motivated students who have graduated with a high class rank from a high school with superior academic standards to achieve a high GPA at the end of first year. On the other hand, students who lack motivation or who have achieved only moderate success in high school are not expected to do so well.

- Rank in high-school class
- High school's overall rating
- High school GPA

This problem is of a fairly general nature. You are interested in a random variable Y (GPA) that is related to a number of independent variables. The objective is to create a *prediction equation* that expresses Y as a function of these independent variables. Then, if you can measure the independent variables, you can substitute these values into the prediction equation and obtain the prediction for Y—the student's university GPA in our example. But which variables should you use as predictors? How strong is their relationship to Y? How do you construct a good prediction equation for Y as a function of the selected predictor variables? We will answer these questions in the next two chapters.

In this chapter, we restrict our attention to the simple problem of predicting Y as a linear function of a single predictor variable X. This problem was originally addressed in Chapter 3 in the discussion of *bivariate data.* Remember that we used the equation of a straight line to describe the relationship between X and Y and we described the strength of the relationship using the correlation coefficient r. We rely on some of these results as we revisit the subject of linear regression and correlation.

A SIMPLE LINEAR PROBABILISTIC MODEL AND LEAST SQUARES

Consider the problem of trying to predict the value of a response y based on the value of an independent variable x. The best-fitting line of Chapter 3,

$$Y = a + bX$$

was based on a *sample* of n bivariate observations drawn from a larger *population* of measurements. The line that describes the relationship between Y and X in the *population* is similar to, but not the same as, the best-fitting line from the *sample.* How can you construct a **population model** to describe the relationship between a random variable Y and a related independent variable X?

You begin by assuming that the variable of interest, Y, is *linearly* related to an independent variable X. To describe the linear relationship, you can use the **deterministic model**

$$Y = \alpha + \beta X$$

where α is the y-intercept—the value of y when $x = 0$—and β is the slope of the line, defined as the change in y for a one-unit change in x, as shown in Figure 12.1. This model describes a deterministic relationship between the variable of interest Y, sometimes called the **response variable,** and the independent variable X, often called the **predictor variable.** That is, the linear equation determines an exact value of y when the value of x is given. Is this a realistic model for an experimental situation? Consider the following example.

FIGURE 12.1
The y-intercept and slope for a line

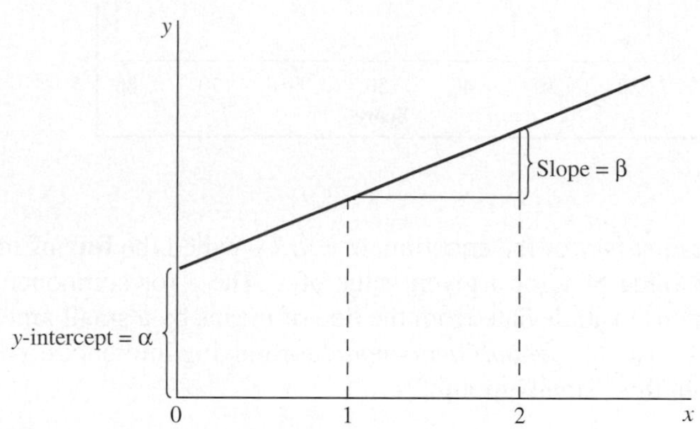

slope = change in y for a 1-unit increase in x
y-intercept = value of y when $x = 0$

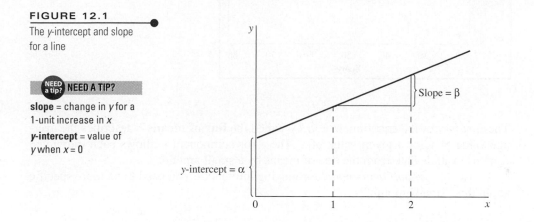

Building a Scatterplot

Table 12.1 displays the mathematics achievement test scores for a random sample of $n = 10$ Grade 12 students along with their final calculus grades. A bivariate plot of these scores and grades is given in Figure 12.2. You can use the **Building a Scatterplot** applet to refresh your memory as to how this plot is drawn. Notice that the points *do not lie exactly on a line* but rather seem to be deviations about an underlying line. A simple way to modify the deterministic model is to add a **random error component** to explain the deviations of the points about the line. A particular response Y is described using the **probabilistic model**

$$Y = \alpha + \beta X + \epsilon$$

TABLE 12.1 **Mathematics Achievement Test Scores and Final Calculus Grade for Grade 12 Students from High School**

Student	Mathematics Achievement Test Score	Final Calculus Grade
1	39	65
2	43	78
3	21	52
4	64	82
5	57	92
6	47	89
7	28	73
8	75	98
9	34	56
10	52	75

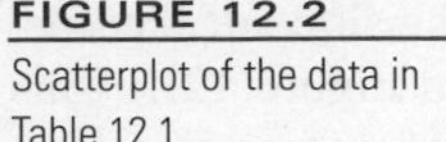

FIGURE 12.2
Scatterplot of the data in Table 12.1

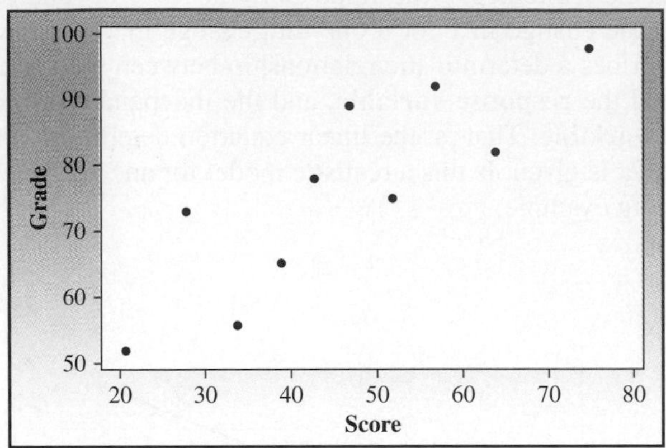

The first part of the equation, $\alpha + \beta X$—called the **line of means**—describes the average value of y for a given value of x. The error component ϵ allows each individual response y to deviate from the line of means by a small amount.

To use this *probabilistic model* for making inferences, you need to be more specific about this "small amount," ϵ.

ASSUMPTIONS ABOUT THE RANDOM ERROR ϵ

Assume that the values of ϵ satisfy these conditions:

- Are independent in the probabilistic sense
- Have a mean of 0 and a common variance equal to σ^2
- Have a normal probability distribution

These assumptions about the random error ϵ are shown in Figure 12.3 for three fixed values of x—say, x_1, x_2, and x_3. Notice the similarity between these assumptions and the assumptions necessary for the tests in Chapters 10 and 11. We will revisit these assumptions later in this chapter and provide some diagnostic tools for you to use in checking their validity.

FIGURE 12.3
Linear probabilistic model

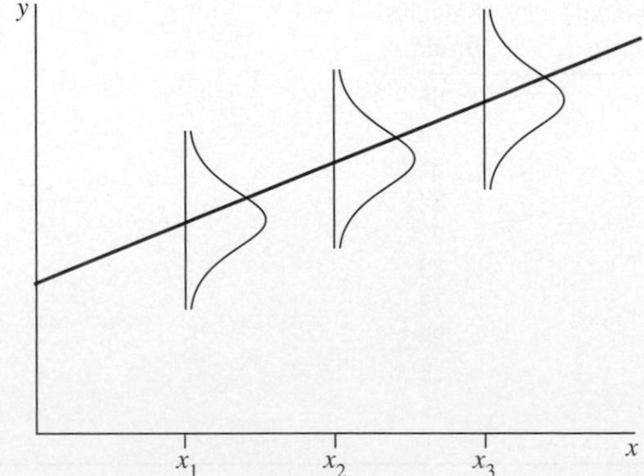

Remember that this model is created for a population of measurements that is generally unknown to you. However, you can use sample information to estimate the values of α and β, which are the coefficients of the line of means, $E(Y) = \alpha + \beta X$. These estimates are used to form the best-fitting line for a given set of data, called the **least squares line** or **regression line.** We review how to calculate the intercept and the slope of this line in the next section.

The Method of Least Squares

The statistical procedure for finding the best-fitting line for a set of bivariate data does mathematically what you do visually when you move a ruler until you think you have minimized the vertical distances, or deviations, from the ruler to a set of points. The formula for the best-fitting line is

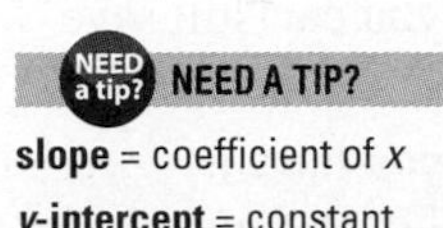

slope = coefficient of x
***y*-intercept** = constant term

$$\hat{y} = a + bx$$

where a and b are the estimates of the intercept and slope parameters α and β, respectively. The fitted line for the data in Table 12.1 is shown in the **Method of Least Squares** applet, Figure 12.4. The blue vertical lines in Figure 12.4 drawn from the prediction line to each point (x_i, y_i) represent the deviations of the points from the line.

FIGURE 12.4
Method of Least Squares applet

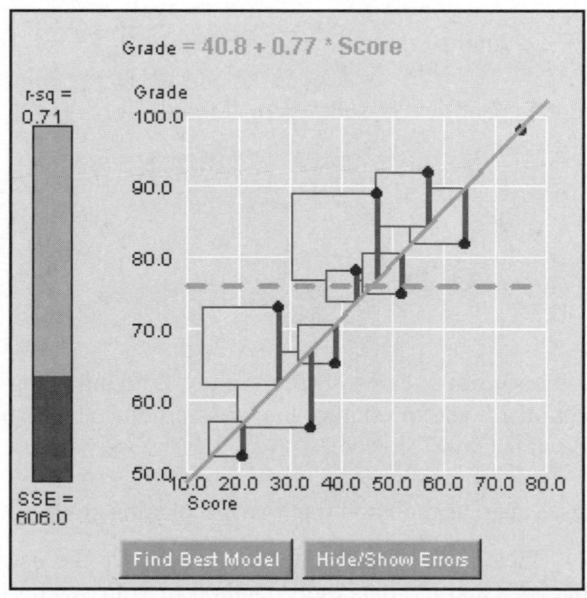

Method of Least Squares

To minimize the distances from the points to the fitted line, you can use the **principle of least squares.**

PRINCIPLE OF LEAST SQUARES

The line that minimizes the sum of squares of the deviations of the observed values of y from those predicted is the **best-fitting line.** The sum of squared deviations is commonly called the **sum of squares for error** (SSE) and is defined as

$$\text{SSE} = \Sigma(y_i - \hat{y}_i)^2 = \Sigma(y_i - a - bx_i)^2$$

Look at the regression line and the data points in Figure 12.4. SSE is the sum of the squared distances represented by the area of the light orange squares in Figure 12.4.

Finding the values of a and b, the estimates of α and β, uses differential calculus, which is beyond the scope of this text. Rather than derive their values, we will simply present formulas for calculating the values of a and b—called the **least-squares estimators** of α and β. We will use notation that is based on the **sums of squares** for the variables in the regression problem, which are similar in form to the sums of squares used in Chapter 11. These formulas look different from the formulas presented in Chapter 3, but they are in fact algebraically identical!

You should use the data entry method for your scientific calculator to enter the sample data.

- If your calculator has only a one-variable statistics function, you can still save some time in finding the necessary sums and sums of squares.
- If your calculator has a two-variable statistics function, or if you have a graphing calculator, the calculator will automatically store all of the sums and sums of squares as well as the values of a, b, and the correlation coefficient r.
- Make sure you consult your calculator manual to find the easiest way to obtain the least squares estimators.

LEAST-SQUARES ESTIMATORS OF α AND β

$$b = \frac{S_{xy}}{S_{xx}} \quad \text{and} \quad a = \bar{y} - b\bar{x}$$

where the quantities S_{xy} and S_{xx} are defined as

$$S_{xy} = \Sigma(x_i - \bar{x})(yi - \bar{y}) = \Sigma x_i y_i - \frac{(\Sigma x_i)(\Sigma y_i)}{n}$$

and

$$S_{xx} = \Sigma(x_i - \bar{x})^2 = \Sigma x_i^2 - \frac{(\Sigma x_i)^2}{n}$$

Notice that the sum of squares of the x-values is found using the computing formula given in Section 2.3 and the sum of the cross-products is the numerator of the *covariance* defined in Section 3.4.

EXAMPLE 12.1 Find the least-squares prediction line for the calculus grade data in Table 12.1.

SOLUTION Table 12.2 provides the calculations for the data in Table 12.1. Use the data in Table 12.2 and the data entry method in your scientific calculator to find the following sums of squares:

$$S_{xx} = \Sigma x_i^2 - \frac{(\Sigma x_i)^2}{n} = 23{,}634 - \frac{(460)^2}{10} = 2474$$

$$S_{xy} = \Sigma x_i y_i - \frac{(\Sigma x_i)(\Sigma y_i)}{n} = 36{,}854 - \frac{(460)(760)}{10} = 1894$$

$$\bar{y} = \frac{\Sigma y_i}{n} = \frac{760}{10} = 76 \qquad \bar{x} = \frac{\Sigma x_i}{n} = \frac{460}{10} = 46$$

TABLE 12.2 • Calculations for the Data in Table 12.1

	y_i	x_i	x_i^2	x_iy_i	y_i^2
	65	39	1521	2535	4225
	78	43	1849	3354	6084
	52	21	441	1092	2704
	82	64	4096	5248	6724
	92	57	3249	5244	8464
	89	47	2209	4183	7921
	73	28	784	2044	5329
	98	75	5625	7350	9604
	56	34	1156	1904	3136
	75	52	2704	3900	5625
Sum	760	460	23,634	36,854	59,816

Then

$$b = \frac{S_{xy}}{S_{xx}} = \frac{1894}{2474} = 0.76556 \quad \text{and} \quad a = \bar{y} - b\bar{x} = 76 - (0.76556)(46) = 40.78424$$

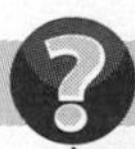
NEED A TIP?

You can predict y for a given value of x by substituting x into the equation to find $\hat{y}$.

The least-squares regression line is then

$$\hat{y} = a + bx = 40.78424 + 0.76556x$$

The graph of this line is shown in Figure 12.4. It can now be used to predict y for a given value of x—either by referring to Figure 12.4 or by substituting the proper value of x into the equation. For example, if a student scored $x = 50$ on the achievement test, the student's predicted calculus grade is (using full decimal accuracy)

$$\hat{y} = a + b(50) = 40.78424 + (0.76556)(50) = 79.06$$

NEED TO KNOW

How to Ensure That Calculations Are Correct

- Be careful of rounding errors. Carry at least six significant figures, and round off only in reporting the end result.
- Use a scientific or graphing calculator to do all the work for you. Most of these calculators will calculate the values for a and b if you enter the data properly.
- Use a computer software program if you have access to one.
- Always plot the data and graph the line. If the line does not fit through the points, you have probably made a mistake!

12.3 AN ANALYSIS OF VARIANCE FOR LINEAR REGRESSION

In Chapter 11, you used the analysis of variance procedures to divide the total variation in the experiment into portions attributed to various factors of interest to the experimenter. In a regression analysis, the response y is related to the independent variable x. Hence, the total variation in the response variable y, given by

$$\text{Total SS} = S_{yy} = \Sigma(y_i - \bar{y})^2 = \Sigma y_i^2 - \frac{(\Sigma y_i)^2}{n}$$

Alternatively, it can be shown that Total SS can be is divided into two portions:

- SSR (sum of squares for regression) measures the amount of variation explained by using the regression line with one independent variable x
- SSE (sum of squares for error) measures the "residual" variation in the data that is not explained by the independent variable x

so that

$$\text{Total SS} = \text{SSR} + \text{SSE}$$

For a particular value of the response y_i, you can visualize this breakdown in the variation using the vertical distances illustrated in Figure 12.5. You can see that SSR is the sum of the squared deviations of the differences between the estimated response without using $x(\bar{y})$ and the estimated response using x (the regression line, $\hat{y}$); SSE is the sum of the squared differences between the regression line ($\hat{y}$) and the point y.

FIGURE 12.5

Deviations from the fitted line

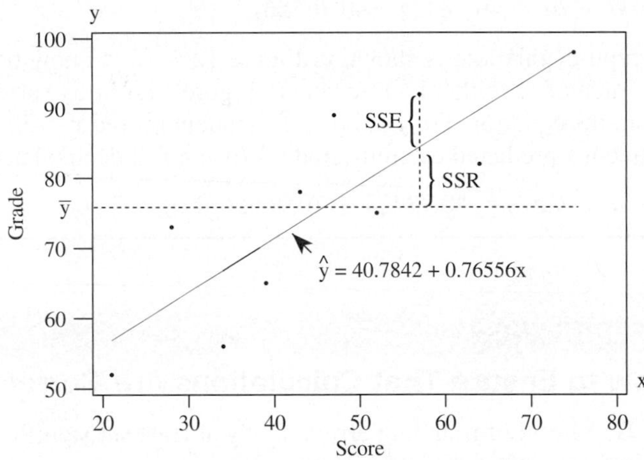

It is not too hard to show algebraically that

$$\text{SSR} = \Sigma(\hat{y}_i - \bar{y}_i)^2 = \Sigma(a + bx_i - \bar{y})^2 = \Sigma(\bar{y} - b\bar{x} + bx_i - \bar{y})^2 = b^2\Sigma(x_i - \bar{x})^2$$

$$= \left(\frac{S_{xy}}{S_{xx}}\right)^2 S_{xx} = \frac{(S_{xy})^2}{S_{xx}}$$

Since Total SS = SSR + SSE, you can complete the partition by calculating

$$\text{SSE} = \text{Total SS} - \text{SSR} = S_{yy} - \frac{(S_{xy})^2}{S_{xx}}$$

Remember from Chapter 11 that each of the various sum of squares, when divided by the appropriate **degrees of freedom,** provides an estimate of the variation in the experiment.

These estimates are called **mean squares**—MS = SS/*df*—and are displayed in an ANOVA table.

In examining the degrees of freedom associated with each of these sums of squares, notice that the total degrees of freedom for n measurements is $(n - 1)$. Since estimating the regression line, $\hat{y} = a + bx_i = \bar{y} - b\bar{x} + bx_i$, involves estimating *one additional* parameter β, there is *one* degree of freedom associated with SSR, leaving $(n - 2)$ degrees of freedom with SSE.

As with all ANOVA tables we have discussed, the mean square for error,

$$\text{MSE} = s^2 = \frac{\text{SSE}}{n - 2}$$

is an unbiased estimator of the underlying variance σ^2. The analysis of variance table is shown in Table 12.3.

TABLE 12.3 Analysis of Variance for Linear Regression

Source	*df*	SS	MS
Regression	1	$\frac{(S_{xy})^2}{S_{xx}}$	MSR
Error	$n - 2$	$S_{yy} - \frac{(S_{xy})^2}{S_{xx}}$	MSE
Total	$n - 1$	S_{yy}	

For the data in Table 12.1, you can calculate

$$\text{Total SS} = S_{yy} = \Sigma y_i^2 - \frac{(\Sigma y_i)^2}{n} = 59{,}816 - \frac{(760)^2}{10} = 2056$$

$$\text{SSR} = \frac{(S_{xy})^2}{S_{xx}} = \frac{(1894)^2}{2474} = 1449.9741$$

so that

$$\text{SSE} = \text{Total SS} - \text{SSR} = 2056 - 1449.9741 = 606.0259$$

and

$$\text{MSE} = \frac{\text{SSE}}{n - 2} = \frac{606.0259}{8} = 75.7532$$

The analysis of variance table, part of the *linear regression output* generated by *MINITAB*, is the lower shaded section in the printout in Figure 12.6(a). The first two lines give the equation of the least-squares line, $\hat{y} = 40.8 + 0.766x$. The least-squares estimates a and b are given with greater accuracy in the column labelled "Coef." You can find instructions for generating this output in the "Technology Today" section at the end of this chapter.

FIGURE 12.6(a)
MINITAB output for the data in Table 12.1

Regression Analysis: y versus x

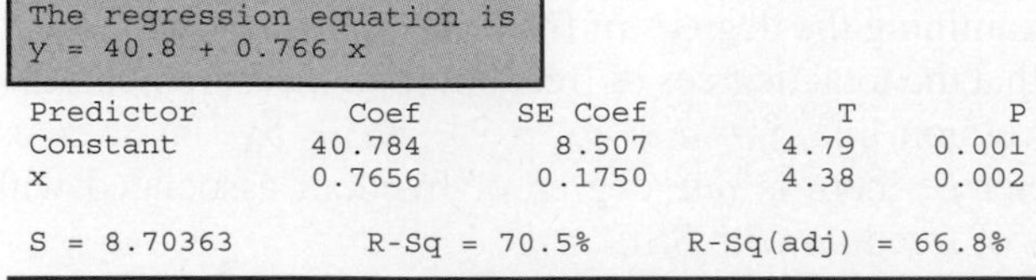

```
The regression equation is
y = 40.8 + 0.766 x

Predictor       Coef    SE Coef       T       P
Constant      40.784      8.507    4.79   0.001
x             0.7656     0.1750    4.38   0.002

S = 8.70363        R-Sq = 70.5%      R-Sq(adj) = 66.8%

Analysis of Variance
Source            DF        SS        MS       F       P
Regression         1    1450.0    1450.0   19.14   0.002
Residual Error     8     606.0      75.8
Total              9    2056.0
```

FIGURE 12.6(b)
Excel output for the data in Table 12.1

SUMMARY OUTPUT

Regression Statistics	
Multiple R	0.8398
R Square	0.7052
Adjusted R Square	0.6684
Standard Error	8.7036
Observations	10

ANOVA

	df	*SS*	*MS*	*F*	*Significance F*
Regression	1	1449.974	1449.974	19.141	0.002
Residual	8	606.026	75.753		
Total	9	2056			

	Coefficients	*Standard Error*	*t Stat*	*P-value*	*Lower 95%*	*Upper 95%*	*Lower 95.0%*
Intercept	40.784	8.507	4.794	0.001	21.167	60.401	21.167
Score	0.766	0.175	4.375	0.002	0.362	1.169	0.362

The computer outputs also give some information about the variation in the experiment. Each of the least-squares estimates, a and b, has an associated standard error, labelled "SE Coef" in Figure 12.6(a) and "Standard Error" in Figure 12.6(b). In the middle of the *MINITAB* output, you will find the best unbiased estimate of σ is $s = \sqrt{\text{MSE}} = \sqrt{75.7532} = 8.70363$—which measures the **residual error,** the unexplained or "leftover" variation in the experiment. This same measure is found in the top portion of the *Excel* output, labelled "Standard Error." It will not surprise you to know that the t and F statistics and their p-values found in the printouts are used to test statistical hypotheses. We explain these entries in the next section.

12.3 EXERCISES

BASIC TECHNIQUES

12.1 Graph the line corresponding to the equation $y = 2x + 1$ by graphing the points corresponding to $x = 0$, 1, and 2. Give the y-intercept and slope for the line.

12.2 Graph the line corresponding to the equation $y = -2x + 1$ by graphing the points corresponding to $x = 0$, 1, and 2. Give the y-intercept and slope for the line. How is this line related to the line $y = 2x + 1$ of Exercise 12.1?

12.3 Give the equation and graph for a line with y-intercept equal to 3 and slope equal to -1.

12.4 Give the equation and graph for a line with y-intercept equal to -3 and slope equal to 1.

12.5 What is the difference between deterministic and probabilistic mathematical models?

12.6 You are given five points with these coordinates:

x	−2	−1	0	1	2
y	1	1	3	5	5

a. Use the data entry method on your scientific or graphing calculator to enter the $n = 5$ observations. Find the sums of squares and cross-products, S_{xx}, S_{xy}, and S_{yy}.

b. Find the least-squares line for the data.

c. Plot the five points and graph the line in part b. Does the line appear to provide a good fit to the data points?

d. Construct the ANOVA table for the linear regression.

12.7 Six points have these coordinates:

x	1	2	3	4	5	6
y	5.6	4.6	4.5	3.7	3.2	2.7

a. Find the least-squares line for the data.

b. Plot the six points and graph the line. Does the line appear to provide a good fit to the data points?

c. Use the least-squares line to predict the value of y when $x = 3.5$.

d. Fill in the missing entries in the *MINITAB* analysis of variance table.

MINITAB ANOVA table for Exercise 12.7

```
Analysis of Variance

Source            DF        SS       MS
Regression         *       ***   5.4321
Residual Error     *    0.1429      ***
Total              *    5.5750
```

12.8 Six points have these coordinates:

x	1	2	3	4	5	6
y	9.7	6.5	6.4	4.1	2.1	1.0

a. Find the least-squares line for the data.

b. Plot the six points and graph the line. Does the line appear to provide a good fit to the data points?

c. Use the least-squares line to predict the value of y when $x = 3.5$.

d. Fill in the missing entries in the *Excel* analysis of variance table.

ANOVA

```
              df        SS        MS
Regression     *       ***   49.7286
Residual       *    1.7848       ***
Total          *   51.5133
```

APPLICATIONS

12.9 Professor Asimov Professor Isaac Asimov was one of the most prolific writers of all time. Prior to his death, he wrote nearly 500 books during a 40-year career. In fact, as his career progressed, he became even more productive in terms of the number of books written within a given period of time.[1] The data give the time in months required to write his books in increments of 100:

Number of books, x	100	200	300	400	490
Time in months, y	237	350	419	465	507

a. Assume that the number of books x and the time in months y are linearly related. Find the least-squares line relating y to x.

b. Plot the time as a function of the number of books written using a scatterplot, and graph the least-squares line on the same paper. Does it seem to provide a good fit to the data points?

c. Construct the ANOVA table for the linear regression.

Data set EX1210 **12.10 A Chemical Experiment** Using a chemical procedure called *differential pulse polarography*, a chemist measured the peak current generated (in microamperes) when a solution containing a given amount of nickel (in parts per billion) is added to a buffer:[2]

x = Ni (ppb)	y = Peak Current (mA)
19.1	0.095
38.2	0.174
57.3	0.256
76.2	0.348
95	0.429
114	0.500
131	0.580
150	0.651
170	0.722

a. Use the data entry method for your calculator to calculate the preliminary sums of squares and cross-products, S_{xx}, S_{yy}, and S_{xy}.

b. Calculate the least-squares regression line.

c. Plot the points and the fitted line. Does the assumption of a linear relationship appear to be reasonable?

d. Use the regression line to predict the peak current generated when a solution containing 100 ppb of nickel is added to the buffer.

e. Construct the ANOVA table for the linear regression.

Data set EX1211 **12.11 Sleep Deprivation** A study was conducted to determine the effects of sleep deprivation on people's ability to solve problems without sleep. A total of 10 subjects participated in the study, two at each of five sleep deprivation levels—8, 12, 16, 20, and 24 hours. After his or her specified sleep deprivation period, each subject was administered a set of simple addition problems, and the number of errors was recorded. These results were obtained:

Number of errors, y	8, 6	6, 10	8, 14
Number of hours without sleep, x	8	12	16

Number of errors, y	14, 12	16, 12
Number of hours without sleep, x	20	24

a. How many pairs of observations are in the experiment?

b. What are the total number of degrees of freedom?

c. Complete the *MINITAB* printout.

MINITAB output for Exercise 12.11

Regression Analysis: y versus x

```
The regression equation is
y = 3.00 + 0.475 x

Predictor      Coef      SE Coef          T        P
Constant      3.000        2.127       1.41    0.196
x               ***       0.1253       3.79    0.005

S = 2.24165     R-Sq = 64.2%      R-Sq(adj) = 59.8%

Analysis of Variance
Source           DF       SS        MS        F        P
Regression       **   72.200    72.200    14.37    0.005
Residual Error   **      ***     5.025
Total            **      ***
```

d. What is the least-squares prediction equation?

e. Use the prediction equation to predict the number of errors for a person who has not slept for 10 hours.

12.12 Sleep Deprivation II

Refer to the data given in the sleep deprivation experiment in Exercise 12.11. Answer the questions posed in parts a, b, d, and e of that exercise by completing the following *Excel* printout:

ANOVA

```
             df    SS     MS        F   Significance F
Regression   **  72.2   72.2 14.36816         0.005308
Residual     **   ***  5.025
Total        **   ***
```

	Coefficients	Standard Error	t Stat	P-value
Intercept	3	2.126617	1.410691	0.196016
X	0.475	0.125312	3.790535	0.005308

Data set EX1213 **12.13 Global Warming?** The following table shows annual mean global surface temperature anomaly for the period 1972 to 2001 provided by the Global Historical Climate Network (GHCN):[3]

Year	Temperature (°C) Anomaly	Year	Temperature (°C) Anomaly
1972	−0.334	1987	0.337
1973	0.113	1988	0.418
1974	−0.260	1989	0.328
1975	−0.050	1990	0.638
1976	−0.357	1991	0.455
1977	0.127	1992	0.234
1978	−0.099	1993	0.133
1979	−0.006	1994	0.512
1980	0.077	1995	0.513
1981	0.390	1996	0.142
1982	−0.047	1997	0.437
1983	0.304	1998	0.852
1984	−0.063	1999	0.587
1985	−0.093	2000	0.445
1986	0.180	2001	0.600

Source: National Climate Data Center

a. Which of the two variables is the independent variable and which is the dependent variable? Explain your choice.

b. Use a scatterplot to plot the data. Is the assumption of a linear relationship between year and temperature anomaly reasonable?

c. Assuming that year and temperature anomaly are linearly related, calculate the least-squares regression line.

d. Plot the line on the scatterplot in part b. Does the line fit through the data points?

Data set EX1214 **12.14 How Long Is It?** How good are you at estimating? To test a subject's ability to estimate sizes, he was shown 10 different objects and asked to estimate their length or diameter. The object was then measured, and the results were recorded in the table below:

Object	Estimated (cm)	Actual (cm)
Pencil	17.78	15.24
Dinner plate	24.13	26.04
Book 1	19.05	17.15
Cellphone	10.16	10.8
Photograph	36.83	40.0
Toy	9.53	12.7
Belt	106.68	105.41
Clothespin	6.99	9.53
Book 2	25.4	23.5
Calculator	8.89	12.07

a. Find the least-squares regression line for predicting the actual measurement as a function of the estimated measurement.

b. Plot the points and the fitted line. Does the assumption of a linear relationship appear to be reasonable?

Data set EX1215 **12.15 Test Interviews** Of two personnel evaluation techniques available, the first requires a two-hour test interview, while the second can be completed in less than an hour. The scores for each of the 15 individuals who took both tests are given in the next table:

Applicant	Test 1 (x)	Test 2 (y)
1	75	38
2	89	56
3	60	35
4	71	45
5	92	59
6	105	70
7	55	31
8	87	52
9	73	48
10	77	41
11	84	51
12	91	58
13	75	45
14	82	49
15	76	47

a. Construct a scatterplot for the data. Does the assumption of linearity appear to be reasonable?

b. Find the least-squares line for the data.

c. Use the regression line to predict the score on the second test for an applicant who scored 85 on Test 1.

12.16 Test Interviews, continued Refer to Exercise 12.15. Construct the ANOVA table for the linear regression relating y, the score on Test 2, to x, the score on Test 1.

Data set EX1217 **12.17 Armspan and Height** Leonardo da Vinci (1452–1519) drew a sketch of a man, indicating that a person's armspan (measuring across the back with your arms outstretched to make a "t") is roughly equal to the person's height. To test this claim, we measured eight people with the following results:

Person	1	2	3	4	5	6	7	8
Armspan (cm)	172.7	158.1	165.1	176.5	172.7	175.3	157.5	153.0
Height (cm)	175.3	157.5	165.1	177.8	170.2	170.2	160.0	157.5

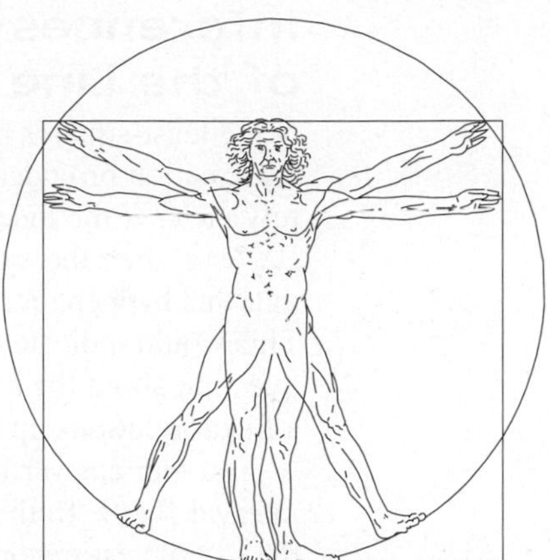

a. Draw a scatterplot for armspan and height. Use the same scale on both the horizontal and vertical axes. Describe the relationship between the two variables.

b. If da Vinci is correct, and a person's armspan is roughly the same as the person's height, what should the slope of the regression line be?

c. Calculate the regression line for predicting height based on a person's armspan. Does the value of the slope b confirm your conclusions in part b?

d. If a person has an armspan of 157.5 cm, what would you predict the person's height to be?

Data set EX1218 **12.18 Strawberries** The following data were obtained in an experiment relating the dependent variable, y (texture of strawberries), with x (coded storage temperature).

x	−2	−2	0	2	2
y	4.0	3.5	2.0	0.5	0.0

a. Find the least-squares line for the data.

b. Plot the data points and graph the least-squares line as a check on your calculations.

c. Construct the ANOVA table.

12.4 TESTING THE USEFULNESS OF THE LINEAR REGRESSION MODEL

In considering linear regression, you may ask two questions:

- Is the independent variable X useful in predicting the response variable Y?
- If so, how well does it work?

This section examines several statistical tests and measures that will help you reach some answers. Once you have determined that the model is working, you can then use the model for predicting the response y for a given value of x.

Inferences Concerning β, the Slope of the Line of Means

Is the least-squares regression line useful? That is, is the regression equation that uses information provided by x substantially better than the simple predictor $\bar{y}$ that does not rely on x? If the independent variable x is *not useful* in the population model $Y = \alpha + \beta X + \epsilon$, then the value of y does not change for different values of x. The only way that this happens for all values of x is when the slope β of the line of means equals 0. This would indicate that the relationship between y and x is not linear, so that the initial question about the usefulness of the independent variable X can be restated as: Is there a linear relationship between X and Y?

You can answer this question by using either a test of hypothesis or a confidence interval for β. Both of these procedures are based on the sampling distribution of b, the sample estimator of the slope β. It can be shown that, if the assumptions about the random error ϵ are valid, then the estimator b has a normal distribution in repeated sampling with mean

$$E(b) = \beta$$

and standard error given by

$$\text{SE} = \sqrt{\frac{\sigma^2}{S_{xx}}}$$

where σ^2 is the variance of the random error ϵ. Since the value of σ^2 is estimated with $s^2 = \text{MSE}$, you can base inferences on the statistic given by

$$t = \frac{b - \beta}{\sqrt{\text{MSE}/S_{xx}}}$$

which has a t distribution with $df = (n - 2)$, the degrees of freedom associated with MSE.

TEST OF HYPOTHESIS CONCERNING THE SLOPE OF A LINE

1. Null hypothesis: $H_0 : \beta = 0$
2. Alternative hypothesis:

One-Tailed Test	**Two-Tailed Test**
$H_a : \beta > 0$ (or $\beta < 0$)	$H_a : \beta \neq 0$

3. Test statistic: $t = \dfrac{b - 0}{\sqrt{\text{MSE}/S_{xx}}}$

   When the assumptions given in Section 12.2 are satisfied, the test statistic will have a Student's t distribution with $(n - 2)$ degrees of freedom.

4. Rejection region: Reject H_0 when

One-Tailed Test	**Two-Tailed Test**
$t > t_\alpha$ (or $t < -t_\alpha$ when the alternative hypothesis is $H_a : \beta < 0$)	$t > t_{\alpha/2}$ or $t < -t_{\alpha/2}$

or when p-value $< \alpha$

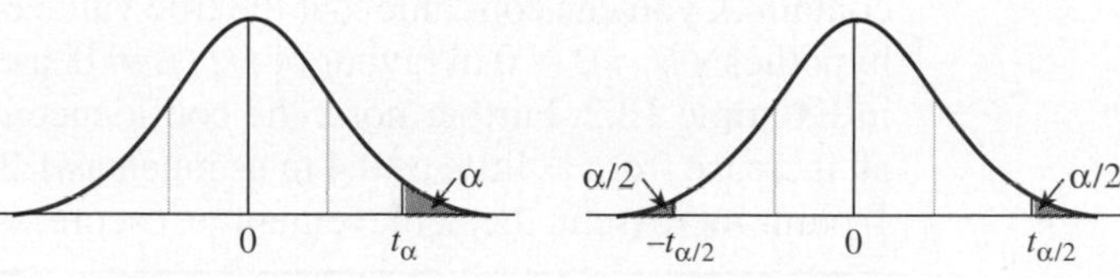

The values of t_α and $t_{\alpha/2}$ can be found using Table 4 in Appendix I. Use the values of t corresponding to $(n - 2)$ degrees of freedom.

EXAMPLE 12.2

Determine whether there is a significant linear relationship between the calculus grades and test scores listed in Table 12.1. Test at the 5% level of significance.

SOLUTION The hypotheses to be tested are

$$H_0 : \beta = 0 \qquad \text{versus} \qquad H_a : \beta \neq 0$$

and the observed value of the test statistic is calculated as

$$t = \frac{b - 0}{\sqrt{\text{MSE}/S_{xx}}} = \frac{0.7656 - 0}{\sqrt{75.7532/2474}} = 4.38$$

with $(n - 2) = 8$ degrees of freedom. With $\alpha = 0.05$, you can reject H_0 when $t > 2.306$ or $t < -2.306$. Since the observed value of the test statistic falls into the rejection region, H_0 is rejected and you can conclude that there is a significant linear relationship between the calculus grades and the test scores for the population of Grade 12 students.

Another way to make inferences about the value of β is to construct a confidence interval for β and examine the range of possible values for β.

A (1 – α)100% CONFIDENCE INTERVAL FOR β

$$b \pm t_{\alpha/2}(\text{SE})$$

where $t_{\alpha/2}$ is based on $(n - 2)$ degrees of freedom and

$$\text{SE} = \sqrt{\frac{s^2}{S_{xx}}} = \sqrt{\frac{\text{MSE}}{S_{xx}}}$$

EXAMPLE 12.3

Find a 95% confidence interval estimate of the slope β for the calculus grade data in Table 12.1.

SOLUTION Substituting previously calculated values into

$$b \pm t_{0.025}\sqrt{\frac{\text{MSE}}{S_{xx}}}$$

you have

$$0.766 \pm 2.306\sqrt{\frac{75.7532}{2474}}$$

$$0.766 \pm 0.404$$

The resulting 95% confidence interval is 0.362 to 1.17. Since the interval does not contain 0, you can conclude that the true value of β is not 0, and you can reject the null hypothesis $H_0 : \beta = 0$ in favour of $H_a : \beta \neq 0$, a conclusion that agrees with the findings in Example 12.2. Furthermore, the confidence interval estimate indicates that there is an increase from as little as 0.4 to as much as 1.2 points in a calculus test score for each 1-point increase in the achievement test score.

If you are using computer software to perform the regression analysis, you will find the t statistic and its p-value on the printout. In the second section of the *MINITAB* output in Figure 12.7(a), you will find the least-squares estimate b of the slope in the line marked "*x*," along with its standard error "SE Coef," the calculated value of the test statistic "T" used for testing the hypothesis $H_0 : \beta = 0$ and its p-value "P." You will find the same information in the last line of the *Excel* output in Figure 12.7(b), along with the upper and lower confidence limits of a 95% confidence interval for β. The F test for significant regression, $H_0 : \beta = 0$, has a p-value of $P = 0.002$, and the null hypothesis is rejected, as in Example 12.2. There is a significant linear relationship between x and y.

FIGURE 12.7(a)
MINITAB output for the calculus grade data

NEED A TIP?

Look for the standard error of b in the column marked "SE Coef" on the *MINITAB* output and "Standard Error" on the *Excel* output.

Regression Analysis: y versus x

```
The regression equation is
y = 40.8 + 0.766 x

Predictor      Coef     SE Coef       T        P
Constant     40.784       8.507    4.79    0.001
x            0.7656      0.1750    4.38    0.002

S = 8.70363   R-Sq = 70.5%   R-Sq(adj) = 66.8%

Analysis of Variance
Source          DF       SS       MS       F       P
Regression       1   1450.0   1450.0   19.14   0.002
Residual Error   8    606.0     75.8
Total            9   2056.0
```

FIGURE 12.7(b)
Excel output for the calculus grade data

SUMMARY OUTPUT

Regression Statistics	
Multiple R	0.8398
R Square	0.7052
Adjusted R Square	0.6684
Standard Error	8.7036
Observations	10

ANOVA

	df	*SS*	*MS*	*F*	*Significance F*
Regression	1	1449.974	1449.974	19.141	0.002
Residual	8	606.026	75.753		
Total	9	2056			

	Coefficients	*Standard Error*	*t Stat*	*P-value*	*Lower 95%*	*Upper 95%*
Intercept	40.784	8.507	4.794	0.001	21.167	60.401
x	0.766	0.175	4.375	0.002	0.362	1.169

ANOVA F tests are always one-tailed (upper-tail).

The Analysis of Variance F Test

The analysis of variance portion of the printout in Figure 12.7(a) shows an F statistic given by

$$F = \frac{\text{MSR}}{\text{MSE}} = 19.14$$

with 1 numerator degree of freedom and $(n - 2) = 8$ denominator degrees of freedom. This is an *equivalent test statistic* that can also be used for testing the hypothesis $H_0 : \beta = 0$. Notice that, within rounding error, the value of F is equal to t^2 with the identical p-value. In this case, if you use five-decimal-place accuracy prior to rounding, you find that $t^2 = (0.76556/0.17498)^2 = (4.37513)^2 = 19.14175 \approx 19.14 = F$ as given in the printout. This is no accident and results from the fact that the square of a t statistic with *df* degrees of freedom has the same distribution as an F statistic with 1 numerator and *df* denominator degrees of freedom. The F test is a more general test of the usefulness of the model and can be used when the model has more than one independent variable.

Measuring the Strength of the Relationship: The Coefficient of Determination

How well does the regression model fit? To answer this question, you can use a measure related to the *correlation coefficient r*, introduced in Chapter 3. Remember that

$$r = \frac{s_{xy}}{s_x s_y} = \frac{S_{xy}}{\sqrt{S_{xx}S_{yy}}} \quad \text{for} \quad -1 \le r \le 1$$

where s_{xy}, s_x, and s_y were defined in Chapter 3 and the various sums of squares were defined in Section 12.3.

The sum of squares for regression, SSR, in the analysis of variance measures the portion of the total variation, Total SS $= S_{yy}$, that can be explained by the regression of y on x. The remaining portion, SSE, is the "unexplained" variation attributed to random error. One way to measure the strength of the relationship between the response variable y and the predictor variable x is to calculate the **coefficient of determination**—the proportion of the total variation y that is explained by the linear regression of y on x. For the calculus grade data, this proportion is equal to

NEED A TIP?

On computer printouts, r^2 is often given as a **percentage** rather than a proportion.

$$\frac{\text{SSR}}{\text{Total SS}} = \frac{1450}{2056} = 0.705 \quad \text{or} \quad 70.5\%$$

Since Total SS $= S_{yy}$ and SSR $= \dfrac{(S_{xy})^2}{S_{xx}}$, you can write

$$\frac{\text{SSR}}{\text{Total SS}} = \frac{(S_{xy})^2}{S_{xx}S_{yy}} = \left(\frac{S_{xy}}{\sqrt{S_{xx}S_{yy}}}\right)^2 = r^2$$

Therefore, the coefficient of determination, which was calculated as SSR/Total SS, is simply the square of the correlation coefficient r. It is the entry labelled "R-Sq" in Figure 12.7(a) and "R Square" in Figure 12.7(b).

Remember that the analysis of variance table isolates the variation due to regression (SSR) from the total variation in the experiment. Doing so reduces the amount of *random variation* in the experiment, now measured by SSE rather than Total SS. In this context, the **coefficient of determination r^2** can be defined as follows:

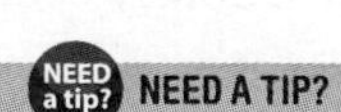

NEED A TIP?

r^2 is called "R-Sq" on the *MINITAB* printout and "R Square" on the *Excel* printout.

DEFINITION The **coefficient of determination r^2** can be interpreted as the percent reduction in the total variation in the experiment obtained by using the regression line $\hat{y} = a + bx$, instead of ignoring x and using the sample mean $\bar{y}$ to predict the response variable y.

For the calculus grade data, a reduction of $r^2 = 0.705$ or 70.5% is substantial. The regression model is working very well!

Interpreting the Results of a Significant Regression

Once you have performed the t or F test to determine the significance of the linear regression, you must interpret your results carefully. The slope β of the line of means is estimated based on data from only a particular region of observation. Even if you do not reject the null hypothesis that the slope of the line equals 0, it does not necessarily mean that y and x are unrelated. It may be that you have committed a Type II error—falsely declaring that the slope is 0 and that x and y are unrelated.

Fitting the Wrong Model

It may happen that y and x are perfectly related in a non-linear way, as shown in Figure 12.8. Here are three possibilities:

FIGURE 12.8

Curvilinear relationship

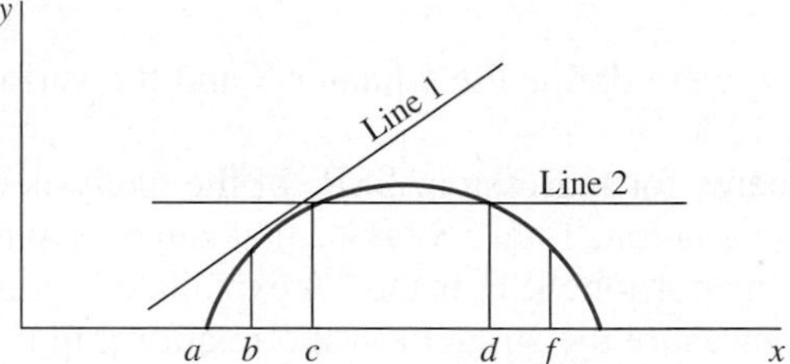

- If observations were taken only within the interval $b < x < c$, the relationship would appear to be linear with a positive slope.
- If observations were taken only within the interval $d < x < f$, the relationship would appear to be linear with a negative slope.
- If the observations were taken over the interval $c < x < d$, the line would be fitted with a slope close to 0, indicating no linear relationship between y and x.

For the example shown in Figure 12.8, no straight line accurately describes the true relationship between x and y, which is really a *curvilinear relationship.* In this case, we have chosen the *wrong model* to describe the relationship. Sometimes this type of mistake can be detected using residual plots, the subject of Section 12.6.

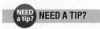

NEED A TIP?
It is dangerous to try to predict values of y outside of the range of the fitted data.

Extrapolation

One serious problem is to apply the results of a linear regression analysis to values of x that are *not included* within the range of the fitted data. This is called **extrapolation** and can lead to serious errors in prediction, as shown for line 1 in Figure 12.8. Prediction results would be good over the interval $b < x < c$ but would seriously overestimate the values of y for $x > c$.

Causality

When there is a significant regression of y and x, it is tempting to conclude that x *causes* y. However, it is possible that one or more unknown variables that you have not even measured and that are not included in the analysis may be causing the observed relationship. In general, the statistician reports the results of an analysis but leaves conclusions concerning causality to scientists and investigators who are experts in these areas. These experts are better prepared to make such decisions.

12.4 EXERCISES

BASIC TECHNIQUES

12.19 Refer to Exercise 12.6. The data are reproduced below:

x	−2	−1	0	1	2
y	1	1	3	5	5

a. Do the data present sufficient evidence to indicate that y and x are linearly related? Test the hypothesis that $\beta = 0$ at the 5% level of significance.

b. Use the ANOVA table to calculate $F = \text{MSR/MSE}$. Verify that the square of the t statistic used in part a is equal to F.

c. Compare the two-tailed critical value for the t test in part a with the critical value for F with $\alpha = 0.05$. What is the relationship between the critical values?

12.20 Refer to Exercise 12.19. Find a 95% confidence interval for the slope of the line. What does the phrase "95% confident" mean?

12.21 Refer to Exercise 12.7. The data, along with the *MINITAB* analysis of variance table, are reproduced below:

x	1	2	3	4	5	6
y	5.6	4.6	4.5	3.7	3.2	2.7

MINITAB ANOVA table for Exercise 12.21

Regression Analysis: y versus x

```
Analysis of Variance
Source          DF      SS       MS        F       P
Regression       1   5.4321   5.4321   152.10   0.000
Residual Error   4   0.1429   0.0357
Total            5   5.5750
```

a. Do the data provide sufficient evidence to indicate that y and x are linearly related? Use the information in the *MINITAB* printout to answer this question at the 1% level of significance.

b. Calculate the coefficient of determination r^2. What information does this value give about the usefulness of the linear model?

12.22 Refer to Exercise 12.8. The data, along with the *Excel* analysis of variance table, are reproduced below:

x	1	2	3	4	5	6
y	9.7	6.5	6.4	4.1	2.1	1.0

Excel ANOVA table for Exercise 12.22

ANOVA

	df	*SS*	*MS*	*F*	*Significance F*
Regression	1	49.72857	49.72857	111.45	0.000
Residual	4	1.78476	0.4462		
Total	5	51.51333			

a. Do the data provide sufficient evidence to indicate that y and x are linearly related? Use the information in the printout to answer this question at the 5% level of significance.

b. Calculate the coefficient of determination r^2. What information does this value give about the usefulness of the linear model?

APPLICATIONS

Data set EX1223 **12.23 Air Pollution** An experiment was designed to compare several different types of air pollution monitors.[4] The monitor was set up, and then exposed to different concentrations of ozone, ranging between 15 and 230 parts per million (ppm) for periods of 8 to 72 hours. Filters on the monitor were then analyzed, and the amount (in micrograms) of sodium nitrate (NO_3) recorded by the monitor was measured. The results for one type of monitor are given in the table:

Ozone, x (ppm/h)	0.8	1.3	1.7	2.2	2.7	2.9
NO_3, y (μg)	2.44	5.21	6.07	8.98	10.82	12.16

a. Find the least-squares regression line relating the monitor's response to the ozone concentration.

b. Do the data provide sufficient evidence to indicate that there is a linear relationship between the ozone concentration and the amount of sodium nitrate detected?

c. Calculate r^2. What does this value tell you about the effectiveness of the linear regression analysis?

Data set **12.24 Cricket (the insect, not the game)** EX1224 Crickets make their chirping sounds by rapidly sliding one wing over the other. The faster they move their wings, the higher the chirping sound that is produced. Scientists have noticed that crickets move their wings faster in warm temperatures than in cold temperatures. Therefore, by listening to the pitch of the chirp of crickets, it is possible to tell the temperature of the air. In a 1948 book called *The Song of Insects,* George W. Pierce, a Harvard physics professor, presented real data relating the number of chirps per second for striped ground crickets to the temperature in degrees Fahrenheit. Apparently the number of chirps represents some kind of average since it is given to the nearest tenth.

The table below gives the recorded pitch (in vibrations per second) of a cricket chirping recorded at 15 different temperatures.[5] The variables are chirps (the number of chirps per second) and temperature (the temperature in degrees Fahrenheit).

Chirps per Second	Temperature (°F)
20.0	88.6
16.0	71.6
19.8	93.3
18.4	84.3
17.1	80.6
15.5	75.2
14.7	69.7
17.1	82.0
15.4	69.4
16.2	83.3
15.0	79.6
17.2	82.6
16.0	80.6
17.0	83.5
14.4	76.3

a. Convert the temperature from Fahrenheit to Celsius by using the formula C/5= (F–32)/9.

b. If you want to estimate the number of chirps per second based on the temperature, which variable is the response variable and which is the independent predictor variable?

c. Assume that there is a linear regression between temperature (in Celsius) and chirps. Calculate the least-squares regression line describing number of chirps as a linear function of temperature.

d. Plot the data points and regression line. Does it appear that the line fits the data?

e. Use the appropriate statistical tests and measures to explain the usefulness of the regression model for predicting number of chirps.

 **12.25 Gestation Times and Longevity** EX1225 The table below shows the gestation time in days and the average longevity in years for a variety of mammals in captivity:[6]

Animal	Gestation (days)	Avg Longevity (yrs)
Baboon	187	20
Bear (black)	219	18
Bison	285	15
Cat (domestic)	63	12
Elk	250	15
Fox (red)	52	7
Goat (domestic)	151	8
Gorilla	258	20
Horse	330	20
Monkey (rhesus)	166	15
Mouse (meadow)	21	3
Pig (domestic)	112	10
Puma	90	12
Sheep (domestic)	154	12
Wolf (maned)	63	5

a. If you want to estimate the average longevity of an animal based on its gestation time, which variable is the response variable and which is the independent predictor variable?

b. Assume that there is a linear relationship between gestation time and longevity. Calculate the least-squares regression line describing longevity as a linear function of gestation time.

c. Plot the data points and the regression line. Does it appear that the line fits the data?

d. Use the appropriate statistical tests and measures to explain the usefulness of the regression model for predicting longevity.

12.26 Professor Asimov, continued Refer to the data in Exercise 12.9, relating x, the number of books written by Professor Isaac Asimov, to y, the number of months he took to write his books (in increments of 100). The data are reproduced here:

Number of books, x	100	200	300	400	490
Time in months, y	237	350	419	465	507

a. Do the data support the hypothesis that $\beta = 0$? Use the p-value approach, bounding the p-value using Table 4 of Appendix I. Explain your conclusions in practical terms.

b. Use the ANOVA table in Exercise 12.9, part c, to calculate the coefficient of determination r^2. What percentage reduction in the total variation is achieved by using the linear regression model?

c. Plot the data or refer to the plot in Exercise 12.9, part b. Do the results of parts a and b indicate that the model provides a good fit for the data? Are there any assumptions that may have been violated in fitting the linear model?

12.27 Sleep Deprivation III Refer to the sleep deprivation experiment described in Exercise 12.11 and data set EX1211. The data and the *MINITAB* and *Excel* printouts are reproduced here:

Number of errors, y	8, 6	6, 10	8, 14
Number of hours without sleep, x	8	12	16

Number of errors, y	14, 12	16, 12
Number of hours without sleep, x	20	24

MINITAB ANOVA table for Exercise 12.27

Regression Analysis: y versus x

```
The regression equation is
y = 3.00 + 0.475 x

Predictor      Coef   SE Coef      T       P
Constant      3.000     2.127   1.41   0.196
x            0.4750    0.1253   3.79   0.005

S = 2.24165   R-Sq = 64.2%   R-Sq(adj) = 59.8%

Analysis of Variance

Source          DF       SS       MS       F       P
Regression       1   72.200   72.200   14.37   0.005
Residual Error   8   40.200    5.025
Total            9  112.400
```

Excel output for Exercise 12.27

ANOVA

	df	*SS*	*MS*	*F*	*Significance F*
Regression	1	72.2	72.2	14.368	0.005
Residual	8	40.2	5.025		
Total	9	112.4			

	Coefficients	*Standard Error*	*t Stat*	*P- value*	*Lower 95%*	*Upper 95%*
Intercept	3	2.1266	1.4107	0.1960	−1.9040	7.9040
x	0.475	0.1253	3.7905	0.0053	0.1860	0.7640

a. Do the data present sufficient evidence to indicate that the number of errors is linearly related to the number of hours without sleep? Identify the two test statistics in the printout that can be used to answer this question.

b. Would you expect the relationship between y and x to be linear if x varied over a wider range (say, $x = 4$ to $x = 48$)?

c. How do you describe the strength of the relationship between y and x?

d. What is the best estimate of the common population variance σ^2?

e. Find a 95% confidence interval for the slope of the line.

12.28 Strawberries II The following data (Exercise 12.18 and data set EX1218) were obtained in an experiment relating the dependent variable, y (texture of strawberries), with x (coded storage temperature). Use the information from Exercise 12.18 to answer the following questions:

x	−2	−2	0	2	2
y	4.0	3.5	2.0	0.5	0.0

a. What is the best estimate of σ^2, the variance of the random error ϵ?

b. Do the data indicate that texture and storage temperature are linearly related? Use $\alpha = 0.05$.

c. Calculate the coefficient of determination, r^2.

d. Of what value is the *linear* model in increasing the accuracy of prediction as compared to the predictor, $\bar{y}$?

Data set EX1229 **12.29 Systolic Blood Pressure (SBP) and Body Mass Index (BMI)** In the case study of Chapter 1 we provided the blood pressure data on 500 persons (both male and females). Here we take a random sample of 40 persons. In the first column the blood pressure (y) and in the second column body mass index (x) were recorded for each randomly selected person. The data and *MINITAB* printout are shown in the following table:[7]

Systolic Blood Pressure	Body Mass Index	Systolic Blood Pressure	Body Mass Index
185	32.68430	194	41.04970
174	18.00509	186	49.77252
147	15.09444	191	25.18267
174	12.03111	121	23.66558
130	20.56960	130	22.56072
160	37.03880	152	23.72454
123	18.61939	127	16.24771
136	21.10426	136	32.11230
182	22.72159	148	32.64552
111	26.42051	130	27.64572
105	20.62475	100	28.20940
224	36.56926	124	12.09412
221	35.26174	147	31.57406
135	18.00972	146	18.34757
122	16.17305	132	26.97432
113	24.19442	176	25.01860
144	25.84728	163	20.07109
135	34.27116	149	23.58075
158	32.18126	130	22.17864
153	40.88436	130	20.92983

Regression Analysis: y versus x

```
The regression equation is
y = 109 + 1.55 x

Predictor      Coef     SE Coef      T        P
Constant     108.74     13.89      7.83    0.000
BMI          1.5453     0.5128     3.01    0.005

S = 26.8692     R-Sq = 19.3%    R-Sq(adj) = 17.2%

Analysis of Variance
Source          DF      SS          MS         F       P
Regression       1     6555.3     6555.3     9.08    0.005
Residual Error  38    27434.3      722.0
Total           39    33939.6
```

a. Construct a scatterplot of the data. Does the assumption of linearity appear to be reasonable?

b. What is the equation of regression line used for predicting SBP as the function of BMI?

c. Do the data present sufficient evidence to indicate that SBP measurement is linearly related to the BMI? Use $\alpha = 0.01$.

d. Find a 99% confidence interval for the slope of the regression line.

12.30 Systolic Blood Pressure (SBP) and Body Mass Index (BMI), continued Refer to Exercise 12.29.

a. Use the *MINITAB* printout to find the value of the coefficient of determination, r^2. Show that $r^2 = \text{SSR/Total SS}$.

b. What percentage reduction in the total variation is achieved by the linear regression model?

12.31 Armspan and Height II In Exercise 12.17 (data set EX1217), we measured the armspan and height of eight people with the following results:

Person	1	2	3	4	5	6	7	8
Armspan (cm)	172.7	158.1	165.1	176.5	172.7	175.3	157.5	153.0
Height (cm)	175.3	157.5	165.1	177.8	170.2	170.2	160.0	157.5

a. Does the data provide sufficient evidence to indicate that there is a linear relationship between armspan and height? Test at the 5% level of significance.

b. Construct a 95% confidence interval for the slope of the line of means, β.

c. If Leonardo da Vinci is correct, and a person's armspan is roughly the same as the person's height, the slope of the regression line is approximately equal to 1. Is this supposition confirmed by the confidence interval constructed in part b? Explain.

12.5 DIAGNOSTIC TOOLS FOR CHECKING THE REGRESSION ASSUMPTIONS

Even though you have determined—using the t-test for the slope (or the ANOVA F test) and the value of r^2—that x is useful in predicting the value of y, the results of a regression analysis are valid only when the data satisfy the necessary regression assumptions.

REGRESSION ASSUMPTIONS

- The relationship between Y and X must be linear, given by the model

  $$Y = \alpha + \beta X + \epsilon$$

- The values of the random error term ϵ (1) are independent, (2) have a mean of 0 and a common variance σ^2, independent of X, and (3) are normally distributed.

Since these assumptions are quite similar to those presented in Chapter 11 for an analysis of variance, it should not surprise you to find that the **diagnostic tools** for checking these assumptions are the same as those we used in that chapter. These tools involve the analysis of the **residual error,** the unexplained variation in each observation once the variation explained by the regression model has been removed.

Dependent Error Terms

The error terms are often dependent when the observations are collected at regular time intervals. When this is the case, the observations make up a **time series** whose error terms are correlated. This in turn causes bias in the estimates of model parameters. Time series data should be analyzed using time series methods.

Residual Plots

The other regression assumptions can be checked using **residual plots,** which are fairly complicated to construct by hand but easy to use once a computer has graphed them for you!

In simple linear regression, you can use the **plot of residuals versus fit** to check for a constant variance as well as to make sure that the linear model is in fact adequate. This plot should be free of any patterns. It should appear as a random scatter of points about 0 on the vertical axis with approximately the same vertical spread for all values of $\hat{y}$. One property of the residuals is that they sum to 0 and therefore have a sample mean of 0. The plot of the residuals versus fit for the calculus grade example is shown in Figure 12.9. There are no apparent patterns in this residual plot, which indicates that the model is adequate for these data.

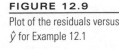

FIGURE 12.9
Plot of the residuals versus $\hat{y}$ for Example 12.1

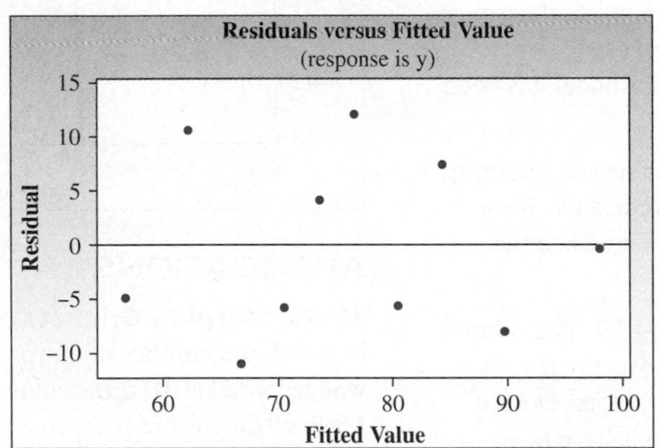

Recall from Chapter 11 that the **normal probability plot** is a graph that plots the residuals against the expected value of that residual if it had come from a normal distribution. When the residuals are normally distributed or approximately so, the plot should appear as a straight line, sloping upward. If the residuals have a standard normal distribution, then the line slopes upward at a 45° angle. The normal probability plot for the residuals in Example 12.1 is given in Figure 12.10. With the exception of the fourth and fifth plotted points, the remaining points appear to lie approximately on a straight line. This plot is not unusual and does not indicate underlying non-normality. The most serious violations of the normality assumption usually appear in the tails of the distribution because this is where the normal distribution differs most from other types of distributions with a similar mean and measure of spread. Hence, curvature in either or both of the two ends of the normal probability plot is indicative of non-normality.

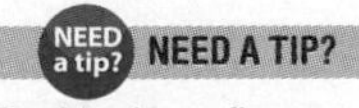

Residuals vs. fits ⇔ random scatter

Normal plot ⇔ straight line, sloping up

FIGURE 12.10
Normal probability plot of residuals for Example 12.1

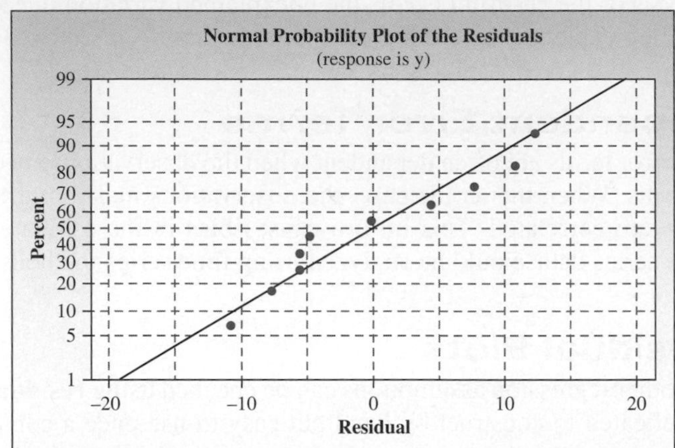

12.5 EXERCISES

BASIC TECHNIQUES

12.32 What diagnostic plot can you use to determine whether the data satisfy the normality assumption? What should the plot look like for normal residuals?

12.33 What diagnostic plot can you use to determine whether the incorrect model has been used? What should the plot look like if the correct model has been used?

12.34 What diagnostic plot can you use to determine whether the assumption of equal variance has been violated? What should the plot look like when the variances are equal for all values of x?

12.35 Refer to the data in Exercise 12.7. The normal probability plot and the residuals versus fitted values plots generated by *MINITAB* are shown here. Does it appear that any regression assumptions have been violated? Explain.

MINITAB output for Exercise 12.35

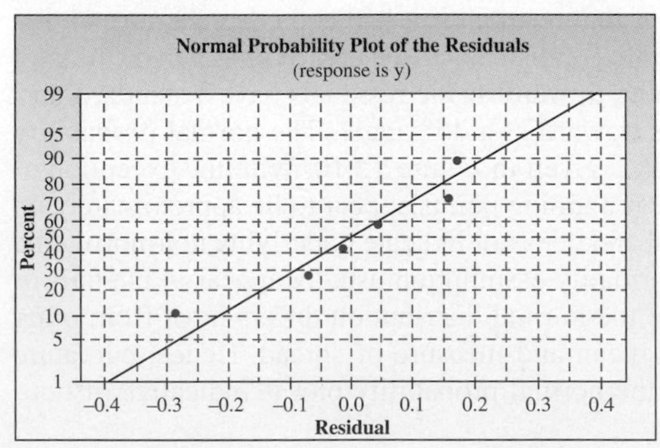

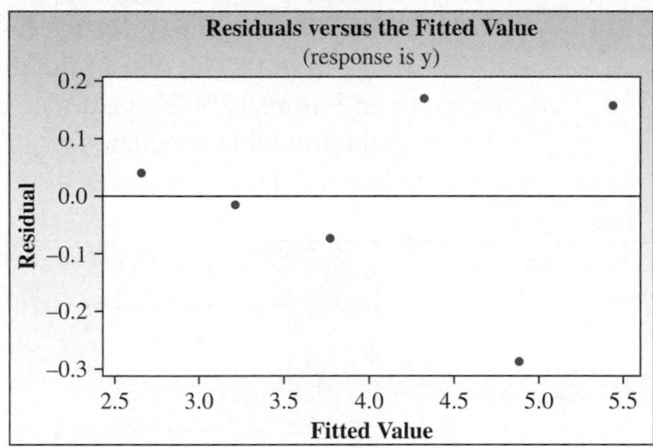

APPLICATIONS

12.36 Chirping Crickets Refer to Exercise 12.24, in which the number of chirps per second for a cricket was recorded at 10 different temperatures. Use the *MINITAB* diagnostic plots to comment on the validity of the regression assumptions.

MINITAB output for Exercise 12.36

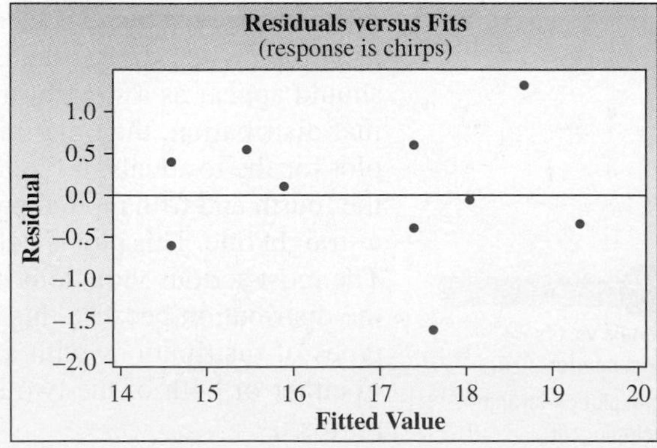

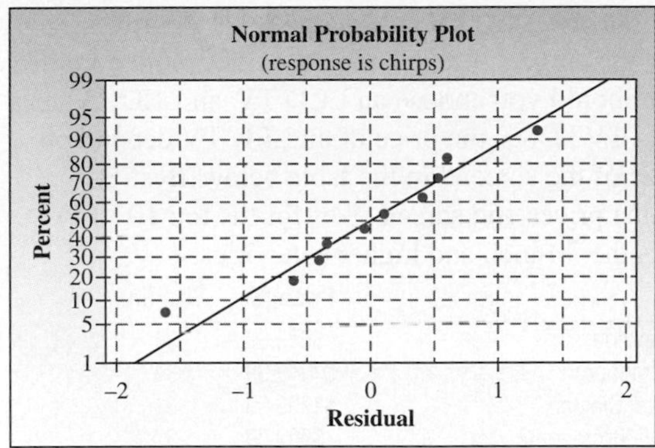

12.37 Air Pollution Refer to Exercise 12.23, in which an air pollution monitor's response to ozone was recorded for several different concentrations of ozone. Use the *MINITAB* residual plots to comment on the validity of the regression assumptions.

MINITAB output for Exercise 12.37

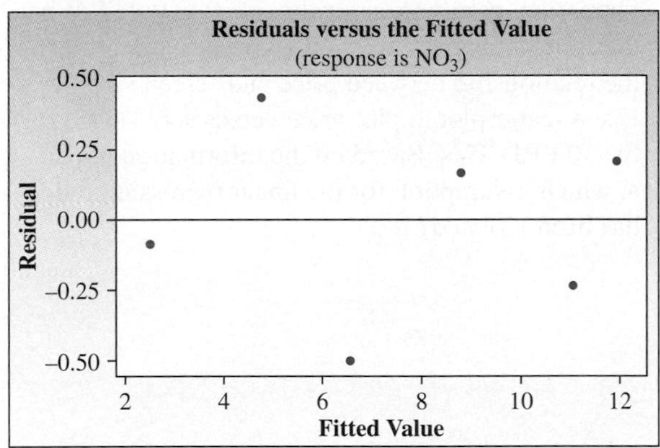

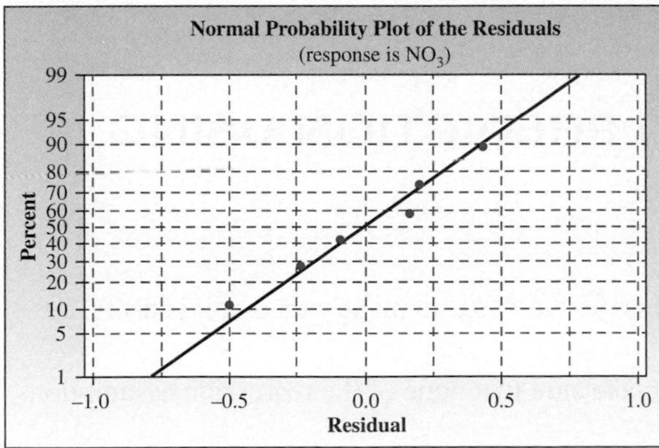

12.38 Professor Asimov, again Refer to Exercise 12.9, in which the number of books x written by Isaac Asimov are related to the number of months y he took to write them. A plot of the data is shown.

a. Can you see any pattern other than a linear relationship in the original plot?

b. The value of r^2 for these data is 0.959. What does this tell you about the fit of the regression line?

c. Look at the accompanying diagnostic plots for these data. Do you see any pattern in the residuals? Does this suggest that the relationship between number of months and number of books written is something other than linear?

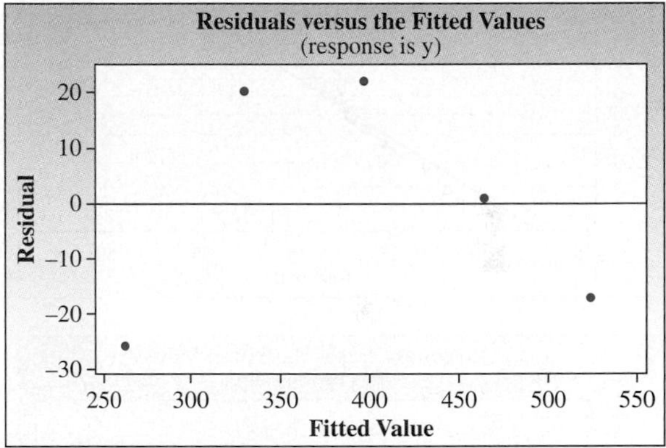

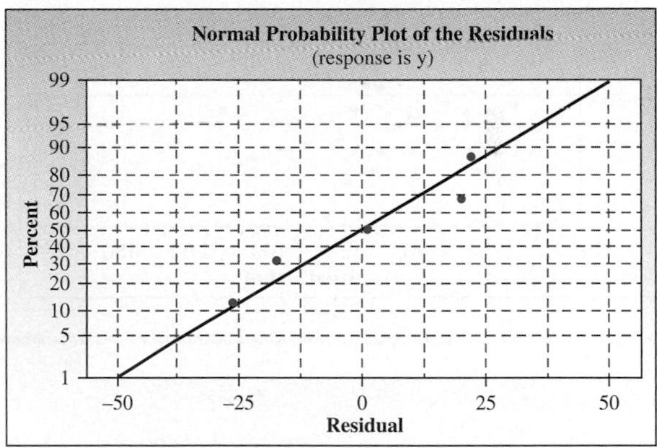

12.39 Systolic Blood Pressure (SBP) and Body Mass Index (BMI), again Refer to Exercise 12.29. The *MINITAB* printout is reproduced here.

Regression Analysis: y versus x

```
The regression equation is
y = 109 + 1.55 x

Predictor   Coef     SE Coef     T      P
Constant    108.74   13.89       7.83   0.000
BMI         1.5453   0.5128      3.01   0.005

S = 26.8692    R-Sq = 19.3%    R-Sq(adj) = 17.2%

Analysis of Variance
Source           DF   SS        MS       F      P
Regression       1    6555.3    6555.3   9.08   0.005
Residual Error   38   27434.3   722.0
Total            39   33989.6
```

a. What assumptions must be made about the distribution of the random error, ϵ?

b. What is the best estimate of σ^2, the variance of the random error, ϵ?

c. Use the diagnostic plots for these data to comment on the validity of the regression assumptions.

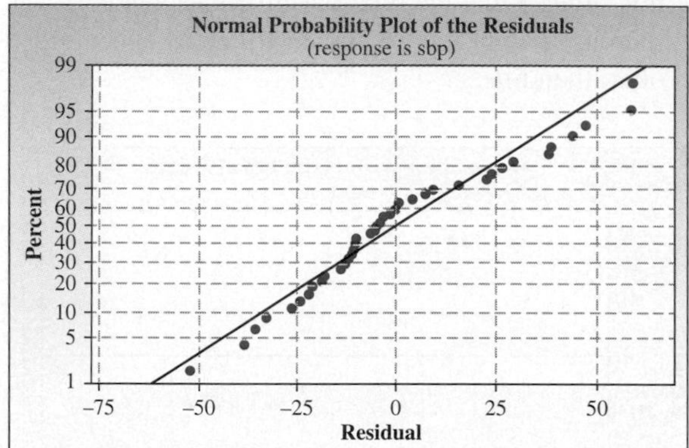

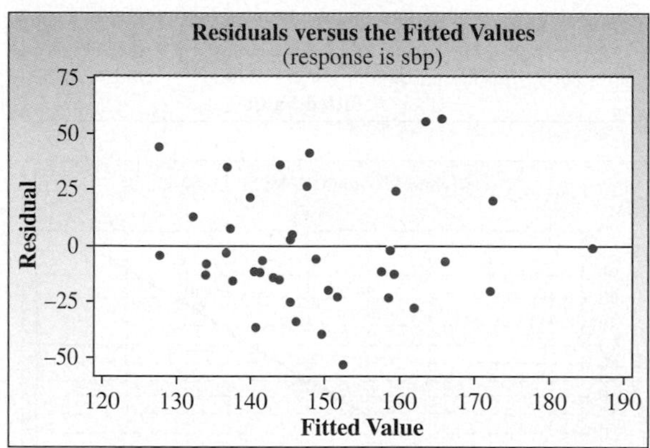

12.40 LED TVs As technology improves, EX1240 the choice of televisions becomes more complicated. Should you choose an LCD TV, an LED TV, or a plasma TV? Does the price of an LED TV depend on the size of the screen? In the table below, Best Buy[8] gives the prices and screen sizes for the top 10 LED TVs in the 30-inch and higher categories.

Brand	Price ($)	Size (in.)
LG UltraWide	$649.99	34
Dell UltraSharp	$1349.99	34
LG Digital Cinema	$1299.99	31
BenQ Widescreen 32	$999.99	32
BenQ Widescreen 55	$6499.99	55
LG 1080p	$599.95	47
HP Z30i	$1429.99	30
LG Widescreen Commercial-Grade	$1599.95	47
Samsung Ultra	$1749.99	31.5
HP Envy	$499.99	32

a. Suppose that we assume that the relationship between size and price is linear, and perform a linear regression, resulting in a value of $r^2 = 0.027$. What does the value of r^2 tell you about the strength of the relationship between price and screen size?

b. Use a scatterplot to plot price versus screen size for the 10 LED TVs. Based on the information in part a, which assumption for the linear regression model has been violated?

12.6 ESTIMATION AND PREDICTION USING THE FITTED LINE

You have now done the following:

- Tested the fitted regression line, $\hat{y} = a + bx$, to make sure that it is useful for prediction
- Used the diagnostic tools to make sure that none of the regression assumptions have been violated

You are ready to use the line for one of its two purposes:

- Estimating the average value of y for a given value of x
- Predicting a particular value of y for a given value of x

The sample of n pairs of observations have been chosen from a population in which the average value of Y is related to the value of the predictor variable X by the **line of means,**

$$E(Y) = \alpha + \beta X$$

an unknown line, shown as a broken line in Figure 12.11. Remember that for a fixed value of X—say, x_0—the particular values of Y deviate from the line of means. These values of Y have a normal distribution with mean equal to $\alpha + \beta x_0$ and variance σ^2, as shown in Figure 12.11.

FIGURE 12.11

Distribution of y for $x = x_0$

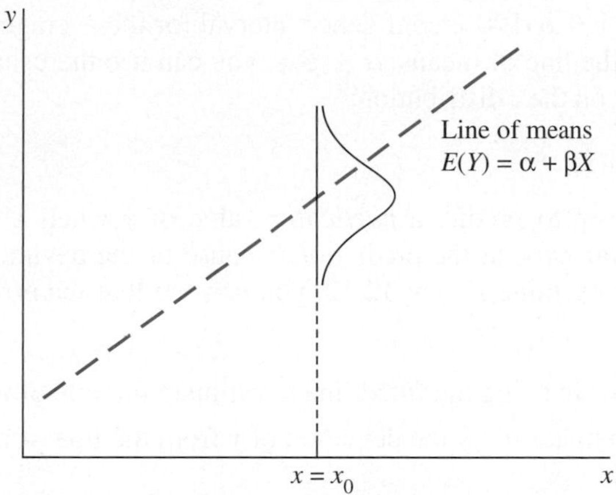

Since the computed values of a and b vary from sample to sample, each new sample produces a different regression line $\hat{y} = a + bx$, which can be used either to estimate the line of means or to predict a particular value of y. Figure 12.12 shows one of the possible configurations of the fitted line (blue), the unknown line of means (black), and a particular value of y (the blue dot).

How far will our estimator $\hat{y} = a + bx_0$ be from the quantity to be estimated or predicted? This depends, as always, on the variability in our estimator, measured by its **standard error.** It can be shown that

$$\hat{y} = a + bx_0$$

FIGURE 12.12

Error in estimating $E(y)$ and in predicting y

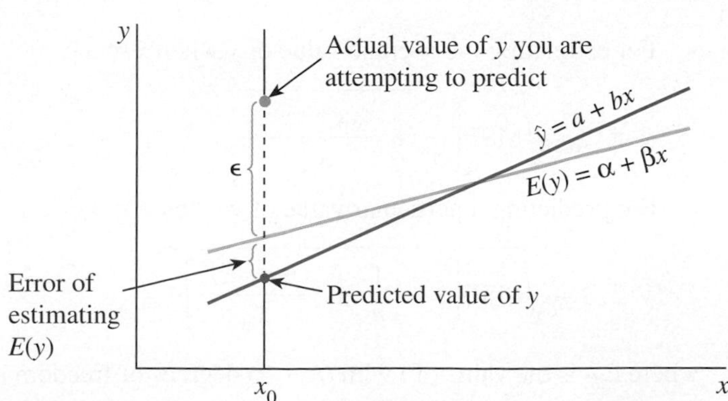

the estimated value of y when $x = x_0$, is an unbiased estimator of the line of means, $\alpha + \beta x_0$, and that $\hat{y}$ is normally distributed with the standard error of $\hat{y}$ estimated by

$$\text{SE}(\hat{y}) = \sqrt{\text{MSE}\left(\frac{1}{n} + \frac{(x_0 - \bar{x})^2}{S_{xx}}\right)}$$

Estimation and testing are based on the statistic

$$t = \frac{\hat{y} - E(y)}{\text{SE}(\hat{y})}$$

which has a t distribution with $(n - 2)$ degrees of freedom.

To form a $(1 - \alpha)100\%$ confidence interval for the average value of y when $x = x_0$, measured by the line of means, $\alpha + \beta x_0$, you can use the usual form for a confidence interval based on the t distribution:

NEED A TIP?

For a given value of x, the prediction interval is always wider than the confidence interval.

$$\hat{y} \pm t_{\alpha/2}\text{SE}(\hat{y})$$

If you choose to predict a *particular* value of y when $x = x_0$, however, there is some additional error in the prediction because of the deviation of y from the line of means. If you examine Figure 12.12, you can see that the error in prediction has two components:

- The error in using the fitted line to estimate the line of means
- The error caused by the deviation of y from the line of means, measured by σ^2

The variance of the difference between y and $\hat{y}$ is the sum of these two variances and forms the basis for the standard error of $(y - \hat{y})$ used for prediction:

$$\text{SE}(y - \hat{y}) = \sqrt{\text{MSE}\left[1 + \frac{1}{n} + \frac{(x_0 - \bar{x})^2}{S_{xx}}\right]}$$

and the $(1 - \alpha)100\%$ prediction interval is formed as

$$\hat{y} \pm t_{\alpha/2}\text{SE}(y - \hat{y})$$

$(1 - \alpha)100\%$ CONFIDENCE AND PREDICTION INTERVALS

- For estimating the average value of y when $x = x_0$:

$$\hat{y} \pm t_{\alpha/2}\sqrt{\text{MSE}\left[\frac{1}{n} + \frac{(x_0 - \bar{x})^2}{S_{xx}}\right]}$$

- For predicting a particular value of y when $x = x_0$:

$$\hat{y} \pm t_{\alpha/2}\sqrt{\text{MSE}\left[1 + \frac{1}{n} + \frac{(x_0 - \bar{x})^2}{S_{xx}}\right]}$$

where $t_{\alpha/2}$ is the value of t with $(n - 2)$ degrees of freedom and area $\alpha/2$ to its right.

EXAMPLE 12.4

Use the information in Example 12.1 to estimate the average calculus grade for students whose achievement score is 50, with a 95% confidence interval.

SOLUTION The point estimate of $E(y|x_0 = 50)$, the average calculus grade for students whose achievement score is 50, is

$$\hat{y} = 40.78424 + 0.76556(50) = 79.06$$

The standard error of $\hat{y}$ is

$$\sqrt{\text{MSE}\left[\frac{1}{n} + \frac{(x_0 - \bar{x})^2}{S_{xx}}\right]} = \sqrt{75.7532\left[\frac{1}{10} + \frac{(50 - 46)^2}{2474}\right]} = 2.840$$

and the 95% confidence interval is

$$79.06 \pm 2.306(2.840)$$

$$79.06 \pm 6.55$$

Our results indicate that the average calculus grade for students who score 50 on the achievement test will lie between 72.51 and 85.61.

EXAMPLE 12.5

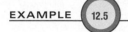

A student took the achievement test and scored 50 but has not yet taken the calculus test. Using the information in Example 12.1, predict the calculus grade for this student with a 95% prediction interval.

SOLUTION The predicted value of y is $\hat{y} = 79.06$, as in Example 12.4. However, the error in prediction is measured by $\text{SE}(y - \hat{y})$, and the 95% prediction interval is

$$79.06 \pm 2.306\sqrt{75.7532\left[1 + \frac{1}{10} + \frac{(50 - 46)^2}{2474}\right]}$$

$$79.06 \pm 2.306(9.155)$$

$$79.06 \pm 21.11$$

or from 57.95 to 100.17. The prediction interval is *wider* than the confidence interval in Example 12.4 because of the extra variability in predicting the actual value of the response y.

One particular point on the line of means is often of interest to experimenters, the ***y*-intercept** α—the average value of y when $x_0 = 0$.

EXAMPLE 12.6

Prior to fitting a line to the calculus grade–achievement score data, you may have thought that a score of 0 on the achievement test would predict a grade of 0 on the calculus test. This implies that we should fit a model with α equal to 0. Do the data support the hypothesis of a 0 intercept?

SOLUTION You can answer this question by constructing a 95% confidence interval for the y-intercept α, which is the average value of y when $x = 0$. The estimate of α is

$$\hat{y} = 40.784 + 0.76556(0) = 40.784 = a$$

and the 95% confidence interval is

$$\hat{y} \pm t_{\alpha/2}\sqrt{\text{MSE}\left[\frac{1}{n} + \frac{(x_0 - \bar{x})^2}{S_{xx}}\right]}$$

$$40.784 \pm 2.306\sqrt{75.7532\left[\frac{1}{10} + \frac{(0-46)^2}{2474}\right]}$$

$$40.784 \pm 19.617$$

or from 21.167 to 60.401, an interval that does not contain the value $\alpha = 0$. Hence, it is unlikely that the y-intercept is 0. You should include a non-zero intercept in the model $Y = \alpha + \beta X + \epsilon$.

For this special situation in which you are interested in testing or estimating the y-intercept α for the line of means, the inferences involve the sample estimate a. The test for a 0 intercept is given in Figure 12.13 in the shaded line labelled "Constant." The coefficient given as 40.784 is a, with standard error given in the column labelled "SE Coef" as 8.507, which agrees with the value calculated in Example 12.6. The value of t = 4.79 is found by dividing a by its standard error with p-value = 0.001.

FIGURE 12.13

Portion of the *MINITAB* output for Example 12.6

```
Predictor        Coef     SE Coef       T       P
Constant       40.784       8.507    4.79   0.001
x              0.7656      0.1750    4.38   0.002
```

You can see that it is quite time-consuming to calculate these estimation and prediction intervals by hand. Moreover, it is difficult to maintain accuracy in your calculations. Fortunately, computer programs can perform these calculations for you. The *MINITAB* regression command provides an option for either estimation or prediction when you specify the necessary value(s) of x. The printout in Figure 12.14 gives the values of $\hat{y} = 79.06$ labelled "Fit"; the standard error of $\hat{y}$, SE($\hat{y}$), labelled "SE Fit"; the *confidence interval* for the average value of y when $x = 50$, labelled "95% CI"; and the prediction interval for y when $x = 50$, labelled "95% PI."

FIGURE 12.14

MINITAB option for estimation and prediction

```
Predicted Values for New Observations
New Obs     Fit    SE Fit        95% CI              95% PI
1         79.06      2.84   (72.51,   85.61)   (57.95,  100.17)

Values of Predictors for New Observations
New Obs      x
1         50.0
```

The confidence bands and prediction bands generated by *MINITAB* for the calculus grades data are shown in Figure 12.15. Notice that in general the confidence bands are narrower than the prediction bands for every value of the achievement test score x. Certainly you would expect predictions for an individual value to be much more variable than estimates of the average value. Also notice that the bands seem to get wider as the value of x_0 gets farther from the mean $\bar{x}$. This is because the standard errors used in the confidence and prediction intervals contain the term $(x_0 - \bar{x})^2$, which gets larger as the two values diverge. In practice, this means that estimation and prediction are more accurate when x_0 is near the centre of the range of the x-values. You can locate the calculated confidence and prediction intervals when $x = 50$ in Figure 12.15.

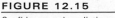

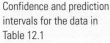

FIGURE 12.15
Confidence and prediction intervals for the data in Table 12.1

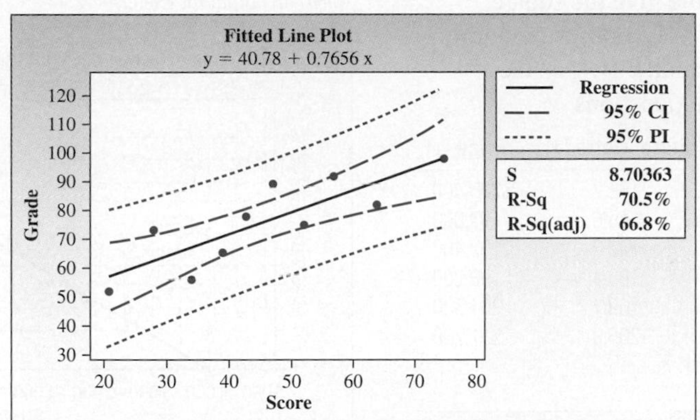

12.6 EXERCISES

BASIC TECHNIQUES

12.41 Refer to Exercise 12.6.

a. Estimate the average value of y when $x = 1$, using a 90% confidence interval.

b. Find a 90% prediction interval for some value of y to be observed in the future when $x = 1$.

12.42 Refer to Exercise 12.7. Portions of the *MINITAB* printout are shown here.

MINITAB output for Exercise 12.42

Regression Analysis: y versus x

```
The regression equation is
y = 6.00 - 0.557 x

Predictor       Coef     SE Coef        T         P
Constant      6.0000      0.1759    34.10     0.000
x           -0.55714     0.04518   -12.33     0.000

Predicted Values for New Observations
New Obs     Fit    SE Fit       95.0% CI            95.0% PI
1        4.8857    0.1027   (4.6006,  5.1708)  (4.2886,  5.4829)
2        1.5429    0.2174   (0.9392,  2.1466)  (0.7430,  2.3427) X
X denotes a point that is an outlier in the predictors.

Values of Predictors for New Observations
New Obs        x
1           2.00
2           8.00
```

a. Find a 95% confidence interval for the average value of y when $x = 2$.

b. Find a 95% prediction interval for some value of y to be observed in the future when $x = 2$.

c. The last line in the third section of the printout indicates a problem with one of the fitted values. What value of x corresponds to the fitted value $\hat{y} = 1.5429$? What problem has the *MINITAB* program detected?

APPLICATIONS

Data set EX1243 **12.43 What to Buy?** A marketing research experiment was conducted to study the relationship between the length of time necessary for a buyer to reach a decision and the number of alternative package designs of a product presented. Brand names were eliminated from the packages to reduce the effects of brand preferences. The buyers made their selections using the manufacturer's product descriptions on the packages as the only buying guide. The length of time necessary to reach a decision was recorded for 15 participants in the marketing research study:

Length of decision time, y (sec)	5, 8, 8, 7, 9	7, 9, 8, 9, 10	10, 11, 10, 12, 9
Number of alternatives, x	2	3	4

a. Find the least-squares line appropriate for these data.

b. Plot the points and graph the line as a check on your calculations.

c. Calculate s^2.

d. Do the data present sufficient evidence to indicate that the length of decision time is linearly related to the number of alternative package designs? (Test at the $\alpha = 0.05$ level of significance.)

e. Find the approximate p-value for the test and interpret its value.

f. If they are available, examine the diagnostic plots to check the validity of the regression assumptions.

g. Estimate the average length of time necessary to reach a decision when three alternatives are presented, using a 95% confidence interval.

Data set EX1244 **12.44 Housing Prices** If you try to rent an apartment or buy a house, you find that real estate representatives establish apartment rents and house prices on the basis of square footage of heated floor space. The data in the table give the square footages and sales prices of $n = 12$ houses randomly selected from those sold in a small city. Use the *MINITAB* printout to answer the questions:

Square Metre (x)	Price (y)	Square Metre (x)	Price (y)
135.6	$188,700	183.7	$205,400
195.9	209,300	149.6	197,000
161.9	201,400	142.2	192,400
139.3	191,100	163.4	198,200
173.2	202,400	169.2	204,300
222.1	214,900	205.9	211,700

Plot of data for Exercise 12.44

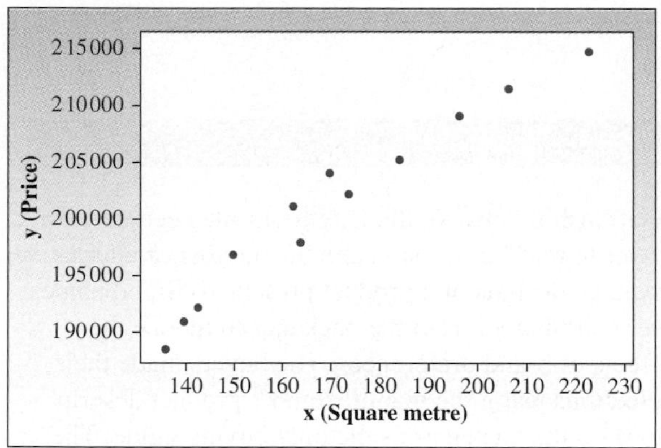

MINITAB output for Exercise 12.44

Regression Analysis: y versus x

```
The regression equation is
y = 151206 + 295 x

Predictor    Coef      SE Coef      T        P
Constant     151206    3389         44.62    0.000
x            294.99    19.68        14.99    0.000

S = 1792.72      R-Sq = 95.7%      R-Sq(adj) = 95.3%

Predicted Values for New Observations
New Obs    Fit       SE Fit         95.0% CI               95.0% PI
1          199989    526       (198817, 201161)    (195826, 204151)
2          206018    602       (204676, 207360)    (201804, 210232)

Values of Predictors for New Observations
New
Obs     C1
1       165
2       186
```

a. Can you see any pattern other than a linear relationship in the original plot?

b. The value of r^2 for these data is 0.957. What does this tell you about the fit of the regression line?

c. Look at the accompanying diagnostic plots for these data. Do you see any pattern in the residuals? Does this suggest that the relationship between price and square metres is something other than linear?

MINITAB output for Exercise 12.44

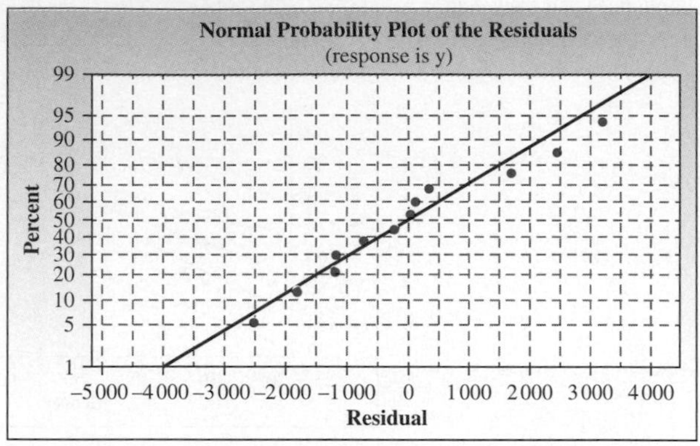

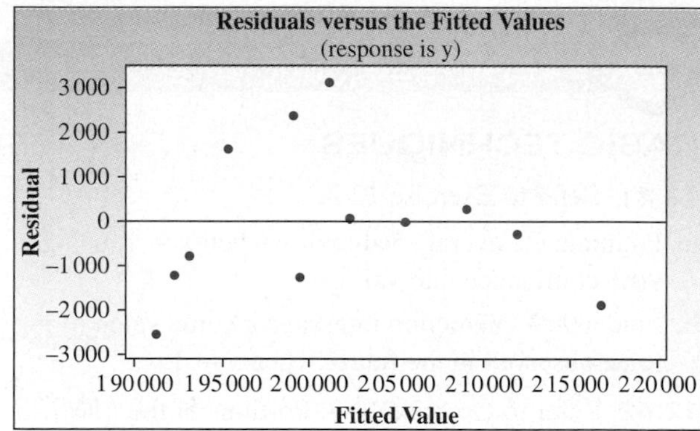

12.45 Housing Prices II Refer to Exercise 12.44 and data set EX1244.

a. Estimate the average increase in the price for an increase of 1 m^2 for houses sold in the city. Use a 99% confidence interval. Interpret your estimate.

b. A real estate salesperson needs to estimate the average sales price of houses with a total of 185.8 m^2 of heated space. Use a 95% confidence interval and interpret your estimate.

c. Calculate the price per square metre for each house and then calculate the sample mean. Why is this estimate of the average cost per square metre not equal to the answer in part a? Should it be? Explain.

d. Suppose that a house with 165.4 m^2 of heated floor space is offered for sale. Construct a 95% prediction interval for the price at which the house will sell.

12.46 Strawberries III The following data (Exercises 12.18 and 12.28) were obtained in an experiment relating the dependent variable, y (texture of strawberries), with x (coded storage temperature):

x	−2	−2	0	2	2
y	4.0	3.5	2.0	0.5	0.0

a. Estimate the expected strawberry texture for a coded storage temperature of $x = 1$. Use a 99% confidence interval.

b. Predict the particular value of y when $x = 1$ with a 99% prediction interval.

c. At what value of x will the width of the prediction interval for a particular value of y be a minimum, assuming n remains fixed?

Data set EX1247 **12.47 Stride Rate** One measure of form for a runner is stride rate, defined as the number of steps per second. A runner is considered to be efficient if the stride rate is close to optimum. The stride rate is related to speed; the greater the speed, the greater the stride rate.

In a study of some top female runners, researchers measured the stride rate for different speeds. The following table gives the average stride rate of these women versus the speed (in metres per second):[9]

Speed	4.83	5.15	5.33	5.68	6.09	6.42	6.74
Stride rate	3.05	3.12	3.17	3.25	3.36	3.46	3.55

a. What is the least-squares line relating the stride rate to speed?

b. What proportion of the total variation is explained by the regression of stride rate (y) on the speed (x)?

c. If they are available, examine the diagnostic plots to check the validity of the regression assumptions.

12.48 Stride Rate, continued Refer to Exercise 12.47.

a. Estimate the average stride rate if the speed is 5.8 metres per second using a 95% confidence interval.

b. Predict the actual number of the stride rate if the speed is 5.8 metres per second using a 95% prediction interval.

c. Would it be advisable to use the least-squares line from Exercise 12.47 to predict stride rate if the speed is 2 metres per second? Explain.

Data set EX1249 **12.49 Drew Brees** The number of passes completed and the total number of passing yards for Drew Brees, quarterback for the New Orleans Saints, were recorded for the 16 regular games in the 2010 football season.[10] He had no games in week 10 and no data was reported.

Week	Completions	Total Yards
1	27	237
2	28	254
3	30	365
4	33	275
5	24	279
6	21	263
7	37	356
8	34	305
9	27	253
10	–	–
11	29	382
12	23	352
13	24	313
14	25	221
15	29	267
16	25	302
17	22	196

a. What is the least-squares line relating the total passing yards to the number of pass completions for Drew Brees?

b. What proportion of the total variation is explained by the regression of total passing yards (y) on the number of pass completions (x)?

c. If they are available, examine the diagnostic plots to check the validity of the regression assumptions.

12.50 Drew Brees, continued Refer to Exercise 12.49.

a. Estimate the average number of passing yards for games in which Brees throws 20 completed passes using a 95% confidence interval.

b. Predict the actual number of passing yards for games in which Brees throws 20 completed passes using a 95% confidence interval.

c. Would it be advisable to use the least-squares line from Exercise 12.49 to predict Brees' total number of passing yards for a game in which he threw only five completed passes? Explain.

12.7 CORRELATION ANALYSIS

In Chapter 3, we introduced the *correlation coefficient* as a measure of the strength of the linear relationship between two variables. The correlation coefficient, r—formally called the **Pearson product moment sample coefficient of correlation**—is defined next.

PEARSON PRODUCT MOMENT COEFFICIENT OF CORRELATION

$$r = \frac{s_{xy}}{s_x s_y} = \frac{S_{xy}}{\sqrt{S_{xx}S_{yy}}} \quad \text{for} \quad -1 \le r \le 1$$

NEED A TIP?

r is always between −1 and +1

The variances and covariance can be found by direct calculation, by using a calculator with a two-variable statistics capacity, or by using a statistical package such as *MINITAB*. The variances and covariance are calculated as

$$s_{xy} = \frac{S_{xy}}{n-1} \qquad s_x^2 = \frac{S_{xx}}{n-1} \qquad s_y^2 = \frac{S_{yy}}{n-1}$$

and use S_{xy}, S_{xx}, and S_{yy}, the same quantities used in regression analysis earlier in this chapter. In general, when a sample of n individuals or experimental units is selected and two variables are measured on each individual or unit so that *both variables are random*, the correlation coefficient r is the appropriate measure of linearity for use in this situation.

EXAMPLE 12.7

The heights and weights of $n = 10$ offensive backfield football players are randomly selected from a county's football all-stars. Calculate the correlation coefficient for the heights (in centimetres) and weights (in kilograms) given in Table 12.4.

TABLE 12.4 **Heights and Weights of $n = 10$ Backfield All-Stars**

Player	Height, x	Weight, y
1	185.4	84.1
2	180.3	79.5
3	190.5	90.9
4	182.9	95.5
5	182.9	86.4
6	190.5	88.6
7	170.2	68.2
8	175.3	77.3
9	180.3	81.8
10	175.3	79.5

SOLUTION You should use the appropriate data entry method of your scientific calculator to verify the calculations for the sums of squares and cross-products:

$$S_{xy} = 328 \qquad S_{xx} = 60.4 \qquad S_{yy} = 2610$$

using the calculational formulas given earlier in this chapter. Then

$$r = \frac{328}{\sqrt{(60.4)(2610)}} = 0.8261$$

or $r = 0.83$. This value of r is fairly close to 1, the largest possible value of r, which indicates a fairly strong positive linear relationship between height and weight.

There is a direct relationship between the calculational formulas for the correlation coefficient r and the slope of the regression line b. Since the numerator of both quantities is S_{xy}, both r and b have the same sign. Therefore, the correlation coefficient has these general properties:

- When $r = 0$, the slope is 0, and there is no linear relationship between x and y.
- When r is positive, so is b, and there is a positive linear relationship between x and y.
- When r is negative, so is b, and there is a negative linear relationship between x and y.

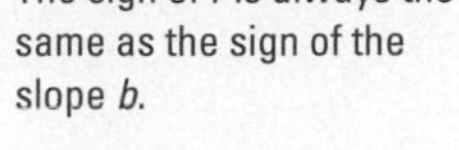
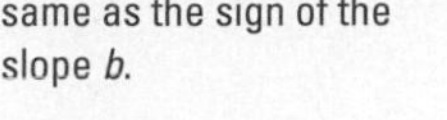

NEED A TIP?

The sign of r is always the same as the sign of the slope b.

ONLINE APPLET

Exploring Correlation

In Section 12.4, we showed that

$$r^2 = \frac{\text{SSR}}{\text{Total SS}} = \frac{\text{Total SS} - \text{SSE}}{\text{Total SS}}$$

In this form, you can see that r^2 can never be greater than 1, so that $-1 \leq r \leq 1$. Moreover, you can see the relationship between the random variation (measured by SSE) and r^2.

- If there is no random variation and all the points fall on the regression line, then $\text{SSE} = 0$ and $r^2 = 1$.
- If the points are randomly scattered and there is no variation explained by regression, then $\text{SSR} = 0$ and $r^2 = 0$.

Figure 12.16 shows four typical scatterplots and their associated correlation coefficients. Notice that in scatterplot (d) there appears to be a curvilinear relationship between x and y, but r is approximately 0, which reinforces the fact that r is a measure of a *linear* (not *curvilinear*) relationship between two variables.

Consider a population generated by measuring two random variables on each experimental unit. In this *bivariate* population, the **population correlation coefficient** ρ (Greek letter rho) is calculated and interpreted as it is in the sample. In this situation, the experimenter can test the hypothesis that there is no correlation between the variables x and y using a test statistic that is *exactly equivalent* to the test of the slope β in Section 12.4. The test procedure is shown next.

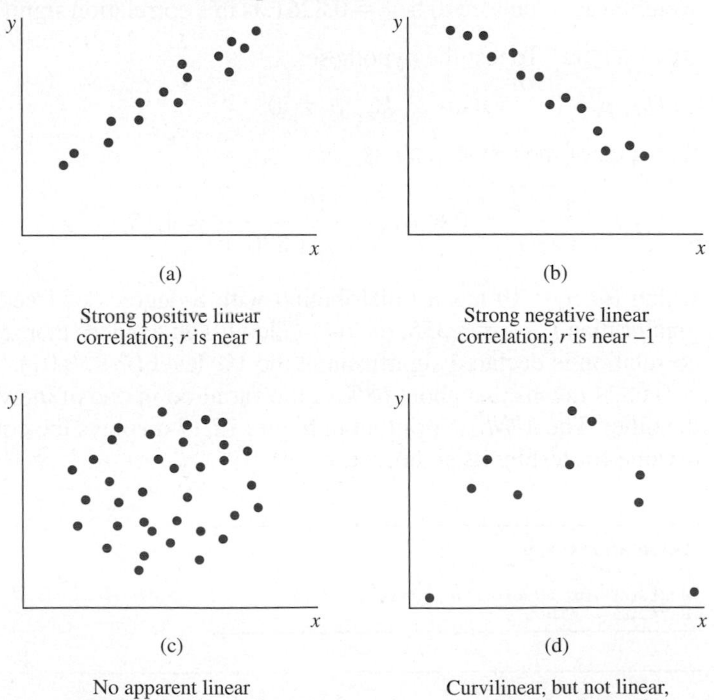
y
x
(a)
Strong positive linear correlation; r is near 1
y
x
(b)
Strong negative linear correlation; r is near –1
y
x
(c)
No apparent linear correlation; r is near 0
y
x
(d)
Curvilinear, but not linear, correlation; r is near 0

FIGURE 12.16 Some typical scatterplots with approximate values of r

TEST OF HYPOTHESIS CONCERNING THE CORRELATION COEFFICIENT ρ

1. Null hypothesis: $H_0 : \rho = 0$
2. Alternative hypothesis:

One-Tailed Test	**Two-Tailed Test**
$H_a : \rho > 0$	$H_a : \rho \neq 0$
(or $\rho < 0$)	

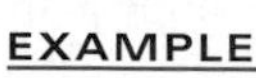

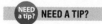

NEED A TIP?

You can prove that

$$t = r\sqrt{\frac{n-2}{1-r^2}} = \frac{b-0}{\sqrt{MSE/S_{xx}}}$$

3. Test statistic: $t = r\sqrt{\dfrac{n-2}{1-r^2}}$

When the assumptions given in Section 12.2 are satisfied, the test statistic will have a Student's t distribution with $(n - 2)$ degrees of freedom.

4. Rejection region: Reject H_0 when

One-Tailed Test	**Two-Tailed Test**
$t > t_\alpha$	$t > t_{\alpha/2}$ or $t < -t_{\alpha/2}$
(or $t < -t_\alpha$ when the alternative hypothesis is $H_a : \rho < 0$)	

or when p-value $< \alpha$

The values of t_α and $t_{\alpha/2}$ can be found using Table 4 in Appendix I. Use the values of t corresponding to $(n - 2)$ degrees of freedom.

EXAMPLE 12.8

Refer to the height and weight data in Example 12.7. The correlation of height and weight was calculated to be $r = 0.8261$. Is this correlation significantly different from 0?

SOLUTION To test the hypotheses

NEED A TIP?

The t value and p-value for testing $H_0 : \rho = 0$ will be identical to the t and p-value for testing $H_0 : \beta = 0$.

$$H_0 : \rho = 0 \quad \text{versus} \quad H_a : \rho \neq 0$$

the value of the test statistic is

$$t = r\sqrt{\frac{n-2}{1-r^2}} = 0.8261\sqrt{\frac{10-2}{1-(.8261)^2}} = 4.15$$

which for $n = 10$ has a t distribution with 8 degrees of freedom. Since this value is greater than $t_{0.005} = 3.355$, the two-tailed p-value is less than $2(0.005) = 0.01$, and the correlation is declared significant at the 1% level ($P < 0.01$). The value $r^2 = 0.8261^2 = 0.6824$ means that about 68% of the variation in one of the variables is explained by the other. The *MINITAB* printout in Figure 12.17 displays the correlation r and the exact p-value for testing its significance.

FIGURE 12.17
MINITAB output for Example 12.8

Correlations: x, y

```
Pearson correlation of x and y = 0.826
P-Value = 0.003
```

If the linear coefficients of correlation between y and each of two variables x_1 and x_2 are calculated to be 0.4 and 0.5, respectively, it does not follow that a predictor using both variables will account for $[(0.4)^2 + (0.5)^2] = 0.41$, or a 41% reduction in the sum of squares of deviations. Actually, x_1 and x_2 might be highly correlated and therefore contribute virtually the same information for the prediction of y.

Finally, remember that r is a measure of **linear correlation** and that x and y could be perfectly related by some curvilinear function when the observed value of r is equal to 0. The problem of estimating or predicting y using information given by several independent variables, $x_1, x_2, \ldots, x_k$, is the subject of Chapter 13.

12.7 EXERCISES

BASIC TECHNIQUES

12.51 How does the coefficient of correlation measure the strength of the linear relationship between two variables y and x?

12.52 Describe the significance of the algebraic sign and the magnitude of r.

12.53 What value does r assume if all the data points fall on the same straight line in these cases?

a. The line has positive slope.

b. The line has negative slope.

12.54 You are given these data:

x	−2	−1	0	1	2
y	2	2	3	4	4

a. Plot the data points. Based on your graph, what will be the sign of the sample correlation coefficient?

b. Calculate r and r^2 and interpret their values.

12.55 You are given these data:

x	1	2	3	4	5	6
y	7	5	5	3	2	0

a. Plot the six points on graph paper.

b. Calculate the sample coefficient of correlation r and interpret.

c. By what percentage was the sum of squares of deviations reduced by using the least-squares predictor $\hat{y} = a + bx$ rather than $\hat{y}$ as a predictor of y?

12.56 Reverse the slope of the line in Exercise 12.55 by reordering the y observations, as follows:

x	1	2	3	4	5	6
y	0	2	3	5	5	7

Repeat the steps of Exercise 12.55. Notice the change in the sign of r and the relationship between the values of r^2 of Exercise 12.55 and this exercise.

APPLICATIONS

Data set EX1257 **12.57 Lobster** The table gives the numbers of *Octolasmis tridens* and *O. lowei* barnacles on each of 10 lobsters.[11] Does it appear that the barnacles compete for space on the surface of a lobster?

Lobster Field Number	*O. tridens*	*O. lowei*
A061	645	6
A062	320	23
A066	401	40
A070	364	9
A067	327	24
A069	73	5
A064	20	86
A068	221	0
A065	3	109
A063	5	350

a. If they do compete, do you expect the number x of *O. tridens* and the number y of *O. lowei* barnacles to be positively or negatively correlated? Explain.

b. If you want to test the theory that the two types of barnacles compete for space by conducting a test of the null hypothesis "the population correlation coefficient ρ equals 0," what is your alternative hypothesis?

c. Conduct the test in part b and state your conclusions.

Data set EX1258 **12.58 Social Skills Training** A social skills training program was implemented with seven mildly challenged students in a study to determine whether the program caused improvement in pre/post measures and behaviour ratings. For one such test, the pre- and posttest scores for the seven students are given in the table:[12]

Subject	Pretest	Posttest
Earl	101	113
Ned	89	89
Jasper	112	121
Charlie	105	99
Tom	90	104
Susie	91	94
Lori	89	99

Source: Gregory K. Torrey, S.F. Vasa, J.W. Maag, and J.J. Kramer, "Social Skills Interventions Across School Settings: Case Study Reviews of Students with Mild Disabilities," *Psychology in the Schools* 29 (July 1992):248. Republished with permission of John Wiley & Sons, Inc.; permission conveyed through Copyright Clearance Center, Inc.

a. What type of correlation, if any, do you expect to see between the pre- and posttest scores? Plot the data. Does the correlation appear to be positive or negative?

b. Calculate the correlation coefficient, r. Is there a significant positive correlation?

12.59 Hockey G.W. Marino investigated the variables related to a hockey player's ability to make a fast start from a stopped position.[13] In the experiment, each skater started from a stopped position and attempted to move as rapidly as possible over a 6-metre distance. The correlation coefficient r between a skater's stride rate (number of strides per second) and the length of time to cover the 6-metre distance for the sample of 69 skaters was -0.37.

a. Do the data provide sufficient evidence to indicate a correlation between stride rate and time to cover the distance? Test using $\alpha = 0.05$.

b. Find the approximate p-value for the test.

c. What are the practical implications of the test in part a?

12.60 Hockey II Refer to Exercise 12.59. Marino calculated the sample correlation coefficient r for the stride rate and the average acceleration rate for the 69 skaters to be 0.36. Do the data provide sufficient evidence to indicate a correlation between stride rate and average acceleration for the skaters? Use the p-value approach.

Data set EX1261 **12.61 Geothermal Power** Geothermal power is an important source of energy. Since the amount of energy contained in 0.45 kg of water is a function of its temperature, you might wonder whether water obtained from deeper wells contains more energy per 0.45 kg. The data in the table are reproduced from an article on geothermal systems by A.J. Ellis:[14]

Location of Well	Average (max.) Drill Hole Depth (m)	Average (max.) Temperature (°C)
El Tateo, Chile	650	230
Ahuachapan, El Salvador	1000	230
Namafjall, Iceland	1000	250
Larderello (region), Italy	600	200
Matsukawa, Japan	1000	220
Cerro Prieto, Mexico	800	300
Wairakei, New Zealand	800	230
Kizildere, Turkey	700	190
The Geysers, United States	1500	250

Is there a significant positive correlation between average maximum drill hole depth and average maximum temperature?

12.62 Cheese, Please! The demand for healthy foods that are low in fat and calories has resulted in a large number of "low-fat" or "fat-free" products. The table shows the number of calories and the amount of sodium (in milligrams) per slice for five different brands of fat-free cheese:

Brand	Sodium (mg)	Calories
Kraft Fat-Free Singles	300	30
Ralphs Fat-Free Singles	300	30
Borden Fat Free	320	30
Healthy Choice Fat Free	290	30
Smart Beat	180	25

a. Should you use the methods of linear regression analysis or correlation analysis to analyze the data? Explain.

b. Analyze the data to determine the nature of the relationship between sodium and calories in fat-free cheese slices. Use any statistical tests that are appropriate.

Data set EX1263 **12.63 Body Temperature and Heart Rate** Is there any relationship between these two variables? To find out, we randomly selected 12 people from a data set constructed by Allen Shoemaker (*Journal of Statistics Education*) and recorded their body temperature and heart rate:[15]

Person	1	2	3	4	5	6
Temperature (degrees Celsius)	35.7	36.3	37.2	37.2	37.2	36
Heart rate (beats per minute)	70	68	80	75	79	75

Person	7	8	9	10	11	12
Temperature (degrees Celsius)	36.9	36.9	37.1	37.1	37.3	37.4
Heart rate (beats per minute)	74	84	73	84	66	68

a. Find the correlation coefficient r, relating body temperature to heart rate.

b. Is there sufficient evidence to indicate that there is a correlation between these two variables? Test at the 5% level of significance.

Data set EX1264 **12.64 Cricket by Chance** One measure of form for a bowler is the average run rate, defined as the number of runs given per wicket taken. A bowler is considered to be efficient if the run rate is close to optimum. The data shown in the following table provide information for 10 cricket bowlers from various countries (based on all matches up to and including Test 1825):[16]

Name	No. of Overs Completed	Runs Given	No. of Wickets Taken	Average Run Rate	Team/ Country
K.D. Mills	87.3	261	14	18.64	New Zealand
C.R.D. Fernando	104.0	312	13	24.00	Sri Lanka
S.C.G. MacGill	125.5	436	20	21.80	Australia
S.E. Bond	131.4	471	18	26.16	New Zealand
L.E. Plunkett	139.0	476	14	34.00	England
Shahid Nazir	159.0	518	17	30.47	Pakistan
J.H Kallis	155.1	491	15	32.73	South Africa
Z. Khan	170.3	633	18	35.16	India
S.I. Mahmood	175.3	685	19	36.05	England
D.W. Steyn	181.4	734	24	30.58	South Africa

a. Plot the average run rate versus number of overs completed. Does it appear that there is any relationship between run rate and number of completed overs?

b. Is there a significant correlation between number of completed overs and run rate? Test at the 5% level of significance.

c. Do you think that the relationship between these two variables would be different if we had looked at all the bowlers, rather than a few selected ones? Any suggestions?

Data set EX1265 **12.65 Baseball Stats** Does a team's batting average depend in any way on the number of home runs hit by the team? The data in the table show the number of team home runs and the overall team batting average for eight selected major league teams for the 2010 season:[17]

Team	Total Home Runs	Team Batting Average
Atlanta Braves	139	0.258
Baltimore Orioles	133	0.259
Boston Red Sox	211	0.268
Chicago White Sox	177	0.268
Houston Astros	108	0.247
LA Dodgers	120	0.252
Philadelphia Phillies	166	0.260
Seattle Mariners	101	0.236

a. Plot the points using a scatterplot. Does it appear that there is any relationship between total home runs and team batting average?

b. Is there a significant positive correlation between total home runs and team batting average? Test at the 5% level of significance.

c. Do you think that the relationship between these two variables would be different if we had looked at the entire set of major league franchises?

CHAPTER REVIEW

Key Concepts and Formulas

I. A Linear Probabilistic Model

1. When the data exhibit a linear relationship, the appropriate model is $Y = \alpha + \beta X + \epsilon$.
2. The random error ϵ has a normal distribution with mean 0 and variance σ^2.

II. Method of Least Squares

1. Estimates a and b, for a and b, are chosen to minimize SSE, the sum of squared deviations about the regression line, $\hat{y} = a + bx$.
2. The least-squares estimates are $b = S_{xy}/S_{xx}$ and $a = \bar{y} - b\bar{x}$

III. Analysis of Variance

1. Total SS = SSR + SSE, where Total SS = S_{yy} and SSR = $(S_{xy})^2/S_{xx}$.
2. The best estimate of σ^2 is MSE = SSE/$(n - 2)$.

IV. Testing, Estimation, and Prediction

1. A test for the significance of the linear regression—$H_0: \beta = 0$—can be implemented using one of two test statistics:

$$t = \frac{b}{\sqrt{\text{MSE}/S_{xx}}} \quad \text{or} \quad F = \frac{\text{MSR}}{\text{MSE}}$$

2. The strength of the relationship between x and y can be measured using

$$r^2 = \frac{\text{SSR}}{\text{Total SS}}$$

which gets closer to 1 as the relationship gets stronger.

3. Use residual plots to check for non-normality, inequality of variances, or an incorrectly fit model.
4. Confidence intervals can be constructed to estimate the intercept α and slope β of the regression line and to estimate the average value of Y, $E(Y)$, for a given value of X.
5. Prediction intervals can be constructed to predict a particular observation, y, for a given value of x. For a given x, prediction intervals are always wider than confidence intervals.

V. Correlation Analysis

1. Use the correlation coefficient to measure the relationship between x and y when both variables are random:

$$r = \frac{S_{xy}}{\sqrt{S_{xx}S_{yy}}}$$

2. The sign of r indicates the direction of the relationship; r near 0 indicates no linear relationship, and r near 1 or -1 indicates a strong linear relationship.
3. A test of the significance of the correlation coefficient uses the statistic

$$t = r\sqrt{\frac{n-2}{1-r^2}}$$

and is identical to the test of the slope β.

Linear Regression Procedures—*Microsoft Excel*

In Chapter 3, we used some of the linear regression procedures available in *Excel* to obtain a scatterplot of the data and the least-squares regression line and to calculate the correlation coefficient r for a bivariate data set. Now that you have studied the testing and estimation techniques for a simple linear regression analysis, more *Excel* options are available to you.

EXAMPLE 12.9

Refer to Table 12.1, in which the relationship between x = mathematics achievement test score and y = final calculus grade was studied.

Student	Mathematics Achievement Test Score (x)	Final Calculus Grade (y)
1	39	65
2	43	78
3	21	52
4	64	82
5	57	92
6	47	89
7	28	73
8	75	98
9	34	56
10	52	75

1. Enter the values for x and y into columns A and B of an *Excel* spreadsheet.
2. Use **Data ▶ Data Analysis ▶ Regression** to generate the dialogue box in Figure 12.18(a). Highlight or type in the cell ranges for the x and y values and check "Labels" if necessary.
3. If you click "Confidence Level," *Excel* will calculate confidence intervals for the regression estimates, a and b. Enter a cell location for the **Output Range** and click **OK.**

4. The output will appear in the selected cell location, and should be adjusted using **Format ▶ AutoFit Column Width** on the **Home** tab in the **Cells** group while it is still highlighted. You can decrease the decimal accuracy if you like, using .00→.0 on the **Home** tab in the **Number** group (see Figure 12.18(b)).
5. The output in Figure 12.18(b) can also be found in Figures 12.6(b) and 12.7(b), with its interpretation found in Sections 12.3 and 12.4 of the text.

FIGURE 12.18(a)(b)
a) *MINITAB* screen shot
b) *MINITAB* screen shot

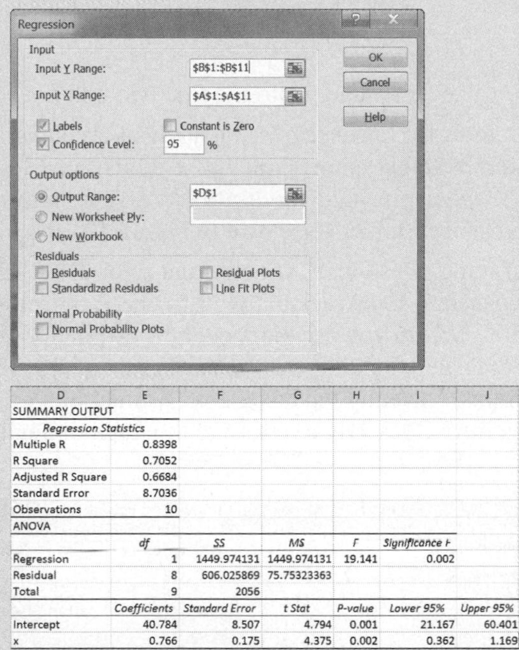

D	E	F	G	H	I	J
SUMMARY OUTPUT						
Regression Statistics						
Multiple R	0.8398					
R Square	0.7052					
Adjusted R Square	0.6684					
Standard Error	8.7036					
Observations	10					
ANOVA						
	df	*SS*	*MS*	*F*	*Significance F*	
Regression	1	1449.974131	1449.974131	19.141	0.002	
Residual	8	606.025869	75.75323363			
Total	9	2056				
	Coefficients	*Standard Error*	*t Stat*	*P-value*	*Lower 95%*	*Upper 95%*
Intercept	40.784	8.507	4.794	0.001	21.167	60.401
x	0.766	0.175	4.375	0.002	0.362	1.169

NOTE: *Excel* does not provide options for estimation and prediction or for the test of significant correlation in Section 12.7. The diagnostic plots which can be generated in *Excel* are not the same plots as we have discussed in Section 12.5 and will not be discussed in this section.

Linear Regression Procedures—*MINITAB*

In Chapter 3, we used some of the linear regression procedures available in *MINITAB* to obtain a graph of the best-fitting least-squares regression line and to calculate the correlation coefficient r for a bivariate data set. Now that you have studied the testing and estimation techniques for a simple linear regression analysis, more *MINITAB* options are available to you.

EXAMPLE 12.10

Refer to Table 12.1, in which the relationship between x = mathematics achievement test score and y = final calculus grade was studied.

Student	Mathematics Achievement Test Score (x)	Final Calculus Grade (y)
1	39	65
2	43	78
3	21	52
4	64	82
5	57	92
6	47	89
7	28	73
8	75	98
9	34	56
10	52	75

1. Enter the values for x and y into the first two columns of a *MINITAB* worksheet.
2. The main tools for linear regression analysis are generated using **Stat ▶ Regression ▶ Regression.** (You will use this same sequence of commands in Chapter 13 when you study *multiple regression analysis.*) The dialogue box for the Regression command is shown in Figure 12.19(a).
3. Select **y** for the "Response" variable and **x** for the "Predictor" variable. You can now generate some residual plots to check the validity of your regression assumptions before you use the model for estimation or prediction. Choose **Graphs** to display the dialogue box in Figure 12.19(b). We have used **Regular** residual plots, checking the boxes for "Normal plot of residuals" and "Residuals versus fits." Click **OK** to return to the main dialogue box.

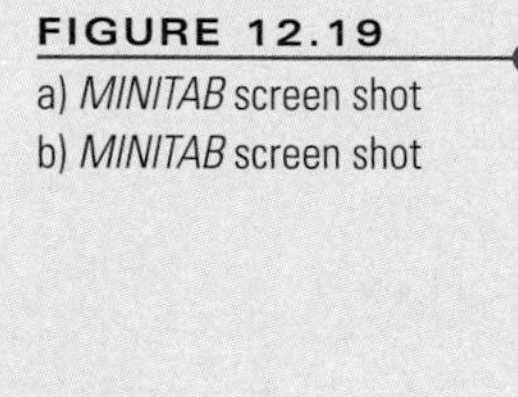

FIGURE 12.19

a) *MINITAB* screen shot
b) *MINITAB* screen shot

(a)

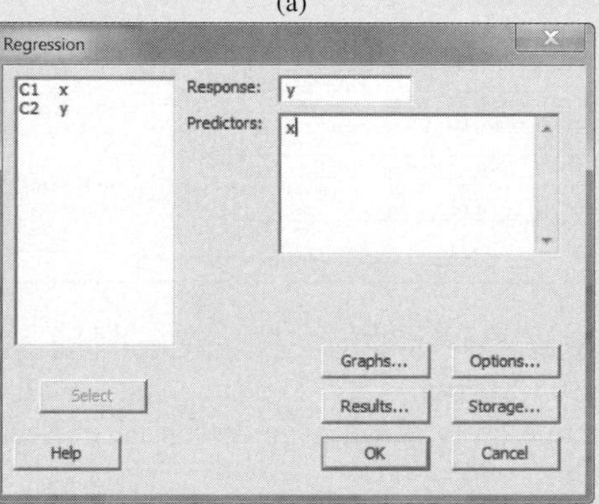

(b)

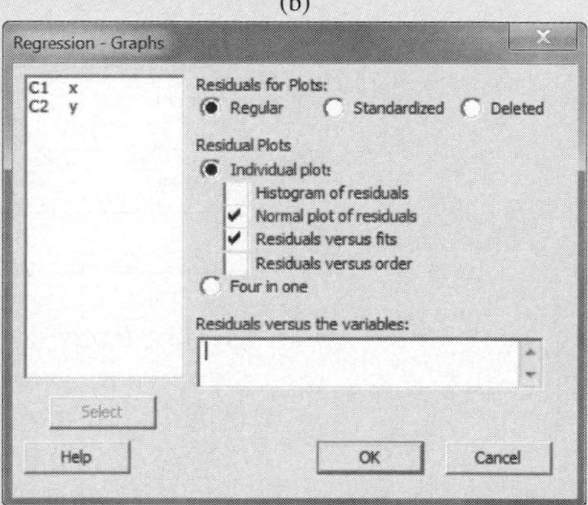

4. If you now choose **Options,** you can obtain confidence and prediction intervals for either of these cases:
   - A single value of x (typed in the box marked "Prediction intervals for new observations").
   - Several values of x stored in a column (say, C3) of the worksheet.
5. Enter the value **50** in Figure 12.20(a) to match the output given in Figure 12.13. When you click **OK** twice, the regression output is generated as shown in Figure 12.20(b). The two diagnostic plots will appear in separate graphics windows.

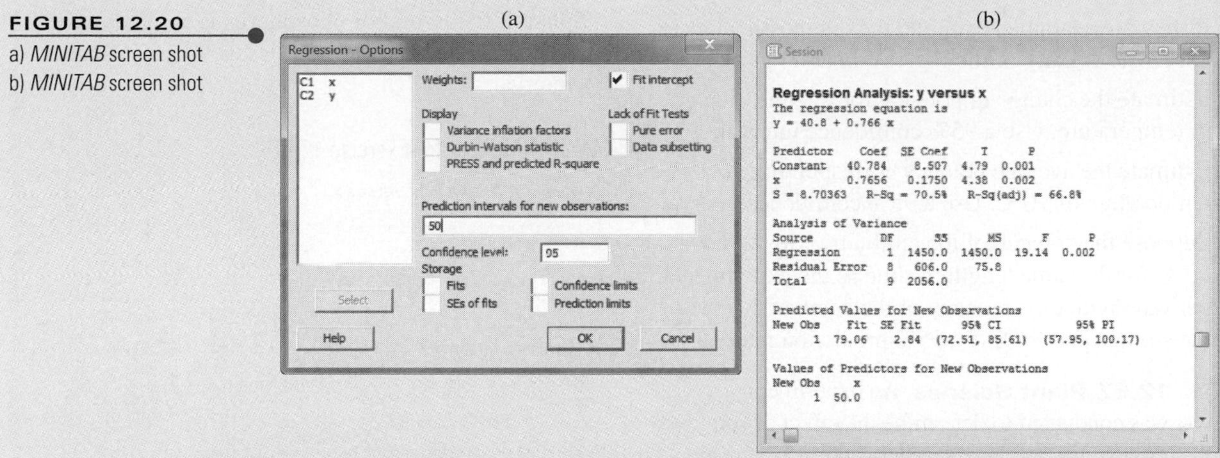

FIGURE 12.20
a) *MINITAB* screen shot
b) *MINITAB* screen shot

6. If you wish, you can now plot the data points, the regression line, and the upper and lower confidence and prediction limits (see Figure 12.14) using **Stat ▶ Regression ▶ Fitted Line Plot.** Select y and x for the response and predictor variables and click "Display confidence interal" and "Display prediction interval" in the **Options** dialogue box. Make sure that **Linear** is selected as the "Type of Regression Model," so that you will obtain a linear fit to the data.
7. Recall that in Chapter 3, we used the command **Stat ▶ Basic Statistics ▶ Correlation** to obtain the value of the correlation coefficient r. Make sure that the box marked "Display p-values" is checked. The output for this command (using the test/grade data) is shown in Figure 12.21. Notice that the p-value for the test of $H_0: \rho = 0$ is identical to the p-value for the test of $H_0: \beta = 0$ because the tests are exactly equivalent!

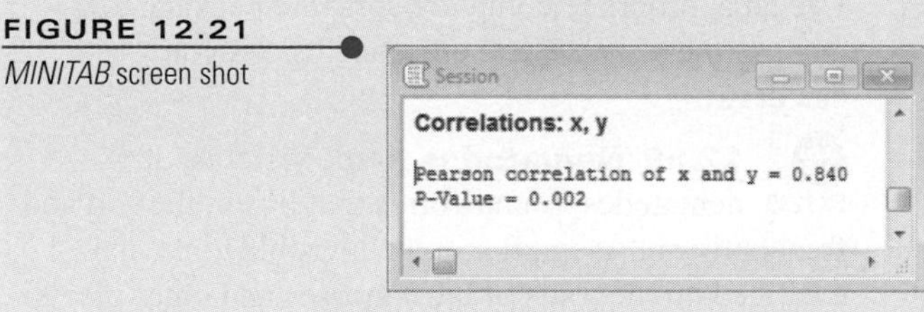

FIGURE 12.21
MINITAB screen shot

Supplementary Exercises

Data set EX1266 **12.66 Potency of an Antibiotic** An experiment was conducted to observe the effect of an increase in temperature on the potency of an antibiotic. Three 28-gram portions of the antibiotic were stored for equal lengths of time at each of these temperatures: −1.1°C, 10°C, 21.1°C, and 32.2°C. The potency readings observed at each temperature of the experimental period are listed here:

Potency readings, y	38, 43, 29	32, 26, 33	19, 27, 23	14, 19, 21
Temperature, x	−1.1°	10°	21.1°	32.2°

Use an appropriate computer program to answer these questions:

a. Find the least-squares line appropriate for these data.

b. Plot the points and graph the line as a check on your calculations.

c. Construct the ANOVA table for linear regression.

d. If they are available, examine the diagnostic plots to check the validity of the regression assumptions.

e. Estimate the change in potency for a 1-unit change in temperature. Use a 95% confidence interval.

f. Estimate the average potency corresponding to a temperature of 10°C. Use a 95% confidence interval.

g. Suppose that a batch of the antibiotic was stored at 10°C for the same length of time as the experimental period. Predict the potency of the batch at the end of the storage period. Use a 95% prediction interval.

Data set EX1267 **12.67 Plant Science** An experiment was conducted to determine the effect of soil applications of various levels of phosphorus on the inorganic phosphorus levels in a particular plant. The data in the table represent the levels of inorganic phosphorus in micromoles (μmol) per gram dry weight of sudan grass roots grown in the greenhouse for 28 days, in the absence of zinc. Use the *MINITAB* output to answer the questions.

Phosphorus Applied, x	Phosphorus in Plant, y
0.5 μmol	204
	195
	247
	245
0.25 μmol	159
	127
	95
	144
0.10 μmol	128
	192
	84
	71

a. Plot the data. Do the data appear to exhibit a linear relationship?

b. Find the least-squares line relating the plant phosphorus levels y to the amount of phosphorus applied to the soil x. Graph the least-squares line as a check on your answer.

c. Do the data provide sufficient evidence to indicate that the amount of phosphorus present in the plant is linearly related to the amount of phosphorus applied to the soil?

d. Estimate the mean amount of phosphorus in the plant if 0.20 μmol of phosphorus is applied to the soil, in the absence of zinc. Use a 90% confidence interval.

MINITAB output for Exercise 12.67

Regression Analysis: y versus x

```
The regression equation is
y = 80.9 + 271 x

Predictor        Coef     SE Coef        T        P
Constant        80.85       22.40     3.61    0.005
x              270.82       68.31     3.96    0.003

S = 39.0419     R-Sq = 61.1%     R-Sq(adj) = 57.2%

Predicted Values for New Observations
New Obs     Fit    SE Fit       95.0% CI             95.0% PI
1         135.0      12.6    (112.1,   157.9)   (60.6,     209.4)

Values of Predictors for New Observations
New Obs       x
1         0.200
```

Data set EX1268 **12.68 Track Stats!** An experiment was conducted to investigate the effect of a training program on the length of time for a typical male university student to complete the 100-metre dash. Nine students were placed in the program. The reduction y in time to complete the 100-metre dash was measured for three students at the end of two weeks, for three at the end of four weeks, and for three at the end of six weeks of training. The data are given in the table:

Reduction in time, y (sec)	1.6, 0.8, 1.0	2.1, 1.6, 2.5	3.8, 2.7, 3.1
Length of training, x (wk)	2	4	6

Use an appropriate computer software package to analyze these data. State any conclusions you can draw.

Data set EX1269 **12.69 Nematodes** Some varieties of nematodes, roundworms that live in the soil and frequently are so small as to be invisible to the naked eye, feed on the roots of lawn grasses and other plants. This pest, which is particularly troublesome in warm climates, can be treated by the application of nematicides. Data collected on the percent kill of nematodes for various rates of application (dosages given in kg per hectare of active ingredient) are as follows:

Rate of application, x	2	3	4	5
Percent kill, y	50, 56, 48	63, 69, 71	86, 82, 76	94, 99, 97

MINITAB diagnostic plots for Exercise 12.69

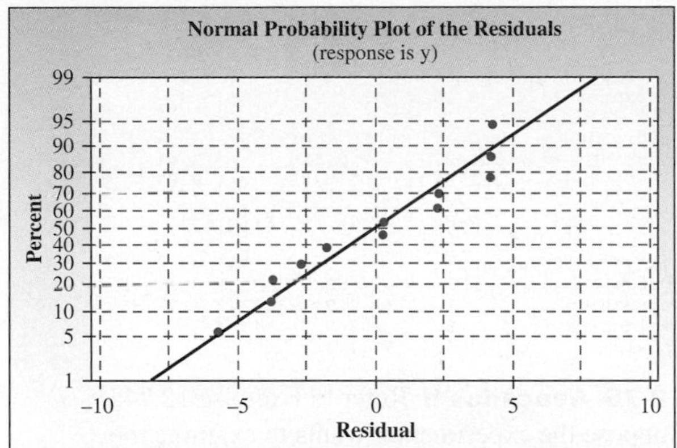

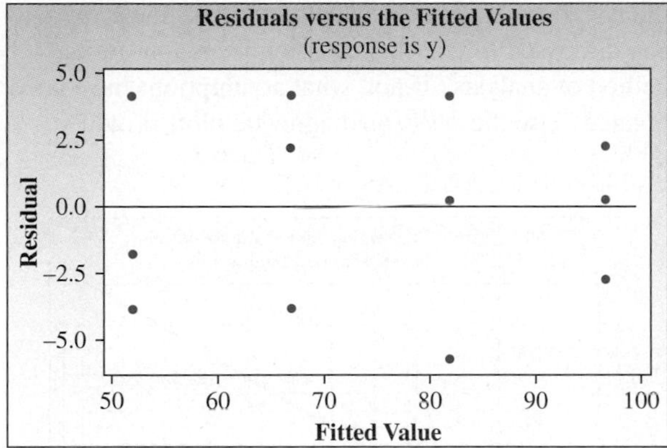

Use an appropriate computer printout to answer these questions:

a. Calculate the coefficient of correlation r between rates of application x and percent kill y.

b. Calculate the coefficient of determination r^2 and interpret.

c. Fit a least-squares line to the data.

d. Suppose you wish to estimate the mean percent kill for an application of 1.82 kg of the nematicide per hectare. What do the diagnostic plots generated by *MINITAB* tell you about the validity of the regression assumptions? Which assumptions may have been violated? Can you explain why?

12.70 Knee Injuries Athletes and others suffering the same type of injury to the knee often require anterior and posterior ligament reconstruction. In order to determine the proper length of bone-patellar tendon-bone grafts, experiments were done using three imaging techniques to determine the required length of the grafts, and these results were compared to the actual length required. A summary of the results of a simple linear regression analysis for each of these three methods is given in the following table:[18]

Imaging Technique	Coefficient of Determination, r^2	Intercept	Slope	p-value
Radiographs	0.80	−3.75	1.031	<0.0001
Standard MRI	0.43	20.29	0.497	0.011
3-Dimensional MRI	0.65	1.80	0.977	<0.0001

a. What can you say about the significance of each of the three regression analyses?

b. How would you rank the effectiveness of the three regression analyses? What is the basis of your decision?

c. How do the values of r^2 and the p-values compare in determining the best predictor of actual graft lengths of ligament required?

Data set EX1271 **12.71 Global Warming II?** Refer to Exercise 12.13 and data set EX1213.

a. Use an appropriate program to analyze the relationship between year and temperature anomaly.

b. Explain all pertinent details of your analysis.

12.72 How Long Is It II? Refer to Exercise 12.14 and data set EX1214 regarding a subject's ability to estimate sizes. The table that follows gives the actual and estimated lengths of the specified objects:

Object	Estimated (cm)	Actual (cm)
Pencil	17.78	15.24
Dinner plate	24.13	26.04
Book 1	19.05	17.15
Cellphone	10.16	10.8
Photograph	36.83	40.0
Toy	9.53	12.7
Belt	106.68	105.41
Clothespin	6.99	9.53
Book 2	25.4	23.5
Calculator	8.89	12.07

a. Use an appropriate program to analyze the relationship between the actual and estimated lengths of the listed objects.

b. Explain all pertinent details of your analysis.

Data set EX1273 **12.73 Tennis, Anyone?** If you play tennis, you know that tennis racquets vary in their physical characteristics. The data in the accompanying table give measures of bending stiffness and twisting stiffness as measured by engineering tests for 12 tennis racquets:

Racquet	Bending Stiffness, x	Twisting Stiffness, y
1	419	227
2	407	231
3	363	200
4	360	211
5	257	182
6	622	304
7	424	384
8	359	194
9	346	158
10	556	225
11	474	305
12	441	235

a. If a racquet has bending stiffness, is it also likely to have twisting stiffness? Do the data provide evidence that x and y are positively correlated?

b. Calculate the coefficient of determination r^2 and interpret its value.

Data set EX1274 **12.74 Avocado Research** Movement of avocados into the United States from certain areas is prohibited because of the possibility of bringing fruit flies into the country with the avocado shipments. However, certain avocado varieties supposedly are resistant to fruit fly infestation before they soften as a result of ripening. The data in the table resulted from an experiment in which avocados ranging from one to nine days after harvest were exposed to Mediterranean fruit flies. Penetrability of the avocados was measured on the day of exposure, and the percentage of the avocado fruit infested was assessed.

Days after Harvest	Penetrability	Percentage Infected
1	0.91	30
2	0.81	40
4	0.95	45
5	1.04	57
6	1.22	60
7	1.38	75
9	1.77	100

Use the *MINITAB* printout of the regression of percentage infected (y) on days after harvest (x) to analyze the relationship between these two variables. Explain all pertinent parts of the printout and interpret the results of any tests.

MINITAB output for Exercise 12.74

Regression Analysis: Percent versus x

```
The regression equation is
Percent = 18.4 + 8.18 x

Predictor      Coef    SE Coef       T       P
Constant     18.427      5.110    3.61   0.015
x            8.1768     0.9285    8.81   0.000

S = 6.35552    R-Sq = 93.9%    R-Sq(adj) = 92.7%

Analysis of Variance
Source           DF      SS      MS       F      P
Regression        1  3132.9  3132.9   77.56  0.000
Residual Error    5   202.0    40.4
Total             6  3334.9
```

12.75 Avocados II Refer to Exercise 12.74. Suppose the experimenter wants to examine the relationship between the penetrability and the number of days after harvest. Does the method of linear regression discussed in this chapter provide an appropriate method of analysis? If not, what assumptions have been violated? Use the *MINITAB* diagnostic plots provided.

MINITAB diagnostic plots for Exercise 12.75

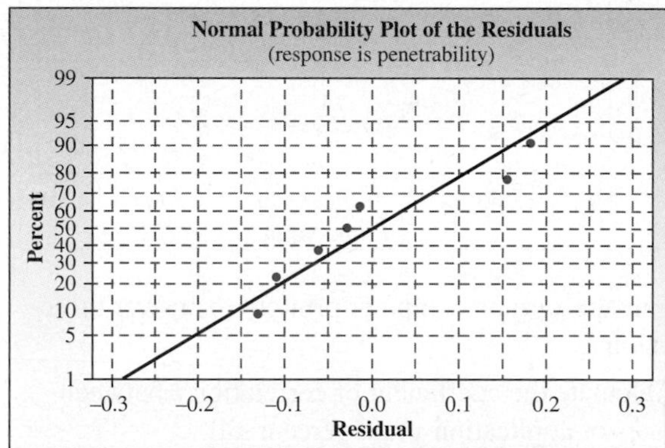

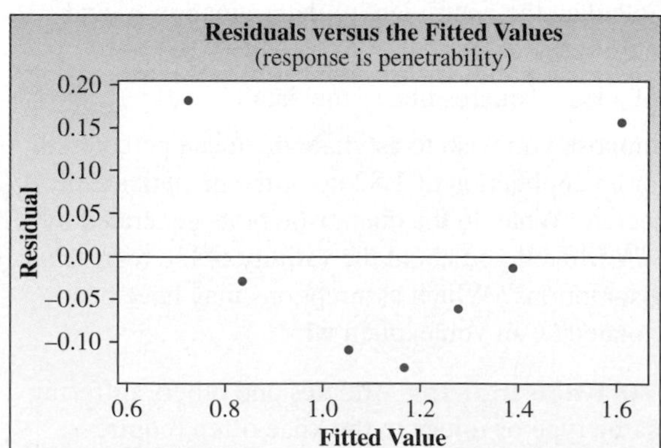

Data set EX1276 **12.76 Metabolism and Weight Gain** Why is it that one person may tend to gain weight, even if he or she eats no more and exercises no less than a slim friend? Recent studies suggest that the factors that control metabolism may depend on your genetic makeup. One study involved 11 pairs of identical twins fed about 1000 calories per day more than needed to maintain initial weight. Activities were kept constant, and exercise was minimal. At the end of 100 days, the changes in body weight (in kg) were recorded for the 22 twins.[19] Is there a significant positive correlation between the changes in body weight for the twins? Can you conclude that this similarity is caused by genetic similarities? Explain.

Pair	Twin A	Twin B
1	4.2	7.3
2	5.5	6.5
3	7.1	5.7
4	7.0	7.2
5	7.8	7.9
6	8.2	6.4
7	8.2	6.5
8	9.1	8.2
9	11.5	6.0
10	11.2	13.7
11	13.0	11.0

Data set EX1277 **12.77 Movie Reviews** How many days can a movie run and still make a reasonable profit? The data that follow show the number of days in release (x) and the gross to date (y) for the top 12 movies on March 30, 2016.[20]

Movie	Gross to Date ($ millions)	Days in Release
Batman v Superman: Dawn of Justice	201	6
Zootopia	253	27
My Big Fat Greek Wedding 2	24	6
Miracles from Heaven	38	15
The Divergent Series: Allegiant	50	13
10 Cloverfield Lane	58	20
Deadpool	351	48
London Has Fallen	57	27
Star Wars (Episode VII): The Force Awakens	934	104
Whiskey Tango Foxtrot	22	27
Kung Fu Panda 3	140	62
The Perfect Match	9	20

Source: © Nash Information Services, LLC.

a. Plot the points in a scatterplot. Does it appear that the relationship between x and y is linear? How would you describe the direction and strength of the relationship?

b. Calculate the value of r^2. What percentage of the overall variation is explained by using the linear model rather than $\bar{y}$ to predict the response variable y?

c. What is the regression equation? Do the data provide evidence to indicate that x and y are linearly related? Test using a 5% significance level.

d. Given the results of parts b and c, is it appropriate to use the regression line for estimation and prediction? Explain your answer.

12.78 In addition to increasingly large bounds on error, why should an experimenter refrain from predicting y for values of x outside the experimental region?

12.79 If the experimenter stays within the experimental region, when will the error in predicting a particular value of y be maximum?

Data set EX1280 **12.80 Oatmeal, Anyone?** An agricultural experimenter, investigating the effect of the amount of nitrogen x applied in 45 kg per hectare on the yield of oats y measured in 35 litres per hectare, collected the following data:

x	1	2	3	4
y	22	38	57	68
	19	41	54	65

a. Find the least-squares line for the data.

b. Construct the ANOVA table.

c. Is there sufficient evidence to indicate that the yield of oats is linearly related to the amount of nitrogen applied? Use $\alpha = 0.05$.

d. Predict the expected yield of oats with 95% confidence if 285 kg of nitrogen per hectare are applied.

e. Estimate the average increase in yield for an increase of 115 kg of nitrogen per hectare with 99% confidence.

f. Calculate r^2 and explain its significance in terms of predicting y, the yield of oats.

Data set EX1281 **12.81 Fresh Roses** A horticulturalist devised a scale to measure the freshness of roses that were packaged and stored for varying periods of time before transplanting. The freshness measurement y and the length of time in days that the rose is packaged and stored before transplanting x are given below:

x	5	10	15	20	25
y	15.3	13.6	9.8	5.5	1.8
	16.8	13.8	8.7	4.7	1.0

a. Fit a least-squares line to the data.

b. Construct the ANOVA table.

c. Is there sufficient evidence to indicate that freshness is linearly related to storage time? Use $\alpha = 0.05$.

d. Estimate the mean rate of change in freshness for a one-day increase in storage time using a 98% confidence interval.

e. Estimate the expected freshness measurement for a storage time of 14 days with a 95% confidence interval.

f. Of what value is the linear model in reference to $\bar{y}$ in predicting freshness?

Data set EX1282 **12.82 Lexus, Inc.** The Lexus GX is a mid-size sport utility vehicle (SUV) sold in North American and Eurasian markets. The GX 470 was introduced in 2002 (as a 2003 model) and was later upgraded with a new off-road suspension system. The sales of the Lexus GX 470 from its inception until 2009 are given in the table:[21]

Calendar Year	Total Sales (U.S.)
2002	2,190
2003	31,376
2004	35,420
2005	34,339
2006	25,454
2007	23,035
2008	15,759
2009	6,235

a. Plot the data using a scatterplot. How would you describe the relationship between year and sales of the Lexus GX 470?

b. Even though the scatterplot in part a might indicate differently, assume that the relationship between year and sales is linear. Find the least-squares regression line relating the sales of the Lexus GX 470 to the year being measured.

c. Is there sufficient evidence to indicate that sales are linearly related to year? Use $\alpha = 0.05$.

d. Examine the diagnostic plots shown below. What can you conclude about the validity of the regression assumptions?

Diagnostic plots for Exercise 12.82

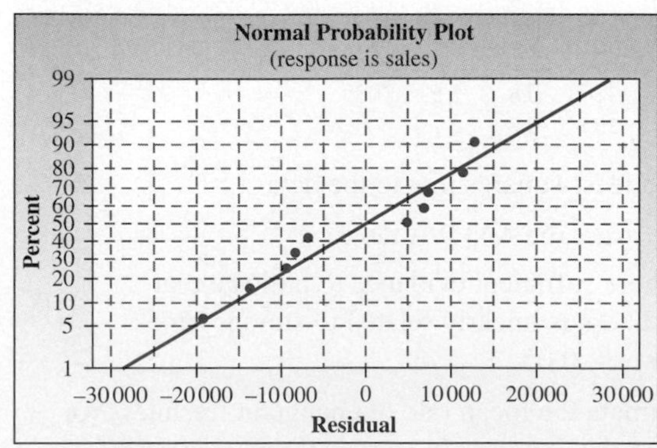

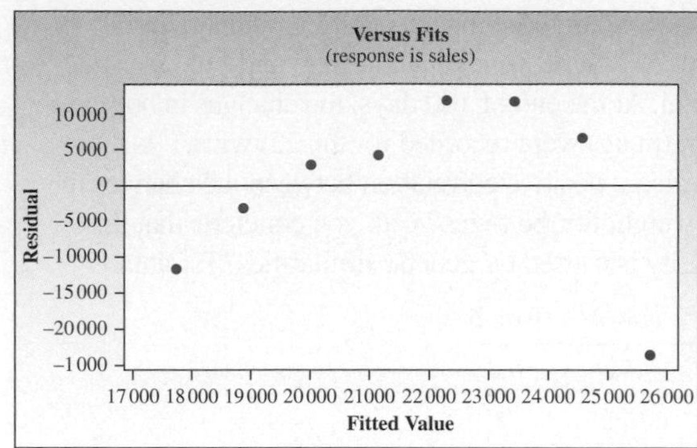

e. Based on your conclusions in part d, is it advisable to predict the 2010 sales using the regression line from part b? Explain.

Data set EX1283 **12.83 Tim Hortons** Here is some nutritional data for a sampling of Tim Hortons beverages. The nutritional information for all Tim Hortons products can be found on the company website, www.timhortons.com/nutrition and in French at www.timhortons/valeurnutritives.[22]

Product	Calories (g)	Fat (g)	Carb. (g)	Fibre	Protein
Original Blend Coffee	0	0	0	0	0
Dark Roast Coffee	0	0	0	0	0
Decaffeinated Coffee	0	0	0	0	0
Original Blend Coffee (1 Cream, 1 Sugar)	70	3.5	9	0	1
Steeped Tea made with Whole Leaf	0	0	0	0	0
Specialty Tea	0	0	0	0	0
Tea (1 Milk, 1 Sugar)	40	0.4	8	0	1
Hot Chocolate	240	6	45	2	2
White Hot Chocolate	260	9	42	0	1
Apple Cider	210	0	52	0	0
Latte	80	0.2	12	0	8
Flavoured Latte	130	0.2	26	0	7
Chocolate Dream Latte	180	5	26	0	7
Cappuccino	60	0.2	10	0	6
French Vanilla Cappuccino	250	8	41	1	4
English Toffee Cappuccino	240	7	41	2	4
Café Mocha	190	8	29	1	1
Iced Capp©™ (Cream)	250	11	33	0	2
Iced Capp©™ (Milk)	150	1.5	32	0	3
Iced Capp©™ (Chocolate Milk)	160	0.5	36	0	3
Flavoured Iced Capp™ (Cream)	250	11	34	0	2
Iced Coffee (Cream)	120	7	12	0	1
Iced Coffee (Milk)	70	1	12	0	2
Iced Coffee (Chocolate Milk)	80	0.5	16	0	2
Flavoured Iced Coffee (Cream)	120	7	13	0	1

Note: All Tim Hortons trademarks referenced herein are owned by Tim Hortons. Used with permission.

Use the appropriate statistical methods to analyze the relationships between some of the nutritional variables given in the table. Write a summary report explaining any conclusions that you can draw from your analysis.

CASE STUDY 

Data set Canadian Economy

Are Foreign Companies "Buying Up the Canadian Economy"?

Canada's for sale! We're losing our sovereignty! Dire statements such as these frequently appear in the media when news breaks about a well-known Canadian company being purchased by a foreign competitor, with the implication that there will soon be no Canadian-owned companies left. In the following case study, we explore the relationship between the percentage of foreign ownership and year, using a simple linear regression analysis. One response has been for the Canadian government to try to regulate takeovers, making it more difficult for foreign companies to directly invest in Canada. At other times, the government has sought to encourage foreign investment by changing or rescinding regulations. For example, the Foreign Investment Review Agency was created in 1975 to monitor and regulate foreign takeovers in Canada. In 1985, after a change of government, the Foreign Investment Review Agency was replaced with a new agency, Investment Canada, whose mandate was intended to be less restrictive. Have these changes had any effect? The data in the table represent the percentage of commercial assets in non-financial corporations under foreign control (y) for the years 1975–2004. To simplify the analysis, we have coded the year using the coded variable x = year − 1975.[23]

Percentage of Assets under Foreign Control in Non-financial Corporations, 1975–2004

Year	x (year minus 1975)	y (percent assets)	Year	x (year minus 1975)	y (percent assets)
1975	0	30.2	1990	15	23.7
1976	1	28.4	1991	16	23.6
1977	2	28.3	1992	17	23.9
1978	3	26.7	1993	18	23.8
1979	4	26.8	1994	19	23.6
1980	5	25.3	1995	20	25.1
1981	6	23.4	1996	21	25.4
1982	7	22.6	1997	22	25.9
1983	8	22.3	1998	23	26.9
1984	9	22.2	1999	24	25.3
1985	10	21.4	2000	25	25.5
1986	11	21.5	2001	26	28.8
1987	12	22.5	2002	27	28.7
1988	13	23.3	2003	28	28.7
1989	14	23.6	2004	29	28.5

1. Using a scatterplot, plot the data for the years 1975–1985. Does there appear to be a linear relationship between the percentage of foreign ownership and the year?
2. Use a computer software package to find the least-squares line for predicting the percentage of foreign ownership as a function of year for the years 1975–1985.
3. Is there a significant linear relationship between the percentage of foreign ownership and year?
4. Use the computer program to predict the percentage of foreign ownership with 95% confidence intervals for the years 2002, 2003, and 2004.
5. Now look at the actual data points for those years. Do the predictions obtained in step 4 provide accurate estimates of the actual values observed in these years? Explain.
6. Add the data for 1986–2004 to your database, and recalculate the regression line. What effect have the new data points had on the slope? What is the effect on SSE?
7. Given the form of the scatterplot, does it appear that a straight line provides an accurate model for the data? What other type of model might be more appropriate? (Use residual plots to help answer the question.)

PROJECTS

Project 12: Aspen Mixedwood Forests in Alberta[24]

Boreal mixedwood forests have been commercially logged for conifers, transected by roads and seismic cutlines, and regionally fragmented by agricultural practices. Most recently, Alberta's public mixedwood forests have been allocated for harvest of trembling aspen for pulp, paper, and oriented strand board products. With relevance to commercial forestry, this report describes the structure and biodiversity of aspen-dominated mixedwood forests of fire origin. The boreal mixedwood forest in Canada extends from southwestern Manitoba, through the central and northern parts of the prairie provinces, and into northeastern British Columbia.

A scientist is studying the relationship between x = diameter at breast height (DBH) (in centimetres) of aspen mixedwood forest in Alberta and y = canopy height (in metres).

The data are as follows:

x	5.1	7.9	7.1	7.4	15.8	16.4	15.9	13.9	37.0	29.8	45.0	37.0
y	7.1	11.0	9.5	9.2	18.7	19.9	18.8	15.1	25.5	22.0	23.4	26.7

a. Develop a scatterplot for these data.

b. What does the scatterplot developed in the previous question indicate about the relationship between the two variables?

c. Calculate the correlation coefficient r. Is there a significant positive correlation between x and y? Perform an appropriate statistical test.

d. Use a statistical software package of your choice and report the regression analysis results.

e. What is the equation of the least-squares regression line?

f. What is the average change in canopy height with an increase of one centimetre in DBH?

g. How do you describe the strength of the relationship between y and x?

h. Find and interpret the coefficient of determination.

i. What percentage of the total variation in y can be explained by the simple linear regression model?

j. Is the simple linear regression model useful for predicting the canopy height of the trees from a given diameter in breast height of the trees? Use the p-value approach to test the usefulness of the linear regression model at the 0.05 level of significance.

k. Use the F test to test the hypotheses $H_0 : \beta = 0$ vs. $H_a : \beta \neq 0$ at $\alpha = 0.05$.

l. Compare the two-tailed critical value at $\alpha = 0.025$ for the t test with the critical value for the F statistic at $\alpha = 0.05$. What is the relationship between the two values?

m. Predict the tree canopy height of a 42-centimetre diameter at breast height.

n. Would you use the least-squares prediction equation line to find the estimated canopy height for a diameter at breast height of 52 centimetres? Why or why not?

o. Find a 95% confidence interval for the slope of the line.

p. Estimate the change in canopy height for a 1-unit change in diameter at breast height. Use a 99% confidence interval. Compare this interval with the interval in part o.

q. What is the best estimate of the common population variance σ^2 ?

r. Use the predicted values and the actual values of y to calculate the residuals.

s. Plot the residuals against the predicted values $\hat{y}$. Does it appear that the constant variance regression assumption has been violated? Explain. What conclusion, if any, can be drawn from the plot?

t. Draw a histogram of the residuals. Does it appear that the errors are normally distributed? Explain.

u. Use the residuals to compute the standardized residuals.

v. Do there appear to be any outliers in the data? Justify your answer.

w. Consider the normal probability plot of the residuals. What conclusion can be drawn from the plot?

x. Find a 95% confidence interval for the mean value of y (canopy height) when diameter at breast height is 10.

y. Compute a 95% prediction interval for the canopy height when diameter at breast height is 10.

z. Which interval in the previous two questions is narrower: the confidence interval estimate of the expected value of y or the prediction interval of y for the same given value of x (10 centimetres) and same confidence level? Why?

13

Multiple Regression Analysis

Songquan Deng/Shutterstock.com

GENERAL OBJECTIVES

In this chapter, we extend the concepts of linear regression and correlation to a situation where the average value of a random variable Y is related to several independent variables—$x_1, x_2, \ldots, x_k$—in models that are more flexible than the straight-line model of Chapter 12. With *multiple regression analysis*, we can use the information provided by the independent variables to fit various types of models to the sample data, to evaluate the usefulness of these models, and finally to estimate the average value of y or predict the actual value of y for given values of $x_1, x_2, \ldots, x_k$.

CHAPTER INDEX

"Buying Up the Canadian Economy"—Another Look

The case study in Chapter 12 examined the effects of government regulation on foreign ownership of assets in Canada using a simple linear regression. However, the relationship was not really linear, and our predictions were far from accurate. We re-examine the same data at the end of this chapter, using the multiple regression analysis.

13.1 INTRODUCTION

Multiple linear regression is an extension of simple linear regression to allow for more than one independent variable. That is, instead of using only a single independent variable x to explain the variation in Y, you can simultaneously use several independent (or predictor) variables. By using more than one independent variable, you should do a better job of explaining the variation in Y and hence be able to make more accurate predictions.

For example, a company's regional sales Y of a product might be related to three factors:

- x_1: The amount spent on television advertising
- x_2: The amount spent on newspaper advertising
- x_3: The number of sales representatives assigned to the region

A researcher would collect data measuring the variables y, x_1, x_2, and x_3 and then use these sample data to construct a prediction equation relating Y to the three predictor variables. Of course, several questions arise, just as they did with simple linear regression:

- How well does the model fit?
- How strong is the relationship between Y and the predictor variables?
- Have any important assumptions been violated?
- How good are estimates and predictions?

The methods of **multiple regression analysis**—which are almost always done with a computer software program—can be used to answer these questions. This chapter provides a brief introduction to multiple regression analysis and the difficult task of model building—that is, choosing the correct model for a practical application.

13.2 THE MULTIPLE REGRESSION MODEL

The **general linear model** for a multiple regression analysis describes a particular response Y using the model given next.

General Linear Model and Assumptions

$$Y = \beta_0 + \beta_1 x_1 + \beta_2 x_2 + \cdots + \beta_k x_k + \epsilon$$

where

- Y is the **response variable** that you want to predict.
- $\beta_0, \beta_1, \beta_2, \cdots, \beta_k$ are unknown constants.
- $x_1, x_2, \cdots, x_k$ are independent **predictor variables** that are measured without error.
- ϵ is the random error, which allows each response to deviate from the average value of y by the amount ϵ. You must assume that the values of ϵ (1) are independent; (2) have a mean of 0 and a common variance σ^2 for any set $x_1, x_2, \cdots, x_k$; and (3) are normally distributed.

When these assumptions about ϵ are met, the *average* value of y for a given set of values $x_1, x_2, \cdots, x_k$ is equal to the *deterministic* part of the model:

$$E(Y) = \beta_0 + \beta_1 x_1 + \beta_2 x_2 + \cdots + \beta_k x_k$$

You will notice that the multiple regression model and assumptions are *very similar* to the model and assumptions used for linear regression. It will probably not surprise you that the testing and estimation procedures are also extensions of those used in Chapter 12.

Multiple regression models are very flexible and can take many forms, depending on the way in which the independent variables $x_1, x_2, \cdots, x_k$ are entered into the model. We begin with a simple multiple regression model, explaining the basic concepts and procedures with an example. As you become more familiar with the multiple regression procedures, we increase the complexity of the examples, and you will see that the same procedures can be used for models of different forms, depending on the particular application.

EXAMPLE 13.1

Suppose you want to relate a random variable Y to two independent variables x_1 and x_2. The multiple regression model is

$$Y = \beta_o + \beta_1 x_1 + \beta_2 x_2 + \epsilon$$

with the mean value of y given as

$$E(Y) = \beta_0 + \beta_1 x_1 + \beta_2 x_2$$

This equation is a three-dimensional extension of the **line of means** from Chapter 12 and traces a **plane** in three-dimensional space (see Figure 13.1). The constant β_0 is called the **intercept**—the average value of y when x_1 and x_2 are both 0. The coefficients β_1 and β_2 are called the **partial slopes** or **partial regression coefficients**. The partial slope β_i (for $i = 1$ or 2) measures the change in y for a one-unit increase in x_i when *all other independent variables are held constant.* The value of the partial regression coefficient—say, β_1—with x_1 and x_2 in the model is generally *not* the same as the slope when you fit a line with x_1 alone. These coefficients are the unknown constants, which must be estimated using sample data to obtain the prediction equation.

NEED A TIP?

Instead of x and y plotted in two-dimensional space, y and $x_1, x_2, \ldots x_k$ have to be plotted in $(k + 1)$ dimensions.

FIGURE 13.1

Plane of means for Example 13.1

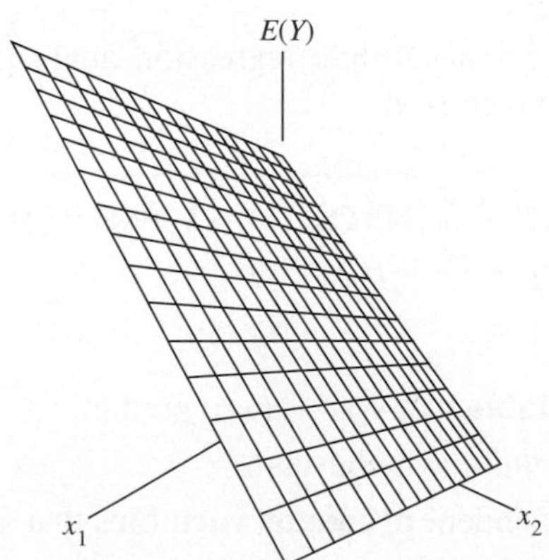

13.3 A MULTIPLE REGRESSION ANALYSIS

A multiple regression analysis involves estimation, testing, and diagnostic procedures designed to fit the multiple regression model

$$E(Y) = \beta_0 + \beta_1 x_1 + \beta_2 x_2 + \cdots + \beta_k x_k$$

to a set of data. Because of the complexity of the calculations involved, these procedures are almost always implemented with a regression program from one of several computer software packages. All give similar output in slightly different forms. We follow the basic patterns set in simple linear regression, beginning with an outline of the general procedures and illustrated with an example.

The Method of Least Squares

The prediction equation

$$\hat{y} = b_0 + b_1x_1 + b_2x_2 + \cdots + b_kx_k$$

is the line that minimizes SSE, the sum of squares of the deviations of the observed values y from the predicted values $\hat{y}$. These values are calculated using a regression program.

EXAMPLE  13.2

How do real estate agents decide on the asking price for a newly listed condominium? A computer database in a small community contains the listed selling price Y (in thousands of dollars), the amount of living area x_1 (in hundreds of square metres), and the numbers of floors x_2, bedrooms x_3, and bathrooms x_4, for $n = 15$ randomly selected condos currently on the market. The data are shown in Table 13.1.

TABLE 13.1 **Data on 15 Condominiums**

Observation	List Price, y	Living Area, x_1	Floors, x_2	Bedrooms, x_3	Baths, x_4
1	69.0	0.56	1	2	1
2	118.5	0.93	1	2	2
3	116.5	0.93	1	3	2
4	125.0	1.02	1	3	2
5	129.9	1.21	1	3	1.7
6	135.0	1.21	2	3	2.5
7	139.9	1.21	1	3	2
8	147.9	1.58	2	3	2.5
9	160.0	1.77	2	3	2
10	169.9	1.67	1	3	2
11	134.9	1.21	1	4	2
12	155.0	1.67	1	4	2
13	169.9	1.58	2	4	3
14	194.5	1.86	2	4	3
15	209.9	1.95	2	4	3

The multiple regression model is

$$E(Y) = \beta_0 + \beta_1x_1 + \beta_2x_2 + \beta_3x_3 + \beta_4x_4$$

which can be fit using either the *MINITAB* or *Microsoft Excel* software packages. You can find instructions for generating this output in the "Technology Today" section at the end of this chapter. The first portion of the *MINITAB* regression output is shown in Figure 13.2(a). You will find the fitted regression equation in the first two lines of the printout:

$$\hat{y} = 18.8 + 67.5x_1 - 16.2x_2 - 2.67x_3 + 30.3x_4$$

The partial regression coefficients are shown with slightly more accuracy in the second section of the *MINITAB* printout; a similar output generated by *Excel* is shown in

Figure 13.2(b). The columns list the name given to each independent predictor variable, its estimated regression coefficient, its standard error, and the t- and p-values that are used to test its significance *in the presence of all the other predictor variables.* We explain these tests in more detail in a later section.

FIGURE 13.2(a)

A portion of the *MINITAB* printout for Example 13.2

Regression Analysis: List Price versus Square Metres, Number of Floors, Bedrooms, Baths

```
The regression equation is
List Price = 18.8 + 67.5 Square Metres - 16.2 Number of Floors
           - 2.67 Bedrooms + 30.3 Baths

Predictor             Coef     SE Coef        T        P
Constant            18.763       9.207     2.04    0.069
Square Metres       67.488       7.806     8.65    0.000
Number of Floors   -16.203       6.212    -2.61    0.026
Bedrooms            -2.673       4.494    -0.59    0.565
Baths               30.271       6.849     4.42    0.001
```

FIGURE 13.2(b)

A portion of the *Excel* printout for Example 13.2

Predictor	*Coefficients*	*Standard Error*	*t stat*	*P-value*	*Lower 95%*	*Upper 95%*
Constant	18.763	9.207	2.04	0.069	−1.75	39.27
Square Metres	67.488	7.806	8.65	0.000	50.09	84.88
Number of Floors	−16.203	6.212	−2.61	0.026	−30.04	−2.36
Bedrooms	−2.673	4.494	−0.59	0.565	−12.65	7.34
Baths	30.271	6.849	4.42	0.001	15.01	45.53

The Analysis of Variance for Multiple Regression

The analysis of variance divides the total variation in the response variable Y,

$$\text{Total SS} = \Sigma y_i^2 - \frac{(\Sigma y_i)^2}{n}$$

into two portions:

- SSR (sum of squares for regression) measures the amount of variation explained by using the regression equation.
- SSE (sum of squares for error) measures the residual variation in the data that is not explained by the independent variables.

so that

$$\text{Total SS} = \text{SSR} + \text{SSE}$$

The **degrees of freedom** for these sums of squares are found using the following argument. There are $(n - 1)$ total degrees of freedom. Estimating the regression line requires estimating k unknown coefficients; the constant b_0 is a function of $\bar{y}$ and the other estimates. Hence, there are k regression degrees of freedom, leaving $(n - 1) - k$ degrees of freedom for error. As in previous chapters, the mean squares are calculated as $\text{MS} = \text{SS}/df$.

The ANOVA table for the real estate data in Table 13.1 is shown in the second portion of the *MINITAB* printout in Figure 13.3(a) and the lower section of the *Excel* printout in Figure 13.3(b). There are $n = 15$ observations and $k = 4$ independent predictor variables. You can verify that the total degrees of freedom, $(n - 1) = 14$, is divided into $k = 4$ for regression and $(n - k - 1) = 10$ for error.

FIGURE 13.3(a)
A portion of the *MINITAB* printout for Example 13.2

```
S = 6.84930         R-Sq = 97.1%       R-Sq(adj) = 96.0%
Analysis of Variance
Source              DF        SS         MS         F        P
Regression           4    15913.0     3978.3     84.80    0.000
Residual Error      10      469.1       46.9
Total               14    16382.2

Source               DF      Seq SS
Square Metres         1     14829.3
Number of Floors      1         0.9
Bedrooms              1       166.4
Baths                 1       916.5
```

FIGURE 13.3(b)
A portion of the *Excel* printout for Example 13.2

SUMMARY OUTPUT

Regression Statistics	
Multiple R	0.986
R Square	0.971
Adjusted R Square	0.960
Standard Error	6.849
Observations	15

ANOVA

	df	*SS*	*MS*	*F*	*Significance F*
Regression	4	15913.0	3978.3	84.80	0.000
Residual	10	469.1	46.9		
Total	14	16382.2			

The best estimate of the random variation σ^2 in the experiment—the variation that is unexplained by the predictor variables—is as usual given by

$$s^2 = \text{MSE} = \frac{\text{SSE}}{n-k-1} = 46.9$$

from the ANOVA table. The first line of Figure 13.3 (a) and fourth line of Figure 13.3(b) also show $s = \sqrt{s^2} = 6.84930$ using computer accuracy. The computer uses these values internally to produce test statistics, confidence intervals, and prediction intervals, which we discuss in subsequent sections.

The second section of Figure 13.3(a) and last section of Figure 13.3(b) show a decomposition of SSR = 15,913.0 in which the conditional contribution of each predictor variable *given the variables already entered into the model* is shown for the order of entry that you specify in your regression program. For the real estate example, the *MINITAB* program entered the variables in this order: square metres, then numbers of floors, bedrooms, and baths. These conditional or **sequential sums of squares** each account for one of the $k = 4$ regression degrees of freedom. It is interesting to notice that the predictor variable x_1 alone accounts for 14,829.3/15,913.0 = 0.932 or 93.2% of the total variation explained by the regression model. However, if you change the order of entry, another variable may account for the major part of the regression sum of squares!

Testing the Usefulness of the Regression Model

Recall in Chapter 12 that you tested to see whether Y and X were linearly related by testing $H_0 : \beta = 0$ with either a t test or an equivalent F test. In multiple regression, there is more than one *partial slope*—the *partial regression coefficients.* The t and F tests are no longer equivalent.

The Analysis of Variance F Test

Is the regression equation that uses information provided by the predictor variables $x_1, x_2, \cdots, x_k$ substantially better than the simple predictor $\bar{y}$ that does not rely on any of the x-values? This question is answered using an overall F test with the hypotheses:

NEED A TIP?

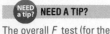

The overall F test (for the significance of the model) in multiple regression is one-tailed.

$$H_0: \beta_1 = \beta_2 = \ldots = \beta_k = 0$$

versus

$$H_a: \text{at least one of } \beta_1, \beta_2, \ldots, \beta_k \text{ is not } 0$$

The test statistic is found in the ANOVA table (Figure 13.3) as

$$F = \frac{\text{MSR}}{\text{MSE}} = \frac{3978.3}{46.9} = 84.80$$

which has an F distribution with $df_1 = k = 4$ and $df_2 = (n - k - 1) = 10$. Since the exact p-value, $P = 0.000$, is given in the printout, you can declare the regression to be highly significant. That is, at least one of the predictor variables is contributing significant information for the prediction of the response variable Y.

The Coefficient of Determination, R^2

NEED A TIP?

MINITAB printouts report R^2 as a percentage rather than a proportion.

How well does the regression model fit? The regression printout provides a statistical measure of the strength of the model in the **coefficient of determination, R^2**—the proportion of the total variation that is explained by the regression of y on $x_1, x_2, \ldots, x_k$—defined as

$$R^2 = \frac{\text{SSR}}{\text{Total SS}} = \frac{15{,}913.0}{16{,}382.2} = 0.971 \text{ or } 97.1\%$$

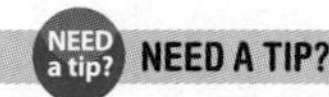

NEED A TIP?

R^2 is the multivariate equivalent of r^2, used in linear regression.

The coefficient of determination is sometimes called **multiple R^2** and is found in the first line of Figure 13.3(a), labelled "R-Sq," and in the second line of Figure 13.3(b), labelled "R Square." Hence, for the real estate example, 97.1% of the total variation has been explained by the regression model. The model fits very well!

It may be helpful to know that the value of the F statistic is related to R^2 by the formula

$$F = \frac{R^2/k}{(1-R^2)/(n-k-1)}$$

so that when R^2 is large, F is large, and vice versa.

Interpreting the Results of a Significant Regression

NEED A TIP?

You can show that $F = \frac{\text{MSR}}{\text{MSE}} = \frac{R^2/k}{(1 - R^2)/(n - k - 1)}$

Testing the Significance of the Partial Regression Coefficients

Once you have determined that the model is useful for predicting Y, you should explore the nature of the "usefulness" in more detail. Do all of the predictor variables add important information for prediction in *the presence of other predictors already in the model?* The individual t tests in the first section of the regression printout are designed to test the hypotheses

$$H_0: \beta_i = 0 \text{ versus } H_a: \beta_i \neq 0$$

for each of the partial regression coefficients, *given that the other predictor variables are already in the model.* These tests are based on the Student's t statistic given by

$$t = \frac{b_i - \beta_i}{SE(b_i)}$$

which has $df = (n - k - 1)$ degrees of freedom. The procedure is identical to the one used to test a hypothesis about the slope β in the simple linear regression model.†

NEED A TIP?

Test for the significance of the individual coefficient β_i, using t tests.

Figure 13.4 shows the t test statistic and p-values from the upper portion of the *MINITAB* printout and the lower section of the *Excel* printout. By examining the p-values in the last column, you can see that all the variables *except* x_3, the number of bedrooms, add very significant information for predicting y, **even with all the other independent variables already in the model.** Could the model be any better? It may be that x_3 is an unnecessary predictor variable. One option is to remove this variable and refit the model with a new set of data!

FIGURE 13.4(a)

A portion of the *MINITAB* printout for Example 13.2

```
Predictor              Coef       SE Coef        T        P
Constant             18.763         9.207     2.04    0.069
Square Metres        67.488         7.806     8.65    0.000
Number of Floors    -16.203         6.212    -2.61    0.026
Bedrooms             -2.673         4.494    -0.59    0.565
Baths                30.271         6.849     4.42    0.001
```

FIGURE 13.4(b)

A portion of the *Excel* printout for Example 13.2

	Coefficients	*Standard Error*	*t Stat*	*P-value*
Intercept	18.763	9.207	2.04	0.069
Square Metres	67.488	7.806	8.65	0.000
Number of Floors	–16.203	6.212	–2.61	0.026
Bedrooms	–2.673	4.494	–0.59	0.565
Baths	30.271	6.849	4.42	0.001

The Adjusted Value of R^2

Notice from the definition of R^2 = SSR/Total SS that its value can never decrease with the addition of more variables into the regression model. Hence, R^2 can be artificially inflated by the inclusion of more and more predictor variables.

An alternative measure of the strength of the regression model is adjusted for degrees of freedom by using mean squares rather than sums of squares:

$$R^2(\text{adj}) = \left(1 - \frac{\text{MSE}}{\text{Total SS}/(n-1)}\right)100\%$$

NEED A TIP?

Use R^2(adj) for comparing one or more possible models.

For the real estate data in Figure 13.3,

$$R^2(\text{adj}) = \left(1 - \frac{46.9}{16{,}382.2/14}\right)100\% = 96.0\%$$

is found in the first line of the *MINITAB* printout in Figure 13.3 and the third line of the *Excel* printout. The value "R-Sq(adj) = 96.0%" or "Adjusted R Square = 0.960" represents the percentage of variation in the response Y explained by the independent variables, corrected for degrees of freedom. The adjusted value of R^2 is mainly used to compare two or more regression models that use different numbers of independent predictor variables.

†Some packages use the t statistic just described, whereas others use the equivalent F statistic ($F = t^2$), since the square of a t statistic with v degrees of freedom is equal to an F statistic with 1 df in the numerator and v degrees of freedom in the denominator.

Checking the Regression Assumptions

Before using the regression model for its main purpose—estimation and prediction of *Y*—you should look at computer-generated **residual plots** to make sure that all the regression assumptions are valid. The *normal probability plot* and the *plot of residuals versus fit* are shown in Figure 13.5 for the real estate data. There appear to be three observations that do not fit the general pattern. You can see them as outliers in both graphs. These three observations should probably be investigated; however, they do not provide strong evidence that the assumptions are violated.

FIGURE 13.5

MINITAB diagnostic plots

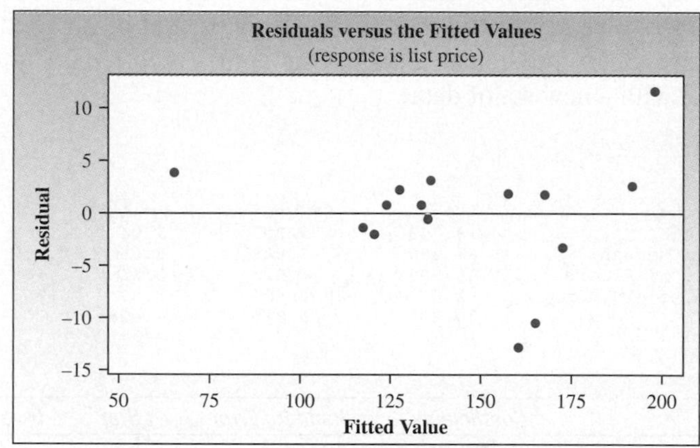

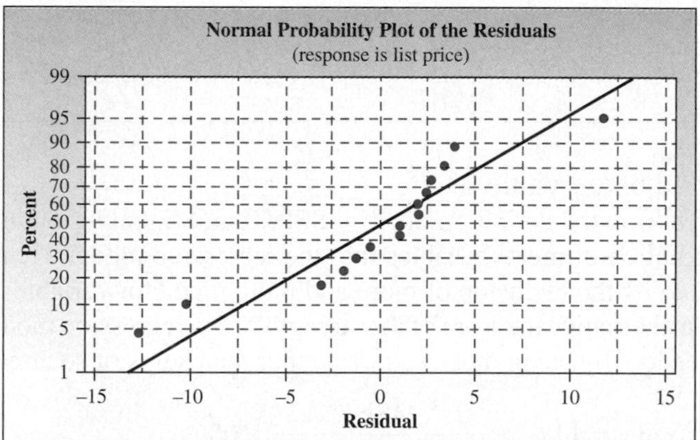

Using the Regression Model for Estimation and Prediction

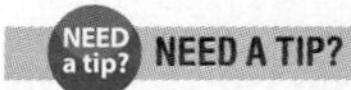

NEED A TIP?

For given values of x_1, x_2, ... x_k, the prediction interval will **always** be wider than the confidence interval.

Finally, once you have determined that the model is effective in describing the relationship between *Y* and the predictor variables $x_1, x_2, \dots, x_k$, the model can be used for these purposes:

- Estimating the average value of *y*—*E*(*Y*)—for given values of $x_1, x_2, \dots, x_k$
- Predicting a particular value of *y* for given values of $x_1, x_2, \dots, x_k$

The values of $x_1, x_2, \dots, x_k$ are entered into the computer, and the computer generates the fitted value $\hat{y}$ together with its estimated standard error and the confidence and prediction intervals. Remember that the prediction interval is *always wider* than the confidence interval.

Let's see how well our prediction works for the real estate data, using another house from the computer database—a house with 93 square metres of living area, one floor, three bedrooms, and two baths, which was listed at $121,500. The printout in Figure 13.6 shows the confidence and prediction intervals for these values. The actual value falls within both intervals, which indicates that the model is working very well!

FIGURE 13.6
Confidence and prediction intervals for Example 13.2

```
Predicted Values for New Observations
New Obs     Fit      SE Fit        95% CI              95% PI
1         117.78      3.11    (110.86,  124.70)  (101.02,  134.54)

Values of Predictors for New Observations
New     Square    Number of
Obs     Metres       Floors     Bedrooms     Baths
1         0.93         1.00         3.00      2.00
```

13.4 A POLYNOMIAL REGRESSION MODEL

In Section 13.3, we explained in detail the various portions of the multiple regression printout. When you perform a multiple regression analysis, you should use a step-by-step approach:

1. Obtain the fitted prediction model.
2. Use the analysis of variance F test and R^2 to determine how well the model fits the data.
3. Check the t tests for the partial regression coefficients to see which ones are contributing significant information in the presence of the others.
4. If you choose to compare several different models, use R^2(adj) to compare their effectiveness.
5. Use computer-generated residual plots to check for violation of the regression assumptions.

NEED A TIP?

A quadratic equation is $y = a + bx + cx^2$. The graph forms a **parabola**.

Once all of these steps have been taken, you are ready to use your model for estimation and prediction.

The predictor variables $x_1, x_2, \ldots, x_k$ used in the general linear model do not have to represent *different* predictor variables. For example, if you suspect that one independent variable x affects the response Y, but that the relationship is *curvilinear* rather than *linear*, then you might choose to fit a **quadratic model:**

$$Y = \beta_0 + \beta_1 x + \beta_2 x^2 + \epsilon$$

The quadratic model is an example of a second-order model because it involves a term whose exponents sum to 2 (in this case, x^2).† It is also an example of a **polynomial model**—a model that takes the form

$$Y = a + bx + cx^2 + dx^3 + \cdots$$

To fit this type of model using the multiple regression program, observed values of y, x, and x^2 are entered into the computer, and the printout can be generated as in Section 13.3.

†The *order* of a term is determined by the sum of the exponents of variables making up that term. Terms involving x_1 or x_2 are first-order. Terms involving x_1^2, x_2^2, or x_1x_2 are second-order.

EXAMPLE 13.3

In a study of variables that affect productivity in the retail grocery trade, W.S. Good uses value added per work-hour to measure the productivity of retail grocery outlets.[1] He defines "value added" as "the surplus [money generated by the business] available to pay for labour, furniture and fixtures, and equipment." Data consistent with the relationship between value added per work-hour Y and the size x of a grocery outlet described in Good's article are shown in Table 13.2 for 10 fictitious grocery outlets. Choose a model to relate Y to x.

TABLE 13.2 Data on Store Size and Value Added

Store	Value Added per Work-Hour, y	Size of Store (thousand square metres), x
1	$4.08	1.95
2	3.40	1.11
3	3.51	2.34
4	3.09	0.97
5	2.92	2.87
6	1.94	0.63
7	4.11	1.82
8	3.16	1.34
9	3.75	2.32
10	3.60	1.77

SOLUTION You can investigate the relationship between Y and x by looking at the plot of the data points in Figure 13.7. The graph suggests that productivity, Y, increases as the size of the grocery outlet, x, increases until an optimal size is reached. Above that size, productivity tends to decrease. The relationship appears to be *curvilinear*, and a quadratic model

$$E(Y) = \beta_0 + \beta_1 x + \beta_2 x^2$$

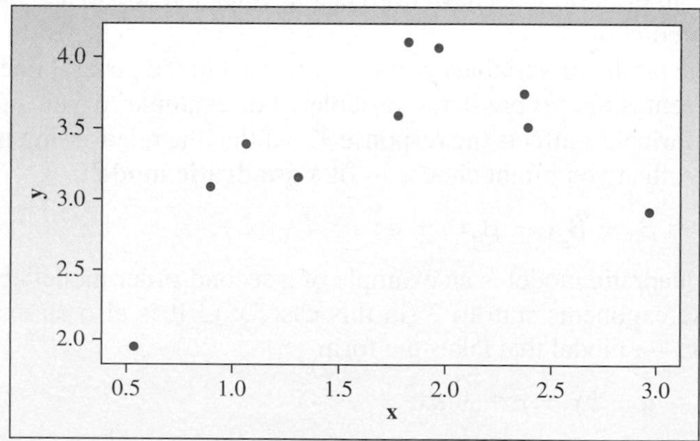

FIGURE 13.7 Plot of store size x and value added y for Example 13.3

may be appropriate. Remember that, in choosing to use this model, we are not saying that the true relationship is quadratic, but only that it may provide more accurate estimations and predictions than, say, a linear model.

EXAMPLE 13.4

Refer to the data on grocery retail outlet productivity and outlet size in Example 13.3. *MINITAB* was used to fit a quadratic model to the data and to graph the quadratic prediction curve, along with the plotted data points. Discuss the adequacy of the fitted model.

SOLUTION From the printout in Figure 13.8, you can see that the regression equation is

$$\hat{y} = -0.159 + 4.219x - 1.100x^2$$

The graph of this quadratic equation together with the data points is shown in Figure 13.9.

FIGURE 13.8

MINITAB printout for Example 13.4

```
Regression Analysis: y versus x, x-sq
The regression equation is
y = - 0.1594 + 4.219 x - 1.100 x**2

Predictor     Coef      SE Coef       T        P
Constant      -0.1594   0.5006        -0.32    0.760
x             4.2187    0.6244        6.76     0.000
x-sq          -1.1001   0.1778        -6.19    0.000

S = 0.250298        R-Sq = 87.9%       R-Sq(adj) = 84.5%

Analysis of Variance
Source        DF        SS          MS          F       P
Regression    2         3.19889     1.59945     25.53   0.001
Error         7         0.43855     0.06265
Total         9         3.63744

Sequential Analysis of Variance
Source        DF    SS          F       P
Linear        1     0.80032     2.26    0.171
Quadratic     1     2.39858     38.29   0.000
```

NEED A TIP?

Look at the computer printout and find the labels for "Predictor." This will tell you what variables have been used in the model.

FIGURE 13.9

Fitted quadratic regression line for Example 13.4

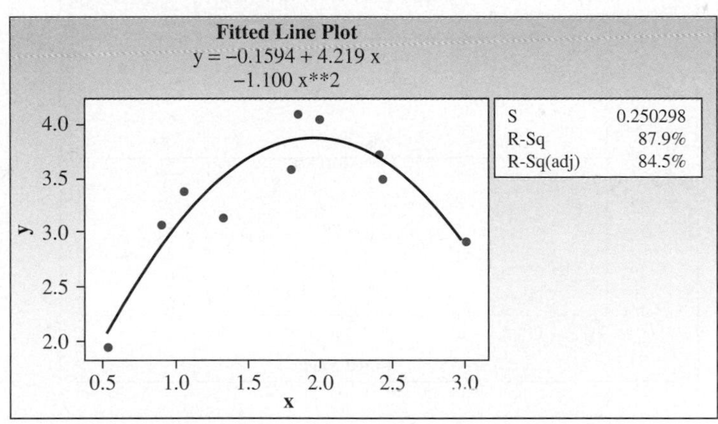

To assess the adequacy of the quadratic model, the test of

$$H_0: \beta_1 = \beta_2 = 0$$

versus

$$H_0: \text{Either } \beta_1 \text{ or } \beta_2 \text{ is not } 0$$

is given in the printout as

$$F = \frac{\text{MSR}}{\text{MSE}} = 25.53$$

with p-value = 0.001. Hence, the overall fit of the model is highly significant. Quadratic regression accounts for $R^2 = 87.9\%$ of the variation in Y [$R^2(\text{adj}) = 84.5\%$].

From the t tests for the individual variables in the model, you can see that both b_1 and b_2 are highly significant with p-values equal to 0.000. Notice from the sequential sum of squares section that the sum of squares for linear regression is 0.8003, with an additional sum of squares of 2.3986 when the quadratic term is added. It is apparent that the simple linear regression model is inadequate in describing the data.

One last look at the residual plots generated by *MINITAB* in Figure 13.10 ensures that the regression assumptions are valid. Notice the relatively linear appearance of the normal plot and the relative scatter of the residuals versus fits. The quadratic model provides accurate predictions for values of x that lie *within the range of the sampled values of x.*

FIGURE 13.10
MINITAB diagnostic plots for Example 13.4

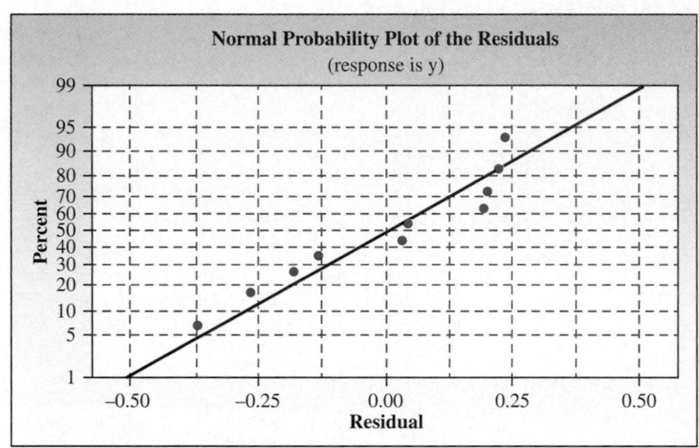

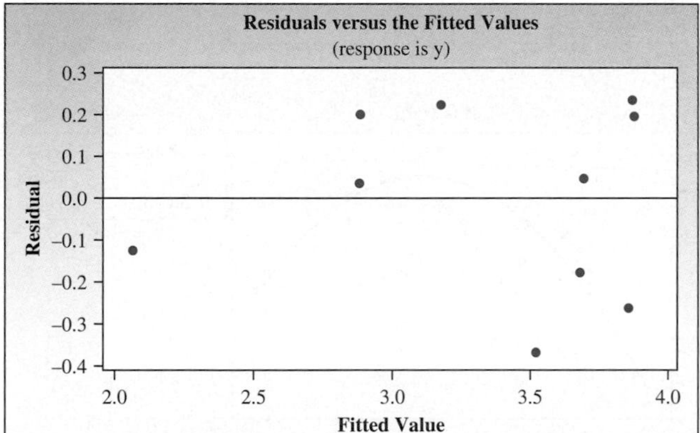

13.4 EXERCISES

BASIC TECHNIQUES

13.1 Suppose that $E(Y)$ is related to two predictor variables, X_1 and X_2, by the equation

$$E(Y) = 3 + x_1 - 2x_2$$

a. Graph the relationship between $E(Y)$ and x_1 when $x_2 = 2$. Repeat for $x_2 = 1$ and for $x_2 = 0$.

b. What relationship do the lines in part a have to one another?

13.2 Refer to Exercise 13.1.

a. Graph the relationship between $E(Y)$ and x_2 when $x_1 = 0$. Repeat for $x_1 = 1$ and for $x_1 = 2$.

b. What relationship do the lines in part a have to one another?

c. Suppose, in a practical situation, you want to model the relationship between $E(Y)$ and two predictor variables x_1 and x_2. What is the implication of using the first-order model $E(Y) = \beta_0 + \beta_1 x_1 + \beta_2 x_2$?

13.3 Suppose that you fit the model

$$E(Y) = \beta_0 + \beta_1 x_1 + \beta_2 x_2 + \beta_3 x_3$$

to 15 data points and found F equal to 57.44.

a. Do the data provide sufficient evidence to indicate that the model contributes information for the prediction of y? Test using a 5% level of significance.

b. Use the value of F to calculate R^2. Interpret its value.

13.4 The computer output for the multiple regression analysis for Exercise 13.3 provides this information:

$b_0 = 1.04$ $\quad b_1 = 1.29$

$\mathrm{SE}(b_1) = 0.42$

$b_2 = 2.72$ $\quad b_3 = 0.41$

$\mathrm{SE}(b_2) = 0.65$ $\quad \mathrm{SE}(b_3) = 0.17$

a. Which, if any, of the independent variables x_1, x_2, and x_3 contribute information for the prediction of y?

b. Give the least-squares prediction equation.

c. On the same sheet of graph paper, graph y versus x_1 when $x_2 = 1$ and $x_3 = 0$ and when $x_2 = 1$ and $x_3 = 0.5$. What relationship do the two lines have to each other?

d. What is the practical interpretation of the parameter β_1?

13.5 Suppose that you fit the model

$$E(Y) = \beta_0 + \beta_1 x + \beta_2 x^2$$

to 20 data points and obtained the accompanying *MINITAB* printout.

MINITAB output for Exercise 13.5

Regression Analysis: y versus x, x-sq

```
The regression equation is
y = 10.6 + 4.44 x - 0.648 x-sq

Predictor        Coef    SE Coef       T       P
Constant      10.5638     0.6951   15.20   0.000
x              4.4366     0.5150    8.61   0.000
x-sq         -0.64754    0.07988   -8.11   0.000

S = 1.191     R-Sq = 81.5%     R-Sq(adj) = 79.3%

Analysis of Variance
Source            DF        SS        MS       F       P
Regression         2   106.072    53.036   37.37   0.000
Residual Error    17    24.128     1.419
Total             19   130.200
```

a. What type of model have you chosen to fit the data?

b. How well does the model fit the data? Explain.

c. Do the data provide sufficient evidence to indicate that the model contributes information for the prediction of y? Use the p-value approach.

13.6 Refer to Exercise 13.5.

a. What is the prediction equation?

b. Graph the prediction equation over the interval $0 \le x \le 6$.

13.7 Refer to Exercise 13.5.

a. What is your estimate of the average value of y when $x = 0$?

b. Do the data provide sufficient evidence to indicate that the average value of y differs from 0 when $x = 0$?

13.8 Refer to Exercise 13.5.

a. Suppose that the relationship between $E(Y)$ and x is a straight line. What would you know about the value of β_2?

b. Do the data provide sufficient evidence to indicate curvature in the relationship between y and x?

13.9 Refer to Exercise 13.5. Suppose that Y is the profit for some business and x is the amount of capital invested, and you know that the rate of increase in profit for a unit increase in capital invested can only decrease as x increases. You want to know whether the data provide sufficient evidence to indicate a decreasing rate of increase in profit as the amount of capital invested increases.

a. The circumstances described imply a one-tailed statistical test. Why?

b. Conduct the test at the 1% level of significance. State your conclusions.

APPLICATIONS

Data set EX1310 **13.10 University Textbooks** A publisher of university textbooks conducted a study to relate profit per text Y to cost of sales x over a six-year period when its sales force (and sales costs) were growing rapidly. These inflation-adjusted data (in thousands of dollars) were collected:

Profit per text, y	16.5	22.4	24.9	28.8	31.5	35.8
Sales cost per text, x	5.0	5.6	6.1	6.8	7.4	8.6

Expecting profit per book to rise and then plateau, the publisher fitted the model $E(Y) = \beta_0 + \beta_1 x + \beta_2 x^2$ to the data.

Excel output for Exercise 13.10

SUMMARY OUTPUT

Regression Statistics	
Multiple R	0.9978
R Square	0.9955
Adjusted R Square	0.9925
Standard Error	0.5944
Observations	6

ANOVA

	df	SS	MS	F	Significance F
Regression	2	234.995	117.478	332.528	0.000
Residual	3	1.060	0.353		
Total	5	236.015			

	Coefficients	Standard Error	t Stat	P-value
Intercept	-44.192	8.287	-5.333	0.013
x	16.334	2.490	6.560	0.007
x-sq	-0.820	0.182	-4.494	0.021

a. Plot the data points. Does it look as though the quadratic model is necessary?

b. Find s on the printout. Confirm that

$$s = \sqrt{\frac{\text{SSE}}{n - k - 1}}$$

c. Do the data provide sufficient evidence to indicate that the model contributes information for the prediction of y? What is the p-value for this test, and what does it mean?

d. How much of the regression sum of squares is accounted for by the quadratic term? The linear term?

e. What sign would you expect the actual value of β_2 to have? Find the value of β_2 in the printout. Does this value confirm your expectation?

f. Do the data indicate a significant curvature in the relationship between y and x? Test at the 5% level of significance.

g. What conclusions can you draw from the accompanying residual plots?

MINITAB plots for Exercise 13.10

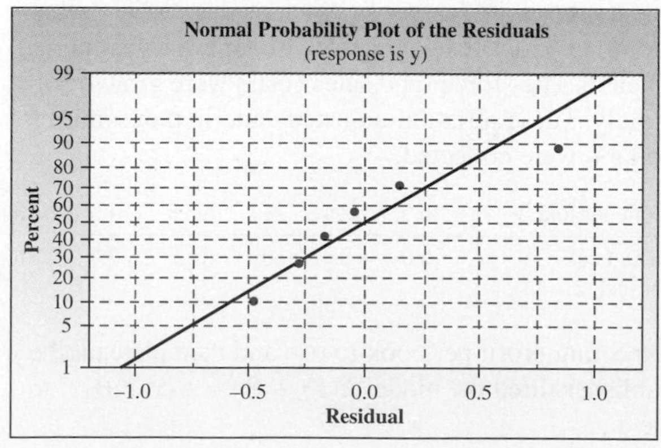

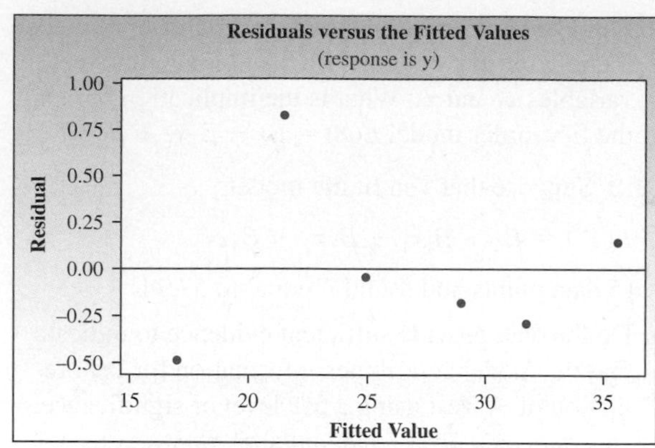

13.11 University Textbooks II Refer to Exercise 13.10.

a. Use the values of SSR and Total SS to calculate R^2. Compare this value with the value given in the printout.

b. Calculate R^2(adj). When would it be appropriate to use this value rather than R^2 to assess the fit of the model?

c. The value of R^2(adj) was 95.7% when a simple linear model was fit to the data. Does the linear or the quadratic model fit better?

Data set EX1312 **13.12 Who Is a Vegan?** A vegan is a vegetarian who does not eat meat or any animal-derived ingredient such as eggs and dairy products. There are many vegan recipes online. A food analyst collected data for 12 different vegan dishes. The data in the table record a flavour and texture score (between 0 and 100) along with the price, number of calories, amount of fat (g), and amount of sodium (mg) per serving. Some of these dishes try to mimic the taste of meat, while others do not. The *MINITAB* printout shows the regression of the taste score y on the four predictor variables: price, calories, fat, and sodium.

Dish	Score, y	Price, x_1	Calories, x_2	Fat, x_3	Sodium, x_4
1	70	91	110	4	310
2	45	68	90	0	420
3	43	92	80	1	280
4	41	75	120	5	370
5	39	88	90	0	410
6	30	67	140	4	440
7	68	73	120	4	430
8	56	92	170	6	520
9	40	71	130	4	180
10	34	67	110	2	180
11	30	92	100	1	330
12	26	95	130	2	340

MINITAB output for Exercise 13.12

Regression Analysis: y versus x1, x2, x3, x4

```
The regression equation is
y = 59.8 + 0.129 x1 - 0.580 x2 + 8.50 x3 + 0.0488 x4

Predictor        Coef     SE Coef       T       P
Constant        59.85       35.68    1.68   0.137
x1             0.1287      0.3391    0.38   0.716
x2            -0.5805      0.2888   -2.01   0.084
x3              8.498       3.472    2.45   0.044
x4            0.04876     0.04062    1.20   0.269

S = 12.7199    R-Sq = 49.9%    R-Sq(adj) = 21.3%

Analysis of Variance
Source            DF        SS      MS      F       P
Regression         4    1128.4   282.1   1.74   0.244
Residual Error     7    1132.6   161.8
Total             11    2261.0

Source    DF    Seq SS
x1         1      11.2
x2         1      19.6
x3         1     864.5
x4         1     233.2
```

a. Comment on the fit of the model using the statistical test for the overall fit and the coefficient of determination, R^2.

b. If you wanted to refit the model by eliminating one of the independent variables, which one would you eliminate? Why?

13.13 Who Is a Vegan? II Refer to Exercise 13.12. A command in the *MINITAB* regression menu provides output in which R^2 and R^2(adj) are calculated for all possible subsets of the four independent variables. The printout is provided here.

MINITAB output for Exercise 13.13

Best Subsets Regression: y versus x1, x2, x3, x4

```
Response is y

                   R-Sq   Mallows             x x x x
Vars    R-Sq      (adj)      C-p         s    1 2 3 4

   1    17.0        8.7      3.6    13.697        X
   1     6.9        0.0      5.0    14.506          X
   2    37.2       23.3      2.8    12.556      X X
   2    20.3        2.5      5.1    14.153        X X
   3    48.9       29.7      3.1    12.020      X X X
   3    39.6       16.9      4.4    13.066    X X X
   4    49.9       21.3      5.0    12.720    X X X X
```

a. If you had to compare these models and choose the best one, which model would you choose? Explain.

b. Comment on the usefulness of the model you chose in part a. Is your model valuable in predicting a taste score based on the chosen predictor variables?

Data set EX1314 **13.14 Air Pollution** An experiment was designed to compare several different types of air pollution monitors.[2] Each monitor was set up and then exposed to different concentrations of ozone, ranging between 15 and 230 parts per million (ppm), for periods of 8 to 72 hours. Filters on the monitor were then analyzed, and the response of the monitor was measured. The results for one type of monitor showed a linear pattern (see Exercise 12.37). The results for another type of monitor are listed in the table:

Ozone (ppm/h), x	0.06	0.12	0.18	0.31	0.57	0.65	0.68	1.29
Relative fluorescence density, y	8	18	27	33	42	47	52	61

a. Plot the data. What model would you expect to provide the best fit to the data? Write the equation of that model.

b. Use a computer software package to fit the model from part a.

c. Find the least-squares regression line relating the monitor's response to the ozone concentration.

d. Does the model contribute significant information for the prediction of the monitor's response based on ozone exposure? Use the appropriate p-value to make your decision.

e. Find R^2 on the printout. What does this value tell you about the effectiveness of the multiple regression analysis?

Data set EX1315 **13.15 Corporate Profits** In order to study the relationship of advertising and capital investment with corporate profits, the following data, recorded in units of \$100,000, were collected for 10 medium-sized firms in the same year. The variable y represents profit for the year, x_1 represents capital investment, and x_2 represents advertising expenditures:

y	x_1	x_2	y	x_1	x_2
15	25	4	1	20	0
16	1	5	16	12	4
2	6	3	18	15	5
3	30	1	13	6	4
12	29	2	2	16	2

a. Using the model

$$Y = \beta_0 + \beta_1 x + \beta_2 x_2 + \epsilon$$

and an appropriate computer software package, find the least-squares prediction equation for these data.

b. Use the overall F test to determine whether the model contributes significant information for the prediction of y. Use $\alpha = 0.01$.

c. Does advertising expenditure x_2 contribute significant information for the prediction of y, given that x_1 is already in the model? Use $\alpha = 0.01$.

d. Calculate the coefficient of determination, R^2. What percentage of the overall variation is explained by the model?

13.16 Lexus, Inc. In Exercise 12.82, we presented sales data for the Lexus GX, a mid-size sport utility vehicle (SUV) sold in North American and Eurasian markets. The sales of the Lexus GX 470 from its inception until 2009 are given in the table:[3]

Data set EX1316

Calendar Year	Total Sales (United States)
2002	2,190
2003	31,376
2004	35,420
2005	34,339
2006	25,454
2007	23,035
2008	15,759
2009	6,235

a. Plot the data. What model would you expect to provide the best fit to the data? Write the equation of that model.

b. Use a computer software package to fit the model from part a.

c. Find the least-squares prediction equation relating the sales of the Lexus GX 470 to the year of production.

d. Does the model contribute significant information for the prediction of sales based on the year of production? Use the appropriate p-value to make your decision.

e. Find R^2 on the printout. What does this value tell you about the effectiveness of the multiple regression analysis?

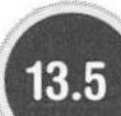

13.5 USING QUANTITATIVE AND QUALITATIVE PREDICTOR VARIABLES IN A REGRESSION MODEL

One reason multiple regression models are very flexible is that they allow for the use of both *qualitative* and *quantitative* predictor variables. For the multiple regression methods used in this chapter, the response variable Y *must be quantitative*, measuring a numerical random variable that has a normal distribution (according to the assumptions of Section 13.2). However, each independent predictor variable can be either a quantitative variable *or* a qualitative variable, whose levels represent qualities or characteristics and can only be categorized.

Quantitative and qualitative variables enter the regression model in different ways. To make things more complicated, we can allow a combination of different types of variables in the model, *and* we can allow the variables to *interact*, a concept that may be familiar to you from the *factorial experiment* of Chapter 11. We consider these options one at a time.

A **quantitative variable** x can be entered as a linear term, x, or to some higher power such as x^2 or x^3, as in the quadratic model in Example 13.3. When more than one quantitative variable is necessary, the interpretation of the possible models becomes more complicated. For example, with two quantitative variables x_1 and x_2, you could use a **first-order model** such as

$$E(Y) = \beta_0 + \beta_1 x_1 + \beta_2 x_2$$

which traces a plane in three-dimensional space (see Figure 13.1). However, it may be that one of the variables—say, x_2—is not related to Y in the same way when $x_1 = 1$ as it is when $x_1 = 2$. To allow x_2 to behave differently depending on the value of x_1, we add an **interaction term,** x_1x_2, and allow the two-dimensional plane to *twist.* The model is now a **second-order model:**

$$E(Y) = \beta_0 + \beta_1 x_1 + \beta_2 x_2 + \beta_3 x_1 x_2$$

The models become complicated quickly when you allow curvilinear relationships *and* interaction for the two variables. One way to decide on the type of model you need is

NEED A TIP?

Enter **quantitative** variables as:

- a single x
- a higher power, x^2 or x^3
- an interaction with another variable.

to plot some of the data—perhaps Y versus x_1, Y versus x_2, and Y versus x_2 for various values of x_1.

In contrast to quantitative predictor variables, **qualitative predictor variables** are entered into a regression model through **dummy** or **indicator variables.** For example, in a model that relates the mean salary of a group of employees to a number of predictor variables, you may want to include the employee's ethnic background. If each employee included in your study belongs to one of three ethnic groups—say, A, B, or C—you can enter the qualitative variable "ethnicity" into your model using two *dummy variables:*

$$x_1 = \begin{cases} 1 & \text{if group B} \\ 0 & \text{if not} \end{cases} \qquad x_2 = \begin{cases} 1 & \text{if group C} \\ 0 & \text{if not} \end{cases}$$

Look at the effect these two variables have on the model $E(Y) = \beta_0 + \beta_1 x_1 + \beta_2 x_2$: For employees in group A,

$$E(Y) = \beta_0 + \beta_1(0) + \beta_2(0) = \beta_0$$

for employees in group B,

$$E(Y) = \beta_0 + \beta_1(1) + \beta_2(0) = \beta_0 + \beta_1$$

and for those in group C,

$$E(Y) = \beta_0 + \beta_1(0) + \beta_2(1) = \beta_0 + \beta_2$$

The model allows a different average response for each group. β_1 measures the difference in the average responses between groups B and A, while β_2 measures the difference between groups C and A.

NEED A TIP?

Qualitative variables are entered as dummy variables—one fewer than the number of categories or levels.

When a qualitative variable involves k categories or levels, $(k - 1)$ dummy variables should be added to the regression model. This model may contain other predictor variables—quantitative or qualitative—as well as cross-products (**interactions**) of the dummy variables with other variables that appear in the model. As you can see, the process of model building—deciding on the appropriate terms to enter into the regression model—can be quite complicated. However, you can become more proficient at model building, gaining experience with the chapter exercises. The next example involves one quantitative and one qualitative variable that interact.

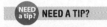

EXAMPLE 13.5

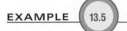

A study was conducted to examine the relationship between university salary Y, the number of years of experience of the faculty member, and the gender of the faculty member. If you expect a straight-line relationship between mean salary and years of experience for both men and women, write the model that relates mean salary to the two predictor variables: years of experience (quantitative) and gender of the professor (qualitative).

SOLUTION Since you may suspect the mean salary lines for women and men to be different, your model for mean salary $E(Y)$ may appear as shown in Figure 13.11. A straight-line relationship between $E(Y)$ and years of experience x_1 implies the model

$$E(Y) = \beta_0 + \beta_1 x_1 \qquad \text{(graphs as a straight line)}$$

The qualitative variable "gender" involves $k = 2$ categories, men and women. Therefore, you need $(k - 1) = 1$ dummy variable, x_2, defined as

$$x_2 = \begin{cases} 1 & \text{if a man} \\ 0 & \text{if a woman} \end{cases}$$

FIGURE 13.11
Hypothetical relationship for mean salary $E(Y)$, years of experience (x_1), and gender (x_2) for Example 13.5

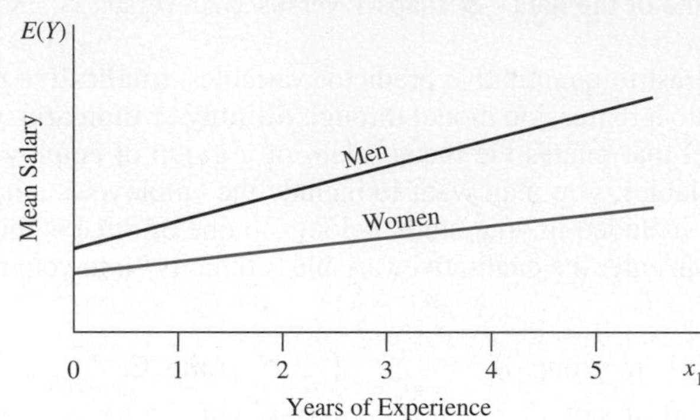

and the model is expanded to become

$$E(Y) = \beta_0 + \beta_1 x_1 + \beta_2 x_2$$

The fact that the slopes of the two lines may differ means that the two predictor variables **interact;** that is, the change in $E(Y)$ corresponding to a change in x_1 depends on whether the professor is a man or a woman. To allow for this interaction (difference in slopes), the interaction term $x_1 x_2$ is introduced into the model. The complete model that characterizes the graph in Figure 13.11 is

$$E(Y) = \beta_0 + \underbrace{\beta_1 x_1}_{\text{years of experience}} + \overbrace{\beta_2 x_2}^{\text{dummy variable for gender}} + \underbrace{\beta_3 x_1 x_2}_{\text{interaction}}$$

where

$$x_1 = \text{Years of experience}$$

$$x_2 = \begin{cases} 1 & \text{if a man} \\ 0 & \text{if a woman} \end{cases}$$

You can see how the model works by assigning values to the dummy variable x_2. When the faculty member is a woman, the model is

$$E(Y) = \beta_0 + \beta_1 x_1 + \beta_2(0) + \beta_3 x_1(0) = \beta_0 + \beta_1 x_1$$

which is a straight line with slope β_1 and intercept β_0. When the faculty member is a man, the model is

$$E(Y) = \beta_0 + \beta_1 x_1 + \beta_2(1) + \beta_3 x_1(1) = (\beta_0 + \beta_2) + (\beta_1 + \beta_3)x_1$$

which is a straight line with slope $(\beta_1 + \beta_3)$ and intercept $(\beta_0 + \beta_2)$. The two lines have *different slopes and different intercepts*, which allows the relationship between salary Y and years of experience x_1 to behave differently for men and women.

EXAMPLE 13.6

Random samples of six female and six male assistant professors were selected from among the assistant professors in a university. The data on salary and years of experience are shown in Table 13.3. Note that both samples contained two professors with three years of experience, but no male professor had two years of experience. Interpret the output of the *Excel* regression printout in Figure 13.12 and graph the predicted salary lines.

TABLE 13.3 **Salary versus Gender and Years of Experience**

Years of Experience, X_1	Salary for Men, Y($)	Salary for Women, Y($)
1	60,710	59,510
2	—	60,440
3	63,160	61,340
3	63,210	61,760
4	64,140	62,750
5	65,760	63,200
5	65,590	—

SOLUTION The *Excel* regression printout for the data in Table 13.3 is shown in Figure 13.12. You can use a step-by-step approach to interpret this regression analysis, beginning with the fitted prediction equation $\hat{y} = 58{,}593 + 969x_1 + 866.71x_2 + 260.13x_1x_2$. By substituting $x_2 = 0$ or 1 into this equation, you get two straight lines—one for women and one for men—to predict the value of y for a given x_1. These lines are

Women: $\hat{y} = 58{,}593 + 969x_1$

Men: $\hat{y} = 59{,}459.71 + 1229.13x_1$

and are graphed in Figure 13.13.

FIGURE 13.12 *Excel* output for Example 13.6

SUMMARY OUTPUT

Regression Statistics	
Multiple R	0.9962
R Square	0.9924
Adjusted R Square	0.9895
Standard Error	201.3438
Observations	12

ANOVA

	df	*SS*	*MS*	*F*	*Significance F*
Regression	3	42108777.03	14036259.01	346.238	0.000
Residual	8	324314.64	40539.330		
Total	11	42433091.67			

	Coefficients	*Standard Error*	*t Stat*	*P-value*
Intercept	58593	207.9470	281.7689	0.000
x_1	969	63.6705	15.2190	0.000
x_2	866.710	305.2568	2.8393	0.022
x_1x_2	260.130	87.0580	2.9880	0.017

Next, consider the overall fit of the model using the analysis of variance F test. Since the observed test statistic in the ANOVA portion of the printout is $F = 346.238$ with p-value ("Significance F") equal to 0.000, you can conclude that at least one of the predictor variables is contributing information for the prediction of y. The strength of this model is further measured by the coefficient of determination, $R^2 = 99.24\%$. You can see that the model appears to fit very well.

FIGURE 13.13

A graph of the faculty salary prediction lines for Example 13.6

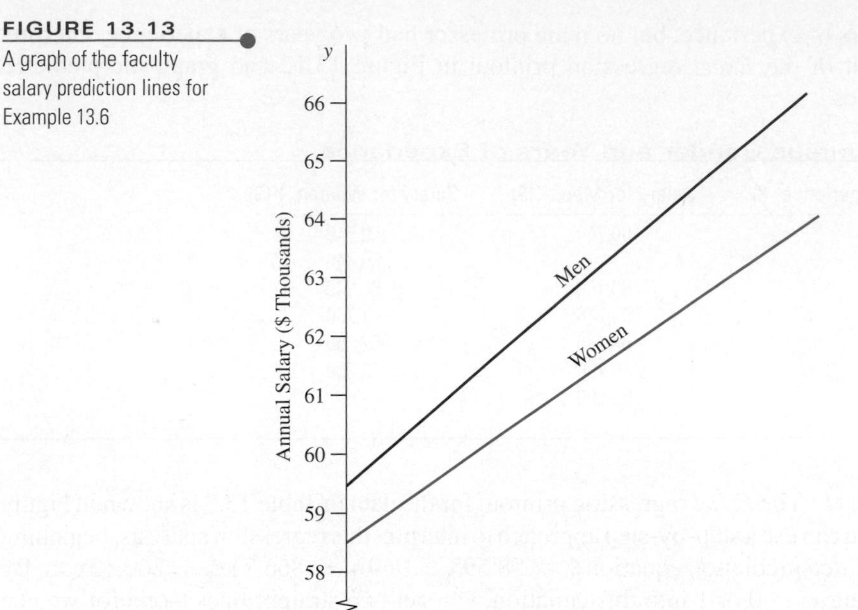

To explore the effect of the predictor variables in more detail, look at the individual t tests for the three predictor variables. The p-values for these tests—0.000, 0.022, and 0.017, respectively—are all significant, which means that all of the predictor variables add significant information to the prediction *with the other two variables already in the model.* Finally, check the diagnostic plots to make sure that there are no strong violations of the regression assumptions. These plots, which behave as expected for a properly fit model, are shown in Figure 13.14.

FIGURE 13.14

Diagnostic plots for Example 13.6

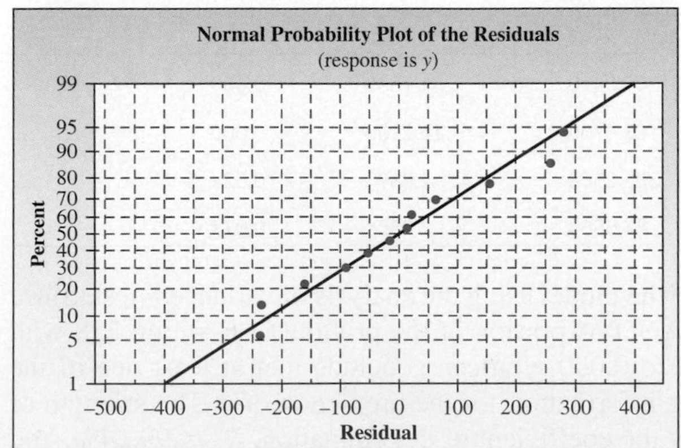

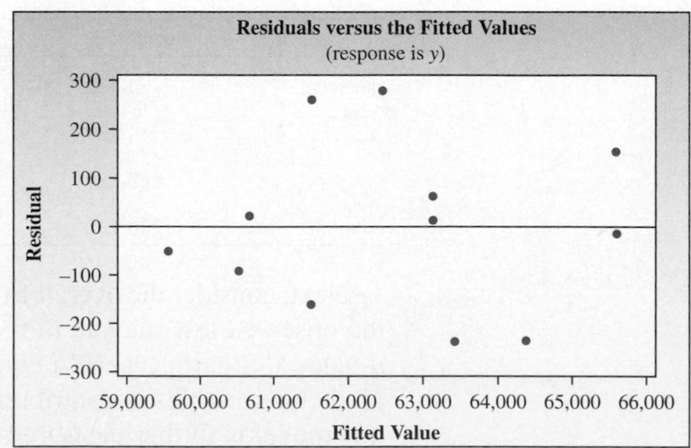

EXAMPLE 13.7

Refer to Example 13.6. Do the data provide sufficient evidence to indicate that the annual rate of increase in male junior faculty salaries exceeds the annual rate of increase in female junior faculty salaries? That is, do the data provide sufficient evidence to indicate that the slope of the men's faculty salary line is greater than the slope of the women's faculty salary line?

SOLUTION Since β_3 measures the difference in slopes, the slopes of the two lines will be identical if $\beta_3 = 0$. Therefore, you want to test the null hypothesis

$$H_0: \beta_3 = 0$$

—that is, the slopes of the two lines are identical—versus the alternative hypothesis

$$H_a: \beta_3 > 0$$

—that is, the slope of the men's faculty salary line is greater than the slope of the women's faculty salary line.

The calculated value of t corresponding to β_3, shown in the row labelled "x1x2" in Figure 13.12, is 2.988. Since the *Excel* regression output provides p-values for two-tailed significance tests, the p-value in the printout, 0.017, is *twice* what it would be for a one-tailed test. For this one-tailed test, the p-value is $0.017/2 = 0.0085$, and the null hypothesis is rejected. There is sufficient evidence to indicate that the annual rate of increase in men's faculty salaries exceeds the rate for women.[†]

13.5 EXERCISES

BASIC TECHNIQUES

13.17 Production Yield Suppose you wish to predict production yield y as a function of several independent predictor variables. Indicate whether each of the following independent variables is qualitative or quantitative. If qualitative, define the appropriate dummy variable(s).

a. The prevailing interest rate in the area

b. The price per kilogram of one item used in the production process

c. The plant (A, B, or C) at which the production yield is measured

d. The length of time that the production machine has been in operation

e. The shift (night or day) in which the yield is measured

13.18 Suppose $E(Y)$ is related to two predictor variables x_1 and x_2 by the equation

$$E(Y) = 3 + x_1 - 2x_2 + x_1x_2$$

a. Graph the relationship between $E(Y)$ and x_1 when $x_2 = 0$. Repeat for $x_2 = 2$ and for $x_2 = -2$.

b. Repeat the instructions of part a for the model

$$E(Y) = 3 + x_1 - 2x_2$$

c. Note that the equation for part a is exactly the same as the equation in part b except that we have added the term x_1x_2. How does the addition of the x_1x_2 term affect the graphs of the three lines?

d. What flexibility is added to the first-order model $E(Y) = \beta_0 + \beta_1x_1 + \beta_2x_2$ by the addition of the term $\beta_3x_1x_2$, using the model $E(Y) = \beta_0 + \beta_1x_1 + \beta_2x_2 + \beta_3x_1x_2$?

13.19 A multiple linear regression model involving one qualitative and one quantitative independent variable produced this prediction equation:

$$\hat{y} = 12.6 + 0.54x_1 - 1.2x_1x_2 + 3.9x_2^2$$

a. Which of the two variables is the quantitative variable? Explain.

b. If x_1 can take only the values 0 or 1, find the two possible prediction equations for this experiment.

c. Graph the two equations found in part b. Compare the shapes of the two curves.

[†]If you want to determine whether the data provide sufficient evidence to indicate that male faculty members start at higher salaries, you would test $H_0: \beta_2 = 0$ versus the alternative hypothesis $H_a: \beta_2 > 0$.

APPLICATIONS

Data set EX1320 **13.20 Less Red Meat!** Canadians are very vocal about their attempts to improve personal well-being by "eating right and exercising more." One desirable dietary change is to reduce the intake of red meat and to substitute poultry or fish. Researchers tracked beef and chicken consumption, Y (in annual kilograms per person) and found the consumption of beef declining and the consumption of chicken increasing over a period of seven years. A summary of their data is shown in the table:

Year	Beef	Chicken
1	85	37
2	89	36
3	76	47
4	76	47
5	68	62
6	67	74
7	60	79

Consider fitting the following model, which allows for simultaneously fitting two simple linear regression lines:

$$E(Y) = \beta_0 + \beta_1 x_1 + \beta_2 x_2 + \beta_3 x_1 x_2$$

where y is the annual meat (either beef or chicken) consumption per person per year,

$$x_1 = \begin{cases} 1 \text{ if beef} \\ 0 \text{ if chicken} \end{cases} \quad \text{and } x_2 = \text{Year}$$

MINITAB output for Exercise 13.20

Regression Analysis: y versus x1, x2, x1x2

```
The regression equation is
y = 23.6 + 69.0 x1 + 7.75 x2 - 12.3 x1x2

Predictor        Coef    SE Coef        T      P
Constant       23.571      3.522     6.69  0.000
x1             69.000      4.981    13.85  0.000
x2             7.7500     0.7875     9.84  0.000
x1x2          -12.286      1.114   -11.03  0.000

S = 4.16705     R-Sq = 95.4%     R-Sq(adj) = 94.1%

Analysis of Variance
Source           DF       SS       MS       F      P
Regression        3   3637.9   1212.6   69.83  0.000
Residual Error   10    173.6     17.4
Total            13   3811.5

Source    DF   Seq SS
x1         1   1380.1
x2         1    144.6
x1x2       1   2113.1

Predicted Values for New Observations
New
Obs     Fit  SE Fit       95% CI            95% PI
1     56.29    3.52  (48.44,  64.13)  (44.13,  68.44)

Values of Predictors for New Observations
New Obs      x1      x2    x1x2
1          1.00    8.00    8.00
```

MINITAB diagnostic plots for Exercise 13.20

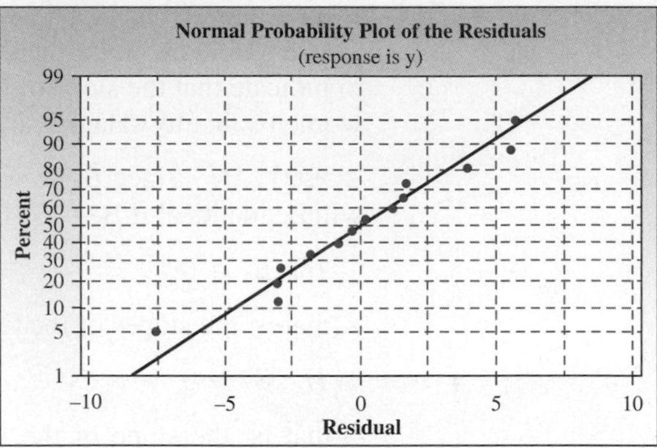

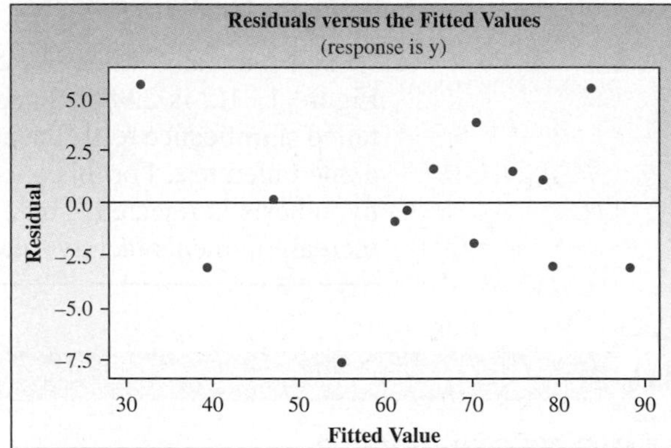

a. How well does the model fit? Use any relevant statistics and diagnostic tools from the printout to answer this question.

b. Write the equations of the two straight lines that describe the trend in consumption over the period of seven years for beef and for chicken.

c. Use the prediction equation to find a point estimate of the average per-person beef consumption in year 8. Compare this value with the value labelled "Fit" in the printout.

d. Use the printout to find a 95% confidence interval for the average per-person beef consumption in year 8. What is the 95% prediction interval for the per-person beef consumption in year 8? Is there any problem with the validity of the 95% confidence level for these intervals?

Data set EX1321 **13.21 Cotton versus Cucumber** In Exercise 11.63, you used the analysis of variance procedure to analyze a 2 × 3 factorial experiment in which each factor–level combination was replicated five times. The experiment involved the number of eggs laid by caged female whiteflies on two different plants at three different temperature levels. Suppose that several of the whiteflies died before the experiment was completed, so that the number of replications was no longer the same for each treatment. The analysis of variance formulas of Chapter 11 can no longer be used, but the experiment *can* be analyzed using a multiple regression analysis. The results of this **2 × 3 factorial experiment with unequal replications** are shown in the table:

Cotton			Cucumber		
21.1°C	25°C	27.8°C	21.1°C	25°C	27.8°C
37	34	46	50	59	43
21	54	32	53	53	62
36	40	41	25	31	71
43	42		37	69	49
31			48	51	

a. Write a model to analyze this experiment. Make sure to include a term for the interaction between plant and temperature.

b. Use a computer software package to perform the multiple regression analysis.

c. Do the data provide sufficient evidence to indicate that the effect of temperature on the number of eggs laid is *different* depending on the type of plant?

d. Based on the results of part c, do you suggest refitting a different model? If so, rerun the regression analysis using the new model and analyze the printout.

e. Write a paragraph summarizing the results of your analyses.

Data set EX1322 **13.22 Blood Pressure Again!** Refer to Exercise 12.29. The systolic blood pressure (SBP) for randomly selected 40 males and females are shown below, along with several other independent variables:[4]

Systolic Blood Pressure y	Body Mass Index x_1	Age x_2	Gender x_3	Married x_4	Smoke x_5
185	32.6843	54	F (female)	N (No)	Y (Yes)
174	18.00509	54	F	Y	Y
147	15.09444	34	M (male)	Y	N
174	12.03111	59	F	Y	Y
130	20.5696	74	M	Y	Y
160	37.0388	42	F	Y	Y
123	18.61939	24	M	Y	Y
136	21.10426	30	F	Y	Y
182	22.72159	57	M	N	Y
111	26.42051	60	M	N	N
105	20.62475	43	F	N	Y
224	36.56926	23	M	N	Y
221	35.26174	58	M	N	Y
135	18.00972	63	F	Y	Y
122	16.17305	61	M	N	Y
113	24.19442	18	F	N	N
144	25.84728	43	F	N	Y
135	34.27116	40	F	N	N
158	32.18126	42	F	N	Y
153	40.88436	25	M	N	N
194	41.0497	37	F	N	N
186	49.77252	36	F	Y	Y
191	25.18267	55	M	Y	N
121	23.66558	41	F	Y	Y
130	22.56072	38	M	Y	Y
152	23.72454	25	M	Y	N
127	16.24771	33	M	Y	Y
136	32.1123	49	M	N	Y
148	32.64552	38	F	N	N
130	27.64572	31	F	N	N
100	28.2094	33	F	N	Y
124	12.09412	45	F	N	N
147	31.57406	55	F	Y	Y
146	18.34757	25	M	N	N
132	26.97432	57	M	N	N
176	25.0186	40	M	N	N
163	20.07109	48	F	Y	Y
149	23.58075	39	M	N	N
130	22.17864	64	F	Y	N
130	20.92983	50	M	N	N

The variables are defined as

y = systolic blood pressure
x_1 = body mass index
x_2 = age
x_3 = 1 if gender is male, 0 otherwise
x_4 = 1 if the person is married, 0 otherwise
x_5 = 1 if the person smokes, 0 otherwise

The *MINITAB* printout for a first-order regression model is given below.

Regression Analysis: y versus x1, x2, x3, x4, x5

```
The regression equation is
y = 66.9 + 1.89 x1 + 0.474 x2 + 13.4 x3 + 7.80 x4 + 5.30 x5

Predictor     Coef      SE Coef    T       P
Constant      66.92     24.47      2.73    0.010
x1            1.8944    0.5440     3.48    0.001
x2            0.4743    0.3417     1.39    0.174
x3            13.472    8.763      1.53    0.135
x4            7.797     9.439      0.83    0.415
x5            5.296     9.245      0.57    0.571

S = 26.5727     R-Sq = 29.4%     R-Sq(adj) = 19.0%

Analysis of Variance
Source            DF    SS         MS        F       P
Regression         5    9981.8     1996.4    2.83    0.031
Residual Error    34    24007.8    706.1
Total             39    33989.6
```

Best Subsets Regression: y versus x1, x2, x3, x4, x5

```
Response is y
                  R-Sq    Mallows                  x x x x x
Vars    R-Sq     (adj)      C-p            S       1 2 3 4 5

  1     87.9      85.8    132.7       22.596               X
  1     84.5      81.9    170.7       25.544           X
  2     97.4      96.4     27.1       11.423       X         X
  2     94.6      92.4     58.8       16.512           X     X
  3     99.0      98.2     11.8       8.1361       X     X   X
  3     98.9      98.2     11.9       8.1654       X       X X
  4     99.8      99.6      4.0       3.8656       X     X X X
  4     99.0      97.8     12.8       8.9626       X X X     X
  5     99.8      99.4      6.0       4.7339       X X X X X
```

a. What is the model that has been fit to this data? What is the least-squares prediction equation?

b. How well does the model fit? Use any relevant statistics from the printout to answer this question.

c. Which, if any, of the independent variables are useful in predicting the SBP, given the other independent variables already in the model? Explain.

d. Use the values of R^2 and R^2(adj) in the printout to choose the best model for prediction. Would you be confident in using the chosen model for predicting the SBP score for next year based on a model containing similar variables? Explain.

13.23 Particle Board A quality control engineer is interested in predicting the strength of particle board Y as a function of the size of the particles x_1 and two types of bonding compounds. If the basic response is expected to be a quadratic function of particle size, write a linear model that incorporates the qualitative variable "bonding compound" into the predictor equation.

Data set EX1324 **13.24 Construction Projects** In a study to examine the relationship between the time required to complete a construction project and several pertinent independent variables, an analyst compiled a list of four variables that might be useful in predicting the time to completion. These four variables were size of the contract, x_1 (in \$1000 units), number of workdays adversely affected by the weather x_2, number of subcontractors involved in the project x_4, and a variable x_3 that measured the presence ($x_3 = 1$) or absence ($x_3 = 0$) of a workers' strike during the construction. Fifteen construction projects were randomly chosen, and each of the four variables as well as the time to completion were measured:

y	x_1	x_2	x_3	x_4
29	60	7	0	7
15	80	10	0	8
60	100	8	1	10
10	50	14	0	5
70	200	12	1	11
15	50	4	0	3
75	500	15	1	12
30	75	5	0	6
45	750	10	0	10
90	1200	20	1	12
7	70	5	0	3
21	80	3	0	6
28	300	8	0	8
50	2600	14	1	13
30	110	7	0	4

An analysis of these data using a first-order model in x_1, x_2, x_3, and x_4 produced the following printout. Give a complete analysis of the printout and interpret your results. What can you say about the apparent contribution of x_1 and x_2 in predicting y?

Excel printout for Exercise 13.24

SUMMARY OUTPUT

Regression Statistics	
Multiple R	0.9204
R Square	0.8471
Adjusted R Square	0.7859
Standard Error	11.8450
Observations	15

ANOVA

	df	*SS*	*MS*	*F*	*Significance F*
Regression	4	7770.297	1942.574	13.846	0.000
Residual	10	1403.036	140.304		
Total	14	9173.333			

	Coefficients	*Standard Error*	*tStat*	*P-value*
Intercept	-1.589	11.656	-0.136	0.894
x1	-0.008	0.006	-1.259	0.237
x2	0.675	1.000	0.675	0.515
x3	28.013	11.371	2.463	0.033
x4	3.489	1.935	1.803	0.102

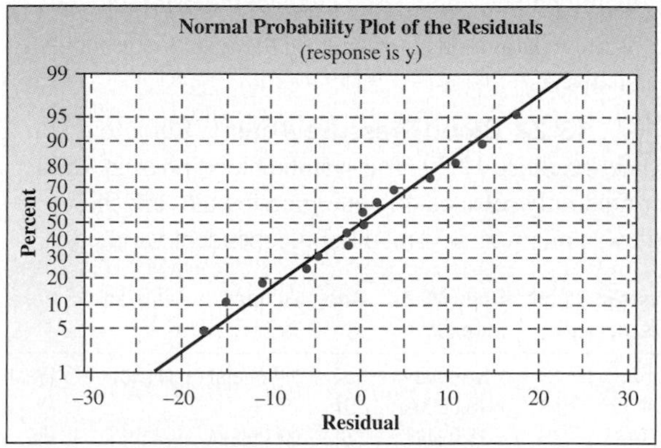

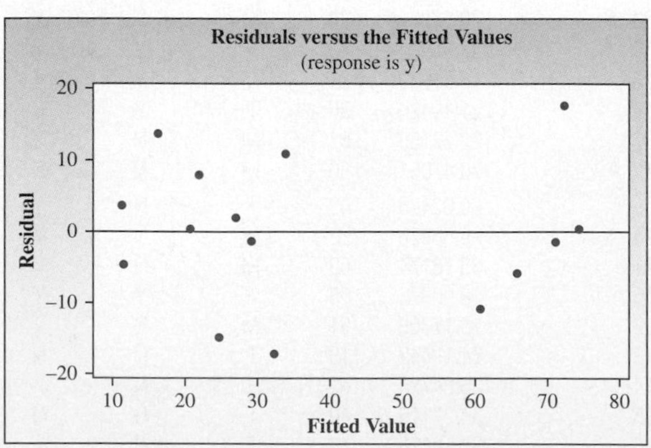

13.6 TESTING SETS OF REGRESSION COEFFICIENTS

In the preceding sections, you have tested the complete set of partial regression coefficients using the F test for the overall fit of the model, and you have tested the partial regression coefficients individually using the Student's t test. Besides these two important tests, you might want to test hypotheses about some subsets of these regression coefficients.

For example, suppose a company suspects that the demand Y for some product could be related to as many as five independent variables, x_1, x_2, x_3, x_4, and x_5. The cost of obtaining measurements on the variables x_3, x_4, and x_5 is very high. If, in a small pilot study, the company could show that these three variables contribute little or no information for the prediction of y, they can be eliminated from the study at great savings to the company.

If all five variables, x_1, x_2, x_3, x_4, and x_5, are used to predict y, the regression model would be written as

$$Y = \beta_0 + \beta_1 x_1 + \beta_2 x_2 + \beta_3 x_3 + \beta_4 x_4 + \beta_5 x_5 + \epsilon$$

However, if x_3, x_4, and x_5 contribute no information for the prediction of y, then they would not appear in the model—that is, $\beta_3 = \beta_4 = \beta_5 = 0$—and the reduced model would be

$$Y = \beta_0 + \beta_1 x_1 + \beta_2 x_2 + \epsilon$$

Hence, you want to test the null hypothesis

$$H_0 : \beta_3 = \beta_4 = \beta_5 = 0$$

—that is, the independent variables x_3, x_4, and x_5 contribute no information for the prediction of y—versus the alternative hypothesis

$$H_a : \text{at least one of the parameters } \beta_3, \beta_4, \text{ or } \beta_5 \text{ differs from } 0$$

—that is, at least one of the variables x_3, x_4, or x_5 contributes information for the prediction of y. Thus, in deciding whether the complete model is preferable to the reduced model in predicting demand, you are led to a test of hypothesis about a set of three parameters, β_3, β_4, and β_5.

A test of hypothesis concerning a set of model parameters involves two models:

Model 1 (reduced model)

$$E(Y) = \beta_0 + \beta_1 x_1 + \beta_2 x_2 + \cdots + \beta_r x_r$$

Model 2 (complete model)

$$E(Y) = \underbrace{\beta_0 + \beta_1 x_1 + \beta_2 x_2 + \cdots + \beta_r x_r}_{\text{terms in model 1}} + \underbrace{\beta_{r+1} x_{r+1} + \beta_{r+2} x_{r+2} + \cdots + \beta_k x_k}_{\text{additional terms in model 2}}$$

Suppose you fit both models to the data set and calculated the sum of squares for error for both regression analyses. If model 2 contributes more information for the prediction of y than model 1, then the errors of prediction for model 2 should be smaller than the corresponding errors for model 1, and SSE_2 should be smaller than SSE_1. In fact, the greater the difference between SSE_1 and SSE_2, the greater is the evidence to indicate that model 2 contributes more information for the prediction of y than model 1.

The test of the null hypothesis

$$H_0 : \beta_{r+1} = \beta_{r+2} = \cdots = \beta_k = 0$$

versus the alternative hypothesis

H_a : At least one of the parameters $\beta_{r+1}, \beta_{r+2}, \ldots, \beta_k$ differs from 0

uses the test statistic

$$F = \frac{(\text{SSE}_1 - \text{SSE}_2)/(k - r)}{\text{MSE}_2}$$

where F is based on $df_1 = (k - r)$ and $df_2 = n - (k + 1)$. Note that the $(k - r)$ parameters involved in H_0 are those added to model 1 to obtain model 2. The numerator degrees of freedom df_1 always equals $(k - r)$, the number of parameters involved in H_0. The denominator degrees of freedom df_2 is the number of degrees of freedom associated with the sum of squares for error, SSE_2, for the complete model.

The rejection region for the test is identical to the rejection region for all of the analysis of variance F tests—namely,

$$F > F_\alpha$$

EXAMPLE 13.8

Refer to the real estate data of Example 13.2 that relate the listed selling price y to the square metres of living area x_1, the number of floors x_2, the number of bedrooms x_3, and the number of bathrooms, x_4. The realtor suspects that square metres of living area is the most important predictor variable and that the other variables might be eliminated from the model without loss of much prediction information. Test this claim with $\alpha = 0.05$.

SOLUTION The hypothesis to be tested is

$$H_0: \beta_2 = \beta_3 = \beta_4 = 0$$

versus the alternative hypothesis that at least one of β_2, β_3, or β_4 is different from 0. The **complete model 2,** given as

$$y = \beta_0 + \beta_1 x_1 + \beta_2 x_2 + \beta_3 x_3 + \beta_4 x_4 + \epsilon$$

was fitted in Example 13.2. A portion of the *MINITAB* printout from Figure 13.3(a) is reproduced in Figure 13.15 along with a portion of the *MINITAB* printout for the simple linear regression analysis of the **reduced model 1,** given as

$$y = \beta_0 + \beta_1 x_1 + \epsilon$$

FIGURE 13.15

Portions of the *MINITAB* regression printouts for (a) complete and (b) reduced models for Example 13.8

Regression Analysis: (a) List Price versus Square Feet, Number of Floors, Bedrooms and Baths

```
S = 6.84930    R-Sq = 97.1%    R-Sq(adj) = 96.0%

Analysis of Variance
Source          DF        SS        MS       F       P
Regression       4   15913.0    3978.3   84.80   0.000
Residual Error  10     469.1      46.9
Total           14   16382.2
```

Regression Analysis: (b) List Price versus Square Feet

```
S = 10.9294    R-Sq = 90.5%    R-Sq(adj) = 89.8%

Analysis of Variance
Source          DF      SS      MS       F       P
Regression       1   14829   14829  124.14   0.000
Residual Error  13    1553     119
Total           14   16382
```

Then $SSE_1 = 1553$ from Figure 13.15(b) and $SSE_2 = 469.1$ and $MSE_2 = 46.9$ from Figure 13.15(a). The test statistic is

$$F = \frac{(SSE_1 - SSE_2)/(k - r)}{MSE_2}$$

$$= \frac{(1553 - 469.1)/(4 - 1)}{46.9} = 7.70$$

The critical value of F with $\alpha = 0.05$, $df_1 = 3$, and $df_2 = n - (k + 1) = 15 - (4 + 1) = 10$ is $F_{0.05} = 3.71$. Hence, H_0 is rejected. There is evidence to indicate that at least one of the three variables, number of floors, bedrooms, or bathrooms, is contributing significant information for predicting the listed selling price.

13.7 INTERPRETING RESIDUAL PLOTS

Once again, you can use residual plots to discover possible violations in the assumptions required for a regression analysis. There are several common patterns you should recognize because they occur frequently in practical applications.

The variance of some types of data changes as the mean changes:

- Poisson data exhibit variation that *increases* with the mean.
- Binomial data exhibit variation that *increases* for values of p from 0.0 to 0.5, and then *decreases* for values of p from 0.5 to 1.0.

Residual plots for these types of data have a pattern similar to that shown in Figure 13.16.

FIGURE 13.16

Plots of residuals against $\hat{y}$

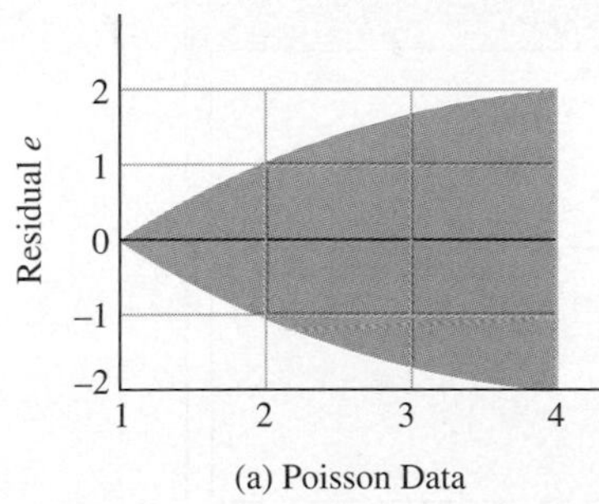

(a) Poisson Data

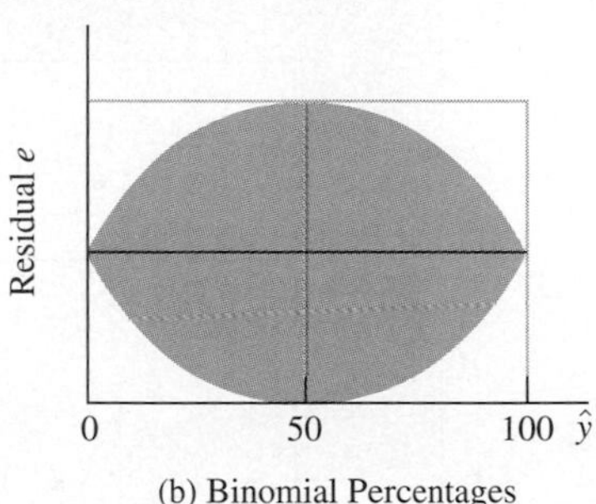

(b) Binomial Percentages

If the range of the residuals increases as $\hat{y}$ increases and you know that the data are measurements on Poisson variables, you can stabilize the variance of the response by running the regression analysis on $y^* = \sqrt{y}$. Or if the percentages are calculated from binomial data, you can use the arcsin transformation, $y^* = \sin^{-1}\sqrt{y}$.[†]

[†]In Chapter 11 and earlier chapters, we represented the response variable by the symbol x. In the chapters on regression analysis, Chapters 12 and 13, the response variable is represented by the symbol y.

Even if you are not sure why the range of the residuals increases as $\hat{y}$ increases, you can still use a transformation of y that affects larger values of y more than smaller values—say, $y^* = \sqrt{y}$ or $y^* = \ln y$. These transformations have a tendency both to stabilize the variance of y^* and to make the distribution of y^* more nearly normal when the distribution of y is highly skewed.

Plots of the residuals versus the fits $\hat{y}$ or versus the individual predictor variables often show a pattern that indicates you have chosen an incorrect model. For example, if $E(Y)$ and a single independent variable x are linearly related—that is,

$$E(Y) = \beta_0 + \beta_1 x$$

and you fit a straight line to the data, then the observed y-values should vary in a random manner about $\hat{y}$, and a plot of the residuals against x will appear as shown in Figure 13.17.

FIGURE 13.17
Residual plot when the model provides a good approximation to reality

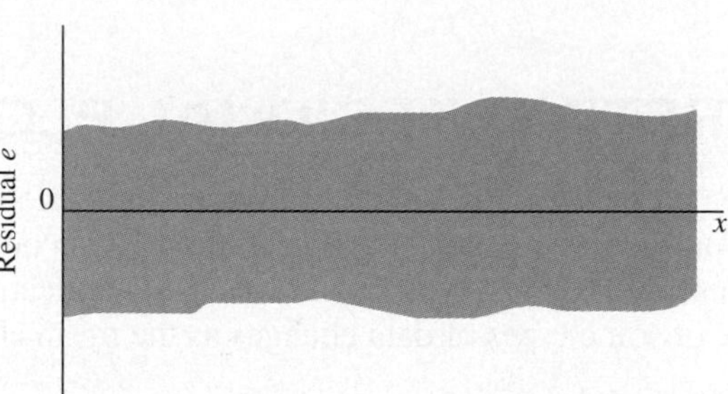

In Example 13.3, you fit a quadratic model relating productivity Y to store size x. If you had incorrectly used a linear model to fit these data, the residual plot in Figure 13.18 would show that the unexplained variation exhibits a curved pattern, which suggests that there is a quadratic effect that has not been included in the model.

FIGURE 13.18
Residual plot for linear fit of store size and productivity data in Example 13.3

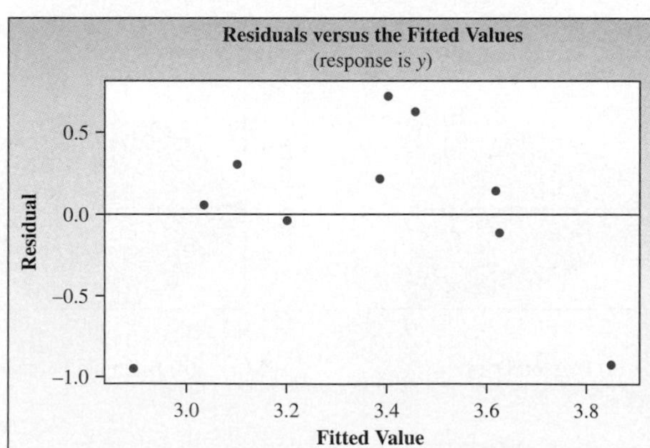

For the data in Example 13.6, the residuals of a linear regression of salary with years of experience x_1 without including gender, x_2, would show one distinct set of positive residuals corresponding to the men and a set of negative residuals corresponding to the women (see Figure 13.19). This pattern signals that the "gender" variable was not included in the model.

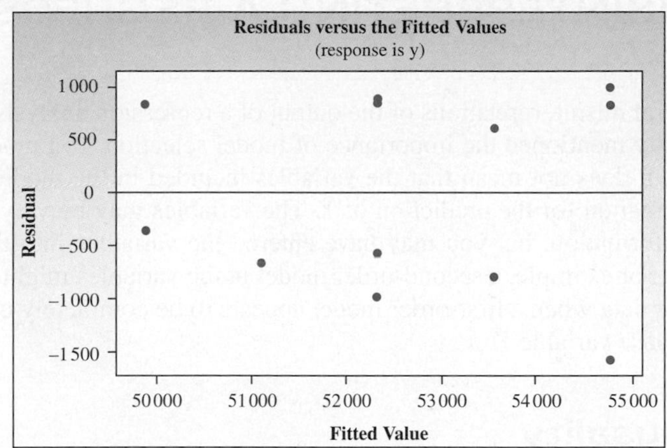

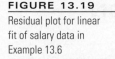

FIGURE 13.19
Residual plot for linear fit of salary data in Example 13.6

Unfortunately, not all residual plots give such a clear indication of the problem. You should examine the residual plots carefully, looking for non-randomness in the pattern of residuals. If you can find an explanation for the behaviour of the residuals, you may be able to modify your model to eliminate the problem.

13.8 STEPWISE REGRESSION ANALYSIS

Sometimes there are a large number of independent predictor variables that *might* have an effect on the response variable Y. For example, try to list all the variables that might affect a first-year university student's GPA:

- Grades in high-school courses, high-school GPA
- Major, number of units carried, number of courses taken
- Work schedule, marital status, commute or live on campus

Which of this large number of independent variables should be included in the model? Since the number of terms could quickly get unmanageable, you might choose to use a procedure called a **stepwise regression analysis,** which is implemented by computer and is available in most statistical packages.

Suppose you have data available on Y and a number of possible independent variables, $x_1, x_2, \dots, x_k$. A stepwise regression analysis fits a variety of models to the data, adding and deleting variables as their significance in the presence of the other variables is either *significant* or *non-significant,* respectively. Once the program has performed a sufficient number of iterations and no more variables are significant when added to the model, and none of the variables in the model are non-significant when removed, the procedure stops.

A stepwise regression analysis is an easy way to locate some variables that contribute information for predicting Y, but it is not foolproof. Since these programs always fit first-order models of the form

$$E(Y) = \beta_0 + \beta_1 x_1 + \beta_2 x_2 + \dots + \beta_k x_k$$

they are not helpful in detecting *curvature* or *interaction* in the data. The stepwise regression analysis is best used as a preliminary tool for identifying which of a large number of variables should be considered in your model. You must then decide how to enter these variables into the actual model you will use for prediction.

13.9 MISINTERPRETING A REGRESSION ANALYSIS

Several misinterpretations of the output of a regression analysis are common. We have already mentioned the importance of model selection. If a model does not fit a set of data, it does not mean that the variables included in the model contribute little or no information for the prediction of Y. The variables may be very important contributors of information, but you may have entered the variables into the model in the wrong way. For example, a second-order model in the variables might provide a very good fit to the data when a first-order model appears to be completely useless in describing the response variable Y.

Causality

You must be careful not to conclude that changes in x *cause* changes in Y. This type of **causal relationship** can be detected only with a *carefully designed experiment.* For example, if you randomly assign experimental units to each of two levels of a variable x—say, $x = 5$ and $x = 10$—and the data show that the mean value of Y is larger when $x = 10$, then you can say that the change in the level of x caused a change in the mean value of Y. But in most regression analyses, in which the experiments are not designed, there is no guarantee that an important predictor variable—say, x_1—caused Y to change. It is quite possible that some variable that is not even in the model causes *both* Y and x_1 to change.

Multicollinearity

Neither the size of a regression coefficient nor its t-value indicates the importance of the variable as a contributor of information. For example, suppose you intend to predict Y, a university student's calculus grade, based on x_1 = high-school mathematics average and x_2 = score on mathematics aptitude test. Since these two variables contain much of the same or **shared information,** it will not surprise you to learn that, once one of the variables is entered into the model, the other contributes very little additional information. The individual t-value is small. If the variables were entered in the reverse order, however, you would see the size of the t-values reversed.

The situation described above is called **multicollinearity,** and occurs when two or more of the predictor variables are highly correlated with one another. When multicollinearity is present in a regression problem, it can have these effects on the analysis:

- The estimated regression coefficients will have large standard errors, causing imprecision in confidence and prediction intervals.
- Adding or deleting a predictor variable may cause significant changes in the values of the other regression coefficients.

How can you tell whether a regression analysis exhibits multicollinearity? Look for these clues:

- The value of R^2 is large, indicating a good fit, but the individual t-tests are nonsignificant.
- The signs of the regression coefficients are contrary to what you would intuitively expect the contributions of those variables to be.
- A matrix of correlations, generated by computer, shows you which predictor variables are highly correlated with each other and with the response Y.

Figure 13.20 displays the matrix of correlations generated for the real estate data from Example 13.2. The first column of the matrix shows the correlations of each predictor variable with the response variable Y. They are all significantly non-zero, but the first variable, x_1 = living area, is the most highly correlated. The last three columns of the matrix show significant correlations between all but one pair of predictor variables. This is a strong indication of multicollinearity. If you try to eliminate one of the variables in the model, it may drastically change the effects of the other three! Another clue can be found by examining the coefficients of the prediction line,

```
ListPrice = 18.8 + 67.5 Square Metres - 16.2
Number of Floors - 2.67 Bedrooms + 30.3 Baths
```

FIGURE 13.20
Correlation matrix for the real estate data in Example 13.2

Correlations: List Price, Square Metres, Number of Floors, Bedrooms, Baths

```
                 ListPrice   SqMetres    Numflrs     Bdrms
Square Metres      0.951
                   0.000

Number of Fl       0.605       0.630
                   0.017       0.012

Bedrooms           0.746       0.711      0.375
                   0.001       0.003      0.168

Baths              0.834       0.720      0.760       0.675
                   0.000       0.002      0.001       0.006

Cell Contents: Pearson Correlation
               P-Value
```

You would expect more floors and bedrooms to increase the list price, but their coefficients are negative.

Since multicollinearity exists to some extent in all regression problems, you should think of the individual terms as *information contributors,* rather than try to measure the practical importance of each term. The primary decision to be made is whether a term contributes sufficient information to justify its inclusion in the model.

13.10 STEPS TO FOLLOW WHEN BUILDING A MULTIPLE REGRESSION MODEL

The ultimate objective of a multiple regression analysis is to develop a model that will accurately predict Y as a function of a set of predictor variables $x_1, x_2, \ldots, x_k$. The step-by-step procedure for developing this model was presented in Section 13.4 and is restated next with some additional detail. If you use this approach, what may appear to be a complicated problem can be made simpler. As with any statistical procedure, your confidence will grow as you gain experience with multiple regression analysis in a variety of practical situations.

1. Select the predictor variables to be included in the model. Since some of these variables may contain shared information, you can reduce the list by running a stepwise regression analysis (see Section 13.8). Keep the number of predictors small enough to be effective yet manageable. Be aware that the number of observations in your data set must exceed the number of terms in your model; the greater the excess, the better!
2. Write a model using the selected predictor variables. If the variables are qualitative, it is best to begin by including interaction terms. If the variables

are quantitative, it is best to start with a second-order model. Unnecessary terms can be deleted later. Obtain the fitted prediction model.

3. Use the analysis of variance F test and R^2 to determine how well the model fits the data.
4. Check the t tests for the partial regression coefficients to see which ones are contributing significant information in the presence of the others. If some terms appear to be non-significant, consider deleting them. If you choose to compare several different models, use R^2(adj) to compare their effectiveness.
5. Use computer-generated residual plots to check for violation of the regression assumptions.

CHAPTER REVIEW

Key Concepts and Formulas

I. The General Linear Model

1. $Y = \beta_0 + \beta_1 x_1 + \beta_2 x_2 + \cdots + b_k x_k + \epsilon$
2. The random error ϵ has a normal distribution with mean 0 and variance σ^2.

II. Method of Least Squares

1. Estimates $b_0, b_1, \ldots, b_k$, for $\beta_0, \beta_1, \ldots, \beta_k$, are chosen to minimize SSE, the sum of squared deviations about the regression line, $\hat{y} = b_0 + b_1 x_1 + b_2 x_2 + \cdots + b_k x_k$.
2. Least-squares estimates are produced by computer.

III. Analysis of Variance

1. Total SS = SSR + SSE, where Total SS = S_{yy}. The ANOVA table is produced by computer.
2. Best estimate of σ^2 is

$$\text{MSE} = \frac{\text{SSE}}{n - k - 1}$$

IV. Testing, Estimation, and Prediction

1. A test for the significance of the regression, $H_0: \beta_1 = \beta_2 = \cdots = \beta_k = 0$, can be implemented using the analysis of variance F test:

$$F = \frac{\text{MSR}}{\text{MSE}}$$

2. The strength of the relationship between x and y can be measured using

$$R^2 = \frac{\text{SSR}}{\text{Total SS}}$$

which gets closer to 1 as the relationship gets stronger.

3. Use residual plots to check for non-normality, inequality of variances, and an incorrectly fit model.
4. Significance tests for the partial regression coefficients can be performed using the Student's t test:

$$t = \frac{b_i - \beta_i}{\text{SE}(b_i)} \quad \text{with error } df = n - k - 1$$

5. Confidence intervals can be generated by computer to estimate the average value of Y, $E(Y)$, for given values of $x_1, x_2, \ldots, x_k$. Computer-generated prediction intervals can be used to predict a particular observation Y for given values of $x_1, x_2, \ldots, x_k$. For given $x_1, x_2, \ldots, x_k$, prediction intervals are always wider than confidence intervals.

V. Model Building

1. The number of terms in a regression model cannot exceed the number of observations in the data set and should be considerably less!
2. To account for a curvilinear effect in a *quantitative* variable, use a second-order polynomial model. For a cubic effect, use a third-order polynomial model.
3. To add a *qualitative* variable with k categories, use $(k - 1)$ dummy or indicator variables.
4. There may be interactions between two quantitative variables or between a quantitative and qualitative variable. Interaction terms are entered as $\beta x_i x_j$.
5. Compare models using R^2(adj).

TECHNOLOGY TODAY

Multiple Regression Procedures—*Microsoft Excel*

The procedure for performing a multiple regression analysis in *Excel* is identical to the linear regression procedure described in the "Technology Today" section in Chapter 12, except that the range of the x-variables will cover more than one column. You might want to review that section before continuing.

EXAMPLE 13.9

Suppose that a response variable Y is related to four predictor variables, x_1, x_2, x_3, and x_4, so that $k = 4$.

1. Enter the observed values of y and each of the $k = 4$ predictor variables into the first $(k + 1)$ columns of an *Excel* spreadsheet. (NOTE: In order for the multiple regression analysis to work properly, there must be a column of values for each independent predictor variable x_i in your model, and the x columns **must be adjacent to each other.**)
2. Use **Data ▶ Data Analysis ▶ Regression** to generate the dialogue box, highlighting or typing in the cell ranges for the x_i and y values and check "Labels" if necessary.
3. If you click "Confidence Level," *Excel* will calculate confidence intervals for the regression estimates, b_0, b_1, b_2, b_3, and b_4. Enter a cell location for the **Output Range** and click **OK** to generate the regression output.

NOTE: *Excel* does not provide options for estimation and prediction. Also, the diagnostic plots which can be generated in *Excel* are not the same plots as we have discussed in Section 13.3 and will not be discussed in this section.

Multiple Regression Procedures—*MINITAB*

The procedure for performing a multiple regression analysis in *MINITAB* is similar to the linear regression procedure described in the "Technology Today" section in Chapter 12, except that the range of the x-variables will cover more than one column. You might want to review that section before continuing.

EXAMPLE 13.10

Suppose that a response variable Y is related to four predictor variables, x_1, x_2, x_3, and x_4, so that $k = 4$.

1. Enter the observed values of y and each of the $k = 4$ predictor variables into the first $(k + 1)$ columns of a *MINITAB* worksheet. Once this is done, the main inferential tools for multiple regression analysis are generated using **Stat ▶ Regression ▶ Regression.** The dialogue box for the **Regression** command is shown in Figure 13.21(a).
2. Select **y** for the Response variable and $\mathbf{x_1, x_2, \ldots, x_k}$ for the Predictor variables. You may now choose to generate some residual plots to check the validity of your regression assumptions before you use the model for estimation or prediction. Choose **Graphs** to display the dialogue box for residual plots, and choose the appropriate diagnostic plot.

FIGURE 13.21
a) *MINITAB* screen shot
b) *MINITAB* screen shot

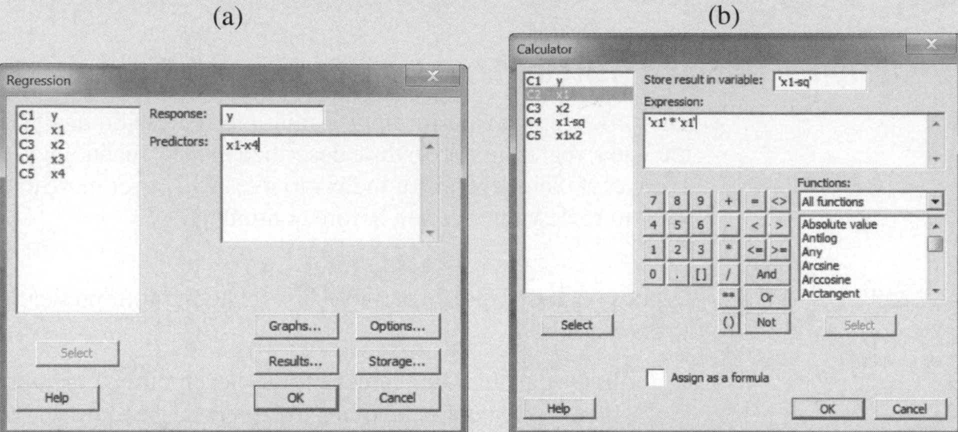

3. Once you have verified the appropriateness of your multiple regression model, you can choose **Options** and obtain confidence and prediction intervals for either of these cases:
   - A single set of values $x_1, x_2, \dots, x_k$ (typed in the box marked "Prediction intervals for new observations")
   - Several sets of values $x_1, x_2, \dots, x_k$ stored in k columns of the worksheet

   When you click **OK** twice, the regression output is generated.
4. The only difficulty in performing the multiple regression analysis using *MINITAB* might be properly entering the data for your particular model. If the model involves polynomial terms or interaction terms, the **Calc ▶ Calculator** command will help you. For example, suppose you want to fit the model

   $$E(Y) = \beta_0 + \beta_1 x_1 + \beta_2 x_2 + \beta_3 x_1^2 + \beta_4 x_1 x_2$$

   You will need to enter the observed values of y, x_1, and x_2 into the first three columns of the *MINITAB* worksheet. Name column C4 "x1-sq" and name C5 "x1x2." You can now use the Calculator dialogue box shown in Figure 13.21(b) to generate these two columns. In the **Expression** box, select **x1 * x1** or **x1 ** 2** and store the results in **C4** (x1-sq). Click **OK.** Similarly, to obtain the data for C5, select **x1 * x2** and store the results in **C5** (x1x2). Click **OK.** You are now ready to perform the multiple regression analysis.
5. If you are fitting either a quadratic or a cubic model in one variable x, you can now plot the data points, the polynomial regression curve, and the upper and lower confidence and prediction limits using **Stat ▶ Regression ▶ Fitted line Plot.** Select y and x for the Response and Predictor variables, and click "Display confidence interval" and "Display prediction interval" in the **Options** dialogue box. Make sure that **Quadratic** or **Cubic** is selected as the "Type of Regression Model," so that you will get the proper fit to the data.
6. Recall that in Chapter 12, you used **Stat ▶ Basic Statistics ▶ Correlation** to obtain the value of the correlation coefficient r. In multiple regression analysis, the same command will generate a matrix of correlations, one for each pair of variables in the set $Y, x_1, x_2, \dots, x_k$. Make sure that the box marked "Display p-values" is checked. The p-values will provide information on the significant correlation between a particular pair, in the presence of all the other variables in the model, and they are identical to the p-values for the individual t tests of the regression coefficients.

Supplementary Exercises

Data set EX1325 **13.25 Biotin Intake in Chicks** Groups of 10-day-old chicks were randomly assigned to seven treatment groups in which a basal diet was supplemented with 0, 50, 100, 150, 200, 250, or 300 micrograms/kilogram (μg/kg) of biotin. The table gives the average biotin intake (x) in micrograms per day and the average weight gain (y) in grams per day:[5]

Added Biotin	Biotin Intake, x	Weight Gain, y
0	0.14	8.0
50	2.01	17.1
100	6.06	22.3
150	6.34	24.4
200	7.15	26.5
250	9.65	23.4
300	12.50	23.3

In the *MINITAB* printout, the second-order polynomial model

$$E(Y) = \beta_0 + \beta_1 x + \beta_2 x^2$$

is fitted to the data. Use the printout to answer the questions.

a. What is the fitted least-squares line?

b. Find R^2 and interpret its value.

c. Do the data provide sufficient evidence to conclude that the model contributes significant information for predicting Y?

d. Find the results of the test of $H_0: \beta_2 = 0$. Is there sufficient evidence to indicate that the quadratic model provides a better fit to the data than a simple linear model does?

e. Do the residual plots indicate that any of the regression assumptions have been violated? Explain.

MINITAB output for Exercise 13.25

Regression Analysis: y versus x, x-sq

```
The regression equation is
y = 8.59 + 3.82 x - 0.217 x-sq

Predictor      Coef    SE Coef       T       P
Constant      8.585      1.641    5.23   0.006
x            3.8208     0.5683    6.72   0.003
x-sq       -0.21663    0.04390   -4.93   0.008

S = 1.83318    R-Sq = 94.4%    R-Sq(adj) = 91.5%

Analysis of Variance
Source          DF      SS       MS       F       P
Regression       2  224.75   112.37   33.44   0.003
Residual Error   4   13.44     3.36
Total            6  238.19

Source     DF    Seq SS
x           1    142.92
x-sq        1     81.83
```

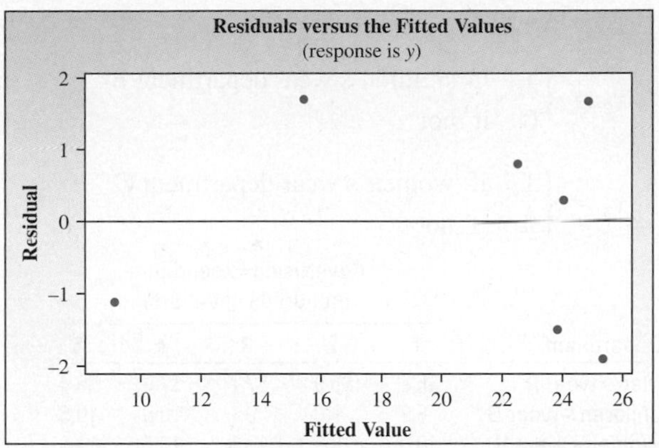

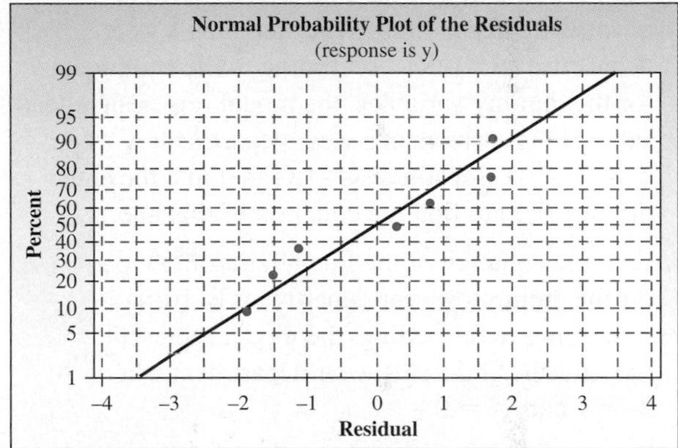

Data set EX1326 **13.26 Advertising and Sales** A department store conducted an experiment to investigate the effects of advertising expenditures on the weekly sales for its men's wear, children's wear, and women's wear departments. Five weeks for observation were randomly selected from each department, and an advertising budget x_1 (in hundreds of dollars) was assigned for each. The weekly sales (in thousands of dollars) are shown in the accompanying table for each of the 15 one-week sales periods. If we expect weekly sales $E(Y)$ to be linearly related to advertising expenditure x_1, and if we expect the slopes of the lines corresponding to the three departments to differ, then an appropriate model for $E(Y)$ is

$$E(Y) = \beta_0 + \underbrace{\beta_1 x_1}_{\substack{\text{quantitative}\\\text{variable}\\\text{"advertising}\\\text{expenditure"}}} + \underbrace{\beta_2 x_2 + \beta_3 x_3}_{\substack{\text{dummy variables}\\\text{used to introduce}\\\text{the qualitative}\\\text{variable "department"}\\\text{into the model}}} + \underbrace{\beta_4 x_1 x_2 + \beta_5 x_1 x_3}_{\substack{\text{interaction terms that}\\\text{introduce differences}\\\text{in slopes}}}$$

where

x_1 = Advertising expenditure

$$x_2 = \begin{cases} 1 & \text{if children's wear department B} \\ 0 & \text{if not} \end{cases}$$

$$x_3 = \begin{cases} 1 & \text{if women's wear department C} \\ 0 & \text{if not} \end{cases}$$

	Advertising Expenditure (hundreds of dollars)				
Department	1	2	3	4	5
Men's wear A	\$5.2	\$5.9	\$7.7	\$7.9	\$9.4
Children's wear B	8.2	9.0	9.1	10.5	10.5
Women's wear C	10.0	10.3	12.1	12.7	13.6

a. Find the equation of the line relating $E(Y)$ to advertising expenditure x_1 for the men's wear department A. (HINT: According to the coding used for the dummy variables, the model represents mean sales $E(Y)$ for the men's wear department A when $x_2 = x_3 = 0$. Substitute $x_2 = x_3 = 0$ into the equation for $E(Y)$ to find the equation of this line.)

b. Find the equation of the line relating $E(Y)$ to x_1 for the children's wear department B. (HINT: According to the coding, the model represents $E(Y)$ for the children's wear department when $x_2 = 1$ and $x_3 = 0$.)

c. Find the equation of the line relating $E(Y)$ to x_1 for the women's wear department C.

d. Find the difference between the intercepts of the $E(Y)$ lines corresponding to the children's wear B and men's wear A departments.

e. Find the difference in slopes between $E(Y)$ lines corresponding to the women's wear C and men's wear A departments.

f. Refer to part e. Suppose you want to test the null hypothesis that the slopes of the lines corresponding to the three departments are equal. Express this as a test of hypothesis about one or more of the model parameters.

13.27 Advertising and Sales, continued Refer to Exercise 13.26. Use a computer software package to perform the multiple regression analysis and obtain diagnostic plots if possible.

a. Comment on the fit of the model, using the analysis of variance F test, R^2, and the diagnostic plots to check the regression assumptions.

b. Find the prediction equation, and graph the three department sales lines.

c. Examine the graphs in part b. Do the slopes of the lines corresponding to the children's wear B and men's wear A departments appear to differ? Test the null hypothesis that the slopes do not differ ($H_0: \beta_4 = 0$) versus the alternative hypothesis that the slopes are different.

d. Are the interaction terms in the model significant? Use the methods described in Section 13.5 to test $H_0: \beta_4 = \beta_5 = 0$. Do the results of this test suggest that the fitted model should be modified?

e. Write a short explanation of the practical implications of this regression analysis.

Data set EX1328 **13.28 Demand for Utilities** Utility companies, which must plan the operation and expansion of electricity generation, are vitally interested in predicting customer demand over both short and long periods of time. A short-term study was conducted to investigate the effect of mean monthly daily temperature x_1 and cost per kilowatt-hour x_2 on the mean daily consumption (in kilowatt-hours, kWh) per household. The company expected the demand for electricity to rise in cold weather (due to heating), fall when the weather was moderate, and rise again when the temperature rose and there was need for air conditioning. They expected demand to decrease as the cost per kilowatt-hour increased, reflecting greater attention to conservation. Data were available for two years, a period in which the cost per kilowatt-hour x_2 increased owing to the increasing cost of fuel. The company fitted the model

$$E(Y) = \beta_0 + \beta_1 x_1 + \beta_2 x_1^2 + \beta_3 x_2 + \beta_4 x_1 x_2 + \beta_5 x_1^2 x_2$$

to the data shown in the table. The *MINITAB* printout for this multiple regression problem is also provided.

Price per kWh, x_2	Daily Temperature and Consumption	Mean Daily Consumption (kWh) Per Household					
8¢	Mean daily temperature (°C), x_1	−0.6	1.1	3.9	5.6	8.3	13.3
		16.7	18.9	20.0	21.7	23.9	25.6
	Mean daily consumption, y	55	49	46	47	40	43
		41	46	44	51	62	73
10¢	Mean daily temperature, x_1	0.0	2.2	3.9	5.6	8.9	13.3
		16.7	18.9	20.0	22.2	23.9	26.1
	Mean daily consumption, y	50	44	42	42	38	40
		39	44	40	44	50	55

MINITAB output for Exercise 13.28

Regression Analysis: y versus x1, x2, x1-sq, x1x2, x1sqx2

```
The regression equation is
y = 77.2 - 7.31 x1 - 2.80 x2 + 0.363 x1-sq
        + 0.540 x1x2 - 0.0283 x1sqx2

Predictor     Coef      SE Coef    T       P
Constant      77.17     12.48      6.18    0.000
x1            -7.313    2.477      -2.95   0.009
x2            -2.803    1.392      -2.01   0.059
x1-sq         0.36324   0.09519    3.82    0.001
x1x2          0.5396    0.2753     1.96    0.066
x1sqx2        -0.02829  0.01054    -2.68   0.015

S = 2.91114      R-Sq = 89.8%      R-Sq(adj) = 87.0%

Analysis of Variance
Source          DF    SS        MS       F       P
Regression      5     1346.08   269.22   31.77   0.000
Residual Error  18    152.55    8.47
Total           23    1498.63

Source    DF   Seq SS
x1        1    140.99
x2        1    203.70
x1-sq     1    883.10
x1x2      1    57.20
x1sqx2    1    61.08

Unusual Observations
Obs   x1     y        Fit      SE Fit   Residual   St Resid
9     20.0   44.000   49.605   1.105    -5.605     -2.08R
12    25.0   73.000   67.798   2.017    5.202      2.48R

R denotes an observation with a large standardized residual.
```

a. Do the data provide sufficient evidence to indicate that the model contributes information for the prediction of mean daily kilowatt-hour consumption per household? Test at the 5% level of significance.

b. Graph the curve depicting $\hat{y}$ as a function of temperature x_1 when the cost per kilowatt-hour is $x_2 = 8¢$. Construct a similar graph for the case when $x_2 = 10¢$ per kilowatt-hour. Are the consumption curves different?

c. If cost per kilowatt-hour is unimportant in predicting use, then you do not need the terms involving x_2 in the model. Therefore, the null hypothesis

$H_0 : x_2$ does not contribute information for the prediction of Y

is equivalent to the null hypothesis $H_0 : \beta_3 = \beta_4 = \beta_5 = 0$ (if $\beta_3 = \beta_4 = \beta_5 = 0$, the terms involving x_2 disappear from the model). The *MINITAB* printout, obtained by fitting the reduced model

$$E(Y) = \beta_0 + \beta_1 x_1 + \beta_2 x_1^2$$

to the data, is shown here. Use the methods of Section 13.5 to determine whether price per kilowatt-hour x_2 contributes significant information for the prediction of Y.

MINITAB output for Exercise 13.28

Regression Analysis: y versus x1, x1sq

```
The regression equation is
y = 52.1 - 2.46 x1 + 0.108 x1sq
Predictor   Coef      SE Coef    T       P
Constant    52.110    2.231      23.36   0.000
x1          -2.4570   0.4413     -5.57   0.000
x1sq        0.10805   0.01690    6.39    0.000

S = 4.68487      R-Sq = 69.2%      R-Sq(adj) = 66.3%

Analysis of Variance
Source          DF   SS        MS       F       P
Regression      2    1037.72   518.86   23.64   0.000
Residual Error  21   460.91    21.95
Total           23   1498.63

Source   DF   Seq SS
x1       1    140.99
x1-sq    1    896.72

Unusual Observations
Obs   x1     y        Fit      SE Fit   Residual   St Resid
12    25.6   73.000   60.020   2.243    12.980     3.16R

R denotes an observation with a large standardized residual.
```

d. Compare the values of R^2(adj) for the two models fit in this exercise. Which of the two models would you recommend?

13.29 Mercury Concentration in Dolphins

Data set EX1329 Because dolphins (and other large marine mammals) are considered to be the top predators in the marine food chain, the heavy metal concentrations in striped dolphins were measured as part of a marine pollution study. The concentration of mercury, the heavy metal reported in this study, is expected to differ in males and females because the mercury in a female is apparently transferred to her offspring during gestation and nursing. This study involved 28 males between the ages of 0.21 and 39.5 years, and 17 females between the ages of 0.80 and 34.5 years. For the data in the table,

x_1 = Age of the dolphin (in years)

$$x_2 = \begin{cases} 0 & \text{if female} \\ 1 & \text{if male} \end{cases}$$

Y = Mercury concentration (in micrograms/gram) in the liver

y	x_1	x_2	y	x_1	x_2
1.70	0.21	1	481.00	22.50	1
1.72	0.33	1	485.00	24.50	1
8.80	2.00	1	221.00	24.50	1
5.90	2.20	1	406.00	25.50	1
101.00	8.50	1	252.00	26.50	1
85.40	11.50	1	329.00	26.50	1
118.00	11.50	1	316.00	26.50	1
183.00	13.50	1	445.00	26.50	1
168.00	16.50	1	278.00	27.50	1
218.00	16.50	1	286.00	28.50	1
180.00	17.50	1	315.00	29.50	1
264.00	20.50	1			

y	x_1	x_2	y	x_1	x_2
241.00	31.50	1	142.00	17.50	0
397.00	31.50	1	180.00	17.50	0
209.00	36.50	1	174.00	18.50	0
314.00	37.50	1	247.00	19.50	0
318.00	39.50	1	223.00	21.50	0
2.50	0.80	0	167.00	21.50	0
9.35	1.58	0	157.00	25.50	0
4.01	1.75	0	177.00	25.50	0
29.80	5.50	0	475.00	32.50	0
45.30	7.50	0	342.00	34.50	0
101.00	8.05	0			
135.00	11.50	0			

a. Write a second-order model relating Y to x_1 and x_2. Allow for curvature in the relationship between age and mercury concentration, and allow for an interaction between gender and age.

Use a computer software package to perform the multiple regression analysis. Refer to the printout to answer these questions.

b. Comment on the fit of the model, using relevant statistics from the printout.

c. What is the prediction equation for predicting the mercury concentration in a female dolphin as a function of her age?

d. What is the prediction equation for predicting the mercury concentration in a male dolphin as a function of his age?

e. Does the quadratic term in the prediction equation for females contribute significantly to the prediction of the mercury concentration in a female dolphin?

f. Are there any other important conclusions that you feel were not considered regarding the fitted prediction equation?

Data set EX1330 **13.30 Cricket by Chance** Is it true that the more balls a cricketer plays the more runs you make? The following table provides the related information about some of the cricket superstars from four countries for the year 2006. The data shown in this table give the total number of runs scored and total balls faced in given cricket matches for players of four countries:[6]

2006 Calendar Year Test Best Innings Batting Strike Rates

Players	Runs Scored	Balls Faced	Country
A. Symonds	55	47	Australia
A. Symonds	29	26	Australia
S.K. Warne	36	31	Australia
S.K. Warne	25	22	Australia
S.K. Warne	36	33	Australia
C.H. Gayle	34	28	West Indies
C.H. Gayle	30	27	West Indies
C.H. Gayle	30	30	West Indies

2006 Calendar Year Test Best Innings Batting Strike Rates

Players	Runs Scored	Balls Faced	Country
Shahid Afridi	60	46	Pakistan
Shahid Afridi	103	80	Pakistan
Shahid Afridi	156	128	Pakistan
S.M. Pollock	32	20	South Africa
S.M. Pollock	26	21	South Africa
S.M. Pollock	40	41	South Africa
S.M. Pollock	44	47	South Africa

Use a computer package to analyze the data with a multiple regression analysis. Comment on the fit of the model, the significant variables, any interactions that exist, and any regression assumptions that may have been violated. Summarize your results in a report, including printouts and graphs if possible.

Data set EX1331 **13.31 Digital Video Recorders** Digital video recorders (DVR) allow you to record and view programs at whatever time you request. Programs can be paused while you go to the kitchen for a snack, and commercials can be eliminated. The sales of DVRs skyrocketed in 2004–2005, as shown in the following table:[7]

Year	2000	2001	2002	2003	2004	2005
DVRs (millions)	0.35	0.88	2.50	5.70	11.50	20.20

a. Plot the predicted number of DVRs (Y) as a function of the year (x) using a scatterplot. Describe the nature of the relationship.

b. What model would you use to predict Y as a function of x? Explain.

c. Using the model from part b and a computer software package, find the least-squares regression equation for predicting the DVR market penetration—that is, the number of DVRs installed in homes—as a function of year.

d. Does the model contribute information for the prediction of y? Test using $\alpha = 0.01$.

e. What is the value of R^2? What does this tell you about the fit of the model?

f. If they are available, examine the residual plots for the analysis. What conclusions can you draw?

Data set EX1332 **13.32 Women in the Workforce** A study was done to track the participation of Canadian adult women in the workforce over an almost 30-year period.[8]

Y = participation (percent of adult women in the workforce)

x_1 = Total fertility rate (tfr): expected births to a cohort of 1000 women at current age–specific fertility rates.

x_2 = Men's average weekly wages

x_3 = Women's average weekly wages

x_4 = Per capita consumer debt, in constant dollars

x_5 = Percent of the active workforce working 34 hours per week or less

Year	y	x_1	x_2	x_3	x_4	x_5
1946	25.3	3748	25.35	14.05	18.18	10.28
1947	24.4	3996	26.14	14.61	28.33	9.28
1948	24.2	3725	25.11	14.23	30.55	9.51
1949	24.2	3750	25.45	14.61	35.81	8.87
1950	23.7	3669	26.79	15.26	38.39	8.54
1951	24.2	3682	26.33	14.58	26.52	8.84
1952	24.1	3845	27.89	15.66	45.65	8.6
1953	23.8	3905	29.15	16.3	52.99	5.49
1954	23.6	4047	29.52	16.57	54.84	6.67
1955	24.3	4043	32.05	17.99	65.53	6.25
1956	25.1	4092	32.98	18.33	72.56	6.32
1957	26.2	4168	32.25	17.64	69.49	7.3
1958	26.6	4073	32.52	18.16	71.71	8.65
1959	26.9	4100	33.95	18.58	78.89	8.8
1960	27.9	4119	34.63	18.95	84.99	9.39
1961	29.1	4159	35.14	18.78	87.71	10.23
1962	29.9	4134	34.49	18.74	95.31	10.77
1963	29.8	4017	35.99	19.71	104.4	10.84
1964	30.9	3886	36.68	20.06	116.8	11.7
1965	32.1	3467	37.96	20.94	130.99	12.33
1966	33.2	3150	38.68	21.2	135.25	12.18
1967	34.5	2879	39.65	21.95	142.93	13.67
1968	35.1	2681	41.2	22.68	155.47	13.82
1969	36.1	2563	42.44	23.75	165.04	14.91
1970	36.9	2571	42.02	25.63	164.53	15.52
1971	37	2503	45.32	26.79	169.63	15.47
1972	37.9	2302	45.61	27.51	190.62	15.85
1973	40.1	1931	45.59	27.35	209.6	15.4
1974	40.6	1875	48.06	29.64	216.66	16.23
1975	42.2	1866	46.12	29.33	224.34	16.71

a. Use a program of your choice to find the correlation matrix for the variables under study. Which variables appear to be highly correlated with Y, participation?

b. Write a model to describe the dependent variable, the participation of women in the workforce, as a function of the variables $x_1 = \text{tfr}$, $x_2 = \text{menWages}$, $x_3 = \text{womenWages}$, $x_4 = \text{debt}$, and $x_5 = \text{part time}$.

c. Use a regression program of your choice to fit the full model using all of the predictors. What proportion of the variation in Y is explained by regression? Does this convey the impression that the model adequately explains the inherent variability in Y?

d. Which variables appear to be good predictors of Y? How might you refine the model in light of these results? Use these variables in refitting the model. What proportion of the variation is explained by this refitted model? Comment on the adequacy of this reduced model in comparison to the full model.

Data set EX1333 **13.33 Quality Control** A manufacturer recorded the number of defective items (Y) produced on a given day by each of 10 machine operators and also recorded the average output per hour (x_1) for each operator and the time in weeks from the last machine service (x_2):

y	x_1	x_2
13	20	3.0
1	15	2.0
11	23	1.5
2	10	4.0
20	30	1.0
15	21	3.5
27	38	0
5	18	2.0
26	24	5.0
1	16	1.5

The printout that follows resulted when these data were analyzed using the *MINITAB* package using the model:

$$E(Y) = \beta_0 + \beta_1 x_1 + \beta_2 x_2$$

MINITAB output for Exercise 13.33

Regression Analysis: y versus x1, x2

```
The regression equation is
y = -28.4 + 1.46 x1 + 3.84 x2

Predictor       Coef    SE Coef         T      P
Constant    -28.3906     0.8273    -34.32  0.000
x1           1.46306    0.02699    -54.20  0.000
x2            3.8446     0.1426     26.97  0.000

S = 0.548433     R-Sq = 99.8%     R-Sq(adj) = 99.7%

Analysis of Variance

Source           DF      SS      MS        F      P
Regression        2  884.79  442.40  1470.84  0.000
Residual Error    7    2.11    0.30
Total             9  886.90

Source  DF   Seq SS
x1       1   666.04
x2       1   218.76
```

a. Interpret R^2 and comment on the fit of the model.

b. Is there evidence to indicate that the model contributes significantly to the prediction of y at the $\alpha = 0.01$ level of significance?

c. What is the prediction equation relating $\hat{y}$ and x_1 when $x_2 = 4$?

d. Use the fitted prediction equation to predict the number of defective items produced for an operator whose average output per hour is 25 and whose machine was serviced three weeks ago.

e. What do the residual plots tell you about the validity of the regression assumptions?

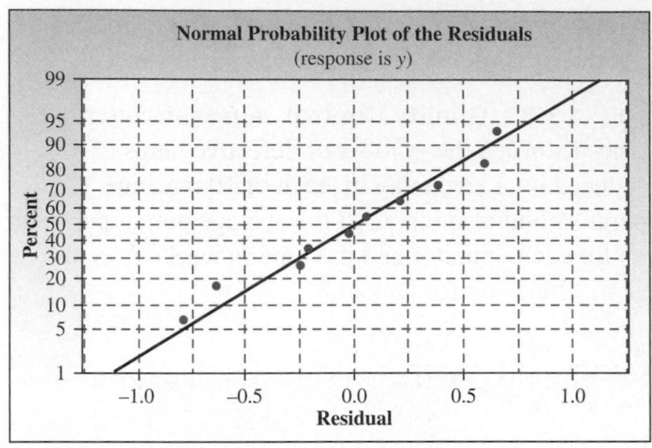

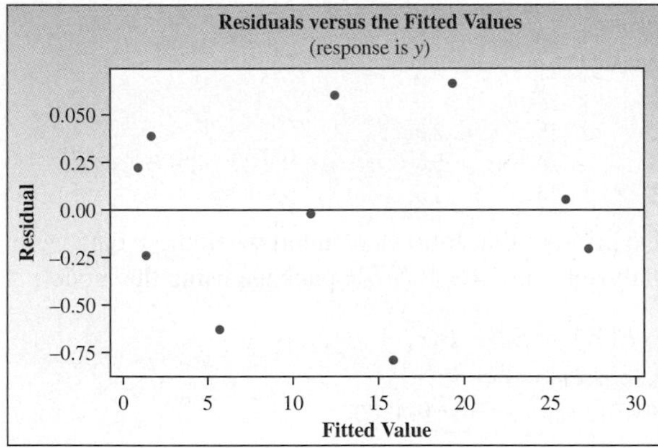

13.34 Metal Corrosion and Soil Acids In an investigation to determine the relationship between the degree of metal corrosion and the length of time the metal is exposed to the action of soil acids, the percentage of corrosion and exposure time were measured weekly:

EX1334

y	0.1	0.3	0.5	0.8	1.2	1.8	2.5	3.4
x	1	2	3	4	5	6	7	8

The data were fitted using the quadratic model, $E(Y) = \beta_0 + \beta_1 x + \beta_2 x^2$, with the following results.

Excel output for Exercise 13.34

SUMMARY OUTPUT

Regression Statistics	
Multiple R	0.9645
R Square	0.9303
Adjusted R Square	0.9187
Standard Error	0.3311
Observations	8

ANOVA

	df	SS	MS	F	Significance F
Regression	1	8.777	8.777	80.052	0.000
Residual	6	0.658	0.110		
Total	7	9.435			

	Coefficients	Standard Error	t Stat	P-value
Intercept	-0.732	0.258	-2.838	0.030
x	0.457	0.051	8.947	0.000

a. What percentage of the total variation is explained by the quadratic regression of Y on x?

b. Is the regression on x and x^2 significant at the $\alpha = 0.05$ level of significance?

c. Is the linear regression coefficient significant when x_2 is in the model?

d. Is the quadratic regression coefficient significant when x_1 is in the model?

e. The data were fitted to a linear model without the quadratic term with the results that follow. What can you say about the contribution of the quadratic term when it is included in the model?

Excel output for Exercise 13.34

SUMMARY OUTPUT

Regression Statistics	
Multiple R	0.9993
R Square	0.9985
Adjusted R Square	0.9979
Standard Error	0.0530
Observations	8

ANOVA

	df	SS	MS	F	Significance F
Regression	2	9.421	4.710	1676.610	0.000
Residual	5	0.014	0.003		
Total	7	9.435			

	Coefficients	Standard Error	t Stat	P-value
Intercept	-1.589	0.074	2.656	0.045
x1	-0.100	0.038	-2.652	0.045
x-sq	0.062	0.004	15.138	0.000

f. The plot of the residuals from the linear regression model in part e shows a specific pattern. What is the term in the model that seems to be missing?

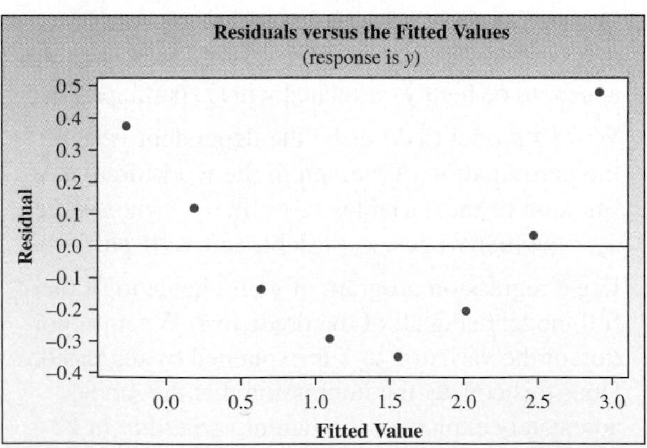

13.35 Managing your Money A particular savings and loan corporation is interested in determining how well the amount of money in family

EX1335

savings accounts can be predicted using the three independent variables—annual income, number in the family unit, and area in which the family lives. Suppose that there are two specific areas of interest to the corporation. The following data were collected, where

Y = Amount in all savings accounts

x_1 = Annual income

x_2 = Number in family unit

x_3 = 0 if in Area 1; 1 if not

Both Y and x_1 were recorded in units of \$1000.

y	x_1	x_2	x_3
0.5	19.2	3	0
0.3	23.8	6	0
1.3	28.6	5	0
0.2	15.4	4	0
5.4	30.5	3	1
1.3	20.3	2	1
12.8	34.7	2	1
1.5	25.2	4	1
0.5	18.6	3	1
15.2	45.8	2	1

The following computer printout resulted when the data were analyzed using *MINITAB*.

Regression Analysis: y versus x1, x2, x3

```
The regression equation is
y = -3.11 + 0.503 x1 - 1.61 x2 - 1.15 x3

Predictor      Coef    SE Coef       T       P
Constant     -3.112      3.600   -0.86   0.421
x1          0.50314    0.07670    6.56   0.001
x2          -1.6126     0.6579   -2.45   0.050
x3           -1.155      1.791   -0.64   0.543

S = 1.89646    R-Sq = 92.2%     R-Sq(adj) = 88.4%

Analysis of Variance
Source          DF        SS      MS      F      P
Regression       3   256.621  85.540  23.78  0.001
Residual Error   6    21.579   3.597
Total            9   278.200

Source  DF   Seq SS
x1       1  229.113
x2       1   26.012
x3       1    1.496
```

a. Interpret R^2 and comment on the fit of the model.

b. Test for a significant regression of Y on x_1, x_2, and x_3 at the 5% level of significance.

c. Test the hypothesis $H_0 : \beta_3 = 0$ against $H_a : \beta_3 \neq 0$ using $\alpha = 0.05$. Comment on the results of your test.

d. What can be said about the utility of x_3 as a predictor variable in this problem?

CASE STUDY  "Buying Up the Canadian Economy"—Another Look

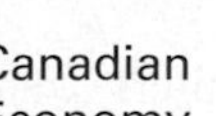 Canadian Economy

The case study in Chapter 12 examined the effects of government regulation on foreign ownership of assets in Canada. For example, the Foreign Investment Review Agency was created in 1975 to monitor and regulate foreign takeovers in Canada. In 1985, after a change of government, the Foreign Investment Review Agency was replaced with a new agency, Investment Canada, whose mandate was intended to be less restrictive. Did this change have any effect? The data in the table represent the percentage of commercial assets in non-financial corporations under foreign control (y) for the years 1975–2004. To simplify the analysis, we have coded the year using the coded variable x = year – 1975.

By examining a scatterplot of the data, you will find that the percentage of foreign ownership of Canadian assets does not appear to follow a linear relationship over time, but rather exhibits a curvilinear response. The question, then, is to decide whether a second-, third-, or higher-order model adequately describes the data.[9]

Percentage of Assets under Foreign Control in Non-financial Corporations, 1975–2004

Year	x (year minus 1975)	y (percent assets)	Year	x (year minus 1975)	y (percent assets)
1975	0	30.2	1990	15	23.7
1976	1	28.4	1991	16	23.6
1977	2	28.3	1992	17	23.9
1978	3	26.7	1993	18	23.8
1979	4	26.8	1994	19	23.6
1980	5	25.3	1995	20	25.1

1981	6	23.4	1996	21	25.4
1982	7	22.6	1997	22	25.9
1983	8	22.3	1998	23	26.9
1984	9	22.2	1999	24	25.3
1985	10	21.4	2000	25	25.5
1986	11	21.5	2001	26	28.8
1987	12	22.5	2002	27	28.7
1988	13	23.3	2003	28	28.7
1989	14	23.6	2004	29	28.5

1. Plot the data and sketch what you consider to be the best-fitting linear, quadratic, and cubic models.
2. Find the residuals using the fitted linear regression model. Does there appear to be any pattern in the residuals when plotted against x? What model do the residuals indicate would produce a better fit?
3. What is the increase in R^2 when you fit a quadratic rather than a linear model? Is the coefficient of the quadratic term significant? Is the fitted quadratic model significantly better than the fitted linear model? Plot the residuals from the fitted quadratic model. Does there seem to be any apparent pattern in the residuals when plotted against x?
4. What is the increase in R^2 when you compare the fitted cubic with the fitted quadratic model? Is the fitted cubic model significantly better than the fitted quadratic? Are there any patterns in a plot of the residuals versus x? What proportion of the variation in the response Y is not accounted for by fitting a cubic model? Should any higher-order polynomial model be considered? Why or why not?

PROJECTS

Project 13: Aspen Mixedwood Forests in Canada, Part 2

Boreal mixedwood forests have been commercially logged for conifers, transected by roads and seismic cutlines, and regionally fragmented by agricultural practices. Most recently, Alberta's public mixedwood forests have been allocated for harvest of trembling aspen for pulp, paper, and oriented strand board products. With relevance to commercial forestry, this report describes the structure and biodiversity of aspen-dominated mixedwood forests of fire-origin. The boreal mixedwood forest in Canada extends from south-western Manitoba, through the central and northern parts of the Prairie provinces, and into north-eastern British Columbia.[10]

Suppose the following data is collected from boreal mixedwood forests in northern Saskatchewan. Use the data in the following table to fit a linear regression model to predict Age using the rest of the variables:

Age (y)	Diameter at Breast Height (x_1)	Canopy Height (x_2)	Stem Density (x_3)
28	8.38	13.42	5213
24	8.18	10.11	4079
26	8.11	11.61	3214
27	7.58	10.01	4854
24	6.36	12.14	2681
27	7.98	10.42	5108
26	6.76	10.94	5049
25	7.18	7.18	2403
24	6.69	8.46	4662

26	8.40	10.33	4256
27	7.98	14.40	2342
27	4.93	10.32	4533
27	6.36	8.29	4175
29	5.10	4.02	2383
25	6.45	13.25	3512
26	7.38	6.70	3481
27	7.33	8.06	5729
28	5.92	9.55	4722
26	6.22	7.35	3279
25	9.50	12.36	4187
27	5.83	6.44	3363
28	5.72	8.60	5005
26	6.65	10.71	3852
26	5.65	11.27	4275
25	6.53	7.59	3851

a. Plot the data and see if a linear regression model can be fit to predict age from the other variables.

b. Fit linear regression models considering each independent variable separately and test for their significance.

c. Build a model taking into account all three independent variables and their interactions. Use the overall F test to determine whether the model contributes significant information for the prediction of y. Use $\alpha = 0.05$.

d. Carry out three separate tests with a significance level of 0.05 to decide if x_1, x_2, and x_3 are significant.

e. Now fit a model with only x_1, x_2, and x_1x_2, (interaction effect) to predict age. How does this model perform compare to the one in (part c)?

f. How would you interpret the regression parameters β_1, β_2, β_3, with reference to the model in (part e)?

g. Does the effect of the quadratic term on predicting age in (part e) appear to be useful? Use $\alpha = 0.05$. How well does the model fit? Use any relevant statistics and diagnostic tools to answer this question.

h. What do the residual plots tell you about the validity of the regression assumptions in (part e)?

i. Based on the model in (part e), which of the predictors make the most significant contribution in predicting the dependent variable Y? Justify your answer.

j. Is there any significant interaction effect of the independent variables on age in the model considered in (part e)?

k. Suppose you suspect that independent variable x_2 affects the response Y, but the relationship is curvilinear. Add the quadratic term that is x_2^2 in the model described in (part e). Use p-value approach to test the effect of the quadratic term on predicting age.

l. Which model would you prefer for prediction of age among the above three models considered? Use the best model to obtain a point prediction of age that has DBH = 8, Canopy Height = 15, and Stem Density = 4900. Also obtain a 95% confidence interval for the prediction.

m. Suppose we wish to investigate the value of regressor variable x_3 given that the regressors x_1 and x_2, and their interactions (x_1x_2) are in the model. Perform the appropriate test of hypothesis to determine whether the reduced model is adequate. Use $\alpha = 0.05$.

14

Analysis of Categorical Data

GENERAL OBJECTIVES

Many types of surveys and experiments result in qualitative rather than quantitative response variables, so that the responses can be classified but not quantified. Data from these experiments consist of the count or number of observations that fall into each of the response categories included in the experiment. In this chapter, we are concerned with methods for analyzing categorical data.

CHAPTER INDEX

NEED TO KNOW

How to Determine the Appropriate Number of Degrees of Freedom

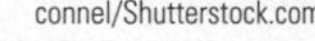

connel/Shutterstock.com

Can a Marketing Approach Improve Library Services?

How do you rate your library? Is the atmosphere friendly, dull, or too quiet? Is the library staff helpful? Are the signs clear and unambiguous? The modern consumer-led approach to marketing, in general, involves the systematic study by organizations of their customers' wants and needs in order to improve their services or products. In the case study at the end of this chapter, we examine the results of a study to explore the attitudes of young adults toward the services provided by libraries.

14.1 A DESCRIPTION OF THE EXPERIMENT

Many experiments result in measurements that are *qualitative* or *categorical* rather than *quantitative*; that is, a *quality* or *characteristic* (rather than a numerical value) is measured for each experimental unit. You can summarize this type of data by creating a list of the categories or characteristics and reporting a **count** of the number of measurements that fall into each category. Here are a few examples:

- People can be classified into five income brackets.
- A mouse can respond in one of three ways to a stimulus.
- An M&M'S® candy can have one of six colours.
- An industrial process manufactures items that can be classified as "acceptable," "second quality," or "defective."

These are some of the many situations in which the data set has characteristics appropriate for the **multinomial experiment.**

THE MULTINOMIAL EXPERIMENT

- The experiment consists of n identical trials.
- The outcome of each trial falls into one of k categories.
- The probability that the outcome of a single trial falls into a particular category—say, category i—is p_i and remains constant from trial to trial. This probability must be between 0 and 1, for each of the k categories, and the sum of all k probabilities is $\Sigma p_i = 1$.
- The trials are independent.
- The experimenter counts the *observed* number of outcomes in each category, written as $O_1, O_2, \ldots, O_k$, with $O_1 + O_2 + \ldots + O_k = n$.

You can visualize the multinomial experiment by thinking of k boxes or **cells** into which n balls are tossed. The n tosses are independent, and on each toss the chance of hitting the ith box is the same. However, this chance can vary from box to box; it might be easier to hit box 1 than box 3 on each toss. Once all n balls have been tossed, the number in each box or **cell**—$O_1, O_2, \ldots, O_k$—is counted.

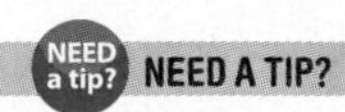
NEED A TIP?

The multinomial experiment is an extension of the *binomial experiment.* For a binomial experiment, $k = 2$.

You have probably noticed the similarity between the *multinomial experiment* and the *binomial experiment* introduced in Chapter 5. In fact, when there are $k = 2$ categories, the two experiments are identical, except for notation. Instead of p and q, we write p_1 and p_2 to represent the probabilities for the two categories, "success" and "failure." Instead of x and $(n - x)$, we write O_1 and O_2 to represent the observed number of "successes" and "failures."

When we presented the binomial random variable, we made inferences about the binomial parameter p (and by default, $q = 1 - p$) using large-sample methods based on the z statistic. In this chapter, we extend this idea to make inferences about the *multinomial parameters,* $p_1, p_2, \ldots, p_k$, using a different type of statistic. This statistic, whose approximate sampling distribution was derived by a British statistician named Karl Pearson in 1900, is called the **chi-square** (or sometimes **Pearson's chi-square**) **statistic.**

14.2 PEARSON'S CHI-SQUARE STATISTIC

Suppose that $n = 100$ balls are tossed at the cells (boxes) and you know that the probability of a ball falling into the first box is $p_1 = 0.1$. How many balls would you *expect* to fall into the first box? Intuitively, you would expect to see $100(0.1) = 10$ balls in the first box. This should remind you of the average or expected number of successes, $\mu = np$, in the binomial experiment. In general, the expected number of balls that fall into cell i—written as E_i—can be calculated using the formula

$$E_i = np_i$$

for any of the cells $i = 1, 2, \ldots, k$.

Now suppose that you *hypothesize* values for each of the probabilities $p_1, p_2, \ldots, p_k$ and calculate the expected number for each category or cell. If your hypothesis is correct, the actual *observed cell counts,* O_i, should not be too different from the *expected cell counts,* $E_i = np_i$. The larger the differences, the more likely it is that the hypothesis is incorrect. The *Pearson chi-square statistic* uses the differences $(O_i - E_i)$ by first squaring these differences to eliminate negative contributions, and then forming a *weighted* average of the squared differences.

PEARSON'S CHI-SQUARE TEST STATISTIC

$$\chi^2 = \Sigma \frac{(O_i - E_i)^2}{E_i}$$

summed over all k cells, with $E_i = np_i$.

Although the mathematical proof is beyond the scope of this book, it can be shown that when n is large, χ^2 has an approximate **chi-square probability distribution** in repeated sampling. If the hypothesized expected cell counts are correct, the differences $(O_i - E_i)$ are small and χ^2 is close to 0. But, if the hypothesized probabilities are incorrect, large differences $(O_i - E_i)$ result in a *large* value of χ^2. You should use a **right-tailed statistical test** and look for an unusually large value of the test statistic.

NEED A TIP?

The Pearson's chi-square tests are always upper-tailed tests.

The chi-square distribution was used in Chapter 10 to make inferences about a single population variance σ^2. Like the F distribution, its shape is not symmetric and depends on a specific number of **degrees of freedom.** Once these degrees of freedom are specified, you can use Table 5 in Appendix I to find critical values or to bound the p-value for a particular chi-square statistic.

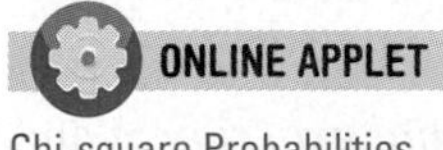

ONLINE APPLET

Chi-square Probabilities

The appropriate degrees of freedom for the chi-square statistic vary depending on the particular application you are using. Although we will specify the appropriate degrees of freedom for the applications presented in this chapter, you should use the general rule given next for determining degrees of freedom for the chi-square statistic.

NEED TO KNOW

How to Determine the Appropriate Number of Degrees of Freedom

1. Start with the number of *categories* or cells in the experiment.
2. Subtract one degree of freedom for each linear restriction on the cell probabilities. You will always lose one *df* because $p_1 + p_2 + \cdots + p_k = 1$.

3. Sometimes the expected cell counts cannot be calculated directly but must be estimated using the sample data. Subtract one degree of freedom for every independent population parameter that must be estimated to obtain the estimated values of E_i.

We begin with the simplest applications of the chi-square test statistic—the **goodness-of-fit** test.

14.3 TESTING SPECIFIED CELL PROBABILITIES: THE GOODNESS-OF-FIT TEST

The simplest hypothesis concerning the cell probabilities specifies a numerical value for each cell. The expected cell counts are easily calculated using the hypothesized probabilities, $E_i = np_i$, and are used to calculate the observed value of the χ^2 test statistic. For a multinomial experiment consisting of k categories or cells, the test statistic has an approximate χ^2 distribution with $df = (k - 1)$.

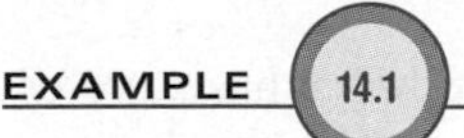

EXAMPLE 14.1 A researcher designs an experiment in which a rat is attracted to the end of a ramp that divides, leading to doors of three different colours. The researcher sends the rat down the ramp $n = 90$ times and observes the choices listed in Table 14.1. Does the rat have (or acquire) a preference for one of the three doors?

TABLE 14.1 Rat's Door Choices

	Door		
	Green	Red	Blue
Observed count (O_i)	20	39	31

SOLUTION If the rat has no preference in the choice of a door, you would expect in the long run that the rat would choose each door an equal number of times. That is, the null hypothesis is

$$H_o: p_1 = p_2 = p_3 = \frac{1}{3}$$

versus the alternative hypothesis

$$H_a: \text{At least one } p_i \text{ is different from } \frac{1}{3}$$

where p_i is the probability that the rat chooses door i, for $i = 1, 2,$ and 3. The expected cell counts are the same for each of the three categories—namely, $np_i = 90(1/3) = 30$. The chi-square test statistic can now be calculated as

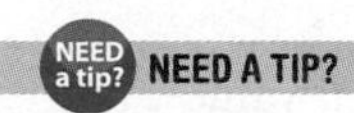

NEED A TIP?

The rejection region and p-value are in the upper tail of the chi-square distribution.

$$\chi^2 = \Sigma\frac{(O_i - E_i)^2}{E_i}$$

$$= \frac{(20 - 30)^2}{30} + \frac{(39 - 30)^2}{30} + \frac{(31 - 30)^2}{30} = 6.067$$

For this example, the test statistic has $(k - 1) = 2$ degrees of freedom because the only linear restriction on the cell probabilities is that they must sum to 1. Hence, you can use Table 5 in Appendix I to find bounds for the right-tailed p-value. Since the observed value, $\chi^2 = 6.067$, lies between $\chi^2_{0.050} = 5.99$ and $\chi^2_{0.025} = 7.38$, the *p*-value is between 0.025 and 0.050. The researcher would report the results as significant at the 5% level ($P < 0.05$), meaning that the null hypothesis of no preference is rejected. There is sufficient evidence to indicate that the rat has a preference for one of the three doors.

What more can you say about the experiment once you have determined statistically that the rat has a preference? Look at the data to see where the differences lie.

The blue door was chosen only a little more than one-third of the time:

$$\frac{31}{90} = 0.344$$

However, the sample proportions for the other two doors are quite different from one-third. The rat chooses the green door least often—only 22% of the time:

$$\frac{20}{90} = 0.222$$

The rat chooses the red door most often—43% of the time:

$$\frac{39}{90} = 0.433$$

You would summarize the results of the experiment by saying that the rat has a preference for the red door. Can you conclude that the preference is *caused* by the door colour? The answer is no—the cause could be some other physiological or psychological factor that you have not yet explored. Avoid declaring a *causal* relationship between colour and preference!

EXAMPLES 14.2

Statistics Canada reported on the proportions of Canadian university teachers/professors by religion. According to Statistics Canada, the proportion of Roman Catholic (R), Christian (C), no religion (N), Jewish (J), Muslim (M), and other (O) professors in Canadian universities are 0.288, 0.282, 0.297, 0.046, 0.024, and 0.064, respectively. To determine whether or not the actual population proportions fit this set of reported probabilities, a random sample of 200 Canadian professors was selected and their religions were recorded. The observed and expected cell counts are shown in Table 14.2. The expected cell counts are calculated as $E_i = 200\,p_i$. Test the goodness of fit of these proportions.

TABLE 14.2 **Counts of Professors' Religious Affiliations**

	R	C	N	J	M	O
Observed (O_i)	59	60	60	8	5	8
Expected (E_i)	57.6	56.4	59.4	9.2	4.8	12.8

SOLUTION The hypothesis to be tested is determined by the model probabilities:

NEED A TIP?
Degrees of freedom for a simple goodness-of-fit test: $df = k - 1$

$$H_o: p_1 = 0.288;\ p_2 = 0.282;\ p_3 = 0.297;\ p_4 = 0.046;\ p_5 = 0.024;\ p_6 = 0.064$$

versus

$$H_a: \text{at least one of the six probabilities is different from the specified value}$$

Then

$$\chi^2 = \Sigma\frac{(O_i - E_i)^2}{E_i}$$

$$= \frac{(59 - 57.6)^2}{57.6} + \ldots + \frac{(8 - 12.8)^2}{12.8} = 2.234$$

From Table 5 in Appendix I, indexing $df = (k - 1) = 5$, you can find that the observed value of the test statistic is less than $\chi^2_{0.100} = 9.23$, so that the p-value is greater than 0.10. You do not have sufficient evidence to reject H_0; that is, you cannot declare that the true proportions are *different* from those reported earlier. The results are non-significant (NS).

You can find instructions in the "Technology Today" section at the end of this chapter that allow you to use *MINITAB* (versions 15 or 16) to perform the chi-square goodness-of-fit test and generate the results.

Notice the difference in the goodness-of-fit hypothesis compared to other hypotheses that you have tested. In the goodness-of-fit test, the researcher uses the null hypothesis to specify the model he believes to be *true,* rather than a model he hopes to prove false! When you could not reject H_0 in Example 14.2, the results were as expected. Be careful, however, when you report your results for goodness-of-fit tests. You cannot declare with confidence that the model is absolutely correct without reporting the value of β for some practical alternatives.

14.3 EXERCISES

BASIC TECHNIQUES

14.1 List the characteristics of a multinomial experiment.

14.2 Use Table 5 in Appendix I to find the value of χ^2 with the following area α to its right:

a. $\alpha = 0.05, df = 3$ **b.** $\alpha = 0.01, df = 8$
c. $\alpha = 0.005, df = 15$ **d.** $\alpha = 0.01, df = 11$

14.3 Give the rejection region for a chi-square test of specified probabilities if the experiment involves k categories in these cases:

a. $k = 7, \alpha = 0.05$ **b.** $k = 10, \alpha = 0.01$
c. $k = 14, \alpha = 0.005$ **d.** $k = 3, \alpha = 0.01$

14.4 Use Table 5 in Appendix I to bound the p-value for a chi-square test:

a. $\chi^2 = 4.29, df = 5$ **b.** $\chi^2 = 20.62, df = 6$
c. $\chi^2 = 0.81, df = 5$ **d.** $\chi^2 = 25.40, df = 13$

14.5 Suppose that a response can fall into one of $k = 5$ categories with probabilities $p_1, p_2, \ldots, p_5$ and that $n = 300$ responses produced these category counts:

Category	1	2	3	4	5
Observed count	47	63	74	51	65

a. Are the five categories equally likely to occur? How would you test this hypothesis?

b. If you were to test this hypothesis using the chi-square statistic, how many degrees of freedom would the test have?

c. Find the critical value of χ^2 that defines the rejection region with $\alpha = 0.05$.

d. Calculate the observed value of the test statistic.

e. Conduct the test and state your conclusions.

14.6 Suppose that a response can fall into one of $k = 3$ categories with probabilities $p_1 = 0.4$, $p_2 = 0.3$, and $p_3 = 0.3$, and $n = 300$ responses produce these category counts:

Category	1	2	3
Observed count	130	98	72

Do the data provide sufficient evidence to indicate that the cell probabilities are different from those specified for the three categories? Find the approximate p-value and use it to make your decision.

APPLICATIONS

14.7 Your Favourite Lane A highway with four lanes in each direction was studied to see whether drivers prefer to drive on the inside lanes. A total of 1000 automobiles were observed during heavy early-morning traffic, and the number of cars in each lane was recorded:

Lane	1	2	3	4
Observed count	294	276	238	192

Do the data present sufficient evidence to indicate that some lanes are preferred over others? Test using $\alpha = 0.05$. If there are any differences, discuss the nature of the differences.

14.8 Peonies A peony plant with red petals was crossed with another plant having streaky petals. A geneticist states that 75% of the offspring from this cross will have red flowers. To test this claim, 100 seeds from this cross were collected and germinated, and 58 plants had red petals. Use the chi-square goodness-of-fit test to determine whether the sample data confirm the geneticist's prediction.

14.9 Heart Attacks on Mondays Do you hate Mondays? Researchers from Germany have provided another reason for you: They concluded that the risk of a heart attack for a working person may be as much as 50% greater on Monday than on any other day.[1] The researchers kept track of heart attacks and coronary arrests over a period of five years among 330,000 people who lived near Augsburg, Germany. In an attempt to verify their claim, you survey 200 working people who had recently had heart attacks and record the day on which their heart attacks occurred:

Day	Observed Count
Sunday	24
Monday	36
Tuesday	27
Wednesday	26
Thursday	32
Friday	26
Saturday	29

Do the data present sufficient evidence to indicate that there is a difference in the incidence of heart attacks depending on the day of the week? Test using $\alpha = 0.05$.

14.10 Mortality Statistics Medical statistics show that deaths due to four major diseases—call them A, B, C, and D—account for 15%, 21%, 18%, and 14%, respectively, of all non-accidental deaths. A study of the causes of 308 non-accidental deaths at a hospital gave the following counts:

Disease	A	B	C	D	Other
Deaths	43	76	85	21	83

Do these data provide sufficient evidence to indicate that the proportions of people dying of diseases A, B, C, and D at this hospital differ from the proportions accumulated for the population at large?

14.11 Schizophrenia Research has suggested a link between the prevalence of schizophrenia and birth during particular months of the year in which viral infections are prevalent. Suppose you are working on a similar problem and you suspect a linkage between a disease observed in later life and month of birth. You have records of 400 cases of the disease, and you classify them according to month of birth. The data appear in the table. Do the data present sufficient evidence to indicate that the proportion of cases of the disease per month varies from month to month? Test with $\alpha = 0.05$.

Month	Jan	Feb	Mar	Apr	May	June
Births	38	31	42	46	28	31

Month	July	Aug	Sept	Oct	Nov	Dec
Births	24	29	33	36	27	35

14.12 Snap Peas Suppose you are interested in following two independent traits in snap peas—seed texture (S = smooth, s = wrinkled) and seed colour (Y = yellow, y = green)—in a second-generation cross of heterozygous parents. Mendelian theory states that the number of peas classified as smooth and yellow, wrinkled and yellow, smooth and green, and wrinkled and green should be in the ratio 9:3:3:1. Suppose that 100 randomly selected snap peas have 56, 19, 17, and 8 in these respective categories. Do these data indicate that the 9:3:3:1 model is correct? Test using $\alpha = 0.01$.

14.13 M&M'S® The Mars, Inc. website reports the following percentages of the various colours of its M&M'S® milk chocolate candies:[2]

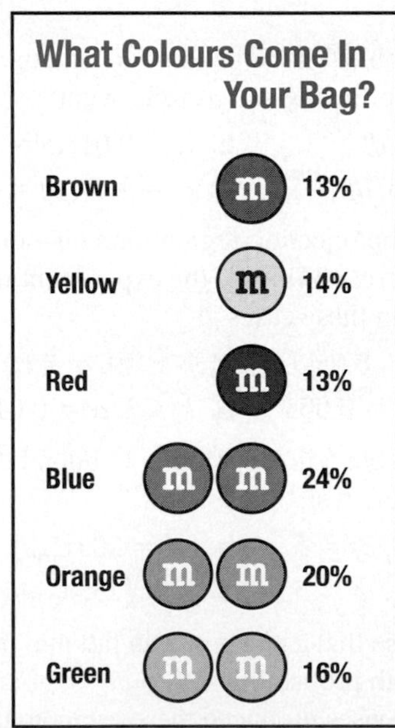

Note: M&M'S® is a registered trademark of Mars, Incorporated. These trademarks are used with permission. Mars, Incorporated is not associated with Cengage Learning, Inc. or Nelson Education Ltd. The images of the M&M'S® mark and M&M'S® candies are printed with permission of Mars, Incorporated.

A 400-gram bag of milk chocolate M&M'S® is randomly selected and contains 70 brown, 72 yellow, 61 red, 118 blue, 108 orange, and 85 green candies. Do the data substantiate the percentages reported by Mars, Inc.? Use the appropriate test and describe the nature of the differences, if there are any.

14.14 M&M'S® Peanut The percentage of various colours are different for Mars M&M'S® peanut variety, as reported by the Mars, Inc. website:[3]

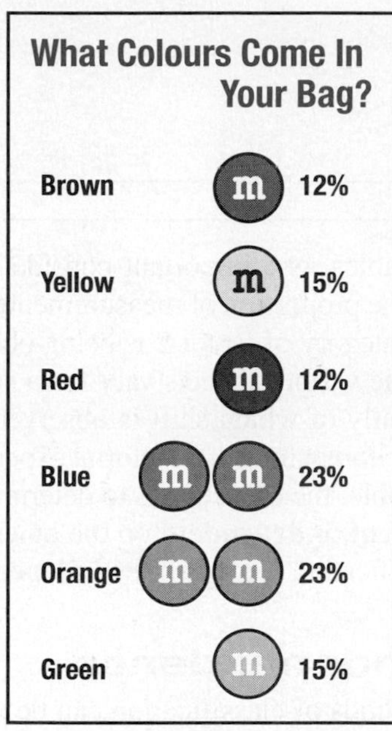

Note: M&M'S® is a registered trademark of Mars, Incorporated. These trademarks are used with permission. Mars, Incorporated is not associated with Cengage Learning, Inc. or Nelson Education Ltd. The images of the M&M'S® mark and M&M'S® candies are printed with permission of Mars, Incorporated.

A 400-gram bag of peanut M&M'S® is randomly selected and contains 70 brown, 87 yellow, 64 red, 115 blue, 106 orange, and 85 green candies. Do the data substantiate the percentages reported by Mars, Inc.? Use the appropriate test and describe the nature of the differences, if there are any.

14.15 Admission Standards Previous enrollment records at a large university indicate that of the total number of persons who apply for admission, 60% are admitted unconditionally, 5% are admitted on a trial basis, and the remainder are refused admission. Of 500 applications to date for the coming year, 329 applicants have been admitted unconditionally, 43 have been admitted on a trial basis, and the remainder have been refused admission. Do these data indicate a departure from previous admission rates? Test using $\alpha = 0.05$.

14.4 CONTINGENCY TABLES: A TWO-WAY CLASSIFICATION

In some situations, the researcher classifies an experimental unit according to *two qualitative variables* to generate *bivariate data,* which we discussed in Chapter 3.

- A defective piece of furniture is classified according to the type of defect and the production shift during which it was made.
- A professor is classified by professional rank and the type of university (public or private) at which the professor works.
- A patient is classified according to the type of preventive flu treatment received and whether or not the patient contracted the flu during the winter.

When two *categorical variables* are recorded, you can summarize the data by counting the observed number of units that fall into each of the various intersections of category levels. The resulting counts are displayed in an array called a **contingency table.**

EXAMPLE 14.3

A total of $n = 309$ furniture defects were recorded and the defects were classified into four types: A, B, C, or D. At the same time, each piece of furniture was identified by the production shift in which it was manufactured. These counts are presented in a contingency table in Table 14.3.

TABLE 14.3 **Contingency Table**

	Shift			
Type of Defects	1	2	3	Total
A	15	26	33	74
B	21	31	17	69
C	45	34	49	128
D	13	5	20	38
Total	94	96	119	309

When you study data that involves two variables, one important consideration is the *relationship between the two variables.* Does the proportion of measurements in the various categories for factor 1 depend on which category of factor 2 is being observed? For the furniture example, do the proportions of the various defects vary from shift to shift, or are these proportions the same, independently of which shift is observed? You may remember a similar phenomenon called *interaction* in the $a \times b$ factorial experiment from Chapter 11. In the analysis of a contingency table, the objective is to determine whether or not one method of classification is **contingent** or **dependent** on the other method of classification. If not, the two methods of classification are said to be **independent.**

NEED A TIP?

With two-way classifications, we do not test hypotheses about specific probabilities. We test whether the two methods of classification are independent.

The Chi-Square Test of Independence

The question of independence of the two methods of classification can be investigated using a test of hypothesis based on the chi-square statistic. These are the hypotheses:

H_0 : The two methods of classification are independent.

H_a : The two methods of classification are dependent.

Suppose we denote the observed cell count in row i and column j of the contingency table as O_{ij}. If you knew the expected cell counts ($E_{ij} = np_{ij}$) under the null hypothesis of independence, then you could use the chi-square statistic to compare the observed and expected counts. However, the expected values are not specified in H_0, as they were in previous examples.

To explain how to estimate these expected cell counts, we must revisit the concept of *independent events* from Chapter 4. Consider p_{ij}, the probability that an observation falls into row i and column j of the contingency table. If the rows and columns are independent, then

$$\begin{aligned} p_{ij} &= P(\text{observation falls in row } i \text{ and column } j) \\ &= P(\text{observation falls in row } i) \times P(\text{observation falls in column } j) \\ &= p_i p_j \end{aligned}$$

where p_i and p_j are the **unconditional** or **marginal probabilities** of falling into row i or column j, respectively. If you could obtain proper estimates of these marginal probabilities, you could use them in place of p_{ij} in the formula for the expected cell count.

Fortunately, these estimates do exist. In fact, they are exactly what you would intuitively choose:

NEED A TIP?

Degrees of freedom for an $r \times c$ contingency table: $df = (r - 1)(c - 1)$

- To estimate a row probability, use $\hat{p}_i = \dfrac{\text{Total observations in row } i}{\text{Total number of observations}} = \dfrac{r_i}{n}$
- To estimate a column probability, use $\hat{p}_j = \dfrac{\text{Total observations in column } j}{\text{Total number of observations}} = \dfrac{c_j}{n}$

The estimate of the expected cell count for row i and column j follows from the independence assumption.

ESTIMATED EXPECTED CELL COUNT

$$\hat{E}_{ij} = n\left(\frac{r_i}{n}\right)\left(\frac{c_j}{n}\right) = \frac{r_i c_j}{n}$$

where r_i is the total for row i and c_j is the total for column j.

The chi-square test statistic for a contingency table with r rows and c columns is calculated as

$$\chi^2 = \Sigma \frac{(O_{ij} - \hat{E}_{ij})^2}{\hat{E}_{ij}}$$

and can be shown to have an approximate chi-square distribution with

$$df = (r - 1)(c - 1)$$

If the observed value of χ^2 is too large, then the null hypothesis of independence is rejected.

EXAMPLE 14.4

Refer to Example 14.3. Do the data present sufficient evidence to indicate that the type of furniture defect varies with the shift during which the piece of furniture is produced?

SOLUTION The estimated expected cell counts are shown in parentheses in Table 14.4. For example, the estimated expected count for a type C defect produced during the second shift is

$$\hat{E}_{32} = \frac{r_3 c_2}{n} = \frac{(128)(96)}{309} = 39.77$$

TABLE 14.4 **Observed and Estimated Expected Cell Counts**

	Shift			
Type of Defects	1	2	3	Total
A	15 (22.51)	26 (22.99)	33 (28.50)	74
B	21 (20.99)	31 (21.44)	17 (26.57)	69
C	45 (38.94)	34 (39.77)	49 (49.29)	128
D	13 (11.56)	5 (11.81)	20 (14.63)	38
Total	94	96	119	309

You can now use the values shown in Table 14.4 to calculate the test statistic as

$$\begin{aligned}\chi^2 &= \Sigma \frac{(O_{ij} - \hat{E}_{ij})^2}{\hat{E}_{ij}} \\ &= \frac{(15 - 22.51)^2}{22.51} + \frac{(26 - 22.99)^2}{22.99} + \cdots + \frac{(20 - 14.63)^2}{14.63} \\ &= 19.18\end{aligned}$$

When you index the chi-square distribution in Table 5 of Appendix I with

$$df = (r - 1)(c - 1) = (4 - 1)(3 - 1) = 6$$

the observed test statistic is greater than $\chi^2_{0.005} = 18.5476$, which indicates that the p-value is less than 0.005. You can reject H_0 and declare the results to be highly significant ($P < 0.005$). There is sufficient evidence to indicate that the proportions of defect types vary from shift to shift.

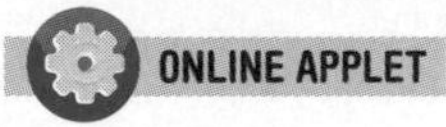

Chi-square Test of Independence

The next obvious question you should ask involves the nature of the relationship between the two classifications. Which shift produces more of which type of defect? As with the factorial experiment in Chapter 11, once a dependence (or interaction) is found, you must look within the table at the relative or *conditional* proportions for each level of classification. For example, consider shift 1, which produced a total of 94 defects. These defects can be divided into types using the *conditional proportions* for this sample shown in the first column of Table 14.5. If you follow the same procedure for the other two shifts, you can then compare the distributions of defect types for the three shifts, as shown in Table 14.5.

Now compare the three sets of proportions (each sums to 1). It appears that shifts 1 and 2 produce defects in the same general order—types C, B, A, and D from most to least—though in differing proportions. Shift 3 shows a different pattern—the most type C defects again but followed by types A, D, and B in that order. Depending on which type of defect is the most important to the manufacturer, each shift should be cautioned separately about the reasons for producing too many defects.

TABLE 14.5 **Conditional Probabilities for Types of Defect within Three Shifts**

	Shift		
Types of Defects	1	2	3
A	$\frac{15}{94} = 0.16$	$\frac{26}{96} = 0.27$	$\frac{33}{119} = 0.28$
B	$\frac{21}{94} = 0.22$	$\frac{31}{96} = 0.32$	$\frac{17}{119} = 0.14$
C	$\frac{45}{94} = 0.48$	$\frac{34}{96} = 0.35$	$\frac{49}{119} = 0.41$
D	$\frac{13}{94} = 0.14$	$\frac{5}{96} = 0.05$	$\frac{20}{119} = 0.17$
Total	1.00	1.00	1.00

NEED TO KNOW

How to Determine the Appropriate Number of Degrees of Freedom

Remember the general procedure for determining degrees of freedom:

1. Start with $k = rc$ categories or cells in the contingency table.
2. Subtract one degree of freedom because all of the rc cell probabilities must sum to 1.
3. You had to estimate $(r - 1)$ row probabilities and $(c - 1)$ column probabilities to calculate the estimated expected cell counts. (The last one of the row and column probabilities is determined because the *marginal* row and column probabilities must also sum to 1.) Subtract $(r - 1)$ and $(c - 1)$ *df*.

The total degrees of freedom for the $r \times c$ contingency table are

$$df = rc - 1 - (r - 1) - (c - 1) = rc - r - c + 1 = (r - 1)(c - 1)$$

EXAMPLE 14.5

A survey was conducted to evaluate the effectiveness of a new flu vaccine that had been administered in a small community. The vaccine was provided free of charge in a two-shot sequence over a period of two weeks. Some people received the two-shotsequence, some appeared for only the first shot, and others received neither. A survey of 1000 local residents the following spring provided the information shown in Table 14.6. Do the data present sufficient evidence to indicate that the vaccine was successful in reducing the number of flu cases in the community?

TABLE 14.6 **2 × 3 Contingency Table**

	No Vaccine	One Shot	Two Shots	Total
Flu	24	9	13	46
No flu	289	100	565	954
Total	313	109	578	1000

SOLUTION The success of the vaccine in reducing the number of flu cases can be assessed in two parts:

- If the vaccine is successful, the proportions of people who get the flu should vary, depending on which of the three treatments they received.
- Not only must this dependence exist, but the proportion of people who get the flu should decrease as the amount of flu prevention treatment increases—from zero to one to two shots.

The first part can be tested using the chi-square test with these hypotheses:

H_0 : No relationship between treatment and incidence of flu

H_a : Incidence of flu depends on amount of flu treatment

NEED A TIP? Use the value of χ^2 and the p-value from the printout to test the hypothesis of independence.

As usual, computer software packages can eliminate all of the tedious calculations and, if the data are entered correctly, provide the correct output containing the observed value of the test statistic and its p-value. Such a printout, generated by *MINITAB*, is shown in Figure 14.1. You can find instructions for generating this printout in the "Technology Today" section at the end of this chapter. The observed value of the test statistic,

FIGURE 14.1 *MINITAB* output for Example 14.5

Chi-Square Test: No Vaccine, One Shot, Two Shots

```
Expected counts are printed below observed counts
Chi-Square contributions are printed below expected counts

       No Vaccine One Shot Two Shots   Total
    1          24        9        13      46
            14.40     5.01     26.59
            6.404    3.169     6.944

    2         289      100       565     954
           298.60   103.99    551.41
            0.309    0.153     0.335

Total         313      109       578    1000

Chi-Sq =  17.313, DF = 2, P-Value = 0.000
```

$\chi^2 = 17.313$, has a p-value of 0.000 and the results are declared highly significant. That is, the null hypothesis is rejected. There is sufficient evidence to indicate a relationship between treatment and incidence of flu.

What is the nature of this relationship? To answer this question, look at Table 14.7, which gives the *incidence* of flu in the sample for each of the three treatment groups. The answer is obvious. The group that received two shots was less susceptible to the flu; only one flu shot does not seem to decrease the susceptibility!

TABLE 14.7 **Incidence of Flu for Three Treatments**

No Vaccine	One Shot	Two Shots
$\frac{24}{313} = 0.08$	$\frac{9}{109} = 0.08$	$\frac{13}{578} = 0.02$

14.4 EXERCISES

BASIC TECHNIQUES

14.16 Calculate the value and give the number of degrees of freedom for χ^2 for these contingency tables:

a.

	Columns			
Rows	1	2	3	4
1	120	70	55	16
2	79	108	95	43
3	31	49	81	140

b.

	Columns		
Rows	1	2	3
1	35	16	84
2	120	92	206

14.17 Suppose that a consumer survey summarizes the responses of $n = 307$ people in a contingency table that contains three rows and five columns. How many degrees of freedom are associated with the chi-square test statistic?

14.18 A survey of 400 respondents produced these cell counts in a 2×3 contingency table:

	Columns			
Rows	1	2	3	Total
1	37	34	93	164
2	66	57	113	236
Total	103	91	206	400

a. If you wish to test the null hypothesis of "independence"—that the probability that a response falls in any one row is independent of the column it falls in—and you plan to use a chi-square test, how many degrees of freedom will be associated with the χ^2 statistic?

b. Find the value of the test statistic.

c. Find the rejection region for $\alpha = 0.01$.

d. Conduct the test and state your conclusions.

e. Find the approximate p-value for the test and interpret its value.

14.19 Gender Differences Male and female respondents to a questionnaire on gender differences were categorized into three groups according to their answers on the first question:

	Group 1	Group 2	Group 3
Men	37	49	72
Women	7	50	31

Use the *MINITAB* printout to determine whether there is a difference in the responses according to gender. Explain the nature of the differences, if any exist.

MINITAB output for Exercise 14.19

Chi-Square Test: Group 1, Group 2, Group 3

```
Expected counts are printed below observed counts
Chi-Square contributions are printed below expected
counts

            Group 1   Group 2   Group 3    Total
     1           37        49        72      158
              28.26     63.59     66.15
              2.703     3.346     0.517

     2            7        50        31       88
              15.74     35.41     36.85
              4.853     6.007     0.927

Total            44        99       103      246

Chi-Sq =  18.352, DF = 2, P-Value = 0.000
```

APPLICATIONS

14.20 Same-Sex Legislation Do you think the next government should let same-sex legislation stand or should it repeal? In 2005, the Strategic Counsel, on behalf of the *Globe and Mail*/CTV polling program, conducted a survey by telephone among a national sample of 1000 adult Canadians 18 years of age or older. The survey indicated that majority of respondents are not in favour of repealing the legislation.

Suppose we randomly select 100 Canadians in each of three subpopulations and record the number who said *following the next election the government in power should attempt to repeal the legislation*. Do the data indicate a significant difference in the proportion among the subpopulations?

	Ontario	Quebec	Rest of Canada
Yes	42	31	44
No	58	69	56

14.21 Anxious Infants A study was conducted by Joseph Jacobson and Diane Wille to determine the effect of early child care on infant-mother attachment patterns.[4] In the study, 93 infants were classified as either "secure" or "anxious" using the Ainsworth strange situation paradigm. In addition, the infants were classified according to the average number of hours per week that they spent in child care. The data are presented in the table:

	Low (0–3 hours)	Moderate (4–19 hours)	High (20–54 hours)
Secure	24	35	5
Anxious	11	10	8

a. Do the data provide sufficient evidence to indicate that there is a difference in attachment pattern for the infants depending on the amount of time spent in child care? Test using $\alpha = 0.05$.

b. What is the approximate p-value for the test in part a?

14.22 Spending Patterns Is there a difference in the spending patterns of Grade 12 students depending on their gender? A study to investigate this question focused on 196 employed Grade 12 students. Students were asked to classify the amount of their earnings that they spent on their car during a given month:

	None or Only a Little	Some	About Half	Most	All or Almost All
Male	73	12	6	4	3
Female	57	15	11	9	6

A portion of the *MINITAB* printout is given here. Use the printout to analyze the relationship between spending patterns and gender. Write a short paragraph explaining your statistical conclusions and their practical implications.

Partial *MINITAB* output for Exercise 14.22

Chi-Square Test: None, Some, Half, Most, All

```
Chi-Sq = 6.696, DF = 4, P-Value = 0.153
2 cells with expected counts less than 5.
```

14.23 Waiting for a Prescription How long do you wait to have your prescriptions filled? According to a survey, "about 3 in 10 Canadians wait more than 20 minutes to have a prescription filled." Suppose a comparison of waiting times for pharmacies in supermarkets and pharmacies in drugstores produced the following results:

Waiting Time	Supermarket	Drugstore
$\leq$ 15 minutes	75	119
16–20 minutes	44	21
$>$ 20 minutes	21	37
Don't know	10	23

a. Is there sufficient evidence to indicate that there is a difference in waiting times for pharmacies in supermarkets and pharmacies in drugstores? Use $\alpha = 0.01$.

b. If we consider only if the waiting time is more than 20 minutes, is there a significant difference in waiting times between pharmacies in supermarkets and pharmacies in drugstores at the 1% level of significance?

Data set EX1424 **14.24 The JFK Assassination** On the 40th anniversary of JFK's assassination, a FOX News poll showed most Americans disagree with the government's conclusions about the killing. The *Warren Commission* found that Lee Harvey Oswald acted alone when he shot Kennedy, but many Americans are not so sure. Do you think that we know all the facts about the assassination of President John F. Kennedy or do you think there was a cover-up? Here are the results from a poll of 900 registered voters countrywide:[5]

	We Know All the Facts	There Was a Cover-Up	(Not Sure)
Democrats	42	309	31
Republicans	64	246	46
Independents	20	115	27

a. Do these data provide sufficient evidence to conclude that there is a difference in voters' opinions about a possible cover-up depending on the political affiliation of the voter? Test using $\alpha = 0.05$.

b. If there is a significant difference in part a, describe the nature of these differences.

Data set EX1425 **14.25 Telecommuting** As an alternative to flextime, many companies allow employees to do some of their work at home. Individuals in a random sample of 300 workers were classified according to salary and number of workdays per week spent at home:

	Workdays at Home Per Week		
Salary	Less Than One	At Least One, but Not All	All at Home
Under $25,000	38	16	14
$25,000 to $49,999	54	26	12
$50,000 to $74,999	35	22	9
Above $75,000	33	29	12

a. Do the data present sufficient evidence to indicate that salary is dependent on the number of workdays spent at home? Test using $\alpha = 0.05$.

b. Use Table 5 in Appendix I to approximate the p-value for this test of hypothesis. Does the p-value confirm your conclusions from part a?

Data set EX1426 **14.26 Telecommuting II** An article addressed the same telecommuting issue (Exercise 14.25) in a slightly different way. It concluded that "people who work exclusively at home tend to be older and better educated than those who have to leave home to report to work."[6] Use the data below based on random samples of 300 workers each to either support or refute these conclusions. Use the appropriate test of hypothesis, and explain why you either agree or disagree with the *article's* conclusions. Note that "Mixed" workers are those who reported working at home at least one full day in a typical week.

	Workers		
Age	Non-Home	Mixed	Home
15–34	73	23	12
35–54	85	40	23
55 and over	22	12	10

	Workers		
Education	Non-Home	Mixed	Homes
Less than H.S. diploma	23	3	5
H.S. graduate	54	12	11
Some college/university	53	24	14
College/university degree or more	41	42	18

14.5 COMPARING SEVERAL MULTINOMIAL POPULATIONS: A TWO-WAY CLASSIFICATION WITH FIXED ROW OR COLUMN TOTALS

An $r \times c$ contingency table results when each of n experimental units is counted as falling into one of the rc cells of a multinomial experiment. Each cell represents a pair of category levels—row level i and column level j. Sometimes, however, it is not advisable to use this type of experimental design—that is, to let the n observations fall where they may. For example, suppose you want to study the opinions of families about their income levels—say, low, medium, and high. If you randomly select $n = 1200$ families for your survey, you may not find any who classify themselves as low-income families! It might be better to decide ahead of time to survey 400 families in each income level. The resulting data will still appear as a two-way classification, but the column totals are fixed in advance.

EXAMPLE 14.6

In another flu prevention experiment like the one described in Example 14.5, the experimenter decides to search the clinic records for 300 patients in each of the three treatment categories: no vaccine, one shot, and two shots. The $n = 900$ patients will then be surveyed regarding their winter flu history. The experiment results in a 2×3 table with the column totals fixed at 300, shown in Table 14.8. By fixing the column totals, the experimenter no longer has a multinomial experiment with $2 \times 3 = 6$ cells. Instead, there are three separate binomial experiments—call them 1, 2, and 3—each with a given probability p_j of contracting the flu and q_j of not contracting the flu. (Remember that for a binomial population, $p_j + q_j = 1$.)

TABLE 14.8 • **Cases of Flu for Three Treatments**

	No Vaccine	One Shot	Two Shots	Total
Flu				r_1
No flu				r_2
Total	300	300	300	n

Suppose you used the chi-square test to test for the independence of row and column classifications. If a particular treatment (column level) does not affect the incidence of flu, then each of the three binomial populations should have the same incidence of flu so that $p_1 = p_2 = p_3$ and $q_1 = q_2 = q_3$.

The 2×3 classification in Example 14.6 describes a situation in which the chi-square test of independence is equivalent to a test of the equality of $c = 3$ binomial proportions. Tests of this type are called **tests of homogeneity** and are used to compare several binomial populations. If there are *more than two* row categories with fixed column totals, then the test of independence is equivalent to a test of the equality of c sets of multinomial proportions.

You do not need to be concerned about the theoretical equivalence of the chi-square tests for these two experimental designs. Whether the columns (or rows) are fixed or not, the test statistic is calculated as

$$\chi^2 = \Sigma \frac{(O_{ij} - \hat{E}_{ij})^2}{\hat{E}_{ij}} \qquad \text{where} \quad \hat{E}_{ij} = \frac{r_i c_j}{n}$$

which has an approximate chi-square distribution in repeated sampling with $df = (r - 1)(c - 1)$.

NEED TO KNOW

How to Determine the Appropriate Number of Degrees of Freedom

Remember the general procedure for determining degrees of freedom:

1. Start with the rc cells in the two-way table.
2. Subtract one degree of freedom for each of the c multinomial populations, whose column probabilities must add to one—a total of c *df*.
3. You had to estimate $(r - 1)$ row probabilities, but the column probabilities are fixed in advance and did not need to be estimated. Subtract $(r - 1)$ *df*.

The total degrees of freedom for the $r \times c$ (fixed-column) table are

$$rc - c - (r - 1) = rc - c - r + 1 = (r - 1)(c - 1)$$

EXAMPLE 14.7

A survey of voter sentiment was conducted in four mid-city political wards to compare the fractions of voters who favour candidate A. Random samples of 200 voters were polled in each of the four wards with the results shown in Table 14.9. The values in parentheses in the table are the expected cell counts. Do the data present sufficient evidence to indicate that the fractions of voters who favour candidate A differ in the four wards?

TABLE 14.9 **Voter Opinions in Four Wards**

	Ward				
	1	2	3	4	Total
Favour A	76 (59)	53 (59)	59 (59)	48 (59)	236
Do not favour A	124 (141)	147 (141)	141 (141)	152 (141)	564
Total	200	200	200	200	800

SOLUTION Since the column totals are fixed at 200, the design involves four binomial experiments, each containing the responses of 200 voters from each of the four wards. To test the equality of the proportions who favour candidate A in all four wards, the null hypothesis

$$H_0 : p_1 = p_2 = p_3 = p_4$$

is equivalent to the null hypothesis

$$H_0 : \text{Proportion favouring candidate A is independent of ward}$$

and will be rejected if the test statistic χ^2 is too large. The observed value of the test statistic, $\chi^2 = 10.722$, and its associated p-value, 0.013, are shown in Figure 14.2. The results are significant ($P < 0.025$); that is, H_0 is rejected and you can conclude that there is a difference in the proportions of voters who favour candidate A among the four wards.

FIGURE 14.2
MINITAB output for Example 14.7

Chi-Square Test: Ward 1, Ward 2, Ward 3, Ward 4,

```
Expected counts are printed below observed counts
Chi-Square contributions are printed below expected counts

        Ward 1   Ward 2   Ward 3   Ward 4    Total
    1       76       53       59       48      236
         59.00    59.00    59.00    59.00
         4.898    0.610    0.000    2.051

    2      124      147      141      152      564
        141.00   141.00   141.00   141.00
         2.050    0.255    0.000    0.858

Total      200      200      200      200      800

Chi-Sq = 10.722 DF = 3, P-Value = 0.013
```

What is the nature of the differences discovered by the chi-square test? To answer this question, look at Table 14.10, which shows the sample proportions who favour candidate A in each of the four wards. It appears that candidate A is doing best in the first ward and worst in the fourth ward. Is this of any *practical significance* to the candidate? Possibly a more important observation is that the candidate does not have a plurality of voters in any of the four wards. If this is a two-candidate race, candidate A needs to increase campaigning!

TABLE 14.10 **Proportions in Favour of Candidate A in Four Wards**

Ward 1	Ward 2	Ward 3	Ward 4
76/200 = 0.38	53/200 = 0.27	59/200 = 0.30	48/200 = 0.24

14.5 EXERCISES

BASIC TECHNIQUES

14.27 Random samples of 200 observations were selected from each of three populations, and each observation was classified according to whether it fell into one of three mutually exclusive categories:

Population	Category 1	2	3	Total
1	108	52	40	200
2	87	51	62	200
3	112	39	49	200

You want to know whether the data provide sufficient evidence to indicate that the proportions of observations in the three categories depend on the population from which they were drawn.

a. Give the value of χ^2 for the test.

b. Give the rejection region for the test for $\alpha = 0.01$.

c. State your conclusions.

d. Find the approximate p-value for the test and interpret its value.

14.28 Suppose you wish to test the null hypothesis that three binomial parameters p_A, p_B, and p_C are equal versus the alternative hypothesis that at least two of the parameters differ. Independent random samples of 100 observations were selected from each of the populations. The data are shown in the table:

	Population A	B	C	Total
Successes	24	19	33	76
Failures	76	81	67	224
Total	100	100	100	300

a. Write the null and alternative hypotheses for testing the equality of the three binomial proportions.

b. Calculate the test statistic and find the approximate p-value for the test in part a.

c. Use the approximate p-value to determine the statistical significance of your results. If the results are statistically significant, explore the nature of the differences in the three binomial proportions.

APPLICATIONS

14.29 Overweight Young Canadians The Canadian Community Health Survey of 2004 deals with overweight/obesity rates by ethnic origin among children aged 2 to 17 (excluding territories).[7] Suppose the results of a similar survey are shown in the following table:

	Overweight/Obese	Not Overweight/Obese
White	189	811
Black	301	699
Southeast/East Asian	117	883
Off-reserves Aboriginal	231	769
Other	162	838

Is there a significant difference in the proportion of overweight/obese children by ethnic origin? Use $\alpha = 0.01$.

14.30 Diseased Chickens A particular poultry disease is thought to be non-communicable. To test this theory, 30,000 chickens were randomly partitioned into three groups of 10,000. One group had no contact with diseased chickens, one had moderate contact, and the third had heavy contact. After a six-month period, data were collected on the number of diseased chickens in each group of 10,000. Do the data provide sufficient evidence to indicate a dependence between the amount of contact between diseased and non-diseased fowl and the incidence of the disease? Use $\alpha = 0.05$.

	No Contact	Moderate Contact	Heavy Contact
Disease	87	89	124
No disease	9,913	9,911	9,876
Total	10,000	10,000	10,000

EX1431 **14.31 Wealth and Education Levels** Does education really make a difference in how much money you will earn? Researchers randomly selected 100 people from each of three income categories—"marginally rich," "comfortably rich," and "super rich"—and then recorded their educational attainment, as in the table:[8]

	Marginally Rich ($70–99 K)	Comfortably Rich ($100–249 K)	Super Rich ($250 K or more)
No college/university	32	20	23
Some college/university	13	16	1
College/university degree	43	51	60
Postgraduate study/degree	12	13	16

a. Describe the multinomial experiments whose proportions are being compared in this experiment.

b. Do these data indicate that the level of wealth is affected by educational attainment? Test at the 1% level of significance.

c. Based on the results of part b, describe the practical nature of the relationship between level of wealth and educational attainment.

14.32 Deep-Sea Research W.W. Menard has conducted research involving manganese nodules, a mineral-rich concoction found abundantly on the deep-sea floor.[9] In one portion of his report, Menard provides data relating the magnetic age of the earth's crust to the "probability of finding manganese nodules." The table gives the number of samples of the earth's core and the percentage of those that contain manganese nodules for each of a set of magnetic-crust ages. Do the data provide sufficient evidence to indicate that the probability of finding manganese nodules in the deep-sea earth's crust is dependent on the magnetic-age classification?

Age	Number of Samples	Percentage with Nodules
Miocene—recent	389	5.9
Oligocene	140	17.9
Eocene	214	16.4
Paleocene	84	21.4
Late Cretaceous	247	21.1
Early and Middle Cretaceous	1120	14.2
Jurassic	99	11.0

Data set EX1433 **14.33 How Big Is the Household?** A local chamber of commerce surveyed 120 households in its city—40 in each of three types of residence (apartment, duplex, or single residence)—and recorded the number of family members in each of the households. The data are shown in the table:

	Type of Residence		
Family Members	Apartment	Duplex	Single Residence
1	8	20	1
2	16	8	9
3	10	10	14
4 or more	6	2	16

Is there a significant difference in the family size distributions for the three types of residence? Test using $\alpha = 0.01$. If there are significant differences, describe the nature of these differences.

Data set EX1434 **14.34 Birth Control versus Abortion** Mexican adults remain opposed to pregnancy termination, according to a poll conducted by Consulta Mitofsky. Only 32.1% of respondents agree with abortion. Further, 93.7% of respondents support the use of a condom as a means of preventing a pregnancy, 87.3% agree with birth control pills, and 60.7% back the morning-after pill.[10] Suppose a similar survey is conducted in Chile and the results are given in the table below:

	Condom	Birth Control Pills	Morning-After Pill	Abortion	Other
Agree	481	423	321	144	233
Disagree	519	577	679	856	767

Do the data provide sufficient evidence to indicate that the same proportion of adults in Mexico agree with the various methods meant to prevent or terminate a pregnancy?

Data set EX1435 **14.35 Evolution: Pro or Con?** According to a poll by the Pew Research Center, 55% of young adults (ages 18–29) believe that evolution is the best explanation for the development of human life.[11] When the data are further categorized by whether or not the responders had a religious affiliation, this proportion changed for those not having a religious affiliation. The data that follow reflect the results of this poll:

	Religious Affiliation	Not Affiliated	Total
Yes	47	152	199
No	53	98	151
Total	100	250	350

a. Do the data indicate that the proportion of young adults who believe that evolution provides the best answer to the development of human life differ for those with a religious affiliation versus those without a religious affiliation? Use $\alpha = 0.05$.

b. If significant differences exist, explain what changes appear to have taken place when religious affiliation is included in the categorization.

14.6 THE EQUIVALENCE OF STATISTICAL TESTS

Remember that when there are only $k = 2$ categories in a multinomial experiment, the experiment reduces to a *binomial experiment* where you record the number of successes x (or O_1) in n (or $O_1 + O_2$) trials. Similarly, the data that result from *two binomial experiments* can be displayed as a two-way classification with $r = 2$ and $c = 2$, so that the chi-square test of *homogeneity* can be used to compare the two binomial proportions, p_1 and p_2. For these two situations, we have presented statistical tests for the binomial proportions based on the z statistic of Chapter 9:

- ***One sample:*** $z = \dfrac{\hat{p} - p_0}{\sqrt{\dfrac{p_0 q_0}{n}}}$

$k = 2$

Successes	Failures

- ***Two samples:*** $z = \dfrac{\hat{p}_1 - \hat{p}_2}{\sqrt{\hat{p}\hat{q}\left(\dfrac{1}{n_1} + \dfrac{1}{n_2}\right)}}$

$r = c = 2$

Sample 1	**Sample 2**
Successes	Successes
Failures	Failures

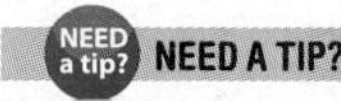

NEED A TIP?

The one- and two-sample binomial tests from Chapter 9 are equivalent to chi-square tests—$z^2 = \chi^2$.

Why are there two different tests for the same statistical hypothesis? Which one should you use? For these two situations, you can use *either* the z test *or* the chi-square test, and you will obtain identical results. For either the one- or two-sample test, we can prove algebraically that

$$z^2 = \chi^2$$

so that the test statistic z will be the square root (either positive or negative, depending on the data) of the chi-square statistic. Furthermore, we can show theoretically that the same relationship holds for the critical values in the z and chi-square tables in Appendix I, which produces *identical p-values* for the two equivalent tests. To test a one-tailed alternative hypothesis such as $H_0: p_1 > p_2$, first determine whether $\hat{p}_1 - \hat{p}_2 > 0$, that is, if the difference in sample proportions has the appropriate sign. If so, the appropriate critical value of χ^2 from Table 5 will have one degree of freedom, a right-tail area of 2α. For example, the critical χ^2 value with 1 *df* and $\alpha = 0.05$ will be $\chi^2_{0.10} = 2.70554 = 1.645^2$.

In summary, you are free to choose the test (z or χ^2) that is most convenient. Since most computer packages include the chi-square test, and most do not include the large-sample z tests, the chi-square test may be preferable to you!

14.7 OTHER APPLICATIONS OF THE CHI-SQUARE TEST

The application of the chi-square test for analyzing count data is only one of many classification problems that result in multinomial data. Some of these applications are quite complex, requiring complicated or calculationally difficult procedures for estimating the expected cell counts. However, several applications are used often enough to make them worth mentioning.

- **Goodness-of-fit tests:** You can design a goodness-of-fit test to determine whether data are consistent with data drawn from a particular probability distribution—possibly the normal, binomial, Poisson, or other distributions. The cells of a sample frequency histogram correspond to the k cells of a multinomial experiment. Expected cell counts are calculated using the probabilities associated with the hypothesized probability distribution.
- **Time-dependent multinomials:** You can use the chi-square statistic to investigate the rate of change of multinomial (or binomial) proportions over time. For example, suppose that the proportion of correct answers on a 100-question exam is recorded for a student, who then repeats the exam in each of the next four weeks. Does the proportion of correct responses increase over time? Is learning taking place? In a process monitored by a quality control plan, is there a positive trend in the proportion of defective items as a function of time?
- **Multidimensional contingency tables:** Instead of only two methods of classification, you can investigate a dependence among three or more classifications. The two-way contingency table is extended to a table in more than two dimensions. The methodology is similar to that used for the $r \times c$ contingency table, but the analysis is a bit more complex.
- **Log-linear models:** Complex models can be created in which the logarithm of the cell probability ($\ln p_{ij}$) is some linear function of the row and column probabilities.

Most of these applications are rather complex and might require that you consult a professional statistician for advice before you conduct your experiment.

In all statistical applications that use *Pearson's chi-square statistic,* assumptions must be satisfied in order that the test statistic have an approximate chi-square probability distribution.

ASSUMPTIONS

- The cell counts $O_1, O_2, \ldots, O_k$ must satisfy the conditions of a multinomial experiment, or a set of multinomial experiments created by fixing either the row or column totals.
- The expected cell counts $E_1, E_2, \ldots, E_k$ should equal or exceed five.

You can usually be fairly certain that you have satisfied the first assumption by carefully preparing and designing your experiment or sample survey. When you calculate the expected cell counts, if you find that one or more is less than five, these options are available to you:

- Choose a larger sample size n. The larger the sample size, the closer the chi-square distribution will approximate the distribution of your test statistic χ^2.
- It may be possible to combine one or more of the cells with small expected cell counts, thereby satisfying the assumption.

Finally, make sure that you are calculating the *degrees of freedom* correctly and that you carefully evaluate the statistical and practical conclusions that can be drawn from your test.

CHAPTER REVIEW

Key Concepts and Formulas

I. The Multinomial Experiment

1. There are n identical trials, and each outcome falls into one of k categories.
2. The probability of falling into category i is p_i and remains constant from trial to trial.
3. The trials are independent, $\Sigma p_i = 1$, and we measure O_i, the number of observations that fall into each of the k categories.

II. Pearson's Chi-Square Statistic

$$\chi^2 = \Sigma \frac{(O_i - E_i)^2}{E_i} \quad \text{where } E_i = np_i$$

which has an approximate chi-square distribution with *degrees of freedom* determined by the application.

III. The Goodness-of-Fit Test

1. This is a one-way classification with cell probabilities specified in H_0.
2. Use the chi-square statistic with $E_i = np_i$ calculated with the hypothesized probabilities.
3. $df = k - 1 -$ (Number of parameters estimated in order to find E_i)
4. If H_0 is rejected, investigate the nature of the differences using the sample proportions.

IV. Contingency Tables

1. A two-way classification with n observations categorized into $r \times c$ cells of a two-way table using two different methods of classification is called a *contingency table.*
2. The test for independence of classification methods uses the chi-square statistic

$$\chi^2 = \Sigma \frac{(O_{ij} - \hat{E}_{ij})^2}{\hat{E}_{ij}}$$

with $\hat{E}_{ij} = \dfrac{r_i c_j}{n}$ and $df = (r - 1)(c - 1)$

3. If the null hypothesis of independence of classifications is rejected, investigate the nature of the dependence using conditional proportions within either the rows or columns of the contingency table.

V. Fixing Row or Column Totals

1. When either the row or column totals are fixed, the test of independence of classifications becomes a test of the homogeneity of cell probabilities for several multinomial experiments.
2. Use the same chi-square statistic as for contingency tables.
3. The large-sample z tests for one and two binomial proportions are special cases of the chi-square statistic.

VI. Assumptions

1. The cell counts satisfy the conditions of a multinomial experiment, or a set of multinomial experiments with fixed sample sizes.
2. All expected cell counts must equal or exceed five in order that the chi-square approximation is valid.

TECHNOLOGY TODAY

The Chi-Square Test—*Microsoft Excel*

The procedure for performing a chi-square test of independence in *Excel* requires that you enter both the observed and the expected cell counts into an *Excel* spreadsheet. If the *raw categorical data* have been stored in the spreadsheet rather than the *observed cell counts,* you may need to tally the data to obtain the cell counts before continuing.

EXAMPLE 14.8

Suppose you have recorded the gender (M or F) and the university status (1Y, 2Y, 3Y, 4Y, Grad) for 100 statistics students, as shown in the table below:

	Status				
Gender	1Y	2Y	3Y	4Y	Grad
F	16	8	8	8	4
M	9	11	12	12	12

1. Enter the observed values into the first five columns of an *Excel* spreadsheet.
2. Calculate (by hand) the 10 estimated expected cell counts and enter them into another range in the spreadsheet.
3. Place your cursor in an empty cell, and use **Formulas ▶ More Functions ▶ Statistical ▶ CHISQ.TEST** (**CHITEST** in earlier versions of *Excel*) to generate the dialogue box in Figure 14.3. Highlight or type in the cell ranges for the observed and expected cell counts.

FIGURE 14.3

MINITAB screen shot

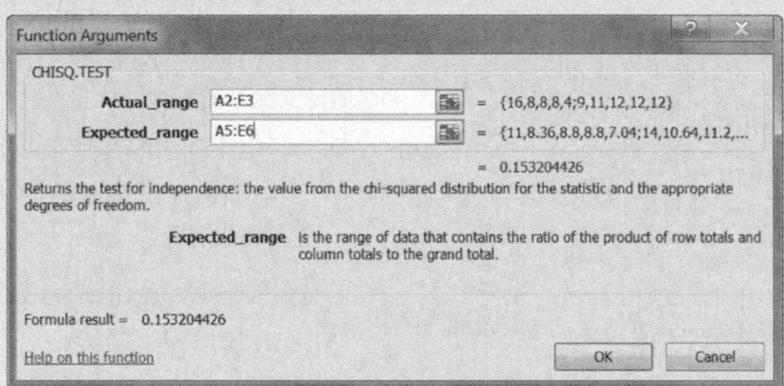
Function Arguments
CHISQ.TEST
Actual_range A2:E3
= {16,8,8,8,4;9,11,12,12,12}
Expected_range A5:E6
= {11,8.36,8.8,8.8,7.04;14,10.64,11.2,...
= 0.153204426
Returns the test for independence: the value from the chi-squared distribution for the statistic and the appropriate degrees of freedom.
Expected_range is the range of data that contains the ratio of the product of row totals and column totals to the grand total.
Formula result = 0.153204426
Help on this function
OK
Cancel

4. When you click **OK,** *Excel* will calculate the p-value associated with the chi-square test of independence. For this data, the large p-value (0.153) indicates a non-significant result. There is insufficient evidence to indicate that a student's gender is dependent on class status.

NOTE: *Excel* does not provide a single command to allow you to perform the chi-square goodness-of-fit test; however, you could manually create formulas in *Excel* to perform this test and obtain the appropriate p-value.

The Chi-Square Test—*MINITAB*

Several procedures are available in the *MINITAB* package for analyzing categorical data. The appropriate procedure depends on whether the data represent a one-way classification (a single multinomial experiment) or a two-way classification or contingency table. If the *raw categorical data* have been stored in the *MINITAB* worksheet rather than the *observed cell counts,* you may need to tally or cross-classify the data to obtain the cell counts before continuing.

EXAMPLE 14.9

Suppose you have recorded the gender (M or F) and the university status (1Y, 2Y, 3Y, 4Y, Grad) for 100 statistics students. The *MINITAB* worksheet would contain two columns of 100 observations each. Each row would contain an individual's gender in column 1 and university status in column 2.

1. To obtain the observed cell counts (O_{ij}) for the 2×5 contingency table, use **Stat ▶ Tables ▶ Cross Tabulation and Chi-Square** to generate the dialogue box shown in Figure 14.4(a).

FIGURE 14.4
a) *MINITAB* screen shot
b) *MINITAB* screen shot

(a)

Cross Tabulation and Chi-Square

Categorical variables:
For rows: Gender
For columns: Status
For layers:
Frequencies are in: (optional)
Display
✔ Counts
Row percents
Column percents
Total percents
Select
Chi-Square... Other Stats...
Options...
Help OK Cancel

(b)

Cross Tabulation - Chi-Square

Display
✔ Chi-Square analysis
✔ Expected cell counts
Raw residuals
Standardized residuals
Adjusted residuals
Each cell's contribution to the Chi-Square statistic
Help OK Cancel

2. Under "Categorical Variables," select "Gender" for the row variable and "Status" for the column variable. Leave the boxes marked "For Layers" and "Frequencies are in:" blank. Make sure that the square labelled "Display Counts" is checked.
3. Click the **Chi-Square . . .** button to display the dialogue box in Figure 14.4(b). Check the boxes for "Chi-Square Analysis" and "Expected Cell Counts." Click **OK** twice. This sequence of commands not only tabulates the contingency table but also performs the chi-square test of independence and displays the results in the Session window shown in Figure 14.5. For the gender/college status data, the large p-value ($P = .153$) indicates a nonsignificant result. There is insufficient evidence to indicate that a student's gender is dependent on class status.

FIGURE 14.5
MINITAB screen shot

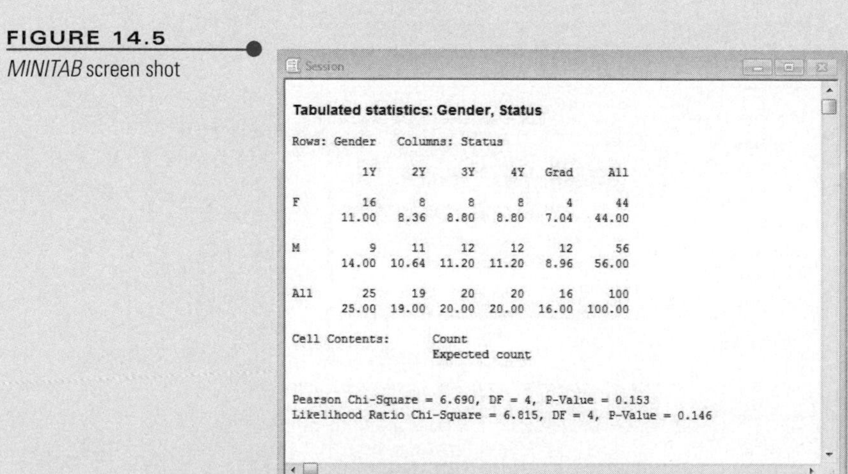

Session

Tabulated statistics: Gender, Status

Rows: Gender Columns: Status

	1Y	2Y	3Y	4Y	Grad	All
F	16	8	8	8	4	44
	11.00	8.36	8.80	8.80	7.04	44.00
M	9	11	12	12	12	56
	14.00	10.64	11.20	11.20	8.96	56.00
All	25	19	20	20	16	100
	25.00	19.00	20.00	20.00	16.00	100.00

Cell Contents: Count
Expected count

Pearson Chi-Square = 6.690, DF = 4, P-Value = 0.153
Likelihood Ratio Chi-Square = 6.815, DF = 4, P-Value = 0.146

4. If the observed cell counts in the contingency table have already been tabulated, simply enter the counts into c columns of the *MINITAB* worksheet, use **Stat ▶ Tables ▶ Chi-Square Test (Two-Way Table in Worksheet),** and select the appropriate columns before clicking **OK.** For the gender/university status data, you can enter the counts into columns C3–C7 as shown in Figure 14.6. The resulting output will be labelled differently but will look exactly like the output in Figure 14.5.

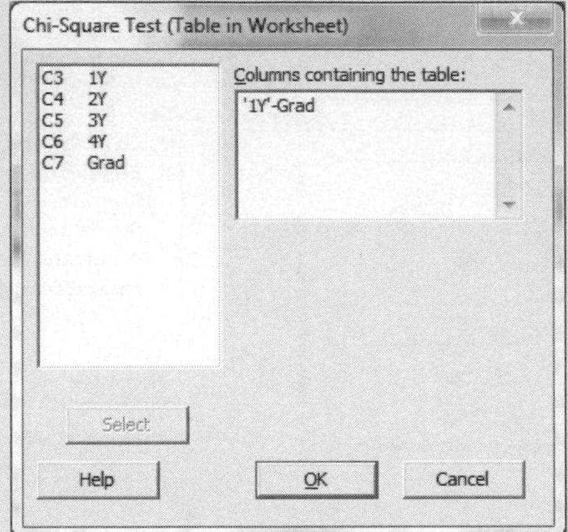

FIGURE 14.6
MINITAB screen shot

EXAMPLE 14.10

A simple test of a single multinomial experiment can be set up by considering whether the proportions of male and female statistics students are the same—that is, $p_1 = 0.5$ and $p_2 = 0.5$.

1. In *MINITAB 15* or *16*, use **Stat ▶ Tables ▶ Chi-Square Goodness-of-Fit Test (One Variable)** to display the dialogue box in Figure 14.7. If you have raw categorical data in a column, click the "Categorical data:" button and enter the "Gender" column in the cell. If you have summary values of observed counts for each category, choose "Observed counts." Then enter the column containing the observed counts or type the observed counts for each category.

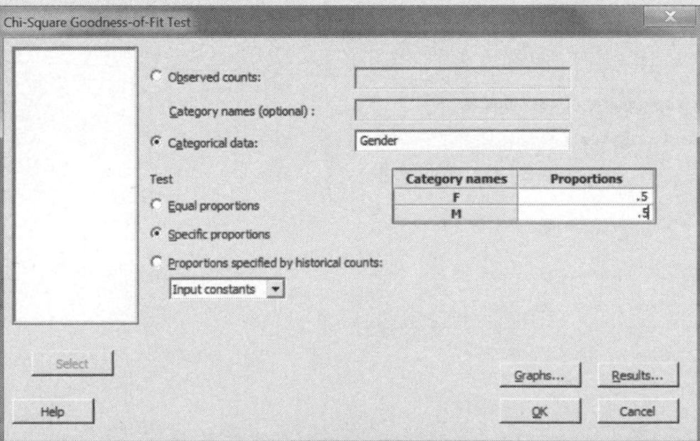

FIGURE 14.7
MINITAB screen shot

2. For this test, we can choose "Equal proportions" to test $H_0: p_1 = p_2 = .5$. When you have different proportions for each category, use "Specific proportions." You can store the proportions for each category in a column, choose "Input column" and enter the column. If you want to type the proportion for each category, choose "Input constants" and type the proportions for the corresponding categories. Click **OK.**
3. The resulting output will include several graphs along with the values for O_i and E_i for each category, the observed value of the test statistic, $\chi^2 = 1.44$, and its p-value = 0.230, which is not significant. There is insufficient evidence to indicate a difference in the proportion of male and female statistics students.

NOTE: If you are using a previous version of *MINITAB,* you will have to determine the observed and expected cell counts, and enter them into separate columns in the worksheet. Then use **Calc ▶ Calculator** and the expression **SUM((‘O’– ‘E’)**2/‘E’)** to calculate the observed value of the test statistic.

Supplementary Exercises

Starred (*) exercises are optional.

14.36 Floor Polish A manufacturer of floor polish conducted a consumer preference experiment to see whether a new floor polish A was superior to those produced by four competitors, B, C, D, and E. A sample of 100 housekeepers viewed five patches of flooring that had received the five polishes, and each indicated the patch that he or she considered superior in appearance. The lighting, background, and so on were approximately the same for all five patches. The results of the survey are listed here:

Polish	A	B	C	D	E
Frequency	27	17	15	22	19

Do these data present sufficient evidence to indicate a preference for one or more of the polished patches of floor over the others? If one were to reject the hypothesis of no preference for this experiment, would this imply that polish A is superior to the others? Can you suggest a better way of conducting the experiment?

14.37 Physical Fitness in Canada A survey was conducted to investigate the interest of middle-aged adults in physical fitness programs in British Columbia, Alberta, Ontario, and Quebec. The objective of the investigation was to determine whether adult participation in physical fitness programs varies from one region of Canada to another. A random sample of people were interviewed in each state and these data were recorded:

	British Columbia	Alberta	Ontario	Quebec
Participate	46	63	108	121
Do not participate	149	178	192	179

Do the data indicate a difference in adult participation in physical fitness programs from one province to another? If so, describe the nature of the differences.

14.38 Fatal Accidents Accident data were analyzed to determine the numbers of fatal accidents for automobiles of three sizes. The data for 346 accidents are as follows:

	Small	Medium	Large
Fatal	67	26	16
Not fatal	128	63	46

Do the data indicate that the frequency of fatal accidents is dependent on the size of automobiles? Write a short paragraph describing your statistical results and their practical implications.

Data set EX1439 **14.39 Life under the EU** Angus Reid Global Monitor reported that many adults in Britain and France are disappointed with the effect of the European Union (EU) in their countries, according to a poll by Harris Interactive published in the *Financial Times.*[12] According to the poll, 52% of Britons and 50% of French respondents say that life has become worse. Suppose, in an online survey with 5000 adults in Britain, France, Germany, Italy, and Spain, the following question was asked: *“Since your country became part of the European Union (EU) has life in your country become better/same or has it become worse?”* The results are shown in the following table:

	Britain	France	Italy	Spain	Germany
Better/same	475	485	531	660	562
Worse	525	515	469	340	438

a. Does this data provide sufficient evidence to conclude that there is a difference in peoples’ opinion from one country to another? Test using $\alpha = 0.05$.

b. If there is a significant difference in part a, describe the nature of these differences.

EX1440 **14.40 Discovery-based Teaching** Two biology instructors set out to evaluate the effects of discovery-based teaching compared to the standard lecture-based teaching approach in the laboratory.[13] The standard lecture-based approach provided a list of instructions to follow at each step of the laboratory exercise, whereas the discovery-based approach asked questions rather than providing directions and used small group reports to decide the best way to proceed in reaching the laboratory objective. One evaluation of the techniques involved written appraisals of both techniques by students at the end of the course. The comparison of the number of positive and negative responses for both techniques is given in the following table:

Group	Positive Evaluations	Negative Evaluations	Total
Discovery	37	11	48
Control	31	17	48

a. Is there a significant difference in the proportion of positive responses for each of the teaching methods? Use $\alpha = 0.05$. If so, how would you describe this difference?

b. What is the approximate p-value for the test in part a?

14.41 Flower Colour and Shape A botanist performs a secondary cross of petunias involving independent factors that control leaf shape and flower colour, where the factor A represents red colour, a represents white colour, B represents round leaves, and b represents long leaves. According to the Mendelian model, the plants should exhibit the characteristics AB, Ab, aB, and ab in the ratio 9:3:3:1. Of 160 experimental plants, the following numbers were observed:

AB	Ab	aB	ab
95	30	28	7

Is there sufficient evidence to refute the Mendelian model at the $\alpha = 0.01$ level?

14.42 Salmonella Is your holiday turkey safe? A federal survey found that 19.6% of turkeys are contaminated with the salmonella bacteria.[14] Use the table that follows to determine if there is a significant difference in the contamination rate at three processing plants. One hundred turkeys were randomly selected from each of the processing lines at these three plants.

Plant	Salmonella Present	Sample Size
1	42	100
2	23	100
3	22	100

Is there a significant difference in the rate of salmonella contamination among these three processing plants? If there is a significant difference, describe the nature of these differences. Use $\alpha = 0.01$.

14.43 An Arthritis Drug A study to determine the effectiveness of a drug (serum) for arthritis resulted in the comparison of two groups, each consisting of 200 arthritic patients. One group was inoculated with the serum; the other received a placebo (an inoculation that appears to contain serum but actually is non-active). After a period of time, each person in the study was asked to state whether his or her arthritic condition had improved. These are the results:

	Treated	Untreated
Improved	117	74
Not improved	83	126

You want to know whether these data present sufficient evidence to indicate that the serum was effective in improving the condition of arthritic patients.

a. Use the chi-square test of homogeneity to compare the proportions improved in the populations of treated and untreated subjects. Test at the 5% level of significance.

b. Test the equality of the two binomial proportions using the two-sample z test of Section 9.6. Verify that the squared value of the test statistic $z^2 = \chi^2$ from part a. Are your conclusions the same as in part a?

14.44 Parking at the University A survey was conducted to determine student, faculty, and administration attitudes about a new university parking policy. The distribution of those favouring or opposing the policy is shown in the table. Do the data provide sufficient evidence to indicate that attitudes about the parking policy are independent of student, faculty, or administration status?

	Student	Faculty	Administration
Favour	252	107	43
Oppose	139	81	40

14.45* The chi-square test used in Exercise 14.43 is equivalent to the two-tailed z test of Section 9.6 provided α is the same for the two tests. Show algebraically that the chi-square test statistic χ^2 is the square of the test statistic z for the equivalent test.

14.46 Fitting a Binomial Distribution You can use a goodness-of-fit test to determine whether all of the criteria for a binomial experiment have actually been met in a given application. Suppose that an experiment consisting of four trials was repeated 100 times. The number of repetitions on which a given number of successes was obtained is recorded in the table:

Possible Results (number of successes)	Number of Times Obtained
0	11
1	17
2	42
3	21
4	9

Estimate p (assuming that the experiment was binomial), obtain estimates of the expected cell frequencies, and test for goodness of fit. To determine the appropriate number of degrees of freedom for χ^2, note that p was estimated by a linear combination of the observed frequencies.

14.47 Antibiotics and Infection Infections sometimes occur when blood transfusions are given during surgical operations. An experiment was conducted to determine whether the injection of antibodies reduced the probability of infection. An examination of the records of 138 patients produced the data shown in the table. Do the data provide sufficient evidence to indicate that injections of antibodies affect the likelihood of transfusional infections? Test by using $\alpha = 0.05$.

	Infection	No Infection
Antibody	4	78
No antibody	11	45

14.48 German Manufacturing Canadian labour unions have traditionally been content to leave the management of the company to managers and corporate executives. But in Europe, worker participation in management decision making is an accepted idea that is continually spreading. To study the effect of worker participation in managerial decision making, 100 workers were interviewed in each of two separate German manufacturing plants. One plant had active worker participation in managerial decision making; the other did not. Each selected worker was asked whether he or she generally approved of the managerial decisions made within the firm. The results of the interviews are shown in the table:

	Participation	No Participation
Generally approve	73	51
Do not approve	27	49

a. Do the data provide sufficient evidence to indicate that approval or disapproval of management's decisions depends on whether workers participate in decision making? Test by using the χ^2 test statistic. Use $\alpha = 0.05$.

b. Do these data support the hypothesis that workers in a firm with participative decision making more generally approve of the firm's managerial decisions than those employed by firms without participative decision making? Test by using the z test presented in Section 9.6. This problem requires a one-tailed test. Why?

14.49 Three Entrances An occupant-traffic study was conducted to aid in the remodelling of an office building that contains three entrances. The choice of entrance was recorded for a sample of 200 persons who entered the building. Do the data in the table indicate that there is a difference in preference for the three entrances? Find a 95% confidence interval for the proportion of persons favouring entrance 1.

Entrance	1	2	3
Number entering	83	61	56

Data set EX1450 **14.50 Graduate Teaching Assistants** Graduate students' responsibilities are often related to their roles as teaching assistants or research assistants. As part of a larger study, K.M. McGoldrick and her colleagues investigated the level of preparation of economics graduate students for their teaching-related duties for students at "top-tier" and those at "second-tier" schools.[15] The responses to the question "Are you satisfied with the level of preparation you have had for your teaching-related duties?" follow:

	Top-Tier	Second-Tier
I am very satisfied	85	197
I am somewhat satisfied	102	171
I am unsatisfied	22	29
Total	209	397

a. Is there a significant difference in the responses to the question between students at "top-tier" schools compared to those at "second-tier" schools?

b. If significant, describe the nature of the differences in response for graduate students at "top-tier" versus "second-tier" schools.

14.51 Publish or Perish In the academic world, students and their faculty advisors often collaborate on research papers, producing works in which publication credit can take several forms. In theory, the first authorship of a student's paper should be given to the student unless the input from the faculty advisor was substantial. In an attempt to see whether this is, in fact, the case, authorship credit was studied for different levels of faculty input and two objectives (dissertation versus non-degree research). The frequency of author assignment decisions for published dissertations is shown in the table as assigned by 60 faculty members and 161 students:[16]

Faculty Respondents

Authorship Assignment	High Input	Medium Input	Low Input
Faculty first author, student mandatory second author	4	0	0
Student first author, faculty mandatory second author	15	12	3
Student first author, faculty courtesy second author	2	7	7
Student sole author	2	3	5

Student Respondents

Authorship Assignment	High Input	Medium Input	Low Input
Faculty first author, student mandatory second author	19	6	2
Student first author, faculty mandatory second author	19	41	27
Student first author, faculty courtesy second author	3	7	31
Student sole author	0	3	3

Source: Costa, M.M., & M. Gatz. (1992). "Determination of Authorship Credit in Published Dissertations." *Psychological Science,* Vol. 3(6), pp. 354–357. Copyright © 1992 by Association for Psychological Science. Reprinted by permission of SAGE Publications, Inc.

a. Is there sufficient evidence to indicate a dependence between the authorship assignment and the input of the faculty advisor as judged by faculty members? Test using $\alpha = 0.01$.

b. Is there sufficient evidence to indicate a dependence between the authorship assignment and the input of the faculty advisor as judged by students? Test using $\alpha = 0.01$.

c. If there is a dependence in the two classifications from parts a and b, does it appear from looking at the data that students are more likely to assign a higher authorship to their faculty advisors than the advisors themselves?

d. Have any of the assumptions necessary for the analysis used in parts a and b been violated? What affect might this have on the validity of your conclusions?

14.52 Are You a Good Driver? How would you rate yourself as a driver? According to a survey, most Albertans think they are good drivers but have little respect for others' driving ability. The data show the distribution of opinions according to gender for two different questions, the first rating themselves as drivers and the second rating others as drivers. We assume that there were 100 men and 100 women in the surveyed group.

Rating Self as a Driver

Gender	Excellent	Good	Fair
Male	43	48	9
Female	44	53	3

Rating Others as Drivers

Gender	Excellent	Good	Fair	Poor/Very Poor
Male	4	42	41	13
Female	3	48	35	14

a. Is there sufficient evidence to indicate that there is a difference in the self-ratings between male and female drivers? Find the approximate p-value for the test.

b. Is there sufficient evidence to indicate that there is a difference in the ratings of other drivers between male and female drivers? Find the approximate p-value for the test.

c. Have any of the assumptions necessary for the analysis used in parts a and b been violated? What affect might this have on the validity of your conclusions?

14.53 Vehicle Colours Each model year seems to introduce new colours and different hues for a wide array of vehicles, from luxury cars to full-size or intermediate models, to compacts and sports cars, to light trucks. However, white and silver/grey continue to make the top five or six colours across all of these categories of vehicles. The top six colours and their percentage of the market share for compact/sports cars are shown in the following table:[17]

Colour	Silver	Black	Grey	Blue	Red	White
Percent	19	17	17	15	12	12

To verify the figures, a random sample consisting of 250 compact/sports cars was taken and the colour of the vehicles recorded. The sample provided the following counts for the categories given above: 52, 43, 48, 41, 32, and 19, respectively.

a. Is any category missing in the classification? How many vehicles belong to that category?

b. Is there sufficient evidence to indicate that our percentages of the colours for compact/sports cars differ from those given? Find the approximate p-value for the test.

Data set EX1454 **14.54 Vehicle Colours, again** Refer to Exercise 14.53. The researcher wants to see if there is a difference in the colour distributions for compact/sports cars versus full/intermediate cars.[18] Another random sample of 250 full/intermediate cars was taken and the colour of the vehicles was recorded. The table below shows the results for both compact/sports and full/intermediate cars:

Colour	Silver	Black	Grey	Blue	Red	White
Compact/ Sports	52	43	48	41	32	19
Full/ Intermediate	50	33	37	32	27	38

Do the data indicate that there is a difference in the colour distributions depending on the type of vehicle? Use $\alpha = 0.05$. (HINT: Remember to include a column called "Other" for cars that do not fall into one of the six categories shown in the table.)

Data set EX1455 **14.55 Good Tasting Medicine** Pfizer Canada Inc. is a pharmaceutical company that makes azithromycin, an antibiotic in a cherry-flavoured suspension used to treat bacterial infections in children. To compare the taste of their product with three competing medications, Pfizer tested 50 healthy children and 20 healthy adults. Among other taste-testing measures, they recorded the number of tasters who rated each of the four antibiotic suspensions as the best tasting.[19] The results are shown in the table. Is there a difference in the perception of the best taste between adults and children? If so, what is the nature of the difference, and why is it of practical importance to the pharmaceutical company?

	Flavour of Antibiotic			
	Banana	Cherry*	Wild Fruit	Strawberry-Banana
Children	14	20	7	9
Adults	4	14	0	2

*Azithromycin produced by Pfizer Canada Inc.

14.56 Funny Cards When you choose a greeting card, do you always look for a humorous card, or does it depend on the occasion? A comparison sponsored by two of the nation's leading manufacturers of greeting cards indicated a slight difference in the proportions of humorous designs made for three different occasions: Father's Day, Mother's Day, and Valentine's Day.[20] To test the accuracy of their comparison, random samples of 500 greeting cards purchased at a local card store in the week prior to each holiday were entered into a computer database, and the results in the table were obtained. Do the data indicate that the proportions of humorous greeting cards vary for these three holidays? (HINT: Remember to include a tabulation for all 1500 greeting cards.)

Holiday	Father's Day	Mother's Day	Valentine's Day
Percent humorous	20	25	24

Data set EX1457 **14.57 Rugby Injuries** Knee injuries are a major problem for athletes in many contact sports. However, athletes who play certain positions are more prone to knee injuries than other players, and their injuries tend to be more severe. The prevalence and patterns of knee injuries among women collegiate rugby players were investigated using a sample questionnaire, to which 42 rugby clubs responded.[21] A total of 76 knee injuries were classified by type as well as the position (forward or back) of the player.

	Type of Knee Injury				
Position	Meniscal Tear	MCL Tear	ACL Tear	Patella Dislocation	PCL Tear
Forward	13	14	7	3	1
Back	12	9	14	2	1

MINITAB output for Exercise 14.57

Chi-Square Test: Men Tear, MCL Tear, ACL Tear, Patella, PCL Tear

```
Expected counts are printed below observed counts
Chi-Square contributions are printed below expected
counts

    Men Tear MCL Tear ACL Tear  Patella PCL Tear  Total
1         13       14        7        3        1     38
       12.50    11.50    10.50     2.50     1.00
       0.020    0.543    1.167    0.100    0.000

2         12        9       14        2        1     38
       12.50    11.50    10.50     2.50     1.00
       0.020    0.543    1.167    0.100    0.000

Total     25       23       21        5        2     76

Chi-Sq = 3.660, DF = 4, P-Value = 0.454
4 cells with expected counts less than 5.0
```

a. Use the *MINITAB* printout to determine whether there is a difference in the distribution of injury types for rugby backs and forwards. Have any of the assumptions necessary for the chi-square test been violated? What effect will this have on the magnitude of the test statistic?

b. The investigators report a significant difference in the proportion of MCL tears for the two positions ($P < 0.05$) and a significant difference in the proportion of ACL tears ($P < 0.05$), but indicate that all other injuries occur with equal frequency for the two positions. Do you agree with those conclusions? Explain.

Data set EX1458 **14.58 Favourite Fast Foods** The number of Canadians who visit fast-food restaurants regularly has grown steadily over the past decade. For this reason, marketing experts are interested in the *demographics* of fast-food customers. Is a customer's preference for a fast-food chain affected by the age of the customer? If so, advertising might need to target a particular age group. Suppose a random sample of 500 fast-food customers aged 16 and older was selected, and their favourite fast-food restaurants along with their age groups were recorded, as shown in the table:

Age Group	McDonald's	Burger King	Wendy's	Other
16–21	75	34	10	6
21–30	89	42	19	10
30–49	54	52	28	18
50+	21	25	7	10

Use an appropriate method to determine whether or not a customer's fast-food preference is dependent on age. Write a short paragraph presenting your statistical conclusions and their practical implications for marketing experts.

14.59 Catching a Cold Is your chance of getting a cold influenced by the number of social contacts you have? A recent study showed that the more social relationships you have, the *less susceptible* you are to colds.[22] A group of 276 healthy men and women were grouped according to their number of relationships (such as parent, friend, church member, neighbour). They were then exposed to a virus that causes colds. An adaptation of the results is shown in the table:

	Number of Relationships		
	Three or Fewer	Four or Five	Six or More
Cold	49	43	34
No cold	31	57	62
Total	80	100	96

a. Do the data provide sufficient evidence to indicate that susceptibility to colds is affected by the number of relationships you have? Test at the 5% significance level.

b. Based on the results of part a, describe the nature of the relationship between the two categorical variables: cold incidence and number of social relationships. Do your observations agree with the conclusions of the study?

EX1460 **14.60 Crime and Educational Achievement** A criminologist studying criminal offenders who have a record of one or more arrests is interested in knowing whether the educational achievement level of the offender influences the frequency of arrests. He has classified his data using four educational level classifications:

A: Completed grade 6 or less
B: Completed grade 7, 8, or 9
C: Completed grade 10, 11, or 12
D: Education beyond grade 12

The contingency table shows the number of offenders in each educational category, along with the number of times they have been arrested:

	Educational Achievement			
Number of Arrests	*A*	*B*	*C*	*D*
1	55	40	43	30
2	15	25	18	22
3 or more	7	8	12	10

Do the data present sufficient evidence to indicate that the number of arrests is dependent on the educational achievement of a criminal offender? Test using $\alpha = 0.05$.

14.61 More Business on the Weekends A department store manager claims that her store has twice as many customers on Fridays and Saturdays than on any other day of the week (the store is closed on Sundays). That is, the probability that a customer visits the store Friday is 2/8, the probability that a customer visits the store Saturday is 2/8, while the probability that a customer visits the store on each of the remaining weekdays is 1/8. During an average week, the following numbers of customers visited the store:

Day	Number of Customers
Monday	95
Tuesday	110
Wednesday	125
Thursday	75
Friday	181
Saturday	214

Can the manager's claim be refuted at the $\alpha = 0.05$ level of significance?

CASE STUDY

Can a Marketing Approach Improve Library Services?

Libraries

Carole Day and Del Lowenthal studied the responses of young adults in their evaluation of library services.[23] Of the $n = 200$ young adults involved in the study, $n_1 = 152$ were students and $n_2 = 48$ were non-students. The table presents the percents and numbers of favourable responses for each group to seven questions in which the atmosphere, staff, and design of the library were examined:

Question		Student Favourable	$n_1 = 152$	Non-student Favourable	$n_2 = 48$	$P(\chi^2)$
3	Libraries are friendly	79.6%	121	56.2%	27	<0.01
4	Libraries are dull	77	117	58.3	28	<0.05
5	Library staff are helpful	91.4	139	87.5	42	NS
6	Library staff are less helpful to teenagers	60.5	92	45.8	22	<0.01
7	Libraries are so quiet they feel uncomfortable	75.6	115	52.05	25	<0.01
11	Libraries should be more brightly decorated	29	44	18.8	9	NS
13	Libraries are badly signposted	45.4	69	43.8	21	NS

The entry in the last column labelled $P(\chi^2)$ is the *p*-value for testing the hypothesis of no difference in the proportion of students and non-students who answer each question favourably. Hence, each question gives rise to a 2×2 contingency table.

1. Perform a test of homogeneity for each question and verify the reported *p*-value of the test.

2. Questions 3, 4, and 7 are concerned with the atmosphere of the library; questions 5 and 6 are concerned with the library staff; and questions 11 and 13 are concerned with the library design. How would you summarize the results of your analyses regarding these seven questions concerning the image of the library?

3. With the information given, is it possible to do any further testing concerning the proportion of favourable versus unfavourable responses for two or more questions simultaneously?

PROJECTS

Project 14-A: Child Safety Seat Survey, Part 3

Canada has a Road Safety Vision of having the safest roads in the world. Yet the leading cause of death of Canadian children remains vehicle crashes. In 2006, a national child safety seat survey was conducted by an AUTO21 research team in collaboration with Transport Canada to empirically measure Canada's progress towards achieving Road Safety Vision 2010.[24] Child seat use was observed in parking lots and nearby intersections in 200 randomly selected sites across Canada. The following table provides a classification of a subset of children in the survey by age groups and type of restraint device they were using at the time of the survey:

Table: Cross-Tabulation of Age Group by Restraint Type

Category	Types of Restraint				
Age Group	**Rear-Facing Infant Seat**	**Forward-Facing Infant Seat**	**Booster Seat**	**Seat Belt Only**	**Total**
Infant (0–1 year)	181	52	1	0	**234**
Toddler (1–4 years)	49	483	117	3	**652**
School (4–9 years)	0	98	450	325	**873**
Older (>9 years)	0	0	16	627	**643**
Total	**230**	**633**	**584**	**955**	**2402**

a. Which statistical technique is appropriate to describe the above data?

b. What is the sampling distribution of the test statistic for a goodness-of-fit test with these categories?

c. Suppose you wish to test whether the types of restraint are related to age group.

(i) State the appropriate hypotheses.

(ii) Which test statistic will you use to test the hypotheses in part (i)? What will be the sampling distribution of your test statistic?

(iii) At a glance on the table, what difficulty, if any, do you expect to encounter in calculating the test statistics for the hypotheses in part (i)?

(iv) Write a short report explaining the issues in performing a formal test on the hypothesis in part (i). In other words, what are the statistical issues?

(v) What remedies will you suggest to alleviate the problems you came across in calculating the test statistic? Make some necessary recommendations to the researcher for a future survey.

d. Let us consider the following adjusted data with fixed column total in the following table:

Category	Types of Restraint			
Age Group	Forward-Facing Infant Seat (A)	Booster Seat (B)	Seat Belt Only (C)	Total
Toddler (1–4 years)	483	250	50	**783**
School (4–9 years)	117	350	550	**1017**
Total	**600**	**600**	**600**	**1800**

Suppose you wish to test the null hypothesis that the three binomial parameters p_A, p_B, and p_C are equal versus the alternative hypothesis that at least two of the parameters differ.

(i) Write the null and alternative hypotheses for testing the equality of the three binomial proportions.

(ii) Calculate the test statistic and find the approximate p-value for the test in the previous question.

(iii) Use the approximate p-value to determine the statistical significance of your results. If the results are statistically significant, explore the nature of the differences in the three binomial proportions.

(iv) If there are more than two row categories in a contingency table with fixed-total c columns, then the test of independence is equivalent to a test of the equality of c sets of multinomial proportions. Justify your answer.

Project 14-B: The Dating Strategies

Social/behavioural scientists are usually interested in knowing what are the important issues for Canadian singles when it comes to dating. The print/digital media also has a keen interest in broadcasting these and other related issues.

Suppose that in an online survey the following question was posed and respondents were given three response categories.

Would financial status affect your decision as to whether or not you would be interested in pursuing a relationship with someone?

A: Matter a bit

B: A huge matter

C: Does not matter at all

The results of the survey are summarized in the following table:

	Responses		
Provinces	A	B	C
Alberta	0.71	0.16	0.13
British Columbia	0.67	0.15	0.18
Ontario	0.69	0.18	0.13
Quebec	0.64	0.10	0.26

Generally speaking, online surveys are not scientific. This type of sampling plan is one form of a convenience sample—a sample that can be easily and simply obtained without random selection. We discussed this issue in Chapter 7, page 273, that all sampling plans used for making inferences must involve randomization. We need to re-emphasize here that non-random samples can be described but cannot be used for making inferences.

For this statistical reason, a social scientist from Quebec finds these results dubious and decided to conduct a scientific survey as opposed to the online survey. The results of her survey are displayed in the following table. The table lists the number of responses for each category in selected provinces:

	Responses		
Provinces	A	B	C
Alberta	140	20	40
British Columbia	130	30	30
Ontario	270	65	55
Quebec	190	50	55

In the following questions treat the percentages from the online survey as if they were the true population proportions and use the scientist's survey as your sample data.

a. Test whether there has been a change from the online percentages in Quebec. Use $\alpha = 0.05$.

b. State the null and alternative hypotheses to test whether the distribution of the response to the question "does not matter at all" agrees with the corresponding distribution from the online survey. State the hypotheses to be tested and use $\alpha = 0.01$ to interpret your results.

c. Do the current data support that the proportions in category "A" have not changed compared to the online survey? Use $\alpha = 0.025$.

d. Based on her results, the social scientist in Quebec believes that the proportions in category "C" are the same across provinces and she would like to test her beliefs.

   (i) State the null and alternative hypotheses.

   (ii) Compute the value of the test statistic.

   (iii) If you were to test this hypothesis using the chi-square statistic, how many degrees of freedom would the test statistic have? What is the rejection region?

   (iv) Test the hypotheses at $\alpha = 0.05$, and write your conclusion.

e. Suppose you wish to analyze the relationship between responses to the question "Would financial status affect your decision as to whether or not you would be interested in pursuing a relationship with someone?" and the province of residence of the respondent. Write a short paragraph explaining your statistical conclusions and their practical implications for the researcher.

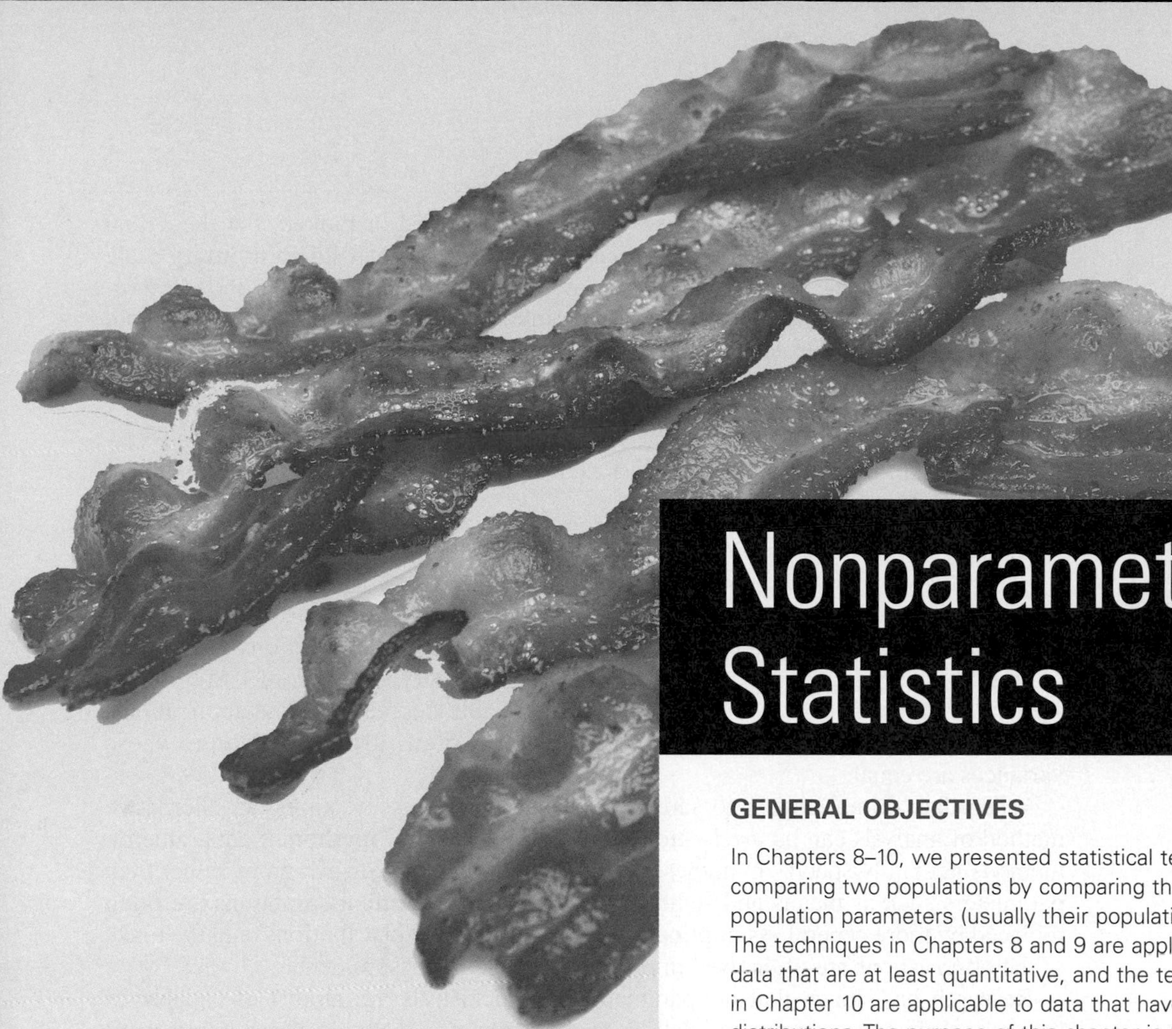

Joe Gough/Shutterstock.com

15

Nonparametric Statistics

GENERAL OBJECTIVES

In Chapters 8–10, we presented statistical techniques for comparing two populations by comparing their respective population parameters (usually their population means). The techniques in Chapters 8 and 9 are applicable to data that are at least quantitative, and the techniques in Chapter 10 are applicable to data that have normal distributions. The purpose of this chapter is to present several statistical tests for comparing populations for the many types of data that do not satisfy the assumptions specified in Chapters 8–10.

CHAPTER INDEX

How's Your Cholesterol Level?

What is your cholesterol level? Many of us have become more health conscious in the last few years as we read the nutritional labels on the food products we buy and choose foods that are low in fat and cholesterol and high in fibre. The case study at the end of this chapter involves a taste-testing experiment to compare three types of egg substitutes, using nonparametric techniques.

INTRODUCTION

Some experiments generate responses that can be ordered or ranked, but the actual value of the response cannot be measured numerically except with an arbitrary scale that you might create. It may be that you are able to tell only whether one observation is larger than another. Perhaps you can rank a whole set of observations without actually knowing the exact numerical values of the measurements. Here are a few examples:

- The sales abilities of four sales representatives are ranked from best to worst.
- The edibility and taste characteristics of five brands of raisin bran are rated on an arbitrary scale of 1 to 5.
- Five automobile designs are ranked from most appealing to least appealing.

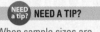

NEED A TIP?

When sample sizes are small and the original populations are not normal, use nonparametric techniques.

How can you analyze these types of data? The small-sample statistical methods presented in Chapters 10–13 are valid only when the sampled populations are normal or approximately so. Data that consist of ranks or arbitrary scales from 1 to 5 *do not satisfy the normality assumption,* even to a reasonable degree. In some applications, the techniques are valid if the samples are randomly drawn from populations whose variances are equal.

When data do not appear to satisfy these and similar assumptions, an alternative method of analysis can be used—**nonparametric statistical method.** Nonparametric methods generally specify hypotheses in terms of population distributions rather than parameters such as means and standard deviations. Parametric assumptions are often replaced by more general assumptions about the population distributions, and the ranks of the observations are often used in place of the actual measurements.

Research has shown that nonparametric statistical tests are almost as capable of detecting differences among populations as the parametric methods of preceding chapters when normality and other assumptions are satisfied. They may be, and often are, *more* powerful in detecting population differences when these assumptions are not satisfied. For this reason, some statisticians advocate the use of nonparametric procedures in preference to their parametric counterparts.

We will present nonparametric methods appropriate for comparing two or more populations using either independent or paired samples. We will also present a measure of association that is useful in determining whether one variable increases as the other increases or whether one variable decreases as the other increases.

THE WILCOXON RANK SUM TEST: INDEPENDENT RANDOM SAMPLES

In comparing the means of two populations based on independent samples, the pivotal statistic was the difference in the sample means. If you are not certain that the assumptions required for a two-sample t test are satisfied, one alternative is to replace the values of the observations by their ranks and proceed as though the ranks were the actual observations. Two different nonparametric tests use a test statistic based on these sample ranks:

- Wilcoxon rank sum test
- Mann-Whitney U test

They are *equivalent* in that they use the same sample information. The procedure that we will present is the Wilcoxon rank sum test, which is based on the sum of the ranks of the sample that has the smaller sample size.

Assume that you have n_1 observations from population 1 and n_2 observations from population 2. The null hypothesis to be tested is that the two population distributions are identical versus the alternative hypothesis that the population distributions are different. These are the possibilities for the two populations:

- If H_0 is true and the observations have come from the same or identical populations, then the observations from both samples should be randomly mixed when jointly ranked from small to large. The sum of the ranks of the observations from sample 1 should be similar to the sum of the ranks from sample 2.
- If, on the other hand, the observations from population 1 tend to be smaller than those from population 2, then these observations would have the smaller ranks because most of these observations would be smaller than those from population 2. The sum of the ranks of these observations would be "small."
- If the observations from population 1 tend to be larger than those in population 2, these observations would be assigned larger ranks. The sum of the ranks of these observations would tend to be "large."

For example, suppose you have $n_1 = 3$ observations from population 1—2, 4, and 6—and $n_2 = 4$ observations from population 2—3, 5, 8, and 9. Table 15.1 shows seven observations ordered from small to large.

TABLE 15.1 **Seven Observations in Order**

Observation	x_1	y_1	x_2	y_2	x_3	y_3	y_4
Data	2	3	4	5	6	8	9
Rank	1	2	3	4	5	6	7

The smallest observation, $x_1 = 2$, is assigned rank 1; the next smallest observation, $y_1 = 3$, is assigned rank 2; and so on. The *sum of the ranks* of the observations from sample 1 is $1 + 3 + 5 = 9$, and the **rank sum** from sample 2 is $2 + 4 + 6 + 7 = 19$. How do you determine whether the rank sum of the observations from sample 1 is significantly small or significantly large? This depends on the probability distribution of the sum of the ranks of one of the samples. Since the ranks for $n_1 + n_2 = N$ observations are the first N integers, the sum of these ranks can be shown to be $N(N + 1)/2$. In this simple example, the sum of the $N = 7$ ranks is $1 + 2 + 3 + 4 + 5 + 6 + 7 = 7(8)/2$ or 28. Hence, if you know the rank sum for one of the samples, you can find the other by subtraction. In our example, notice that the rank sum for sample 1 is 9, whereas the second rank sum is $(28 - 9) = 19$. This means that only one of the two rank sums is needed for the test. To simplify the tabulation of critical values for this test, you should use the rank sum from the smaller sample as the test statistic. What happens if two or more observations are equal? Tied observations are assigned the average of the ranks that the observations would have had if they had been slightly different in value.

To implement the Wilcoxon rank sum test, suppose that independent random samples of size n_1 and n_2 are selected from populations 1 and 2, respectively. Let n_1 represent the *smaller* of the two sample sizes, and let T_1 represent the sum of the ranks of the observations in sample 1. If population 1 lies to the left of population 2, T_1 will be "small." T_1 will be "large" if population 1 lies to the right of population 2.

FORMULAS FOR THE WILCOXON RANK SUM STATISTIC (FOR INDEPENDENT SAMPLES)

Let

$$T_1 = \text{Sum of the ranks for the first sample}$$
$$T_1^* = n_1(n_1 + n_2 + 1) - T_1$$

T_1^* is the value of the rank sum for n_1 if the observations had been ranked from *large to small.* (It is *not* the rank sum for the second sample.) Depending on the nature of the alternative hypothesis, one of these two values will be chosen as the test statistic, T.

Table 7 in Appendix I can be used to locate *critical values* for the test statistic for four different values of one-tailed tests with $\alpha = 0.05, 0.025, 0.01$, and 0.005. To use Table 7 for a two-tailed test, the values of α are doubled—that is, $\alpha = 0.10, 0.05, 0.02$, and 0.01. The tabled entry gives the value of a such that $P(T \le a) \le \alpha$. To see how to locate a critical value for the Wilcoxon rank sum test, suppose that $n_1 = 8$ and $n_2 = 10$ for a one-tailed test with $\alpha = 0.05$. You can use Table 7(a), a portion of which is reproduced in Table 15.2. Notice that the table is constructed assuming that $n_1 \le n_2$. It is for this reason that we designate the population with the smaller sample size as population 1. Values of n_1 are shown across the top of the table, and values of n_2 are shown down the left side. The entry—$a = 56$, shaded—is the critical value for rejection of H_0. The null hypothesis of equality of the two distributions should be rejected if the observed value of the test statistic T is less than or equal to 56.

TABLE 15.2 A Portion of the 5% Left-Tailed Critical Values, Table 7 in Appendix I

	n_1						
n_2	2	3	4	5	6	7	8
3	—	6					
4	—	6	11				
5	3	7	12	19			
6	3	8	13	20	28		
7	3	8	14	21	29	39	
8	4	9	15	23	31	41	51
9	4	10	16	24	33	43	54
10	4	10	17	26	35	45	56

THE WILCOXON RANK SUM TEST

Let n_1 denote the smaller of the two sample sizes. This sample comes from population 1. The hypotheses to be tested are

H_0 : The distributions for populations 1 and 2 are identical
versus one of three alternative hypotheses:

H_a : The distributions for populations 1 and 2 are different (a two-tailed test)

H_a : The distribution for population 1 lies to the left of that for population 2 (a left-tailed test)

H_a : The distribution for population 1 lies to the right of that for population 2 (a right-tailed test)

1. Rank all $n_1 + n_2$ observations from small to large.
2. Find T_1, the rank sum for the observations in sample 1. This is the test statistic for a left-tailed test.
3. Find $T_1^* = n_1(n_1 + n_2 + 1) - T_1$, the sum of the ranks of the observations from population 1 if the assigned ranks had been reversed from large to small. (The value of T_1^* is not the sum of the ranks of the observations in sample 2.) This is the test statistic for a right-tailed test.
4. The test statistic for a two-tailed test is T, the *minimum* of T_1 and T_1^*.
5. H_0 is rejected if the observed test statistic is less than or equal to the critical value found using Table 7 in Appendix I.

We illustrate the use of Table 7 with the next example.

EXAMPLE 15.1

The wing stroke frequencies of two species of Euglossine bees were recorded for a sample of $n_1 = 4$ *Euglossa mandibularis* Friese (species 1) and $n_2 = 6$ *Euglossa imperialis* Cockerell (species 2).[1] The frequencies are listed in Table 15.3. Can you conclude that the distributions of wing strokes differ for these two species? Test using $\alpha = 0.05$.

TABLE 15.3 **Wing Stroke Frequencies for Two Species of Bees**

Species 1	Species 2
235	180
225	169
190	180
188	185
	178
	182

SOLUTION You first need to rank the observations from small to large, as shown in Table 15.4.

TABLE 15.4 **Wing Stroke Frequencies Ranked from Small to Large**

Data	Species	Rank
169	2	1
178	2	2
180	2	3
180	2	4
182	2	5
185	2	6
188	1	7
190	1	8
225	1	9
235	1	10

The hypotheses to be tested are

H_0 : The distributions of the wing stroke frequencies are the same for the two species

versus

H_a : The distributions of the wing stroke frequencies differ for the two species

Since the sample size for individuals from species 1, $n_1 = 4$, is the smaller of the two sample sizes, you have

$$T_1 = 7 + 8 + 9 + 10 = 34$$

and

$$T_1^* = n_1(n_1 + n_2 + 1) - T_1 = 4(4 + 6 + 1) - 34 = 10$$

For a two-tailed test, the test statistic is $T = 10$, the smaller of $T_1 = 34$ and $T_1^* = 10$.

For this two-tailed test with $\alpha = 0.05$, you can use Table 7(b) in Appendix I with $n_1 = 4$ and $n_2 = 6$. The critical value of T such that $P(T \leq a) \leq \alpha/2 = 0.025$ is 12, and you should reject the null hypothesis if the observed value of T is 12 or less. Since the observed value of the test statistic—$T = 10$—is less than 12, you can reject the hypothesis of equal distributions of wing stroke frequencies at the 5% level of significance.

A *MINITAB* printout of the Wilcoxon rank sum test (called Mann–Whitney by *MINITAB*) for these data is given in Figure 15.1. You will find instructions for generating this output in the "Technology Today" section at the end of this chapter. Notice that the rank sum of the first sample is given as W = 34.0, which agrees with our calculations. With a reported p-value of 0.0142 calculated by *MINITAB*, you can reject the null hypothesis at the 5% level.

FIGURE 15.1
Printout for Example 15.1

Mann-Whitney Test and CI: Species 1, Species 2

```
           N    Median
Species 1  4    207.50
Species 2  6    180.00

Point estimate for ETA1-ETA2 is 30.50
95.7 Percent CI for ETA1-ETA2 is (5.99,56.01)
W = 34.0
Test of ETA1 = ETA2 vs ETA1 not = ETA2 is significant at 0.0142
The test is significant at 0.0139 (adjusted for ties)
```

Normal Approximation for the Wilcoxon Rank Sum Test

Table 7 in Appendix I contains critical values for sample sizes of $n_1 \leq n_2 = 3, 4, \ldots, 15$. Provided n_1 is not too small,† approximations to the probabilities for the Wilcoxon rank sum statistic T can be found using a normal approximation to the distribution of T. It can be shown that the mean and variance of T are

$$\mu_T = \frac{n_1(n_1 + n_2 + 1)}{2} \quad \text{and} \quad \sigma_T^2 = \frac{n_1 n_2(n_1 + n_2 + 1)}{12}$$

†Some researchers indicate that the normal approximation is adequate for samples as small as $n_1 = n_2 = 4$.

The distribution of

$$z = \frac{T - \mu_T}{\sigma_T}$$

is approximately normal with mean 0 and standard deviation 1 for values of n_1 and n_2 as small as 10.

If you try this approximation for Example 15.1, you get

$$\mu_T = \frac{n_1(n_1 + n_2 + 1)}{2} = \frac{4(4 + 6 + 1)}{2} = 22$$

and

$$\sigma_T^2 = \frac{n_1 n_2(n_1 + n_2 + 1)}{12} = \frac{4(6)(4 + 6 + 1)}{12} = 22$$

The p-value for this test is $2P(T \geq 34)$. If you use a 0.5 correction for continuity in calculating the value of z because n_1 and n_2 are both small,[†] you have

$$z = \frac{T - \mu_T}{\sigma_T} = \frac{(34 - 0.5) - 22}{\sqrt{22}} = 2.45$$

The p-value for this test is

$$2P(T \geq 34) \approx 2p(z \geq 2.45) = 2(0.0071) = 0.0142$$

the value reported on the *MINITAB* printout in Figure 15.1.

THE WILCOXON RANK SUM TEST FOR LARGE SAMPLES: $n_1 \geq 10$ AND $n_2 \geq 10$

1. Null hypothesis: H_0 : The population distributions are identical.
2. Alternative hypothesis: H_a : The two population distributions are not identical (a two-tailed test). Or H_a : The distribution of population 1 is shifted to the right (or left) of the distribution of population 2 (a one-tailed test).
3. Test statistic: $z = \dfrac{T - n_1(n_1 + n_2 + 1)/2}{\sqrt{n_1 n_2(n_1 + n_2 + 1)/12}}$
4. Rejection region:
   a. For a two-tailed test, reject H_0 if $z > z_{\alpha/2}$ or $z < -z_{\alpha/2}$.
   b. For a one-tailed test in the right tail, reject H_0 if $z > z_\alpha$.
   c. For a one tailed test in the left tail, reject H_0 if $z < -z_\alpha$.

Or reject H_0 if p-value $< \alpha$.

Tabulated values of z are found in Table 3 of Appendix I.

EXAMPLE 15.2

An experiment was conducted to compare the strengths of two types of kraft papers: one a standard kraft paper of a specified weight and the other the same standard kraft paper treated with a chemical substance. Ten pieces of each type of paper, randomly selected from production, produced the strength measurements shown in Table 15.5. Test the null hypothesis of no difference in the distributions of strengths for the two types of paper versus the alternative hypothesis that the treated paper tends to be stronger (i.e.,

[†]Since the value of $T = 34$ lies to the right of the mean 22, the subtraction of 0.5 in using the normal approximation takes into account the lower limit of the bar above the value 34 in the probability distribution of T.

its distribution of strength measurements is shifted to the right of the corresponding distribution for the untreated paper).

TABLE 15.5 Strength Measurements (and Their Ranks) for Two Types of Paper

Standard 1	Treated 2
1.21 (2)	1.49 (15)
1.43 (12)	1.37 (7.5)
1.35 (6)	1.67 (20)
1.51 (17)	1.50 (16)
1.39 (9)	1.31 (5)
1.17 (1)	1.29 (3.5)
1.48 (14)	1.52 (18)
1.42 (11)	1.37 (7.5)
1.29 (3.5)	1.44 (13)
1.40 (10)	1.53 (19)
Rank sum	$T_1 = 85.5$
	$T_1^* = n_1(n_1 + n_2 + 1) - T_1 = 210 - 85.5 = 124.5$

SOLUTION Since the sample sizes are equal, you are at liberty to decide which of the two samples should be sample 1. Choosing the standard treatment as the first sample, you can rank the 20 strength measurements, and the values of T_1 and T_1^* are shown at the bottom of the table. Since you want to detect a shift in the standard (1) measurements to the left of the treated (2) measurements, you conduct a left-tailed test:

H_0 : No difference in the strength distributions

H_a : Standard distribution lies to the left of the treated distribution

and use $T = T_1$ as the test statistic, looking for an unusually small value of T.

To find the critical value for a one-tailed test with $\alpha = 0.05$, index Table 7(a) in Appendix I with $n_1 = n_2 = 10$. Using the tabled entry, you can reject H_0 when $T \leq 82$. Since the observed value of the test statistic is $T = 85.5$, you are not able to reject H_0. There is insufficient evidence to conclude that the treated kraft paper is stronger than the standard paper.

To use the normal approximation to the distribution of T, you can calculate

$$\mu_T = \frac{n_1(n_1 + n_2 + 1)}{2} = \frac{10(21)}{2} = 105$$

and

$$\sigma_T^2 = \frac{n_1 n_2(n_1 + n_2 + 1)}{12} = \frac{10(10)(21)}{12} = 175$$

with $\sigma_T = \sqrt{175} = 13.23$. Then

$$z = \frac{T - \mu_T}{\sigma_T} = \frac{85.5 - 105}{13.23} = -1.47$$

The one-tailed p-value corresponding to $z = -1.47$ is

$$p\text{-value} = P(z \leq -1.47) = 0.0708$$

which is larger than $\alpha = 0.05$. The conclusion is the same. You cannot conclude that the treated kraft paper is stronger than the standard paper.

When should the Wilcoxon rank sum test be used in preference to the two-sample unpaired t test? The two-sample t test performs well if the data are normally distributed with equal variances. If there is doubt concerning these assumptions, a normal probability plot could be used to assess the degree of non-normality, and a two-sample F test of sample variances could be used to check the equality of variances. If these procedures indicate either non-normality or inequality of variance, then the Wilcoxon rank sum test is appropriate.

15.2 EXERCISES

BASIC TECHNIQUES

15.1 Suppose you want to use the Wilcoxon rank sum test to detect a shift in distribution 1 to the right of distribution 2 based on samples of size $n_1 = 6$ and $n_2 = 8$.

a. Should you use T_1 or T_1^* as the test statistic?

b. What is the rejection region for the test if $\alpha = 0.05$?

c. What is the rejection region for the test if $\alpha = 0.01$?

15.2 Refer to Exercise 15.1. Suppose the alternative hypothesis is that distribution 1 is shifted either to the left or to the right of distribution 2.

a. Should you use T_1 or T_1^* as the test statistic?

b. What is the rejection region for the test if $\alpha = 0.05$?

c. What is the rejection region for the test if $\alpha = 0.01$?

15.3 Observations from two random and independent samples, drawn from populations 1 and 2, are given here. Use the Wilcoxon rank sum test to determine whether population 1 is shifted to the left of population 2.

Sample 1	1	3	2	3	5
Sample 2	4	7	6	8	6

a. State the null and alternative hypotheses to be tested.

b. Rank the combined sample from smallest to largest. Calculate T_1 and T_1^*.

c. What is the rejection region for $\alpha = 0.05$?

d. Do the data provide sufficient evidence to indicate that population 1 is shifted to the left of population 2?

15.4 Independent random samples of size $n_1 = 20$ and $n_2 = 25$ are drawn from non-normal populations 1 and 2. The combined sample is ranked and $T_1 = 252$. Use the large-sample approximation to the Wilcoxon rank sum test to determine whether there is a difference in the two population distributions. Calculate the p-value for the test.

15.5 Suppose you wish to detect a shift in distribution 1 to the right of distribution 2 based on sample sizes $n_1 = 12$ and $n_2 = 14$. If $T_1 = 193$, what do you conclude? Use $\alpha = 0.05$.

APPLICATIONS

15.6 Alzheimer's Disease In some tests of healthy, elderly men, a new drug has restored their memory almost to that of young people. It will soon be tested on patients with Alzheimer's disease, the fatal brain disorder that destroys the mind. According to Dr. Gary Lynch, the drug, called ampakine CX-516, accelerates signals between brain cells and appears to significantly sharpen memory.[2] In a preliminary test on students in their early 20s and on men aged 65 to 70, the results were particularly striking. After being given mild doses of this drug, the 65- to 70-year-old men scored nearly as high as the young people. The accompanying data are the numbers of nonsense syllables recalled after 5 minutes for 10 men in their 20s and 10 men aged 65 to 70. Use the Wilcoxon rank sum test to determine whether the distributions for the number of nonsense syllables recalled are the same for these two groups.

20s	3	6	4	8	7	1	1	2	7	8
65–70s	1	0	4	1	2	5	0	2	2	3

15.7 Alzheimer's, continued Refer to Exercise 15.6. Suppose that two more groups of 10 men each are tested on the number of nonsense syllables they can remember after 5 minutes. However, this time the 65- to 70-year-old men are given a mild dose of ampakine CX-516. Do the data provide sufficient evidence to conclude that this drug improves memory in men aged 65 to 70 compared with that of 20-year-olds? Use an appropriate level of α.

20s	11	7	6	8	6	9	2	10	3	6
65–70s	1	9	6	8	7	8	5	7	10	3

15.8 Dissolved O_2 Content The observations in the table are dissolved oxygen contents in water. The higher the dissolved oxygen content, the greater the ability of a river, lake, or stream to support aquatic life. In this experiment, a pollution-control inspector suspected that a river community was releasing semitreated sewage into a river. To check this theory, five randomly selected specimens of river water were selected at a location above the town and another five below. These are the dissolved oxygen readings (in parts per million):

Above town	4.8	5.2	5.0	4.9	5.1
Below town	5.0	4.7	4.9	4.8	4.9

a. Use a one-tailed Wilcoxon rank sum test with $\alpha = 0.05$ to confirm or refute the theory.

b. Use a Student's t test (with $\alpha = 0.05$) to analyze the data. Compare the conclusion reached in part a.

Data set EX1509 **15.9 Eye Movement** In an investigation of the visual scanning behaviour of deaf children, measurements of eye movement were taken on nine deaf and nine hearing children. The table gives the eye-movement rates and their ranks (in parentheses). Does it appear that the distributions of eye-movement rates for deaf children and hearing children differ?

	Deaf Children	Hearing Children
	2.75 (15)	0.89 (1)
	2.14 (11)	1.43 (7)
	3.23 (18)	1.06 (4)
	2.07 (10)	1.01 (3)
	2.49 (14)	0.94 (2)
	2.18 (12)	1.79 (8)
	3.16 (17)	1.12 (5.5)
	2.93 (16)	2.01 (9)
	2.20 (13)	1.12 (5.5)
Rank sum	126	45

Data set EX1510 **15.10 Comparing NFL Quarterbacks** How does Aaron Rodgers, quarterback for the 2011 Super Bowl winners, the Green Bay Packers, compare to Drew Brees, quarterback for the 2010 Super Bowl winners, the New Orleans Saints? The table below shows the number of completed passes for each athlete during the 2010 NFL football season:[3]

Aaron Rodgers			Drew Brees		
19	21	7	27	37	25
19	15	25	28	34	29
34	27	19	30	27	35
12	22		33	29	22
27	26		24	23	
18	21		21	24	

Use the Wilcoxon rank sum test to analyze the data and test to see whether the population distributions for the number of completed passes differ for the two quarterbacks. Use $\alpha = 0.05$.

Data set EX1511 **15.11 Weights of Turtles** The weights of turtles caught in two different lakes were measured to compare the effects of the two lake environments on turtle growth. All the turtles were the same age and were tagged before being released into the lakes. The weights for $n_1 = 10$ tagged turtles caught in lake 1 and $n_2 = 8$ caught in lake 2 are listed here:

Lake	Weight (grams)									
1	399.7	430.9	394.1	411.1	416.7	391.2	396.9	456.4	360.0	433.7
2	345.9	368.8	399.7	385.6	351.5	337.4	354.4	391.2		

Do the data provide sufficient evidence to indicate a difference in the distributions of weights for the tagged turtles exposed to the two lake environments? Use the Wilcoxon rank sum test with $\alpha = 0.05$ to answer the question.

Data set EX1512 **15.12 Chemotherapy** Cancer treatment by means of chemicals—chemotherapy—kills both cancerous and normal cells. In some instances, the toxicity of the cancer drug—that is, its effect on normal cells—can be reduced by the simultaneous injection of a second drug. A study was conducted to determine whether a particular drug injection reduced the harmful effects of a chemotherapy treatment on the survival time for rats. Two randomly selected groups of 12 rats were used in an experiment in which both groups, call them A and B, received the toxic drug in a dose large enough to cause death, but in addition, group B received the antitoxin, which was to reduce the toxic effect of the chemotherapy on normal cells. The test was terminated at the end of 20 days, or 480 hours. The survival times for the two groups of rats, to the nearest four hours, are shown in the table. Do the data provide sufficient evidence to indicate that rats receiving the antitoxin tend to survive longer after chemotherapy than those not receiving the antitoxin? Use the Wilcoxon rank sum test with $\alpha = 0.05$.

Chemotherapy Only A	Chemotherapy plus Drug B
84	140
128	184
168	368
92	96
184	480
92	188
76	480
104	244
72	440
180	380
144	480
120	196

THE SIGN TEST FOR A PAIRED EXPERIMENT

The sign test is a fairly simple procedure that can be used to compare two populations when the samples consist of paired observations. This type of experimental design is called the **paired-difference** or **matched pairs** design, which you used to compare the average wear for two types of tires in Section 10.5. In general, for each pair, you measure whether the first response—say, A—exceeds the second response—say, B. The test statistic is *x,* the number of times that A exceeds B in the *n* pairs of observations.

When the two population distributions are identical, the probability that A exceeds B is $p = 0.5$, and *x,* the number of times that A exceeds B, has a *binomial* distribution. Only pairs without ties are included in the test. Hence, you can test the hypothesis of identical population distributions by testing $H_0 : P = 0.5$ versus either a one- or two-tailed alternative. Critical values for the rejection region or exact *p*-values can be found using the cumulative binomial tables in Appendix I.

THE SIGN TEST FOR COMPARING TWO POPULATIONS

1. Null hypothesis: H_0 : The two population distributions are identical and $P(\text{A exceeds B}) = p = 0.5$.
2. Alternative hypothesis:
   a. H_a : The population distributions are not identical and $p \neq 0.5$.
   b. H_a : The population of A measurements is shifted to the right of the population of B measurements and $p > 0.5$.
   c. H_a : The population of A measurements is shifted to the left of the population of B measurements and $p < 0.5$.
3. Test statistic: For *n,* the number of pairs with no ties, use *x,* the number of times that (A − B) is positive.
4. Rejection region:
   a. For the two-tailed test $H_a : P \neq 0.5$, reject H_0 if $x \leq x_L$ or $x \geq x_U$, where $P(x \leq x_L) \leq \alpha/2$ and $P(x \geq x_U) \leq \alpha/2$ for *x* having a binomial distribution with $p = 0.5$.
   b. For $H_a : P > 0.5$, reject H_0 if $x \geq x_U$ with $P(x \geq x_U) \leq \alpha$.
   c. For $H_a : P < 0.5$, reject H_0 if $x \leq x_L$ with $P(x \leq x_L) \leq \alpha$.

Or calculate the *p*-value and reject H_0 if the *p*-value $< \alpha$.

One problem that may occur when you are conducting a sign test is that the measurements associated with one or more pairs may be equal and therefore result in **tied observations.** When this happens, delete the tied pairs and reduce *n,* the total number of pairs. The following example will help you understand how the sign test is constructed and used.

EXAMPLE 15.3

The numbers of defective electrical fuses produced by two production lines, A and B, were recorded daily for a period of 10 days, with the results shown in Table 15.6. The response variable, the number of defective fuses, has an exact binomial distribution with a large number of fuses produced per day. Although this variable will have an approximately normal distribution, the plant supervisor would prefer a quick-and-easy statistical test to determine whether one production line tends to produce more defectives than the other. Use the sign test to test the appropriate hypothesis.

TABLE 15.6 **Defective Fuses from Two Production Lines**

Day	Line A	Line B	Sign of Difference
1	170	201	−
2	164	179	−
3	140	159	−
4	184	195	−
5	174	177	−
6	142	170	−
7	191	183	+
8	169	179	−
9	161	170	−
10	200	212	−

SOLUTION For this *paired-difference* experiment, x is the number of times that the observation for line A exceeds that for line B in a given day. If there is no difference in the distributions of defectives for the two production lines, then p, the proportion of days on which A exceeds B, is 0.5, which is the hypothesized value in a test of the binomial parameter p. Very small or very large values of x, the number of times that A exceeds B, are contrary to the null hypothesis.

Since $n = 10$ and the hypothesized value of p is 0.5, Table 1 of Appendix I can be used to find the exact p-value for the test of

$$H_0 : P = 0.5 \text{ versus } H_a : p \neq 0.5$$

The observed value of the test statistic—which is the number of "plus" signs in the table—is $x = 1$, and the p-value is calculated as

$$p\text{-value} = 2P(x \leq 1) = 2(0.011) = 0.022$$

The fairly small p-value $= 0.022$ allows you to reject H_0 at the 5% level. There is significant evidence to indicate that the number of defective fuses is not the same for the two production lines; in fact, line B produces more defectives than line A. In this example, the sign test is an easy-to-calculate rough tool for detecting faulty production lines and works perfectly well to detect a significant difference using only a minimum amount of information.

Normal Approximation for the Sign Test

When the number of pairs n is large, the critical values for rejection of H_0 and the approximate p-values can be found using a normal approximation to the distribution of x, which was discussed in Section 6.4. Because the binomial distribution is perfectly symmetric when $p = 0.5$, this approximation works very well, even for n as small as 10.

For $n \geq 25$, you can conduct the sign test by using the z statistic,

$$z = \frac{x - np}{\sqrt{npq}} = \frac{x - 0.5n}{0.5\sqrt{n}}$$

as the test statistic. In using z, you are testing the null hypothesis $p = 0.5$ versus the alternative $p \neq 0.5$ for a two-tailed test or versus the alternative $p > 0.5$ (or $p < 0.5$) for a one-tailed test. The tests use the familiar rejection regions of Chapter 9.

SIGN TEST FOR LARGE SAMPLES: $n \geq 25$

1. Null hypothesis: $H_0 : p = 0.5$ (one treatment is not preferred to a second treatment)
2. Alternative hypothesis: $H_a : p \neq 0.5$, for a two-tailed test (NOTE: We use the two-tailed test as an example. Many analyses might require a one-tailed test.)
3. Test statistic: $z = \dfrac{x - 0.5n}{0.5\sqrt{n}}$
4. Rejection region: Reject H_0 if $z \geq z_{\alpha/2}$ or $z \leq -z_{\alpha/2}$, where $z_{\alpha/2}$ is the z-value from Table 3 in Appendix I corresponding to an area of $\alpha/2$ in the upper tail of the normal distribution.

EXAMPLE 15.4

A production superintendent claims that there is no difference between the employee accident rates for the day versus the evening shifts in a large manufacturing plant. The number of accidents per day is recorded for both the day and evening shifts for $n = 100$ days. It is found that the number of accidents per day for the evening shift x_E exceeded the corresponding number of accidents on the day shift x_D on 63 of the 100 days. Do these results provide sufficient evidence to indicate that more accidents tend to occur on one shift than on the other or, equivalently, that $P(x_E > x_D) \neq 1/2$?

SOLUTION This study is a paired-difference experiment, with $n = 100$ pairs of observations corresponding to the 100 days. To test the null hypothesis that the two distributions of accidents are identical, you can use the test statistic

$$z = \frac{x - 0.5n}{0.5\sqrt{n}}$$

where x is the number of days in which the number of accidents on the evening shift exceeded the number of accidents on the day shift. Then for $\alpha = 0.05$, you can reject the null hypothesis if $z \geq 1.96$ or $z \leq -1.96$. Substituting into the formula for z, you get

$$z = \frac{x - 0.5n}{0.5\sqrt{n}} = \frac{63 - (0.5)(100)}{0.5\sqrt{100}} = \frac{13}{5} = 2.60$$

Since the calculated value of z exceeds $z_{\alpha/2} = 1.96$, you can reject the null hypothesis. The data provide sufficient evidence to indicate a difference in the accident rate distributions for the day versus evening shifts.

When should the sign test be used in preference to the paired t test? When only the *direction* of the difference in the measurement is given, *only* the sign test can be used. On the other hand, when the data are quantitative and satisfy the normality and constant variance assumptions, the paired t test should be used. A normal probability plot can be used to assess normality, while a plot of the residuals $(d_i - \bar{d})$ can reveal large deviations that might indicate a variance that varies from pair to pair. When there are doubts about the validity of the assumptions, statisticians often recommend that both tests be performed. If both tests reach the same conclusions, then the parametric test results can be considered to be valid.

15.3 EXERCISES

BASIC TECHNIQUES

15.13 Suppose you wish to use the sign test to test $H_a : p > 0.5$ for a paired-difference experiment with $n = 25$ pairs.

a. State the practical situation that dictates the alternative hypothesis given.

b. Use Table 1 in Appendix I to find values of α ($\alpha < 0.15$) available for the test.

15.14 Repeat the instructions of Exercise 15.13 for $H_a : p \neq 0.5$.

15.15 Repeat the instructions of Exercises 15.13 and 15.14 for $n = 10$, 15, and 20.

Data set EX1516 **15.16** A paired-difference experiment was conducted to compare two populations. The data are shown in the table. Use a sign test to determine whether the population distributions are different.

	Pairs						
Population	1	2	3	4	5	6	7
1	8.9	8.1	9.3	7.7	10.4	8.3	7.4
2	8.8	7.4	9.0	7.8	9.9	8.1	6.9

a. State the null and alternative hypotheses for the test.

b. Determine an appropriate rejection region with $\alpha \approx 0.01$.

c. Calculate the observed value of the test statistic.

d. Do the data present sufficient evidence to indicate that populations 1 and 2 are different?

APPLICATIONS

Data set EX1517 **15.17 Property Values** In Exercise 10.53, you compared the property evaluations of two tax assessors, A and B. Their assessments for eight properties are shown in the table:

Property	Assessor A	Assessor B
1	76.3	75.1
2	88.4	86.8
3	80.2	77.3
4	94.7	90.6
5	68.7	69.1
6	82.8	81.0
7	76.1	75.3
8	79.0	79.1

a. Use the sign test to determine whether the data present sufficient evidence to indicate that one of the assessors tends to be consistently more conservative than the other; that is, $P(x_A > x_B) \neq 1/2$. Test by using a value of α near 0.05. Find the p-value for the test and interpret its value.

b. Exercise 10.53 uses the t statistic to test the null hypothesis that there is no difference in the mean property assessments between assessors A and B. Check the answer (in the answer section) for Exercise 10.53 and compare it with your answer to part a. Do the test results agree? Explain why the answers are (or are not) consistent.

Data set EX1518 **15.18 Gourmet Cooking** Two gourmets, A and B, rated 22 meals on a scale of 1 to 10. The data are shown in the table. Do the data provide sufficient evidence to indicate that one of the gourmets tends to give higher ratings than the other? Test by using the sign test with a value of α near 0.05.

Meal	A	B	Meal	A	B
1	6	8	12	8	5
2	4	5	13	4	2
3	7	4	14	3	3
4	8	7	15	6	8
5	2	3	16	9	10
6	7	4	17	9	8
7	9	9	18	4	6
8	7	8	19	4	3
9	2	5	20	5	4
10	4	3	21	3	2
11	6	9	22	5	3

a. Use the binomial tables in Appendix I to find the exact rejection region for the test.

b. Use the large-sample z statistic. (NOTE: Although the large-sample approximation is suggested for $n \geq 25$, it works fairly well for values of n as small as 15.)

c. Compare the results of parts a and b.

15.19 Lead Levels in Blood A study reported in the *American Journal of Public Health (Science News)*—the first to follow blood lead levels in law-abiding handgun hobbyists using indoor firing ranges—documents a significant risk of lead poisoning.[4] Lead exposure measurements were made on 17 members of a law enforcement trainee class before, during, and after a three-month period of firearm instruction at an owned indoor firing range. No trainee had elevated blood lead levels before the training, but 15 of the 17 ended their training with blood lead levels "elevated." If the use of an indoor firing range causes no increase in blood lead levels, then p, the probability that a person's blood lead level increases, is less than or equal to 0.5. If, however, use of the indoor firing range causes an increase in a person's blood lead levels, then $p > 0.5$. Use the sign test to determine whether using an indoor firing range has the effect of increasing a person's blood lead level with $\alpha = 0.05$. (HINT: The normal approximation to binomial probabilities is fairly accurate for $n = 17$.)

EX1520 **15.20 Recovery Rates** Clinical data concerning the effectiveness of two drugs in treating a particular disease were collected from 10 hospitals. The numbers of patients treated with the drugs varied from one hospital to another. You want to know whether the data present sufficient evidence to indicate a higher recovery rate for one of the two drugs.

a. Test using the sign test. Choose your rejection region so that α is near 0.05.

b. Why might it be inappropriate to use the Student's t test in analyzing the data?

	Drug A				Drug B		
Hospital	Number in Group	Number Recovered	Percentage Recovered	Hospital	Number in Group	Number Recovered	Percentage Recovered
1	84	63	75.0	1	96	82	85.4
2	63	44	69.8	2	83	69	83.1
3	56	48	85.7	3	91	73	80.2
4	77	57	74.0	4	47	35	74.5
5	29	20	69.0	5	60	42	70.0
6	48	40	83.3	6	27	22	81.5
7	61	42	68.9	7	69	52	75.4
8	45	35	77.8	8	72	57	79.2
9	79	57	72.2	9	89	76	85.4
10	62	48	77.4	10	46	37	80.4

15.4 A COMPARISON OF STATISTICAL TESTS

The experiment in Example 15.3 is designed as a paired-difference experiment. If the assumptions of normality and constant variance, σ_d^2, for the differences were met, would the sign test detect a shift in location for the two populations as efficiently as the paired t test? Probably not, since the t test uses much more information than the sign test. It uses not only the sign of the difference, but also the actual values of the differences. In this case, we would say that the sign test is not as *efficient* as the paired t test. However, the sign test might be more efficient if the usual assumptions were not met.

When two different statistical tests can *both* be used to test a hypothesis based on the same data, it is natural to ask, Which is better? One way to answer this question would be to hold the sample size n and α constant for both procedures and compare β, the probability of a Type II error. Statisticians, however, prefer to examine the **power** of a test.

DEFINITION Power $= 1 - \beta = P(\text{reject } H_0 \text{ when } H_a \text{ is true})$

Since β is the probability of failing to reject the null hypothesis when it is false, the **power** of the test is the probability of rejecting the null hypothesis when it is false and some specified alternative is true. It is the probability that the test will do what it was designed to do—that is, detect a departure from the null hypothesis when a departure exists.

Probably the most common method of comparing two test procedures is in terms of the relative efficiency of a pair of tests. **Relative efficiency** is the ratio of the sample sizes for the two test procedures required to achieve the same α and β for a given alternative to the null hypothesis.

In some situations, you may not be too concerned whether you are using the most powerful test. For example, you might choose to use the sign test over a more powerful competitor because of its ease of application. Thus, you might view tests as microscopes that are used to detect departures from an hypothesized theory. One need not know the exact power of a microscope to use it in a biological investigation, and the same applies to statistical tests. If the test procedure detects a departure from the null hypothesis, you are delighted. If not, you can reanalyze the data by using a more powerful microscope (test), or you can increase the power of the microscope (test) by increasing the sample size.

15.5 THE WILCOXON SIGNED-RANK TEST FOR A PAIRED EXPERIMENT

A signed-rank test proposed by F. Wilcoxon can be used to analyze the paired-difference experiment of Section 10.5 by considering the paired differences of two treatments, 1 and 2. Under the null hypothesis of no differences in the distributions for 1 and 2, you would expect (on the average) half of the differences in pairs to be negative and half to be positive; that is, the expected number of negative differences between pairs would be $n/2$ (where n is the number of pairs). Furthermore, it follows that positive and negative differences of equal absolute magnitude should occur with equal probability. If one were to order the differences according to their absolute values and rank them from smallest to largest, the expected rank sums for the negative and positive differences would be equal. Sizable differences in the sums of the ranks assigned to the positive and negative differences would provide evidence to indicate a shift in location between the distributions of responses for the two treatments, 1 and 2.

If distribution 1 is shifted to the right of distribution 2, then more of the differences are expected to be positive, and this results in a small number of negative differences. Therefore, to detect this one-sided alternative, use the rank sum T^-—the sum of the ranks of the negative differences—and reject the null hypothesis for significantly small values of T^-. Along these same lines, if distribution 1 is shifted to the left of distribution 2, then more of the differences are expected to be negative, and the number of positive differences is small. Hence, to detect this one-sided alternative, use T^+—the sum of the ranks of the positive differences—and reject the null hypothesis if T^+ is significantly small.

CALCULATING THE TEST STATISTIC FOR THE WILCOXON SIGNED-RANK TEST

1. Calculate the differences $(x_1 - x_2)$ for each of the n pairs. Differences equal to 0 are eliminated, and the number of pairs, n, is reduced accordingly.
2. Rank the **absolute values** of the differences by assigning 1 to the smallest, 2 to the second smallest, and so on. Tied observations are assigned the average of the ranks that would have been assigned with no ties.
3. Calculate the **rank sum** for the **negative** differences and label this value T^-. Similarly, calculate T^+, the **rank sum** for the **positive** differences.

For a **two-tailed test,** use the **smaller of these two quantities T as a test statistic** to test the null hypothesis that the two population relative frequency histograms are identical. The smaller the value of T, the greater is the weight of evidence favouring rejection of the null hypothesis. **Therefore, you will reject the null hypothesis if T is less than or equal to some value—say, T_0.**

To detect the **one-sided alternative,** that **distribution 1 is shifted to the right of distribution 2, use the rank sum T^-** of the negative differences and reject the null hypothesis for small values of T^-—say, $T^- \le T_0$. If you wish to detect a **shift of distribution 2 to the right of distribution 1, use the rank sum T^+** of the positive differences as a test statistic and reject the null hypothesis for small values of T^+—say, $T^+ \le T_0$.

The probability that T is less than or equal to some value T_0 has been calculated for a combination of sample sizes and values of T_0. These probabilities, given in Table 8 in Appendix I, can be used to find the rejection region for the T test.

An abbreviated version of Table 8 in Appendix I is shown in Table 15.7. Across the top of the table you see the number of differences (the number of pairs) n. Values of α for a one-tailed test appear in the first column of the table. The second column gives values of α for a two-tailed test. Table entries are the critical values of T. You will recall that the critical value of a test statistic is the value that locates the boundary of the rejection region.

For example, suppose you have $n = 7$ pairs and you are conducting a two-tailed test of the null hypothesis that the two population relative frequency distributions are identical. Checking the $n = 7$ column of Table 15.7 and using the second row (corresponding to $\alpha = 0.05$ for a two-tailed test), you see the entry 2 (shaded). This value is T_0, the critical value of T. As noted earlier, the smaller the value of T, the greater is the evidence to reject the null hypothesis. Therefore, you will reject the null hypothesis for all values of T less than or equal to 2. The rejection region for the Wilcoxon signed-rank test for a paired experiment is always of the form: Reject H_0 if $T \le T_0$, where T_0 is the critical value of T. The rejection region is shown symbolically in Figure 15.2.

TABLE 15.7 **An Abbreviated Version of Table 8 in Appendix I; Critical Values of T**

One-Sided	Two-Sided	$n = 5$	$n = 6$	$n = 7$	$n = 8$	$n = 9$	$n = 10$	$n = 11$
$\alpha = 0.050$	$\alpha = 0.10$	1	2	4	6	8	11	14
$\alpha = 0.025$	$\alpha = 0.05$		1	2	4	6	8	11
$\alpha = 0.010$	$\alpha = 0.02$			0	2	3	5	7
$\alpha = 0.005$	$\alpha = 0.01$				0	2	3	5

One-Sided	Two-Sided	$n = 12$	$n = 13$	$n = 14$	$n = 15$	$n = 16$	$n = 17$
$\alpha = 0.050$	$\alpha = 0.10$	17	21	26	30	36	41
$\alpha = 0.025$	$\alpha = 0.05$	14	17	21	25	30	35
$\alpha = 0.010$	$\alpha = 0.02$	10	13	16	20	24	28
$\alpha = 0.005$	$\alpha = 0.01$	7	10	13	16	19	23

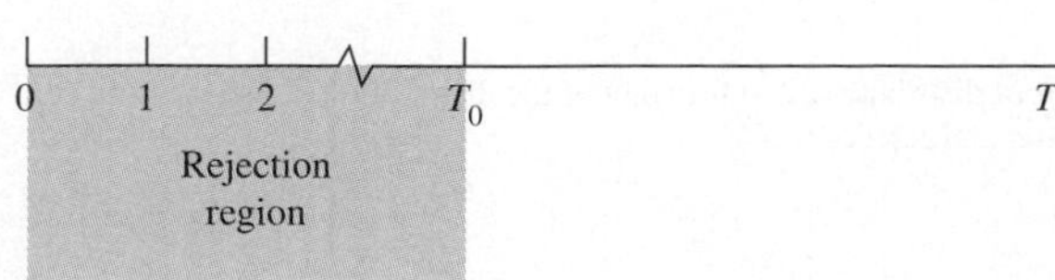

FIGURE 15.2 Rejection region for the Wilcoxon signed-rank test for a paired experiment (reject H_0 if $T \le T_0$)

WILCOXON SIGNED-RANK TEST FOR A PAIRED EXPERIMENT

1. Null hypothesis: H_0 : The two population relative frequency distributions are identical.
2. Alternative hypothesis: H_a : The two population relative frequency distributions differ in location (a two-tailed test). Or H_a : the population 1 relative frequency distribution is shifted to the right of the relative frequency distribution for population 2 (a one-tailed test).
3. Test statistic:
   a. For a two-tailed test, use T, the smaller of the rank sum for positive differences and the rank sum for negative differences.
   b. For a one-tailed test (to detect the alternative hypothesis described above), use the rank sum T^- of the negative differences.
4. Rejection region:
   a. For a two-tailed test, reject H_0 if $T \leq T_0$, where T_0 is the critical value given in Table 8 in Appendix I.
   b. For a one-tailed test (to detect the alternative hypothesis described above), use the rank sum T^- of the negative differences. Reject H_0 if $T^- \leq T_0$.†

(NOTE: It can be shown that $T^+ + T^- = \dfrac{n(n+1)}{2}$.)

EXAMPLE 15.5

An experiment was conducted to compare the densities of cakes prepared from two different cake mixes, A and B. Six cake pans received batter A, and six received batter B. Expecting a variation in oven temperature, the experimenter placed an A and a B cake side by side at six different locations in the oven. Test the hypothesis of no difference in the population distributions of cake densities for two different cake batters.

SOLUTION The data (density in grams per cubic centimetre) and differences in density for six pairs of cakes are given in Table 15.8. The box plot of the differences in Figure 15.3 shows fairly strong skewing and a very large difference in the right tail, which indicates that the data may not satisfy the normality assumption. The sample of differences is too small to make valid decisions about normality. In this situation, Wilcoxon's signed-rank test may be the prudent test to use.

As with other nonparametric tests, the null hypothesis to be tested is that the two population frequency distributions of cake densities are identical. The alternative hypothesis, which implies a two-tailed test, is that the distributions are different. Because the amount of data is small, you can conduct the test using $\alpha = 0.10$. From Table 8 in Appendix I, the critical value of T for a two-tailed test, $\alpha = 0.10$, is $T_0 = 2$. Hence, you can reject H_0 if $T \leq 2$.

†To detect a shift of distribution 2 to the right of the distribution 1, use the rank sum T^+ of the positive differences as the test statistic and reject H_0 if $T^+ \leq T_0$.

TABLE 15.8 **Densities of Six Pairs of Cakes**

x_A	x_B	Difference $(x_A - x_B)$	Rank
0.234	0.223	0.010	2
0.176	0.208	−0.031	5
0.170	0.194	−0.024	4
0.244	0.263	−0.019	3
0.227	0.234	−0.007	1
0.249	0.282	−0.033	6

FIGURE 15.3
Box plot of differences for Example 15.5

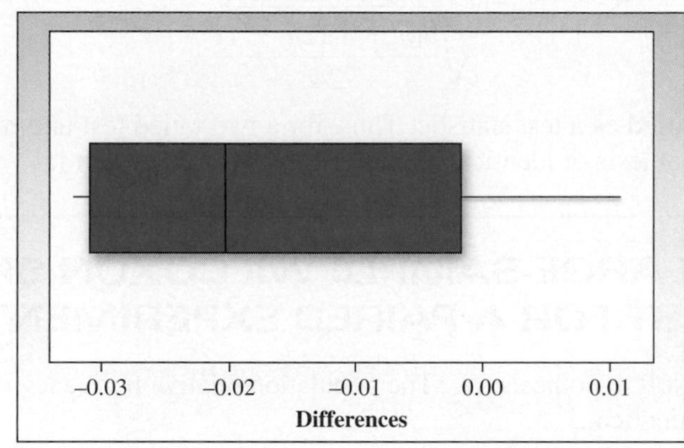

The differences $(x_1 - x_2)$ are calculated and ranked according to their absolute values in Table 15.8. The sum of the positive ranks is $T^+ = 2$, and the sum of the negative ranks is $T^- = 19$. The test statistic is the smaller of these two rank sums, or $T = 2$. Since $T = 2$ falls in the rejection region, you can reject H_0 and conclude that the two population frequency distributions of cake densities differ.

A *MINITAB* printout of the Wilcoxon signed-rank test for these data is given in Figure 15.4. You will find instructions for generating this output in the "Technology Today" section at the end of this chapter. You can see that the value of the test statistic agrees with the other calculations, and the p-value indicates that you can reject H_0 at the 10% level of significance.

FIGURE 15.4
MINITAB printout for Example 15.5

Wilcoxon Signed Rank Test: Difference

```
Test of median = 0.000000 versus median not = 0.000000

                  N for  Wilcoxon              Estimated
              N    Test  Statistic        P       Median
Difference    6       6        2.0    0.093     -0.01100
```

Normal Approximation for the Wilcoxon Signed-Rank Test

Although Table 8 in Appendix I has critical values for n as large as 50, T^+, like the Wilcoxon signed-rank test, will be approximately normally distributed when the null hypothesis is true and n is large—say, 25 or more. This enables you to construct a large-sample z test, where

$$E(T^+) = \frac{n(n+1)}{4}$$

$$\sigma^2_{T^+} = \frac{n(n+1)(2n+1)}{24}$$

Then the z statistic

$$z = \frac{T^+ - E(T^+)}{\sigma_{T^+}} = \frac{T^+ - \dfrac{n(n+1)}{4}}{\sqrt{\dfrac{n(n+1)(2n+1)}{24}}}$$

can be used as a test statistic. Thus, for a two-tailed test and $\alpha = 0.05$, you can reject the hypothesis of identical population distributions when $|z| \geq 1.96$.

A LARGE-SAMPLE WILCOXON SIGNED-RANK TEST FOR A PAIRED EXPERIMENT: $n \geq 25$

1. Null hypothesis: H_0: The population relative frequency distributions 1 and 2 are identical.
2. Alternative hypothesis: H_a: The two population relative frequency distributions differ in location (a two-tailed test). Or H_a: the population 1 relative frequency distribution is shifted to the right (or left) of the relative frequency distribution for population 2 (a one-tailed test).
3. Test statistic: $z = \dfrac{T^- - [n(n+1)/4]}{\sqrt{[n(n+1)(2n+1)/24]}}$
4. Rejection region: Reject H_0 if $z > z_{\alpha/2}$ or $z < -z_{\alpha/2}$ for a two-tailed test. For a one-tailed test, place all of α in one tail of the z distribution. To detect a shift in distribution 1 to the right of distribution 2, reject H_0 when $z > z_\alpha$. To detect a shift in the opposite direction, reject H_0 if $z < -z_\alpha$.

Tabulated values of z are given in Table 3 in Appendix I.

15.5 EXERCISES

BASIC TECHNIQUES

15.21 Suppose you wish to detect a difference in the locations of two population distributions based on a paired-difference experiment consisting of $n = 30$ pairs.

a. Give the null and alternative hypotheses for the Wilcoxon signed-rank test.

b. Give the test statistic.

c. Give the rejection region for the test for $\alpha = 0.05$.

d. If $T^+ = 249$, what are your conclusions? (NOTE: $T^+ + T^- = n(n+1)/2$.)

15.22 Refer to Exercise 15.21. Suppose you wish to detect only a shift in distribution 1 to the right of distribution 2.

a. Give the null and alternative hypotheses for the Wilcoxon signed-rank test.

b. Give the test statistic.

c. Give the rejection region for the test for $\alpha = 0.05$.

d. If $T^+ = 249$, what are your conclusions? (NOTE: $T^+ + T^- = n(n+1)/2$.)

15.23 Refer to Exercise 15.21. Conduct the test using the large-sample z test. Compare your results with the nonparametric test results in Exercise 15.22, part d.

15.24 Refer to Exercise 15.22. Conduct the test using the large-sample z test. Compare your results with the nonparametric test results in Exercise 15.21, part d.

15.25 Refer to Exercise 15.16 and data set EX1516. The data in this table are from a paired-difference experiment with $n = 7$ pairs of observations:

	Pairs						
Population	1	2	3	4	5	6	7
1	8.9	8.1	9.3	7.7	10.4	8.3	7.4
2	8.8	7.4	9.0	7.8	9.9	8.1	6.9

a. Use Wilcoxon's signed-rank test to determine whether there is a significant difference between the two populations.

b. Compare the results of part a with the result you got in Exercise 15.16. Are they the same? Explain.

APPLICATIONS

15.26 Property Values II In Exercise 15.17, you used the sign test to determine whether the data provided sufficient evidence to indicate a difference in the distributions of property assessments for assessors A and B.

a. Use the Wilcoxon signed-rank test for a paired experiment to test the null hypothesis that there is no difference in the distributions of property assessments between assessors A and B. Test by using a value of α near 0.05.

b. Compare the conclusion of the test in part a with the conclusions derived from the t test in Exercise 10.53 and the sign test in Exercise 15.17. Explain why these test conclusions are (or are not) consistent.

Data set EX1527 **15.27 Machine Breakdowns** The number of machine breakdowns per month was recorded for nine months on two identical machines, A and B, used to make wire rope:

Month	A	B
1	3	7
2	14	12
3	7	9
4	10	15
5	9	12
6	6	6
7	13	12
8	6	5
9	7	13

a. Do the data provide sufficient evidence to indicate a difference in the monthly breakdown rates for the two machines? Test by using a value of α near 0.05.

b. Can you think of a reason the breakdown rates for the two machines might vary from month to month?

15.28 Gourmet Cooking II Refer to the comparison of gourmet meal ratings in Exercise 15.18, and use the Wilcoxon signed-rank test to determine whether the data provide sufficient evidence to indicate a difference in the ratings of the two gourmets. Test by using a value of α near 0.05. Compare the results of this test with the results of the sign test in Exercise 15.18. Are the test conclusions consistent?

Data set EX1529 **15.29 Traffic Control** Two methods for controlling traffic, A and B, were used at each of $n = 12$ intersections for a period of one week, and the numbers of accidents that occurred during this time period were recorded. The order of use (which method would be employed for the first week) was selected in a random manner. You want to know whether the data provide sufficient evidence to indicate a difference in the distributions of accident rates for traffic control methods A and B.

	Method			Method	
Intersection	A	B	Intersection	A	B
1	5	4	7	2	3
2	6	4	8	4	1
3	8	9	9	7	9
4	3	2	10	5	2
5	6	3	11	6	5
6	1	0	12	1	1

a. Analyze using a sign test.

b. Analyze using the Wilcoxon signed-rank test for a paired experiment.

Data set EX1530 **15.30 Jigsaw Puzzles** Eight people were asked to perform a simple puzzle-assembly task under normal conditions and under stressful conditions. During the stressful time, a mild shock was delivered to subjects 3 minutes after the start of the experiment and every 30 seconds thereafter until the task was completed. Blood pressure readings were taken under both conditions. The data in the table are the highest readings during the experiment. Do the data present sufficient evidence to indicate higher blood pressure readings under stressful conditions? Analyze the data using the Wilcoxon signed-rank test for a paired experiment.

Subject	Normal	Stressful
1	126	130
2	117	118
3	115	125
4	118	120
5	118	121
6	128	125
7	125	130
8	120	120

Data set EX1531 **15.31 Images and Word Recall** A psychology class performed an experiment to determine whether a recall score in which instructions to form images of 25 words were given differs from an initial recall score for which no imagery instructions were given. Twenty students participated in the experiment with the results listed in the table:

Student	With Imagery	Without Imagery	Student	With Imagery	Without Imagery
1	20	5	11	17	8
2	24	9	12	20	16
3	20	5	13	20	10
4	18	9	14	16	12
5	22	6	15	24	7
6	19	11	16	22	9
7	20	8	17	25	21
8	19	11	18	21	14
9	17	7	19	19	12
10	21	9	20	23	13

a. What three testing procedures can be used to test for differences in the distribution of recall scores with and without imagery? What assumptions are required for the parametric procedure? Do these data satisfy these assumptions?

b. Use both the sign test and the Wilcoxon signed-rank test to test for differences in the distributions of recall scores under these two conditions.

c. Compare the results of the tests in part b. Are the conclusions the same? If not, why not?

15.6 THE KRUSKAL–WALLIS *H* TEST FOR COMPLETELY RANDOMIZED DESIGNS

Just as the Wilcoxon rank sum test is the nonparametric alternative to Student's *t* test for a comparison of population means, the Kruskal–Wallis *H* test is the nonparametric alternative to the analysis of variance *F* test for a completely randomized design. It is used to detect differences in locations among more than two population distributions based on independent random sampling.

The procedure for conducting the Kruskal–Wallis *H* test is similar to that used for the Wilcoxon rank sum test. Suppose you are comparing *k* populations based on independent random samples n_1 from population 1, n_2 from population 2, …, n_k from population *k,* where

$$n_1+n_2 + \cdots + n_k = n$$

The first step is to rank all *n* observations from the smallest (rank 1) to the largest (rank *n*). Tied observations are assigned a rank equal to the average of the ranks they would have received if they had been nearly equal but not tied. You then calculate the rank sums $T_1, T_2, \ldots, T_k$ for the *k* samples and calculate the test statistic

$$H = \frac{12}{n(n+1)} \Sigma \frac{T_i^2}{n_i} - 3(n+1)$$

which is proportional to $\Sigma n_i(\overline{T}_i - \overline{T})^2$, the sum of squared deviations of the rank means about the grand mean $\overline{T} = n(n + 1)/2n = (n + 1)/2$. The greater the differences in locations among the k population distributions, the larger is the value of the H statistic. Thus, you can reject the null hypothesis that the k population distributions are identical for large values of H.

How large is large? It can be shown (proof omitted) that when the sample sizes are moderate to large—say, each sample size is equal to five or larger—and when H_0 is true, the H statistic will have approximately a chi-square distribution with $(k - 1)$ degrees of freedom. Therefore, for a given value of α, you can reject H_0 when the H statistic exceeds χ^2_α (see Figure 15.5).

FIGURE 15.5

Approximate distribution of the *H* statistic when H_0 is true

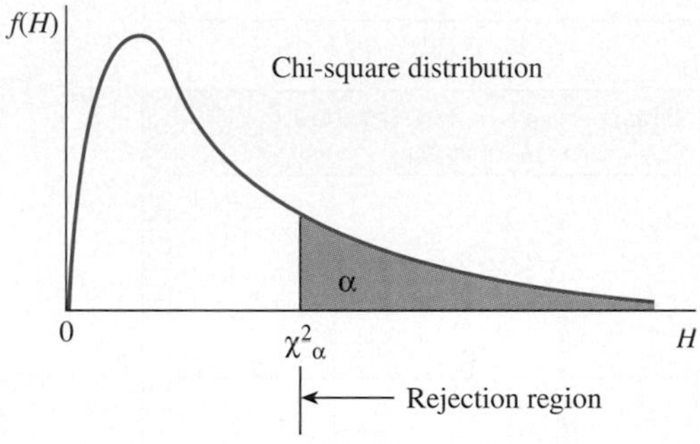

EXAMPLE 15.6

The data in Table 15.9 were collected using a completely randomized design. They are the achievement test scores for four different groups of students, each group taught by a different teaching technique. The objective of the experiment is to test the hypothesis of no difference in the population distributions of achievement test scores versus the alternative that they differ in location; that is, at least one of the distributions is shifted above the others. Conduct the test using the Kruskal–Wallis H test with $\alpha = 0.05$.

TABLE 15.9 **Test Scores (and Ranks) from Four Teaching Techniques**

	1	2	3	4
	65 (3)	75 (9)	59 (1)	94 (23)
	87 (19)	69 (5.5)	78 (11)	89 (21)
	73 (8)	83 (17.5)	67 (4)	80 (14)
	79 (12.5)	81 (15.5)	62 (2)	88 (20)
	81 (15.5)	72 (7)	83 (17.5)	
	69 (5.5)	79 (12.5)	76 (10)	
		90 (22)		
Rank sum	$T_1 = 63.5$	$T_2 = 89$	$T_3 = 45.5$	$T_4 = 78$

SOLUTION Before you perform a nonparametric analysis on these data, you can use a one-way analysis of variance to provide the two plots in Figure 15.6. It appears that technique 4 has a smaller variance than the other three and that there is a marked deviation in the right tail of the normal probability plot. These deviations could be considered minor and either a parametric or nonparametric analysis could be used.

FIGURE 15.6

A normal probability plot and a residual plot following a one-way analysis of variance for Example 15.6

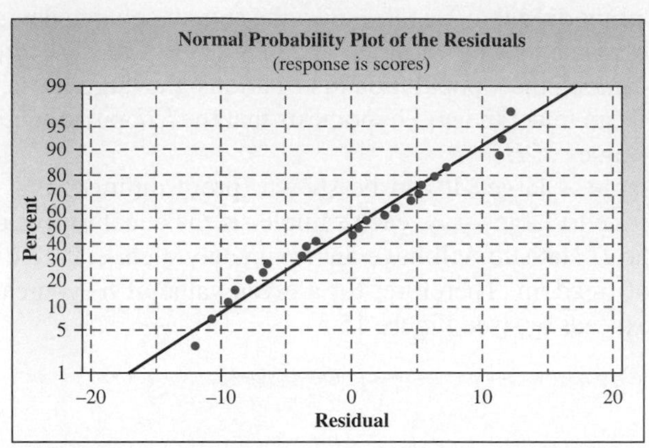

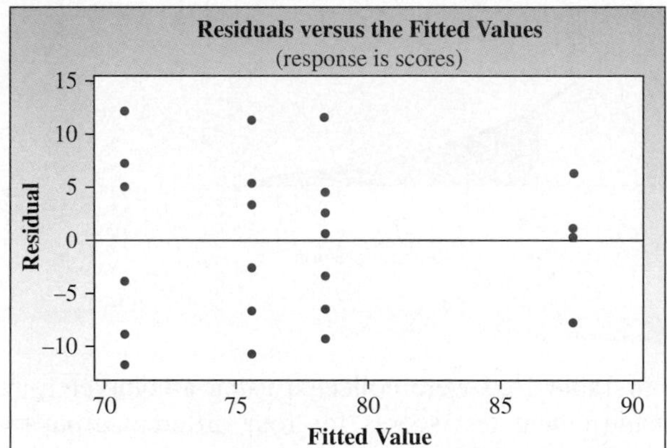

In the Kruskal–Wallis H test procedure, the first step is to rank the $n = 23$ observations from the smallest (rank 1) to the largest (rank 23). These ranks are shown in parentheses in Table 15.9. Notice how the ties are handled. For example, two observations at 69 are tied for rank 5. Therefore, they are assigned the average 5.5 of the two ranks (5 and 6) that they would have occupied if they had been slightly different. The rank sums T_1, T_2, T_3, and T_4 for the four samples are shown in the bottom row of the table. Substituting rank sums and sample sizes into the formula for the H statistic, you get

$$H = \frac{12}{n(n+1)} \Sigma \frac{T_i^2}{n_i} - 3(n+1)$$

$$= \frac{12}{23(24)} \left[\frac{(63.5)^2}{6} + \frac{(89)^2}{7} + \frac{(45.5)^2}{6} + \frac{(78)^2}{4} \right] - 3(24)$$

$$= 79.775102 - 72 = 7.775102$$

The rejection region for the H statistic for $\alpha = 0.05$ includes values of $H \geq \chi^2_{0.05}$, where $\chi^2_{0.05}$ is based on $(k - 1) = (4 - 1) = 3$ *df*. The value of χ^2 given in Table 5 in Appendix I is $\chi^2_{0.05} = 7.81473$. The observed value of the H statistic, $H = 7.775102$, does not fall into the rejection region for the test. Therefore, there is insufficient evidence to indicate differences in the distributions of achievement test scores for the four teaching techniques.

A *MINITAB* printout of the Kruskal–Wallis H test for these data is given in Figure 15.7. Notice that the p-value, 0.051, is only slightly greater than the 5% level necessary to declare statistical significance.

FIGURE 15.7
MINITAB printout for the Kruskal–Wallis test for Example 15.6

Kruskal–Wallis Test: Scores versus Techniques

```
Kruskal-Wallis Test on Scores

Techniques   N    Median   Ave Rank       Z
1            6     76.00       10.6   -0.60
2            7     79.00       12.7    0.33
3            6     71.50        7.6   -1.86
4            4     88.50       19.5    2.43
Overall     23                 12.0

H = 7.78  DF = 3  P = 0.051
H = 7.79  DF = 3  P = 0.051 (adjusted for ties)

* NOTE * One or more small samples
```

In passing, we would like to remark that Examples 15.6 and 15.7 illustrates that parametric methods are more statistically powerful than non-parametric methods if parametric methods are acceptable.

EXAMPLE 15.7

Compare the results of the analysis of variance F test and the Kruskal–Wallis H test for testing for differences in the distributions of achievement test scores for the four teaching techniques in Example 15.6.

SOLUTION The *MINITAB* printout for a one-way analysis of variance for the data in Table 15.9 is given in Figure 15.8. The analysis of variance shows that the F test for testing for differences among the means for the four techniques is significant at the 0.028 level. The Kruskal–Wallis H test did not detect a shift in population distributions at the 0.05 level of significance. Although these conclusions seem to be far apart, the test results do not differ strongly. The p-value $= 0.028$ corresponding to $F = 3.77$, with $df_1 = 3$ and $df_2 = 19$, is slightly less than 0.05, in contrast to the p-value $= 0.051$ for $H = 7.78$, $df = 3$, which is slightly greater than 0.05. Someone viewing the p-values for the two tests would see little difference in the results of the F and H tests. However, if you adhere to the choice of $\alpha = 0.05$, you cannot reject H_0 using the H test.

FIGURE 15.8
MINITAB printout for Example 15.7

One-way ANOVA: Scores versus Techniques

```
Source      DF      SS      MS     F      P
Techniques   3   712.6   237.5  3.77  0.028
Error       19  1196.6    63.0
Total       22  1909.2
```

THE KRUSKAL–WALLIS *H* TEST FOR COMPARING MORE THAN TWO POPULATIONS: COMPLETELY RANDOMIZED DESIGN (INDEPENDENT RANDOM SAMPLES)

1. Null hypothesis: H_0 : The k population distributions are identical.
2. Alternative hypothesis: H_a : At least two of the k population distributions differ in location.
3. Test statistic: $H = \dfrac{12}{n(n+1)} \Sigma \dfrac{T_i^2}{n_i} - 3(n+1)$

THE KRUSKAL–WALLIS *H* TEST FOR COMPARING MORE THAN TWO POPULATIONS: COMPLETELY RANDOMIZED DESIGN (INDEPENDENT RANDOM SAMPLES) *(continued)*

where

n_i = Sample size for population i
T_i = Rank sum for population i
n = Total number of observations
$= n_1 + n_2 + \ldots + n_k$

4. Rejection region for a given α: $H > \chi^2_\alpha$ with $(k - 1)$ df

Assumptions

- All sample sizes are greater than or equal to 5.
- Ties take on the average of the ranks that they would have occupied if they had not been tied.

The Kruskal–Wallis *H* test is a valuable alternative to a one-way analysis of variance when the normality and equality of variance assumptions are violated. Again, normal probability plots of residuals and plots of residuals per treatment group are helpful in determining whether these assumptions have been violated. Remember that a normal probability plot should appear as a straight line with a positive slope; residual plots per treatment groups should exhibit the same spread above and below the 0 line.

15.6 EXERCISES

BASIC TECHNIQUES

Data set EX1532 **15.32** Three treatments were compared using a completely randomized design. The data are shown in the table:

Treatment		
1	2	3
26	27	25
29	31	24
23	30	27
24	28	22
28	29	24
26	32	20
	30	21
	33	

Do the data provide sufficient evidence to indicate a difference in location for at least two of the population distributions? Test using the Kruskal–Wallis *H* statistic with $\alpha = 0.05$.

Data set EX1533 **15.33** Four treatments were compared using a completely randomized design. The data are shown here:

Treatment			
1	2	3	4
124	147	141	117
167	121	144	128
135	136	139	102
160	114	162	119
159	129	155	128
144	117	150	123
133	109		

Do the data provide sufficient evidence to indicate a difference in location for at least two of the population distributions? Test using the Kruskal–Wallis *H* statistic with $\alpha = 0.05$.

APPLICATIONS

15.34 Swampy Sites II Exercise 11.13 presents data (see data set EX1113) on the rates of growth of vegetation at four swampy underdeveloped sites. Six plants were randomly selected at each of the four sites to be used in the comparison. The data are the mean leaf length per plant (in centimetres) for a random sample of 10 leaves per plant:

Location	Mean Leaf Length (cm)					
1	5.7	6.3	6.1	6.0	5.8	6.2
2	6.2	5.3	5.7	6.0	5.2	5.5
3	5.4	5.0	6.0	5.6	4.9	5.2
4	3.7	3.2	3.9	4.0	3.5	3.6

a. Do the data present sufficient evidence to indicate differences in location for at least two of the distributions of mean leaf length corresponding to the four locations? Test using the Kruskal–Wallis *H* test with $\alpha = 0.05$.

b. Find the approximate *p*-value for the test.

c. You analyzed this same set of data in Exercise 11.13 using an analysis of variance. Find the *p*-value for the *F* test used to compare the four location means in Exercise 11.13.

d. Compare the *p*-values in parts b and c and explain the implications of the comparison.

15.35 Heart Rate and Exercise Exercise 11.58 presented data (data set EX1158) on the heart rates for samples of 10 men randomly selected from each of four age groups. Each man walked a treadmill at a fixed grade for a period of 12 minutes, and the increase in heart rate (the difference before and after exercise) was recorded (in beats per minute). The data are shown in the table:

	10–19	20–39	40–59	60–69
	29	24	37	28
	33	27	25	29
	26	33	22	34
	27	31	33	36
	39	21	28	21
	35	28	26	20
	33	24	30	25
	29	34	34	24
	36	21	27	33
	22	32	33	32
Total	309	275	295	282

a. Do the data present sufficient evidence to indicate differences in location for at least two of the four age groups? Test using the Kruskal–Wallis *H* test with $\alpha = 0.01$.

b. Find the approximate *p*-value for the test in part a.

c. Since the *F* test in Exercise 11.58 and the *H* test in part a are both tests to detect differences in location of the four heart-rate populations, how do the test results compare? Compare the *p*-values for the two tests and explain the implications of the comparison.

Data set EX1536 **15.36 pH Levels in Water** A sampling of the acidity of rain for 10 randomly selected rainfalls was recorded at three different locations in Canada: The prairies, the Atlantic region, and the West Coast. The pH readings for these 30 rainfalls are shown in the table. (NOTE: pH readings range from 0 to 14; 0 is acid, 14 is alkaline. Pure water falling through clean air has a pH reading of 5.7.)

Prairies	Atlantic	West Coast
4.45	4.60	4.55
4.02	4.27	4.31
4.13	4.31	4.84
3.51	3.88	4.67
4.42	4.49	4.28
3.89	4.22	4.95
4.18	4.54	4.72
3.95	4.76	4.63
4.07	4.36	4.36
4.29	4.21	4.47

a. Do the data present sufficient evidence to indicate differences in the levels of acidity in rainfalls in the three different locations? Test using the Kruskal–Wallis *H* test.

b. Find the approximate *p*-value for the test in part a and interpret it.

Data set EX1537 **15.37 Advertising Campaigns** The results of an experiment to investigate product recognition for three advertising campaigns were reported in Example 11.15. The responses were the percentage of 400 adults who were familiar with the newly advertised product. The normal probability plot indicated that the data were not approximately normal and another method of analysis should be used. Is there a significant difference among the three population distributions from which these samples came? Use an appropriate nonparametric method to answer this question.

Campaign		
1	2	3
0.33	0.28	0.21
0.29	0.41	0.30
0.21	0.34	0.26
0.32	0.39	0.33
0.25	0.27	0.31

15.7 THE FRIEDMAN F_r TEST FOR RANDOMIZED BLOCK DESIGNS

The Friedman F_r test, proposed by Nobel Prize–winning economist Milton Friedman, is a nonparametric test for comparing the distributions of measurements for k treatments laid out in b blocks using a randomized block design. The procedure for conducting the test is very similar to that used for the Kruskal–Wallis H test. The first step in the procedure is to rank the k treatment observations within each block. Ties are treated in the usual way; that is, they receive an average of the ranks occupied by the tied observations. The rank sums $T_1, T_2, \cdots, T_k$ are then obtained and the test statistic

$$F_r = \frac{12}{bk(k+1)} \Sigma T_i^2 - 3b(k+1)$$

is calculated. The value of the F_r statistic is at a minimum when the rank sums are equal—that is, $T_1 = T_2 = \cdots = T_k$—and increases in value as the differences among the rank sums increase. When either the number k of treatments or the number b of blocks is larger than five, the sampling distribution of F_r can be approximated by a chi-square distribution with $(k - 1)$ *df*. Therefore, as for the Kruskal–Wallis H test, the rejection region for the F_r test consists of values of F_r for which

$$F_r > \chi^2_\alpha$$

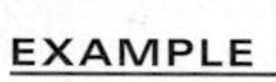

EXAMPLE 15.8

Suppose you wish to compare the reaction times of people exposed to six different stimuli. A reaction time measurement is obtained by subjecting a person to a stimulus and then measuring the time until the person presents some specified reaction. The objective of the experiment is to determine whether differences exist in the reaction times for the stimuli used in the experiment. To eliminate the person-to-person variation in reaction time, four persons participated in the experiment and each person's reaction time (in seconds) was measured for each of the six stimuli. The data are given in Table 15.10 (ranks of the observations are shown in parentheses). Use the Friedman F_r test to determine whether the data present sufficient evidence to indicate differences in the distributions of reaction times for the six stimuli. Test using $\alpha = 0.05$.

TABLE 15.10 Reaction Times to Six Stimuli

	Stimulus					
Subject	A	B	C	D	E	F
1	0.6 (2.5)	0.9 (6)	0.8 (5)	0.7 (4)	0.5 (1)	0.6 (2.5)
2	0.7 (3.5)	1.1 (6)	0.7 (3.5)	0.8 (5)	0.5 (1.5)	0.5 (1.5)
3	0.9 (3)	1.3 (6)	1.0 (4.5)	1.0 (4.5)	0.7 (1)	0.8 (2)
4	0.5 (2)	0.7 (5)	0.8 (6)	0.6 (3.5)	0.4 (1)	0.6 (3.5)
Rank sum	$T_1 = 11$	$T_2 = 23$	$T_3 = 19$	$T_4 = 17$	$T_5 = 4.5$	$T_6 = 9.5$

SOLUTION In Figure 15.9, the plot of the residuals for each of the six stimuli reveals that stimuli 1, 4, and 5 have variances somewhat smaller than the other stimuli. Furthermore, the normal probability plot of the residuals reveals a change in the slope of the line following the first three residuals, as well as curvature in the upper portion of the plot. It appears that a nonparametric analysis is appropriate for these data.

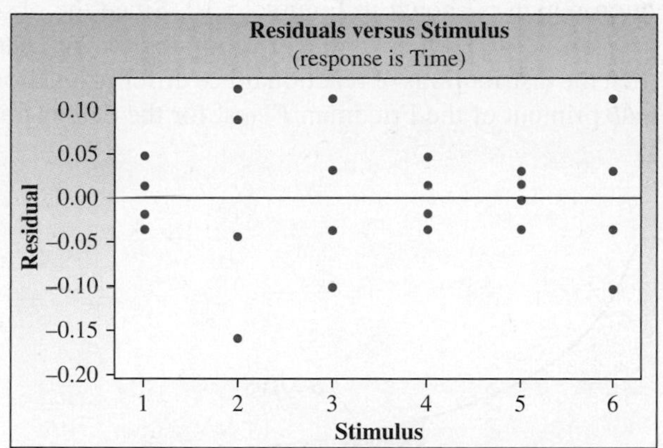

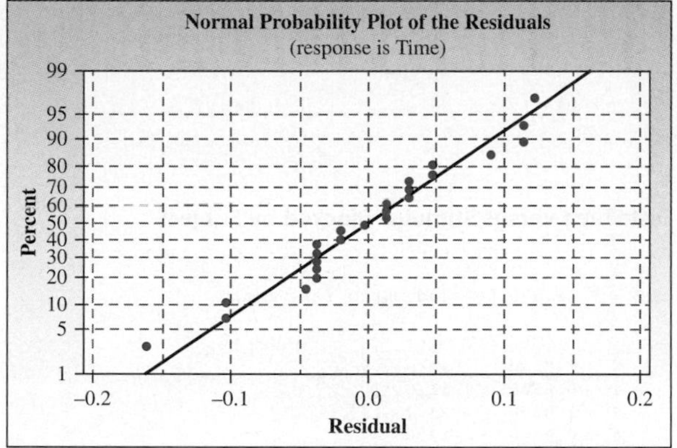

FIGURE 15.9
A plot of treatments versus residuals and a normal probability plot of residuals for Example 15.8

You wish to test

H_0 : The distributions of reaction times for the six stimuli are identical

versus the alternative hypothesis

H_a : At least two of the distributions of reaction times for the six stimuli differ in location

Table 15.10 shows the ranks (in parentheses) of the observations within each block and the rank sums for each of the six stimuli (the treatments). The value of the F_r statistic for these data is

$$F_r = \frac{12}{bk(k+1)}\Sigma T_i^2 - 3b(k+1)$$
$$= \frac{12}{(4)(6)(7)}[(11)^2 + (23)^2 + (19)^2 + \cdots + (9.5)^2] - 3(4)(7)$$
$$= 100.75 - 84 = 16.75$$

Since the number $k = 6$ of treatments exceeds 5, the sampling distribution of F_r can be approximated by a chi-square distribution with $(k - 1) = (6 - 1) = 5$ *df*. Therefore, for $\alpha = 0.05$, you can reject H_0 if

$$F_r > \chi^2_{0.05} \qquad \text{where } \chi^2_{0.05} = 11.0705$$

This rejection region is shown in Figure 15.10. Since the observed value $F_r = 16.75$ exceeds $\chi^2_{0.05} = 11.0705$, it falls in the rejection region. You can therefore reject H_0 and conclude that the distributions of reaction times differ in location for at least two stimuli. The *MINITAB* printout of the Friedman F_r test for the data is given in Figure 15.11.

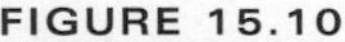

FIGURE 15.10
Rejection region for Example 15.8

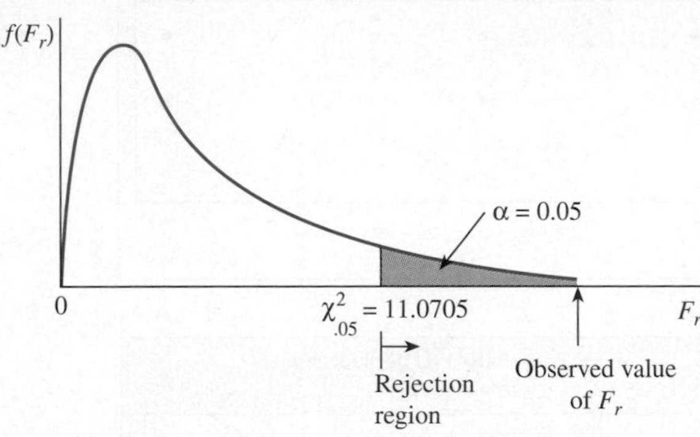

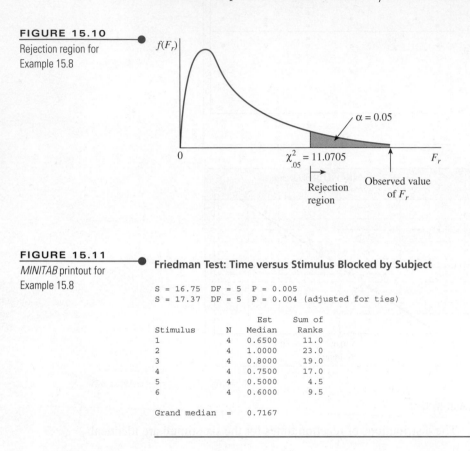

FIGURE 15.11
MINITAB printout for Example 15.8

Friedman Test: Time versus Stimulus Blocked by Subject

```
S = 16.75  DF = 5  P = 0.005
S = 17.37  DF = 5  P = 0.004 (adjusted for ties)

                     Est    Sum of
Stimulus       N  Median     Ranks
1              4  0.6500      11.0
2              4  1.0000      23.0
3              4  0.8000      19.0
4              4  0.7500      17.0
5              4  0.5000       4.5
6              4  0.6000       9.5

Grand median   =  0.7167
```

EXAMPLE 15.9

Find the approximate p-value for the test in Example 15.8.

SOLUTION Consulting Table 5 in Appendix I with 5 *df*, you find that the observed value of $F_r = 16.75$ exceeds the table value $\chi^2_{0.005} = 16.7496$. Hence, the p-value is very close to, but slightly less than, 0.005.

THE FRIEDMAN F_r TEST FOR A RANDOMIZED BLOCK DESIGN

1. Null hypothesis: H_0 : The k population distributions are identical.
2. Alternative hypothesis: H_a : At least two of the k population distributions differ in location.
3. Test statistic: $F_r = \dfrac{12}{bk(k+1)}\sum T_i^2 - 3b(k+1)$

where

b = Number of blocks

k = Number of treatments

T_i = Rank sum for treatment i, $i = 1, 2, \ldots, k$

4. Rejection region: $F_r > \chi^2_\alpha$, where χ^2_α is based on $(k - 1)$ df

Assumption: Either the number k of treatments or the number b of blocks is greater than five.

15.7 EXERCISES

BASIC TECHNIQUES

Data set EX1538 **15.38** A randomized block design is used to compare three treatments in six blocks:

Block	Treatment 1	2	3
1	3.2	3.1	2.4
2	2.8	3.0	1.7
3	4.5	5.0	3.9
4	2.5	2.7	2.6
5	3.7	4.1	3.5
6	2.4	2.4	2.0

a. Use the Friedman F_r test to detect differences in location among the three treatment distributions. Test using $\alpha = 0.05$.

b. Find the approximate p-value for the test in part a.

c. Perform an analysis of variance and give the ANOVA table for the analysis.

d. Give the value of the F statistic for testing the equality of the three treatment means.

e. Give the approximate p-value for the F statistic in part d.

f. Compare the p-values for the tests in parts a and d, and explain the practical implications of the comparison.

Data set EX1539 **15.39** A randomized block design is used to compare four treatments in eight blocks:

Block	Treatment 1	2	3	4
1	89	81	84	85
2	93	86	86	88
3	91	85	87	86
4	85	79	80	82
5	90	84	85	85
6	86	78	83	84
7	87	80	83	82
8	93	86	88	90

a. Use the Friedman F_r test to detect differences in location among the four treatment distributions. Test using $\alpha = 0.05$.

b. Find the approximate p-value for the test in part a.

c. Perform an analysis of variance and give the ANOVA table for the analysis.

d. Give the value of the F statistic for testing the equality of the four treatment means.

e. Give the approximate p-value for the F statistic in part d.

f. Compare the p-values for the tests in parts a and d, and explain the practical implications of the comparison.

APPLICATIONS

Data set EX1540 **15.40 Supermarket Prices** In a comparison of the prices of items at five supermarkets, six items were randomly selected and the price of each was recorded for each of the five supermarkets. The objective of the study was to see whether the data indicated differences in the levels of prices among the five supermarkets. The prices (in dollars) are listed in the table:

Item	Loblaws	IGA	Zehrs	Food Basics	No Frills
Celery	0.33	0.34	0.69	0.59	0.58
Colgate toothpaste	1.28	1.49	1.44	1.37	1.28
Campbell's beef soup	1.05	1.19	1.23	1.19	1.10
Crushed pineapple	0.83	0.95	0.95	0.87	0.84
Spaghetti	0.68	0.79	0.83	0.69	0.69
Heinz ketchup	1.41	1.69	1.79	1.65	1.49

a. Does the distribution of the prices differ from one supermarket to another? Test using the Friedman F_r test with $\alpha = 0.05$.

b. Find the approximate p-value for the test and interpret it.

Data set EX1541 **15.41 Toxic Chemicals** An experiment was conducted to compare the effects of three toxic chemicals, A, B, and C, on the skin of rats. One-centimetre squares of skin were treated with the chemicals and then scored from 0 to 10 depending on the degree of irritation. Three adjacent 1-centimetre squares were marked on the backs of eight rats, and each of the three chemicals was applied to each rat. Thus, the experiment was blocked on rats to eliminate the variation in skin sensitivity from rat to rat.

Rats							
1	2	3	4	5	6	7	8
B	A	A	C	B	C	C	B
5	9	6	6	8	5	5	7
A	C	B	B	C	A	B	A
6	4	9	8	8	5	7	6
C	B	C	A	A	B	A	C
3	9	3	5	7	7	6	7

a. Do the data provide sufficient evidence to indicate a difference in the toxic effects of the three chemicals? Test using the Friedman F_r test with $\alpha = 0.05$.

b. Find the approximate p-value for the test and interpret it.

Data set EX1542 **15.42 Good Tasting Medicine** In a study of the palatability of antibiotics in children, Dr. Doreen Matsui and colleagues used a voluntary sample of healthy children to assess their reactions to the taste of four antibiotics.[5] The children's response was measured on a 10-centimetre (cm) visual analogue scale incorporating the use of faces, from sad (low score) to happy (high score). The minimum score was 0 and the maximum was 10. For the accompanying data (simulated from the results of Matsui's report), each of five children was asked to taste each of four antibiotics and rate them using the visual (faces) analogue scale from 0 to 10 cm.

	Antibiotic			
Child	1	2	3	4
1	4.8	2.2	6.8	6.2
2	8.1	9.2	6.6	9.6
3	5.0	2.6	3.6	6.5
4	7.9	9.4	5.3	8.5
5	3.9	7.4	2.1	2.0

a. What design is used in collecting these data?

b. Using an appropriate statistical package for a two-way classification, produce a normal probability plot of the residuals as well as a plot of residuals versus antibiotics. Do the usual analysis of variance assumptions appear to be satisfied?

c. Use the appropriate nonparametric test to test for differences in the distributions of responses to the tastes of the four antibiotics.

d. Comment on the results of the analysis of variance in part b compared with the nonparametric test in part c.

15.8 RANK CORRELATION COEFFICIENT

In the preceding sections, we used ranks to indicate the relative magnitude of observations in nonparametric tests for the comparison of treatments. We will now use the same technique in testing for a relationship between two ranked variables. Two common rank correlation coefficients are the **Spearman r_s** and the **Kendall τ**. We will present the Spearman r_s because its computation is identical to that for the sample correlation coefficient r of Chapters 3 and 12.

Suppose eight elementary school science teachers have been ranked by a judge according to their teaching ability and all have taken a "national teachers' examination." The data are listed in Table 15.11. Do the data suggest an agreement between the judge's ranking and the examination score? That is, is there a correlation between ranks and test scores?

TABLE 15.11 **Ranks and Test Scores for Eight Teachers**

Teacher	Judge's Rank	Examination Score
1	7	44
2	4	72
3	2	69
4	6	70
5	1	93
6	3	82
7	8	67
8	5	80

The two variables of interest are rank and test score. The former is already in rank form, and the test scores can be ranked similarly, as shown in Table 15.12. The ranks for tied observations are obtained by averaging the ranks that the tied observations would have had if no ties had been observed. The Spearman rank correlation coefficient r_s is calculated by using the ranks of the paired measurements on the two variables x and y in the formula for r (see Chapter 12).

TABLE 15.12 **Ranks of Data in Table 15.11**

Teacher	Judge's Rank, x_i	Test Rank, y_i
1	7	1
2	4	5
3	2	3
4	6	4
5	1	8
6	3	7
7	8	2
8	5	6

SPEARMAN'S RANK CORRELATION COEFFICIENT

$$r_s = \frac{S_{xy}}{\sqrt{S_{xx}S_{yy}}}$$

where x_i and y_i represent the ranks of the ith pair of observations and

$$S_{xy} = \sum (x_i - \bar{x})(y_i - \bar{y}) = \sum x_i y_i - \frac{(\sum x_i)(\sum y_i)}{n}$$

$$S_{xx} = \sum (x_i - \bar{x})^2 = \sum x_i^2 - \frac{(\sum x_i)^2}{n}$$

$$S_{yy} = \sum (y_i - \bar{y})^2 = \sum y_i^2 - \frac{(\sum y_i)^2}{n}$$

When there are no ties in either the x observations or the y observations, the expression for r_s algebraically reduces to the simpler expression

$$r_s = 1 - \frac{6\sum d_i^2}{n(n^2 - 1)} \quad \text{where } d_i = (x_i - y_i)$$

If the number of ties is small in comparison with the number of data pairs, little error results in using this shortcut formula.

EXAMPLE 15.10

Calculate r_s for the data in Table 15.12.

SOLUTION The differences and squares of differences between the two rankings are provided in Table 15.13. Substituting values into the formula for r_s, you have

$$r_s = 1 - \frac{6\sum d_i^2}{n(n^2 - 1)}$$

$$= 1 - \frac{6(144)}{8(64 - 1)} = -0.714$$

TABLE 15.13 Differences and Squares of Differences for the Teacher Ranks

Teacher	x_i	y_i	d_i	d_i^2
1	7	1	6	36
2	4	5	−1	1
3	2	3	−1	1
4	6	4	2	4
5	1	8	−7	49
6	3	7	−4	16
7	8	2	6	36
8	5	6	−1	1
Total				144

The Spearman rank correlation coefficient can be used as a test statistic to test the hypothesis of no association between two populations. You can assume that the n pairs of observations (x_i, y_i) have been randomly selected and, therefore, no association between the populations implies a random assignment of the n ranks within each sample. Each random assignment (for the two samples) represents a simple event associated with the experiment, and a value of r_s can be calculated for each. Thus, it is possible to calculate the probability that r_s assumes a large absolute value due solely to chance and thereby suggests an association between populations when none exists.

The rejection region for a two-tailed test is shown in Figure 15.12. If the alternative hypothesis is that the correlation between x and y is negative, you would reject H_0 for negative values of r_s that are close to -1 (in the lower tail of Figure 15.12). Similarly, if the alternative hypothesis is that the correlation between x and y is positive, you would reject H_0 for large positive values of r_s (in the upper tail of Figure 15.12).

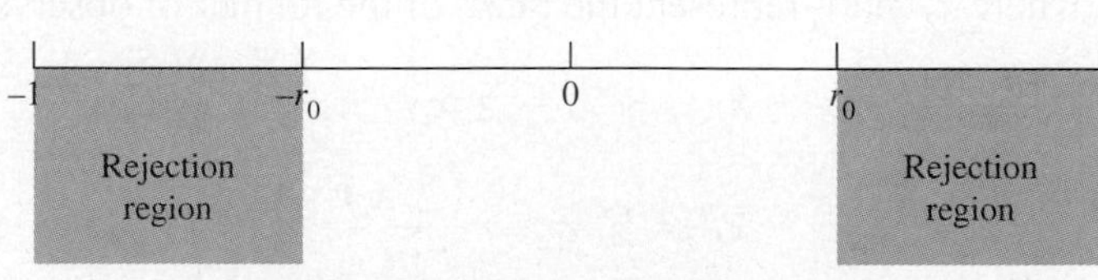

FIGURE 15.12 Rejection region for a two-tailed test of the null hypothesis of no association, using Spearman's rank correlation test

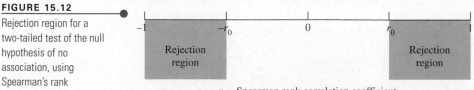

The critical values of r_s are given in Table 9 in Appendix I. An abbreviated version is shown in Table 15.14. Across the top of Table 15.14 (and Table 9 in Appendix I) are the recorded values of α that you might wish to use for a one-tailed test of the null hypothesis of no association between x and y. The number of rank pairs n appears at the left side of the table. The table entries give the critical value r_0 for a one-tailed test. Thus, $P(r_s \geq r_0) = \alpha$.

For example, suppose you have $n = 8$ rank pairs and the alternative hypothesis is that the correlation between the ranks is positive. You would want to reject the null hypothesis of no association for only large positive values of r_s, and you would use a one-tailed test. Referring to Table 15.14 and using the row corresponding to $n = 8$ and the column for $\alpha = 0.05$, you read $r_0 = 0.643$. Therefore, you can reject H_0 for all values of r_s greater than or equal to 0.643.

The test is conducted in exactly the same manner if you wish to test only the alternative hypothesis that the ranks are negatively correlated. The only difference is that you would reject the null hypothesis if $r_s \leq -0.643$. That is, you use the negative of the tabulated value of r_0 to get the lower-tail critical value.

TABLE 15.14 **An Abbreviated Version of Table 9 in Appendix I, for Spearman's Rank Correlation Test**

n	$\alpha = 0.05$	$\alpha = 0.025$	$\alpha = 0.01$	$\alpha = 0.005$
5	0.900	—	—	—
6	0.829	0.886	0.943	—
7	0.714	0.786	0.893	—
8	0.643	0.738	0.833	0.881
9	0.600	0.683	0.783	0.833
10	0.564	0.648	0.745	0.794
11	0.523	0.623	0.736	0.818
12	0.497	0.591	0.703	0.780
13	0.475	0.566	0.673	0.745
14	0.457	0.545		
15	0.441	0.525		
16	0.425			
17	0.412			
18	0.399			
19	0.388			
20	0.377			

To conduct a two-tailed test, you reject the null hypothesis if $r_s \geq r_0$ or $r_s \leq -r_0$. The value of α for the test is double the value shown at the top of the table. For example, if $n = 8$ and you choose the 0.025 column, you will reject H_0 if $r_s \geq 0.738$ or $r_s \leq -0.738$. The α-value for the test is $2(0.025) = 0.05$.

SPEARMAN'S RANK CORRELATION TEST

1. Null hypothesis: H_0 : There is no association between the rank pairs.
2. Alternative hypothesis: H_a : There is an association between the rank pairs (a two-tailed test). Or H_a : The correlation between the rank pairs is positive or negative (a one-tailed test).
3. Test statistic: $r_s = \dfrac{S_{xy}}{\sqrt{S_{xx}S_{yy}}}$

   where x_i and y_i represent the ranks of the ith pair of observations.
4. Rejection region: For a two-tailed test, reject H_0 if $r_s \geq r_0$ or $r_s \leq -r_0$, where r_0 is given in Table 9 in Appendix I. Double the tabulated probability to obtain the value of α for the two-tailed test. For a one-tailed test, reject H_0 if $r_s \geq r_0$ (for an upper-tailed test) or $r_s \leq -r_0$ (for a lower-tailed test). The α-value for a one-tailed test is the value shown in Table 9 in Appendix I.

EXAMPLE 15.11 Test the hypothesis of no association between the populations for Example 15.10.

SOLUTION The critical value of r_s for a one-tailed test with $\alpha = 0.05$ and $n = 8$ is 0.643. You may assume that a correlation between the judge's rank and the teachers' test scores could not possibly be positive. (A low rank means good teaching and should be associated with a high test score if the judge and the test measure teaching ability.) The alternative hypothesis is that the **population rank correlation coefficient** ρ_s is less than 0, and you are concerned with a one-tailed statistical test. Thus, α for the test is the tabulated value for 0.05, and you can reject the null hypothesis if $r_s \le -0.643$.

The calculated value of the test statistic, $r_s = -0.714$, is less than the critical value for $\alpha = 0.05$. Hence, the null hypothesis is rejected at the $\alpha = 0.05$ level of significance. It appears that some agreement does exist between the judge's rankings and the test scores. However, it should be noted that this agreement could exist when *neither* provides an adequate yardstick for measuring teaching ability. For example, the association could exist if both the judge and those who constructed the teachers' examination had a completely erroneous, but similar, concept of the characteristics of good teaching.

What exactly does r_s measure? Spearman's correlation coefficient detects not only a linear relationship between two variables but also any other monotonic relationship (either y increases as x increases or y decreases as x increases). For example, if you calculated r_s for the two data sets in Table 15.15, both would produce a value of $r_s = 1$ because the assigned ranks for x and y in both cases agree for all pairs (x, y). It is important to remember that a significant value of r_s indicates a relationship between x and y that is either increasing or decreasing, but is not necessarily linear.

TABLE 15.15 Twin Data Sets with $r_s = 1$

x	$y = x^2$	x	$y = \log_{10}(x)$
1	1	10	1
2	4	100	2
3	9	1000	3
4	16	10,000	4
5	25	100,000	5
6	36	1,000,000	6

15.8 EXERCISES

BASIC TECHNIQUES

15.43 Give the rejection region for a test to detect positive rank correlation if the number of pairs of ranks is 16 and you have these α-values:

a. $\alpha = 0.05$ **b.** $\alpha = 0.01$

15.44 Give the rejection region for a test to detect negative rank correlation if the number of pairs of ranks is 12 and you have these α-values:

a. $\alpha = 0.05$ **b.** $\alpha = 0.01$

15.45 Give the rejection region for a test to detect rank correlation if the number of pairs of ranks is 25 and you have these α-values:

a. $\alpha = 0.05$ **b.** $\alpha = 0.01$

15.46 The following paired observations were obtained on two variables x and y:

x	1.2	0.8	2.1	3.5	2.7	1.5
y	1.0	1.3	0.1	−0.8	−0.2	0.6

a. Calculate Spearman's rank correlation coefficient r_s.

b. Do the data present sufficient evidence to indicate a correlation between x and y? Test using $\alpha = 0.05$.

APPLICATIONS

Data set EX1547 **15.47 Rating Political Candidates** A political scientist wished to examine the relationship between the voter image of a conservative political candidate and the distance (in kilometres) between the residences of the voter and the candidate. Each of 12 voters rated the candidate on a scale of 1 to 20:

Voter	Rating	Distance
1	12	75
2	7	165
3	5	300
4	19	15
5	17	180
6	12	240
7	9	120
8	18	60
9	3	230
10	8	200
11	15	130
12	4	130

a. Calculate Spearman's rank correlation coefficient r_s.

b. Do these data provide sufficient evidence to indicate a negative correlation between rating and distance?

Data set EX1548 **15.48 Competitive Running** Is the number of years of competitive running experience related to a runner's distance running performance? The data on nine runners, obtained from the study by Scott Powers and colleagues, are shown in the table:[6]

Runner	Years of Competitive Running	10-Kilometre Finish Time (min)
1	9	33.15
2	13	33.33
3	5	33.50
4	7	33.55
5	12	33.73
6	6	33.86
7	4	33.90
8	5	34.15
9	3	34.90

a. Calculate the rank correlation coefficient between years of competitive running x and a runner's finish time y in the 10-kilometre race.

b. Do the data provide sufficient evidence to indicate a rank correlation between y and x? Test using $\alpha = 0.05$.

Data set EX1549 **15.49 Tennis Racquets** The data shown in the accompanying table give measures of bending stiffness and twisting stiffness as determined by engineering tests on 12 tennis racquets:

Racquet	Bending Stiffness	Twisting Stiffness
1	419	227
2	407	231
3	363	200
4	360	211
5	257	182
6	622	304
7	424	384
8	359	194
9	346	158
10	556	225
11	474	305
12	441	235

a. Calculate the rank correlation coefficient r_s between bending stiffness and twisting stiffness.

b. If a racquet has bending stiffness, is it also likely to have twisting stiffness? Use the rank correlation coefficient to determine whether there is a significant positive relationship between bending stiffness and twisting stiffness. Use $\alpha = 0.05$.

15.50 Student Ratings A school principal suspected that a teacher's attitude toward a Grade 1 student depended on his original judgment of the child's ability. The principal also suspected that much of that judgment was based on the Grade 1 student's IQ score, which was usually known to the teacher. After three weeks of teaching, a teacher was asked to rank the nine children in his class from 1 (highest) to 9 (lowest) as to his opinion of their ability. Calculate r_s for these teacher–IQ ranks:

Teacher	1	2	3	4	5	6	7	8	9
IQ	3	1	2	4	5	7	9	6	8

15.51 Student Ratings, continued Refer to Exercise 15.50. Do the data provide sufficient evidence to indicate a positive correlation between the teacher's ranks and the ranks of the IQs? Use $\alpha = 0.05$.

Data set EX1552 **15.52 Art Critics** Two art critics each ranked 10 paintings by contemporary (but anonymous) artists in accordance with their appeal to the respective critics. The ratings are shown in the table. Do the critics seem to agree on their ratings of contemporary art? That is, do the data provide sufficient evidence to indicate a positive correlation between critics A and B? Test by using α value of a near 0.05.

Painting	Critic A	Critic B
1	6	5
2	4	6
3	9	10
4	1	2
5	2	3
6	7	8
7	3	1
8	8	7
9	5	4
10	10	9

Data set EX1553 **15.53 Rating Tobacco Leaves** An experiment was conducted to study the relationship between the ratings of a tobacco leaf grader and the moisture content of the tobacco leaves. Twelve leaves were rated by the grader on a scale of 1 to 10, and corresponding readings of moisture content were made:

Leaf	Grader's Rating	Moisture Content
1	9	0.22
2	6	0.16
3	7	0.17
4	7	0.14
5	5	0.12
6	8	0.19
7	2	0.10
8	6	0.12
9	1	0.05
10	10	0.20
11	9	0.16
12	3	0.09

Calculate r_s. Do the data provide sufficient evidence to indicate an association between the grader's ratings and the moisture contents of the leaves?

Data set EX1554 **15.54 Social Skills Training** A social skills training program was implemented with seven mildly challenged students in a study to determine whether the program caused improvements in pre/post measures and behaviour ratings. For one such test, the pre- and posttest scores for the seven students are given in the table:

Student	Pretest	Posttest
Earl	101	113
Ned	89	89
Jasper	112	121
Charlie	105	99
Tom	90	104
Susie	91	94
Lori	89	99

a. Use a nonparametric test to determine whether there is a significant positive relationship between the pre- and posttest scores.

b. Do these results agree with the results of the parametric test in Exercise 12.58?

15.9 SUMMARY

The nonparametric tests presented in this chapter are only a few of the many nonparametric tests available to experimenters. The tests presented here are those for which tables of critical values are readily available.

Nonparametric statistical methods are especially useful when the observations can be rank ordered but cannot be located exactly on a measurement scale. Also, nonparametric methods are the only methods that can be used when the sampling designs have been correctly adhered to, but the data are not or cannot be assumed to follow the prescribed one or more distributional assumptions.

We have presented a wide array of nonparametric techniques that can be used when either the data are not normally distributed or the other required assumptions are not met. One-sample procedures are available in the literature; however, we have concentrated on analyzing two or more samples that have been properly selected using random and independent sampling as required by the design involved. The nonparametric analogues of the parametric procedures presented in Chapters 10–14 are straightforward and fairly simple to implement:

- The Wilcoxon rank sum test is the nonparametric analogue of the two-sample t test.

- The sign test and the Wilcoxon signed-rank tests are the nonparametric analogues of the paired-sample t test.
- The Kruskal–Wallis H test is the rank equivalent of the one-way analysis of variance F test.
- The Friedman F_r test is the rank equivalent of the randomized block design two-way analysis of variance F test.
- Spearman's rank correlation r_s is the rank equivalent of Pearson's correlation coefficient.

These and many more nonparametric procedures are available as alternatives to the parametric tests presented earlier. It is important to keep in mind that when the assumptions required of the sampled populations are relaxed, our ability to detect significant differences in one or more population characteristics is decreased.

CHAPTER REVIEW

Key Concepts and Formulas

I. Nonparametric Methods

1. These methods can be used when the data cannot be measured on a quantitative scale, or when
2. The numerical scale of measurement is arbitrarily set by the researcher, or when
3. The parametric assumptions such as normality or constant variance are seriously violated.

II. Wilcoxon Rank Sum Test: Independent Random Samples

1. Jointly rank the two samples. Designate the smaller sample as sample 1. Then

$$T_1 = \text{Rank sum of sample 1}$$

$$T_1^* = n_1(n_1 + n_2 + 1) - T_1$$

2. Use T_1 to test for population 1 to the left of population 2. Use T_1^* to test for population 1 to the right of population 2. Use the smaller of T_1 and T_1^* to test for a difference in the locations of the two populations.
3. Table 7 of Appendix I has critical values for the rejection of H_0.
4. When the sample sizes are large, use the normal approximation:

$$\mu_T = \frac{n_1(n_1 + n_2 + 1)}{2}$$

$$\sigma_T^2 = \frac{n_1 n_2(n_1 + n_2 + 1)}{12}$$

$$z = \frac{T - \mu_T}{\sigma_T}$$

III. Sign Test for a Paired Experiment

1. Find x, the number of times that observation A exceeds observation B for a given pair.
2. To test for a difference in two populations, test $H_0 : P = 0.5$ versus a one- or two-tailed alternative.
3. Use Table 1 of Appendix I to calculate the p-value for the test.
4. When the sample sizes are large, use the normal approximation:

$$z = \frac{x - 0.5n}{0.5\sqrt{n}}$$

IV. Wilcoxon Signed-Rank Test: Paired Experiment

1. Calculate the differences in the paired observations. Rank the *absolute values* of the differences. Calculate the rank sums T^+ and T^- for the positive and negative differences, respectively. The test statistic T is the smaller of the two rank sums.
2. Table 8 in Appendix I has critical values for the rejection of H_0 for both one- and two-tailed tests.
3. When the sample sizes are large, use the normal approximation:

$$z = \frac{T^+ - [n(n + 1)/4]}{\sqrt{[n(n + 1)(2n + 1)]/24}}$$

V. Kruskal–Wallis H Test: Completely Randomized Design

1. Jointly rank the n observations in the k samples. Calculate the rank sums, T_i = rank sum of sample i, and the test statistic

$$H = \frac{12}{n(n+1)} \sum \frac{T_i^2}{n_i} - 3(n+1)$$

2. If the null hypothesis of equality of distributions is false, H will be unusually large, resulting in a one-tailed test.
3. For sample sizes of five or greater, the rejection region for H is based on the chi-square distribution with $(k - 1)$ degrees of freedom.

VI. The Friedman F_r Test: Randomized Block Design

1. Rank the responses within each block from 1 to k. Calculate the rank sums, $T_1, T_2, \cdots, T_k$, and the test statistic

$$F_r = \frac{12}{bk(k+1)} \sum T_i^2 - 3b(k+1)$$

2. If the null hypothesis of equality of treatment distributions is false, F_r will be unusually large, resulting in a one-tailed test.
3. For block sizes of five or greater, the rejection region for F_r is based on the chi-square distribution with $(k - 1)$ degrees of freedom.

VII. Spearman's Rank Correlation Coefficient

1. Rank the responses for the two variables from smallest to largest.
2. Calculate the correlation coefficient for the ranked observations:

$$r_s = \frac{S_{xy}}{\sqrt{S_{xx}S_{yy}}} \quad \text{or} \quad r_s = 1 - \frac{6\sum d_i^2}{n(n^2-1)}$$

if there are no ties

3. Table 9 in Appendix I gives critical values for rank correlations significantly different from 0.
4. The rank correlation coefficient detects not only significant linear correlation but also any other monotonic relationship between the two variables.

Nonparametric Procedures

Although these are not options for nonparametric procedures in *Excel*, many nonparametric procedures are available in the *MINITAB* package, including most of the tests discussed in this chapter. The dialogue boxes are all familiar to you by now, and we will discuss the tests in the order presented in the chapter.

To implement the Wilcoxon rank sum test for two independent random samples, enter the two sets of sample data into two columns (say, C1 and C2) of the *MINITAB* worksheet. The dialogue box in Figure 15.13 is generated using **Stat** ▶ **Nonparametrics** ▶ **Mann-Whitney.** Select C1 and C2 for the **First** and **Second Samples,** and indicate the appropriate confidence coefficient (for a confidence interval) and alternative hypothesis. Clicking **OK** will generate the output in Figure 15.1.

The sign test *and* the Wilcoxon signed-rank test for paired samples are performed in exactly the same way, with a change only in the last command of the sequence. Even the dialogue boxes are identical! Enter the data into two columns of the *MINITAB* worksheet (we used the cake mix data in Example 15.5). Before you can implement either test, you must generate a column of differences using **Calc** ▶ **Calculator,** as shown in Figure 15.14. Use **Stat** ▶ **Nonparametrics** ▶ **1-Sample Sign** or **Stat** ▶ **Nonparametrics** ▶ **1-Sample Wilcoxon** to generate the appropriate dialogue box

FIGURE 15.13

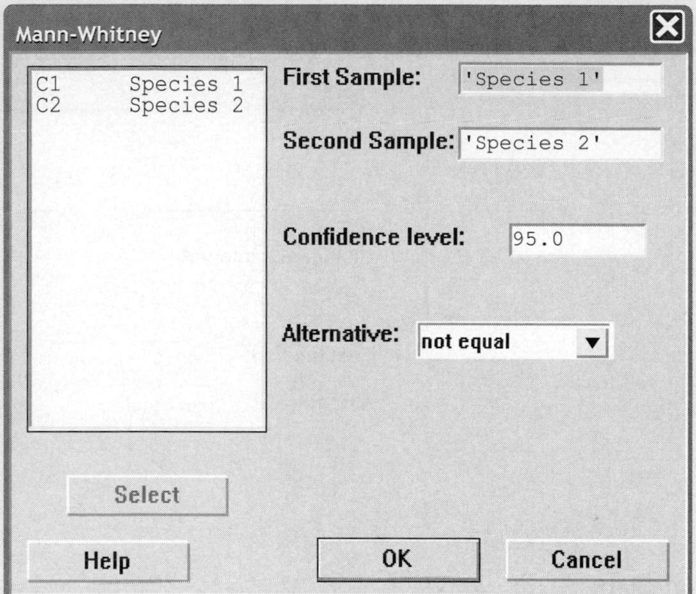

FIGURE 15.14

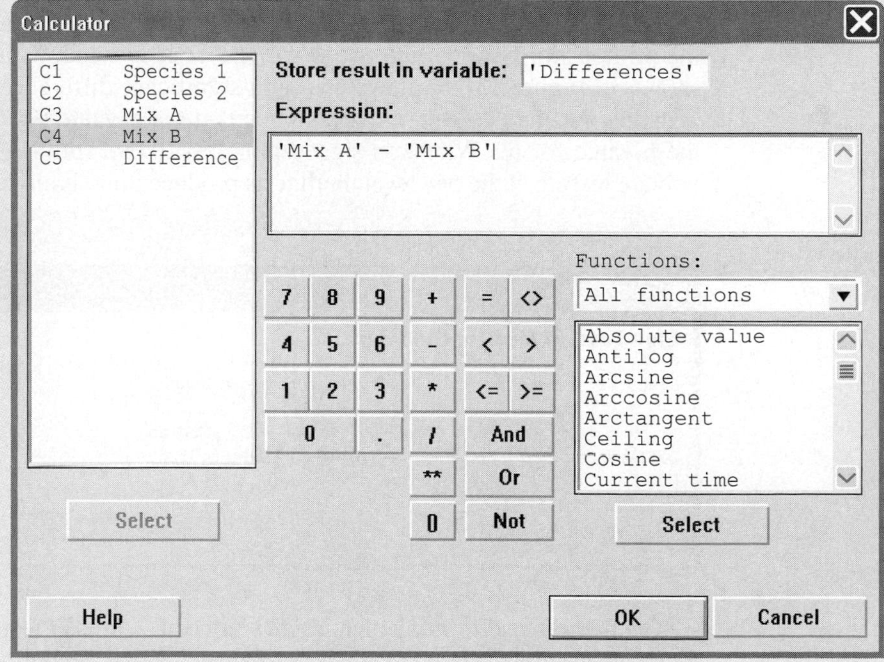

shown in Figure 15.15. Remember that the median is the value of a variable such that 50% of the values are smaller and 50% are larger. Hence, if the two population distributions are the same, the median of the differences will be 0. This is equivalent to the null hypothesis

$$H_0 : P(\text{positive difference}) = P(\text{negative difference}) = 0.5$$

FIGURE 15.15

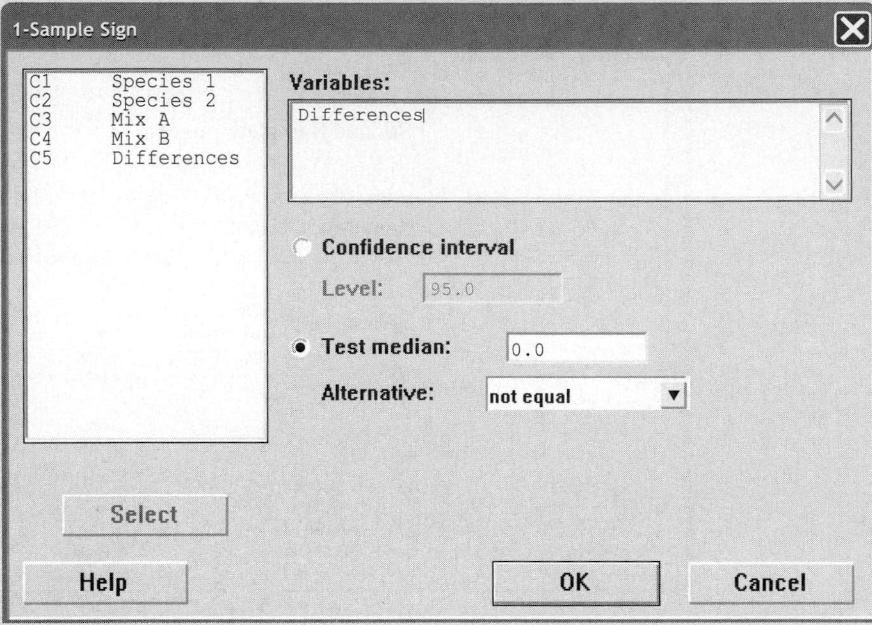

used for the sign test. Select the column of differences for the Variables box, and select that the test median equals 0 with the appropriate alternative. Click **OK** to obtain the printout for either of the two tests. The Session window printout for the sign test, shown in Figure 15.16, indicates a non-significant difference in the distributions of densities for the two cake mixes. Notice that the p-value (0.2188) is not the same as the p-value for the Wilcoxon signed-rank test (0.093 from Figure 15.4). However, if you are testing at the 5% level, both tests produce non-significant differences.

FIGURE 15.16

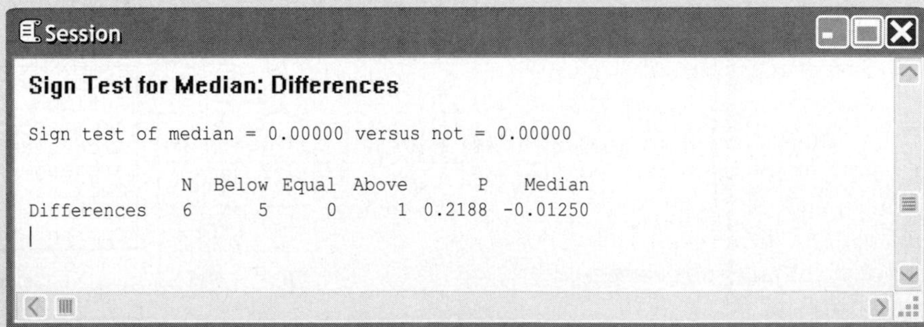

The procedures for implementing the Kruskal–Wallis H test for k independent samples and Friedman's F_r test for a randomized block design are identical to the procedures used for their parametric equivalents. Review the methods described in the "Technology Today" section in Chapter 11. Once you have entered the data as explained in that section, the commands **Stat ▶ Nonparametrics ▶ Kruskal–Wallis** or **Stat ▶ Nonparametrics ▶ Friedman** will generate a dialogue box in which you specify the Response column and the Factor column, or the Response column, the Treatment column, and the Block column, respectively. Click **OK** to obtain the outputs for these nonparametric tests.

Finally, you can generate the nonparametric rank correlation coefficient r_s if you enter the data into two columns and rank the data using **Data ▶ Rank.** For example, the data on judge's rank and test scores were entered into columns C6 and C7 of our *MINITAB* worksheet. Since the judge's ranks are already in rank order, we need only to rank C7 by selecting 'Exam Score' and storing the ranks in C8 (see 'Rank (y)' in Figure 15.17). The commands **Stat ▶ Basic Statistics ▶ Correlation** will now produce the rank correlation coefficient when C6 and C8 are selected. However, the *p*-value that you see in the *output does not* produce exactly the same test as the critical values in Table 15.14. You should compare your value of r_s with the tabled value to check for a significant association between the two variables.

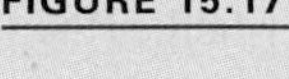
FIGURE 15.17

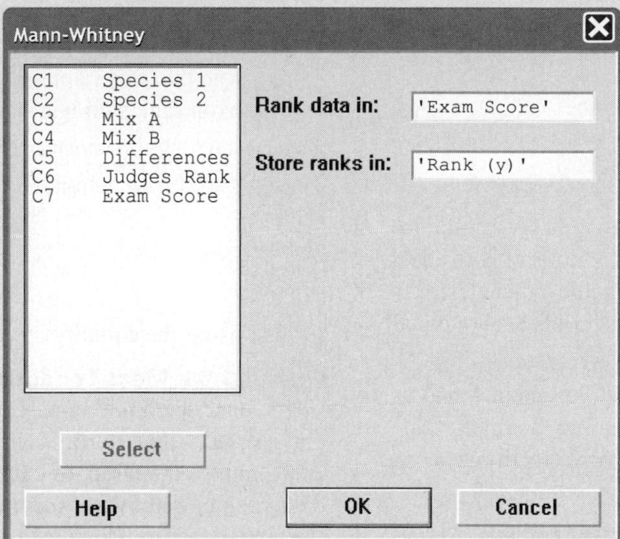

Supplementary Exercises

15.55 Response Times An experiment was conducted to compare the response times for two different stimuli. To remove natural person-to-person variability in the responses, both stimuli were presented to each of nine subjects, thus permitting an analysis of the differences between stimuli *within* each person. The table lists the response times (in seconds):

Data set EX1555

Subject	Stimulus 1	Stimulus 2
1	9.4	10.3
2	7.8	8.9
3	5.6	4.1
4	12.1	14.7
5	6.9	8.7
6	4.2	7.1
7	8.8	11.3
8	7.7	5.2
9	6.4	7.8

a. Use the sign test to determine whether sufficient evidence exists to indicate a difference in the mean response times for the two stimuli. Use a rejection region for which $\alpha \leq 0.05$.

b. Test the hypothesis of no difference in mean response times using Student's *t* test.

15.56 Response Times, continued Refer to Exercise 15.55. Test the hypothesis that no difference exists in the distributions of response times for the two stimuli, using the Wilcoxon signed-rank test. Use a rejection region for which α is as near as possible to the α achieved in Exercise 15.55, part a.

15.57 Identical Twins To compare two middle schools, A and B, in academic effectiveness, an experiment was designed requiring

Data set EX1557

the use of 10 sets of identical twins, each twin having just completed Grade 6. In each case, the twins in the same set had obtained their schooling in the same classrooms at each grade level. One child was selected at random from each pair of twins and assigned to school A. The remaining children were sent to school B. Near the end of Grade 8, a certain achievement test was given to each child in the experiment. The test scores are shown in the table:

Twin Pair	School A	School B
1	67	39
2	80	75
3	65	69
4	70	55
5	86	74
6	50	52
7	63	56
8	81	72
9	86	89
10	60	47

a. Test (using the sign test) the hypothesis that the two schools are the same in academic effectiveness, as measured by scores on the achievement test, versus the alternative that the schools are not equally effective.

b. Suppose it was known that middle school A had a superior faculty and better learning facilities. Test the hypothesis of equal academic effectiveness versus the alternative that school A is superior.

15.58 Identical Twins II Refer to Exercise 15.57. What answers are obtained if Wilcoxon's signed-rank test is used in analyzing the data? Compare with your earlier answers.

Data set EX1559 **15.59 Paper Brightness** The coded values for a measure of brightness in paper (light reflectivity), prepared by two different processes, are given in the table for samples of nine observations drawn randomly from each of the two processes. Do the data present sufficient evidence to indicate a difference in the brightness measurements for the two processes? Use both a parametric and a nonparametric test and compare your results.

Process	Brightness								
A	6.1	9.2	8.7	8.9	7.6	7.1	9.5	8.3	9.0
B	9.1	8.2	8.6	6.9	7.5	7.9	8.3	7.8	8.9

15.60 Precision Instruments Assume (as in the case of measurements produced by two well-calibrated measuring instruments) the means of two populations are equal. Use the Wilcoxon rank sum statistic for testing hypotheses concerning the population variances as follows:

a. Rank the combined sample.

b. Number the ranked observations "from the outside in"; that is, number the smallest observation 1, the largest 2, the next-to-smallest 3, the next-to-largest 4, and so on. This sequence of numbers induces an ordering on the symbols A (population A items) and B (population B items). If $\sigma_A^2 > \sigma_B^2$, one would expect to find a preponderance of A's near the first of the sequences, and thus a relatively small "sum of ranks" for the A observations.

c. Given the measurements in the table produced by well-calibrated precision instruments A and B, test at near the $\alpha = 0.05$ level to determine whether the more expensive instrument B is more precise than A. (Note that this implies a one-tailed test.) Use the Wilcoxon rank sum test statistic.

Instrument A	Instrument B
1060.21	1060.24
1060.34	1060.28
1060.27	1060.32
1060.36	1060.30
1060.40	

d. Test using the equality of variance F test.

Data set EX1561 **15.61 Meat Tenderizers** An experiment was conducted to compare the tenderness of meat cuts treated with two different meat tenderizers, A and B. To reduce the effect of extraneous variables, the data were paired by the specific meat cut, by applying the tenderizers to two cuts taken from the same steer, by cooking paired cuts together, and by using a single judge for each pair. After cooking, each cut was rated by a judge on a scale of 1 to 10, with 10 corresponding to the most tender meat. The data are shown for a single judge. Do the data provide sufficient evidence to indicate that one of the two tenderizers tends to receive higher ratings than the other? Would a Student's t test be appropriate for analyzing these data? Explain.

	Tenderizer	
Cut	A	B
Shoulder roast	5	7
Chuck roast	6	5
Rib steak	8	9
Brisket	4	5
Club steak	9	9
Round steak	3	5
Rump roast	7	6
Sirloin steak	8	8
Sirloin tip steak	8	9
T-bone steak	9	10

EX1562 **15.62 Interviewing Job Prospects** A large corporation selects university graduates for

employment using both interviews and a psychological achievement test. Interviews conducted at the home office of the company are far more expensive than the tests that can be conducted on campus. Consequently, the personnel office was interested in determining whether the test scores were correlated with interview ratings and whether tests could be substituted for interviews. The idea was not to eliminate interviews but to reduce their number. To determine whether the measures were correlated, 10 prospects were ranked during interviews and tested. The paired scores are as listed here:

Subject	Interview Rank	Test Score
1	8	74
2	5	81
3	10	66
4	3	83
5	6	66
6	1	94
7	4	96
8	7	70
9	9	61
10	2	86

Calculate the Spearman rank correlation coefficient r_s. Rank 1 is assigned to the candidate judged to be the best.

15.63 Interviews, continued Refer to Exercise 15.62. Do the data present sufficient evidence to indicate that the correlation between interview rankings and test scores is less than 0? If this evidence does exist, can you say that tests can be used to reduce the number of interviews?

15.64 Word Association Experiments A comparison of reaction times for two different stimuli in a psychological word-association experiment produced the accompanying results when applied to a random sample of 16 people:

Stimulus	Reaction Time (sec)							
1	1	3	2	1	2	1	3	2
2	4	2	3	3	1	2	3	3

Do the data present sufficient evidence to indicate a difference in mean reaction times for the two stimuli? Use an appropriate nonparametric test and explain your conclusions.

Data set EX1565 **15.65 Math and Art** The table gives the scores of a group of 15 students in mathematics and art. Use Wilcoxon's signed-rank test to determine whether the median scores for these students differ significantly for the two subjects.

Student	Math	Art	Student	Math	Art
1	22	53	9	62	55
2	37	68	10	65	74
3	36	42	11	66	68
4	38	49	12	56	64
5	42	51	13	66	67
6	58	65	14	67	73
7	58	51	15	62	65
8	60	71			

15.66 Math and Art, continued Refer to Exercise 15.65. Compute Spearman's rank correlation coefficient for these data and test H_0 : no association between the rank pairs at the 10% level of significance.

15.67 Yield of Wheat Exercise 11.66 presented an analysis of variance of the yields of five different varieties of wheat, observed on one plot each at each of six different locations (see data set EX1166). The data from this randomized block design are listed here:

	Location					
Varieties	1	2	3	4	5	6
A	35.3	31.0	32.7	36.8	37.2	33.1
B	30.7	32.2	31.4	31.7	35.0	32.7
C	38.2	33.4	33.6	37.1	37.3	38.2
D	34.9	36.1	35.2	38.3	40.2	36.0
E	32.4	28.9	29.2	30.7	33.9	32.1

a. Use the appropriate nonparametric test to determine whether the data provide sufficient evidence to indicate a difference in the yields for the five different varieties of wheat. Test using $\alpha = 0.05$.

b. Exercise 11.66 presented a computer printout of the analysis of variance for comparing the mean yields for the five varieties of wheat. How do the results of the analysis of variance F test compare with the test in part a? Explain.

15.68 Learning to Sell In Exercise 11.59, you compared the numbers of sales per trainee after completion of one of four different sales training programs (see data set EX1159). Six trainees completed training program 1, eight completed 2, and so on. The numbers of sales per trainee are shown in the table:

	Training Program			
	1	2	3	4
	78	99	74	81
	84	86	87	63
	86	90	80	71
	92	93	83	65
	69	94	78	86
	73	85		79
		97		73
		91		70
Total	482	735	402	588

a. Do the data present sufficient evidence to indicate that the distribution of number of sales per trainee differs from one training program to another? Test using the appropriate nonparametric test.

b. How do the test results in part a compare with the results of the analysis of variance F test in Exercise 11.59?

15.69 Pollution from Chemical Plants In Exercise 11.64, you performed an analysis of variance to compare the mean levels of effluents in water at four different industrial plants (see data set EX1164). Five samples of liquid waste were taken at the output of each of four industrial plants. The data are shown in the table:

Plant	Polluting Effluents (400 g/4 L of waste)				
A	1.65	1.72	1.50	1.37	1.60
B	1.70	1.85	1.46	2.05	1.80
C	1.40	1.75	1.38	1.65	1.55
D	2.10	1.95	1.65	1.88	2.00

a. Do the data present sufficient evidence to indicate a difference in the levels of pollutants for the four different industrial plants? Test using the appropriate nonparametric test.

b. Find the approximate p-value for the test and interpret its value.

c. Compare the test results in part a with the analysis of variance test in Exercise 11.64. Do the results agree? Explain.

15.70 AIDS Research Scientists have shown that a newly developed vaccine can shield rhesus monkeys from infection by a virus closely related to the AIDS-causing human immunodeficiency virus (HIV). In their work, Ronald C. Resrosiers and his colleagues gave each of $n = 6$ rhesus monkeys five inoculations with the simian immunodeficiency virus (SIV) vaccine. One week after the last vaccination, each monkey received an injection of live SIV. Two of the six vaccinated monkeys showed no evidence of SIV infection for as long as a year and a half after the SIV injection.[7] Scientists were able to isolate the SIV virus from the other four vaccinated monkeys, although these animals showed no sign of the disease. Does this information contain sufficient evidence to indicate that the vaccine is effective in protecting monkeys from SIV? Use $\alpha = 0.10$.

Data set EX1571 **15.71 Heavy Metal** An experiment was performed to determine whether there is an accumulation of heavy metals in plants that were grown in soils amended with sludge and whether there is an accumulation of heavy metals in insects feeding on those plants.[8] The data in the table are cadmium concentrations (in μg/kg) in plants grown under six different rates of application of sludge for three different harvests. The rates of application are the treatments. The three harvests represent time blocks in the two-way design.

Rate	Harvest 1	Harvest 2	Harvest 3
Control	162.1	153.7	200.4
1	199.8	199.6	278.2
2	220.0	210.7	294.8
3	194.4	179.0	341.1
4	204.3	203.7	330.2
5	218.9	236.1	344.2

a. Based on the *MINITAB* normal probability plot and the plot of residuals versus rates, are you willing to assume that the normality and constant variance assumptions are satisfied?

MINITAB residual plots for Exercise 15.71

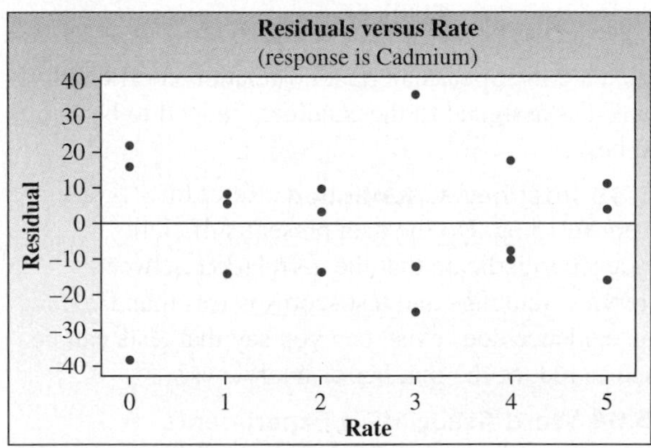

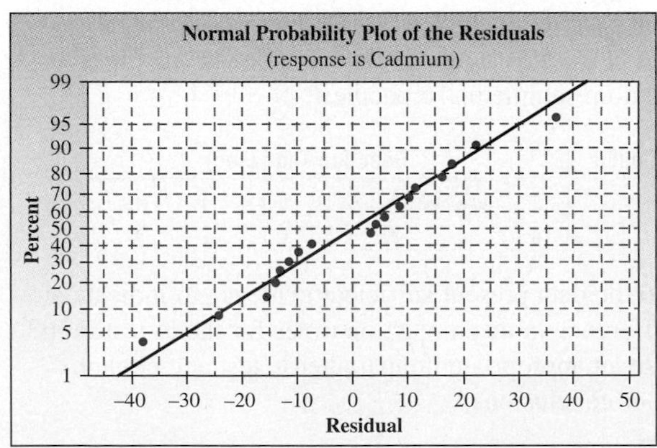

b. Using an appropriate method of analysis, analyze the data to determine whether there are significant differences among the responses due to rates of application.

Data set EX1572 **15.72** Refer to Exercise 15.71. The data in this table are the cadmium concentrations found in aphids that fed on the plants grown in soil amended with sludge:

	Harvest		
Rate	1	2	3
Control	16.2	55.8	65.8
1	16.9	119.4	181.1
2	12.7	171.9	184.6
3	31.3	128.4	196.4
4	38.5	182.0	163.7
5	20.6	191.3	242.8

a. Use the *MINITAB* normal probability plot of the residuals and the plot of residuals versus rates of application to assess whether the assumptions of normality and constant variance are reasonable in this case.

b. Based on your conclusions in part a, use an appropriate statistical method to test for significant differences in cadmium concentrations for the six rates of application.

MINITAB residual plots for Exercise 15.72

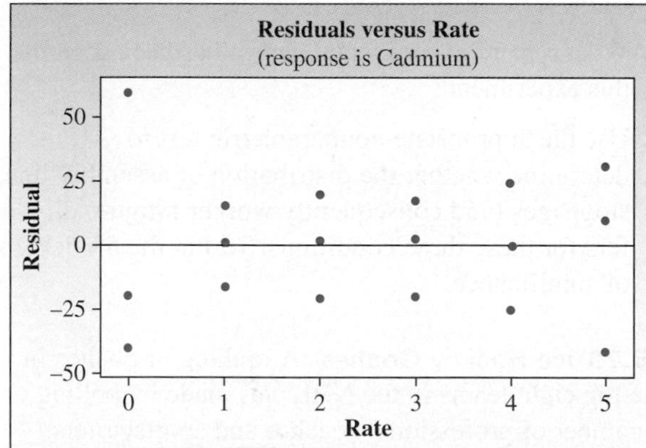

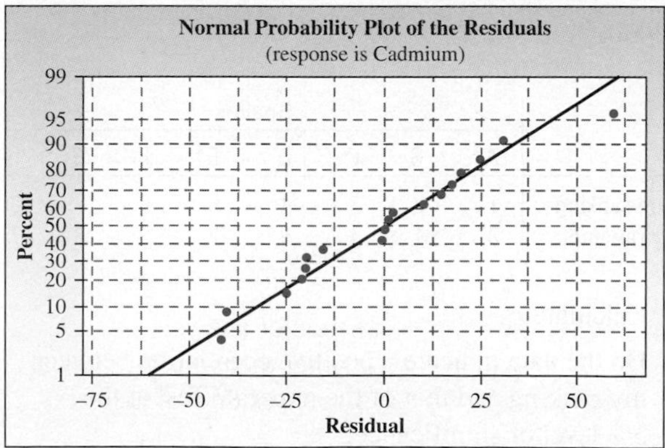

15.73 Rating Teaching Applicants Before filling several new teaching positions at the high school, the principal formed a review board consisting of five teachers who were asked to interview the 12 applicants and rank them in order of merit. Seven of the 12 applicants held university degrees but had limited teaching experience. Of the remaining five applicants, all had university degrees and substantial experience. The review board's rankings are given in the table:

Limited Experience	Substantial Experience
4	1
6	2
7	3
9	5
10	8
11	
12	

Do these rankings indicate that the review board considers experience a prime factor in the selection of the best candidates? Test using $\alpha = 0.05$.

Data set EX1574 **15.74 Contaminants in Chemicals** A manufacturer uses a large amount of a certain chemical. Since there are just two suppliers of this chemical, the manufacturer wishes to test whether the percentage of contaminants is the same for the two sources against the alternative that there is a difference in the percentages of contaminants for the two suppliers. Data from independent random samples are shown below:

Supplier			
1		2	
0.86	0.65	0.55	0.58
0.69	1.13	0.40	0.16
0.72	0.65	0.22	0.07
1.18	0.50	0.09	0.36
0.45	1.04	0.16	0.20
1.41	0.41	0.26	0.15

a. Use the Wilcoxon rank sum test to determine whether there is a difference in the contaminant percentages for the two suppliers. Use $\alpha = 0.05$.

b. Use the large-sample approximation to the Wilcoxon rank sum test to determine whether there is a difference in the contaminant percentages for the two suppliers. Use $\alpha = 0.05$. Compare your conclusions to the conclusions from part a.

15.75 Lighting in the Classroom The productivity of 35 students was observed and measured both before and after the installation of new lighting in their classroom. The productivity of 21 of the 35 students was observed to have improved, whereas the productivity of the others appeared to show no perceptible gain as a result of the new lighting. Use the normal approximation to the sign test to determine whether or not the new lighting was effective in increasing student productivity at the 5% level of significance.

Data set EX1576 **15.76 Reducing Cholesterol** A drug was developed for reducing cholesterol levels in heart patients. The cholesterol levels before and after drug treatment were obtained for a random sample of 25 heart patients with the following results:

Patient	Before	After	Patient	Before	After
1	257	243	14	210	217
2	222	217	15	263	243
3	177	174	16	214	198
4	258	260	17	392	388
5	294	295	18	370	357
6	244	236	19	310	299
7	390	383	20	255	258
8	247	233	21	281	276
9	409	410	22	294	295
10	214	216	23	257	227
11	217	210	24	227	231
12	340	335	25	385	374
13	364	343			

a. Use the sign test to determine whether or not this drug reduces the cholesterol levels of heart patients. Use $\alpha = 0.01$.

b. Use the Wilcoxon signed-rank test to test the hypothesis in part a at the 1% level of significance. Are your conclusions the same as those in part a?

Data set EX1577 **15.77 Legos** The time required for kindergarten children to assemble a specific Lego creation was measured for children who had been instructed for four different lengths of time. Four children were randomly assigned to each instructional group, but two were eliminated during the experiment because of sickness. The length of time (in minutes) to assemble the Lego creation was recorded for each child in the experiment:

Training Period (hours)			
0.5	1.0	1.5	2.0
8	9	4	4
14	7	6	7
9	5	7	5
12		8	

Use the Kruskal–Wallis H Test to determine whether there is a difference in the distribution of times for the four different lengths of instructional time. Use $\alpha = 0.01$.

Data set EX1578 **15.78 Worker Fatigue** To investigate methods of reducing fatigue among employees whose jobs involve a monotonous assembly procedure, 12 randomly selected employees were asked to perform their usual job under each of three trial conditions. As a measure of fatigue, the experimenter used the number of assembly line stoppages during a four-hour period for each trial condition:

	Conditions		
Employee	1	2	3
1	31	22	26
2	20	15	23
3	26	21	18
4	31	22	32
5	12	16	18
6	22	29	34
7	28	17	26
8	15	9	12
9	41	31	46
10	19	19	25
11	31	34	41
12	18	11	21

a. What type of experimental design has been used in this experiment?

b. Use the appropriate nonparametric test to determine whether the distribution of assembly line stoppages (and consequently worker fatigue) differs for these three conditions. Test at the 5% level of significance.

15.79 Ice Hockey Goalies A ranking of goalies in the top eight teams in the NHL was made by polling a number of professional coaches and sportswriters. This "true ranking" is shown below, together with "my ranking":

	Goalies							
	A	B	C	D	E	F	G	H
True ranking	1	2	3	4	5	6	7	8
My ranking	3	1	4	5	2	8	6	7

a. Calculate r_s.

b. Do the data indicate a positive correlation between my ranking and that of the experts? Test at the 5% level of significance.

CASE STUDY 

 Eggs

How's Your Cholesterol Level?

As consumers become more and more interested in eating healthy foods, many "light," "fat-free," and "cholesterol-free" products are appearing in the marketplace. One such product is the frozen egg substitute, a cholesterol-free product that can be used in cooking and baking in many of the same ways that regular eggs can—though not all. Some consumers even use egg substitutes for Caesar salad dressings and other recipes calling for raw eggs because these products are pasteurized and thus eliminate worries about bacterial contamination.

Unfortunately, the products currently on the market exhibit strong differences in both flavour and texture when tasted in their primary preparation as scrambled eggs. Five panelists, all experts in nutrition and food preparation, were asked to rate each of three egg substitutes on the basis of taste, appearance, texture, and whether they would buy the product.[9] The judges tasted the three egg substitutes and rated them on a scale of 0 to 20. The results, shown in the table, indicate that the highest rating, by 23 points, went to ConAgra's Healthy Choice Egg Product, which the tasters unanimously agreed most closely resembled eggs as they come from the hen. The second-place product, Morningstar Farms' Scramblers, struck several tasters as having an "oddly sweet flavour . . . similar to carrots." Finally, none of the tasters indicated that they would be willing to buy Fleishmann's Egg Beaters, which was described by the testers as "watery," "slippery," and "unpleasant." Oddly enough, these results are contrary to a similar taste test done four years earlier, in which Egg Beaters were considered better than competing egg substitutes.

Taster	Healthy Choice	Scramblers	Egg Beaters
Dan Bowe	16	9	7
John Carroll	16	7	8
Donna Katzl	14	8	4
Rick O'Connell	15	16	9
Roland Passot	13	11	2
Total	74	51	30

Source: Republished with permission of *San Francisco Chronicle*. From Karola Saekel, "Egg Substitutes Range in Quality," *San Francisco Chronicle*, 10 February 1993, p. 8.

1. What type of design has been used in this taste-testing experiment?
2. Do the data satisfy the assumptions required for a parametric analysis of variance? Explain.
3. Use the appropriate nonparametric technique to determine whether there is a significant difference between the average scores for the three brands of egg substitutes.

PROJECTS 

Project 15-A: Air-Conditioning Makes You Gain Weight[10]

A report published in the *International Journal of Obesity* claims the "biggie size" fries and lack of exercise aren't the only factors to blame for North America's battle of the bulge. Researchers say some aspects of modern life, including air-conditioning, lack of sleep, and exposure to chemicals, may be making people pudgy.

With air-conditioning, your body doesn't have to work as hard to keep cool, so it burns fewer calories. The calories you burn just lying in bed doing nothing is called your basal metabolic rate (BMR)—about 70 percent of the calories you burn in a day can be attributed to BMR.

To investigate the effect of air-conditioning on basal metabolism, eight females from Vancouver who used air-conditioning during the hours of night sleep (Group A), and six females who didn't use air-conditioning during the hours of night sleep (Group B), were selected and their basal metabolism recorded as shown below:

Group A	1150	1088	1090	1027	1150	1145	1199	950
Group B	1086	998	1011	987	1101	1099	1125	

a. Use the Wilcoxon rank sum test to determine whether the distributions for metabolism measurements are the same for both groups. Use $\alpha = 0.05$.

b. Use the Wilcoxon rank sum test to determine whether the metabolism measurements for Group A are significantly higher than those of Group B. Use $\alpha = 0.05$.

c. Can you use the normal approximation for the Wilcoxon rank sum test in the above two questions? Why or why not? How about using t test? Explain.

Project 15-B: Does Drinking Water Increase Metabolism?

Having the appropriate amount of water will increase or boost metabolism and regulate body temperature. The basal metabolism of 10 females was examined and recorded before and after drinking the appropriate amount of water (as displayed in the following table) to see if we can conclude that drinking an appropriate amount of water was helpful in increasing metabolism.

After (y)	1150	1088	1090	1027	1151	1145	1199	950	1179	998
Before (x)	1105	1077	1020	989	1120	1099	1145	907	1131	988

a. State the null and alternative hypotheses.

b. Describe what the test statistic is for the sign test. What is the value of the test statistic in this problem?

c. Which test is appropriate for this situation?

d. Is this a one-tailed test or a two-tailed test? Find the rejection region for $\alpha = 0.10$.

e. Using $\alpha = 0.10$, can we conclude drinking an appropriate amount of water was helpful in increasing metabolism?

f. At what level of significance could we reject H_0?

g. Use the large-sample z statistic for testing the hypothesis in part (a) using $\alpha = 0.10$. (Note: Although the large-sample approximation is suggested for $n \geq 25$, it works fairly well for values of n as small as 10.)

h. Compare the results of questions (e) and (g).

i. State the null and alternative hypotheses if the Wilcoxon signed-rank test is used on the data to see if we can conclude that drinking an appropriate amount of water was helpful in increasing metabolism.

j. Describe the test statistic for the Wilcoxon signed-rank test. What is the value of the test statistic in this problem?

k. Determine the rejection region for the Wilcoxon signed-rank test at the 5% significance level.

l. Using the Wilcoxon signed-rank test, can we conclude that drinking an appropriate amount was helpful in increasing metabolism?

m. Find the p-value for the Wilcoxon signed-rank test.

n. Use large-sample Wilcoxon signed-rank test for testing the hypothesis in part (i).

o. Compare the results of parts (e), (l) and (n).

p. Calculate Spearman's rank correlation coefficient r_s.

q. Do the data present sufficient evidence to indicate a correlation between x and y? Test using $\alpha = 0.05$.

Project 15-C: Increase Your Overall Muscle Mass and Boost Your Metabolism

You can increase your basal metabolic rate (BMR) by implementing one of the following strategies:

Strategy 1: Having the appropriate amount of water will increase or boost metabolism and regulate body temperature.

Strategy 2: The more frequently you eat, the greater your metabolism will become to burn fat and calories. Aim for six to eight small meals with protein, veggies, and good fats.

Strategy 3: Ingesting complete protein, whole wheat, and spicy foods increases BMR.

Strategy 4: Cold and hot hydrotherapy boosts or increases BMR.

Strategy 5: Caffeine, nicotine, and some fat-burning supplements will increase metabolism.

A registered dietician took a random sample of six females. These strategies were compared using a completely randomized design. The data are shown in the table below:

Strategy 1	Strategy 2	Strategy 3	Strategy 4	Strategy 5
998	1001	877	1129	1009
1055	955	1151	1098	905
975	1099	1173	1012	1191
1123	987	890	981	993
1074	1134	1089	1034	1143
1101	1195	1188	995	1159

The dietician was interested in knowing whether the data provide sufficient evidence to indicate a difference in location for at least two of the population distributions.

a. Do the data provide sufficient evidence to indicate a difference in location for at least two of the five strategies? Test using the Kruskal-Wallis H statistic with $\alpha = 0.05$.

b. What is the most accurate statement that can be made about the p-value of this test?

c. Explain how to use the p-value for testing the hypotheses.

d. Perform an analysis of variance and give the ANOVA table for the analysis, and using statistical software to approximate the p-value for the F statistic in testing the equality of the means test using $\alpha = 0.05$.

e. Compare the p-values for the tests in parts (a) and (d).

Another registered dietician also planned to evaluate five different strategies to assess their effectiveness in increasing or boosting BMR. However, this dietician consulted a statistician about the design of her study. Because she was looking for an effective method of increasing BMR, the statistician suggested she should divide the subjects into groups according to their age. These age groups represent relatively homogeneous experimental units within which the comparison of different ways to increase the BMR may be made. Within each age group, the five participating females were randomly assigned to the five strategies and measurements of BMR were recorded. The data is summarized in the following table:

Age	Strategy 1	Strategy 2	Strategy 3	Strategy 4	Strategy 5
15–19	1187	1103	997	1029	1207
20–24	1310	1278	1251	1198	1305
25–29	1287	1191	1287	1121	1256
30–34	1223	1167	1234	1109	1245
35–39	1250	1289	1223	1244	1230
40–44	1312	11324	1288	1295	1159

a. What experimental design did the statistician use?

b. The dietician wasn't sure whether the assumptions for the usual analysis of variance were valid, so she decided to use a nonparametric procedure. Use the appropriate method to determine whether the BMR is the same for all the strategies. Use $\alpha = 0.05$.

c. Can we conclude that all the strategies are equally effective?

d. What is the observed significance level of this test?

e. The dietician would like to perform an analysis of variance anyway. Give the ANOVA table for the analysis, using statistical software to approximate the p-value for the F statistic in testing the equality of the five treatment means. Test using $\alpha = 0.05$.

f. Compare the p-values for the tests in parts (b) and (e).

Appendix I
Tables

CONTENTS

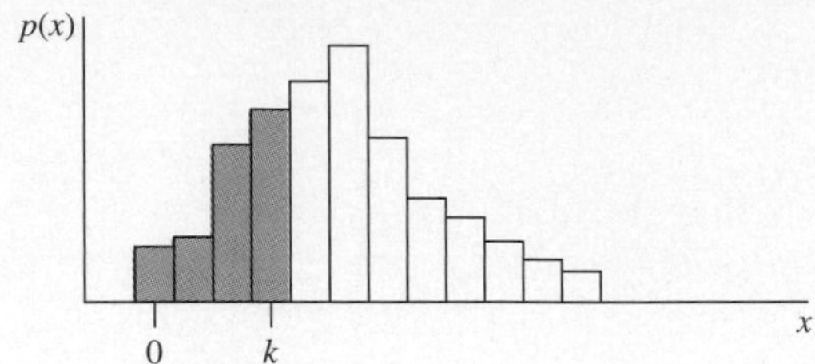

TABLE 1 Cumulative Binomial Probabilities
Tabulated values are $P(X \leq k) = p(0) + p(1) + \cdots + p(k)$
(Computations are rounded at the third decimal place.)

$n = 2$

k	0.01	0.05	0.10	0.20	0.30	0.40	0.50	0.60	0.70	0.80	0.90	0.95	0.99	k
0	0.980	0.902	0.810	0.640	0.490	0.360	0.250	0.160	0.090	0.040	0.010	0.002	0.000	0
1	1.000	0.998	0.990	0.960	0.910	0.840	0.750	0.640	0.510	0.360	0.190	0.098	0.020	1
2	1.000	1.000	1.000	1.000	1.000	1.000	1.000	1.000	1.000	1.000	1.000	1.000	1.000	2

$n = 3$

k	0.01	0.05	0.10	0.20	0.30	0.40	0.50	0.60	0.70	0.80	0.90	0.95	0.99	k
0	0.970	0.857	0.729	0.512	0.343	0.216	0.125	0.064	0.027	0.008	0.001	0.000	0.000	0
1	1.000	0.993	0.972	0.896	0.784	0.648	0.500	0.352	0.216	0.104	0.028	0.007	0.000	1
2	1.000	1.000	0.999	0.992	0.973	0.936	0.875	0.784	0.657	0.488	0.271	0.143	0.030	2
3	1.000	1.000	1.000	1.000	1.000	1.000	1.000	1.000	1.000	1.000	1.000	1.000	1.000	3

$n = 4$

k	0.01	0.05	0.10	0.20	0.30	0.40	0.50	0.60	0.70	0.80	0.90	0.95	0.99	k
0	0.961	0.815	0.656	0.410	0.240	0.130	0.062	0.026	0.008	0.002	0.000	0.000	0.000	0
1	0.999	0.986	0.948	0.819	0.652	0.475	0.312	0.179	0.084	0.027	0.004	0.000	0.000	1
2	1.000	1.000	0.996	0.973	0.916	0.821	0.688	0.525	0.348	0.181	0.052	0.014	0.001	2
3	1.000	1.000	1.000	0.998	0.992	0.974	0.938	0.870	0.760	0.590	0.344	0.185	0.039	3
4	1.000	1.000	1.000	1.000	1.000	1.000	1.000	1.000	1.000	1.000	1.000	1.000	1.000	4

TABLE 1 *(continued)*

$n = 5$

k	0.01	0.05	0.10	0.20	0.30	0.40	0.50	0.60	0.70	0.80	0.90	0.95	0.99	k
0	0.951	0.774	0.590	0.328	0.168	0.078	0.031	0.010	0.002	0.000	0.000	0.000	0.000	0
1	0.999	0.977	0.919	0.737	0.528	0.337	0.188	0.087	0.031	0.007	0.000	0.000	0.000	1
2	1.000	0.999	0.991	0.942	0.837	0.683	0.500	0.317	0.163	0.058	0.009	0.001	0.000	2
3	1.000	1.000	1.000	0.993	0.969	0.913	0.812	0.663	0.472	0.263	0.081	0.023	0.001	3
4	1.000	1.000	1.000	1.000	0.998	0.990	0.969	0.922	0.832	0.672	0.410	0.226	0.049	4
5	1.000	1.000	1.000	1.000	1.000	1.000	1.000	1.000	1.000	1.000	1.000	1.000	1.000	5

$n = 6$

k	0.01	0.05	0.10	0.20	0.30	0.40	0.50	0.60	0.70	0.80	0.90	0.95	0.99	k
0	0.941	0.735	0.531	0.262	0.118	0.047	0.016	0.004	0.001	0.000	0.000	0.000	0.000	0
1	0.999	0.967	0.886	0.655	0.420	0.233	0.109	0.041	0.011	0.002	0.000	0.000	0.000	1
2	1.000	0.998	0.984	0.901	0.744	0.544	0.344	0.179	0.070	0.017	0.001	0.000	0.000	2
3	1.000	1.000	0.999	0.983	0.930	0.821	0.656	0.456	0.256	0.099	0.016	0.002	0.000	3
4	1.000	1.000	1.000	0.998	0.989	0.959	0.891	0.767	0.580	0.345	0.114	0.033	0.001	4
5	1.000	1.000	1.000	1.000	0.999	0.996	0.984	0.953	0.882	0.738	0.469	0.265	0.059	5
6	1.000	1.000	1.000	1.000	1.000	1.000	1.000	1.000	1.000	1.000	1.000	1.000	1.000	6

$n = 7$

k	0.01	0.05	0.10	0.20	0.30	0.40	0.50	0.60	0.70	0.80	0.90	0.95	0.99	k
0	0.932	0.698	0.478	0.210	0.082	0.028	0.008	0.002	0.000	0.000	0.000	0.000	0.000	0
1	0.998	0.956	0.850	0.577	0.329	0.159	0.062	0.019	0.004	0.000	0.000	0.000	0.000	1
2	1.000	0.996	0.974	0.852	0.647	0.420	0.227	0.096	0.029	0.005	0.000	0.000	0.000	2
3	1.000	1.000	0.997	0.967	0.874	0.710	0.500	0.290	0.126	0.033	0.003	0.000	0.000	3
4	1.000	1.000	1.000	0.995	0.971	0.904	0.773	0.580	0.353	0.148	0.026	0.004	0.000	4
5	1.000	1.000	1.000	1.000	0.996	0.981	0.938	0.841	0.671	0.423	0.150	0.044	0.002	5
6	1.000	1.000	1.000	1.000	1.000	0.998	0.992	0.972	0.918	0.790	0.522	0.302	0.068	6
7	1.000	1.000	1.000	1.000	1.000	1.000	1.000	1.000	1.000	1.000	1.000	1.000	1.000	7

$n = 8$

k	0.01	0.05	0.10	0.20	0.30	0.40	0.50	0.60	0.70	0.80	0.90	0.95	0.99	k
0	0.923	0.663	0.430	0.168	0.058	0.017	0.004	0.001	0.000	0.000	0.000	0.000	0.000	0
1	0.997	0.943	0.813	0.503	0.255	0.106	0.035	0.009	0.001	0.000	0.000	0.000	0.000	1
2	1.000	0.994	0.962	0.797	0.552	0.315	0.145	0.050	0.011	0.001	0.000	0.000	0.000	2
3	1.000	1.000	0.995	0.944	0.806	0.594	0.363	0.174	0.058	0.010	0.000	0.000	0.000	3
4	1.000	1.000	1.000	0.990	0.942	0.826	0.637	0.406	0.194	0.056	0.005	0.000	0.000	4
5	1.000	1.000	1.000	0.999	0.989	0.950	0.855	0.685	0.448	0.203	0.038	0.006	0.000	5
6	1.000	1.000	1.000	1.000	0.999	0.991	0.965	0.894	0.745	0.497	0.187	0.057	0.003	6
7	1.000	1.000	1.000	1.000	1.000	0.999	0.996	0.983	0.942	0.832	0.570	0.337	0.077	7
8	1.000	1.000	1.000	1.000	1.000	1.000	1.000	1.000	1.000	1.000	1.000	1.000	1.000	8

TABLE 1 *(continued)*

$n = 9$

	p													
k	0.01	0.05	0.10	0.20	0.30	0.40	0.50	0.60	0.70	0.80	0.90	0.95	0.99	*k*
0	0.914	0.630	0.387	0.134	0.040	0.010	0.002	0.000	0.000	0.000	0.000	0.000	0.000	0
1	0.997	0.929	0.775	0.436	0.196	0.071	0.020	0.004	0.000	0.000	0.000	0.000	0.000	1
2	1.000	0.992	0.947	0.738	0.463	0.232	0.090	0.025	0.004	0.000	0.000	0.000	0.000	2
3	1.000	0.999	0.992	0.914	0.730	0.483	0.254	0.099	0.025	0.003	0.000	0.000	0.000	3
4	1.000	1.000	0.999	0.980	0.901	0.733	0.500	0.267	0.099	0.020	0.001	0.000	0.000	4
5	1.000	1.000	1.000	0.997	0.975	0.901	0.746	0.517	0.270	0.086	0.008	0.001	0.000	5
6	1.000	1.000	1.000	1.000	0.996	0.975	0.910	0.768	0.537	0.262	0.053	0.008	0.000	6
7	1.000	1.000	1.000	1.000	1.000	0.996	0.980	0.929	0.804	0.564	0.225	0.071	0.003	7
8	1.000	1.000	1.000	1.000	1.000	1.000	0.998	0.990	0.960	0.866	0.613	0.370	0.086	8
9	1.000	1.000	1.000	1.000	1.000	1.000	1.000	1.000	1.000	1.000	1.000	1.000	1.000	9

$n = 10$

	p													
k	0.01	0.05	0.10	0.20	0.30	0.40	0.50	0.60	0.70	0.80	0.90	0.95	0.99	*k*
0	0.904	0.599	0.349	0.107	0.028	0.006	0.001	0.000	0.000	0.000	0.000	0.000	0.000	0
1	0.996	0.914	0.736	0.376	0.149	0.046	0.011	0.002	0.000	0.000	0.000	0.000	0.000	1
2	1.000	0.988	0.930	0.678	0.383	0.167	0.055	0.012	0.002	0.000	0.000	0.000	0.000	2
3	1.000	0.999	0.987	0.879	0.650	0.382	0.172	0.055	0.011	0.001	0.000	0.000	0.000	3
4	1.000	1.000	0.998	0.967	0.850	0.633	0.377	0.166	0.047	0.006	0.000	0.000	0.000	4
5	1.000	1.000	1.000	0.994	0.953	0.834	0.623	0.367	0.150	0.033	0.002	0.000	0.000	5
6	1.000	1.000	1.000	0.999	0.989	0.945	0.828	0.618	0.350	0.121	0.013	0.001	0.000	6
7	1.000	1.000	1.000	1.000	0.998	0.988	0.945	0.833	0.617	0.322	0.070	0.012	0.000	7
8	1.000	1.000	1.000	1.000	1.000	0.998	0.989	0.954	0.851	0.624	0.264	0.086	0.004	8
9	1.000	1.000	1.000	1.000	1.000	1.000	0.999	0.994	0.972	0.893	0.651	0.401	0.096	9
10	1.000	1.000	1.000	1.000	1.000	1.000	1.000	1.000	1.000	1.000	1.000	1.000	1.000	10

$n = 11$

	p													
k	0.01	0.05	0.10	0.20	0.30	0.40	0.50	0.60	0.70	0.80	0.90	0.95	0.99	*k*
0	0.895	0.569	0.314	0.086	0.020	0.004	0.000	0.000	0.000	0.000	0.000	0.000	0.000	0
1	0.995	0.898	0.697	0.322	0.113	0.030	0.006	0.001	0.000	0.000	0.000	0.000	0.000	1
2	1.000	0.985	0.910	0.617	0.313	0.119	0.033	0.006	0.001	0.000	0.000	0.000	0.000	2
3	1.000	0.998	0.981	0.839	0.570	0.296	0.113	0.029	0.004	0.000	0.000	0.000	0.000	3
4	1.000	1.000	0.997	0.950	0.790	0.533	0.274	0.099	0.022	0.002	0.000	0.000	0.000	4
5	1.000	1.000	1.000	0.988	0.922	0.754	0.500	0.246	0.078	0.012	0.000	0.000	0.000	5
6	1.000	1.000	1.000	0.998	0.978	0.901	0.726	0.467	0.210	0.050	0.003	0.000	0.000	6
7	1.000	1.000	1.000	1.000	0.996	0.971	0.887	0.704	0.430	0.161	0.019	0.002	0.000	7
8	1.000	1.000	1.000	1.000	0.999	0.994	0.967	0.881	0.687	0.383	0.090	0.015	0.000	8
9	1.000	1.000	1.000	1.000	1.000	0.999	0.994	0.970	0.887	0.678	0.303	0.102	0.005	9
10	1.000	1.000	1.000	1.000	1.000	1.000	1.000	0.996	0.980	0.914	0.686	0.431	0.105	10
11	1.000	1.000	1.000	1.000	1.000	1.000	1.000	1.000	1.000	1.000	1.000	1.000	1.000	11

TABLE 1 *(continued)*

$n = 12$

	p													
k	0.01	0.05	0.10	0.20	0.30	0.40	0.50	0.60	0.70	0.80	0.90	0.95	0.99	*k*
0	0.886	0.540	0.282	0.069	0.014	0.002	0.000	0.000	0.000	0.000	0.000	0.000	0.000	0
1	0.994	0.882	0.659	0.275	0.085	0.020	0.003	0.000	0.000	0.000	0.000	0.000	0.000	1
2	1.000	0.980	0.889	0.558	0.253	0.083	0.019	0.003	0.000	0.000	0.000	0.000	0.000	2
3	1.000	0.998	0.974	0.795	0.493	0.225	0.073	0.015	0.002	0.000	0.000	0.000	0.000	3
4	1.000	1.000	0.996	0.927	0.724	0.438	0.194	0.057	0.009	0.001	0.000	0.000	0.000	4
5	1.000	1.000	0.999	0.981	0.882	0.665	0.387	0.158	0.039	0.004	0.000	0.000	0.000	5
6	1.000	1.000	1.000	0.996	0.961	0.842	0.613	0.335	0.118	0.019	0.001	0.000	0.000	6
7	1.000	1.000	1.000	0.999	0.991	0.943	0.806	0.562	0.276	0.073	0.004	0.000	0.000	7
8	1.000	1.000	1.000	1.000	0.998	0.985	0.927	0.775	0.507	0.205	0.026	0.002	0.000	8
9	1.000	1.000	1.000	1.000	1.000	0.997	0.981	0.917	0.747	0.442	0.111	0.020	0.000	9
10	1.000	1.000	1.000	1.000	1.000	1.000	0.997	0.980	0.915	0.725	0.341	0.118	0.006	10
11	1.000	1.000	1.000	1.000	1.000	1.000	1.000	0.998	0.986	0.931	0.718	0.460	0.114	11
12	1.000	1.000	1.000	1.000	1.000	1.000	1.000	1.000	1.000	1.000	1.000	1.000	1.000	12

$n = 15$

	p													
k	0.01	0.05	0.10	0.20	0.30	0.40	0.50	0.60	0.70	0.80	0.90	0.95	0.99	*k*
0	0.860	0.463	0.206	0.035	0.005	0.000	0.000	0.000	0.000	0.000	0.000	0.000	0.000	0
1	0.990	0.829	0.549	0.167	0.035	0.005	0.000	0.000	0.000	0.000	0.000	0.000	0.000	1
2	1.000	0.964	0.816	0.398	0.127	0.027	0.004	0.000	0.000	0.000	0.000	0.000	0.000	2
3	1.000	0.995	0.944	0.648	0.297	0.091	0.018	0.002	0.000	0.000	0.000	0.000	0.000	3
4	1.000	0.999	0.987	0.836	0.515	0.217	0.059	0.009	0.001	0.000	0.000	0.000	0.000	4
5	1.000	1.000	0.998	0.939	0.722	0.403	0.151	0.034	0.004	0.000	0.000	0.000	0.000	5
6	1.000	1.000	1.000	0.982	0.869	0.610	0.304	0.095	0.015	0.001	0.000	0.000	0.000	6
7	1.000	1.000	1.000	0.996	0.950	0.787	0.500	0.213	0.050	0.004	0.000	0.000	0.000	7
8	1.000	1.000	1.000	0.999	0.985	0.905	0.696	0.390	0.131	0.018	0.000	0.000	0.000	8
9	1.000	1.000	1.000	1.000	0.996	0.966	0.849	0.597	0.278	0.061	0.002	0.000	0.000	9
10	1.000	1.000	1.000	1.000	0.999	0.991	0.941	0.783	0.485	0.164	0.013	0.001	0.000	10
11	1.000	1.000	1.000	1.000	1.000	0.998	0.982	0.909	0.703	0.352	0.056	0.005	0.000	11
12	1.000	1.000	1.000	1.000	1.000	1.000	0.996	0.973	0.873	0.602	0.184	0.036	0.000	12
13	1.000	1.000	1.000	1.000	1.000	1.000	1.000	0.995	0.965	0.833	0.451	0.171	0.010	13
14	1.000	1.000	1.000	1.000	1.000	1.000	1.000	1.000	0.995	0.965	0.794	0.537	0.140	14
15	1.000	1.000	1.000	1.000	1.000	1.000	1.000	1.000	1.000	1.000	1.000	1.000	1.000	15

TABLE 1 *(continued)*

$n = 20$

	p													
k	0.01	0.05	0.10	0.20	0.30	0.40	0.50	0.60	0.70	0.80	0.90	0.95	0.99	*k*
0	0.818	0.358	0.122	0.012	0.001	0.000	0.000	0.000	0.000	0.000	0.000	0.000	0.000	0
1	0.983	0.736	0.392	0.069	0.008	0.001	0.000	0.000	0.000	0.000	0.000	0.000	0.000	1
2	0.999	0.925	0.677	0.206	0.035	0.004	0.000	0.000	0.000	0.000	0.000	0.000	0.000	2
3	1.000	0.984	0.867	0.411	0.107	0.016	0.001	0.000	0.000	0.000	0.000	0.000	0.000	3
4	1.000	0.997	0.957	0.630	0.238	0.051	0.006	0.000	0.000	0.000	0.000	0.000	0.000	4
5	1.000	1.000	0.989	0.804	0.416	0.126	0.021	0.002	0.000	0.000	0.000	0.000	0.000	5
6	1.000	1.000	0.998	0.913	0.608	0.250	0.058	0.006	0.000	0.000	0.000	0.000	0.000	6
7	1.000	1.000	1.000	0.968	0.772	0.416	0.132	0.021	0.001	0.000	0.000	0.000	0.000	7
8	1.000	1.000	1.000	0.990	0.887	0.596	0.252	0.057	0.005	0.000	0.000	0.000	0.000	8
9	1.000	1.000	1.000	0.997	0.952	0.755	0.412	0.128	0.017	0.001	0.000	0.000	0.000	9
10	1.000	1.000	1.000	0.999	0.983	0.872	0.588	0.245	0.048	0.003	0.000	0.000	0.000	10
11	1.000	1.000	1.000	1.000	0.995	0.943	0.748	0.404	0.113	0.010	0.000	0.000	0.000	11
12	1.000	1.000	1.000	1.000	0.999	0.979	0.868	0.584	0.228	0.032	0.000	0.000	0.000	12
13	1.000	1.000	1.000	1.000	1.000	0.994	0.942	0.750	0.392	0.087	0.002	0.000	0.000	13
14	1.000	1.000	1.000	1.000	1.000	0.998	0.979	0.874	0.584	0.196	0.011	0.000	0.000	14
15	1.000	1.000	1.000	1.000	1.000	1.000	0.994	0.949	0.762	0.370	0.043	0.003	0.000	15
16	1.000	1.000	1.000	1.000	1.000	1.000	0.999	0.984	0.893	0.589	0.133	0.016	0.000	16
17	1.000	1.000	1.000	1.000	1.000	1.000	1.000	0.996	0.965	0.794	0.323	0.075	0.001	17
18	1.000	1.000	1.000	1.000	1.000	1.000	1.000	0.999	0.992	0.931	0.608	0.264	0.017	18
19	1.000	1.000	1.000	1.000	1.000	1.000	1.000	1.000	0.999	0.988	0.878	0.642	0.182	19
20	1.000	1.000	1.000	1.000	1.000	1.000	1.000	1.000	1.000	1.000	1.000	1.000	1.000	20

TABLE 1 *(continued)*

$n = 25$

	p													
k	0.01	0.05	0.10	0.20	0.30	0.40	0.50	0.60	0.70	0.80	0.90	0.95	0.99	*k*
0	0.778	0.277	0.072	0.004	0.000	0.000	0.000	0.000	0.000	0.000	0.000	0.000	0.000	0
1	0.974	0.642	0.271	0.027	0.002	0.000	0.000	0.000	0.000	0.000	0.000	0.000	0.000	1
2	0.998	0.873	0.537	0.098	0.009	0.000	0.000	0.000	0.000	0.000	0.000	0.000	0.000	2
3	1.000	0.966	0.764	0.234	0.033	0.002	0.000	0.000	0.000	0.000	0.000	0.000	0.000	3
4	1.000	0.993	0.902	0.421	0.090	0.009	0.000	0.000	0.000	0.000	0.000	0.000	0.000	4
5	1.000	0.999	0.967	0.617	0.193	0.029	0.002	0.000	0.000	0.000	0.000	0.000	0.000	5
6	1.000	1.000	0.991	0.780	0.341	0.074	0.007	0.000	0.000	0.000	0.000	0.000	0.000	6
7	1.000	1.000	0.998	0.891	0.512	0.154	0.022	0.001	0.000	0.000	0.000	0.000	0.000	7
8	1.000	1.000	1.000	0.953	0.677	0.274	0.054	0.004	0.000	0.000	0.000	0.000	0.000	8
9	1.000	1.000	1.000	0.983	0.811	0.425	0.115	0.013	0.000	0.000	0.000	0.000	0.000	9
10	1.000	1.000	1.000	0.994	0.902	0.586	0.212	0.034	0.002	0.000	0.000	0.000	0.000	10
11	1.000	1.000	1.000	0.998	0.956	0.732	0.345	0.078	0.006	0.000	0.000	0.000	0.000	11
12	1.000	1.000	1.000	1.000	0.983	0.846	0.500	0.154	0.017	0.000	0.000	0.000	0.000	12
13	1.000	1.000	1.000	1.000	0.994	0.922	0.655	0.268	0.044	0.002	0.000	0.000	0.000	13
14	1.000	1.000	1.000	1.000	0.998	0.966	0.788	0.414	0.098	0.006	0.000	0.000	0.000	14
15	1.000	1.000	1.000	1.000	1.000	0.987	0.885	0.575	0.189	0.017	0.000	0.000	0.000	15
16	1.000	1.000	1.000	1.000	1.000	0.996	0.946	0.726	0.323	0.047	0.000	0.000	0.000	16
17	1.000	1.000	1.000	1.000	1.000	0.999	0.978	0.846	0.488	0.109	0.002	0.000	0.000	17
18	1.000	1.000	1.000	1.000	1.000	1.000	0.993	0.926	0.659	0.220	0.009	0.000	0.000	18
19	1.000	1.000	1.000	1.000	1.000	1.000	0.998	0.971	0.807	0.383	0.033	0.001	0.000	19
20	1.000	1.000	1.000	1.000	1.000	1.000	1.000	0.991	0.910	0.579	0.098	0.007	0.000	20
21	1.000	1.000	1.000	1.000	1.000	1.000	1.000	0.998	0.967	0.766	0.236	0.034	0.000	21
22	1.000	1.000	1.000	1.000	1.000	1.000	1.000	1.000	0.991	0.902	0.463	0.127	0.002	22
23	1.000	1.000	1.000	1.000	1.000	1.000	1.000	1.000	0.998	0.973	0.729	0.358	0.026	23
24	1.000	1.000	1.000	1.000	1.000	1.000	1.000	1.000	1.000	0.996	0.928	0.723	0.222	24
25	1.000	1.000	1.000	1.000	1.000	1.000	1.000	1.000	1.000	1.000	1.000	1.000	1.000	25

TABLE 2 Cumulative Poisson Probabilities
Tabulated values are $P(X \leq k) = p(0) + p(1) + \cdots + p(k)$
(Computations are rounded at the third decimal place.)

	μ										
k	0.1	0.2	0.3	0.4	0.5	0.6	0.7	0.8	0.9	1.0	1.5
0	0.905	0.819	0.741	0.670	0.607	0.549	0.497	0.449	0.407	0.368	0.223
1	0.995	0.982	0.963	0.938	0.910	0.878	0.844	0.809	0.772	0.736	0.558
2	1.000	0.999	0.996	0.992	0.986	0.977	0.966	0.953	0.937	0.920	0.809
3		1.000	1.000	0.999	0.998	0.997	0.994	0.991	0.987	0.981	0.934
4				1.000	1.000	1.000	0.999	0.999	0.998	0.996	0.981
5							1.000	1.000	1.000	0.999	0.996
6										1.000	0.999
7											1.000

	μ										
k	2.0	2.5	3.0	3.5	4.0	4.5	5.0	5.5	6.0	6.5	7.0
0	0.135	0.082	0.055	0.033	0.018	0.011	0.007	0.004	0.003	0.002	0.001
1	0.406	0.287	0.199	0.136	0.092	0.061	0.040	0.027	0.017	0.011	0.007
2	0.677	0.544	0.423	0.321	0.238	0.174	0.125	0.088	0.062	0.043	0.030
3	0.857	0.758	0.647	0.537	0.433	0.342	0.265	0.202	0.151	0.112	0.082
4	0.947	0.891	0.815	0.725	0.629	0.532	0.440	0.358	0.285	0.224	0.173
5	0.983	0.958	0.916	0.858	0.785	0.703	0.616	0.529	0.446	0.369	0.301
6	0.995	0.986	0.966	0.935	0.889	0.831	0.762	0.686	0.606	0.563	0.450
7	0.999	0.996	0.988	0.973	0.949	0.913	0.867	0.809	0.744	0.673	0.599
8	1.000	0.999	0.996	0.990	0.979	0.960	0.932	0.894	0.847	0.792	0.729
9		1.000	0.999	0.997	0.992	0.983	0.968	0.946	0.916	0.877	0.830
10			1.000	0.999	0.997	0.993	0.986	0.975	0.957	0.933	0.901
11				1.000	0.999	0.998	0.995	0.989	0.980	0.966	0.947
12					1.000	0.999	0.998	0.996	0.991	0.984	0.973
13						1.000	0.999	0.998	0.996	0.993	0.987
14							1.000	0.999	0.999	0.997	0.994
15								1.000	0.999	0.999	0.998
16									1.000	1.000	0.999
17											1.000

TABLE 2 *(continued)*

	μ								
k	7.5	8.0	8.5	9.0	9.5	10.0	12.0	15.0	20.0
0	0.001	0.000	0.000	0.000	0.000	0.000	0.000	0.000	0.000
1	0.005	0.003	0.002	0.001	0.001	0.000	0.000	0.000	0.000
2	0.020	0.014	0.009	0.006	0.004	0.003	0.001	0.000	0.000
3	0.059	0.042	0.030	0.021	0.015	0.010	0.002	0.000	0.000
4	0.132	0.100	0.074	0.055	0.040	0.029	0.008	0.001	0.000
5	0.241	0.191	0.150	0.116	0.089	0.067	0.020	0.003	0.000
6	0.378	0.313	0.256	0.207	0.165	0.130	0.046	0.008	0.000
7	0.525	0.453	0.386	0.324	0.269	0.220	0.090	0.018	0.001
8	0.662	0.593	0.523	0.456	0.392	0.333	0.155	0.037	0.002
9	0.776	0.717	0.653	0.587	0.522	0.458	0.242	0.070	0.005
10	0.862	0.816	0.763	0.706	0.645	0.583	0.347	0.118	0.011
11	0.921	0.888	0.849	0.803	0.752	0.697	0.462	0.185	0.021
12	0.957	0.936	0.909	0.876	0.836	0.792	0.576	0.268	0.039
13	0.978	0.966	0.949	0.926	0.898	0.864	0.682	0.363	0.066
14	0.990	0.983	0.973	0.959	0.940	0.917	0.772	0.466	0.105
15	0.995	0.992	0.986	0.978	0.967	0.951	0.844	0.568	0.157
16	0.998	0.996	0.993	0.989	0.982	0.973	0.899	0.664	0.221
17	0.999	0.998	0.997	0.995	0.991	0.986	0.937	0.749	0.297
18	1.000	0.999	0.999	0.998	0.996	0.993	0.963	0.819	0.381
19		1.000	0.999	0.999	0.998	0.997	0.979	0.875	0.470
20			1.000	1.000	0.999	0.998	0.988	0.917	0.559
21					1.000	0.999	0.994	0.947	0.644
22						1.000	0.997	0.967	0.721
23							0.999	0.981	0.787
24							0.999	0.989	0.843
25							1.000	0.994	0.888
26								0.997	0.922
27								0.998	0.948
28								0.999	0.966
29								1.000	0.978
30									0.987
31									0.992
32									0.995
33									0.997
34									0.999
35									0.999
36									1.000

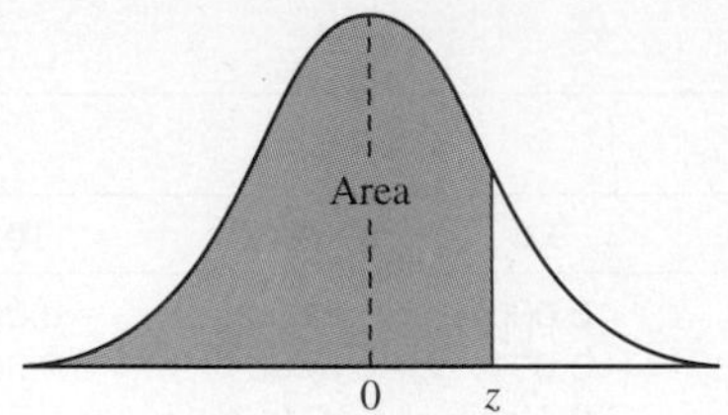

TABLE 3 **Areas under the Normal Curve**

z	0.00	0.01	0.02	0.03	0.04	0.05	0.06	0.07	0.08	0.09
−3.4	0.0003	0.0003	0.0003	0.0003	0.0003	0.0003	0.0003	0.0003	0.0003	0.0002
−3.3	0.0005	0.0005	0.0005	0.0004	0.0004	0.0004	0.0004	0.0004	0.0004	0.0003
−3.2	0.0007	0.0007	0.0006	0.0006	0.0006	0.0006	0.0006	0.0005	0.0005	0.0005
−3.1	0.0010	0.0009	0.0009	0.0009	0.0008	0.0008	0.0008	0.0008	0.0007	0.0007
−3.0	0.0013	0.0013	0.0013	0.0012	0.0012	0.0011	0.0011	0.0011	0.0010	0.0010
−2.9	0.0019	0.0018	0.0017	0.0017	0.0016	0.0016	0.0015	0.0015	0.0014	0.0014
−2.8	0.0026	0.0025	0.0024	0.0023	0.0023	0.0022	0.0021	0.0021	0.0020	0.0019
−2.7	0.0035	0.0034	0.0033	0.0032	0.0031	0.0030	0.0029	0.0028	0.0027	0.0026
−2.6	0.0047	0.0045	0.0044	0.0043	0.0041	0.0040	0.0039	0.0038	0.0037	0.0036
−2.5	0.0062	0.0060	0.0059	0.0057	0.0055	0.0054	0.0052	0.0051	0.0049	0.0048
−2.4	0.0082	0.0080	0.0078	0.0075	0.0073	0.0071	0.0069	0.0068	0.0066	0.0064
−2.3	0.0107	0.0104	0.0102	0.0099	0.0096	0.0094	0.0091	0.0089	0.0087	0.0084
−2.2	0.0139	0.0136	0.0132	0.0129	0.0125	0.0122	0.0119	0.0116	0.0113	0.0110
−2.1	0.0179	0.0174	0.0170	0.0166	0.0162	0.0158	0.0154	0.0150	0.0146	0.0143
−2.0	0.0228	0.0222	0.0217	0.0212	0.0207	0.0202	0.0197	0.0192	0.0188	0.0183
−1.9	0.0287	0.0281	0.0274	0.0268	0.0262	0.0256	0.0250	0.0244	0.0239	0.0233
−1.8	0.0359	0.0351	0.0344	0.0336	0.0329	0.0322	0.0314	0.0307	0.0301	0.0294
−1.7	0.0446	0.0436	0.0427	0.0418	0.0409	0.0401	0.0392	0.0384	0.0375	0.0367
−1.6	0.0548	0.0537	0.0526	0.0516	0.0505	0.0495	0.0485	0.0475	0.0465	0.0455
−1.5	0.0668	0.0655	0.0643	0.0630	0.0618	0.0606	0.0594	0.0582	0.0571	0.0559
−1.4	0.0808	0.0793	0.0778	0.0764	0.0749	0.0735	0.0722	0.0708	0.0694	0.0681
−1.3	0.0968	0.0951	0.0934	0.0918	0.0901	0.0885	0.0869	0.0853	0.0838	0.0823
−1.2	0.1151	0.1131	0.1112	0.1093	0.1075	0.1056	0.1038	0.1020	0.1003	0.0985
−1.1	0.1357	0.1335	0.1314	0.1292	0.1271	0.1251	0.1230	0.1210	0.1190	0.1170
−1.0	0.1587	0.1562	0.1539	0.1515	0.1492	0.1469	0.1446	0.1423	0.1401	0.1379
−0.9	0.1841	0.1814	0.1788	0.1762	0.1736	0.1711	0.1685	0.1660	0.1635	0.1611
−0.8	0.2119	0.2090	0.2061	0.2033	0.2005	0.1977	0.1949	0.1922	0.1894	0.1867
−0.7	0.2420	0.2389	0.2358	0.2327	0.2296	0.2266	0.2236	0.2206	0.2177	0.2148
−0.6	0.2743	0.2709	0.2676	0.2643	0.2611	0.2578	0.2546	0.2514	0.2483	0.2451
−0.5	0.3085	0.3050	0.3015	0.2981	0.2946	0.2912	0.2877	0.2843	0.2810	0.2776
−0.4	0.3446	0.3409	0.3372	0.3336	0.3300	0.3264	0.3228	0.3192	0.3156	0.3121
−0.3	0.3821	0.3783	0.3745	0.3707	0.3669	0.3632	0.3594	0.3557	0.3520	0.3483
−0.2	0.4207	0.4168	0.4129	0.4090	0.4052	0.4013	0.3974	0.3936	0.3897	0.3859
−0.1	0.4602	0.4562	0.4522	0.4483	0.4443	0.4404	0.4364	0.4325	0.4286	0.4247
−0.0	0.5000	0.4960	0.4920	0.4880	0.4840	0.4801	0.4761	0.4721	0.4681	0.4641

TABLE 3 *(continued)*

z	0.00	0.01	0.02	0.03	0.04	0.05	0.06	0.07	0.08	0.09
0.0	0.5000	0.5040	0.5080	0.5120	0.5160	0.5199	0.5239	0.5279	0.5319	0.5359
0.1	0.5398	0.5438	0.5478	0.5517	0.5557	0.5596	0.5636	0.5675	0.5714	0.5753
0.2	0.5793	0.5832	0.5871	0.5910	0.5948	0.5987	0.6026	0.6064	0.6103	0.6141
0.3	0.6179	0.6217	0.6255	0.6293	0.6331	0.6368	0.6406	0.6443	0.6480	0.6517
0.4	0.6554	0.6591	0.6628	0.6664	0.6700	0.6736	0.6772	0.6808	0.6844	0.6879
0.5	0.6915	0.6950	0.6985	0.7019	0.7054	0.7088	0.7123	0.7157	0.7190	0.7224
0.6	0.7257	0.7291	0.7324	0.7357	0.7389	0.7422	0.7454	0.7486	0.7517	0.7549
0.7	0.7580	0.7611	0.7642	0.7673	0.7704	0.7734	0.7764	0.7794	0.7823	0.7852
0.8	0.7881	0.7910	0.7939	0.7967	0.7995	0.8023	0.8051	0.8078	0.8106	0.8133
0.9	0.8159	0.8186	0.8212	0.8238	0.8264	0.8289	0.8315	0.8340	0.8365	0.8389
1.0	0.8413	0.8438	0.8461	0.8485	0.8508	0.8531	0.8554	0.8577	0.8599	0.8621
1.1	0.8643	0.8665	0.8686	0.8708	0.8729	0.8749	0.8770	0.8790	0.8810	0.8830
1.2	0.8849	0.8869	0.8888	0.8907	0.8925	0.8944	0.8962	0.8980	0.8997	0.9015
1.3	0.9032	0.9049	0.9066	0.9082	0.9099	0.9115	0.9131	0.9147	0.9162	0.9177
1.4	0.9192	0.9207	0.9222	0.9236	0.9251	0.9265	0.9279	0.9292	0.9306	0.9319
1.5	0.9332	0.9345	0.9357	0.9370	0.9382	0.9394	0.9406	0.9418	0.9429	0.9441
1.6	0.9452	0.9463	0.9474	0.9484	0.9495	0.9505	0.9515	0.9525	0.9535	0.9545
1.7	0.9554	0.9564	0.9573	0.9582	0.9591	0.9599	0.9608	0.9616	0.9625	0.9633
1.8	0.9641	0.9649	0.9656	0.9664	0.9671	0.9678	0.9686	0.9693	0.9699	0.9706
1.9	0.9713	0.9719	0.9726	0.9732	0.9738	0.9744	0.9750	0.9756	0.9761	0.9767
2.0	0.9772	0.9778	0.9783	0.9788	0.9793	0.9798	0.9803	0.9808	0.9812	0.9817
2.1	0.9821	0.9826	0.9830	0.9834	0.9838	0.9842	0.9846	0.9850	0.9854	0.9857
2.2	0.9861	0.9864	0.9868	0.9871	0.9875	0.9878	0.9881	0.9884	0.9887	0.9890
2.3	0.9893	0.9896	0.9898	0.9901	0.9904	0.9906	0.9909	0.9911	0.9913	0.9916
2.4	0.9918	0.9920	0.9922	0.9925	0.9927	0.9929	0.9931	0.9932	0.9934	0.9936
2.5	0.9938	0.9940	0.9941	0.9943	0.9945	0.9946	0.9948	0.9949	0.9951	0.9952
2.6	0.9953	0.9955	0.9956	0.9957	0.9959	0.9960	0.9961	0.9962	0.9963	0.9964
2.7	0.9965	0.9966	0.9967	0.9968	0.9969	0.9970	0.9971	0.9972	0.9973	0.9974
2.8	0.9974	0.9975	0.9976	0.9977	0.9977	0.9978	0.9979	0.9979	0.9980	0.9981
2.9	0.9981	0.9982	0.9982	0.9983	0.9984	0.9984	0.9985	0.9985	0.9986	0.9986
3.0	0.9987	0.9987	0.9987	0.9988	0.9988	0.9989	0.9989	0.9989	0.9990	0.9990
3.1	0.9990	0.9991	0.9991	0.9991	0.9992	0.9992	0.9992	0.9992	0.9993	0.9993
3.2	0.9993	0.9993	0.9994	0.9994	0.9994	0.9994	0.9994	0.9995	0.9995	0.9995
3.3	0.9995	0.9995	0.9995	0.9996	0.9996	0.9996	0.9996	0.9996	0.9996	0.9997
3.4	0.9997	0.9997	0.9997	0.9997	0.9997	0.9997	0.9997	0.9997	0.9997	0.9998

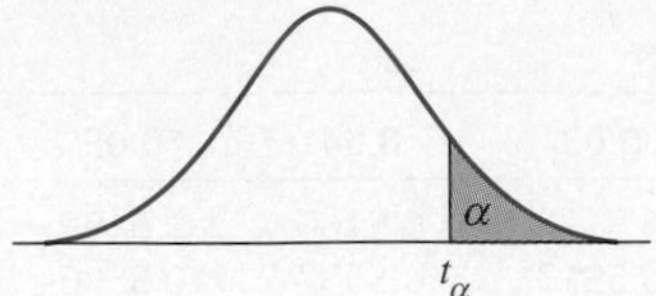

TABLE 4
Critical Values of *t*

df	$t_{0.100}$	$t_{0.050}$	$t_{0.025}$	$t_{0.010}$	$t_{0.005}$	*df*
1	3.078	6.314	12.706	31.821	63.657	1
2	1.886	2.920	4.303	6.965	9.925	2
3	1.638	2.353	3.182	4.541	5.841	3
4	1.533	2.132	2.776	3.747	4.604	4
5	1.476	2.015	2.571	3.365	4.032	5
6	1.440	1.943	2.447	3.143	3.707	6
7	1.415	1.895	2.365	2.998	3.499	7
8	1.397	1.860	2.306	2.896	3.355	8
9	1.383	1.833	2.262	2.821	3.250	9
10	1.372	1.812	2.228	2.764	3.169	10
11	1.363	1.796	2.201	2.718	3.106	11
12	1.356	1.782	2.179	2.681	3.055	12
13	1.350	1.771	2.160	2.650	3.012	13
14	1.345	1.761	2.145	2.624	2.977	14
15	1.341	1.753	2.131	2.602	2.947	15
16	1.337	1.746	2.120	2.583	2.921	16
17	1.333	1.740	2.110	2.567	2.898	17
18	1.330	1.734	2.101	2.552	2.878	18
19	1.328	1.729	2.093	2.539	2.861	19
20	1.325	1.725	2.086	2.528	2.845	20
21	1.323	1.721	2.080	2.518	2.831	21
22	1.321	1.717	2.074	2.508	2.819	22
23	1.319	1.714	2.069	2.500	2.807	23
24	1.318	1.711	2.064	2.492	2.797	24
25	1.316	1.708	2.060	2.485	2.787	25
26	1.315	1.706	2.056	2.479	2.779	26
27	1.314	1.703	2.052	2.473	2.771	27
28	1.313	1.701	2.048	2.467	2.763	28
29	1.311	1.699	2.045	2.462	2.756	29
∞	1.282	1.645	1.960	2.326	2.576	∞

Source: Maxine Merrington, "Table of Percentage Points of the *t*-Distribution," *Biometrika*, Vol. 32 (3–4), 1942: 300. Reproduced by permission of Oxford University Press on behalf of the *Biometrika* Trustees.

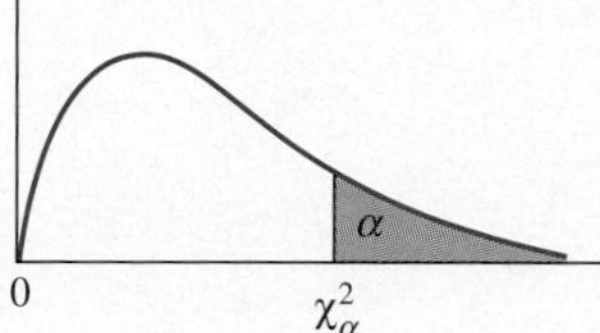

TABLE 5
Critical Values of Chi-Square

df	$\chi^2_{0.995}$	$\chi^2_{0.990}$	$\chi^2_{0.975}$	$\chi^2_{0.950}$	$\chi^2_{0.900}$
1	0.0000393	0.0001571	0.0009821	0.0039321	0.0157908
2	0.0100251	0.0201007	0.0506356	0.102587	0.210720
3	0.0717212	0.114832	0.215795	0.351846	0.584375
4	0.206990	0.297110	0.484419	0.710721	1.063623
5	0.411740	0.554300	0.831211	1.145476	1.61031
6	0.675727	0.872085	1.237347	1.63539	2.20413
7	0.989265	1.239043	1.68987	2.16735	2.83311
8	1.344419	1.646482	2.17973	2.73264	3.48954
9	1.734926	2.087912	2.70039	3.32511	4.16816
10	2.15585	2.55821	3.24697	3.94030	4.86518
11	2.60321	3.05347	3.81575	4.57481	5.57779
12	3.07382	3.57056	4.40379	5.22603	6.30380
13	3.56503	4.10691	5.00874	5.89186	7.04150
14	4.07468	4.66043	5.62872	6.57063	7.78953
15	4.60094	5.22935	6.26214	7.26094	8.54675
16	5.14224	5.81221	6.90766	7.96164	9.31223
17	5.69724	6.40776	7.56418	8.67176	10.0852
18	6.26481	7.01491	8.23075	9.39046	10.8649
19	6.84398	7.63273	8.90655	10.1170	11.6509
20	7.43386	8.26040	9.59083	10.8508	12.4426
21	8.03366	8.89720	10.28293	11.5913	13.2396
22	8.64272	9.54249	10.9823	12.3380	14.0415
23	9.26042	10.19567	11.6885	13.0905	14.8479
24	9.88623	10.8564	12.4011	13.8484	15.6587
25	10.5197	11.5240	13.1197	14.6114	16.4734
26	11.1603	12.1981	13.8439	15.3791	17.2919
27	11.8076	12.8786	14.5733	16.1513	18.1138
28	12.4613	13.5648	15.3079	16.9279	18.9392
29	13.1211	14.2565	16.0471	17.7083	19.7677
30	13.7867	14.9535	16.7908	18.4926	20.5992
40	20.7065	22.1643	24.4331	26.5093	29.0505
50	27.9907	29.7067	32.3574	34.7642	37.6886
60	35.5346	37.4848	40.4817	43.1879	46.4589
70	43.2752	45.4418	48.7576	51.7393	55.3290
80	51.1720	53.5400	57.1532	60.3915	64.2778
90	59.1963	61.7541	65.6466	69.1260	73.2912
100	67.3276	70.0648	74.2219	77.9295	82.3581

Source: "Tables of the Percentage Points of the χ^2-Distribution," *Biometrika Tables for Statisticians*, Vol. 1, 3rd ed. edited by E.S. Pearson and H.O. Hartley (Cambridge University Press, 1966). Reproduced by permission of Oxford University Press on behalf of the *Biometrika* Trustees.

TABLE 5 *(continued)*

$\chi^2_{0.100}$	$\chi^2_{0.050}$	$\chi^2_{0.025}$	$\chi^2_{0.010}$	$\chi^2_{0.005}$	*df*
2.70554	3.84146	5.02389	6.63490	7.87944	1
4.60517	5.99147	7.37776	9.21034	10.5966	2
6.25139	7.81473	9.34840	11.3449	12.8381	3
7.77944	9.48773	11.1433	13.2767	14.8602	4
9.23635	11.0705	12.8325	15.0863	16.7496	5
10.6446	12.5916	14.4494	16.8119	18.5476	6
12.0170	14.0671	16.0128	18.4753	20.2777	7
13.3616	15.5073	17.5346	20.0902	21.9550	8
14.6837	16.9190	19.0228	21.6660	23.5893	9
15.9871	18.3070	20.4831	23.2093	25.1882	10
17.2750	19.6751	21.9200	24.7250	26.7569	11
18.5494	21.0261	23.3367	26.2170	28.2995	12
19.8119	22.3621	24.7356	27.6883	29.8194	13
21.0642	23.6848	26.1190	29.1413	31.3193	14
22.3072	24.9958	27.4884	30.5779	32.8013	15
23.5418	26.2962	28.8485	31.9999	34.2672	16
24.7690	27.8571	30.1910	33.4087	35.7185	17
25.9894	28.8693	31.5264	34.8053	37.1564	18
27.2036	30.1435	32.8523	36.1908	38.5822	19
28.4120	31.4104	34.1696	37.5662	39.9968	20
29.6151	32.6705	35.4789	38.9321	41.4010	21
30.8133	33.9244	36.7807	40.2894	42.7956	22
32.0069	35.1725	38.0757	41.6384	44.1813	23
33.1963	36.4151	39.3641	42.9798	45.5585	24
34.3816	37.6525	40.6465	44.3141	46.9278	25
35.5631	38.8852	41.9232	45.6417	48.2899	26
36.7412	40.1133	43.1944	46.9630	49.6449	27
37.9159	41.3372	44.4607	48.2782	50.9933	28
39.0875	42.5569	45.7222	49.5879	52.3356	29
40.2560	43.7729	46.9792	50.8922	53.6720	30
51.8050	55.7585	59.3417	63.6907	66.7659	40
63.1671	67.5048	71.4202	76.1539	79.4900	50
74.3970	79.0819	83.2976	88.3794	91.9517	60
85.5271	90.5312	95.0231	100.425	104.215	70
96.5782	101.879	106.629	112.329	116.321	80
107.565	113.145	118.136	124.116	128.299	90
118.498	124.342	129.561	135.807	140.169	100

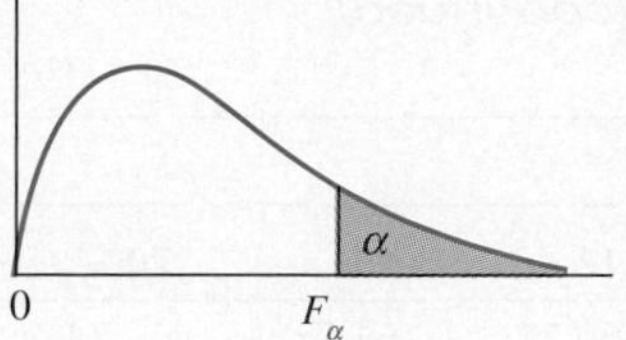

TABLE 6 Percentage Points of the *F* Distribution

		df_1								
df_2	α	1	2	3	4	5	6	7	8	9
1	0.100	39.86	49.50	53.59	55.83	57.24	58.20	58.91	59.44	59.86
	0.050	161.4	199.5	215.7	224.6	230.2	234.0	236.8	238.9	240.5
	0.025	647.8	799.5	864.2	899.6	921.8	937.1	948.2	956.7	963.3
	0.010	4052	4999.5	5403	5625	5764	5859	5928	5982	6022
	0.005	16211	20000	21615	22500	23056	23437	23715	23925	24091
2	0.100	8.53	9.00	9.16	9.24	9.29	9.33	9.35	9.37	9.38
	0.050	18.51	19.00	19.16	19.25	19.30	19.33	19.35	19.37	19.38
	0.025	38.51	39.00	39.17	39.25	39.30	39.33	39.36	39.37	39.39
	0.010	98.50	99.00	99.17	99.25	99.30	99.33	99.36	99.37	99.39
	0.005	198.5	199.0	199.2	199.2	199.3	199.3	199.4	199.4	199.4
3	0.100	5.54	5.46	5.39	5.34	5.31	5.28	5.27	5.25	5.24
	0.050	10.13	9.55	9.28	9.12	9.01	8.94	8.89	8.85	8.81
	0.025	17.44	16.04	15.44	15.10	14.88	14.73	14.62	14.54	14.47
	0.010	34.12	30.82	29.46	28.71	28.24	27.91	27.64	27.49	27.35
	0.005	55.55	49.80	47.47	46.19	45.39	44.84	44.43	44.13	43.88
4	0.100	4.54	4.32	4.19	4.11	4.05	4.01	3.98	3.95	3.94
	0.050	7.71	6.94	6.59	6.39	6.26	6.16	6.09	6.04	6.00
	0.025	12.22	10.65	9.98	9.60	9.36	9.20	9.07	8.98	8.90
	0.010	21.20	18.00	16.69	15.98	15.52	15.21	14.98	14.80	14.66
	0.005	31.33	26.28	24.26	23.15	22.46	21.97	21.62	21.35	21.14
5	0.100	4.06	3.78	3.62	3.52	3.45	3.40	3.37	3.34	3.32
	0.050	6.61	5.79	5.41	5.19	5.05	4.95	4.88	4.82	4.77
	0.025	10.01	8.43	7.76	7.39	7.15	6.98	6.85	6.76	6.68
	0.010	16.26	13.27	12.06	11.39	10.97	10.67	10.46	10.29	10.16
	0.005	22.78	18.31	16.53	15.56	14.94	14.51	14.20	13.96	13.77
6	0.100	3.78	3.46	3.29	3.18	3.11	3.05	3.01	2.98	2.96
	0.050	5.99	5.14	4.76	4.53	4.39	4.28	4.21	4.15	4.10
	0.025	8.81	7.26	6.60	6.23	5.99	5.82	5.70	5.60	5.52
	0.010	13.75	10.92	9.78	9.15	8.75	8.47	8.26	8.10	7.98
	0.005	18.63	14.54	12.92	12.03	11.46	11.07	10.79	10.57	10.39
7	0.100	3.59	3.26	3.07	2.96	2.88	2.83	2.78	2.75	2.72
	0.050	5.59	4.74	4.35	4.12	3.97	3.87	3.79	3.73	3.68
	0.025	8.07	6.54	5.89	5.52	5.29	5.12	4.99	4.90	4.82
	0.010	12.25	9.55	8.45	7.85	7.46	7.19	6.99	6.84	6.72
	0.005	16.24	12.40	10.88	10.05	9.52	9.16	8.89	8.68	8.51
8	0.100	3.46	3.11	2.92	2.81	2.73	2.67	2.62	2.59	2.56
	0.050	5.32	4.46	4.07	3.84	3.69	3.58	3.50	3.44	3.39
	0.025	7.57	6.06	5.42	5.05	4.82	4.65	4.53	4.43	4.36
	0.010	11.26	8.65	7.59	7.01	6.63	6.37	6.18	6.03	5.91
	0.005	14.69	11.04	9.60	8.81	8.30	7.95	7.69	7.50	7.34
9	0.100	3.36	3.01	2.81	2.69	2.61	2.55	2.51	2.47	2.44
	0.050	5.12	4.26	3.86	3.63	3.48	3.37	3.29	3.23	3.18
	0.025	7.21	5.71	5.08	4.72	4.48	4.32	4.20	4.10	4.03
	0.010	10.56	8.02	6.99	6.42	6.06	5.80	5.61	5.47	5.35
	0.005	13.61	10.11	8.72	7.96	7.47	7.13	6.88	6.69	6.54

Source: A portion of "Tables of percentage points of the inverted beta (*F*) distribution," *Biometrika*, vol. 33 (1943) by M. Merrington and C.M. Thompson and from Table 18 of *Biometrika Tables for Statisticians*, Vol. 1, edited by E.S. Pearson and H.O. Hartley (Cambridge University Press, 1954). Reproduced by permission of Oxford University Press on behalf of the *Biometrika* Trustees.

TABLE 6 *(continued)*

df_1											
10	12	15	20	24	30	40	60	120	∞	α	df_2
60.19	60.71	61.22	61.74	62.00	62.26	62.53	62.79	63.06	63.33	0.100	1
241.9	243.9	245.9	248.0	249.1	250.1	251.2	252.2	253.3	254.3	0.050	
968.6	976.7	984.9	993.1	997.2	1001	1006	1010	1014	1018	0.025	
6056	6106	6157	6209	6235	6261	6287	6313	6339	6366	0.010	
24224	24426	24630	24836	24940	25044	25148	25253	25359	25465	0.005	
9.39	9.41	9.42	9.44	9.45	9.46	9.47	9.47	9.48	9.49	0.100	2
19.40	19.41	19.43	19.45	19.45	19.46	19.47	19.48	19.49	19.50	0.050	
39.40	39.41	39.43	39.45	39.46	39.46	39.47	39.48	39.49	39.50	0.025	
99.40	99.42	99.43	99.45	99.46	99.47	99.47	99.48	99.49	99.50	0.010	
199.4	199.4	199.4	199.4	199.5	199.5	199.5	199.5	199.5	199.5	0.005	
5.23	5.22	5.20	5.18	5.18	5.17	5.16	5.15	5.14	5.13	0.100	3
8.79	8.74	8.70	8.66	8.64	8.62	8.59	8.57	8.55	8.53	0.050	
14.42	14.34	14.25	14.17	14.12	14.08	14.04	13.99	13.95	13.90	0.025	
27.23	27.05	26.87	26.69	26.60	26.50	26.41	26.32	26.22	26.13	0.010	
43.69	43.39	43.08	42.78	42.62	42.47	42.31	42.15	41.99	41.83	0.005	
3.92	3.90	3.87	3.84	3.83	3.82	3.80	3.79	3.78	3.76	0.100	4
5.96	5.91	5.86	5.80	5.77	5.75	5.72	5.69	5.66	5.63	0.050	
8.84	8.75	8.66	8.56	8.51	8.46	8.41	8.36	8.31	8.26	0.025	
14.55	14.37	14.20	14.02	13.93	13.84	13.75	13.65	13.56	13.46	0.010	
20.97	20.70	20.44	20.17	20.03	19.89	19.75	19.61	19.47	19.32	0.005	
3.30	3.27	3.24	3.21	3.19	3.17	3.16	3.14	3.12	3.10	0.100	5
4.74	4.68	4.62	4.56	4.53	4.50	4.46	4.43	4.40	4.36	0.050	
6.62	6.52	6.43	6.33	6.28	6.23	6.18	6.12	6.07	6.02	0.025	
10.05	9.89	9.72	9.55	9.47	9.38	9.29	9.20	9.11	9.02	0.010	
13.62	13.38	13.15	12.90	12.78	12.66	12.53	12.40	12.27	12.14	0.005	
2.94	2.90	2.87	2.84	2.82	2.80	2.78	2.76	2.74	2.72	0.100	6
4.06	4.00	3.94	3.87	3.84	3.81	3.77	3.74	3.70	3.67	0.050	
5.46	5.37	5.27	5.17	5.12	5.07	5.01	4.96	4.90	4.85	0.025	
7.87	7.72	7.56	7.40	7.31	7.23	7.14	7.06	6.97	6.88	0.010	
10.25	10.03	9.81	9.59	9.47	9.36	9.24	9.12	9.00	8.88	0.005	
2.70	2.67	2.63	2.59	2.58	2.56	2.54	2.51	2.49	2.47	0.100	7
3.64	3.57	3.51	3.44	3.41	3.38	3.34	3.30	3.27	3.23	0.050	
4.76	4.67	4.57	4.47	4.42	4.36	4.31	4.25	4.20	4.14	0.025	
6.62	6.47	6.31	6.16	6.07	5.99	5.91	5.82	5.74	5.65	0.010	
8.38	8.18	7.97	7.75	7.65	7.53	7.42	7.31	7.19	7.08	0.005	
2.54	2.50	2.46	2.42	2.40	2.38	2.36	2.34	2.32	2.29	0.100	8
3.35	3.28	3.22	3.15	3.12	3.08	3.04	3.01	2.97	2.93	0.050	
4.30	4.20	4.10	4.00	3.95	3.89	3.84	3.78	3.73	3.67	0.025	
5.81	5.67	5.52	5.36	5.28	5.20	5.12	5.03	4.95	4.86	0.010	
7.21	7.01	6.81	6.61	6.50	6.40	6.29	6.18	6.06	5.95	0.005	
2.42	2.38	2.34	2.30	2.28	2.25	2.23	2.21	2.18	2.16	0.100	9
3.14	3.07	3.01	2.94	2.90	2.86	2.83	2.79	2.75	2.71	0.050	
3.96	3.87	3.77	3.67	3.61	3.56	3.51	3.45	3.39	3.33	0.025	
5.26	5.11	4.96	4.81	4.73	4.65	4.57	4.48	4.40	4.31	0.010	
6.42	6.23	6.03	5.83	5.73	5.62	5.52	5.41	5.30	5.19	0.005	

TABLE 6 *(continued)*

		df_1								
df_2	α	1	2	3	4	5	6	7	8	9
10	0.100	3.29	2.92	2.73	2.61	2.52	2.46	2.41	2.38	2.35
	0.050	4.96	4.10	3.71	3.48	3.33	3.22	3.14	3.07	3.02
	0.025	6.94	5.46	4.83	4.47	4.24	4.07	3.95	3.85	3.78
	0.010	10.04	7.56	6.55	5.99	5.64	5.39	5.20	5.06	4.94
	0.005	12.83	9.43	8.08	7.34	6.87	6.54	6.30	6.12	5.97
11	0.100	3.23	2.86	2.66	2.54	2.45	2.39	2.34	2.30	2.27
	0.050	4.84	3.98	3.59	3.36	3.20	3.09	3.01	2.95	2.90
	0.025	6.72	5.26	4.63	4.28	4.04	3.88	3.76	3.66	3.59
	0.010	9.65	7.21	6.22	5.67	5.32	5.07	4.89	4.74	4.63
	0.005	12.23	8.91	7.60	6.88	6.42	6.10	5.86	5.68	5.54
12	0.100	3.18	2.81	2.61	2.48	2.39	2.33	2.28	2.24	2.21
	0.050	4.75	3.89	3.49	3.26	3.11	3.00	2.91	2.85	2.80
	0.025	6.55	5.10	4.47	4.12	3.89	3.73	3.61	3.51	3.44
	0.010	9.33	6.93	5.95	5.41	5.06	4.82	4.64	4.50	4.39
	0.005	11.75	8.51	7.23	6.52	6.07	5.76	5.52	5.35	5.20
13	0.100	3.14	2.76	2.56	2.43	2.35	2.28	2.23	2.20	2.16
	0.050	4.67	3.81	3.41	3.18	3.03	2.92	2.83	2.77	2.71
	0.025	6.41	4.97	4.35	4.00	3.77	3.60	3.48	3.39	3.31
	0.010	9.07	6.70	5.74	5.21	4.86	4.62	4.44	4.30	4.19
	0.005	11.37	8.19	6.93	6.23	5.79	5.48	5.25	5.08	4.94
14	0.100	3.10	2.73	2.52	2.39	2.31	2.24	2.19	2.15	2.12
	0.050	4.60	3.74	3.34	3.11	2.96	2.85	2.76	2.70	2.65
	0.025	6.30	4.86	4.24	3.89	3.66	3.50	3.38	3.29	3.21
	0.010	8.86	6.51	5.56	5.04	4.69	4.46	4.28	4.14	4.03
	0.005	11.06	7.92	6.68	6.00	5.56	5.26	5.03	4.86	4.72
15	0.100	3.07	2.70	2.49	2.36	2.27	2.21	2.16	2.12	2.09
	0.050	4.54	3.68	3.29	3.06	2.90	2.79	2.71	2.64	2.59
	0.025	6.20	4.77	4.15	3.80	3.58	3.41	3.29	3.20	3.12
	0.010	8.68	6.36	5.42	4.89	4.56	4.32	4.14	4.00	3.89
	0.005	10.80	7.70	6.48	5.80	5.37	5.07	4.85	4.67	4.54
16	0.100	3.05	2.67	2.46	2.33	2.24	2.18	2.13	2.09	2.06
	0.050	4.49	3.63	3.24	3.01	2.85	2.74	2.66	2.59	2.54
	0.025	6.12	4.69	4.08	3.73	3.50	3.34	3.22	3.12	3.05
	0.010	8.53	6.23	5.29	4.77	4.44	4.20	4.03	3.89	3.78
	0.005	10.58	7.51	6.30	5.64	5.21	4.91	4.69	4.52	4.38
17	0.100	3.03	2.64	2.44	2.31	2.22	2.15	2.10	2.06	2.03
	0.050	4.45	3.59	3.20	2.96	2.81	2.70	2.61	2.55	2.49
	0.025	6.04	4.62	4.01	3.66	3.44	3.28	3.16	3.06	2.98
	0.010	8.40	6.11	5.18	4.67	4.34	4.10	3.93	3.79	3.68
	0.005	10.38	7.35	6.16	5.50	5.07	4.78	4.56	4.39	4.25
18	0.100	3.01	2.62	2.42	2.29	2.20	2.13	2.08	2.04	2.00
	0.050	4.41	3.55	3.16	2.93	2.77	2.66	2.58	2.51	2.46
	0.025	5.98	4.56	3.95	3.61	3.38	3.22	3.10	3.01	2.93
	0.010	8.29	6.01	5.09	4.58	4.25	4.01	3.84	3.71	3.60
	0.005	10.22	7.21	6.03	5.37	4.96	4.66	4.44	4.28	4.14
19	0.100	2.99	2.61	2.40	2.27	2.18	2.11	2.06	2.02	1.98
	0.050	4.38	3.52	3.13	2.90	2.74	2.63	2.54	2.48	2.42
	0.025	5.92	4.51	3.90	3.56	3.33	3.17	3.05	2.96	2.88
	0.010	8.18	5.93	5.01	4.50	4.17	3.94	3.77	3.63	3.52
	0.005	10.07	7.09	5.92	5.27	4.85	4.56	4.34	4.18	4.04
20	0.100	2.97	2.59	2.38	2.25	2.16	2.09	2.04	2.00	1.96
	0.050	4.35	3.49	3.10	2.87	2.71	2.60	2.51	2.45	2.39
	0.025	5.87	4.46	3.86	3.51	3.29	3.13	3.01	2.91	2.84
	0.010	8.10	5.85	4.94	4.43	4.10	3.87	3.70	3.56	3.46
	0.005	9.94	6.99	5.82	5.17	4.76	4.47	4.26	4.09	3.96

TABLE 6 *(continued)*

df_1											
10	12	15	20	24	30	40	60	120	∞	α	df_2
2.32	2.28	2.24	2.20	2.18	2.16	2.13	2.11	2.08	2.06	0.100	10
2.98	2.91	2.85	2.77	2.74	2.70	2.66	2.62	2.58	2.54	0.050	
3.72	3.62	3.52	3.42	3.37	3.31	3.26	3.20	3.14	3.08	0.025	
4.85	4.71	4.56	4.41	4.33	4.25	4.17	4.08	4.00	3.91	0.010	
5.85	5.66	5.47	5.27	5.17	5.07	4.97	4.86	4.75	4.64	0.005	
2.25	2.21	2.17	2.12	2.10	2.08	2.05	2.03	2.00	1.97	0.100	11
2.85	2.79	2.72	2.65	2.61	2.57	2.53	2.49	2.45	2.40	0.050	
3.53	3.43	3.33	3.23	3.17	3.12	3.06	3.00	2.94	2.88	0.025	
4.54	4.40	4.25	4.10	4.02	3.94	3.86	3.78	3.69	3.60	0.010	
5.42	5.24	5.05	4.86	4.76	4.65	4.55	4.44	4.34	4.23	0.005	
2.19	2.15	2.10	2.06	2.04	2.01	1.99	1.96	1.93	1.90	0.100	12
2.75	2.69	2.62	2.54	2.51	2.47	2.43	2.38	2.34	2.30	0.050	
3.37	3.28	3.18	3.07	3.02	2.96	2.91	2.85	2.79	2.72	0.025	
4.30	4.16	4.01	3.86	3.78	3.70	3.62	3.54	3.45	3.36	0.010	
5.09	4.91	4.72	4.53	4.43	4.33	4.23	4.12	4.01	3.90	0.005	
2.14	2.10	2.05	2.01	1.98	1.96	1.93	1.90	1.88	1.85	0.100	13
2.67	2.60	2.53	2.46	2.42	2.38	2.34	2.30	2.25	2.21	0.050	
3.25	3.15	3.05	2.95	2.89	2.84	2.78	2.72	2.66	2.60	0.025	
4.10	3.96	3.82	3.66	3.59	3.51	3.43	3.34	3.25	3.17	0.010	
4.82	4.64	4.46	4.27	4.17	4.07	3.97	3.87	3.76	3.65	0.005	
2.10	2.05	2.01	1.96	1.94	1.91	1.89	1.86	1.83	1.80	0.100	14
2.60	2.53	2.46	2.39	2.35	2.31	2.27	2.22	2.18	2.13	0.050	
3.15	3.05	2.95	2.84	2.79	2.73	2.67	2.61	2.55	2.49	0.025	
3.94	3.80	3.66	3.51	3.43	3.35	3.27	3.18	3.09	3.00	0.010	
4.60	4.43	4.25	4.06	3.96	3.86	3.76	3.66	3.55	3.44	0.005	
2.06	2.02	1.97	1.92	1.90	1.87	1.85	1.82	1.79	1.76	0.100	15
2.54	2.48	2.40	2.33	2.29	2.25	2.20	2.16	2.11	2.07	0.050	
3.06	2.96	2.86	2.76	2.70	2.64	2.59	2.52	2.46	2.40	0.025	
3.80	3.67	3.52	3.37	3.29	3.21	3.13	3.05	2.96	2.87	0.010	
4.42	4.25	4.07	3.88	3.79	3.69	3.58	3.48	3.37	3.26	0.005	
2.03	1.99	1.94	1.89	1.87	1.84	1.81	1.78	1.75	1.72	0.100	16
2.49	2.42	2.35	2.28	2.24	2.19	2.15	2.11	2.06	2.01	0.050	
2.99	2.89	2.79	2.68	2.63	2.57	2.51	2.45	2.38	2.32	0.025	
3.69	3.55	3.41	3.26	3.18	3.10	3.02	2.93	2.84	2.75	0.010	
4.27	4.10	3.92	3.73	3.64	3.54	3.44	3.33	3.22	3.11	0.005	
2.00	1.96	1.91	1.86	1.84	1.81	1.78	1.75	1.72	1.69	0.100	17
2.45	2.38	2.31	2.23	2.19	2.15	2.10	2.06	2.01	1.96	0.050	
2.92	2.82	2.72	2.62	2.56	2.50	2.44	2.38	2.32	2.25	0.025	
3.59	3.46	3.31	3.16	3.08	3.00	2.92	2.83	2.75	2.65	0.010	
4.14	3.97	3.79	3.61	3.51	3.41	3.31	3.21	3.10	2.98	0.005	
1.98	1.93	1.89	1.84	1.81	1.78	1.75	1.72	1.69	1.66	0.100	18
2.41	2.34	2.27	2.19	2.15	2.11	2.06	2.02	1.97	1.92	0.050	
2.87	2.77	2.67	2.56	2.50	2.44	2.38	2.32	2.26	2.19	0.025	
3.51	3.37	3.23	3.08	3.00	2.92	2.84	2.75	2.66	2.57	0.010	
4.03	3.86	3.68	3.50	3.40	3.30	3.20	3.10	2.99	2.87	0.005	
1.96	1.91	1.86	1.81	1.79	1.76	1.73	1.70	1.67	1.63	0.100	19
2.38	2.31	2.23	2.16	2.11	2.07	2.03	1.98	1.93	1.88	0.050	
2.82	2.72	2.62	2.51	2.45	2.39	2.33	2.27	2.20	2.13	0.025	
3.43	3.30	3.15	3.00	2.92	2.84	2.76	2.67	2.58	2.49	0.010	
3.93	3.76	3.59	3.40	3.31	3.21	3.11	3.00	2.89	2.78	0.005	
1.94	1.89	1.84	1.79	1.77	1.74	1.71	1.68	1.64	1.61	0.100	20
2.35	2.28	2.20	2.12	2.08	2.04	1.99	1.95	1.90	1.84	0.050	
2.77	2.68	2.57	2.46	2.41	2.35	2.29	2.22	2.16	2.09	0.025	
3.37	3.23	3.09	2.94	2.86	2.78	2.69	2.61	2.52	2.42	0.010	
3.85	3.68	3.50	3.32	3.22	3.12	3.02	2.92	2.81	2.69	0.005	

TABLE 6 *(continued)*

df_2	α	df_1 1	2	3	4	5	6	7	8	9
21	0.100	2.96	2.57	2.36	2.23	2.14	2.08	2.02	1.98	1.95
	0.050	4.32	3.47	3.07	2.84	2.68	2.57	2.49	2.42	2.37
	0.025	5.83	4.42	3.82	3.48	3.25	3.09	2.97	2.87	2.80
	0.010	8.02	5.78	4.87	4.37	4.04	3.81	3.64	3.51	3.40
	0.005	9.83	6.89	5.73	5.09	4.68	4.39	4.18	4.01	3.88
22	0.100	2.95	2.56	2.35	2.22	2.13	2.06	2.01	1.97	1.93
	0.050	4.30	3.44	3.05	2.82	2.66	2.55	2.46	2.40	2.34
	0.025	5.79	4.38	3.78	3.44	3.22	3.05	2.93	2.84	2.76
	0.010	7.95	5.72	4.82	4.31	3.99	3.76	3.59	3.45	3.35
	0.005	9.73	6.81	5.65	5.02	4.61	4.32	4.11	3.94	3.81
23	0.100	2.94	2.55	2.34	2.21	2.11	2.05	1.99	1.95	1.92
	0.050	4.28	3.42	3.03	2.80	2.64	2.53	2.44	2.37	2.32
	0.025	5.75	4.35	3.75	3.41	3.18	3.02	2.90	2.81	2.73
	0.010	7.88	5.66	4.76	4.26	3.94	3.71	3.54	3.41	3.30
	0.005	9.63	6.73	5.58	4.95	4.54	4.26	4.05	3.88	3.75
24	0.100	2.93	2.54	2.33	2.19	2.10	2.04	1.98	1.94	1.91
	0.050	4.26	3.40	3.01	2.78	2.62	2.51	2.42	2.36	2.30
	0.025	5.72	4.32	3.72	3.38	3.15	2.99	2.87	2.78	2.70
	0.010	7.82	5.61	4.72	4.22	3.90	3.67	3.50	3.36	3.26
	0.005	9.55	6.66	5.52	4.89	4.49	4.20	3.99	3.83	3.69
25	0.100	2.92	2.53	2.32	2.18	2.09	2.02	1.97	1.93	1.89
	0.050	4.24	3.39	2.99	2.76	2.60	2.49	2.40	2.34	2.28
	0.025	5.69	4.29	3.69	3.35	3.13	2.97	2.85	2.75	2.68
	0.010	7.77	5.57	4.68	4.18	3.85	3.63	3.46	3.32	3.22
	0.005	9.48	6.60	5.46	4.84	4.43	4.15	3.94	3.78	3.64
26	0.100	2.91	2.52	2.31	2.17	2.08	2.01	1.96	1.92	1.88
	0.050	4.23	3.37	2.98	2.74	2.59	2.47	2.39	2.32	2.27
	0.025	5.66	4.27	3.67	3.33	3.10	2.94	2.82	2.73	2.65
	0.010	7.72	5.53	4.64	4.14	3.82	3.59	3.42	3.29	3.18
	0.005	9.41	6.54	5.41	4.79	4.38	4.10	3.89	3.73	3.60
27	0.100	2.90	2.51	2.30	2.17	2.07	2.00	1.95	1.91	1.87
	0.050	4.21	3.35	2.96	2.73	2.57	2.46	2.37	2.31	2.25
	0.025	5.63	4.24	3.65	3.31	3.08	2.92	2.80	2.71	2.63
	0.010	7.68	5.49	4.60	4.11	3.78	3.56	3.39	3.26	3.15
	0.005	9.34	6.49	5.36	4.74	4.34	4.06	3.85	3.69	3.56
28	0.100	2.89	2.50	2.29	2.16	2.06	2.00	1.94	1.90	1.87
	0.050	4.20	3.34	2.95	2.71	2.56	2.45	2.36	2.29	2.24
	0.025	5.61	4.22	3.63	3.29	3.06	2.90	2.78	2.69	2.61
	0.010	7.64	5.45	4.57	4.07	3.75	3.53	3.36	3.23	3.12
	0.005	9.28	6.44	5.32	4.70	4.30	4.02	3.81	3.65	3.52
29	0.100	2.89	2.50	2.28	2.15	2.06	1.99	1.93	1.89	1.86
	0.050	4.18	3.33	2.93	2.70	2.55	2.43	2.35	2.28	2.22
	0.025	5.59	4.20	3.61	3.27	3.04	2.88	2.76	2.67	2.59
	0.010	7.60	5.42	4.54	4.04	3.73	3.50	3.33	3.20	3.09
	0.005	9.23	6.40	5.28	4.66	4.26	3.98	3.77	3.61	3.48
30	0.100	2.88	2.49	2.28	2.14	2.05	1.98	1.93	1.88	1.85
	0.050	4.17	3.32	2.92	2.69	2.53	2.42	2.33	2.27	2.21
	0.025	5.57	4.18	3.59	3.25	3.03	2.87	2.75	2.65	2.57
	0.010	7.56	5.39	4.51	4.02	3.70	3.47	3.30	3.17	3.07
	0.005	9.18	6.35	5.24	4.62	4.23	3.95	3.74	3.58	3.45

TABLE 6 *(continued)*

df_1											
10	12	15	20	24	30	40	60	120	∞	α	df_2
1.92	1.87	1.83	1.78	1.75	1.72	1.69	1.66	1.62	1.59	0.100	21
2.32	2.25	2.18	2.10	2.05	2.01	1.96	1.92	1.87	1.81	0.050	
2.73	2.64	2.53	2.42	2.37	2.31	2.25	2.18	2.11	2.04	0.025	
3.31	3.17	3.03	2.88	2.80	2.72	2.64	2.55	2.46	2.36	0.010	
3.77	3.60	3.43	3.24	3.15	3.05	2.95	2.84	2.73	2.61	0.005	
1.90	1.86	1.81	1.76	1.73	1.70	1.67	1.64	1.60	1.57	0.100	22
2.30	2.23	2.15	2.07	2.03	1.98	1.94	1.89	1.84	1.78	0.050	
2.70	2.60	2.50	2.39	2.33	2.27	2.21	2.14	2.08	2.00	0.025	
3.26	3.12	2.98	2.83	2.75	2.67	2.58	2.50	2.40	2.31	0.010	
3.70	3.54	3.36	3.18	3.08	2.98	2.88	2.77	2.66	2.55	0.005	
1.89	1.84	1.80	1.74	1.72	1.69	1.66	1.62	1.59	1.55	0.100	23
2.27	2.20	2.13	2.05	2.01	1.96	1.91	1.86	1.81	1.76	0.050	
2.67	2.57	2.47	2.36	2.30	2.24	2.18	2.11	2.04	1.97	0.025	
3.21	3.07	2.93	2.78	2.70	2.62	2.54	2.45	2.35	2.26	0.010	
3.64	3.47	3.30	3.12	3.02	2.92	2.82	2.71	2.60	2.48	0.005	
1.88	1.83	1.78	1.73	1.70	1.67	1.64	1.61	1.57	1.53	0.100	24
2.25	2.18	2.11	2.03	1.98	1.94	1.89	1.84	1.79	1.73	0.050	
2.64	2.54	2.44	2.33	2.27	2.21	2.15	2.08	2.01	1.94	0.025	
3.17	3.03	2.89	2.74	2.66	2.58	2.49	2.40	2.31	2.21	0.010	
3.59	3.42	3.25	3.06	2.97	2.87	2.77	2.66	2.55	2.43	0.005	
1.87	1.82	1.77	1.72	1.69	1.66	1.63	1.59	1.56	1.52	0.100	25
2.24	2.16	2.09	2.01	1.96	1.92	1.87	1.82	1.77	1.71	0.050	
2.61	2.51	2.41	2.30	2.24	2.18	2.12	2.05	1.98	1.91	0.025	
3.13	2.99	2.85	2.70	2.62	2.54	2.45	2.36	2.27	2.17	0.010	
3.54	3.37	3.20	3.01	2.92	2.82	2.72	2.61	2.50	2.38	0.005	
1.86	1.81	1.76	1.71	1.68	1.65	1.61	1.58	1.54	1.50	0.100	26
2.22	2.15	2.07	1.99	1.95	1.90	1.85	1.80	1.75	1.69	0.050	
2.59	2.49	2.39	2.28	2.22	2.16	2.09	2.03	1.95	1.88	0.025	
3.09	2.96	2.81	2.66	2.58	2.50	2.42	2.33	2.23	2.13	0.010	
3.49	3.33	3.15	2.97	2.87	2.77	2.67	2.56	2.45	2.33	0.005	
1.85	1.80	1.75	1.70	1.67	1.64	1.60	1.57	1.53	1.49	0.100	27
2.20	2.13	2.06	1.97	1.93	1.88	1.84	1.79	1.73	1.67	0.050	
2.57	2.47	2.36	2.25	2.19	2.13	2.07	2.00	1.93	1.85	0.025	
3.06	2.93	2.78	2.63	2.55	2.47	2.38	2.29	2.20	2.10	0.010	
3.45	3.28	3.11	2.93	2.83	2.73	2.63	2.52	2.41	2.29	0.005	
1.84	1.79	1.74	1.69	1.66	1.63	1.59	1.56	1.52	1.48	0.100	28
2.19	2.12	2.04	1.96	1.91	1.87	1.82	1.77	1.71	1.65	0.050	
2.55	2.45	2.34	2.23	2.17	2.11	2.05	1.98	1.91	1.83	0.025	
3.03	2.90	2.75	2.60	2.52	2.44	2.35	2.26	2.17	2.06	0.010	
3.41	3.25	3.07	2.89	2.79	2.69	2.59	2.48	2.37	2.25	0.005	
1.83	1.78	1.73	1.68	1.65	1.62	1.58	1.55	1.51	1.47	0.100	29
2.18	2.10	2.03	1.94	1.90	1.85	1.81	1.75	1.70	1.64	0.050	
2.53	2.43	2.32	2.21	2.15	2.09	2.03	1.96	1.89	1.81	0.025	
3.00	2.87	2.73	2.57	2.49	2.41	2.33	2.23	2.14	2.03	0.010	
3.38	3.21	3.04	2.86	2.76	2.66	2.56	2.45	2.33	2.21	0.005	
1.82	1.77	1.72	1.67	1.64	1.61	1.57	1.54	1.50	1.46	0.100	30
2.16	2.09	2.01	1.93	1.89	1.84	1.79	1.74	1.68	1.62	0.050	
2.51	2.41	2.31	2.20	2.14	2.07	2.01	1.94	1.87	1.79	0.025	
2.98	2.84	2.70	2.55	2.47	2.39	2.30	2.21	2.11	2.01	0.010	
3.34	3.18	3.01	2.82	2.73	2.63	2.52	2.42	2.30	2.18	0.005	

TABLE 6 *(continued)*

df_2	α	df_1 1	2	3	4	5	6	7	8	9
40	0.100	2.84	2.44	2.23	2.09	2.00	1.93	1.87	1.83	1.79
	0.050	4.08	3.23	2.84	2.61	2.45	2.34	2.25	2.18	2.12
	0.025	5.42	4.05	3.46	3.13	2.90	2.74	2.62	2.53	2.45
	0.010	7.31	5.18	4.31	3.83	3.51	3.29	3.12	2.99	2.89
	0.005	8.83	6.07	4.98	4.37	3.99	3.71	3.51	3.35	3.22
60	0.100	2.79	2.39	2.18	2.04	1.95	1.87	1.82	1.77	1.74
	0.050	4.00	3.15	2.76	2.53	2.37	2.25	2.17	2.10	2.04
	0.025	5.29	3.93	3.34	3.01	2.79	2.63	2.51	2.41	2.33
	0.010	7.08	4.98	4.13	3.65	3.34	3.12	2.95	2.82	2.72
	0.005	8.49	5.79	4.73	4.14	3.76	3.49	3.29	3.13	3.01
120	0.100	2.75	2.35	2.13	1.99	1.90	1.82	1.77	1.72	1.68
	0.050	3.92	3.07	2.68	2.45	2.29	2.17	2.09	2.02	1.96
	0.025	5.15	3.80	3.23	2.89	2.67	2.52	2.39	2.30	2.22
	0.010	6.85	4.79	3.95	3.48	3.17	2.96	2.79	2.66	2.56
	0.005	8.18	5.54	4.50	3.92	3.55	3.28	3.09	2.93	2.81
∞	0.100	2.71	2.30	2.08	1.94	1.85	1.77	1.72	1.67	1.63
	0.050	3.84	3.00	2.60	2.37	2.21	2.10	2.01	1.94	1.63
	0.025	5.02	3.69	3.12	2.79	2.57	2.41	2.29	2.19	2.11
	0.010	6.63	4.61	3.78	3.32	3.02	2.80	2.64	2.51	2.41
	0.005	7.88	5.30	4.28	3.72	3.35	3.09	2.90	2.74	2.62

TABLE 6 *(continued)*

df_1											
10	12	15	20	24	30	40	60	120	∞	α	df_2
1.76	1.71	1.66	1.61	1.57	1.54	1.51	1.47	1.42	1.38	0.100	40
2.08	2.00	1.92	1.84	1.79	1.74	1.69	1.64	1.58	1.51	0.050	
2.39	2.29	2.18	2.07	2.01	1.94	1.88	1.80	1.72	1.64	0.025	
2.80	2.66	2.52	2.37	2.29	2.20	2.11	2.02	1.92	1.80	0.010	
3.12	2.95	2.78	2.60	2.50	2.40	2.30	2.18	2.06	1.93	0.005	
1.71	1.66	1.60	1.54	1.51	1.48	1.44	1.40	1.35	1.29	0.100	60
1.99	1.92	1.84	1.75	1.70	1.65	1.59	1.53	1.47	1.39	0.050	
2.27	2.17	2.06	1.94	1.88	1.82	1.74	1.67	1.58	1.48	0.025	
2.63	2.50	2.35	2.20	2.12	2.03	1.94	1.84	1.73	1.60	0.010	
2.90	2.74	2.57	2.39	2.29	2.19	2.08	1.96	1.83	1.69	0.005	
1.65	1.60	1.55	1.48	1.45	1.41	1.37	1.32	1.26	1.19	0.100	120
1.91	1.83	1.75	1.66	1.61	1.55	1.50	1.43	1.35	1.25	0.050	
2.16	2.05	1.94	1.82	1.76	1.69	1.61	1.53	1.43	1.31	0.025	
2.47	2.34	2.19	2.03	1.95	1.86	1.76	1.66	1.53	1.38	0.010	
2.71	2.54	2.37	2.19	2.09	1.98	1.87	1.75	1.61	1.43	0.005	
1.60	1.55	1.49	1.42	1.38	1.34	1.30	1.24	1.17	1.00	0.100	∞
1.83	1.75	1.67	1.57	1.52	1.46	1.39	1.32	1.22	1.00	0.050	
2.05	1.94	1.83	1.71	1.64	1.57	1.48	1.39	1.27	1.00	0.025	
2.32	2.18	2.04	1.88	1.79	1.70	1.59	1.47	1.32	1.00	0.010	
2.52	2.36	2.19	2.00	1.90	1.79	1.67	1.53	1.36	1.00	0.005	

TABLE 7 Critical Values of T for the Wilcoxon Rank Sum Test, $n_1 \le n_2$

TABLE 7(a)
5% Left-Tailed Critical Values

	n_1													
n_2	2	3	4	5	6	7	8	9	10	11	12	13	14	15
3	—	6												
4	—	6	11											
5	3	7	12	19										
6	3	8	13	20	28									
7	3	8	14	21	29	39								
8	4	9	15	23	31	41	51							
9	4	10	16	24	33	43	54	66						
10	4	10	17	26	35	45	56	69	82					
11	4	11	18	27	37	47	59	72	86	100				
12	5	11	19	28	38	49	62	75	89	104	120			
13	5	12	20	30	40	52	64	78	92	108	125	142		
14	6	13	21	31	42	54	67	81	96	112	129	147	166	
15	6	13	22	33	44	56	69	84	99	116	133	152	171	192

TABLE 7(b)
2.5% Left-Tailed Critical Values

	n_1													
n_2	2	3	4	5	6	7	8	9	10	11	12	13	14	15
4	—	—	10											
5	—	6	11	17										
6	—	7	12	18	26									
7	—	7	13	20	27	36								
8	3	8	14	21	29	38	49							
9	3	8	14	22	31	40	51	62						
10	3	9	15	23	32	42	53	65	78					
11	3	9	16	24	34	44	55	68	81	96				
12	4	10	17	26	35	46	58	71	84	99	115			
13	4	10	18	27	37	48	60	73	88	103	119	136		
14	4	11	19	28	38	50	62	76	91	106	123	141	160	
15	4	11	20	29	40	52	65	79	94	110	127	145	164	184

Source: Adapted from "An Extended Table of Critical Values for the Mann-Whitney (Wilcoxon) Two-Sample Statistics" by Roy C. Milton, *Journal of the American Statistical Association*, Volume 59, Issue 307 (September 1964), pp. 925–934. Reprinted by permission of the publisher (Taylor & Francis Ltd, http://www.tandfonline.com).

TABLE 7(c)
1% Left-Tailed Critical Values

	n_1													
n_2	2	3	4	5	6	7	8	9	10	11	12	13	14	15
3	—	—												
4	—	—	—											
5	—	—	10	16										
6	—	—	11	17	24									
7	—	6	11	18	25	34								
8	—	6	12	19	27	35	45							
9	—	7	13	20	28	37	47	59						
10	—	7	13	21	29	39	49	61	74					
11	—	7	14	22	30	40	51	63	77	91				
12	—	8	15	23	32	42	53	66	79	94	109			
13	3	8	15	24	33	44	56	68	82	97	113	130		
14	3	8	16	25	34	45	58	71	85	100	116	134	152	
15	3	9	17	26	36	47	60	73	88	103	120	138	156	176

TABLE 7(d)
0.5% Left-Tailed Critical Values

	n_1												
n_2	3	4	5	6	7	8	9	10	11	12	13	14	15
3	—												
4	—	—											
5	—	—	15										
6	—	10	16	23									
7	—	10	16	24	32								
8	—	11	17	25	34	42							
9	6	11	18	26	35	45	56						
10	6	12	19	27	37	47	58	71					
11	6	12	20	28	38	49	61	73	87				
12	7	13	21	30	40	51	63	76	90	105			
13	7	13	22	31	41	53	65	79	93	109	125		
14	7	14	22	32	43	54	67	81	96	112	129	147	
15	8	15	23	33	44	56	69	84	99	115	133	151	171

TABLE 8
Critical Values of T for the Wilcoxon Signed-Rank Test, $n = 5$ to $n = 50$

One-Sided	Two-Sided	$n = 5$	$n = 6$	$n = 7$	$n = 8$	$n = 9$	$n = 10$
$\alpha = 0.050$	$\alpha = 0.10$	1	2	4	6	8	11
$\alpha = 0.025$	$\alpha = 0.05$		1	2	4	6	8
$\alpha = 0.010$	$\alpha = 0.02$			0	2	3	5
$\alpha = 0.005$	$\alpha = 0.01$				0	2	3
One-Sided	**Two-Sided**	$n = 11$	$n = 12$	$n = 13$	$n = 14$	$n = 15$	$n = 16$
$\alpha = 0.050$	$\alpha = 0.10$	14	17	21	26	30	36
$\alpha = 0.025$	$\alpha = 0.05$	11	14	17	21	25	30
$\alpha = 0.010$	$\alpha = 0.02$	7	10	13	16	20	24
$\alpha = 0.005$	$\alpha = 0.01$	5	7	10	13	16	19
One-Sided	**Two-Sided**	$n = 17$	$n = 18$	$n = 19$	$n = 20$	$n = 21$	$n = 22$
$\alpha = 0.050$	$\alpha = 0.10$	41	47	54	60	68	75
$\alpha = 0.025$	$\alpha = 0.05$	35	40	46	52	59	66
$\alpha = 0.010$	$\alpha = 0.02$	28	33	38	43	49	56
$\alpha = 0.005$	$\alpha = 0.01$	23	28	32	37	43	49
One-Sided	**Two-Sided**	$n = 23$	$n = 24$	$n = 25$	$n = 26$	$n = 27$	$n = 28$
$\alpha = 0.050$	$\alpha = 0.10$	83	92	101	110	120	130
$\alpha = 0.025$	$\alpha = 0.05$	73	81	90	98	107	117
$\alpha = 0.010$	$\alpha = 0.02$	62	69	77	85	93	102
$\alpha = 0.005$	$\alpha = 0.01$	55	68	68	76	84	92
One-Sided	**Two-Sided**	$n = 29$	$n = 30$	$n = 31$	$n = 32$	$n = 33$	$n = 34$
$\alpha = 0.050$	$\alpha = 0.10$	141	152	163	175	188	201
$\alpha = 0.025$	$\alpha = 0.05$	127	137	148	159	171	183
$\alpha = 0.010$	$\alpha = 0.02$	111	120	130	141	151	162
$\alpha = 0.005$	$\alpha = 0.01$	100	109	118	128	138	149
One-Sided	**Two-Sided**	$n = 35$	$n = 36$	$n = 37$	$n = 38$	$n = 39$	
$\alpha = 0.050$	$\alpha = 0.10$	214	228	242	256	271	
$\alpha = 0.025$	$\alpha = 0.05$	195	208	222	235	250	
$\alpha = 0.010$	$\alpha = 0.02$	174	186	198	211	224	
$\alpha = 0.005$	$\alpha = 0.01$	160	171	183	195	208	
One-Sided	**Two-Sided**	$n = 40$	$n = 41$	$n = 42$	$n = 43$	$n = 44$	$n = 45$
$\alpha = 0.050$	$\alpha = 0.10$	287	303	319	336	353	371
$\alpha = 0.025$	$\alpha = 0.05$	264	279	295	311	327	344
$\alpha = 0.010$	$\alpha = 0.02$	238	252	267	281	297	313
$\alpha = 0.005$	$\alpha = 0.01$	221	234	248	262	277	292
One-Sided	**Two-Sided**	$n = 46$	$n = 47$	$n = 48$	$n = 49$	$n = 50$	
$\alpha = 0.050$	$\alpha = 0.10$	389	408	427	446	466	
$\alpha = 0.025$	$\alpha = 0.05$	361	379	397	415	434	
$\alpha = 0.010$	$\alpha = 0.02$	329	345	362	380	398	
$\alpha = 0.005$	$\alpha = 0.01$	307	323	339	356	373	

Source: F. Wilcoxon and R. A. Wilcox, "Some Rapid Approximate Statistical Procedures," *Annals of the New York Academy of Sciences*, March 1950, Vol. 52, pp. 808–814.

TABLE 9
Critical Values of Spearman's Rank Correlation Coefficient for a One-Tailed Test

n	$\alpha = 0.05$	$\alpha = 0.025$	$\alpha = 0.01$	$\alpha = 0.005$
5	0.900	—	—	—
6	0.829	0.886	0.943	—
7	0.714	0.786	0.893	—
8	0.643	0.738	0.833	0.881
9	0.600	0.683	0.783	0.833
10	0.564	0.648	0.745	0.794
11	0.523	0.623	0.736	0.818
12	0.497	0.591	0.703	0.780
13	0.475	0.566	0.673	0.745
14	0.457	0.545	0.646	0.716
15	0.441	0.525	0.623	0.689
16	0.425	0.507	0.601	0.666
17	0.412	0.490	0.582	0.645
18	0.399	0.476	0.564	0.625
19	0.388	0.462	0.549	0.608
20	0.377	0.450	0.534	0.591
21	0.368	0.438	0.521	0.576
22	0.359	0.428	0.508	0.562
23	0.351	0.418	0.496	0.549
24	0.343	0.409	0.485	0.537
25	0.336	0.400	0.475	0.526
26	0.329	0.392	0.465	0.515
27	0.323	0.385	0.456	0.505
28	0.317	0.377	0.448	0.496
29	0.311	0.370	0.440	0.487
30	0.305	0.364	0.432	0.478

Source: E.G. Olds, "Distribution of Sums of Squares of Rank Differences for Small Samples," *Annals of Mathematical Statistics* 9 (2), June 1938, pp. 133–48. Permission conveyed through Copyright Clearance Center, Inc.

TABLE 10 Random Numbers

Line	Column 1	2	3	4	5	6	7	8	9	10	11	12	13	14
1	12208	24645	51954	52367	70041	29436	63165	08837	07502	10149	08179	53274	23126	26100
2	60949	33052	09897	89471	34478	43445	29109	03613	22700	14166	17272	76291	68803	14145
3	10125	46170	58787	52307	07665	64727	08345	54683	54353	45493	86548	85100	34150	82828
4	97118	40064	59700	93934	29985	51335	02930	56894	30512	98814	14599	47940	67920	94497
5	84033	61125	82315	31512	12114	36381	93301	35910	32733	91754	63426	97761	62300	10876
6	07820	41513	28543	16453	33233	74608	60331	89813	76943	97832	94710	88549	44635	43695
7	00234	43696	42735	17786	4025	87100	07395	52091	00832	57361	34279	05350	42153	23364
8	89302	37564	50840	47507	58074	49310	82937	44872	11847	89562	59876	49140	30278	53204
9	98696	07951	87790	13918	65342	40593	97785	85330	75532	78929	19289	87944	27536	38304
10	73539	43436	22704	08757	14239	46112	46914	77138	59535	01510	81367	35527	49944	21629
11	55103	84942	79545	20065	61265	23526	15461	40037	46003	59195	61448	16268	60561	66288
12	41045	12160	64217	40769	95304	46459	96419	10368	24136	01103	05745	05804	42592	12293
13	23850	58229	94169	73127	36686	57658	72752	80822	37910	03454	95931	20323	15948	78285
14	20681	97532	1344	57739	87295	77618	10523	94581	30300	52127	38482	42538	11380	31201
15	98632	42711	69112	22573	40208	19594	80180	08816	39764	36276	51626	73602	57803	91804
16	58814	83477	36429	53798	23731	43189	33745	24033	92000	62483	29257	23497	75123	16747
17	89540	72335	50722	61442	97766	41744	21239	92366	33292	10990	99628	39529	59671	80331
18	70689	90419	11713	97741	62132	71642	55798	04331	14179	34346	17864	86670	00577	16375
19	94733	23533	53497	00448	59701	64661	76086	79356	63094	18723	06144	81742	83766	23162
20	14527	70109	58790	82196	66411	52538	63906	14490	83687	91191	77030	74205	93421	90513
21	32981	71474	55418	95983	08363	06588	69382	11771	25139	05836	78280	82278	92911	96114
22	43827	90926	19156	56860	88768	23931	60455	46556	73438	90761	53112	94316	25130	36214
23	14927	58208	51337	34109	83005	54569	38943	91520	32909	84737	18982	37375	49478	15282
24	78604	76419	35608	02927	56196	24941	15074	50191	19461	40018	23677	83597	93612	42621
25	70574	75824	95331	99994	45032	04159	76949	60108	07441	98042	49363	27057	64660	37556
26	59572	11754	23401	87063	14034	88071	14214	68390	10690	20344	24333	29879	49597	03075
27	51903	32650	81090	38650	11784	52192	54856	50690	95780	33949	66752	94679	43455	36137
28	04929	38887	57405	00142	93085	03152	80218	92384	95467	32945	63971	80999	83369	66273
29	95442	76741	16848	80125	85165	27545	85072	60222	87099	00810	69387	33099	49006	85377
30	40106	23109	75067	02874	38151	88117	91020	29804	25349	86190	29144	57430	05126	87101
31	20217	25571	49253	77636	53454	29944	02473	64865	61403	76255	78138	99569	44798	38249
32	81586	37012	14302	34070	00623	51557	52423	08260	42281	03890	86205	93117	58934	93110
33	98927	37863	18714	52289	40415	38142	43133	35412	23899	59451	29702	86955	11168	31625
34	53661	17418	66400	76324	79650	09008	37545	57393	34060	03850	67013	99180	78207	60786
35	07438	23455	39141	79868	94619	18887	94506	80669	20762	21644	49262	68451	28250	69767
36	60931	85536	88331	68211	02498	70301	34052	68591	65700	83555	83159	95480	11289	41200
37	94206	94875	07193	66610	65756	83072	51551	96665	89114	28879	26503	55640	98991	19279
38	85065	39321	95550	50484	40264	19032	12600	40136	43247	18763	92825	41267	22449	83895
39	06574	23024	89779	77864	6637	43879	14396	76785	03111	38054	68752	30534	68588	33238
40	14662	56351	95099	05032	67453	76276	87241	04756	68574	96571	39186	38407	61146	09731
41	38343	59916	16059	64027	85331	04382	38797	19971	77118	76606	76124	35913	26863	42644
42	99746	97427	10227	56118	68397	06584	19744	00759	48757	71813	87959	96068	27320	16333
43	36698	54171	51751	46513	10278	48772	86874	89069	75512	95393	77252	56276	34909	52380
44	61486	68800	92472	01328	25032	21933	38166	47956	88500	59925	49268	34991	13019	56103
45	31573	09581	79484	60562	33439	99539	38912	12079	01397	83189	22685	23921	45539	36426
46	42853	44830	81141	44382	36115	37740	34097	67680	65471	34396	66419	84451	11451	55660
47	95202	01956	48319	74650	92863	31305	98992	07191	13636	62264	54002	00285	98411	13004
48	12127	76887	15700	99324	06540	37115	28802	07917	50920	49182	53084	11269	66174	83705
49	27955	77084	70487	60423	24686	51553	22787	39335	73740	24673	09572	51042	11231	73109
50	32874	31879	58394	91948	00970	83700	04273	72675	99867	29157	48266	93447	45136	10419

TABLE 10 *(continued)*

	Column													
Line	1	2	3	4	5	6	7	8	9	10	11	12	13	14
51	43146	28783	81783	35301	48029	59230	92843	98504	13483	90181	46171	73922	56502	39017
52	44220	58294	65513	69045	75794	04184	35613	44393	60792	73666	55718	55854	42609	18881
53	33214	75877	09488	15466	22035	62519	23847	28528	44125	78312	95481	84012	90492	46523
54	36569	79997	85531	64156	30212	87711	13808	06597	35833	60542	36286	89726	47755	93879
55	46718	58503	73964	44048	58750	90197	18556	24204	27809	68620	55102	43181	91886	78902
56	55241	18081	26016	53193	26699	80368	68025	15633	63865	54476	56018	92884	85640	14450
57	26823	23849	48932	77788	23825	68374	3884	71509	59254	94395	28552	31644	18452	52541
58	84565	21197	38208	84431	56637	06623	32668	70791	20063	85306	04773	44423	68724	21659
59	69809	49234	26403	57225	26679	92637	34374	21376	89597	77749	4450	21744	90562	45272
60	68701	61671	91698	06285	94994	27448	29580	18911	14041	28510	71764	89145	44107	85752
61	67395	48305	03269	39920	38337	51033	02302	02730	65997	34801	53108	18233	92695	31093
62	18632	33352	30818	94502	28843	91668	28986	15003	48138	29686	08685	69674	22428	50617
63	02494	73310	50247	26158	21293	62810	77067	41875	56248	12340	99552	77654	71050	62647
64	48228	10660	80117	27259	37371	48345	22423	68160	79138	05397	50859	08601	71597	87422
65	88327	25662	36727	58289	03024	59840	46226	90771	66483	72239	95391	69131	02703	96426
66	02276	23422	72667	34204	56279	47400	63879	31291	07037	33455	45742	52178	71973	75122
67	08663	78086	95643	86488	65661	23577	72539	71029	54634	91067	84541	31579	36301	98755
68	10987	22530	91237	16663	20960	31000	58226	47410	91894	40466	18033	63198	24365	47012
69	64913	98762	80306	56098	87383	76114	20587	37796	81583	30616	51327	72196	56275	30700
70	92171	82087	00603	37130	53023	15211	88168	51560	66280	84631	77421	77005	39304	86339
71	67323	49348	60421	74047	2655	52763	82312	51594	46014	69094	37945	83466	88469	63059
72	11437	02265	49941	42996	64641	04395	44249	96484	08931	73306	16659	66067	27309	88737
73	49481	18292	93867	25435	62138	79514	63630	99645	14270	67513	02160	27091	75152	30448
74	20218	04852	04035	62227	60414	12812	23879	45465	76239	42718	95114	26681	43480	48166
75	52448	57276	22268	64976	91241	91887	16898	54702	65230	57866	32391	65348	50585	28385
76	06951	91603	93567	04376	19305	10206	96028	45598	94348	33204	42138	23588	31097	36717
77	80030	19449	84673	52684	31045	97738	26207	78072	95051	40694	84672	90130	91469	37334
78	18868	55012	96510	53722	66271	10454	93038	49863	62479	56587	66689	02715	52951	78191
79	47473	64079	58696	15803	42279	80389	50441	35124	97747	00336	10298	13498	78531	72712
80	28047	44350	25963	28427	65803	14775	74303	36997	38398	50174	38700	34281	65296	02315
81	53914	76762	14512	91658	49418	50997	07990	74252	71907	40196	25533	93814	82036	27418
82	68559	58758	19550	47614	20635	59647	18910	94815	96448	96142	79153	72998	51926	93164
83	52390	51394	11369	47813	93600	84009	80838	51375	05658	50987	05192	03600	37167	47759
84	64694	29958	14745	12798	24079	00484	28255	83819	01244	40100	10825	79559	91218	41610
85	80711	70335	18568	54770	83115	92577	79212	87339	14664	33251	12458	73663	68658	79653
86	19788	50566	48171	57327	76273	52553	50896	98605	72532	27948	36024	15932	32809	52590
87	08175	82442	64597	67137	57196	05833	67853	80887	82354	48306	54521	31500	17322	33723
88	02151	58882	99383	51519	26381	81617	62015	61263	62593	82932	94394	25542	35977	91643
89	07098	30062	40289	37560	93494	15023	76676	39224	06604	84039	49605	78247	82111	47386
90	00513	48707	86149	03782	29615	56325	42864	50495	37694	45148	96021	92898	93682	63206
91	19452	03270	87476	82110	19584	17927	12223	23225	58748	02157	52454	27417	47706	95488
92	79511	22274	21873	08155	51009	94820	09171	10334	11388	90769	35789	46408	06904	20482
93	37921	39095	96221	67712	00130	07156	03610	41023	50599	21150	42241	39979	62635	70948
94	12191	98435	97777	23786	23987	87770	84440	85462	76652	18981	53147	57842	06544	60431
95	66951	07460	26091	40404	75069	56519	54951	14793	89851	58848	66663	52160	87441	14406
96	78134	48437	36059	04162	76856	74138	10632	51849	85121	22896	00617	00567	88799	23961
97	68730	08199	58971	14461	63523	68277	17728	31832	30995	24511	13455	10226	12391	39554
98	34336	41241	48877	39251	17796	30313	65311	61826	64233	25668	93416	34124	16752	78131
99	80950	90475	66033	63371	33724	14982	43108	27847	00462	53751	01127	85419	32927	28781
100	42390	80434	48371	60163	09971	75724	18343	90382	41362	33772	44030	29599	39597	46081

TABLE 11(a) Percentage Points of the Studentized Range, $q_{0.05}(k, df)$; Upper 5% Points

df	*k* = 2	3	4	5	6	7	8	9	10	11
1	17.97	26.98	32.82	37.08	40.41	43.12	45.40	47.36	49.07	50.59
2	6.08	8.33	9.80	10.88	11.74	12.44	13.03	13.54	13.99	14.39
3	4.50	5.91	6.82	7.50	8.04	8.48	8.85	9.18	9.46	9.72
4	3.93	5.04	5.76	6.29	6.71	7.05	7.35	7.60	7.83	8.03
5	3.64	4.60	5.22	5.67	6.03	6.33	6.58	6.80	6.99	7.17
6	3.46	4.34	4.90	5.30	5.63	5.90	6.12	6.32	6.49	6.65
7	3.34	4.16	4.68	5.06	5.36	5.61	5.82	6.00	6.16	6.30
8	3.26	4.04	4.53	4.89	5.17	5.40	5.60	5.77	5.92	6.05
9	3.20	3.95	4.41	4.76	5.02	5.24	5.43	5.59	5.74	5.87
10	3.15	3.88	4.33	4.65	4.91	5.12	5.30	5.46	5.60	5.72
11	3.11	3.82	4.26	4.57	4.82	5.03	5.20	5.35	5.49	5.61
12	3.08	3.77	4.20	4.51	4.75	4.95	5.12	5.27	5.39	5.51
13	3.06	3.73	4.15	4.45	4.69	4.88	5.05	5.19	5.32	5.43
14	3.03	3.70	4.11	4.41	4.64	4.83	4.99	5.13	5.25	5.36
15	3.01	3.67	4.08	4.37	4.60	4.78	4.94	5.08	5.20	5.31
16	3.00	3.65	4.05	4.33	4.56	4.74	4.90	5.03	5.15	5.26
17	2.98	3.63	4.02	4.30	4.52	4.70	4.86	4.99	5.11	5.21
18	2.97	3.61	4.00	4.28	4.49	4.67	4.82	4.96	5.07	5.17
19	2.96	3.59	3.98	4.25	4.47	4.65	4.79	4.92	5.04	5.14
20	2.95	3.58	3.96	4.23	4.45	4.62	4.77	4.90	5.01	5.11
24	2.92	3.53	3.90	4.17	4.37	4.54	4.68	4.81	4.92	5.01
30	2.89	3.49	3.85	4.10	4.30	4.46	4.60	4.72	4.82	4.92
40	2.86	3.44	3.79	4.04	4.23	4.39	4.52	4.63	4.73	4.82
60	2.83	3.40	3.74	3.98	4.16	4.31	4.44	4.55	4.65	4.73
120	2.80	3.36	3.68	3.92	4.10	4.24	4.36	4.47	4.56	4.64
∞	2.77	3.31	3.63	3.86	4.03	4.17	4.29	4.39	4.47	4.55

Source: A portion of "Tables of percentage points" from *Biometrika Tables for Statisticians*, Vol. 1, 3rd ed., edited by E.S. Pearson and H.O. Hartley (Cambridge University Press, 1966). Reproduced by permission of Oxford University Press on behalf of the *Biometrika* Trustees.

TABLE 11(a)
(continued)

k									
12	13	14	15	16	17	18	19	20	*df*
51.96	53.20	54.33	55.36	56.32	57.22	58.04	58.83	59.56	1
14.75	15.08	15.38	15.65	15.91	16.14	16.37	16.57	16.77	2
9.95	10.15	10.35	10.52	10.69	10.84	10.98	11.11	11.24	3
8.21	8.37	8.52	8.66	8.79	8.91	9.03	9.13	9.23	4
7.32	7.47	7.60	7.72	7.83	7.93	8.03	8.12	8.21	5
6.79	6.92	7.03	7.14	7.24	7.34	7.43	7.51	7.59	6
6.43	6.55	6.66	6.76	6.85	6.94	7.02	7.10	7.17	7
6.18	6.29	6.39	6.48	6.57	6.65	6.73	6.80	6.87	8
5.98	6.09	6.19	6.28	6.36	6.44	6.51	6.58	6.64	9
5.83	5.93	6.03	6.11	6.19	6.27	6.34	6.40	6.47	10
5.71	5.81	5.90	5.98	6.06	6.13	6.20	6.27	6.33	11
5.61	5.71	5.80	5.88	5.95	6.02	6.09	6.15	6.21	12
5.53	5.63	5.71	5.79	5.86	5.93	5.99	6.05	6.11	13
5.46	5.55	5.64	5.71	5.79	5.85	5.91	5.97	6.03	14
5.40	5.49	5.57	5.65	5.72	5.78	5.85	5.90	5.96	15
5.35	5.44	5.52	5.59	5.66	5.73	5.79	5.84	5.90	16
5.31	5.39	5.47	5.54	5.61	5.67	5.73	5.79	5.84	17
5.27	5.35	5.43	5.50	5.57	5.63	5.69	5.74	5.79	18
5.23	5.31	5.39	5.46	5.53	5.59	5.65	5.70	5.75	19
5.20	5.28	5.36	5.43	5.49	5.55	5.61	5.66	5.71	20
5.10	5.18	5.25	5.32	5.38	5.44	5.49	5.55	5.59	24
5.00	5.08	5.15	5.21	5.27	5.33	5.38	5.43	5.47	30
4.90	4.98	5.04	5.11	5.16	5.22	5.27	5.31	5.36	40
4.81	4.88	4.94	5.00	5.06	5.11	5.15	5.20	5.24	60
4.71	4.78	4.84	4.90	4.95	5.00	5.04	5.09	5.13	120
4.62	4.68	4.74	4.80	4.85	4.89	4.93	4.97	5.01	∞

TABLE 11(b) Percentage Points of the Studentized Range, $q_{0.01}(k, df)$; Upper 1% Points

	k									
df	2	3	4	5	6	7	8	9	10	11
1	90.03	135.0	164.3	185.6	202.2	215.8	227.2	237.0	245.6	253.2
2	14.04	19.02	22.29	24.72	26.63	28.20	29.53	30.68	31.69	32.59
3	8.26	10.62	12.17	13.33	14.24	15.00	15.64	16.20	16.69	17.13
4	6.51	8.12	9.17	9.96	10.58	11.10	11.55	11.93	12.27	12.57
5	5.70	6.98	7.80	8.42	8.91	9.32	9.67	9.97	10.24	10.48
6	5.24	6.33	7.03	7.56	7.97	8.32	8.61	8.87	9.10	9.30
7	4.95	5.92	6.54	7.01	7.37	7.68	7.94	8.17	8.37	8.55
8	4.75	5.64	6.20	6.62	6.96	7.24	7.47	7.68	7.86	8.03
9	4.60	5.43	5.96	6.35	6.66	6.91	7.13	7.33	7.49	7.65
10	4.48	5.27	5.77	6.14	6.43	6.67	6.87	7.05	7.21	7.36
11	4.39	5.15	5.62	5.97	6.25	6.48	6.67	6.84	6.99	7.13
12	4.32	5.05	5.50	5.84	6.10	6.32	6.51	6.67	6.81	6.94
13	4.26	4.96	5.40	5.73	5.98	6.19	6.37	6.53	6.67	6.79
14	4.21	4.89	5.32	5.63	5.88	6.08	6.26	6.41	6.54	6.66
15	4.17	4.84	5.25	5.56	5.80	5.99	6.16	6.31	6.44	6.55
16	4.13	4.79	5.19	5.49	5.72	5.92	6.08	6.22	6.35	6.46
17	4.10	4.74	5.14	5.43	5.66	5.85	6.01	6.15	6.27	6.38
18	4.07	4.70	5.09	5.38	5.60	5.79	5.94	6.08	6.20	6.31
19	4.05	4.67	5.05	5.33	5.55	5.73	5.89	6.02	6.14	6.25
20	4.02	4.64	5.02	5.29	5.51	5.69	5.84	5.97	6.09	6.19
24	3.96	4.55	4.91	5.17	5.37	5.54	5.69	5.81	5.92	6.02
30	3.89	4.45	4.80	5.05	5.24	5.40	5.54	5.65	5.76	5.85
40	3.82	4.37	4.70	4.93	5.11	5.26	5.39	5.50	5.60	5.69
60	3.76	4.28	4.59	4.82	4.99	5.13	5.25	5.36	5.45	5.53
120	3.70	4.20	4.50	4.71	4.87	5.01	5.12	5.21	5.30	5.37
∞	3.64	4.12	4.40	4.60	4.76	4.88	4.99	5.08	5.16	5.23

Source: From *Biometrika Tables for Statisticians*, Vol. 1, 3rd ed., edited by E.S. Pearson and H.O. Hartley (Cambridge University Press, 1966). Reproduced by permission of Oxford University Press on behalf of the *Biometrika* Trustees.

TABLE 11(b)
(continued)

k									
12	13	14	15	16	17	18	19	20	*df*
260.0	266.2	271.8	277.0	281.8	286.3	290.0	294.3	298.0	1
33.40	34.13	34.81	35.43	36.00	36.53	37.03	37.50	37.95	2
17.53	17.89	18.22	18.52	18.81	19.07	19.32	19.55	19.77	3
12.84	13.09	13.32	13.53	13.73	13.91	14.08	14.24	14.40	4
10.70	10.89	11.08	11.24	11.40	11.55	11.68	11.81	11.93	5
9.48	9.65	9.81	9.95	10.08	10.21	10.32	10.43	10.54	6
8.71	8.86	9.00	9.12	9.24	9.35	9.46	9.55	9.65	7
8.18	8.31	8.44	8.55	8.66	8.76	8.85	8.94	9.03	8
7.78	7.91	8.03	8.13	8.23	8.33	8.41	8.49	8.57	9
7.49	7.60	7.71	7.81	7.91	7.99	8.08	8.15	8.23	10
7.25	7.36	7.46	7.56	7.65	7.73	7.81	7.88	7.95	11
7.06	7.17	7.26	7.36	7.44	7.52	7.59	7.66	7.73	12
6.90	7.01	7.10	7.19	7.27	7.35	7.42	7.48	7.55	13
6.77	6.87	6.96	7.05	7.13	7.20	7.27	7.33	7.39	14
6.66	6.76	6.84	6.93	7.00	7.07	7.14	7.20	7.26	15
6.56	6.66	6.74	6.82	6.90	6.97	7.03	7.09	7.15	16
6.48	6.57	6.66	6.73	6.81	6.87	6.94	7.00	7.05	17
6.41	6.50	6.58	6.65	6.72	6.79	6.85	6.91	6.97	18
6.34	6.43	6.51	6.58	6.65	6.72	6.78	6.84	6.89	19
6.28	6.37	6.45	6.52	6.59	6.65	6.71	6.77	6.82	20
6.11	6.19	6.26	6.33	6.39	6.45	6.51	6.56	6.61	24
5.93	6.01	6.08	6.14	6.20	6.26	6.31	6.36	6.41	30
5.76	5.83	5.90	5.96	6.02	6.07	6.12	6.16	6.21	40
5.60	5.67	5.73	5.78	5.84	5.89	5.93	5.97	6.01	60
5.44	5.50	5.56	5.61	5.66	5.71	5.75	5.79	5.83	120
5.29	5.35	5.40	5.45	5.49	5.54	5.57	5.61	5.65	∞

Data Sources

Introduction

1. http://www.electionalmanac.com/ea/canada-election-polls/
2. Andrew Heard, "Canada Election Pollsters Success," http://www.sfu.ca/~aheard/elections/poll-results.html
3. http://www.ipsos-na.com/news-polls/pressrelease.aspx?id=3341
4. http://www.statcan.gc.ca/ca-ra2006/articles/snapshot-portrait-eng.htm
5. "Hot News: 98.6 Not Normal," *The Press-Enterprise* (Riverside, CA), 23 September 1992; Adapted from A. Mackowiak, S.S. Wasserman, and M.M. Levine, "A Critical Appraisal of 98.6 Degrees F, The Upper Limit of the Normal Body Temperature, and Other Legacies of Carl Reinhold August Wunderlich," *Journal of the American Medical Association* (268):1578–1580.

Chapter 1

1. http://www.socialbakers.com/facebook-statistics/canada
2. http://www.autotrader.ca/newsfeatures/20160106/canadas-25-best-selling-cars-in-2015/#ODmVogSLo8F9kJSJ.97
3. http://www.advancededucation.gov.ab.ca/news/2003/September/nr-PostSecPays.asp
4. Canadian complaints against airlines: http://www.otc-cta.gc.ca/eng/report-released-air-travel-complaints-canada; U.S. complaints: http://airconsumer.ost.dot.gov/reports/2004/0402atcr.doc; U.S. airline passengers: http://airconsumer.ost.dot.gov/reports/2004/0402atcr.doc; Air Canada passengers: http://publications.gc.ca/collections/Collection-R/Statcan/51-004-XIB/0050451-004-XIB.pdf
5. http://en.wikipedia.org/wiki/List_of_Canadian_Prime_Ministers_by_age
6. http://www.infoplease.com/ipsa/A0758956.html © Gerry Brown, ESPN Almanac.
7. "Major Religions of the World Ranked by Number of Adherents," http://www.adherents.com/Religions_By_Adherents.html, 6 July 2010.
8. Robert P. Wilder, D. Brennan, and D.E. Schotte, "A Standard Measure for Exercise Prescription for Aqua Running,"*The American Journal of Sports Medicine* 21, no. 1 (1993):45.
9. Written by Elke Town for *Professionally Speaking/Pour parler profession* magazines, Ontario College of Teachers. *Professionally Speaking*, March 2007, p. 33.
10. Source: Library of Parliament/Bibliotheque du Parlement, "Electoral Results by Party," http://www.lop.parl.gc.ca/ParlInfo/compilations/electionsandridings/ResultsParty.aspx
11. https://www.currentresults.com/Weather/Canada/Cities/precipitation-annual-average.php
12. https://en.wikipedia.org/wiki/Queen's_Plate
13. OECD Broadband Statistics and OECD Telecommunications Database 2005.
14. Elections Canada. Calculations and adaptation rest with the authors.
15. "Dealing with Mugabe's Diamonds," *Time Magazine,* 5 July 2010, p. 14.
16. A. Tubb, A.J. Parker, and G. Nickless, "The Analysis of Romano-British Pottery by Atomic Absorption Spectrophotometry," *Archaeometry* 22, 2 (August 1989):153.
17. EKOS Research Associates; http://www.ekospolitics.com/articles/6nov2006background.pdf
18. Mercer Human Resource Consulting, Cost of Living Survey – Worldwide Ranking 2004; http://www.uwo.ca/theory/FinancialAssistance/CostofLiving.pdf; http://www.immigration-quebec.gouv.qc.ca/en/choose-quebec/quality-life/cost-living/cost-index.html
19. Consumer's Association of Canada: http://www.consumer.ca/1542
20. http://www.skyscan.ca/SampleData.htm © 1999–2005 by Sky Scan, Edmonton, Alberta, Canada. Used by permission of Skyscan.ca

21. A. Azzalini and A.W. Bowman, "A Look at Some Data on the Old Faithful Geyser," *Applied Statistics* (1990):57.
22. Open Gov't License: http://open.canada.ca/en/open-government-licence-canada.
23. Sarah Janssen, ed., *The World Almanac and Book of Facts 2011* (New York, NY: Infobase Publishing, 2011).
24. William A. McGeveran, Jr., ed., *The World Almanac and Book of Facts, 2004* (New York: St. Martin's Press, 2004). Copyright © 2003 World Almanac Education Group. All Rights Reserved.
25. http://www.timhortons.com/
26. P.A. Mackowiak, S.S. Wasserman, and M.M. Levine, "A Critical Appraisal of 98.6 Degrees F, the Upper Limit of the Normal Body Temperature, and Other Legacies of Carl Reinhold August Wunderlich," *Journal of the American Medical Association* (268): 1578–1580; and Allen L. Shoemaker, "What's Normal? Temperature, Gender, and Heart Rate," *Journal of Statistics Education* (1996).
27. Allen L. Shoemaker, "What's Normal? Temperture, Gender, and Heart Rate," *Journal of Statistics Education* (1996).

Chapter 2

1. Brittany Darwell, "Survey suggests Facebook advertising benchmarks: $0.80 CPC, 0.041 percent CTR," *SocialTimes*, April 4, 2012, http://www.adweek.com/socialtimes/survey-suggests-facebook-advertising-benchmarks-0-80-cpc-0-014-percent-ctr/277429
2. Adapted from http://www.fraserinstitute.ca/shared/readmore.asp?sNav=pb&id=894. Source: Table 1: "Estimated Average Automobile Insurance Premiums, 2004–05, by Province, Straight Dollars," from *The False Promise of Government Auto Insurance,* p. 5, by Brett J. Skinner, February 2007. www.fraserinstitute.org
3. http://abclocal.go.com/wls/story?section=nation_world&id=5107136
4. "Birth Order and the Baby Boom," *American Demographics* (Trend Cop), March 1997, p. 10.
5. http://www.timhortons.com/
6. Prices from http://www.bestbuy.ca/ as of April 2016.
7. CBC News, "Cold or flu? It's not just in your head," January 12, 2009, http://www.cbc.ca/news/technology/cold-or-flu-it-s-not-just-in-your-head-1.822772
8. James Brady, "Rich Eisen eyeing record 40-yard dash at NFL Combine," SB Nation, February 23, 2015, http://www.sbnation.com/nfl/2015/2/23/8080037/nfl-combine-2015-rich-eisen-40-yard-dash Permission courtesy of Vox Media, Inc.
9. A. Tubb, A.J. Parker, and G. Nickless, "The Analysis of Romano-British Pottery by Atomic Absorption Spectrophotometry," *Archaeometry* 22, 2 (August 1989): 153.
10. Data from Numbeo.com, http://www.numbeo.com/cost-of-living/city_result.jsp?country=Canada&city=Regina
11. A. Azzalini and A.W. Bowman, "A Look at Some Data on the Old Faithful Geyser," *Applied Statistics* (1990):57.
12. M. Mendoza, B. Poblete, and C. Castillo, "Twitter Under Crisis: Can we trust what we RT?" Paper presented at *1st Workshop on Social Media Analytics (SOMA '10)*, July 25, 2010, Washington, D.C., http://chato.cl/papers/mendoza_poblete_castillo_2010_twitter_terremoto.pdf
13. http://www.hockeydb.com/ihdb/stats/players.html
14. "CFS Report—Fuel Price Update—February 2015," http://www.njc-cnm.gc.ca/doc.php?did=622&lang=eng
15. http://www.hockeydb.com/ihdb/stats/players.html
16. Library of Parliament/Bibliotheque du Parlement, "Electoral Results by Party," http://www.lop.parl.gc.ca/ParlInfo/compilations/electionsandridings/ResultsParty.aspx
17. "comScore Ranks the Top 50 U.S. Digital Media Properties for August 2014," news release, comScore, September 18, 2014, http://www.comscore.com/Insights/Market-Rankings/comScore-Ranks-the-Top-50-US-Digital-Media-Properties-for-August-2014
18. Data from Numbeo.com, http://www.numbeo.com/cost-of-living/city_result.jsp?country=Canada&city=Calgary

19. Allen L. Shoemaker, "What's Normal? Temperature, Gender, and Heart Rate," *Journal of Statistics Education* (1996).
20. Data provided by gasbuddy.com
21. http://www.hockeydb.com/ihdb/stats/players.html
22. http://www.thestrategiccounsel.com/our_news/polls/2007-01-17%20GMCTV%20Jan%2011-14%20f.pdf

Chapter 3

1. Adapted from Michael J. Weiss, "The New Summer Break," *American Demographics,* August 2001, p. 55.
2. Statistics Canada http://www.statcan.gc.ca/tables-tableaux/sum-som/l01/cst01/econ09a-eng.htm
3. Statistics Canada: http://www.statcan.gc.ca/daily-quotidien/150217/dq150217c-eng.htm
4. "U.S. Facebook Audience," *EContent Magazine,* Vol. 33, No. 2, March 2010, p. 4.
5. https://www.cmhc-schl.gc.ca/en/corp/about/cahoob/data/data_001.cfm
6. Gregory K. Torrey, S.F. Vasa, J.W. Maag, and J.J. Kramer, "Social Skills Interventions Across School Settings: Case Study Reviews of Students with Mild Disabilities," *Psychology in the Schools* 29 (July 1992):248.
7. Adapted from http://www.jama.ca/ar/2006/index.asp
8. www.bestbuy.ca, July 2015
9. Stellan Ohlsson, "The Learning Curve for Writing Books: Evidence from Professor Asimov," *Psychological Science* 3, no. 6 (1992):380–382.
10. "Summery Scoreboard," *Entertainment Weekly* #614, 14 September 2001, p. 13.
11. http://www.tbs-sct.gc.ca/fcsi-rscf/cen.aspx?Language=EN&sid=wu312154445220&dataset=prov&sort=name; http://www40.statcan.ca/l01/cst01/phys01.htm
12. "Aaron Rodgers #12 QB," http://sports.espn.go.com/nfl/players, 18 March 2011.
13. A. Tubb, A.J. Parker, and G. Nickless, "The Analysis of Romano-British Pottery by Atomic Absorption Spectrophotometry," *Archaeometry* 22, 2 (August 1989): 153.
14. OECD Broadband Statistics and OECD Telecommunications Database 2005.
15. Matthew Mendelsohn and Richard Nadeau, "The Religious Cleavage and the Media in Canada," *Canadian Journal of Political Science* 30, 1 (March 1997): 129–146.
16. Excerpt: "Travel by Canadians to foreign countries, top 15 countries visited (2013)," from the Statistics Canada website: http://www.statcan.gc.ca/tables-tableaux/sum-som/l01/cst01/arts37a-eng.htm
17. Canadian complaints against airlines: http://www.otc-cta.gc.ca/eng/report-released-air-travel-complaints-canada; U.S. complaints: http://airconsumer.ost.dot.gov/reports/2004/0402atcr.doc; U.S. airline passengers: http://airconsumer.ost.dot.gov/reports/2004/0402atcr.doc; Air Canada passengers: http://publications.gc.ca/collections/Collection-R/Statcan/51-004-XIB/0050451-004-XIB.pdf
18. Total Payroll: USA Today Salaries Databases, Hockey, 2005–06: http://www.usatoday.com/sports/nhl/bruins/salaries/2005/team/all/; Performance Measures: nhl.com, Standings, 2005–2006: http://www.nhl.com/

Chapter 4

1. M. Duggen, N.B. Ellison, C. Lampe, A. Lenhart, and M. Madden, "Social Media Update 2014," Pew Research Center, January 9, 2015, http://www.pewinternet.org/2015/01/09/social-media-update-2014/
2. S. Thakore, Z. Ismail, S. Jarvis, E. Payne, S. Keetbaas, R. Payne, and L. Rothenburg, "The Perceptions and Habits of Alcohol Consumption and Smoking among Canadian Medical Students," *Academic Psychiatry* 33, 3 (May–June 2009): 193–97.
3. http://www.therecord.com/news/canada/article/306689—canadians-found-to-be-huge-online-users
4. http://www.princanada.com/survey-says-47-of-canadians-use-twitter/
5. http://articles.nydailynews.com/2010-02-16/entertainment/27056462_1_new-poll-web-users-internet; http://www.datingsitesreviews.com/staticpages/index.php?page=online-dating-industry-facts-statistics; http://www.bbc.co.uk/pressoffice/pressreleases/stories/2010/02_february/13/poll.shtml

6. http://www.tdsb.on.ca/Portals/0/AboutUs/Research/2011StudentCensus.pdf
7. http://www.ctv.ca/CTVNews/Canada/20101013/motherhood-gap-101012/#ixzz1olocglXz; http://www.td.com/document/PDF/economics/special/td-economics-special-bc1010-career-interrupted.pdf
8. http://www.hockey-reference.com/leaders/shot_pct_active.html
9. http://www40.statcan.ca/101/cst01/labor12.htm?sdi=employment
10. http://www40.statcan.ca/101/cst01/demo10a.htm?sdi=age
11. http://globalnews.ca/news/2426553/canadians-support-for-taking-in-syrian-refugees-is-increasing-ipsos-poll/
12. http://www.ipsos-na.com/news/pressrelease.cfm?id=3383
13. Adapted from David L. Wheeler, "More Social Roles Means Fewer Colds," *Chronicle of Higher Education* XLIII, no. 44 (July 11, 1997): A13.
14. P.D. Franklin, R.A. Lemon, and H.S. Barden, "Accuracy of Imaging the Menisci on an In Office, Dedicated, Magnetic Resonance Imaging Extremity System," *The American Journal of Sports Medicine* 25, no. 3 (1997): 382.
15. Mya Frazier, "The Reality of the Working Woman," *Ad Age Insights*—White Paper, Spring, 2010: http://adage.com/images/bin/pdf/aa_working_women_whitepaper_web.pdf, p. 4.
16. Michael Crichton, *Congo* (New York: Knopf, 1980).

Chapter 5

1. The Canadian Press. Data taken from "Cancer Rate Drops with Plastic Surgery: But Suicide Rate Rises with Cosmetic Procedures," *The Windsor Star*, 10 October 2006, p. A10.
2. http://www.vegasinsider.com/nhl/odds/futures/
3. https://ca.sports.yahoo.com/nba/players/4886/
4. http://www.oxfordseminars.ca/LSAT/lsat_profiles.php, March 2015.
5. http://usatoday.com/snapshot/life/2001-06-11-potter.htm, 9 October 2001.
6. "Percentage of Major League Sports Players Born Outside the USA," http://www.usatoday.com/news/snapshot.htm, 6 July 2010.
7. Christy Fisher, "The Not-So-Great American Road Trip," *American Demographics,* May 1997, p. 47.
8. http://www.ipsos-na.com/news-polls/pressrelease.aspx?id=2981
9. G.M. Mullet, "Simeon Poisson and the NHL," *The American Statistician* 31 (1977): 8–12.
10. http://www.windsorstar.com
11. L.D. Williams, P.S. Hamilton, B.W. Wilson, and M.D. Estock, "An Outbreak of *Escherichia coli 01257:H7* involving Long Term Shedding and Person-to-Person Transmission in a Child Care Center," *Environmental Health,* May 1997, p. 9.
12. http://www.ipsos-na.com/news/pressrelease.cfm?id=3382
13. Data adapted from *American Demographics,* May 1997, p. 32.
14. "Call It in the Air," *The Press-Enterprise* (Riverside, CA), 19 October 1992.
15. Mark A. Atkinson, "Diet, Genetics, and Diabetes," *Food Technology* 51, no. 3 (March 1997), p. 77.
16. "SquareTrade's Report on Simple vs. Smartphone Reliability," http://bgr.com/2010/11/09/square-trade-apple-motorola-htc-make-most-reliable-smartphones/
17. "Most Popular Chocolate," http://usatoday.com/news/snapshot.htm?section=L&label=2006-10-27-wash
18. http://www.ipsos-na.com/news/pressrelease.cfm?id=3013
19. http://www.ekospolitics.com/articles/25July2005background.pdf
20. http://www.ekospolitics.com/articles/22Sept2005Background.pdf
21. The Canadian Press. Data taken from "Cancer Rate Drops with Plastic Surgery: But Suicide Rate Rises with Cosmetic Procedures," *The Windsor Star*, 10 October 2006, p. A10.
22. http://www.theglobeandmail.com/life/health-and-fitness/health/conditions/child-abuse-victims-more-likely-to-develop-cancer/article572644/
23. http://www.phac-aspc.gc.ca/cd-mc/cancer/fs-fi/cancer-child-enfant/index-eng.php

Chapter 6

1. L. Jimenez and P. Morreale, "Mobile Assistive Technology: Mapping and Integration," Proceedings of the 17th International Conference on Human Computer Interaction (HCI), Lecture Notes in Computer Science series, Springer Computer Science, 2015.
2. Adapted from A.M. Goodman and A.R. Ennos, "The Response of Field-grown Sunflower and Maize to Mechanical Support," *Annals of Botany* 79 (1997): 703.
3. "Medical Encyclopedia: Pulse," *Medline Plus: Trusted Health Information for You,* http://www.nlm.nih.gov/medlineplus/ency/article/003399.htm#Normal%20Values, 2 April 2004.
4. http://www.pagina12.com.ar
5. http://elcomercio.pe/
6. "Pepsico (PEP)," http://www.pepsico.com/docs/album/default-document-library/pepsico-2014-annual-report_final.pdf?sfvrsn=0
7. Philip A. Altman and D.S. Dittmer, *The Biology Data Book,* 2nd ed., Vol. I (Bethesda, MD: Federation of American Societies for Experimental Biology, 1964), p. 137.
8. Allen L. Shoemaker, "What's Normal? Temperature, Gender, and Heart Rate," *Journal of Statistics Education* (1996).
9. AboutKidsHealth, "About Premature Babies," http://www.aboutkidshealth.ca
10. https://secure.cihi.ca/free_products/too_early_too_small_en.pdf

Chapter 7

1. The Queen's Printer for Ontario, 2007, http://www.attorneygeneral.jus.gov.on.ca/english/courts/jury/geninfo.asp
2. Alice Park, "Omega-3 May Reduce Heart Risks Less Than Thought," http://www.time.com/time/health/article/0,8599,2014603,00.html, 30 August 2010.
3. "Chlorinated Water Byproduct, Rat Cancer Linked," *The Press-Enterprise* (Riverside, CA), 18 June 1997, p. A-6.
4. Chris Gilberg, J.L. Cos, H. Kashima, and K. Eberle, "Survey Biases: When Does the Interviewer's Race Matter?" *Chance* (Fall 1996): 23.
5. P.D. Franklin, R.A. Lemon, and H.S. Barden, "Accuracy of Imaging the Menisci on an In-Office, Dedicated, Magnetic Resonance Imaging Extremity System," *The American Journal of Sports Medicine* 25, no. 3 (1997): 382.
6. Liz Szabo, "Study: Tai Chi Could Ease Fibromyalgia Pain," *The Press-Enterprise* (Riverside, CA), 19 August 2010, p. 4D.
7. "New Drug a Bit Better Than Aspirin," *The Press-Enterprise* (Riverside, CA), 14 November 1996, p. A-14.
8. http://www.thestrategiccouncil.com/our_news/polls/0805%20GMCTV%20August%20Poll.pdf; http://rc-rc.ca/wp-content/uploads/2015/11/Can_speaks_eng.pdf
9. http://www.statcan.gc.ca/pub/84f0210x/2007000/t015-eng.htm
10. http://www.statcan.gc.ca/tables-tableaux/sum-som/l01/cst01/labr69g-eng.htm
11. Allen L. Shoemaker, "What's Normal? Temperature, Gender, and Heart Rate," *Journal of Statistics Education* (1996).
12. Nicola Maffulli, V. Testa, G. Capasso, and A. Sullo, "Calcific Insertional Achilles Tendinopathy," *The American Journal of Sports Medicine* 32, no. 1 (January/February 2004): 174.
13. http://www.statcan.gc.ca/tables-tableaux/sum-som/l01/cst01/health74b-eng.htm
14. "Must Have Accessories on Family Road Trips," http://www.usatoday.com/news/snapshot.htm, 25 August 2010.
15. American Public Health Association. *American Journal of Public Health* 94, no. 9 (September 2004): 1555–1559, http://www.ajph.org/cgi/content/abstract/94/9/1555
16. "Views Mixed on Role of Internet in Education," *The Press-Enterprise* (Riverside, CA) 21 August 2001, p. A-7.
17. http://us.mms.com/us/about/products/milkchocolate, 3 March 2004.
18. T. Lobstein, L. Baur, and R. Uauy, Obesity in Children and Young People: A Crisis in Public Health, *Obesity Reviews* 2004, 5 (Suppl. 1): 4–85.
19. http://www.statcan.gc.ca/tables-tableaux/sum-som/l01/cst01/health83b-eng.htm

20. Adam Fernandez, "Nuts About You," *American Demographics* 26, no. 1 (February 2004): 14.
21. P.C. Karalekas, Jr., C R. Ryan, and F.B. Taylor, "Control of Lead, Copper, and Iron Pipe Corrosion in Boston," *American Water Works Journal,* February, 1983.
22. *Science News* 136 (19 August 1989):124.
23. "Half in Study Were Abused As Adults," *The Press-Enterprise* (Riverside, CA), 15 July 1997, p. D-4.
24. Catherine M. Santaniello and R.E. Koning, "Are Radishes Really Allelopathic to Lettuce?" *The American Biology Teacher* 58, no. 2 (February 1996): 102.
25. http://www.thestrategiccounsel.com/our_news/polls/0905%20GMCTV%20Sept%20Poll.pdf
26. Reprinted with permission from Journal of Quality Technology © 1991 American Society for Quality. J. Hackl, *Journal of Quality Technology,* April 1991.
27. Daniel Seligman, "The Road to Monte Carlo," *Fortune,* 15 April 1985.

Chapter 8

1. *Science News* 136 (19 August 1989):124.
2. "Hotels for any Budget," *Consumer Reports,* June 2010, and "Hotel Ratings," http://consumerreports.org/cro/magazine-archive/2010/june/shopping/hotels/ratings/index.htm, 2 September 2010.
3. Laurie Lucas, "It's Elementary, Mister," *The Press-Enterprise* (Riverside, CA), 28 May 1997, p. D-1.
4. http://www.ekos.ca/admin/articles/14Nov2005/UpDown.pdf
5. http://www.angus-reid.com/wp-content/uploads/2010/11/2010.11.08_Melting_CAN.pdf
6. "Caught in the Middle," *American Demographics,* July 2001, p. 14–15.
7. http://www.ctf-fce.ca/en/projects/MERP/summaryfindings.pdf
8. Alison Stein Wellner, "A New Cure for Shoppus Interuptus," *The Marketing Tool Directory,* 2002.
9. Allen L. Shoemaker, "What's Normal? Temperature, Gender, and Heart Rate," *Journal of Statistics Education* (1996).
10. http://www.angus-reid.com/wp-content/uploads/2010/11/2010.11.08_Melting_CAN.pdf
11. http://www.elections.ca/eca/eim/article_search/article.asp?id=105&lang=e&frmPageSize=&textonly=false
12. William H. Leonard, Barbara J. Speziale, and John E. Pernick, "Performance Assessment of a Standards-Based High School Biology Curriculum," *The American Biology Teacher* 63, no. 5 (2001): 310–316.
13. Ibid.
14. "Hotels for any Budget," *Consumer Reports,* June 2010 and "Hotel Ratings," http://www.consumerreports.org/cro/magazine-archive/2010/june/shopping/hotels/overview/index.htm, 2 September 2010.
15. Shoemaker, "What's Normal? Temperature, Gender, and Heart Rate."
16. http://www.ipsos-na.com/news/pressrelease.cfm?id=3395
17. David L. Wheeler, "More Social Roles Means Fewer Colds," *Chronicle of Higher Education* XLIII, no. 44 (11 July 1997): A13.
18. Adapted from "Toplines: To the Moon?" Rebecca Gardyn, ed., *American Demographics,* August 2001, p. 9.
19. "Generation Next: A Snapshot," www.pewtrusts.org/, 9 November 2006; and "The American Freshman: National Norms for Fall 2005," http://www.heri.ucla.edu/tfsPublications.php, 9 November 2006.
20. Reed Abelson, "A Survey of Wall St. Finds Women Disheartened," *The New York Times on the Web,* http://www.nytimes.com, July 26, 2001.
21. *Ecological Economics* 42 (2002):185–199.

22. Adapted from A.M. Goodman and A.R. Ennos, "The Responses of Field-grown Sunflower and Maize to Mechanical Support," *Annals of Botany* 79 (1997): 703.
23. http://www.newswire.ca/en/releases/archive/January2007/03/c2880.html; http://www.decima.ca/en/inthenews/
24. G. Wayne Marino, "Selected Mechanical Factors Associated with Acceleration in Ice Skating," *Research Quarterly for Exercise and Sport* 54, no. 3 (1983).
25. http://www.cbc.ca/news/canada/average-student-debt-difficult-to-pay-off-delays-life-milestones-1.2534974
26. Andrew Parkin and Matthew Mendelsohn, "Table 5: Proud to Be Canadian," *The CRIC Papers: A New Canada: An Identity Shaped by Diversity*, October 2003, p. 11, http://130.15.161.246:82/docs/Government%20surveys/cric/cricgmnc03-eng/cricpapers_oct3003.pdf
27. http://www.ohqc.ca/pdfs/ohqcsuccess_stories-2.2.6may12.pdf. Used with permission from the Ontario Health Quality Council. Disclaimer: North York General Hospital did not use statistical rigor to assess the significance of the improvement or interventions designed to reduce ambulatory patient wait time. Although these improvements have been felt the results are not statistically relevant and purely anecdotal at this time. Data integrity issues and inappropriate sample size precluded appropriate statistical analysis.

Chapter 9

1. http://www.statcan.gc.ca/pub/84f0210x/2007000/t015-eng.htm
2. http://www.angus-reid.com/wp-content/uploads/2010/11/2010.11.08_Melting_CAN.pdf
3. Jan Pergl, Irena Perglova, Per Pysek, and Hansjorg Dietz, 2006. "Population Age Structure and Reproductive Behavior of the Monocarpic Perennial *Heraculaneum Mantegazzianum* (Apiaceae) in its Native and Invaded Distribution Ranges," *American Journal of Botany,* 93(7):1018–1028.
4. Adapted from Pamela Paul, "Coming Soon: More Ads Tailored to Your Tastes," *American Demographics,* August 2001, p. 28.
5. Allen L. Shoemaker, "What's Normal? Temperature, Gender, and Heart Rate," *Journal of Statistics Education* (1996).
6. "Hot News: 98.6 Not Normal," *The Press-Enterprise* (Riverside, CA), 23 September 1992.
7. Nicola Maffulli, V. Testa, G. Capasso, and A. Sullo, "Calcific Insertional Achilles Tendinopathy," *The American Journal of Sports Medicine* 32, no. 1 (January/February 2004): 174.
8. http://allthingsd.com/20120927/instagram-beats-twitter-in-daily-mobile-users-for-the-first-time-data-says/
9. http://data.worldbank.org/indicator/SP.DYN.LE00.IN
10. "Hotels for any Budget," *Consumer Reports*, June 2010, and "Hotel Ratings," http://www.consumerreports.org/cro/magazine-archive/2010/june/shopping/hotels/overview/index.htm, 2 February 2016.
11. Ibid
12. Lance Wallace and Terrence Slonecker, "Ambient Air Concentrations of Fine ($PM_{2.5}$) Manganese in the U.S. National Parks and in California and Canadian Cities: The Possible Impact of Adding MMT to Unleaded Gasoline," *Journal of the Air and Waste Management Association* 47 (June 1997):642–651.
13. Shoemaker, "What's Normal? Temperature, Gender, and Heart Rate."
14. "It Pays to Buy Store Brands," *Consumer Reports*, October 2009, and http://www.consumerreports.org/cro/magazine-archive/october-2009/shopping/buying-store-brands/overview/buying-store-brands-ov.htm, 16 February 2016.
15. http://www.thestrategiccounsel.com/our-news/polls/1005%20GMCTV%20October%20Report%20-%Avian%20Flu%20NAFTA%20Final.pdf
16. http://www.elections.ca/eca/eim/article_search/article.asp?id=124&lang=e&frmPageSize=&textonly=false
17. https://www.canadianveterinarians.net/documents/canadian-pet-population-figures-cahi-2014

18. Liz Szabo, "Study: Tai Chi Could Ease Fibromyalgia Pain," *The Press-Enterprise* (Riverside, CA), 19 August 2010, p. 4D.
19. Paul, "Coming Soon: More Ads Tailored to Your Tastes."
20. Denise Grady, "Study Finds Alzheimer's Danger in Hormone Therapy," *The Press-Enterprise* (Riverside, CA), 28 May 2003.
21. *Heart Healthy Women* website, http://www.hearthealthywomen.org/patients/medications/blood_thinners__aspirin_6.html
22. Adapted from "Toplines: To the Moon?" Rebecca Gardyn, ed., *American Demographics,* August 2001, p. 9.
23. Loren Hill, *Bassmaster,* September/October 1980.
24. Charles Dickey, "A Strategy for Big Bucks," *Field and Stream,* October 1990.
25. *Science News* 136 (19 August 1989): 124.
26. Copyright © 2016 Ipsos in Canada. http://www.ipsos-na.com/news/pressrelease.cfm?id=3393
27. Copyright © 2016 Ipsos in Canada. http://www.ipsos-na.com/news/pressrelease.cfm?id=3396
28. The Strategic Counsel, "Does Canada accept the right number of immigrants per year?" From the August Survey for the *Globe and Mail* and CTV: Immigration Terrorism and National Security, August 7, 2005, p. 14. http://www.thestrategiccounsel.com
29. Kurt Grote, T.L. Lincoln, and J.G. Gamble, "Hip Adductor Injury in Competitive Swimmers," *The American Journal of Sports Medicine* 32, no. 1 (January/February 2004): 104.
30. McElhaney et al., "A Placebo-Controlled Trial of a Proprietary Extract of North American Ginseng (CVT-E002) to Prevent Acute Respiratory Illness in Institutionalized Older Adults," *Journal of the American Geriatrics Society* 52 no. 1 (2004): 13–19.
31. February 25, 2006, the *Vancouver Sun* published an article in which two professors from the University of British Columbia criticized the claims
32. http://www.ejcancer.com/issue/S0959-8049%2809%29X0005-5
33. http://www.bellaonline.com/articles/art20257.asp; https://en.wikipedia.org/wiki/Preferred_walking_speed

Chapter 10

1. http://quitsmoking.about.com/od/chemicalsinsmoke/a/tar_in_cigs.htm; http://www.imperialtobaccocanada.com/groupca/sites/imp_7vsh6j.nsf/vwPagesWebLive/DO7VXN23?opendocument&SKN=1
2. http://ibdcrohns.about.com/od/diagnostictesting/p/testrbc.htm
3. W.B. Jeffries, H.K. Voris, and C.M. Yang, "Diversity and Distribution of the Pedunculate Barnacles *Octolasmis* Gray, 1825 Epizoic on the Scyllarid Lobster *Thenus orientalis* (Lund, 1793)," *Crustaceana* 46, no. 3 (1984).
4. http://www.hockeydb.com/ihdb/stats/players.html
5. Wendy K. Baell and E. H. Wertheim, "Predictors of Outcome in the Treatment of Bulimia Nervosa," *British Journal of Clinical Psychology* 31 (1992): 330–332.
6. "L.A. Heart Data." Adapted from data found at http://www-unix.oit.umass.edu/~statdata/statdata/data/laheart.dat
7. Jan D. Lindhe, "Clinical Assessment of Antiplaque Agents," *Compendium of Continuing Education in Dentistry,* Suppl. 5 (1984).
8. Susan J. Beckham, W.A. Grana, P. Buckley, J.E. Breasile, and P.L. Claypool, "A Comparison of Anterior Compartment Pressures in Competitive Runners and Cyclists," *The American Journal of Sports Medicine* 21, no. 1 (1992): 36.
9. Michael A. Brehm, J.S. Buguliskis, D.K. Hawkins, E.S. Lee, D. Sabapathi, and R.A. Smith, "Determining Differences in Efficacy of Two Disinfectants Using *t*-tests," *The American Biology Teacher* 58, no. 2 (February 1996): 111.
10. http://www.hockeydb.com/ihdb/stats/players.html
11. "Aaron Rodgers #12 QB," http://sports.espn.go.com/nfl/players, and "Drew Brees #9 QB," http://sports.espn.go.com/nfl/players, 18 March 2011.

12. A. Tubb, A.J. Parker, and G. Nickless, "The Analysis of Romano-British Pottery by Atomic Absorption Spectrophotometry," *Archaeometry* 22, 2 (August 1989): 153.
13. The Fraser Institute. Table 1: "Estimated Average Automobile Insurance Premiums, 2004–05, by Province, Straight Dollars," from *The False Promise of Government Auto Insurance*, p. 5, by Brett J. Skinner, February 2007. http://www.fraserinstitute.ca
14. Beckham et al., "A Comparison of Anterior Compartment Pressures."
15. "2010 College-Bound Seniors: Total Group Profile Report," http://research.collegeboard.org/programs/sat/data/archived/cb-seniors-2010
16. "Aaron Rodgers #12 QB," http://sports.espn.go.com/nfl/players, and "Ben Roethlisberger #7 QB," http://sports.espn.go.com/nfl/players, 18 March 2011.
17. Beckham et al., "A Comparison of Anterior Compartment Pressures."
18. Carlos E. Macellari, "Revision of Serpulids of the Genus *Rotularia (Annelida)* at Seymour Island (Antarctic Peninsula) and Their Value in Stratigraphy," *Journal of Paleontology* 58, no. 4 (1984).
19. T.M. Casey, M.L. May, and K.R. Morgan, "Flight Energetics of Euglossine Bees in Relation to Morphology and Wing Stroke Frequency," *Journal of Experimental Biology* 116 (1985).
20. Karl J. Niklas and T.G. Owens, "Physiological and Morphological Modifications of *Plantago Major (Plantaginaceae)* in Response to Light Conditions," *American Journal of Botany* 76, no. 3 (1989): 370–382.
21. "Expensive Bumper Repair: Latest Crash Tests," *Consumers' Research,* April 2001, pp. 20–21.
22. "TicketNetwork Advisory—Lakers vs. Celtics NBA Finals Ticket Prices Lower Than 2008," http://www.marketwire.com/press-release/TicketNetwork-Advisory-Lakers-vs-Celtics-NBA-Finals-Ticket-Prices-Lower-Than-2008-1269109.htm, 1 June 2010.
23. "KFC: Too Finger-Lickin' Good?" *Good Housekeeping* Saavy Consumer Product Tests: http://www.prnewswire.com/news-releases/kfc-too-finger-lickin-good—from-the-august-2002-issue-of-good-housekeeping—76280657.html
24. John Fetto, "Shop Around the Clock," *American Demographics* 25, no. 7 (September 2003): 18.
25. Adapted from Donald L. Barlett and James B. Steele, "Why We Pay So Much for Drugs," *Time,* 2 February 2004, p. 44.
26. Adapted from Peter Quinby, James Hodson, and Mike Henry, "Forest Landscape Baselines: Evaluating Track Plate Independence and Bait Type for Detecting Marten in Temagami, Ontario," 25, Table 2, p. 3, 2005. http://www.ancientforest.org/wp-content/uploads/2014/12/flb25.pdf
27. S.K. Martin, "Feeding Ecology of American Martens and Fishers," *Martens, Sables, and Fishers: Biology and Conservation* (Cornell University Press: Ithaca, N.Y., 1994), pp. 297–315.
28. http://www.idf.org/signs-and-symptoms-diabetes
29. http://www.mayoclinic.org/diseases-conditions/diabetes/expert-blog/prediabetes/bgp-20056566

Chapter 11

1. H.F. Barsam and Z.M. Simutis, "Computer-Based Graphics for Terrain Visualization Training," *Human Factors,* vol. 26, no. 6, 1984, pp. 659–665. Copyright © 1984 by the Human Factors and Ergonomics Society. Reprinted by permission of SAGE Publications, Inc.
2 http://www.payscale.com/research/CA/Job=Flight_Attendant/Hourly_Rate; http://www1.salary.com/CA/Ontario/Flight-Attendant-salary.html
3. Ciril Rebetez, Mireille Betrancourt, Mirweis Sangrin, and Pierre Dillenbourg, "Learning from Animation Enabled by Collaboration," *Intstructional Science* 35, no. 5 (September 2010): 471–485.
4. From January 30 2001, "Bumper crash tests: Hyundai Elantra performs best" and September 13, 2000 "FLIMSY BUMPERS ON SUVs FAIL TO RESIST DAMAGE; FOUR OF FIVE TESTED EARN LOWEST POSSIBLE RATING ARLINGTON." Copyright © 2001 Consumers Research. Reprinted by permission.

5. Russell R. Pate, Chia-Yih Wang, Marsha Dowda, Stephen W. Farrell, and Jennifer R. O'Neill, "Cardiorespiratory Fitness Levels Among U.S. Youth 12 to 19 Years of Age," *Archives of Pediatric Adolescent Medicine* 160 (October 2006): 1005–1011.
6. Adapted from "Average Salary for Men and Women Faculty, by Category, Affiliation, and Academic Rank 2002–2003," *Academe: Bulletin of the American Association of University Professors* (March–April 2003): 37.
7. A. Tubb, A.J. Parker, and G. Nickless, "The Analysis of Romano-British Pottery by Atomic Absorption Spectrophotometry," *Archaeometry* 22, 2 (August 1989): 153.
8. http://www.statcan.ca/cgi-bin/downpub/listpub.cgi?catno=75-001-XIE2005102; http://www.statcan.ca/bsolc/english/bsolc?catno=81-595-M2006046
9. Data taken from the published fee schedules of the cities listed, April 2007.
10. www.newswire.ca/news-releases/r-e-p-e-a-t----hard-to-shake-globe-and-mail-series-exposes-the-pervasive-health-risks-associated-with-canadas-excessive-salt-consumption-537866071.html; http://www.heartandstroke.com/site/c.ikIQLcMWJtE/b.3484241/k.6D9D/Healthy_living__Salt.htm; from *The Globe and Mail* series, "Hard to Shake," a four-part investigative series, starting Saturday, June 20, 2009; © 2015, Heart and Stroke Foundation of Canada. Reproduced with the permission of the Heart and Stroke Foundation of Canada. www.heartandstroke.ca.

Chapter 12

1. Stellan Ohlsson, "The Learning Curve for Writing Books: Evidence from Professor Asimov," *Psychological Science* 3, no. 6 (1992): 380–382.
2. Daniel C. Harris, *Quantitative Chemical Analysis,* 3rd ed. (New York: Freeman, 1991).
3. http://www.ncdc.noaa.gov/oa/climate/research/ghcn/ghcngrid.html#Data
4. Adapted from J. Zhou and S. Smith, "Measurement of Ozone Concentrations in Ambient Air Using a Badge-Type Passive Monitor," *Journal of the Air & Waste Management Association* 47 (June 1997): 697.
5. George W. Pierce, *The Song of Insects* (1948), http://exploringdata.cqu.edu.au/stories.htm#chirps
6. Sarah Janssen, Ed., *The World Almanac and Book of Facts 2011* (New York, NY: Infobase Publishing), 2011.
7. http://www.math.yorku.ca/Who/Faculty/Ng/ssc2003/CaseStudiesMain.htm
8. www.bestbuy.ca
9. R.C. Nelson, C.M. Brooks, and N.L. Pike, "Biomechanical Comparison of Male and Female Runners," in P. Milvy (ed.), *The Marathon: Physiological, Medical, Epistemological, and Psychological Studies* (New York Academy of Sciences, 1977), pp. 793–807.
10. "Drew Brees #9 QB," http://sports.espn.go.com/nfl/players, 22 March 2011.
11. W.B. Jeffries, H.K. Voris, and C.M. Yang, "Diversity and Distribution of the Pedunculat Barnacles *Octolasmis* Gray, 1825 Epizoic on the Scyllarid Lobster, *Thenus orientalis* (Lund, 1793)," *Crustaceana* 46, no. 3 (1984).
12. Gregory K. Torrey, S.F. Vasa, J.W. Maag, and J.J. Kramer, "Social Skills Interventions across School Settings: Case Study Reviews of Students with Mild Disabilities," *Psychology in the Schools* 29 (July 1992): 248.
13. G. Wayne Marino, "Selected Mechanical Factors Associated with Acceleration in Ice Skating," *Research Quarterly for Exercise and Sport* 54, no. 3 (1983).
14. A.J. Ellis, "Geothermal Systems," *American Scientist,* September/October 1975.
15. Allen L. Shoemaker, "What's Normal? Temperature, Gender, and Heart Rate," *Journal of Statistics Education* (1996).
16. http://usa.cricinfo.com/db/stats/by_calendar/2000S/2006/test_bowl_best_sr_2006.html
17. http://espn.go.com/mlb/stats/team/_/stat/batting/year/2010/seasontype/2, 5 November 2010.
18. David R. McAllister et al., "A Comparison of Preoperative Imaging Techniques for Predicting Patellar Tendon Graft Length before Cruciate Ligament Reconstruction," *The American Journal of Sports Medicine* 20(4): 461–465.

19. Henry Gleitman, *Basic Psychology,* 4th ed. (New York: Norton, 1996).
20. http://www.the-numbers.com/daily-box-office-chart, 30 March 2016.
21. "Lexus GX," http://en.wikipedia.org/wiki/Lexus_GX, 27 September 2010.
22. http://www.timhortons.com/nutrition and http://www.timhortons/valeurnutritives
23. J. R. Baldwin and G. Gellatly, *Global Links: Long-Term Trends in Foreign Investment and Foreign Control in Canada, 1960 to 2000.* Micro-economic Analysis Division, 18th Floor, R.H. Coats Building, Ottawa, K1A 0T6. Telephone: 1-800-263-1136. Catalogue no. 11-622-MIE–No. 008. ISSN: 1705-6896. ISBN: 0-662-42013-6; For years 2001–2004, CANSIM Table 179-0004: Corporations Returns Act (CRA), major financial variables, annual, 1999 to 2004.
24. http://publications.gc.ca/collections/Collection/Fo46-12-363E.pdf

Chapter 13

1. W.S. Good, "Productivity in the Retail Grocery Trade," *Journal of Retailing* 60, no. 3 (1984).
2. Adapted from J. Zhou and S. Smith, "Measurement of Ozone Concentrations in Ambient Air Using a Badge-Type Passive Monitor," *Journal of the Air & Waste Management Association* (June 1997): 697.
3. "Lexus GX," http://en.wikipedia.org/wiki/Lexus_GX, 27 September 2010.
4. http://www.math.yorku.ca/Who/Faculty/Ng/ssc2003/CaseStudiesMain.htm
5. R. Blair and R. Miser, "Biotin Bioavailability from Protein Supplements and Cereal Grains for Growing Broiler Chickens," *International Journal of Vitamin and Nutrition Research* 59 (1989): 55–58.
6. http://usa.cricinfo.com/db/Stats/by_calendar/2000S/2006/test_bat_highest_sr_2006.html
7. Jeff Howe, "Total Control," *American Demographics,* July 2001, p. 30.
8. B. Fox, *Women's Domestic Labour and their Involvement in Wage Work*, unpublished doctoral dissertation, 1980, p. 449.
9. J.R. Baldwin and G. Gellatly, *Global Links: Long-Term Trends in Foreign Investment and Foreign Control in Canada, 1960 to 2000*. Micro-economic Analysis Division, 18th Floor, R. H. Coats Building, Ottawa, K1A 0T6. Telephone: 1 800 263-1136. Catalogue no. 11-622-MIE–No. 008. ISSN: 1705-6896. ISBN: 0-662-42013-6; For years 2001–2004, CANSIM Table 179-0004: Corporations Returns Act (CRA), major financial variables, annual, 1999 to 2004.
10. http://dsp-psd.pwgsc.gc.ca/Collection/Fo46-12-363E.pdf

Chapter 14

1. Daniel Q. Haney, "Mondays May Be Hazardous," *The Press-Enterprise* (Riverside, CA), 17 November 1992, p. A16.
2. M&M'S® is a registered trademark of Mars, Incorporated. These trademarks are used with permission. Mars, Incorporated is not associated with Cengage Learning, Inc. or Nelson Education Ltd. The images of the M&M'S® mark and M&M'S® Brand candies are printed with permission of Mars, Incorporated.
3. M&M'S® is a registered trademark of Mars, Incorporated. These trademarks are used with permission. Mars, Incorporated is not associated with Cengage Learning, Inc. or Nelson Education Ltd. The images of the M&M'S® mark and M&M'S® Brand candies are printed with permission of Mars, Incorporated.
4. Adapted from Linda Schmittroth, ed., *Statistical Record of Women Worldwide* (Detroit and London: Gale Research, 1991).
5. Adapted from Dana Blanton, "Poll: Most Believe 'Cover-Up' of JFK Assassination Facts," http://www.foxnews.com/story/0,2933,102511,00.html, 10 February 2004.
6. "No Shows," *American Demographics,* 25, no. 9 (November 2003): 11.
7. Statistics Canada, Catalogue no. 82-620-MWE.
8. Adapted from Rebecca Piirto Heath, "Life on Easy Street," *American Demographics,* April 1997, p. 33.

9. W.W. Menard, "Time, Chance and the Origin of Manganese Nodules," *American Scientist,* September/October, 1976.
10. "Mexicans Support Birth Control, Not Abortion," http://www.angus-reid.com/polls/index.cfm/fuseaction/viewItem/itemID/15192
11. "Religion among the Millennials: Less Religiously Active Than Older Americans, But Fairly Traditional In Other Ways," http://pewforum.org/Age/Religion-Among-the-Millennials.aspx, 17 February 2010.
12. Harris Interactive/Financial Times: "Britons, French Upset with Life Under EU" from http://www.angus-reid.com/polls/index.cfm/fuseaction/viewItem/itemID/15181
13. Thomas Lord and Terri Orkwiszewski, "Moving from Didactic to Inquiry-Based Instruction in a Science Laboratory," *American Journal of Primatolgy,* 68 (October 2006).
14. http://www.inspection.gc.ca/english/anima/meavia/mmopmmhv/chap19/baseline-e.pdf
15. Kim Marie McGoldrick, Gail Hoyt, and David Colander, "The Professional Development of Graduate Students for Teaching Activities: The Students' Perspective," *Journal of Economic Education*, 41, no. 2 (2010): 194–201.
16. M. Martin Costa and M. Gatz, "Determination of Authorship Credit in Published Dissertations," *Psychological Science* 3, no. 6 (1992): 54.
17. Sarah Janssen, Ed., *The World Almanac and Book of Facts 2011* (New York, NY: Infobase Publishing), 2011.
18. Sarah Janssen, Ed., *The World Almanac and Book of Facts 2011* (New York, NY: Infobase Publishing), 2011.
19. Doreen Matsui, R. Lim, T. Tschen, and M.J. Rieder, "Assessment of the Palatability of b-Lactamase-Resistant Antibiotics in Children," *Archives of Pediatric Adolescent Medicine* 151 (June 1997): 599.
20. "Every Dad Has His Day," *Time,* 16 June 1997, p. 16.
21. Andrew S. Levy, M.J. Wetzler, M. Lewars, and W. Laughlin, "Knee Injuries in Women Collegiate Rugby Players," *The American Journal of Sports Medicine* 25, no. 3 (1997): 360.
22. Adapted from David L. Wheeler, "More Social Roles Means Fewer Colds," *Chronicle of Higher Education* XLIII, no. 44 (July 11, 1997): A13.
23. Carole Day and Del Lowenthal, "The Use of Open Group Discussions in Marketing Library Services to Young Adults," *British Journal of Educational Psychology* 62 (1992):324–340. The British Journal of Educational Psychology by BRITISH PSYCHOLOGICAL SOCIETY; TRAINING COLLEGE ASSOCIATION (GREAT BRITAIN). Reproduced with permission of SCOTTISH ACADEMIC PRESS, in the format Book via Copyright Clearance Center.
24. T. Yi Wen, A. Snowdon, S.E. Ahmed, and A.A. Hussein. (2009). *Accommodating Nonrespondents in the Canadian National Child Safety Seat Survey.* © Her Majesty the Queen in Right of Canada, represented by the Minister of Transport (2010). This information has been reproduced with the permission of Transport Canada.

Chapter 15

1. T.M. Casey, M.L. May, and K.R. Morgan, "Flight Energetics of Euglossine Bees in Relation to Morphology and Wing Stroke Frequency," *Journal of Experimental Biology* 116 (1985).
2. "Alzheimer's Test Set for New Memory Drug," *The Press-Enterprise* (Riverside, CA), 18 November 1997, p. A-4.
3. "Aaron Rodgers #12 QB," http://sports.espn.go.com/nfl/players, and "Drew Brees #9 QB," http://sports.espn.go.com/nfl/players, 18 March 2011.
4. *Science News* 136 (August 1989):126.
5. Doreen Matsui, R. Lim, T. Tschen, and M.J. Rieder, "Assessment of the Palatability of b-Lactamase-Resistant Antibiotics in Children," *Archives of Pediatric Adolescent Medicine* 151 (1997): 559–601.
6. Scott K. Powers and M.B. Walker, "Physiological and Anatomical Characteristics of Outstanding Female Junior Tennis Players," *Research Quarterly for Exercise and Sport* 53, no. 2 (1983).

7. *Science News,* 1989, p. 116.
8. G. Merrington, L. Winder, and I. Green, "The Uptake of Cadmium and Zinc by the Birdcherry Oat Aphid *Rhopalosiphum Padi (Homoptera:Aphididae)* Feeding on Wheat Grown on Sewage Sludge Amended Agricultural Soil," *Environmental Pollution* 96, no. 1 (1997): 111–114.
9. Karola Sakekel, "Egg Substitutes Range in Quality," *San Francisco Chronicle,* 10 February 1993, p. 8. Copyright © 1993 San Francisco Chronicle. Reprinted by permission.
10. http://www.citynews.ca/2006/06/29/air-conditioning-makes-you-gain-weight/

Answers to Selected Exercises

Chapter 1

1.9 **a–b.** Survival times, *quantitative continuous* variable
c. Population of survival times for all patients having a particular type of cancer and having undergone a particular type of radiotherapy.
d–e. Problem with sampling: If all patients having cancer and radiotherapy are sampled, some may still be living and cannot be measure survival times. Need alternative population to sample.

1.11 **a–b.** 50 people constitute experimental unit; category, qualitative variable measured.
e. Yes, depends on the order of presentation; unimportant.
f. 0.78
g. 72%

1.13 **a.** Voter opinions on the conflict between Islam and the West.
b. All adults from 27 countries.
c. Category "Other" for the other 15% of the responses.
d. Answers will vary.

1.15 **a–b.** Educational attainment; qualitative variable.
c. Percentage of Aboriginal and non-Aboriginal population in each educational attainment category.
f. 6% of Aboriginal population; significant gap.

1.17 **a.** Age of Facebook users.
b. Quantitative.
c. Percentage of the Facebook users in different age groups.
d. Yes; percentages add up to 100%.
f. Bar chart, easier to follow; pie chart visually interesting.
g. Gender, type of device being, Internet connection speed.

1.19 **a.** For $n = 5$, between 8 and 10.
c. 0.86
d. 0.66
e. Stem and leaf plot has a more peaked mound-shaped distribution than relative frequency histogram because of the smaller number of groups.

1.21 **b.** Same as the proportion of "2," or 0.30.
c. Same as the proportion of "0" and "1," or 0.70.
d. 30
e. Relatively symmetric, mound-shaped distribution; no outliers.

1.23 Yes, since time decreases each successive day.

1.25 **a. Stem and Leaf Plot: Scores**

```
Stem and leaf of Scores N = 20
Leaf Unit = 1.0
  2 5 57
  5 6 123
  8 6 578
  9 7 2
 (2) 7 56
  9 8 24
  7 8 6679
  3 9 134
```

b–c. Not mound-shaped, bimodal with two peaks; might indicate two different student groups.

1.27 **a.** Bar chart.
c. Average salary increases as educational level increases.

1.29 **b.** Pareto chart.
c. The larger the airline, the more difficult it may be to serve passengers without any complaints. Other variables: size of airline, number of passengers served, airfare.

1.31 **a.** Skewed right; yes.
c. Two graphs convey the same information.

1.33 **a.** Answers will vary.
b. Stem and Leaf Plot: Age

```
Stem and leaf of Age N = 22
Leaf Unit = 1.0
1 3 9
1 4
8 4 5666778
11 5 124
11 5 579
8 6 01
6 6 5669
2 7 04
```

c. Clark, Mulroney, Harper, Campbell, and Meighen; all Conservative.

1.35 **b.** 20%

1.59 **a.** Height of each bar increases as the price increases, but not in the correct proportion to the actual prices. Price scale starts at \$8.50. Height of "Alberta Highest = \$10.79" looks more than double of "Alberta Lowest = \$9.50," the actual price is only about 14% higher.

1.61 Answers may vary.

1.63 **a.** **Stem and Leaf Plot: Megawatts**

```
Stem and leaf of Megawatts N = 20
Leaf Unit = 1000
10 0 2222233333
10 0 444
7 0 666
4 0 8
3 1 0
2 1 2
1 1
1 1
1 1 8
```

b. Skewed right.

1.65 **a.** Skewed right.
c. Cities with larger populations may have a larger number of Tim Hortons shops.

1.67 **a.** Demand for all three increased.
b. Demands are different for each produce during each quarter.
c. Yes. In Q3 and Q4, the demand of each produce increases. In Q2, as demand of corn and okra goes down, the demand of wheat increases.

Chapter 2

2.5 **a.** $\bar{x} = 1082.3$
b. $\bar{x} = 1102.5$
c. The average premium cost in different provinces is not as important to the consumer as the average cost for a variety of consumers in his or her geographical area.

2.7 **a.** **Stem and Leaf Plot: Wealth**

```
Stem and leaf of Wealth N = 20
Leaf Unit = 1.0
5 1 78888
(8) 2 00001234
7 2 66
5 3 23
3 3
3 4
3 4 9
2 5 2
1 5 6
```

b. $\bar{x} = 26.82$; $m = 21.75$
c. Mean is strongly affected by outliers; the median would be a better measure of centre.

2.9 **a.** $\bar{x} = 4.04$
b. Position of median = 7.5; median = 4.13
c. Skewed left.

2.11 **a.** $\bar{x} = 215$
b. Position of median = 5.5; median = 212.5
c. No unusually large or small observations to affect the value of the mean; mean or average time on task

2.13 **a.** $\bar{x} = 1667.98$
b. Median = 1324.99
c. Average cost as important as other variables: picture quality, size, lowest cost for the best quality, etc.

2.15 **a.** $\bar{x} = 6.25$
b. No, data set is small for mode; no, median will not provide any new meaningful information.
c. A plot of time against the year would be appropriate; most improvement is from 2005 to 2006, largest difference in time.
d. 40-yard dash time will be less than 2014.

2.19 **a.** 5.17
b. 5
c. $s = 2.64$
d. Range is larger than standard deviation; range measures variability from extreme values, standard deviation measures variability of every value in the data set from the mean, standard deviation is more reliable.

2.21 **a.** 119.96
b. 250.85
c. 35.25

2.25 **a.** $\mu \pm 3\sigma$; contains at least 8/9 of the measurements.
b. $\mu \pm 2\sigma$; contains at least 3/4 of the measurements.
c. At most 1/4 of the measurements can be less than 65.

2.27 **a.** **Stem and Leaf Plot: Weight**

```
Stem and leaf of Weight N = 27
Leaf Unit = 0.010
1 7 5
2 8 3
6 8 7999
8 9 23
13 9 66789
```

```
13 10
(3) 10 688
11 11 2244
7 11 788
4 12 4
3 12 8
2 13
2 13 8
1 14 1
```

b. $\bar{x} = 0.052$; $s = 0.166$

c.

k	$\bar{x} \pm ks$	Interval	Number in Interval	Percentage
1	1.052 ± 0.166	0.866 to 1.218	21	78%
2	1.052 ± 0.332	0.720 to 1.384	26	96%
3	1.052 ± 0.498	0.554 to 1.550	27	100%

2.29 a. Stem and Leaf Plot: Method 1, Method 2

```
Stem and leaf of Method 1 N = 10
Leaf Unit = 0.00010
1 10 0
3 11 00
4 12 0
(4) 13 0000
2 14 0
1 15 0

Stem-and-leaf of Method 2 N = 10
Leaf Unit = 0.00010
1 11 0
3 12 00
5 13 00
5 14 0
4 15 00
2 16 0
1 17 0
```

b. *Method 1*: $\bar{x} = 0.0125$, $s = 0.00151$; *Method 2*: $\bar{x} = 0.0138$, $s = 0.00193$; confirm the conclusions of part a.

2.31 a.

k	$\bar{x} \pm ks$	Interval	Number in Interval	Percentage
1	4.586 ± 2.892	1.694 to 7.478	43	61%
2	4.586 ± 5.784	−1.198 to 10.370	70	100%
3	4.586 ± 8.676	−4.090 to 13.262	70	100%

b. Part a does not agree with Empirical Rule; shape of the distribution is not mound-shaped, but flat.

2.33 a. $\mu - \sigma = -4$
b. Approximately 34%.
c–d. Latter is impossible; approximate values given by the Empirical Rule are not accurate, distribution cannot be mound-shaped.

2.35 a. $s \approx 1.0625$
b. $s = 1.351$; fairly close to the approximate value.

2.37 a. The data is spread out; the variance and standard deviation will be large.
b. Very little variance in average number of tweets; the variance in the average number of tweets as a function of followees starts to increase significantly after about 500 followees. Similarly, the variance for the average number of tweets as a function of followers starts to increase significantly at approximately 300 followers.

2.39 a. $s = 0.0810$
b. Range $= 0.225$; range about three times as big as standard deviation.
c. $s \approx 0.0563$; approximation not very accurate.
d. $s^2 = 0.00657$; subtracting (or adding) 0.5 (or any constant) does affect the sample variance.

2.41 $s = 1.27$

2.43 a. Min $= 0$; $Q_1 = 3.25$; m $= 5.5$; $Q_3 - 6.75$; max $= 8$; $IQR = 3.5$
b. $\bar{x} = 4.75$; $s = 9.0287$
c. Smaller observation: z-score $= -1.94$; largest observation, z-score $= 1.32$; since neither z-score exceeds 2 in absolute value, none of the observations are unusually small or large.

2.45 Lower and upper fence: 16 and 32; $x = 12$ is an outlier.

2.47 Lower and upper fence: –31.88 and 87.13; $x = 181$ is an outlier.

2.53 a. Skewed right.
b. $m = 108{,}697$, $\bar{x} = 121{,}248$; mean > median implies skewed right.
c. Lower fence and upper fence: –22,532.63 and 247,666.38

Chapter 3

3.15 $y = -24901.91 + 12.69x$

3.35 b. Most of the data points are clustered together in the lower region; strong positive linear relationship.
c. 0.999; very strong positive correlation.

3.39 a. Positive, linear, and moderate.
b. 0.959
c. $y = 71.97 + 0.0153x$

Chapter 4

4.67 a. 0.75
b. 0.7
c. 0.35
d. 0.47
e. 0.46
f. 0.46
g. 0.76
h. 0.275

4.89 a. $p(0) = 0.0973, p(1) = 0.3428,$
$p(2) = 0.4024, p(3) = 0.1575$
c. $p(1) = 0.3428$
d. $\mu = 1.62; \sigma = 0.8632$

Chapter 5

5.1

k	0	1	2	3	4	5	6	7	8
$P(x \le k)$	0.000	0.001	0.011	0.058	0.194	0.448	0.745	0.942	1.000

The Problem	List the Values of x	Write the Probability	Rewrite the Probability	Find the Probability
Three or less	0, 1, 2, 3	$P(x \le 3)$	not needed	0.058
Three or more	3, 4, 5, 6, 7, 8	$P(x \ge 3)$	$1 - P(x \le 2)$	0.989
More than three	4, 5, 6, 7, 8	$P(x > 3)$	$1 - P(x \le 3)$	0.942
Fewer than three	0, 1, 2	$P(x < 3)$	$P(x \le 2)$	0.011
Between 3 and 5 (inclusive)	3, 4, 5	$P(3 \le x \le 5)$	$P(x \le 5) - P(x \le 2)$	0.437
Exactly three	3	$P(x = 3)$	$P(x \le 3) - P(x \le 2)$	0.047

5.3 not binomial; dependent trials; p varies from trial to trial

5.5 a. 0.2965 b. 0.8145 c. 0.1172
d. 0.3670

5.7 a. 0.097 b. 0.329 c. 0.671
d. 2.1 e. 1.212

5.9 $p(0) = 0.000; p(1) = 0.002; p(2) = 0.015;$
$p(3) = 0.082; p(4) = 0.246; p(5) = 0.393;$
$p(6) = 0.262$

5.11 a. 0.251 b. 0.618 c. 0.367 d. 0.633
e. 4 f. 1.549

5.13 a. 0.901 b. 0.015 c. 0.002 d. 0.998

5.15 a. 0.748 b. 0.610 c. 0.367 d. 0.966
e. 0.656

5.17 a. 1; 0.99 b. 90; 3 c. 30; 4.58
d. 70; 4.58 e. 50; 5

5.19 a. 0.9568 b. 0.957 c. 0.9569
d. $\mu = 2; \sigma = 1.342$ e. 0.7455; 0.9569; 0.9977 f. yes; yes

5.21 No; the variable is not the number of successes in n trials. Instead, the number of trials n is variable.

5.23 a. 1.000 b. 0.997 c. 0.086

5.25 a. 0.098 b. 0.991 c. 0.098 d. 0.138
e. 0.430 f. 0.902

5.27 a. .999
b. .054
c. .004

5.29 600, 420, 20.4939; Tchebysheff's Theorem concludes that at least .96 of the measurements lie within 4.88σ of the mean (i.e., 600 ± 100). Therefore, at most .04 of the measurements are less than 500 or greater than 700. Since the distribution is fairly mound-shaped and symmetric, .02 of the measurements are greater than 700.

5.31 Since $\mu = 5$ and $\sigma = 1.58$, the mouse should choose the red door between 1.84 to 8.16 (roughly 2 to 8) times if there is no colour preference. If more than 8 or less than 2 times, the unusual results might suggest a colour preference.

5.33 a. 10, 2.449
b. 5.102 to 14.898; values of x in the range $6 \le x \le 14$
c. .937; this value agrees with Tchebysheff's Theorem and also with the Empirical Rule.

5.35 a. .196
b. .059
c. .059

5.37 a. .0498
b. .1494
c. .8008

5.39 a. 0.135335 b. 0.27067 c. 0.593994
d. 0.036089

5.41 a. 0.677 b. 0.6767 c. yes

5.43 a. 0.0067 b. 0.1755 c. 0.560

5.45 a. 0.257 b. 0.713 c. 0.287 d. 0.918

5.47 $P(x > 5) = 0.017$; unlikely

5.49 a. 0.6 b. 0.5143 c. 0.0714

5.51 a. 0
b. .9762
c. .2381

5.53 a. .5357
b. .0179
c. .1786

5.55 **a.** $p(x) = \dfrac{C_x^2 C_{2-x}^3}{C_2^5}$ for $x = 0,1,2$
b. .8, .36

5.61 **a.** 1/8, 3/8, 3/8, 1/8
c. 1.5, .866
d. 3/4, 1; Consistent with both.

5.63 **a.** 1
b. .588
c. .021

5.65 **a.** $p(x) = \dfrac{1}{10}$ for $x = 0, 1, 2, ..., 9$
b. 3/10
c. 7/10

5.67 **a.** $p(x) = C_x^{15}(.8)^x(.2)^{15-x}$ for $x = 0, 1, 2, ..., 15$
b. .018
c. .982
d. 9

5.69 **a.** .005
b. .021
c. .021

5.71 .017; less than 80%

5.73 **a.** .3
b. 2.29129
c. 1.09
d. not likely

5.75 **a.** .018
b. .433
c. .371
d. Approximately 95% of the values of x should lie in the interval 0 to 8.

5.77 **a.** .5
b. 10
c. .006
d. .456

5.79 **a.** Yes, 5.
b. .265
c. .742
d. .032

5.81 **a.** .022935
b. .291639
c. .95818

5.83 **a.** .1252
b. .3323
c. .9955

5.85 **a.** .999
b. 0
c. .230

5.87 **a.** .031
b. .887
c. .048

5.89 **a.** .033
b. .764

5.91 .183940

5.93 Approximately 95% of the values of x should lie in the interval between 1202 and 1246.

5.95 .214, .214

5.99 **a.** .9606
b. .9994

Chapter 6

6.51 **a.** 0.0004
b. 0.5626
c. 0.9986
d. No; higher than claimed.

6.53 **a.** 0.1841
b. 0.9904
c. Approximately 1

6.55 **a.** 0.9500
b. 0.0250
c. 0.0250

6.57 **a.** 1.96
b. 2.33
c. 0.90
d. 0.99

6.59 **a.** 0.9788
b. 0.5873
c. Approximately 0

6.63 **a.** 0.3227
b. 0.1586

6.65 0

6.67 z_0 represents 75th percentile and $-z_0$ corresponds to 25th percentile.

6.69 Range of faculty ages should be approximately from 25 to 65; distribution skewed to the right.

6.71 5.065

6.73 0.0401

6.75 $z_0 = 1.28$ or $x_0 = 85.36$

6.77 –1.26; no reason to doubt the claim.

6.79 226.7

6.79 **a.** 141
b. 0.0401

6.81 0.9474

6.83 0.3557

6.85 **a.** lower quartile: 269.96; upper quartile 286.04
b. 180; unusual

6.87 **a.** Approximately 0
b. 0.6026
c. Not random; survey results are not reliable.

6.89 **a.** 0.2743
b. 37.429

6.91 **a.** 1.27
b. 0.1020

Chapter 7

7.51 No change.

7.53 No change.

7.55 **a.** 0.058
b. 0.0041
c. 0.0505

Chapter 8

8.15 **a.** $\hat{p} = 0.792$
b. SE = 0.0371; MOE = 0.0726

8.17 $\bar{x} = 4.2$; MOE = 0.339

8.19 **a.** $\hat{p} = 0.68$; MOE = 0.0578
b. $\hat{p} = 0.48$; MOE = 0.0619

8.21 $\hat{p} = 0.16$; MOE = 0.045

8.23 **a.** 0.54
b. MOE = 0.0308

8.25 $\bar{x} = 19.3$; MOE = 1.86

8.27 **a.** (0.797, 0.883)
b. (21.469, 22.331)

8.29 $\hat{p} = 0.877$; (0.846, 0.908)

8.31 **a.** 3.92
b. 2.772
c. 1.96

8.33 **a.** 3.29
b. 5.16
c. As the confidence coefficient increases, so does the width of the confidence interval.

8.63 **a.** (−0.221, 0.149)
b. Since $p_1 - p_2 = 0$ is in the confidence interval, it is possible that $p_1 = p_2$. Cannot conclude that there is a difference in the proportion.

8.81 $\mu < 5636.99$

8.83 $n \geq 2134.2$ or $n_1 = n_2 = 2135$

8.85 **a.** $(\mu_1 - \mu_2) > 2.52$
b. Since the difference in the means is positive, conclude that there has been a decrease in the average per-capita beef consumption over the last 10 years.

8.87 $n \geq 97$

8.107 Answers will vary.

8.109 $n \geq 65$

8.111 $0.059 < (p_1 - p_2) < 0.141$

8.113 **a.** $\bar{x} \pm 1.96\, s/\sqrt{69}$
b. $1.922 < \mu < 1.984$

8.115 **a.** $5.543 < \mu < 5.963$
b. $n \geq 530$

8.117 $0.568 < p < 0.738$

8.119 $\bar{x} = 21.6$, MOE = 0.49

8.121 $n \geq 246$

Chapter 9

9.11 **a.** p-value = 0.0032 is less than or equal than 5% significance level; null hypothesis is reject, results are statistically significant.
b. p-value = 0.1868 is greater than 5% significance level; null hypothesis is not reject, results are not statistically significant.
c. p-value = 0.1867 is greater than 5% significance level; null hypothesis is not reject, results are not statistically significant.

9.21 **a.** H_0: $\mu = 170$
b. H_a: $\mu \neq 170$
c,d. $z = 2.19$ with p-value = 0.0286. Null hypothesis is rejected. Evidence to indicate a difference in the mean.
e. Answers may vary.

9.27 **a.** one-tailed, H_0: $\mu_1 - \mu_2 = 0$ versus H_a: $\mu_1 - \mu_2 > 0$
b. $z \approx 2.074$; reject H_0. Evidence to indicate that $\mu_1 - \mu_2 > 0$.

9.29 **a.** two-tailed, H_0: $\mu_1 - \mu_2 = 0$ versus H_a: $\mu_1 - \mu_2 \neq 0$, $z = -2.26$ with p-value = 0.0238. Null hypothesis is rejected. Evidence to indicate a difference in the mean.
b. $-3.55 < (\mu_1 - \mu_2) < -0.25$
c. Since the value $\mu_1 - \mu_2 = 5$ or $\mu_1 - \mu_2 = -5$ is not in the confidence interval in part b, it is not likely that the difference will be more than 5 ppm, and hence the statistical

significance of the difference is not of practical importance to the engineers.

9.31 **a.** two-tailed, H_0: $\mu_1 - \mu_2 = 0$ versus H_a: $\mu_1 - \mu_2 \neq 0$
b. $z \approx -3.75$; H_0 is rejected. Evidence to indicate that there is a difference in the mean.
c. p-value ≈ 0.0000, null hypothesis can be rejected at the 1% level. Sufficient evidence to conclude that $\mu_1 - \mu_2 \neq 0$.

9.33 **a.** two-tailed, H_0: $\mu_1 - \mu_2 = 0$ versus H_a: $\mu_1 - \mu_2 \neq 0$, $z \approx -3.18$ with p-value $= 0.0014$. Null hypothesis is rejected. Evidence to indicate a difference in the mean.
b. $-3.01 < (\mu_1 - \mu_2) < -0.71$; Since the value $\mu_1 - \mu_2 = 0$ does not fall in the interval in part b, it is not likely that $\mu_1 = \mu_2$. There is evidence to indicate that the means are different, confirming the conclusion in part a.

9.35 **a.** two-tailed, H_0: $\mu_1 - \mu_2 = 0$ versus H_a: $\mu_1 - \mu_2 \neq 0$, $z \approx -1.27$ with p-value $= 0.2040$. Null hypothesis is not rejected. Insufficient evidence to indicate a difference in the mean.
b. Since p-value $= 0.2040$, reject H_0 neither at the 5% level nor at the 1% level. Not statistically significant.

Chapter 10

10.1 Sample must be randomly selected. Population from which sampling must be normally distributed.

10.3 Slope of t-distribution is a normal distribution. If an assumption is violated, the statistic does not change significantly.

10.5 As sample size n increases beyond 50, the variability of t decreases.

Chapter 11

11.1

Source	df
Treatments	5
Error	54
Total	59

11.3 **a.** (2.731, 3.409) **b.** (0.07, 1.03)

11.5 **a.**

Source	df	SS	MS	F
Treatments	3	339.8	113.267	16.98
Error	20	133.4	6.67	
Total	23			

b. $df_1 = 3$ and $df_2 = 20$ **c.** $F > 3.10$
d. yes, $F = 16.98$ **e.** p-value < 0.005; yes

11.7 **a.** CM = 103.142857; Total SS = 26.8571
b. SST = 14.5071; MST = 7.2536
c. SSE = 12.3500; MSE = 1.1227
d. Analysis of Variance

Source	DF	SS	MS	F	P
Trts	2	14.51	7.25	6.46	0.014
Error	11	12.35	1.12		
Total	13	26.86			

f. $F = 6.46$; reject H_0 with $0.01 < p$-value < 0.025

11.9 **a.** (1.95, 3.65) **b.** (0.27, 2.83)

11.11 **a.** (67.86, 84.14) **b.** (55.82, 76.84)
c. (−3.629, 22.963) **d.** no, they are not independent

11.13 **a.** Each observation is the mean length of ten leaves.
b. yes, $F = 57.38$ with p-value $= 0.000$
c. reject H_0; $t = 12.09$
d. (1.810, 2.924)

11.15

Analysis of Variance for Percent

Source	DF	SS	MS	F	P
Method	2	0.0000041	0.0000021	16.38	0.000
Error	12	0.0000015	0.0000001		
Total	14	0.0000056			

11.17 **a.** completely randomized design
b.

Source	DF	SS	MS	F	P
State	3	3272.2	1090.7	26.44	0.000
Error	16	660.0	41.3		
Total	19	3932.2			

c. $F = 26.44$; reject H_0 with p-value $= 0.000$

11.19 sample means must be independent; equal sample sizes

11.21 **a.** $1.878s$ **b.** $2.1567s$

11.23 $\bar{x}_1$ $\bar{x}_2$ $\bar{x}_3$ $\bar{x}_4$

11.25 **a.** no; $F = 0.60$ with p-value $= 0.562$
b. no differences

11.27 **a.** yes; $F = 8.55$, p-value $= 0.005$
b. (−157.41, −28.59) **c.** $\bar{x}_3$ $\bar{x}_1$ $\bar{x}_2$

11.29

Source	df	SS	MS	F
Treatments	2	11.4	5.70	4.01
Blocks	5	17.1	3.42	2.41
Error	10	14.2	1.42	
Total	17	42.7		

11.31 (−3.833, −0.767)

11.33 **a.** yes; $F = 19.19$ **b.** yes; $F = 135.75$ **c.** $\underline{\bar{x}_1 \quad \bar{x}_3 \quad \bar{x}_4 \quad \bar{x}_2}$ **d.** $(-5.332, -2.668)$ **e.** yes

11.35 **a.** 7 **b.** 7 **c.** 5 **e.** yes; $F = 9.68$ **f.** yes; $F = 8.59$

11.37

Two-way ANOVA: y versus Blocks, Chemicals

```
Analysis of Variance for y
Source      DF        SS       MS       F       P
Blocks       2    7.1717   3.5858   40.21   0.000
Chemical     3    5.2000   1.7333   19.44   0.002
Error        6    0.5350   0.0892
Total       11   12.9067
```

11.39 **a.** yes; $F = 10.06$ **b.** yes; $F = 10.88$ **c.** $\omega = 4.35$ **d.** $(1.12, 5.88)$

11.41

Two-way ANOVA: Cost versus Estimator, Job

```
Analysis of Variance for Cost
Source      DF       SS       MS       F       P
Estimator    2   10.862    5.431    7.20   0.025
Job          3   37.607   12.536   16.61   0.003
Error        6    4.528    0.755
Total       11   52.997
```

11.45 **b–c.** Yes, $F = 6.67$ and the rejection region is $F > 3.63$. **d.** Since $F = 6.67$ lies between $F_{0.01}$ and $F_{0.005}$, $0.005 < p\text{-value} < 0.01$ **e.** differences should be explored individually

11.47 **a.** no, $F = 1.21$ with p-value $= 0.362$
b. Both A and B are significant.
c. Plot does not indicate a significant difference in the behaviour of the mean responses for the two different locations.
d. $(-31, 0.01, -1.99)$

11.51 **a.** A 2×4 factorial experiment with $r = 5$ replications; Two factors: Gender and School, one at two levels and one at four levels.
b.

Source	*df*	SS	MS	*F*
G	1	6200.1	6200.100	2.09
Sc	3	246725.8	82241.933	27.75
G × Sc	3	10574.9	3524.967	1.19
Error	32	94825.6	2963.300	
Total	39	358326.4		

c. There is insufficient evidence to indicate interaction between gender and schools.
d. There is a small difference between the average scores for male and female students at schools 1 and 2, but no difference at the other two schools. The interaction is not significant.
e. There is a significant effect due to schools.

11.53 **a.**

Source	*df*	SS	MS	*F*
A	1	4489	4489	117.49
B	1	132.25	132.25	3.46
A × B	1	56.25	56.25	1.47
Error	12	458.5	38.2083	
Total	15	5136		

b. The interaction term is not significant.
c. Factor B (Situation) is not significant.
d. Factor A (Training) is highly significant.
e. The response is much higher for the supervisors who have been trained; very little change in the response for the two different situations; no interaction between the two factors.

11.55 Significant differences are observed between treatments A and C, B and C, C and E, and D and E.

11.57 **a.** There is a significant difference in the effect of the five stimuli.
b. The ranked means:

E	A	B	D	C
.525	.7	.8	1.025	1.05

(E and A underlined together; A and B underlined together; D and C underlined together)

c. The block differences are significant; blocking has been effective.

11.59 Answers will vary.

11.61 **a.** There is insufficient evidence to indicate an interaction.
b. p-value $> .10$
c. There is evidence that factor A affects the response.
d. Factor B affects the response.

11.63 **a.** A 2×3 factorial and a two-way analysis of variance is generated.
b. There is insufficient evidence to suggest that the effect of temperature is different depending on the type of plant.
c. The temperature appears to have a quadratic effect on the number of eggs laid in both cotton and cucumber. However, the treatment means are higher overall for the cucumber plants.
d. $-22.56 < (\mu_{Cotton} - \mu_{Cucumber}) < -5.84$

11.65 **a.** Randomized block design, with weeks representing blocks and stores as treatments.

b. Two-way ANOVA: Total versus Week, Store

```
Source DF       SS       MS     F       P
  Week  3   571.71  190.570  8.27  0.003
 Store  4   684.64  171.159  7.43  0.003
 Error 12   276.38   23.032
 Total 19  1532.73
S = 4.799 R-Sq = 81.97%
R-Sq(adj) = 71.45%
```

c. There is a significant difference in the average weekly totals for the five supermarkets.

d. The ranked means:

1	2	3	4	5
240.23	249.19	252.18	254.87	256.99

11.67 **a.** Randomized block design with type of crash (block) and bumpers on different cars (treatments).

b. There is evidence of significant difference in the average cost due to the type of automobile.

c. There is insufficient evidence to indicate a difference in the average cost due to the type of crash.

d. The ranked means:

1	3	4	2	5
209.75	370.25	637.75	653.5	690.25

6	7	8
1030.5	1294.25	2265.25

11.71 **a.** Randomized block design, with telephone companies as treatments and cities as blocks.

b. Two-way ANOVA: Score versus City, Carrier

```
 Source DF      SS       MS      F      P
   City  3  55.688  18.5625   3.88  0.049
Carrier  3 285.688  95.2292  19.90  0.000
  Error  9  43.063   4.7847
  Total 15 384.438
S = 2.187 R-Sq = 88.80%
R-Sq(adj) = 81.33%
```

c. The there is a significant difference in the average satisfaction scores for the four carriers.

d. There is a significant difference in the average satisfaction scores for the four carriers.

Chapter 12

12.1 y-intercept = 1, slope = 2

12.3 $y = 3 - x$

12.7 **a.** $\hat{y} = 6.00 - 0.557x$ **c.** 4.05

d.

```
Analysis of Variance
Source          DF      SS      MS
Regression       1  5.4321  5.4321
Residual Error   4  0.1429  0.0357
Total            5  5.5750
```

12.13 **a.** y = Temperature Anomaly; x = Year
b. yes **c.** $\hat{y} = -53.972 + 0.0273x$
d. yes

12.15 **a.** yes **b.** $\hat{y} = -11.665 + 0.755x$
c. $\hat{y} = 52.51$

12.17 **a.** strong positive linear relationship
b. approximately 1
c. $\hat{y} = 31.103 + 0.815x$ **d.** $\hat{y} = 159.4655$

12.19 **a.** yes, $t = 5.20$ **b.** $F = 27.00$
c. $t_{0.025} = 3.182$; $F_{0.05} = 10.13$

12.21 **a.** yes, $F = 152.10$ with p-value = 0.000
b. $r^2 = 0.974$

12.23 **b.** y =number of chirps, x = temperature
c. $\hat{y} = 6.472 + 0.38147x$
e. $t = 5.47$, $r^2 = 0.697$

12.27 **a.** yes; $t = 3.79$ and $F = 14.37$ with p-value = 0.005 **b.** no **c.** $r^2 = 0.642$
d. MSE = 5.025 **e.** (0.186, 0.764)

12.29 **a.** yes **b.** $\hat{y} = 108.74 + 1.5453x$
c. yes; $t = 3.01$ and $F = 9.08$ with p-value = 0.005 **d.** (0.2243, 2.8663)

12.31 **a.** yes; reject H_0, $t = 7.11$
b. (0.5346, 1.0954) **c.** yes; the value $\beta = 1$ is contained in the interval

12.33 plot residuals versus fit; random scatter of points, free of patterns

12.35 no

12.37 The two plots behave as expected if the regression assumptions have been satisfied.

12.39 **a.** The random error ε must have a normal distribution with mean 0 and a common variance σ^2, independent of x.
b. The best estimate of σ^2 is 722.0
c. There does not appear to be any extreme violations of the regression assumptions.

12.41 **a.** $3.259 < E(y) < 5.141$.
b. $2.24 < y < 6.16$

12.43 **a.** $4.3 + 1.5x$
c. 1.53
d. There is sufficient evidence to indicate that the independent variable x does help in predicting values of the dependent variable y.
e. 0.01
f. The regression assumptions are valid.
g. $8.11 < E(y) < 9.49$

12.47 **a.** $1.769 + .263x$
b. 0.998
c. There are no strong violations of assumptions.

12.49 **a.** $156.135 + 4.844x$
b. 0.163
c. There are no strong violations of assumptions.

12.51 The coefficient of correlation provides a meaningful measure of the strength of the linear relationship between two variables, y and x based on its sign and magnitude. If r is positive, then the least squares line slopes upward to the right. Similarly, if r is negative, the line slopes downward to the right. Values of r close to ± 1 indicate a strong correlation.

12.53 **a.** $+1$
b. -1

12.55 **b.** -0.982
c. 96.47%

12.57 **a.** Negatively correlated; When x is large, y should be small if the barnacles compete for space on the lobster's surface.
b. $H_a: \rho < 0$
c. Reject H_0. There is evidence of negative correlation.

12.67 **a.** yes **b.** $\hat{y} = 80.85 + 270.82x$
c. yes; $t = 3.96$ with p-value $= 0.003$
d. (112.1, 157.9)

12.69 **a.** $r = 0.980$ **b.** $r^2 = 0.961$
c. $\hat{y} = 21.9 + 15.0x$ **d.** variance is not constant for all x

12.71

Regression Analysis: Temperature Anomaly versus Year

```
The regression equation is
Temperature Anomaly = -54.0 + 0.0273 Year
Predictor        Coef    SE Coef      T       P
Constant      -53.972      7.944  -6.79   0.000
Year         0.027279   0.003999   6.82   0.000

S = 0.189592  R-Sq = 62.4%  R-Sq(adj) = 61.1%
Analysis of Variance
Source          DF      SS      MS      F      P
Regression       1  1.6724  1.6724  46.53  0.000
Residual Error  28  1.0065  0.0359
Total           29  2.6789
```

12.73 **a.** no; $t = 2.066$ with p-value > 0.05
b. $r^2 = 0.299$

12.75 No; variance is not constant for all x.

12.79 at the extremes of the experimental region

12.81 **a.** $\hat{y} = 20.47 - 0.758x$
b.

```
Source          DF      SS      MS       F      P
Regression       1  287.28  287.28  493.40  0.000
Residual Error   8    4.66    0.58
Total            9  291.94
```

c. reject H_0, $t = -22.21$
d. $(-0.86, -0.66)$ **e.** (9.296, 10.420)
f. $r^2 = 0.984$

Chapter 13

13.1 **b.** parallel lines

13.3 **a.** yes, $F = 57.44$ with p-value < 0.005
b. $R^2 = 0.94$

13.5 **a.** quadratic
b. $R^2 = 0.815$; relatively good fit
c. yes, $F = 37.37$ with p-value $= 0.000$

13.7 **a.** $b_0 = 10.5638$
b. yes, $t = 15.20$ with p-value $= 0.000$

13.9 **b.** $t = -8.11$ with p-value $= 0.000$; reject H_0: $\beta_2 = 0$ in favour of H_a: $\beta_2 < 0$

13.11 **a.** $R^2 = 0.9955$ **b.** $R^2(\text{adj}) = 99.3\%$
c. The quadratic model fits slightly better.

13.13 **a.** Use variables x_2, x_3, and x_4. **b.** no

13.15 **a.** $\hat{y} = -8.177 + 292x_1 + 4.434x_2$
b. Reject H_0, $F = 16.28$ with p-value $= 0.002$. The model contributes significant information for the prediction of y. **c.** yes; $t = 5.54$ with p-value $= 0.001$ **d.** $R^2 = 0.823$; 82.3%

13.17 **a.** quantitative **b.** quantitative
c. qualitative; $x_1 = 1$ if plant B, 0 otherwise; $x_2 = 1$ if plant C, 0 otherwise
d. quantitative **e.** qualitative; $x_1 = 1$ if day shift, 0 if night shift

13.19 **a.** x_2 **b.** $\hat{y} = 12.6 + 3.9x_2^2$ or $\hat{y} = 13.14 - 1.2x_2 + 3.9x_2^2$

13.21 **a.** $y = \beta_0 + \beta_1 x_1 + \beta_2 x_2 + \beta_3 x_1 x_2 + \varepsilon$ with $x_2 = 1$ if cucumber, 0 if cotton **c.** No; the test for interaction yields $t = 0.63$ with p-value $= 0.533$. **d.** yes

13.23 $y = \beta_0 + \beta_1 x_1 + \beta_2 x_1^2 + \beta_3 x_2 + \beta_4 x_1 x_2 + \beta_5 x_1^2 x_2 + \varepsilon$

13.25 **a.** $\hat{y} = 8.585 + 3.8208x - 0.21663x^2$ **b.** $R^2 = 0.944$ **c.** yes; $F = 33.44$ **d.** yes; $t = -4.93$ with p-value $= 0.008$ **e.** no

13.27 **b.** $\hat{y} = 4.10 + 1.04x_1 + 3.53x_2 + 4.76x_3 - 0.43x_1x_2 - 0.08x_1x_3$ **c.** yes; $t = -2.61$ with p-value $= 0.028$ **d.** no; $F = 3.86$; Consider eliminating the interaction terms.

13.29 **a.** $y = \beta_0 + \beta_1 x_1 + \beta_2 x_2 + \beta_3 x_1^2 + \beta_4 x_1 x_2 + \beta_5 x_1^2 x_2 + \varepsilon$ **b.** $F = 25.85$; $R^2 = 0.768$ **c.** $\hat{y} = 4.51 + 6.394x_1 + 0.1318x_1^2$ **d.** $\hat{y} = -46.34 + 23.458x_1 - 0.3707x_1^2$ **e.** no; $t = 0.78$ with p-value $= 0.439$

13.31 **a.** curvilinear relationship **b.** $y = \beta_0 + \beta_1 x + \beta_2 x^2 + \varepsilon$ **c.** $\hat{y} = 4114749 - 4113.4x + 1.02804x^2$ **d.** yes; $F = 542.11$ with p-value $= 0.000$ **e.** $R^2 = 0.997$; very good fit

13.35 **a.** 99.8% of the total variation is accounted for by using x and x^2 in the model. **b.** Reject H_0. The model provides valuable information for the prediction of y. **c.** $\hat{y} = -28.3906 + 1.463x_1 + 3.8446(4)$ $\hat{y} = -13.0122 + 1.463x_1$ **d.** 19.7197 **e.** No violations of the regression assumptions.

Chapter 14

14.3 **a.** $\chi^2 > 12.59$ **b.** $\chi^2 > 21.666$ **c.** $\chi^2 > 29.8194$ **d.** $\chi^2 > 5.99$

14.5 **a.** H_0: $p_1 = p_2 = p_3 = p_4 = p_5 = 1/5$ **b.** 4 **c.** 9.4877 **d.** $\chi^2 = 8.00$ **e.** do not reject H_0

14.7 yes, $\chi^2 = 24.48$; drivers tend to prefer the inside lanes.

14.9 no, $\chi^2 = 3.63$

14.11 no, $\chi^2 = 13.58$

14.13 yes; do not reject H_0, $\chi^2 = 1.247$

14.15 yes; reject H_0, $\chi^2 = 28.386$

14.17 8

14.19 reject H_0, $\chi^2 = 18.352$ with p-value $= 0.000$

14.21 **a.** yes; $\chi^2 = 7.267$ **b.** $0.025 < p\text{-value} < 0.05$

14.23 **a.** yes; reject H_0; $\chi^2 = 20.937$ **b.** no; $\chi^2 = 1.255$

14.25 **a.** no; do not reject H_0, $\chi^2 = 6.447$ **b.** p-value > 0.10; yes

14.27 **a.** $\chi^2 = 10.597$ **b.** $\chi^2 > 13.2767$ **c.** do not reject H_0 **d.** $0.025 < p\text{-value} < 0.05$

14.29 yes; $\chi^2 = 122.6$

14.31 **a.** Each income category represents a multinomial population in which we measure education levels. **b.** yes; $\chi^2 = 19.172$

14.33 yes; reject H_0, $\chi^2 = 36.499$

14.35 **a.** There is sufficient evidence to indicate that proportion of young adults who believe that evolution is the best explanation for the development of human life differs depending on religious affiliation. **b.** The percentage who believe that evolution is the best explanation for the development of human life would be higher.

14.37 There is a difference in the proportions for the four provinces.

14.39 **a.** There is sufficient evidence to indicate that there is a difference in peoples' opinion from one country to another.

14.43 **a.** The serum is effective. **b.** Reject the null hypothesis of no difference and conclude that the serum is effective.

14.49 $.347 < p_1 < .483$.

14.51 **b.** Reject H_0 in both cases. **c.** Answers will vary. **d.** Some cells have expected cell counts less than 5. This is a violation of the assumptions necessary for this test, and results should thus be viewed with caution.

14.53 **a.** Since the percentages do not add to 100%, there is a category missing; 15. **b.** We cannot reject H_0. There is insufficient evidence to suggest a difference from the given percentages.

14.55 There is insufficient evidence to indicate a difference in perception of the best taste between adults and children. If the company intends to use a flavour as a marketing tool, the cherry flavour does not seem to provide an incentive to buy this product.

14.57 **a.** The results are not significant; H_0 is not rejected, and there is insufficient evidence to indicate a difference in the distribution of injury types for rugby forwards and backs. Since the assumption that E_i is greater than or equal to five for each cell has been violated. **b.** The p-value for the Chi-square test is greater than the .05 claimed by the researcher, and does not suggest a significant difference.

14.59 **a.** The results are highly significant and we conclude that there is a difference in the susceptibility to colds depending on the number of relationships you have.

14.61 Yes; H_0 is rejected and the manager's claim is refuted.

Chapter 15

15.1 **a.** T_1^* **b.** $T \leq 31$ **c.** $T \leq 27$

15.3 **a.** H_0: population distributions are identical; H_a: population 1 shifted to the left of population 2. **b.** $T_1 = 16$; $T_1^* = 39$ **c.** $T \leq 19$ **d.** yes; reject H_0

15.5 do not reject H_0; $z = -1.59$

15.7 do not reject H_0; $T = 102$

15.9 yes; reject H_0; $T = 45$

15.11 yes; reject H_0; $T = 44$

15.13 **b.** $\alpha = 0.002, 0.007, 0.022, 0.054, 0.115$

15.15 one-tailed: $n = 10$: $\alpha = 0.001, 0.011, 0.055$; $n = 15$: $\alpha = 0.004, 0.018, 0.059$; $n = 20$: $\alpha = 0.001, 0.006, 0.021, 0.058, 0.132$; two-tailed: $n = 10$: $\alpha = 0.002, 0.022, 0.110$; $n = 15$: $\alpha = 0.008, 0.036, 0.118$; $n = 20$: $\alpha = 0.002, 0.012, 0.042, 0.116$

15.17 **a.** $H_0: p = \frac{1}{2}$; $H_a: p \neq \frac{1}{2}$; rejection region: $\{0, 1, 7, 8\}$; $x = 6$; do not reject H_0 at $\alpha = 0.07$; p-value $= 0.290$

15.19 $z = 3.15$; reject H_0

15.21 **b.** $T = \min\{T^+, T^-\}$ **c.** $T \leq 137$ **d.** do not reject H_0

15.23 do not reject H_0; $z = -0.34$

15.25 **a.** reject H_0; $T = 1.5$ **b.** Results do not agree.

15.27 **a.** no; $T = 6.5$

15.29 **a.** do not reject H_0; $x = 8$ **b.** do not reject H_0; $T = 14.5$

15.31 **a.** paired difference test, sign test, Wilcoxon signed-rank test **b.** reject H_0 with both tests; $x = 0$ and $T = 0$

15.33 yes, $H = 13.90$

15.35 **a.** no; $H = 2.63$ **b.** p-value > 0.10 **c.** p-value > 0.10

15.37 no; $H = 2.54$ with p-value > 0.10

15.39 **a.** reject H_0; $F_r = 21.19$ **b.** p-value $<$ 0.005 **d.** $F = 75.43$ **e.** p-value $<$ 0.005 **f.** Results are identical.

15.41 **a.** do not reject H_0; $F_r = 5.81$ **b.** $0.05 < p$-value < 0.10

15.43 **a.** $r_s \geq 0.425$ **b.** $r_s \geq 0.601$

15.45 **a.** $|r_s| \geq 0.400$ **b.** $|r_s| \geq 0.526$

15.47 **a.** -0.593 **b.** yes

15.49 **a.** $r_s = 0.811$ **b.** yes

15.51 yes

15.53 yes, $r_s = 0.9118$

15.55 **a.** do not reject H_0; $x = 2$ **b.** do not reject H_0; $t = -1.646$

15.57 **a.** do not reject H_0; $x = 7$ **b.** do not reject H_0; $x = 7$

15.59 Do not reject H_0 with the Wilcoxon rank sum test ($T = 77$) or the paired difference test ($t = 0.30$).

15.61 do not reject H_0 using the sign test ($x = 2$); no

15.63 yes; $r_s = -0.845$

15.65 reject H_0; $T = 14$

15.67 **a.** reject H_0; $F_r = 20.13$ **b.** The results are the same.

15.69 **a.** reject H_0; $H = 9.08$ **b.** $0.025 < p$-value < 0.05 **c.** The results are the same.

15.71 **a.** no **b.** Significant differences among the responses to the three rates of application; $F_r = 10.33$ with p-value $= 0.006$.

15.73 $T = 19$. $T_{0.05} = 21$ ($T_{0.01} = 18$.); reject H_0

15.75 $z = 1.285 < z_{0.05} = 1.645$. Lighting not effective.

15.77 $H = 7.43$ $df = 3$ p-value $= 0.059$. No significant difference.

15.79 **a.** $r_s = 0.738$. **b.** p-value $= 0.025 < 0.05$; Yes, there is a positive correlation.

Index

Note: Following a page number, "f" denotes a figure and a "t" denotes a table.